CONCRETE CONSTRUCTION ENGINEERING
HANDBOOK

Editor-in-Chief
Dr. Edward G. Nawy, P.E., C. Eng.
Distinguished Professor
Rutgers—The State University of New Jersey
New Brunswick, New Jersey

CRC Press
Boca Raton   New York

Library of Congress Cataloging-in-Publication Data

Concrete construction engineering handbook / Edward G. Nawy.
    p. cm.
Includes bibliographical references and index.
ISBN 0-8493-2666-4 (alk. paper)
1. Concrete construction.   I. Nawy, Edward G.
TA681.C743 1997
624.1'834—dc21
DNLM/DLC
for Library of Congress
                                                              97-7652
                                                               CIP

No claim to original U.S. Government works
International Standard Book Number 0-8493-2666-4
Library of Congress Card Number 97-7652
Printed in the United States of America 1 2 3 4 5 6 7 8 9 0
Printed on acid-free paper

Contents

Preface

A great need has existed for an in-depth handbook on concrete construction engineering and technology which can assist the constructor in making correct technical judgements in the various areas of constructed systems. This Handbook is intended to fill this very need. All chapters treat the particular subjects with sufficient detail and depth of discussion, a feature that is lacking in any comparable texts. Also, each chapter culminates with an extensive list of selected references for the user to consult beyond the scope of the Handbook.

The topics covered are state-of-the-art statements on what the design engineer and the constructor should know about concrete as the most versatile material of this century and for the next millennium. These topics can be grouped into four categories:

1. The latest advances in engineered concrete materials including concrete constituents, high performance concretes, design of mixtures for both normal strength and high strength concretes, and special concrete applications including architectural concrete.
2. Reinforced concrete construction including recommendations on the vast array of types of constructed facilities, long-term effects on behavior and performance including creep and shrinkage, construction loading effects, formwork and falsework proportioning and automation in construction.
3. Specialized construction such as prestressed concrete construction in buildings and transportation facilities, construction and proportioning of structures in seismic zones, masonry construction, heavy concrete construction including roller-compacted concrete and concrete marine structures such as offshore platforms concrete. For seismic design, it presents the latest provisions of the 1997 Uniform Building Code (UBC).
4. Design recommendations for high performance including deflection reduction in buildings, proportioning of concrete structural elements by the latest ACI 318-95 Building Code, prefabricated precasting, geotechnical and foundation engineering, nondestructive evaluation of long-term structural performance, and structural concrete repair, retrofit and rehabilitation.

This Handbook is the only publication in this category which has in a single chapter a summary of all the design expressions in accordance with the latest ACI 318-95 Building Code for flexure, shear, torsion, compression, long-term effects, long columns and development of reinforcement. Both PI (in-lb) and SI formats are included. In this manner a design office can at a glance go through all the latest requirements for structural concrete.

Twenty-nine contributors produced the twenty-seven chapters of the Handbook. These authors are the leading authorities in the field with a combined professional practice of at least 800 years. All of them are national or international leaders in research, design and construction.

This Handbook should enable the designers, the constructors, the educators, and the field personnel together to produce the best and most durably engineered constructed facilities. It is for these professionals that this Handbook is written with the hope that the wealth of the most up-to-date knowledge embodied in this comprehensive work will provide, in the next millennium, vastly better, more efficient, and longer enduring constructed concrete.

I consider myself lucky to have had the chance to work with such outstanding experts in developing this Handbook. My acknowledgment and thanks are given to all the authors who, busy as they are, have brought to the profession in the chapters presented herein the vast experience gained from their many years of engineering and construction practice.

Acknowledgment and thanks are due to the American Concrete Institute for permitting unrestricted use by the various authors of their vast technical resource publications and to Prentice Hall and to Addison Wesley Longman for permitting me to extensively use material covered in my three textbooks with them.

I wish to extend my special thanks to Mr. Tim Pletscher, Acquiring Editor of CRC Press, who through valuable input, deliberations, and cooperation has made the Handbook a reality. Also, my gratitude to Ms. Nora Konopka, Associate Editor, who worked on the Handbook's early development stages, Ms. Gerry Jaffe, Production Editor, who has meticulously overseen the correct development of its contents, the staff of CRC Press for the hard work involved and to Mr. Andrew Wilson, Project Manager, TechBooks, for his expert management of this massive reference work.

Last but not least, acknowledgment is due to my wife, Rachel, who has had enduring patience and gave unlimited support while I was totally immersed in the development of the Handbook.

Edward G. Nawy
Rutgers University
Piscataway, New Jersey

Editor-in-Chief

Edward G. Nawy, distinguished professor, Department of Civil and Environmental Engineering at Rutgers, the State University of New Jersey, is internationally recognized for his extensive research work in the fields of reinforced and prestressed concrete, particularly in the areas of serviceability and crack control. He has practiced civil and structural engineering in excess of forty years and has been on the faculty of Rutgers University almost as long, having served also as chairman and graduate director for two terms. He also served two terms on the Board of Governors and one term on the Board of Trustees of the University.

His work has been published in technical journals worldwide, over 150 technical papers, and he has been editor of several Special Publication volumes of the American Concrete Institute since 1972. He is the author of several textbooks: *Simplified Reinforced Concrete* (1986), *Reinforced Concrete—A Fundamental Approach* (1st ed., 1985, 3rd ed., 1996) and *Prestressed Concrete—A Fundamental Approach* (1st ed., 1989, 2nd ed., 1996), all published by Prentice–Hall as well as translated versions in Spanish, Chinese, South Korean and Malaysian languages, and *High Strength High Performance Concrete* (1996) published by Addison Wesley Longman. He is also the author of major chapters in several Handbooks including the *Handbook of Structural Concrete* published by McGraw Hill and the *Engineering Handbook* published by CRC Press.

Dr. Nawy is a charter Fellow of the American Concrete Institute, Fellow of the American Society of Civil Engineers, Fellow of the Institution of Civil Engineers (London) and member of the Precast/Prestressed Concrete Institute. He has chaired several committees of the American Concrete Institute including being the founding chairman of ACI Committee 224 on Cracking, chairman of ACI Committee 435 on Deflection of Structures, member of ACI–ASCE Joint Committee on Slabs, member of ACI Committee 340 on the Design Handbook including the chairmanship of its Subcommittee on Two-Way Slabs and Plates, and member of the Technical Activities Committee of the Precast/Prestressed Concrete Institute.

He is a recipient of several major awards, including the Henry L. Kennedy Award of the American Concrete Institute and the Honorary Professorship of the Nanjing Institute of Technology, Nanjing,

China; is a licensed Professional Engineer in the States of New York, New Jersey, Pennsylvania, California and Florida; Chartered Civil Engineer in the United Kingdom and the Commonwealth; Program Evaluator for the National Accreditation Board for Engineering and Technology (ABET); Panelist for the National Science Foundation, Washington, D.C.; University Representative to the Transportation Research Board, Washington, D.C., and chairman of the TRB Committee A2E03 on Materials, National Research Council. He has been engineering consultant to agencies throughout the United States particularly in areas of structures and materials forensic engineering. He has been listed in *Who's Who in America* since 1967, in *Who's Who in the World* and in several other major standard reference works.

Contributors

John Albinger
President, VANS Materials, Inc., Chicago, Illinois. Expert in mixture design and in high-strength, high-performance concrete. (Chapter 6, Part B co-author)

Florian G. Barth, P.E.
Principal, FBA, Inc., Consulting Engineers, Belmont, California. Expert in the design of post-tensioned concrete structures and in the design and rehabilitation of buildings in seismic regions. (Chapters 12 and 26)

Dr. Terry O. Blackburn, P.E.
Principal and Director, Structural Department, Schoor DePalma–Engineers and Design Professionals, Manalapan, New Jersey. Expert in design and rehabilitation of reinforced and prestressed concrete structures. (Chapter 10)

Dr. Nicholas J. Carino
Research Structural Engineer, Structures Division at the National Institute of Standards and Technology (NIST), Gaithersburg, Maryland. Expert in nondestructive test methods for concrete. (Chapter 19)

Walter L. Dickey, P.E.
Consulting Engineer, Los Angeles, California. Expert in Masonry Construction and has been consulting since the 1930s. Author of *Reinforced Masonry Design* with R. Schneider. (Chapter 21)

M.J. Dickey
Engineering Consultant, Los Angeles, California. Author and consultant on masonry design and construction. (Chapter 21 co-author)

Russell S. Fling, P.E.
Consulting Structural Engineer, Columbus, Ohio. Has been in engineering practice since 1949, has published numerous papers and a textbook and was president of the American Concrete Institute. (Chapter 9)

Sidney Freedman
Director of Architectural Precast Concrete Services at the Precast/Prestressed Concrete Institute, Chicago, Illinois. Expert in architectural precast concrete since the 1960s and author of the PCI Manual on this subject. (Chapter 20)

Dr. Ben C. Gerwick, Jr., P.E.
Chairman, Ben C. Gerwick, Inc., San Francisco, California and Professor Emeritus of Civil Engineering, University of California at Berkeley. An international authority on prestressed concrete and former president of FIP. (Chapter 11)

Dr. S.K. Ghosh
Director, Engineering Services, Codes and Standards, Portland Cement Association, Skokie, Illinois. Expert in structural design, particularly in the areas of serviceability and author of the PCA Manuals on the ACI318 Building Code. (Chapter 8)

Dr. Manjriker Gunaratne, P.E.
Associate Professor, University of South Florida, Tampa. Specialized in various areas of geotechnical engineering, including foundation design, numerical modeling and soil stabilization. (Chapter 14)

Dr. George C. Hoff, D.Eng.
Engineering Consultant, Mobil Technology Company, Dallas, Texas. Expert on offshore concrete platforms and marine structures and on concrete behavior under severe conditions. He was president of the American Concrete Institute. (Chapter 13)

Mark B. Hogan, P.E.
Vice President of Engineering, National Concrete Masonry Association, Herndon, Virginia. He is an active member of committees in several professional societies, including ACI/TMS Committee 216 on Fire Resistance and Fire Protection of Structures. (Chapter 24)

Dr. David W. Johnston, P.E.
Professor of Civil Engineering, North Carolina State University, Raleigh, North Carolina since 1977. He is an active member of several committees of the American Concrete Institute and is a Fellow of both the American Concrete Institute and the American Society of Civil Engineers. (Chapter 7)

Allan R. Kenney, P.E.
President, Precast Systems Consultants, Inc. Venice, Florida. Expert on Architectural Concrete. He was co-editor of the PCI Handbook on this subject. (Chapter 20)

Steven H. Kosmatka, P.E.
Director, Construction Information Services, Portland Cement Association, Skokie, Illinois. Was formerly Manager of Research and Development, P.C.A. and is an active member of several committees of the ACI, including the Concrete Research Council and ASTM committees on cement and concrete. (Chapter 5)

Dr. Raghavan Kunigahalli
Research and Development Coordinator, Bentley Systems, Inc., Exton, Pennsylvania. Expert in computer-aided design and construction and computer automation. (Chapter 17 co-author)

Dr. V. Mohan Malhotra, P.Eng.
Program Manager, Advanced Concrete Technology Program, CANMET, Ottawa, Canada. Expert on concrete behavior, long-term effects on concrete behavior, particularly in the area of use of mineral admixtures. (Chapter 2)

Dr. Richard A. Miller, P.E.
Associate Professor of Civil Engineering, University of Cincinnati, Cincinnati, Ohio. Expert in nondestructive testing and evaluation of existing bridges. (Chapter 23)

Dr. Sidney Mindess, P.Eng.
Professor of Civil Engineering and Associate Vice President Academic, University of British Columbia, Vancouver, Canada. Expert in concrete materials behavior and in composites. (Chapter 1)

Jaime Moreno
President, Cement Technology Corp., Chicago, Illinois. Expert on cements, blending of cements, high-performance concretes and efficient mixture proportioning. (Chapter 6, Part B)

Walid M. Naja, S.E.
Principal, W.N. Structural Engineers, Belmont, California. Expert in the design and construction of concrete structures in seismically active regions, particularly low to mid-rise buildings. (Chapter 26 co-author)

Dr. Edward G. Nawy, P.E., C.Eng.
Distinguished Professor, Civil Engineering, Rutgers University—The State University of New Jersey. Expert in concrete structures and materials and author of several textbooks on reinforced and prestressed concrete design and on concrete materials and technology. (Chapters 4, 6, 22 and 27)

Dr. Randall W. Poston, P.E.
Principal of Whitlock Dalrymple Poston & Associates, Inc., Manassas, Virginia. Expert in structural design and forensic engineering, both in reinforced and prestressed concrete and in retrofit of concrete building and transportation structures. (Chapter 16)

John M. Scanlon

Senior Consultant, Wiss, Janney, Elstner Associates, Inc. and formerly Chief, Concrete Technology Division, US Army Waterways Experiment Station, Vicksburg, Mississippi. Expert in concrete behavior and technology. (Chapter 15)

Ernest K. Schrader, P.E.

President, Schrader Consulting Engineers, Walla Walla, Washington. Expert in all phases of roller-compacted concrete, mass concrete, deformation of concrete, and structural joints. (Chapter 18)

Dr. Miroslaw J. Skibniewski

Professor of Civil Engineering, Purdue University, West Lafayette, Indiana. Expert in construction automation and in technology management. (Chapter 17 co-author)

Michael M. Sprinkel

Research Manager, Virginia Transportation Research Council, Charlottesville, Virginia. Expert since the 1970s in materials and construction, particularly in public works. (Chapter 25)

Dr. David P. Whitney

Research Operations Manager, Construction Materials Research Group, University of Texas, Austin, Texas. Expert in construction materials with particular emphasis on polymers and other additives. (Chapter 3)

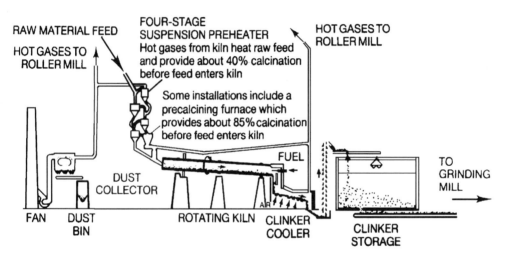

Schematic of Portland cement manufacturing process (Courtesy Prentice Hall and E.G. Nawy).

1

Concrete Constituent Materials

by
Sidney Mindess, P.Eng.
Professor, Civil Engineering and Associate Vice President Academic, University of British Columbia, Vancouver, Canada. Expert in concrete materials behavior and in composites.

1.1 Introduction

Portland cement concrete is a composite material, made by combining cement, supplementary cementing materials, aggregates, water, and chemical admixtures in suitable proportions and allowing the resulting mixture to set and harden over time. Since hardened concrete is a relatively brittle material with a low tensile strength, steel reinforcing bars and sometimes discontinuous fibers are used in structural concrete to provide some tensile load-bearing capacity and to increase the toughness of the material. In this chapter, we will deal with some of the basic constituents: cements, aggregates, water, steel reinforcement, and fiber reinforcement. Chemical admixtures and supplementary cementing materials (often referred to as mineral admixtures) will be covered in Chapter 2.

It must be emphasized that merely choosing the appropriate constituent materials for a particular concrete is a necessary, but not sufficient, condition for the production of high-quality concrete. The materials must be proportioned correctly, and the concrete must then be mixed, placed, and cured properly (Chapter 6). In addition, there must be careful quality control of every part of the concrete-making process. This requires full cooperation among the materials or ready-mixed-concrete supplier, the engineer, and the contractor.

1.2 Portland Cement

Portland cement is by far the most important member of the family of **hydraulic cements**, that is, cements which harden through chemical interaction with water. The first patent for "Portland" cement was taken out in England in 1824 by Joseph Aspdin, though it was probably not a true Portland cement; the first true Portland cements were produced about 20 years later. Since then, many improvements have been made to cement production, leading to the sophisticated, though common, cements that are now so widely available.

1.2.1 Manufacture of Portland Cement

The manufacture of Portland cement is, in principle, a simple process that relies upon the use of cheap and abundant raw materials. In short, an intimate mixture of limestone ($CaCO_3$) and clay or silt (iron-bearing aluminosilicates), to which a small amount of iron oxide (Fe_2O_3) and sometimes quartz (SiO_2) are added, is heated in a kiln to a temperature of between 1400°C and 1600°C; in this temperature range, the materials react chemically to form calcium silicates, calcium aluminates, and calcium aluminoferrites.

The cement production process is shown schematically in Figure 1.1. The raw materials, which are ground to a fineness of less than about 75 μm, are introduced at the top end of an inclined rotary

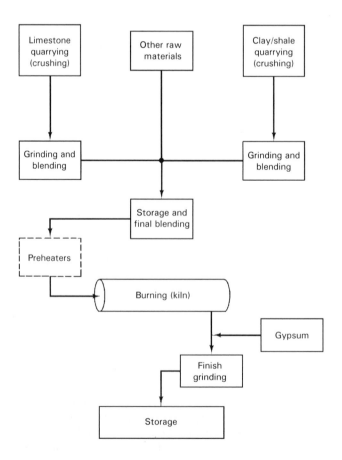

FIGURE 1.1   Schematic outline of Portland cement production.

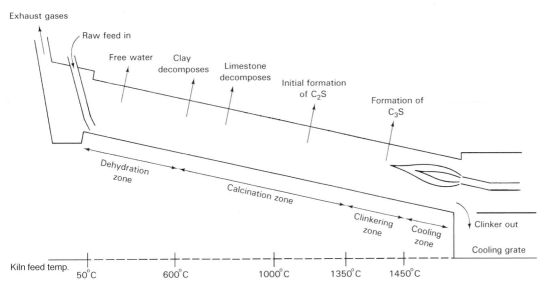

FIGURE 1.2   Schematic outline of reactions in a typical dry-process rotary cement kiln.

kiln, as shown schematically in Figure 1.2. The kiln is heated by fuel (natural gas, oil, or pulverized coal) that is injected and burnt at the kiln's lower end, with the hot gases passing up through the kiln. Thus, in a period ranging from about 20 min to 2.5 h depending on the kiln design, the raw ingredients see increasingly higher temperatures as they pass through the kiln, and a complex series of chemical reactions takes place. In the hottest part of the kiln (the **clinkering zone**), the calcium aluminates and ferrites, which have already formed, melt, and calcium silicates form in this liquid phase. As the charge in the kiln moves through the final few feet, its temperature drops rapidly, and it emerges from the kiln as **clinker**, dark colored nodules, about 6 mm to 50 mm in diameter. This is then cooled and is finally interground with gypsum ($CaSO_4 \cdot 2H_2O$), to a particle size of about 10 μm or less. The gypsum is added to control the early hydration reactions of the cement.

Because of the nature of the raw materials, there are a number of impurities that are incorporated into the cement, including MgO (from the limestone), Na_2O and K_2O (from the clay), and SO_3 (from the fuel). These have little effect on the mechanical properties of the cement but may have significant effects on its durability, as will be discussed later.

The chemical composition of Portland cement is customarily reported in terms of the **oxides** of the various elements that are present, using the shorthand notation given in Table 1.1. Using this notation, the typical compound composition of ordinary Portland cement may be given as shown in Table 1.2.

TABLE 1.1   Shorthand Notation for the Oxides in Portland Cement

Oxide	Shorthand Notation	Common Name	Typical Weight Percent in Ordinary Cement
CaO	C	Lime	63
SiO_2	S	Silica	22
Al_2O_3	A	Alumina	6
Fe_2O_3	F	Ferric oxide	2.5
MgO	M	Magnesia	2.5
K_2O	K ⎫	Alkalis	0.6
Na_2O	N ⎭		0.4
SO_3	$\overline{S}$	Sulfur trioxide	2.0
CO_2	$\overline{C}$	Carbon dioxide	—
H_2O	H	Water	—

TABLE 1.2 Typical Compound Composition of Ordinary Portland Cement

Chemical Formula	Shorthand Notation	Chemical Name	Weight Percent
$3CaO \cdot SiO_2$	C_3S	Tricalcium silicate	50
$2CaO \cdot SiO_2$	C_2S	Dicalcium silicate	25
$3CaO \cdot Al_2O_3$	C_3A	Tricalcium aluminate	12
$4CaO \cdot Al_2O_3 \cdot Fe_2O_3$	C_4AF	Tetracalcium aluminoferrite	8
$CaSO_4 \cdot 2H_2O$	$C\bar{S}H_2$	Calcium sulfate dihydrate (gypsum)	3.5

TABLE 1.3 Contribution of Cement Compounds to the Hydration of Portland Cement

Compound	Reaction Rate	Heat Liberated	Contribution to Strength
C_3S	Moderate	High	High
C_2S	Slow	Low	Low initially, high later
$C_3A + C\bar{S}H_2$	Fast	Very high	Low
$C_4AF + C\bar{S}H_2$	Moderate	Moderate	Low

TABLE 1.4 Approximate Chemical Compositions of the Principal Types of Portland Cement

ASTM Designation	Common Name	Percent by Weight					Fineness*
		C_3S	C_2S	C_3A	C_4AF	$C\bar{S}H_2$	
Type I	Ordinary	50	25	12	8	5	350
Type II	Modified	45	30	7	12	5	350
Type III	High-early strength	60	15	10	8	5	450
Type IV[†]	Low heat	25	50	5	12	4	300
Type V	Sulfate resistant	40	40	4	10	4	350

*Blaine fineness, measured in m^2/kg.

[†]Now rarely produced; replaced with blends of Portland cement and fly ash.

The characteristics of these compounds when cement is hydrated are indicated in Table 1.3. It can be seen that the two calcium silicates are primarily responsible for the strength that the cement will develop upon hydration.

By making relatively small changes in the relative proportions of raw materials, one can bring about relatively large changes in the relative proportions of the principal compounds of Portland cement. In North America, this has led to the specification of five types of Portland cement, as indicated in Table 1.4.

It is thus possible, to a considerable degree, to "tailor make" cements for particular applications, as long as the quantities required are sufficiently large to be economically feasible. For instance, special cements have been formulated for very high strength concretes and for particular durability considerations.

The hydration reactions of Portland cement do not involve the complete dissolution of the cement grains; rather, the reactions take place between water and the exposed surfaces of the cement particles. As a result, the fineness of the cement will have a considerable effect on its rate of reaction, since this will determine the surface area exposed to water. Clearly, more finely ground cements will hydrate more rapidly, but they give rise to higher rates of heat liberation during hydration, the consequences of which will be discussed later.

1.2.2 Hydration of Portland Cement

The hydration reactions that take place between finely ground Portland cement and water are highly complex, since the individual cement grains vary in size and composition. As a consequence, the resulting hydration products are also not uniform; their chemical composition and microstructural characteristics vary not only with time but also with their location within the concrete.

The basic characteristics of the hydration of Portland cement may be described as follows:

(i) As long as the individual cement grains remain separated from each other by water, the cement paste remains fluid.
(ii) The products of the hydration reactions occupy a greater volume than that occupied by the original cement grains.
(iii) As the hydration products begin to intergrow, setting occurs.
(iv) As the hydration reactions continue, additional bonds are formed between the cement grains, leading to strengthening of the system.

1.2.2.1 Chemistry of Hydration

The principal products of the hydration reactions, which are primarily responsible for the strength of concrete, are the calcium silicate hydrates that make up most of the hydrated cement. They are formed from the reactions between the two calcium silicates and water. Using the shorthand notation of Tables 1.1 and 1.2, these reactions may be written as

$$2C_3S + 7H \rightarrow C_3S_2H_4 + 3CH$$
$$2C_2S + 5H \rightarrow C_3S_2H_4 + CH \tag{1.1}$$

In reality, calcium silicate hydrate is a largely amorphous material that does not have the precise composition indicated in Eq. (1.1). It is thus more often referred to merely as C-S-H, so that no specific formula is implied. The reactions of Eq. (1.1) are highly exothermic. These reactions, and the others described below, occur first on the surfaces of the finely divided cement; as the surface layers react, water must diffuse through the hydration products to reach still unhydrated material for the reactions to proceed. The reactions will continue, at an ever-decreasing rate, either until all of the water available for hydration is used up or until all of the space available for the hydration products is filled.

In the absence of the gypsum that is interground with the Portland cement clinker, the C_3A would react very rapidly with the water, leading to early setting (within a very few minutes) of the cement, which is of course highly undesirable. In the presence of gypsum, however, a layer of **ettringite** forms on the surface of the C_3A particles, slowing down the subsequent hydration:

$$C_3A \quad + 3C\bar{S}H_2 + 26H \rightarrow C_6A\bar{S}_3H_{32}$$
$$\text{tricalcium aluminate} + \text{gypsum} + \text{water} \rightarrow \text{ettringite} \tag{1.2}$$

As the gypsum becomes depleted by this reaction, the ettringite and the C_3A react further:

$$C_6A\bar{S}_3H_{32} + \quad 2C_3A \quad + 4H \rightarrow \quad 3C_4A\bar{S}H_{12}$$
$$\text{ettringite} + \text{tricalcium aluminate} + \text{water} \rightarrow \text{monosulfoaluminate} \tag{1.3}$$

The monosulfoaluminate is thus the stable phase in concrete.

The ferrite phase is much less reactive than the C_3A, and so it does not combine with much of the gypsum. Its reaction may be written as

$$C_4AF + \quad 2CH \quad + 18H \rightarrow \quad C_4(A, F)H_{12} \quad + \quad C_2(A, F)H_8$$
$$\text{ferrite} + \quad \text{calcium} \quad + \text{water} \rightarrow \quad \text{tricalcium} \quad + \text{ferric-alumina} \tag{1.4}$$
$$\text{hydroxide} \qquad\qquad \text{aluminate hydrate} \qquad \text{hydroxide}$$

(A, F) means that A and F occur interchangeably in the so-called hexagonal hydrates, but the ratio A/F need not be the same as that in the parent compounds. These hydrates derive their name from the fact that they tend to occur in thin, hexagonal plates. The tetracalcium aluminate hydrate is structurally related to monosulfoaluminate; the ferric-alumina hydroxide is amorphous.

It must be emphasized again that the chemical formulae presented in Eqs. (1.1) to (1.4) are only approximate. There may be as much as 5% of various other "impurities" in the raw materials used to make cement (K_2O, Na_2O, MgO, etc.), and these atoms also find their way into the structure of the hydration compounds, and so the pure phases represented above are rarely found in that form. While in general, this has little effect on the mechanical properties of hardened cement or concrete, the impurities may have a considerable effect on the durability of the concrete and on interactions between the cement and modern chemical admixtures.

1.2.2.2 Development of Hydration Products

The hydration reactions described above occur at quite different rates, and so the rates of strength development and the final strengths achieved by the various hydration products vary widely (Figure 1.3). Most of the strength comes from the hydration of the calcium silicates. The C_3S hydrates more rapidly than the C_2S, and so is responsible for most of the early strength gain. The aluminate and ferrite phases hydrate quickly but contribute little to strength.

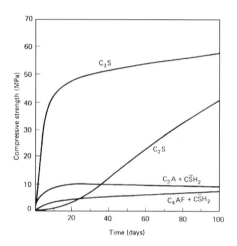

FIGURE 1.3 Compressive strength development of the pure cement compounds.

The course of the hydration of Portland cement is best described by reference to Figure 1.4 in which the hydration process is divided into five stages on the basis of the amount of heat being liberated. The first stage lasts only a few minutes; the heat liberated is due mostly to the wetting and early dissolution of the cement grains. In the second, or induction stage, which may last for several hours, there is very little hydration activity, and the cement paste remains fluid. The beginning of the hydration of C_3S marks the start of the third stage, during which both initial set and final set occur, owing to the prod-

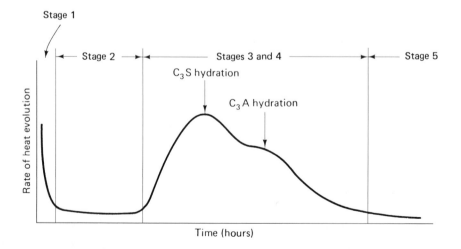

FIGURE 1.4 Rate of heat evolution during the hydration of Portland cement.

uction of the hydration products and the development of a solid microstructural "skeleton." Stage four is marked by the hydration of the C_3A after the depletion of the gypsum. Finally, in stage five, there is a slowing rate of reaction, as long as water is present, during which the skeleton developed in stage three is filled in and densified by additional hydration products.

1.2.2.3 Mechanical Properties of Hydration Products

What determines the mechanical properties of the hardened cement is not so much the chemical details of the hydration reactions, but the physical microstructure that is developed as a result of these reactions. As a continuous matrix of C-S-H is formed, there is a reduction in the porosity of the system, and it is this reduction in porosity that is largely responsible for the gain in strength with an increasing degree of hydration. Of course, in addition to the C-S-H, the hardened matrix also contains the still unhydrated residues of the cement grains, relatively large crystals of calcium hydroxide, and monosulfoaluminate crystallites, but the latter two are more important for durability than for strength considerations. Here, we will focus on the resultant porosity of the system.

Pores may exist in hydrated Portland cement over a wide range of sizes. They may generally be classified into the following size ranges:

micropores: <2.5 nm
mesopores: 2.5 nm to 100 nm (0.1 μm)
macropores: >100 nm

However, if we adopt the simplified model of pore structure first suggested by Powers (1958), it is possible to relate the strength of the hardened paste to its porosity. Powers subdivided pores into two types: gel pores, with a diameter of <10 nm, are an intrinsic part of the microstructure of the hardened paste. Capillary pores (>10 nm in diameter) represent the spaces in the hardened paste that were originally filled with mixing water and have not been completely filled by the various hydration products. The larger the amount of mixing water used, therefore, the greater the volume of capillary pores; the volume of gel pores is largely independent of the amount of mixing water. It is possible to calculate the volume fraction of the pores, and the solid fraction, in terms of two parameters: the original water/cement (w/c) ratio and the degree of hydration (α), which is the fraction of cement that is hydrated and ranges from 0 to 1. The following equations were originally determined empirically by Powers, and are still often used:

$$\text{volume of total hydration products} = 0.68\alpha \text{ cm}^3/\text{g of original cement}$$
$$\text{(including gel pores)} \tag{1.5}$$

$$\text{volume of unhydrated cement} = 0.32(1 - \alpha) \text{ cm}^3/\text{g of original cement} \tag{1.6}$$

$$\text{volume of capillary pores} = [\text{w/c} - 0.36\alpha] \text{ cm}^3/\text{g of original cement} \tag{1.7}$$

$$\text{volume of gel pores} = 0.16\alpha \text{ cm}^3/\text{g of original cement} \tag{1.8}$$

$$\text{capillary porosity (relative volume of capillary pores)} = \frac{\text{w/c} - 0.36\alpha}{\text{w/c} + 0.32} \tag{1.9}$$

These volume relationships can be seen more clearly in Figure 1.5, in terms of the degree of hydration and w/c ratio. From the above, it may be seen that the w/c ratio essentially controls the capillary porosity, which in turn controls the permeability and strength of the hardened paste. This is the basis of the w/c ratio "law" on which most mix design procedures are based. To produce high-strength, low-permeability concretes, it is thus necessary to use a low w/c ratio and to ensure a high degree of hydration by following proper curing procedures.

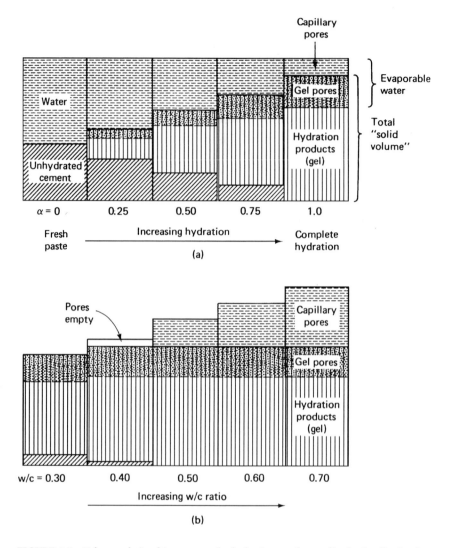

FIGURE 1.5 Volume relationships among the hydration products of hydrating Portland cement pastes: (a) Constant w/c ratio = 0.50; (b) Increasing w/c ratio ($\alpha = 1.0$).

1.3 Modified Portland Cements

Increasingly, modern concretes contain a blend of Portland cement and other cementitious materials. When other materials are added to Portland cement at the time at which the concrete is batched, they are referred to as **mineral admixtures**, which are described in detail in Chapter 2. However, there are also hydraulic cements that are produced either by forming other compounds during the burning process or by adding other materials to the clinker and then intergrinding them. The common types of such modified cements are described in the following sections.

1.3.1 Portland Pozzolan Cements

Portland pozzolan cements are blends of Portland cement and a pozzolanic material (see Chapter 2). The role of the pozzolan is to react slowly with the calcium hydroxide that is liberated during cement hydration. This tends to reduce the heat of hydration and the early strength, but can increase the ultimate strength of the material. These cements tend to be more resistant to sulfate attack and to the alkali-aggregate reaction.

1.3.2 Portland Blast-Furnace Slag Cements

Ground granulated blast-furnace slag (GGBFS), which is a byproduct of the iron and steel industry, is composed largely of lime, silica, and alumina and thus is a potentially cementitious material. In order for it to hydrate, however, it must be activated by the addition of other compounds. When the GGBFS is to be activated by lime, the lime is most easily supplied by the hydration of the Portland cement itself. Slags may be present in proportions ranging from 25% to 90%. They react slowly to form C-S-H, that is, the same product that results from the hydration of the calcium silicates. In general, because they react more slowly than Portland cement, slag cements have both lower heats of hydration and lower rates of strength gain. On the other hand, they have an enhanced resistance to sulfate attack. When the GGBFS is to be activated with calcium sulfate ($CaSO_4$), together with a small amount of lime or Portland cement, the material is known as **supersulfated** cement. This cement is available mostly in Europe, where it is used for its lower heat of hydration and its resistance to sulfate attack.

1.3.3 Expansive Cements

Expansive cements were developed to try to offset the drying shrinkage that concrete undergoes. This is particularly important when the concrete is restrained against contraction or when it is to be cast against mature concrete in repair situations. In both cases, severe cracking may occur as a result of the shrinkage. Expansive cements are based on the formation of large quantities of ettringite during the first few days of hydration. However, they are little used today, in large part because it is very difficult to control (or predict) the amount of expansion that will take place for a particular concrete formulation.

1.4 High-Alumina Cement

There are a number of non-Portland inorganic cements available, but by far the most important is high-alumina cement (also known as calcium-aluminate cement). It was developed originally for its sulfate-resistant properties but was subsequently used structurally because of its high rate of strength gain. However, structural problems because of the loss of strength that can occur in certain circumstances, including several disastrous structural collapses, have limited its use.

High alumina cement (HAC) is about 60% CA, 10% C_2S, and 5–20% C_2AS (gehlenite), with 10–25% various minor constituents. When this material hydrates, much depends on the temperature:

$$CA + H \quad \begin{array}{l} \xrightarrow{<10^\circ C} CA + H_{10} \\ \xrightarrow{10-25^\circ C} C_2AH_8 + AH_3 \\ \xrightarrow{>25^\circ C} C_3AH_6 + 2AH_3 \end{array} \qquad (1.10)$$

These reactions take place rapidly, so that HAC reaches about 75% of its ultimate strength in one day. Unfortunately, C_2AH_8 and CAH_{10} will be transformed to C_3AH_6 at temperatures above $30^\circ C$, particularly in moist conditions. This leads to a considerable loss in strength, because C_3AH_6 has a smaller solid volume than the other two calcium aluminate hydrates, causing a large increase in

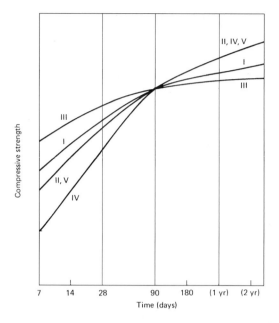

FIGURE 1.6 Relative compressive strengths of concretes made with different cements. Adapted from *Concrete Manual*, 8th ed., US Bureau of Reclamation, Denver, Colorado, 1975.

porosity. This loss in strength can be minimized if a low w/c ratio (<0.40) is used, but because of difficulties in predicting or controlling strength loss, HAC is now not used structurally. Rather, it is used primarily for refractory purposes because of its good high-temperature properties.

1.5 Performance of Different Cements in Concrete

The compositions of each of the five ASTM types of cements may vary widely from cement to cement, due to variations in locally available raw materials, kiln design, burning conditions, and so on. Their fineness may also be quite variable. As a result, their cementitious properties may also vary widely. In some regions, for instance, it may not be possible to find a commercially available cement for the production of very high strength (>100 MPa) concrete. However, since most cements contain about 75% by weight of calcium silicates and undergo the same hydration reactions, though perhaps at different rates, their ultimate performances in concrete are similar, as shown in Figure 1.6. Here, the major differences are in the rate of strength gain and in the heat of hydration.

It should be noted that the major differences between cements are more likely to be manifested in their durability and their compatibility with various admixtures, rather than in their strength. This is because durability and admixture compatibility are much more affected by the so-called "minor" constituents in cement, which can be very different, rather than by the strength-determining calcium silicates.

1.6 Water

Although the water itself is often not considered when dealing with materials that go into production of concrete, it is an important ingredient. Typically, 150 to 200 kg/m^3 of water is used. The old rule of thumb for water quality is: "If you can drink it, you can use it in concrete." However, good-quality

TABLE 1.5 Tolerable Levels of Some Impurities in Mixing Water

Impurity	Maximum Concentration, ppm	Remarks
Suspended matter (turbidity)	2000	Silt, clay, organic matter
Algae	500–1000	Entrain air
Carbonates	1000	Decrease setting times
Bicarbonates	400–1000	400 ppm for bicarbonates of Ca or Mg
Sodium sulfate	10,000	May increase early strength, but reduce
Magnesium sulfate	40,000	later strength
Sodium chloride	20,000	Decrease setting times, increase early
Calcium chloride	50,000	strength, reduce ultimate strength, and
Magnesium chloride	40,000	may lead to corrosion of reinforcing steel
Sugar	500	Affects setting behavior

concrete can be made with water that is not really potable. Indeed, more bad concrete is made by using too much drinkable water than by using the right amount of undrinkable water. The tolerable limits for various common impurities in mixing water are given in Table 1.5. When in question, the suitability of the water is determined by comparing the strength of concrete made with the suspect water to the strength of concrete made with a known "good" water.

1.7 Water/Cement Ratio

For brittle ceramic materials, including cementitious systems, the strength has been found to be inversely proportional to the porosity. Often, an exponential equation is used to relate strength to porosity, for example

$$f_c = f_{c0}e^{-kp} \qquad (1.11)$$

where f_c is the strength, f_{c0} is the "intrinsic" strength at zero porosity, p is the porosity, and k is a constant that depends on the particular system. Equations such as this do not consider the pore-size distribution, the pore shape, and whether the pores are empty or filled with water; they thus are a gross simplification of the true strength versus porosity relationship. Nonetheless, for ordinary concretes for the same degree of cement hydration, the strength does indeed depend primarily upon the porosity. Since the porosity, in turn, depends mostly upon the original w/c ratio, mix proportioning for normal-strength concretes is based to a large extent on the w/c ratio "law" articulated by Abrams in 1919: "For given materials, the strength depends only on one factor—the ratio of water to cement." This can be expressed as

$$f_c = \frac{K_1}{K_2^{(w/c)}} \qquad (1.12)$$

where K_1 and K_2 are constants, and (w/c) is the water/cement ratio by weight.

In fact, of course, given the variability in raw materials from concrete to concrete, the w/c ratio law is really a family of relationships for different mixtures. As stated by Gilkey (1961a),

> For a given cement and acceptable aggregates, the strength that may be developed by a workable, properly placed mixture of cement, aggregate, and water (under the same mixing, curing and, testing conditions) is influenced by the: (a) ratio of cement to mixing water, (b) ratio of cement to aggregate; (c) grading, surface texture, shape, strength, and stiffness of aggregate particles; and (d) maximum size of aggregate.

Thus, in some cases, simple reliance on the w/c ratio law may lead to serious errors.

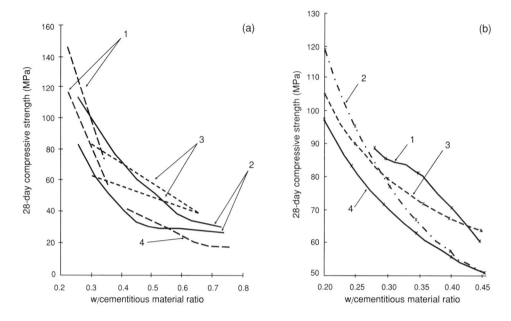

FIGURE 1.7 Water/cementitious material versus strength relationships obtained by different investigators. (a) 1, after Aitcin (1992); 2, after Fiorato (1989); 3, after Cook (1989); 4, after Canadian Portland Cement Association (1991). (b) 1, after Addis and Alexander (1990); 2, after Hattori (1979); 3, Ordinary Portland cement, after Suzuki (1987); 4, High early strength cement, after Suzuki (1987).

It should be noted that many modern concretes contain one or more mineral admixtures, which are, in themselves, cementitious to a greater or lesser degree. Therefore it is becoming more common to use the term **water/cementitious material** ratio to reflect this fact rather than the simpler water/cement ratio.

Nonetheless, for "ordinary" concretes, the w/c ratio law works well for a given set of raw materials, since the aggregate strength is generally much greater than the paste strength. However, for high-strength concretes, in which the strength-limiting factor may be the aggregate strength, or the strength of the interfacial zone between the cement and the aggregate, the w/c ratio law is more problematic. While it is, of course, necessary to use very low w/c ratios in order to achieve very high strengths, the w/c ratio versus strength relationship is not as straightforward as it is for normal concretes. Figure 1.7 shows a variety of w/cementitious material versus strength relationships obtained by a number of different investigators. It can be seen that there is a great deal of scatter in the results. In addition, the range of strengths for a given w/c ratio increases as the w/c ratio decreases, leading to the conclusion that for these concretes, the w/c ratio is not by itself a very good predictor of strength; a different w/c ratio law must be determined for every different set of materials.

1.8 Aggregates

Aggregates make up about 75% of the volume of concrete, and so their properties have a large influence on the properties of the concrete. Aggregates are granular materials, most commonly natural gravels and sands or crushed stone, though occasionally synthetic materials such as slags or expanded clays or shales are used. Most aggregates have specific gravities in the range of 2.6 to 2.7, though both heavyweight and lightweight aggregates are sometimes used for special concretes, as described later. The role of the aggregate is to provide much better dimensional stability and wear resistance; without aggregates large castings of neat cement paste would essentially self-destruct upon drying. Also, being cheaper than Portland cement, aggregates lead to the production of more economical concretes.

In general, aggregates are much stronger than the cement paste, and so (except for very high strength concretes) their exact mechanical properties are not considered to be of much importance. Similarly, they are also assumed to be completely inert in a cement matrix, though this is not always true, as will be seen in the discussion of the alkali-aggregate reaction. For ordinary concretes, the most important aggregate properties are the particle grading (or particle size distribution), shape, and porosity and possible reactivity with the cement. Of course, all aggregates should be clean, that is, free of impurities such as salt, clay, dirt, or foreign matter. As a matter of convenience, aggregates are generally divided into two size ranges: **coarse** aggregate, which is the fraction of material retained on a No. 4 (4.75 mm) sieve, and **fine** aggregate, which is the fraction passing the #4 sieve but retained on the No. 100 (0.15 mm) sieve.

1.8.1 Particle Shape and Texture

Ideally, in order to minimize the amount of cement paste required to provide adequate workability of the fresh concrete, aggregate particles for ordinary concrete should be roughly equidimensional with relatively smooth surfaces, such as most natural sands and gravels. Where natural sands and gravels are unavailable, crushed stone may be used. Crushed stone tends to have a rougher surface and to be more angular in shape. As a result, it tends to require rather more cement paste for workability. Whether using natural gravels or crushed stone, however, either flat or elongated particles should be avoided, as they will lead to workability and finishability problems.

1.8.2 Particle Grading

The **particle-size distribution** in a sample of aggregate, referred to as the **grading**, is generally expressed in terms of the cumulative percentage of particles passing (or retained on) a specific series of sieves. These distributions are most commonly shown graphically as **grading curves**. Examples of such curves are given in Figure 1.8, which shows the usual North American grading limits for fine aggregate and for a particular maximum size (38.1 mm) of coarse aggregate. Such grading limits

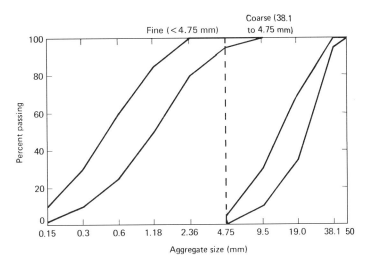

FIGURE 1.8   ASTM grading limits for fine aggregate and for coarse aggregate with a maximum particle size of 38.1 mm.

TABLE 1.6 Calculation of Fineness Modulus

Sieve Size	Weight Retained	Wt% Retained	Cumulative Wt% Retained
#4	30	3	3
#8	100	10	13
#16	195	19.5	32.5
#30	200	20	52.5
#50	260	26	78.5
#100	215	21.5	100.0
			$\sum = 279.5$

Note: Sample weight 1000 g. FM = (279.5/100) = 280.

have been determined empirically; they are intended to provide a fairly dense packing of aggregate particles, again to minimize the cement paste requirement. There is, however, no "ideal" aggregate grading that can be derived theoretically; in practice, one can provide good concrete with quite a range of aggregate gradings.

While the continuous type of grading described in Figure 1.8 is the most common, other types of grading are sometimes used for special purposes. For instance, **gap-grading** refers to a grading in which one or more of the intermediate size fractions is omitted. This is sometimes convenient when it is necessary to blend different aggregates to achieve a suitable grading. Such concretes are also prone to segregation of the fresh concrete. **No-fines** concrete is a special case of gap-graded concrete in which the fine aggregate (<4.75 mm) is omitted entirely. This produces a porous, lighter weight concrete, which may be made, for instance, to allow water to drain through it.

For fine aggregates, the particle-size distribution tends to be described by a single number, the **fineness modulus**, which is defined as

$$\mathrm{FM} = \frac{\sum(\text{cumulative \% retained on standard sieves})}{100} \qquad (1.13)$$

where the standard sieve sizes are No. 100 (0.15 mm), No. 50 (0.30 mm), No. 30 (0.60 mm), No. 16 (1.18 mm), No. 8 (2.36 mm), and No. 4 (4.75 mm). An example of the calculation of the fineness modulus is shown in Table 1.6. Normally, the FM should fall between 2.3 and 3.1 (higher values imply a coarser material). The value of the FM is required for mix design purposes. Also see Chapter 6 of this handbook, which discusses the proportioning of concrete mixtures.

1.8.3 Aggregate Moisture Content

Aggregates can hold water in two ways: absorbed within the aggregate porosity, or held on the particle surface as a moisture film. Thus, depending on the relative humidity, recent weather conditions, and location within the aggregate stockpile, aggregate particles can have a variable moisture content. However, for the purposes of mix proportioning, it is necessary to know how much water the aggregate will absorb from the mix water or how much extra water the aggregate might contribute.

Figure 1.9 schematically represents four different moisture states:

(i) ***Oven-dry (OD).*** All moisture is removed by heating the aggregates in an oven at 105°C to constant weight.

(ii) ***Air-dry (AD).*** No surface moisture, but the pores may be partially full.

(iii) ***Saturated surface dry (SSD).*** All pores are full, but the surface is completely dry.

(iv) ***Wet.*** All pores are full, and there is a water film on the surface.

Of these four states, only two (OD and SSD) correspond to well-defined moisture conditions, and so either can be used as a reference point for calculating the moisture contents. In the following,

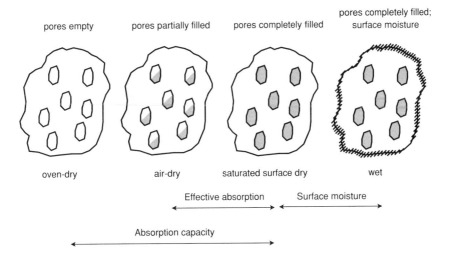

FIGURE 1.9 Moisture states of aggregate.

the SSD state will be so used. Now, to determine how much water the aggregate may add to or take from the mixing water, three further quantities must be defined.

(i) The **absorption capacity** (AC) represents the maximum amount of water the aggregates can absorb. This is, from Figure 1.9, the difference between the SSD and OD states, expressed as a percentage of the OD weight:

$$AC = \frac{W_{SSD} - W_{OD}}{W_{OD}} \times 100\% \qquad (1.14)$$

where W represents weight. It should be noted that, for most common aggregates, the absorption capacities are of the order of 0.5–2.0%. Absorption capacities greater than 2% are often an indication that the aggregates may have potential durability problems.

(ii) **Effective absorption** (EA) refers to the amount of water required for the aggregate to go from the AD to the SSD state:

$$EA = \frac{W_{SSD} - W_{AD}}{W_{SSD}} \times 100\% \qquad (1.15)$$

To calculate the weight of water absorbed (W_{abs}) by the aggregate in the concrete mix

$$W_{abs} = (EA)\, W_{agg} \qquad (1.16)$$

(iii) **Surface moisture** (SM) represents water in excess of the SSD state, held on the aggregate surface:

$$SM = \frac{W_{wet} - W_{SSD}}{W_{SSD}} \times 100\% \qquad (1.17)$$

Thus the extra water added to the concrete from the wet aggregates will be

$$W_{add} = (SM)\, W_{agg} \qquad (1.18)$$

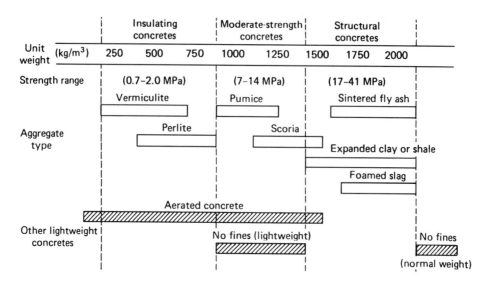

FIGURE 1.10   Classification of lightweight concretes.

1.8.4 Lightweight Aggregates

Lightweight aggregates, which can be either natural or synthetic materials, are characterized by a high internal porosity. While ordinary concrete has a unit weight of about 2,300 kg/m^3, lightweight concretes with unit weights as low as 120 kg/m^3 can be produced, though of course they are accompanied by a significant decrease in concrete strength. Natural lightweight aggregates include pumice, scoria, and tuff. However, most lightweight aggregates are synthetically produced. The most common such lightweight aggregates are made from expanded clay, shale, or slate. The raw material is either crushed to the desired size (or ground and then pelletized) and then heated to 1000°–1200°C. At these temperatures, the material will **bloat** (or puff up) from the rapid generation of gas produced by the combustion of the small amounts of organic material that these particles generally contain. (The process is similar to that in popping popcorn.) Other materials, such as volcanic glass (perlite), calcium silicate glasses (slags), or vermiculite can similarly be bloated. Lightweight aggregates tend to be angular and irregular in shape and can also be quite variable. They will also tend to have high porosities, leading to a considerable potential for absorbing water from the mix. Hence, mix design with lightweight aggregates is much more of a "trial-and-error" procedure than with normal-weight aggregates.

Lightweight concretes made with these aggregates may be classified as shown in Figure 1.10. The properties of different types of lightweight concrete are described in Table 1.7. It should be noted that despite the high porosity and relative weakness of the aggregate there is no problem in reaching

TABLE 1.7   Properties of Some Lightweight Concretes

Type of Lightweight Concrete	Type of Aggregate	Aggregate Density, kg/m^3	Concrete Density, kg/m^3
Aerated	—	—	400–600
Partially compacted	Expanded vermiculite and perlite	5–240	400–1150
	Foamed slag	480–960	960–1500
	Sintered pulverized-fuel ash	640–960	1100–1300
	Expanded clay or shale	560–1040	950–1200
Structural lightweight aggregate concrete	Foamed slag	480–960	1650–2050
	Sintered pulverized-fuel ash	640–960	1350–1750
	Expanded clay or shale	560–1040	1350–1850

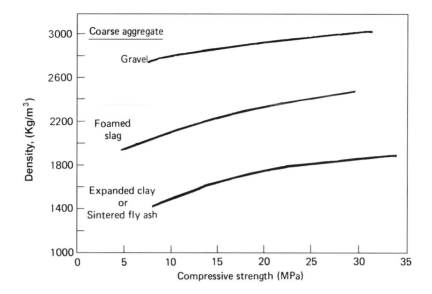

FIGURE 1.11 Average density versus compressive strength relationships. Adapted from A. Short and W. Kinniburgh, *Lightweight Concrete*, 3rd ed., Applied Science Publishers, London, 1978.

strengths as high as 40 MPa. To do this, lower w/c ratios are required for lightweight concretes than are required for ordinary concretes of the same strength.

Although there is a general relationship between strength and density of the concrete, it can be seen from Figure 1.11 that this relationship depends on the particular aggregate used.

1.8.5 Heavyweight Aggregates

Heavyweight aggregates are used to make heavyweight concretes, with units weights ranging from about 2900 to 6000 kg/m^3. Such concretes are used primarily for radiation shielding, though they are sometimes used to make counterweights as well. Natural heavyweight aggregates include materials such as goethite, limonite, barite, illmenite, magnetite and hematite, with specific gravities (SG) ranging from about 3.5 up to about 5.3, leading to concretes with unit weights up to about 4100 kg/m^3. For higher unit weight concretes, synthetic materials such as ferrophosphorous (SG 5.8–6.8) or scrap iron steel punchings (SG 6.2–7.8) can be used, with resulting concrete unit weights of up to 6100 kg/m^3.

High-density aggregates are good attenuators of gamma rays and of fast neutrons, hence their use in radiation shielding. However, while heavyweight concretes can be proportioned in much the same way as ordinary concretes, the aggregates tend to be harsh and have a tendency to segregate from the rest of the mix. As a result, both higher than usual cement contents and a higher ratio of fine to coarse aggregates are recommended.

1.8.6 Aggregate Durability

Though it is generally assumed that aggregates are inert in concrete, this is often not true, and so their durability must also be considered.

1.8.6.1 Soundness

Soundness refers to the ability of aggregates to withstand cyclic volume changes due to wetting and drying, or freezing and thawing, without deteriorating. Rocks that are susceptible to wetting and drying cycles are rare, and so soundness generally refers to freeze-thaw resistance. Aggregates will deteriorate if high internal stresses are developed when water absorbed within the aggregate freezes. This deterioration, in turn, depends on the size, porosity, permeability, and the degree of saturation of the aggregate. For most aggregates, the critical size above which unsoundness develops is greater than the maximum size normally used in concrete. However, for some sedimentary rocks (chert, graywacke, sandstone, shale, and poorly consolidated limestone), the critical size can be less than 25 mm. The most susceptible rocks are those that have a relatively high absorption (porosity), greater than 2%, combined with a very fine pore structure (low permeability) so that the freezing water cannot easily be expelled from the aggregate. If aggregates are unsound, this can lead to surface "pop-outs," and to D-cracking in pavements and slabs.

1.8.6.2 Alkali-Aggregate Reaction

A large number of failures have resulted from expansions caused by the reactions between certain types of siliceous aggregates and the alkalis (K_2O and Na_2O) contained in the cement. The types of rocks that are likely to participate in these reactions include siliceous limestones, cherts, shale, flint, volcanic glasses, opaline rocks, quartzite, sandstone, and some granites and schists. In some cases, these expansive reactions, which are first manifested by extensive surface cracking, occur within a few years after the concrete has been cast. In other cases, however, damage may not appear until 15 to 20 years after construction.

The factors that control the rate and extent of the alkali-aggregate reaction are (i) the nature of the reactive silica, (ii) the amount of reactive silica, (iii) the particle size of the reactive material, (iv) the amount of alkali available, and (v) the amount of available moisture. Regardless of the specifics of the cement and the aggregate, however, the alkali-silica reaction progresses through the steps indicated in Table 1.8. Step 1 is controlled by the alkali content of the cement. Any real control of the reaction potential must be taken at this step, either by avoiding reactive aggregates or by keeping the alkali content of the cement quite low ($Na_2O + 0.6\ K_2O \leq 0.60$). Step 2 depends on the exact form of the silica and determines the rate at which the reaction will take place. It is at Step 3 that the damage occurs as the alkali silicate gel imbibes water and swells, cracking the matrix. Step 4 mostly produces the visible signs of the reaction. It should be noted that Steps 3 and 4 will not occur if the concrete is kept dry.

If low-alkali cements are not available in a particular region (because the raw materials used to produce the cement are themselves too high in alkali content), a useful strategy is often to replace some of the cement with a pozzolanic material (such as fly ash) that contains no alkalis at all, in order to reduce the total alkali content of the cementitious materials to an acceptable level. Though pozzolanic materials contain silica in a reactive form, the silica is very finely divided and reacts very quickly with the alkalis. Harmful expansions do not occur because the resulting alkali-silicate gel is distributed throughout the cementitious matrix, and so the expansions are distributed. The worst case is to have relatively small volumes of reactive aggregates in large pieces scattered throughout the concrete; this localizes and concentrates the expansions, leading to severe cracking.

TABLE 1.8   The Alkali-Silica Reaction

Reaction Step	Reaction	Consequences
1	Release of alkali ions during cement hydration	Increased alkalinity of water in the pores (pore solution)
2	Initial hydrolysis of the reactive silicates in the alkaline pore solution: $K(Na)OH + SiO_2 \rightarrow K_2O(Na_2O)\text{-}SiO_2\text{-}H_2O$, amorphous alkali silicate gel	Aggregate integrity destroyed
3	Alkali silicate gel swells as it imbibes water	Localized cracking
4	Liquefaction of alkali silicate gel as it imbibes more moisture	Liquid gel is expelled through the cracks

1.8.6.3 Alkali-Carbonate Reaction

Expansive reactions can also occur between the alkalis in the cement and certain dolomitic lime-stones ($MgCO_3/CaCO_3$) that contain some clay. Not all dolomitic limestones are subject to this reaction; those that are have the following features: (i) very small crystals of $MgCO_3$, (ii) presence of considerable fine-grained calcite, (iii) abundant interstitial clay, and (iv) dolomite and calcite crystals uniformly distributed in the clay matrix. The expansive reaction is

$$CM\overline{C}_2 + NH \text{ (or KH)} \rightarrow MH + C\overline{C} + N\overline{C}$$

$$\text{dolomite} + \text{from cement} \rightarrow \text{brucite} + \text{calcite} + \text{soluble carbonates}$$

(1.19)

However, this reaction is still not fully understood.

The alkali-carbonate reaction can also be controlled by keeping the alkali content of the cementitious material very low (<0.40%). Unlike the case of the alkali-silica reaction, however, pozzolanic materials in this case cannot control the alkali-carbonate reaction.

1.9 Reinforcement

1.9.1 General

Plain concrete is a brittle material, with low tensile strength and strain capacities. Hence reinforcement has to be used to balance this deficiency. Main bar reinforcement is used in tensile zones to enhance the capacity of concrete elements so that the structural beam or slab will be able to withstand high loads and deform adequately without brittle failure. Chapter 27 of this handbook details the ACI 318-95 design equations for proportioning reinforced concrete members.

Another method that can increase the tensile capacity of plain concrete within a structural bar-reinforced section is the use of fiber reinforcement. There has been a steady increase since the early 1960s in the use of **fiber-reinforced concrete** (FRC) and, lately, in the use of **fiber-reinforced plastics** (FRP).

1.9.2 Fiber Reinforcement

The role of the fibers is not to increase strength, though modest strength increases may occur. Rather, the role of discrete, discontinuous, randomly oriented fibers is to bridge the cracks that develop in concrete either as it is subjected to environmental changes (such as drying) or as it is loaded. If the fibers are strong enough, stiff enough, present in sufficient quantity, and develop sufficient bonds with the cementitious matrix, they will serve to keep the crack widths small and will permit the FRC to withstand significant stresses over a relatively large strain capacity in the post-cracking (or strain softening) stage. In other words, by bridging the cracks that develop, the fibers can provide a considerable amount of post-cracking "ductility." This is often referred to as **toughness**. The toughness may be defined in terms of the area under the load versus the deflection curve.

The stress is transferred from the matrix to the fibers in two ways:

1) At early stages of loading, the stress is transmitted from the matrix to the fibers through an elastic shear-transfer mechanism, which, however, has only a small effect on the limit of proportionality or first crack stress of the FRC.

2) After the initial cracking, and at more advanced stages of loading, there is debonding along the fiber-matrix interface, and frictional slip becomes the process controlling the stress transfer. The degree of frictional slip affects both the ultimate strength and strain of the FRC. At this stage, the fibers may increase the strength of the FRC by transferring loads and stresses across the cracked matrix. In addition, and more importantly, they may increase the toughness of the FRC by providing energy absorption mechanisms, related to the pull-out processes of

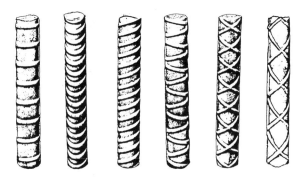

FIGURE 1.12   Various forms of ASTM-approved bars.

the fibers. Chapter 22 of this handbook gives an in-depth treatment of this subject for both FRC and FPC concretes.

1.9.3 Steel Reinforcement

Typically, steel reinforcement used in concrete has yield strengths in the range of 275 to 550 MPa (40,000 to 80,000 psi). For all these steels, the modulus of elasticity is in the range of 200×10^6 MPa (29×10^6 psi). The bar sizes in the United States range from No. 3 to No. 18 deformed bars. In slabs, the use of welded wire fabric is more prevalent.

Tables 1.9 and 1.10a and 1.10b give the properties of most commonly used bars for concrete reinforcement. Table 1.11 gives standard wire fabric reinforcement for one-way and two-way slabs and plates. These tables are derived from the ASTM Standards. The bar deformations shown in Figure 1.12 are in accordance with ASTM Standard A616.

One of the problems resulting from permeability of concrete to moisture is the corrosion of the reinforcement. Corrosion can lead to a reduction in the effective cross-sectional area of the steel and a spalling of the concrete above the steel, both of which can lead to severe damage or even failure of a structure or structural element.

TABLE 1.9   Reinforcement Grades and Strengths

1982 Standard Type	Minimum Yield Point or Yield Strength, f_y, psi	Ultimate Strength f_u, psi
Billet steel (A615)		
Grade 40	40,000	70,000
Grade 60	60,000	90,000
Axle steel (A617)		
Grade 40	40,000	70,000
Grade 60	60,000	90,000
Low-alloy steel (A706)		
Grade 60	60,000	80,000
Deformed wire		
Reinforced	75,000	85,000
Fabric	70,000	80,000
Smooth wire		
Reinforced	70,000	80,000
Fabric	65,000, 56,000	75,000, 70,000

Source: Nawy, E.G. 1996. *Reinforced Concrete—A Fundamental Approach*, 840 pp. Prentice Hall, Upper Saddle River, NJ.

TABLE 1.10a   Weight, Area, and Perimeter of Individual Bars

Bar Designation Number	Weight Per Foot, lb	1982 Standard Nominal Dimensions		
		Diameter, d_b, in (mm)	Cross-Sectional Area, A_b, in^2	Perimeter, in
3	0.376	0.375 (9)	0.11	1.178
4	0.668	0.500 (13)	0.20	1.571
5	1.043	0.625 (16)	0.31	1.963
6	1.502	0.750 (19)	0.44	2.356
7	2.044	0.875 (22)	0.60	2.749
8	2.670	1.000 (25)	0.79	3.142
9	3.400	1.128 (28)	1.00	3.544
10	4.303	1.270 (31)	1.27	3.990
11	5.313	1.410 (33)	1.56	4.430
14	7.65	1.693 (43)	2.25	5.32
18	13.60	2.257 (56)	4.00	7.09

TABLE 1.10b   ASTM Standard Metric Reinforcing Bars

Bar Size Designation Number	Nominal Dimensions		
	Mass, kg/m	Diameter, mm	Area, mm^2
10 M	0.785	11.3	100
15 M	1.570	16.0	200
20 M	2.355	19.5	300
25 M	3.925	25.2	500
30 M	5.495	29.9	700
35 M	7.850	35.7	1000
45 M	11.775	43.7	1500
55 M	19.625	56.4	2500

Source: Nawy, E.G. 1996. *Reinforced Concrete—A Fundamental Approach*, 840 pp. Prentice Hall, Upper Saddle River, NJ.

Note: ASTM A615M Grade 300 is limited to size No. 5, 10 M through No. 20 M, otherwise grades 400 or 500 MPa for all the sizes. Check availability with local suppliers for No. 45 M and 55 M.

The mechanisms of corrosion in steel are by now well known. An isolated steel bar will rust spontaneously (Figure 1.13) because different areas of the bar may have different electrochemical potentials and thus set up anode-cathode pairs, with corrosion occurring in localized anodic areas. In high-quality concrete, however, steel corrosion should not occur, even in the presence of moisture and oxygen. This is because the highly alkaline concrete (pH of about 12 to 12.5) causes a passive oxide film to form on the surface of the steel, which prevents corrosion. Only when the pH falls below 11 will the passive iron-oxide layer be destroyed, permitting rusting to occur. There are two ways in which the pH may be reduced:

1) The calcium hydroxide that is responsible for the high pH can be converted to calcium carbonate by atmospheric carbonation. For good concrete, the depth of carbonation rarely exceeds 25 mm, and so adequate cover (25 mm to 40 mm) over the reinforcement should provide adequate protection. For high-permeability concretes, or very severe exposures, a thicker cover must be used. Of course, if there are cracks in the concrete, perhaps due to drying shrinkage, CO_2 may penetrate to the steel and initiate corrosion. However, it has been found empirically that if the crack widths at the surface of the concrete can be kept below 0.025 mm, by suitable reinforcing details, then the rate of CO_2 diffusion will be slow enough to avoid extensive corrosion.

TABLE 1.11   Standard Wire Reinforcement

W&D Size		U.S. Customary			Area in^2/ft of Width for Various Spacings						
		Nominal Diameter,	Nominal Area,	Nominal Weight,	Center-to-Center Spacing, in						
Smooth	Deformed	in	in^2	lb/ft	2	3	4	6	8	10	12
W31	D31	0.628	0.310	1.054	1.86	1.24	0.93	0.62	0.465	0.372	0.31
W30	D30	0.618	0.300	1.020	1.80	1.20	0.90	0.60	0.45	0.366	0.30
W28	D28	0.597	0.280	0.952	1.68	1.12	0.84	0.56	0.42	0.336	0.28
W26	D26	0.575	0.260	0.934	1.56	1.04	0.78	0.52	0.39	0.312	0.26
W24	D24	0.553	0.240	0.816	1.44	0.96	0.72	0.48	0.36	0.288	0.24
W22	D22	0.529	0.220	0.748	1.32	0.88	0.66	0.44	0.33	0.264	0.22
W20	D20	0.504	0.200	0.680	1.20	0.80	0.60	0.40	0.30	0.24	0.20
W18	D18	0.478	0.180	0.612	1.08	0.72	0.54	0.36	0.27	0.216	0.18
W16	D16	0.451	0.160	0.544	0.96	0.64	0.48	0.32	0.24	0.192	0.16
W14	D14	0.422	0.140	0.476	0.84	0.56	0.42	0.28	0.21	0.168	0.14
W12	D12	0.390	0.120	0.408	0.72	0.48	0.36	0.24	0.18	0.144	0.12
W11	D11	0.374	0.110	0.374	0.66	0.44	0.33	0.22	0.165	0.132	0.11
W10.5		0.366	0.105	0.357	0.63	0.42	0.315	0.21	0.157	0.126	0.105
W10	D10	0.356	0.100	0.340	0.60	0.40	0.30	0.20	0.15	0.12	0.10
W9.5		0.348	0.095	0.323	0.57	0.38	0.285	0.19	0.142	0.114	0.095
W9	D9	0.338	0.090	0.306	0.54	0.36	0.27	0.18	0.135	0.108	0.09
W8.5		0.329	0.085	0.289	0.51	0.34	0.255	0.17	0.127	0.102	0.085
W8	D8	0.319	0.080	0.272	0.48	0.32	0.24	0.16	0.12	0.096	0.08
W7.5		0.309	0.075	0.255	0.45	0.30	0.225	0.15	0.112	0.09	0.075
W7	D7	0.298	0.070	0.238	0.42	0.28	0.21	0.14	0.105	0.084	0.07
W6.5		0.288	0.065	0.221	0.39	0.26	0.195	0.13	0.097	0.078	0.065
W6	D6	0.276	0.060	0.204	0.36	0.24	0.18	0.12	0.09	0.072	0.06
W5.5		0.264	0.055	0.187	0.33	0.22	0.165	0.11	0.082	0.066	0.055
W5	D5	0.252	0.050	0.170	0.30	0.20	0.15	0.10	0.075	0.06	0.05
W4.5		0.240	0.045	0.153	0.27	0.18	0.135	0.09	0.067	0.054	0.045
W4	D4	0.225	0.040	0.136	0.24	0.16	0.12	0.08	0.06	0.048	0.04
W3.5		0.211	0.035	0.119	0.21	0.14	0.105	0.07	0.052	0.042	0.035
W3		0.195	0.030	0.102	0.18	0.12	0.09	0.06	0.045	0.036	0.03
W2.9		0.192	0.029	0.098	0.174	0.116	0.087	0.058	0.043	0.035	0.029
W2.5		0.178	0.025	0.085	0.15	0.10	0.075	0.05	0.037	0.03	0.025
W2		0.159	0.020	0.068	0.12	0.08	0.06	0.04	0.03	0.024	0.02
W1.4		0.135	0.014	0.049	0.084	0.056	0.042	0.028	0.021	0.017	0.014

Source: Nawy, E.G. 1996. *Reinforced Concrete—A Fundamental Approach*, 840 pp. Prentice Hall, Upper Saddle River, NJ .

2) Chloride ions have the capacity to destroy the passive oxide layer even at high pH, and relatively little chloride is required to initiate corrosion. Chloride ions may enter concrete from $CaCl_2$ used as an accelerating admixture, from seawater, and from the deicing salts commonly used on bridge decks and pavements. This may be particularly severe for prestressing steel, since in this case **stress corrosion** may occur. What ever the source of the Cl^- ions, once a critical concentration of chlorides develops at the level of the steel (0.6 to 1.2 kg Cl^-/m^3), corrosion will occur.

Several strategies to combat steel corrosion have been developed, but corrosion remains one of the principal concrete construction problems. The most common techniques at this time (apart from making more-impermeable high-performance concrete) are the following:

1) Suppression of electrochemical corrosion by using **cathodic protection**. In

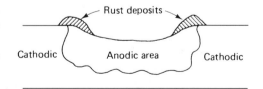

FIGURE 1.13   Schematic of rusting in reinforcing steel.

this technique, an electric current is applied to the rebars (which must be connected electrically) in the opposite direction to the current flow in spontaneous corrosion. This makes the steel cathodic, preventing corrosion. This technique has been used successfully but is difficult to apply.

2) Various corrosion inhibiting admixtures are now available commercially, which serve to maintain the passive oxide film.

3) Epoxy-coated rebars are becoming more and more common. There are two problems with this technique: (i) if the bars are scratched during handling, or are bent too severely, the epoxy coating may be damaged, exposing the steel to corrosion; (ii) the epoxy-coated bars develop substantially less bond with the concrete. If this is not recognized, severe structural cracking can occur. New design techniques are now being developed for epoxy coated bars to take the reduction in bond into account. More efficient deformation patterns are also being developed.

1.10 Durability Considerations

If concrete is properly designed for the environment to which it is to be exposed, and is properly placed and cured, it should last for many decades without costly repairs. This is particularly the case for the modern generation of high-performance concretes. However, concrete is potentially vulnerable to both chemical attack and physical attack. Several of these durability problems (alkali-aggregate reaction and corrosion of reinforcing steel) have already been discussed, while freeze-thaw durability will be discussed in Chapter 4 of this handbook. Here, the major consideration is primarily chemical attack.

The single parameter that has the largest influence on durability is the porosity of the concrete (governed by the w/c ratio), since damage due to surface attack of concrete occurs slowly. A dense, well-cured concrete has a permeability similar to that of low-porosity rocks, despite the much higher porosity of the paste, because in such concretes the porosity is discontinuous. This means that, instead of the bulk flow that would occur through continuous capillary pores, water can flow only by a much slower molecular diffusion process through micropores. Permeability can be further reduced by the use of silica fume or blast-furnace slag, which help to create an even more discontinuous pore structure.

1.10.1 Leaching and Efflorescence

Efflorescence consists of salts that are leached out of the concrete by water and then are crystallized on the concrete surface by the subsequent evaporation of the water. Efflorescence due to the leaching of calcium hydroxide from concrete, and its subsequent carbonation (to $CaCO_3$) on the concrete surface, is frequently seen as white deposits on the underside of floor slabs in parking garages. Efflorescence itself is primarily an aesthetic rather than a durability problem, but it does indicate that there is substantial leaching in the concrete, which will lead to an increase in porosity and permeability.

1.10.2 Sulfate Attack

Sulfates are often present in groundwaters and in seawater. Chemical attack by sulfates occurs when sulfate ions penetrate into the concrete and react to form gypsum. This is followed by the formation of ettringite, which is accompanied by a volume expansion sufficient to cause cracking of the concrete. This process can continue to complete destruction of the concrete. The principal methods of preventing sulfate attack are to reduce the permeability of the concrete by reducing the w/c ratio and/or by using a slag or pozzolanic mineral admixture. The use of a sulfate-resistant cement (Type V, with $\leq 5\%$ C_3A) will prevent the damaging formation of ettringite.

1.10.3 Acid Attack

Since concrete is an alkaline material (pH $\cong$ 12.5), it is susceptible to attack by acids, such as might occur in industrial wastes or near some mining operations. The problem is particularly severe when the acidity is high (pH < 4) and there is water flow. Any acid that forms soluble calcium salts on contact with concrete is particularly aggressive (hydrochloric, nitric, acetic, sulfuric, and lactic acids). However, for acids that form an insoluble salt (phosphoric, tannic acids), the precipitation of their salts within the capillary pores will slow or even prevent further deterioration.

References

Addis, B.J. and Alexander, M.G. 1990. A method of proportioning trial mixes for high-strength concrete. *High Strength Concrete, Second International Symposium*, ACI SP-121, pp. 287–308. American Concrete Institute, Farmington Hills, MI.

Canadian Portland Cement Association. 1991. *Design and Control of Concrete Mixtures*, 213 pp. CPCA, Ottawa, Canada.

Cook, J.E. 1989. 10,000 psi concrete. *Concrete Int.* 11(10):67–75.

Fiorato, A.E. 1989. PCA research on high-strength concrete. *Concrete Int.* 11(4):44–50.

Gilkey, H.J. 1961a. Water–cement ratio versus strength—Another look. *ACI J.* 57(10):1287–1312.

Gilkey, H.J. 1961b. Discussion of water–cement ratio versus strength—Another look. *ACI J.* 58(6):1851–1878.

Hattori, K. 1979. Experiences with Mighty superplasticizer in Japan. *Superplasticizers in Concrete*, ACI SP-62, pp. 37–66. American Concrete Institute, Farmington Hills, MI.

Nawy, E. G. 1996. *Reinforced Concrete—A Fundamental Approach*, 840 pp. Prentice Hall, Upper Saddle River, NJ.

Powers, T.C. 1958. Structure and physical properties of hardened Portland cement paste. *J. Am. Ceram. Soc.* 4(1):1–6.

Short, A. and Kinniburgh, W. 1978. *Lightweight Concrete*, 3rd ed. Applied Sceince Publishers, London.

Suzuki, T. 1987. Experimental studies on high strength superplasticized concrete. *Utilization of High Strength Concrete, Symposium Proceedings*, pp. 53–54. Tapis Publishers, Trondheim, Norway.

(b)

(a)

(a) Society Center, Cleveland, Ohio Composite Frame (Courtesy Portland Cement Association); (b) Scotia Plaza, Toronto. 68 story office tower, 1000 psi silica fume concrete (Courtesy Portland Cement Association).

2

Mineral Admixtures

by
V.M. Malhotra, P.Eng.
Program Manager, Advanced Concrete Technology Program, CANMET, Ottawa, Canada. Expert on concrete behavior, long-term effects on concrete behavior, particularly in the area of use of mineral admixtures.

2.1 Fly Ash

2.1.1 Introduction

Fly ash is a byproduct of the combustion of pulverized coal in thermal power plants. A dust-collection system removes the fly ash, as a fine particulate residue, from combustion gases before they are discharged into the atmosphere (Figure 2.1).

The types and relative amounts of incombustible matter in the coal used determine the chemical composition of fly ash. More than 85% of most fly ashes comprise chemical compounds and glasses formed from the elements silicon, aluminum, iron, calcium, and magnesium. Generally, fly ash

0-8493-2666-4/97/$0.00+$.50

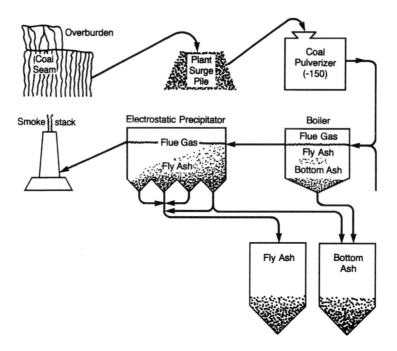

FIGURE 2.1   A schematic of a fossil fuel plant.

from the combustion of subbituminous coals contains more calcium and less iron than fly ash from bituminous coal; also, fly ash from subbituminous coals contains very little unburned carbon. Plants that operate only intermittently (peak-load stations) and that burn bituminous coals produce the largest percentage of unburned carbon. Fly-ash particles are typically spherical, ranging in diameter from <1 μm up to 150 μm.

Fly ashes exhibit pozzolanic activity. The American Society for Testing and Materials (ASTM) [ASTM, 1975] defines a pozzolan as "a siliceous or siliceous and aluminous material which in itself possesses little or no cementitious value but which will, in finely divided form and in the presence of moisture, chemically react with calcium hydroxide at ordinary temperature to form compounds possessing cementitious properties." Fly ashes contain metastable alumino-silicates that will react with calcium ions, in the presence of moisture, to form calcium silicate hydrates.

The term fly ash was first used in the electrical power industry around 1930. The first comprehensive data on the use of fly ash in concrete in North America were reported by Davis et al. (1937). The first major practical application was reported in 1948 with the publication of the United States Bureau of Reclamation's data on the use of fly ash in the construction of the Hungry Horse Dam. Worldwide acceptance of fly ash as a component of concrete slowly followed these early efforts, but interest was particularly noticeable in the wake of the rapid increases in energy costs (and hence cement costs) that occurred during the 1970s.

In recent years, it has become evident that fly ashes differ in significant and definable ways that reflect their combustion and, to some extent, their origin. The ASTM recognizes two general classes of fly ash:

- Class C, normally produced from lignite or subbituminous coals; and
- Class F, normally produced from bituminous coals.

In recent years, several publications have become available that discuss in detail the properties and use of fly ash in concrete [ASTM, 1978; Berry and Malhotra, 1978; Malhotra and Mehta, 1996; Malhotra and Ramezanianpour, 1994], and Table 2.1 shows the estimated production and use of coal ash in major coal-using countries [Malhotra and Ramezanianpour, 1994].

TABLE 2.1   Coal Ash Production and Use in Major Coal-Using Countries

Country	Fly Ash, kt/year	Coarse Ash, kt/year	Total Ash, kt/year	Use, kt/year	Percentage Use	Year
Australia	7,050	850	7,900	800	10	1990
Belgium	930	160	1,090	795	73	1989
Canada	3,830	1,420	5,250	1,575	30	1987
France	2,200	405	2,605	1,300	50	1987
Germany	7,480	4,120	11,600	6,465	56	1989
Italy	1,300	135	1,435	900	63	1988
Japan	3,480	445	3,925	1,920	49	1989
Spain	7,390	1,305	8,695	1,220	14	1987
United Kingdom	9,950	2,590	12,540	6,120	49	1989
United States	48,430	16,750	65,190	15,895	24	1989
China	—	—	62.500	16,200	26	1989
Czechosolvakia	—	—	18,100	1,400	8	1989
East Germany (former GDR)	—	—	19,100	7,200	38	1989
Hungary	—	—	4,100	1,100	27	1987
India	—	—	40,000	6,750	17	1991
Poland	—	—	29,500	4,500	15	1989
Romania	—	—	27,000	700	3	1989
Former Soviet Union	—	—	125,000	11,500	9	1989
Others	—	—	116,470	3,660	3	1989

Source: Malhotra, V.M. and Ramezanianpour, A.A. 1994. Fly ash in concrete. *CANMET, MSL 94-45 (IR)*, 307 pp. Energy, Mines and Resources Canada, Ottawa.

2.1.2 Physical, Chemical, and Mineralogical Properties of Fly Ash

2.1.2.1 Physical Properties

Fly ash is a fine-grained material consisting mostly of spherical, glassy particles. Some ashes also contain irregular or angular particles. The size of particles varies depending on the sources. Some ashes may be finer or coarser than Portland cement particles. Figures 2.2 and 2.3 show the scanning electron microscope (SEM) micrographs of polished sections of subbituminous and lignite fly ashes [Malhotra, 1983]. Figure 2.4 shows a secondary electron SEM image of bituminous fly-ash particles. Some of these particles appear to be solid, whereas other larger particles appear to be portions of thin, hollow spheres containing many smaller particles.

2.1.2.2 Fineness

Dry- and wet-sieving methods are commonly used to measure the fineness of fly ashes. ASTM designation C 311-77 recommends determining the amount of the sample retained after it is wet sieved on a 45-μm sieve, in accordance with ASTM method C 430, except that a representative sample of the fly ash or natural pozzolan is substituted for hydraulic cement in the determination. Dry sieving

FIGURE 2.2   SEM micrograph of a subbituminous ash. (Backscattered electron image of a polished section of the dispersed sample.) From Malhotra and Ramezanianpour (1994).

on a 45-μm sieve can be performed according to a method established at Canada Center for Mineral and Energy Technology (CANMET) [Malhotra and Wallace, 1993]. Table 2.2 shows the fineness of 11 fly ashes as determined by wet and dry sieving.

Particle-size distribution of fly ash can be determined by various means, such as X-ray sedigraph, laser particle-size analyzer, and coulter counter. In some cases, agglomeration of a number of small particles may form a large particle. In most cases, fly ashes contain particles of >1-μm diameter. Mehta (1994), using an X-ray sedimentation technique, reported particle-size distribution data for several U.S. fly ashes. Mehta found that high-calcium fly ashes were finer than low-calcium fly ashes, and he related this difference to the presence of larger amounts of alkali sulphates in high-calcium fly ashes.

2.1.2.3 Specific Surface

The specific surface of fly ash, which is the area of a unit of mass, can be measured by various techniques. The most common technique is the Blaine specific-surface method, which measures the resistance of compacted particles to air flow. ASTM C 204 describes this method for measurement of the surface area of Portland cement.

2.1.2.4 Specific gravity

The specific gravity of hydraulic cements is determined according to ASTM C 188. This method can also be used to determine the specific gravity of fly ashes. If fly ashes contain water-soluble compounds, the use of a nonaqueous solvent, instead of water, is recommended.

The specific gravity of different fly ashes varies over a wide range. In the CANMET investigation of 11 fly ashes [Carette and Malhotra, 1986], the specific gravity ranged from a low value of 1.90 for a subbituminous ash to a high value of 2.96 for an iron-rich bituminous ash. Three subbituminous ashes had a comparatively low specific gravity of ~2.0, and this suggested that hollow particles, such as cenospheres or plerospheres, were present in significant proportions in the three ashes.

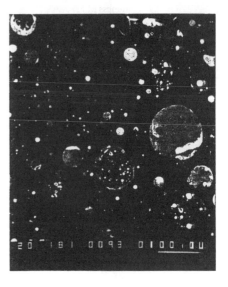

FIGURE 2.3 SEM micrograph of a lignite fly ash. (Backscattered electron image of a polished section of the dispersed sample.) From Malhotra and Ramezanianpour (1994).

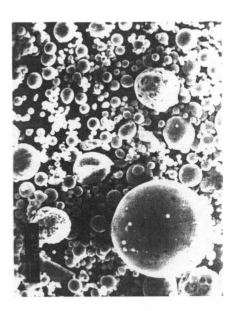

FIGURE 2.4 SEM micrograph of bituminous fly ash. (Secondary electron image of the sample.) From Malhotra and Ramezanianpour (1994).

In general, the physical characteristics of fly ashes vary over a significant range, corresponding to their source. Attempts have been made to correlate the physical properties of different fly ashes. In one CANMET investigation, no apparent relationship was found between type of fly ash and fineness, as determined by percentage retained on a 45-μm sieve. Fineness is probably influenced more by factors such as coal combustion and ash collection and classification than by the nature of the coal itself [Carette and Malhotra, 1986]. Similarly, the type of fly ash showed no apparent influence on the specific surface as measured by the

TABLE 2.2 Fineness of Fly Ashes

Fly Ash Source	Type of Coal*	Specific (Le Chatelier Method)	Fineness (% retained on 45-μm sieve) Wet Sieving[†]	Dry Sieving (Alpine Jet)	Blaine Specific Surface, m^2/kg
1	B	2.53	17.3 (14.9)	12.3	289
2	B	2.58	14.7 (12.7)	10.2	312
3	B	2.88	25.2 (21.7)	18.0	127
4	B	2.96	19.2 (16.6)	14.0	198
5	B	2.38	21.2 (18.3)	16.1	448
6	B	2.22	40.7 (35.1)	30.3	303
7	SB	1.90	33.2 (28.7)	26.4	215
8	SB	2.05	19.4 (16.7)	14.3	326
9	SB	2.11	46.0 (39.7)	33.0	240
10	L	2.38	24.9 (21.5)	18.8	286
11	L	2.53	2.7 (2.4)	2.5	581

Source: Malhotra, V.M. and Ramezanianpour, A.A. 1994. Fly ash in concrete. *CANMET, MSL 94-45 (IR)*, 307 pp. Energy, Mines and Resources Canada, Ottawa.
*B, bituminous; SB, subbituminous; L, lignite.
[†]Values in parentheses do not include sieve correction factor.

TABLE 2.3 Chemical Composition of Fly Ashes

Fly-Ash Source	Type of Coal[†]	Chemical Composition, wt%* SiO$_2$	Al$_2$O$_3$	Fe$_2$O$_3$	CaO	MgO	Na$_2$O	K$_2$O	TiO$_2$	P$_2$O$_3$	MnO	BaO	SO$_3$	LOI[‡]
1	B	47.1	23.0	20.4	1.21	1.17	0.54	3.16	0.85	0.16	0.78	0.07	0.67	2.88
2	B	44.1	21.4	26.8	1.95	0.99	0.56	2.32	0.80	0.25	0.12	0.07	0.96	0.70
3	B	35.5	12.5	44.7	1.89	0.63	0.10	1.75	0.56	0.59	0.12	0.04	0.75	0.75
4	B	38.3	12.8	39.7	4.49	0.43	0.14	1.54	0.59	1.54	0.20	0.04	1.34	0.88
5	B	45.1	22.2	15.7	3.77	0.91	0.58	1.52	0.98	0.32	0.32	0.12	1.40	9.72
6	B	48.0	21.5	10.6	6.72	0.96	0.56	0.86	0.91	0.26	0.36	0.21	0.52	6.89
7	SB	55.7	20.4	4.61	10.7	1.53	4.65	1.00	0.43	0.41	0.50	0.75	0.38	0.44
8	SB	55.6	23.1	3.48	12.3	1.21	1.67	0.50	0.64	0.13	0.56	0.47	0.30	0.29
9	SB	62.1	21.4	2.99	11.0	1.76	0.30	0.72	0.65	0.10	0.69	0.33	0.16	0.70
10	L	46.3	22.1	3.10	13.3	3.11	7.30	0.78	0.78	0.44	0.13	1.18	0.80	0.65
11	L	44.5	21.1	3.38	12.9	3.10	6.25	0.80	0.94	0.66	0.17	1.22	7.81	0.82

Source: Carette, G.G. and Malhotra, V.M. 1986. Characterization of Canadian fly ashes and their relative performance in concrete. *CANMET Report 86-6E.* Energy, Mines and Resources Canada, Ottawa.
*By inductively coupled argon plasma (ICAP) technique, except for Na$_2$O, K$_2$O, SO$_3$, and LOI.
[†]B, bituminous; SB, Subbituminous; L, Lignite.
[‡]105–750°C.

Blaine technique. Moreover, except in one or two cases, there was very little relationship between the specific surface as measured by the Blaine technique and the fineness as determined by percentage retained on a 45-μm sieve.

2.1.3 Chemical and Mineralogical Composition

2.1.3.1 Chemical Composition

Several authors have reported the chemical composition of various fly ashes produced in North America. Carette and Malhotra (1986) in their study of 11 Canadian fly ashes indicated a wide range of chemical composition (Table 2.3). Manz et al. (1989) examined 19 North American lignite fly ashes, and their data are given in Table 2.4. The results of CANMET investigations [Carette and Malhotra, 1986] and the data reported by Manz et al. (1989), on bituminous, subbituminous,

TABLE 2.4   Chemical Analyses for North American Lignite Fly Ashes

Fly Ash	SiO$_2$	Al$_2$O$_3$	Fe$_2$O$_3$	Sum	CaO	MgO	SO$_3$	Na$_2$O	K$_2$O	Available Alkalis	Loss on Ignition (LOI)
							Bulk Chemical Analysis, wt%				
				North Dakota and Montana Lignite							
81–271	25.7	15.0	9.2	49.9	26.8	7.2	8.8	—	—	1.2	0.3
81–560	30.2	12.5	4.6	47.3	23.6	7.9	9.6	7.3	0.6	5.2	1.8
82–179	42.1	12.0	8.1	62.2	18.5	5.0	4.1	8.0	1.2	3.8	0.4
83–275	45.6	15.5	7.3	68.4	20.3	5.0	1.9	1.0	1.7	0.8	0.1
85–352	39.6	14.0	10.7	64.3	15.9	5.7	2.6	4.4	1.4	2.2	0.1
87–139	27.9	10.7	9.9	48.5	21.6	5.5	12.3	5.4	1.5	3.7	1.4
86–305	35.2	20.3	6.3	61.8	25.0	6.8	1.1	0.2	0.5	0.2	0.3
				Saskatchewan Lignite							
85–147	50.4	21.4	3.5	75.3	11.6	3.0	0.5	6.6	0.9	3.1	0.5
86.805	46.4	24.5	4.9	75.8	13.7	4.0	0.6	0.9	1.6	0.7	0.1
87–144	47.9	21.9	4.9	74.7	13.3	2.9	1.1	6.1	1.0	2.9	0.1
				Texas and Louisiana Lignite							
87–146	50.3	20.2	5.5	76.0	14.4	4.0	0.7	0.9	1.2	—	0.2
87–147	57.9	26.3	3.9	88.1	9.6	2.1	0.4	0.0	0.4	0.3	0.3
87–154	62.3	20.9	2.2	85.3	6.1	0.7	0.5	4.1	2.1	5.5	0.2
87–155	52.2	18.0	10.5	80.7	11.9	2.5	1.3	0.2	0.4	0.5	0.1
87–156	55.5	18.6	4.3	78.4	7.0	0.8	0.3	0.6	0.9	0.3	0.1
87–159	57.5	20.6	7.0	85.1	9.1	2.6	0.2	0.4	1.4	0.3	0.1
87–219	62.0	20.1	2.0	84.1	6.9	1.2	0.6	0.9	0.9	1.5	0.4
87–239	48.9	18.5	21.8	89.1	7.3	2.6	0.5	0.4	0.9	0.8	0.1
87–157	52.8	23.6	8.9	85.3	9.5	2.7	0.4	1.1	0.8	1.6	0.0

Source: Manz, O.E., McCarthy, G.J., Stevenson, R.J., Dockter, B.A., Hassett, D.G., and Thedchanamoorthy, A. 1989. Characterization and classification of North American lignite fly ashes for use in Concrete. *Proc. 3rd CANMET/ACI Int. Conf. Use of Fly Ash, Silica Fume, Slag, and Other Mineral Byproducts in Concrete,* Supplementary Papers, pp. 16–32.

Note: ASTM C 618 specification limits: Class F fly ash: SiO$_2$ + Al$_2$O$_3$ + Fe$_2$O$_2$ = 70%; SO$_3$ = 5% max.; LOI = 5% max. Class C fly ash: SiO$_2$ + Al$_2$O$_3$ + Fe$_2$O$_2$ = 50%; SO$_3$ = 5% max.; LOI = 5% max.

and lignite ashes obtained from various coal-powered plants in North America show significant differences in the chemical composition of fly ashes.

2.1.3.2 Mineralogical Composition

In general, both the type and source of fly ash influence its mineralogical composition. Owing to the rapid cooling of burned coal in the power plant, fly ashes consist of noncrystalline particles ($\leq$90%), or glass, and a small amount of crystalline material. Depending on the system of burning, some unburned coal may be collected with ash particles.

In addition to a substantial amount of glassy material, each fly ash may contain one or more of the four major crystalline phases: quartz, mullite, magnetite, and hematite. In subbituminous fly ashes, the crystalline phases may include C$_3$A, C$_4$A$_3\overline{S}$, calcium sulphate, and alkali sulphates [Mehta, 1989]. Table 2.5 shows mineralogical composition of some selected fly ashes.

The reactivity of fly ashes is related to the noncrystalline phase or glass. The reasons for the high reactivity of high-calcium fly ashes may partially lie in the chemical composition of the glass. Mehta (1989) pointed out that the composition of glass in low-calcium fly ashes is different from that in high-calcium fly ashes.

2.1.4 Proportioning Concretes Containing Fly Ash

In most applications, the objective of using fly ash in concrete is to achieve one or more of the following benefits:

- reducing the cement content to reduce costs,

TABLE 2.5 Mineralogical Composition of Some Selected Fly Ashes

Fly-Ash Source	Type of Coal*	Phase Composition, %					
		Glass	Quartz	Mullite	Magnetite	Hematite	LOI (%)
1	B	72.1	4.0	12.6	6.2	1.6	3.5
4	B	70.1	3.2	3.3	17.2	4.7	1.5
5	B	55.6	6.2	19.8	5.6	3.1	9.7
6	B	54.2	8.3	23.5	4.4	2.1	7.5
7	SB	90.2	2.9	6.1	—	—	0.8
8	SB	83.9	4.1	10.2	—	1.4	0.4
9	SB	79.8	8.7	11.5	—	—	0.8
10	L	94.5	4.6	—	—	—	0.9

Source: Carette, G.G. and Malhotra, V.M. 1986. Characterization of Canadian fly ashes and their relative performance in concrete. *CANMET Report 86-6E.* Energy, Mines and Resources Canada, Ottawa.

*B, Bituminous; SB, subbituminous; L, lignite.

- obtaining reduced heat of hydration,
- improving workability,
- attaining required levels of strength in concrete at ages >90 d, and
- improving durability.

The properties of any particular fly ash will greatly affect the properties of the concrete in which it is used. The mixture-proportioning method can minimize the effects that the inclusion of different fly ashes has on concrete performance.

In practice, fly ash can be introduced into concrete in one of two ways:

- A blended cement containing fly ash may be used in place of Portland cement.
- Fly ash may be introduced as an additional component at the concrete-mixing stage.

The use of blended cement is the simpler of the two, as it is free from the complication of batching additional materials and may ensure more uniform control. The relative proportions of fly ash and cement are predetermined, and this limits the range of mixture proportions.

The addition of fly ash at the concrete-mixing stage is flexible and allows for more complete exploitation of the qualities of fly ash as a component of concrete. It does, however, demand that the unique properties of fly ash be considered in determining the proportions of the mixture. In current trends fly ash plays more than one role in concrete. In freshly mixed concrete, it generally acts as a fine aggregate and to some degree may reduce the demand for water. In hardened concrete, because of the pozzolanic nature of fly ash, it becomes a component of the cementitious matrix and influences strength and durability. Thus the use of fly ash in a concrete introduces a number of complexities regarding proportioning, if the accepted relationships between workability, strength, and water/cement ratio are taken into account. Two common assumptions are made when selecting an approach to mixture proportioning of fly-ash concrete:

- Fly ash usually reduces the strength of concrete at early ages.
- For equal workability, concrete incorporating fly ash usually requires less water than concrete containing only Portland cement.

Neither assumption is universally true, and both are influenced by the presence of other common concrete components. However, both assumptions have strongly influenced the approach to mixture proportioning of fly-ash concrete.

As with any other type of concrete, the mixture proportions for a fly-ash concrete can be selected either by reference to some standard concrete (excluding fly ash) or on the basis of the ways in which all the concrete components (including fly ash) will behave in fresh and hardened states.

Throughout the more than 40 years fly ash has been used in concrete, common practice has been to use some plain concrete as a standard of comparison for the mixture proportions of fly-ash concretes. Similarly, the properties of both fresh and hardened concrete usually have been compared with those of a reference concrete. Thus fly ash has been generally thought of as replacing cement, rather than as a component that complements the functions of the cement, sand, or water. Recently, there has been a trend toward considering the components of fly-ash concrete as a whole and treating it as a unique material without reference to an equivalent plain-concrete mixture.

2.1.5 Influence of Fly Ash on the Setting Time of Portland Cement Concrete

The rate at which concrete sets during the first few hours after mixing is expressed as the initial and final setting time and is determined by some form of penetrometer test. Fly ash may be expected to influence the rate of hardening of cement for a number of reasons:

- The ash may be cementitious (high calcium).
- Fly ash may contain sulphates that react with cement in the same way as the gypsum added to Portland cement does.
- The fly ash–cement mortar may contain less water as a consequence of the presence of fly ash, and this will influence the rate of stiffening.
- The ash may absorb surface-active agents added to modify the rheology (water reducers) of concrete, and again this influences the stiffness of the mortar.
- Fly-ash particles may act as nuclei for crystallization of cement hydration products.

There seems to be general agreement in the literature that low-calcium fly ashes retard the setting of cement. In experiments conducted at CANMET [Carette and Malhotra, 1986], the data show that all but 2 of the 11 ashes significantly increase both the initial and final setting times.

2.1.6 Effect of Fly Ash on Workability, Water Requirement, and Bleeding of Fresh Concrete

The small size and the essentially spherical form of low-calcium fly-ash particles influence the rheological properties of cement pastes, causing a reduction in the water required or an increase in workability compared with that of an equivalent paste without fly ash. As Davis et al. (1937) noted, fly ash differs from other pozzolans, which usually increase the water requirement of concrete mixtures. The improved workability allows a reduction in the amount of water used in concrete. According to Owens (1979), the major factor influencing the effects of ash on the workability of concrete is the proportion of coarse material ($>45\,\mu$m) in the ash. Owens has shown that, for example, substitution of 50% by mass of the cement with fine particulate fly ash can reduce the water requirement by 25%. A similar substitution using ash with 50% of the material larger than $45\,\mu$m has no effect on the water requirement.

2.1.7 Effects of Fly Ash on Air Entrainment in Fresh Concrete

Cycles of freezing and thawing are extremely destructive to water-saturated concretes that are not properly proportioned. Concrete will be frost-resistant if it is made with sound, coarse aggregate and is properly protected until some maturity is attained.

To obtain the number of correctly spaced air voids in hardened concrete necessary for frost resistance, an air-entraining admixture (AEA) is added (at a prescribed dosage) to the concrete during mixing. Two attributes are important: the AEA must produce the required volume of air bubbles of the desired size and spacing in the concrete and it must do so in a manner that allows the air content to remain stable while the concrete is mixed, transported, and placed.

The use of some fly ashes causes an increase in the quantity of AEA required to produce a given level of air entrainment in fresh concrete. Larson (1994), writing on the use of fly ash in air-entrained concrete and reviewing the work of other investigators, concluded that the primary effect of fly ash was on the AEA requirement rather than on the air entrainment as such.

Gebler and Klieger (1983) examined 10 different fly ashes representing a range of chemical and physical properties. Carbon content was 0.14–4.19%, total organics were 0.09–1.04%, CaO was 1.2–9.0%, and fineness (as a percentage retained on a 45-μm sieve) was 11.24–38.45%.

Concretes were proportioned by simple replacement of 25% of the cement by fly ash (by mass). All mixtures were proportioned to have 75 $\pm$ 25-mm slump and 6 $\pm$ 1% air. Neutralized vinsol resin was the only AEA used.

The AEA requirement as a percentage of that for the control concrete (for 6% air content) showed the following results:

- For ashes containing >10% CaO, the range of AEA requirements was 126–173%.
- For ashes containing <10% CaO, the range of AEA requirements was 170–553%.

Gebler and Klieger (1983, p. 107) offered the following summary of the findings and conclusions relevant to air entrainment in fresh concrete:

- Generally, concretes containing Class C fly ash require less air-entraining admixture than those concretes with Class F fly ash. All concretes with fly ash required more air-entraining admixture than the Portland cement concretes without fly ash.
- Plastic concretes containing Class C fly ash tended to lose less air than concretes with Class F ash.
- As the air-entraining admixture requirement increases for a concrete containing fly ash, the air loss increases.
- Air contents in plastic concrete containing Class F fly ash were reduced as much as 59%, 90 min after completion of mixing.
- As the organic matter content, carbon content, and loss-on-ignition of fly ash increase, the air-entraining admixture requirement increases, as does the loss of air in plastic concrete.
- Generally, as the total alkalis in fly ash increase, the air-entraining admixture requirement decreases.
- As the specific gravity of a fly ash increases, the retention of air in concrete increases. Concrete containing a fly ash that has a high lime content (Class C fly ash) and less organic matter tends to be less vulnerable to loss of air.
- Generally, as the SO_3 content of fly ash increases, the retained air in concrete increases.

2.1.8 Effects of Fly Ash on Properties of Hardened Concrete

2.1.8.1 Strength Development in Fly-Ash Concrete

As discussed earlier, the main factors determining strength in concrete are the amount of cement used and the water/cement ratio. In practice, these are established as a compromise between the need for workability in the freshly mixed state, strength and durability in the hardened state, and cost. The degree and manner in which fly ash affects workability are major factors in its influence on strength development. As was shown earlier, a fly ash that permits a reduction in the total water requirement in concrete will generally present no problems in selection of mixture proportions and permit any rate of strength development.

Many variables influence the strength development of fly-ash concrete; the most important are the following:

- the properties of the fly ash,
- chemical composition,
- particle size,

- reactivity, and
- temperature and other curing conditions.

2.1.8.2 Effect of Fly-Ash Type on Concrete Strength

The first difference among fly ashes is that some are cementitious even in the absence of Portland cement; these are the so-called ASTM Class C, or high-calcium, fly ashes, usually produced at power plants that burn subbituminous or lignitic coals.

In general, the rate of strength development in concretes tends to be only marginally affected by high-calcium fly ashes. Concrete incorporating high-calcium fly ashes can be made on an equal-weight or equal-volume replacement basis without any significant effect on strength at early ages.

Yuan and Cook (1983) examined the strength development of concretes with and without high-calcium fly ash (CaO = 30.3 wt%). The data from their research are shown in Figures 2.5 and 2.6. Using a simple replacement method of mixture proportioning (Table 2.6), they found the rate of strength development of fly-ash concrete to be comparable to that of the control concrete, with or without air entrainment.

Low-calcium fly ashes, the so-called ASTM Class F ashes, were the first to be examined for use in concrete. Most of what has been written on the behavior of fly-ash concrete examines concretes that use Class F ashes. In addition, the ashes used in much of the early work came from older power plants and were coarse in particle size, contained unburned fuel, and were often relatively inactive pozzolans. Used in concrete and proportioned by simple replacement, these ashes showed

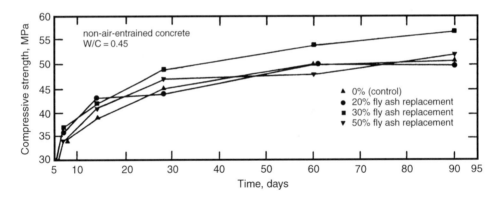

FIGURE 2.5 Compressive-strength development of non-air-entrained concretes containing high-calcium fly ash. From Yuan and Cook (1983).

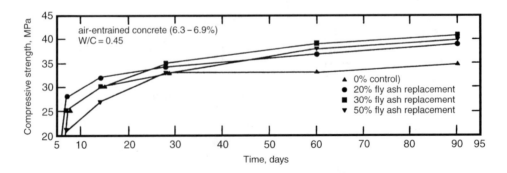

FIGURE 2.6 Compressive-strength development of air-entrained concrete containing high-calcium fly ash. From Yuan and Cook (1983).

TABLE 2.6 Mixture Designations, Proportions, and Properties of Concrete Incorporating High-Calcium Fly Ash

	Mixture Designation			
	C1	C2	C3	C4
Proportions, kg/m³				
Cement	387	309	272	196
Fly ash	0	77	117	196
Cement + fly ash	387	386	389	392
Water	145	145	146	147
Coarse aggregate	1146	1144	1153	1160
Fine aggregate	701	690	678	654
Properties				
Slump, cm	3	9	12	21
Air content, %	2.1	1.9	1.9	1.4
Unit weight, kg/m³	2377	2364	2364	2352
Fly ash as percentage of cement	0	20	30	50

Source: Yuan, R.L. and Cook, J.E. 1983. Study of a Class C fly ash concrete. *Proc., 1st Int. Conf. Use of Fly Ash, Silica Fume, Slag, and Other Mineral Byproducts in Concrete.* V.M. Malhotra, ed. American Concrete Institute, Spec. Pub. SP-79, pp. 307–319.

exceptionally slow rates of strength development. This has led to the erroneous view that fly ash reduces strength at all ages.

Gebler and Klieger (1986) evaluated the effect of ASTM Class F and Class C fly ashes from 10 different sources on the compressive-strength development of concretes under different curing conditions, including effects of low temperature and moisture availability. Their tests indicated that concrete containing fly ash had the potential to produce satisfactory compressive-strength development. The influence of the class of fly ash on the long-term compressive strength of concrete was not significant. In general, compressive-strength development of concretes containing Class F fly ash was more susceptible to low curing temperature than concretes with Class C fly ash or the control concretes.

Gebler and Klieger concluded that Class F fly-ash concretes required more initial moist curing for long-term, air-cured compressive-strength development than did concretes containing Class C fly ashes or the control concretes.

2.1.8.3 Effects of Temperature and Curing Regime on Strength Development in Fly-Ash Concretes

When concrete made with Portland cement is cured at temperatures >30°C, there is an increase in strength at early ages but a marked decrease in strength in the mature concrete.

Concretes containing fly ash and control concretes behave significantly differently. Figure 2.7 shows the general way in which the temperature, maintained during early ages of curing, influences the 28-d strength of concrete [Williams and Owens, 1982].

2.1.8.4 Effect of Fly-Ash on Elastic Properties of Concrete

Published data indicate that fly ash has little influence on the elastic properties of concrete.

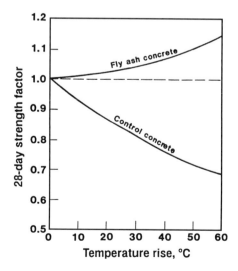

FIGURE 2.7 Effect of temperature rise during curing on the compressive-strength development of concretes. From Williams and Owens (1982).

Abdun-Nur (1961) made the following observation:

The modulus of elasticity of fly ash concrete is lower at early ages, and higher at later ages. In general, fly ash increases the modulus of elasticity of concrete when concretes of the same strength with and without fly ash are compared.

2.1.8.5 Effect of Fly Ash on Creep Properties of Concrete

Data on creep of fly-ash concrete are limited. Ghosh and Timusk (1981) examined bituminous fly ashes of different carbon contents and fineness values in concretes at nominal strength levels of 20, 35, and 55 MPa (water/cement ratio of 1.0, 0.4, and 0.2, respectively). Each concrete was proportioned for equivalent strength at 28 days. Fly-ash concretes showed less creep in the majority of specimens than the reference concretes. This was attributed to a relatively higher rate of strength gain after the time of loading for the fly-ash concretes than for the reference concretes.

Yuan and Cook (1983) investigated creep of high-strength concrete containing a high-calcium fly ash and showed that concrete containing 30 and 50% fly ash exhibited more creep than either the control concrete or a concrete with 20% fly ash.

2.1.8.6 Effect of Fly Ash on Volume Changes of Concrete

It has been generally reported that the use of fly ash in normal proportions does not significantly influence the drying shrinkage of concrete. Typical of the conclusions of most researchers in this area are those made by Davis et al. (1937), who commented as follows:

For masses of ordinary thickness, such as are normally found in highway slabs and in the walls and frames of buildings, the drying shrinkage at the exposed surfaces of concrete up to the age of one year for fly-ash cements is about the same as, or somewhat less than, that for corresponding Portland cement. At a short distance from the exposed surface the drying shrinkage up to the age of one year is substantially less for concretes containing corresponding Portland cements.

For very thin sections and for cements of normal fineness, the drying shrinkage of concretes containing finely ground high-early-strength cements may be somewhat reduced by the use of fly ash.

2.1.9 Effects of Fly Ash on the Durability of Concrete

Increasingly, concrete is being selected for use as a construction material in aggressive or potentially aggressive environments. Concrete structures have always been exposed to the action of sea water. In modern times, the demands placed on concrete in marine environments have increased greatly, as concrete structures are used in arctic, temperate, and tropical waters to contain and support the equipment, people, and products of oil and gas exploration and production. Concrete structures are used to contain nuclear reactors and must be capable of containing gases and vapors at elevated temperatures and pressures under emergency conditions. Concrete is increasingly being placed in contact with sulfate and acidic waters. In all of these instances, the use of fly ash as a concrete material plays a role, and an understanding of its effect on concrete durability is essential to its correct and economical application.

2.1.9.1 Effects of Fly Ash on Permeability of Concrete

A number of investigations have studied the influence of fly ash on the relative permeability of concrete pipes containing fly ash substituted for cement in amounts of 30–50%. In a study by Davis (1954), permeability tests were made on 150×150-mm cylinders at the ages of 28 d and 6 mo. The results of these tests are shown in Table 2.7.

It is clear from these data that the permeability of the concrete was directly related to the quantity of hydrated cementitious material at any given time. After 28 d of curing, at which time little pozzolanic activity would have occurred, the fly-ash concretes were more permeable than the control concretes.

TABLE 2.7 Relative Permeability of Concretes With and Without Fly Ash

Fly Ash		W/(C + F)	Relative Permeability, %	
Type	% by Weight	by Weight	28 d	6 mo
None	—	0.75	100	26
Chicago	30	0.70	220	5
	60	0.65	1410	2
Cleveland	30	0.70	320	5
	60	0.69	1880	7

Source: Davis, R.E. 1954. Pozzolanic materials—With special reference to their use in concrete pipe. Technical Memo. American Concrete Pipe Association.

Note: W/(C + F), water/(cement + fly ash) ratio.

At 6 mo, this was reversed. Considerable imperviousness had developed, presumably as a result of the pozzolanic reaction of fly ash.

Short and Page (1982) reported on the diffusion of chloride ions in solution into Portland blended cement pastes and found the following values of diffusion coefficient, D_c, for different cement types:

Type of Cement	D_c Value ($\times 10^{-9}$ cm^2/s)
Normal Portland	44.7
Sulphate-resisting	100.0
Fly ash/Portland	14.7
Slag/Portland	4.1

It was concluded from these data that slag and fly-ash cements were more effective in limiting chloride diffusion in pastes than were normal or sulfate-resisting cements.

2.1.9.2 Effects of Fly Ash on Carbonation of Concrete

In moist conditions, calcium hydroxide, and to a lesser degree, calcium silicates and aluminates in hydrated Portland cement react with carbon dioxide from the atmosphere to form calcium carbonate. The process, termed carbonation, occurs in all Portland cement concretes. The rate at which concrete carbonates is determined by its permeability, the degree of saturation, and the mass of calcium hydroxide available for reaction. Well-compacted and properly cured concrete, at a low water/cement ratio will be sufficiently impermeable to resist the advance of carbonation beyond the first few millimeters.

If carbonation progresses into a mass of concrete, two deleterious consequences may follow: shrinkage may occur, and carbonation of concrete immediately adjacent to steel reinforcement may reduce the resistance of steel to corrosion.

Nagataki et al. (1986) reported the long-term results of experiments carried out since 1969 that investigate the depth of carbonation in concrete with and without fly ash. The authors concluded that the initial curing period affects the carbonation of concrete cured indoors; hence it is necessary for fly-ash concrete to have a longer curing period in water at early ages. The carbonation of concrete cured outdoors is not affected by the initial curing period in water, provided it is cured in water for a period of 7 d.

2.1.9.3 Effects of Fly Ash on the Durability of Concrete Subjected to Repeated Cycles of Freezing and Thawing

It is generally accepted, other criteria also being met, that air entrainment renders concrete frost resistant. Fly ashes, in common with other finely divided mineral components in concrete, tend to cause an increase in the quantity of admixture required to obtain specified levels of entrained air in concrete. In some instances, the stability of the air or the rate of air loss from fresh concrete is also affected. In general, the observed effects of fly ash on freezing and thawing durability support the

TABLE 2.8   Mass of Scaling Residue After 50 Freezing and Thawing Cycles—Series I

Time of Moist Curing, days	Time of Air Drying, weeks	Mass of Scaling Residue, kg/m^3								
		W/(C + F) = 0.35, Percentage of Fly Ash			W/(C + F) = 0.45, Percentage of Fly Ash			W/(C + F) = 0.55, Percentage of Fly Ash		
		0	20	30	0	20	30	0	20	30
3	3	0.195	0.504	0.184	0.149	0.160	0.206	0.123	0.282	0.321
	4	0.122	0.237	0.208	0.178	0.200	0.680	0.126	0.281	0.638
	5	0.076	0.128	0.143	0.091	0.243	0.634	1.131	0.734	0.354
	6	0.100	0.074	0.158	0.105	0.306	0.263	0.129	0.255	0.226
7	3	0.154	0.047	0.371	0.135	0.212	0.362	0.160	0.335	0.426
	4	0.147	0.092	0.265	0.158	0.448	0.209	0.172	0.312	0.885
	5	0.098	0.108	0.151	0.114	0.180	0.199	0.119	0.238	0.396
	6	0.192	0.038	0.223	0.169	0.268	0.177	0.118	0.370	0.562
14	3	0.139	0.670	1.094	0.188	0.264	0.409	0.517	0.895	0.705
	4	0.144	0.158	0.449	0.126	0.198	0.202	0.131	0.636	0.625
	5	0.174	0.066	0.493	0.135	0.319	0.839	0.162	0.811	0.613
	6	0.168	0.064	0.189	0.117	0.293	0.463	0.286	0.728	0.814

Source: Bilodeau, A., Carette, G.G., Malhotra, V.M., and Langley, W.S. 1991. Influence of curing and drying on salt scaling resistance of fly ash concrete. *Proc. 2nd CANMET/ACI Int. Conf. Durability of Concrete*, V.M. Malhotra, ed. American Concrete Institute Spec. Publ. SP-126, Vol. 1, pp. 201–228.

Note: Each value represents the average of results from two slabs.

view expressed by Larson (1994):

> Fly ash has no apparent ill effects on the air voids in hardened concrete. When a proper volume of air is entrained, characteristics of the void system meet generally accepted criteria.

Klieger and Gebler (1987) also evaluated the durability of concretes containing ASTM Class F and Class C fly ashes. Their test results indicated that air-entrained concretes, with or without fly ash, that were moist cured at 23°C generally showed good resistance to freezing and thawing. For specimens cured at a low temperature (4.4°C), air-entrained concretes with Class F fly ash showed slightly less resistance to freezing and thawing than similar concretes made with Class C fly ash.

Bilodeau et al. (1991), in an investigation carried out at CANMET, determined the scaling resistance of concrete incorporating fly ashes. Water/cement + fly ash ratios of 0.35, 0.45, and 0.55 were used. Concrete without fly ash and concretes containing 20 and 30% fly ash as replacement by mass for cement were made. Bilodeau et al.'s results showed that the concrete containing ≥30% fly ash performed satisfactorily under the scaling test with minor exceptions (Table 2.8).

Carette and Langley (1990) studied the performance of fly ash concrete subjected to 50 freezing and thawing cycles in the presence of deicing salts. They concluded that the incorporation of fly ash in concrete mixtures with ≤30% replacement of Portland cement did not show significant difference in salt-scaling resistance in the presence of a 4% calcium chloride solution when examined by visual rating of surface deterioration. In the measurement of weight loss due to surface deterioration, which they believed was a meaningful way to assess surface deterioration, concretes containing fly ash showed greater weight loss than control concrete. Carette and Langley stated that the surface scaling appeared not to be sensitive to the length of time that specimens were moist cured or air dried subsequent to initial moist curing, at least within the periods investigated.

2.1.9.4 Abrasion and Erosion of Fly-Ash Concrete

Under many circumstances, concrete is subjected to wear by attrition, scraping, or the sliding action of vehicles, ice, and other objects. When water flows over concrete surfaces, erosion may occur. In general, regardless of the type of test performed, the abrasion resistance of concrete is usually found to be proportional to its compressive strength. Similarly, at constant slump, resistance to erosion improves with increased cement content and strength. It may be anticipated that fly-ash concrete that is incompletely or inadequately cured may show reduced resistance to abrasion.

Carrasquillo (1987) examined the abrasion resistance of concretes containing no fly ash, 35% ASTM Class C fly ash, or 35% ASTM Class F fly ash. Specimens tested were cast from concretes having similar strengths, air contents, and cementitious materials contents. The abrasion resistance of concrete containing Class C fly ash was greater than that of concrete containing Class F fly ash or no fly ash. The latter two exhibited approximately equal abrasion resistance; measurement was based on the depth of wear.

Naik et al. (1992) carried out an investigation of the compressive strength and abrasion resistance of concrete containing ASTM Class C fly ash. They proportioned concrete mixtures to have cement replacement in the range of 15–70 wt% fly ash. The water/cementitious materials ratio varied from 0.31 to 0.37. Their results showed that the abrasion resistance of concrete containing $\leq30\%$ fly ash was similar to that of the control concrete. However, abrasion resistance of concretes containing $>40\%$ fly ash was lower than that of control concrete without fly ash.

2.1.9.5 Effects of Fly Ash on Sulfate Resistance of Concrete

In 1967, Dikeou (1970) reported the results of sulfate-resistance studies on 30 concrete mixtures made with Portland cement, Portland–fly-ash cement, or fly ash. From this work, it was concluded that all of the 12 fly ashes tested greatly improved sulfate resistance.

Kalousek et al. (1972) studied the requirements of concretes for long-term service in a sulfate environment. From their study, they drew the following conclusions:

- Eighty-four percent of ASTM types V and II cement concretes without pozzolan showed a life expectancy of <50 years.
- Certain pozzolans very significantly increased the life expectancy of concrete exposed to 2.1% sodium sulfate solution. Fly ashes meeting present-day specifications were prominent among the group of pozzolans showing the greatest improvements.
- Concretes for long-term survival in a sulfate environment should be made with high-quality pozzolans and a sulfate-resisting cement. The pozzolan should not increase significantly, but should preferably decrease, the amount of water required.
- Cement to be used in making sulfate-resisting concrete with pozzolan of proven performance should have a maximum C_3A content of 6.5% and maximum C_4AF content of 12%. Restrictions of cements to those meeting present-day specifications for type V cement does not appear justified.

The fly-ash samples examined by Dikeou (1970) and those examined by Kalousek et al. (1972) originated from bituminous coals.

Dunstan (1976) reported the results of experiments on 13 concrete mixtures made with fly ashes from lignite or subbituminous coal sources. On the basis of this work, he concluded that lignite and subbituminous fly-ash concrete generally exhibited reduced resistance to sulfate attack. The *Concrete Manual* published by the U.S. Bureau of Reclamation gives options for cementitious materials for producing sulfate-resistant concretes [Pierce, 1982; Bureau of Reclamation, 1981].

2.1.9.6 Effects of Fly Ash on Alkali-Aggregate Reactions in Concrete

Shortly after Stanton (1940) discovered that alkali-aggregate reactions (AAR) caused expansion and damage in some concretes, he reported that the effects could be reduced by adding finely ground reactive materials to the concrete mixture. Subsequently, a variety of natural and artificial pozzolans and mineral admixtures, including fly ash, were found to be effective in reducing the damage caused by AAR. The effectiveness of fly ash (and other mineral admixtures) in reducing expansion due to AAR appears to be limited to reactions involving siliceous aggregates. A form of AAR, known as alkali-carbonate reaction, has been reported [Poole, 1981] and has been shown to be relatively unresponsive to the addition of pozzolans [Swenson and Gillott, 1960].

The role of fly ashes in reducing expansion by AAR can be summarized as follows:

- There are substantial published data to show that low-calcium fly ashes with alkali contents of less than about 4% are effective in reducing expansion caused by alkali-silica reactions when the fly ashes are used at a replacement level in the range of 25–30%. High-volume fly ash concrete is very effective in this regard.
- The use of high-calcium ashes has received less attention; hence the background information relevant to their use is less well developed. If they are to be used, there is some indication that effective replacement levels may be higher than those for low-calcium ashes.
- The mechanism and details of the control of expansion caused by alkali-silica reactions are not fully understood. Much research remains before a satisfactory understanding can be developed.

2.1.9.7 Effects of Fly Ash on the Corrosion of Reinforcing Steel in Concrete

Recently, an issue of concern has been the corrosion of steel reinforcement in fly-ash concrete structures exposed to chloride ions from deicing salts or sea water.

If the concrete cover over steel reinforcement is sufficiently thick and impermeable, it will normally provide adequate protection against corrosion. The protective effect of the concrete cover is of both a physical and a chemical nature and functions in three ways:

- It provides an alkaline medium in the immediate vicinity of the steel surface.
- It offers a physical and chemical barrier to the ingress of moisture, oxygen, carbon dioxide, chlorides, and other aggressive agents.
- It provides an electrically resistive medium around the steel members.

Under alkaline conditions (pH higher than $\sim$11.5), a protective oxide film will form on a steel surface, rendering it immune to further corrosion.

When concrete carbonates and the depth of carbonation reaches the steel-concrete boundary, passivation may be reduced and corrosion may occur if sufficient oxygen and moisture reach the metal surface. Chlorides or other ions may also undermine the protective effect of passivation and encourage corrosion.

The Réunion internationale des laboratoires d'essais et de recherches sur les matériaux et les constructions (RILEM) Technical Committee on Corrosion of Steel in Concrete (1974) made the following statements, which give perspective to this issue:

- "The efficiency of the (concrete) cover in preventing corrosion is dependent on many factors which collectively are referred to as its 'quality'. In this context, the 'quality' implies impermeability and a high reserve of alkalinity which satisfies both the physical needs and chemical requirements of the concrete cover. If the concrete is permeable to atmospheric gases or lean in cement, corrosion of the reinforcement can be anticipated and good protection should be attempted by the use of dense aggregate and a well-compacted mix with a reasonably low water/cement ratio."
- "If chloride corrosion is excepted, it is now usually agreed that carbonation of concrete cover is the essential condition for corrosion of reinforcement."

As discussed in Effects of Fly Ash on Carbonation of Concrete above, the issue of carbonation of fly-ash concrete has received some attention in recent years. However, it is our belief that the carbonation of fly-ash concrete is not a matter of concern, provided attention is paid to obtaining adequate impermeability in the concrete mass.

2.1.9.8 Effects of Fly Ash on Concrete Exposed to Sea Water

Exposure of concrete to the marine environment subjects it to an array of severely aggressive factors, including most of those discussed in the preceding sections of this chapter.

Concrete in tidal zones is the most severely attacked, subjected as it is to alternating wetting and drying, wave action, abrasion by sand and debris, frequent freezing and thawing cycles, and corrosion of reinforcement–all occurring in a chemically aggressive medium. Permanently immersed concrete is less severely affected.

Very little direct observation of fly-ash concrete in sea water has been reported in the literature, although research in this area is in progress [Malhotra et al., 1980].

In 1978, CANMET [Malhotra et al., 1988, 1992] started a long-term project on marine-environment performance of concretes incorporating supplementary cementing materials. Test specimens were exposed to repeated cycles of wetting and drying and up to approximately 100 freezing and thawing cycles per year.

Even under exposure to severe marine conditions, concretes incorporating 25% fly ash from a bituminous source were in satisfactory condition after 15 years. The only exceptions were the specimens with a water/cement + fly ash ratio of 0.60. It was concluded that fly-ash concrete at 25% cement replacement level (by mass) can be satisfactory under such conditions of exposure, provided the water/cementitious materials ratio is ≤0.50.

Whereas permeability is considered to be the major factor affecting the durability of concrete in sea water, it is evident that fly ash has the potential to contribute to a number of aspects of concrete durability in the marine environment. It is clear also that this is an aspect of fly-ash concrete behavior that is greatly in need of research.

2.2 Blast-Furnace Slag

2.2.1 Ground, Granulated, or Pelletized Blast-Furnace Slag

Blast-furnace slag is a byproduct of iron manufacture. When it is rapidly quenched with water to a glassy state and finely ground, it develops the property of latent hydraulicity. Most of the slags so produced are, in themselves, cementitious materials to a certain degree, whereas others become so in the presence of activators such as Portland cement and calcium sulfate. Their performance in concrete is, however, independent of whether they are inherently cementitious or not. Figure 2.8 shows a flow chart for the production of pig iron and blast-furnace slag [Kim, 1975]. The technology for utilization of granulated blast-furnace slags in concrete is now well established worldwide.

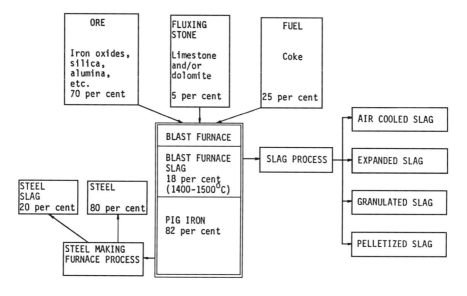

FIGURE 2.8   Flow chart showing production of pig iron and blast-furnace slag. From Kim (1975).

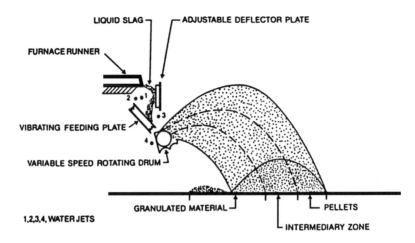

FIGURE 2.9   Diagram of a typical slag pelletizer. From Hooton (1987a).

In the early 1970s, a pelletizing process to produce glassy slag was introduced in southern Ontario. This process uses considerably less water than granulation techniques. The molten slag is first expanded by treatment with water sprays, and the material is then passed over a rotating finned drum. The semimolten material is then thrown into the air for cooling and pelletization (Figure 2.9). A number of pelletization plants are in operation worldwide [Hooton, 1987a].

2.2.2 Mixture Proportions and Properties of Fresh Concrete Incorporating Blast-Furnace Slag

2.2.2.1 Mixture Proportions

The proportions of ground, granulated, or ground, pelletized, blast-furnace slag[1] used in concrete depend upon the job requirements. In normal ready-mixed concrete operations, in which the primary aim is to conserve cement, the usual proportions vary from 25 to 50% by weight of cement on a cement-replacement basis. However, if the purpose is to enhance some aspect of concrete durability, for example, sulfate resistance, then the slag content is at least 50% of the total cementitious material. As each slag has a unique chemical composition, glass content, and fineness, it is necessary to perform exploratory laboratory investigations with the cement, aggregates, and chemical admixtures to be used at a project to determine the correct percentage of slag to be incorporated into concrete. This aspect cannot be overemphasized.

The specific gravity of slags ranges from 2.85 to 2.95, compared with 3.15 for Portland cements. Thus a given replacement of cement by slag on a weight basis results in a higher volume of paste in a concrete mixture. This result is of little consequence at lower percentages of cement replacement. If, however, 50 to 75% cement replacements are being considered, this will affect the rheology of the concrete mixtures and may allow some increase in the volume of coarse aggregate to be used, especially in mixtures incorporating higher amounts of cement.

2.2.2.2 Time of Setting

The incorporation of slag as a replacement for Portland cement in concrete normally results in increased setting time. Final setting time can be delayed up to several hours depending upon the ambient temperature, concrete temperature, and mixture proportions. At temperatures lower

---

[1]Granulated slag implies that the slag is granulated by rapid-water quenching of the molten slag, whereas pelletized slag implies that the granulation is achieved by a pelletizing process. Hereafter, these are referred to either as granulated or pelletized slag, or only as slag when reference is made to both types.

TABLE 2.9 Data on Time of Setting for Air-Entrained Concrete Incorporating Granulated Slag

Properties	Control, No Slag	Slag, Content, %			Control, No Slag	Slag, Content, %		
		40	50	65		40	50	65
Water/(cement + slag)	0.40	0.40	0.40	0.40	0.55	0.55	0.55	0.55
Cement factor (cement and slag), kg/m^3	413	435	419	408	272	290	245	301
Fine aggregate/coarse aggregate	33/67	33/67	33/67	33/67	46/54	46/54	46/54	46/54
Air content, %	5.4	3.4	4.5	5.0	4.3	3.5	6.0	4.9
Unit weight, kg/m^3	2330	2350	2320	2310	2345	2355	2310	2295
Slump, mm	75	75	95	90	70	15	50	75
Air-entraining admixture mL/kg cement	4.4	4.4	9.0	9.5	8.4	4.6	8.2	8.2
Initial set at 21.1°C (70°F), h : min	4 : 06	4.02	4 : 31	4 : 30	4 : 32	5 : 02	5 : 10	5 : 21
Final set at 21.1°C (70°F), h : min	5 : 34	5 : 40	6 : 29	7 : 04	7 : 03	6 : 40	8 : 10	8 : 09
Initial set at 32.2°C (70°F), h : min	3 : 30	—	3 : 45	—	—	—	—	—
Final set at 32.2°C (70°F), h : min	4 : 30	—	4 : 50	—	—	—	—	—

Source: Hogan, F.J. and Meusel, J.W. 1981. The evaluation for durability and strength development of ground granulated blast furnace slag. *ASTM Cement, Concrete, Aggregates* 3(1):40–52.

than 23°C, considerable retardation in setting time can be expected for slag concretes compared with control concrete, which has serious implications in winter concreting. At higher temperatures (>30°C), there is little or no change in the setting time of slag concrete as compared to control concrete. Data by Hogan and Meusel (1981) on initial and final setting for concrete incorporating granulated slag are shown in Table 2.9.

2.2.2.3 Bleeding

Few published data are available on the bleeding of slag concretes. Slags are generally ground to a higher fineness than normal Portland cement and, therefore, a given mass of slag has a higher surface area than the corresponding mass of Portland cement. As the bleeding of concrete is governed by the ratio of the surface area of solids to the volume of water, in all likelihood the bleeding of slag concrete will be lower than that of the corresponding control concrete. The slags now available in Canada and the United States have fineness, as measured by the Blaine method, greater than 4500 cm^2/g, compared with that of about 3000 cm^2/g for Portland cement. Thus in concrete in which a given mass of Portland cement is replaced by an equivalent mass of slag, bleeding should not be a problem.

2.2.2.4 Dosage of Air-Entraining Admixtures

The dosage requirement of an air-entraining admixture to entrain a given volume of air in slag concrete increases with increasing amounts of slag. The increased demand for the admixture is, once again, probably due to the higher total surface area of the slag particles compared with that of the Portland cement particles. For example, in one investigation reported by Malhotra (1979), the admixture dosage needed to entrain about 5% air increased from 177 mL/m^3 for the control concrete to 562 mL/m^3 for a concrete mixture incorporating 65% pelletized slag. The water/cement + slag ratio was 0.30.

2.2.2.5 Rates of Slump Loss and Entrained Air Loss

Meusel and Rose (1979) have shown that the rate of slump loss of concrete incorporating granulated slag at 50% cement replacement was comparable to that of the control concrete.

2.2.3 Properties of Hardened Concrete

2.2.3.1 Color

Concrete incorporating slags is generally lighter in color than normal Portland cement concrete, owing to the lighter color of slags. When concrete is tested for compression or flexure, the interior of the broken specimens exhibit a deep blue-green color. After sufficient exposure to air, the color disappears. The degree of color, which results from the reaction of sulfides in the slag with other compounds in cement, depends upon the percentage of slag used, curing conditions, and the rate of oxidation.

2.2.3.2 Curing

The rate and degree of hydration of cement paste, and consequently its strength, are affected significantly by lack of proper curing because of the slow formation of strength-producing hydrates. This effect becomes more pronounced when the paste incorporates high percentages of slag. Thus, to ensure proper strength and durability of concretes incorporating high percentages of slag (>30%), it is important that they be given more curing than concretes without slag. Such extended curing is especially important during winter concreting in Canada and the northern United States. The increase in curing time would depend upon the ambient temperature, the concrete temperature, the type and amount of cement used, and the percentage of cement replacement.

2.2.3.3 Compressive Strength

The compressive-strength development of slag concrete depends primarily upon the type, fineness, activity index, and proportions of slag used in concrete mixtures. Other factors that affect the performance of slag in concrete are the water/cementitious materials ratio and the type of cement used. In general, the strength development of concrete incorporating slags is slow at one to five days compared with that of the control concrete. Between 7 and 28 d, the strength approaches that of the control concrete, and beyond this period, the strength of slag concrete exceeds the strength of the control concrete. Figures 2.10 and 2.11 show compressive-strength development with age for granulated slag concrete for water/cement + slag [W/C + S] ratios of 0.40 and 0.55. It is noted that the highest strength gain at 28 d was for concrete with a slag content of 40% cement replacement.

Malhotra et al. (1985) reported investigations in which small amounts of condensed silica fume were added to pelletized slag concrete to increase the early-age strength. Figure 2.11 illustrates the

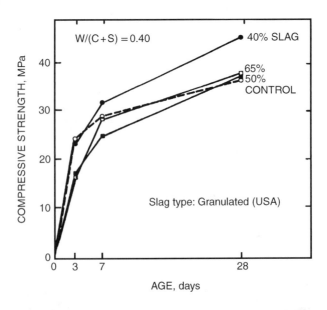

FIGURE 2.10   Age versus compressive-strength relationship for air-entrained concrete: W/(C+S) = 0.40. From Hogan and Meusel (1981).

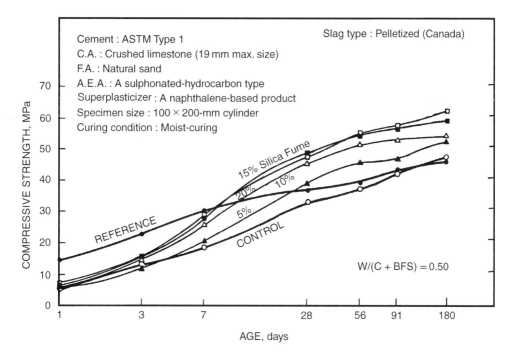

FIGURE 2.11 Age versus compressive-strength relationship for concrete incorporating condensed silica fume and pelletized slag, water/cement + blast-furnace slag [W/(C + BFS)] = 0.50. From Malhotra et al. (1985).

strength development of concrete from 1 to 180 d. The authors concluded:

The low early-age strength of portland cement concrete incorporating blast-furnace slag can be increased by the incorporation of condensed silica fume. The gain in strength is, generally, directly proportional to the percentage of the fume used.

At three days, the increase in strength is generally marginal, especially for concrete with high W/(C + BFS). However, at the age of 14 days and beyond, with minor exceptions, the loss in compressive strength of concrete due to the incorporation of BFS can be fully compensated for with a given percentage of condensed silica fume, regardless of the W/(C + BFS). This is also true for the flexural strength.

The continuing increase in strength at 56, 91 and 180 days of the concrete incorporating BFS and condensed silica fume indicates that sufficient lime (liberated during the hydration of portland cement) is present at these ages for the cementitious reaction to continue.

2.2.3.4 Flexural Strength

In general, at seven days and beyond, the flexural strength of concrete incorporating slag is comparable to, or greater than, the corresponding strength of control concrete; however, in one instance, the reverse was reported for a water/cement + slag ratio of 0.38. The increased flexural strength of slag concrete is probably due to the stronger bonds in the cement/slag/aggregate system because of the shape and surface texture of the slag particles.

2.2.3.5 Young's Modulus of Elasticity

According to Stutterheim (1960), at the same strength level, there is little, if any, difference between the modulus of elasticity of the control concrete and that containing a granulated slag of South African origin. No published data are available on Young's modulus of elasticity of slags currently available in North America. Investigations performed by Nakamura et al. (1986) on a Japanese slag showed no significant difference between the values of the Young's modulus of elasticity of concrete incorporating granulated slag and that of the control concrete.

2.2.3.6 Drying Shrinkage

Hogan and Meusel (1981) have shown that drying shrinkage of concrete incorporating granulated slag is more than that of control concrete. The increase in shrinkage is attributed to increased paste volume in concrete when slag is used as replacement for Portland cement on an equal weight basis because of the lower specific gravity of the slag. This finding may or may not be true for other slags, and further research is needed to confirm this. Fulton (1974a) has suggested that the shrinkage of concrete incorporating granulated slag can be reduced by taking advantage of improved workability to increase the aggregate/cement ratio or by reducing the water/cement ratio of concrete.

2.2.3.7 Creep

There are few published data on creep of concrete incorporating currently available North American slags. The available data from South Africa and Japan are conflicting [Fulton, 1974a]. This conflict is due primarily to the fineness of slags used, methods of tests, age of testing, humidity conditions, and the stress/strength ratio employed. For example, it has been shown that fineness of cement significantly affects the creep strains [Fulton, 1974a]. Bamforth (1980) has reported limited data on creep strains for concretes with and without fly ash and granulated slags, loaded to a constant stress/strength ratio of 0.25. He found that for concretes loaded at an age greater than 24 h, the effects of fly ash and slag significantly reduced the magnitude of the creep.

Neville and Brooks (1975) have shown that when creep tests are performed at ordinary room temperature and humidity conditions (i.e., 20°C and 60% relative humidity) on test specimens which have been loaded after moist curing for 28 d, the total creep of the concrete incorporating a slag from a British source was greater than that of the control concrete, though not significantly so. The rationale for this finding may be that under such test conditions, the rate of gain of strength of the slag concrete is lower than that of the control concrete.

2.2.3.8 Permeability

The permeability of concrete depends mainly upon the permeability of the cement paste, which, in turn, depends upon its pore-size distribution. Using mercury-intrusion techniques, several investigators [Manmohan and Mehta, 1981; Mehta, 1983] have shown that incorporating granulated slags in cement paste helps transform large pores into smaller pores, resulting in decreased permeability of the matrix and, hence, of the concrete. The exact mechanism by which the pore refinement occurs in hydrated slag-cement matrix is, however, not fully understood. Detailed data on the comparison of permeability of concrete with and without slags are not available, although it is an observed phenomenon that granulated slag concretes incorporating slags at up to 75% cement replacement have performed satisfactorily when exposed to sea water [Wiebenga, 1980].

2.2.4 Durability of Concrete Incorporating Blast-Furnace Slag

It is believed that the increased durability of Portland cement concrete incorporating blast-furnace slag results from a finer pore structure and a reduction in easily leached calcium hydroxide in the hardened cement paste. Subsequently, the volume previously occupied by calcium hydroxide is filled in with hydration products, resulting in a less-permeable material. Permeability controls the physical and chemical processes of degradation caused by the action of migrating water; therefore permeability to water determines the rate of deterioration.

2.2.4.1 Resistance to sulfate attack

Sulfates attack concrete and affect its coherence and strength. Concrete's resistance to sulfate attack is improved by partially replacing Portland cement with ground, granulated blast-furance slag. In

Germany, France, and the Netherlands, cements with a high blast-furnace slag content have been used for many years and are considered appropriate for use in a high-sulfate environment [DIN 1164, 1978; NEN 3550, 1979].

Hogan and Meusel (1981) carried out a study that showed high resistance to sulfate attack when the granulated slag proportion exceeded 50% of the total cementitious material; ASTM Type II cements were used.

Results of studies carried out by Frearson (1986) confirm that ordinary Portland cements and blends of both ordinary and sulfate-resisting Portland cement containing lower levels of granulated slag replacement have inferior resistance to sulfate attack. Sulfate resistance increased with granulated slag content, and a mortar with 70% slag content was found to have a resistance superior to mortars containing sulfate-resisting Portland cements alone. Also, the influence of slag content on sulfate resistance was found to be more significant than the water/cement ratio in the mixtures investigated.

According to Ludwig (1989), the cements exhibiting resistance to sulfate attack are

Portland cement with C_3A content ≤ 3 wt%,
Portland cement with $\leq 70\%$ slag content, and
Nonstandard cements such as high-alumina and super-sulfated cements.

Bakker (1983) found that slag concretes with a high slag content display an increased resistance to sulfates because of the low permeability of the concrete to different ions and water, as shown by the diffusion coefficient values in Table 2.10.

Where granulated slag is used in sufficient quantities, several changes occur that improve resistance to sulfate attack. These changes include the following:

The C_3A content of the mixture is proportionally reduced depending on the percentage of slag used. However, Lea (1970) reported that increased sulfate resistance not only depends on the C_3A content of Portland cement alone but also on the Al_2O_3 content of the granulated slag. Lea further reported that sulfate resistance increased where the alumina content of the slag is less than 11%, regardless of the C_3A content of the Portland cement when blends with 20–50% granulated slags were used.

Through the reduction of soluble $Ca(OH)_2$ in the formation of calcium silicate hydrates (CSH), the environment for the formation of ettringite is reduced.

TABLE 2.10 Diffusion Coefficients of Na, K, and Cl Ions in Hardened Cement Paste and Mortars Made with Portland and Blast-Furnace Cement After Different Hardening Times

Diffused Ion	Hardening Time, days	W/C	D_m (10^{-8}cm^2/s) OPC	BFC	Slag Content, %
Na$^+$	3	0.50	7.02	1.44	75
	14	0.50	2.38	0.10	75
	28	0.55	1.47	0.05	60
	28	0.60	3.18	0.05	60
	28	0.65	4.73	0.06	60
K$^+$	3	0.50	11.38	2.10	75
	14	0.50	3.58	0.21	75
Cl$^-$	28	0.55	3.57	0.12	60
	28	0.60	6.21	0.23	60
	28	0.65	8.53	0.41	60
	5	0.50	5.08	0.42	75
	103	0.50	2.96	0.04	75
	60	0.50	4.47	0.41	75

Source: Bakker, R.F.M. 1983. Permeability of blended cement concrete. *Proc. 2nd CANMET/ACI Int. Conf. Fly Ash, Silica Fume Slag and Natural Pozzolans in Concrete,* ACI Spec. Publ. SP-79, pp. 589–605.

Note: W/C, water/cement ratio; OPC, Portland cement; BFC, blast-furnace cement. D_m, diffusion coefficient.

Resistance to sulfate attack is greatly dependent on the permeability of the concrete or cement paste. The formation of CSH in pore spaces, usually occupied by alkalis and $Ca(OH)_2$, reduces the permeability of the paste and prevents the intrusion of aggressive sulfates.

Mehta (1981) tested pastes incorporating natural pozzolans, rice-husk ash, and granulated slag. The 28-day-old paste of the blended cement containing 70% blast-furnace slag showed excellent resistance to sulfate attack. There were hardly any large pores present in the hydrated paste, although the total porosity (pores > 45 Å) was the highest among all the cements tested. The direct relationship between sulfate resistance of a cement and the slope of its pore-size distribution plot in the range 500–45 Å probably shows that the presence of a large number of fine pores is associated with improved sulfate resistance of the material (Figure 2.12). Although the total porosity of the cement containing 30% slag was considerably less than the cement containing 70% slag, the former was not found to be sulfate resistant. On the basis of the test results, Mehta proposed that the chemical resistance of blended Portland cements results mainly from the process of pore refinement, which is associated with the pozzolanic reactions involving the removal of $Ca(OH)_2$.

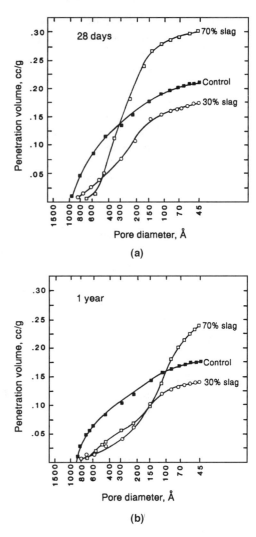

FIGURE 2.12 Pore-size distribution of hydrated cements containing 30 or 70% granulated blast-furnace slag [Mehta, 1981].

2.2.4.2 Resistance to Sea Water

The action of sulfate in sea water on concrete is rather similar to that of sulfate-bearing ground water, but in the former case the attack is not accompanied by expansion of the concrete. The absence of expansion is partly due to the presence of a large quantity of chlorides in the sea water, which inhibit the expansion; gypsum and calcium sulphoaluminate are more soluble in a chloride solution than in water, so they either do not form or are leached out by the sea water.

Regourd et al. (1977) studied mortar cubes that had been exposed to sea water since 1904 at the port of La Rochelle, France. They concluded that all Portland slag cements with a slag content >60% perform well in sea water. In the case of lower slag content, the $MgSO_4$ reacts with the $Ca(OH)_2$ from C_3S and C_2S hydration and produces gypsum. The gypsum reacts with the aluminates to form expansive ettringite.

Mehta (1989), on the other hand, supports the idea that the deterioration of concrete by sea water is not characterized by expansion, but rather is affected by erosion or loss of the solid constituents from the mass. Mehta proposes that ettringite expansion is suppressed in environments where $(OH)^-$ ions have been replaced by Cl^- ions.

In sea water, well-cured concretes containing large amounts of granulated slag or pozzolan usually outperform reference concretes containing only Portland cement, partly because the former contain less uncombined $Ca(OH)_2$ after curing. In permeable concretes, the normal amount of CO_2 present in sea water is sufficient to decompose the cementitious products. The presence of calcium silico-carbonate (thaumasite), calcium carboaluminate hydrate (hydrocalumite) and calcium carbonate (aragonite) have been reported in cement pastes derived from deteriorated concretes exposed to sea water for long periods.

2.2.4.3 Reduction of Expansion Due to Alkali-Silica Reaction

In concrete containing reactive siliceous aggregates, slag cements are preferable to Portland cements, which are rich in alkalis [Regourd, 1980]. Research undertaken at the Concrete Research Institute of the Dutch Cement Industry and by other investigators [Smolczyk, 1974, 1975] confirms that the reason for the high resistance of concretes incorporating slags to the alkali-silica reaction is the low permeability of these concretes to different ions and to water. The low permeability is due not only to the amount of gel formed but also to the locality where the gel is precipitated, that is, the gel can block a pore when Portland cement and slag grains are close to each other.

The potential alkali-aggregate reactivity for combinations of Portland cement and granulated slag was investigated by Hogan and Meusel (1981) using ASTM Test C 227; the aggregate used was Pyrex glass, known to be highly reactive. The data indicate that the expansion of mortar bars made with slag-cement mixtures and Pyrex glass is significantly less than for bars made with Portland cement alone. The cement used for these tests had an alkali content of 0.51% sodium oxide equivalent, which conforms to the ASTM C 150 specification for Portland cement requirement for low-alkali cement.

Suppression of the alkali-aggregate reaction by the addition of slag was cited by Mather (1965), who suggested that an alkali limit for Portland slag cements, which have a performance equal to that of 0.60% Na_2O for Portland cements alone, could be as high as 1.20% Na_2O equivalent.

2.2.4.4 Resistance to Repeated Cycles of Freezing and Thawing

Many studies have been published in which granulated blast-furnace slag has been used as partial replacement for Portland cement in concrete subjected to repeated cycles of freezing and thawing [Mather, 1957; Klieger and Isberner, 1967; Fulton, 1974]. Results of these studies indicate that when mortar or concrete made with granulated slag and Portland cement were tested in comparison with Type I and Type II cements, their resistance to freezing and thawing (ASTM C 666, Procedure A) was essentially the same, provided the concrete was air-entrained.

Malhotra (1983b) reported results of tests performed in an automatic unit capable of performing eight freezing and thawing cycles per day (ASTM C 666, Procedure B). The percentage of slag used as replacement for normal Portland cement varied from 25 to 65 wt % of cement. Initial measurements were taken at 14 d. After about every 100 cycles, the specimens were measured, weighed, and tested by resonant frequency and by the ultrasonic-pulse velocity method. The test was terminated at 700 freezing and thawing cycles. Durability of the exposed concrete prisms was determined from weight, length, resonant frequency, and pulse velocity of the test prisms before and after the freezing and thawing cycling, and relative durability factors (ASTM C 666) were calculated. The test results (Table 2.11) indicate that regardless of the water/cement + slag [W/(C + S)] ratio and whether the concretes were air-entrained or air-entrained and superplasticized, these specimens performed excellently in freezing and thawing tests, with relative durability factors greater than 91%.

2.2.5 Carbonation

Concrete exposed to air will partially release its free water from the layers next to the surface. During evaporation, the pore water in the concrete is replaced by air, and reactions between the CO_2 of the atmosphere and the alkali compounds of the concrete take place. This process between the CO_2 of the atmosphere and the hydration products of the hardened cement paste is called carbonation. The properties of the concrete, as well its protective properties against corrosion of reinforcing steel, are affected by these reactions.

TABLE 2.11 Summary of Freeze-Thaw Test Results for Concrete Series B and D

| | | | Summary of Freeze-Thaw Test Results | | | | | | | | | |
| | | | At Zero Cycles | | | | At Completion of 700 Cycles | | | | | |
Mix Series	W/C + S*	Type of Mix	Weight, kg	Length,† mm	Longitudinal Resonant Frequency, Hz	Pulse Velocity, m/s	Weight, kg	Length, mm	Longitudinal Resonant Frequency, Hz	Pulse Velocity, %	Durability Factor, %	Relative Durability Factor, %
B	0.38	Control + AEA	8.703	2.89	5150	4717	8.693	2.90	5200	4747	102	100
		Control + AEA + SP	8.499	2.70	5150	4684	8.486	2.72	5138	4661	99	97
		25% slag + AEA	8.697	3.00	5300	4788	8.673	3.05	5225	4788	97	95
		25% slag + AEA + SP	8.540	2.96	5125	4684	8.517	3.01	5100	4656	99	97
		65% slag + AEA	8.622	2.74	5140	4684	8.626	2.91	4950	4568	93	91
		65% slag + AEA + SP	8.302	1.59	5025	4589	8.302	1.68	4875	4531	94	92
D	0.56	Control + AEA	8.331	2.56	5000	4568	8.299	2.56	5010	4600	100	—
		Control + AEA + SP	8.443	2.76	4980	4568	8.394	2.76	4980	4504	100	100
		25% slag + AEA	8.451	2.85	5000	4573	8.416	2.88	5000	4606	100	100
		25% slag + AEA + SP	8.544	2.83	5040	4639	8.483	2.91	5050	4622	100	100
		65% slag + AEA	8.465	2.61	4950	4546	‡	2.88	§	§	—	59
		65% slag + AEA + SP	8.471	2.52	4930	4563	‡	2.75	§	§	70	

Source: Malhotra, V.M. 1983b. Strength and durability characteristics of concrete incorporating a pelletized blast furnace slag. *Proc. 1st Int. Conf. Use of Fly Ash, Silica Fume, Slag and Other Mineral By products in Concrete.* V.M. Malhotra, ed., American Concrete Institute Spec. Pub. SP-79. Vol. 2, pp. 892–921 and 923–931.

*Water/cement + slag ratio.
† Gauge length of 345 mm should be added to this value to arrive at the exact length.
‡ Prisms failed at the end of 533 freeze-thaw cycles when the resonant frequency was 3840 Hz.
§ Prisms failed at the end of 450 freeze-thaw cycles when the resonant frequency was 4150 Hz.

In general, in well-compacted low water/cement ratio slag concrete, carbonation is not a problem. However, if concrete incorporates large percentages of slag, is not cured properly, and has a high water/cement + slag ratio, the depth of carbonation will exceed that for normal Portland cement concrete.

Steel in the presence of high concentrations of hydroxyl does not corrode. Bird (1969) stated that this passivity is the result of the formation of a protective film of gamma ferric oxide on the surface of the steel. As long as this protective film is maintained by a high pH, and is not disrupted by aggressive substances, complete protection of the steel against corrosion is assured. Carbonation can reduce the pH to an extent determined by the permeability of the concrete. Hamada (1968) and Meyer (1968) seem to agree that carbonation proceeds more rapidly in concretes incorporating slag than in those made with ordinary portland cement. This finding has been disputed by Schröder and Smolczyk (1968), who point out that comparative tests should be based on specimens of equal initial permeability rather than on specimens of equal age.

2.3 Silica Fume

2.3.1 Production of Silica Fume

Silica fume is a byproduct of the manufacture of silicon or of various silicon alloys produced in submerged electric-arc furnaces. The type of alloy produced and the composition of quartz and coal, the two major components used in the submerged-electric arc furnace, greatly influence the chemical composition of silica fume [Malhotra et al., 1987]. Most of the published data on the use of silica fume in cement and concrete are related to silica fume collected during the production of a silicon alloy containing at least 75% silicon.

2.3.1.1 Forms of Silica Fume

There are several forms in which silica fume is available commercially both in North America and in Europe. The available forms are discussed in the following sections.

As-Produced Silica Fume. This refers to silica fume as collected in dedusting systems known as bag houses. In this form, the material is very fine and has a bulk density of about 200 to 300 kg/m^3, compared with 1500 kg/m^3 for Portland cement [Malhotra et al., 1987]. As-produced silica fume is available in bags or in bulk. Because of its extreme fineness, this form does pose handling problems; in spite of this, the material can be and has been transported and handled like Portland cement.

Compacted Silica Fume. This form of silica fume has a bulk density ranging from 500 to 700 kg/m^3 and is considerably easier to handle than as-produced silica fume. In order to produce the compacted form, the as-produced silica fume is placed in a silo, and compressed air is blown in from the bottom of the silo. This causes the particles to tumble, and in doing so they agglomerate. The heavier agglomerates fall to the bottom of the silo and are removed at intervals. The air compaction of the as-produced silica fume is designed so, that the agglomerates produced are rather weak and quickly break down during concrete mixing. Mechanical means have also been used to produce compacted silica fume.

Water-Based Silica Fume Slurry. In order to overcome the handling and transporting problems associated with as-produced silica fume, several companies are marketing water-based slurries of silica fume, which contain about 40 to 60% solid particles. Typically, these slurries have a density of about 1300 kg/m^3. Some slurries may contain chemical admixtures such as superplasticizers, water reducers, and retarders. One such product known as Force 10,000 has been successfully marketed in North America.

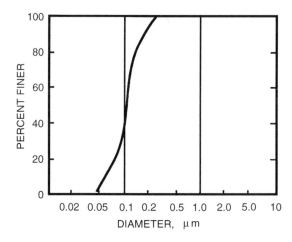

FIGURE 2.13   Typical particle size distribution of silica fume. From Hjorth (1982).

2.3.2 Physical and Chemical Characteristics of Silica Fume

2.3.2.1 Physical Characteristics of Silica Fume

Silica fume varies in color from pale to dark grey. The carbon content and, to a lesser extent, the iron content seems to have preponderant influence on the color of silica fume. The bulk specific weight is of the order of 200 kg/m^3 and is 500 kg/m^3 when compacted.

The specific gravity of silica fume is about 2.20. The particles have a wide range of sizes, but they are perfectly spherical (Figure 2.13). The mean particle diameter is 0.1 μm, compared with 10 μm for particles of cement. The specific surface of silica fume ranges from 13,000 to 30,000 m^2/kg as measured by the nitrogen adsorption technique; the values for Portland cement are 300 to 400 m^2/kg as measured by the Blaine method.

Table 2.12 gives values of fineness, specific surface, pozzolanic activity index, and specific gravity for silica fume from the production of silicon and ferrosilicon alloy [Malhotra et al., 1987].

TABLE 2.12   Physical Characteristics of Silica Fume

Physical Characteristics	Nebesar and Carette (1986) Results		Pistilli et al. (1984) Results	
	Si*	FeSi-75%*	Mixture of Si & FeSi-75%[†]	FeSi-75%[‡]
Fineness by 45 μm sieve, % passing	94.6	98.2	94.0	96.3
Specific surface area, m^2/kg	20,000[§]	17,200[§]	3,750[‖]	5,520[‖]
Pozzolanic activity index with Portland cement, %	102.8	96.5	91.9	95.3
Water requirement, %	138.8	139.2	140.1	144.5
Pozzolanic activity index with lime, MPa	8.9	—	7.0	9.1
Specific gravity	—	—	2.27	2.26

Source: Malhotra, V.M., Ramachandran, V.S., Feldman, R.F., and Aitcin, P.C. 1987. *Condensed Silica Fume in Concrete.* CRC Press Inc., Boca Raton, FL.

*Twenty-four samples.
[†]Thirty-two samples.
[‡]Thirty samples.
[§]Nitrogen adsorption.
[‖]Blaine permeability.

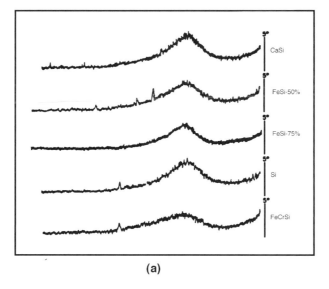

(a)

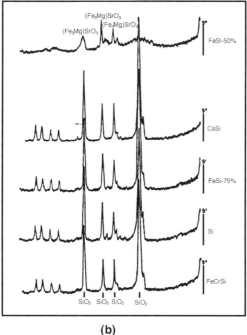

(b)

FIGURE 2.14 X-ray diffractograms of different types of silica fume for concrete with water/cement + silica fume [W/(C + SF)] of 0.64. (a) Before heating; (b) after heating at 1100°C. From Carette and Malhotra (1983).

    X-ray diffractograms of samples of different types of silica fume show them to be vitreous. All the diffractograms exhibit a very wide scattering peak centered at about 4.4 Å, the most important peak of cristobalite (Figure 2.14). When heated to 1100°C, silica fume crystallizes in the form of cristobalite, except the FeSi-50% type, which crystallizes as enstatite, most probably due to the presence of a high amount of iron and magnesium oxide.

TABLE 2.13   Chemical Characteristics of Silica Fume

Chemical Composition	Nebesar and Carette (1986) Results*		Pistilli et al. (1984) Results	
	Si-SF	FeSi-75%-SF	Mixture of Si and FeSi-75%[†]	FeSi-75%[‡]
SiO_2	93.7[§]	93.2	92.1	91.4
Al_2O_3	0.28	0.3	0.25	0.57
Fe_2O_3	0.58	1.1	0.79	3.86
CaO	0.27	0.44	0.38	0.73
MgO	0.25	1.08	0.35	0.44
Na_2O	0.02	0.10	0.17	0.20
K_2O	0.49	1.37	0.96	1.06
S	0.20	0.22	—	—
SO_3	—	—	0.36	0.36[‖]
C	3.62	1.92	—	—
LOI	4.4	3.1	3.20	2.62[‖]

Source: Malhotra, V.M., Ramachandran, V.S., Feldman, R.F., and Aitcin, P.C. (1987) *Condensed Silica Fume in Concrete.* CRC Press, Boca Raton, FL.

*Twenty-four samples.
[†]Thirty-two samples.
[‡]Six Samples.
[§]Calculated.
[‖]Thirty samples.

2.3.2.2 Chemical Composition of Silica Fume

The chemical composition of silica fume from different furnaces is given in Table 2.13. A change in the type of alloy manufactured may cause changes in the characteristics of the silica fume produced. Therefore it is important that concrete plants using silica fume know of any changes in the source of raw materials used for the furnace or changes in the nature of the alloy being produced by a plant.

2.3.3 Physical and Chemical Mechanisms in the Cement-Silica Fume System

2.3.3.1 Physical Mechanisms

There are several physical mechanisms by which silica fume enhances the properties of concrete. These include increasing the strength of the bond between the paste and aggregate by reducing the size of the CH crystals in the region by (a) providing nucleation sites for the CH crystals so that they are smaller and more randomly oriented and (b) reducing the thickness of the weaker transition zone [Detwiler and Mehta, 1989; Monteiro and Mehta 1986]. Physical mechanisms also include increasing the density of composite system by the filler packing effect and by providing a more refined pore structure [Sellevold and Nilsen, 1987; Bache, 1981; Hjorth, 1982]. For the above mechanisms to take place, it is essential that silica-fume particles be well dispersed in a concrete mixture. To achieve this, the use of high-range water reducers (superplasticizers) becomes almost mandatory.

2.3.3.2 Chemical Mechanisms and Pozzolanic Reactions

Malhotra et al. (1987) have reported extensively on the chemical reactions involved in the cement-silica fume-water system. In the presence of Portland cement, the basic reaction, known as the pozzolanic reaction, involves combining finely divided amorphous silica with lime to form a calcium silicate hydrate.

2.3.4 Properties of Fresh Concrete

2.3.4.1 Color

In general, the color of silica-fume concrete is darker than that of conventional concrete. This color difference is more evident on the surface of wet-hardened concrete and in fresh concrete. Also, in concretes incorporating high levels of silica fume that has high carbon content, the dark color is more pronounced. In investigations of silica-fume concrete made with dark-colored fume, it was observed that the color difference disappeared when concrete specimens were stored in the laboratory environment for extended periods. Presumably, this difference in color was neutralized by drying and perhaps by carbonation.

2.3.4.2 Water Demand

Owing to its spherical, small particles, silica fume fills the pores between larger grains of cement and gives a better particle-size distribution, leading to a decrease in water demand in silica-fume concrete. However, the high specific area of the silica-fume particles tends to increase water demand, giving a net effect of increased water demand compared to Portland cement concrete with the same level of workability. Figure 2.15 shows that the increase in water demand is almost directly proportional to the amount of silica fume used in concretes that have an initial water/cement ratio of 0.64 [Carette and Malhotra, 1983]. Superplasticizers or high-range water reducers can be used in silica-fume concretes to reduce the water demand.

In a recent study, using high-resolution nuclear magnetic resonance (NMR) in combination with thermal analysis (DTA/TG), Justness et al. (1992) investigated the pozzolanicity of condensed silica fume in cement pastes. Their results confirmed that condensed silica fume is a very reactive pozzolan. The conversion rate of condensed silica fume to hydration products after 3 d of curing was higher than for cement at the same age.

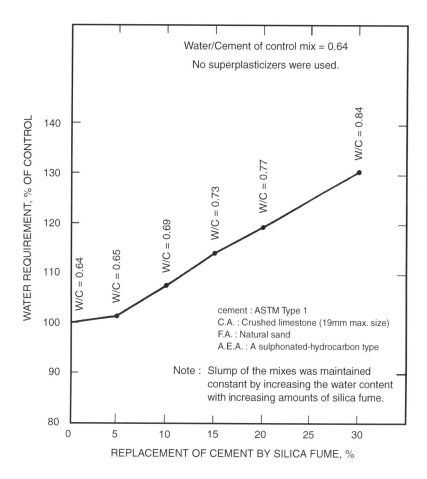

FIGURE 2.15 Relation between water requirement and dosage of silica fume for concrete with a W/(C + SF) of 0.64. From Carette and Malhotra (1983).

In a parallel study, Sellevold and Justnes (1992) studied the decrease in relative humidity and chemical shrinkage during hydration for sealed cement pastes. The Portland cement pastes incorporated 0, 8, and 16% condensed silica fume and had water/cement + silica fume ratios of 0.2, 0.3, and 0.40, respectively. They found that the relative humidity (RH) decreased rapidly during the first two weeks and reached about 78% RH after more than a year for the lowest water/cement + silica fume ratio pastes; the value for the highest water/cement + silica fume ratio paste was about 87% RH. According to the authors, the decrease in RH is the main cause of the increased cracking tendency in high-strength concretes that incorporate silica fume.

2.3.4.3 Bleeding

Bleeding of silica fume concretes is generally lower than that of plain Portland cement concrete. The extremely fine silica-fume particles attach themselves to the cement particles, reducing the channels for bleeding and leaving very little free water available in the fresh concrete.

2.3.4.4 Workability

When compared to Portland cement concrete, silica-fume concrete is more cohesive and resistant to segregation owing to the increase in the number of solid-to-solid contact points. Silica-fume concrete also tends to lose slump rapidly, and a higher initial slump than that of a conventional concrete is often required. However, in lean concretes incorporating less than 300 kg/m^3 cement, workability appears to improve when silica fume is added.

2.3.4.5 Air Entrainment

Because of the extremely high surface area of the silica fume, the dosage of air-entraining admixture required to produce a certain volume of air in silica fume concretes increases considerably with increasing silica-fume dosage. The presence of carbon in the silica fume adds to the increase in air-entraining admixture demand. It has been reported that entrainment of more than 5% air is difficult in concretes that incorporate high amounts of silica fume, even in the presence of a superplasticizer [Carette and Malhotra, 1983].

2.3.4.6 Shrinkage Cracking

Shrinkage cracking occurs in fresh concrete under curing conditions that cause a net removal of water from exposed concrete surfaces, thus creating tensile stresses beyond the low early-age tensile strength capacity of concretes. As concretes containing silica fume show little or no bleeding, thus allowing very little water to rise to the surface, the risk of cracking is high in fresh concrete. Shrinkage cracking can be a very serious problem under curing conditions of elevated temperatures, low humidity, and high winds, which allow rapid evaporation of water from freshly placed concrete. Johansen (1980) and Sellevold (1984) reported that fresh concrete is most vulnerable to shrinkage cracking as it approaches initial set. To overcome this problem, the surface of concrete should be protected from evaporation by covering it with plastic sheets or wet burlap, or by using curing compounds and evaporation-retarding admixtures.

2.3.4.7 Setting Time

Ordinary mixtures (with 250 to 300 kg/m^3 cement) that incorporate small amounts of silica fume, up to 10% by weight of cement, exhibit no significant difference in setting times compared with conventional concretes. As silica fume is invariably used in concretes in combination with water reducers and superplasticizers, the effect of silica fume on setting of concrete is masked by the effects of the admixtures. Investigations by Bilodeau (1985) have shown that the addition of 5 to 10% silica fume to superplasticized and nonsuperplasticized concretes had a negligible effect on the setting time of concrete. However, in concrete with a water/cementitious-materials ratio of 0.40 and 15% silica fume, there was a noticeable delay in setting time. The high dosage of superplasticizer used because of the high silica-fume content in the concrete could have contributed to the setting delay [Bilodeau, 1985].

2.3.5 Properties of Hardened Concrete

2.3.5.1 Compressive Strength

It is well recognized that silica fume can contribute significantly to the compressive-strength development of concrete. This is because of the filler effect and the excellent pozzolanic properties of the material, which translate into a stronger transition zone at the paste-aggregate interface.

The extent to which silica fume contributes to the development of compressive strength depends on various factors such as percentage of silica fume, water/cementitious-materials ratio W/(C + SF), cementitious materials content, cement composition, type and dosage of superplasticizer, temperature, curing conditions, and age.

Superplasticizing admixtures play an important role in ensuring an optimum strength development of silica-fume concrete. The water demand of silica-fume concrete is directly proportional to the amount of silica fume (used as a percentage replacement for Portland cement) if the slump of concrete is to be kept maintained constant by increasing the water content rather than by using a superplasticizer. In such instances, the increase in the strength of silica-fume concrete over that of control concrete is largely offset by the higher water demand, especially for high silica-fume content at early ages. In general, the use of superplasticizer is a prerequisite in order to achieve proper dispersion of the silica fume in concrete and to fully utilize the strength potential of the fume. In fact, many important applications of silica fume in concrete depend strictly upon its utilization in conjunction with superplasticizing admixtures.

Silica-fume concretes have compressive-strength development patterns that are generally different from those of Portland cement concretes. The strength-development characteristics of these concretes are somewhat similar to those of fly-ash concrete, except that the results of the pozzolanic reactions of the former are evident at earlier ages. This is due to the fact that silica fume is a very fine material with a very high amorphous silica content. The main contribution of silica fume to concrete strength development at normal temperatures takes place between the ages of about 3 and 28 d. The overall strength-development patterns can vary according to concrete proportions and composition and are also affected by the curing conditions.

Carette and Malhotra (1992) have reported investigations dealing with the short- and long-term strength development of silica-fume concrete under conditions of both continuous water curing and dry curing after an initial moist-curing period of 7 d. Their investigations covered superplasticized concretes incorporating 0 and 10% silica fume as a replacement by weight for Portland cement and W/C + SF ranging between 0.25 and 0.40. As expected, the major contributions of silica fume to the strength took place prior to 28 d; the largest gains in strength of the silica-fume concrete over the control concrete were recorded at the ages of 28 and 91 d. However, this gain progressively diminished with age. For concretes with a W/C + SF of 0.30 and 0.40, it largely disappeared at later ages. Under air-drying conditions, the strength-development pattern was found to be significantly different from that of water-cured concretes up to the age of about 91 d; thereafter, however, air-drying clearly had some adverse effect on the strength development of both types of concrete. The effect was generally more severe for silica-fume concrete, where some reduction in strength was recorded between the ages of 91 days and 3.5 years, especially for concretes with a W/C + SF of 0.30 and 0.40. These trends of strength reduction have not yet been clearly explained, but they appear to stabilize at later ages and therefore are probably of little practical significance.

Curing temperatures have also been shown to affect significantly the strength development of silica-fume concrete. This aspect has been examined in some detail by several investigators in Scandinavia. In general, these investigations have indicated that the pozzolanic reaction of silica fume is very sensitive to temperature, and elevated-temperature curing has a greater strength-accelerating effect on silica-fume concrete than on comparable Portland cement concrete.

The dosage of silica fume is obviously an important parameter influencing the compressive strength of silica-fume concrete. For general construction, the optimum dosage generally varies between 7 and 10%; however, in specialized situations, up to 15% silica fume has been incorporated successfully in concrete.

2.3.5.2 Young's Modulus of Elasticity

Based on the data published by various investigators, there appears to be no significant differences between the Young's modulus of elasticity, E, of concrete with and without silica fume. Malhotra et al. (1985) have reported data on the Young's modulus of elasticity of Portland cement/blast-furnace slag/silica-fume concrete. They found that regardless of the various percentages of silica fume and water/cementitious materials ratios, there is no significant difference between the E values obtained at 28 d.

2.3.5.3 Creep

The published data on the creep strain of silica-fume concrete are sparse; Bilodeau et al. (1989) reported an investigation on the mechanical properties, creep, and drying shrinkage of high-strength concretes incorporating fly ash, slag, and silica fume. The cementitious-material content of the concretes was about 530 kg/m^3, and the water/cementitious material ratio was 0.22. In addition to the reference concrete, concretes with silica fume at 7 and 12% cement replacements, fly ash at 25% cement replacement, slag at 40% cement replacement, a combination of 7% silica fume and 25% fly ash, and 7% silica fume and 40% slag were investigated. Specimens of these concretes were subjected to creep loading after 35 d of moist curing and an applied stress equivalent to 35% of the compressive strength. After one year, the creep strains of the reference, 7% silica fume, and 12% silica fume concretes were 1505×10^{-6}, 713×10^{-6}, and 836×10^{-6}, respectively. For concrete with 7% silica fume and 40% slag, the creep strain measured was the lowest at 641×10^{-6}. The pore structure and minimal quantity of free water in concrete with the highest amount of supplementary materials could have resulted in these very low creep values, about 40% of the creep of the reference concrete.

2.3.5.4 Permeability

The incorporation of supplementary cementing materials such as fly ash, slag, silica fume, and natural pozzolans in concrete results in fine pore structure and changes to the aggregate-paste interface, leading to a decrease in permeability. This decrease is much higher in concretes incorporating silica fume, owing to its high pozzolanicity. The filler and pozzolanic activity of the silica fume, as well as the virtual elimination of bleeding, improve the interfacial zone through pore refinement.

Plante and Bilodeau (1989) reported that the addition of 8% silica fume significantly reduced the penetration of chloride ions into concrete. With increasing cementitious materials content and decreasing water/cementitious material ratios, the chloride-ion penetration was reduced further. At a W/C + SF of 0.21 and 500 kg/m^3 cement and 40 kg/m^3 silica fume, the chloride-ion penetration was found to be 196 coulombs at 28 d, compared to 1246 coulombs for the reference concrete. This reduction is primarily due to the refined pore structure and increased density of the matrix.

2.3.6 Durability Aspects

2.3.6.1 Carbonation

Silica fume, like other pozzolanic materials, reduces the $Ca(OH)_2$ content of the concrete, and this promotes a faster rate of carbonation. On the other hand, this effect may be offset by the more impermeable nature of silica-fume concrete, which tends to impede the ingress of CO_2 into concrete. The net effect can be somewhat variable, as it depends on various factors such as silica-fume content, water/cementitious materials ratio, and curing conditions, all of which can have a determinant influence on the ultimate $Ca(OH)_2$ content and permeability of the concrete.

Vennesland and Gjorv (1983) reported that using up to 20% silica fume as an addition to cement in concrete, in combination with the use of a plasticizer, reduced the rate of carbonation. The concrete specimens were initially moist cured for a period of 7 d prior to air storage at 60% relative humidity.

Carette and Malhotra (1992) compared the rate of carbonation of silica-fume concrete with that of reference Portland cement concrete up to the age of 3.5 years. The test specimens had been initially moist cured for 7 d before being stored under ambient room drying conditions. At the water/cementitious material ratio of 0.25, they found that all concretes remained free of any noticeable carbonation during a 3.5 year period, whereas at a W/C + SF of 0.40, all specimens exhibited signs of carbonation, the effect being slightly more marked for the silica-fume concrete. In general, however, it is agreed that carbonation is not a problem in adequately cured, high-quality low water/cement ratio concrete, and this also applies to silica-fume concrete.

2.3.6.2 Chemical Resistance

The chemical attack on concrete causes destructive expansion and decomposition of the cement paste, leading to severe deterioration. A permeating solvent as innocuous as water can result in the leaching of calcium hydroxide liberated from the hydration of cement. The ingress of chemicals and acids into concrete allows them to react with calcium hydroxide to form water-soluble salts that leach out of concrete, increasing the permeability of concrete and allowing further ingress of chemicals. Sulfates react with calcium hydroxide, forming ettringite, which causes expansion and cracking of the concrete. Silica-fume concretes have better resistance to chemicals than comparable Portland cement concretes owing to the depletion of calcium hydroxide liberated during the hydration of Portland cement by means of pozzolanic reaction with silica fume, which thus reduces the amount of lime available for leaching, and also owing to the decrease in permeability resulting from the refined pore structure of the mortar phase of the concrete.

Mehta (1985) compared the chemical resistance of low water/cement ratio (0.33 to 0.35) concretes exposed to solutions of 1% hydrochloric acid, 1% sulphuric acid, 1% lactic acid, 5% acetic acid, 5% ammonium sulfate, and 5% sodium sulfate. Specimens of a reference concrete, latex-modified concrete, and silica-fume concrete with 15% silica fume by weight of cement were used, and the criteria for failure was 25% weight loss when fully submerged in the above solutions. The investigation showed that concrete incorporating silica fume better resisted chemical attacks than did the other two types of concrete. The only exception was the silica-fume concrete specimens immersed in ammonium sulfate solution, which performed poorly. This was attributed to the ability of the ammonium salts to decompose the calcium silicate hydrate in the hydrated cement paste.

2.3.6.3 Freezing and Thawing Resistance

Extensive laboratory and field experience in Canada and the United States has shown that for satisfactory performance of concrete under repeated cycles of freezing and thawing, the cement paste should be protected by incorporating air bubbles, 10 to 100 μm in size, using an air-entraining admixture. Briefly, the most important parameters concerning the entrainment of air in concrete are the air content, bubble-spacing factor, and specific surface. For satisfactory freezing and thawing resistance, it is recommended that air-entrained concrete should have bubble-spacing factor $(\overline{L})$ values of less than 200 μm and specific surface (α) greater than 24 mm^{-1}. Usually, fresh Portland cement concrete incorporating between 4 to 7% entrained air by volume will yield the above values of $\overline{L}$ and α.

Several investigators have performed studies on the freezing and thawing resistance of silica-fume concrete. These include, among others, Sorensen (1983), Gjorv (1983), Carette and Malhotra (1983), Malhotra (1984), Yamato et al. (1986), Hooton (1987b), Hammer and Sellevold (1990), Virtanen (1985), Pigeon et al. (1986), and Batrakov et al. (1992).

In one CANMET investigation, the freezing and thawing resistance of non-air-entrained and air-entrained concrete incorporating various percentages of silica fume was compared (Malhotra,

1984). The study led to the following conclusions:

Non-air-entrained concrete: Non-air-entrained concrete, regardless of the W/(C + SF), and irrespective of the amount of condensed silica fume, shows very low durability factors and excessive expansion when tested in accordance with ASTM C 666 (Procedure A or B). The concrete appears to show somewhat increasing distress with increasing amounts of fume. Therefore the use of non-air-entrained condensed silica-fume concrete is not recommended when it is to be subjected to repeated cycles of freezing and thawing.

Air-entrained concrete: Air-entrained concrete, regardless of the W/(C + SF) and containing up to 15% condensed silica fume as partial replacement for cement, performs satisfactorily when tested in accordance with ASTM C 666 Procedures A and B. However, concrete incorporating 30% of the fume and a W/(C + SF) of 0.42 performs very poorly (durability factors less than 10) irrespective of the procedure used. This is probably due to the hardened concrete having high values of $\overline{L}$ or due to the high amount of condensed silica fume in concrete, resulting in a very dense cement matrix system that, in turn, might have adversely affected the movement of water. It was difficult to entrain more than 5% air in the above type of concrete and this amount of air may or may not provide satisfactory $\overline{L}$ values in hardened concrete for durability purposes. The users are therefore asked to exercise caution when using high percentages of condensed silica fume as replacement for Portland cement in concretes with W/(C+SF) of the order of 0.40 if these concretes are to be subjected to repeated cycles of freezing and thawing.

2.3.6.4 Frost Resistance in the Presence of Deicing Salts

Limited published data are available on the deicing salt scaling resistance of silica-fume concrete. Gagne et al. (1990) have reported that the frost resistance of concrete incorporating 6% silica fume as an addition to cement is satisfactory in the procedure of deicing salt scaling tests (ASTM C 672). This was true for both air-entrained and non-air-entrained concrete. The W/(C + SF) of the concretes ranged from 0.23 to 0.30, and the concretes were air dried for 28 d before being subjected to the deicing salt scaling test.

Using the ASTM C 672 tests, Bilodeau and Carette (1989) investigated the deicing salt scaling resistance of Portland cement concrete and concretes incorporating 8% silica fume as replacement for cement. In their investigation, superplasticized and nonsuperplasticized concretes were made, with water/cementitious materials ratios ranging from 0.40 to 0.65, all having adequate air-void parameters. They reported that, in general, all concretes performed satisfactorily under the action of deicing salts, though the silica-fume concrete exhibited a slightly inferior performance when compared to that of Portland cement concrete. In particular, silica-fume concrete with a W/(C + SF) of 0.60 showed appreciable weight loss after 50 cycles.

2.3.6.5 Role of Silica Fume in Reducing Expansion Due to Alkali-Silica Reaction

A number of laboratory investigations have indicated that silica fume, like other pozzolans, is effective in reducing the above-mentioned expansions due to alkali-silica reactions. However, the percentage of silica fume to be incorporated into concrete would depend upon the type of reactive aggregate, the exposure conditions, the alkali and silica contents, the silica fume used, the type of cement used, and the water/cementitious materials ratio of the mixture. Published data indicate that the percentage replacement of cement by silica fume may range from about 10 to 15% [Durand, 1991]. CANMET-funded studies have indicated that silica fume is not effective in controlling expansions due to alkali-carbonate reactions [Chen and Sunderman, 1990].

At present, no significant well-documented data are available as to the long-term effectiveness of silica fume in controlling alkali-silica expansions in actual field structures. Based upon current knowledge, it is recommended that those contemplating the use of silica fume to control expansion due to alkali-silica reactions perform accelerated tests in the laboratory, using the materials to be employed on a job site, to determine the percentage of silica fume to be used as replacement for cement.

2.4 Highly Reactive Metakaolin

Highly reactive metakaolin has recently become available as a very active pozzolanic material for use in concrete. Unlike fly ash, slag, or silica fume, this material is not a by product but is manufactured from a high-purity kaolin clay by calcination at temperatures in the region of 700 to 800°C [Caldarone et al., 1994]. The material, ground to an average particle size of 1 to 2 μm, is white in color. In 1994, a plant was commissioned in Atlanta to produce the material on a commercial scale.

2.4.1 Chemical and Mineralogical Composition

Unlike silica fume, which contains more than 85% SiO_2, highly reactive metakaolin contains equal proportions of SiO_2 and Al_2O_3 by mass. A typical oxide analysis is given below:

SiO_2	Al_2O_3	Fe_2O_3	CaO	MgO	Alkalies	Loss on Ignition
51	40	1	2	0.1	0.5	2

The highly reactive metakaolin derives its reactivity from the combination of two factors, namely, a totally noncrystalline structure and a high surface area.

As far as mineralogical character is concerned, like silica fume, metakaolin is composed essentially of noncrystalline aluminosilicate (Si-Al-O) phase. Occasionally, a small amount of crystalline impurities may be present, that is, 1 to 2% of quartz, feldspar, or titania. The material has a specific surface of about 20 m^2/g and a specific gravity of 2.5.

2.4.2 Properties of Fresh Concrete

According to the limited published data, the initial and final setting time of concrete incorporating 10% of metakoalin by mass are comparable to those of control concrete and concrete incorporating 10% silica fume. Also, the air-entraining admixture demand is similar to that of silica-fume concrete.

Because of its very high specific surface, the bleeding of concrete incorporating metakolin is negligible.

According to the data published by Zhang and Malhotra (1995), the autogenous temperature rise of concrete made with metakaolin was higher than that of the control and silica-fume concrete at ages up to 6 d, indicating high reactivity of the material (Figure 2.16).

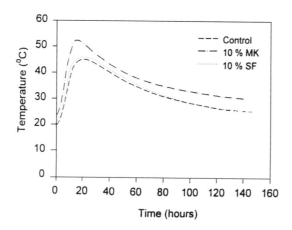

FIGURE 2.16 Autogenous temperature rise in control, silica fume, and metakaolin concretes.

TABLE 2.14 Mechanical Properties of Hardened Concrete

					Strength, MPa								
		Silica	W/C or	Unit	Compressive*						Splitting-tensile[†]	Flexura[‡]	E Modulus[§] GPa
Mix	MK Content, %	Fume Content, %	W/C + MK or W/C + SF	Weight, kg/m³	1 day	3 days	7 days	28 days	90 days	180 days	28 days	28 days	28 days
CO	0	0	0.40	2350	20.9	25.5	28.9	36.4	42.5	44.2	2.7	6.3	29.6
MK10	10	—	0.40	2330	25.0	32.9	37.9	39.9	43.0	46.2	3.1	7.4	32.0
SF10	—	10	0.40	2320	23.2	28.6	34.7	44.4	48.0	50.2	2.8	7.0	31.1

Source: Zhang, M.H., and Malhotra, V.M. 1995. Characteristics of a thermally activated alumina-silicate pozzolanic material and its use in concrete. *Cement Concrete Res.* 25(8):1713–1725.

Note: MK, Metakaolin; W/C, water/cementitous material ratio; SF, silica fume.

* Average of three 102 × 203-mm cylinders.
[†] Average of two 152 × 305-mm cylinders.
[‡] Average of two 102 × 76 × 406-mm prims.
[§] Average of two 152 × 305-mm cylinders.

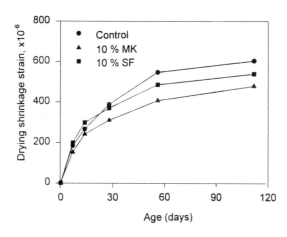

FIGURE 2.17 Drying shrinkage of control silica fume and metakaolin concrete.

2.4.3 Mechanical Properties of Hardened Concrete

2.4.3.1 Strength Development

Table 2.14 gives some data on strength development and Young's modulus for concrete made with metakaolin; also, data on the control concrete and silica fume are included for comparison purposes. The faster strength development of the metakaolin concrete at early ages as compared with the silica-fume concrete is probably due to the faster rate of hydration, as discussed earlier.

2.4.3.2 Drying Shrinkage

Zhang and Malhotra (1995) investigated the drying shrinkage of concrete incorporating metakaolin, and the data are shown in Figure 2.17, together with data for control and silica-fume concrete. The concretes were exposed to drying shrinkage after 7 d of initial curing in lime-saturated water. The metakaolin concrete had a lower drying shrinkage strain compared with that of the control and silica-fume concrete. After 112 d of drying at a relative humidity of 50%, the metakaolin concrete had a drying shrinkage strain of 427×10^{-6} compared with 596×10^{-6} for the control concrete.

2.4.4 Durability Aspects of Hardened Concrete

Air-entrained concrete incorporating 10% metakaolin by mass of cement has high resistance to the penetration of chloride ions and excellent durability in regard to repeated cycles of freezing and thawing. Tables 2.15 and 2.16 show some test results on these above aspects of durability [Zhang and Malhotra, 1995].

TABLE 2.15 Resistance of Concrete to Chloride-Ion Penetration

Mix	Type of Concrete	W/C or W/C + MK or W/C + SF	Unit Weight, kg/m³	28-d Compressive Strength, MPa	Resistance to Chloride-ion Penetration, coulombs	
					28 days	90 days
CO-D	Control	0.40	2320	36.5	3175	1875
MK10-D	10% MK	0.40	2330	37.9	390	300
SF10-D	10% SF	0.40	2310	42.8	410	360

Source: Zhang, M.H. and Malhotra, V. M. 1995. Characteristics of a thermally activated alumina-silicate pozzolanic material and its use in concrete. *Cement Concrete Res.* 25(8):1713–1725.

Note: W/C, water/cementitious materials ratio; MK, metakaolin; SF, silica fume.

TABLE 2.16 Summary of Test Results After 300 Cycles of Freezing and Thawing

Mix	Type of Concrete	28-d Compressive Strength, MPa	Air Content		Specific Surface, mm⁻¹	Spacing Factor L, mm	Change at the End of 300 Freezing/Thawing Cycles, %				Durability Factor, %	Residual Flexural Strength, %
			Fresh Concrete, %	Hardened Concrete, %	mm^{-1}	mm	W	L	PV	RF	%	%
CO-D	Control	36.5	5.8	6.6	21.2	0.15	0.08	0.006	−0.55	−0.84	98.3	85
MK10-D	10% MK	37.9	5.1	4.9	17.9	0.22	0.09	−0.003	0	0.16	100.3	89
SF-10-D	10% SF	42.8	5.8	5.0	22.2	0.17	0.12	0.001	0.19	0.47	100.9	96

Source: Zhang, M.H. and Malhotra, V.M. 1995. Characteristics of a thermally activated alumina-silicate pozzolanic material and its use in concrete. *Cement Concrete Res.* 25(8):1713–1725.

Note: W, weight; L, length; PV, pulse velocity; RF, resonant frequency.

Limited data on the deicing salt scaling resistance of the metakaolin concrete indicates that its performance in the ASTM deicing salt scaling test (ASTM C 672) is comparable to that of silica-fume concrete but somewhat inferior to that of plain Portland cement concrete.

References

Abdun-Nur, E.A. 1961. Fly ash in concrete: An evaluation. *Highway Research Bulletin 284.* Highway Research Board, Washington, DC.

American Society for Testing Materials. 1975. ASTM C 595. In: Annual Book of ASTM Standards, Part 13, Standard Specification for Blended Hydraulic Cements. ASTM, Philadelphia, PA. pp. 353.

American Society for Testing and Materials. 1978. ASTM C 618-78. Specification for Fly Ash and Raw or Calcined Natural Pozzolan for Use as a Mineral Admixture in Portland Cement Concrete. ASTM, Philadelphia, PA.

Bache, H.H. 1981. Densified cement/ultra fine particle-based materials. Paper presented at 2nd International Conference on Superplasticizers in Concrete, Ottawa, Canada.

Bakker, R.F.M. 1983. Permeability of blended cement concrete. *Proc. 1st Int. Conf. Use of Fly Ash, Silica Fume, Slag and Other Mineral Byproducts in Concrete*, V.M. Malhotra, ed. American Concrete Institute Spec. Pub. SP-79. Vol. 1, pp. 589–605.

Bamforth, P.B. 1980. *In-situ* measurements of the effects of partial Portland cement replacement using fly ash, ground granulated blast furnace slag, on the performance of mass concrete. *Proc., Inst. Civ. Eng.* Part 2, 777–800, September.

Batrakov, V.G., Kaprielov, S.S., and Sheinfeld, A.V. 1992. Influence of different types of silica fume having varying silica content on the microstructure and properties of concrete. *ACI SP132.* V.M. Malhotra, ed., pp. 943–963. American Concrete Institute, Detroit, MI.

Berry, E.E. and Malhotra, V.M. 1978. Fly ash for use in concrete, Part II, A critical review of the effects of fly ash on the properties of concrete. *CANMET Report 78-16.* Energy, Mines and Resources Canada, Ottawa.

Bilodeau, A. 1985. Influence des fumées de silice sur le ressuage et le temps de prise du béton. *CANMET Rep. MRP/MSL 85-22 (TR)*, 11 pp. Energy, Mines and Resources Canada, Ottawa.

Bilodeau, A. and Carette, G.G. 1989. Resistance of condensed silica fume concrete to the combined action of freezing and thawing cycling and deicing salts. *American Concrete Institute Spec. Pub. SP-114*, Vol. 2. V.M. Malhotra, ed., pp. 945–970.

Bilodeau, A., Carette, G.G., and Malhotra, V.M. 1989. Mechanical properties of non air-entrained, high-strength concrete incorporating supplementary cementing materials. *Division Report MSL 89-129*, 30 pp. CANMET, Energy, Mines and Resources Canada, Ottawa.

Bilodeau, A., Carette, G.G., Malhotra, V.M., and Langley, W.S. 1991. Influence of curing and drying on salt scaling resistance of fly ash concrete. *Proc. 2nd CANMET/ACI Int. Conf. Durability of Concrete*, V.M. Malhotra, ed. American Concrete Institute Spec. Publ. SP-126, Vol. 1, pp. 201–228.

Bird, C.E. 1969. Corrosion of steel in reinforced concrete in marine atmosphere. In *Concrete Technology*, F.S. Fulton, ed. Portland Cement Institute, Johannesburg.

Bureau of Reclamation. 1981. *Concrete Manual*, 8th ed., pp. 11. Bureau of Reclamation, Denver, CO.

Caldarone, M.A., Gruber, K.A., and Burg, R.G. 1994. High-reactivity metakaolin: A new generation mineral admixture. *ACI Concrete Int.* 16(11):37–40.

Carette, G.G. and Langley, W.S. 1990. Evaluation of deicing salt scaling of fly ash concrete. *Proc. Int. Workshop Alkali-Aggregate Reactions in Concrete: Occurrences, Testing and Control*, Halifax, NS, 1990. CANMET, Ottawa, Canada.

Carette, G.G. and Malhotra, V.M. 1983. Mechanical properties, durability and drying shrinkage of Portland cement concrete incorporating silica fume. *ASTM J. Cement, Concrete Aggregates* 5(1):3–13.

Carette, G.G. and Malhotra, V.M. 1986. Characterization of Canadian fly ashes and their relative performance in concrete. *CANMET Report 86-6E*. Energy, Mines and Resources Canada, Ottawa.

Carette, G.G. and Malhotra, V.M. 1992. Long-term strength development of silica fume concrete. *Proc. 4th Int. Conf. Fly Ash, Slag and Silica Fume in Concrete*, V.M. Malhotra, ed. American Concrete Institute Spec. Publ. 132. Vol. 2, pp. 1017–1044.

Carrasquillo, P. 1987. Durability of concrete containing fly ash for use in highway applications. *Proc., Katharine and Bryant Mather Int. Conf. Concrete Durability*, J.M. Scanlon, ed. American Concrete Institute Spec. Publ. SP-100, Vol. 1, pp. 843–861.

Chen, H. and Suderman, R.W. 1990. The effectiveness of Canadian supplementary cementing materials in reducing alkali-aggregate reactivity. *Final Report*. DSS Contract Number 0SQ83-00215, 245 pp. CANMET, Energy, Mines and Resources Canada, Ottawa, Canada.

Davis, R.E. 1954. Pozzolanic materials—With special reference to their use in concrete pipe. Technical Memo. American Concrete Pipe Association, Irving, TX.

Davis, R.E., Carlson, R.W., Kelly, J.W., and Davis, H.E. 1937. Properties of cements and concretes containing fly ash. *J. Am. Concrete Inst.* 33:577–612.

Detwiler, R.J. and Mehta, P.K. 1989. Chemical and physical effects of silica fume on the mechanical behaviour of concrete. *ACI Mat. J.* 86(6):609–614.

Dikeou, J.T. 1970. Fly ash increases resistance of concrete to sulphate attack. *Water Resour. Tech. Pub. Res. Rep. 23*. Bureau of Reclamation, Denver, CO.

DIN 1164, Part 1. 1978. *Portland, Einsenportland Hochofen und Trabzement* (Portland, Blast Furnace and Pozzolanic Cements).

Dunstan, E.R. 1976. Performance of lignite and sub-bituminous fly ash in concrete—A progress report. *Report REC-ERC-76-1*. Bureau of Reclamation, Denver, CO.

Durand, B. 1991. Preventive measures against alkali-aggregate reactions. In *Course Manual: Petrography and Alkali-Aggregate Reactivity*, pp. 399–489. CANMET, Energy, Mines and Resources Canada.

Frearsen, J.P.H. 1986. Sulfate resistance of combination of Portland cement and ground granulated blast furnace slag. *Proc. 2nd Int. Conf. Fly Ash, Silica Fume, Slag and Natural Pozzolans in Concrete*, V.M. Malhotra, ed. American Concrete Institute Spec. Pub. SP-91. Vol. 2, pp. 1495–1524.

Fulton, F.S. 1974a. *The Properties of Portland Cement Containing Milled Granulated Blast Furnace Slag*, 78 pp. The Portland Cement Institute, Johannesburg.

Fulton, F.S. 1974b. *The Properties of Portland Cement Containing Milled Granulated Blast Furnace Slag*, Portland Cement Institute Monograph. Portland Cement Institute, Johannesburg.

Gagne, R., Pigeon, M., and Aitcin, P.C. 1990. Deicer salt scaling resistance of high performance concrete. *American Concrete Institute Spec. Pub. SP-122*, pp. 29–44.

Gebler, S. and Klieger, P. 1983. Effect of fly ash on the air-void stability of concrete. *Proc. 1st Int. Conf. Use of Fly Ash, Silica Fume, Slag, and Other Mineral Byproducts in Concrete*, V.M. Malhotra, ed. American Concrete Institute SP-79, pp. 103–142.

Gebler, S. and Klieger, P. 1986. Effect of fly ash on physical properties of concrete. *Proc. 2nd CANMET/ACI Int. Conf. Use of Fly Ash, Silica Fume, Slag, and Natural Pozzolan in Concrete*, V.M. Malhotra, ed., American Concrete Institute Spec. Pub. SP-91, pp. 1–50.

Ghosh, R.S. and Timusk, J. 1981. Creep of fly ash concrete. *J. Am. Concrete Inst.*, 78:351–357.

Gjorv, O.E. 1983. Durability of concrete containing condensed silica fume. *American Concrete Institute Spec. Pub. SP-79*, V.M. Malhotra, ed., pp. 695–708.

Hamada, M. 1968. Neutralization (carbonation) of concrete and corrosion of reinforcing steel. *Proc. Fifth Int. Symp. Chem. Cement*, pp. 343–369. Tokyo, Japan.

Hammer, T.A. and Sellevold, E.J. 1990. Frost resistance of high-strength concrete. *American Concrete Institute Spec. Pub. SP-121*, pp. 457–487.

Hjorth, L. 1982. Silica fumes as additions to concrete. In *Characterization and Performance Prediction of Cement and Concrete*, pp. 165. Engineering Foundation, Henniker.

Hogan, F.J. and Meusel, J.W. 1981. The evaluation for durability and strength development of ground granulated blast furnace slag. *ASTM Cement, Concrete, Aggregates* 3(1): 40–52.

Hooton, R.D. 1987a. The reactivity and hydration products of blast-furnace slag. In *Supplementary Cementing Materials*, V.M. Malhotra, ed., pp. 247–330. CANMET, Ottawa, Ontario.

Hooton, R.D. 1987b. Some aspects of durability with condensed silica fume in pastes, mortars, and concretes. Paper presented at CANMET Sponsored International Workshop on Condensed Silica Fume in Concrete Montreal.

Johansen, R. 1980. Risstendens ved plastisk svinn. *Report STF 65 A80016*, FCB/SINTEF. The Norwegian Institute of Technology, Trondheim, Norway.

Justnes, H., Sellevold, E.J., and Lundevall, G. 1992. High-strength concrete binders, Part A, Reactivity and composition of cement pastes with and without condensed silica fume. *ACI SP132*. V.M. Malhotra, ed., pp. 873–889. American Concrete Institute, Detroit, MI.

Kalousek, G.L., Porter, L.C., and Benton, E.J. 1972. Concrete for long-term service in sulphate environment. *Cement Concrete Res.* 2:79–89.

Kim, C.S. 1975. *Waste and Secondary Product Utilization in Highway Construction*, M.S. Thesis, 256 pp. McMaster University, Hamilton, Ontario.

Klieger, P. and Gebler, S. 1987. Fly ash and concrete durability. *Proc. Katharine and Bryant Mather Int. Conf. Concrete Durability*, J.M. Scanlon, ed. American Concrete Institute Spec. Pub. SP-100, Vol. 1, pp. 1043–1069.

Klieger, P. and Isberner, A. 1967. Laboratory studies of blended cement—Portland blast furnace slag cement. *J. PCA Res. Dev. Lab.* 9(3):35.

Larson, T.D. 1994. Air entrainment and durability aspects of fly ash concrete. *Proc. Am. Soc. Test. Mater.* 64:866–886.

Lea, F.M. 1970. *The Chemistry of Cement and Concrete*, E. Arnold Ltd., London.

Malhotra, V.M. 1979. Strength and durability characteristics of concrete incorporating a pelletized blast furnace slag. *ACI SP-79*, V.M. Malhotra, ed., pp. 891–922. American Concrete Association, Detroit, MI.

Malhotra, V.M., ed. 1983a. *Proc. 1st Int. Conf. Use of Fly Ash, Silica Fume, Slag, and Other Mineral Byproducts in Concrete*. American Concrete Institute, Spec. Pub. SP-79. 1182 pp.

Malhotra, V.M. 1983b. Strength and durability characteristics of concrete incorporating a pelletized blast furnace slag. *Proc. 1st Int. Conf. Use of Fly Ash, Silica Fume, Slag and Other Mineral By products in Concrete*, V.M. Malhotra, ed. American Concrete Institute Spec. Pub. SP-79. Vol. 2, pp. 892–921 and 923–931.

Malhotra, V.M. 1984. Mechanical properties and freezing and thawing resistance of non air-entrained and air-entrained condensed silica fume concrete using ASTM Test C 666 Procedure A and B. *Division Report MRP/MSL 84-153 (OP&J)*, CANMET, Energy, Mines and Resources Canada, Ottawa.

Malhotra, V.M. and Mehta, P.K. 1996. Pozzolanic and Cementitious Materials. Gordon and Breach Publishers, Philadelphia, PA.

Malhotra, V.M. and Ramezanianpour, A.A. 1994. Fly ash in concrete. *CANMET, MSL 94-45 (IR)*, 307 pp. Energy, Mines and Resources Canada, Ottawa.

Malhotra, V.M. and Wallace, G.G. 1993. A new method for determining fineness of cement. *Mines Branch Investigation Report 63-119*, Energy, Mines and Resources Canada, Ottawa.

Malhotra, V.M., Carette, G.G., and Bremner, T.W. 1980. Durability of concrete containing granulated blast furnace slag or fly ash or both in marine environment. *CANMET Report 80-18E*, Energy, Mines and Resources Canada, Ottawa.

Malhotra, V.M., Carette, G.G., and Aitcin, P.C. 1985. Mechanical properties of portland cement concrete incorporating blast furnace slag and condensed silica fume. *Proc. RILEM/ACI Sympo.*

Tech. Concrete When Pozzolans, Slags and Chemical Admixtures are Used, Monterrey, Mexico, pp. 395–414.

Malhotra, V.M., Ramachandran, V.S., Feldman, R.F., and Aitcin, P.C. 1987. *Condensed Silica Fume in Concrete,* CRC Press, Boca Raton, FL.

Malhotra, V.M., Carette, G.G., and Bremner, T.W. 1988. Current status of CANMET's studies on the durability of concrete containing supplementary cementing materials in marine environment. *Proc. 2nd CANMET/ACI Int. Conf. Performance of Concrete in Marine Environment,* V.M. Malhotra, ed. American Concrete Institute Spec. Publ. SP-109, pp. 31–72.

Malhotra, V.M., Carette, G.G., and Bremner, T.W. 1992. CANMET investigations dealing with the performance of concrete containing supplementary cementing materials at Treat Island, Maine. *CANMET Report MSL 744,* Energy, Mines and Resources, Canada, Ottawa.

Manmohan, D. and Mehta, P.K. 1981. Influence of pozzolanic, slag, and chemical admixtures on pore size distribution and permeability of hardened cement paste. *ASTM Cement, Concrete, Aggregates* 3(1):63–67.

Manz, O.E., McCarthy, G.J., Stevenson, R.J., Dockter, B.A., Hassett, D.G., and Thedchanamoorthy, A. 1989. Characterization and classification of North American lignite fly ashes for use in Concrete. *3rd CANMET/ACI Int. Conf. Use of Fly Ash, Silica Fume, Slag, and Other Mineral Byproducts in Concrete,* Supplementary Papers, pp. 16–32.

Mather, B. 1957. Laboratory test of Portland blast furnace slag cements. *J. ACI* 54(13):205–232.

Mather, B. 1965. Investigation of Portland blast-furnace slag cement. *Technical Report 6-4555.* Supplementary data. U.S. Army Engineers Waterways Experimental Station, Vicksburg, Mississippi.

Mehta, P.K. 1981. Sulfate resistance of blended Portland cements containing pozzolans and granulated blast furnace Slag. *Proc. Fifth Int. Symp. Concrete Technol.,* R. Rivera, ed. University of N.L., Monterrey, Mexico, pp. 35–50.

Mehta, P.K. 1983. Pozzolanic and cementitious by products as mineral admixtures for concrete—A critical review. *Proc. 1st CANMET/ACI Int. Conf. Fly Ash Silica Fume, Slag and Natural Pozzolans in Concrete,* V.M. Malhotra, ed. American Concrete Institute Spec. Publ. SP-79, Vol. 1, pp. 1–46.

Mehta, P.K. 1985. Durability of low water-to-cement ratio concretes containing latex or silica fume as admixtures. *Proc. RILEM-ACI Symp. Technol. Concrete When Pozzolans, Slags, and Chemical Admixtures are Used,* Monterrey, Mexico, pp. 325–340.

Mehta, P.K. 1989. Pozzolanic and cementitious by-products in concrete—Another look. *Proc., 3rd CANMET/ACI Int. Conf. Use of Fly Ash, Silica Fume, Slag, and Other Mineral By products in Concrete,* Malhotra, ed. American Concrete Institute, Spec. Publ. SP-114, Vol. 1, pp. 1–43.

Mehta, P.K. 1994. Testing and correlation of fly ash properties with respect to pozzolanic behaviour. *Report CS3314.* Electric Power Research Institute, Palo Alto, CA.

Meusel, J.W. and Rose, J.H. 1979. Production of blast furnace slag at Sparrows Point, and the workability and strength potential of concrete incorporating the slag. *ACI SP-79,* V.M. Malhotra, ed., pp. 867–890.

Meyer, A. 1968. Investigations on the carbonation of concrete. *Proc. Fifth Int. Symp. Chem. Cement,* pp. 394–401.

Monteiro, P.J.M., and Mehta, P.K. 1986. Improvement of the aggregate-cement paste transition zone by grain refinement of hydration products. *Proc. 8th Int. Conf. Chem. Cements,* Vol. 3, Rio de Janeiro, pp. 433–437.

Nagataki, S., Ohga, H., and Kim, E.K. 1986. Effect of curing conditions on the carbonation of concrete with fly ash and the corrosion of reinforcement in long term tests. *Proc., 2nd CANMET/ACI Int. Conf. Use of Fly Ash, Silica Fume, Slag, and Natural Pozzolans in Concrete,* V.M. Malhotra, ed. American Concrete Institute, Spec. Pub. SP-91, pp. 521–540.

Naik, T.R., Singh, S.A., and Hossein, M.M. 1992. Abrasion resistance of high-volume fly ash concrete systems. EPRI Report. Electric Power Research Institute, Palo Alto, CA.

Nakamura, N., Sakai, M., Koibuchi, K., and Iijima, Y. 1986. Properties of high strength concrete incorporating very finely ground granulated blast-furnace slag. *Proc. 2nd CANMET/ACI Int. Conf. Fly Ash, Silica Fume, Slag and Natural Pozzolans in Concrete*, V.M. Malhotra, ed. American Concrete Institute Spec. Publ. SP-91, pp. 1361–1380.

Nebesar, B., and Carette, G.G. 1986. Variations in the chemical composition, surface area, fineness and pozzolanic activity of a Canadian silica fume. *ASTM Cement Concrete and Aggregates*, 8(1):42–46.

NEN 3550. 1979. *Netherlands Standard for Cement*.

Neville, A.M., and Brooks, J.J. 1975. Time dependent behaviour of Cemsave concrete. *Concrete*, 9(3):36-39.

Owens, P.L. 1979. Fly ash and its usage in concrete. Concrete: *J. Concrete Soc.* 13:21–26.

Pierce, J.S. 1982. Use of fly ash in combating sulphate attack in concrete. *Proc. 6th Int. Symp. Fly Ash Utilization*, DOE/METC/82-52, pp. 208–231. U.S. Department of the Environment, Washington, DC.

Pigeon, M., Pleau, R., and Aitcin, P.C. 1986. Freeze-thaw durability of concrete with and without silica fume in ASTM C666 (Procedure A) test method: Internal cracking versus scaling. *ASTM J. Cement, Concrete Aggregates* 8(2):76–85.

Pistilli, N.F., Rau, G., and Cechner, R. 1984. The variability of condensed silica fume from a Canadian source and its influence on the properties of Portland cement concrete. *ASTM Cement Concrete and Aggregates*, 6(2):120–124.

Plante, P. and Bilodeau, A. 1989. Rapid chloride ion permeability test data on concretes incorporating supplementary cementing materials. *American Concrete Institute Spec. Pub. SP 114*, Vol. 1, V.M. Malhotra, ed., pp. 625–644.

Poole, A.B. 1981. Alkali-carbonate reactions in concrete. Paper presented at, 5th International Conference on Alkali-Aggregate Reaction in Concrete, Cape Town, South Africa, Mar. 30–Apr. 3, 1981.

Regourd, M. 1980. Structure and behaviour of slag Portland cement hydrates. *Proc Seventh Int. Cong. Chem. Cement*, Vol. I, III-2 10:14. Paris, France.

Regourd, M., Hornain, H., and Montureux, B. 1977. Résistance à l'eau de mer des ciments au laitiers. *Silicates Indust.* 1:19–27.

RILEM Technical Committee 12-CRC. 1974. Corrosion of reinforcement and prestressing tendons—A state of the art report. *Matér. Const.* 9:187–206.

Schröder, F. and Smolczyk, H.G. 1968. Carbonation and protection from steel corrosion. *Proc. Fifth Int. Symp. Chem. Cement*, pp. 188–191.

Sellevold, E.J. 1984. Review: Microsilica in concrete. *Project Report No. 08037-EJS TJJ*. Norwegian Building Research Institute, Oslo, Norway.

Sellevold, E.J. and Justnes, H. 1992. High-strength concrete binders, Part B, Non-evaporable water, self-desiccation and porosity of cement pastes with and without condensed silica fume. *ACI SP132*, V.M. Malhotra, ed., pp. 891–902. American Concrete Institute, Detroit, MI.

Sellevold, E.J. and Nilsen, T. 1987. Condensed silica fume in concrete: A world review. In Supplementary Cementing Materials for Concrete. *CANMET Publ. SP 86-8E*, V.M. Malhotra, ed., pp. 167–229. Energy, Mines and Resources Canada, Ottawa.

Short, N.R., and Page, C.L. 1982. The diffusion of chloride ions through Portland and blended cement pastes. *Silicates Indust.* 47:237–240.

Smolczyk, H.G. 1974. Slag cements and alkali-reactive aggregates. *Proc. Sixth Int. Cong. Chem. Cement*, Suppl. Pap., Sect. III. Moscow, Russia.

Smolczyk, H.G. 1975. Investigation on the diffusion of Na ion in concrete. Paper presented at Symposium on Alkali-Aggregate Reaction. Reykjavik.

Sorensen, E.V. 1983. Freezing and thawing resistance of condensed silica fume (microsilica) concrete exposed to deicing salts. *Proc. 1st CAMET/ACI Inst. Conf. Fly Ash Silica Fume Slag and Natural*

Pozzolans in Concrete, V.M. Malhotra, ed. American Concrete Institute Spec. Pub. SP-79, pp. 709–718.

Stanton, T.E. 1940. Expansion of concrete through reaction between cement and aggregate. Proc. Amer. Soc. Civ. Eng., Vol. 66, p. 1781.

Stutterheim, N. 1960. Properties and uses of high-magnesia Portland slag cement concretes. *J. ACI* 31(10):1027–1045.

Swenson, E.G. and Gillott, J.E. 1960. Characteristics of Kingston carbonate rock reaction. Bulletin 275. Highway Research Board, Washington, DC.

Vennesland, O. and Gjorv, O.E. 1983. Silica concrete-protection against corrosion of embedded steel. *American Concrete Institute Spec. Pub. SP-79*, V.M. Malhotra, ed., pp. 719–729.

Virtenan, J. 1985. Mineral by products and freeze-thaw resistance of concrete. *Publication No. 22:85*, pp. 231–251. Dansk Betonforening, Copenhagen, Denmark.

Wiebenga, J.G. 1980. Durability of concrete structures along the North Sea coast of the Netherlands. *ACI SP-65*, V.M. Malhotra, ed., pp. 437–452. American Concrete Institute, Detroit, MI.

Williams, J.T. and Owens, P.L. 1982. The implications of a selected grade of United Kingdom pulverized fuel ash on the engineering design and use in structural concrete. *Proc., Int. Symp. Use of PFA in Concrete*, J.G. Cabrera and A.R. Cusens, eds., pp. 301–313. Department of Civil Engineering, University of Leeds, Leeds, U.K.

Yamato, T., Emoto, Y., and Soeda, M. 1986. Strength and freezing and thawing resistance of concrete incorporating condensed silica fume. *American Concrete Institute Spec. Pub. SP-91*, V.M. Malhotra, ed., pp. 1095–1117.

Yuan, R.L. and Cook, J.E. 1983. Study of a Class C fly ash concrete. *Proc., 1st Int. Conf. Use of Fly Ash, Silica Fume, Slag, and Other Mineral Byproducts in Concrete*, V.M. Malhotra, ed. American Concrete Institute, Spec. Pub. SP-79, pp. 307–319.

Zhang, M.H. and Malhotra, V. M. 1995. Characteristics of a thermally activated alumina-silicate pozzolanic material and its use in concrete. *Cement Concrete Res.* 25(8):1713–1725.

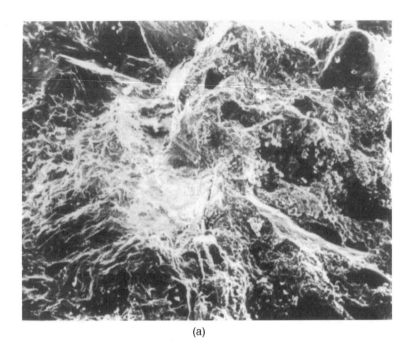

(a)

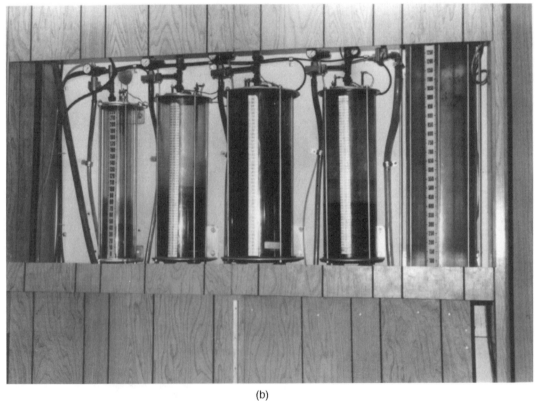

(b)

(a) Scanning electron micrograph of a polymer concrete fracture surface (Nawy et al., Courtesy Dr. Edward G. Nawy, Rutgers University). (b) Chemical admixtures analysis (Courtesy Portland Cement Association).

3

Chemical Admixtures

by
David P. Whitney
Research Operations Manager, Construction Materials Research Group, University of Texas, Austin, Texas. Expert in construction materials with particular emphasis on polymers and other additives.

3.1 Introduction to Chemical Admixtures

Chemical admixtures have been developed in order to improve work times, workability, strength, and durability of Portland cement concrete. Most efforts have centered around improving the properties of concrete with minimal technological investment or skills on the part of the contractors' labor forces. This has resulted in construction cost reductions and has sometimes provided remedies for unexpected problems during construction.

The function of each admixture fulfills a specific need, so each was developed independently of the other. Later, some were combined for ease of addition during the batching process. Common chemical admixture definitions and specifications are discussed in ASTM C494 and in the ACI Manual of Concrete Practice 212.3R. and 212.4R.

Retarders were discovered and developed to allow for longer working times with minimal effect on the final cure strength. In the heat of summer, this allowed for better finishing and better quality concrete.

Accelerators were discovered and developed that started the critical portion of the hydration process much sooner and at lower temperatures, when they were added to a batch design originally intended for use at higher temperatures. It was observed that air entrainment improved the concrete's resistance to freezing and thawing by introducing many well-distributed tiny bubbles that act as pressure relief mechanisms whenever water in concrete pores expanded and contracted through the freezing point.

Water reducers, sometimes called plasticizers, were developed to enable finishers to place and work the concrete with much less water and thus produce higher strength cured concrete. After this, high range water reducers were developed as superplasticizers to adjust the plasticity of low-water concrete to a consistency that allowed the concrete to be pumped up to higher elevations without significantly affecting its strength or durability.

Later, organic polymers such as latex and epoxies were developed to modify the concrete matrix in such a way as to improve the bond of the cured concrete with a given substrate or to reduce the permeability of the cured matrix. Often, a stronger matrix resulted. Monomer systems have also been used to impregnate cured Portland cement concrete. Monomers fill the small pores, capillaries, and voids with a liquid that quickly hardens, resulting in a less porous, higher modulus, and more chemical-resistant cured concrete.

Finally, a polymer-bound product was developed that omitted the Portland cement and water from the matrix altogether. This product, in which sand and coarse aggregate are bound together by a polymer, is called polymer concrete and is used in many commercial applications today.

All these admixtures have been refined to provide designers and builders with concrete options that allow concrete to be adapted to a wider variety of applications and ambient conditions. It is estimated that one or more chemical admixtures, not including air-entraining agents, are present in 80% of the concrete placed today, and that figure rises to almost 100% when air-entraining agents are included [Whiting et al., 1994].

The most commercially important chemical admixtures are described in the Manual of Concrete Practice [ACI Committee 212, 1991] and ASTM C494, Specifications for Chemical Admixtures in Concrete (1992). In order to better understand the recommended usage for various applications of these chemical admixtures in concrete, a review of each functional category is presented in this chapter.

It would be simpler if each of the admixtures worked independently in the matrix, but the performance of each is often affected by the presence of another. For this reason, known reactions and interactions will be discussed wherever appropriate for specific materials.

3.2 Retarding Admixtures

Retarding admixtures were discovered, used, and developed to slow down the initial set of the concrete whenever elevated ambient temperatures shortened the working times beyond the practical limitations of normal placement and finishing operations. Retarders are specified in ASTM C494 as Type B admixtures and may be used in varying proportions and in combination with some other admixtures, so that, as temperature increases, higher doses of the admixture may be used to obtain a uniform setting time [ACI 305R, 1989]. Simple retarders typically consist of one of four relatively inexpensive materials: lignin, borax, sugars, and tartaric acids or tartaric salts.

Retarders serve best to compensate for the unwanted acceleration of working times due to temperature or other admixture side effects. They also are used to extend the working time needed for complicated or high-volume placements and for retarding the set of concrete at a surface where an exposed aggregate finish is desired.

Retarding admixtures interfere with the critical chemical reactions of the fastest hydrating cement reactant groups, C_3A and C_3S [Collepardi, 1984]. These reactants normally initiate the hydration process in early stages of concrete formation. Eventually, the hydration process accelerates due to another, initially slower, reaction group, and the heat of reaction allows the hydration to continue at a normal rate until completion. Typical retardation effects are significant for the first 24 to 72 hours.

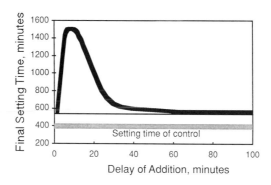

FIGURE 3.1 Effect of delayed addition of retarding admixture on its retarding power.

The goal for retarders is to affect the concrete in its plastic state, while the properties of hardened concrete show little or no negative effect. Improvements to fresh concrete's properties include extended times for workability and set, better workability with less water, frequently, an increase in air content, and some delay in cure time. Retarders may cause critical physical properties to be reduced if used in excess. Changes in the properties of hardened concrete due to retarders usually relate to delayed early strength development, which may indicate increases in early (especially plastic) shrinkage, cracking, and creep [Daughterty and Kowalewski, 1976].

Dosage rates vary considerably depending upon the needs of concrete application, ambient and material temperatures, the retarder type and concentration, cement type and content, and the presence of other admixtures in the concrete, but it generally is best to use the least amount necessary to produce the required properties. This is typically between 2 and 7 fl oz per 100 lb cement [ACI 211.1-4, 1989], introduced with a portion of the mixing water a few minutes after the first addition of water (Figure 3.1). This is recommended in order to ensure the most efficient and uniform dispersion throughout the batch. For determining dosages, manufacturers' recommendations are a good place to start, but trial batches used under the expected field conditions, followed by a complete evaluation of effects on all critical properties should be conducted.

It should be noted that retardation may be a side effect of other chemical admixtures that are specified to adjust other properties in the matrix. If the effect is desirable and predictable, as in many water-reducing admixtures, the material is marketed as such, for instance, water reducing and retarding admixtures. Because of improved technology, many retarders today are designed to serve double duty as water reducers also. This added functionality is discussed in Section 3.3. If a retardation side effect is predictable but not desired, the simple addition of a compatible accelerator may be all that is needed. If, however, the side effect retardation is neither desired nor predictable, an incompatibility may exist between the cement or other admixtures and the corrupting admixture [Johnston, 1987; Dodson and Hayden, 1989]. Every effort must be made to determine the faulty ingredient and replace it with a more compatible choice.

3.3 Water-Reducing Admixtures

Water-reducing agents, or plasticizers, are added to provide workability in a freshly mixed concrete matrix while using significantly lower amounts of mix water. They provide lubricity in coarse mixes that would normally require more paste or more water in the paste. According to ASTM C494, these admixtures are classified as Type A and must allow at least a 5% reduction

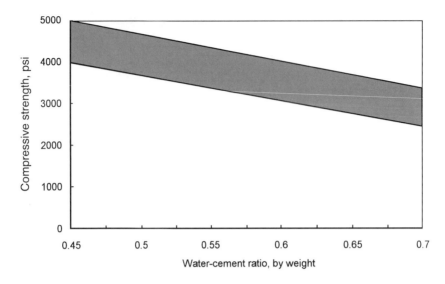

FIGURE 3.2   Typical relationship between 28-d compressive strength and water/cement ratio for a wide variety of air-entrained concretes using Type I cement. *Source:* Kosmatka, S.H. and Panarese, W.C. 1988. *Design and Control of Concrete Mixtures,* 13th ed. Portland Cement Association, Skokie, IL.

in water without changing the consistency or reducing the strength from that of a control batch having a higher water content without the admixture. They often are able to reduce the water demand 10% and even as much as 15%, resulting in the strength and durability benefits derived from lowering the water-to-cement ratio (Figure 3.2). This class of admixtures is typically made from relatively inexpensive lignosulfonates, hydroxylated carboxylic acids, or carbohydrates.

Several theories exist as to what mechanisms are responsible for the reduction in water demand in the plastic concrete matrix, but all agree that the improvement is mainly due to the chemical and physical effects of the water reducing admixtures (WRAs) on the surface of the hydrating cement particles. Deflocculating and dispersing the cement particles is the net result, making better use of the available water for more uniform lubrication and hydration. Also, since many WRAs entrain as much as 2% air, increased lubricity is in part due to the added air bubbles.

Many WRAs retard set times, and sometimes an accelerator is used for compensation. When WRAs are combined with an accelerator into one admixture, it is classified as Type E under ASTM C494. Of course, retardation may be desired for higher-temperature concreting conditions. When the natural tendency of WRAs to retard hydration is not adequate for the desired application, additional retarders are added. When the retarder is included with the WRA into a single admixture, the result is classified as Type D under ASTM C494.

Many WRAs are associated with higher shrinkage rates and faster slump loss, even though the water/cement ratio is reduced. Bleeding properties, too, are sometimes affected by the choice of WRA. To overcome these tendencies, WRAs are often added at the batch plant along with much less of the more expensive, but more efficient, high-range water reducers described in the next section.

Considerations for usage of WRA are based on economics, and strategies fall into the following three main categories [Collepardi, 1984]:

1. To reduce the water/cement ratio for higher strengths and more durability while maintaining workability and cement content.

2. To reduce water and cement for the purpose of reducing shrinkage and heat development in massive placements. Workability, strength, and durability are maintained at a comparative level with the control batch having more water or paste.
3. To keep water and cement the same and maintain strength and durability, but WRA to improve workability.

Efficient dosage rates vary with the chemical composition of the WRA, individual batch designs, cement types, other admixtures, environmental conditions [Kosmatka and Panarese, 1988], and end product needs. Manufacturers' guidelines typically recommend 2 to 7 fl oz/100 lb cement be added to the mix water. These recommendations should be used as a starting point for several trial batches that are closely monitored for critical properties under field conditions in both the fresh and hardened states.

3.4 High-Range Water-Reducing Admixtures

High-range water reducers (HRWRs) are also known as superplasticizers, superfluidizers, and super water reducers because they are more efficient than conventional WRAs for improving workability and flow of concrete mixes. They were developed for use in situations where the addition of WRA required for the desired slump or flow resulted in unacceptable reductions of other critical properties. Changing the chemistry enabled developers to produce an admixture that allowed contractors to place highly workable, pumpable, or even flowing concrete that had higher strengths, greater durability, and less shrinkage when the concrete mix was properly designed. The specifications for superplasticizers are detailed in ASTM C494 as Type F for high-range water reduction with normal set times or Type G for high-range water reduction with retarded setting time. ASTM C1017 (1992) specifies chemical admixtures for use in flowing concrete.

HRWRs are typically from one of four chemical groups: sulfonated melamine-formaldehyde condensate (SMF), sulfonated naphthalene-formaldehyde condensate (SNF), modified lignosulfonates (MLS), and others, which may include sulfonic acid esters or carbohydrate esters.

HRWRs deflocculate and disperse the cement particles in a similar manner but much more efficiently than WRAs. Superplasticizers can reduce water demand in the matrix by as much as 30%, and since HRWRs can be added into the transit mixer at the plant and again at the job site, workability can be customized at the site for specific application needs regardless of transit-time slump loss [Fisher, 1994] (Figure 3.3).

HRWRs are often referred to as first-, second-, and third-generation superplasticizers.

1. The first-generation superplasticizers are primarily anionic materials that create negative charges on the cement particles, resulting in reduced friction due to the particles repelling

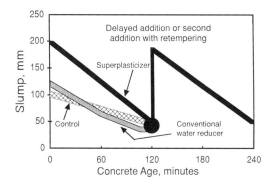

FIGURE 3.3 Effect of superplasticizers on slump loss.

each other. These HRWRs have no effect on the hydration process and, because of their short workability times, are normally added at the job site. Their chemistry allows a reduction in water of 20 to 30%.

2. The second-generation superplasticizers, which are normally added at the batch plant, coat the cement particles with a thixatropic material that lubricates the mix, allows lower water/cement ratios, and adds a measure of control in the hydration process. This allows second-generation HRWRs to be used at higher concrete temperatures, reducing or eliminating the need for ice. Water demand is typically reduced 20 to 30%, and the workability time is extended.

3. The third-generation superplasticizers coat the cement particles and are added at the batch plant, just like second-generation HRWRs. Third-generation superplasticizers offer the same advantages as second-generation types offer plus offer the added bonus of maintaining initial setting characteristics that are similar to normal concrete while producing a highly plastic mix at an extremely low water/cement ratio. Second- and third-generation superplasticizers are relatively expensive, typically costing an additional $5 to $6 per yard of concrete, but they have proven to be cost effective in applications including hot-weather concreting, wall placements, bucket and crane placements, slabs on grade, and pumped concrete [Guennewig, 1988].

Strategies for the use of HRWRs include those listed in section 3.3. Application-related considerations may dictate that HRWR be used and require a choice to be made as to which type to use and how much is needed to achieve the proper concrete placement. Determining superplasticizer dosage can be a relatively complex task that has impacts on cost, rheology of the fresh concrete, and mechanical properties at early ages. The optimal dosage is highly dependent upon the saturation concentration, that is, the highest ratio of the mass of HRWR solids to the mass of cementitious materials over which any higher dosage will not significantly improve the workability of the paste [Gagné et al., 1996]. This ratio is usually 0.8 to 1.2% and is affected by type and quality of superplasticizer, finess of cement, C_3A content, type and content of sulfates, and the speed and shearing action of the mixer employed [Baalbaki, 1990].

Ability to flow through congested reinforcement and into otherwise inaccessible areas and the need for minimal vibration make superplasticized concrete a natural choice in any applications where such problems occur. The ease by which it is pumped and placed makes it a good choice for floors, structural and foundation slabs, pavements, bridges, and roof decks. Contractors and precasters prefer superplasticized concrete for faster strength gain, improved surface details, and pigment dispersion in architectural applications. HRWRs are used in sprayed concrete applications and underwater tremie piped placements because of the increased fluidity and higher strengths. The concept of high-performance concrete has been implemented primarily through the use of engineered concrete mix designs that are made possible by including HRWR to significantly reduce water/cement ratios and by adding high-surface fine materials like silica fume and flyash. Because of this, higher and earlier strengths and increased durability are commonly expected and achieved.

Because of the regularly recorded successes of concrete applications where HRWRs were used, the tendency is to think of HRWRs as a forgiving panacea for the problems associated with concrete. This tendency should be carefully considered. Good batch designs that consider the level of added HRWR are critically important to avoid serious segregation and excessive bleeding. Typically, in normal mixes made flowable through the addition of HRWRs, sand needs to be increased by approximately 5%. Surface finishes can become mottled and irregular when too much HRWR is used. Water/cement ratios are critical to strength and durability, but evaporation and surface absorption keep the practical lower limits well above the theoretical minimum required for complete hydration. Even though HRWRs enable concrete to be made with the lowest water/cement ratios, care must be exercised to ensure that enough water is present throughout the entire hydration process to fully hydrate all of the cement in the matrix.

As with WRAs, HRWRs tend to retard concrete setting times, but may be compensated for or overridden by the addition of accelerating admixtures or further retarded by retarding admixtures. When the retarder is combined with a superplasticizer, the resulting admixture is classified by ASTM C494 as Type G.

Slump loss is a problem of particular importance in high slump or flowable concretes utilizing HRWRs. The rate of slump loss may be affected by the type and dosage of HRWR, other admixtures, the order of addition, type and brand of cement, the concrete temperature, and the concert batch design. Concrete that has HRWR added at the batch plant will tend to experience moderate to rapid slump loss, unless special slump-loss-control admixtures are also added. Normally, higher dosage rates tend to slow down the rate of slump loss, and cement with higher levels of C_3A, as found in Type I and III cements, exhibit a more rapid rate of slump loss. Higher concrete temperatures tend to accelerate the rate of slump loss also [ACI Committee 212, 1993], but adherence to ACI Committee 305 (1989) hot-weather-concreting procedures can minimize it.

Entrained air content is initially increased by the addition of HRWR, but redosing flowing concrete causes loss of entrainment with each dose. Also, air bubbles may have a tendency to be larger and coalesce in flowing concrete.

Shrinkage and creep are generally as good or better than that of control mixes, but flowing concrete may exhibit more shrinkage depending upon water and cement content and choice of HRWR.

Durability is generally better for concrete modified with HRWR because strength is higher, cement particles are more uniformly hydrated, and permeability is lower. Resistance to sulfate attack and abrasion should be better, and resistance to salt scaling and resistance to corrosion of reinforcing steel are comparable with control batches containing no superplasticizer [ACI Committee 212, 1993].

As always, evaluation of trial batches cured under field conditions is highly recommended.

3.5 Accelerating Admixtures

Accelerators today are frequently used to offset retardation effects from other admixtures, although overcoming weather-induced retardation due to colder temperatures at the job site is probably the primary application. Through the use of accelerators contractors, can place concrete at much lower temperatures than would be practical without them [ACI 306R-88, 1988]. Accelerating admixtures are also commonly used to speed up normal set and cure times for earlier service than possible with an unaccelerated mix design. This last application is seen most often in concrete repair mix designs and in prestressed or precast applications, where time delays cost customers significant amounts of money and cause inconvenience.

Accelerating admixtures should conform to ASTM C 494, Type C. These materials are usually calcium chloride or closely related salts because of their availability, relatively low costs, and more than a century of documented usage. Other materials such as silicates, flourosilicates, thiocyanates, alkali hydroxides, calcium formate, calcium nitrate, calcium thiosulfate, potassium carbonate, sodium chloride, aluminum chloride, and various organic compounds, including triethanolamine, are also used as accelerating admixtures [ACI 517.2R, 1987]. Calcium nitrite is frequently recommended and commercially available as an accelerator that also inhibits corrosion of reinforcing steel wherever exposure to salt from deicing materials or sea water is likely to contribute to corrosion problems, although lignosulfonate, benzoates, phosphates, hypophosphates, chromates, fluorides, calgon, and alkali nitrates have been promoted as corrosion inhibitors, also.

Because of the long-term usage of calcium-chloride-based accelerating admixtures, many of their effects on concrete are well known and highly touted in manufacturers' literature. They should conform to ASTM D98 requirements and be sampled and tested according to the procedures in ASTM D345 (1990). On the beneficial side, calcium chlorides are known to reduce initial and final setting times, workability of the mix is usually improved by accelerators, and the use of calcium

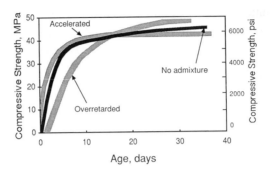

FIGURE 3.4   Effect of set-modifying admixtures on
strength development of concrete.

chloride may slightly increase the air content of the matrix with average sized voids. They also tend
to reduce the bleeding rate and bleeding capacity of concrete.

The effects of accelerators on hardened concrete are as important as they are for the fresh mix. The
benefits are demonstrated by accelerated strength development in both compression (Figure 3.4)
and flexural modes, although less so in the latter. The modulus of elasticity also increases at a
faster rate. Abrasion resistance and erosion resistance are improved with the use of accelerating
admixtures, as is pore structure because porosity is reduced. Frost resistance in concrete is better
at early ages when calcium chloride accelerators are used, but performance declines with time, and
resistance is actually worse at later ages.

Other disadvantages to the use of common calcium chloride accelerating admixtures include slight
increases in drying shrinkage and some increases, from 20 to 35%, in creep that may occur because
rapid strength gain is usually obtained at the expense of reduced ultimate strength. (Figure 3.4)
Also, these accelerators often increase alkali-silica reactions while decreasing resistance to sulfate
attack. Additionally, they are known to increase the rate of freeze-thaw scaling and of corrosion of
steel reinforcement, except for those accelerating admixtures known as corrosion inhibitors, that
is, calcium nitrite and sodium thiocyanate [Nmai et al., 1994]. Accelerated concrete is also usually
darker in color than unaccelerated concrete.

With these effects on properties in mind, the Portland Cement Association (PCA) recommends
that calcium chloride and other admixtures containing soluble chlorides should not be used in the
following:

1. prestressed concrete, owing to potential corrosion hazards;
2. concrete containing embedded aluminum, owing to serious corrosion problems in humid
   environments;
3. concrete subjected to alkali-silica reaction or exposed to soil or water containing sulfates;
4. floor slabs intended to receive dry-shake metallic finishes;
5. hot ambient conditions or hot constituent materials like aggregates or water; or
6. massive concrete placements.

PCA also recommends caution when using calcium chloride in the following applications:

1. concrete subjected to steam curing, because of possible expansion problems with delayed
   ettringite formation;
2. concrete containing embedded dissimilar metals, especially when electrically connected to
   reinforcing steel; and
3. concrete slabs supported on galvanized steel forms [Kosmatka and Panarese, 1988].

The ACI 318, Committee on Building Code Requirements for Reinforced Concrete (1989) has recommended maximum allowable levels of chloride ions for corrosion protection in steel reinforced concrete. These levels are shown in Table 3.1.

Calcium chlorides are used on a regular basis with predictable performance. Recommended usage rates are very dependent upon mix design, temperatures, and exposure conditions. As with any admixture, though, no more should be used in the batching than is necessary to accomplish its intended purpose. Even then it is not generally used at rates higher than 2% of the weight of cement [Kosmatka and Panarese, 1988], and care to ascertain the complete dissolution of all the salt into the mixing water before addition to the rest of the matrix must be made to avoid popouts and serious local staining problems. Laboratory evaluations of field-cured trial batches are recommended before full scale placements are made.

TABLE 3.1 ACI 318 Maximum Chloride-Ion Content for Corrosion Protection

Type of Application	Maximum Water-Soluble Chloride Ion (Cl^-) in Concrete, Percent by Weight of Cement
Prestressed concrete	0.06
Reinforced concrete exposed to chloride in service	0.15
Reinforced concrete that will be dry or protected from moisture in service	1.00
Other reinforced concrete construction	0.30

Source: ACI 318-95/318R-95 Building Code and Commentary.

3.5.1 Accelerating Corrosion Inhibiting Admixtures

When designers are concerned with corrosion of reinforcing steel because of shallow concrete cover, corrosion inhibiting admixtures are often prescribed for the batch design. This class of materials interferes with the galvanic process by scavenging electrons or protons near the steel.

The following are the three general categories of corrosion inhibitors:

1. ***Anodic or active.*** This means that inhibitors actively interfere with the corrosion process by occupying overactive electrons, oxidizing rapidly dissolving ferrous oxide ions into an insoluble protective coating of ferric oxide on their way to the surface of the anodic steel reinforcement. Calcium nitrite is the primary material in this category, although sodium nitrite, sodium benzoate, and sodium chromate are also used outside the United States.

2. ***Cathodic or passive.*** Acting as proton acceptors, inhibitors indirectly interfere with the corrosion process by slowing down the cathodic reactions and impeding the corrosion current. They are normally highly alkaline materials that render the ferrous ions at the surface of the reinforcing steel insoluble in water and unable to participate in the corrosion current. The most commonly used materials in this category are sodium hydroxide, sodium bicarbonate, and ammonium hydroxide, although use of organic materials like substituted forms of aniline and mercaptobenzothiazole has been reported.

3. ***Mixed or passive-active.*** This might be preferable over either of the previous types, since it attacks the problem at both ends of the reaction, interfering with the corrosion process before it can begin. These materials can be in the form of more complicated organic molecules that are attracted to and tie up both cathodic and anodic sites, forming salts as they accept either the available electrons or protons. This category may also be a physical mixture of two simpler compounds one of which affects the cathodic reaction site while the other affects the anodic site.

Corrosion inhibitors are typically used at dosage rates of 1 to 4%, based on the weight of cement.

Concrete properties are affected by the use of corrosion-inhibiting admixtures. Inorganic salts like calcium nitrite are known accelerators, but organic corrosion inhibitors may retard set times. Workability may be improved at levels up to 2% but begins to degenerate above that. In hardened concrete, strengths are often slightly diminished, and the alkali-silica reaction may be aggravated by corrosion inhibitors if reactive aggregates are present in the matrix.

3.6 Air-Entraining Admixtures

In the 1930s air-entrained concrete was developed, and today its use is recommended for nearly every commercial application. Air-entraining agents are provided already ground into the cement (air-entrained cement) or as admixtures whose addition can be adjusted for individual batch design needs.

Because air-entraining agents provide extremely small and well-dispersed air bubbles in the paste, they act as localized stress reducers in the cured matrix. This is advantageous in concrete that is exposed to moisture and, especially, to wet deicing chemicals during freezing and thawing conditions (Figure 3.5). Since the bubble voids provide room for microscopically localized expansions, resistance to damage from alkali-silica reactions and sulfate attack is enhanced as well.

The trade-off is that air is compressible and not strength enhancing, and therefore some loss in strength of the concrete will result. A good rule of thumb is that every 1% increase in air content reduces the strength of a well-designed concrete 2–4% [Kosmatka and Panarese, 1988] (Figure 3.6). Generally, manufacturers recommend between 0.3 and 2.0 ml/kg of cement, depending upon the specific air-entraining admixture, batch conditions, and the amount of air required. Manufacturers' recommendations for a specific air-entraining agent should be used as a guideline, but property evaluations should be made on several small batches before deciding on the optimum quantity needed to produce the air content required for the batch design (Table 3.2).

Air-entraining admixtures are members of a class of chemicals called surface-active substances, or surfactants. These surfactants are made of molecules that have a polar (water-attracted) head and a nonpolar (water-repulsed) tail. If the head is negatively charged, the surfactant is called anionic and is represented by carboxylates, sulfonates, and sufate esters. When the head is positively charged, the molecule is known as cationic, which is represented by substituted ammonium-ion products. Nonionic surfactants are made of molecules with uncharged polar heads, and they are generally polyoxyethylenated compounds. Commercial air-entraining agents for concrete are inexpensive

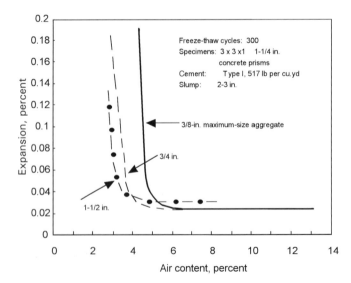

FIGURE 3.5 Relationship between air content and expansion of concrete test specimens during 300 cycles of freezing and thawing for various maximum aggregate sizes. *Source:* Kosmatka, S.H. and Panarese, W.C. 1988. *Design and Control of Concrete Mixtures*, 13th ed. Portland Cement Association, Skokie, IL.

TABLE 3.2   Properties of Typical Air-Entraining Admixtures

Name Brand	Manufacturer	Active Ingredient	Dosage	Sp. Gr.
Protex regular	Protex Industries	Neutral vinusol resin	0.3–1.0	1.044
Darex AEA	WR Grace & Co.	Organic acid salts	0.65–1.95	1.00–1.05
Airex "D"	Mulco, Inc.	Sulfonated HC salt	1.5–1.85	1.01–1.03
Plastair	Sika Chemical Corp.	Vinusol resin	1.4	—
Plastade	Sternson	Coconut acid amide	0.6–1.9	1.0

Sources: Dolch, W.L. 1984. Air-entraining admixtures. In *Concrete Admixtures Handbook Properties, Science, and Technology,* V.S. Ramachandran, ed., pp. 269–302. Noyes Publications, Park Ridge, NJ.

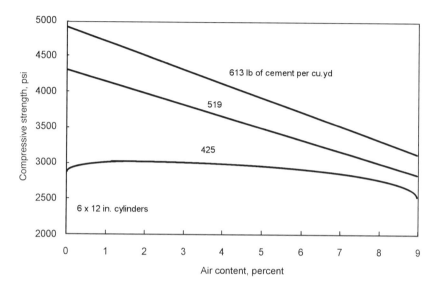

FIGURE 3.6   Relationship between air content and 28-d compressive strength for concrete at three constant cement contents. Water content was reduced with increased air content to maintain a constant slump. *Source:* Kosmatka, S.H. and Panarese, W.C. 1988. *Design and Control of Concrete Mixtures,* 13th ed. Portland Cement Association, Skokie, IL.

and have been proven to work well over time. They come from surfactants that can be categorized into the following seven groups [Dolch, 1984].

1. *Vinusol.*   The most widely used type of air entraining admixture, it is the byproduct of distillation and extraction of pine stumps. The left-over insoluble residue is neutralized in sodium hydroxide, resulting in the soluble solution used to produce entrained air in concrete.

2. *Synthetic detergents.*   These are normally alkyl aryl sulfonates and come from petroleum-based (typically C_{12}) residues that are condensed with benzene, then sulfonated and neutralized to obtain the soluble salt.

3. *Salts of sulfonated lignin.*   These are byproducts of the paper industry. These tend to be relatively poor air-entraining agents but are used extensively in concrete as water reducers and retarders.

4. *Salts of petroleum acids.*   These are the leftovers from petroleum refineries. Sludge left after the extraction of white oils using sulfuric acid contains water-soluble sulfonates that are then neutralized with sodium hydroxide.

5. *Salts of proteinaceous materials.*   These are products of the animal- and hide-processing industries. They consist of salts from a complex mixture of carboxylic and amino acids. This group does not produce many commercial air-entraining admixtures.

6. ***Fatty and resinous acids and their salts.*** These are produced from various sources like production of vegetable oil, coconut oil, and tall oil, another byproduct of the paper industry.
7. ***Organic salts of sulfonated hydrocarbons.*** These are the same water-soluble sulfonates described in group 4, except that they are neutralized with triethanolamine instead of sodium hydroxide.

When determining air content, an observant technician would note that even concrete without air-entraining admixtures contains some air. This air, however, is air that has become entrapped during the mixing process. Entrapped air is evident as much larger irregularly spaced voids that do nothing to resist freeze-thaw stresses.

Entrained air has a significant influence on many of the properties of both fresh and cured concrete. In fresh concrete it is known to reduce water demand and the tendency toward bleeding, as well as reducing plastic shrinkage. It increases slump and workability. In hardened concrete, entrained air improves deicer scaling resistance and freeze-thaw resistance, although small and predictable reductions in compressive, flexural, and bond strengths are to be expected (Table 3.3).

If air is determined to be excessive, very small quantities of defoaming or air-detraining agents may be used, as recommended by the manufacturer. Defoaming agents are typically composed of silicones, esters of carbonic acid (water-insoluble), octyl alcohol, dibutyl phthalate, or tributyl phosphate. These materials are very effective, but too much may hurt concrete properties. Care must be made to use only enough defoamer/detrainer to bring concrete air into the specified range.

All air-entraining admixtures that are to be added at the time of mixing should conform to ASTM C260-95 specifications. Air-entraining cements should conform to ASTM C150-94 and C595-94a using ASTM C226-93 specifications for air-entraining additives in air-entraining cements.

TABLE 3.3 Effect of Entrained Air on Concrete

Properties	Effect
	Plastic Concrete
Bleeding	Significant reduction of bleeding
Plastic shrinkage	Reduction
Slump	Increases: 25mm more slump per 0.5–1% more air
Unit weight	Decreases as air content increases
Water demand (for equal slump)	Decreases 3–6 kg/m³ per 1% increase in air
Workability	Increases as air increases
	Hardened Concrete
Abrasion	Insignificant; only as it relates to strength and E
Absorption	Insignificant
Alkali-silica reactivity	Concrete expansion decreases as air increases
Bond to steel	Decreases as air increases
Compressive strength	Typically decreases strength 2–6% per 1% air increase; unusually harsh or lean mixes may gain strength.
Creep	Insignificant; only as it relates to strength and E
Deicer scaling	Significantly more resistant as air increases
Fatigue	Insignificant; only as it relates to strength and E
Flexural strength	Decreases strength 2–4% per 1% increase in air
Freeze-thaw resistance	Water-saturated freeze-thaw resistance improves with added air
Heat of hydration	Insignificant
Modulus of elasticity	E decreases 724–1380 MPa (1.05×10^5–2.00×10^5 lb/in²) per 1% air increase
Permeability	Minimal; if higher air means lower w/c, then lower permeability results

Source: Kosmatka, S.H. and Panarese, W.C. 1988. *Design and Control of Concrete Mixtures*, 13th ed. Portland Cement Association, Skokie, IL.

Note: E, modulus of elasticity; w/c, water/cement ratio.

TABLE 3.4 Total Target Air Content for Concrete

Nominal Maximum Aggregate Size, in	Air Content, %*		
	Severe Exposure[†]	Moderate Exposure[†]	Mild Exposure[†]
3/8	7 1/2	6	4 1/2
1/2	7	5 1/2	4
3/4	6	5	3 1/2
1	6	4 1/2	3
1 1/2	5 1/2	4 1/2	2 1/2
2[‡]	5	4	2
3[‡]	4 1/2	3 1/2	1 1/2

Source: Kosmatka, S.H. and Panarese, W.C. 1988. *Design and Control of Concrete Mixtures,* 13th ed. Portland Cement Association, Skokie, Ill.

*Project specifications often allow the air content of the delivered concrete to be within −1 to +2 percentage points of the table target values.

[†]Severe exposure is an environment in which concrete is exposed to wet freeze-thaw conditions, deicers, or other aggressive agents. Moderate exposure is an environment in which concrete is exposed to freezing but will not be continually moist, will not be exposed to water for long periods before freezing, and will not be in contact with deicers or aggressive chemicals. Mild exposure is an environment in which concrete is not exposed to freezing conditions, deicers, or aggressive agents.

[‡]These air contents apply to total mix, as for the preceding aggregate sizes. When testing these concretes, however, aggregate larger than 1 1/2 in is removed by hand-picking or sieving and air content is determined on the minus 1 1/2 in fraction of mix. (Tolerance on air content as delivered applies to this value.) Air content of the total mix is computed from the value determined on the minus 1 1/2 in fraction.

Typically, total air content needs are dependent upon anticipated exposure conditions versus the strength and quality of the matrix mortar and coarse aggregates. Table 3.4 illustrates target air content considerations for different sized coarse aggregates.

3.7 Antifreezing Admixtures

As the name implies, this category of admixtures is employed to allow most types of concrete construction work and precasting to take place at freezing and well-below freezing temperatures. These admixtures are sometimes used in conjunction with external energy and heat sources, but are often used without them, even under bitterly cold environmental conditions.

Antifreezing admixtures work by lowering the freezing point of the water in fresh concrete [Brook et al., 1988; Ratinov and Rosenberg, 1984]. This is accomplished by dissolving salts or mixing in higher-molecular-weight alcohols, ammonia, or carbamide into the mix water. For this reason, dosage rates are based upon the amount of mixing water in the given batch design. Most antifreezes tend to seriously retard the set and cure properties of the matrix, and, of course, the freezing temperatures do as well.

It becomes obvious then, that at the same time the water is kept unfrozen and thus available for hydration, acceleration of the hydration process is also important and must be designed so as to work with the antifreezing component. These two active agents, antifreeze and accelerator, are often combined into one complex multicomponent additive. This type of additive normally consists of inexpensive combinations of potash and calcium chloride.

A third group of antifreeze admixture that is used occasionally promotes a highly accelerated hydration process and utilizes the much higher earlier exothermic temperatures in the concrete mass to push the curing process, even though ambient temperatures are too low to allow normal curing. This group includes ferric sulfate and aluminum sulfate.

Dosage rates for the various materials may differ significantly even for the same antifreeze admixture. This is because admixture effect is influenced by many things, including the temperatures of forms, air, aggregates, and mix water, concrete mass/geometry, construction technology, and type and brand of cement used.

3.8 Antiwashout Admixtures

This class of chemical admixture was developed as a viscosity-modifying admixture that could improve the rheological properties of cement paste. It has proved to significantly improve the cohesiveness of concrete placed under water, where the exposed matrix is in jeopardy of being diluted and segregated or washed away by the surrounding water. It is most commonly used underwater in large placements and in repairs. This type of admixture is also used to produce self-leveling concrete and wherever extreme congestion due to reinforcement configurations or unusual geometry of the forms requires a very fluid, cohesive concrete that resists bleeding and segregation [Khayat, 1996].

The disadvantages of these specialty admixtures are the typical reductions in strength and modulus of elasticity. Depending upon the base concrete batch design, water/cement ratio, and type and dosage rate of antiwash admixture (AWA), compressive strength has been measured from 75 to 100% of a control mix without the admixture. Flexural strengths have been reported at 84 to 100%, and modulus of elasticity has been measured at 80 to 100% of the control batch.

The two most commonly used AWAs are based upon welan gum or on hydroxypropyl methyl cellulose. Other AWAs are made from variations of related microbial saccharides like welan gum or different cellulose-based polymers like hydroxyethyl cellulose and hydroxyethyl methyl cellulose.

The mechanisms that enable this admixture to work include the attachment of its long molecules to water molecules. This inhibits free displacement of water by heavier mix constituents. The long chains can slip past each other under conditions of high shear such as mixing and pumping and rapid flow, but when the moving concrete slows down, the chains intermesh and the matrix appears to become much more viscous. This minimizes segregation and bleeding. Workability is usually better also, since the constituents of the mix are better dispersed, even with higher additions of HRWR.

Set time, cure time, shrinkage, and creep are not significantly affected by the presence of AWAs themselves but may be influenced by the addition of higher levels of HRWR associated with the use of AWAs. Electric current passed through cured permeability test specimens is lower in concrete containing AWAs, indicating reduced chloride ion permeability.

Dosage levels for AWAs are totally dependent upon the application needs and constraints, but AWA function seems to be based upon availability of free water. Trial batches must be evaluated for all important properties before concrete containing AWA is placed.

Known interactions with other admixtures include higher air-entraining admixture demand when higher levels of HRWR are needed. Acceptable entrainment levels are easily attainable but require trial-batch evaluations. Hydroxypropyl methyl cellulose tends to entrap air unless a deaerating agent is also added into the mixer. Therefore careful adherence to a specific mixing procedure may be more important than with other batch designs.

3.9 Shrinkage-Reducing Admixtures

Shrinkage reducing or shrinkage-compensating admixtures expand the concrete by the same amount that normal drying shrinkage contracts it. The net change in length of hardened concrete should be small enough to prevent shrinkage cracks. These admixtures can be used to great advantage in slabs, bridge decks, structures, and repair work where cracking can lead to steel reinforcement corrosion problems.

The typical materials used for shrinkage compensation in concrete are based on calcium sulfoaluminate or calcium aluminate and calcium oxide. Some loss in properties is typical with the introduction of these antishrinkage agents. Any ill effects are minimized, however, by the use of HRWR, which keeps the workability good while reducing the water content. Usage rates vary with batch designs and water content, but typically range from 8 to 25%.

ACI 223, Standard Practice for the Use of Shrinkage-Compensating Concrete specifies the best methods for utilization of this admixture.

3.10 Polymer Modifiers and Binder Systems

This section discusses the different types of polymers that are used in concrete. Some are used as polymerizing admixtures for comatrix formation. Some impregnate hydrated concrete with liquids that polymerize into hard plastics. And some even form polymer systems that entirely substitute for the cement and water paste in a polymer-concrete matrix.

3.10.1 Polymer-Modified Concrete

Polymer-modified concrete (PMC), also referred to as latex-modified concrete, is the result of adding higher molecular weight polymers to concrete batch designs for the purposes of improved adhesion, higher chemical resistance, lower permeability, lower drying shrinkage, improved tensile strength, or accelerated curing. Different chemical families and physical forms of polymers have been tried with varying degrees of success, but latex, acrylic, and epoxy additives are the most often used commercially. They are available as powdered or liquid forms of resins, monomers, or emulsions, and their uses include concretes and mortars for flooring, ship decks, bridge deck overlays, repair, anticorrosive coatings, and adhesives.

Improvements over properties of normal concrete or mortar depend upon the polymer-phase formation and cement hydration, forming an interpenetrating network of polymer and hydrated cement phases. The resulting monolithic matrix exhibits properties better than either the hydraulic cement phase or the polymer phase is capable of by itself.

Initially, the mixing process disperses the admixture into the fresh concrete matrix. As hydration begins and free water is lost, membranes of polymer begin forming either through water loss or independent polymerization. These membranes adhere to major portions of the hydrating cement-particle surfaces. As the process continues, membranes interconnect and hydration progresses somewhat independently until both constituents of the matrix have cured (Figure 3.7). This modified matrix tends to arrest propagating microcracking due to local tensile or impact insults to the concrete. The mechanics of this crack arresting seem to depend upon the higher tensile strengths of the polymer network and the better bond of the polymer to aggregates and hydration product.

Because of significantly higher costs, these polymer modifiers are considered to be premium admixtures for concrete with special needs. The gains in tensile and flexural strength are primarily a function of polymer/cement ratio rather than water/cement ratio.

Although mixing and placement of PMC is similar to that of normal concrete, there are some important differences, and finishing and curing may be very different. Initial set times for PMC may be much more sensitive to ambient, substrate, and concrete constituents' temperatures, although normally set times are delayed in PMCs. Because of sensitivity to temperatures and unwanted additional air content from overmixing, it is generally recommended that latex-modified concrete be mixed at the jobsite. With latex in particular, too much entrained air necessitates the addition of an antifoaming agent if it is not already included in the admixture. Initially, workability may be better than unmodified concrete, but as the polymer phase progresses and the surface begins to dry, finishing operations may tear the sticky or crusty surface because of polymer adhesion. Styrene-butadiene resin (SBR) concretes should be wet cured for 24 to 48 h in order to permit concrete to gain strength before the latex is allowed to cure.

Bleeding and segregation are reduced in most PMCs because of the hydrophilic nature of the polymer modifiers used. Placement and finishing tools must be cleaned thoroughly and immediately after each use, and reuseable formwork must be carefully coated with a special form release that releases from the latex or epoxy as well as the cement paste. All joints should be formed, in-

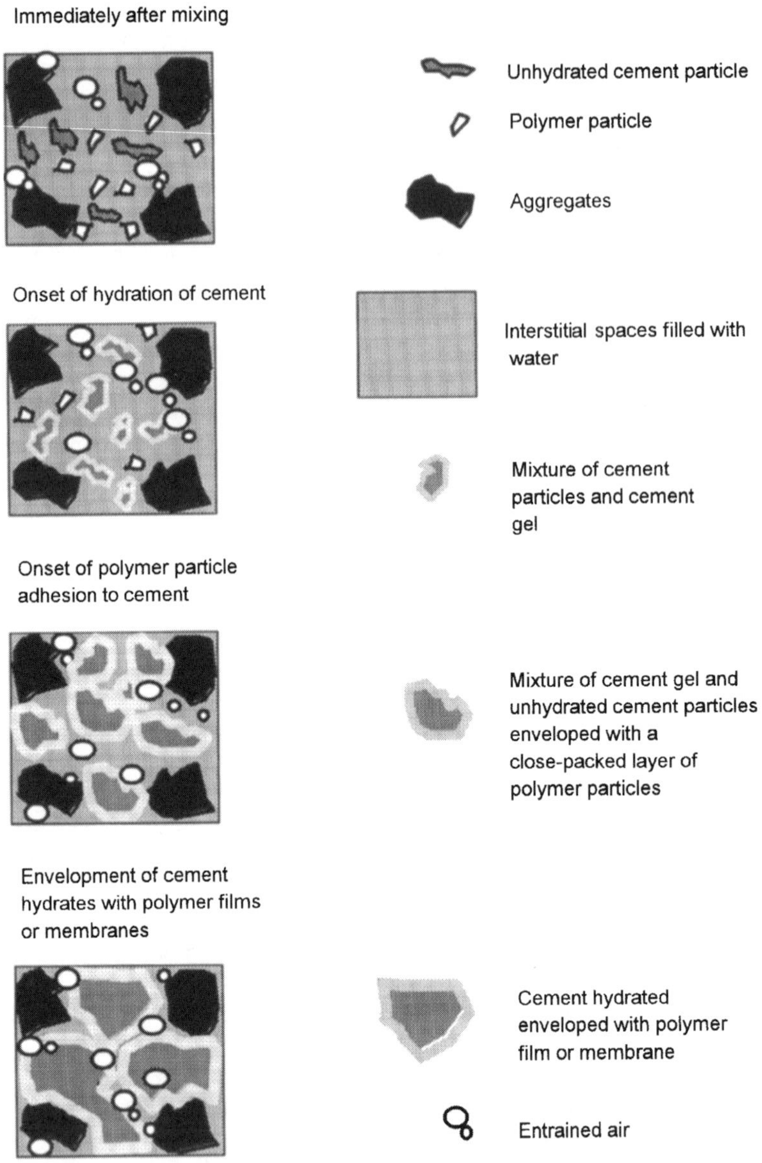

Immediately after mixing

Onset of hydration of cement

Onset of polymer particle
adhesion to cement

Envelopment of cement
hydrates with polymer films
or membranes

Unhydrated cement particle

Polymer particle

Aggregates

Interstitial spaces filled with
water

Mixture of cement
particles and cement
gel

Mixture of cement gel and
unhydrated cement particles
enveloped with a
close-packed layer of
polymer particles

Cement hydrated
enveloped with polymer
film or membrane

Entrained air

FIGURE 3.7 Simplified formation model for a latex-cement comatrix. *Source:* Ohama, Y. 1973. Study on properties and mix proportioning of polymer modified mortars for buildings. In *Report of the Building Research Institute, No. 65.* Building Research Institute, Tokyo, Japan.

cluding construction joints and control joints. Joints sawed early for controlled cracking are not recommended [ACI 548 Committee, 1992].

It is important to anticipate changes in hardened concrete properties. Shrinkage is dependent upon the batch design and the choice and amount of polymer modifier. Some PMCs exhibit the same or less shrinkage and some exhibit more than expected from normal concrete. Creep properties from SBR latex modified concrete are typically lower than for normal concrete, but may be significantly higher for, epoxy modified concrete depending upon the polymer loading in the matrix. The coefficient of thermal expansion for PMCs is minimally affected by the polymer since it is such a small constituent in the matrix compared to the aggregates. Although significant increases in tensile strength, flexural strength, and bond strength may be expected, compressive strengths do not necessarily increase. Cured PMC may exhibit lower water absorption and water or water vapor permeability because larger pores are filled with polymer (Figures 3.8, 3.9, and 3.10). However, many polymers reemulsify, and those may reduce the strength of the matrix when wet. Epoxy-modified concretes are more suitable for constantly wet conditions [Popovics, 1985].

PMC strengths are better at ambient temperatures below 100°F. At typical ambient temperatures, PMC will remain higher in tensile strength than conventional concrete, but the strength benefits

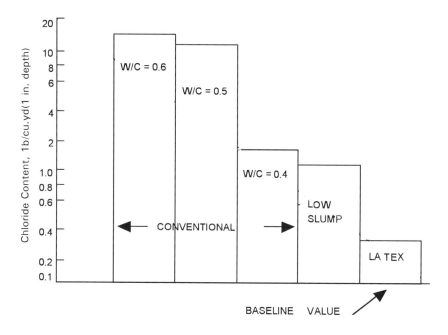

FIGURE 3.8   FHWA test results of 90-d ponding test. From Dow Chemical Company.

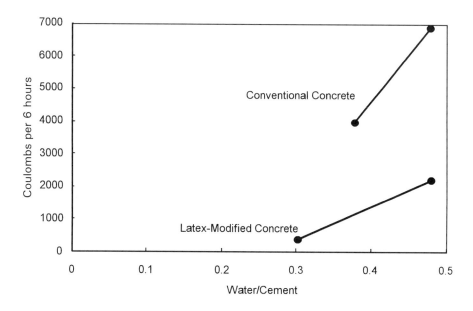

FIGURE 3.9   Permeability versus water-cement ratio. From Dow Chemical Company.

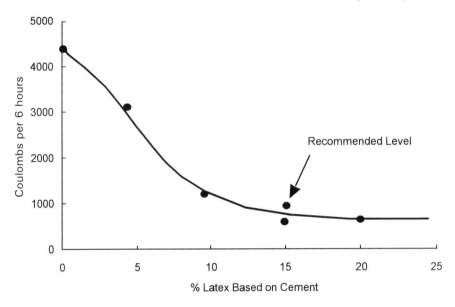

FIGURE 3.10   Effect of latex level on permeability. From Dow Chemical Company.

decrease with an increase in temperature. Abrasion resistance, frost resistance, chemical resistance, and the bond to concrete or steel substrates are normally better in PMC than in unmodified concrete.

Epoxy-modified concrete typically has a polymer/cement ratio of 20% for water-reducible resins and more than 30% for the others (even more than 50% for some resins) [Ohama, 1984]. This makes epoxy-modified concretes more expensive than latex modified, but all strengths are typically higher, including compressive strength, and these systems can be cured in water, where as latex-modified concrete usually must be allowed to dry after 48 h of moist curing.

Other commercially available polymers such as polyvinyl acetates, water-soluble unsaturated polyester resins, methyl cellulose, and polyurethanes are used to modify concrete for specific applications, but latexes (particularly SBR and acrylic) and epoxies are the two that share the most of the market.

PMC is used most often in protective coatings like shotcrete [Shorn, 1985], overlays [Irvin, 1989], and large surface-area repairs [Smoak, 1985] because these applications can take cost-effective advantage of the better adhesion, better abrasion resistance, and lower permeability afforded the matrix through the use of polymer modifiers. Also the crews working with these applications generally tend to be better trained for the special batching, placing, finishing, and curing needs of PMCs.

Job-specific materials, usage, and batch-design recommendations from polymer-modifier manufacturers should be solicited and closely followed to make trial batches under field conditions, and must be tested for critical properties before large-scale use of PMC is implemented.

3.10.2 Polymer Impregnated Concrete

Polymer impregnated concrete (PIC) results from filling the pores of cured and dried concrete with a polymerizing liquid that cures in concrete pores to a hardened plastic interpenetrating network that is much stronger than the PMC system. This process is done to render the concrete relatively impermeable and much more resistant to abrasion, scaling, cavitation (in spillways), and corrosion. Additionally, the relatively incompressible plastic in the pores significantly increases the strength and modulus and reduces creep in the concrete. Although it has shown many promising property improvements for concrete in the laboratory, it has been used only minimally in industry or the field. This is because the process is fairly cost intensive to start up and is probably only justifiable

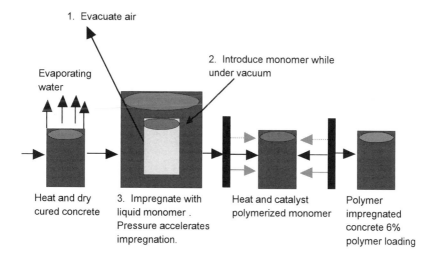

FIGURE 3.11 Schematic for the method of impregnating concrete with polymer. Courtesy of the American Concrete Institute; ACI Committee 548, 1992, Guide for Use of Polymers in Concrete.

on a volume basis. Also, the monomers most often used are volatile, odorous, flammable, and hazardous to personnel when used outside a plant facility, and this may be more than a contractor is willing to deal with.

Commercially, PIC is normally thought of in terms of partial-depth impregnation, where fully hydrated concrete surface is soaked with a low-viscosity, good-wetting liquid prepolymeric resin or monomer until the desired depth of surface saturation is reached. The liquid in the concrete is then quickly polymerized with the application of heat or microwave radiation. This has been done successfully on bridge decks [Fowler et al., 1973], vertical walls of dam outlets [Fowler and Paul, 1975], and dam spillways [Smoak, 1985].

Another method is known as full-depth impregnation. In this method, the dried concrete element is immersed in a bath of polymerizable liquid until it is saturated throughout. At this point, the element is removed from the liquid (or the liquid is drained from the bath or chamber) and cured under hot water, steam, or radiant energy. In the laboratory, vacuum and pressure have been used to speed up the impregnation process, but this process has been used in very few commercial applications (Figure 3.11).

The impregnation and curing process is difficult and costly to perform under field conditions. The emerging benefits of other sytems of admixtures such as the mineral types discussed in Chapter 2 and the chemical types presented in the preceding and following sections have rendered the use of polymer-impregnated concrete unjustifiable except in very specialized cases [Nawy, 1996].

3.10.3 Polymer Concrete

Polymer concrete (PC) is a specialty concrete that has no hydraulic cement binder. Instead the sand, coarse aggregates, and fillers are bound solely by a polymer. The liquid monomer or resin is blended with the inorganic materials to form a very workable mix that can be designed to finish like normal concrete. Through the use of a chemical hardener system that is blended into the liquid prepolymer or monomer binder, the concrete hardens, usually much more quickly than conventional Portland cement concrete. Typically, within hours PC strengths are very high; tensile and flexural strengths are much higher than in ordinary Portland cement concrete, while the modulus of elasticity is somewhat lower.

The coefficient of thermal expansion is typically 3 to 10 times that of normal concrete. Shrinkage and creep are usually significantly higher than for normal concrete. Although optimizing the aggregate blend will mitigate some of the thermal expansion, shrinkage, and creep problems by reducing the polymer content, bond and tensile properties may be reduced when resin content falls below 10 to 11% of the matrix by weight.

PC costs more than normal concrete, and for that reason it is recommended for use in circumstances where normal Portland cement concrete does not perform as well. Such conditions frequently include sites where external or internal chemical attack is problematic in normal concrete. Sulfate attack and alkali-silica reactions do not happen in PC since it is practically impermeable to water, and no freeze-thaw, sulfate-conversion, or alkali reactions with aggregate can be present when no water enters the matrix. This allows users to incorporate into PC many aggregates that would not be chemically acceptable for use in normal concrete.

The bond to normal concrete is also very good, so PC is often used as a durable patching material [Nawy et al., 1977; Nawy, 1981; Kudlapur et al., 1989; Kudlapur and Nawy, 1990], even at sub-freezing temperatures and as a tough, impact-resistant, and chemically protective coating for concrete floors, bridge decks, chemical drains, or foundations for chemical pumps or tanks. Specific chemical resistance is normally a function of the type of polymer binder employed. These binders may be epoxies, polyesters, acrylics, urethanes, styrene, or furan resins. Although most PC systems require dry aggregates for good cohesion and dry substrate surfaces for adhesion, many epoxy grouts and concretes are especially suited to wet and even underwater repairs or placements. Good repair preparation and placement procedures are presented in the ACI Manual of Concrete Practice by ACI Committee 546 (1980) on repairing concrete.

Safety in storing, handling, placing, finishing, curing, and disposal of the polymers, hardeners, peroxides, and cleaning solvents is especially important in these organic-polymer systems. Manufacturers' material safety data sheets must be obtained, carefully read, and followed. Many of these materials can cause problems to lungs, eyes, skin, internal organs, and the environment. A formal training session on safe handling practices for chemicals, cured PC waste materials, and all related equipment should be mandatory for all workers and supervisors involved with the handling of these materials [ACI Committee 548, 1994]. Training topics should include protective apparel, breathing contamination protection, cross-chemical contamination, spark and heat avoidance, proper washing and neutralization methods for people and equipment, and acceptable disposal practices.

3.11 Conclusion

It should be noted that all of the many admixtures are designed to enhance the properties of concrete, but they are not meant to substitute for proper concrete design, batching, transport, and finishing practices. It is often more cost effective to change the mixture proportions or the aggregate than to use higher quantities of admixtures; therefore it is recommended that a cost analysis be done on both the proposed and alternate batch designs.

Any change in admixtures or their quantities should be verified in trial-batch evaluations for strength and any other critical properties before being delivered to any job site. Trial batches should be mixed with all intended admixtures and cured under expected field conditions before evaluation.

References

ACI Committee 201. 1982. *Guide to Durable Concrete*, ACI 201.2R. American Concrete Institute, Farmington Hills, MI.

ACI Committee 211. 1989. *Standard Practice for Selecting Proportions for Normal, Heavyweight, and Mass Concrete*, ACI 211.1-89. American Concrete Institute, Farmington Hills, MI.

ACI Committee 212. 1991. *Chemical Admixtures for Concrete*, ACI 212.3R-91. American Concrete Institute, Farmington Hills, MI.

ACI Committee 212. 1993. *Guide for the Use of High-Range Water-Reducing Admixtures (Superplasticizers) in Concrete*, 212.4R-93. American Concrete Institute, Farmington Hills, MI.

ACI Committee 223. 1993. *Standard Practice for the Use of Shrinkage Compensating Concrete*, ACI 223. American Concrete Institute, Farmington Hills, MI.

ACI Committee 305. 1989. *Hot Weather Concreting*, 305R-89. American Concrete Institute, Farmington Hills, MI.

ACI Committee 306. 1988. *Cold Weather Concreting*, 306R-88. American Concrete Institute, Farmington Hills, MI.

ACI Committee 318. 1989. *Building Code Requirements for Reinforced Concrete*, ACI Report 318. American Concrete Institute, Farmington Hills, MI.

ACI Committee 517. 1987. *Accelerating Curing of Concrete at Atmospheric Pressure-State of the Art*, 517.2R-87. American Concrete Institute, Farmington Hills, MI.

ACI Committee 546. 1980. *Guide for Repair of Concrete Bridge Superstructures*, 546.1R-80. American Concrete Institute, Farmington Hills, MI.

ACI Committee 548. 1992. *Guide for Use of Polymers in Concrete*, 548.1R-92, pp. 1–33. American Concrete Institute, Farmington Hills.

ACI Committee 548. 1994. *Guide for Polymer Concrete Overlays*, 548.5R-94. American Concrete Institute, Farmington Hills, MI.

ASTM C 150-94. Standard Specification for Portland Cement. In *Annual Book of ASTM Standards*, Vol. 04.01, pp. 125–129. American Society for Testing and Materials, Philadelphia, PA.

ASTM C 226-93. Standard Specification for Air-Entraining Additions for Use in Manufacture of Air-Entraining Portland Cement. In *Annual Book of ASTM Standards*, Vol. 04.01, pp. 170–173. American Society for Testing and Materials, Philadelphia, PA.

ASTM C 260-95. Standard Specification for Air-Entraining Admixtures for Concrete. In *Annual Book of ASTM Standards*, Vol. 04.02, pp. 153–155. American Society for Testing and Materials, Philadelphia, PA.

ASTM D 98-93. Standard Specification for Calcium Chloride. In *Annual Book of ASTM Standards*, Vol. 04.02, pp. 20–25. American Society for Testing and Materials, Philadelphia, PA.

ASTM D 345-90. Test Method for Sampling Calcium Chloride for Roads and Structural Applications. In *Annual Book of ASTM Standards*, Vol. 4.03, pp. 63–64. American Society for Testing and Materials, Philadelphia, PA.

ASTM C 494-92. Standard Specification for Chemical Admixtures for Concrete. In *Annual Book of ASTM Standards*, Vol. 04.02, pp. 251–259. American Society for Testing and Materials, Philadelphia, PA.

ASTM C 595-94a. Standard Specification for Blended Hydraulic Cement. In *Annual Book of ASTM Standards*, Vol. 04.01, pp. 284–289. American Society for Testing and Materials, Philadelphia, PA.

ASTM C 1017-92. Standard Specification for Chemical Admixtures for Use Producing Flowing Concrete, In *Annual Book of ASTM Standards*, Vol. 04.02, pp. 498–505. American Society for Testing and Materials, Philadelphia, PA.

Baalbaki, M. 1990. Practical means for estimating superplasticizer dosage: Determining the saturation point. In *Superplasticizers—Report of the Canadian Network of Center of Excellence in High Performance Concrete*, pp. 49–57. University of Sherbrooke.

Brook, J.W., Factor, D.F., Kinney, F.D., and Sarter, A.K. 1988. Cold weather admixture. *Concrete Int.*

Collepardi, M. 1984. Water reducers/retarders. In *Concrete Admixtures Handbook Properties, Science, and Technology*, V.S. Ramachandran, ed., pp. 116–210. Noyes Publications, Park Ridge, NJ.

Collepardi, M. 1994. Superplasticizer and air-entraining agents: State of the art and future needs. In *Proc. V. Mohan Malhotra Symp.*, ACI SP-144, P.K. Mehta, ed., pp. 399–416. American Concrete Institute, Farmington Hills.

Daugherty, K.E. and Kowalewski, M.J., Jr. 1976. Use of Admixtures Placed at High Temperatures. Report 564-10-20. Transportation Research Board.

Dodson, V.H. and Hayden, T.D. 1989. Another look at the Portland cement/chemical admixture incompatibility problem. *Cement, Concrete, and Aggregates*, 11(1):

Dolch, W.L. 1984. Air-entraining admixtures. In *Concrete Admixtures Handbook Properties, Science, and Technology*, V.S. Ramachandran, ed., pp. 269–302. Noyes Publications, Park Ridge, NJ.

Dow Chemical Company. *A Handbook on Portland Cement Concrete and Mortar Containing Styrene/Butadiene Latex*, Midland, MI.

Fisher, T.S. 1994. A contractor's guide to superplasticizers. *Concrete Const.* 39(7):547–550.

Fowler, D.W. 1983. Polymers in Concrete. In *Handbook of Structural Concrete*, pp. 8.1–8.32. McGraw–Hill, New York.

Fowler, D.W. and Paul, D.R. 1975. Polymer Impregnation of Concrete Walls. Research Report to U.S. Army District Engineers. Walla Walla Corps of Engineers, Walla Walla, WA.

Fowler, D.W., Houston, J.T., and Paul, D.R. 1973. Polymer-Impregnated Concrete Surface Treatment for Highway Bridge Decks. *ACI SP-40*, pp. 93–117. American Concrete Institute, Farmington Hills, MI.

Gagné, R., Boisvert, A., and Pigeon, M. 1996. Effect of superplasticizer dosage on mechanical properties, permeability, and freeze-thaw durability of high strength concrete with and without silica fume. *ACI Mater. J.* 93(2):111–120.

Guennewig, T. 1988. Cost effective use of superplasticizers. *Concrete Int.*

Irvin, B.D. 1989. Application of styrene-butadiene latex modified Portland cement concrete overlays in parking structure repair and rehabilitation. In *Polymers in Concrete: Advances and Applications*, P. Mendis and C. McClaskey, eds., pp. 1–14. American Concrete Institute, Farmington Hills, MI.

Johnston, C.D. 1987. Admixture cement incompatibility: A case history. *Concrete Int.*

Khayat, K.H. 1996. Effects of antiwashout admixtures on properties of hardened concrete. *ACI Mater. J.* 93(2):134–146.

Kosmatka, S.H. and Panarese, W.C. 1988. *Design and Control of Concrete Mixtures*, 13th ed. Portland Cement Association, Skokie, IL.

Kudlapur, P., Hanaor A., Balaguru, P.N., and Nawy, E.G. 1989. Evaluation of cold weather patching materials. *ACI Mater. J.* 86(1):36–44.

Kudlapur, P. and Nawy, E.G. 1990. Shear interaction of high strength two-layered concretes at early ages placed in sub-freezing temperatures. *Trans. No. 1284*, pp. 32–52. Transportation Research Board, National Research Council, Washington, DC.

Mindess, S. and Young, J.F. 1981. *Concrete*. Prentice Hall, Englewood Cliffs, NJ.

Nawy, E.G. 1996. *Fundamentals of High Strength High Performance Concrete*, pp. 350. Addison Wesley–Longman, New York and London.

Nawy, E.G., Ukadike, M.M., and Sauer, J.A. 1977. High strength field modified polymer concrete, *Proc. ASCE J. Struct. Div. ST 12*, pp. 2307–2322.

Nawy, E.G. 1981. *Shear Transfer Behavior in Concrete and Polymer Modified Concrete Two-Layered Systems with Application to Infrastructure Rehabilitaion and New Designs*, ACI SP-89-4, pp. 51–90. American Concrete Institute, Farmington Hills, MI.

Nmai, C.K. 1995. Corrosion inhibiting admixtures: Passive, passive-active versus active systems. In *Advances in Concrete Technology, Proc. 2nd CANMET/ACI Int. Symp.*, M. Malhotra, ed., pp. 565–585. American Concrete Institute, Farmington Hills, MI.

Nmai, C.K., Bury, M.A., and Farzam, H. 1994. Corrosion evaluation of a sodium thiocyanate-based admixture. *Concrete Int.* 16(4):22–25.

Ohama, Y. 1973. Study on properties and mix proportioning of polymer modified mortars for buildings. In *Report of the Building Research Institute, No. 65*, Building Research Institute, Tokyo, Japan.

Ohama, Y. 1984. Polymer Modified Mortars and Concretes. In *Concrete Admixtures Handbook Properties, Science, and Technology*, V.S. Ramachandran, ed., pp. 116–210. Noyes Publications, Park Ridge, NJ.

Popovics, S. 1985. Modification of Portland cement concrete with epoxy as admixture. In *Polymer Concrete Uses Materials and Properties*, J. Dikeou and D.W. Fowler, eds., pp. 207–229. American Concrete Institute, Farmington Hills, MI.

Ratinov, V.B. and Rosenberg, T.I. 1984. Antifreezing admixtures. In *Concrete Admixtures Handbook Properties, Science, and Technology*, V.S. Ramachandran, ed., pp. 116–210. Noyes Publications, Park Ridge, NJ.

Shorn, H. 1985. Epoxy modified shotcrete. In *Polymer Concrete Uses Materials and Properties*, J. Dikeou and D.W. Fowler, eds., pp. 249–260. American Concrete Institute, Farmington Hills, MI.

Smoak, W.G. 1985. Polymer impregnation and polymer concrete repairs at Grand Coulee Dam. In *Polymer Concrete Uses Materials and Properties*, J. Dikeou and D.W. Fowler, eds., pp. 43–49. American Concrete Institute, Farmington Hills, MI.

Whiting, D., Nagi, M., Okamoto, P., Yu, T., Peshkin, D., Smith, K., Darter, M., Clifton, J., and Kaetzel, L. 1994. *SHRP-C-373 Optimization of Highway Technology*, Strategic Highway Research Program, National Research Council, Washington, DC.

Laboratory test on long-term deterioration of concrete prisms (Courtesy Portland Cement Association).

4

Long Term Effects and Serviceability

by
Edward G. Nawy, D.Eng., P.E., C.Eng.
Distinguished Professor, Civil Engineering, Rutgers University—The State University of New Jersey. Expert in concrete structures and materials.

4.1 Creep Deformations in Concrete

Creep or lateral material flow is the increase in strain with time due to sustained load. Initial deformation due to load is the **elastic strain**, while the additional strain due to the same sustained load is the **creep strain**. This practical assumption is quite acceptable since the initial recorded deformation includes few time-dependent effects. Figure 4.1 illustrates the increase in creep strain with time, and as in the case of shrinkage, it can be seen that the rate of creep decreases with time. Creep cannot be measured directly but is determined only by deducting elastic strain and shrinkage strain from the total deformation. Although shrinkage and creep are not independent phenomena, it can be assumed that superposition of strains is valid; hence

$$\text{total strain } (\epsilon_t) = \text{elastic strain } (\epsilon_e) + \text{creep } (\epsilon_c) + \text{shrinkage } (\epsilon_{sh})$$

An example [Nawy, 1996c; Ross, 1937] of the relative numerical values of strain due to elastic strain creep, and shrinkage is presented for a normal concrete specimen subjected to 900 psi in

0-8493-2666-4/97/$0.00+$.50

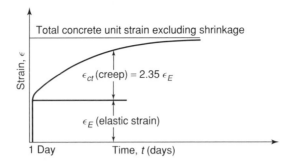

FIGURE 4.1 Long-term creep stress-time curve [Nawy, 1996a,c].

compression:

Immediate elastic strain, ϵ_e	$= 250 \times 10^{-6}$ in/in
Shrinkage strain after 1 yr, ϵ_{sh}	$= 500 \times 10^{-6}$ in/in
Creep strain after 1 yr, ϵ_c	$= 750 \times 10^{-6}$ in/in
	$\epsilon_t = 1500 \times 10^{-6}$ in/in

These relative values illustrate that stress-strain relationships for short-term loading in normally reinforced or plain concrete elements lose their significance and the effects of long-term loadings become dominant on the behavior of a structure. In cases of large heavily reinforced columns in buildings, elastic strain can be a more significant component of the total strain.

Figure 4.2 shows a three-dimensional model of the three types of strain discussed that result from sustained compressive stress and shrinkage. Since creep is time dependent, this model has to be such that its orthogonal axes are deformation, stress, and time.

Numerous tests have indicated that creep deformation is proportional to the applied stress, but the proportionality is valid only for low stress levels. The upper limit of the relationship cannot be determined accurately but can vary between 0.2 and 0.5 of the ultimate strength f_c'. This range in

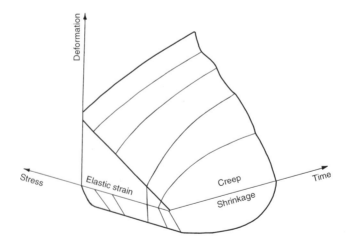

FIGURE 4.2 Three-dimensional model of time-dependent structural behavior. *Source:* Nawy, E.G. © 1996a. *Reinforced Concrete*, 3rd ed. Reprinted by permission of Prentice Hall, Upper Saddle River, NJ.

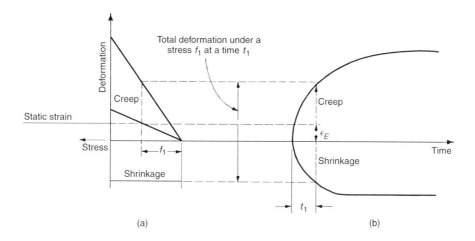

FIGURE 4.3   Two-dimensional section of deformations in Figure 4.2. [Nawy, 1996a,c]. (a) Section parallel to stress-deformation plane; (b) Section parallel to deformation-time plane.

the limit of the proportionality is expected owing to the large number of microcracks that exist at about 40% of the ultimate load.

Figure 4.3a shows a section of the three-dimensional model presented in Figure 4.2 parallel to the plane that contains the stress and deformation axes at time t_1. The figure indicates that both elastic and creep strains are linearly proportional to applied stress. In a similar manner, Figure 4.3b illustrates a section parallel to the plane that contains the time and strain axes at a stress f_1; hence it shows the familiar creep time and shrinkage time relationships.

As in the case of shrinkage, creep is not completely reversible. If a specimen is unloaded after a period of being under a sustained load, an immediate elastic recovery is obtained that is less than the strain precipitated on loading. The instantaneous recovery is followed by a gradual decrease in strain, called **creep recovery**. The extent of the recovery depends on the age of the concrete when loaded; older concretes present higher creep recoveries, while residual strains or deformations become frozen in the structural element, as shown in Figure 4.4.

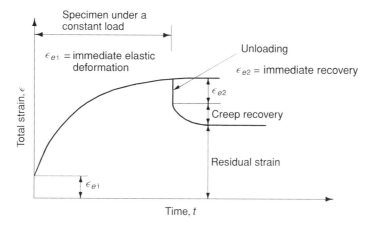

FIGURE 4.4   Creep recovery versus time. *Source:* Nawy, E.G. © 1996a. *Rein-forced Concrete*, 3rd ed. Reprinted by permission of Prentice Hall, Upper Saddle River, NJ.

4.1.1 Creep Effects

As in shrinkage, creep increases the deflection of beams and slabs and causes loss of prestress in prestressed elements. In addition, the initial eccentricity of a reinforced concrete column increases with time owing to creep, resulting in the transfer of the compressive load from the concrete to the steel in the concrete section.

Once the steel yields, additional load has to be carried by the concrete. Consequently, the resisting capacity of the column is reduced and the curvature of the column increases further, resulting in overstress in the concrete and leading to failure. Similar behavior occurs in axially loaded columns.

4.1.2 Rheological Models

Rheological models are mechanical devices that portray the general deformation behavior and flow of materials under stress. A model is basically composed of elastic springs and ideal dashpots denoting stress, elastic strain, delayed elastic strain, irrecoverable strain, and time. The springs represent the proportionality between stress and strain, and the dashpots represent the proportionality of stress to the rate of strain. A spring and a dashpot in parallel form a Kelvin unit, and in series they form a Maxwell unit.

Two rheological models will be discussed: the Burgers model and the Ross model. The Burgers model, in Figure 4.5, is shown since it can approximately simulate the stress-strain-time behavior of concrete at the limit of proportionality, with some limitations. This model simulates the instantaneous recoverable strain (a), the delayed recoverable elastic strain in the spring (b), and the irrecoverable time-dependent strain in dashpots (c and d). The weakness in this model is that it continues to deform at a uniform rate as long as the load is sustained by the Maxwell dashpot—a behavior not similar to concrete, where creep reaches a limiting value with time, as shown in Figure 4.1.

A modification in the form of the Ross rheological model [Ross, 1958] in Figure 4.6 can eliminate this deficiency. In this model, A represents the Hookian direct proportionality of the stress-to-strain element, D represents the Newtonian element, and B and C are the elastic springs that can transmit the applied load $P(t)$ to the enclosing cylinder walls by direct friction. Since each coil has a defined frictional resistance, only those coils whose resistance equals the applied load $P(t)$ are displaced;

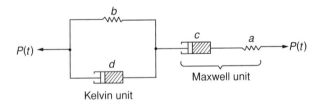

FIGURE 4.5 Burger's rheological model. *Source:* Nawy, E.G. © 1996a. *Reinforced Concrete*, 3rd ed. Reprinted by permission of Prentice Hall, Inc., Upper Saddle River, NJ.

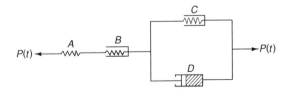

FIGURE 4.6 Ross rheological model. *Source:* Nawy, E.G. © 1996a. *Reinforced Concrete*, 3rd ed. Reprinted by permission of Prentice Hall, Upper Saddle River, NJ.

the others remain unstressed, symbolizing irrecoverable deformation in concrete. As the load continues to increase, it overcomes the spring resistance of unit B, pulling the spring from the dashpot and signifying failure in a concrete element. More rigorous models have been used, such as Roll's model, to assist in predicting the creep strains. Mathematical expressions for such predictions can be very rigorous. One convenient expression from Ross defines creep C under load after a time interval t as follows:

$$C = \frac{t}{a + bt} \tag{4.1}$$

where a and b are constants that can be determined from tests.

As will be discussed in Sections 4.2, Creep Prediction from Nonstandard Conditions, this model seems to represent the creep deformation of concrete and is the background for the ACI Code equations for creep.

4.2 Creep Prediction

4.2.1 Creep Prediction for Standard Conditions

Creep and shrinkage are interrelated phenomena because of the similarity of the variables affecting both, including the forms of their strain-time curves, as seen in Figure 4.3. The ACI [ACI Committee 209, 1992] proposes a similar format for expressing both creep and shrinkage behavior.

The expression for creep is as follows:

$$C_t = \frac{t^\alpha}{a + t^\alpha} C_u \tag{4.2}$$

where a and α are experimental constants and t, in days, is the duration of loading.

Work by Branson (1971, 1977) formed the basis for Eqs. (4.2) and (4.3) in a simplified creep evaluation. The additional strain ϵ_{cu} due to creep can be defined as

$$\epsilon_{cu} = \rho_u f_{ci} \tag{4.3}$$

where

ρ_u = unit creep coefficient, generally called **specific creep**

f_{ci} = stress intensity in the structural member corresponding to initial unit strain ϵ_{ci}

$$C_u = \frac{\epsilon_{cu}}{\epsilon_{ci}} = \rho_u E_c \tag{4.4}$$

If C_u is the ultimate creep coefficient. An average value of $C_u \simeq 2.35$.

Branson's model, verified by extensive tests, relates the creep coefficient C_t at any time to the ultimate creep coefficient for standard conditions as follows:

$$C_t = \frac{t^{0.6}}{10 + t^{0.6}} C_u \tag{4.5}$$

or alternatively

$$\rho_t = \frac{t^{0.6}}{10 + t^{0.6}} \tag{4.6}$$

where t is the time in days during which the load is applied. Standard conditions are summarized in Table 4.1 for both creep and shrinkage [ACI Committee 209, 1992].

TABLE 4.1   Standard Conditions for Creep and Shrinkage Factors

Parameter		Factors	Variable Considered	Standard Conditions
Concrete	Concrete	Cement paste content	Type of cement	Type I and III
(creep and	composition	Water/cement ratio	Slump	2.7 in (70 mm)
shrinkage)		Mixture proportions	Air content	≤6%
		Aggregate characteristics	Fine aggregate percentage	50%
		Degree of compaction	Cement content	470 to 752 lb/yd^3
				(279 to 446 kg/m^3)
	Initial curing	Length of initial curing	Moist cured	7 d
			Steam cured	1–3 d
		Curing temperature	Moist cured	73.4 ± 4°F (23 ± 2°C)
			Steam cured	≤212°F (≤100°C)
		Curing humidity	Relative humidity	≥95
Member	Environment	Concrete temperature	Concrete temperature	73.4 ± 4°F (23 ± 2°C)
geometry and		Concrete water content	Ambient relative humidity	40%
environment				
(creep and	Geometry	Size and shape	Volume/surface ratio (v/s),	v/s = 1.5 in (38 mm)
shrinkage)			or Minimum thickness	6 in (150 mm)
Loading	Loading	Concrete age at load	Moist cured	7 d
(only creep)	history	Application	Steam cured	1–3 d
		Duration	Sustained load	Sustained load
		Duration of unloading	—	—
		period		
		Number of unloading cycles		
	Stress	Type of stress and	Compressive stress	Axial compression
	conditions	distribution across the		
		section		
		Stress/strength ratio	Stress/strength ratio	≤0.50

Source: ACI Committee 209, 1992. *Prediction of Creep, Shrinkage, and Temperature Effects in Concrete Structures*, ACI 209R-92, pp. 1–47. American Concrete Institute, Farmington Hills, MI.

4.2.2 Factors Affecting Creep

Creep is greatly affected by concrete constituents. The coarse-aggregate modulus affects the creep strain level. But the cementitious paste and its shear-friction interaction with the aggregate are constituents that significantly influence the time-dependant load-induced strain. Other factors are environmental effects. A summary of these factors follows:

1. **Sustained load.**   Creep as a result of sustained load is proportional to the sustained stress and is recoverable up to 30–50% of the ultimate strain.
2. **Water/cementitious materials ratio [$W/(C + P)$].**   The higher this ratio, the larger the creep, as seen in Figure 4.7, which relates specific creep to the W/(C + P) ratio [Mindess and Young, 1981].
3. **Aggregate modulus and aggregate/paste ratio.**   For a constant paste-volume content, an increase in aggregate volume decreases creep. As an example, an increase from 65 to 75% lowered creep by 10% [Neville, 1995]. This behavior is the same whether the coarse aggregate is natural stone or lightweight artificial aggregate.
4. **Age at time of loading.**   The older the concrete at the time of loading, the smaller the induced creep strain for the same load level.
5. **Relative humidity.**   Reconditioning the concrete at a lower relative humidity before applying the sustained external load reduces the resulting creep strain. If creep is considered in two categories: drying creep and wetting creep, the creep strain develops irrespective of the direction of change [Mindess and Young, 1981], provided that the exposure is above 40%.
6. **Temperature.**   Creep increases with increase in temperature if the concrete is maintained at elevated temperatures while under sustained load. It increases in a linear manner up to

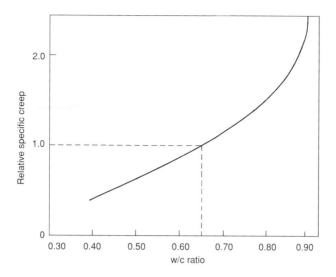

FIGURE 4.7 Water/cement ratio effect on the relative specific creep. *Source:* Mindess, S. and Young, J.F. 1981. *Concrete.* Prentice Hall, Upper Saddle River, NJ.

a temperature of 175°F (80°C). Its value at this temperature level is almost three times the creep value at ambient temperatures.

7. **Concrete member size.** Creep strain decreases with increasing thickness of the concrete member.
8. **Reinforcement.** Creep effects are reduced by use of reinforcement in the compressive zones of concrete members.

4.2.3 Creep Prediction for Nonstandard Conditions

As the standard conditions for creep described in Table 4.1 change, corrective modifying multipliers have to be applied to the ultimate creep coefficient C_u in Eq. (4.5).

If the average ultimate creep C_u is 2.35 for standard conditions, it has to be adjusted by a multiplier γ_{CR} so that

$$C_u = 2.35\gamma_{CR} \qquad (4.7)$$

Here, γ_{CR} has component coefficients that account for the change in conditions enumerated in the preceding section. ACI Committee 209 recommends, in detailed tabular form, the various component coefficients for the γ_{CR} multiplier [ACI Committee 209, 1992]. These are generally based on Branson's (1977) studies. Tabulated values are given in graphical form [ACI Committee 435, 1995; Branson, 1977; Meyers and Thomas, 1983] in Figure 4.8 for the multiplier as follows:

$$\gamma_{CR} = K_h^c K_d^c K_s^c K_f^c K_{ac}^c K_{to}^c$$

where

$\gamma_{CR} = 1$ for standard conditions
K_h^c = relative humidity factor
K_d^c = minimum member thickness factor
K_s^c = concrete consistency factor

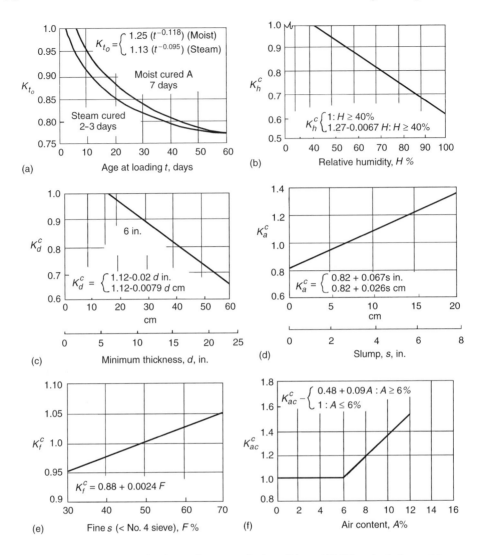

FIGURE 4.8 Creep correction factors for nonstandard conditions, ACI 209 method. *Source:* Meyers, B.L. and Thomas, E.W. 1983. Elasticity, shrinkage, creep and thermal movement of concrete. In *Handbook of Structural Concrete*, Chapt. 11, Kong et al., ed., pp. 11.1–11.33. McGraw–Hill, New York.

K_f^c = fine aggregate content factor
K_{ac}^c = air-content factor
K_{to}^c = age of concrete at load application factor

4.3 Shrinkage in Concrete

4.3.1 General Shrinkage Behavior

In general, there are two types of shrinkage: *plastic shrinkage* and *drying shrinkage*; carbonation shrinkage another form of shrinkage.

Plastic shrinkage occurs during the first few hours after placing fresh concrete in forms. Exposed surfaces such as floor slabs are more easily affected by exposure to dry air because of their large contact surface. In such cases, moisture evaporates from the concrete surface faster than it is replaced

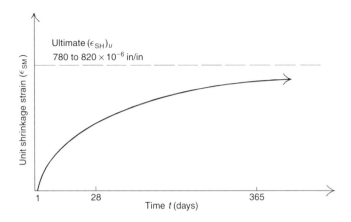

FIGURE 4.9 Concrete shrinkage versus time curve. *Source:* Nawy, E.G. © 1996a. *Reinforced Concrete*, 3rd ed. Reprinted by permission of Prentice Hall, Upper Saddle River, NJ.

by bleed water from the lower layers. **Drying shrinkage**, on the other hand, occurs after the concrete has already attained its final set and a good portion of the chemical hydration process in the cement gel has been accomplished.

Drying shrinkage is the decrease in the volume of a concrete element when it loses moisture by evaporation. The opposite phenomenon, that is, volume increase through water absorption, is termed swelling. In other words, shrinkage and swelling represent water movement out of or into the gel structure of a concrete specimen and is caused by the difference in humidity or saturation levels between the specimen and the surroundings, irrespective of the external load.

Shrinkage is not a completely reversible process. If a concrete unit is saturated with water after having fully shrunk, it will not expand to its original volume. Figure 4.9 relates the increase in shrinkage strain ϵ_{sh} with time. The rate decreases with time since older concretes are more resistant to environmental effects and consequently undergo less shrinkage, so that the shrinkage strain becomes almost asymptotic with time.

Several factors affect the magnitude of drying shrinkage:

1. *Aggregate.* Aggregate acts to restrain the shrinkage of cement paste; hence concretes with high aggregate content are less vulnerable to shrinkage. In addition, the degree of restraint of a given concrete is determined by the properties of the aggregates: those with a high modulus of elasticity or with rough surfaces are more resistant to the shrinkage process. See Figures 4.10 and 4.11 [Mindess and Young, 1981].

2. *Water/cementitious materials ratio.* The higher the water/cementitious materials ratio, the higher the shrinkage effects. Figure 4.12 is a typical plot relating shrinkage to aggregate content and, significantly, to the water/cement ratio.

3. *Size of the concrete element.* Both the rate and total magnitude of shrinkage decrease with an increase in the volume of the concrete element. However, the duration of shrinkage is longer for larger members since more time is needed for drying to reach the internal regions. It is possible that 1 yr may be needed for the drying process to begin at a depth of 10 in from the exposed surface, and 10 yr to begin at 24 in below the external surface; large members may never dry out completely.

4. *Ambient conditions of the medium.* The relative humidity of the medium greatly affects the magnitude of shrinkage; the rate of shrinkage is lower at high relative humidity. Temperature is another factor, in that shrinkage becomes stabilized at low temperatures.

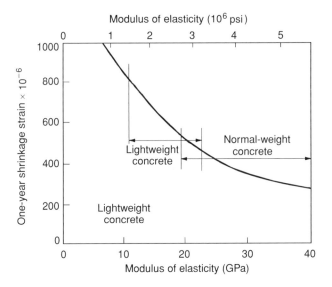

FIGURE 4.10 Aggregate modulus effect on shrinkage strain. *Source:* Mindness, S. and Young, J.F. 1981. *Concrete.* pp. 671. Prentice Hall, Upper Saddle River, NJ.

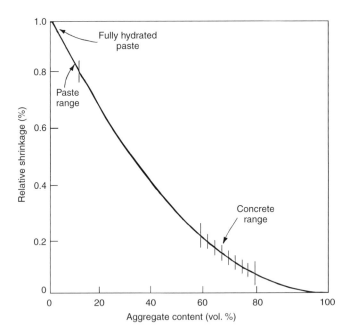

FIGURE 4.11 Aggregate content effect on drying shrinkage. *Source:* Mindness, S. and Young, J.F. 1981. *Concrete.* pp. 671. Prentice Hall, Upper Saddle River, NJ.

5. ***Amount of reinforcement.***   Reinforced concrete shrinks less than plain concrete; the relative difference is a function of the reinforcement percentage.
6. ***Admixtures.***   This effect varies depending on the type of admixture. An accelerator such as calcium chloride, used to accelerate the hardening and setting of the concrete, increases the shrinkage. Pozzolans can also increase the drying shrinkage, whereas air-entraining agents have little effect.

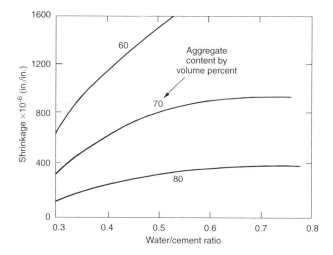

FIGURE 4.12 Water/cement ratio and aggregate content effect on shrinkage. *Source:* Nawy, E.G. © 1996a *Reinforced Concrete*, 3rd ed. Reprinted by permission of Prentice Hall, Upper Saddle River, NJ.

7. ***Type of cement.*** Rapid-hardening cement shrinks somewhat more than other types, while shrinkage-compensating cements minimize or eliminate shrinkage cracking when used with restraining reinforcement.

8. ***Carbonation.*** Carbonation shrinkage is caused by the reaction between carbon dioxide (CO_2) present in the atmosphere and that present in the cement paste. The amount of combined shrinkage from carbonation and drying varies according to the sequence of carbonation and drying processes. If both phenomena take place simultaneously, less shrinkage occurs. The process of carbonation, however, is dramatically reduced at relative humidities below 50%.

4.3.2 Shrinkage Prediction for Standard Conditions

The mathematical model for shrinkage prediction in Eq. (4.3) is

$$(\epsilon_{SH})_t = \frac{t^\beta}{b + t^\beta}(\epsilon_{SH})_u \tag{4.8}$$

where β is a constant and t, in days, is the amount of time after curing that the concrete hardened. The value of the ultimate shrinkage strain at the standard conditions defined in Table 4.1 has the following range:

$$(\epsilon_{SH})_u = 415 \times 10^{-6} \text{ to } 1070 \times 10^{-6} \text{ in/in} \quad (\text{mm/mm})$$

An average value of $(\epsilon_{SH})_u$, as recommended by ACI Committee 209 (1992) is as follows:

Moist cured for 7 d

$$(\epsilon_{SH})_u = 800 \times 10^{-6} \text{ in/in} \quad (\text{mm/mm})$$

Steam cured for 1–3 d

$$(\epsilon_{SH})_u = 730 \times 10^{-6} \text{ in/in} \quad (\text{mm/mm})$$

A common sufficiently accurate average shrinkage strain in standard conditions for both moist-cured and steam-cured concretes which can be used [ACI Committee 209, 1992] is:

$$(\epsilon_{SH})_u = 780 \times 10^{-6} \text{ in/in} \quad (\text{mm/mm})$$

The constant b in the mathematical model of Eq. (4.8) is $b = 35$ for 7-d moist-cured specimens and $b = 55$ for 1–3 d steam-cured specimens. Hence the shrinkage-strain prediction expressions for standard conditions become

Shrinkage after 7 d of moist curing

$$(\epsilon_{SH})_t = \frac{t}{35 + t}(\epsilon_{SH})_u \tag{4.9a}$$

where t is the age of concrete in days after curing.

Shrinkage after 1–3 d of steam curing

$$(\epsilon_{SH})_t = \frac{t}{55 + t}(\epsilon_{SH})_u \tag{4.9b}$$

4.3.3 Shrinkage Prediction for Nonstandard Conditions

As the standard conditions for shrinkage described in Table 4.1 change, corrective modifying multipliers have to be applied to the ultimate value of the shrinkage strain $(\epsilon_{SH})_u$ in Eqs. (4.8) and (4.9).

If γ_{SH} is the shrinkage adjusting-multiplier, the average ultimate shrinkage strain for nonstandard conditions becomes

$$(\epsilon_{SH})_u = 780 \times 10^{-6}\gamma_{SH} \tag{4.10}$$

or

$$(\epsilon_{SH})_{u,n} = \gamma_{SH}(\epsilon_{SH})_u \tag{4.11}$$

where $(\epsilon_{SH})_{u,n}$ is the average ultimate strain for nonstandard conditions. Hence, for nonstandard conditions, Eqs. (4.9a) and (4.9b), respectively, become

$$(\epsilon_{SH})_t = \frac{t}{35 + t}\gamma_{SH}(\epsilon_{SH})_u \tag{4.12a}$$

and

$$(\epsilon_{SH})_t = \frac{t}{55 + t}\gamma_{SH}(\epsilon_{SH})_u \tag{4.12b}$$

The multiplier γ_{SH} has component coefficients that account for the change in conditions enumerated in the General Shrinkage Behavior section. ACI Committee 209 recommends, in detailed tabular form, the various component coefficients for the γ_{SH} multiplier [ACI Committee 435, 1995]. These are generally based on Branson's (1977) studies. The tabulated values [ACI Committee 435, 1995; Branson, 1977; Meyers and Thomas, 1983] are given in graphical form in Figure 4.13 for

$$\gamma_{SH} = K_H^s K_d^s K_s^s K_F^s K_B^s K_{AC}^s \tag{4.13}$$

where

$\gamma_{SH} = 1$ for standard conditions
$K_H^s = $ relative humidity-factor

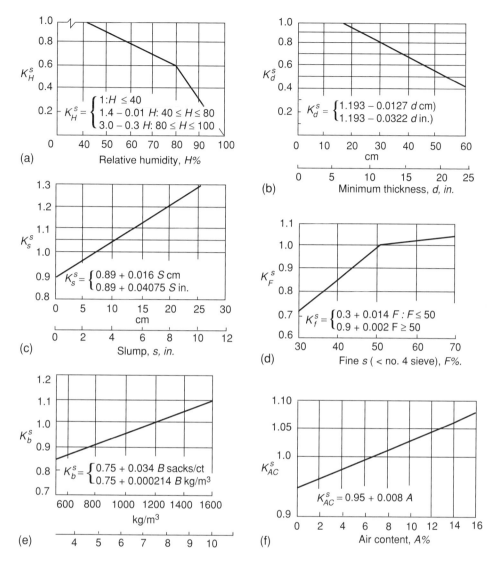

FIGURE 4.13 Shrinkage correction factors for nonstandard conditions, ACI method. *Source:* Meyers, B.L. and Thomas, E.W. 1983. Elasticity, shrinkage, creep and thermal movement of concrete. In *Handbook of Structural Concrete*, Chapt. 11, Kong et al., ed., pp. 11.1–11.33. McGraw–Hill, New York.

K_d^s = minimum member thickness factor
K_s^s = slump factor
K_F^s = fine aggregate content factor
K_B^s = cement-content factor
K_{AC}^s = air-content factor

Values of these factors are given in Figure 4.13.

4.3.4 Alternate Method for Shrinkage Prediction in Prestressed Concrete Elements

For standard conditions, the Prestressed Concrete Institute stipulates average value of nominal ultimate shrinkage strain $(\epsilon_{SH})_u = 820 \times 10^{-6}$ in/in (mm/mm). If ϵ_{SH} is the shrinkage strain after

TABLE 4.2   Values of K_{SH} for Post-tensioned Members

	Time from End of Moist Curing to Application of Prestress, Days							
	1	3	5	7	10	20	30	60
K_{SH}	0.92	0.85	0.80	0.70	0.73	0.64	0.58	0.45

Source: Precast/Prestress Concrete Institute (PCI).

adjusting for relative humidity (RH) at a volume-to-surface ratio, v/s, the shrinkage strain is

$$\epsilon_{SH} = 8.2 \times 10^{-6} K_{SH}\left(1 - 0.06\frac{V}{S}\right)(100 - RH) \tag{4.14}$$

where $K_{SH} = 1.0$ for pretensioned members. Table 4.2 gives the values of K_{SH} for post-tensioned members. Adjustment of shrinkage losses for standard conditions as a function of time t in days is made using Eqs. (4.9a) and (4.9b) for standard conditions and Eqs. (4.12a) and (4.12b) for nonstandard conditions. RH = Relative humidity in percent V/S = Volume to surface ratio (in)

4.4 Strength and Elastic Properties of Concrete versus Time

4.4.1 Cylinder Compressive Strength f'_c

Cylinder compressive strength increases with time as the cement hydration reaction progresses in the presence of water. As a function of time, the developing strength [Mindess and Young, 1981] is

$$(f'_c)_t = \frac{t}{\alpha/\beta + t}(f'_c)_u \tag{4.15}$$

where

α/β = age of concrete, in days, at which one half of the ultimate (in time) compressive strength
of concrete ($f'_c)_u$ is reached
t = age of concrete in days

The range of α and β for normal weight, sand lightweight, and all lightweight concrete is

$$\alpha = 0.05 - 9.25$$
$$\beta = 0.67 - 0.98$$

These constants are a function of the type of cement and the type of curing applied. Typical values for α/β and the time strength ratios are given in Table 4.3.

4.4.2 Modulus of Rupture f_r and Tensile Strength f'_t

The modulus of rupture f_r can be expressed as

$$f_r = g_r\sqrt{w(f'_c)_t} \tag{4.16}$$

g_r has a range of 0.6 to 1.00, with an average value of 0.65 (in SI units, this range is 0.012–0.021, with an average of 0.0135 MPa for f_r); w is the unit weight of the concrete in pounds per cubic foot

TABLE 4.3 Values of Constant α/β and Time Strength Ratio $(f_c')_t/(f_c')_u$ at a Given Age

Type of Curing	Cement Type	Constant α/β	Days							Years		Ultimate (in time)
			3	7	14	21	28	56	91	1	10	
Moist cured	I	$\alpha/\beta = 4.71$	0.39	0.60	0.75	0.82	0.86	0.92	0.95	0.99	1.0	1.0
	III	$\alpha/\beta = 2.5$	0.54	0.74	0.85	0.89	0.92	0.96	0.97	0.99	1.0	1.0
Steam cured	I	$\alpha/\beta = 1.50$	0.74	0.87	0.93	0.95	0.96	0.98	0.99	1.0	1.0	1.0
	III	$\alpha/\beta = 0.71$	0.81	0.91	0.95	0.97	0.97	0.99	0.99	1.0	1.0	1.0

Source: ACI Committee 209. 1992. *Prediction of Creep, Shrinkage, and Temperature Effects in Concrete Structures*, ACI 209R-92, pp. 1–47. American Concrete Institute, Farmington Hills, MI.

for f_r in psi or kg/m^3 for f_r in megapascals. Hence Eq. (4.16) becomes

$$f_r(\text{psi}) = 0.65\sqrt{wf_c'} \qquad (4.17a)$$

and

$$f_r(\text{MPa}) = 0.013\sqrt{wf_c'} \qquad (4.17b)$$

Equation (4.17) is applicable for concrete strengths up to 12,000 psi (83 MPa). For normal-weight concrete, $w = 145$ lb/ft^3 (2320 kg/m^3); Eq. (4.17) becomes

$$f_r(\text{psi}) = 7.5\sqrt{f_c'} \qquad (4.18a)$$

$$f_r(\text{MPa}) = 0.60\sqrt{f_c'} \qquad (4.18b)$$

ACI Committee 363 (1992) on high-strength concrete [Nawy, 1996a] recommends higher values for the modulus of rupture, as follows, for normal-weight concrete:

$$f_r(\text{psi}) = 11.7\sqrt{f_c'} \qquad (4.19a)$$

$$f_r(\text{MPa}) = 0.94\sqrt{f_c'} \qquad (4.19b)$$

The tensile splitting strength f_t' as recommended in ACI Committee 363 (1992) and ACI (1997) for normal-weight concrete of a compressive-strength range up to 12,000 psi (83 MPa) is

$$f_t'(\text{psi}) = 7.4\sqrt{f_c'} \qquad (4.20a)$$

$$f_t'(\text{MPa}) = 0.59\sqrt{f_c'} \qquad (4.20b)$$

4.4.3 Modulus of Elasticity, E_c

The modulus of elasticity of concrete is strongly influenced by the concrete materials and mix proportions used. An increase in compressive strength is accompanied by an increase in the modulus, since the slope of the ascending branch of the stress-strain diagram becomes steeper. For concretes with densities in the range of 90 to 155 lb/ft^3 (1440–2320 kg/m^3), based on the secant modulus at $0.45 f_c'$ intercept and compressive strength up to 6000 psi (42 MPa)

$$E_c(\text{psi}) = 33w^{1.5}\sqrt{f_c'} \qquad (4.21a)$$

$$E_c(\text{MPa}) = 0.0143w_c^{1.5}\sqrt{f_c'} \qquad (4.21b)$$

As the strength of the concrete increases beyond 6000 psi, the measured value of E_c increases at a slower rate such that the value expressed in Eq. (4.21) underestimates the actual value of the modulus. The value of the modulus for a compressive strength range of 6000–12,000 psi (42–83 MPa) [Nilson, 1985] can be predicted by

$$E_c(\text{psi}) = \left(40{,}000\sqrt{f_c'} + 1.0 \times 10^6\right)\left(\frac{w_c}{145}\right)^{1.5} \tag{4.22a}$$

$$E_c(\text{MPa}) = \left(3.32\sqrt{f_c'} + 6895\right)\left(\frac{w_c}{2320}\right)^{1.5} \tag{4.22b}$$

Figure 4.14 from Nilson (1985) gives the best fit for E_c versus f_c' for high-strength concretes. Deviations from the predicted values are highly sensitive to properties of the coarse aggregate such as size, porosity, and hardness. When very high strength concretes [20,000 psi (140 MPa) or higher] are used in major structures or when deformation is critical, E_c should be determined from actual field cylinder test values and the 0.45 f_c' intercept in the resulting stress-strain diagram.

Long-term effects on the modulus of elasticity can be viewed in terms of the gain in the compressive strength $(f_c')_t$ such that

$$E_{ct} = E_c\sqrt{[(f_c')_t / f_c']} \tag{4.23}$$

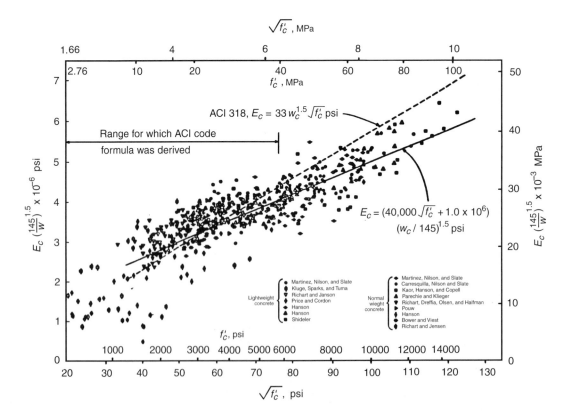

FIGURE 4.14 Modulus of elasticity versus concrete strength. *Source:* Courtesy of American Concrete Institute; Nilson, A.H. 1985. Design implications of current research on high strength concrete, *ACI SP-87-7.*

where $(f'_c)_t$ is equal to compressive strength at later ages and f'_c is equal to 28 d compressive strength.

4.5 Serviceability Long-Term Considerations

In concrete structural members, serviceability is evaluated by cracking and deflection behavior. Creep and shrinkage effects on cracking and deflection are well established. Both deflections and crack widths increase with time. As a section cracks, its gross moment of inertia is reduced, resulting in reduced stiffness, and hence larger deformations and deflections. The crack width, w, and the cracking moment, M_{cr}, are the principal parameters, together with the contribution of the reinforcement (compressive reinforcement in the case of deflection), that determine the long-term behavior of structural elements and systems.

4.5.1 Cracking Moment M_{cr} and Effective Moment of Inertia I_e

4.5.1.1 Reinforced Concrete Beams

Tension cracks develop when externally imposed loads cause bending moments in excess of the cracking moment, M_{cr}. As a result, tensile stresses in the concrete at the tensile extreme fibers exceed the modulus, f_r, of the concrete. The cracking moment for a noncracked section can be computed from the basic flexural formula

$$M_{cr} = \frac{f_r I_g}{Y_t} \tag{4.24a}$$

$$M_{cr} = \frac{f_r}{S_t} \tag{4.24b}$$

where

f_r = modulus of rupture
I_g = gross moment of inertia
y_t = distance form the neutral axis to the extreme *tension* fibers
S_t = section modulus to the extreme tension fibers

Cracks develop at several sections along a member length. At those sections where the modulus of rupture, f_r, is exceeded, cracks develop and the moment of inertia is reduced to a cracked moment, I_{cr}. At other sections' where cracks did not develop, I_g is used for evaluating the sections' stiffness.

Branson's work (1977), the basis of the ACI 318 Code, proposes using the effective moment of inertia I_e for cracked sections as follows,

$$I_e = \left(\frac{M_{cr}}{M_g}\right)^3 I_g + \left[1 - \left(\frac{M_{cr}}{M_a}\right)^3\right] I_{cr} \le I_g \tag{4.25}$$

where

M_{cr} = cracking moment
M_a = maximum moment at the stage at which deflections are being considered
I_g = gross moment of inertia of the section
I_{cr} = moment of inertia of the cracked transformed section

The two moments, I_g and I_{cr}, are based on the assumption of bilinear load-deflection behavior, as

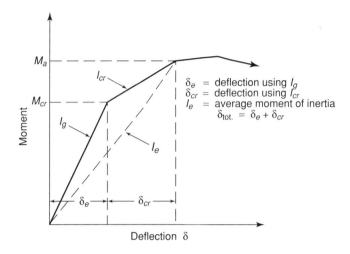

FIGURE 4.15 Bilinear moment of inertia diagram. *Source:* Courtesy of American Concrete Institute; ACI Committee 435. 1995. *Control of Deflection in Concrete Structures.*

seen in Figure 4.15 [Nawy, 1996a]. The cracked moment of inertia I_{cr} is

$$I_{cr} = nA_s d(1 - 1.6\sqrt{n\rho}) \tag{4.26}$$

where

n = modular ratio = E_s/E_c
$\rho = A_s/b\,d$
d = effective depth

Eq. (4.25) can be rewritten as follows:

$$I_e = I_{cr} + \left(\frac{M_{cr}}{M_a}\right)^3 (I_g - I_{cr}) \leq I_g \tag{4.27}$$

Continuous Members. For continuous beams, ACI 318-95 stipulates that I_e may be taken as the average value obtained from Eqs. (4.25) or (4.26) for the critical positive and negative moment sections. For prismatic sections, I_e may be taken as the value obtained at midspan for continuous spans. If the designer chooses to average the effective moments of inertia, I_e, the following expression is used

$$I_e = 0.5 I_{e(m)} + 0.25(I_{e1} + I_{e2}) \tag{4.28}$$

where m, 1, and 2 refer to midspan and the two beam ends, respectively.

Improved results for continuous prismatic members can, however, be obtained using a weighted average as follows [ACI Committee 435, 1995] for beams that are continuous on both ends

$$I_e = 0.70 I_{e(m)} + 0.15(I_{e1} + I_{e2}) \tag{4.29a}$$

Or for beams continuous on one end

$$I_e = 0.85 I_{e(m)} + 0.15(I_{e1}) \tag{4.29b}$$

4.5.1.2 Prestressed Concrete Beams

The effective moment of inertia, I_e, in Eqs. (4.25) or (4.27) is based on different moment levels for M_{cr} and M_a in the case of prestressed concrete beams because of the initial compressive stress imposed by the prestressing force. The $(\frac{M_{cr}}{M_a})$ value is defined by

$$\left(\frac{M_{cr}}{M_a} \right) = \left(1 - \frac{f_{TL} - f_r}{f_L} \right) \tag{4.30}$$

where

$$f_r = 7.5\lambda \sqrt{f_c'}$$

(here, $\lambda = 1.0$, normal concrete; $\lambda = 0.85$ sand lightweight concrete; $\lambda = 0.75$ all lightweight concrete)

where

f_{TL} = *total* calculated stress in the member
f_L = calculated stress owing to *live load*
M_{cr} = moment due to that portion of the *unfactored live-load* moment that causes cracking
M_a = maximum *unfactored* live-load moment
y_t = distance from the neutral axis to the tensile face

In prestressed beams that are partially prestressed by the addition of mild steel reinforcement

$$I_{cr} = \left(n_p A_{ps} d_p^2 + n_s A_s d^2 \right) \left[\left(1 - 1.6 \sqrt{n_p \rho_p + n_s \rho} \right) \right] \tag{4.31}$$

4.5.1.3 Effect of Compression Reinforcement

Compression reinforcement in reinforced flexural members and nontensioned reinforcement in prestressed flexural members tend to offset the movement of the neutral axis caused by creep [ACI Committee 209, 1992]. A reverse movement toward the tensile fibers can thus result.

A multiplier λ has to be used to account for increase in deflection, as required in the ACI 318 Building Code

$$\lambda = \frac{\xi}{1 + 50\rho'} \tag{4.32}$$

where

ξ = time-dependent factor for the long-term increase in deflection obtained from Figure 4.16 [ACI Committee 435, 1995].
$\rho' = A_s'/b\,d$
A_s' = area of compression reinforcement (in square inches)

Nilson (1985) suggested that two modifying factors should be applied to Eq. (4.32): the material modifier μ_m to be applied to ξ and the section modifier μ_s to be applied to ρ'. Both μ_m and μ_s have a value of one or less. Combining the two multipliers, without significant loss in accuracy,

Eq. (4.32) becomes

$$\lambda = \frac{\mu\xi}{1 + 50\mu\rho'} \tag{4.33}$$

with the range of μ value as follows within the 6000 to 9500 psi (42–66 MPa) compressive strength tests conducted:

$$\mu \geq 0.7$$
$$\mu \leq (1.3 - 0.00005\, f_c') \leq 1.0 \tag{4.34a}$$

or, megapascals for f_c'

$$\mu \leq (1.3 - 0.0072\, f_c') \leq 1.0 \tag{4.34b}$$

Further evaluations are needed for cases where the concrete strength is higher than 12,000 psi (83 MPa).

4.5.2 Flexural Crack Width Development

External load results in direct and bending stresses, which cause flexural, bond, and diagonal tension cracks. Immediately after the tensile stress in the concrete exceeds its tensile strength at a particular location, any internal microcracks that might have formed begin to propagate into macrocracks. These cracks develop into macrocracks that propagate to the external fiber zones of the element.

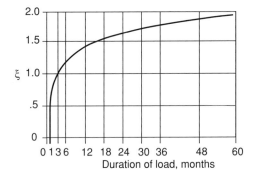

FIGURE 4.16   Multipliers for long-term deflection. *Source:* Nawy, E.G. © 1996a. *Reinforced Concrete*, 3rd ed. Reprinted by permission of Prentice Hall, Upper Saddle River, NJ.

Immediately after full development of the first crack in a reinforced concrete element, the stress in the concrete at the cracking zone is reduced to zero and is assumed by the reinforcement [Nawy, 1996a,c]. The distribution of ultimate bond stress μ, longitudinal stress in the concrete f_t, and longitudinal tensile stress f_s in the reinforcement can be schematically represented, as shown in Figure 4.17.

Crack width is primarily a function of the deformation of reinforcement between the two adjacent cracks, 1 and 2, in Figure 4.17 if the small concrete tensile strain along the crack interval a_c is neglected. The crack width is thus a function of crack spacing up to the load level at which no more cracks develop, which leads to stabilization of the crack spacing as in Figure 4.18.

The major parameters affecting the development and characteristics of cracks are percentage of reinforcement, bond characteristics and size of bar, concrete cover, and the concrete area in tension. On this basis, one can propose the following mathematical model:

$$w = \alpha a_c^\beta \epsilon_s^\gamma \tag{4.35}$$

where w is maximum crack width and α, β, and γ are nonlinearity constants. Crack spacing a_c is a function of the factors to be subsequently discussed and is inversely proportional to bond strength

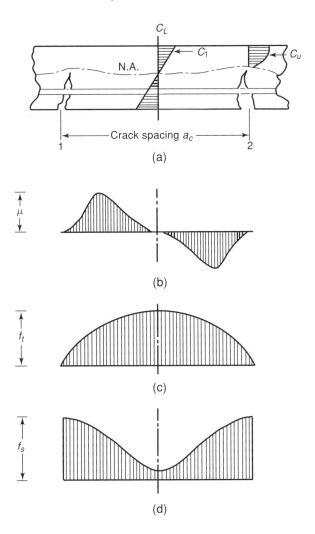

FIGURE 4.17   Schematic stress distribution between two flexural cracks [Nawy, 1996a; Nawy and Blair, 1971].

and active steel ratio (steel percentage in terms of the concrete area in tension). Here, ϵ_s is the strain in the reinforcement induced by external load.

The basic mathematical model in Eq. (4.35), with the appropriate experimental values of the constants α, β, and γ, can be derived for a particular type of structural member. Such a member can be a one-dimensional element such as a beam, a two-dimensional structure such as a two-way slab, or a three-dimensional member such as a shell or circular tank wall. Hence it is expected that different forms or expressions apply for evaluation of macrocracking behavior of different structural elements consistent with their fundamental structural behavior [ACI Committee 318, 1996; ACI Committee 224, 1980; CEB-FIP, 1990; Gergely and Lutz, 1968; Nawy, 1972a,b; Nawy, 1994; Nawy and Blair, 1971].

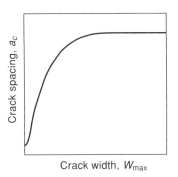

FIGURE 4.18   Schematic variation of crack width with crack spacing [Nawy, 1996a; Nawy and Blair, 1971].

4.5.2.1 Reinforced Concrete Beams and One-Way Slabs

The requirements for crack control in beams and thick one-way slabs [10 in (250 mm) or thicker] in the ACI Building Code [ACI Committee 318, 1996] are based on the statistical analysis of maximum crack width data from a number of sources. On the basis of the analysis, the following general conclusions were reached [Gergely and Lutz, 1968; Nawy, 1972a,b; Nawy, 1994; Nawy and Blair, 1971]:

1. The steel stress is the most important variable.
2. The thickness of the concrete cover is an important variable.
3. The area of concrete surrounding each reinforcing bar is an important geometric variable.
4. Bar diameter is not a major variable.
5. The bottom crack width is influenced by the amount of strain gradient from the level of the steel to the tension face of the beam.

The simplified expression relating crack width to steel stress is given as follows [Gergely and Lutz, 1968].

$$w_{max} \text{ (in)} = 0.076\beta f_s \sqrt[3]{d_c A} \times 10^{-3} \qquad (4.36)$$

where

f_s = reinforcing steel stress, ksi
A = area of concrete symmetrical with reinforcing steel divided by number of bars, in^2
d_c = thickness of concrete cover measured from extreme tension fiber to center of bar or wire closest thereto, inches
$\beta = h_2/h_1$, where h_1 is the distance from the neutral axis to the reinforcing steel, in; h_2 is the distance from the neutral axis to the extreme concrete tensile surface.

In the ACI Code [ACI Committee 318, 1996], when the design yield strength f_y for tension reinforcement exceeds 40 ksi (276 MPa), cross sections of maximum positive and negative moment have to be so proportioned that the quantity z given by Eq. (4.37)

$$z = f_s \sqrt[3]{d_c A} \qquad (4.37)$$

does not exceed 175 kips/in (30 MN/m) for interior exposure and 145 kips/in (25 MN/m) for exterior exposure. Calculated stress in the reinforcement at service load f_s, (kips/in^2) is computed as the moment divided by the product of steel area and internal moment arm. In lieu of such computations, it is allowable to take f_s as 60% of the specified yield strength f_y.

When the strain, ϵ_s, in the steel reinforcement is used instead of stress, f_s, Eq. (4.37) becomes

$$w = 2.2\beta\epsilon_s \sqrt[3]{d_c A} \qquad (4.38)$$

Equation (4.38) is valid in any system of measurement.

The cracking behavior in thick one-way slabs is similar to that in shallow beams. For one-way slabs that have a clear concrete cover in the range of 1 in (25.4 mm), Eq. (4.38) can be adequately applied if $\beta = 1.25$–1.35 is used.

4.5.2.2 Prestressed Concrete Beams

Crack Spacing. Primary cracks form in the region of maximum bending moment when the external load reaches the cracking load. Sometimes, in post-tensioned parking garage elements, cracks form in the draped region before forming at the maximum moment region. They can also form at debonded tendon locations. As loading is increased, additional cracks will form, and the number of cracks will be stabilized when the stress in the concrete no longer exceeds its tensile strength at further locations, regardless of load increase. This condition is important, as it essentially produces the absolute minimum crack spacing that can occur at high steel stresses, termed the **stabilized minimum crack spacing**. The maximum possible crack spacing under this stabilized condition is twice the minimum, termed the **stabilized maximum crack spacing**. Hence the stabilized **mean** crack spacing a_{cs} is deduced as the mean value of the two extremes.

The total tensile force T transferred from the steel to the concrete over the stabilized mean crack spacing [Nawy, 1990, 1994, 1996b] can be defined as

$$T = \gamma\, a_{cs}\, \mu \sum o \tag{4.39a}$$

where

$\gamma =$ a factor reflecting the distribution of bond stress
$\mu =$ maximum bond stress that is a function of $\sqrt{f_c'}$
$\sum o =$ sum of reinforcing elements' circumferences

Figure 4.19 illustrates the forces that cause the formation of the stabilized crack.

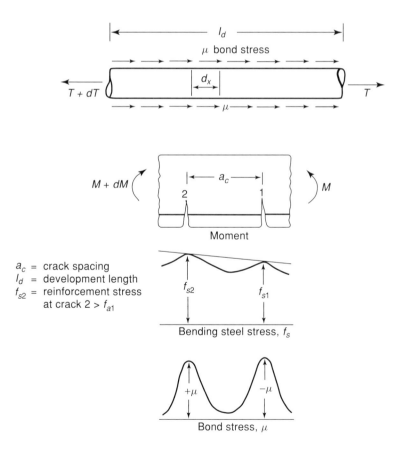

$a_c =$ crack spacing
$l_d =$ development length
$f_{s2} =$ reinforcement stress
    at crack 2 $> f_{a1}$

FIGURE 4.19 Force and stress distribution in stabilized crack in a prestressed Beam. *Source:* Courtesy of American Concrete Institute; Nawy, E.G. 1990. Flexural cracking behavior of partially prestressed pretensioned and post-tensioned beam—State-of-the-art. In *Cracking in Prestressed Concrete Structures.*

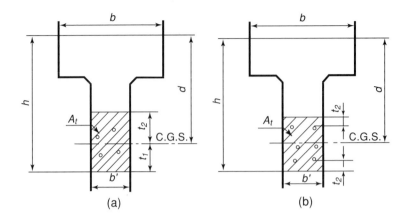

FIGURE 4.20   Effective concrete area in tension. [Nawy, 1990, 1996a]. (a) Even reinforcement distribution; (b) noneven reinforcement distribution.

The resistance R of the concrete area in tension A_t, can be defined as

$$R = A_t f_t' \qquad (4.39b)$$

where f_t' is the tensile splitting strength of the concrete. By equating Eqs. (4.39a) and (4.39b), the following expression for a_{cs} is obtained, where c is a constant to be developed from the tests:

$$a_{cs} = c \frac{A_t f_t'}{\sum o \sqrt{f_c'}} \qquad (4.40a)$$

The concrete stretched area, namely, the concrete area A_t that is under tension for both the evenly distributed and nonevenly distributed reinforcing elements, is illustrated in Figure 4.20. With a mean value of $f_t'/\sqrt{f_c'} = 7.95$, the mean stabilized crack spacing becomes

$$a_{cs} = 1.20 \frac{A_t}{\sum o} \qquad (4.40b)$$

Crack Width.   If Δf_s is the net stress in the prestressed tendon or the magnitude of the tensile stress in normal steel at any crack width load level in which the decompression load (decompression here means $f_c' = 0$ at the level of the reinforcing steel) is taken as the reference point (Nawy, 1990, 1996b), then for the prestressed tendon

$$\Delta f_s = f_{nt} - f_d \qquad (4.41)$$

where

     f_{nt} = stress in the prestressing steel at any load beyond the decompression load
     f_d = stress in the prestressing steel corresponding to the decompression load

The unit strain $\epsilon_s = \Delta f_s / E_s$. It is logical to disregard as insignificant unit strains in the concrete caused by temperature, shrinkage, and elastic shortening effects. The maximum crack width as defined in Eq. (4.35) can therefore be taken as

$$w_{max} = k a_{cs} \epsilon_s^\alpha \qquad (4.42a)$$

or

$$w_{\max} = k' a_{cs} (\Delta f_s)^{\alpha} \qquad (4.42b)$$

where k' is a constant in terms of constant k.

Expression for Pretensioned Beams. Equation (4.42b) is rewritten in terms of Δf_s to give the maximum crack width at the reinforcement level as follows:

$$w_{\max} \text{ (in)} = 5.85 \times 10^{-5} \frac{A_t}{\sum o}(\Delta f_s) \qquad (4.43a)$$

where A_t = square inches, $\sum o$ = inches and Δf_s = kips/in^2

$$w_{\max} \text{ (mm)} = 8.48 \times 10^{-5} \frac{A_t}{\sum o}(\Delta f_s) \qquad (4.43b)$$

(where A_t = square centimeters, $\sum o$ = cm, and Δf_s = MPa). The maximum crack width (inches) at the tensile face of the concrete is

$$w'_{\max} = 5.85 \times 10^{-5} R_i \frac{A_t}{\sum o}(\Delta f_s) \qquad (4.43c)$$

where R_i is the distance ratio equal to h_2 / h_1 with h_2 being the distance from the neutral axis to the extreme tension fibers and h_1 the distance from the neutral axis to the reinforcement centroid.

A plot of the pretensioned beams test data and the best fit expression for Eq. (4.43a) is given in Figure 4.21 with a 40% spread, which is reasonable in view of the randomness of crack development.

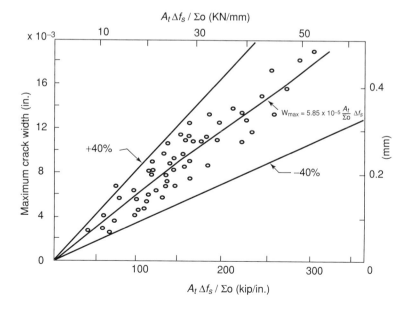

FIGURE 4.21   Linearized maximum crack width versus $(A_t/\sum o)\Delta f_s$ for pretensioned beams [Nawy, 1990, 1994, 1996a,c].

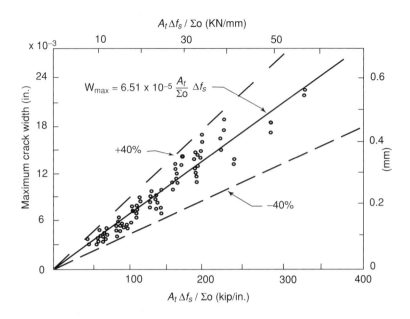

FIGURE 4.22   Linearized maximum crack width versus $(A_t/\sum o)\Delta f_s$ for post-tensioned beams [Nawy, 1990, 1994, 1996a,c].

Expressions for Post-tensioned Beams.   The expression developed for the crack width in post-tensioned bonded beams that contain mild steel reinforcement is

$$w_{\max} \text{ (in)} = 6.51 \times 10^{-5}\frac{A_t}{\sum o}(\Delta f_s) \tag{4.44a}$$

$$w_{\max} \text{ (mm)} = 9.44 \times 10^{-5}\frac{A_t}{\sum o}(\Delta f_s) \tag{4.44b}$$

for the width at the reinforcement level closest to the tensile face. At the tensile face, the crack width for the post-tensioned beams becomes

$$w_{\max} \text{ (in)} = 6.51 \times 10^{-5} R_i\frac{A_t}{\sum o}(\Delta f_s) \tag{4.44c}$$

For nonbonded beams, the factor 6.51 in Eqs. (4.44a) and (4.44c) becomes 6.83. A plot of the data and the best fit expression for Eq. (4.44a) is given in Figure 4.22.

A typical plot of the effect of the various steel percentages on the crack spacing at various stress levels Δf_s is given in Figure 4.23. It can be seen from this plot that crack spacing stabilizes at a net stress level range of 30 to 36 kips/in² (207–248 MPa).

Cracking of High-Strength Prestressed Beams.   Analysis of recent work at Rutgers University on the cracking behavior of pretensioned and nonbonded post-tensioned beams having cylinder compressive strengths in the range of 10,200 psi to 14,200 psi (70.3 to 97.9 MPa) have resulted in the following expression for crack width at the reinforcement level of pretensioned members:

$$w_{\max} \text{ (in)} = 2.75 \times 10^{-5}\frac{A_t}{\sum o}(\Delta f_s) \tag{4.45a}$$

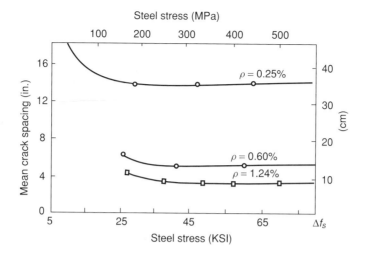

FIGURE 4.23 Effect of steel percentage on mean crack spacing in pre-stressed beams [Nawy, 1990, 1994, 1996a,c].

or

$$w_{\max} \text{ (mm)} = 4.0 \times 10^{-5} \frac{A_t}{\sum o} (\Delta f_s) \qquad (4.45b)$$

The factor 2.75 is an average of values from the following statistical expression [Nawy, 1994] for a reduction multiplier λ_r of $w_{\max}$ in Eq. (4.43) such that

$$\lambda_r = \frac{2}{(0.75 + 0.06\sqrt{f'_c})\sqrt{f'_c}} \qquad (4.45c)$$

This reduced crack width due to use of high-strength concrete is expected in view of the increased bond interaction between the concrete and the reinforcement.

Other Work on Cracking in Prestressed Concrete. After analyzing various investigators results (Harajli and Naaman, 1989; Naaman and Siriaksorn, 1979), Naaman produced the following modified expression for partially prestressed pretensioned members:

$$w_{\max} \text{ (in)} = \left(42 + 5.58 \frac{A_t}{\sum o} (\Delta f_s)\right) \times 10^{-5} \qquad (4.46)$$

This expression is very close to Eq. (4.43) by this author. If plotted against results of the various researchers, work, it gives a best fit as shown in Figure 4.24.

4.5.2.3 Two-Way Supported Slabs and Plates

Flexural crack control is essential in structural floors, most of which are under two-way action. Cracks at service-load and overload conditions can be serious in floors such as those in office buildings, schools, parking garages, industrial buildings, and other floors where the design service load and overload levels exceed loads in normal-size apartment building panels. Such cracks can only lead to detrimental effects on the integrity of the total structure, particularly in adverse environmental conditions.

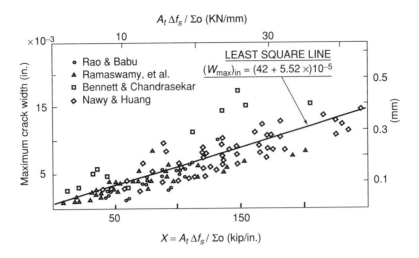

FIGURE 4.24 Reinforcement stress versus crack width (best fit data of several investigators). *Source:* Nawy, E.G. 1994. Cracking of concrete: ACI and CEB approaches. In *Proc. CANMET Int. Symp. Advances in Concrete Technol.*, 2nd ed., V.M. Malhotra, ed., pp. 203–242. Canada Center for Mineral and Energy Technology (CANMET), Ottawa.

Flexural Cracking Mechanism and Fracture Hypothesis. Flexural cracking behavior in concrete structural floors under two-way action is significantly different from that in one-way members. Crack control equations for beams underestimate crack widths developed in two-way slabs and plates and do not tell the designer how to space reinforcement. Cracking in two-way slabs and plates is primarily controlled by the steel stress level and the **spacing** of reinforcement in two perpendicular directions. In addition, the clear concrete cover in two-way slabs and plates is nearly constant [3/4 in (20 mm) for interior exposure], whereas it is a major variable in the crack control equations for beams.

The results from extensive tests on slabs and plates by Nawy et al. [Nawy (1972a, 1994); Nawy and Blair (1971)] demonstrate this difference in behavior in a fracture hypothesis on crack development and propagation in two-way plate action. Nawy's work also conclusively demonstrates that surface deformations of individual reinforcing elements have little effect for arresting the generation of cracks or controlling crack type or width in a two-way-action slab or plate. In another conclusion, one may assume that the scale effect on two-way-action cracking behavior is insignificant, since the cracking grid is a reflection of the reinforcement grid if the preferred orthogonal narrow cracking widths develop.

Therefore, to control cracking in two-way-action floors, the major parameter to be considered is the reinforcement spacing in the two perpendicular directions. Concrete cover has only a minor effect, since the cover is usually small with a constant value of 0.75 in (20 mm). Maximum spacing of the reinforcement in both orthogonal directions should not exceed 12 in (30 cm) in any structural floor.

Crack control equation. The basic Eq. (4.35) for relating crack width to strain in the reinforcement is

$$w = \alpha a_c^{\beta} \epsilon_s^{\gamma}$$

The effect of the tensile strain in the concrete between the cracks is neglected as insignificant. The parameter a_c is the crack spacing, ϵ_s is the unit strain in the reinforcement, and α, β' and γ are constants evaluated by tests. The mathematical model in Eq. (4.35) and the statistical analysis of the data of 90 slabs tested to failure, gives the following equation [ACI Committee 224, 1980; Nawy and Blair, 1971] for serviceability requirements for crack control:

$$w \text{ (in)} = K\beta f_s \sqrt{G_I} \tag{4.47}$$

Using SI units, the expression becomes

$$w_{max} \text{ (mm)} = 0.145 k\beta f_s \sqrt{G_I} \tag{4.48}$$

where f_s is in megapascals and all the terms for the grid index G_I in Eq. (4.49) are in millimeters. $G_I = d_{b1}s_2/\rho_{t1}$ is termed the grid index that defines the reinforcement distribution in two-way action slabs and plates. It can be transformed in Eq. (4.47) to

$$G_I = \frac{s_1 s_2 d_c}{d_{b1}} \frac{8}{\pi} \tag{4.49}$$

where

K = fracture coefficient, having a value of $K = 2.8 \times 10^{-5}$ for uniformly loaded restrained two-way action square slabs and plates. For concentrated loads or reactions, or when the ratio of short to long span is less than 0.75 but larger than 0.5, a value of $K = 2.1 \times 10^{-5}$ is applicable. For a span-aspect ratio of 0.5, $K = 1.6 \times 10^{-5}$. Units of coefficient K are in square inch per pound

β = ratio of the distance from the neutral axis to the tensile face of the slab to the distance from the neutral axis to the centroid of the reinforcement grid (to simplify the calculations use $\beta = 1.25$, although it varies between 1.20 and 1.35)

f_s = actual average service load reinforcement stress level, or 40% of the design yield strength, f_s, ksi

d_{b1} = diameter of the reinforcement in direction 1 closest to the concrete outer fibers, inches

s_1 = spacing of the reinforcement in perpendicular direction 1, inches, closest to the tensile face

s_2 = pacing of the reinforcement in perpendicular direction 2, inches

1 = direction of the reinforcement closest to the outer concrete fibers; this is the direction for which the crack control check is to be made.

ρ_{t1} = active steel ratio in direction 1

$$\rho_{t1} = \frac{\text{area of steel } A_s \text{ per foot width}}{12(d_{b1} + 2c_1)}$$

where c_1 is the clear concrete cover measured from the tensile face of the concrete to the nearest edge of the reinforcing bar in direction 1.

w = crack width at the face of the concrete caused by flexural load, inches.

Subscripts one and two pertain to the directions of reinforcement. Detailed values of the fracture coefficients for various boundary conditions are given in Table 4.4. A graphical solution of Eq. (4.47) is given in Figure 4.25 for $f_y = 60$ ksi (414 MPa) and $f_s = 40\%$, $f_y = 40$ ksi (165.5 MPa) for rapid determination of the reinforcement size and spacing needed for crack control.

Since cracking in two-way slabs and plates is primarily controlled by the grid intersections of the reinforcement, concrete strength is not of major consequence. Hence the value of crack width in two-way action predicted by Eq. (4.47) should not be significantly affected if higher-strength concretes are used in excess of 6000 psi (41.4 MPa). It has to be pointed out that in two-way normal-slab floors, the use of much higher strengths is not justified in economical terms.

Tolerable crack widths in concrete structures. The maximum crack width that a structural element should be permitted to develop depends on the particular function of the element and the

TABLE 4.4   Fracture Coefficients for Slabs and Plates

Loading Type*	Slab Shape	Boundary Condition†	Span Ratio,‡ S/L	Fracture Coefficient 10^{-5} K
A	Square	4 edges r	1.0	2.1
A	Square	4 edges s	1.0	2.1
B	Rectangular	4 edges r	0.5	1.6
B	Rectangular	4 edges r	0.7	2.2
B	Rectangular	3 edges r, 1 edge h	0.7	2.3
B	Rectangular	2 edges r, 2 edges r	0.7	2.7
B	Square	4 edges r	1.0	2.8
B	Square	3 edges r, 1 edge h	1.0	2.9
B	Square	2 edges r, 2 edges h	1.0	4.2

Sources: Nawy, E.G. and Blair, K.W. 1971. Further studies on flexural crack control in structural slab systems. In *Cracking, Deflection, and Ultimate Load of Concrete Slab Systems*, ACI SP-30, E.G. Nawy, ed., pp. 1–41. American Concrete Institute, Farmington Hills, MI.

*Loading type: A, concentrated; B, uniformly distributed.
†Boundary condition: r, restrained; s, simply supported; h, hinged.
‡Span ratio S/L:S, clear short span; L, clear long span.

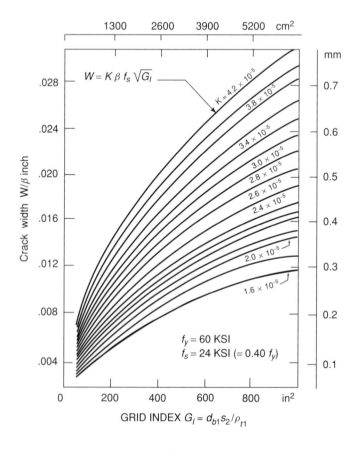

FIGURE 4.25   Crack control reinforcement distribution in two-way action slabs and plates [Nawy, 1994, 1996a; Nawy and Blair, 1971].

environmental conditions to which the structure may be subjected. Table 4.5 is a reasonable guide to the tolerable average crack widths in concrete structures under various normally encountered environmental conditions. Its values are in close agreement with the Comité Eurointernational du Beton recommendations [CEB-FIP, 1990] for most exposure conditions.

The crack-control equation and guidelines presented are important not only for the control of corrosion in reinforcement but also for deflection control. The reduction of the stiffness EI of the two-way slabs or plates due to orthogonal cracking when the permissible crack widths in Table 4.5 are exceeded can lead to both short- and long-term excessive deflection. Deflection values that are several times greater than those anticipated in the design, including deflection due to construction loading, can be reasonably controlled through camber and control of the flexural-crack width in the slab or plate. Proper selection of reinforcement spacing s_1 and s_2 in the perpendicular directions, as discussed in this section, that does *not exceed* 12 in (30 cm) center to center can maintain good serviceability performance of a slab system under normal and reasonable overload conditions.

TABLE 4.5 Maximum Tolerable Flexural-Crack Widths

	Crack Width	
Exposure Condition	in	mm
Dry air or protective membrane	0.016	0.40
Humidity, moist air, soil	0.012	0.30
Deicing chemicals	0.007	0.18
Seawater and seawater spray; wetting and drying	0.006	0.15
Water-retaining structures (excluding nonpressure pipes)	0.004	0.10

Sources: ACI Committee 224. 1980. Control of cracking in concrete structures. *Proc. ACI. J.* 20(10):35–76. Nawy, E.G. 1996a. *Reinforced Concrete—A Fundamental Approach*, 3rd ed., pp. 838. Prentice Hall, Upper Saddle River, NJ.

Long-term effects on cracking. In most cases, the magnitude of crack widths increases in long-term exposure and long-term loading. Increase in crack width can vary considerably in cases of cyclic loading, such as in bridges. However, width increases at decreasing rate with time. In most cases, a doubling of crack width after several years under sustained loading is not unusual.

4.5.2.4 Cracking in Prestressed Concrete Circular Tanks

Circular prestressed tanks are cylindrical shell elements of very large diameter in relation to their height. Hence, with respect to flexural cracking, it is possible to treat the wall of a tank in a manner similar to the treatment of two-way action plates. Vessay and Preston (1978) modified Nawy and Blair's (1971) expressions for two-way action slabs and plates (Nawy and Blair, 1971) so that the maximum crack width can be defined as

$$w_{max} \text{ (in)} = 4.1 \times 10^{-6} \epsilon_{ct} E_{ps} \sqrt{G_I} \qquad (4.50)$$

Using SI units, the expression becomes

$$w_{max} \text{ (mm)} = 0.6 \times 10^{-6} \epsilon_{ct} E_{ps} \sqrt{G_I} \qquad (4.51)$$

where E_{ps} is in MPa and the dimensions of all the parameters of the grid index G_I are in millimeters.

ϵ_{ct} = tensile surface strain in the concrete = $(\lambda_t f_p)/(E_{ps})$
f_p = actual stress in the steel
f_{pi} = initial prestress before losses
$\lambda_t = f_p/f_{pi}$
G_I = grid index = $\dfrac{s_1 s_2 d_c}{d_{b1}} \dfrac{8}{\pi}$
d_{b1} = diameter of steel in direction 1
s_1 = spacing of the reinforcement in direction 1 closest to the tensile face

s_2 = spacing of the reinforcement in direction 2

d_c = concrete cover to center of steel, inches.

Note that w_{max} = 0.004 in (0.1 mm) should be the limit of crack width for liquid-retaining tanks.

Acknowledgments

This Chapter is based on extensive material taken with permission from *Fundamentals of High Strength High Performance Concrete* by E.G. Nawy, 1996, Addison Wesley Longman, London and California; *Reinforced Concrete—A Fundamental Approach* (3rd. Ed.) and *Prestressed Concrete—A Fundamental Approach* (2nd. Ed.), both by E.G. Nawy, 1996, Prentice Hall, Upper Saddle River, New Jersey and Committee Reports and Standards of the American Concrete Institute, Farmington Hills, Michigan.

References

ACI Committee 209. 1992. *Prediction of Creep, Shrinkage, and Temperature Effects in Concrete Structures*, ACI 209R-92, pp. 1–47. American Concrete Institute, Farmington Hills, MI.

ACI Committee 363. 1992. *State-of-the Art Report on High Strength Concrete*, ACI Report 363R-92, pp. 1–55. American Concrete Institute, Farmington Hills, MI.

ACI Committee 435. 1995. *Control of Deflection in Concrete Structures*, ACI 435 Committee Report, pp. 77. American Concrete Institute, Farmington Hills, MI.

ACI Committee 318. 1996. *Building Code Requirements for Structural Concrete*, ACI 318-95; and *Commentary*, ACI 318R-95. American Concrete Institute, Farmington Hills, MI.

ACI 1997. *Manual of Concrete Practice*, Vols. 1–5. American Concrete Institute, Farmington Hills, MI, in press.

ACI Committee 224. 1980. Control of cracking in concrete structures. *Proc. ACI. J.* 20(10):35–76.

Branson, D.E. 1971. Compression steel effects on long term deflections, *Proc. J. Am. Concrete Inst.* 68:555–559.

Branson, D.E. 1977. *Deformation of Concrete Structures*, pp. 546. McGraw–Hill, New York.

CEB-FIP 1990. *Model Code for Concrete Structures*, Vols. 1, 2, and 3, CEB-FIP, Paris.

Gergely, P. and Lutz, L.A. 1968. Maximum crack width in reinforced concrete flexural members. In *Causes, Mechanism, and Control of Cracking in Concrete*, ACI SP-20, E.G. Nawy, ed., pp. 87–117. American Concrete Institute, Farmington Hills, MI.

Harajli, M.H. and Naaman, A.E. 1989. Cracking in partially prestressed beams under static and fatigue loading. In *ACI SP-113*, pp. 29–56. American Concrete Institute, Farmington Hills, MI.

Meyers, B.L. and Thomas, E.W. 1983. Elasticity, shrinkage, creep and thermal movement of concrete, In *Handbook of Structural Concrete*, Chapt. 11, Kong et al., ed., pp. 11.1–11.33. McGraw–Hill, New York.

Mindess, S. and Young, J.F. 1981. *Concrete*, pp. 671. Prentice Hall, Upper Saddle River, NJ.

Naaman, A.E. and Siriaksorn, A. 1979. Serviceability based design of partially prestressed beams, Part I—Analysis. *Proc. PCI J.* 24(2):64–89.

Nawy, E.G. 1972a. Crack control through reinforcement distribution in two-way acting slabs and plates. *Proc. ACI J.* 69(4):217–219.

Nawy, E.G. 1972b. Crack control in beams reinforcement with bundled bars. *Proc. ACI J.* 69:637–639.

Nawy, E.G. 1990. Flexural cracking behavior of partially prestressed pretensioned and post-tensioned beams—State-of-the-Art. In *Cracking in Prestressed Concrete Structures*, ACI SP-113, pp. 1–42. American Concrete Institute, Farmington Hills, MI.

Nawy, E.G. 1994. Cracking of concrete: ACI and CEB approaches. In *Proc. CANMET Int. Symp. Advances Concrete Technol*, 2nd ed., V.M. Malhotra, ed., pp. 203–242. Canada Center for Mineral and Energy Technology (CANMET), Ottawa.

Nawy, E.G. 1996a. *Reinforced Concrete—A Fundamental Approach*, 3rd ed., pp. 838. Prentice Hall, Upper Saddle River, NJ.

Nawy, E.G. 1996b. *Prestressed Concrete—A Fundamental Approach*, 2nd ed., pp. 810. Prentice Hall, Upper Saddle River, NJ.

Nawy, E.G. 1996c. *Fundamentals of High Strength High Performance Concrete*, pp. 350. Addison Wesley Longman, London and California.

Nawy, E.G. and Blair, K.W. 1971. Further studies on flexural crack control in structural slab systems. In *Cracking, Deflection, and Ultimate Load of Concrete Slab Systems*, ACI SP-30, E.G. Nawy, ed., pp. 1–41. American Concrete Institute, Farmington Hills, MI.

Neville, A.M. 1995. *Properties of Concrete*, 4th ed., pp. 844. Addison Wesley Longman, London and California.

Nilson, A.H. 1985. Design implications of current research on high strength concrete. *ACI SP-87-7*, pp. 85–118. American Concrete Institute, Farmington Hills, MI.

Ross, A.D. 1937. Creep concrete data. *Proc. Insti. Struct. Eng.* 15:314–326.

Ross, A.D. 1958 The elasticity, creep and shrinkage of concrete. *Proc. Conf. Non-Metallic Brittle Mater.*, pp. 157–174. Interscience, London.

Shah, S.P. and McGarry, F.J. 1971. Griffith fracture criteria and concrete, *Proc., J. of Eng. Mech. Div. ASCE.* 47(EM6):1663–1676.

Vessey, J.V. and Preston, R.L. 1978. A Critical review of code requirements for circular prestressed concrete reservoir's, *FIP Bull*, pp. 6. Federation International PreContrainte, Paris.

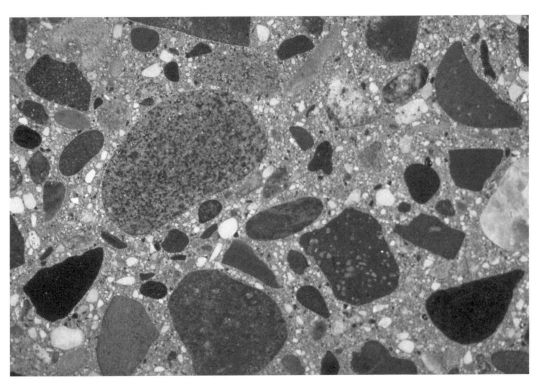

Uniform distribution of ingredients in high performance concrete (Courtesy Portland Cement Association).

5

Properties and Performance of Normal-Strength and High-Strength Concrete

by
Steven H. Kosmatka, P.E.
Director, Construction Information Services, Portland Cement Association, Skokie, Illinois. Formerly Manager of Research and Development, P.C.A. and active member of several committees of the ACI including the Concrete Research Council and ASTM committees on cement and concrete.

5.1 Introduction

Portland cement concrete is a simple material in appearance but has a very complex internal nature. Despite its internal complexity, concrete's versatility, durability, and economy has made it the world's most-used construction material. This can be seen in the variety of structures for which it is used, from highways and bridges to buildings and dams (Figure 5.1).

Concrete is a mixture of portland cement, water, and aggregates, with or without admixtures. Portland cement and water form a paste that hardens owing to chemical reactions between the cement and water. The paste acts as a glue, binding the aggregates, composed of sand and gravel or crushed stone, into a solid rocklike mass. The quality of the paste and aggregates dictates the engineering properties of concrete construction. Paste qualities are directly related to the amount of water used in relation to the amount of cement. The less water used, the better the quality of the concrete. Reduced water content results in improved strength and durability and reduced permeability and shrinkage. Fine and coarse aggregates make up 60–75% of the total volume of the concrete (Figure 5.2); therefore selection of aggregate is important. Aggregates must be of adequate strength, be resistant to exposure conditions, and be durable.

(a)

(b)

(c)

(d)

FIGURE 5.1   Concrete used in a variety of applications. (a) Highways; (b) bridges, (c) buildings, and (d) dams. *Source:* Courtesy of the Portland Cement Association.

Concrete is often discussed as being of "normal strength" or "high strength." Normal-strength concrete typically has a compressive strength of between 3,000 psi and 6,000 psi (20 MPa to 40 MPa). In this chapter, high-strength concrete refers to concrete with a compressive strength of between 6,000 and 20,000 psi (40 MPa and 140 MPa). The distinction between normal- and high-strength concrete has changed through history as concrete technology has advanced. One hundred years ago, 4,000-psi (28 MPa) concrete would have been considered a high-strength concrete. Today concretes for special applications can achieve compressive strengths over 120,000 psi (800 MPa)—these are called reactive powder con-

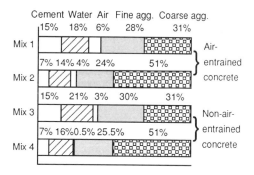

FIGURE 5.2 Range of proportions of materials used in concrete. *Source:* Courtesy of the Portland Cement Association; Kosmatka, S.H. and Panarese, W.C. 1992. *Design and Control of Concrete Mixture.*

cretes. These very exotic materials will not be addressed owing to their limited use in general construction. However, for simplicity, normal- and high-strength concretes (up to 20,000 psi) will be discussed in tandem as they use the same materials and have related or similar properties [Farny and Panarase, 1994; Kosmatka and Panarese, 1992; Nawy, 1996].

Normal-strength concrete is used in most construction applications. Only a small percentage of concrete structures use high-strength concrete. High-strength concrete is used for reduced weight, creep, or permeability; improved durability; or where architectural considerations require smaller load-carrying elements. The Two Union Square building in Seattle has the highest strength concrete used in a major building—almost 20,000 psi (140 MPa) in the columns (Figure 5.3). In this case,

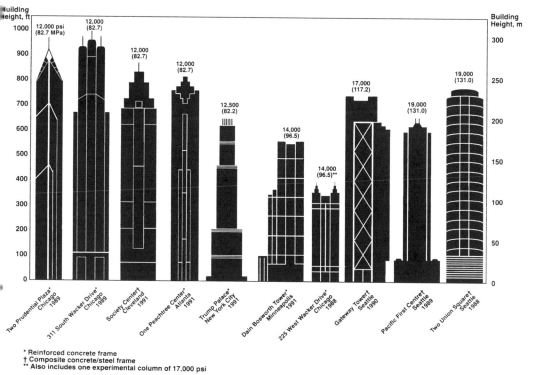

* Reinforced concrete frame
† Composite concrete/steel frame
** Also includes one experimental column of 17,000 psi

FIGURE 5.3 High-strength concrete in buildings in 1980s–1990s. *Source:* Courtesy of the Portland Cement Association; Farny, J.A. and Panarese, W.C. 1994. *High-Strength Concrete.*

high strength was needed to minimize creep. In other high-rise buildings, high-strength concrete helps achieve more efficient floor plans through smaller vertical members.

5.2 Workability, Bleeding, and Consolidation

Freshly mixed concrete should be plastic or in a semifluid state that can be molded by hand or by mechanical means. In a plastic concrete mixture, all the particles of sand and coarse aggregate are encased and held in suspension. The ingredients should not segregate or separate during transport or handling. A uniform distribution of aggregate particles helps control segregation. After the concrete hardens, it becomes a homogeneous mixture of all the components. Concrete of plastic consistency should not crumble, but flow sluggishly without segregation (Figure 5.4). Concrete consistency is measured by the slump test, ASTM C 143 (Figure 5.5). Slumps of 1 to 3 in (25 to 75 mm) are used for pavements and slabs. Slumps of 3 to 5 in (75 to 125 mm) are used for columns and walls. Higher-slump concretes, made with plasticizers, are used in thin

FIGURE 5.4   Workable concrete should flow sluggishly into place without segregation. *Source:* Courtesy of the Portland Cement Association; Kosmatka, S.H. and Panarese, W.C. 1992. *Design and Control of Concrete Mixture.*

FIGURE 5.5   Consistency slump test of concrete. Courtesy of the Portland Cement Association.

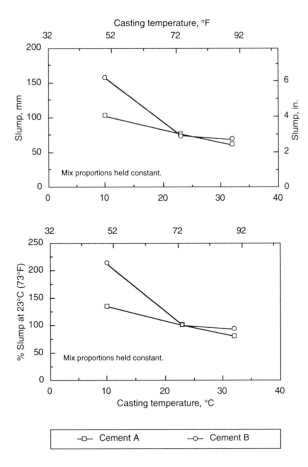

FIGURE 5.6 Slump characteristics versus casting temperature for two different concretes. *Source:* Courtesy of the Portland Cement Association; Burg, R.G. 1996a. *The Influence of Casting and Curing Temperatures on the Properties of Fresh and Hardened Concrete.*

applications, locations with large amounts of reinforcing steel, or where concrete needs to be essentially self-consolidating.

(a) **Workability.** This means the ease of placing, consolidating, and finishing freshly mixed concrete. Concrete should be workable but should not segregate or bleed excessively. Normal-strength concrete usually has good workability as long as concrete ingredients are used in proper proportions and an adequate aggregate gradation is used. However, high-strength concrete is often sticky and difficult to handle and place, even with the aid of plasticizers, owing to its high cement content. Concrete temperature also affects workability. Figure 5.6 illustrates that increases in temperature for an established mix reduce slump and workability, resulting in the need to adjust the mix for environmental conditions. As concrete mixture temperature is increased from 73°F (23°C), slump will decrease approximately 0.8 in for each 20°F temperature increase (20 mm for each 10°C increase).

(b) **Bleeding.** This is the development of a layer of water at the top of freshly placed concrete that is caused by settlement of solid particles of cement and aggregate and the simultaneous upward migration of water (Figure 5.7). A small amount of bleeding is normal and will not

FIGURE 5.7   Bleed water on freshly placed concrete surface. *Source:* Courtesy of the Portland Cement Association; Kosmatka, S.H. 1994. *Bleeding.*

FIGURE 5.8   Proper vibration facilitates concrete placement even in heavily reinforced members. *Source:* Courtesy of the Portland Cement Association; Kosmatka, S.H. and Panarese, W.C. 1992. *Design and Control of Concrete Mixture.*

affect the durability or strength of the concrete. However, excessive bleeding can result in high water/cement ratios that can cause durability and surface-strength problems. Finishing should be performed after the presence of bleed water is gone. Use of a properly graded aggregate and a proper dosage of cementitious materials will reduce bleeding to an acceptable level. High-strength concrete essentially does not bleed owing to its low water content and large amount of cementitious materials. Air-entrained concrete also has little to no bleed water.

To achieve desired strength and durability, concrete must be consolidated to form a homogeneous mass without the presence of large voids. Internal and external vibration of concrete, using vibrators, allows stiff, low-slump mixtures to be properly densified. The use of mechanical vibration provides an economical, practical method to quickly consolidate concrete without detrimentally affecting its properties (Figure 5.8).

5.3 Mixing, Transporting, and Placing Concrete

All concrete should be mixed thoroughly until it is uniform in appearance and all ingredients are evenly distributed. If concrete has been adequately mixed, samples taken from different portions of a batch will have essentially the same unit weight, air content, slump, and strength. Concrete is sometimes mixed at a job site in a stationary mixer or paving mixer, and other times it is mixed in central mixers at ready-mix plants. Ready-mixed concrete can be manufactured by any of the following methods:

1. Centrally mixed concrete is completely mixed in a stationary mixer and then delivered in a truck agitator or in a truck mixer at agitating speed to the job site.
2. Shrink-mixed concrete is partially mixed in a stationary mixer and completed in a truck mixer.
3. Truck-mixed concrete is mixed completely in a truck mixer.

Once concrete is transported to a job site, it is conveyed by a variety of methods including belt conveyors, buckets, shoots, cranes, pumps, wheelbarrows, and other equipment (Figure 5.9). Concrete should be conveyed in such a manner that it is not allowed to dry out, be delayed, or allowed to segregate before it is placed. Table 5.1 describes the advantages of various methods and equipment for transporting and handling concrete.

Concrete should be deposited continuously as near as possible to its final position. Concrete should be placed in horizontal layers of uniform thickness, each layer being thoroughly consolidated before the next is placed. The rate of placement should be rapid enough that a layer of concrete is not yet set when a new layer is placed upon it. Drop shoots should be used to prevent segregation and spattering of mortar on reinforcement and forms in wall placements.

Once concrete is placed into a form or layer, it should be consolidated or compacted into the mold so that it forms in and around embedded items and reinforcement and removes entrapped air. Consolidation is usually accomplished by mechanical methods. Use of internal vibrators is the most common method of consolidating concrete. When concrete is vibrated, it behaves like a liquid; it settles in the forms by the action of gravity, and large entrapped air voids rise to the surface. Workers must be careful to make sure that the vibrator is inserted into the concrete at a uniform distribution within the vibrator's radius of action to properly consolidate all the concrete. Even highly fluid, plasticized mixes with slumps over 7 1/2 in (190 mm) need some vibration for proper consolidation.

(a)   (b)

FIGURE 5.9 (a) Truck-mounted pump and boom and (b) a conveyer belt can conveniently move concrete to the desired location. *Source:* Courtesy of the Portland Cement Association; Kosmatka, S.H. and Panarese, W.C. 1992. *Design and Control of Concrete Mixture.*

TABLE 5.1   Methods and Equipment for Transporting and Handling Concrete

Equipment	Type and Range of Work for Which Equipment is Best Suited	Advantages	Points to Watch for
Mobile batcher mixers	Used for intermittent production of concrete at jobsite.	A combined materials transporter and mobile batching and mixing system for quick, precise proportioning of specified concrete. One-man operation.	Trouble-free operation requires a preventive maintenance program equipment. Materials must be identical to those in original mix design.
Nonagitating trucks	Used to transport concrete on short hauls over smooth roadways.	Capital cost of nonagitating equipment is lower than that of truck agitators or mixers.	Concrete slump should be limited. Possibility of segregation. Height needed for high lift of truck body discharge.
Pneumatic guns (shortcrete)	Used where concrete is to be placed in difficult locations and where thin sections and large areas are needed.	Ideal for placing concrete in free-form shapes, for repairing and strengthening buildings, for protective coatings, and thin linings.	Quality of work depends on skill of those using equipment. Only experienced nozzlemen should be employed.
Pumps	Used to convey concrete directly from central discharge point at jobsite to formwork or to secondary discharge point.	Pipelines take up little space and can be readily extended. Delivers concrete in continuous stream. Pump can move concrete both vertically and horizontally. Mobile pumps can be delivered when necessary to small or large projects. Stationary pump booms provide continuous concrete for tall building construction.	Constant supply of freshly mixed concrete is needed with average consistency and without any tendency to segregate. Care must be taken in operating pipeline to ensure an even flow and to clean out at conclusion of each operation. Pumping vertically, around bends and through flexible hose will considerably reduce the maximum pumping distance.
Screw spreaders	Used for spreading concrete over flat areas, as in pavements.	With a screw spreader a batch concrete discharged from bucket or truck can be quickly spread over a wide area to a uniform depth. The spread concrete has good uniformity of compaction before vibration is used for final compaction.	Screw spreaders are usually used as part of a paving train. They should be used for spreading before vibration is applied.
Tremies	For placing concrete underwater.	Can be used to funnel concrete down through the water into the foundation or other part of the structure being cast.	Precautions are needed to ensure that the tremie discharge end is always buried in fresh concrete, so that a seal is preserved between water and concrete mass. Diameter should be 10 to 12 in unless press is available. Concrete mixture needs more cement, 700 lb per cubic yard, and greater slump, 6 to 9 in, because concrete must flow and consolidate without any vibration.
Truck agitators	Used to transport concrete for all uses in pavements, structures, and buildings. Haul distances must allow discharge of concrete within 1 1/2 hours, but limit may be waived under certain circumstances.	Truck agitators usually operate from central mixing plants where quality concrete is produced under controlled conditions. Discharge from agitators is well controlled. There is uniformity and homogeneity of concrete on discharge.	Timing of deliveries to suit job organization. Concrete crew and equipment must be ready onsite to handle concrete

TABLE 5.1 (*continued*)

Equipment	Type and Range of Work for Which Equipment is Best Suited	Advantages	Points to Watch for
Truck mixers	Used to transport concrete for all uses in pavements, structures, and buildings. Haul distances must allow discharge of concrete within 1 1/2 hours, but limit may be waived under certain circumstances.	No central mixing plant needed, only a batching plant, since concrete is completely mixed in truck mixer. Discharge is same as for truck agitator.	Timing of deliveries to suit job organization. Concrete crew and equipment must be ready onsite to handle concrete. Control of concrete quality is not as good as with central mixing.
Wheelbarrows and buggies	For short flat hauls on all types of onsite concrete construction, especially where accessibility to work area is restricted.	Very versatile and therefore ideal inside and on jobsites where placing conditions are constantly changing.	Slow and labor intensive.
Belt conveyors	For conveying concrete horizontally or to a higher or lower level. Usually used between main discharge point and secondary discharge point.	Belt conveyors have adjustable reach, traveling diverter, and variable speed both forward and reverse. Can place large volumes of concrete quickly when access is limited.	End-discharge arrangements needed to prevent segregation, leave no mortar on return belt. In adverse weather (hot, windy) long reaches of belt need cover.
Belt conveyors mounted on truck mixers	For conveying concrete to a lower, horizontal, or higher level.	Conveying equipment arrives with the concrete. Adjustable reach and variable speed.	End-discharge arrangments needed to prevent segregation, leave no mortar on return belt.
Buckets	Used with cranes, cableways, and helicopters for construction of buildings and dams. Convey concrete directly from central discharge point to formwork or to secondary discharge point.	Enable full versatility of cranes, cableways, and helicopters to be exploited. Clean discharge. Wide range of capacities.	Select bucket capacity to conform to size of the concrete batch and capacity of placing equipment. Discharge should be controllable.
Chutes	For conveying concrete to lower level, usually below ground level, on all types of concrete construction.	Low cost and easy to maneuver. No power required, gravity does most of the work.	Slopes range between 1 to 2 and 1 to 3 and chutes must be adequately supported in all positions. Arrange for discharge at end (downpipe) to prevent segregation.
Cranes	The right tool for work above ground level.	Can handle concrete, reinforcing steel, formwork, and sundry items in high-rise, concrete-framed buildings.	Has only one hook. Careful scheduling between trades and operations are needed to keep it busy.
Dropchutes	Used for placing concrete in vertical forms of all kinds. Some chutes are one piece, others are assembled from loosely connected segments.	Dropchutes direct concrete into formwork and carry it to bottom of forms without segregation. Their use avoids spillage of grout and concrete on the form sides, which is harmful when off-the-form surfaces are specified. They also will prevent segregation of coarse particles.	Dropchutes should have sufficiently large, splayed-top openings into which concrete can be discharged without spillage. The cross section of dropchute should be chosen to permit inserting into the formwork without interfering with reinforcing steel.

Source: Courtesy of the Portland Cement Association; Kosmatka, S.H. and Panarese, W.C. 1992. *Design and Control of Concrete Mixture.*

5.4 Permeability

Concrete's durability, corrosion resistance, and resistance to chemical attack is directly related to permeability. If a substance cannot enter concrete, it cannot damage it. Concrete permeability is a function of the permeability of the paste and aggregate. Decreased permeability improves concrete's resistance to saturation, sulfate attack, chemical attack and chloride penetration. Paste permeability has the greatest influence on concrete permeability. Paste permeability is directly related to the water/cement ratio and the degree of hydration or length of moist-curing. A low water/cement ratio and an adequate moist curing period result in concrete with low permeability. The permeability of mature, good-quality normal-strength concrete, is approximately 1×10^{-10} cm/s. Figure 5.10 illustrates the relationship between water/cement ratio and permeability. Supplementary cementing materials such as fly ash and, especially, silica fume reduce permeability.

High-strength concrete usually has very low permeability as a result of its low water/cement ratio and the common use of silica fume. High-strength concretes often have water permeability coefficients from 1×10^{-11} to 1×10^{-13} cm/s. Low permeability and inherent resistance to chlorides make high-strength concrete ideal for structures such as bridges and parking decks that are exposed to deicers or seawater.

A variety of methods are used to determine the permeability of concrete. Some of the common methods are chloride ponding test (AASHTO T259), electrical conductance—rapid chloride permeability test (ASTM C 1202), water permeability (Army Corps CRD-C 163), air permeability (SHRP 2031), and volume of permeable voids (ASTM C 642). Table 5.2 illustrates the relationship between different permeability methods on normal and high-strength concretes of different water/cement ratios. Table 5.3 provides conversions for various permeability units. The first step in achieving a low permeability is to specify a water/cement ratio of 0.4 or less and then consider using a supplementary cementing material or blended hydraulic cement.

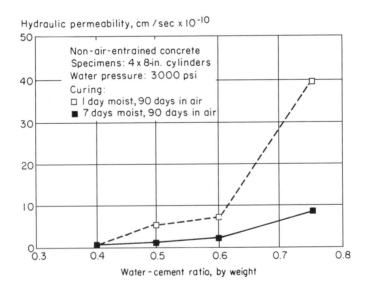

FIGURE 5.10 Water permeability of concrete as influenced by water/cement ratio and curing. *Source:* Courtesy of American Concrete Institute; Whiting, D. 1988. Permeability of selected concretes. In *Permeability of Concrete.*

TABLE 5.2   Various Permeability Tests on Concrete

Mix	W/C	Cure Time	RCPT, Coulombs	90-day Ponding, % Cl	Permeability Hydraulic, μDarcys[†]	Air, μDarcys[†]	Porosity, %	Volume Permeable Voids, %
1	0.26	1 day	44	0.013	[‡]	37	8.3	6.3
		7 days	65	0.013	[‡]	29	7.5	6.2
2	0.28	1 day	942	0.017	[‡]	28	9.1	8.1
		7 days	852	0.022	[‡]	33	8.8	8.0
3	0.4	1 day	3897	0.062	0.030	130	11.3	11.4
		7 days	3242	0.058	0.027	120	11.3	12.2
4	0.5	1 day	5703	0.103	0.560	120	12.4	13.0
		7 days	4315	0.076	0.200	170	12.5	12.7
5	0.6	1 day	5911	0.104	0.740	200	13.0	12.8
		7 days	4526	0.077	0.230	150	12.7	12.5
6	0.75	1 day	7065	0.112	4.100	270	13.0	14.2
		7 days	5915	0.085	0.860	150	13.0	13.3
Coefficient of Variation, %			7.0	12.9	20.9	14.0	2.5	2.4

Source: Courtesy of American Concrete Institute; Whiting, D. 1988. Permeability of selected concretes. In *Permeability of Concrete.*

Note: W/C = water/cement ratio, RCPT = rapid chloride permeability test.

[†]To convert from μDarcys to m^2 multiply by 9.87×10^{-7}.

[‡]Permeability too small to measure.

TABLE 5.3   Conversion Table for Permeability Units

	Darcy	Millidarcy	cm/s	m^2	Meinzers	ft/d
Darcy	1	1000	9.68×10^{-4}	9.87×10^{-13}	20.50	2.75
Millidarcy	10^{-3}	1	9.68×10^{-7}	9.87×10^{-16}	2.05×10^{-2}	2.75×10^{-3}
cm/s	1.03×10^3	1.03×10^6	1	1.03×10^{-9}	2.12×10^4	2.84×10^3
m^2	1.01×10^{12}	1.01×10^{15}	9.71×10^8	1	20.7×10^{12}	2.78×10^{12}
Meinzers	4.88×10^{-2}	48.78	4.72×10^{-5}	4.83×10^{-14}	1	1.34×10^{-1}
ft/day	3.64×10^{-1}	3.64×10^2	3.52×10^{-4}	3.60×10^{-13}	7.46	1

Source: Courtesy of American Concrete Institute; Whiting, D. 1988. Permeability of selected concretes. In *Permeability of Concrete.*

Note: To convert from units in leftmost column to units at top, multiply by indicated factor. Conversions given are appropriate for cases of saturated, steady-state flow. For units associated with diffusion processes or with relative, empirical test procedures, no direct conversions are available.

5.5 Carbonation

Carbonation is related to permeability. Carbonation of portland cement concrete paste occurs on all exposed concrete surfaces. Carbonation is a chemical reaction in which carbon dioxide in air or water reacts with compounds in the hardened cement paste to form carbonates, primarily calcium carbonate. The reaction begins immediately upon exposure of the concrete to air. The rate of carbonation is largely a function of paste permeability, temperature, relative humidity, and the concentration of carbon dioxide in the air. Paste permeability is controlled by the water/cement ratio, amount of moist curing, and the degree of "densification" of the exposed surface.

Carbonation increases drying shrinkage and lowers the alkalinity (pH) of concrete. The inherent high alkalinity of concrete, pH greater than 12.5, protects embedded reinforcing steel from corrosion by causing a passive and noncorroding protective oxide film to form on the steel surface.

By reducing the pH, carbonation destroys the protective film and, in the presence of moisture and oxygen, allows steel to corrode. Engineers specify a certain amount of protective concrete cover over reinforcing steel, assuming that the small depth of carbonated concrete will fall far short of the steel. In effect, the designer assumes that the concrete will not carbonate down to the depth of the steel during the expected life of the structure. The depth of carbonation is usually of little importance, less than 0.5 in (13 mm) after 10 years for good-quality portland-cement concrete, and the depth of cover required by building codes provides an adequate factor of safety against corrosion [Campbell et al., 1991].

5.6 Early Age Characteristics and Strength

Concrete is primarily known for its massive compressive strength. Concrete gains strength through the reaction between cement and water called hydration. Portland cement is primarily a calcium-silicate cement. The calcium silicate combines with water and forms calcium silicate hydrate, which is responsible for the primary engineering properties of concrete, such as strength and dimensional stability.

During the first few hours, the cement in the concrete slowly reacts with water, giving the concrete some early strength. This period of early hardening is called the setting time. Initial set is defined as the time at which concrete has attained 500 psi (3.5 MPa); final set is defined as the time at which it reaches 4000 psi (27.6 MPa), based on penetration resistance. The contractor should be aware of the setting properties of a concrete to properly control finishing operations on slabs and pavements. Temperature has one of the greatest effects on setting time. Figure 5.11 illustrates how an increase in temperature decreases set time and a decrease in temperature increases set time. Setting time can be expected to change approximately 30% for each 10°F change (50% for each 10°C change) in temperature from a base temperature of 73°F (23°C).

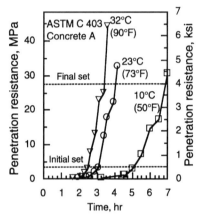

FIGURE 5.11   Temperature effect on set time. *Source:* Courtesy of the Portland Cement Association; Burg, R.G. 1996b. The influence of concrete casting and curing temperatures. In *Concrete Technology Today.*

The compressive strength of concrete is directly related to the water/cement ratio. A decrease in water/cement ratio results in higher strength. Table 5.4 illustrates a conservative relationship

TABLE 5.4   Typical Relationship Between Water/Cement Ratio and Compressive Strength of Concrete

Compressive Strength at 28 Days, psi*	Water/Cement Ratio by Weight	
	Non-Air-Entrained Concrete	Air-Entrained Concrete
6000	0.41	—
5000	0.48	0.40
4000	0.57	0.48
3000	0.68	0.59
2000	0.82	0.74

Note: Strength is based on 6 × 12-in cylinders moist-cured 28 days at 73.4°F ± 3°F in accordance with Section 9b of ASTM C31. Relationship assumes maximum size of aggregate about 3/4 in to 1 in.

*Values are estimated average strengths for concrete.

TABLE 5.5 Examples of Trial Mixes for Air-Entrained Concrete of Medium Consistency, 3–4 in Slump (English Units)

Water-Cement Ratio, lb per lb	Maximum Size of Aggregate, in	Air Content, %	Water, lb/yd³ of Concrete	Cement, lb/yd³ of Concrete	With Fine Sand, Fineness Modulus = 2.50			With Coarse Sand, Fineness Modulus = 2.90		
					Fine Aggregate, Percent of Total Aggregate	Fine Aggregate, lb/yd³ of Concrete	Coarse Aggregate, lb/yd³ of Concrete	Fine Aggregate, Percent of Total Aggregate	Fine Aggregate, lb/yd³ of Concrete	Coarse Aggregate, lb/yd³ of Concrete
0.40	3/8	7.5	340	850	50	1250	1260	54	1360	1150
	1/2	7.5	325	815	41	1060	1520	46	1180	1400
	3/4	6	300	750	35	970	1800	39	1090	1680
	1	6	285	715	32	900	1940	36	1010	1830
	1 1/2	5	265	665	29	870	2110	33	990	1990
0.45	3/8	7.5	340	755	51	1330	1260	56	1440	1150
	1/2	7.5	325	720	43	1140	1520	47	1260	1400
	3/4	6	300	665	37	1040	1800	41	1160	1680
	1	6	285	635	33	970	1940	37	1080	1830
	1 1/2	5	265	590	31	930	2110	35	1050	1990
0.50	3/8	7.5	340	680	53	1400	1260	57	1510	1150
	1/2	7.5	325	650	44	1200	1520	49	1320	1400
	3/4	6	300	600	38	1100	1800	42	1220	1680
	1	6	285	570	34	1020	1940	38	1130	1830
	1 1/2	5	265	530	33	980	2110	36	1100	1990
0.55	3/8	7.5	340	620	54	1450	1260	53	1560	1150
	1/2	7.5	325	590	45	1250	1520	49	1370	1400
	3/4	6	300	545	39	1140	1800	43	1260	1680
	1	6	285	520	35	1060	1940	39	1170	1830
	1 1/2	5	265	480	33	1030	2110	37	1150	1990
0.60	3/8	7.5	340	565	54	1490	1260	58	1600	1150
	1/2	7.5	325	540	46	1290	1520	50	1410	1400
	3/4	6	300	500	40	1180	1800	44	1300	1680
	1	6	285	475	36	1100	1940	40	1210	1830
	1 1/2	5	265	440	33	1060	2110	37	1180	1990
0.65	3/8	7.5	340	525	55	1530	1260	59	1640	1150
	1/2	7.5	325	500	47	1330	1520	51	1450	1400
	3/4	6	300	460	40	1210	1800	44	1330	1680
	1	6	285	440	37	1130	1940	40	1240	1830
	1 1/2	5	265	410	34	1090	2110	38	1210	1990
0.70	3/8	7.5	340	485	55	1560	1260	59	1670	1150
	1/2	7.5	325	465	47	1360	1520	51	1480	1400
	3/4	6	300	430	41	1240	1800	45	1360	1680
	1	6	285	405	37	1160	1940	41	1270	1830
	1 1/2	5	265	380	34	1110	2110	38	1230	1990

Source: Courtesy of the Portland Cement Association; Kosmatka, S.H. and Panarese, W.C. 1992. *Design and Control of Concrete Mixtures.*

between strength and water/cement ratio, and Tables 5.5 and 5.6 illustrate typical mix designs for concretes with various water/cement ratios. The compressive strength of concrete increases with age as long as an appropriate moisture content and temperature are available (Figures 5.12 and 5.13). The compressive strength of concrete at 7 d is approximately 75% of the 28-d strength for concrete cast and cured at 73°F (23°C). Later-age strength of concrete cast and cured at 50°F (10°C) often equals or exceeds that of the same concrete cast and cured at 73°F (23°C). For most concretes cast and cured at 90°F (32°C), 3-d compressive strength will be approximately 70% of 28-d compressive strength of concrete cast and cured at 90°F (32°C). The effect of casting and curing at 90°F (32°C) roughly results in a 3-d compressive strength similar to 7-d strength for concrete cast and cured at 73°F (23°C).

TABLE 5.6 Example of Trial Mixes for Air-Entrained Concrete of Medium Consistency, 80–100 mm Slump (Metric Units)

Water/ Cement Ratio, kg per kg	Nominal Maximum Size of Aggregate, mm	Air Content, %,	Water, kg/m³ of Concrete	Cement, kg/m³ of Concrete	With Fine Sand. Fineness Modulus = 2.50			With Coarse Sand Fineness Modulus = 2.90		
					Fine Aggregate, Percent of Total Aggregate	Fine, Aggregate, kg/m³ of Concrete	Coarse Aggregate, kg/m³ of Concrete	Fine Aggregate, Percent of Total Aggregate	Fine Aggregate, kg/m³ of Concrete	Coarse Aggregate, kg/m³ of Concrete
0.40	10	7.5	202	505	50	744	750	54	809	684
	14	7.5	194	485	41	630	904	46	702	833
	20	6	178	446	35	577	1071	39	648	1000
	40	5	158	395	29	518	1255	33	589	1184
0.45	10	7.5	202	450	51	791	750	56	858	684
	14	7.5	194	428	43	678	904	47	750	833
	20	6	178	395	37	619	1071	41	690	1000
	40	5	158	351	31	553	1225	35	625	1184
0.50	10	7.5	202	406	53	833	750	57	898	684
	14	7.5	194	387	44	714	904	49	785	833
	20	6	178	357	38	654	1071	42	726	1000
	40	5	158	315	32	583	1225	36	654	1184
0.55	10	7.5	202	369	54	862	750	58	928	684
	14	7.5	194	351	45	744	904	49	815	833
	20	6	178	324	40	702	1071	43	750	1000
	40	5	158	286	33	613	1225	37	684	1184
0.60	10	7.5	202	336	54	886	750	58	952	684
	14	7.5	194	321	46	768	904	50	839	833
	20	6	178	298	40	720	1071	44	773	1000
	40	5	158	262	33	631	1225	37	702	1184
0.65	10	7.5	202	312	55	910	750	59	976	684
	14	7.5	194	298	47	791	904	51	863	833
	20	6	178	274	40	720	1071	44	791	1000
	40	5	158	244	34	649	1225	38	720	1184
0.70	10	7.5	202	288	55	928	750	59	994	684
	14	7.5	194	277	47	809	904	51	880	883
	20	6	178	256	41	738	1071	45	809	1000
	40	5	158	226	34	660	1225	38	732	1184

Outdoor concrete continues to gain strength as moisture is maintained by natural humidity or rainfall. Strength gain for different concretes under different environmental conditions is illustrated in Figures 5.14 to 5.16. Figure 5.14 also illustrates the effect of water/cement ratio. Compressive strength is usually specified at the age of 28 d; however, depending on the project, ages of 3 and 7 d can also be specified. For high-strength concrete, ages of 56 or 90 d may be specified. For general-use concrete, a compressive-strength between 20 MPa and 35 MPa (3000 psi and 6000 psi) is used. Compressive, strength development of high-strength concrete is illustrated in Figure 5.17 and corresponds to the mixes in Table 5.7. The rate of strength gain for high-strength concrete versus that of normal-strength concrete is illustrated in Figure 5.18. Compressive strength is usually determined by casting and testing 6 in by 12 in (150 mm by 300 mm) cylinders; however, 4 in by 8 in (100 mm by 200 mm) cylinders are commonly used for high-strength concrete and sometimes for normal-strength concrete. For high-strength concrete, 4 in by 8 in (100 mm by 200 mm) cylinders generally have strengths within about 1% of 6 in by 12 in (150 mm by 300 mm) cylinders.

Increase in strength with age continues as long as any unhydrated cement is still present, the relative humidity in the concrete is approximately 80% or higher, and the concrete temperature is favorable.

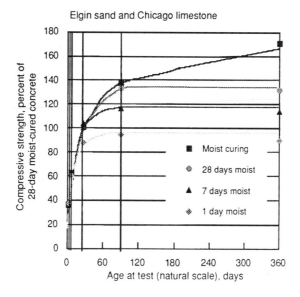

FIGURE 5.12   Increase in concrete strength with age as affected by curing moisture levels. *Source:* Courtesy of the Portland Cement Association.

TABLE 5.7   Example of High Strength Concretes, SI (English Units) (See Figures 5.17, 5.22, 5.24, 5.26, 5.30.)

Parameter, Units per Cubic Yard (Cubic Meter)	Mix Number				
	1	2	3	4	5
Cement Type I, lb (kg)	950 (564)	800 (475)	820 (487)	950 (564)	800 (475)
Silica fume, lb[†] (kg)	—	40 (24)	80 (47)	150 (89)	125 (74)
Fly ash, lb (kg)	—	100 (59)	—	—	175 (104)
Coarse agg. SSD, lb[‡] (kg)	1800 (1068)	1800 (1068)	1800 (1068)	1800 (1068)	1800 (1068)
Fine agg. SSD, lb (kg)	1090 (647)	1110 (659)	1140 (676)	1000 (593)	1000 (593)
HRWR type F, fl oz (L)	300 (11.60)	300 (11.60)	290 (11.22)	520 (20.11)	425 (16.44)
Retarder type D, fl oz (L)	29 (1.12)	27 (1.05)	25 (0.97)	38 (1.46)	39 (1.50)
Total water, lb[§] (kg)	267 (158)	270 (160)	262 (155)	242 (144)	254 (151)
Water/cement ratio	0.28	0.34	0.32	0.26	0.32
Water/cementitious materials ratio	0.28	0.29	0.29	0.22	0.23

Source: Courtesy of the Portland Cement Association; Burg, R.G. and Ost, B.W. 1994. *Properties of Commercially Available High-Strength Concretes (Including Three-Year Data).*

Note: HRWR, high-range water reducer.

[*]As reported by ready-mix supplier.

[†]Dry weight.

[‡]Maximum aggregate size 12.5 mm.

[§]Mass of total water in mix including admixtures.

In order to maintain this increase in strength, concrete must be properly cured. Curing means not only that a favorable temperature be present, but also that moisture loss will not be permitted or extra water will be provided at the surface. Use of a wet burlap or plastic covering for 7 d or use of a curing compound usually provides adequate curing for normal- and high-strength concrete.

Although concrete is very strong in compression, it is weak in tensile strength. Tensile strength is about 8 to 12% of the compressive strength. Flexural strength (modulus of rupture) is 5 to 7.5 times (0.7 to 0.8 times for metric units) the square root of the compressive strength. Shear strength

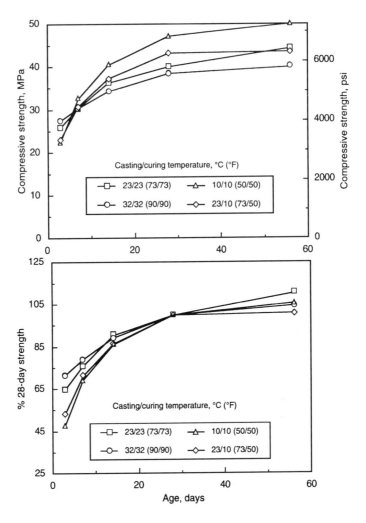

FIGURE 5.13   Strength Gain vs. Age for Different Casting and Curing Temperatures. *Source:* Courtesy of the Portland Cement Association; Burg, R.G. 1996a. *The Influence of Casting and Curing Temperatures on the Properties of Fresh and Hardened Concrete.*

is about 20% of the compressive strength. Modulus of elasticity ranges from 2 to 6 million psi or 57,000 times the square root of the compressive strength, in English units (14,000 to 41,000 MPa), and can be estimated as 5,000 times the square root of the compressive strength in metric units. Figures 5.19 to 5.21 illustrate these properties. Table 5.8 lists the flexural and tensile strengths for high-strength concretes cited in Table 5.7.

5.7 Density

Normal-weight concrete, as is used in pavements and bridges, has a density of 140 to 150 lb/ft^3 (2240 to 2400 kg/m^3). The density of concrete varies with the relative density of the aggregate, the amount of air present in the paste, and the amount of water and cement in the mixture. The density of non-air-entrained high-strength concrete is often over 150 lb/ft^3 (2400 kg/m^3). The high-strength concretes in Table 5.7 use a crushed dolomite coarse aggregate and have densities ranging from 152 to 155 lb/ft^3 (2430 to 2490 kg/m^3) and air contents ranging from 0.7 to 1.6%.

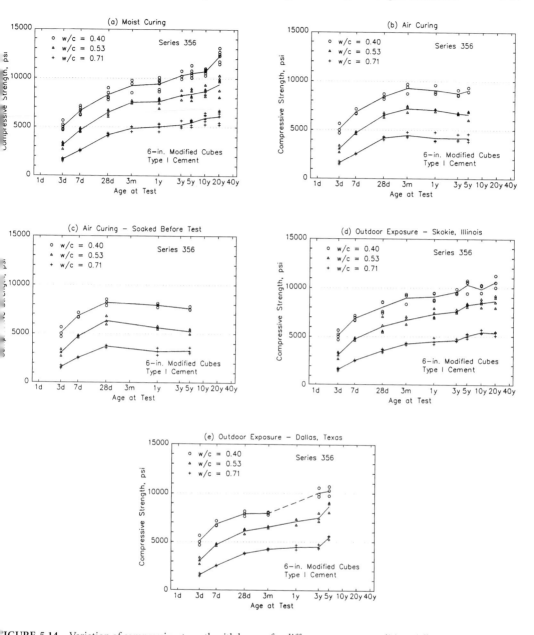

FIGURE 5.14 Variation of compressive strength with log age for different exposure conditions (all specimens received 7-d moist curing prior to ultimate exposure). *Source:* Courtesy of the Portland Cement Association; Wood, S.L. 1992. *Evaluation of the Long-Term Properties of Concrete.*

Heavyweight concretes, used for radiation shielding, can have densities approaching 400 lb/ft³ (6400 kg/m³).

5.8 Abrasion Resistance

Structures such as pavements and bridge decks are subjected to constant abrasion. Therefore concrete in these applications must have a high degree of abrasion resistance. Abrasion resistance is directly related to the compressive strength of the concrete. The type of aggregate and surface finish also have a strong influence on abrasion resistance. A hard aggregate, such as a granite, provides more abrasion resistance than a soft limestone aggregate.

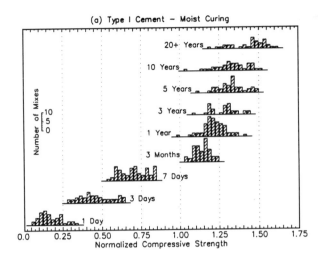

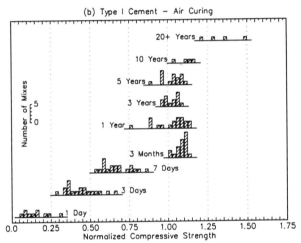

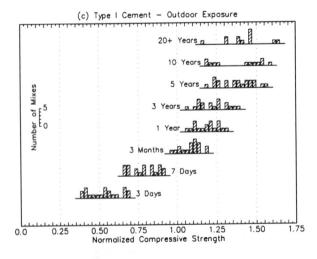

FIGURE 5.15 Distribution of compressive strength normalized to 28 d for Type I cement concrete (all specimens received a 7-d moist curing prior to ultimate exposure). *Source:* Courtesy of the Portland Cement Association; Wood, S.L. 1992. *Evaluation of the Long-Term Properties of Concrete.*

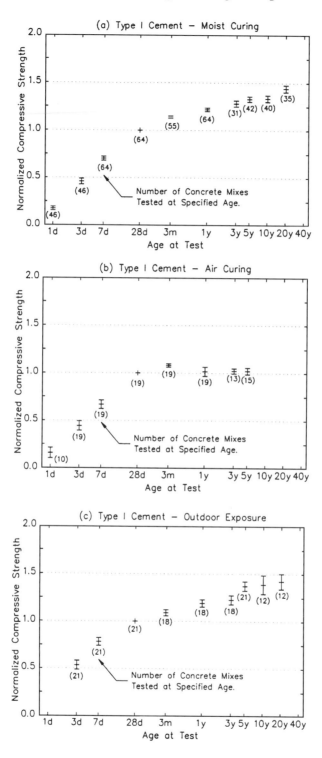

FIGURE 5.16 Ninety-five percent confidence intervals for mean normalized compressive strength of concrete made with Type I cement (7 d in moist room followed by indoor dry air exposure at 50% humidity, significantly reduced long-term strength gain). *Source:* Courtesy of the Portland Cement Association; Wood, S.L. 1992. *Evaluation of the Long-Term Properties of Concrete.*

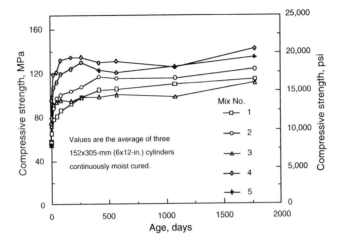

FIGURE 5.17 Strength development for five high-strength concrete mixes over a five-year period. Mixes correspond to Table 5.7. *Source:* Courtesy of the Portland Cement Association; Detwiler, R.J. 1996. Long-term strength tests of high-strength concrete. In *Concrete Technology Today.*

TABLE 5.8　Relative Flexural and Tensile Strengths of High-Strength Concrete at 91 d (Compressive Strengths in Figure 5.17) (See Mixes in Table 5.7)

Mix	Cure	Modulus of Rupture, psi*	Modulus of Rupture, MPa	Tensile Strength, psi[†]	Tensile Strength, MPa
1	Moist	1370	9.4	925	6.4
1	Air	890	6.1	445	3.1
2	Moist	1470	10.1	860	5.9
2	Air	770	5.3	630	4.3
3	Moist	1440	9.9	720	5.0
3	Air	810	5.6	710	4.9
4	Moist	1990	13.7	1050	7.2
4	Air	980	6.8	940	6.5
5	Moist	1390	9.6	950	6.6
5	Air	940	6.5	670	4.6

　Source: Courtesy of the Portland Cement Association; Burg, R.G. and Ost, B.W. 1994. *Properties of Commercially Available High-Strength Concretes (Including Three-Year Data).*
　*Values are averages of three 6 × 6 × 30-in (152 × 152 × 762-mm) specimens.
　[†]Values are averages of three 6 × 12-in (152 × 305-mm) specimens.

5.9 Volume Change and Crack Control

Concrete changes slightly in volume for various reasons. Understanding the nature of these changes is useful in planning concrete work and preventing cracks from forming. If concrete is free to move, normal volume changes have very little consequence; but because concrete in service is usually restrained by foundations, subgrades, reinforcement, or connecting elements, significant stresses can develop. As the concrete shrinks, tensile stresses develop that can exceed the tensile strength of the concrete, resulting in crack formation.

　The primary factors affecting volume change are temperature and moisture changes. Concrete expands slightly as temperature rises and contracts as temperature falls. The average value for the coefficient of thermal expansion of concrete is about 5.5 millionths per degree Fahrenheit (10 millionths per degree Celsius). This amounts to a length change of 0.66 in for 100 ft (5 mm for a 10-m length) of concrete subjected to a rise or fall of 100°F (50°C). The thermal coefficient of

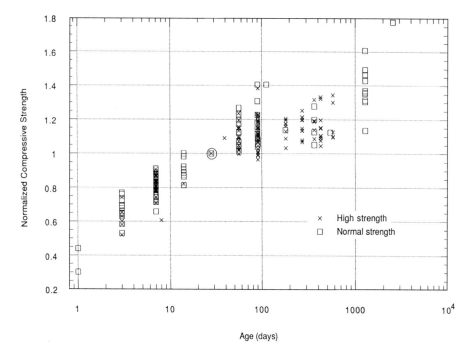

FIGURE 5.18 Compressive strength versus time for normal and high-strength concretes cast in the 1980s and 1990s, normalized to 28-d strength. *Source:* Courtesy of the Portland Cement Association; Lange, D.A. 1994. *Long-Term Strength Development of Concrete.*

expansion for steel is about 6.5 millionths per degree Fahrenheit (12 millionths per degree Celsius), comparable to that of concrete. The coefficient for reinforced concrete can be assumed to be 6 millionths per degree Fahrenheit (11 millionths per degree Celsius).

Concrete expands slightly with a gain in moisture and contracts with a loss in moisture. The drying shrinkage of normal and high-strength concrete specimens ranges from 400 to 800 millionths when exposed to air at a 50% relative humidity. Figure 5.22 illustrates the drying-shrinkage characteristics of high-strength concretes in Table 5.7. Concrete with a unit shrinkage of 550 millionths shortens by about the same amount as a thermal contraction caused by a decrease in temperature of 100°F (55°C). The shrinkage of reinforced concrete is less than that for plain concrete owing to the restraint offered by the reinforcement. Reinforced concrete structures with normal amounts of reinforcement have a drying shrinkage in the range of 200–300 millionths. The amount of shrinkage is directly related to the amount of water in the concrete. Higher water content results in higher shrinkage. Specimen size also has an effect. Larger specimens shrink less than small specimens. Cement type and content have little effect on drying shrinkage. Drying shrinkage can also be reduced by using aggregates that do not have high drying-shrinkage properties such as quartz, granite, or limestone. Some chemical admixtures may increase drying shrinkage.

Drying-shrinkage is an inherent and unavoidable property of concrete; therefore properly positioned reinforcing steel is used to reduce crack widths, or joints are used to predetermine or control the location of cracks. Thermal stress due to fluctuations in temperature can also cause cracking, particularly at early ages.

5.10 Deformation and Creep

Concrete deforms a small amount at initial loading. When concrete is loaded, the deformation caused by the load can be divided into two parts: a deformation that occurs immediately, such

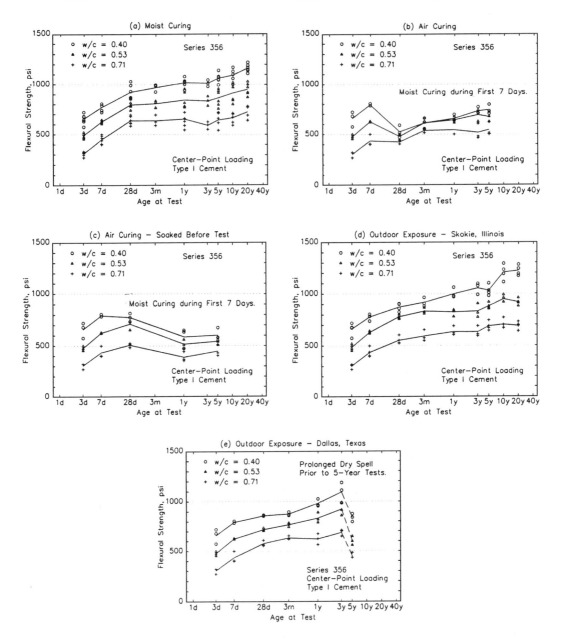

FIGURE 5.19 Variation of concrete flexural strength with log of age. *Source:* Courtesy of the Portland Cement Association; Wood, S.L. 1992. *Evaluation of the Long-Term Properties of Concrete.*

as elastic strain, and a time-dependent deformation that begins immediately but continues at a decreasing rate for as long as the concrete is loaded. This latter is called creep. The amount of creep is dependent on the magnitude of the stress, the age and strength of the concrete when the stress is applied, and the length of time the concrete is stressed.

Figure 5.23 illustrates the combined effects of elastic and creep strains. Creep is of little concern for normal concrete buildings, pavements, and bridges; however, creep should be considered when designing for very tall buildings or very long bridges. High-strength concrete is sometimes used to control creep by keeping load values low. The specific creep of high-strength concretes in Table 5.7 is illustrated in Figure 5.24.

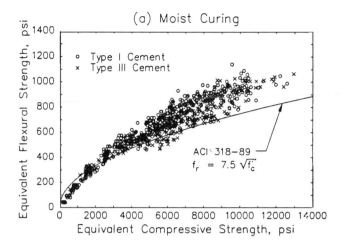

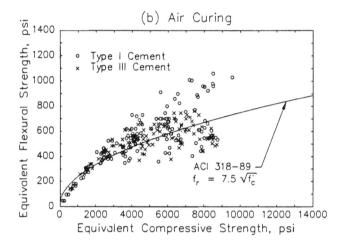

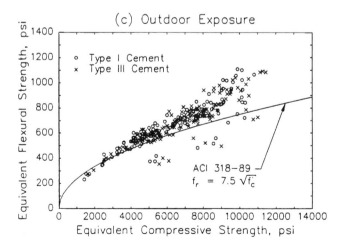

FIGURE 5.20   Flexural strength versus compressive strength (such a relationship is often used in paving projects where cylinders are used in place of flexural beams). *Source:* Courtesy of the Portland Cement Association; Wood, S.L. 1992. *Evaluation of the Long-Term Properties of Concrete.*

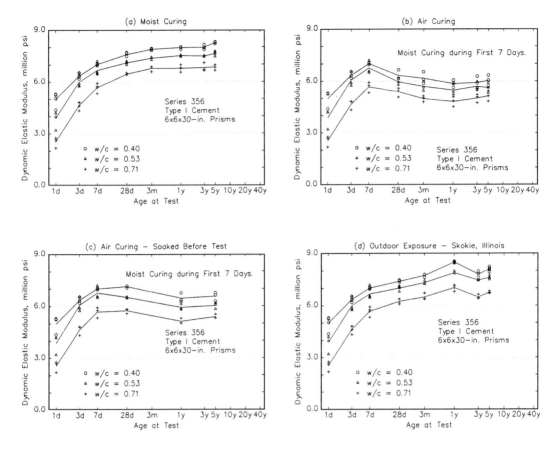

FIGURE 5.21 Dynamic modulus of elasticity of concrete versus age. *Source:* Courtesy of the Portland Cement Association; Wood, S.L. 1992. *Evaluation of the Long-Term Properties of Concrete.*

5.11 Concrete Ingredients

Normal- and high-strength concretes use a variety of cements, aggregates, mineral admixtures, and chemical admixtures. These materials are reviewed briefly in this section.

5.11.1 Portland Cements

Portland cements are hydraulic cements, that is, they set and harden by reacting with water. This reaction, called hydration, causes water and cement to combine to form a stonelike mass. Portland cement was invented in 1824 by an English mason, Joseph Aspdin, who named his product portland cement because it produced a concrete that was of the same color as natural stone from the Isle of Portland in the English Channel.

Portland cement is produced by combining appropriate proportions of lime, iron, silica, and alumina and heating them. These raw ingredients are fed into a kiln that heats the ingredients to temperatures from 2600°F to 3000°F (1450°C to 1650°C) and chemically changes the raw materials into cement clinker. The clinker is cooled and then pulverized. During this operation, a small amount of gypsum is added to control the setting of the cement. The finished pulverized product is portland cement. Portland cement is essentially a calcium silicate cement. Powdered coal, oil, natural gas, or other materials are used as fuel for the kiln. A detailed discussion of cements and their chemistry is given in Chapter 1 of this handbook.

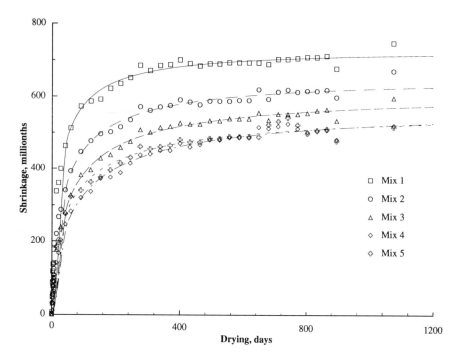

FIGURE 5.22   Drying shrinkage of $3 \times 3 \times 11.25$-in ($76 \times 76 \times 286$-mm) prisms of high-strength concrete after 28-d moist curing. *Source:* Courtesy of the Portland Cement Association; Burg, R.G. and Ost, B.W. 1994. *Properties of Commercially Available High-Strength Concretes (Including Three-Year Data).*

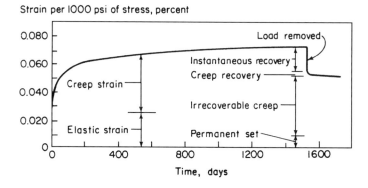

FIGURE 5.23   Combined elastic-creep strain versus time (cylinders loaded at 8 d immediately after removal from fog room, then stored at 70°F and 5% relative humidity; applied stress = 25% compressive strength at 8 d). *Source:* Courtesy of the Portland Cement Association; Kosmatka, S.H. and Panarese, W.C. 1992. *Design and Control of Concrete Mixtures.*

The American Society for Testing and Materials (ASTM) Standard C 150, Specification for portland cement, defines the following types of portland cement:

Type I	General portland cement
Type II	Moderate sulfate-resistant cement
Type III	High-early-strength cement
Type IV	Low heat of hydration cement
Type V	High-sulfate-resistant cement

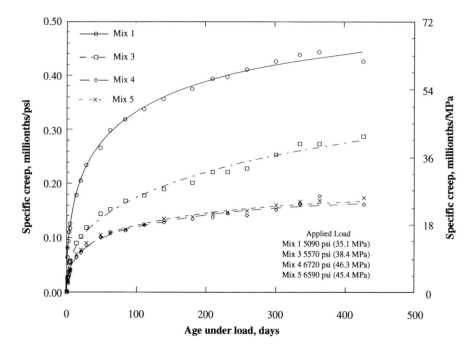

FIGURE 5.24 Specific creep of high-strength concrete in 6 × 12-in (150 × 300-mm) cylinders. *Source:* Courtesy of the Portland Cement Association; Burg, R.G. and Ost, B.W. 1994. *Properties of Commercially Available High-Strength Concretes (Including Three-Year Data).*

Types I, II, and III may also be designated as being air entraining. Type I portland cement is a general cement suitable for all uses where the special properties of other cements are not required. It is commonly used in pavements, buildings, bridges, and precast concrete products.

Type II portland cement is used where precaution against moderate sulfate attack is important, as in bridge or pavement structures, or where sulfate concentrations in ground water or soil are higher than normal, but not too severe. Type II cement can also be specified to generate less heat than Type I cement. This moderate heat of hydration is helpful when placing massive structures such as piers, heavy abutments, and retaining walls. Type II cement may be specified when water-soluble sulfate in soil is between 0.1 and 0.2% or when the sulfate content in water is between 150 and 1500 ppm.

Type III portland cement provides strength at an early age. Figure 5.25 illustrates the earlier strength gain of Type III cement over that of other cement types. It is chemically similar to Type I cement except that the particles are ground finer to increase the rate of hydration. It is commonly used in fast-track paving or when a concrete structure must be put into service as soon as possible, as in bridge deck repair.

Type IV portland cement is used where the rate and amount of heat generated from hydration must

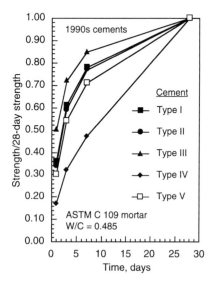

FIGURE 5.25 Strength gain curves for ASTM C 109 mortar cubes for different 1990s cement types. *Source:* Courtesy of the Portland Cement Association; PCA 1996. Portland cement: Past and present characteristics. In *Concrete Technology Today.*

be minimized. This low heat of hydration cement is intended for massive structures such as gravity dams. Type IV cement is rarely available.

Type V portland cement is used in concrete exposed to very severe sulfate exposures. Type V cement is used when concrete is exposed to soil with a water-soluble sulfate content of 0.2% and higher or to water with over 1500 ppm of sulfate. The high sulfate resistance of Type V cement is attributed to its low tricalcium aluminate content.

Cement Types I and II are commonly available and represent about 90% of the cement shipped in the United States, while Type III and white cement (discussed in the following section) represent only 4% of the cement produced. Type IV cement is manufactured only when specified for massive structures and is therefore not readily available. Type V is only available in specific severe-sulfate environments.

5.11.2 White Portland Cement

White portland cement has essentially the same chemistry as regular gray portland cement except that it is white in color. White portland cement is made of select raw materials that contain a negligible amount of iron and other dark metal oxides. White portland cement is usually used for architectural purposes such as noise barrier walls and decorative concrete.

5.11.3 Blended Hydraulic Cements

Blended hydraulic cements are produced by blending two or more types of cementitious material. Primary blending materials are portland cement, ground granulated blast-furnace slag, fly ash, natural pozzolans, and silica fume. These cements are commonly used in the same manner as portland cements. Blended hydraulic cements conform to the requirements of ASTM C 595 or C 1157. ASTM C 595 cements are as follows: Type IS—portland blast-furnace slag cement, Type IP and Type P—portland-pozzolan cement, Type S—slag cement, Type I (PM)—pozzolan-modified portland cement, and Type I (SM)—slag-modified portland cement. The blast-furnace slag content of Type IS is between 25% and 70% by mass. The pozzolan content of Types IP and P is between 15% and 40% of the blended cement by mass. Type I (PM) contains less than 15% pozzolan. Type S contains at least 70% slag by mass. Type I (SM) contains less than 25% slag by mass. These blended cements may also be designated as air-entraining, moderate sulfate-resistant, or as moderate or low heat of hydration.

ASTM C 1157 blended hydraulic cements include the following: Type GU—blended hydraulic cement for general construction use, Type HE—high-early strength cement, Type MS—moderate sulfate resistant cement, Type HS—high sulfate resistant cement, Type MH—moderate heat of hydration cement, and Type LH—low heat of hydration cement. These cements can also be designated for low reactivity (option R) with alkali-reactive aggregates. There are no restrictions on the composition of the C 1157 cements. The manufacturer can optimize ingredients, such as pozzolans and slags, to optimize particular concrete properties.

The most common blended cements available are Types IP and IS. The United States uses a relatively small amount of blended cement compared with countries in Europe or Asia. However, this may change as consumer demands for products with specific properties and as environmental and energy concerns change.

5.11.4 Expansive Cements

Expansive cements are hydraulic cements that expand slightly during the early hardening period following setting. They meet the requirements of ASTM C 845 in which they are designated Type E-1. Although three varieties of expansive cement are designated in the standard as K, M, and S, only K is available in the United States. Type E-1 (K) contains portland cement, anhydrous tetracalcium trialuminosulfate, calcium sulfate, and uncombined calcium oxide (lime). Expansive cement is used to make shrinkage-compensating concrete that is used (1) to compensate for volume decrease due to drying shrinkage, (2) to induce tensile stress in reinforcement, and (3)to stabilize long-term dimensions of posttensioned concrete structures. One of the major advantages of using expansive cement is in the control and reduction of drying-shrinkage cracks. In recent years, shrinkage-compensating concrete has been of particular interest in bridge deck construction, where crack development must be minimized.

5.11.5 Supplementary Cementing Materials—Mineral Admixtures

Supplementary cementing materials, also called mineral admixtures, contribute to the properties of hardened concrete through hydraulic or pozzolanic activity. Typical examples are natural pozzolans, fly ash, ground granulated blast-furnace slag, and silica fume, which can be used individually with portland or blended cement or in different combinations. These materials react chemically with calcium hydroxide released from the hydration of portland cement to form cement compounds.

Fly ash is a finely divided residue that results from the combustion of pulverized coal and is carried from the combustion chamber of the furnace by exhaust gases. Commercially available fly ash is a byproduct of thermal power generating stations.

Blast-furnace slag, or iron blast-furnace slag, is a nonmetallic product consisting essentially of silicates, aluminosilicates of calcium, and other compounds that form in a molten condition simultaneously with iron in the blast furnace.

Silica fume, also called condensed silica fume and microsilica, is a finely divided residue resulting from the production of elemental silicon or ferrosilicon alloys that is carried from the furnace by the exhaust gases. Silica fume, with or without fly ash or slag, is usually used to make high-strength concrete. The following is a summary of the specifications and classes of supplementary cementing materials.

Ground Granulated Iron Blast-Furnace Slag—ASTM C 989.

> Grade 80—Slags with a low activity index
> Grade 100—Slags with a moderate activity index
> Grade 120—Slags with a high activity index

Fly Ash and Natural Pozzolan—ASTM C 618.

> Class N—Raw or calcined natural pozzolans including diatomaceous earth, opaline cherts, shales, tuffs, volcanic ashes, and some calcined clays and shales
> Class F—Fly ash with pozzolanic properties
> Class C—Fly ash with pozzolanic and cementitious properties

Silica Fume—ASTM C 1240.

A detailed discussion of mineral admixtures is given in Chapter 2 of this handbook.

5.11.5.1 Effects of Supplementary Cementing Materials on Properties of Freshly Mixed Concrete

Concrete mixes containing fly ash or slag will generally require less water, about 1 to 10%, for a given slump than concrete containing only portland cement. Silica fume requires more water unless a water reducer is used.

The amount of air-entraining admixture required to obtain a specified air content is normally greater when fly ash or silica fume is used. Class C fly ash generally requires less air-entraining admixture than Class F fly ash and tends to lose less air with mixing time. Ground slag has about the same effect as portland cement on air-entrainment dosage. Fly ashes with high carbon contents require higher air-entraining dosages and also tend to lose more air with mixing or agitation.

Fly ash and ground slag will generally improve the workability of concrete of equal slump and strength. Silica fume may reduce workability. Fly ash and ground slag can reduce the amount of heat of hydration in concrete for structures where lower heat development is required. The use of fly ash, natural pozzolans, and ground slag will generally retard the setting time of concrete. For example, in one study, fly ash caused the initial set to be retarded by 10 to 55 min and the final set to be retarded by 5 to 130 min for concrete incorporating 10 different fly ashes [Gebler and Klieger, 1986]. The degree of retardation depends on the amount of portland cement, the water requirement, the type of supplementary cementing material, and the concrete temperature. Finishability and pumping are generally improved with these materials.

Fly ash is commonly used in the United States at a dosage of 15 to 25% by mass of cementitious material. High-calcium Class C ashes have been used at higher dosages for particular situations. Ground slag commonly averages about 40% of the cementing material in the mix. Silica fume is typically used at a 5 to 10% dosage rate by weight of cementitious material. The dosage of a supplementary material depends on the particular concrete property the user is trying to improve or affect. Therefore it is very important to determine the optimal amount of the mineral admixture to use. An overdose or underdose can be harmful or not have the desired effect. Also, these mineral admixtures react differently with different cements.

The effects of temperature and moisture on the setting properties and strength development of concretes containing finely divided mineral admixtures are similar to the effects on concrete made with only portland cement; however, the curing time may need to be longer for some materials.

Relatively high dosages of silica fume can make a very cohesive concrete with very little bleeding. With little to no bleed water available at the surface for evaporation, plastic cracking can readily develop on hot days. Therefore special precautions to minimize evaporation during early curing are essential when using silica fume in concrete.

5.11.5.2 Effects on Hardened Concrete

Supplementary cementing materials contribute to the strength gain of concrete by providing additional cementitious compounds. However, the rate of strength gain of concrete containing these materials often differs from the strength gain of concrete that uses portland cement as the only cementitious material. Owing to the lower rate of hydration when using some of these materials, the early strength gain can be lower than that of comparable concrete without such materials. Yet, other materials perform in a manner very similar to portland cement. Drying shrinkage and creep are little affected by the presence of normal dosages of supplementary cementing materials. The permeability of concrete containing these mineral admixtures is usually reduced with adequate curing. Silica fume is frequently used in bridges to minimize chloride penetration into concrete.

5.11.5.3 Effects on Durability

The effect on durability of normal dosages of supplementary cementing materials is usually insignificant in regard to freeze-thaw resistance and deicer scaling. Depending on the dosage and the specific mineral admixture, alkali-silica reactivity and sulfate resistance may be improved, not affected, or even aggravated. These materials will usually improve the corrosion resistance of embedded steel because they reduce permeability and help keep chloride ions out of the concrete. However, some materials may slightly increase the depth of carbonation of concrete. Carbonation reduces the alkalinity of the concrete that is needed to protect the embedded steel. Normal dosages of supplementary materials have an insignificant effect on carbonation.

5.11.6 Mixing Water for Concrete

Almost any natural water that is drinkable can be used as mixing water for concrete. However, some waters that are not fit for consumption may be suitable for concrete. Of particular concern are the levels of chloride, sulfate, and alkalies, as these can affect the durability of the concrete. Chapter 3 of Kosmatka and Panarese (1992) provides guidance concerning the use of waters containing alkali carbonates, chlorides, sulfates, acids, oils, and other materials and provides guidance as to allowable levels of contamination. Sea water containing salt can be suitable for nonreinforced concrete, but it should not be used for steel-reinforced concrete owing to corrosion concerns.

5.11.7 Aggregates for Concrete

The importance of using the right type and quality of aggregates cannot be overemphasized since the fine and coarse aggregates occupy between 60 and 75% of the concrete volume and strongly influence the concrete's freshly mixed and hardened properties, mix proportions, and economy. Fine aggregates consist of natural sand or fresh stone with particles smaller than 0.2 in (5 mm). Coarse aggregates consist of a combination of gravel or crushed aggregate with particles predominately larger than 0.2 in (5 mm) and generally between 3/8 and 1 1/2 in (10 and 38 mm). The most common coarse-aggregate sizes are 3/4 in and 1 in (19 mm and 25 mm). High-strength concrete often uses 1/2 in (12 mm) maximum coarse aggregate. Natural gravel and sand are usually dug from pits. Crushed aggregate is produced by crushing quarry rock, boulders, or large-size gravel. Crushed air-cooled blast-furnace slag is also used as a fine or coarse aggregate. About half the coarse aggregates used in portland-cement concrete in the United States are gravels and the remainder are crushed stone. Recycled concrete or crushed waste concrete is also a common source of aggregates and an economic necessity where good aggregates are scarce.

The most commonly used aggregates, such as gravel or crushed stone, produce a concrete with a freshly mixed unit weight of about 140 to 150 lb/ft^3 (2240 to 2400 kg/m^3). Special heavyweight or lightweight aggregates are also available for making concretes with a range of densities. Normal-weight aggregates should meet the requirements of ASTM C 33. This specification limits the amounts of harmful substances and states the requirements for aggregate characteristics such as grading.

Grading is the particle-size distribution of an aggregate and is determined by a sieve analysis. The seven standard ASTM C 33 sieves for fine aggregate have openings ranging from 150 μm to 3/8 in (10 mm). The 13 standard sieves for coarse aggregate have openings ranging from 0.046 to 4 in (1.18 mm to 100 mm). For highway construction, ASTM D 448 lists the 13 sizes in ASTM C 33 plus additional coarse aggregate sizes.

The grading and maximum size of the aggregate affect the relative aggregate proportions as well as cement and water requirements, workability, pumpability, economy, shrinkage, and durability of the concrete. Variations in grading can seriously affect the uniformity of the concrete from batch to batch. In general, aggregates that do not have a large deficiency or excess of any size and have a uniform distribution of particle sizes will produce the most satisfactory results.

The particle shape and surface texture of an aggregate influences the properties of freshly mixed concrete more than the properties of hardened concrete. Rough-textured, angular, elongated particles require more water to produce a workable concrete mixture than do smooth, rounded, compact aggregates. Hence aggregate particles that are angular require more cement to maintain the same water/cement ratio. With satisfactory gradation, both crushed and noncrushed aggregates of the same rock type generally give essentially the same strength for the same cement content and water/cement ratio.

The unit weight, also called the unit mass or bulk density, of an aggregate is the weight of aggregate required to fill a container of a specified unit volume. The volume referred to here is occupied by both the aggregate and the voids between aggregate particles. The approximate bulk density of aggregates commonly used in concrete ranges from about 70 to 110 lb/ft^3 (1120 to 1760 kg/m^3). Most natural aggregates have specific gravities or mass densities of between 2.4 and

2.9 with corresponding solid densities of 150 to 180 lb/ft³ (2400 to 2880 kg/m³). The density of an aggregate used in mixture-proportioning computations (not including voids between particles) is determined by multiplying the specific gravity of the aggregate times the density of water, a value of 62.4 lb/ft³ (1000 kg/m³) is used for the density of water.

The absorption and surface moisture of aggregate should be determined according to ASTM test methods so that the net water content of the concrete can be controlled and correct batch weights determined. The internal structure of an aggregate particle is made up of solid matter and voids that may or may not contain water. Following are the moisture conditions of aggregate:

1. Oven dry—fully absorbent
2. Air dry—dry at the particle surface but containing some interior moisture and thus still somewhat absorbent
3. Saturated surface dry (SSD)—neither absorbing water nor contributing water to the concrete mixture
4. Damp or wet—containing excessive moisture on the surface (free water)

The amount of water used in the concrete mixture must be adjusted for the moisture conditions of the aggregates in order to meet the designated water requirement. If the water content of the concrete mixture is not kept constant, the compressive strength, workability, and other properties will vary from batch to batch. Coarse and fine aggregate will generally have absorption levels in the range of 0.2% to 0.4% and 0.2% to 2%, respectively. Free-water contents will usually range from 0.5% to 2% for coarse aggregate and 2% to 6% for fine aggregate.

Harmful substances that may be present in aggregate include organic impurities, silt, clay, shale, iron oxide, coal, and certain lightweight or soft particles. In addition, certain substances and minerals such as chert may also be reactive to alkalis in the concrete. Certain aggregates such as shales will cause pop-outs at the concrete surface. Most specifications limit the permissible amounts of these substances in aggregates. Aggregates are potentially harmful if they contain compounds known to react chemically with portland-cement concrete and produce significant volume changes in the paste, aggregates, or both; interfere with the normal hydration of the cement; or otherwise have harmful byproducts. Alkali-aggregate reactivity will be discussed in Section 5.18.

5.11.8 Chemical Admixtures for Concrete

Admixtures are ingredients in concrete other than portland cement, water, and aggregates that are added to the mixture immediately before or during mixing. Common chemical admixtures include air-entraining, water-reducing, retarding, accelerating, and superplasticizing admixtures. Finely divided mineral admixtures were discussed in Section 5.11.5. The major reasons for using admixtures are to reduce the cost of concrete construction, achieve certain properties in concrete more effectively than by other means, or to ensure the quality of concrete during the stages of mixing, transporting, placing, or curing in adverse weather conditions.

The most common admixtures are air-entraining admixtures. Air-entraining admixtures are used to purposely entrain microscopic air bubbles in concrete. Air entrainment will dramatically improve the durability of concrete exposed to moisture during freezing and thawing. Entrained air greatly improves concrete's resistance to surface scaling caused by chemical deicers. The workability of fresh concrete is also improved significantly, and segregation and bleeding are reduced or eliminated. Air-entraining admixtures are commonly used to provide between 5 and 8% air content in concrete.

The second most common admixture is the water-reducing admixture. Water-reducing admixtures are used to reduce the quantity of mixing water required to produce concrete of a certain slump, reduce the water/cement ratio, reduce cement content, or increase slump. Typical water reduces reduce the water content by approximately 5% to 10%. High-range water reducers reduce the water content by approximately 12% to 30%.

Retarding admixtures are used to retard the rate of setting of concrete. High temperatures of fresh concrete, around 90°F (32°C) or higher, often cause an increased rate of hardening that makes placing and finishing difficult. One of the most practical methods of counteracting this effect is to reduce the temperature of the concrete by cooling the mix water or aggregates. Retarders are sometimes used to offset the accelerating effect of hot weather on the setting of concrete, delay the initial set of concrete when difficult or unusual conditions of placement occur, or delay the set for special finishing processes, such as exposed aggregate finishes.

An accelerating admixture is used to accelerate the strength development of concrete at an early age. Calcium chloride is the most common accelerating admixture used for nonreinforced concrete. Care must be taken to make sure that the chloride content induced by the calcium chloride does not exceed the maximum chloride content allowed by local building codes. An excess of calcium chloride can cause corrosion in reinforced concrete. Besides accelerating strength gain, calcium chloride causes an increase in drying shrinkage and may slightly darken the concrete. Calcium chloride is not an antifreeze and will not reduce the freezing point of water by more than a few degrees. The amount of calcium chloride added should be no more than necessary to produce the desired results and should in no case exceed 2% by weight of cement in nonreinforced concrete. Nonchloride, noncorrosive admixtures are also available where corrosion is a concern.

Superplasticizers are high-range water reducers that are added to concrete with a low to normal slump and water/cement ratio to make high-slump flowing concrete. Flowing concrete is a highly fluid, workable concrete that can be placed with little or no vibration or compaction and yet is free of excessive bleeding or segregation. Superplasticized flowing concrete is used in thin section placements, areas of closely spaced reinforcing steel, underwater placements, and other applications where a more fluid mixture is more economically handled or placed.

All the admixtures discussed in this section must meet the appropriate ASTM standards. When used at appropriate dosages, these materials should not adversely affect concrete properties. However, an overdose has significant effects on strength, shrinkage, and durability. In-depth details of chemical admixtures are given in Chapter 3 of this handbook.

5.12 Proportioning of Concrete Mixtures

The objective in proportioning concrete mixtures is to determine the most economical and practical combination of readily available materials to produce a concrete that will satisfy the performance requirements for the particular conditions of use. To fulfill these objectives, a properly proportioned concrete mix should possess these followig qualities: 1) acceptable workability of freshly mixed concrete; 2) durability, strength, and uniform appearance of hardened concrete; and 3) economy. Only with proper selection of materials and mixture characteristics can the above qualities be obtained in concrete production. Concrete mixtures should be kept as simple as possible, as an excess number of ingredients often make a concrete mixture difficult to control.

The key to designing a concrete mixture is to be fully aware of the relationship between water/cement ratio and strength and durability. The specified compressive strength at 28-d and durability concerns dictate the water/cement ratio established for a concrete mixture. The water/cement ratio is simply the weight of water divided by the weight of cement. If pozzolans or slag are used, the ratio would include their weights and be referred to as the water/cementitious material ratio. The water/cement ratio can be established by a known relationship to strength or by durability requirements. For example, a concrete structure may require only 3000 psi compressive strength, which would relate to a water/cement ratio of about 0.6; however, if the concrete is exposed to deicers, the maximum water/cement ratio should be 0.45. For corrosion protection or for reinforced concrete exposed to deicers, the maximum water/cement ratio should be 0.40. When designing concrete mixtures, it is important to remember that where durability is concerned, the water/cement ratio should be as low as practical.

Aggregate grading and the nature of the particles such as shape, size, and surface texture all affect the mix design. The maximum size that can be used depends upon the size and shape of the concrete member to be cast as well as on the amount of reinforcing steel present. The maximum size of coarse aggregate should not exceed one-fifth of the narrowest dimension between sides of forms or three fourths of the clear space between individual reinforcing bars.

The designer has to decide whether air entrainment is required. Entrained air must be used in all concrete that will be exposed to freezing and thawing and deicing chemicals. The typical range for the amount of entrained air depends upon the size of the coarse aggregate and the degree of exposure. A typical air content for concrete would range from 5 to 8%. Concrete must also be workable and have an adequate consistency for the job placement conditions. The slump test is a measure of concrete consistency. The slump range for pavements and slabs is between 1 and 3 in (25 and 75 mm). Slumps for wall and other applications can range from 3 to 6 in (75 to 150 mm).

The water content of concrete is influenced by a number of factors: aggregate size and shape, slump, water/cement ratio, air content, cement content and admixtures, and environmental conditions. Increased air content and aggregate size, reduction in water/cement ratio and slump, rounded aggregates, and the use of water-reducing admixtures or fly ash all reduce water demand. On the other hand, increased temperatures, cement contents, water/cement ratio, and aggregate angularity all increase water demand. The cement content is determined from the selected water/cement ratio and water content.

Kosmatka and Panarese (1992) provides the Portland Concrete Association's (PCA) step-by-step procedures for proportioning concrete by absolute volume methods from field or laboratory data. Farny and Panarese (1994) provides guidance for high-strength concrete. Chapter 6 of this handbook presents a detailed treatment including numerical examples of the design of concrete mixtures both for normal- and high-strength concretes using the American Concrete Institutes (ACI) method [ACI 1989, 1991a, 1991b, 1993 and Nawy, 1996].

5.13 Hot and Cold Weather Concreting

All concrete must be properly cured. Curing is maintaining a satisfactory moisture content and temperature in concrete for some definite time period immediately following placing and finishing so that the desired properties of strength and durability may develop. Concrete should be moist cured for seven days at a temperature between 50°F and 80°F (10 to 27°C). Common methods of curing include: ponding, spraying or fogging, use of wet covers, impervious paper, plastic sheets, or membrane-forming curing compounds, or some combination of these. These techniques will properly maintain adequate moisture in the concrete. When hot or cold weather conditions are present, temperature protection must be provided. In hot weather concreting, the concrete temperature should be less than 90°F (32°C). This can be achieved cooling the concrete with ice, cold water, or nitrogen injection when the concrete is batched. High-strength concrete especially needs extra care to minimize plastic cracking during hot weather.

When the temperature approaches freezing, cold-weather concreting practices should be instituted. This may include methods for accelerating the early strength of the concrete such as using extra cement, a lower water/cement ratio, or an accelerator. Cold temperatures can dramatically reduce the setting rate of concrete. Use of insulating blankets or temporary heated enclosures are effective means of providing an adequate temperature for concrete. The natural temperature rise of concrete helps maintain the desired temperature. The temperature rise of high-strength concrete is illustrated in Figure 5.26. The temperature rise of high-strength concrete, being greater than that of normal-strength concrete, also leads to concern about controlling the cracking caused by extreme temperature differences. Figures 5.11 and 5.13 illustrate the effects of temperature on set time and

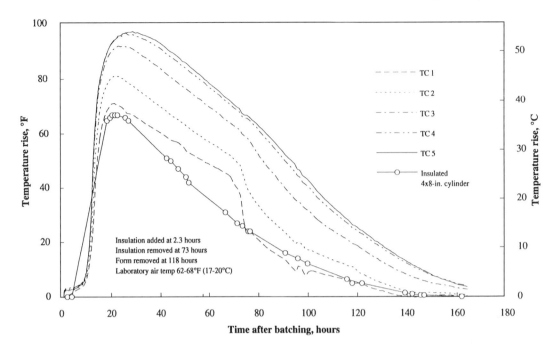

FIGURE 5.26   Temperature rise versus time for high-strength concrete in a 4 ft (1.2 m) Cube (TC1 through TC5 represent temperatures from the outside to the inside of the cube). *Source:* Courtesy of the Portland Cement Association; Burg, R.G. and Ost, B.W. 1994. *Properties of Commercially Available High-Strength Concretes (Including Three-Year Data).*

FIGURE 5.27   Cold weather concreting under suitable preparation conditions. *Source:* Courtesy of the Portland Cement Association; Kosmatka, S.H. and Panarese, W.C. 1992. *Design and Control of Concrete Mixtures.*

strength gain for normal-strength concrete. Concrete temperatures below freezing essentially halt hydration and strength gain. With appropriate precautions, concrete can be successfully placed in the winter, (Figure 5.27).

5.14 Control Tests

Satisfactory concrete construction and performance requires concrete possessing specific properties. To assure these properties are obtained, quality control and acceptance testing are indispensable parts of the construction process. Past experience and sound judgment must be relied on in evaluating tests and assessing their significance in the ultimate performance of concrete.

The most common tests performed on concrete include the slump, strength, temperature, and air-content tests. The number of tests done depends on project conditions and specifications. Strength tests should be done every day at least and should be done no less than once for each 150 yd^3 (115 m^3) of concrete. A 7-d strength cylinder, along with two 28-d cylinders, is often made to provide early indications of strength development. The slump test should be performed in accordance with ASTM C 143. The temperature of concrete should be measured in accordance with ASTM C 1064. The air content should be determined in accordance with ASTM C 231, C 173, or C 138. Strength specimens should be made and cured in accordance with ASTM C 31. Typically, concrete cylinders 6 in in diameter by 12 in high are used for concretes with aggregates no larger than 2 in.

5.15 Freeze-Thaw and Deicer Scaling Resistance

As water freezes in wet concrete, it expands 9%, producing hydraulic pressures in the capillaries and pores of cement paste and aggregate. If the pressure exceeds the tensile strength of the paste or aggregate, the cavity will rupture. Accumulated effects of successive freeze-thaw cycles and disruption of the paste and aggregate eventually cause significant expansion and extensive deterioration of the concrete. The deterioration is visible in the form of cracking, scaling, and crumbling, as seen in Figure 5.28.

The resistance of hardened concrete to freezing and thawing in moist conditions, with or without the presence of deicers, is significantly improved by the use of entrained air. Air entrainment prevents frost damage and scaling and is required for all concretes exposed to freezing and thawing or deicer chemicals. An air content of between 5 and 8% should be specified for normal- and high-strength concrete exposed to wet freezing conditions (Figure 5.29). In the past, it was thought that high-strength concrete did not need to be air entrained because of its low permeability. However, research has shown that in most cases high-strength concrete also needs air entrainment (Figure 5.30). A good air-void system should have a spacing factor of less 0.008 in. Air-entrained concrete should be composed of durable materials and have a low water/portland cement ratio (maximum 0.45), a minimum cement content of 564 lb/yd^3 (335 kg/m^3), proper finishing after bleed water has evaporated from the surface, adequate drainage, a minimum of 7-d moist curing at or above 50°F (10°C), a minimum compressive strength upon frost exposure of 4000 psi (28 MPa), and a minimum 30-d drying period after moist curing. Sealers may also be applied to provide additional protection against the effects of freezing and thawing and deicers. However, a sealer should not be necessary for properly proportioned and placed concrete.

FIGURE 5.28   Frost-damaged concrete with surface scaling.

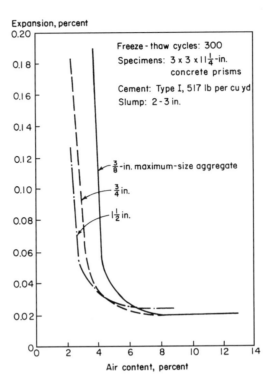

FIGURE 5.29 Air content versus expansion of normal-strength concrete (air content above 5–6%, depending on aggregate size, essentially eliminates frost damage). *Source:* Courtesy of the Portland Cement Association; Klieger, P. 1952. *Studies of the Effect of Entrained Air on the Strength and Durability of Concretes Made with Various Maximum Sizes of Aggregates.*

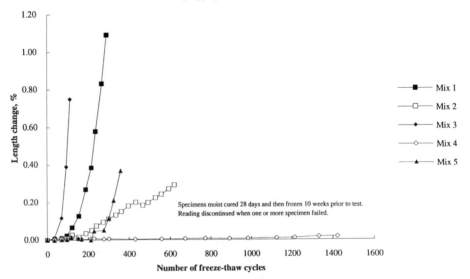

FIGURE 5.30 Length change versus wet freeze-thaw cycles in non-air-entrained high-strength concrete prisms; from Table 5.7 (expansions over 0.10% considered failure; hence most non-air-entrained high-strength concretes are not durable under frost exposure). *Source:* Courtesy of the Portland Cement Association; Burg, R.G. and Ost, B.W. 1994. *Engineering Properties of Commercially Available High-Strength Concretes (Including Three-Year Data).*

5.16 Sulfate-Resistant Concrete

Excessive amounts of sulfates in soil or water can, over 5 to 30 years, attack and destroy concrete that is not properly designed. Sulfates attack concrete by reacting with hydrated compounds in the hardened cement paste, especially calcium aluminate hydrate. Owing to crystallization, these expansive reactions can induce sufficient pressure to disrupt the cement paste, resulting in cracking and disintegration of the concrete. Furthermore, magnesium sulfate solutions can attack concrete by reacting with calcium silicate hydrate, the primary binding material in portland-cement paste. Advanced sulfate attack can turn concrete into a noncohesive mass of rubble, similar to crushed stone or gravel in appearance.

The water/cement ratio (or water/cementitious materials ratio) is the most critical part of a mix design for preventing sulfate attack. The ratio should be no more than the value recommended in Table 5.9 for a particular sulfate exposure and should preferably be less than 0.4 where practical to maximize sulfate resistance. Figures 5.31 and 5.32 clearly illustrate the great importance of using a low water/cementitious materials ratio to protect concrete against the ravages of sulfate attack. Air entrainment is also helpful for improving sulfate resistance when it reduces the water/cement ratio. The cement content for normal-weight and lightweight concrete should be at least 564 lb/yd^3 (335 kg/m^3). When possible, concrete should be allowed to achieve its design strength before it is exposed to sulfate solutions.

TABLE 5.9   Requirements for Concrete Exposed to Sulfate

Sulfate Exposure	Water-Soluble Sulfate (SO$_4$) in Soil, % by Weight	Sulfate (SO$_4$) in Water, ppm	Cement Type*	Maximum Water/Cement Ratio, by Mass
Negligible	0.00–0.10	0–150	Use any cement	—
Moderate†	0.10–0.20	150–1500	II, IP(MS), IS(MS), P(MS), MS, I(PM)(MS), I(SM)(MS)	0.50
Severe	0.20–2.00	1500–10,000	V, HS	0.45
Very severe	Over 2.00	Over 10,000	V or V plus pozzolan‡, HS	0.40

*Cement Types II and V are specified in ASTM C 150 and the remaining types, blended cements, are specified in ASTM C 595 or C 1157.

†Sea water.

‡Pozzolan (ASTM C 618 or C 1240) that has been determined by test or service record to improve sulfate resistance when used in concrete containing Type V cement.

(a)   (b)

FIGURE 5.31   Sulfate attack of concrete beams placed in high-sulfate soil (5-year performance with water/cementitious ratio of [a] 0.50 and [b] 0.39). *Source:* Courtesy of the Portland Cement Association; Stark, D. 1989a. *Durability of Concrete in Sulfate-Rich Soils.*

Selection of the cement type is the second step (after selecting the water/cement ratio) in designing a sulfate-resistant concrete. Table 5.9 provides recommendations for cement types for different sulfate conditions (adapted from requirements originally established by the U.S. Bureau of Reclamation). Cement standards such as ASTM C 150 (Specification for portland Cement), ASTM C 595 (Specification for Blended Hydraulic Cements), and ASTM C 1157 (Performance Specification for Blended Hydraulic Cement) provide specifications for cements that are to be used in concrete exposed to sulfate conditions. Portland cements for sulfate exposures must have low amounts of tricalcium aluminate (C_3A). ASTM C 150 cement Types II and V are intended for use in moderate and severe sulfate exposures, respectively. Type II cement is usually available where needed; however, Type V and the blended cements are available only in certain areas. Mineral admixtures can also be used in sulfate environments, especially in low water/cement ratio concrete; however, their effects on sulfate resistance should be demonstrated by field record or test. Some mineral admixtures are helpful and others are detrimental.

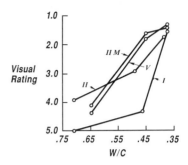

FIGURE 5.32 Sulfate resistance of concrete with various cement types and water/cement ratios (a rating of 1 is no deterioration and a rating of 5 is severe deterioration, namely, mortar loss). *Source:* Courtesy of the Portland Cement Association; Stark, D. 1989a. *Durability of Concrete in Sulfate-Rich Soils.*

For concrete exposed to seawater, cement with a C_3A content up to 10% can be used owing to the ability of chloride in seawater to inhibit expansive sulfate reactions. Usually a Type II cement is specified for use in seawater (see Table 5.9), but some Type I cements will meet this criteria.

An analysis of the sulfate conditions to which concrete will be exposed and proper selection of materials, proportioning, and testing of concrete mixtures will minimize the effects of sulfate attack and ensure the long-term durability of concrete structures.

5.17 Corrosion Protection

Concrete protects embedded steel from corrosion through its highly alkaline nature. The high-pH environment (usually greater than 12.5) causes a passive and noncorroding protective oxide film to form on steel. However, carbonation or the presence of chloride ions from deicers or seawater can destroy or penetrate the film. Once this happens, an electric cell is formed along the steel or between steel bars and the electrochemical process of corrosion begins. Some steel areas along the bar become anodes, discharging current in the electric cell, and in those areas, the iron goes into solution. Steel areas that receive current are cathodes, where hydroxide ions are formed. The iron and hydroxide ions form iron hydroxide, FeOH, which further oxidizes to from rust, or iron oxide. Rusting is an expansive process—rust expands to up to four times its original volume—which induces internal stress and eventual spalling in concrete over reinforcing steel. The cross-sectional area of the steel can also be significantly reduced. Once it starts, the rate of steel corrosion is influenced by the concrete's electrical resistivity, its moisture content, and the rate at which oxygen migrates through the concrete to the steel. Chloride ions alone can also penetrate the passive film on the reinforcement and combine with iron ions to form a soluble iron chloride complex that carries the iron into the concrete where it is later oxidized and forms rust. As little as 0.15% water-soluble chloride (about 0.2% acid soluble) by weight of cement is sufficient to initiate corrosion of embedded steel under some conditions.

Premature deterioration of concrete bridges and parking structures can be a major concrete problem. These structures are often exposed to very severe conditions. Deicer and ocean salts, temperature and moisture changes, freezing and thawing, and abrasion all challenge the life of bridges and other structures. Of these exposure conditions, chloride salts from deicers and ocean spray in marine environments provide the greatest challenge to durability. They can corrode embedded steel, usually causing cracks, spalls, rust stains, and sometimes structural failure.

Deficient corrosion protection has resulted in millions of dollars in repairs over the last two decades. The severity of chloride-induced corrosion and the need for extra protection were not fully realized in the design, construction, and maintenance of early structures. However, field experience and extensive research in recent years have provided new insight into the mechanism of chloride-induced corrosion and methods to cope with the problem. The use of certain protective strategies to prevent or retard steel corrosion can result in bridges with long-term durability, extended structural life, and significantly reduced maintenance and repair costs. In order to reduce the risk of corrosion, the following protective strategies can be used individually or in combination:

1. Cover thickness of 3.5 in (90 mm) or more of concrete over the top of reinforcing steel in compression zones in bridge decks (Note: excessive cover in tension zones exacerbates-surface crack width)
2. Low-slump dense concrete overlay
3. Latex-modified concrete overlay
4. Interlayer membrane/asphaltic concrete systems
5. Epoxy-coated reinforcing steel
6. Corrosion-inhibiting admixtures in concrete
7. Sealers with or without overlay
8. Silica fume or other pozzolans that significantly reduce concrete permeability
9. Low water/cement ratio superplasticized concrete, monolithic or in overlay
10. Cathodic protection
11. Polymer concrete overlay
12. Galvanized reinforcing steel
13. Polymer impregnation
14. Lateral and longitudinal prestressing for crack control
15. Blended cements containing silica fume or other pozzolans to reduce permeability

The purpose of most of these strategies is to delay or stop chloride ions from coming in contact with reinforcing steel. Most of the strategies can also be used alone or in combination for repair and maintenance of older structures showing signs of deterioration.

Studies of the corrosion activity of concretes made with varying water/cement ratios, calcium nitrite corrosion-inhibiting admixtures, silane and methacrylate surface treatments, epoxy-coated rebars and prestressing strands, galvanized reinforcement, and silica-fume admixtures were reported in *Protective Systems for New Prestressed and Substructure Concrete* FHWA/RD-86/193 [Federal Highway Administration, 1987]. The three-year corrosion research project involved 11 protection systems. A summary of some design considerations and research conclusions from this study are as follows:

1. Use as low a water/cement ratio as possible, preferably in the range of 0.32 to 0.44. Concrete permeability and corrosion severity both decrease dramatically as the water/cement ratio is decreased. Long-term chloride permeability to the 1-in deep (25 mm) level in concrete was reduced in this study by about 80% and 95% when the water/cement ratio was reduced from 0.51 to 0.4 and 0.28, respectively.
2. Maximize the clear concrete cover over reinforcement. It should be as large as possible to minimize long-term penetration of water and chloride ions.
3. Silica-fume admixtures and surface sealers (such as silane) both drastically reduced chloride penetration.
4. Corrosion-inhibiting admixtures significantly reduced corrosion severity.
5. Epoxy-coated reinforcing bars and coated prestressing strands were free of corrosion in chloride-laden concrete.

A belt-and-suspenders approach to bridge construction—using more than one protection method simultaneously—can result in significant savings in maintenance costs and produce a structure with a long trouble-free life. Examples of just some double-protection systems would be use of silica fume and a corrosion inhibitor, silica fume and cathodic protection, or latex-modified concrete and extra concrete cover over steel, assuming a low water/cement ratio in each case.

In addition, concrete surfaces must be sloped for adequate drainage, and concrete must be properly proportioned, placed, finished, and cured. Admixtures containing chlorides should be avoided or, at least, the acid-soluble chloride content should be limited to a maximum of 0.08% and 0.20% (preferably less) by weight of cement for prestressed and reinforced concrete, respectively [ACI Committee 222, 1989; Stark, 1989b]. Today, engineers can feel confident that, with the use of good design and construction practices and one or more available corrosion protection systems, a concrete structure can be built to endure even the most severe environment for many years with little maintenance.

5.18 Alkali-Silica Reaction

Most aggregates are chemically stable in hydraulic cement concrete, that is, without deleterious interaction with other concrete ingredients. However, this is not the case for aggregates containing certain siliceous substances that react with soluble alkalies in concrete. Alkali-silica reactivity (ASR) has been reported in structures throughout the world. ASR is usually first identified by surface map-pattern cracks. Very reactive aggregates can induce cracks within a year, whereas slowly reactive aggregates can take over 20 years to induce noticeable cracks (Figure 5.33). Because ASR deterioration is slow, the risk of catastrophic failure is low; however, cracking due to ASR can cause serviceability problems and can exacerbate other deterioration mechanisms such as occur in frost, deicer, or sulfate exposures. Normal-strength and high-strength concretes are both susceptible to ASR.

The basic options for avoiding reactivity problems are relatively simple in concept:

- Do not use potentially reactive aggregates.
- Use fly ash, slag, silica fume, natural pozzolans, or blended hydraulic cement to control reactions.
- Reduce the level of soluble alkalis in the concrete.

Avoiding potentially reactive aggregates is the most effective approach to solving reactivity problems. Unfortunately, with a diminishing supply of proven nonreactive aggregate sources, concrete

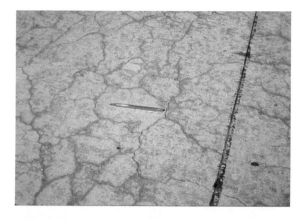

FIGURE 5.33 Cracking of concrete as a result of alkali-silica reactivity.

producers have no alternative but to use potentially reactive aggregate. As this is often the case, other options must be considered. Use of pozzolans, slag, or blended cement is an effective option in areas where appropriate materials are available. Reducing alkalis is commonly equated with using low-alkali cement. However, for a number of reasons, low alkali-cements are not always readily available, nor is reliance on a simple low-alkali cement specification always effective with highly reactive aggregates. The following provides guidance for detecting which aggregates are potentially reactive and methods of controlling ASR based on the current state of the art.

Alkali-silica reactivity is an expansive reaction between reactive forms of silica in aggregate and alkali hydroxides. The alkalies, potassium and sodium, come mostly from cement, though some come from aggregates, pozzolans, admixtures, and mixing water. External sources of alkali from soils, deicers, and industrial processes can also contribute to reactivity. The reaction forms an alkali-silica gel that swells as it draws water from the surrounding paste, inducing pressure, expansion, and cracking of the aggregate and surrounding paste. The reaction can be visualized as a two-step process:

1.   Alkali + silica → gel reaction product

2.   Gel reaction product + moisture → expansion

The outward manifestation of the expansion often results in map-pattern cracks. Crack patterns are affected by restraint conditions and other volume change movements.

Aggregate sources are prone to ASR if the following silica constituents are present: opal (more than 0.5%); chert and chalcedony (more than 3.0%); tridymite and cristobalite (more than 1.0%); strained or microcrystalline quartz (more than 5.0%) as found in granites, granite gneiss, graywackes, argillites, phyllites, siltstones, and natural sand and gravel; and natural volcanic glasses. The aggregate should then be evaluated for ASR in accordance with ASTM C 1260 (rapid mortar bar test) or ASTM C 1293 (concrete prism test). As with other methods, the C 1260 test is an accelerated procedure that is intended to "force" the potential reactions, the high alkali content and temperature are not representative of field conditions. It is an aggregate screening test and does not identify safe cement-aggregate combinations. A variation of C 1260, as described in PCA (1995) and Farny and Kosmatka (1997) can be used to determine the effectiveness of pozzolans or slags to control ASR. The use of ASTM C 1293 comes from demonstrated Canadian practice. Criteria to determine if the aggregate is potentially reactive based on these physical tests follow:

ASTM C1260	Mean Expansion at 14 days:	>0.10%, Potentially reactive
ASTM C1293	Expansion at 1 year:	>0.04%, Potentially reactive

After an aggregate is determined to be potentially reactive by the above tests, special precautions in mix proportioning and mix ingredient selection must be taken. ASR is avoidable by using the following approaches individually or in combination:

5.18.1 Blended Cement

Blended hydraulic cement meeting ASTM C 595 or C 1157 can be used to control ASR as long as it has been demonstrated to be effective in controlling ASR using ASTM C 1260 or C 441 as outlined in PCA (1995).

5.18.2 Supplementary Cementing Materials

Supplementary cementing materials such as fly ash, slag, silica fume, and natural pozzolans can be added to concrete to control ASR (Figure 5.34). The benefits of these materials come from their pozzolanic properties. Fly ash and natural pozzolans should meet ASTM C 618 requirements.

Ground granulated blast-furnace slag should meet ASTM C 989. Silica fume should meet ASTM C 1240. As an incorrect dosage of supplementary cementing materials may not control ASR or can increase potential for ASR, the optimum amount should be determined using ASTM C 1260 as described in PCA (1995) or other methods as described in Farny and Kosmatka (1997). Low-alkali Class F ash is generally more effective than many Class C ashes in reducing ASR.

5.18.3 Controlling Concrete Alkali Content

The concrete alkali content can be limited to a level that has been demonstrated to be effective by the field performance of concretes with the potentially reactive aggregate. This approach assumes that concrete ingredients for future construction are the same as those materials used in the past where detrimental ASR did not occur. Specifications in the United States have traditionally limited the alkali content in concrete through use of low-alkali portland cement. ASTM C 150 defines low-alkali cement as having a maximum alkali content of 0.60% equivalent sodium oxide. However, higher alkali limits have been safely used with certain moderately reactive aggregates. Low-alkali cements are not available

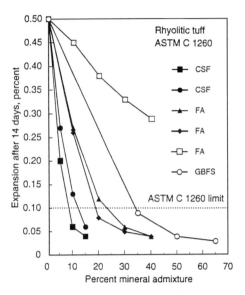

FIGURE 5.34 ASR expansion versus dosage for various mineral admixtures: silica fume (CSF), fly ash (FA), and slag (GBFS); *Source:* Courtesy of American Concrete Institute; Berube, M.A. and Duchesne, J. 1992. Evaluation of testing methods used for assessing the effectiveness of mineral admixtures in suppressing expansion due to alkali-aggregate reaction. In *Fly Ash, Silica Fume, Slag, and Natural Pozzolans in Concrete*, pp. 549–575. American Concrete Institute, Farmington Hills, MI.

in many areas and deleterious reactivity has been observed with certain glassy volcanic aggregates, especially andesite and rhyolite rocks, even when low-alkali cements (alkali contents of 0.35 to 0.6%) were used [Farny and Kosmatka, 1997]. Thus the use of local cements in combination with one or more of the above alternatives is the preferable approach to controlling ASR.

The methods outlined to evaluate the potential for aggregate reaction provide estimates based on accelerated laboratory tests. While these methods are often the only approach available, the best indicator of potential aggregate performance is field history under conditions similar to the intended application. If a proven service record can be documented, the "comfort level" regarding potential performance is increased dramatically. Complete documentation of service history should include condition evaluations of several structures, preferably at least 15 years old, exposed to moisture in service, and made with concrete containing the aggregates in question and cement having an alkali content at least as high (equivalent sodium oxide) as the cement intended for the new construction. Condition evaluations should include petrographic examination of the concrete by ASTM C 856. Requirements for reactivity are

- a potentially reactive aggregate;
- sufficiently available alkali; and
- sufficiently available moisture.

If any of the three requirements are missing, deleterious reactivity will not occur. The approaches listed above for avoiding reactivity focus on the first two requirements. However, the moisture requirement, while difficult to control, should not be ignored. Of particular importance are wetting and drying exposures, as they can concentrate alkalies near the drying surface, inducing reactivity. Design considerations should include steps to minimize available moisture and wet-dry cycles. ASR needs at least an 80% relative humidity (RH) at 73°F (23°C) in the concrete to develop. In some applications, it may be possible to dry out the concrete to less than 80% RH. For example, sealed laboratory concrete with a water/cement ratio of 0.35 or less can self-desiccate to below 80% RH and thereby provides little opportunity for expansion due to ASR. Another environmental factor that affects reactivity is temperature. As with any chemical reaction, the rate of reactivity increases with temperature, and thus structures in warmer exposures are more susceptible to ASR than those in colder climates.

5.19 Related Standards

5.19.1 American Society for Testing and Materials (ASTM)

C 33	Specification for Concrete Aggregates
C 150	Specification for portland Cement
C 227	Test Method for Potential Alkali Reactivity of Cement-Aggregate Combinations (Mortar Bar Method)
C 289	Test Method for Potential Reactivity of Aggregates (Chemical Method)
C 294	Descriptive Nomenclature of Constituents of Natural Mineral Aggregates
C 295	Practice for Petrographic Examination of Aggregates for Concrete
C 441	Test Method for Effectiveness of Mineral Admixtures or Ground Blast-Furnace Slag in Preventing Excessive Expansion of Concrete Due to the Alkali-Silica Reaction
C 586	Test Method for Potential Alkali Reactivity of Carbonate Rocks for Concrete Aggregates (Rock Cylinder Method)
C 595	Specification for Blended Hydraulic Cements
C 618	Specification for Fly Ash and Raw or Calcined Natural Pozzolans for Use as a Mineral Admixture in portland-Cement Concrete
C 823	Practice for Examination and Sampling of Hardened Concrete in Construction
C 856	Practice for Petrographic Examination of Hardened Concrete
C 989	Specification for Ground Granulated Blast-Furnace Slag for Use in Concrete and Mortars
C 1105	Standard Test Method for Length of Change of Concrete Due to Alkali-Carbonate Rock Reaction
C 1157	Performance Specification for Blended Hydraulic Cement
C 1240	Specification for Silica Fume for Use in Hydraulic-Cement Concrete and Mortar Test Method for Accelerated Detection of Potential Deleterious Expansion of Mortar Bars Due to Alkali-Silica Reaction
C 1293	Test Method for Concrete Aggregates by Determination of Length Change of Concrete Due to Alkali-Silica Reaction.

References

ACI Committee 222. 1989. *Corrosion of Metals in Concrete*, ACI 222R-89, 30 pp. American Concrete Institute, Farmington Hills, MI.

ACI Committee 211. 1991a. *Standard Practice for Selecting Proportions for Normal, Heavyweight and Mass Concrete*, ACI 211.1-91, 38 pp. American Concrete Institute, Farmington Hills, MI.

ACI Committee 211. 1991b. *Standard Practice for Selecting Proportions for Structural Lightweight Concrete*, ACI 211. 2-91, 14 pp. American Concrete Institute, Farmington Hills, MI.

ACI Committee 211. 1993. *Guide for Selecting Proportions for High Strength Concrete with Portland Cement and Fly Ash*, ACI 211.4 R-93. American Concrete Institute, Farmington Hills, MI.

Berube, M.A. and Duchesne, J. 1992. Evaluation of testing methods used for assessing the effectiveness of mineral admixtures in suppressing expansion due to alkali-aggregate reaction. In *Fly Ash, Silica Fume, Slag, and Natural Pozzolans in Concrete*, pp. 549–575. American Concrete Institute, Farmington Hills, MI.

Burg, R.G. 1996a. *The Influence of Casting and Curing Temperatures on the Properties of Fresh and Hardened Concrete*, RD113, 20 pp. Portland Cement Association, Skokie, IL.

Burg, R.G. and Farny, J.A. 1996b. The influence of concrete casting and curing temperatures. In *Concrete Technology Today*, PL963, 3 pp. Portland Cement Association, Skokie, IL.

Burg, R.G. and Ost, B.W. 1994. *Engineering Properties of Commercially Available High-Strength Concretes (Including Three-Year Data)*, RD104, 62 pp. Portland Cement Association, Skokie, IL.

Campbell, D.H., Sturm, R.D., and Kosmatka, S.H. 1991. Detecting Carbonation. PL911. *Concrete Technology Today*, 6 pp. Portland Cement Association, Skokie, IL.

Detwiler, R.J. 1996. Long-term strength tests of high-strength concrete. In *Concrete Technology Today*, PL962. Portland Cement Association, Skokie, IL.

Farny, J.A. and Kosmatka, S.H. 1997. *Diagnosis and Control of Alkali-Aggregate Reactions in Concrete*, IS413. Portland Cement Association, Skokie, IL.

Farny, J.A. and Panarese, W.C. 1994. *High-Strength Concrete*, EB114, 58 pp. Portland Cement Association, Skokie, IL.

Federal Highway Administration. 1987. *Protective System for New Prestressed and Substructure Concrete*, FRWA/RD-86/193. Washington, DC.

Gebler, S.H. and Klieger, P. 1986. *Effect of Fly Ash on Some of the Physical Properties of Concrete*, RD089. Portland Cement Association, Skokie, IL.

Klieger, P. 1952. *Studies of the Effect of Entrained Air on the Strength and Durability of Concretes Made with Various Maximum Sizes of Aggregates*, RX040. Portland Cement Association, Skokie, IL.

Kosmatka, S.H. 1994. *Bleeding*, RP328, 28 pp. Portland Cement Association, Skokie, IL.

Kosmatka, S.H. and Panarese, W.C. 1992. *Design and Control of Concrete Mixtures*, EB001, 218 pp. Portland Cement Association, Skokie, IL.

Lange, D.A. 1994. *Long-Term Strength Development of Concrete*, RP326, 36 pp. Portland Cement Association, Skokie, IL.

Nawy, E.G. 1996a. *Fundamentals of High Strength High Performance Concrete*, 350 pp. Addison Wesley Longman, California and London.

Nawy, E.G. 1996b. *Reinforced Concrete—A Fundamental Approach*, 3rd ed., 840 pp. Prentice Hall, Upper Saddle River, NJ.

PCA 1995. *Guide Specification for Concrete Subject to Alkali-Silica Reactions*, IS415, 8 pp. Portland Cement Association, Skokie, IL.

PCA 1996. Portland cement: Past and present characteristics. In *Concrete Technology Today*, PL962, 4 pp. Portland Cement Association, Skokie, IL.

Stark, D. 1989a. *Durability of Concrete in Sulfate-Rich Soils*, RD097, 18 pp. Portland Cement Association, Skokie, IL.

Stark, D. 1989b. *Influence of Design and Materials on Corrosion Resistance of Steel in Concrete*, RD098, 40 pp. Portland Cement Association, Skokie, IL.

Whiting, D. 1988. Permeability of Selected Concretes. In *Permeability of Concrete*, SP108-11, pp. 195–225. American Concrete Institute, Farmington Hills, MI.

Wood, S.L. 1992. *Evaluation of the Long-Term Properties of Concrete*. RD102, 99 pp. Portland Cement Association, Skokie, IL.

(a)

(b)

(a) Hoist delivery and placement of ready mix concrete. (b) Truck delivery of ready mix concrete (Courtesy Portland Cement Association).

6

Part A: Design of Concrete Mixtures

Part A by
Edward G. Nawy, D.Eng., P.E., C.Eng.
Distinguished Professor, Civil Engineering at Rutgers University—The State University of New Jersey. Expert in concrete structures and materials.

Part B by
Jaime Moreno
President, Cement Technology Corp., Chicago. Expert on cements, blending of cements, high-performance concretes, and efficient mixture proportioning.

John Albinger
President, VANS Materials, Inc., Chicago. Expert in mixture design and in high-strength, high-performance concrete.

0-8493-2666-4/97/$0.00+$.50
© 1997 by CRC Press LLC

6.1 General

Design of concrete mixtures involves choosing appropriate proportions of ingredients for particular strengths, long-term qualities, and performance of the concrete produced. Several factors determine these properties. They include the following parameters, which are summarized in Figure 6.1:

1. quality of cement,
2. proportion of cement in relation to water in the mix (water/cementitious ratio),
3. strength and cleanliness of aggregate,
4. interaction or adhesion between cement paste and aggregate,
5. adequate mixing of the ingredients,
6. proper placing, finishing, and compaction of the fresh concrete,
7. curing at a temperature not below 50°F while the placed concrete gains strength, and

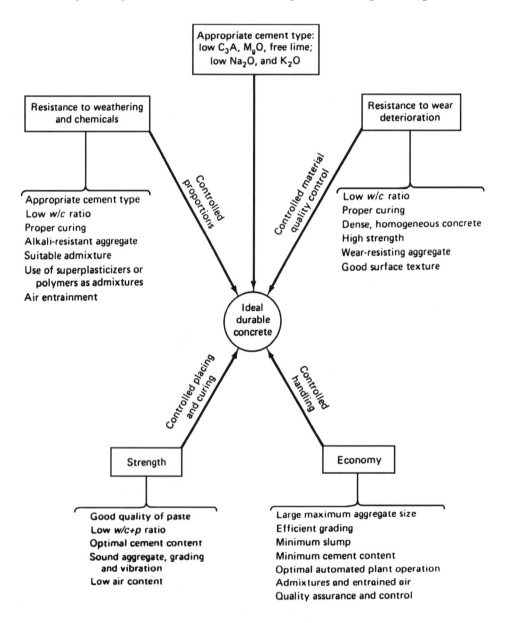

FIGURE 6.1   Principal properties of good concrete. (Nawy, 1996a, 1996b).

8. chloride content not to exceed 0.15% in reinforced concrete exposed to chlorides in service and 1% in dry protected concrete.

A study of these requirements shows that most control actions have to be taken prior to placing the fresh concrete. Since such control is governed by the proportions of ingredients and the mechanical ease or difficulty in handling and placing the concrete, the development of criteria based on the theory of proportioning for each mix should be studied. Most mixture design methods have become essentially only of historical and academic value.

The two universally accepted methods of mixture proportioning for normal-weight and light-weight concrete are the American Concrete Institute's (ACI) methods of proportioning, described in ACI's recommended practices for selecting proportions for normal-weight, heavyweight, and mass concrete and in the recommended practice for selecting proportions for structural light-weight concrete (ASTM, 1993; ACI, 1991; Nawy, 1996a).

6.2 Selection of Constituent Materials

6.2.1 Cement and Other Cementitious Materials

As discussed in Chapter 1, several types of cement are available. The most commonly used is Portland cement. Table 6.1 lists the general composition of the various types of Portland cement. Types I and III fundamentally differ in that type III (high early cement) is considerably finer than type I, and in 7 d achieves the strength that normal cement concrete achieves in 28 d. Other hydraulic cements include blended cements, rapid-setting cements, expansive cements (type K), very high early cements, and exotic cements such as macrodefect-free cement (MDF), densified cement (DSP), perlite cement, and alkali-activated cement.

6.2.2 Normal Weight and Lightweight Aggregate

Aggregates are those parts of the concrete that constitute the bulk of the finished product. They comprise 60% to 80% of the volume of the concrete and have to be graded so that the whole mass of concrete acts as a relatively solid, homogeneous, dense combination, with the smaller particles acting as inert filler for the voids that exist between the larger particles.

Aggregates are of two types:

1. *coarse aggregate:* gravel, crushed stone, or blast-furnace slag, and
2. *fine aggregate:* natural or manufactured sand.

TABLE 6.1 Percentage Composition of Portland Cements

Type of Cement	Component (%)							General Characteristics
	C_3S	C_2S	C_3A	C_4AF	$CaSO_4$	CaO	MgO	
Normal: I	49	25	12	8	2.9	0.8	2.4	All-purpose cement
Modified: II	45	29	6	12	2.8	0.6	3.0	Comparative low heat liberation: used in large structures
High early strength: III	56	15	12	8	3.9	1.4	2.6	High strength in 3 days
Low heat: IV	30	46	5	13	2.9	0.3	2.7	Used in mass concrete dams
Sulfate resisting: V	43	36	4	12	2.7	0.4	1.6	Used in sewers and structures exposed to sulfates

Sources: Nawy, E.G. 1996a. *Reinforced Concrete—A Fundamental Approach*, 3rd ed., 840 pp. Prentice Hall, Upper Saddle River, NJ. Nawy, E.G. 1996b. *Fundamentals of High Strength High Performance Concrete*, 350 pp. Longman U.K., London. Addison–Wesley, New York.

Since the aggregate constitutes the major part of the mix, the more aggregate in the mix the cheaper the concrete, provided that the mix is of reasonable workability for the specific job for which it is used.

Coarse aggregate is classified as such if the smallest particle is greater than 1/4 in (6 mm). Properties of coarse aggregate affect the final strength of hardened concrete and its resistance to disintegration, weathering, and other destructive effects. Mineral coarse aggregates must be clean of organic impurities and must bond well with the cement gel.

Common types of coarse aggregate are as follows.

1. *Natural crushed stone.* This is produced by crushing natural stone or rock from quarries. The rock may be igneous, sedimentary, or metamorphic. Although crushed rock gives higher concrete strength, it is less workable for mixing and placing than are the other types of aggregate.

2. *Natural gravel.* This is produced by the weathering action of running water on the beds and banks of streams. It gives less strength than crushed rock but is more workable.

3. *Artificial coarse aggregates.* These are mainly slag and expanded shale and are frequently used to produce lightweight concrete. They are byproducts of other manufacturing processes such as blast-furnace slag or expanded shale, or pumice for lightweight concrete.

4. *Heavyweight and nuclear-shielding aggregates.* With the specific demands of our atomic age and the hazards of nuclear radiation, owing to the increasing number of atomic reactors and nuclear power stations, special concretes have had to be produced to shield against X-rays, gamma rays, and neutrons. In such concretes, economic and workability considerations are not of prime importance. The main heavy coarse aggregate types are steel punchings, barites, magnetites, and limonites.

   Whereas concrete with ordinary aggregate weighs about 144 lb/ft^3, concrete made with these heavy aggregates weighs from 225 to 330 lb/ft^3. The property of heavyweight radiation-shielding in concrete depends on the density of the compact product rather than on the water/cement ratio criterion. In certain cases, high density is the only consideration, whereas in others both density and strength govern.

5. *Fine aggregate.* This is a smaller filler made of sand. It ranges in size from No. 4 to No. 100 in U.S. standard sieve sizes. A good fine aggregate should always be free of organic impurities, clay, or any deleterious material or excessive filler of size smaller than No. 100 sieve. Preferably, it should have a well-graded combination conforming to the American Society for Testing

TABLE 6.2 Grading Requirements for Aggregates in Normal-Weight Concrete (ASTM C-33)

U.S. Standard Sieve Size	Percent Passing				
	Coarse Aggregate				
	No. 4 to 2 in	No. 4 to 1 1/2 in	No 4 to 1 in	No. 4 to 3/4 in	Fine Aggregate
2 in	95–100	100	—	—	—
1 1/2 in	—	95–100	100	—	—
1 in	25–70	—	95–100	100	—
3/4 in	—	35–70	—	90–100	—
1/2 in	10–30	—	25–60	—	—
3/8 in	—	10–30	—	20–55	100
No. 4	0–5	0–5	0–10	0–10	95–100
No. 8	0	0	0–5	0–5	80–100
No. 16	0	0	0	0	50–85
No. 30	0	0	0	0	25–60
No. 50	0	0	0	0	10–30
No. 100	0	0	0	0	2–10

Source: American Society for Testing and Materials. 1993. *Annual Book of ASTM Standards*, Part 14, *Concrete and Mineral Aggregates*, 834 pp. ASTM, Philadelphia, PA.

TABLE 6.3 Grading Requirements for Aggregates in Lightweight Structural Concrete (ASTM C-330)

			Percentage (by weight) Passing Sieves with Square Openings						
Size Designation	1 in (25.0 mm)	3/4 in (19.0 mm)	1/2 in (12.5 mm)	3/8 in (9.5 mm)	No. 4 (4.75 mm)	No. 8 (2.36 mm)	No. 16 (1.18 mm)	No. 50 (300 μm)	No. 100 (150 μm)
Fine Aggregate									
No. 4 to 0	—	—	—	100	85–100	—	40–80	10–35	5–25
Coarse Aggregate									
1 in to No. 4	95–100	—	25–60	—	0–10	—	—	—	—
3/4 in to No. 4	100	90–100	—	10–50	0–15	—	—	—	—
1/2 in to No. 4	—	100	90–100	40–80	0–20	0–10	—	—	—
3/8 in to No. 8	—	—	100	80–100	5–40	0–20	0–10	—	—
Combined Fine and Coarse Aggregate									
1/2 in to 0	—	100	95–100	—	50–80	—	—	5–20	2–15
3/8 in to 0	—	—	100	90–100	65–90	35–65	—	10–25	5–15

Source: American Society for Testing and Materials. 1993. *Annual Book of ASTM Standards*, Part 14, *Concrete and Mineral Aggregates*, 834 pp. ASTM, Philadelphia, PA.

TABLE 6.4 Grading Requirements for Coarse Aggregate for Aggregate Concrete (ASTM C-637)

	Percentage Passing	
Sieve Size	Grading 1*	Grading 2†
Coarse Aggregate		
2 in (50 mm)	100	
1 1/2 in (37.5 mm)	95–100	100
1 in (25.0 mm)	40–80	95–100
3/4 in (19.0 mm)	20–45	40–80
1/2 in (12.5 mm)	0–10	0–15
3/8 in (9.5 mm)	0–2	0–2
Fine Aggregate		
No. 8 (2.36 mm)	100	
No. 16 (1.18 mm)	95–100	100
No. 30 (600 μm)	55–80	75–95
No. 50 (300 μm)	30–55	45–65
No. 100 (150 μm)	10–30	20–40
No. 200 (75 μm)	0–10	0–10
Fineness modulus	1.30–2.10	1.00–1.60

Note: Reprinted with permission from the American Society for Testing and Materials, Philadelphia, PA.
*for 1 1/2 in (37.5 mm) maximum-size aggregate.
†for 3/4 in (19.0 mm) maximum-size aggregate.

and Materials (ASTM) sieve analysis standards. For radiation-shielding concrete, fine steel shot and crushed iron ore are used as fine aggregate.

The recommended grading of coarse and fine aggregates for normal-weight concrete is presented in Table 6.2 (ASTM C-33). Grading requirements for lightweight aggregate (ASTM C-330) are given in Table 6.3. Table 6.4 gives grading requirements for coarse aggregate for aggregate concrete (ASTM C-637). Data in these tables are used with permission from The American Society for Testings and Materials, Philadelphia, PA. The unit weights of the commonly used aggregates in structural concrete are given in Table 6.5.

6.2.3 Workability-Enhancing and Strength-Enhancing Admixtures

Admixtures are materials other than water, aggregate, or hydraulic cement that are used as ingredients of concrete and that are added to the batch immediately before or during mixing. Their function is to modify the properties of concrete for strength, workability, and long-term performance. The

TABLE 6.5 Unit Weight of Aggregates

Type	Unit Weight of Dry-Rodded Aggregate (lb/ft³)*	Unit Weight of Concrete (lb/ft³)*
Insulating concretes (perlite, vermiculite, etc.)	15–50	20–90
Structural lightweight	40–70	90–110
Normal weight	70–110	130–160
Heavyweight	>135	180–380

Sources: American Society for Testing and Materials. 1993. *Annual Book of ASTM Standards*, Part 14, *Concrete and Mineral Aggregates*, 834 pp. ASTM, Philadelphia, PA. Nawy, E.G. 1996a. *Reinforced Concrete—A Fundamental Approach.* 3rd ed. 840 pp. Prentice Hall, Upper Saddle River, NJ. Nawy, E.G. 1996b. *Fundamentals of High Strength High Performance Concrete.* 350 pp. Longman U.K., London. Addison–Wesley, New York.
*1 lb/ft³ = 16.02 kg/m³.

major types of admixtures can be summarized as follows:

1. accelerating admixtures,
2. air-entraining admixtures,
3. water-reducing admixtures and set-controlling admixtures,
4. finely divided mineral admixtures,
5. admixtures for no-slump concretes,
6. polymers, and
7. superplasticizers.

Details of the various mineral and organic admixtures for strength enhancement are given in Chapters 2 and 3. As presented in Nawy (1996a), the following is a brief outline of these admixtures.

6.2.3.1 Accelerating Admixtures

These admixtures are added to the concrete mixture to reduce the time of setting and accelerate early strength development. The best known are calcium chlorides. Other accel-

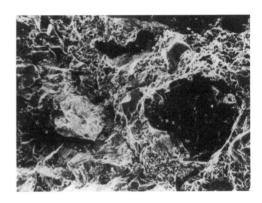

PHOTO 6.1   Electron-microscope photographs of concrete (Test by E.G. Nawy et al.).

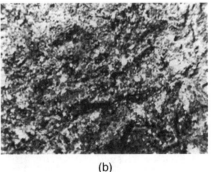

| (a) | (b) |

PHOTO 6.2   Fracture surface in tensile splitting test. (a) Mortar failure, $f_t' = 450$ psi, 3.1 MPa; (b) Aggregate failure, $f_t' = 1550$ psi, 10.7 MPa. From E.G. Nawy, *Fundamentals of High Strength High Performance Concrete*, © 1996. Reprinted with permission of Addison–Wesley, New York.

erating chemicals include a wide range of soluble salts such as chlorides, bromides, carbonates, and silicates, and some other organic compounds such as triethanolamine.

It must be stressed that calcium chlorides should not be used where progressive corrosion of steel reinforcement can occur. The maximum dosage recommended is 0.5% by weight of Portland cement.

6.2.3.2 Air-Entraining Admixtures

These admixtures form minute bubbles that are 1 mm in diameter or smaller in concrete or mortar during mixing; they are used to increase both the workability of the mix during placing and the frost resistance of the finished product. Most air-entraining admixtures are in liquid form, although a few are powders, flakes, or semisolids. The amount of admixture required to obtain a given air content depends on the shape and grading of the aggregate used. The finer the aggregate, the larger the percentage of admixture needed. The amount required is also governed by several other factors such as type and condition of the mixer, use of fly ash or other pozzolans, and degree of agitation of the mix. It can be expected that air entrainment reduces the strength of the concrete. Maintaining cement content and workability, however, offsets the partial reduction of strength because of the resulting reduction in the water/cement ratio.

6.2.3.3 Water-Reducing and Set-Controlling Admixtures

These admixtures increase the strength of the concrete. They also enable reduction of cement content in proportion to the reduction of water content.

Most admixtures of the water-reducing type are water soluble. The water they contain becomes part of the mixing water in the concrete and is added to the total weight of water in the design of the mix. It has to be emphasized that the proportion of mortar to coarse aggregate should always remain the same. Changes in the water content, air content, or cement content are compensated for by changes in the fine-aggregate content so that the volume of the mortar remains the same.

6.2.3.4 Finely Divided Admixtures

These are mineral admixtures used to rectify deficiencies in concrete mixes by providing missing fines from the fine aggregate, improving one or more qualities of the concrete, such as reducing permeability or expansion, and reducing the cost of concrete-making materials. Such admixtures include hydraulic lime, slag cement, fly ash, and raw or calcined natural pozzolan.

6.2.3.5 Admixtures for No-Slump Concrete

No-slump concrete is defined as a concrete with a slump of 1 in (25 mm) or less immediately after mixing. The choice of admixture depends on the desired properties of the finished product such as plasticity, setting time and strength development, freeze-thaw effects, strength, and cost.

PHOTO 6.3   Slump test of fresh concrete (Test by E.G. Nawy et al.).

6.2.3.6 Polymers

These admixtures enable production of concretes of very high strength up to a compressive strength of 15,000 psi or higher and a tensile splitting strength of 1500 psi or higher. Such concretes are generally produced using a polymerizing material through (1) modification of the concrete property by water reduction in the field or (2) impregnation and irradiation under elevated temperature in a laboratory environment.

Polymer concrete (PC) is concrete made through the addition of resin and hardener as admixtures. The principle is to replace part of the mixing water with the polymer so as to attain a high compressive strength and other desired qualities. The optimum polymer/concrete ratio by weight to achieve such high compressive strengths seems to lie within the range of 0.3 to 0.45.

6.2.3.7 Superplasticizers

These admixtures can be termed "high range water reducing chemical admixtures." There are four types of plasticizers:

1. sulfonated melamine formaldehyde condensates, with a chloride content of 0.005% (MSF);
2. sulfonated naphthalene formaldehyde condensates, with negligible chloride content (NSF); and
3. modified lignosulfonates, which contain no chlorides.

The three plasticizers above are made from organic sulfonates and are termed superplasticizers in view of their considerable ability to facilitate reduction of water content in a concrete mix, while simultaneously increasing the slump up to 8 in (206 mm) or more. A dosage of 1% to 2% by weight of cement is advisable. Higher dosages can result in a reduction in compressive strength.

4. other super plasticizers such as sulfonic acid esters or other carbohydrate esters.

A dosage of 1% to 2 1/2% by weight of cement is advisable. Higher dosages can result in a reduction in compressive strength unless the cement content is increased to balance the reduction. It should be noted that superplasticizers exert their action by decreasing the surface tension of water and by equidirectional charging of cement particles. These properties, coupled with the addition of silica fume, help achieve high strength and water reduction in concrete without loss of workability.

6.2.3.8 Silica Fume Admixture Use in High-Strength Concrete

Silica fume is generally accepted as an efficient mineral admixture for high-strength concrete mixes. It is a pozzolanic material that has received considerable attention in both research and application. Silica fume is a byproduct resulting from the use of high-purity quartz with coal in electric arc furnaces during production of silicon and ferrosilicon alloys. Its main constituent, fine spherical particles of silicon dioxide, makes it an ideal cement replacement and simultaneously raises concrete strength. As it is a waste product that is relatively easy to collect, compared to fly ash or slag, silica fume is extensively used worldwide.

Proportions of silica fume in concrete mixes vary from 5% to 30% by weight of the cement, depending on strength and workability requirements. However, water demand is greatly increased as the proportion of silica fume is increased, and high-range water reducers are essential to keep the water/cementitious ratio low in order to produce higher-strength yet workable concrete. Silica fume seems to reach a high early strength in about 3 to 7 d with relatively less increase in strength at 28 d. The strength-development pattern of flexural and tensile splitting strengths is similar to that of compressive-strength gain for silica-fume-added concrete. The addition of silica fume to the mixture can produce significant increases in strength in excess of 20,000 psi, increased modulus of elasticity, and increased flexural strength.

6.3 Mixture Proportioning for High-Performance Normal-Strength Concrete (Compressive Cylinder Strength Limit 6000 psi)

The ACI method of designing mixtures for normal-weight concrete is summarized in Figure 6.2, based on water slump values selected from Tables 6.6, 6.7, and 6.8. One aim of the mixture design is to produce workable concrete that is easy to place in forms. A measure of the degree of consistency and extent of workability is the **slump**. In the slump test, a plastic concrete specimen is formed into a conical metal mold, as described in ASTM Standard C-143. The mold is lifted, leaving the concrete to slump, that is, to spread or drop in height. This drop in height is the slump measure of the degree of workability of the mixture.

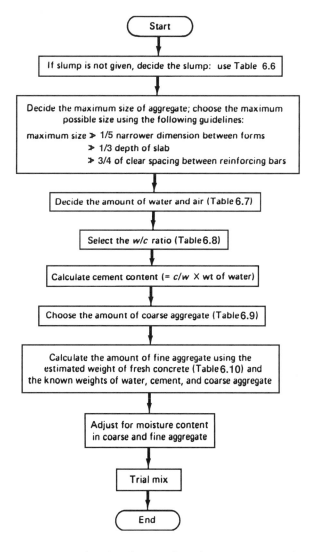

FIGURE 6.2   Flow chart for normal-weight concrete mixture design.   From E.G. Nawy, *Reinforced Concrete*, 3rd ed., © 1996. Reprinted by permission of Prentice Hall, Upper Saddle River, NJ.

TABLE 6.6 Recommended Slumps for Various Types of Construction

Types of Construction	Slump, in*	
	Maximum[†]	Minimum
Reinforced foundation walls and footings	3	1
Plain footings, caissons, and substructure walls	3	1
Beams and reinforced walls	4	1
Building columns	4	1
Pavements and slabs	3	1
Mass concrete	2	1

Sources: ACI Committee 211 *Standard Practice for Selecting Proportions for Normal, Heavyweight, and Mass Concrete.* ACI 211.1-91. 38 pp. American Concrete Institute, Farmington Hills. Nawy, E.G. 1996a. *Reinforced Concrete—A Fundamental Approach.* 3rd ed. 840 pp. Prentice Hall, Upper Saddle River, NJ.

*1 in = 25.4 mm.

[†] May be increased 1 in for methods of consolidation other than vibration.

6.3.1 Example 6.1: Mixture Proportions Design for Normal Strength Concrete (Nawy, 1996a)

Design a concrete mixture using the following details:

> Required strength: 4000 psi (27.6 MPa)
> Type of structure: beam
> Maximum size of aggregate: 3/4 in (18 mm)
> Fineness modulus of sand: 2.6
> Dry-rodded weight of coarse aggregate: 100 lb/ft^3
> Moisture absorption: 3% for coarse aggregate and 2% for fine aggregate

Solution.

> Required slump for beams (Table 6.6): 3 in. Maximum aggregate size (given): 3/4 in. For a slump between 3 and 4 in and a maximum aggregate size of 3/4 in, the weight of water required per cubic yard of concrete (Table 6.7) is 340 lb/yd^3.
> For the specified compression strength, 4000 psi, the w/c ratio (Table 6.8) is 0.57. Table 6.9 is also needed if volumes instead of weights are used in the mix design calculations. Therefore, the amount of cement required per cubic yard of concrete is 340/0.57 = 596.5 lb/yd^3.
> Using a sand fineness value of 2.6 and Table 6.9, the volume of coarse aggregate is 0.64 yd^3. Using the dry-rodded weight of 100 lb/ft^3 for coarse aggregate, the weight of coarse aggregate is $(0.64) \times (27\ \text{ft}^3/\text{yd}^3) \times 100 = 1728\ \text{lb/yd}^3$.
> The estimated weight of fresh concrete for aggregate of 3/4 in is as follows:
> maximum size (Table 6.10) = 3690 lb/yd^3
> weight of sand = [weight of fresh concrete − weights of (water/cement + coarse aggregate)]

$$3960 - 340 - 596.5 - 1728 = 1295.5\ \text{lb}$$

> Net weight of sand to be taken (moisture absorption 2%) = $1.02 \times 1295.5 = 1321.41$ lb
> Net weight of gravel (moisture absorption 3%) = $1.03 \times 1728 = 1779.84$ lb
> Net weight of water = $340 - 0.02 \times 1295.5 - 0.03 \times 1728 = 262.25$ lb

TABLE 6.7 Approximate Mixing Water and Air Content Requirements for Different Slumps and Nominal Maximum Sizes of Aggregates

Slump, in	Water (lb/yd³ of Concrete for Indicated Nominal Maximum Sizes of Aggregate)							
	3/8 in*	1/2 in*	3/4 in*	1 in*	1 1/2 in*	2 in*,†	3 in†,‡	6 in†,‡
	Non-Air-Entrained Concrete							
1 to 2	350	335	315	300	275	260	220	190
3 to 4	385	365	340	325	300	285	245	210
6 to 7	410	385	360	340	315	300	270	—
Approximate amount of entrapped air in non-air-entrained concrete, %	3	2.5	2	1.5	1	0.5	0.3	0.2
	Air-Entrained Concrete							
1 to 2	305	295	280	270	250	240	205	180
3 to 4	340	325	305	295	275	265	225	200
6 to 7	365	345	325	310	290	280	260	-
Recommended average total air content§ (percent for level of exposure)								
Mild exposure	4.5	4.0	3.5	3.0	2.5	2.0	1.5‖,#	1.0‖,#
Moderate exposure	6.0	5.5	5.0	4.5	4.5	4.0	3.5‖,#	3.0‖,#
Extreme exposure**	7.5	7.0	6.0	6.0	5.5	5.0	4.5‖,#	4.0‖,#

Sources: ACI Committee 211. *Standard Practice for Selecting Proportions for Normal, Heavyweight, and Mass Concrete,* ACI 211.1-91, 38 pp. American Concrete Institute, Farmington Hills; Nawy, E.G. 1996a. *Reinforced Concrete—A Fundamental Approach.* 3rd ed. 840 pp. Prentice Hall, Upper Saddle River, NJ.

*These quantities of mixing water are for use in computing cement factors for trial batches. They are maximal for reasonably well-shaped angular coarse aggregates graded within limits of accepted specifications.

†The slump values for concrete containing aggregate larger than 1 1/2 in are based on slump tests made after removal of particles larger than 1 1/2 in by wet screening.

‡These quantities of mixing water are for use in computing cement factors for trial batches when 3-in or 6-in nominal maximum-size aggregate is used. They are average for reasonably well-shaped coarse aggregates, well graded from coarse to fine.

§Additional recommendations for air content and necessary tolerances on air content for control in the field are given in a number of ACI documents, including ACI 201, 345, 318, 301, and 302. ASTM C-94 for ready-mixed concrete also gives air-content limits. The requirements in other documents may not always agree exactly, so in proportioning concrete consideration must be given to selecting an air content that will meet the needs of the job and also meet the applicable specifications.

‖For concrete containing large aggregates that will be wet screened over the 1 1/2 in sieve prior to testing for air content, the percentage of air expected in the 1 1/2 in-minus material should be tabulated in the 1 1/2 in column. However, initial proportioning calculations should include the air content as a percent of the whole.

#When using large aggregate in low-cement-factor concrete, air entrainment need not be detrimental to strength. In most cases, the mixing water requirement is reduced sufficiently to improve the water/cement ratio and thus to compensate for the strength-reducing effect of entrained-air concrete. Generally, therefore, for these large maximum sizes of aggregate, air contents recommended for extreme exposure should be considered even though there may be little or no exposure to moisture and freezing.

**These values are based on the criteria that 9% air is needed in the mortar phase of the concrete. If the mortar volume will be substantially different from that determined in this recommended practice, it may be desirable to calculate the needed air content by taking 9% of the actual mortar volume.

For 1 yd³ of Concrete.

cement 596.5 lb = 600 lb (273 kg)
sand 1321.41 lb = 1320 lb (600 kg)
gravel 1779.84 lb = 1780 lb (810 kg)
water 262.25 lb = 260 lb (120 kg)

6.3.1.1 PCA Method of Design of Concrete Mixtures

The mixture design method of the Portland Cement Association (PCA) is essentially similar to the ACI method. Generally, results would be very close once trial batches are prepared in the laboratory.

TABLE 6.8   Relationship Between Water/Cement Ratio and Compressive Strength of Concrete

Compressive Strength at 28 Days,* psi [†]	Water/Cement Ratio, by weight	
	Non-Air-Entrained Concrete	Air-Entrained Concrete
6000	0.41	—
5000	0.48	0.40
4000	0.57	0.48
3000	0.68	0.59
2000	0.82	0.74

Sources: ACI Committee 211. *Standard Practice for Selecting Proportions for Normal, Heavyweight, and Mass Concrete*, ACI 211.1-91, 38 pp. American Concrete Institute, Farmington Hills; Nawy, E.G. 1996a. *Reinforced Concrete—A Fundamental Approach*, 3rd ed., 840 pp. Prentice Hall, Upper Saddle River, NJ.

*Values are estimated average strengths for concrete containing not more than the percentage of air shown in Table 6.7. For a constant water/cement ratio, the strength of concrete is reduced as the air content is increased.

Strength is based on 6 in × 12 in cylinders moist-cured 28 days at 73.4 ± 3°F(23 ± 1.7°C) in accordance with Section 9(b) of ASTM C-31, "Making and Curing Concrete Compression and Flexure Test Specimens in the Field."

Relationship assumes maximum size of aggregate about *j* to 1 in; for a given source, strength produced for a given water/cement ratio will increase as maximum size of aggregate decreases.

[†] 1000 psi = 6.9 MPa.

TABLE 6.9   Volume of Coarse Aggregate per Unit of Volume of Concrete

Maximum Size of Aggregate, in*	Volume of Dry-Rodded Coarse Aggregate[†] per Unit Volume of Concrete for Different Fineness Moduli of Sand			
	2.40	2.60	2.80	3.00
3/8	0.50	0.48	0.46	0.44
1/2	0.59	0.57	0.55	0.53
3/4	0.66	0.64	0.62	0.60
1	0.71	0.69	0.67	0.65
1 1/2	0.75	0.73	0.71	0.69
2	0.78	0.76	0.74	0.72
3	0.82	0.80	0.78	0.76
6	0.87	0.85	0.83	0.81

Source: American Society for Testing and Materials 1993. *Annual Book of ASTM Standards*, Part 14, *Concrete and Mineral Aggregates*, 834 pp. ASTM, Philadelphia, PA; ACI Committee 211. *Standard Practice for Selecting Proportions for Normal, Heavyweight, and Mass Concrete*, ACI 211.1-91, 38 pp. American Concrete Institute, Farmington Hills; Nawy, E.G. 1996a. *Reinforced Concrete—A Fundamental Approach*, 3rd ed., 840 pp. Prentice Hall, Upper Saddle River, NJ.

*1 in = 25.4 mm.

[†] Volumes are based on aggregates in dry-rodded condition as described in ASTM C-29, "Unit Weight of Aggregate." These volumes are selected from empirical relationships to produce concrete with a degree of workability suitable for usual reinforced construction. For less workable concrete, such as that required for concrete pavement construction, they may be increased about 10%. For more workable concrete, the coarse aggregate content may be decreased up to 10%, provided that the slump and water/cement ratio requirements are satisfied.

TABLE 6.10 First Estimate of Weight of Fresh Concrete

Maximum Size of Aggregate, in*	First Estimate of Concrete Weight,† lb/yd³	
	Non-Air-Entrained Concrete	Air-Entrained Concrete
3/8	3840	3690
1/2	3890	3760
3/4	3960	3840
1	4010	3900
1 1/2	4070	3960
2	4120	4000
3	4160	4040
6	4230	4120

Source: ACI Committee 211. *Standard Practice for Selecting Proportions for Normal, Heavyweight, and Mass Concrete,* ACI 211.1-91, 38 pp. American Concrete Institute, Farmington Hills; Nawy, E.G. 1996a. *Reinforced Concrete—A Fundamental Approach,* 3rd ed., 840 pp. Prentice Hall, Upper Saddle River, NJ.

*1 in = 25.4 mm.

†Values calculated and presented in this table are for concrete of medium richness (550 lb of cement per cubic yard) and medium slump with aggregate specific gravity of 2.7. Water requirements are based on values for 3- to 4-in slump in Table 5.3.2 of ASTM C-143. If desired, the estimated weight may be refined as follows if necessary information is available: for each 10-lb difference in mixing water from Table 5.3.2, values for 3- to 4-in slump, correct the weight per cubic yard 15 lb in the opposite direction; for each 100-lb difference in cement content from 550 lb, correct the weight per cubic yard 10 lb in the same direction: for each 0.1 by which aggregate specific gravity deviates from 2.7, correct the concrete weight 100 lb in the same direction.

The PCA publication *Design and Control of Concrete Mixtures* (1994) gives the details of the method as well as other information on properties of the ingredients.

6.3.2 Estimating Compressive Strength of a Trial Mixture Using the Specified Compressive Strength

The compressive strength for which the trial mixture is designed is not the strength specified by the designer. The mixture should be overdesigned to ensure that the actual structure uses concrete with the specified minimum compressive strength. The extent of mixture overdesign depends on the degree of quality control available in the mixing plant.

ACI Committee 318 (1995) specifies a systematic way of determining the compressive strength for mixture designs using the specified compressive strength, f'_c. The procedure is presented in a self-explanatory flow-chart form in Figure 6.3. The cylinder compressive strength, f'_c is the test result 28 d after casting normal-weight concrete. Mixture design has to be based on an adjusted higher value, the adjusted cylinder compressive strength, f'_{cr}. The f'_{cr} value for which a trial mixture design is calculated depends on the

PHOTO 6.4 Cylinder compressive-strength testing set up.

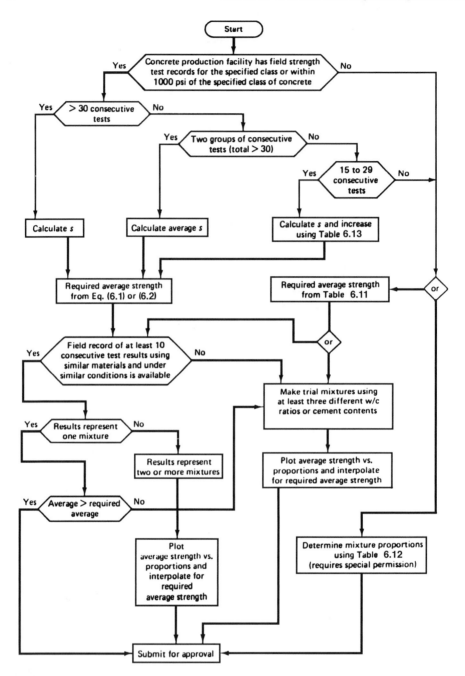

FIGURE 6.3 Flow chart of selection and documentation of concrete properties. E.G. Nawy, *Reinforced Concrete*, 3rd ed., © 1996. Reprinted by permission of Prentice Hall, Upper Saddle River, NJ.

extent of field data available, as shown in the following list.

1. *No cylinder test records available:* If field-strength test records for the specified class (or within 1000 psi of the specified class) of concrete are not available, the trial mixture strength can be calculated by increasing the cylinder compressive strength, f'_c, by a reasonable value,

depending on the spread in values expected in the supplied concrete. Such a spread can be quantified by the standard deviation values represented by values in excess of f'_c in Table 6.11. Table 6.12 can then be used to obtain the water/cement ratio needed for the required cylinder compressive strength value f'_c.

2. *Data are available on more than 30 consecutive cylinder tests:* If more than 30 consecutive test results are available, Eqs. (6.1), (6.2), and (6.3a) can be used to establish the required mixture strength, f'_{cr}, from f'_c. If two groups of consecutive test results with a total of more than 30 are available, f'_{cr} can be obtained using Eqs. (6.1), (6.2), and (6.3b).

3. *Data are available on fewer than 30 consecutive cylinder tests:* If the number of consecutive tests results available is fewer than 30 and more than 15, Eqs. (6.1), (6.2), and (6.3a) should be used in conjunction with Table 6.13. Essentially, the designer should calculate the standard deviation s using Eq. (6.3a), multiply the s value by the magnification factor provided in Table 6.13, and use the magnified s in Eqs. (6.1) and (6.2). In this manner, the expected degree of spread of cylinder test values as measured by the standard deviation, s, is well accounted for.

TABLE 6.11 Required Average Compressive Strength When Data Are Not Available to Establish a Standard Deviation

Specified Compressive Strength, f'_c, psi	Required Average Compressive Strength, f'_{cr}, psi*
Less than 3000	$f'_c + 1000$
3000–5000	$f'_c + 1200$
More than 5000	$f'_c + 1400$

Source: ACI Committee 211. Standard Practice for Selecting Proportions for Normal, Heavyweight, and Mass Concrete, ACI 211.1-91, 38 pp. American Concrete Institute, Farmington Hills; Nawy, E.G. 1996a. Reinforced Concrete—A Fundamental Approach, 3rd ed., 840 pp. Prentice Hall, Upper Saddle River, NJ.

*1000 psi = 6.9 MPa.

6.3.3 Recommended Proportions for Concrete Strength, f'_{cr}

Once the required average strength, f'_{cr}, for mixture design is determined, the actual mixture needed to obtain this strength can be established using either existing field data or a basic trial mixture design.

1. *Use of field data:* Field records of existing f'_{cr} values can be used if at least 10 consecutive test results are available. The test records should cover a period of at least 45 d. The material and conditions of the existing field mix data should be the same as those to be used in the proposed work.

TABLE 6.12 Maximum Permissible Water/Cement Ratios for Concrete When Strength Data From Field Experience or Trial Mixtures Are Not Available

Specified Compressive Strength,* f'_c, psi	Absolute Water/Cement Ratio by Weight	
	Non-Air-Entrained Concrete	Air-Entrained Concrete
2500	0.67	0.54
3000	0.58	0.46
3500	0.51	0.40
4000	0.44	0.35
4500	0.38	‡
5000	‡	‡

Source: ACI Committee 211. Standard Practice for Selecting Proportions for Normal, Heavyweight, and Mass Concrete. ACI 211.1-91. 38 pp. American Concrete Institute, Farmington Hills; Nawy, E.G. 1996a. Reinforced Concrete—A Fundamental Approach. 3rd ed. 840 pp. Prentice Hall, Upper Saddle River, NJ.

*28-day strength. With most materials, the water/cement ratios shown will provide average strengths greater than those calculated using Eqs. 6.1 and 6.2.

†1000 psi = 6.9 MPa.

‡For strengths above 4500 psi for non-air-entrained concrete and 4000 psi for air-entrained concrete, mix proportions should be established using trial mixes.

2. *Trial mixture design:* If field data are not available, trial mixtures should be used to establish the maximum water/cement ratio or minimum cement content necessary for designing a mix that produces a 28-d f'_{cr} value. In this procedure, the following requirements have to be met:

(a) The materials used and age at testing should be the same for the trial mixture and the concrete used in the structure.

(b) At least three water/cement ratios or three cement contents should be tried in the design of the mixture. The trial mixtures should result in the required f'_{cr} value. Three cylinders should be tested for each water/cement ratio and each cement content tried.

(c) The slump and air content should be within ±0.75 in and 0.5% of permissible limits.

(d) A plot should be constructed of the compressive strength at the designated age versus the cement content or water/cement ratio, from which one can then choose the water/cement ratio or the cement content that will give the average f'_{cr} value required.

TABLE 6.13 Modification Factor for Standard Deviation When Fewer Than 30 Tests Are Available

Number of Tests*	Modification Factor for Standard Deviation[†]
Less than 15	Use Table 6.11
15	1.16
20	1.08
25	1.03
30 or more	1.00

Source: ACI Committee 211. *Standard Practice for Selecting Proportions for Normal, Heavyweight, and Mass Concrete,* ACI 211.1-91, 38 pp. American Concrete Institute, Farmington Hills; Nawy, E.G. 1996a. *Reinforced Concrete—A Fundamental Approach,* 3rd ed., 840 pp. Prentice Hall, Upper Saddle River, NJ.

*Interpolate for intermediate number of tests.

[†]Modified standard deviation to be used to determine required average strength f'_{cr} in Eqs. 6.1 and 6.2.

If field test data are available for more than 30 consecutive tests, the trial mixture should be designed for a compressive strength f'_{cr} calculated from

$$f'_{cr} = f'_c + 1.34\,s \tag{6.1}$$

or from

$$f'_{cr} = f'_c + 2.3\,s - 500 \tag{6.2}$$

The larger value of f'_{cr} from Eqs. (6.1) and (6.2) should be used in designing the mixture, with the expectation that the minimum specified compressive strength will be attained. The standard deviation, s, is defined by the expression

$$s = \left[\frac{\sum(f_{ci} - f_c)^2}{(n-1)} \right]^{\frac{1}{2}} \tag{6.3a}$$

where

f'_c = individual strength, and
f'_{cr} = average of the n specimens.

If two test records are used to determine the average strength, the standard deviation becomes

$$s = \left[\frac{(n_1 - 1)(s_1)^2 + (n_2 - 1)(s_2)^2}{(n_1 + n_2 - 2)} \right]^{\frac{1}{2}} \tag{6.3b}$$

where

s_1, s_2 = standard deviations calculated from two test records, 1 and 2, respectively, and
n_1, n_2 = number of tests in each test record, respectively.

If the number of test results available is fewer than 30 and more than 15, the value of s used in Eqs. (6.1) and (6.2) should be multiplied by the appropriate modification factor given in Table 6.13.

6.3.4 Example 6.2: Calculation of Design Strength for a Trial Mixture

Calculate the average compressive strengths for the design of a concrete mixture where the specified compressive strength is 5000 psi (334.5 MPa) and such that (a) the standard deviation obtained using more than 30 consecutive tests is 500 psi (3.45 MPa), (b) the standard deviation obtained using 15 consecutive tests is 450 psi (3.11 MPa); and (c) records of prior cylinder test results are not available.

Solution.

(a) Using Eq. (6.1)

$$f'_{cr} = 5000 + 1.34 \times 500 = 5670 \text{ psi}$$

Using Eq. (6.2)

$$f'_{cr} = 5000 + 2.33 \times 500 - 500 = 5665 \text{ psi}$$

Hence the required trial mix strength is 5670 psi (39.12 MPa).

(b) The standard deviation is equal to 450 psi in 15 tests. From Table 6.13, the modification factor for s is 1.16. Hence the value of the standard deviation to be used in Eqs. (6.1) and (6.2) is $1.16 \times 450 = 522$ psi (3.6 MPa). Using Eq. (6.1)

$$f'_{cr} = 5000 + 1.34 \times 522 = 5700 \text{ psi}$$

Using Eq. (6.2)

$$f'_{cr} = 5000 + 2.33 \times 522 - 500 = 5716 \text{ psi}$$

Hence the required trial mixture strength is 5716 psi (39.44 MPa).

(c) Records of prior test results are not available. Using Table 6.11

$$f'_{cr} = f_c + 1200$$

for 5000 psi concrete. Hence the trial mixture strength is $5000 + 1200 = 6200$ psi (42.78 MPa).

If a mixing plant keeps good records of its cylinder test results over a long period, the required trial mixture strength can be reduced as a result of quality control, hence reducing costs for the owner.

6.4 Mixture Proportioning for High-Performance High-Strength Concrete (Cylinder Compressive Strength Exceeding 6000 psi)

6.4.1 General

By present ACI definitions, high-strength concrete covers concretes whose cylinder compressive strength exceeds 6000 psi (41.4 MPa). Proportioning concrete mixtures is more critical for high-strength concrete than for normal-strength concrete. The procedure is similar to the proportioning process for normal strength concrete[6.x] except that adjustments have to be made for admixtures that replace part of the cement content in the mixture and the need to often use smaller size aggregates in very high strength concretes.

As discussed in Chapters 2 and 3 and in Section 6.5, there are several types of strength-modifying admixtures: high-range water reducers (superplasticizers), fly ash, polymers, silica fume, and blast-furnace slag. However, in mix proportioning for very high strength concrete, isolating the water/cementitious-materials ratio (W/C + P) from the paste/aggregate ratio because of the very low water content can be more effective for arriving at the optimum mixture with fewer trial mixtures and field trial batches. But few other methods are available today. The very low W/C + P ratio required for strengths in the range of 20,000 psi (138 MPa) or higher require major modifications to the standard approach used for mixture proportioning that seems to work well for strengths up to 12,000 psi (83 MPa). The optimum mixture, chosen with minimum trials, has to produce a satisfactory concrete product both in its plastic and its hardened states.

An approach presented in ACI Committee 211, 1993 (*Guide for Selecting Proportions for High Strength Concrete with Portland Cement and Fly Ash*, ACI Report 211.4R-93. American Concrete Institute, Farmington Hills, MI.) is based on mortar volume/stone volume ratio, proportioning the solids in the mortar on the basis of the ratio.

$$\frac{\text{solid sand volume} \pm \text{cementitious solid volume}}{\text{mortar volume}}$$

The ACI standard is well established for fly ash concretes (FAC). Ample mixture-proportioning results are available for polymers. The same is true for silica-fume concretes (SFC) and slag concrete (SC or GGBFSC). They are, however, not established in the form of a standard such as the committee reports of the *ACI Manual of Concrete Practice* (1997). Additionally, several sections in this book list mixture proportions for the types of strength-generating admixtures that have been discussed in this chapter. The computational example in this chapter on fly ash concrete design for strengths up to 12,000 psi (83 MPa) should serve as a systematic step-by-step guide for proportioning mixtures using polymers, silica fume, and granulated blast furnace slag within the example's compressive strength range.

6.4.2 Strength Requirements

The age at test is a governing criteria for selecting mixture proportions. The standard 28-d strength for normal-strength concrete penalizes high-strength concrete since the latter continues gaining

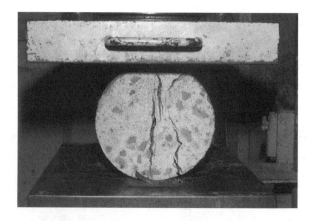

PHOTO 6.5   Tensile splitting strength test to failure.

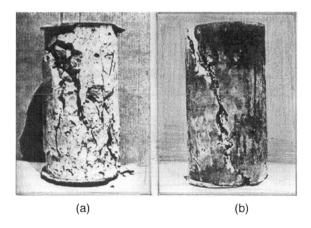

PHOTO 6.6 Cylinder compressive-strength test result. (a) Normal-strength concrete; (b) higher-strength concrete (Tests by E.G. Nawy et al.).

strength after that age. One also has to consider that a structure is subjected to its service load at 60 to 90 days age at the earliest. Consequently, in this case mixture proportioning has to be based on the latter age levels and also on either *field experience* or *laboratory batch trials*. The average compressive field strength results should exceed the specified design compressive strength by a sufficiently high margin so as to reduce the probability of lower test results.

6.4.2.1 Mixture Proportions on the Basis of Field Experiences

If the concrete producer chooses the mixture on the basis of field experience, the required average strength f'_{cr} should be the larger of

$$f'_{cr} = f'_c + 1.34\,s$$

or

$$f'_{cr} = 0.90\,f'_c + 2.33\,s$$

where

f'_c = specified design compressive strength, and
s = sample standard deviation in pounds per square inch from Eq. (6.3).

6.4.2.2 Mixture Proportions on the Basis of Laboratory Trial Batches

In this case, the laboratory trial batches should give

$$f'_{cr} = \frac{(f'_c + 1400)}{0.90} \quad \text{psi} \tag{6.4a}$$

or, in SI units

$$f'_{cr} = \frac{(f'_c + 27.6)}{0.90} \quad \text{MPa} \tag{6.4b}$$

It is important to note that high-strength high-performance concrete requires *special* attention in the selection and control of ingredients in the mixture in order to obtain optimum proportioning and maximum strength. To achieve this aim, care in the choice of the particular cement, admixture brand, dosage rate, mixing procedure, and quality and size of aggregate becomes paramount. Since all the cement does not hydrate, it is advisable that the cement content be kept at a minimum for optimum mixture proportioning.

6.4.3 Selection of Ingredients

6.4.3.1 Cement and Other Cementitious Ingredients

A proper selection of types and source of cement is extremely important. ASTM cement requirements are only minimum requirements, and certain brands are better than others owing to variations in the physical and chemical makeup of various cements. High-strength concrete requires high cementitious materials content, namely, a low water/cementitious-materials ratio $[(W/(C + P)]$, and the fineness of the cementitious materials has a major effect on the workability of the fresh mixture and the strength of the hardened concrete. They contribute to the reduction in water demand and lower the temperature of hydration. Hence a determination has to be made whether to choose Class F fly ash or Class G, silica fume, or granulated slag. The Chapter 2 discussion of these cementitious materials gives guidelines for choosing ingredients.

PHOTO 6.7 Failure patterns of higher-strength concrete compressive test (E.G. Nawy et al.).

6.4.3.2 Coarse Aggregate

Aggregates greatly influence the strength of the hardened concrete, as they comprise the largest segment of all the constituents. Consequently, only hard aggregate should be used for normal weight high-strength concrete so that the aggregate *at least* has the strength of the cement gel. As higher strength is sought, aggregate size should be decreased. It is advisable to limit aggregate size to a 3/4 in (19 mm) maximum size for strengths up to 9000 psi (62 MPa). For higher strengths, a 1/2 in or preferably 3/8 in aggregate should be used (12.7–9.5 mm). In order to achieve very high strengths in the range of 15,000–20,000 psi (103–138 MPa), higher-strength trap rock from selected quarries should be used. Beyond 20,00–30,000 psi strength, aggregate size should not exceed 3/8 in for structural components.

6.4.3.3 Fine Aggregate

A fineness modulus (FM) in the range of 2.5–3.2 is recommended for high-strength concrete to facilitate workability. Lower values result in decreased workability and a higher water demand. The mixing-water demand is dependent on the void ratio in the sand. The basic void ratio is 0.35 and should be adjusted for other void ratios such that the void content, V, in percent can be evaluated from

$$V = \left(1 - \frac{\text{oven-dry rodded unit weight (lb/ft}^3)}{\text{bulk dry specific gravity} \times 62.4}\right) 100 \qquad (6.5a)$$

or in SI units

$$V = \left(1 - \frac{\text{oven-dry rodded unit weight (kg/m}^3)}{\text{bulk dry specific gravity} \times 10^3}\right) 100 \qquad (6.5b)$$

The mixing water has to be adjusted accordingly to account for the change in the basic void ratio such that the mixing-water adjustment (A) would be as follows:

$$A = 8(V - 35) \tag{6.6a}$$

where A is pounds per cubic/yard, or, in SI units

$$A = 4.7(V - 35) \tag{6.6b}$$

where A is kilograms per cubic meter.

6.4.3.4 Workability-Enhancing Chemical Admixtures

High-strength mixtures have a rich cementitious content that requires a high water content, keeping in mind that excessive water reduces the compressive strength of the concrete and affects its long-term performance. Thus water-reducing admixtures become mandatory. High-range water-reducing admixtures (HRWR), discussed in Chapters 2 and 3, are used. These are sometimes called superplasticizers. The dosage rate is usually based on fluid ounce per 100 pounds (45 kg) of total cementitious materials if the HRWRs are in liquid form. If the water-reducing agent is in powdered form, the dosage rate would be on a weight ratio basis. The optimum admixture percentage should be determined on a trial and adjustment basis, as admixtures can reduce the water demand by almost 30 to 35% with a corresponding increase in compressive strength. A slump of 1 to 2 in (25 to 50 mm) is considered adequate. If, however, no HRWR admixtures are used, the slump should be increased to 2 to 4 in (50 to 100 mm). In addition, air-entraining admixtures are used if the concrete is exposed to freezing and thawing cycles in severe environmental conditions. For structural components in building systems, air entraining is unnecessary as these are usually not subjected to the type of frost action that exposed bridge decks or marine platforms endure.

6.4.4 Recommended Proportions

Tables 6.14 to 6.22, adapted from Nawy (1996), ACI 212 (1983), Neville (1996), ACI 221 (1994), Russell (1993), and ACI 211 (1993), recommend the necessary ingredients for proportioning mixes for high-strength concrete.

Where the void content, V, in percent is

$$V = \left(1 - \frac{\text{oven-dry rodded unit weight}}{\text{bulk specific gravity (dry)} \times 62.4}\right) \times 100$$

TABLE 6.14 Required Average Compressive Strength When Data Are Not Available to Establish a Standard Deviation

Specified Strength, f'_{cr}, psi (MPa)	Required Strength f'_{cr}, psi (MPa)
>5000 (34.5)	$f'_c + 1400$ ($f'_c + 9.7$)

TABLE 6.15 Maximum Size Coarse Aggregate

Required concrete Strength, f'_c, psi (MPa)	Maximum Aggregate Size, in (mm)
<9000 (62)	3/4–1 (19–25)
≥9000 (62)	3/8–1/2 (9.5–12.7)

TABLE 6.16 Coarse Aggregate to Concrete Fractional Volume Ratio (Sand Fineness Modulus 2.5–3.2)

Nominal Maximum Size, in (mm)	0.03352	0.004701	3/4 (19)	1 (25)
Fractional Volume of oven-dry rodded coarse aggregate	0.65	0.68	0.72	0.75

TABLE 6.17 Recommended Slump

With HRWR,* in (mm)	NO HRWR, in (mm)
1–2 (25–50) before adding HRWR	2–4 (50–100)

Note: HRWR, high-range water reducer.

*Adjust slump to that desired in the field by adding HRWR.

TABLE 6.18   Mixing-Water Requirement and Air Content of Fresh Concrete Using Sand With 35% Void Ratio—First Trial Water Content

| | Mixing Water, lb/yd^3 (kg/m^3) | | | |
| | Maximum Size Coarse Aggregate, in (mm) | | | |
Slump, in (mm)	3/8 (9.5)	1/2 (12.7)	3/4 (19)	1 (25)
1–2 (25–50)	310 (183)	295 (174)	285 (168)	280 (165)
2–3 (50–75)	320 (189)	310 (183)	295 (174)	290 (171)
3–4 (75–100)	330 (195)	320 (189)	305 (180)	300 (177)
Entrapped Air,* %	3 (2.5)†	2.5 (2.0)	2 (1.5)	1.5 (1.0)

Sources: Nawy, E.G. 1996b. *Fundamentals of High Strength High Performance Concrete.* 350 pp. Longman U.K., London. Addison–Wesley, New York. ACI Committee 211. 1993. *Guide for Selecting Proportions for High Strength Concrete with Portland Cement and Fly Ash.* ACI Report 211.4R-93, American Concrete Institute, Farmington Hills.

Note: lb/yd^3 = 0.59 kg/m^3.

*Adjust mixing-water values for sand-void ratios other than 35%.

†Mixtures using HRWR.

TABLE 6.19   W/(C + P) Ratio for Concrete Without High-Range Water Reducer

| | | (W/C + P) Ratio | | | |
| | | Maximum-Size Coarse Aggregate, in (mm) | | | |
Field Strength,* f'_{cr},† psi (Pa)		3/8 (9.5)	1/2 (12.7)	3/4 (19)	1 (25)
7000	28 day	0.42	0.41	0.40	0.39
(48)	56 day	0.46	0.45	0.44	0.43
8000	28 day	0.35	0.34	0.33	0.33
(55)	56 day	0.38	0.37	0.36	0.35
9000	28 day	0.30	0.29	0.29	0.28
(62)	56 day	0.33	0.32	0.31	0.30
10000	28 day	0.26	0.26	0.25	0.25
(69)	56 day	0.29	0.28	0.27	0.26

*Here, $f'_{cr} = f'_c + 1400$ ($f'_{cr} = f'_c + 9.7$).

†These are average field values; enter into the table 0.9 (required f'_{cr}).

and the mixing-water adjustment, in pounds per cubic yard, is

$$(V - 35) \times 8$$

and in kilograms per cubic meter is

$$(V - 35) \times 4.7$$

6.4.5 Step-by-Step Procedure for Selecting Proportions

The following are the steps necessary in the proportioning selection process.

1. Select the slump and required strength f'_{cr}, (Table 6.14) and,

   (a) based on field experience, the f'_{cr}, that is the larger of $f'_c + 1.34\,s$ or $0.9\,f'_{cr} + 2.33s$; or,

   (b) based on laboratory batching, $(f'_c + 1400)/0.9$.

TABLE 6.20 W/(C + P) Ratio for Concrete With High-Range Water Reducer

Field Strength,* f'_{cr},[†] psi (MPa)		W/(C + P) Ratio			
		Maximum-Size Coarse Aggregate, in (mm)			
		3/8 (9.5)	1/2 (12.7)	3/4 (19)	1 (25)
7000 (48)	28 day	0.50	0.48	0.45	0.43
	56 day	0.55	0.52	0.48	0.46
8000 (55)	28 day	0.44	0.42	0.40	0.38
	56 day	0.48	0.45	0.42	0.40
9000 (62)	28 day	0.38	0.36	0.335	0.34
	56 day	0.42	0.39	0.37	0.36
10,000 (69)	28 day	0.33	0.32	0.31	0.30
	56 day	0.37	0.35	0.33	0.32
11,000 (76)	28 day	0.30	0.29	0.27	0.27
	56 day	0.37	0.31	0.29	0.29
12,000 (83)	28 day	0.27	0.26	0.25	0.25
	56 day	0.30	0.28	0.27	0.26

Note: A comparison of the values contained in Tables 6.19 and 6.20 permits, in particular, the following conclusions:

1. For a given water/cementitious material ratio, the field strength of concrete is greater with the use of HRWR than without it, and this greater strength is reached within a shorter period of time.
2. With the use of HRWR, a given concrete field strength can be achieved in a given period of time using less cementitious material than would be required when not using HRWR.

*These are average field values; enter into the table 0.9 (required f'_{cr}).

[†] $f'_{cr} = f'_c + 1400$ ($f'_{cr} = f'_c + 9.7$).

2. Select the maximum size of aggregate (Table 6.15).
3. Select the optimum coarse-aggregate content (Table 6.16).
4. Estimate the mixing water and air content (Table 6.18).
5. Select the W/(C + P), where C is the cement content and P is the pozzolanic content by weight (Tables 6.19 and 6.20).
6. Compute the necessary content of the cementitious material P. This can be obtained by dividing the amount of mixing water per cubic yard or cubic meter of concrete (Step 4) by the W/(C + P) ratio.
7. Proportion a basic mixture without the cementitious pozzolanic material P.
8. Proportion a companion mixture using the cementitious pozzolanic material P such as fly ash.
9. Produce a trial mix for each of the trial mix proportions designed in Steps 1–8.
10. Adjust the mixture proportions to achieve the required slump by changing the contents and adjusting the HRWR agent rate for several trial mixes.
11. Select the optimum mixture.

6.4.6 Example 6.3: Mixture Proportions Design for High Strength Concrete

Design a high-strength concrete mixture for columns in a multistory structure for 28-d compressive strength of 10,000 psi (69 MPa). A slump of 9 in (229 mm) is required for the workability needed for congested reinforcement in the columns. Do not use an aggregate size exceeding 1/2 in (12.7 mm). Use a high range water reducer (HRWR) to obtain the

TABLE 6.21 Fly Ash Values to Replace Part of the Cement

Type	Replacement % by Weight
ASTM Class F	15–25
ASTM Class C	20–35

9 in slump and a set-retarding admixture. Assume that the ready-mix producer has no prior history with high-strength concrete.

Assume the following sand properties:

Fineness modulus (FM) = 2.90
Bulk specific gravity (over dry) $(BSG)_{dry}$ = 2.59
Absorption based on dry weight (Abs) = 1.1%
Dry rodded unit weight (DRUW) = 103 lb/ft^3 (1620 kg/m^3)
Moisture content in sand = 6.4%

Solution.

1. **Select the slump and required concrete strength.** Because a HRWR agent is used, choose strength on the basis of a 1–2 in slump prior to the addition of HRWR. Also, since the ready-mix producer has no prior history with high-strength concrete, laboratory trial mixtures have to be designed for selection of the optimum proportions. From Eq. (6.4a)

$$f'_{cr} = (f'_c + 1400)/0.90$$
$$= (10,000 + 1400)/0.90$$
$$= 12,670 \text{ psi (87 MPa)}$$

2. **Select the maximum aggregate size.** A crushed limestone graded 1/2 in (12.7 mm) maximum size is selected with BSG_{dry} = 2.76, Abs = 0.70, DRUW = 101 lb/ft^3, and stone moisture content = 0.5%.

3. **Select the optimum coarse aggregate content.** From Table 6.16, the fractional ratio is equal to 0.68. The dry weight of coarse aggregate per cubic yard of concrete is

$$W_{dry} = (\%DRUW) \times (DRUW \times 27)$$
$$= 0.68 \times 101 \times 27 = 1854 \text{ lb (841 kg)}$$

4. **Estimate the mixing water and air content.** From Table 6.18, the first estimate of the required mixing water is 295 lb/yd^3 (174 kg/m^3) of concrete, and the entrapped air content when HRWR is used is 2.0%.

   From Eq. (6.5a), the void content of the sand to be used is

$$V = \left[1 - \frac{103}{2.59 \times 62.4}\right] \times 100 = 36\%$$

   From Eq. (6.6a), the mixing water adjustment, A, is

$$A = 8(V - 35) = 8(36 - 35) = +8 \text{ lb/yd}^3 (4.7 \text{ kg/m}^3)$$

   of concrete.

   Hence the total mixing water, W, is 295 + 8 = 303 lb (138 kg).

5. **Select the water/cementitious materials ratio [W/(C + P)].** The values in Tables 6.19 and 6.20 are average field-strength values. Hence the strength f'_{cr}, for which the W/(C + P) ratio is to be found is

$$f'_{cr} = 0.90 \times 12,670 = 11,400 \text{ psi (77 MPa)}$$

TABLE 6.22 Modification Factor for Standard Deviation When Fewer Than 30 Tests Are Available

Number of Tests*	Modification Factor for Standard Deviation[†]
Less than 15	Use Table 6.14
15	1.16
20	1.08
25	1.03
30 or more	1.00

*Interpolate for intermediate number of tests.
[†]Modified standard deviation to be used to determine required average strength f'_{cr} in Eqs. (6.1) and (6.2.).

From Table 6.20, for a 1/2-in aggregate, the desirable $W/(C + P)$ ratio is 0.272, by interpolation.

6. **Compute the content of cementitious material.** From Step 6, the mixing water, W, is 303 lb; hence,

$$C + P = 303/0.272 = 1114 \text{ lb (505 kg)}$$

7. **Proportion the basic mixture with cement only.** Volumes of all materials per cubic yard, except sand, are as follows:

Cement	$1114 \div (3.15 \times 62.4)$	$= 5.67$ (ft^3)
Stone	$1854 \div (2.76 \times 62.4)$	$= 10.77$
Water	$303 \div 62.4$	$= 4.86$
Air	0.02×27	$= 0.54$
Total*		21.77 ft^3 (0.62 m^3)

*$1 \text{ m}^3 = 35.31 \text{ ft}^3$.

Hence the required volume of sand per cubic yard of concrete is

$$27 - 21.77 = 5.23 \text{ ft}^3$$

Converting the sand volume to weight

$$\text{sand} = 5.23 \times 62.4 \times 2.59 = 845 \text{ lb} (384 \text{ kg})$$

The mixture proportions by weight for the no-fly-ash concrete would be

	lb/yd^3	(kg/m^3)
Cement	1114	(661)
Sand, dry	845	(501)
Stone, dry	1854	(1100)
Water (including 3 oz/cwt* retarding admixture)	303	(180)
Total	4116 lb/yd^3	(2442 kg/yd^3)

*Here cwt = 100 weight of cement.

8. **Proportion companion mixtures using cement and fly ash.** In this case use ASTM Class C fly ash (FA), which has a bulk specific gravity (sg) of 2.64. From Table 6.21, the fly ash replacement is 20–35%. Use four trial mixtures: 20, 25, 30, and 35% levels. For trial mixture 1, the silica fume is $0.20(1114) = 223$ lb, hence the cement is $1114 - 223 = 891$ lb.

In a similar manner, the weights of the cementitious materials would be as shown in the following table.

Mixture	Cement, lb (kg)	Fly Ash, lb (kg)
1	891 (404)	223 (101)
2	835 (379)	279 (126)
3	780 (354)	334 (151)
4	724 (328)	390 (177)

In mixture 1, the volumes of components, except sand, per cubic yard of concrete are

Cement	$891 \div (3.15 \times 62.4) = 4.53 \text{ ft}^3$
FA	$223 \div (2.64 \times 62.4) = 1.35$
Stone*	$= 10.77$
Water* (including 2.5 oz/cwt retarder)	$= 4.86$
Air*	$= 0.54$
Total	22.05 ft^3

* From earlier table.

TABLE 6.23 Mixture Proportion in Example 6.3 Without Moisture Trial Batch Adjustment

Ingredient	Basic Mix Cement Only, lb	Cement + CFA† Mixes, lb			
		#1	#2	#3	#4
Cement	1114	891	835	780	724
Fly ash	0	223	279	334	390
Sand (dry)	845	800	790	781	773
Stone (dry)	1854	1854	1854	1854	1854
Water (+ retarder)	303	303	303	303	303
Total,* lb/yd³ concrete	4116	4071	4061	4052	4044
Total, kg/m³ concrete	2428	2402	2396	2391	2386

*1 lb/yd³ = 1 kg/m³.
†CFA = Fly Ash Class C.

The sand volume is

$$27 - 22.05 = 4.95\,\text{ft}^3 = 4.95 \times 62.4 \times 2.59 = 800\,\text{lb}$$

The mixture proportions by weight for the fly-ash concrete mixture 1 would be

	lb/yd³*	(kg/m³)
Cement	891	(526)
Fly ash	223	(132)
Sand, dry	800	(472)
Stone, dry	1854	(1094)
Water, incl. retarder	303	(179)
Total	4071 lb/yd³	(2402 kg/m³)

*1 lb/yd³ = 0.59 kg/m³.

In a similar manner, the mixture proportions for 25, 30, and 35% fly-ash content are computed to give the companion mixtures in Table 6.23.

9. ***Adjust the trial mixtures for absorbed water content in aggregate.***   From before, the moisture content in sand is 6.4%, and the moisture content in stone is 0.5%. From Table 6.23, corrections in the basic mixture for the wetness of the aggregates are as follows:

wet sand = 845(1 + 0.064) = 899 lb
wet stone = 1854(1 + 0.005) = 1863 lb

From input data, the sand absorption based on dry weight is 1.1% and the stone absorption is 0.7%. Hence the water correction is

$$303 - 845(0.064 - 0.011) - 1854(0.005 - 0.007)$$
$$= 303 - 45 + 4 = 262\,\text{lb}\ (119\,\text{kg})$$

Accordingly, the batch weight of water has to be corrected to account for the excess moisture contributed by

$$\text{aggregates} = \text{total moisture} - \text{aggregate absorbed moisture.}$$

Hence Table 6.23 is modified to give Table 6.24.

TABLE 6.24 Moisture-Adjusted Mixture Proportion in Example 6.3

Ingredient	Basic Mix Cement Only, lb	Cement + CFA[†] Mixes, lb			
		#1	#2	#3	#4
Cement	1114	891	835	780	724
Fly ash	0	223	279	334	390
Sand (wet)	899	851	841	831	823
Stone (wet)	1863	1863	1863	1863	1863
Water (+ retarder)	262	262	262	262	262
Total,* lb/yd³ concrete	4138	4090	4080	4062	
Total, kg/m³ concrete	2441	2413	2407	2401	2397

*1 lb/yd³ = 0.89 kg/m³.
[†]CFA = Fly Ash Class C.

10. **Select the size of the laboratory trial mixture.** The usual size of the trial mixture is 3.0 ft³ (0.085 m³). The reduced batch weights to yield 3.0 ft³ of concrete would be 1/9 of the values tabulated in Table 6.24 to yield the values in the following table.

Ingredient	Basic Mix Cement Only, lb	Cement + CFA Mixes, lb			
		#1	#2	#3	#4
Cement	123.78	99.00	92.78	86.67	80.44
Fly ash	0	24.78	31.00	37.11	43.33
Sand (dry)	99.89	94.56	93.44	92.33	91.44
Stone (dry)	207.00	207.00	207.00	207.00	207.00
Water (+ retarder)	29.11	29.11	29.11	29.11	29.11
Total, lb/3 yd³ concrete	460	455	453	452	451
Total, kg/(1/10) m³ concrete	245.3	242.7	241.6	241.1	240.5

11. **Adjust the trial mixture on the basis of slump observation.**

(a) **Basic mixture:** Assume that the water calculated to produce 1–2 in slump, namely 29.11 lb, was found to be not adequate and has to be increased to 30 lbs per 3 cu. ft., including the 2.5 oz/cwt retarding admixture.

The actual batch weights therefore have to be adjusted so that the actual batch weight for the basic mixture (no fly ash) becomes:

Cement	123.78, lb
Sand	99.89
Stone	207.00
Water	30.00

These values have to be adjusted for moisture correction to the dry weight.

The basic total added water is $30 \times 9 = 270$ lb/yd³. From before, the absorbed water in the aggregates is $45 - 4 = 41$ lb. The actual total water content is $270 + 41 = 311$ lb/yd³ $= 34.56$ lb/ft³.

Cement		= 123.78 lb
Sand	99.89 ÷ 1.064	= 93.88 lb
Stone	207 ÷ 1.005	= 205.97 lb
Batch water	30.00 + 45/9 − 4/9	= 34.56 lb

(b) *Yield of trial batch:*   The actual yield of the trial mixture becomes:

Cement	$123.78 \div (3.15 \times 62.4) = 0.63$
Sand	$93.88 \div (2.59 \times 62.4) = 0.58$
Stone	$205.97 \div (2.76 \times 62.4) = 1.20$
Water	$34.56 \div 62.4 \quad\quad = 0.55$
Air	$0.02 \times 3 \text{ ft}^3 \quad\quad = 0.06$
Total yield volume of trial batch	3.02 ft^3

The yield in pounds per cubic yard of concrete is obtained by multiplying all the previous values by nine and converting the volumes to weights, giving

Cement	$1114 \times \dfrac{3.0}{3.02} = 1107 \text{ lb}$
Sand, dry	$845 \times \dfrac{3.0}{3.02} = 839 \text{ lb}$
Stone	$1854 \times \dfrac{3.0}{3.02} = 1841 \text{ lb}$
Water (including retarder)	$311 \times \dfrac{3.0}{3.02} = 309 \text{ lb}$

The new mixture proportions result in a water/cementitious materials ratio of

$$W/(C + P) = 309/1107 = 0.28$$

versus the desirable ratio of 0.272 previously obtained from Table 6.20. In order to maintain the 0.272 ratio, the weight of cement should be increased to $309/0.272 = 1136 \text{ lb/yd}^3$ of concrete. The increase in volume due to the adjustment of the weight of cement is

$$(1136 - 1107) \div (3.15 \times 62.4) = 0.148 \text{ ft}^3$$

This increase in volume should be adjusted for by the removal of an equal volume of sand. Hence the weight of sand to be removed is $0.148 \times 2.59 \times 62.4 = 23.98 \text{ lb/yd}^3$, say, 18 lb/yd^3. The resulting adjusted mixture proportions become:

Cement	1131 lb
Sand, dry	$839 - 24 = 815$
Stone, dry	1841
Water + 2.5 oz/cwt retarder	309

(c) *Increasing slump to 9 in (229 mm):*   The required slump in this example is 9 in (229 mm). To achieve this value without the addition of water, which will reduce the strength, a high-range water reducer, namely, a plasticizer is used. The dosage recommended by the manufacturer of the HRWR ranged between 8 and 16 oz per 100 lb of cementitious material. Laboratory tests in a laboratory with an ambient temperature of 74°F indicated the following:

an 8 oz dosage produced 5 in slump;

an 11 oz dosage produced 10 in slump; and

a 16 oz dosage produced segregation of the fresh concrete.

In all these cases, a constant dosage rate of a retarding admixture of 2.5 oz/cwt was also added to the mixture along with the mixing water.

The HRWR was added to the mixture about 15 min after initial mixing. It was determined that

1. the mixture with 10 in (255 mm) slump had adequate workability; hence no correction was needed to the coarse-aggregate content;
2. the air content of the HRWR concrete mixture was found to be 1.9%; hence no correction was needed;

TABLE 6.25 Laboratory Final Trial Mixtures

Ingredient, lb	Basic Mix Cement Only, lb	Cement + CFA† Mixes, lb			
		#1 20% CFA	#2 25% CFA	#3 30% CFA	#4 35% CFA
Cement	1131	905	848	792	735
Fly ash	—	226	283	339	396
Sand (dry)	803	754	748	739	731
Stone (dry)	1841	1850	1850	1850	1850
Water (+ retarder)	309	302	298	296	295
Slump, in (mm)	1.00 (25)	1.20 (31)	1.15 (29)	1.50 (38)	1.90 (48)
Retarder, oz/cwt	3.5	2.5	2.0	2.5	2.0
HRWR, oz/cwt	10.00	10.50	11.00	10.25	9.00
Slump, in (mm)	10.00 (250)	10.75 (270)	8.75 (220)	10.50 (270)	9.25 (235)
28 day, psi (MPa)	12,600 (87)	12,400 (85)	12,550 (87)	12,750 (88)	12,250 (84)

Source: Nawy, E.G. 1996b. *Fundamentals of High Strength High Performance Concrete*, 350 pp. Longman U.K., London. Addison–Wesley, New York.

†CFA = Fly Ash Class C.

    3. the 28-d compressive strength of the No. 3 mixture was found to be 12,750 psi, satisfying the required f'_{cr} value of 12,670 psi. (*Note:* It is important to recognize that if additional water at this stage was needed to produce the required slump and workability, then an additional cycle of corrections to actual batches of aggregate would have to be executed in the same manner as in the previous steps.)

12. ***Summary of the trial mixtures laboratory performance.***   Table 6.25 is a summary table of the performance of the five mixtures, namely, the basic no-fly-ash concrete and four concretes with fly-ash contents of 20, 25, 30, and 35% of the total cementitious material. Slump values for no-HRWR mixtures and those with HRWR were measured in the laboratory slump tests.

      In addition, field trials have to be done to verify the choice of laboratory trial mix. In this case, mix 3 from Table 6.25, which gave the highest 28-d compressive strength of 12,750 psi (88 MPa), is closest to the required f'_{cr} of 12,670 psi that gives the average compressive strength f'_c of 10,000 required in this example.

Part B: Applications and Constructability

6.5 Applications and Constructability With an Emphasis on High-Strength High-Performance Concrete

6.5.1 General

High-strength concrete (HSC) has a relative meaning according to the availability and utilization of materials and resources in each geographical area. In Chicago, Seattle, and Texas, concrete strengths of more than 84 MPa (12,000 psi) have been produced, while in other parts of the country

42 MPa (6,000 psi) is considered high. The reasons for such divergence are necessity, availability of materials, and "know-how." It should be understood that these reasons are relative to the type of construction, initiative of the design engineer, commitment of the concrete producer, and quality of local materials.

The permissible margin of error for high-strength concrete is smaller than for normal-strength concrete. Small variations in mixture proportions and deviations from good testing practices can have a greater effect on the strength or perceived strength of HSC than on normal-strength concrete. Preconstruction meetings are advisable to define responsibilities and avoid problems during construction.

6.5.2 Constructability Preparation Process

6.5.2.1 Preconstruction Meetings

Prior to construction, all of the project participants should meet to clarify contract requirements, discuss planned placement conditions, and review the planned inspection and testing programs of the various parties. The effects of time, temperature, placing, curing, and acceptance criteria, and how those criteria will be established should be reviewed. The capabilities of the contractor's work force, inspection staff, and testing and batching facilities should also be reviewed.

The meeting should establish lines of communication and identify responsibilities. It is especially important to review the procedures the inspector will follow when noncompliance with contract requirements is found or suspected. This advance understanding will minimize future disputes and will give all members of the construction team an opportunity to participate in the quality process. Timely and accurate reporting are paramount. Trial production batches should have resulted in a workable mixture, but it may be necessary to make adjustments owing to site conditions such as changing weather. The ready-mix concrete producer is essential to this discussion since the producer is most familiar with and responsible for the product. Only designated individuals, such as the concrete supplier's quality-control personnel, should have the authority to add admixtures or water at the site. No water in excess of the approved mixture proportions should be added to high-strength concrete.

6.5.2.2 Material Selection and Proportioning

Unlike more conventional, lower-strength concrete, established procedures cannot always be used to determine the proportioning of HSC. As will be seen later, modifications to published procedures for mixture design of HSC must be made, although the basic principles remain. Maintaining the lowest possible water/cement ratio, for example, is not merely important but critical (Table 6.26). The basic consideration is still the selection and combination of materials that will produce a quality concrete with the desired workability, ease of placement, strength, and durability.

Because of the innumerable types and grades of aggregate, varying chemical contents and physical characteristic of cements, pozzolans, and chemical admixtures, and the interaction of these materials, arriving at the optimum combination often becomes a matter of trial and error. Caution must be exercised when examining the data of others, inasmuch as conclusions reached may reflect a particular set of materials unlike those in the researcher's area. For this reason, it is suggested that the following data be evaluated in a conceptual rather than quantitative manner. The slump required for pumping, placement, and consolidation normally varies with individual job conditions. If a high-range water reducer (superplasticizer) is not part of the original design, it should be considered in order to facilitate placement.

TABLE 6.26 Suggested Water-Cementitious Material* Ratio

Specified Strength psi (MPa)	Maximum W/C
6000 (41.4)	0.40
8000 (55.2)	0.36
9000 (62.1)	0.34
10000 (69.0)	0.32
12000 (82.8)	0.30
14000 (96.6)	0.28

Note: W/C, water/cementitious material ratio.

*For purposes of calculating the W/C of the mixture, the weight of all pozzolan should be added to the weight of the cement.

6.5.3 Mixture Components

6.5.3.1 Cement

In a study conducted by Blick et al. (1973), five cements of various types were ground and evaluated (Table 6.27). It was found that mortar cubes of brand C, type III, performed best at all ages (Figure 6.4). When it was used in concrete, the results were quite different (Figure 6.5). This phenomenon was further substantiated by tests conducted in the laboratory. In this study, mortar cubes were made with five different types of cement. When 20% of the cement, by mass, was replaced with fly ash, the results were quite different, as seen in Figures 6.6a, 6.6b, and 6.6c. The mortar containing the highest-strength cement showed a loss of strength, while the mortar with the lowest strength cement showed the highest strength gain. Mortar cubes alone might not tell the whole story at the design level, but they are an important means of control after a selection has been made. A stress performance at ages up to 90 d should be evaluated. Limits of the

TABLE 6.27 Chemical and Physical Analysis of Cements for High-Strength Concrete Program

	Brand A, Type I	Brand B, Type II	Brand C, Type I	Brand C, Type II	Brand C, Type III
Composition					
SiO_2, %	21.80	20.60	20.50	21.90	20.40
Al_2O_3, %	5.30	6.10	5.70	4.10	5.50
Fe_2O_3, %	2.00	3.20	2.10	2.90	2.00
CaO, %	65.20	63.20	63.80	65.10	63.60
MgO, %	2.50	2.80	3.60	2.70	3.50
SO_3, %	2.10	2.60	2.50	2.00	3.00
Ignition loss, %	1.10	1.10	2.00	1.00	2.10
Na_2O, %	0.19	0.31	0.11	0.14	0.12
K_2O, %	0.41	0.70	0.17	0.24	0.20
Na_2O equivalent, %	55.00	48.00	55.00	61.00*	56.00
C_3S, %	21.00	23.00	17.00	17.00	17.00
C_2S, %	10.60	10.70	11.50	5.90*	11.20
C_3A, %	6.10	9.70	6.40	8.90	6.10
C_4AF, %	0.33	0.59	0.09	0.15	0.12
Blaine (air permeability)	3670	3780	3550	4220	5400
+325	4.10	8.40	5.60	2.50	0.90
N.C.	25.80	23.60	23.00	24.80	20.60
Setting time vicat	2:35	2:15	1.35	1:55	1:45
False set					
H_2O, %	30.00	30.00	30.00	30.00	33.00
3 min	50.00	50.00	50.00	50.00	50.00
5 min	35.00	50.00	50.00	50.00	50.00
8 min	18.00	50.00	50.00	46.00	45.00
11 min	12.00	50.00	50.00	40.00	39.00
Remix 15 min	11.00	None	None	50.00	50.00
Air content, %	9.10	9.50	7.60	6.60	6.20
H_2O, %	69.00	68.00	68.00	68.00	71.00
Comp. strength					
H_2O, %	50.40	49.00	48.00	47.00	51.40
Flow, %	114.00	106.00	113.00	106.00	108.00
1-Day, psi	1460	1400	1200	1860	2090
3-Day, psi	2930	2870	2700	3280	4420
7-Day, psi	4460	3980	4370	4560	6290
28-Day, psi	6220	5730	6410	6770	7870
63-Day, psi	6870	6320	7310	7850	8320

Source: Blick, R.L. 1973. Some factors influencing high strength concrete. In *Modern Concrete*.

*This cement does not meet Type II specifications for moderate heat of hydration. Maximum percent of C_3S + C_3A = 58, and maximum percent of C_3A = 8% (according to ASTM C 150).

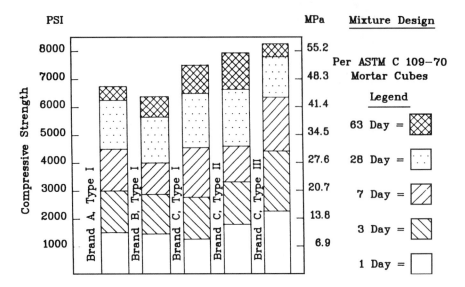

FIGURE 6.4   Effect of various cements on mortar cube compressive strength.

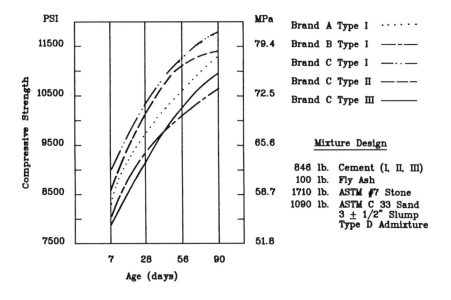

FIGURE 6.5   Effect of various cements on concrete compressive strength.

physical properties of the cement should be established and submitted to the cement producer for compliance.

The optimum cement quantity must be obtained through a series of laboratory trial mixtures. At least three cement quantities ranging from 136 to 409 kg/m³ (600 to 900 lbs/yd³) should be tested. The tests should be conducted at a constant slump if a superplasticizer is not used or at a constant water/cement ratio (W/C) if a superplasticizer is used.

The strength efficiency per pound of cement is influenced by all the variables that affect the strength of concrete. If the cement quantity of a mix is below optimum, higher strength may be obtained by using a larger size aggregate. If the cement quantity is above the optimum, higher strength may be achieved by using a smaller aggregate. After an optimum cement content for a given set of materials is selected, the maximum strength will not be increased by adding additional cement (Figure 6.7). Excessive cement content will also cause the concrete to become sticky and unworkable.

6.5.3.2 Fly Ash

The benefits derived from the use of fly ash vary with its class (ASTM C-618), chemical and physical properties, and quantity and compatibility with the cement. Cook (1982) and others have shown that higher strengths can be achieved using a Class C fly ash with a high calcium oxide content, as shown in Figure 6.8. Amounts of 10 to 15% of fly ash by mass of cement are common, but amounts of 25% and higher should be evaluated if Class C fly ash is being considered. The calcium hydroxide liberated during the hydration of the cement combines with the silica in the fly ash to form additional calcium silicate hydrate gel. This strength gain generally occurs after 14 d. Class F will have more of an effect on later strength, 28 to 90 d. The gains achieved with the use of fly ash cannot be attained through the use of additional cement. Additional benefit is derived mechanically from the fly ash inasmuch as it acts as a filler that optimizes the particle distribution of the cementitious materials (Kiell, 1985).

6.5.3.3 Micro Silica

Micro silica, or silica fume as it is sometimes called, has been used to create high-strength concretes. Its chemistry is similar to both cement and fly ash (Table 6.28). Its super fineness of over 100,000 cm^2/g, in combination with its pozzolanic reactivity, offers strengths in excess of 15,000 psi (103.4 MPa). Quantities of 5% to 15% of cement weight should be evaluated. A superplasticizer must be used in conjunction

(a)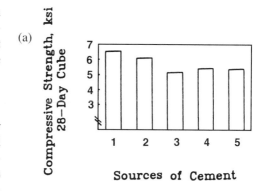

(b)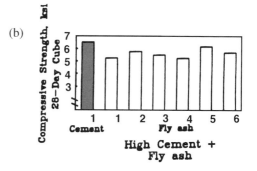

(c)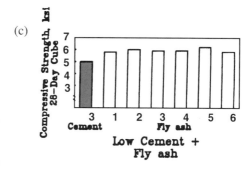

FIGURE 6.6 Effect of compressive strength on various cements and fly ash: (a) source of cement; (b) high cement content plus fly ahs; and, (c) low cement content plus fly ash.

with micro silica because of the negative affect on the slump caused by the fineness of the silica. Micro silica has been used with and without fly ash. The benefit of the fly ash may be lessened when it is used with micro silica, but when attempting to produce an ultra-high-strength concrete, every benefit must be exploited (Figure 6.9).

6.5.3.4 Chemical Admixtures

The use of a normal-set water reducer, retarding water reducer, or a combination of these becomes necessary to efficiently utilize all cementitious materials and to maintain the lowest practical water/cement ratio. Dosages higher than those recommended by the admixture manufacturer have been found to increase strength without detrimental affects (Figure 6.10). The use of an ASTM

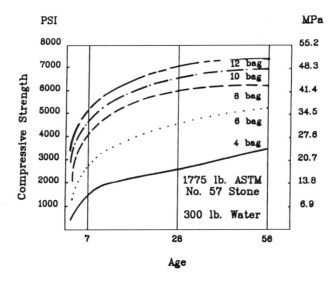

FIGURE 6.7   Cement efficiency versus compressive strength.

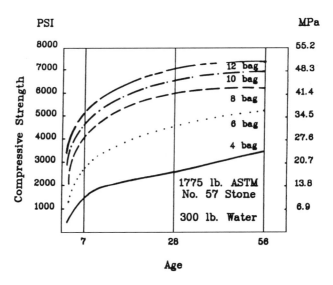

FIGURE 6.8   Effect of CaO on compressive strength (ASTM C-618, Class C fly ash.)

type D retarding water reducer offered strength benefits beyond 56-d strengths. Varying dosages of this material help offset the rapid setting time that might be expected from a mix containing a high amount of cement. During hot weather, this type of control is imperative. The heat of hydration of this type of concrete generates quickly and if left uncontrolled will cause high early strength, lower ultimate strength, and excessive cracking. Factors to be considered when evaluating an admixture are cement and fly ash compatibility, water reduction, setting times, workability, time of addition, and dosages.

Superplasticizers (high-range water reducers) offer the ability to achieve higher strengths than were previously attainable. Water reductions of up to 30% are possible while still maintaining a placeable consistency (Figure 6.11). This additional water reduction was instrumental in increasing our 9000 psi mix to 11,000 psi. Superplasticizers should be used in conjunction with an ASTM type D retarder and can successfully be added at the ready-mix plant and again at the job site if

TABLE 6.28   Chemical Analysis

	Normal Portland Cement, %	Micro Silica, %	Fly Ash Class F, %	Fly Ash Class C, %
Silica (SiO$_2$)	20.13	94–98	49.00	40.40
Calcium Oxide (CaO)	63.44	0.08–0.30	5.00	25.40
Magnesium Oxide (MgO)	2.86	0.30–0.90	1.50	4.70
Ferric Oxide (Fe$_2$O$_3$)	2.96	0.02–0.15	6.00	5.90
Aluminum Oxide (Al$_2$O$_3$)	5.10	0.10–0.40	26.00	17.00
Sulfur Trioxide (SO$_3$)	3.00	—	0.50–0.60	2.75
Potassium Oxide (K$_2$O)	1.12	0.20–0.70	0.80–0.90	0.27
Sodium Oxide (Na$_2$O)	0.30	0.10–0.40	0.25	1.60
Loss on ignition	0.80	0.80–1.50	3.50	0.43
Silicon Carbide (SiC)	—	0.20–0.10		
Carbon (C)	—	0.20–1.30		
C$_4$AF	9.00			
C$_3$S	57.40			
C$_2$S	14.40			
C$_3$A	8.50			
Blaine, cm^{2}g	3782	100,000+		

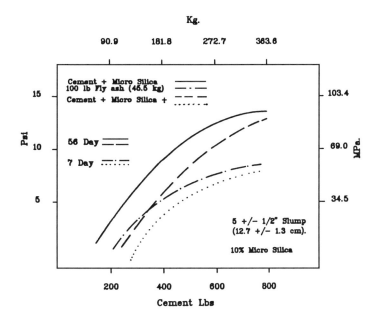

FIGURE 6.9   Benefit of fly ash when used with micro silica.

necessary. Air-entraining agents are not generally used because of the accompanying strength loss and because the type of application normally precludes their use (i.e., caissons, interior columns, and shear walls).

6.5.3.5 Aggregates

Careful consideration should be given to the shape, surface texture, and mineralogy of aggregates. Cubically shaped crushed stone with a rough surface texture appears to produce the highest strength. Smoother faced, uncrushed gravel may be used to produce strengths of up to about 10,000 psi, but it does not have the bond strength needed to produce higher strengths. Tests we conducted showed that gravels produce lower compressive strengths and moduli of elasticity when compared to the same size crushed stone with equal cement content (Figure 6.12).

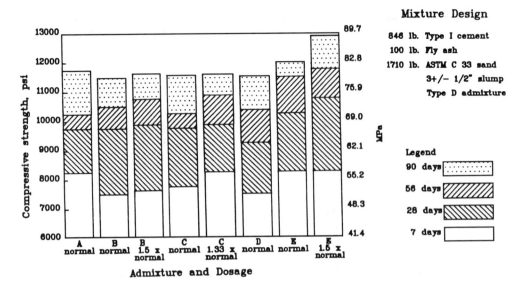

FIGURE 6.10   Effect of chemical admixtures on concrete compressive strength.

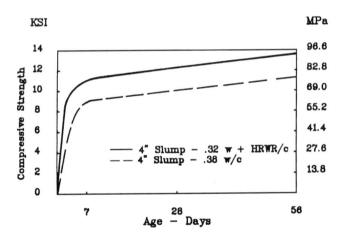

FIGURE 6.11   Compressive strength of high-strength concrete containing a high-range water reducer.

Each strength level will have an optimum size aggregate that will yield the greatest compressive strength. Because of the high volume of cementitious materials in high-strength concrete, the optimum aggregate surface area is necessary to provide maximum bonding. In addition to optimizing the gradation of the coarse aggregate, further benefit can be derived from altering the proportions of coarse to fine aggregate. It was found that the proportions recommended by ACI 211.1-82 do not apply, as seen from the comparative values in Table 6.29.

Well-graded fine aggregates that provide good finishing characteristics in regular-strength concrete are not only unnecessary but may require more water than can be tolerated (Table 6.30). Fine aggregates, with a fineness modulus of 2.5 and under will produce concrete with a sticky consistency, less workability, and lower compressive strength. A fineness modulus of 2.75 to 3.20 appears to produce the highest strengths. As with coarse aggregates, cubically shaped particles in fine aggregate have been found to increase bonding (Kiell, 1985; National Crushed Stone Association, 1975, 1976). Very angular manufactured sands should not be used since they increase water demand.

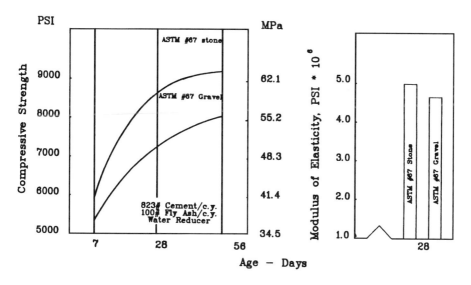

FIGURE 6.12 Compressive strength and modulus of elasticity versus age for various sizes and types of coarse aggregate.

6.5.3.6 Water

Various sources of water have been evaluated, and it appears that the limitations listed in ASTM C-94 are adequate. Water need not be potable. ASTM C-94 states, "If it contains quantities of substances which discolor it or make it smell or taste unusual or objectionable or cause suspicion, it shall not be used unless service records of concrete made with it or other information indicates that it is not injurious to the quality of the concrete."

TABLE 6.29 Volume of Dry-Rodded Coarse Aggregate per Unit Volume of Concrete for 2.80 Fineness Modulus of Sand

Maximum Size of Aggregate	Modified for High Strength	From ACI 211.1-82
3/8"	0.61	0.46
1/2"	0.64	0.55
3/4"	0.66	0.62
1"	0.70	0.67

6.5.4 Materials Control

After the materials are chosen, the consistency of the materials becomes imperative. If variations become excessive, the required average strength may become unattainable. As an example, if the specified strength was 9000 psi and the standard deviation was 697 psi (7% Coefficient of Variation), the overdesign required would be 8.4 MPa (1200 psi). On the other hand, if the standard deviation is 1414 psi (13% C of V), the overdesign would have to be nearly 21 MPa (3000 psi) (Table 6.31).

TABLE 6.30 Fine Aggregate for 7500 psi and Above Concrete (Fineness Modulus, 2.78)

Typical Gradation			
Sieve Size	% Passing	Chemical Analysis	Approx. Percent
3/8 in	100.0	Silicon Dioxide (SiO_2)	35%
No. 4	96.5	Iron Oxide (FeO_3)	
No. 8	81.7	Aluminum Oxide (Al_2O_3)	5% combined total
No. 16	70.9	Magnesium Carbonate	12%
No. 30	53.5	Calcium Carbonate	48%
No. 50	16.5		
No. 100	2.8		

TABLE 6.31 Required Increases in Design Strength as Coefficient of Variation Increases

SS, f_c'	CV 7%		CV 10%		CV 13%	
	SD	DS	SD	DS	SD	DS
6000	465	6700	693	7200	943	7690
7500	581	8400	866	9010	1178	9740
9000	697	10,200	1040	10,920	1414	11,790

Note: SS, specified strength; CV; coefficient of variation; SD, standard deviation; DS, design strength.

TABLE 6.32 Material Testing Schedule

Material	Frequency	Test
Cement	1/week	Mortar Cubes Fineness
Fly ash	1/week	325 Sieve Loss on Ignition
	1/month	Pozzolantic Activity
Admixtures	Each delivery	Specific Gravity
Water	As needed	
Aggregates	1/week	Gradation 200 Wash Specific Grativity

A materials-testing program should be established so that variations can be dealt with before they affect concrete strength (Table 6.32). All suppliers of constituent materials should be made aware of the importance of providing consistent materials. They should also know that their material will be checked on a periodic basis and the results forwarded to them. The handling and storage of materials need not be substantially different from the procedures used for conventional concrete. Proper stockpiling of aggregates, control of moisture, and adequate segregation of cementitious materials are essential.

Table 6.33 gives selected data on some HSC mixes used. However, the information tabulated is intended only as an example and should not replace locally generated data.

6.5.5 Mixing, Transporting, Placing and Curing

6.5.5.1 Mixing

High-strength concrete may be produced in manual, semiautomatic, or automatic plants. It should be noted, however, that total automation almost always improves consistency. Normal batching procedures need not be altered, but a modification in timing may help. Admixtures have been found to be most effective if they are introduced after the cement has been wetted. Caution should be exercised if the concrete is to be truck mixed. Truck mixers that are not maintained or that have worn fins and blades will not thoroughly mix the concrete and will cause unacceptable variations in all the physical characteristics of the concrete. Once in the stationary or truck mixer, the materials should be mixed as follows: 1 min for 1 yd^3 plus 0.25 min for each additional cubic yard, for example, 3.75 m^3 (5 yd^3) = 2 min. The speed at which the concrete is mixed should be in accordance with the mixer manufacturer's recommendations.

Care should be taken so that the temperature of the concrete as delivered does not exceed 42°C (90°F). As with all concrete, higher temperatures mean more water. If the water demand becomes excessive or slump loss becomes rapid, increased dosages of chemical retarders or superplasticizers should be considered. If problems persist, ice or chilled aggregates may be required to lower the temperature. Maintaining a minimum temperature is normally not a problem, in that the concrete is typically used in more massive structural members where lower temperatures are beneficial. Additionally, lower concrete temperature means less total water, higher later strengths, and less cracking.

TABLE 6.33 Composition of Concretes Produced in a Ready-Mix Plant

Mixture Ingredients and Characteristics	Concrete Type				
	Reference	Silica Fume	Fly Ash	Slag + Silica Fume	
Water/cementitious material ratio	0.30	0.30	0.30	0.30	0.25
Ingredients kg/m^3 (lb/yd^3)					
Water	127 (214)	128 (216)	129 (217)	131(221)	128 (216)
Cement ASTM Type II	450 (759)	425 (716)	365 (615)	228 (384)	168 (283)
Silica fume	—	45 (76)	—	45 (76)	54 (91)
Fly ash	—	—	95 (160)	—	—
Slag	—	—	—	183 (308)	320 (539)
Dolomitic limestone					
Coarse aggregate	1100 (1854)	1110 (1871)	1115 (1879)	1100 (1871)	1100 (1854)
Fine aggregate	815 (1374)	810 (1365)	810 (1365)	800 (1349)	730 (1230)
Superplasticizer * l/m^3 (oz/yd^3)	153 (395)	14 (362)	13 (336)	12 (310)	13 (336)
Slump after 45 min mm (in)	110 (4-1/4)	180 (7)	170 (6-3/4)	220 (8-3/4)	210 (8-1/4)
Compressive strength					
at 28 d MPa (psi)	99 (14,360)	110 (15,950)	90 (13,050)	105 (15,230)	114 (16,530)
at 91 d MPa (psi)	109 (15,810)	118 (17,110)	111 (16,100)	121 (17,550)	126 (18,280)
at 1 year MPa (psi)	119 (17,260)	127 (18,420)	125 (18,130)	127 (18,420)	137 (19,870)

*Sodium salt of a naphthalene sulfonate.

6.5.5.2 Transporting

High-strength concrete can be successfully mixed and transported in a number of ways. The quality-assurance/quality-control (QA/QC) personnel must be aware that prolonged mixing will cause slump loss and result in lower workability. Adequate job control must be established to prevent delays. When practical, withholding some of the high-range water-reducing admixture until the truck arrives at the job site or on-site addition of high-range water-reducing admixtures may be desirable. When materials are added at the site, proper mixing is required. Because of the consequences of segregation, QA/QC personnel should pay close attention to on-site mixing and verify that the mixture is uniform. The ACI 304.R Committee Report contains information on proper mixing. Truck mixers should be equipped with a drum revolution counter, and their fins should be in good condition. All mixer trucks used to transport high-strength concrete should be regularly inspected and certified to comply with the NRMCA Inspection Requirements prior to and during their use in this capacity.

6.5.5.3 Placing

High-strength concrete is typically produced with slumps in excess of 200 mm (8 in). Despite their fluid appearance, these mixes require thorough consolidation. All concrete should be consolidated quickly and thoroughly. Standby vibratory equipment is recommended, with at least one standby vibrator for every three vibrators required. The provisions in the ACI 309 Committee Report should be followed for proper consolidation. In construction, different strength concretes are often used within or between different members. QA/QC personnel should be aware of the exact location for each approved mixture.

Often, "mushrooming" is performed over column and shear-wall locations when placing floor slabs. That is, high-strength concrete is mushroomed around those locations to form a cap prior to placing lower-strength concrete around it in the floor. QA/QC personnel should be aware of how far the cap should extend. Since cold joints between the two concretes are not allowed, the inspector should determine that the high-strength mushroom is still plastic enough to blend with the lower-strength slab concrete. When two (or more) concrete mixes are being used in the same pour, it is mandatory that sufficient control be exercised at the point of discharge from each truck to insure that the intended concrete is placed as specified. Planning is necessary to determine the best procedures. Consideration might be given to the use of retarding admixtures.

6.5.5.4 Curing

The full potential strength and durability of high-strength concrete will be fully developed only if it is properly cured for an adequate period prior to being placed in service or subjected to construction loading. Many acceptable methods for curing are available, as presented in the ACI 308 Committee Report. However, high-strength concretes are extremely dense and impermeable. Therefore, appropriate curing methods for various structural elements should be selected in advance. For interior columns, curing is often impractical and durability is not a problem. The period during which the forms are in place may be adequate in such instances. QA/QC personnel should verify that the accepted methods are properly employed in the work.

High-strength concretes usually do not exhibit much bleeding, and without protection from loss of surface moisture, plastic shrinkage cracks may form on exposed surfaces. Protection methods include fog misting, use of evaporation retarders, covering concrete with polyethylene sheeting, or application of a curing compound. Water curing of high-strength concrete is recommended because of the low water/cementitious material ratios employed. Water curing of vertical members is usually impractical, and other curing methods should be employed.

When forms are released or removed at very early ages, typically less than one day, prevention of thermal cracking by providing insulation should be considered, particularly in cold weather.

Each high-strength mixture has heat evolution and dissipation characteristics in the context of its curing environment. Maximum temperature and thermal gradients, and their effects on constructability and long-term design properties should be determined by preconstruction trials. Consideration might be given to computer simulation of the expected thermal history in a structure so that appropriate curing and protection can be performed.

The higher cement contents of high-strength concrete generate a higher heat of hydration and, possibly, thermal gradients in excess of $20°C/m$ ($11°F/ft$), especially in uninsulated mass placements. However, studies have shown that thermal gradients are similar to those for conventional strength concretes. Data have shown that for deep foundations, when the ultimate internal temperature developed during hydration rose to $78°C$ ($172°F$), the in situ strength and stiffness was not adversely affected. The architect/engineer should understand the effects of heat generation in the various structural elements and address them in-project specifications.

Specifications for mass concrete often require that the temperature difference between the concrete interior and surface not exceed $20°C$ ($36°F$). The inspector should monitor and record ambient temperatures and curing temperatures at the surface and center of large concrete components so that the design/construction team can effectively make any adjustments such as mixture design changes or use of insulating forms during the course of the project. Concrete delivered at temperatures exceeding specification limits should be rejected, unless alternate procedures are agreed to at the preconstruction meeting.

6.6 Job-Site Control

Coordination of delivery time between the concrete supplier and contractor is critical. Concrete should be delivered so that minimal waiting is experienced. Delays on the job may result in slump loss and, subsequently, the concrete may require retempering. Water added at the job site can be extremely detrimental to the structural integrity of the concrete and therefore should not be permitted. Any adjustments made in the slump should be made with a high-range water reducer. As always, the amount of water in the admixture must also be considered in the water/cementitious material ratio.

The time allowed from loading to discharge should be limited to 90 min. If the concrete is older but still has a placeable consistency, it may be used. Again the concern is that retempering may be required as the concrete ages.

If a high-range water reducer is being used, slumps should not be used as a basis of acceptance or rejection. The concrete should be used, regardless of the slump, as long as it is placeable or not segregated. The specified water/cement ratio should be the governing factor. The method of placement (pump, bucket, or conveyor), quantity, and spacing of reinforcement will also dictate what slump is necessary.

6.7 Testing

6.7.1 General

Measurement of mechanical properties during construction provides the basic information needed to evaluate whether design considerations are met and the concrete is acceptable. Since high-strength concrete is more sensitive to testing variables than normal-strength concrete, the quality of these measurements is very important. Factors having little or no effect on 21 MPa (3000 psi) concrete can have significant effects on high-strength concrete, especially on compressive strength. Standard, established procedures such as the latest editions of ASTM C-31, 39, and 94 must be followed. Deviation from these procedures cannot be tolerated. Such deviations not only result in nonrepresentative tests but introduce variations that cannot be compensated for.

6.7.2 Sampling

Statistical methods are an excellent means to evaluate high-strength concrete. For statistical procedures to be valid, the data (slump, unit weight, temperature, air content, and strength) should be derived from samples obtained by means of a random sampling plan designed to reduce the possibility that choice will be exercised by the testing technician. Random sampling means that any portion of the material being presented has an equal chance of being selected. The samples should represent the quality of concrete supplied. Therefore composite samples are taken in accordance with ASTM C-172. They should be combined and remixed to ensure uniformity prior to testing the freshly mixed concrete or casting test specimens.

These samples are representative of the quality of concrete delivered to the job site and may not truly represent the quality of the concrete in the structure, which may be affected by site transportation and placing methods. If additional test samples are required to check the quality of the concrete at the point of placement, as in pumped concrete, this should be established at the preconstruction meeting.

6.7.3 Extent of Testing

Tests for air content, unit weight, slump, and temperature should be made on the first truckload each day to establish that batching is adequate. Thereafter, these tests should be performed on a random basis in accordance with project specifications. When visual inspection reveals inconsistent concrete, it should be rejected unless additional tests show it to be acceptable. Such test results should not be counted in the statistical evaluation of the mixture unless they are made on samples that are taken at random.

The architect/engineer will generally take advantage of the fact that high-strength concrete containing fly ash or ground granulated blast-furnace slag develops considerable strength at later ages, that is, at 56 and 90 d. It is not uncommon therefore that more test specimens than normally required are specified. The technician should be prepared to take a large enough sample to properly accommodate the volume required to cast all test specimens. Under no circumstances should the technician use other samples to "top off" test specimens. If the sample is too small, another sample should be taken.

However, only a reasonable number of specimens can be made in a high-quality manner and within the correct time frame from collection of each sample. No more than nine specimens should be made per sample unless sufficient personnel and facilities are available to properly handle a larger number. At least three specimens per test age are recommended, with three held in reserve. Where later ages are specified to enable the use of more economical mixes, it may be desirable to make an early assessment of potential strength by testing early age specimens or specimens with accelerated curing.

6.7.4 Compressive Strength Specimens

Since one of the main interests in high-strength structural concrete is its strength in compression, compressive-strength measurements are of primary concern. The primary function of standard laboratory-cured specimens is to provide assurance that the concrete mix as delivered to the job site has the potential to meet contract-specification requirements. The potential strength and variability of the concrete can be established only by specimens made, cured, and tested under standard conditions. ASTM standards specify a 150-mm (6 in) cylindrical specimen, 300 mm (12 in) long.

This specimen size has evolved over the years from practical considerations, and the architectural/ engineering profession is familiar with the empirical values obtained. However, 100 mm (4 in) by 200 mm (8 in) cylindrical specimens have also been used to successfully determine compressive strength. The use of the smaller cylinder is recommended, provided that the test result is determined in accordance with ASTM C-39. Regardless of specimen size, the size used to determine mix proportions should be consistent with the size specified for acceptance testing and approved by the engineer of record.

6.7.4.1 Mold Type

The choice of mold material can have a significant effect on measured compressive strength. A given consolidation effort is more effective with rigidly constructed molds. Rigid single-use plastic molds with 6-mm (1/4 inch) wall thickness or greater have been successfully used for 70 MPa (10,000 psi) concrete. Steel molds in particular have been found to produce strength results approximately 13% higher than high-quality cardboard molds. Therefore cardboard molds are not recommended. Because of these differences, the mold type used for field specimens should be the same as the mold type used to develop the design mix. Molds should comply with ASTM C-470.

6.7.4.2 Testing Apparatus

The higher compressive loads carried by high-strength concrete put more demands on compression machines. Machine characteristics that may affect the measured strength include calibration accuracy, longitudinal and lateral stiffness, alignment of the machine components, type of platens, and the behavior of the platen spherical seating. Testing machines should meet the requirements of ASTM C-39. On the basis of practical experience, it is recommended that the machine have a load capacity at least 20% in excess of the expected cylinder breaking load, since premature damage and loss of calibration has been known to occur as a result of large numbers of explosive failures at high loads.

6.7.5 Prequalification of Testing Laboratories and Ready-Mix Suppliers

To prequalify a laboratory as well as a ready-mix supplier, both should be examined from two perspectives: how they have performed in the past and how well they are equipped to perform properly in the future. A review of past test data for high-strength concrete analyzed in accordance with ACI 214 will show that in-test variability is a measure of the testing consistency of the laboratory.

The qualifications and experience of technicians and inspectors should be reviewed. The laboratory should be accredited or inspected for conformance to the requirement of ASTM C-1077 by a recognized agency such as the American Association for Laboratory Accreditation (AALA), AASHTO Materials Reference Laboratory (AMRL), National Voluntary Laboratory Accreditation Program (NV-LAP), Cement and Concrete Reference Laboratory (CCRL), or their equivalent.

The architect/engineer should approve the testing laboratory. A comprehensive internal quality control protocol that covers test procedures, the use, care, and calibration of testing equipment, and the checking and reporting procedures to be followed provides evidence of a well-run laboratory. Depending on the results of the review of past and potential performance of the laboratory, some tests of personnel and equipment may or may not be made to conclude the prequalification examinations.

Acknowledgments

Part A of this Chapter (Sections 6.1–6.4) is based on extensive material taken with permission from *Fundamentals of High Strength High Performance Concrete* by E.G. Nawy, 1996, Addison–Wesley, London and California; *Reinforced Concrete—A Fundamental Approach*, 3rd edition, and *Prestressed Concrete—A Fundamental Approach*, 2nd edition both by E.G. Nawy, 1996, Prentice Hall, Upper Saddle River, New Jersey, and Committee Reports and Standards of the American Concrete Institute, Farmington Hills, MI).

References

ACI Committee 221. 1961. Selection and use of aggregate for concrete. *J. Am. Concrete Inst.* 58(5): 513–542.

ACI Committee 212. 1983. Admixtures for concrete. In *ACI Manual of Concrete Practice 1983*, ACI 212.1 R-81, 29 pp. American Concrete Institute, Farmington Hills, MI.

ACI Committee 211. 1993. *Guide for Selecting Proportions for High Strength Concrete with Portland Cement and Fly Ash*, ACI Report 211.4R-93. American Concrete Institute, Farmington Hills, MI.

American Concrete Institute. 1994. *ACI Manual of Concrete Practice 1994, Part I, Materials*, ACI Farmington Hills, MI.

ACI Committee 304. 1989. *Guide for Measuring, Mixing, Transporting and Placing Concrete*, ACI Report 304-89. American Concrete Institute, Farmington Hills, MI.

ACI Committee 308. 1992. *Standard Practice for Curing Concrete*, ACI Report 308-92. American Concrete Institute, Farmington Hills, MI.

ACI Committee 309. 1987. *Guide for Consolidation of Concrete*, ACI Report 309R-87. American Concrete Institute, Farmington Hills, MI.

ACI Committee 318. 1995a. *Building Code Requirements for Structural Concrete*, ACI Standard 318–95. American Concrete Institute, Farmington Hills, MI.

ACI Committee 318. 1995b. *Commentary on Building Code Requirements for Structural Concrete*, ACI Standard 318–95. American Concrete Institute. Farmington Hills, MI.

ACI Committee 211. 1991. *Standard Practice for Selecting Proportions for Normal, Heavyweight, and Mass Concrete*, ACI 211.1-91, 38 pp. American Concrete Institute, Farmington Hills, MI.

ACI Committee 211. 1991. *Standard Practice for Selecting Proportions for Structural Lightweight Concrete*, ACI 211.2-91, 14 pp. American Concrete Institute, Farmington Hills, MI.

Addis, B.J., and Alexander M.G. 1990. A method of proportioning trial mixes for high strength concrete. *Proc. 2nd Int. Symp. High Strength Concrete*, ACI SP-121, pp. 287–308. American Concrete Institute, Farmington Hills, MI.

American Concrete Institute. 1997. *ACI Manual of Concrete Practice*, Vols. 1, 2, 3, 4, and 5. Farmington Hills, MI.

American Society for Testing and Materials. 1978. *Significance of Tests and Properties of Concrete and Concrete Making Materials*, Spec. Tech. Pub. 169B, 882 pp. ASTM, Philadelphia.

American Society for Testing and Materials. 1993. *Annual Book of ASTM Standards*, Part 14, *Concrete and Mineral Aggregates*, 834 pp. ASTM, Philadelphia.

Blick, R.L. 1973. Some factors influencing high strength concrete. In *Modern Concrete.*

Cook, J.E. 1982. Research and applications of high strength concrete using Class C fly ash. In *Concrete International*, American Concrete Institute, Farmington Hills, MI.

Kiell, J. 1985. The effect of reactor products and particle size distribution on the stiffening behavior of cement pastes, mortar, and concrete. Public Paper of the Cement Industry.

National Crushed Stone Association. 1975. Quality concrete with crushed stone aggregate. *Pub. CSC-3*. MD.

National Crushed Stone Association. 1976. Stone sand for Portland cement concrete. *Stone Products Update*, 1st ed. MD.

Nawy, E.G. 1996a. *Reinforced Concrete—A Fundamental Approach*, 3rd ed., 840 pp. Prentice Hall, Upper Saddle River, NJ.

Nawy, E.G. 1996b. *Fundamentals of High Strength High Performance Concrete*, 350 pp. Addison Wesley Longman, London and California.

Neville, A.M. 1996. *Properties of Concrete*, 4th ed., 844 pp. Pitman Books, London.

Portland Cement Association. 1994. *Design and Control of Concrete Mixtures*, 13th ed., Steven H. Kosmatka and William C. Panarese, eds., 212 pp. Skokie, IL.

Russell, H.G. 1993. High strength concrete. In *ACI Compilation 17*, pp. 3. American Concrete Institute, Farmington Hills, MI.

Formwork for concrete wall (Courtesy Portland Cement Association).

7

Design and Construction of Concrete Formwork

by
David W. Johnston P.E., Ph.D.
Professor of civil engineering at North Carolina State University, where he teaches and conducts research in the construction and performance of structures, formwork loadings and design, and bridge management systems. Prior to joining the University in 1977, he worked for eight years as an engineer and partner in consulting firms. He is a Fellow of both the American Concrete Institute and the American Society of Civil Engineers.

0-8493-2666-4/97/$0.00+$.50

7.1 Introduction

Forms are extremely important in concrete construction. They mold the concrete to the required size and shape while controlling its position and alignment (Figure 7.1). Forms are self-supporting structures that are also sufficient to hold the dead load of the reinforcement and fresh concrete and the live load of equipment, workers, and miscellaneous materials (Figure 7.2). In building and designing formwork, three major objectives must be considered:

1. **Quality:** Forms must be designed and built with sufficient stiffness and accuracy so that the size, shape, position, and finish of the cast concrete are attained within the required tolerances.
2. **Safety:** Forms must be built with sufficient strength and factors of safety so that they are capable of supporting all dead and live loads without collapse or danger to workers and to the concrete structure.
3. **Economy:** Forms must be built efficiently, minimizing time and cost in the construction process and schedule for the benefit of both the contractor and the owner.

Economy is important because the costs of formwork often range from 35 to 60% or more of the total cost of the concrete structure. Considering the impact of formwork on total cost, it is critical

FIGURE 7.1   Formwork serves as a mold to define concrete structure shape.

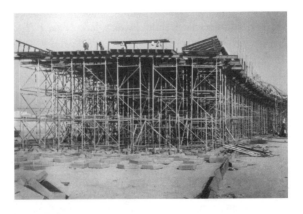

FIGURE 7.2   Formwork elements must support many heavy loads.

that the structural engineer of the facility also design the facility structure for economy of forming, not just for economy of the materials in the finished structure.

Ideally, the builder will achieve maximum economy with no cost to either safety or specified quality. In designing formwork, the construction engineer can reduce costs by carefully considering the materials and equipment to be used; the fabrication, erection and stripping procedures; and the reuse of forms. However, economy measures that result in either formwork failure or poor-quality products that require (often expensive) modification are self-defeating.

Correctly designed formwork will ensure that the concrete maintains the desired size and shape by having the proper dimensions and being rigid enough to hold its shape under the stresses of the concrete. It must be stable and strong in order to keep large sections of concrete in alignment (Figure 7.3). Finally, formwork must be substantially constructed so it can be reused and frequently handled while maintaining its shape. Formwork must remain in place until the concrete is strong enough to carry its own weight. In addition, the surface finish of the concrete is dependent on the contact material of the form.

FIGURE 7.3 Large and complex placements require substantial formwork.

The quality of the formwork itself has a direct impact on safety, accidents, and failures. A floor formwork system filled with wet concrete has its weight at the top and is not inherently stable. As a result, one of the most frequent causes of failure is from effects that induce lateral forces or displacement of supporting elements. Therefore inadequate crossbracing and/or horizontal bracing is one of the most frequently involved factors in formwork failure. Poor bracing can make a minor failure turn into a major disaster, in what might be thought of as a domino effect or a **progressive failure**: a failure at one point in the formwork that can become an extensive collapse through chain reaction. Vibration is one factor that can trigger failure through inadequate bracing. Two other formwork problems are unstable soil under mudsills and shoring that is not plumb: formwork is stable if adequately braced and built so all loads are carried to solid ground through vertical and bracing members.

Regardless of the quality of the formwork, premature removal of the forms or shores, often out of a wish for economy, can result in collapse or sagging. Sagging, while not an immediate problem, can lead to hairline cracks and extensive maintenance problems. Careless reshoring, often involving inadequate size, spacing, or attachment, can also cause damage or collapse. Specific related standards (OSHA, ACI, ASCE, etc.) for formwork will be discussed in Section 7.3.

In addition to optimizing material in the form-design process, there are three major factors to be considered when planning forms that are cost effective:

1. designing and planning for maximum reuse,
2. economical form assembly, and
3. efficient setting and stripping.

Each factor must be balanced with the other two in order to determine the most efficient form design.

In planning for maximum reuse, specifications, rate of concrete strength gain, and local code requirements regarding stripping must be taken into account. The sooner a form can be stripped safely, the more practical it is to reschedule many reuses. In addition, for a minimum of cost, the least number of forms required for a smooth work schedule should be built. For example, the formwork on the outside of a spandrel beam can be stripped sooner than the formwork on the bottom; hence fewer side forms than bottom forms need be built because they can be reused more frequently. It is also important to consider the labor involved in reuse—Does the form need to be disassembled and rebuilt? To create a plan for reuse, the construction engineer needs to make a detailed study of the work flow and construction sequence to determine the practical number of reuses that will result in smooth and efficient construction with the lowest total cost (Figure 7.4). When comparing designs, the construction engineer should calculate the ratio of total contact area of the formed concrete structure to the first-use form contact area for various alternatives, which is a general indication of overall reuse efficiency.

There are several considerations involved in determining an economical form construction:

FIGURE 7.4 Planning is required to achieve maximum overall construction efficiency.

- cost and feasibility of adapting materials on hand versus cost of buying or renting new materials;
- cost of a higher grade of material versus cost of lower grade of material plus labor to improve for required quality and use;
- selection of more expensive materials that provide greater durability and capability for reuse versus less expensive materials that have a shorter use-life;
- building on site versus building in a central shop and shipping to site: this depends on the site itself and space available, the size of project, the distance of shipping, etc.

An estimate of the cost of constructing a particular job-built form can be obtained by determining the quantity of wood materials needed to make one square foot of contact area, while allowing for waste and rejection of some wood, and multiplying by the unit prices of lumber involved. This provides the contractor with an estimate of lumber costs for one square foot of contact area. In addition to lumber, the costs for labor, hardware, miscellaneous materials, handling, and clean up must also be evaluated. The construction engineer should also consider the possibility of using prefabricated forms, either rented or purchased. The advantages of using rented prefabricated forms are reduced risk, no investment cost, and transfer of some management responsibility to the formwork supplier or subcontractor.

The last major factor, efficient setting and stripping practices, has a direct impact on the two already discussed. Reuse of a form is only fully efficient if the form can be stripped and rebuilt without too much labor time or damage to the form. The estimate for constructing a form must take into account the worker hours needed to erect and dismantle the form during each reuse. When calculating time and cost for setting and stripping forms, the contractor should allow for delays from weather, equipment problems, etc., as well as cleaning and other miscellaneous expenses.

In addition to the above elements of cost, planning of formwork operations should consider the overall flow of operations, including the following:

1. *Crew efficiency:* Providing a reasonable schedule creates a smooth daily repetition of the same operation so that the workers can be familiar with their tasks and thus perform efficiently
2. *Concreting:* The ease and speed of pouring the concrete is directly related to the choice of design and the construction schedule
3. *Bar setting and other trades (mechanical, electrical, piping):* Schedules of these activities must be coordinated with the concreting schedule so that all groups can work efficiently

4. ***Cranes and hoists:*** Plan to use appropriate cranes at times when they are not needed for other functions and reduce idle time

7.2 Types of Formwork

Formwork components can be assembled in a wide variety of systems for casting many structural shapes. The terms formwork and falsework are often used in combination. **Formwork** is the total system of support for freshly placed concrete and includes the sheathing that is in contact with the concrete as well as all supporting members, hardware, and necessary bracing. **Falsework** is a temporary structure erected to support work in the process of construction. Falsework may be the temporary support for steel bridge girders, for precast concrete elements to be post-tensioned together, or for many other applications. When this term is used in relation to formwork, the forms are often considered to be the horizontal system of elements directly under heavier concrete placements, such as cast-in-place bridges (Figure 7.2), and the falsework includes the temporary girders, shores or vertical posts, and lateral bracing.

Forms can either be job built or prefabricated. Job-built forms, frequently of wood (Figure 7.5), are most frequently selected where the shape does not conform to the constraints of commercial systems or where the economics are viable. Prefabricated forms can be purchased, rented, or rented with an option to buy. They are usually constructed substantially for the purpose of frequent reuse. These forms can either be ready made or custom made (Figure 7.6). The latter is designed for specialized use, usually on a single job, but is often reused multiple times on that project.

7.2.1 Contact Surface Materials

The material serving as the contact face of forms is known as **sheathing** and sometimes is referred to as lagging or sheeting in specialty applications. Plywood is frequently used for sheathing, but some forms use steel sheet metal, steel plate, fiber-reinforced plastic, paperboard, wood boards, or other materials. A primary characteristic in selection of the sheathing type and grade is the quality of surface required by the specification. For some applications, steel or high-density overlay plywood may be needed. In other cases where a decorative surface is required, the sheathing may be specially treated (wire brushing to expose wood grain, addition of rustication strips, etc.) or provided with a commercial plastic **liner** imprinted or shaped to provide a specified design (Figure 7.7).

Permanent forms are any form that remains in place after the concrete has developed its design strength. The form may or may not become an integral part of the structure. Metal deck forms, the most prevalent permanent form, are made of a ribbed or corrugated steel sheet, usually galvanized

FIGURE 7.5 Wood column form.

to reduce the potential for future rust staining, and are secured to the structure with clips or by welding. The diaphragm created by the attached deck may also contribute to the lateral stability of the supporting members during concrete placement. Metal deck forms are used in floor and roof slabs cast over steel joists or beams, in bridge decks (Figure 7.8), and over pipe trenches and other inaccessible locations where removing wood forms is impractical. Precast concrete deck forms are often used in combination with prestressed concrete bridge girders. Sometimes the deck forms must be temporarily shored at intermediate points to support the loads applied during construction. However, deck forms of adequate section profile can often be selected to span between the permanent structural members and safely support the weights of reinforcement, fresh concrete, and construction live loads. In the latter case, the added cost of the stronger section is often offset by savings in shoring materials and labor.

FIGURE 7.6   Custom-made steel form with integrated access platforms.

7.2.2 Floor-Forming Systems

Floor-forming systems vary somewhat with the configuration of the concrete floor system being cast. Figure 7.9 illustrates the basic wood member arrangements for flat plate floors, most areas of flat slabs, and the slab areas of slab and beam floors. The terms used to describe the members are the same in systems assembled from steel or aluminum members. The **joist** is a horizontal structural member supporting the deck-form sheathing and usually rests on stringers or ledgers. **Stringers** are horizontal structural members usually (in slab forming) supporting joists and resting on vertical supports such as shores.

In one- (pan joist) and two-way (waffle) joist construction, a similar layout is usually adopted. **Pans and domes** (Figure 7.10) are used in concrete joist construction, which is a cast-in-place floor system with a thin slab integral with regularly spaced joists that span to beams and columns. In some cases, the pans or domes are placed on the plywood sheathing; in other cases, the pan or dome edges are supported on wide joists, and the sheathing is omitted. Pans are prefabricated form units,

FIGURE 7.7   Plastic form liner used to create rough ribbed pattern.

FIGURE 7.8 Stay-in-place corrugated galvanized metal deck for bridge-slab form.

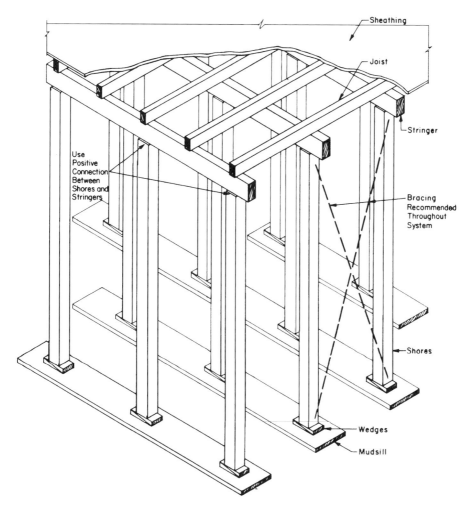

FIGURE 7.9 Typical elevated flat-slab formwork elements. *Source:* Courtesy of American Concrete Institute. Hurd, M.K. 1995. *Formwork for Concrete*, Publication SP-4.

FIGURE 7.10   Fiber-reinforced plastic pan and dome molds for floor formwork.

usually steel or fiberglass, used to form single-direction joists. Domes, also usually made of steel or fiberglass, are square pan forms used in two-way, or waffle, concrete joist construction.

Steel and aluminum joists and stringers are usually flanged shapes. Some aluminum extrusions have special configurations allowing easy connection and incorporating a top channel for a wood nailer. Commercial steel sections fabricated specifically for formwork systems also incorporate connection features. **Horizontal shoring** is formed by adjustable beams, trusses, or combinations of the two that support formwork over clear spans and eliminate numerous vertical supports. Metal or timber support beams are used for small spans. Telescoping shores, steel lattice, plate, or box members are used to support forms in spans of 6 to 30 ft. Heavy-duty horizontal shores, for example, trusses supporting flying-form panels, can span up to 80 ft. The disadvantage of using horizontal shoring is the potential need for special bearing plates to support the high end load on each individual shore.

Flying forms, or table forms, are large crane-handled sections of floor formwork (frequently including supporting truss, beam, or scaffolding frames) that are completely unitized. Such forms can be lowered for clearance under joists or beams, rolled out the face of the building bay, picked by a crane, and reset at the next floor level. By having a large movable unit, the costs of stripping and reassembly are reduced, particularly when a crane is available on site.

7.2.3 Column Forms

Column-form materials tend to vary with the column shape. Wood or steel is often used with square or rectangular columns (Figure 7.11). Round column forms (Figure 7.12), more typically premanufactured in a range of standard diameters, are available in steel, paperboard, and fiber-reinforced plastic. Square and rectangular forms are composed of short-span bending elements contained by external ties or clamps. Round column forms are more structurally efficient since the internal concrete pressures can be resisted by a hoop membrane tension in the form skin with little or no bending induced. Round, single-piece glass fiber reinforced plastic tubes with a single joint can be removed from the column without cutting. They are held together with either bolts or clamps.

Round paperboard tubes are single-use forms that are stripped by unwrapping and then discarded. They can be cut to the exact length needed and sections of the tube can be adapted to making partial column sections (half-round, quarter-round, etc.).

Steel column forms have built-in bracing for short heights so that the only external bracing needed is to keep the column plumb and for taller columns. Both half-round and rectangular panel units are

FIGURE 7.11   Modular steel form for rectangular columns.

available in various section heights that can be connected vertically to form tall columns. Round steel forms are generally used for larger columns and bridge piers and come in diameters ranging from about 14 in to 10 ft.

7.2.4 Wall Forms

Wall forms principally resist the lateral pressures generated by fresh concrete as a liquid or semiliquid material. The pressures can be quite large, certainly many times the magnitude of live loads on permanent floors. Thus wall form design often involves closely spaced and well-supported members as shown in Figure 7.13. As mentioned, the contact surface of the wall form is called sheathing. **Studs** are vertical supporting members to which sheathing is attached. **Wales** are long horizontal members (usually double) used to support the studs. The studs and wales are often wood, steel, or aluminum beamlike elements. Commercial form suppliers are innovative in devising elements as well as hardware for connections. The wall-form members are sometimes oriented with the stud members placed horizontal rather than vertical, and the wales are run vertical.

FIGURE 7.12   Steel round-column form with access scaffolding in preparation.

The wales are in turn supported on washer plates or other bearing devices attached to form ties. A concrete **form tie** is a tensile unit connecting opposite sides of the form and providing a link for equilibrium. Form ties are usually steel, although some fiber-reinforced plastic ties are also available. The ties come in a wide range of types (Figure 7.14) and tension working capacities rated by the manufacturer. Snap ties, loop ties, and flat ties are single-use ties, usually of relatively low capacity (1500 to 3200 lb) that are twisted and snapped off a specified distance back from the concrete surface. Coil ties, she bolts, and he bolts are examples of ties where some parts are left embedded within the cast wall and some parts can be reused. The taper tie, a tapered rod threaded on each end, is completely removed and reused. The tension capacity of heavy ties can range upward to over 60,000 lb.

Some of the ties have built-in provisions for spacing the forms a definite distance apart; this is particularly true of single-use ties if this feature is ordered. An alternate means of maintaining the

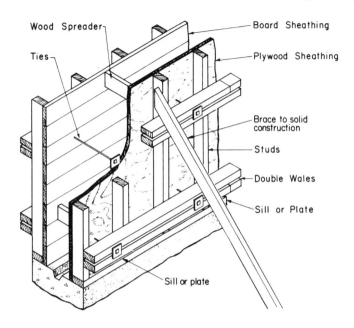

FIGURE 7.13 Typical wall-form components with alternate sheathing materials illustrated. *Source:* Courtesy of American Concrete Institute. Hurd, M.K. 1995. *Formwork for Concrete*, Publication SP-4.

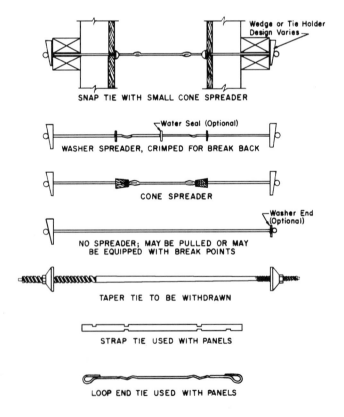

FIGURE 7.14 Examples of form ties. *Source:* Courtesy of American Concrete Institute. Hurd, M.K. 1995. *Formwork for Concrete*, Publication SP-4.

FIGURE 7.15 Ganged forms allow movement by crane and reduce reerection cost.

correct inside distance is by means of a ***spreader***, a strut, usually of wood, inserted inside the forms that can be retrieved with an attached rope or wire when the concrete placement reaches that level.

Strongbacks, frames attached to the back of a form or additional vertical wales placed outside horizontal wales, are sometimes added for strength, to improve alignment, or to assemble a ganged form. **Gang forms** (Figure 7.15) are prefabricated panels joined to make a much larger unit for convenience in erection, stripping, and reuse; they are usually braced with wales, strongbacks, or special lifting hardware. Such units require the use of a crane for stripping and resetting.

Panel forms, sections of form sheathing constructed from boards, plywood, metal, etc. that can be erected and stripped as units, are primarily used in wall construction. They may be adapted for use as slab or column forms. There are four basic types of panel form: unframed plywood, plywood in a metal frame (Figure 7.16), all-metal, and heavier steel frame. The first two are most frequently used for general light construction and erecting walls with heights ranging from 2 to 24 ft. Both are sometimes backed by steel braces. All-metal panel forms can be made of either steel or aluminum. The heavier steel frame panel forms are faced with either wood or plywood and are used for projects involving large pressures or loads.

Slip-forms are forms that move, usually continuously, during placing of the concrete. Movement may be either horizontal or vertical. Slip-forming is like an extrusion process, with the forms acting as moving dies to shape the concrete. In wall forming (Figure 7.17), the slip-form is usually moved vertically at a rate of 6 to 12 in per hour. This method can be economical when constructing concrete cores of tall buildings, tall concrete stacks, and concrete towers.

Climbing forms, or jump forms, are forms that are raised vertically for succeeding lifts of concrete in a given structure, usually supported by anchors embedded in the previous lift. The form is moved only after an entire lift is placed and (partially) hardened; this should not be confused with a slip form that moves during placement of the concrete. Support of the climbing form is usually provided

FIGURE 7.16 Steel-framed plywood panel system with integrated loop-tie anchors.

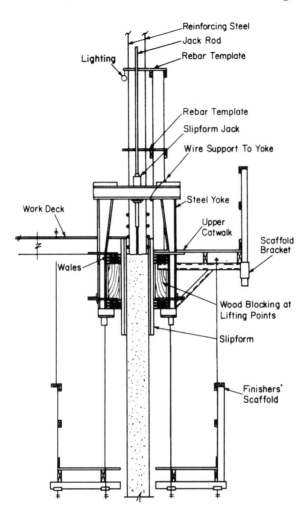

FIGURE 7.17 Typical wall slip-form components. *Source:* Courtesy of American Concrete Institute. Hurd, M.K. 1995. *Formwork for Concrete*, Publication SP-4.

by anchors cast in the previous placements. It is critical that the concrete strength gain at the anchors be sufficient at each stage of the operation to resist the imposed loads.

Although the use of proper support chairs for reinforcement is routinely required in horizontal construction, similar elements are sometimes neglected in wall and column forms. **Side form spacers** are devices similar to chairs that can be used to advantage in wall and column forms to attain correct cover for the reinforcement.

7.2.5 Shoring

Shores are vertical or inclined columnlike compression supports for forms. Shoring systems may be made of wood or metal posts, scaffold-type frames, or various patented members. **Scaffolding** is an elevated platform to support workmen, tools, and materials. In concrete work, heavy duty scaffolding is often adapted to double as shoring.

The simplest type of vertical shore is a 4 × 4 or 6 × 6 piece of lumber with special hardware attached to the top to facilitate joining to the stringers with a minimum of nailing. Metal shore-jack fittings may be placed at the lower end of the shore to allow some adjustment for exact height. Various manufacturers sell all-metal adjustable shores, also known as **jack shores** or simply as jacks, in a number of designs. They are usually available in adjustable heights from 6 to 16 ft and can carry working loads ranging from 2500 to 9000 lb with a safety factor of 2.5, depending on the type and the length of the shore. A third type of vertical support is a device that attaches to a column or bearing wall of the structure. Components of this kind of shore include friction collars and shore brackets that are attached to the support with through bolts or heavy embedded anchors. These attached supports are particularly useful in supporting slab-forming systems.

Basic scaffold-type shoring is made from tubular steel frames. End frames are assembled with diagonal bracing, locking connections, and adjustable bases to create a shoring tower. These may have flat top plates, U-heads, or other upper members for attaching to supported forms. Most scaffold-type shoring has a safe working load between 4000 and 25,000 lb per leg, depending on the height, bracing, and construction of the tower

FIGURE 7.18 Shoring towers used for bridge falsework.

(Figure 7.18). Ultra-high load shoring frames can support up to 100,000 lb of load per leg. Scaffold-type shoring can also be made of tubular aluminum, which has the advantage of being more lightweight than a similar steel frame.

In multistory concrete building construction (Figure 7.4), a process called **shoring and reshoring** is used. The weight of the fresh concrete, reinforcing, forms, and construction live load for an individual floor usually exceeds the design live-load capacity of the floor below. Furthermore, that floor has not gained full 28-d strength since floors are often cast at intervals of 4 to 14 d. By interconnecting several floors with shores and reshores, the loads at the top can be distributed over several floors. When this construction process is engineered and controlled properly, the time-varying loads on all elements (floors, forms, shores, and reshores) are safely within their time-varying strengths. One or more sets of floor forms with shores may be used. After the forms and shores are stripped completely from the lowest form-supported floor, that floor and the cast floors above deflect and must share the equivalent of the support that has been removed. Reshores are then placed under the stripped floor. **Reshores** are shores placed snugly under a concrete floor so that future loads imposed from construction at the highest level can be shared over sufficient floors to carry the dead and live loads safely.

In some multistory construction, this process is varied slightly. **Backshores** are shores placed snugly under a concrete slab or structural member after the original forms and shores have been removed from a small area without allowing the slab or member to deflect or support its own weight or existing construction loads from above. **Preshores** are added shores placed snugly under selected panels of a deck-forming system before any primary (original) shores are removed. Preshores and the panels they support remain in place until the remainder of the bay has been stripped and backshored, a small area at a time.

Centering is a specialized temporary vertical support used in construction of arches, shells, and space structures where the entire temporary support is lowered (**struck** or **decentered**) as a unit to avoid introduction of injurious stresses in any part of the structure.

Shores that are supported on the ground must have an adequate strength and size temporary footing termed, in formwork, a **mudsill** (Figure 7.19). The mudsill may be a plank, wood grillage, or precast pad, depending on the loads and ground conditions.

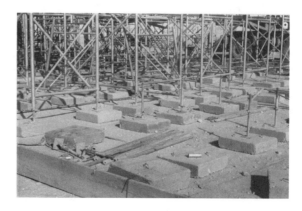

FIGURE 7.19   Precast concrete mudsills distribute bridge falsework load to soil.

FIGURE 7.20   Formwork brace anchored to buried concrete mass.

7.2.6 Bracing and Lacing

A brace is any structural member used to support another, always designed for compression loads and sometimes for tension under special load conditions. In formwork, **diagonal bracing** is a supplementary (not horizontal or vertical) member designed to resist lateral load. Form braces are frequently made of wood or steel. Commercial steel pipe braces in various diameters, wall thicknesses, and load rated for adjustable lengths are popular.

Buckling strength of braces is always a primary design consideration. **Horizontal lacing**, horizontal members attached to shores or braces to reduce their unsupported length, can thus increase the available load capacity.

Both bracing and lacing must be adequately connected at each end. This can be accomplished with bolts, nails, and a variety of commercial devices, depending on the materials involved. When attaching braces to the ground, a buried or above ground concrete mass known as a deadman is sometimes used (Figure 7.20).

7.2.7 Other Forms and Components

The above represents only a brief summary of some of the most typical form elements. Many other configurations need to be considered in many jobs. Beam forms are somewhat like short wall

FIGURE 7.21 Plywood templates secure dowels or anchor bolts at correct position.

forms in that lateral pressures must be resisted; however, they also involve concentrations of vertical load, requiring strong bottom forms and more shoring. The casting of footings sometimes requires forming where the concrete cannot be cast against vertical earth sides. This forming often must be braced from the outside to hold the lateral pressure of the earth if the width of the foundation is large. Supplementary forming elements such as **templates** are often incorporated in foundation forms to precisely locate anchor bolts and dowels (Figure 7.21).

Where concrete elements must be subdivided into two or more placement sections, a form **bulkhead** is usually placed, either as a construction joint or as an end closure. The bulkhead, although involving short spans, must be carefully designed and connected since it must resist the same pressure magnitudes as the faces of the form.

Forms may also incorporate a host of other features. **Chamfer strips** are triangular or curved inserts placed in the inside corner of a form to produce a rounded or beveled corner. These are often specified in rectangular columns and at outside corners of walls. **Cleanouts** are openings sometimes provided at the bottom of wall or column forms for removal of refuse before the concrete is placed, better assuring a good construction joint. There must be a means of closing and supporting the cleanout door to resist concrete pressure. **Wrecking strips** are small pieces or panels fitted into a formwork assembly in such a way that they can be easily removed ahead of the main panels or sections, making it easier to strip the major form components. Various references such as ACI SP-4 [Hurd, 1995] and form-supplier catalogs provide numerous illustrations of formwork details.

7.3 Formwork Standards and Recommended Practices

Ultimately, formwork safety is dependent upon the system in place on individual projects to assure proper and safe design, fabrication, handling, erection, inspection, monitoring, and stripping of the forms and supports (Figure 7.22). As noted earlier, formwork materials and labor are roughly equal in cost to the concrete and reinforcing materials and placing labor. The loads supported by formwork, say for an individual floor, are usually similar in magnitude to the loads supported by the finished structure and are sometimes much greater. Thus it is justifiable for the design and planning of formwork to require the significant time of a professional, just as is required for design of the structure of the facility being built. Savings can accrue from a well planned and designed system.

This section summarizes some of the resources available to the construction engineer to guide the planning and design of formwork. Most of the resources are in the form of guides or recommended

FIGURE 7.22   Guardrails and rated, well-maintained access ladders are components of a safety system.

practices. Except for some provisions of OSHA, there has previously been no uniformly mandated standard in the United States for design of temporary structures such as formwork. However, this is in the process of change as will be noted in the following sections. As in all engineering work, the construction engineer should seek the most recent edition of the following resources. Most standards are updated or reconfirmed on a cycle of six years or less.

7.3.1 American Concrete Institute Recommendations

Prior to about 1958, formwork was designed based upon only limited data and guidance for loads. At that time, recommendations for loads and pressures for the design of formwork became available from the American Concrete Institute. These recommendations have evolved over the years and are available in two often-consulted and periodically updated ACI publications:

> *Guide to Formwork for Concrete*, ACI 347R-94 [ACI, 1994] and
> *Formwork for Concrete*, ACI SP-4, 6th ed. [Hurd, 1995].

The former provides the recommended practice for design and construction of formwork, including recommendations for loads and pressures. The latter is a manual that extensively describes systems and provides design procedures, design aids, and examples. ACI also publishes, through its journals and other monographs, numerous articles on concrete formwork as well as guides for craftspersons involved in formwork fabrication and erection.

ACI Committee 347 recommendations for loads and pressures to be applied in the design of formwork are discussed more thoroughly in Section 7.4 of this chapter.

7.3.2 OSHA Standards

By the early 1970s, legislation to establish the Occupational Health and Safety Act (OSHA) began to have an impact on formwork. OSHA expectations include not only an adequate temporary structure but also emphasize components for worker fall protection (Figures 7.23 and 7.24) and procedures for safe handling and erection of formwork. Sections of the OSHA *Occupational Safety and Health Standards for the Construction Industry*, 29 CFR, Part 1926, particularly affecting formwork planning, design, and execution includes the following requirements:

> Subpart L—Scaffolding
> Subpart M—Fall Protection
> Subpart Q—Concrete and Masonry Construction

FIGURE 7.23 Fall-protection nets.

FIGURE 7.24 Work platform being installed on tall wall form.

Subpart Q contains sections on scope, application, and definitions; general requirements; equipment and tools; cast-in-place concrete; precast concrete; lift-slab operations; and masonry construction. Of primary interest here is the following section on cast-in-place concrete:

§ 1926.703 Requirements for cast-in-place concrete.

(a) General requirements for formwork.

(1) Formwork shall be designed, fabricated, erected, supported, braced, and maintained so that it will be capable of supporting without failure all vertical loads that may reasonably be anticipated to be applied to the formwork. Formwork which is designed, fabricated, erected, supported, braced, and maintained in conformance with the Appendix to this section will be deemed to meet the requirements of this paragraph.

(2) Drawings or plans, including all revisions, for the jack layout, formwork (including shoring equipment), working decks, and scaffolds, shall be available at the jobsite.

(b) Shoring and reshoring.

(1) All shoring equipment (including equipment used in reshoring operations) shall be inspected prior to erection to determine that the equipment meets the requirements specified in the formwork drawings.

(2) Shoring equipment found to be damaged such that its strength is reduced to less than that required by § 1926.703(a)(1) shall not be used for shoring.

(3) Erected shoring equipment shall be inspected immediately prior to, during, and immediately after concrete placement.

(4) Shoring equipment that is found to be damaged or weakened after erection, such that its strength is reduced to less than that required by § 1926.703(a)(1), shall be immediately reinforced.

(5) The sills for shoring shall be sound, rigid, and capable of carrying the maximum intended load.

(6) All base plates, shore heads, extension devices, and adjustment screws shall be in firm contact, and secured when necessary, with the foundation and the form.

(7) Eccentric loads on shore heads and similar members shall be prohibited unless these members have been designed for such loading.

(8) Whenever single post shores are used one on top of another (tiered), the employer shall comply with the following specific requirements in addition to the general requirements for formwork:

    (i) The design of the shoring shall be prepared by a qualified designer and the erected shoring shall be inspected by an engineer qualified in structural design.

    (ii) The single post shores shall be vertically aligned.

    (iii) The single post shores shall be spliced to prevent misalignment.

    (iv) The single post shores shall be adequately braced in two mutually perpendicular directions at the splice level. Each tier shall also be diagonally braced in the same two directions.

(9) Adjustment of single post shores to raise formwork shall not be made after the placement of concrete.

(10) Reshoring shall be erected, as the original forms and shores are removed, whenever the concrete is required to support loads in excess of its capacity.

(c) Vertical slip forms.

(1) The steel rods or pipes on which jacks climb or by which the forms are lifted shall be

    (i) Specifically designed for that purpose; and

    (ii) Adequately braced where not encased in concrete.

(2) Forms shall be designed to prevent excessive distortion of the structure during the jacking operation.

(3) All vertical slip forms shall be provided with scaffolds or work platforms where employees are required to work or pass.

(4) Jacks and vertical supports shall be positioned in such a manner that the loads do not exceed the rated capacity of the jacks.

(5) The jacks or other lifting devices shall be provided with mechanical dogs or other automatic holding devices to support the slip forms whenever failure of the power supply or lifting mechanism occurs.

(6) The form structure shall be maintained within all design tolerances specified for plumbness during the jacking operation.

(7) The predetermined safe rate of lift shall not be exceeded.

(d) Reinforcing steel.

(1) Reinforcing steel for walls, piers, columns, and similar vertical structures shall be adequately supported to prevent overturning and to prevent collapse.

(2) Employers shall take measures to prevent unrolled wire mesh from recoiling. Such measures may include, but are not limited to, securing each end of the roll or turning over the roll.

(e) Removal of formwork.

(1) Forms and shores (except those used for slabs on grade and slip forms) shall not be removed until the employer determines that the concrete has gained sufficient strength to support its weight and superimposed loads. Such determination shall be based on compliance with one of the following:

(i) The plans and specifications stipulate conditions for removal of forms and shores, and such conditions have been followed, or

(ii) The concrete has been properly tested with an appropriate ASTM standard test method designed to indicate the concrete compressive strength, and the test results indicate that the concrete has gained sufficient strength to support its weight and superimposed loads.

(2) Reshoring shall not be removed until the concrete being supported has attained adequate strength to support its weight and all loads in place upon it.

APPENDIX TO § 1926.703(a)(1) General requirements for formwork. (This Appendix is non-mandatory.)

This appendix serves as a non-mandatory guideline to assist employers in complying with the formwork requirements in § 1926.703(a)(1). Formwork which has been designed, fabricated, erected, braced, supported, and maintained in accordance with Sections 6 and 7 of the American National Standard for Construction and Demolition Operations-Concrete and Masonry Work, ANSI A10.9-1983, shall be deemed to be in compliance with the provision of § 1926.703(a)(1).

7.3.3 American National Standards Institute

The standard suggested in the above OSHA provisions as a means for achieving compliance is the American National Standard for Construction and Demolition Operations—Concrete and Masonry Work—Safety Requirements, ANSI A10.9-1983. Section 6, on vertical shoring, has provisions for loads and design, field practices, removal, tubular welded frame shoring, tube and coupler tower shoring, and single-post shores. Section 7, on formwork, has provisions for loads, formwork design, placing and removal of forms, vertical slip forms, flying-deck forms, and horizontal shoring beams. Most of the load and design provisions refer to the ACI Committee 347 *Guide to Formwork for Concrete* as the procedure to follow.

7.3.4 Scaffolding, Shoring, and Forming Institute Guides and Rules

This industry-supported institute publishes many relevant guides, rules, and training aids related to formwork and its support systems. Examples include the following:

Scaffolding Safety Guidelines
Guide to Scaffolding Erection & Dismantling Procedures
Guidelines for Safe Practices for Erecting & Dismantling Frame Shoring
Horizontal Shoring Beam Safety Rules
Single Post Shore Safety Rules
Rolling Shore Bracket Safety Rules
Recommended Steel Frame Shoring Erection Procedures
Guide to Horizontal Shoring Beam Erection Procedures for Stationary Systems
Guidelines for Safety Requirements for Shoring Concrete
Flying Deck Form Safety Rules
Scaffolding Safety Do's and Don'ts Slide Presentation
Shoring Safety Do's and Don'ts Slide Presentation
Forming Safety Do's and Don'ts Slide Presentation

7.3.5 American Society of Civil Engineers Standards Development

At the time of this writing, a committee of the American Society of Civil Engineers is developing a standard for loads imposed on structures during construction, including temporary structures. The proposed standard covers many topics, including loads for formwork, falsework, and shoring. The loads and pressures imposed by concrete, construction interval winds, and other sources are proposed to be defined. The proposed standard is in an advanced draft form at this writing and may become available for use if it successfully completes the standardization process. As with any voluntary consensus standard, it would then be available for groups to specify in individual contracts, or it could be cited as the standard for use by government agencies in rule making.

Although various guides and industry practices have been available for many years for the design of formwork, this standard would make available for the first time in the United States a document that defines loads in mandatory language that could be incorporated in contracts, codes, and safety laws. Thus the reader is advised to seek an update on the status of this standardization effort.

7.3.6 Other Standards and Recommendations for Pressures and Loads

Several international sources can supplement information and methods for calculating loads and pressures on formwork. The Construction Industry Research and Information Association (CIRIA) of England has conducted studies and published a report with recommended methods for calculating concrete lateral pressures on formwork. CIRIA Report 108, *Concrete Pressure on Formwork* [Clear and Harrison, 1985], is noteworthy in that it contains pressure modification factors covering use of various admixtures as well as slag and fly ash. Similarly, the Canadian Standards Association has published a standard, Concrete Formwork, CAN/CSA-S269.3-M92 (1992), covering design loads for formwork.

7.3.7 Information from Formwork Suppliers

Formwork suppliers of course publish literature defining their products and limitations for their use. Data sheets indicate the load ratings of products such as form ties, shores, braces, wales, etc., for reference by form designers of job-built systems.

In addition, many form manufacturers supply pre-engineered formwork systems (Figure 7.25) under rental or purchase arrangements for individual projects or general use. Such systems will come with rules and limitations whose implementation is required by the contractor as part of the rental or sales agreement. It is critical that these requirements be implemented as the normal procedures for use and for training workers involved with the formwork.

7.3.8 General Material Design Standards

Although they do not specifically address formwork, the same criteria used for design of permanent structures in any particular material should also be applied to the design of formwork once the loads and safety factors appropriate for formwork have been determined. Most forms are made of wood, steel, or aluminum, or some combination of these materials. The usual reference design standards for these materials are as follows for

FIGURE 7.25 Pre-engineered wall form system.

allowable stress design:

> *National Design Specification for Wood Construction*, ANSI/NFoPA NDS-1991. Revised 1993. American Forest & Paper Association.
>
> *Manual of Steel Construction—Allowable Stress Design*, 9th ed. 1993. American Institute of Steel Construction.
>
> *Specifications for Aluminum Structures, Construction Manual Series*. 1986. The Aluminum Association.
>
> *Plywood Design Specification*. 1997 ed. American Plywood Association.

The use of load and resistance factor design (LRFD) methods for formwork is currently limited, but the application of this approach is expected to increase. Applicable standards for LRFD design include the following:

> *Standard for Load and Resistance Factor Design (LRFD) for Engineered Wood Construction*, ASCE/AFPA 16-95. 1995. American Forest & Paper Association/American Society of Civil Engineers.
>
> *Manual of Steel Construction—Load and Resistance Factor Design*, 2nd ed. 1993. American Institute of Steel Construction.

Addresses for many of the above sources of standards and specifications are given following the references at the end of this chapter.

7.4 Loads and Pressures on Formwork

The possible loads acting on formwork are many. Vertical loads are usually associated with the dead load of the placed concrete and formwork and the live load of workers and their equipment. Internal pressures on vertical formwork result from the liquid or semiliquid behavior of the fresh concrete. External forces such as wind exert horizontal loads on the forms, requiring bracing systems for lateral form stability.

7.4.1 Vertical Loads

Vertical loads acting on formwork (Figure 7.26) include the self-weight of the forms, the placed reinforcement, the weight of fresh concrete, the weight of the workers, and the weight of placing equipment and tools. The dead load of the concrete is usually estimated at 145 to 150 lb/ft^3 including allowance for normal amounts of reinforcement. In cases where the reinforcement appears to be

FIGURE 7.26 Weights of concrete, reinforcing, and formwork contribute to dead load.

heavy, the materials should be calculated separately to determine the actual unit weight. Adjustments are also made for lightweight concrete densities. An 8 in normal-weight slab thus imposes a dead load of 150 lb/ft^3 × 8 in/12 in/ft or 100 psf on horizontal formwork.

ACI Committee 347 (1994) recommends that horizontal formwork be designed for a minimum vertical live load of 50 psf to allow for workers and their incidental placing tools such as screeds, vibrators, and hoses. When motorized carts or buggies are used, a minimum live load of 75 psf is recommended. Furthermore, it is recommended that the minimum combined total design dead and live load should be no less than 100 psf, or 125 psf if motorized buggies are used.

Formwork self-weight can be calculated from the unit weights and sizes of the components. The weight of the formwork is often much less than the dead load of the concrete and the construction live loads. Thus, during design, an allowance is frequently made for the form components as a superimposed load per square foot. Since forms often weigh 5 to 15 psf, an initial estimate is made in this range based upon experience and then is verified after the members are sized.

7.4.2 Lateral Pressures of Concrete

Vertical formwork, such as that for walls and columns, is subjected to internal lateral pressure from the accumulated depth of concrete placed. In a placement, fresh concrete, at least near the top and sometimes at greater depths, behaves as a liquid during vibration and generates lateral pressures equal to the vertical liquid head. Although concrete is a granular material with internal friction, the fluidization of the concrete resulting from internal vibration at least temporarily creates a liquid state. However, many factors appear to contribute to the lateral pressures being less than a liquid head at depths below the controlled depth of vibration. If the vertical placement rate is slow, the concrete mass below may have time to begin stiffening. If the concrete is warm, this stiffening may begin earlier. Internal concrete granular friction, form friction, migration of pore water, and other factors may also reduce the resulting lateral pressures. Admixtures, different types of cements, cement substitutes, and construction practices also affect the level of lateral pressure.

Tests have indicated that the pressures often have a distribution, as indicated in Figure 7.27, starting as a liquid pressure near the top and reaching a maximum at some lower level. For

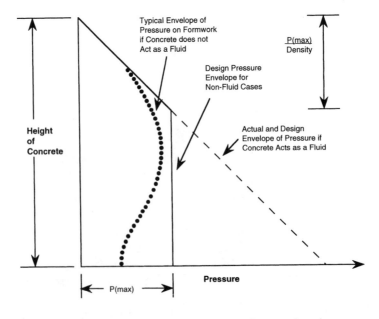

FIGURE 7.27 Typical and assumed distributions of concrete lateral pressure.

simplicity, design practice usually assumes that the maximum pressure is uniform at a conservative value.

ACI Committee 347 (1994) recommends calculation of the pressure magnitude versus depth as a full liquid head

$$p = wh \qquad (7.1)$$

where

p = lateral pressure of concrete, pounds per square foot;
w = unit weight of fresh concrete, pounds per cubic foot; and
h = depth of fluid or plastic concrete, feet.

However, ACI Committee 347 further states that for concrete made with Type I cement that weighs 150 lb/ft^3, contains no pozzolans or admixtures, has a slump of 4 in or less, and has normal internal vibration to a depth of 4 ft or less, the formwork may be designed for a lateral pressure as follows, where R is the rate of vertical placement, feet per hour; and T is the temperature of concrete, in degrees Farenheit.
For walls with placement, R, less than 7 ft/hr, the maximum pressure is

$$p = 150 + 9000 \frac{R}{T} \qquad (7.2)$$

with a maximum of 2000 psf, and a minimum of 600 psf but in no case greater than 150h.
For walls with placement, R, of 7 to 10 ft/hr, the maximum pressure is

$$p = 150 + \frac{43{,}400}{T} + 2800 \frac{R}{T} \qquad (7.3)$$

with a maximum of 2000 psf, and a minimum of 600 psf but in no case greater than 150h.
For column forms, the maximum pressure is

$$p = 150 + 9000 \frac{R}{T} \qquad (7.4)$$

with a maximum of 3000 psf, and a minimum of 600 psf but in no case greater than 150h.

Caution should be used if a concrete form is filled by pumping upward from the base of the form. Not only can a liquid head be developed, but a portion of the pump surcharge pressure is likely to be developed due to friction and drag resistance of the concrete moving upward through the form and reinforcing.

The limitations upon the use of the reduced pressure equations above are very restrictive. Excluding mixes with admixtures, other than Type I cement, cement substitutes such as fly ash and slag, etc., virtually make it necessary to design for a full liquid head under the current ACI 347R-94 (1994) guidance. These limitations have developed owing to insufficient test data to date to extend the range of the equations or modify them. Alternatively, ACI 347R-94 provides that a method based on appropriate experimental data and reference sources such as the CIRIA Report 108 [Clear and Harrison, 1985] may be used to determine the lateral pressure used for form design.

Internal concrete lateral pressures are usually resisted in wall forms by transferring the pressures through beamlike members (plywood, studs, and wales) to tension "ties" linking the two wall-form sides (Figure 7.28). Since tension elements are the most efficient structural members, this is also the most cost-effective solution. The internal pressures in most column forms are transferred to external tie elements on adjacent faces of the form that serve as links between the opposite sides of a square or rectangular column form. Circular columns have a great forming advantage in that the column-form skin can act as a hoop, resisting the pressures in tension.

However, although these are effective methods of resisting internal pressures of the concrete, separate resisting elements must be provided to resist external horizontal loads that tend to overturn wall, column, and slab forms.

7.4.3 Horizontal Loads

Horizontal loads from such forces as wind, seismic activity, cables, inclined supports, inclined dumping of concrete, and starting or stopping of equipment must be resisted by braces and shores. ACI Committee 347 (1994) recommends the following minimum loads for design to prevent lateral collapse of the formwork.

FIGURE 7.28 Tall wall form with internal pressures resisted by ties and wind resisted by external braces.

> For elevated floor formwork in building construction, the horizontal load, w, for design in any direction at each floor line should not be less than 100 lb per linear foot of floor edge or 2% of the total dead load on the form distributed as a uniform load per foot of slab edge, whichever is greater (see Figure 7.29).
>
> For wall forms bracing should be designed to meet the minimum wind load requirements of ANSI A58.1 or the local building code, whichever is more stringent. The minimum wind design pressure, q, should not be less than 15 psf and bracing should be designed for at least $w = 100$ lb per linear foot of wall, applied at the top.

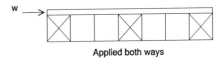

FIGURE 7.29 Schematic bracing of slab formwork.

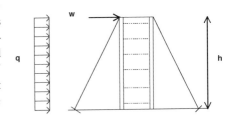

FIGURE 7.30 Schematic bracing of wall formwork.

In Figure 7.30, the minimum lateral load, w, for design of the bracing system would be the greater of $q \times h/2$ or 100 lb/ft.

The formwork designer must also be alert for other conditions such as post-tensioning operations, unusual geometry, or sequence of construction or operations that may create special loads on the forming system.

7.5 Formwork Design Criteria

Formwork components can be designed and constructed in many materials, such as plywood, wood, steel, aluminum, and fiber composites. Frequently, a mixture of materials is used (Figure 7.31). Steel, aluminum, and fiber composites are more likely to be parts of manufactured components or systems

FIGURE 7.31   Wood joists supported on steel beams and steel scaffolding towers.

that are rated or designed by the producer and may be supplied predesigned on a rental basis for the project needs. Forms intended to be job built are often made of wood and require design by the construction engineer associated with the project or by a consultant to the contractor. The examples in this chapter will illustrate the latter case for wood components designed by allowable stress methods.

In order to understand the examples, it is necessary to provide some of the essentials of wood design. The reader undertaking design of formwork in wood is advised to obtain and follow the more comprehensive specifications in the *National Design Specification for Wood Construction* [AFPA-NDS, 1993] and the *Plywood Design Specification* [APA-PDS, 1997].

Most of the lumber used in formwork is surfaced on four sides (S4S) to achieve its final dimensions as shown in Table 7.1. The S4S dimensions are smaller than the nominal sizes referred to in the table. Except for classification purposes, it is the actual dimensions and actual section properties that are used in design. A second set of sizes known as rough lumber (not shown in Table 7.1) has slightly larger dimensions but is still not the full nominal size. Rough lumber sizes are sometimes used in heavy falsework-supporting forms.

Plywood is frequently used as the surface layer of the formwork in contact with the fresh concrete. Plywood has different strengths and stiffness depending upon the direction of its span relative to the direction of the grain in the outer layers. The equivalent section, considering the varying elastic

TABLE 7.1   Example Section Properties of Standard Dressed (S4S) Sawn Lumber*

Nominal Size	Standard Dressed Size (S4S) $b \times d(X\text{-}X)$ $d \times b(Y\text{-}Y)$, in $\times$ in	Area of Section, A, in^2	X-X Axis		Y-Y Axis		Approximate Weight (lb/ft) When Wood Density Equals the Following Weights		
			Section Modulus, S_{xx}, in^3	Moment of Intertia, I_{xx}, in^4	Section Modulus, S_{yy}, in^3	Moment of Inertia, I_{yy}, in^4	25 lb/ft^3	30 lb/ft^3	35 lb/ft^3
× 4	1-1/2 × 3-1/2	5.250	3.063	5.359	1.313	0.984	0.911	1.094	1.276
× 6	1-1/2 × 5-1/2	8.250	7.563	20.80	2.063	1.547	1.432	1.719	2.005
× 8	1-1/2 × 7-1/4	10.88	13.14	47.63	2.719	2.039	1.888	2.266	2.643
× 10	1-1/2 × 9-1/4	13.88	21.39	98.93	3.469	2.602	2.409	2.891	3.372
× 4	3-1/2 × 3-1/2	12.25	7.146	12.51	7.146	12.51	2.127	2.522	2.977
× 6	3-1/2 × 5-1/2	19.25	17.65	48.53	11.23	19.65	3.342	4.010	4.679
× 8	3-1/2 × 7-1/4	25.38	30.66	111.1	14.80	25.90	4.405	5.286	6.168

*Abbreviated list of properties of selected sizes; see [AFPA-NDS, 1993] for additional data.

modulus and strength between parallel-to-grain loading and side-grain loading, is illustrated by the equivalent sections in Figure 7.32. When the grain of
the outer layers is parallel to the span direction, the
strength and stiffness is greatest (Figure 7.33). Many
types of plywood are available. Section properties
for B-B Plyform Class I plywood, one of the most
frequently selected types for moderate reuse in formwork, are given in Table 7.2.

Note that owing to the alternating grain directions
in the plywood veneer layers, conventional methods
for calculating section properties of homogeneous,
isotropic sections do not apply. The section properties given in Table 7.2 have been determined by considering the varying properties in the different layers
as well as the complications of weakness induced by
the tendency of fibers to roll over each other in shear
lateral to the grain, or **rolling shear** (Figure 7.34).
For these reasons, use the listed value of S only in
bending calculations; use I only for deflection calculations, and use Ib/Q, the rolling-shear constant, for
shear calculations.

The basic design values for wood and for plywood
of the species, grades, sizes, and types frequently used
in formwork are listed in Table 7.3. The species and
grades readily available in the area of the project
should always be verified. Contractors also often
have stocks of form lumber for reuse from previous
projects. Such lumber should always be inspected for
defects as the material is assembled and unsuitable
pieces must be rejected.

7.5.1 Adjustment Factors for Lumber Stresses

The AFPA-NDS (1993) provides for adjustment of
the lumber **basic design values** (F), such as those
given in Table 7.3, by a series of multipliers yielding
the **allowable design values** (F') for stress as follows:

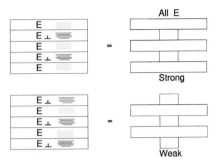

FIGURE 7.32 Plywood equivalent sections recognizing the weakness of lateral modulus.

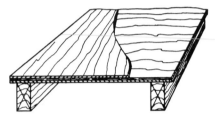

Plywood Used The Weak Way

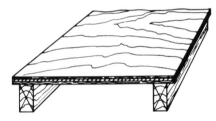

Plywood Used The Strong Way

FIGURE 7.33 Plywood load capacity and
stiffness varies with direction of face grain
span. *Source:* Courtesy of American Concrete Institute. Hurd, M.K. 1995. *Formwork
for Concrete.* Publication SP-4.

Bending

$$F'_b = F_b \times C_D \times C_M \times C_L \times C_F \times C_{fu} \times C_r \times [C_t \times C_V \times C_c \times C_f] \quad (7.5)$$

Shear

$$F'_v = F_v \times C_D \times C_M \times C_H \times [C_t] \quad (7.6)$$

Bearing

$$F'_{c\perp} = F_{c\perp} \times C_M \times C_b \times [C_t] \quad (7.7)$$

TABLE 7.2   Example Effective-Section Properties* for Plywood—12 in Unit Width of B-B Plyform, Class I

Plywood Net Thickness, in	12-in Width, Used with Face Grain Parrallel to Span			12-in Width, Used with Face Grain Perpendicular to Span			Weight, psf
	Moment of Inertia, I, in^4	Effective Section Modulus, S, in^3	Rolling Shear Constant, Ib/Q, in^2	Moment of Inertia, I, in^4	Effective Section Modulus, S, in^3	Rolling Shear Constant, Ib/Q, in^2	
1/2	0.077	0.268	5.153	0.024	0.130	2.739	1.5
5/8	0.130	0.358	5.717	0.038	0.175	3.094	1.8
3/4	0.199	0.455	7.187	0.092	0.306	4.063	2.2
7/8	0.296	0.584	8.555	0.151	0.306	6.028	2.6
1	0.427	0.737	9.374	0.270	0.634	7.014	3.0

*Use listed S value in bending calculations and use I only in deflection calculations. Properties of B-B Plyform, Class I, taken from American Plywood Association (1994). Consult reference or manufacturer for other plywood types.

Compression

$$F'_c = F_c \times C_D \times C_M \times C_F \times C_P \times [C_t]$$ (7.8)

Elastic Modulus

$$E' = E' \times C_M \times [C_t \times C_T]$$ (7.9)

Some of the adjustment factors [in brackets] only apply to glue-laminated lumber (volume factor, C_V; curvature factor, C_C), to truss members (buckling stiffness factor, C_T), when the member depth is >12 in (form factor, C_f), or when the temperature is >100° F (temperature factor, C_t), and thus have only rare uses in formwork. The remaining factors are discussed below.

7.5.2 Load Duration Factor

The adjustment for load duration, C_D, reflects the ability of wood to exhibit increased strength under shorter periods of loading. The following values may be applied for the indicated cumulative durations:

$C_D = 0.9$ load duration > 10 yr
$C_D = 1.0$ 2 mo < load duration ≤ 10 yr
$C_D = 1.15$ 7 days < load duration ≤ 2 mo
$C_D = 1.25$ load duration ≤ 7 days
$C_D = 1.6$ wind/earthquake
$C_D = 2.0$ impact

For most formwork, an adjustment of $C_D = 1.25$ is applied. However, when the components are reused for longer cumulative durations, C_D should be appropriately reduced.

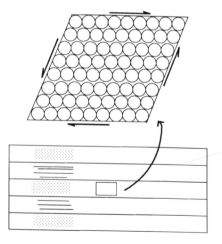

FIGURE 7.34   Grains of wood tend to roll when subjected to lateral shear.

7.5.3 Moisture Factor

Wood gains in strength as it loses moisture in a range below the fiber saturation point (about 30% moisture content). The basic design values are established for lumber that has a moisture content of 19% or less, typical of air-dried lumber. When the exposure is such that the wood moisture content will exceed 19% for an extended period of time, the design values should be multiplied by the C_M values indicated in Table 7.3.

TABLE 7.3 Example Basic Design Values for Visually Graded Dimension Lumber at 19% Maximum Moisture and Plywood Used Wet

Species, Specific Gravity, Grade, and Size Category	Extreme Fiber Bending, F_b, psi	Compression ⊥ to Grain, $F_{c\perp}$, psi	Compression ∥ to Grain, F_c, psi	Horizontal Shear, F_v, psi	Modulus of Elasticity, E, psi
Douglas fir-larch, $G = 0.50$					
No. 2, 2–4 in thick, 2 in and wider	875	625	1300	95	1,600,000
Construction, 2–4 in thick, 2–4 in wide	1000	625	1600	95	1,500,000
Douglas fir-south, $G = 0.49$					
No. 2, 2–4 in thick, 2 in and wider	825	520	1300	90	1,200,000
Construction, 2–4 in thick, 2–4 in wide	925	520	1550	90	1,200,000
Southern pine, $G = 0.55$	[Size-adjusted values]				
No. 2, 2–4 in thick, 2–4 in wide	1500	565	1650	90	1,600,000
No. 2, 2–4 in thick, 5–6 in wide	1250	565	1600	90	1,600,000
No. 2, 2–4 in thick, 8 in wide	1200	565	1500	90	1,600,000
Construction, 2–4 in thick, 4 in wide	1100	565	1800	100	1,500,000
Spruce-pine-fir, $G = 0.42$					
No. 2. 2–4 in thick, 2 in and wider	875	425	1100	70	1,400,000
Construction, 2–4 in thick, 4 in wide	975	425	1350	70	1,300,000
Hem-fir $G = 0.43$					
No. 2, 2–4 in thick, 2 in and wider	850	405	1250	75	1,300,000
Construction, 2–4 in thick, 2–4 in wide	975	405	1500	75	1,300,000
Adjustment factor, C_M, for Moisture content > 19% (Lumber)	0.85[†]	0.67	0.8[‡]	0.97	0.9
Adjustment factor, C_D, for Maximum load duration 7 d or less (Lumber and Plywood)	1.25	—	1.25	1.25	—
Other applicable adjustment Factors for Lumber	Temperature, size,* flat use, beam Stability, and repetitive member	Temperature, bearing length	Temperature, size,* Column Stability	Temperature, shear Stress	Temperature
Plywood Sheathing Used Wet: B-B Plyform, Class I	1545**	Face Bearing 210	—	Rolling shear 57**	1,500,000

Note: From recommendations of the American Forest and Paper Association (1993) and from recommendations of the American Plywood Association (1994) as formatted by Hurd (1995).

*Size adjustments apply to all lumber basic bending and compression parallel to the grain design values except southern pine. The size adjustments are already included in southern pine basic design values. Consult Table 7.4 for details of size adjustments.

**Plywood stresses include an experience factor of 1.3 recommended by the American Plywood Association.

[†]When size-adjusted bending stress is less than or equal to 1150 psi, no moisture adjustment is required.

[‡]When size-adjusted compression parallel is less than or equal to 750 psi, no moisture adjustment is required.

TABLE 7.4   Size and Flat-Use Adjustment Factors for Example Dimension Lumber Grades

| Grade | Width, in | Size Factor (all species except southern pine) | | | Flat-Use Factor | |
| | | Bending Stress, F_b | | Compression, F_C | Bending Stress, F_b | |
		2 and 3 in thick	4 in thick		2 and 3 in thick	4 in thick
No. 1 and No. 2	2 and 3	1.5	1.5	1.15	1.0	—
	4	1.5	1.5	1.15	1.1	1.0
	5	1.4	1.4	1.1	1.1	1.05
	6	1.3	1.3	1.1	1.15	1.05
	8	1.2	1.3	1.05	1.15	1.05
	10	1.1	1.2	1.0	1.2	1.1
	12	1.0	1.1	1.0	1.2	1.1
	14 and wider	0.9	1.0	0.9	1.2	1.1
Construction	2 and 3	1.1	1.1	1.05	1.0	—
	4	1.0	1.0	1.0	1.1	1.0

7.5.4 Shear Factor

The basic design values for horizontal shear stress have been established conservatively, since shrinkage and other cracks in wood can reduce the horizontal shear strength. Such cracks are sometimes found near the cut ends of members. In situations where cracking may be expected or the wood will not be inspected for end cracks during form fabrication, C_H should be taken as 1.0. However, when the amount of cracking is known, the value of C_H may be increased as tabulated in the AFPA-NDS (1993). When end-cracked or end-split members are disallowed by inspection, the value of C_H may be 2.0. For sawn lumber, at locations greater than five times the depth of the member, d, from a cut end (i.e., $>5d$), C_H may be taken as 2.0.

7.5.5 Size Factor

Tests indicate that member overall size affects the failure stress. To account for these variations, the size factor (C_F) as shown in Table 7.4 is applied to the bending and compression basic design values. Note that the size factor does not apply to the basic design values of southern pine, whose basic design values in Table 7.3 are preadjusted to reflect the size effect.

7.5.6 Flat-Use Factor

Lumber loaded on its wide face and bending about its weak axis (Y-Y) exhibits a slightly higher failure strength. To reflect these variations, the flat-use factor, C_{fu}, adjustments in Table 7.4 may be applied to the basic design values for bending stress.

7.5.7 Beam-Stability Factor

The AFPA-NDS (1996) provides equations for determining the beam-stability factor, C_L, an adjustment less than 1.0, when the compression edge of a beam may become unstable. However, for sawn lumber, the AFPA-NDS also provides prescriptive d/b ratios, based on nominal dimensions and lateral support conditions where the member may be assumed to be stable and no reduction for C_L is needed, as follows:

$d/b = 2$ to 1 or less	No lateral support is necessary.
$d/b = 3$ to 1 or 4 to 1	Ends shall be held in position against lateral rotation or displacement by blocking or connection to other members.
$d/b = 5$ to 1	One edge shall be held in line for the entire length.

$d/b = 6$ to 1   Bridging, blocking, or cross bracing shall be installed at intervals
                              not exceeding 8 ft unless both edges are held in line.
$d/b = 7$ to 1   Both edges shall be held in line for the entire length.

The wood bending members in most formwork can be made to meet these support criteria, and thus C_L can be taken as 1.0 in typical formwork situations. In cases where the lateral support is less than indicated above, further evaluation is needed.

7.5.8 Column-Stability Factor

The column-stability factor, C_P, is an adjustment less than 1.0 to reduce the allowable compression stress parallel to the grain when longer columnlike members such as shores or braces may fail in a buckling mode rather than by crushing. The factor is given by

$$C_P = \frac{1 + \left(F_{cE}/F_c^*\right)}{2c} - \sqrt{\left[\frac{1 + \left(F_{cE}/F_c^*\right)}{2c}\right]^2 - \frac{F_{cE}/F_c^*}{c}} \qquad (7.10)$$

in which

F_c^* = tabulated compression design value multiplied by all applicable adjustment factors except C_P
K_{cE} = 0.3 (for visually graded lumber and machine-evaluated lumber)
c = 0.8 (for sawn lumber)
$F_{cE} = K_{cE} E'/(l_e/d)^2$

and l_e/d is the larger of the slenderness ratios about the possible buckling axes. The value of l_e/d cannot normally exceed 50 and l_e is the effective length.

7.5.9 Bearing-Area Factor

The bearing-area factor, C_b, is for the case of bearing perpendicular to the grain of the wood, that is, bearing on the side grain. The bearing factor is normally taken as 1.0. However, if the bearing area is more than 3 in from the end of the member and less than 6 in length as measured along the grain, then the following increase factor may be applied:

$$C_b = \frac{l_b + 0.375}{l_b} \qquad (7.11)$$

where l_b is the length of bearing in inches measured parallel to the grain. Note that in formwork design, unless there is certainty that the bearing area will not be within 3 in of the end of the member, it is usually best to assume $C_b = 1.0$.

7.5.10 Repetitive-Use Factor

The AFPA-NDS (1993) includes a repetitive-use factor, C_r, that may be used to increase the bending design value when there are at least three members spaced not more than 24 in on center, such as joists and studs, and they are joined by a load-distributing member, such as sheathing. The increase is allowed because it is unlikely that normal defects would occur in the repetitive members at the same critical location and the load could be shared if one had a defect. However, in formwork design, ACI SP-4 [Hurd, 1995] suggests that this factor should only be applied to carefully constructed panels whose components are securely nailed or bolted together. The factor values, >1.0, are listed in the AFPA-NDS.

TABLE 7.5   Minimum Safety Factors of Formwork Accessories

Accessory	Safety Factor	Type of Construction
Form tie	2.0	All applications
Form anchor	2.0	Fromwork supporting form weight and concrete pressures only
	3.0	Formwork supporting form weight of forms, concrete, construction live loads, and impact
Form hanger	2.0	All applications
Anchoring inserts used as form ties	2.0	Precast concrete panels when used as formwork

7.5.11 Adjustment Factors for Plywood Stresses

Relative to formwork, the allowable stresses given by the American Plywood Association (1994) are subject to three primary adjustments: load duration, wet use, and experience factors. A load-duration factor similar to the value of C_D for wood is normally applied as listed in Table 7.3. The wet use and experience factors appropriate to formwork applications have been incorporated in the allowable stresses listed for B-B Plyform Class I plywood in Table 7.3. The calculation of plywood deflection can be refined to include both the bending deflection and the shear deflection. In this presentation, for simplicity, only the bending deflection will be considered, but the lower value of plywood elastic modulus will be used to partially compensate for this. Consult American Plywood Association (1994) for procedures for calculating the shear deflection component.

7.5.12 Manufactured Wood Products

Various manufactured wood products are sometimes used in formwork applications. These not only include sheathing but also include structural composite lumber. **Laminated veneer lumber** (LVL) is made of wood-veneer sheet elements 1/4 in thick or less, bonded with an exterior adhesive, with wood fibers oriented along the length of the member. **Parallel strand lumber** (PSL) is made of wood strands having a least dimension of 1/4 in or less and an average length of at least 150 times the least dimension, bonded with exterior adhesive, and with strands oriented along the length of the member. **Laminated strand lumber** (LSL) is made of wood strands with a least dimension of about 1/32 in, approximately 12 in long, parallel to the length of the member, and bonded with an exterior adhesive. The suitability of these materials for exposed formwork should be evaluated for each application. Design information can be obtained from the manufacturers or the manufacturers' associations.

7.5.13 Safety Factors for Formwork Accessories

Various accessories are used in formwork such as form ties, anchors, and hangers. Most of these devices are made of steel and are rated by the manufacturers. The ratings may be listed as either the allowable or ultimate strengths of the devices under certain types of loading. The type of rating (allowable or ultimate) should be carefully determined from the information provided. If ultimate strengths are listed, the allowable strength should be calculated by dividing by an appropriate factor of safety. ACI Committee 347 (1994) has recommended the safety factors listed in Table 7.5 for such accessories.

7.6 Formwork Design

Most components of a form system can be subdivided into members that are primarily bending elements (sheathing, joists, studs, stringers, and wales) and members that are primarily tension or compression elements (shores, braces, etc.). In addition, there are a host of details to design, such as

FIGURE 7.35   Connections at corners, braces, and bulkheads require specific design.

FIGURE 7.36   Wood timber grillage mudsill.

connections (Figure 7.35), hangers, and footings or mudsills (Figure 7.36). The following sections will provide example designs, in wood, of the main members of an elevated floor slab form and a vertical wall form to convey a sense of the procedures involved. After the design is complete, the formwork material specifications, member layout, member sizes, connection details, erection procedures, and use limitations should be conveyed by means of drawings with appropriate notes to the field workers who will fabricate and erect the form.

7.6.1 Determination of Resultants from Loads

The bending members of a wood form system are either single-span or continuous multiple-span elements, usually with bearing supports as illustrated in Figure 7.37. Although the members may sometimes have more than three spans, the benefits from considering more than three span conditions are very limited. Many of the member loads are uniform. In other cases, the loadings may be a series of closely spaced concentrated loads that can often be approximated as a uniform load if there are a sufficient number in a span. Figure 7.37 also provides the formulas for the maximum moments, shears, and deflections for the uniformly loaded one- two- and three-span cases.

It should be noted that the formulas for calculating the maximum shear force are modified to calculate the shear at a distance d from the face of the supporting member, where d is the depth of the member being designed and l_b is the length of bearing at the supporting member. In wood design, the AFPA-NDS provides that loads within a distance d of the face of the support can be

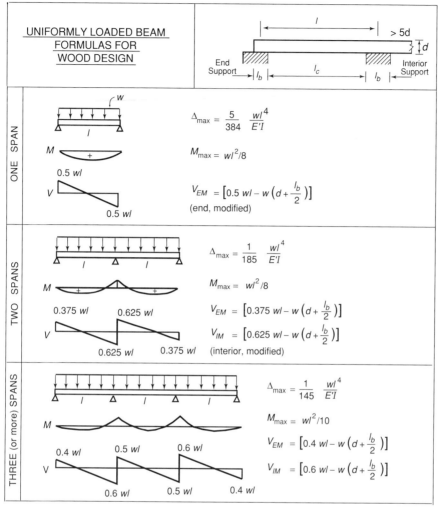

FIGURE 7.37 Beam formulas for one-, two-, and three-span conditions.

neglected when designing for shear if the member is loaded on one face and supported on the opposite face or edge. However, in cases where the member is notched or connection is made in the web, as by bolting, other AFPA-NDS special provisions should be consulted, which, in effect, magnify the shear force used for design.

7.6.2 Fundamental Relations Between Resultants and Stresses

From mechanics of solids, the following relationships apply to the elastic design of wood elements.

Bending of beams or plywood:

$$f_b = \frac{M}{S} \qquad (7.12)$$

Shear of solid rectangular beams:

$$f_v = \frac{3}{2}\frac{V}{bd} \tag{7.13}$$

Shear of plywood:

$$f_v = \frac{VQ}{Ib} = \frac{V}{Ib/Q} \tag{7.14}$$

where Ib/Q is the rolling shear constant.

Bearing and axial compression:

$$f_{c\perp} \quad \text{or} \quad f_c = \frac{P}{A} \tag{7.15}$$

When f is used in the above equations, the actual stress is sought in the calculation from the actual resultant. When F is used, the maximum allowable stress is implied and the maximum resultants are sought. Load-deflection relationships are given in Figure 7.37 as a function of the number of spans.

7.6.3 Basis of Examples

When undertaking a form design, some parameters are frequently known, and others need to be calculated to meet the required strength needs and deflection limitations. For example, the spacings of members may be dictated by the geometry of the area to be formed or the modular needs of the system or plywood facing. For this case, the unknowns become the needed member sizes. In another case, the contractor may have material of a given size that is desired to be used. For the latter case, the unknowns are the maximum spacings of the members needed to limit the loads on each to its allowable value.

In the following elevated slab form example, the spacings will be assumed to be set by job conditions and the members will be the unknowns. In the wall form design example, the member sizes will be assumed to be predetermined by available material and the spacings will be the unknowns.

7.7 Slab-Form Design Example

Assume that an elevated flat plate floor slab of 8 in thickness is to be formed with a top elevation 10 ft above the supporting surface below. The general layout of the plywood sheathing, joists, stringers, and shores is shown in Figure 7.9. Because of the floor layout, the contractor desires to space the joists at 16 in on center, the stringers at 5 ft on center under the joists, and the shores at 5 ft on center under the stringers. The materials are Plyform B-B Class I plywood and No. 2 spruce-pine-fir (SPF) joists, stringers, and shores. It has been determined that the strength gain of the concrete will allow the forms to be stripped in 7 days, followed by installation of reshores. The project specification requires that the formwork element deflections shall be no greater than $l/360$. The plywood will be assumed to be used wet since it is frequently exposed for a lengthy period during rebar placement and in contact with the fresh concrete. However, the lumber elements will be assumed to be relatively sheltered and not exposed to significant moisture for lengthy periods. The plywood will be oriented with its face grain parallel to the span direction, that is, the strong way.

From the earlier section on vertical loads, the distributed pressure, q, for design on a working basis is

Dead load of 8 in concrete slab	100 psf
Construction live load	50 psf
Formwork estimated dead load	5 psf
Total	155 psf

7.7.1 Sheathing Design

The plywood span is the spacing of the joists, and 8 ft long plywood panels can typically have three or more spans. The plywood allowable stresses adjusted for load duration (see Table 7.3) are

$$F_b' = F_b \times C_D = 1545 \text{ psi} \times 1.25 = 1931 \text{ psi}$$

$$F_v' = F_v \times C_D = 57 \text{ psi} \times 1.25 = 71 \text{ psi}$$

$$E' = E = 1,500,000 \text{ psi}$$

Considering the loading to be a uniform load, w, of 1 ft width corresponding to the section properties of the plywood based on 12 in width

$$w = q \times \text{unit width} = 155 \text{ psf} \times 1.0 \text{ ft} = 155 \text{ lb/ft}.$$

For the limitation of flexural bending, the maximum moment, M, is

$$M = \frac{wl^2}{10} = \frac{155 \text{ lb/ft} \times (1.33 \text{ ft})^2}{10} = 27.5 \text{ ft-lb} = 300 \text{ in-lb}$$

and the required section modulus, S, is given by

$$S = \frac{M}{F_b'} = \frac{300 \text{ in-lb}}{1931 \text{ psi}} = 0.171 \text{ in}^3$$

For the shear limitation, the maximum shear force for design is determined at the center of the support since the support width is not yet known.

$$V = 0.62wl = 0.6(155 \text{ lb/ft})(16 \text{ in}/12 \text{ in/ft}) = 124 \text{ lb}$$

$$\frac{Ib}{Q} = \frac{V}{F_v'} = \frac{124 \text{ lb}}{71 \text{ psi}} = 1.74 \text{ in}^3$$

The deflection limit is determined from

$$\frac{l}{360} = \frac{1}{145} \frac{wl^4}{EI}$$

$$I = \frac{360}{145} \frac{155 \text{ lb/ft}}{12 \text{ in/ft}} \frac{(16 \text{ in})^3}{1,500,000 \text{ psi}} = 0.0876 \text{ in}^4$$

From these requirements 1/2 in plywood would be adequate for bending and shear, but 5/8 in is needed for deflection. Select the plywood thickness to be 5/8 in. Since deflection controls, there is no benefit to further refining the shear calculation after the joist width is known.

7.7.2 Joist Design

The joist span is the 5-ft spacing of the stringers, and the joists can be constructed with a three-span continuous arrangement (Figure 7.38). Lateral buckling of the joists will be restrained by nailing the sheathing to the joists. For selection of the joist basic design values (Table 7.3),

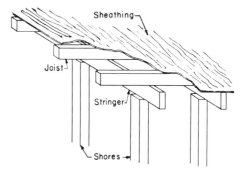

FIGURE 7.38 Slab-formwork layout. *Source:* Courtesy of American Concrete Institute Hurd, M.K., 1995, *Formwork for Concrete*, Publication SP-4.

assume initially that a 2 × 6 joist might work. Allowable stresses for the No. 2 S-P-F joist are as follows:

$$F_b' = F_b \times C_D \times C_M \times C_L \times C_F \times C_r$$
$$= 875 \text{ psi} \times 1.25 \times 1.0 \times 1.0 \times 1.3 \times 1.0 = 1421 \text{ psi}$$

$$F_v' = F_v \times C_D \times C_M \times C_H$$
$$= 70 \text{ psi} \times 1.25 \times 1.0 \times 1.0 = 87.5 \text{ psi} \quad \text{(within } 5d \text{ of the ends)}$$

$$F_v' = F_v \times C_D \times C_M \times C_H$$
$$= 70 \text{ psi} \times 1.25 \times 1.0 \times 2.0 = 175 \text{ psi} \quad (>5d \text{ from the ends)}$$

$$E' = E \times C_M = 1,400,000 \text{ psi} \times 1.0 = 1,400,000 \text{ psi}$$

Considering the loading to be a uniform load, w, of 16 in tributary width corresponding to the spacing of the joists

$$w = q \times \text{unit width} = 155 \text{ psf} \times 1.33 \text{ ft} = 207 \text{ lb/ft}.$$

For the limitation of flexural bending, the maximum moment, M, is

$$M = \frac{wl^2}{10} = \frac{207 \text{ lb/ft} \times (5 \text{ ft})^2}{10} = 517 \text{ ft-lb} = 6210 \text{ in-lb}$$

and the required section modulus, S, is given by

$$S = \frac{M}{F_b'} = \frac{6210 \text{ in-lb}}{1421 \text{ psi}} = 4.37 \text{ in}^3$$

For the shear limitation, the maximum shear force for design is determined at d from the center of the support since the supporting stringer width is not yet known. Both the joist end support and the joist interior support located $>5d$ from a cut end are examined since different shear forces as well as different allowable stresses apply.

At the ends

$$V = 0.4wl - wd = 0.4(207 \text{ lb/ft})(5 \text{ ft}) - 207 \text{ lb/ft}\left(\frac{5.5}{12} \text{ ft.}\right) = 319 \text{ lb}$$

$$bd = \frac{3}{2} \frac{V}{F_v'} = \frac{3 \times 319 \text{ lb}}{2 \times 87.5 \text{ psi}} = 5.47 \text{ in}^2$$

At the interior support

$$V = 0.6wl - wd = 0.6(207 \text{ lb/ft})(5 \text{ ft}) - 207 \text{ lb/ft}\left(\frac{5.5}{12} \text{ ft.}\right) = 526 \text{ lb}$$

$$bd = \frac{3}{2} \frac{V}{F_v'} = \frac{3 \times 526 \text{ lb}}{2 \times 175 \text{ psi}} = 4.51 \text{ in}^2$$

The deflection limit is determined from

$$\frac{l}{360} = \frac{1}{145} \frac{wl^4}{EI}$$

$$I = \frac{360}{145} \frac{207 \text{ lb/ft}}{12 \text{ in/ft}} \frac{(60 \text{ in})^3}{1,400,000 \text{ psi}} = 6.61 \text{ in}^4$$

From these combined requirements, 2×6 joist is the optimum size having the least area. The result also agrees with the size assumption for selecting the allowable stresses and size factor. If the size did not agree, the calculations would be repeated with an improved size assumption. In this case, there is no benefit to further refining the shear calculation after the joist width is known since shear alone is not controlling the size.

7.7.3 Stringer Design

The stringer span is the 5-ft spacing of the shores, and the stringers can be constructed with a three-span continuous arrangement. For selection of the stringer basic design values (Table 7.3), assume initially that a 4×6 stringer might work. Lateral buckling of a stringer of this size would not be a problem since $d/b \leq 2$ to 1. Allowable stresses for the No. 2 SPF stringer are as follows:

$$F_b' = F_b \times C_D \times C_M \times C_L \times C_F \times C_r$$
$$= 875 \text{ psi} \times 1.25 \times 1.0 \times 1.0 \times 1.3 \times 1.0 = 1421 \text{ psi}$$

$$F_v' = F_v \times C_D \times C_M \times C_H$$
$$= 70 \text{ psi} \times 1.25 \times 1.0 \times 1.0 = 87.5 \text{ psi} \quad \text{(within } 5d \text{ of the ends)}$$

$$F_v' = F_v \times C_D \times C_M \times C_H$$
$$= 70 \text{ psi} \times 1.25 \times 1.0 \times 2.0 = 175 \text{ psi} \quad (>5d \text{ from the ends)}$$

$$E' = E \times C_M = 1,400,000 \text{ psi} \times 1.0 = 1,400,000 \text{ psi}$$

The stringer loading is actually a series of concentrated loads from the joists at 16 in on center. However, the starting position of the loads can vary in each span. Owing to the complications of considering many possible starting positions and recognizing that loads within a distance d of the support can be neglected for shear calculations, it is often the practice in formwork to assume a uniform load as being adequately similar [Hurd, 1995]. This assumption works reasonably well when there are three or more equally spaced concentrated loads of equal magnitude in the span. Considering the loading to be a uniform load, w, of 5 ft tributary width corresponding to the spacing of the stringers

$$w = q \times \text{unit width} = 155 \text{ psf} \times 5 \text{ ft} = 775 \text{ lb/ft}.$$

For the limitation of flexural bending, the maximum moment, M, is

$$M = \frac{wl^2}{10} = \frac{775 \text{ lb/ft} \times (5 \text{ ft})^2}{10} = 1938 \text{ ft-lb} = 23,250 \text{ in-lb}$$

and the required section modulus, S, is given by

$$S = \frac{M}{F_b'} = \frac{23,250 \text{ in-lb}}{1421 \text{ psi}} = 16.36 \text{ in}^3$$

For the shear limitation, the maximum shear force for design is determined at d from the face of the support, assuming the shore will be at least 4×4 or have a head of equal or larger size if it is a metal shore. Both the stringer end support and the stringer interior support located $>5d$ from a cut end are examined since different shear forces as well as different allowable stresses apply.

At the ends

$$V = 0.4wl - w\left(d + \frac{l_b}{2}\right)$$

$$= 0.4(775 \text{ lb/ft})(5 \text{ ft}) - 775 \text{ lb/ft}\left(\frac{5.5}{12} + \frac{3.5}{2 \times 12}\text{ft}\right) = 1082 \text{ lb}$$

$$bd = \frac{3}{2}\frac{V}{F'_v}$$

$$= \frac{3 \times 1082 \text{ lb}}{2 \times 87.5 \text{ psi}} = 18.55 \text{ in}^2$$

At the interior support

$$V = 0.6wl - w\left(d + \frac{l_b}{2}\right)$$

$$= 0.6(775 \text{ lb/ft})(5 \text{ ft}) - 775 \text{ lb/ft}\left(\frac{5.5}{12} + \frac{3.5}{2 \times 12}\text{ft}\right) = 1857 \text{ lb}$$

$$bd = \frac{3}{2}\frac{V}{F'_v}$$

$$= \frac{3 \times 1865 \text{ lb}}{2 \times 175 \text{ psi}} = 15.92 \text{ in}^2$$

The deflection limit is determined from

$$\frac{l}{360} = \frac{1}{145}\frac{wl^4}{EI}$$

$$I = \frac{360}{145}\frac{775 \text{ lb/ft}}{12 \text{ in/ft}}\frac{(60 \text{ in})^3}{1,400,000 \text{ psi}} = 24.73 \text{ in}^4$$

From these requirements, a 4 × 6 stringer is the optimum size meeting all the requirements and having the least area. The result also agrees with the size assumption for selecting the allowable stresses and size factor. If the size did not agree, the calculations would be repeated with an improved size assumption.

A check is also needed to make sure that there is adequate bearing area for the joist on the stringer. The tributary load is

$$P = 155 \text{ psf} \times 1.33 \text{ ft} \times 5 \text{ ft} = 1031 \text{ lb}$$

The allowable bearing stress is given by

$$F'_{c\perp} = F_{c\perp} \times C_M \times C_b = 425 \text{ psi} \times 1.0 \times 1.0 = 425 \text{ psi}$$

and the actual stress is

$$f_{c\perp} = \frac{P}{A} = \frac{1031 \text{ lb}}{1.5 \text{ in} \times 3.5 \text{ in}} = 196 \text{ psi} < 425 \text{ psi}$$

7.7.4 Shore Design

To determine the shore unbraced height, the slab thickness and depth of forming elements are subtracted from the floor-to-floor height.

$$l_e = 120 - 8 - 0.625 - 5.5 - 5.5 \text{ in} = 100.4 \text{ in}$$

For design purposes, the shore is assumed to be concentrically loaded, pinned at the top and the bottom, and prevented from translation by an adequate overall form diagonal bracing system. The load to be supported based upon the tributary area is

$$P = 155 \text{ psf} \times 5 \text{ ft} \times 5 \text{ ft} = 3875 \text{ lb}$$

Two stress checks are needed to size the shore: the bearing perpendicular to the grain of the stringer at the upper end of the shore and the compression parallel to the grain of the shore (including the possibility of buckling). Evaluation of bearing first will establish a first trial size of the shore area. The allowable bearing stress is

$$F'_{c\perp} = F_{c\perp} \times C_M \times C_b = 425 \text{ psi} \times 1.0 \times 1.0 = 425 \text{ psi}$$

C_b is assumed to be 1.0 since the bearing length is not yet known and since the shore may sometimes be located at the end of the stringer. The needed bearing area is given by

$$A = \frac{P}{F_{c\perp}} = \frac{3875 \text{ lb}}{425 \text{ psi}} = 9.12 \text{ in}^2$$

A 4×4 shore is adequate for bearing and matches the width of the 4×6 stringer. For this 4×4 shore size, the allowable compression stress is

$$F'_c = F_c \times C_D \times C_M \times C_F \times C_P$$
$$= 1100 \text{ psi} \times 1.25 \times 1 \times 1.15 \times C_P = 1581 \text{ psi} \times C_P = F^*_c \times C_P$$

Since the trial shore is square and braced only at the ends to resist translation, the l_e/d ratios for each buckling axis are equal; thus

$$\frac{l_e}{d} = \frac{100.4 \text{ in}}{3.5 \text{ in}} = 28.68 < 50$$

and

$$F_{cE} = \frac{K_{cE} E'}{(l_e/d)^2} = \frac{0.3 \times 1,400,000 \text{ psi}}{(28.68)^2} = 510 \text{ psi}$$

so that

$$\frac{F_{cE}}{f^*_c} = \frac{510 \text{ psi}}{1581 \text{ psi}} = 0.323$$

and, C_P from equation 7.10 is

$$C_p = \frac{1 + 0.323}{2 \times 0.8} - \sqrt{\left[\frac{1 + 0.323}{2 \times 0.8}\right]^2 - \frac{0.323}{0.8}} = 0.298$$

Thus the allowable stress in compression parallel to the grain for the unbraced length is

$$F'_c = 1581 \text{ psi} \times C_P = 1581 \text{ psi} \times 0.298 = 470 \text{ psi}$$

Since $F'_c > F_{c\perp}$, compression perpendicular to the grain in bearing controls the design and the 4×4 shore is adequate.

In summary, the design requires 5/8 in Plyform spanning the strong way, 2×6 joists at 16 in on center, 4×6 stringers at 5 ft on center, and 4×4 shores at 5 ft on center each way.

7.7.5 Bracing Design Considerations

The slab form must be braced, as a minimum, to resist the horizontal loads recommended by ACI 347-94. When the concrete columns have been placed prior to erection of the floor forms, the column can contribute to lateral stability if the form has an adequate horizontal diaphragm and is tied to the columns. Horizontal cross bracing can be added to improve the diaphragm. Vertical cross bracing in two directions at right angles, in combination with an adequate diaphragm, can also provide a workable system. The braces need not be located in the opening between every pair of shores (Figure 7.29). Often they are located between alternate pairs of shores (and sometimes farther apart) in a well dispersed, symmetrical pattern in each direction. The brace size is usually controlled by the buckling resistance in the compression direction of force, and the analysis and design proceeds in a manner similar to the shore or wall-form brace design. It is critical to provide a connection at the top and bottom ends of the brace that is adequate to resist the forces imposed.

7.8 Wall-Form Design Example

Assume that a form is needed for the placement of a 12 ft high wall. The general layout of the plywood sheathing, studs, and wales is shown in Figure 7.39. Because of the availability of the material from previous projects, the contractor wishes to use 3/4 in B-B Plyform Class I plywood for the sheathing and No. 2 southern pine 2×4 studs and double 2×4 wales. Ties can be ordered to meet project needs. The project specification requires that the formwork element deflections be no greater than $l/240$. The plywood will be assumed to be used wet since it is in contact with the fresh concrete and the lumber elements will be assumed to be exposed to sufficient rain moisture on the job such that the wood-moisture content might rise to above 19%. The normal-weight concrete, made with Type I cement, has a slump <4 in and contains no pozzolans or admixtures. Internal vibration during placement will be controlled and limited to an immersion depth <4 ft. The rate of vertical placement, R, will be limited to 4 ft per hour and the concrete temperature, T, is expected to be above a minimum of 70°F.

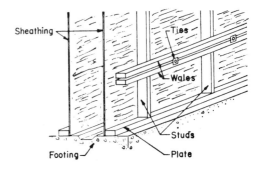

FIGURE 7.39 Wall form components. *Source:* Courtesy of American Concrete Institute. Hurd, M.K. 1995. *Formwork for Concrete*, Publication SP-4.

From the earlier section on lateral pressures, the maximum distributed pressure, p, for design on a working basis is given by Eq. (7.2).

$$p = 150 + 9000\frac{R}{T} = 150 + \frac{9000 \times 4}{70} = 664 \text{ psf}$$

The pressure is assumed to gradually increase with depth at a rate of 150 lb/ft^3/ft until reaching the maximum uniform pressure at a depth of 664 psf/150 lb/ft^3 or 4.4 ft. For purposes of illustrating the form-design procedure, first consider the spacing of members in the uniform pressure region. Since the sizes of the members are predetermined, the unknowns are the maximum allowable spacings of their supports.

7.8.1 Sheathing Design

The plywood span is the spacing of the studs, and the 8 ft long plywood panels can typically have three or more spans. The plywood allowable stresses adjusted for load duration (see Table 7.3) are

$$F_b' = F_b \times C_D = 1545 \text{ psi} \times 1.25 = 1931 \text{ psi}$$
$$F_v' = F_v \times C_D = 57 \text{ psi} \times 1.25 = 71 \text{ psi}$$
$$E' = E = 1,500,000 \text{ psi}$$

Considering the loading to be uniform load, w, of 1-ft width corresponding to the section properties of the plywood based on 12-in width

$$w = p \times \text{unit width} = 664 \text{ psf} \times 1.0 \text{ ft} = 664 \text{ lb/ft}$$

For the limitation of flexural bending, the maximum allowable moment, M, is

$$M = f_b' S = 1931 \text{ psi} \times 0.455 \text{ in}^3 = 878 \text{ in-lb} = 73.2 \text{ ft-lb}$$

and the maximum span, l, is given by

$$M = \frac{wl^2}{10}$$
$$73.2 \text{ ft-lb} = \frac{664 \text{ lb/ft} \times l^2}{10}$$
$$l = 1.05 \text{ ft} = 12.6 \text{ in}$$

For the shear limitation, the support width is known to be 1.5 in so that the maximum shear force for design can be determined at the face of the support.

$$V = F_v'\frac{Ib}{Q} = 71 \text{ psi} \times 7.187 \text{ in}^2 = 510 \text{ lb}$$
$$V = 0.6wl - w\left(\frac{l_b}{2}\right)$$
$$510 \text{ lb} = 0.6(664 \text{ lb/ft})(l) - 664\left(\frac{1.5 \text{ in}/2}{12 \text{ in/ft}}\right)$$
$$l = 1.38 \text{ ft} = 16.6 \text{ in}$$

For the deflection limit, the maximum span is determined from

$$\frac{l}{240} = \frac{1}{145}\frac{wl^4}{EI}$$

$$\frac{l}{240} = \frac{1}{145}\frac{664\,\text{lb/ft}(l)^4}{12\,\text{in/ft} \times 1{,}500{,}000\,\text{psi} \times 0.199\,\text{in}^4}$$

$$l = 1.48\,\text{in}$$

From these requirements, 12.6 in (based on bending) is the maximum span of the plywood. Considering the 8 ft modular length of the plywood panel, provide eight equal spans with the studs spaced at 12 in on center.

7.8.2 Stud Design

The stud span is the spacing of the double wales, and the studs can be constructed with a three-span continuous arrangement. Lateral buckling of the studs will be restrained by nailing the sheathing to the studs. Allowable stresses for the No. 2 southern pine 2 × 4 stud from Table 7.3 are as follows:

$$F_b' = F_b \times C_D \times C_M \times C_L \times C_F \times C_r$$
$$= 1500\,\text{psi} \times 1.25 \times 0.85 \times 1.0 \times 1.0 \times 1.0 = 1593\,\text{psi}$$

$$F_v' = F_v \times C_D \times C_M \times C_H$$
$$= 90\,\text{psi} \times 1.25 \times 0.97 \times 1.0 = 109\,\text{psi} \quad \text{(within } 5d \text{ of the ends)}$$

$$F_v' = F_v \times C_D \times C_M \times C_H$$
$$= 90\,\text{psi} \times 1.25 \times 0.97 \times 2.0 = 218\,\text{psi} \quad (>5d \text{ from the ends)}$$

$$E' = E \times C_M$$
$$= 1{,}600{,}000\,\text{psi} \times 0.9 = 1{,}440{,}000\,\text{psi}$$

Considering the loading as a uniform load, w, of 12 in tributary width corresponding to the spacing of the studs

$$w = p \times \text{unit width} = 664\,\text{psf} \times 1.0\,\text{ft} = 664\,\text{lb/ft}$$

For the limitation of flexural bending, the maximum allowable moment, M, is

$$M = F_b'S = 1593\,\text{psi} \times 3.063\,\text{in}^3 = 4{,}879\,\text{in-lb} = 406.6\,\text{ft-lb}$$

and the maximum allowable span is given by

$$M = \frac{wl^2}{10}$$

$$406.6\,\text{ft-lb} = \frac{664\,\text{lb/ft} \times l^2}{10}$$

$$l = 2.47\,\text{ft} = 29.7\,\text{in}$$

For the shear limitation, the maximum shear force for design is determined at d from the face of the support since the supporting wale width ($2 \times 1.5\,\text{in} = 3\,\text{in}$) is known. Both the joist end support and the joist interior support located $>5d$ from a cut end are examined since different shear forces as well as different allowable stresses apply.

At the ends

$$V = \frac{2}{3}F_v'bd = \frac{2}{3}(109 \text{ psi} \times 1.5 \text{ in} \times 3.5 \text{ in}) = 381 \text{ lb}$$

$$V = 0.4wl - w\left(d + \frac{l_b}{2}\right)$$

$$381 \text{ lb} = 0.4(664 \text{ lb/ft})(l) - 664 \text{ lb/ft}\left(\frac{3.5 \text{ in} + 3.0 \text{ in}/2}{12 \text{ in/ft}}\right)$$

$$l = 2.47 \text{ ft} = 29.7 \text{ in}$$

At the interior support

$$V = \frac{2}{3}F_v'bd = \frac{2}{3}(218 \text{ psi} \times 1.5 \text{ in} \times 3.5 \text{ in}) = 763 \text{ lb}$$

$$V = 0.6wl - w\left(d + \frac{l_b}{2}\right)$$

$$763 \text{ lb} = 0.6(664 \text{ lb/ft})(l) - 664 \text{ lb/ft}\left(\frac{3.5 \text{ in} + 3.0 \text{ in}/2}{12 \text{ in/ft}}\right)$$

$$l = 2.60 \text{ ft} = 31.2 \text{ in}$$

The deflection limit is determined from

$$\frac{l}{240} = \frac{1}{145}\frac{wl^4}{EI}$$

$$\frac{l}{240} = \frac{1}{145}\frac{664 \text{ lb/ft}(l)^4}{12 \text{ in/ft} \times 1{,}440{,}000 \text{ psi} \times 5.359 \text{ in}^4}$$

$$l = 43.8 \text{ in}$$

From these requirements, the spans for bending and shear control the maximum span to be 29.7 in. Use 28 in for layout convenience.

7.8.3 Wale Design

The wale span is the spacing of the form ties, and the wales can be constructed with a three-span continuous arrangement. Lateral buckling of the wale is not a problem since $d/b \leq 2$ to 1. Allowable stresses for the No. 2 southern pine double 2×4 wales are the same as determined above for the studs.

The wale loading is actually a series of concentrated loads from the studs at 12 in on center. However, the starting position of the loads can vary in each span. Owing to the complications of considering many possible starting positions and recognizing that loads within a distance d of the support can be neglected for shear calculations, it is often the practice in formwork to assume a uniform load to be adequately similar to many closely-spaced equal concentrated loads [Hurd, 1995]. This assumption works reasonably well when there are three or more equally-spaced concentrated loads of equal magnitude in the span. Considering the loading to be a uniform load, w, of 2.33 ft tributary width corresponding to the spacing of the wales

$$w = p \times \text{unit width} = 664 \text{ psf} \times 2.33 \text{ ft} = 1549 \text{ lb/ft}$$

For the limitation of flexural bending, the maximum allowable moment, M, is

$$M = F'_b S = 1593 \text{ psi} \times 2 \times 3.063 \text{ in}^3 = 9{,}758 \text{ in-lb} = 813 \text{ ft-lb}$$

and the maximum allowable span is given by

$$M = \frac{wl^2}{10}$$

$$813 \text{ ft-lb} = \frac{1549 \text{ lb/ft} \times l^2}{10}$$

$$l = 2.29 \text{ ft} = 27.4 \text{ in}$$

For the shear limitation, the maximum shear force for design is determined at d from the face of the support assuming that the tie-washer plate will have at least a 3 in length of bearing. Both the wale end support and the wale interior support located $>5d$ from a cut end are examined since different shear forces as well as different allowable stresses apply.

At the ends

$$V = \frac{2}{3} F'_v bd = \frac{2}{3}(109 \text{ psi} \times 2 \times 1.5 \text{ in} \times 3.5 \text{ in}) = 763 \text{ lb}$$

$$V = 0.4wl - w\left(d + \frac{l_b}{2}\right)$$

$$763 \text{ lb} = 0.4(1549 \text{ lb/ft})(l) - 1549 \text{ lb/ft}\left(\frac{3.5 \text{ in} + 3.0 \text{ in}/2}{12 \text{ in/ft}}\right)$$

$$l = 2.27 \text{ ft} = 27.2 \text{ in}$$

At the interior support

$$V = \frac{2}{3} F'_v bd = \frac{2}{3}(218 \text{ psi} \times 2 \times 1.5 \text{ in} \times 3.5 \text{ in}) = 1526 \text{ lb}$$

$$V = 0.6wl - w\left(d + \frac{l_b}{2}\right)$$

$$1526 \text{ lb} = 0.6(1549 \text{ lb/ft})(l) - 1549 \text{ lb/ft}\left(\frac{3.5 \text{ in} + 3.0 \text{ in}/2}{12 \text{ in/ft}}\right)$$

$$l = 2.33 \text{ ft} = 28.0 \text{ in}$$

The deflection limit is determined from

$$\frac{l}{240} = \frac{1}{145} \frac{wl^4}{EI}$$

$$\frac{l}{240} = \frac{1}{145} \frac{1549 \text{ lb/ft}(l)^4}{12 \text{ in/ft} \times 1{,}440{,}000 \text{ psi} \times 2 \times 5.359 \text{ in}^4}$$

$$l = 41.6 \text{ in}$$

From these requirements, the maximum span is 27.2 in and is controlled by shear. Select a wale span of 24 in so that the tie spacing can be coordinated easily with the stud spacing.

A check is also needed to make sure that there is adequate bearing area for the stud on the double wale. The tributary load is

$$P = 664 \text{ psf} \times 1.0 \text{ ft} \times 2.33 \text{ ft} = 1549 \text{ lb}$$

The allowable bearing stress is given by

$$F'_{c\perp} = F_{c\perp} \times C_M \times C_b = 565 \text{ psi} \times 0.67 \times 1.0 = 378 \text{ psi}$$

and the actual stress is

$$f_{c\perp} = \frac{P}{A} = \frac{1549 \text{ lb}}{1.5 \text{ in} \times 3.0 \text{ in}} = 344 \text{ psi} < 378 \text{ psi}$$

7.8.4 Tie Design

Select the tie capacity to exceed the tie force, T, based upon the tributary area supported.

$$T = p \times \text{wale spacing} \times \text{tie spacing} = 664 \text{ lb} \times 2.33 \text{ ft} \times 2.0 \text{ ft} = 3{,}094 \text{ lb}$$

Select a heavy snap tie with a 3200 lb working capacity.

Check the bearing of the 3 in square tie washer on the double wale (Figure 7.40).

$$f_{c\perp} = \frac{P}{A} = \frac{3094 \text{ lb}}{2 \times 1.5 \text{ in} \times 3.0 \text{ in}} = 344 \text{ psi} < 378 \text{ psi}$$

7.8.5 Bracing Design

Braces are needed to keep the form from overturning due to wind or incidental construction loads such as ladders leaning on the form or workers climbing on the form. The effects of an elevated work platform may also have to be considered in some designs. If designing for the minimum recommendations of ACI 347-94, the loadings would include a wind pressure of 15 psf or a line load of 100 lb/linear foot at the top, whichever is greater. The contractor desires to lay out the braces as

FIGURE 7.40 Bearing of form tie bracket on double wood wale.

shown in Figure 7.41, with the braces at 4 ft on center, the brace attached at an elevation of 10 ft on the 12 ft form, and the brace anchored to the ground 7.5 ft from the form. This results in a brace length of 12.5 ft, including the adjustable connectors at each end. For design purposes, the brace is assumed to be concentrically loaded, pinned at the top and the bottom, and prevented from translation by the longitudinal rigidity of the form. The horizontal brace load, H, from the wind is

$$H = 15 \text{ psf} \times 4 \text{ ft} \times 12 \text{ ft} \times 6 \text{ ft}/10 \text{ ft} = 432 \text{ lb}$$

Alternately, the horizontal brace load from the 100 lb/ft line load is

$$H = 100 \text{ lb/ft} \times 4 \text{ ft} \times 12 \text{ ft}/10 \text{ ft} = 480 \text{ lb}$$

The brace should be designed to resist a horizontal component of 480 lb, which results in an axial compression force, P, of

$$P = 480 \text{ lb} \times 12.5 \text{ ft}/7.5 \text{ ft} = 800 \text{ lb}$$

Try a No. 2 southern pine 2 × 4 brace with a bracing strut at midlength in the weak buckling direction. For this 2 × 4 size, the allowable compression stress is

$$F'_c = F_c \times C_D \times C_M \times C_F \times C_P$$
$$= 1650 \text{ psi} \times 1.25 \times 0.8 \times 1.0 \times C_P = 1650 \text{ psi} \times C_P = F^*_c \times C_P$$

Since the trial brace is rectangular, check the l_e/d ratios for each buckling axis.

$$\frac{l_1}{d_1} = \frac{150 \text{ in}}{3.5 \text{ in}} = 42.8 \leq 50$$

$$\frac{l_2}{d_2} = \frac{75 \text{ in}}{1.5 \text{ in}} = 50.0 \leq 50$$

Thus the weak axis controls and $l_e/d = 50$. Substituting to determine C_P

$$F_{cE} = \frac{K_{cE} E'}{(l_e/d)^2} = \frac{0.3 \times 1,400,000 \text{ psi}}{(50.0)^2} = 173 \text{ psi}$$

so that

$$\frac{F_{cE}}{F^*_C} = \frac{173 \text{ psi}}{1650 \text{ psi}} = 0.105$$

and C_P from Eq. (7.10) is

$$C_P = \frac{1 + 0.105}{2 \times 0.8} - \sqrt{\left[\frac{1 + 0.105}{2 \times 0.8}\right]^2 - \frac{0.105}{0.8}} = 0.103$$

Thus the allowable stress in compression parallel to the grain for the unbraced length is

$$F'_c = 1650 \text{ psi} \times C_P = 1650 \text{ psi} \times 0.103 = 170 \text{ psi}$$

$$f'_c = \frac{P}{A} = \frac{800 \text{ lb}}{5.25 \text{ in}^2} = 153 \text{ psi} \leq 170 \text{ psi}$$

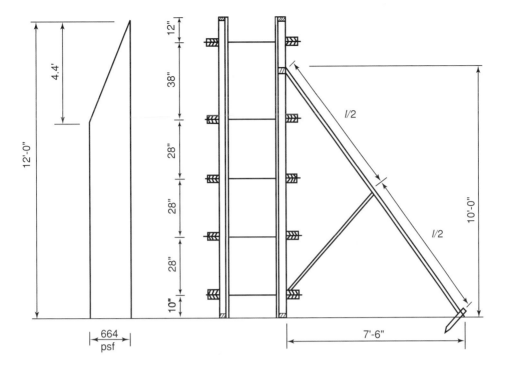

FIGURE 7.41 Layout of wall form members.

The 2 × 4 brace is adequate with the midlength strut. Alternately, a steel-pipe brace rated for the load at the extended length of 12.5 ft could be used. Appropriate connection, adequate for a tension or compression of 800 lb must be made at the top and bottom ends of the brace by bolting, nailing, or a rated commercial hardware device, without inducing significant eccentricities.

A possible layout for the wall form is shown in Figure 7.41. The top and bottom extensions of the studs beyond the wales are usually about 1/3 of the adjacent span up to perhaps 12 in. At the top of the form where the pressures are lower, the span of the studs can be investigated with ramp loading, and the wale spacing can often be increased, sometimes eliminating a level of wales and ties in comparison to the number required for a constant spacing.

Careful consideration of all design details, communicating these requirements to the form craftspersons, and providing knowledgeable verification of execution can lead to safe and economical support of concrete during construction.

References

ACI Committee 347. 1994. *Guide to Formwork for Concrete*, ACI 347R-94, 34 pp. American Concrete Institute, Farmington Hills, MI.

American Forest and Paper Association. 1993. *National Design Specification for Wood Construction.* ANSI/NFoPA NDS-1991. Revised ed. Washington, DC.

American Plywood Association. 1997. *Plywood Design Specification*, 31 pp. Tacoma, WA.

American Plywood Association. 1994. *Concrete Forming*, 26 pp. Tacoma, WA.

Canadian Standards Association 1992. *Concrete Formwork*, CAN/CSA-S269.3-M92. Toronto, Ontario, Canada.

Clear, C. A. and Harrison, T. A. 1985. *Concrete Pressure on Formwork*, Report 108, 32 pp. Construction Industry Research and Information Association, London, England.

Hurd, M.K. 1995. *Formwork for Concrete*, 6th ed. Publication SP-4. American Concrete Institute, Farmington Hills, MI.

Further Information

American Concrete Institute, P.O. Box 9094, Farmington Hills, MI 48333-9094, (810) 848-3700.

American Forest and Paper Association, 1111 19th Street, Seventh Floor, Washington, DC 20036.

American National Standards Institute, 11 West 42 Street, New York, NY 10036, (212) 642-4900.

American Plywood Association, P.O. Box 11700, Tacoma, WA 98411, (206) 565-6600.

Canadian Standards Association, 178 Rexdale Boulevard, Rexdale, Ontario M9W 1R3, Canada.

Construction Industry Research and Information Association, 6 Storey's Gate, Westminster, London SW1P 3AU, England.

Scaffolding, Shoring and Forming Institute, Inc., 1300 Sumner Avenue, Cleveland, OH 44115-2851, (216) 241-7333.

(a)

(b)

(a) Construction loading in high rise buildings (Courtesy Dr. S.K. Ghosh, Portland Cement Association). (b) Ronald McDonald house, New York, under construction (Courtesy New York Construction News and Dr. Edward G. Nawy, Rutgers University).

8

Construction Loading in High-Rise Buildings

by
S.K. Ghosh

Director, Engineering Services, Codes and Standards, Portland Cement Association, Skokie, Illinois. Expert in structural design particularly in the areas of serviceability and author of the PCA Manuals on the ACI 318 Building Code.

8.1 Introduction

Structural formwork and its support system need to be given careful consideration in two different respects: (1) loads that are applied to the formwork and its supports and (2) loads that the formwork and its supports apply to the structure. The first is the subject of Chapter 7 on formwork and falsework; the second is the subject of this chapter.

8.2 Construction Loads

8.2.1 Design Considerations

In the construction of multistory buildings with reinforced concrete floor slabs, a step-by-step sequence of operations is employed. The sequence comprises the steps of setting up shoring on the most recently poured floor, forming the next floor, setting up reinforcement, and concreting the slab. Since the floor below the one being concreted is usually between 7 and 14 days old, it is

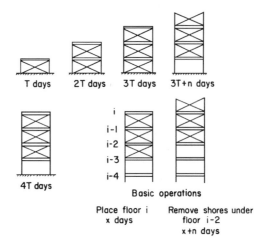

FIGURE 8.1   Construction sequence using three levels of shores.

common practice to leave formwork supports in place between that floor and a small number of floors below it. A typical construction cycle using three levels of shores is illustrated in Figure 8.1.

In discussing construction loads, it is convenient to express them as a factor times the sum of the self-weight of the floor and the dead load of the formwork. The term **floor-loading ratio** is used for this factor.

Progressive slab deflections had been causing trouble in Sweden in the early 1950s and led to the first study by Nielsen (1952) of a rational approach to stripping formwork for floors. Nielsen gave a detailed analysis of the distribution of load between a system of connected shores and floor slabs. The method considered the deformation characteristics of both the slabs and the shores. The maximum load ratio obtained by Nielsen on a slab assuming three levels of shores was 2.56.

Grundy and Kabaila (1963) developed a simplified method of analysis for floor loads based upon the following assumptions:

1. The slabs behave elastically.
2. Initially, the slabs are supported from a completely rigid foundation.
3. The shores supporting the slabs and formwork may be regarded as a continuous uniform elastic support, the elastic properties of which may be expressed by a coefficient K, where K is the load intensity that produces unit deformation of the support.
4. Any added load is distributed among the supporting slabs in proportion to their relative flexural stiffnesses.

Grundy and Kabaila assumed that K was infinite and carried out their analyses assuming a constant flexural stiffness for all connected slabs, as well as a flexural stiffness increasing with age. It was found that the error introduced by assuming equal relative stiffnesses for the floors is not appreciable.

Figure 8.2 represents the construction of a multifloor building using three levels of shores (and forms) for a 7-d casting cycle, with stripping in 5 d. The loads carried by the slabs and the shores, in terms of the loading ratio, are indicated on the figure adjacent to the element concerned. Floors 1, 2, and 3, supported from the ground by stiff shores, cannot deflect and therefore carry no load; all the load is carried by the shores directly to the foundation. At the age of 26 d, the lowest level of shores is removed, allowing all three slabs to deflect and carry their self-weight. The removed shores are placed on the third floor slab and the fourth floor is poured. As all three supporting slabs have equal stiffness, the weight of the newly poured slab is carried equally by the three lower slabs. The absolute maximum load ratio occurs when the shores connecting the supporting assembly with the

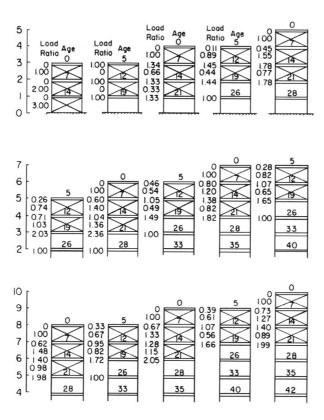

3 Levels of Shoring
Construction Cycle Time, T = 7 days
Time of Removal of Lowest Level of Shores
from Concreting of Top Floor, m = 5 days

FIGURE 8.2 Load ratio versus time for three levels of shores.

ground level are removed; however, the load ratio converges for upper floor levels. For the same structure considered by Nielsen, Grundy and Kabaila obtained an absolute maximum load ratio of 2.36, while the converged value for upper floor levels was 2.00 (Figure 8.2). The most heavily loaded slab is always the last slab that is supported directly from the foundation. The load time histories of the third floor slab and a typical slab are shown in Figure 8.3. Altering the number of shored levels has little effect on the maximum load ratio value. However, by decreasing the number of shored levels, the age of the slab at which the maximum ratio occurs also decreases.

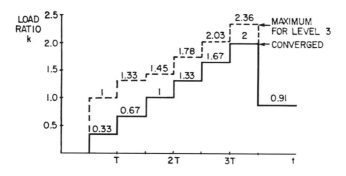

FIGURE 8.3 Load-ratio history for three levels of shores.

TABLE 8.1   Floor-Loading Ratio

Author	Maximum Value, Converged Values in Brackets				Comment
	$m = 2$	$m = 3$	$m = 4$	$m = 5$	
Nielsen (1952)	2.17	2.28×			Values for floor level 2 only
	2.0Δ	2.56+			×Timber props
					+Steel props ($n = 1$)
					Δ Observed
Grundy and	2.25	2.36	2.43		($n = 5$)
Kabaila (1963)	(2.00)	(2.00)	(2.00)		
Beresford (1964)	2.25	(2.00)*	2.45	2.50	*Obtained for rapid hardening, normal
		(2.06)* 2.35			and slow maturing concretes, respectively
		(2.32)			($n = 5$)
Blakey and	2.25	2.3			+Stepwise construction
Beresford (1965)	2.25+				
Beresford (1971)		2.2			($n = 4$)
		1.5Δ			Δ Observed

Source: Wheen, R.J. 1982. An invention to control construction floor loads in tall concrete buildings. *Concrete Int.* 4(5):56–62.

Note: Here m = number of levels of shoring used and n = time in days for removal of lowest levels of shores after concreting top floor.

Beresford (1964) used infinite as well as various finite values of K (see assumption 3 above) and found that the results of floor-load analysis were not appreciably affected.

In addition to variations in moduli of elasticity due to concrete age, cracking of slabs that occurs during construction affects the distribution of load between slabs in the supporting assembly. Sbarounis (1984) reports that incorporating the effects of cracking into the load distribution factors for the supporting slabs reduces the previously calculated maximum-load ratios by approximately 10%.

Blakey and Beresford (1965) recommended a stepped sequence of construction in a system of floors and shores as a means of controlling the construction loads imposed on both the slabs and the shores. The advantage of this method lies in the fact that a young slab is given more time to develop adequate strength before the application of construction load from the casting of a new slab directly above.

Table 8.1, adapted from Wheen (1982), shows clearly that all writers on the subject agree that floor-loading ratios during construction usually exceed values of 2.

The sequence of construction illustrated in Figure 8.1 uses three sets of forms. Economic considerations usually necessitate the removal of formwork as soon as possible for reuse. This necessity has given rise to the widespread practice of reshoring. The reshore technique typically involves using only one level of forms or shores and several levels of reshores. Basically, the forms or shores are removed from beneath a slab, allowing it to deflect and carry its own weight; reshores are then installed, allowing the load during subsequent concrete placement to be shared between the various slabs in the supporting assembly. At the time of installation, reshores carry no significant load. Figure 8.4 illustrates a construction scheme with two levels of shoring and one level of reshoring.

Agarwal and Gardner (1974) in Canada and Marosszeky (1972) in Australia analyzed the loads imposed on slabs using the shoring/reshoring method of construction, utilizing the same simplifying assumptions as used earlier by Grundy and Kabaila (1963). Using two levels of shores and one level of reshores, the slab loads are calculated as shown in Figure 8.5. The slab load-time history using two levels of shores and one level of reshores is shown in Figure 8.6.

Taylor (1967) suggested a technique of stripping formwork to reduce the loads imposed on a slab during construction. By slackening and tightening adjustable shores before each new slab is cast, the loads distributed to the shores and slabs are reduced considerably. With this method, all shores

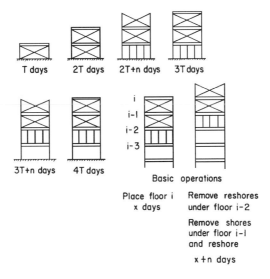

FIGURE 8.4 Construction sequence using two levels of shores and one level of reshores.

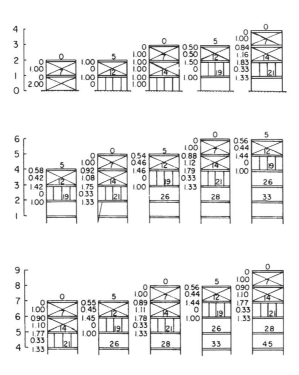

FIGURE 8.5 Load ratio versus time for two levels of shores and one level of reshores.

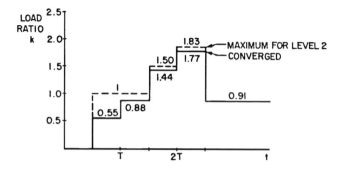

FIGURE 8.6   Load-ratio history for two levels of shores and one level of reshores.

at one level must be slackened simultaneously, thus it requires greater supervision and inspection. Taylor reported that the maximum load ratio was reduced from the 2.36 obtained by Grundy and Kabaila to a value of 1.44. Taylor's method is the same in principle as stripping and immediate reshoring of a slab.

Marosszeky (1972) described complete release and reshoring of a floor slab such that the floor carried its own dead weight at a time $(T-1)$ days, where T is the construction cycle of floors. This reshoring technique produces less construction load on the supporting slabs and props when compared with using undisturbed shores.

Although the load ratios previously defined reflect the slab-plus-formwork dead weights, a construction live load is also present. While the dead load can be estimated with reasonable certainty, the live loads can vary significantly depending upon the construction method used to place concrete. ACI Committee 347 (1988) recommends that the design live load should be at least 50 psf (2.4 kPa) of horizontal projection. A comprehensive study [Fattal, 1983] on construction loads in concrete buildings indicated that when concrete is placed by a bucket, the live load may be as high as 40–50 psf (1.9–2.4 kPa). On the other hand, the American National Standards Institute (ANSI) Standard A10.9 (1983) does not provide a numerical value and leaves the determination of live load to the formwork designer. However, the standard does list factors to be considered in estimating the design load such as the weight of workers, equipment, runways, and impact of concrete. The OSHA safety standard (1972) requires at least 20 psf (1.0 kPa) for live load and the weight of formwork. The Scaffolding and Shoring Institute also recommended (1977) a construction live load of 20 psf (1.0 kPa). According to Lew (1985), there is no apparent basis for such a small magnitude of the live load requirement.

Hurd (1989) suggests a minimum construction live load of 50 psf (2.4 kPa) for designing forms. Lasisi and Ng (1979) presented a method to include the live-load effect. For a typical flat plate structure, assuming a supporting assembly of two shore levels plus one reshore level, a construction live load of 50 psf (2.4 kPa) removed after the casting day, and a constant modulus of elasticity for the connected slabs, the absolute maximum load ratio increased 9% over that predicted by Grundy and Kabaila (1963). Both Agarwal and Gardner (1974) and Sbarounis (1984) accounted for the construction live-load effect by increasing the maximum load carried by the lowest slab in the supporting assembly. Sbarounis recommended additional loads, due to a 50 psf (2.4 kPa) live load, of $55/N$ and $35/N$ psf ($2.6/N$ and $1.7/N$ kPa) for uncracked and cracked slabs, respectively, with N representing the total number of levels in the supporting assembly.

Table 8.2 shows absolute maximum and converged maximum load ratios on slabs and supporting props for various combinations of levels of shores and reshores. Table 8.2 and Figure 8.6 indicate that the use of two levels of shoring and one level of reshoring, rather than three levels of shores, reduces the absolute maximum load ratio from 2.00 to 1.78. This is advantageous in most situations, although with reshoring, the maximum-load ratios come into play at an earlier age than with shoring,

TABLE 8.2 Theoretical Maximum Load Ratios on Floor and Prop for Various Shore/Reshore Combinations

Shore + Reshore	Absolute Maximum Load Ratio		Converged Maximum Load Ratio	
	On Floor Slab	On Prop	On Floor Slab	On Prop
1 + 1	1.50	1.0	1.50	1.0
1 + 2	1.34	1.0	1.34	1.0
1 + 3	1.25	1.0	1.25	1.0
1 + 4	1.20	1.0	1.20	1.0
1 + 5	1.17	1.0	1.17	1.0
2 + 0	2.25	2.0	2.00	1.0
2 + 1	1.83	2.0	1.78	1.11
2 + 2	1.75	2.0	1.67	1.17
2 + 3	1.61	2.0	1.60	1.21
2 + 4	1.60	2.0	1.56	1.25
2 + 5	1.55	2.0	1.53	1.24
3 + 0	2.36	3.0	2.00	1.34
3 + 1	2.10	3.0	1.87	1.37
3 + 2	1.97	3.0	1.80	1.40
3 + 3	1.84	3.0	1.76	1.42
3 + 4	1.77	3.0	1.72	1.43
3 + 5	1.77	3.0	1.70	1.43

Source: Lasisi, M.Y. and Ng, S.F. 1979. Construction loads imposed on high-rise floor slabs. *Concrete Int.* 1(2):24–29.

TABLE 8.3 Comparison of Construction Loads With Service Loads

Construction Loads, psf					Service Loads, psf	
8 in slab	100				8 in slab	100
Formwork	10				Ceiling and mech	15
Subtotal	110				Partitions	20
					Live load	50
		3 Levels of Shores		*2 Levels of Shores Plus 1 Level of Reshores*	Total (after 28 days)	185
Load ratio	1.00 at 5 days	110	1.00 at 5 days	110		
	1.34 at 7 days	147	1.00 at 7 days	110		
	1.45 at 12 days	160	1.50 at 12 days[†]	165		
	1.78 at 14 days	196	1.83 at 14 days	201		
	2.03 at 19 days	223*	1.00 at 19 days	110		
	2.36 at 21 days	260				
	1.00 at 26 days	110				

*Deflection not restrained beyond 19 days.

[†]Allowed to deflect under these construction loads at 12 days. Further deflections partly restrained for the next 7 days.

as should be apparent from Figures 8.2 and 8.5. Table 8.3 presents construction loads for Floor 3, which experiences the absolute maximum load ratio. The construction loads are compared with the design service loads in the table. It is clear that the construction loads are more critical than the design loads. Also important, construction loads act on concrete that has not attained the age at which it is supposed to experience design service loads.

8.2.2 Field Verification

Agarwal and Gardner (1974) used field measurements to check the accuracy of different analysis methods for estimating the shore and slab load distribution. Their investigation consisted of directly

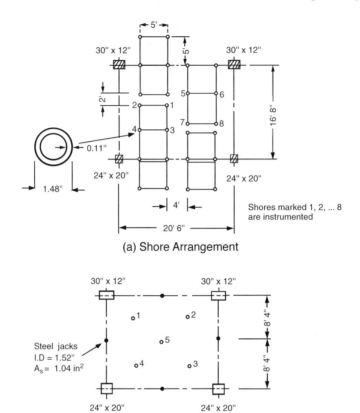

(a) Shore Arrangement

(b) Reshore Arrangement

FIGURE 8.7   (a) Shore and (b) reshore arrangements for Alta Vista Towers. *Source:* Courtesy of American Concrete Institute; Agarwal, R.K. and Gardner, N.J. 1974. Form and shore requirements for multistory flat slab type buildings, *ACI J.* 71(11):559–569.

measuring the load ratios applied to shores and reshores during construction and determining the loads carried by the floor slabs at different stages of construction.

The shore and reshore loads for two buildings were measured [Agarwal and Gardner, 1974]: Alta Vista Towers in Ottawa, Ontario, Canada, and Place du Portage in Hull, Quebec, Canada. Steel shores were used for both structures.

Three levels of shores and four levels of reshores were used to temporarily support the flat slab floors in the construction of the 22-story Alta Vista Towers. Measurement of the shore and reshore loads, for the shore and reshore arrangements shown in Figure 8.7, was limited to the seventh through the thirteenth story.

Three levels of shores and no reshores were used in the construction of the 27-story Place du Portage structure, which is a flat-slab office building. The shoring arrangement is shown in Figure 8.8. Measurements were taken from the nineteenth through the twenty-second stories.

According to Liu et al. (1985a), three factors concerning the field measurement data should be noted:

1. Field measurements were not taken from the ground level, so the actual slab and shore loads of the entire system during construction were not available.

2. Typical shores and reshores chosen for instrumentation were located in the central portions of the slabs (Figures 8.7 and 8.8). Thus the influence of the surrounding boundary beams and columns was less pronounced. From the measured data reported in Agarwal and Gardner (1974), the coefficient of variation of the shore axial force varied from 0.04 to 0.23, depending on the distance of the shore from the boundary line.

3. The values of the modulus of elasticity of concrete E_c and the shore height were not reported in Agarwal and Gardner (1974). Without these, it is not possible to use the measured shore-load data to check other methods further.

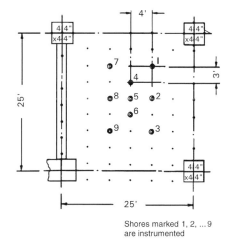

Shores marked 1, 2, ...9 are instrumented

FIGURE 8.8 Shoring arrangements for Place du Portage. *Source:* Courtesy of American Concrete Institute; Agarwal, R.K. and Gardner, N.J. 1974. Form and shore requirements for multistoy flat slab type buildings. *ACI J.* 71(11):559–569.

Comparisons between field measurements for both buildings, and the slab and shore loads given by the simplified analysis of Grundy and Kabaila (1963), are reported in Agarwal and Gardner (1974). In order to check the assumptions of equal slab stiffness originally used by Grundy and Kabaila, the slab and shore loads were calculated by both using equal slab stiffnesses and varying the slab stiffness for each floor to account for an increase in E_c with age. Table 8.4 shows the comparison of field measurements for each story in both buildings to the calculated maximum slab and shore loads.

Good agreement between the predictions of simplified analysis and the field measurements was generally observed. Uses of variable slab stiffnesses in the simplified analysis produced closer agreement: the maximum discrepancy for variable slab stiffness was less than 10%, while for constant

TABLE 8.4 A Comparison Between Field Measurements and Calculated Maximum Slab and Shore Loads

		Maximum Slab Load ÷ Slab Weight					Maximum Shore Load ÷ Slab Weight				
		Field Measmt.	Calc. with E_c Constant	Calc. with E_c Variable	Comparisons		Field Measmt.	Calc. with E_c Constant	Calc. with E_c Variable	Comparisons	
					(2)/(1)	(3)/(1)				(4)/(5)	(4)/(6)
lding	Level	(1)	(2)	(3)	(1)	(1)	(4)	(5)	(6)	(5)	(6)
a	7	1.88	1.72	1.83	0.91	0.97	1.65	1.40	1.58	0.85	0.96
ta	8	1.93	1.70	1.85	0.88	0.96	1.51	1.46	1.57	0.97	1.04
vers	9	1.91	1.72	1.74	0.90	0.91	1.70	1.42	1.58	0.84	0.93
	10	2.02	1.72	1.99	0.85	0.99	1.37	1.43	1.46	1.04	1.07
	11	1.07	1.13	0.99	1.06	0.93	1.68	1.44	1.62	0.89	1.00
	12						1.64	1.43	1.61	0.87	0.98
an					0.92	0.95				0.90	1.00
ndard eviation					0.08	0.03				0.09	0.05
ce du	19	2.11		2.11		1.00	1.48		1.41		0.95
tage	20	1.34		1.30		0.97	1.44		1.44		1.00
	21						1.45		1.41		0.97
an						0.99					0.97
ndard eviation						0.02					0.03

Source: Taylor, P.J. 1967. Effects of formwork stripping time on deflections of flat slabs and plates. *Australian Civil Eng. st.* 8(2):31–35.

slab stiffness it was as high as 16%. Further, it was shown in Agarwal and Gardner (1974) that the calculated results based on either constant or variable slab stiffnesses consistently predicted the correct construction step and the location of the maximum shore and slab loads.

Lasisi and Ng (1979) carried out measurements on five regular floors to investigate peak construction loads on slabs, shores, and reshores in a 15-story flat-slab office building in Ottawa, Canada. One level of shores and two levels of reshores were utilized in construction. The period between the casting of a slab section and the next slab directly above was 10 d on average. The forms were supported on tubular steel shores. Reshores were made of telescopic steel jacks. The average stripping time for a slab in the instrumented section of the building was 5 d. Measurement on this building was commenced during concreting of the seventh floor slab and continued until after the stripping of the eleventh floor. An interior bay of a section of the building was instrumented. Load cells were installed beneath a shore or a reshore at the chosen location; the instrumented reshore on each level was vertically below the load cell placed underneath the scaffold shore.

Shore and reshore loads were calculated by Lasisi and Ng (1979) using Grundy and Kabaila's simplified analysis, taking into consideration the presence of 50 psf (2.4 kPa) of construction live load during concrete placement. As can be seen from Table 8.5, there was reasonably close agreement between measured and theoretical maximum shore loads. However, the loads measured on the reshores were found to be on average considerably less than the corresponding theoretical values.

TABLE 8.5   Measured Construction Loads on Shores and Reshores

Shore or Reshore Resting on Level	Construction Operation		Imposed Loads		
	A (Concreting)	B (After Stripping and Reshoring)	psf	Load Ratios	Theoretical Load Ratios
4	A-7R		50.1	0.40	0.47
5	A-7R		77.0	0.62	0.93
		B-7R	5.0	0.04	0
6	A-8R		13.4	0.11	0.47
	A-7S		154.0	1.23	1.40
		B-7	—	—	—
	A-8R		97.3	0.78	0.93
		B-8R	8.9	0.07	0
7	A-9R		17.6	0.14	0.47
	A-8S		188.8	1.51	1.40
		B-8	—	—	—
	A-9R		106.4	0.85	0.93
		B-9R	21.1	0.17	0
8	A-10R		40.1	0.32	0.47
	A-9S		209.5	1.68	1.40
		B-9	—	—	—
	A-10R		110.9	0.89	0.93
		B-10R	9.4	0.08	0
9	A-11R		28.2	0.23	0.47
	A-10S		135.6	1.09	1.40
		B-10	—	—	—
	A-11R		59.9	0.48	0.93
		B-11R	23.3	0.19	0
10	A-11S		182.5	1.46	1.40
		B-11	—	—	—

Source: Lasisi, M.Y. and Ng, S.F. 1979. Construction loads imposed on high-rise floor slabs. *Concrete Int.* 1(2):24–29.

Note: S = shore; R = reshore.

8.2.3 Refined Analysis

Liu et al. (1985a) carried out refined analysis to determine shore and slab load distribution, considering the actual rigidity of shores and the time-dependent variation in slab stiffness due to concrete maturity. The refined analysis technique [Liu et al., 1985a] treated the shore-slab interaction as a two-dimensional problem and was based on the following assumptions:

1. The slabs behave elastically and their stiffnesses are time dependent.
2. The shores and reshores behave as continuous uniform elastic supports and their axial stiffnesses are finite and time independent.
3. The foundation is rigid.
4. The joints between the shores and slabs are pinned connections.
5. The slab edges are either fixed or simply supported.

Comparison of maximum loads computed by the simplified Grundy-Kabaila method (1963) with those predicted by the refined method indicated that the maximum relative differences in the two methods varied between −5 and +9%. Because the effect of shore stiffness was not considered, the calculated results using the simplified method showed larger errors (deviations from field measurements) than the results predicted using the refined method.

The simplified Grundy-Kabaila method reliably predicted the construction step and location where the maximum slab and shore loads would occur, but generally underestimated the actual load ratios. Consequently, the maximum slab and shore loads predicted by the simplified method could be corrected using a modification coefficient. Liu et al. (1985a) recommended that the value of the modification coefficient for the simplified method vary from 1.05 to 1.10 for design purposes.

Liu et al. (1985b) also used a more realistic three-dimensional (3-D) model to check the approximation of the Grundy-Kabaila simplified method and the two-dimensional (2-D) refined method [Liu et al., 1985a]. A number of simplifying assumptions were made in conducting the refined 3-D analysis. First, the reinforced concrete slabs were assumed to behave elastically; their stiffnesses were taken as time dependent. Slab edges were either free or rotationally fixed (but free to deflect). Second, the vertical deflections of the slabs at the slab-column joints were neglected. Third, the weight and structural details of each floor were assumed to be similar. Fourth, the shores and reshores were presumed to be continuous ideal elastic supports with equal axial stiffnesses. The joints between the shores and the slabs were assumed to be pin-ended connections. Lastly, the foundation was assumed to be rigid and unyielding. The influences of foundation rigidity, column deformation, and slab aspect ratio were examined using the 3-D model.

On the basis of the various factors examined for the 3-D model, the following conclusions and observations were made:

1. The maximum slab loads given by the 2-D refined analysis and the 3-D refined analysis were nearly identical. The maximum shore load increased slightly for the 3-D refined analysis.
2. Variations of the foundation rigidity affected slab displacements more than the maximum shore loads and slab moments. When the rigidity of the foundation decreased, the maximum slab moments and the maximum shore loads decreased.
3. The vertical deformation of columns could be neglected when computing the maximum shore loads and slab moments.
4. A change in the slab aspect ratio produced very little increment in the maximum shore load for slabs with free edges. For a slab with fixed edges, the total increment of the maximum shore load was 3% for the aspect ratios examined (between 0.6 and 1.0).
5. In all cases examined, a modification coefficient that varied from 1.05 to 1.10 could be used to conservatively correct the results of the Grundy-Kabaila simplified method.
6. It was found that nonuniform distribution of shore stiffnesses did not change the prediction of the construction step and location where the maximum slab moments and shore loads occurred.

Stivaros and Halvorsen (1990) questioned the good agreement mentioned earlier between the simplified Grundy-Kabaila method and existing field measurements. They pointed out that in actual building cases investigated [Agarwal and Gardner, 1974; Lasisi and Ng, 1979], steel shoring systems had been used, justifying the Grundy-Kabaila assumption of infinite shoring system stiffness. The fact that the field measurements were observations around the center of the slab and thus ignored the influence of structural continuity contributed to the good agreement between field data and the theoretical values. Another important fact was that field measurements were not taken from the ground level, so the complete loading-unloading construction cycle was not recorded.

Stivaros and Halvorsen (1991) also pointed out that most practical methods for analyzing construction loads have employed single-bay idealizations of frames. This is a definite limitation to the proper evaluation of the punching shear forces that are the critical load effect in most flat-slab systems. A single-bay structural idealization cannot fully assess the maximum shear-force effects at the columns from both direct shear forces and unbalanced moments. Therefore a practical and simple method that took into account the continuity of concrete structures over several bays, as well as the interaction of the floor slabs and the shoring system over several stories, was thought to be needed, [Stivaros and Halvorsen, 1991].

The equivalent frame method (EFM) of analysis was found [Stivaros and Halvorsen, 1990] to be a reasonable procedure for accomplishing the preceding goal. The EFM was first developed as a viable procedure for analyzing reinforced concrete slab systems, particularly for the effects of gravity loads. It has been applied to lateral load analysis of slab systems as well. A computer program was developed [Stivaros and Halvorsen, 1990] to determine the construction load distribution among the slabs and the shoring system during the construction of multistory buildings. A multistory building under construction is modeled according to the assumptions of the equivalent frame method as described in ACI 318 (1995). A schematic representation of an idealized structure is shown in Figure 8.9. The shores and reshores are replaced by an equal number of elastic supports, each having a stiffness equal to the total stiffness of the corresponding shores. It is assumed that the reshores carry no load at the time of installation. The load of the freshly cast concrete is applied to the top-floor shores as concentrated loads. The load is shared among the shores according to their tributary areas. The connection between the shores or reshores and the floor slabs can be reasonably assumed to be pin-jointed. At ground level, shores and reshores are assumed to be supported on a rigid foundation. The EFM concepts are applied to determine the frame member stiffness coefficients. The idealized two-dimensional frame is then analyzed elastically using conventional stiffness analysis procedures to determine the member forces, including the axial load on the shoring system.

A typical floor plan of the equivalent frame used by Stivaros and Halvorsen (1990) for parametric studies is shown in Figure 8.10. The shores and reshores are idealized as series of vertical truss-type elements with a stiffness equivalent to the total stiffness of the shores or reshores on each row. Two levels of shores and one level of reshores were assumed, with a construction rate of one floor per week. Only the slab dead load was considered in analysis.

The loads in slabs were normalized to the self-weight of each slab, and the loads on shores were normalized to the weight of the slab supported by the shores. The construction operations were denoted with a set of two numbers. The first number represented the number of the floor level under construction, and the second indicated the phase of construction (Figure 8.11). The numeral 1 indicated casting of the top floor, 2 denoted the removal of lower level of reshores, and 3 indicated

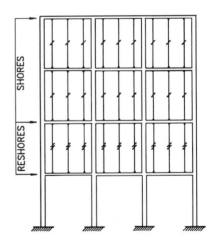

FIGURE 8.9   Structural idealization. From Stivaros and Halvorsen (1991).

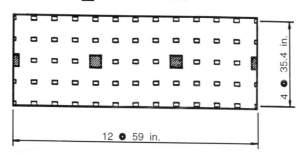

FIGURE 8.10 Typical plan of equivalent frame studied by Stivaros and Halvorsen. *Source:* Courtesy of American Concrete Institute; Stivaros, P.C. and Halvorsen, G.T. 1990. Shoring/reshoring operations for multistory buildings, *ACI Structural J.* 87(5):589–596.

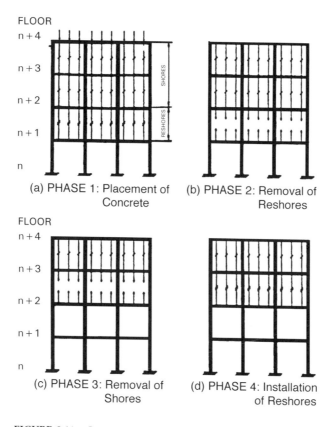

FIGURE 8.11 Construction phases. From Stivaros and Halvorsen (1992).

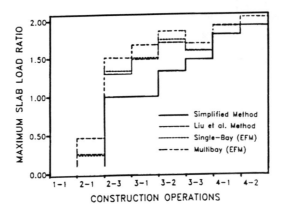

FIGURE 8.12 Comparison of methods of construction load analysis. *Source:* Courtesy of American Concrete Institute; Stivaros, P.C. and Halvorsen, G.T. 1990. Shoring/reshoring operations for multistory buildings. *ACI Structural J.* 87(5):589–596.

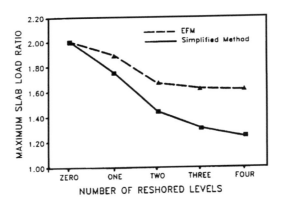

FIGURE 8.13 Comparison of EFM and simplified method: one shored level. *Source:* Courtesy of American Concrete Institute; Stivaros, P.C. and Halvorsen, G.T. 1990. Shoring/reshoring operations for multistory buildings. *ACI Structural J.* 87(5):589–596.

the removal of the lower-level shores. For example, Operation 3-2 referred to the maximum slab or shore loads during the removal of the lower level of reshores after casting the third floor.

The maximum slab loads evaluated using the EFM, both multibay and single-bay models, along with the corresponding results of the simplified Grundy-Kabaila method and results by Liu et al. are plotted in Figure 8.12 with respect to the sequential construction operations. The results reported for the EFM multibay model are those for an interior bay.

Figures 8.13 and 8.14 show the variation of the maximum slab load with respect to the number of reshored levels for both the equivalent frame model and the simplified method for one level and two levels of shoring, respectively. As the figures show, the EFM and the simplified method diverge as the number of reshored levels increases. As the number of shored levels increases (comparing both figures), the diversion of the EFM from the simplified method slows.

Figure 8.14 shows that the combination of two shored levels and one reshored level (the combination used by Liu et al.) is a case where the EFM and the simplified method converge with

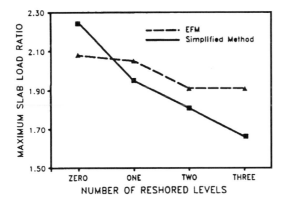

FIGURE 8.14 Comparison of EFM and simplified method: two shored level. *Source:* Courtesy of American Concrete Institute; Stivaros, P.C. and Halvorsen, G.T. 1990. Shoring/reshoring operations for multistory buildings. *ACI Structural Journal* 87(5):589–596.

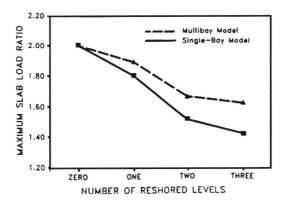

FIGURE 8.15 Comparison of single-bay and multibay models: one-shored level. *Source:* Courtesy of American Concrete Institute; Stivaros, P.C. and Halvorsen, G.T. 1990. Shoring/reshoring operations for multistory buildings. *ACI Structural J.* 87(5):589–596.

a difference of about 5%. The difference increases as the number of reshored levels increases, approaching 15% when three reshored levels are used. Considering the case of the one-shore scheme with four reshored levels from Figure 8.13, the difference between the two methods is about 30%.

Figure 8.15 shows the variation of the maximum slab load with respect to the number of reshored levels for one level of shoring and for both the single-bay and multibay models. As can be seen, the two methods diverge as the number of reshored levels increases. The multibay model always predicts higher maximum slab loads than the single-bay model, with differences as great as 14% in the case of three reshored levels.

The stiffness of the shoring system affects the behavior of both the single-bay and multibay models. Figure 8.16 shows the variation of the maximum slab load of the first floor during the casting of the third floor with respect to the shoring system stiffness for both models. The stiffness

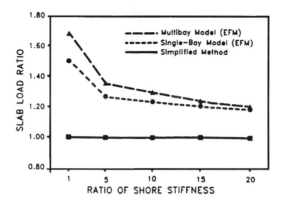

FIGURE 8.16 Comparison of analytical models based on shoring system stiffness. *Source:* Courtesy of American Concrete Institute; Stivaros, P.C. and Halvorsen, G.T. 1990. Shoring/reshoring operations for multistory buildings. *ACI Structural J.* 87(5):589–596.

of the shoring system varies in five increments from the actual value to 20 times the actual value. As the figure shows, the multibay model predicts higher values than the single-bay model. As the stiffness of the shoring system increases, both methods tend to converge. When the stiffness reaches a relatively infinite value, both methods produce almost identical results approaching the simplified method.

The multibay model is more advantageous than the single-bay model because it represents the concrete structure as a more realistic monolithic continuous frame. Furthermore, it can provide information about the total shear force (due to the gravity load and the unbalanced moments) applied to slab-column joints. Considering the advantages that the multibay analysis offers, it is preferable to apply the more realistic continuous multibay model when possible.

To investigate the influence of slab stiffness, the construction example was analyzed for two construction cycles of 3 and 7 d. The percentage of the 28-d concrete strength for the 3- and 7-d cycles were determined using the maturity concept. The age of the slab where the maximum load occurred differed for the two construction schedules. Figure 8.17 shows the maximum slab loads

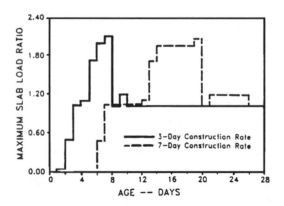

FIGURE 8.17 Comparison of construction rates. *Source:* Courtesy of American Concrete Institute; Stivaros, P.C. and Halvorsen, G.T. 1990. Shoring/reshoring operations for multistory buildings. *ACI Structural J.* 87(5):589–596.

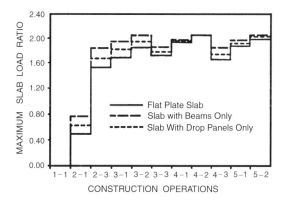

FIGURE 8.18 Comparison of slab types. *Source:* Courtesy of American Concrete Institute; Stivaros, P.C. and Halvorsen, G.T. 1990. Shoring/reshoring operations for multistory buildings. *ACI Structural J.* 87(5):589–596.

with respect to the slab age in days for the 3-and 7-d construction rates. Despite the fact that both schedules predict almost identical maximum slab loads, the ages of the slabs under these loads are different. The maximum load of 2.08 times the slab self-weight for the 3-d schedule occurred on a 7-d-old slab, while the maximum load of 2.06 times the slab self-weight for the 7-d schedule occurred on a 19-d-old slab.

The presence of drop panels around the columns or beams between columns can significantly increase the stiffness of slabs and columns. To determine the influence of drop panels and beams, the construction example was analyzed for two additional conditions. The first condition assumes beams spanning between columns in both directions, with an overall depth of 20 in (510 mm) and width of 12 in (300 mm). The second condition assumes a type of flat-slab construction with square drop panels around columns with depth equal to half the slab thickness of 7.1 in (180 mm) and length equal to one-third the 236 in (6000 mm) span.

Figure 8.18 compares the results of the cases with and without beams or drop panels. The slabs with beams or drop panels share higher loads than the flat plate slab because the beams and the drop panels increase the stiffness of the slab. A similar construction load distribution occurs when the slab thickness, and consequently the slab stiffness, is increased. The degree of influence of beams, drop panels, or the slab thickness on the construction load distribution is totally dependent on the size of these structural members relative to the global size of the equivalent frame as well as on the stiffness of the shoring system. If the stiffness of the shoring system is infinite relative to the stiffness of the supported slabs, the effect of the slab stiffness variation on the load distribution will be minimal.

To determine the influence of the axial shore stiffness on the load distribution, the construction example was investigated by increasing the shoring system stiffness by five increments up to 20 times the actual value. The slab stiffness and the number of shores or reshores per bay were kept constant for this analysis, the results of which are shown in Figure 8.19. As can be seen, the shoring stiffness has a considerable influence on the construction load distribution. The difference in the slab load, for example, is up to 40% during Operation 3-1. Figure 8.19 also incorporates the values predicted by the simplified method and illustrates how the shoring system stiffness affects the slab load during construction.

An important observation from Figure 8.19 is that as the stiffness of the shoring system approaches infinity, the predicted maximum slab loads using the EFM draw near the loads predicted by the simplified method. This clearly shows that the differences of the equivalent frame method and the simplified method are very much due to the shoring system stiffness.

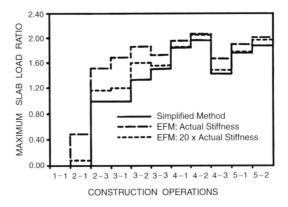

FIGURE 8.19 Comparison of shoring system stiffnesses. *Source:* Courtesy of American Concrete Institute; Stivaros, P.C. and Halvorsen, G.T. 1990. Shoring/reshoring operations for multistory buildings. *ACI Structural J.* 87(5): 589–596.

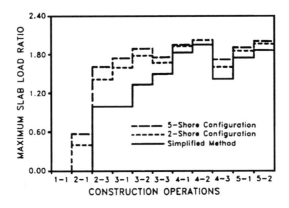

FIGURE 8.20 Comparison of shoring configuration types. *Source:* Courtesy of American Concrete Institute; Stivaros, P.C. and Halvorsen, G.T. 1990. Shoring/reshoring operations for multistory buildings. *ACI Structural J.* 87(5): 589–596.

It has been common practice to consider the shores to be spaced closely enough that shore reactions can be treated as uniformly distributed. This may not be true when a flying truss forming system is used. To determine the effects of various shoring systems, the construction example was investigated using two, three, four, and five shores in each bay. The total stiffness of shores and reshores in each bay was kept constant for all cases and was evenly distributed among the shores and reshores. The slab stiffness was also kept constant. As can be seen from Figure 8.20, the two-shore-per-bay configuration predicts lower values than the five-shore configuration throughout the construction, despite the fact that the total stiffness of the shoring system is the same for both cases. The fewer the number of shores per bay, the lower the construction load the floor slabs share. A comparison between the two-shore and five-shore configurations and the simplified method shows that the two-shore configuration is closer to the simplified method. Thus the simplified method is more suitable for cases with fewer concentrated reaction-type shores per bay. The simplified method underestimates the construction slab loads during the early stages of construction, when a "denser" shoring system is used.

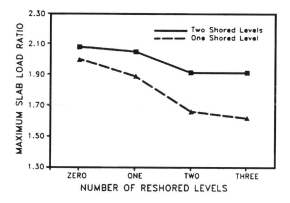

FIGURE 8.21 Influence of the number of reshored levels on the maximum slab load. *Source:* Courtesy of American Concrete Institute; Stivaros, P.C. and Halvorsen, G.T. 1990. Shoring/reshoring operations for multistory buildings. *ACI Structural J.* 87(5):589–596.

Figure 8.21 shows how the maximum slab load changes with respect to the number of reshored levels for combinations with one and two levels of shoring. The maximum slab load decreases at a decreasing rate as the number of reshored levels increases.

According to the simplified method, any applied construction load is distributed equally among the interconnected slabs or in proportion to their relative stiffnesses. As the number of slabs in the shoring system increases, the amount of load that each slab shares decreases. However, when compressible shores or reshores are used, the uppermost floor slabs carry most of the applied construction load, leaving the lower slabs without much load to share. This makes the additional reshored levels redundant and ineffective.

Figure 8.22 shows the relationship between the number of shored levels, along with various combinations of reshores, and the maximum slab load. The figure shows that as the number of the shored levels increases, the maximum slab load also increases. While this is true, the maximum slab load occurs on an older slab, which has more strength, when more shored levels are used.

During the design of the shoring system, it is important to determine the minimum number of shored levels required so that the maximum applied load will occur on a slab that is old enough to

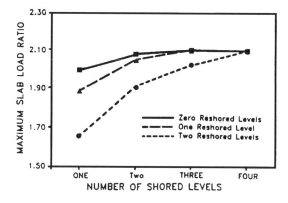

FIGURE 8.22 Influence of the number of shored levels on the maximum slab load. *Source:* Courtesy of American Concrete Institute; Stivaros, P.C. and Halvorsen, G.T. 1990. Shoring/reshoring operations for multistory buildings. *ACI Structural J.* 87(5):589–596.

have developed the necessary strength to withstand this load. Either the number of shored levels or the construction rate can be controlled.

It is also interesting to observe from Figure 8.22 that all reshore combinations, from zero to three reshored levels, tend to converge as the number of shored levels increases. Therefore it is important to determine the maximum number of reshored levels that can cause an appreciable change in the maximum applied construction loads.

8.3 Properties of Concrete at Early Ages

As discussed in detail in Section 8.2, when the usual shore/reshore method of construction is used for multistory buildings, high early-age, short-duration loads are imposed upon the slabs. These loads can be comparable in magnitude to, or even exceed, the design service loads. Also, they are applied to concrete slabs that have not achieved their specified concrete strength. The consequences may very well be structural failure, unless adequate precautions are taken during construction, based on proper evaluation of the construction loads and sound knowledge of the early-age properties of concrete.

Structural failures may be associated with collapse of a structure or part of a structure or structural member, or they may be associated with structural distress that causes a significant reduction in the capability of a structure to function in the way originally intended. The former may be classed as "strength" failures, while the latter may be referred to as "serviceability" failures. Strength failures occur relatively infrequently but can be catastrophic, sometimes leading to loss of life. Serviceability failures occur more frequently and, although not usually catastrophic or life threatening, may result in significant financial losses. Objectionable cracking or excessive floor-slab deflections are examples of serviceability failure.

Heavy construction loads applied to slabs at early ages may be a major contributing factor to strength failures as well as serviceability failures. To understand the strength-related consequences, which may be due to deficiencies of flexural-axial loading strength, shear strength, or bond strength, a knowledge of the compressive strength, tensile strength, shear strength, and bond strength of concrete is necessary. For an understanding of the serviceability related consequences, a knowledge of the modulus of elasticity, shrinkage, and creep of concrete at early ages is essential.

8.3.1 Compressive Strength

Price (1951) and Klieger (1958) have generated considerable information on the development of the compressive strength of concrete under varying temperature and moisture conditions.

Temperature and moisture have pronounced effects on the strength development of concrete. Figure 8.23 [Price, 1951] shows that the development of strength stops at an early age when a concrete specimen is exposed to dry air with no previous moist curing. Concrete exposed to dry air from the time it is placed is about 42% as strong at 6 months as concrete that was continuously moist cured. Specimens cured in water at 70°F were found to be stronger at 28 d than those cured in a fog room at 100% relative humidity. The richer mixes showed up to better advantage than the leaner ones under water curing. The strength of the water-cured specimens was about 10% higher than that of the fog-cured specimens for concrete having water/cement ratios of 0.55 by weight.

Curing temperatures have a marked effect on the strength development of concrete. Test results were obtained by casting and curing concrete specimens at different temperatures, and under such treatment the highest temperatures developed the highest 28-d strength. Other concrete specimens were cured at 70°F after holding the specimens at different casting temperatures for two hours. Such treatment produced opposite results, as the specimens made at the lower temperatures produced the highest 28-d strength. It was speculated that concrete cast at high temperatures is weakened by rapid setting, which is not overcome by subsequent curing at 70°F. Continued curing at higher temperatures for the full 28-d period, as was done for some of the first sets of specimens, accelerated

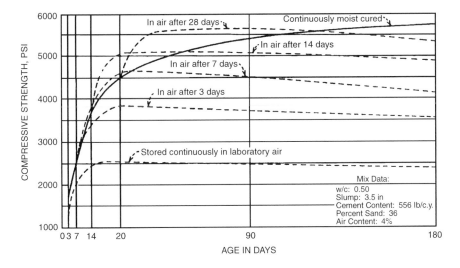

FIGURE 8.23 Compressive strength of concrete dried in laboratory air after preliminary moist curing. *Source:* Courtesy of American Concrete Institute; Price, W.H. 1951. Factors influencing concrete strength. *ACI J.* 22(6):417–432.

the strength development sufficiently to produce the highest strength for the highest temperature. At later ages, however, the specimens made and cured at higher temperatures had lower strengths than those made and cured at lower temperatures.

In Klieger's (1958) investigation, concretes were mixed and placed at temperatures of 40, 55, 73, 90, 105, and 120°F. Specimens were tested at ages 1, 3, 7, and 28 d, 3 months, and 1 year. ASTM Types I, II, and III cements were used. Concretes were made both with and without calcium chloride as an accelerator. An air-entraining agent was added at the mixer to entrain a prescribed amount of air in all of the concretes.

In Part I (73°F and lower), all specimens were continuously moist cured (100% relative humidity) at the mixing and placing temperature for 28 d or less, depending on the test age. After the initial 28-d period, half of the remaining specimens for tests at 3 months and 1 year, were stored at 73°F and 100% relative humidity and the other half at 73°F and 50% relative humidity. Additional concretes were mixed and placed at 40°F and, immediately after placing, the specimens in their molds were stored at 25°F. All surfaces were kept continuously moist (prior to and following removal from the molds at 1 d of age) for 28 d or less, depending upon test age. After 28 d, specimens kept for tests at 3 months and 1 year were treated like those stored at other temperatures.

In Part II (73°F and higher), half of the specimens were moist cured for 7 d at the fabrication temperature, while the remainder of the specimens were moist cured for 28 d at the fabrication temperature. At the end of the 7- and 28-d preliminary curing periods, each half was divided into two groups, one cured at 73°F and 100% relative humidity and the other at 73°F and 50% relative humidity.

For concretes mixed at 40 and 73°F, the net water/cement ratio was held at the value determined to produce the target slump at 55°F. In Part II, the net water/cement ratio for each concrete was such as to produce a certain target slump at a concrete temperature of 90°F. Continuity between Parts I and II was provided by a repetition of the tests for all the concretes at 73°F; there was a small difference in net water/cement ratio between the concretes in Parts I and II.

Two 6 × 6 × 30 in beam specimens were cast for each test age and curing condition. Beams were tested in flexure with load applied at the third points of an 18-in span. Two flexural breaks were obtained for each beam. The two beam ends were tested for compression as 6-in modified cubes. (For the particular aggregate, the ratio of 6 × 12 in cylinder strength to 6-in modified cube strength was taken as 0.93). For each test age, concrete mix, and curing procedure, two beams were tested, yielding four flexural and four compressive test results for averaging. All strength tests were made

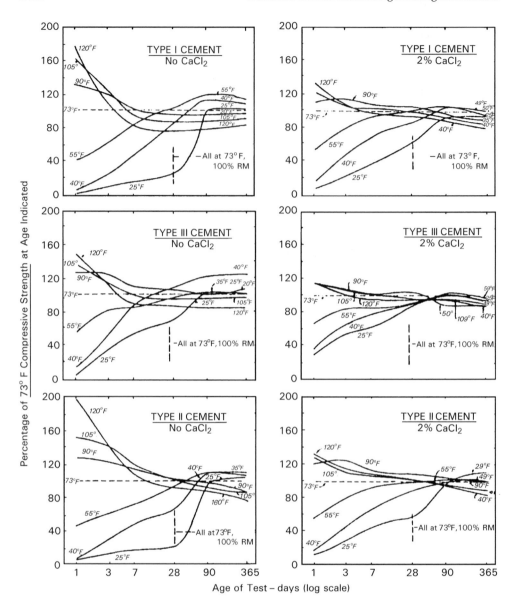

FIGURE 8.24 Effect of temperature on compressive strength of concretes made with different types of cement with and without an accelerator. *Source:* Courtesy of American Concrete Institute; Klieger, P. 1958. Effect of mixing and curing temperature on concrete strength. *ACI J.* 29(12):1063–1081.

at a concrete temperature of 73°F. Specimens stored at temperatures other than 73°F were placed in the 73°F moist room for temperature conditioning 1/2 hr before testing.

Figure 8.24 shows all the compressive strength data for three types of cement with and without accelerator, expressed as percentages of the strengths developed at 73°F for each test age. This figure shows the accelerating effect of temperatures above 73°F on the early-age strengths, with a sacrifice, however, in strength at later ages. On the other hand, concretes placed and cured at temperatures below 73°F, while showing lower strengths at the early ages, show strengths at the later ages in excess of those developed at 73°F. This was true even for concretes mixed and cast at 40°F and stored immediately after casting at 25°F for the first 28 d. For concretes cured initially at low temperatures followed by curing at 73°F, 1-yr strengths close to or exceeding those for the concretes cured continuously moist at 73°F were attained only when moist curing was employed. Air drying during this subsequent 73°F period generally resulted in lower strengths, particularly for concretes made with Type I and Type II cements.

The data in Figure 8.24 are for concretes moist cured the first 28 d. In Part II, a 7-d moist-curing period at the fabrication temperature was included for comparison with the 28-d moist-curing period at the fabrication temperature. The qualitative interpretation of results for this 7-d group was similar to that for the 28-d group.

The following conclusions were drawn by Klieger (1958):

1. At 1, 3, and 7 d, concrete compressive strengths increase with an increase in the initial and curing temperatures of the concrete.
2. Increasing the initial and curing temperatures results in considerably lower compressive strengths at 3 mo and 1 yr.
3. Compressive strengths of concretes made with the three cement types used were influenced in a like manner by temperature; differences were in degree only.
4. These tests indicated that there is a temperature during the early life of concrete, which may be considered optimum with regard to strength at later ages, or more strictly, at comparable degrees of hydration. This temperature is influenced somewhat by cement type. For Types I and II, this temperature appeared to be 55°F; for Type III, it was 40°F.
5. For concretes with calcium chloride added, compressive strength increases due to the accelerator were proportionately greater at early ages and lower temperatures.

ACI Committee 209 (1971) has recommended the following expression for the time-dependent strength of moist-cured (as distinct from steam-cured) concrete using Type I cement:

$$f'_{ct} = \frac{t}{4.00 + 0.85t} f'_c \tag{8.1}$$

where t is the time in days from casting up to loading and f'_c is the 28-d compressive strength of concrete.

The validity of Eq. (8.1) at very early ages needs to be examined in view of the data reported above and a significant volume of European data that is also available [Byfors, 1980; RILEM Commission 42-CEA, 1981].

8.3.2 Tensile Strength and Bond Strength

Price (1951) reported data (Table 8.6) from tests at the Portland Cement Association, showing relationships among the compressive strength, the tensile strength, and the flexural strength (modulus

TABLE 8.6 Comparison of Compressive, Flexural and Tensile Strength of Plain Concrete

Strength of Plain Concrete, psi			Ratio, %		
Compressive	Modulus of Rupture	Tensile	Modulus of Rupture to Compressive Strength	Tensile Strength to Compressive Strength	Tensile Strength to Modulus of Rupture
1000	230	110	23.0	11.0	48
2000	375	200	18.8	10.0	53
3000	485	275	16.2	9.2	57
4000	580	340	14.5	8.5	59
5000	675	400	13.5	8.0	59
6000	765	460	12.8	7.7	60
7000	855	520	12.2	7.4	61
8000	930	580	11.6	7.2	62
9000	1010	630	11.2	7.0	63

Source: Price, W.H. 1951. Factors influencing concrete strength. *ACI J. Proc.* 22(6):417–432.

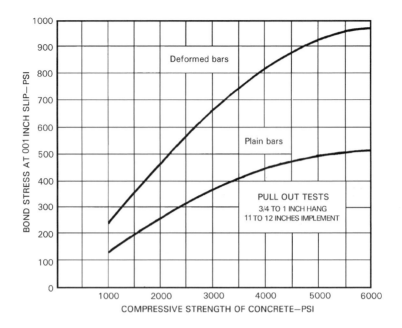

FIGURE 8.25 Variation of bond with strength of concrete. *Source:* Courtesy of American Concrete Institute; Price, W.H. 1951. Factors influencing concrete strength. *ACI J.* 22(6):417–432.

of rupture) of concrete. It was not stated, but may probably be assumed, that the tensile strength reported was the splitting tensile strength. "The ratio of tensile to compressive strength decreases as the compressive strength increases and approaches a constant of about 7% for higher compressive strength. Other published data indicate that the ratio of tensile to compressive strength decreases as the age of concrete increases to the same constant value of 7%" [Price, 1951].

Figure 8.25 from Price (1951) shows in a qualitative way the relationship between bond strength and compressive strength of concrete for plain and deformed bars. The ratio of bond strength to compressive strength decreases as compressive strength increases. The bond strength of the deformed bars is 24% of the compressive strength for 2000 psi concrete and 18% of the compressive strength for 5000 psi concrete. "The relationship of bond to compressive strength apparently is not changed materially by air entrained in the proportions recommended."

Klieger's (1958) flexural strength data for the same concretes as were tested for Figure 8.24 showed that the discussion about compressive strength data in the previous section applied equally to the flexural strength data. Flexural strengths at early ages increased with increase in temperature. At later ages, the effect of temperature was reversed. Concretes made and cured at lower temperatures showed the highest flexural strengths at 1 yr. The optimum temperatures for flexural strength development appeared to be the same as those for compressive strength. The use of calcium chrloride frequently resulted in flexural strengths at later ages that were somewhat lower than for comparable concretes without calcium chrlorides. Maximum reductions noted were on the order of 10%.

In a significant investigation by Gardner and Poon (1976), six series of specimens were cast: three using Type I and three using Type III cement concretes. The specimens for each series were cast from a single batch of ready-mixed concrete. Two series of tests, one made with Type I cement concrete and one with Type III cement concrete, were carried out with the concrete continuously moist cured at 72°F (22°C). Two series of specimens, one made using Type I cement concrete and one with Type III cement concrete, were subjected to extended curing, under wet burlap, at a steady temperature of 55°F (13°C). The remaining two series were subjected to extended curing under wet burlap at a steady temperature of 35°F (2°C).

All specimens were cast at 72°F (22°C). To examine the effect of time on the concrete, specimens were cured at 72°F (22°C) for different periods before exposing them to low temperatures. One-third of the specimens of each series were cured for 1 d, one-third were cured for 3 d, and the remainder were cured for 7 d at 72°F (22°C) before being transferred to the cooler environment.

Each series involved the casting of 160 standard 6 × 12 in (150 × 300 mm) cylinders and 80 bond specimens. Five specimens were tested for compressive strength, five for tensile strength, and five for bond strength at 1, 3, 7, 14, and 28 d and at 3 mo for each of the 14 curing schedules. All concretes had a specified cylinder strength of 4000 psi (28 MPa) and a water/cement ratio of approximately 0.5. Compressive and tensile strengths were determined by the standard compressive strength test and the split-cylinder test, respectively, on 6 × 12 in (150 × 300 mm) cylinders. Bond specimens were made by casting a #6 reinforcing bar into a 6 × 6 in (150 × 150 mm) cylindrical block of concrete to closely approximate the ASTM standard specimen, which is a 6 in (150 mm) concrete cube or a 6 × 6 × 12 in (150 × 150 × 300 mm) block of concrete. Concrete strengths are plotted against the log of time in Figure 8.26 for concretes made with Type I cement and subjected to extended curing at 35°F (2°C).

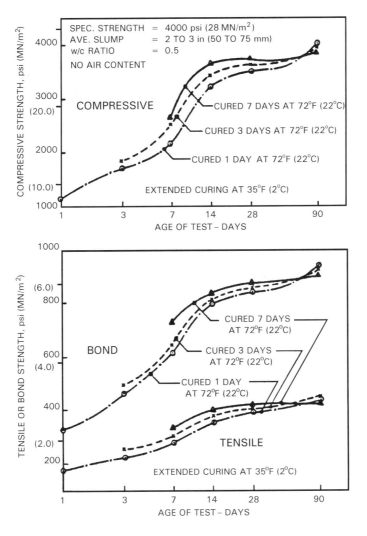

FIGURE 8.26 Variation of concrete strength with temperature for Type I cement concrete cured at 35°F (2°C). *Source:* Courtesy of American Concrete Institute; Gardner, N.J. and Poon, S.M. 1976. Time and temperature effects on tensile, bond, and compressive strength. *ACI J.* 73(7):405–409.

Figure 8.26 was typical of all six batches of concrete. The following conclusions were drawn:

1. Temperature influences the tensile and bond strength development of concrete in much the same manner as it does compressive strength development. Compressive strength, tensile strength, and bond strength are all related, and an increase in one is reflected in the others.
2. The compressive strength, tensile strength, and bond strength of concrete at early ages increases with increased curing temperature. The lower the initial curing temperature the greater the eventual ultimate strength of the concrete, provided curing is continuous.
3. Bond strength and tensile strength are proportional to the 0.8 power of the cylinder strength at the appropriate age. Neither extended curing at temperatures of 35°F (2°C) and 55°F (13°C) nor type of cement appears to have any significant effect on the interrelationship of bond strength or tensile strength and cylinder strength.
4. With respect to construction schedules, there is a 5% to 9% gain in the 7-d and 14-d strengths due to casting and curing Type I cement concretes for 3 d and 7 d, respectively, at 72°F (22°C), compared to casting and curing for 1 d at 72°F (22°C).
5. Type III cement concretes exhibited a small strength gain due to prolonged initial curing at 72°F (22°C) when subjected to extended curing at 35°F (2°C) but no strength gain for prolonged initial curing at 72°F (22°C) when subjected to extended curing at 55°F (13°C).

Lew and Reichard (1978) performed three types of tests:

1. compressive strength tests of cylindrical specimens,
2. splitting tensile tests of cylindrical specimens, and
3. bond-strength tests using cylindrical pullout specimens.

All three types of tests were made on specimens cured at different temperatures [35°F (2°C), 55°F (13°C), and 73°F (22°C)] and tested at nine ages (1, 2, 3, 5, 7, 14, 21, 28, and 42 d).

The combined effect of temperature and time, or "maturity," usually expressed in degree-days (or hours), may be defined as the sum of the product of the increment of age of cure and the difference between curing and some temperature below which no strength gain takes place. The definition can be written as

$$M = \sum (T - T_0) \Delta t \qquad (8.2)$$

where T is temperature of the concrete at any time, T_0 is a datum temperature below which no strength gain of concrete takes place, and Δt is the increment of time. Taking $T_0 = 10°F (-12°C)$ on the basis of prior tests, Eq. (8.2) can be rewritten as

$$M = \sum (T - 10) \Delta t \qquad (8.3)$$

(where T is in degrees Farenheit). Analysis of Lew and Reichard's test data showed that the compressive strength of concrete can be related to maturity expressed in terms of concrete temperature and age of cure. When related to maturity, the elastic modulus data and the splitting tensile strength data of specimens cured at various temperatures could be treated as being from a single group.

The pullout strength, when not governed by the yielding of the bar, could be expressed in terms of maturity, thus allowing the specimens cured at various ages to be treated as a single group as well.

Figure 8.27 from [Lew and Reichard, 1978] shows that, at early ages, the rate of increase in the splitting tensile strength is about the same as that of the compression strength, whereas the rate of increase in the pullout bond strength and the modulus are slightly greater than that of the compressive strength.

Carino and Lew (1982) performed a series of statistical analyses on published data of Gardner and Poon (1976), on published data from the National Bureau of Standards (NBS) [Lew and Reichard, 1978], and on previously unpublished data from NBS. These data were used to determine the best relationship between the splitting tensile strength and the compressive strength of normal-weight

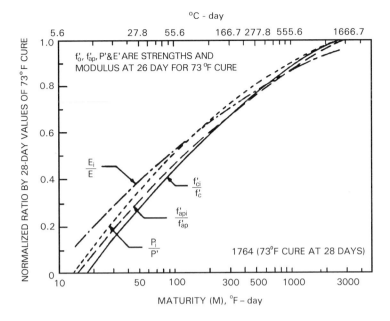

FIGURE 8.27 Mechanical properties normalized with respect to 28-d values versus maturity. *Source:* Courtesy of American Concrete Institute; Lew, H.S. and Reichard, T.W. 1978. Mechanical properties of concrete at early ages. *ACI J.* 75(10):533–542.

concrete, which was then compared with a larger group of published data. The unpublished data were from tests of specimens made with concretes similar to those reported in Lew and Reichard (1978), but which were allowed to cure under ambient, outdoor conditions. Three replicate specimens were tested at ages ranging from 1 to 37 d.

This study clearly indicated that the assumed proportionality of splitting tensile strength to the square root of compressive strength is not the most precise relationship when dealing with a wide range of compressive strengths. The square-root relationship was originally formulated on the basis of tests on mature concrete with strengths generally greater than 2500 psi (17.2 MPa). For low compressive strength, the current ACI formula overestimates splitting tensile strength and underestimates it for high compressive strength. A simple power law of the form $f'_{sp} = (f'_c)^b$ was found to be a more precise representation of data over the full range of concrete strengths. For the Gardner-Poon data and data from the two series of NBS tests mentioned above, the best-fit estimate of b was found to be 0.73. When additional data from four different published references and from other unpublished work at NBS were added to the mix, $f'_{sp} = (f'_c)^{(0.71)}$ seemed to be appropriate for estimating the average expected splitting tensile strength. The function $f'_{sp} = (f'_c)^{(0.71)}$ appeared to give a reasonable lower bound estimate, overestimating about 10% of the data.

8.3.3 Punching Shear Strength

According to Nilson (1991) (who was discussing mature concretes), a reasonable estimate for the split-cylinder strength f'_{sp} is 6 to 7 times the square root of f'_c for normal weight concretes. The true tensile strength f'_t is on the order of 0.5 to 0.7 times f'_{sp} (3 to 5 $\sqrt{f'_c}$) and the flexural tension strength f_r (modulus of rupture) varies from 1.25 to 1.75 times f'_{sp} (7.5 to 12 $\sqrt{f'_c}$). The smaller of the foregoing factors apply to higher-strength concretes and the larger to lower-strength concretes.

The modulus of rupture controls the behavior of flexural members subject to large bending moments and small shear forces; for large shear forces, the concern becomes safety against premature failure due to diagonal tension in the concrete, resulting from combined shear and longitudinal flexural stresses. In flexural members subject to large shear forces and small bending moments, "web-shear" cracks can be expected to form when the diagonal tension stress in the vicinity of the neutral axis becomes equal to the tension strength of the concrete (3 to 5 $\sqrt{f_c'}$). In members subject to large shear forces as well as large bending moments, "flexure-shear" cracks form when the diagonal tension stress at the tip of one or more flexural cracks exceeds the tensile strength of the concrete.

The information presented on the early-age modulus of rupture and splitting cylinder strength of concrete in the previous section is directly relevant to the flexural and diagonal tension failure of beams and one-way slabs. However, as noted, one of the common causes of failure of flat-plate structures during construction is insufficient early-age punching shear strength under relatively high construction loads. Direct knowledge of early-age punching shear strength was thus thought to be desirable and was investigated by Gardner (1960).

A relatively large number of circular flat plates were made and subjected to a centrally applied load through a circular steel plate. All steel was located on the tension side of the slabs, and steel ratios ranged from 1/2 to 5%. Concrete strengths ranged from 2000 to nearly 8000 psi (14 to 56 MPa). Slab thicknesses were 2, 4, and 6 in (50, 100, and 150 mm). An initial set of experiments was undertaken to determine how punch diameter relates to slab thickness, slab outer diameter, and slab support diameter.

For punching shear failure to occur, at least three slab thicknesses are necessary between the punch and the slab support. Further, the slab outer diameter must be sufficient to enable the slab steel to develop any necessary stress; otherwise, the slab will fail in bond, separating at the reinforcement level. All specimens with zero steel or steel ratios less than 1/2% failed in flexure. A few conclusions, given below, were drawn.

1. The steel ratio in the region 3d from the column should be of the order of 0.5 % in each direction, and the spacing should be similar to the effective depth.
2. In all cases, slabs should be detailed so that some of the top (negative-moment tension) steel passes through the column.
3. The cube-root relationship between shear strength and concrete strength is preferable to the square-root relationship currently used by ACI 318 (1995).
4. If the punching shear strength is in doubt, the shear perimeter should be increased by using larger columns or column capitals.

8.3.4 Modulus of Elasticity

Figure 8.28 from Byfors (1980) provides an idea of how the stress-strain relationship of concrete in compression changes with age. The stress and the strain are related to the compressive strength and the corresponding strain, respectively. The arrows in the figure indicate possible unloading. The relationship is practically a straight-line curve at a very early age (5.0 h). It can be seen that the strain is mainly inelastic. At this early stage, concrete shows properties that are similar to those of a claylike material. A few hours later (7.9 h), the stress-strain relationship has a completely different character. One can see a clear transition from the elastic to the inelastic part. After 15.8 h the curve begins to get the appearance that normally characterizes concrete. The continued changes in the appearance of the stress-strain relationship are not, however, as radical any longer. Expressed in absolute values, the changes are nevertheless large.

Modulus of elasticity, also referred to as elastic modulus, Young's modulus, and Young's modulus of elasticity, may be defined in general terms as the ratio of normal stress to corresponding strain for tensile or compressive stress below the proportional limit of the material. Few materials, however, conform to Hooke's law throughout the entire range of stress-strain relations; deviations therefrom are caused by inelastic behavior. If the deviations are significant, the slope of the tangent to the stress-strain curve at the origin, the slope of the tangent to the stress-strain curve at any given stress, the slope of the secant drawn from the origin to any specified point on the stress-strain curve, or the slope of the chord connecting any two specified points on the stress-strain curve, may be considered

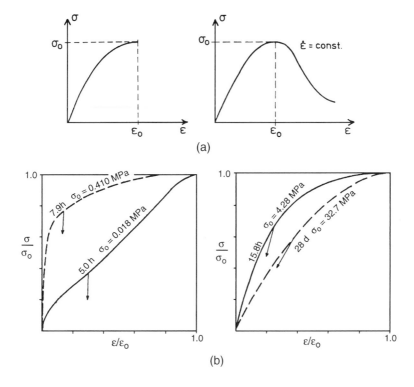

FIGURE 8.28 (a) Basic appearance of the compressive stress-strain relationship under constant stress rate (left) and constant deformation rate (right). (b) Change of shape of compressive stress-strain relationship at early ages [Byfors, 1980].

to be the modulus; in such cases, the modulus is designated, as the initial tangent modulus, the tangent modulus, the secant modulus, or the chord modulus, respectively, and the stress is stated. Normally, the static modulus of elasticity of concrete is determined as the secant modulus based on short-term experiments but with a certain number of repeated loadings at a low stress level. The stress level for the secant's intersection is normally 30–50% of strength. Studies leading to the expression for modulus of elasticity of concrete in ACI 318 are summarized in Pauw (1960), where E_c was defined as the slope of the line drawn from a stress of zero to a compressive stress of $0.45 f_c'$.

According to ACI Committee 209 (1971) the effect of age of concrete at the time of loading on the modulus of elasticity of concrete is accounted for when strength of concrete at the time of loading, rather than at 28 d, is inserted into the modulus of elasticity expression given in ACI 318 (1995):

$$E_{ct} = 33w^{1.5}\sqrt{f_{ct}'} \tag{8.4}$$

where E_{ct} is the time-dependent modulus of elasticity of concrete in pounds per square inch, w is the unit weight of concrete in pounds per cubic feet, and f_{ct}' is the time-dependent concrete strength in pounds per square inch.

Equation (8.4) for normal-weight concrete, when converted to SI units and rounded off, becomes

$$E_{ct} = 5000\sqrt{f_{ct}'} \tag{8.5}$$

where E_{ct}, f_{ct}' are in megapascals. Figure 8.29 shows a comparison of Eq. (8.5) with the experimental information currently available on the modulus of elasticity of concrete at early ages [RILEM Commision 42-CEA, 1981]. Concrete strengths of 1 MPa (145 psi) and less are encountered only

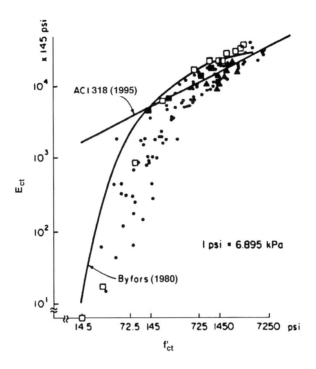

FIGURE 8.29   Modulus of elasticity of concrete at early ages. *Source:* Courtesy of Portland Cement Association; Fintel, M. et al. 1987. Column shortening in tall structures—prediction and compensation. Publication EB108-01D, Portland Cement Association, Skokie, IL.

during the first few hours after concrete is poured and are not of interest here. In the range of practical interest, the ACI equation relating modulus of elasticity and strength agrees quite favorably with the trend of experimental results.

8.3.5 Shrinkage of Unreinforced (Plain) Concrete

Shrinkage of concrete is caused by the evaporation of moisture from the surface. The rate of shrinkage is high at early ages and decreases with an increase in age until the curve becomes asymptotic to the final value of shrinkage. The rate and amount of evaporation and consequently of shrinkage depend greatly upon the relative humidity of the environment, size of the member, and mix proportions of the concrete. In a dry atmosphere, moderate-size members (24-in or 600-mm diameter) may undergo up to half of their ultimate shrinkage within two to four months, while identical members kept in water may exhibit growth instead of shrinkage. In moderate-size members, the inside relative humidity has been measured at 80% after four years of storage in a laboratory at 50% relative humidity.

8.3.5.1 Basic Value of Shrinkage

Let e_s denote the ultimate shrinkage of 6-in (150-mm)-diameter standard cylinders (volume-to-surface or $v:s$ ratio = 1.5 in or 38 mm) moist-cured for 7 d and then exposed to 40% ambient relative humidity. If concrete has been cured for less than 7 d, multiply e_s by a factor linearly varying from 1.2 for 1 d of curing to 1.0 for 7 d of curing [ACI Committee 209, 1971].

   Attempts have been made in the past to correlate e_s with parameters such as concrete strength. In view of available experimental data [Russell and Corley, 1977], it appears that no such correlation may in fact exist. The only possible correlation is probably that between e_s and the water content of a concrete mix [Troxell et al., 1968]. In the absence of specific shrinkage data for concretes to be used in a particular structure, the value of e_s may be taken as between 500×10^{-6} in/in (mm/mm)(low value) and 800×10^{-6} in/in (mm/mm) (high value) [Fintel et al., 1987]. The latter value has been recommended by ACI Committee 209 (1971).

8.3.5.2 Effect of Member Size

Since evaporation occurs from the surface of members, the volume-to-surface ratio of a member has a pronounced effect on the amount of its shrinkage. The amount of shrinkage decreases as the size of the specimen increases.

For shrinkage of members having volume-to-surface ratios different from 1.5, e_s must be multiplied by the following factor:

$$SH_{v:s} = \frac{0.037(v{:}s) + 0.944}{0.177(v{:}s) + 0.734} \tag{8.6}$$

where $v{:}s$ is the volume-to-surface ratio in inches. The metric equivalent of Eq. (8.6) is

$$SH_{v:s} = \frac{0.015(v{:}s) + 0.944}{0.070(v{:}s) + 0.734} \tag{8.6'}$$

where $v{:}s$ is the volume-to-surface ratio in millimeters. As indicated in Fintel et al. (1987), Eq. (8.6) is based on laboratory data (Hansen and Mattock, 1966) and European recommendations [*Recommendations for an International Code of Practice for Reinforced Concrete*; CEB, 1972].

Much of the shrinkage data available in the literature was obtained from tests on prisms of a 3×3-in (75×75-mm) section ($v{:}s = 0.75$ in or 19 mm). According to Eq. (8.6), the size coefficient for prisms of that size is 1.12. Thus shrinkage measured on a prism of a 3×3-in (75×75-mm) section must be divided by 1.12 before the size coefficient given by Eq. (8.6) is applied. It should be cautioned, however, that as the specimen size becomes smaller, the extrapolation to full-size members becomes less accurate.

8.3.5.3 Effect of Relative Humidity

The rate and amount of shrinkage greatly depend upon the relative humidity of the environment. If ambient relative humidity is substantially greater than 40%, e_s must be multiplied by

$$SH_H = \begin{cases} 1.40 - 0.010H & \text{for } 40 \leq H \leq 80 \\ 3.00 - 0.030H & \text{for } 80 \leq H \leq 100 \end{cases} \tag{8.7}$$

where H is the relative humidity in percent. Average annual values of H should probably be used. Maps giving average annual relative humidities for locations around the United States are available [Fintel et al., 1987]. However, if locally measured humidity data are available, they are likely to be more accurate than information such as that included in Fintel et al. (1987) and should be used in conjunction with Eq. (8.7). Equation (8.7) is based on ACI Committee 209 (1971) recommendations. A comparison with European recommendations is shown in Fintel et al. (1987). If shrinkage specimens are stored under job site conditions rather than under standard laboratory conditions, the correction for humidity, as given by Eq. (8.7), should be eliminated.

8.3.5.4 Progress of Shrinkage with Time

Hansen and Mattock (1966) established that the size of a member influences not only the final value of shrinkage but also the rate of shrinkage, which appears to be only logical. Their expression giving

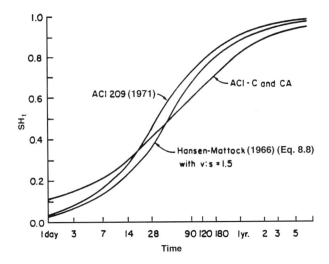

FIGURE 8.30   Progress of shrinkage with time. *Source:* Courtesy of Portland Cement Association; Fintel, M. et al. 1987. Column shortening in tall structures—prediction and compensation. Publication EB108-01D, Portland Cement Association, Skokie, IL.

the progress of shrinkage with time is

$$SH_t = \frac{\epsilon_{st}}{\epsilon_s} = \frac{t}{26.0e^{0.36(v:s)} + t} \tag{8.8}$$

where ϵ_{st} and ϵ_s are shrinkage strains up to time t and time infinity, respectively; and t is measured from the end of moist curing.

Equation (8.8) is compared in Figure 8.30 with the progress of shrinkage curve from *Recommendations for an International Code of Practice for Reinforced Concrete*. Also shown in Figure 8.30 is a comparison of Eq. (8.8) with the progress of shrinkage relationship recommended by ACI Committee 209 (1971). It should be noted that both the ACI and Cement and Concrete Association (C&CA) [*Recommendations for an International Code of Practice for Reinforced Concrete*] and the ACI Committee 209 (1971) relationships are independent of the volume-to-surface ratio.

8.3.6 Creep of Unreinforced (Plain) Concrete

Creep is a time-dependent increment of the strain of a stressed element that continues for many years. The basic phenomenon of creep is not yet conclusively explained. During the initial period following the loading of a structural member, the rate of creep is significant. The rate diminishes as time progresses until it eventually approaches zero.

Creep consists of two components:

1. Basic (or true) creep that occurs under conditions of hygral equilibrium, which means that no moisture movement occurs to or from the ambient medium. In the laboratory, basic creep can be reproduced by sealing a specimen in copper foil or by keeping it in a fog room.
2. Drying creep that results from an exchange of moisture between the stressed member and its environment. Drying creep has its effect only during the initial period under load.

Creep of concrete is very nearly a linear function of stress up to stresses that are about 40% of the ultimate strength. This includes all practical ranges of stresses in columns and walls. Beyond this level, creep becomes a nonlinear function of stress.

For structural engineering practice, it is convenient to consider specific creep, which is defined as the ultimate creep strain per unit of sustained stress.

8.3.6.1 Value of Specific Creep

Specific creep values can be obtained by extrapolating results from a number of laboratory tests performed on samples prepared from the actual mix to be used in a structure. It is obvious that sufficient time for such tests must be allowed prior to the start of construction, since reliability of the prediction improves with the length of time over which creep is actually measured.

A way of predicting basic specific creep (excluding drying creep), without testing, from the modulus of elasticity of concrete at the time of loading was proposed by Hickey (1968) on the basis of long-term creep studies at the Bureau of Reclamation in Denver. A simpler suggestion was made in Fintel et al. (1987), as described below.

Let ϵ_c denote the specific creep (basic plus drying) of 6-in (150-mm)-diameter standard cylinders ($v{:}s = 1.5$ in or 38 mm) exposed to 40% relative humidity following about 7 d of moist curing and loaded at the age of 28 d. In the absence of specific creep data for concretes to be used in a particular structure, the following likely values of ϵ_c may be used:

$$\epsilon_c = 0.003/f_c' \text{ (low value) to } 0.005/f_c' \text{ (high value)} \tag{8.9}$$

where ϵ_c is in inch per inch per kips per square inch if f_c' is in kips per square inch, or in inch per inch per pounds per square inch if f_c' is in pounds per square inch. The metric equivalent of Eq. (8.9) is

$$\epsilon_c = 0.000435/f_c' \text{ (low value) to } 0.000725/f_c' \text{ (high value)} \tag{8.9'}$$

where ϵ_c is in millimeter per millimeter per megapascal if f_c' is in megapascal, or in millimeter per millimeter per kilopascal if f_c' is in kilopascal. The lower end of the proposed range is in accord with specific creep values suggested by Neville (1981). The upper end agrees with laboratory data obtained by testing the concretes used in Water Tower Place in Chicago [Russell and Corley, 1977].

8.3.6.2 Effect of Age of Concrete at Loading

For a given mix of concrete, the amount of creep depends not only on the stress level but also to a great extent on the age of the concrete at the time of loading. Figure 8.31 shows the relationship between creep and age at loading as developed by Comité Europeen du Beton (CEB), using available information from many tests [CEB, 1972]. The coefficient CR_{LA} relates the creep for any age at loading to the creep of a specimen loaded at the age of 28 d. The 28-d creep is used as a basis of comparison, the corresponding CR_{LA} being equal to 1.0.

Figure 8.31 also depicts the following suggested relationship [Fintel et al., 1987] between creep and age at loading:

$$CR_{LA} = 2.3t_{LA}^{-0.25} \tag{8.10}$$

where t_{LA} is the age of concrete at the time of loading, in days. The form of Eq. (8.10) is as suggested by ACI Committee 209 (1971). Equation (8.10) gives better correlation with the CEB mean curve than the corresponding equation suggested by Committee 209. Figure 8.31 also shows comparison with a few experimental results [Lew and Reichard, 1978; Pfeifer et al., 1971]. According to Eq. (8.10), the creep of concrete loaded at 7 d age is 41% higher than that of concrete loaded at 28 d.

8.3.6.3 Effect of Member Size

Creep is less sensitive to member size than shrinkage, since only the drying-creep component of the total creep is affected by the size and shape of members, whereas basic creep is independent of size and shape.

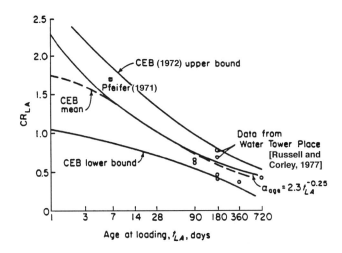

FIGURE 8.31   Creep versus age of concrete at time of loading. *Source:* Courtesy of Portland Cement Association; Fintel, M. et al. 1987. Column shortening in tall structures—prediction and compensation. Publication EB108-01D, Portland Cement Association, Skokie, IL.

For members with volume-to-surface ratios different from 1.5 in or 38 mm, ϵ_c should be multiplied by

$$CR_{v:s} = \frac{0.044(v:s) + 0.934}{0.1(v:s) + 0.85} \qquad (8.11)$$

where *v:s* is the volume-to-surface ratio in inches. The metric equivalent of Eq. (8.11) is

$$CR_{v:s} = \frac{0.017(v:s) + 0.934}{0.039(v:s) + 0.85} \qquad (8.11')$$

where *v:s* is the volume-to-surface ratio in millimeters. As indicated in Fintel et al. (1987), Eq. (8.11) is based on laboratory data [Hansen and Mattock, 1966] and European recommendations [*Recommendations for an International Code of Practice for Reinforced Concrete*; CEB, 1972].

Much of the creep data available in the literature was obtained by testing 6-in (150-mm)-diameter standard cylinders wrapped in foil. The wrapped specimens simulate very large columns. Equation (8.11) yields a value of $CR_{v:s}$ equal to 0.49 for $v:s = 100$. Thus it is suggested that creep data obtained from sealed specimen tests should be multiplied by $2(1/0.49 \cong 2)$ before the modification factor given by Eq. (8.11) is applied to such data.

8.3.6.4 Effect of Relative Humidity

For an ambient relative humidity greater than 40%, ϵ_c should be multiplied by the following factor, as suggested by ACI Committee 209 (1971).

$$CR_H = 1.40 - 0.01H \qquad (8.12)$$

where *H* is the relative humidity in percent. Again, it is suggested that the average annual value of *H* be used.

8.3.6.5 Progress of Creep with Time

The progress of creep relationship recommended by ACI Committee 209 (1971) is given by the following expression:

$$CR_t = \frac{\epsilon_{ct}}{\epsilon_c} = \frac{t^{0.6}}{10 + t^{0.6}} \qquad (8.13)$$

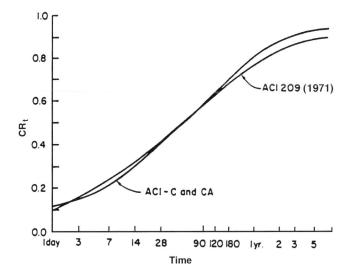

FIGURE 8.32 Progress of creep with time. *Source:* Courtesy of Portland Cement Association; Fintel, M. et al. 1987. Column shortening in tall structures—prediction and compensation. Publication EB108-01D, Portland Cement Association, Skokie, IL.

where ϵ_{ct} is creep strain per unit stress up to time t and t is measured from the time of loading. This above relationship is plotted in Figure 8.32 where it compares well with the creep-versus-time curve suggested in European recommendations [*Recommendations for an International Code of Practice for Reinforced Concrete*].

8.3.6.6 Irreversible Nature of Creep

Another important aspect of creep of concrete needs to be addressed. Figure 8.33 schematically shows the strain history of a concrete specimen, that was stored in air at near 100% relative humidity to eliminate shrinkage and subjected to a sustained axial stress that was removed at 120 d from the time of loading. This instantaneous recovery is followed by a gradual decrease in strain, called creep recovery. The shape of the creep-recovery curve is similar to that of the creep curve, but the recovery approaches its maximum value much more rapidly. The reversal of creep is not complete, so that any sustained application of load, even over only a day, results in a residual deformation [Neville, 1981].

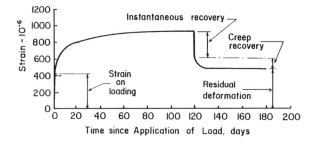

FIGURE 8.33 Creep strain versus time when part of sustained load is removed at a certain time after application *Source:* Courtesy of Portland Cement Association; Fintel, M. et al. 1987. Column shortening in tall structures—prediction and compensation. Publication EB108-01D, Portland Cement Association, Skokie, IL.

8.3.7 Effects of Drying on Flexural Cracking

Hover (1988) pointed out that the resistance to flexural cracking in beams and slabs decreases sharply during periods in which the surface of the concrete is drying, such as the period that immediately follows the removal of forms.

Hover considered a 6-in (150-mm) thick reinforced concrete slab with a concrete compression strength of 4000 psi (28 MPa). The solid lines in Figure 8.34a show the increase in the "allowable" superimposed live load as a function of the amount of reinforcing steel for continuous spans of 14 to 18 ft (4.2 to 5.4 m). The allowable load was determined by subtracting the factored dead load from the ultimate load and dividing the result by an appropriate load factor. The dotted lines in Figure 8.34a indicate the superimposed live load that caused cracking of the same slabs. Note that for lightly reinforced slabs, the cracking load is actually greater than the safe allowable load. This means that cracking will be avoided by complying with the load limitations imposed on the basis of strength calculations. While the allowable load on the basis of strength considerations increases dramatically with an increasing amount of reinforcing steel, the cracking load increases by only 25% or so for an 8-fold increase in the amount of reinforcing steel. For members with the maximum allowable amount of reinforcing steel, the safe load on the basis of strength exceeds the cracking load by a factor of four. One would therefore expect heavily reinforced slabs and beams to crack at loads well below the safe or allowable load calculated on the basis of strength.

In Figure 8.34b it is assumed that the same slabs have a compressive strength of only 3000 psi (21 MPa) at the time of load application. Note that this case corresponds to the above 4000 psi (28 MPa) concrete about 7 d after casting, when the concrete has attained approximately 75% of its expected 28-d strength. Comparison of Figures 8.34a with 8.34b shows an approximate 6% decrease in the allowable load on the basis of strength and an approximate 20% decrease in the cracking load for a compression strength decrease from 4000 to 3000 psi (28 to 21 MPa). Thus low concrete compressive strength has a more significant effect on resistance to cracking than on the strength of the slabs considered in these examples.

Calculation of the cracking load in the above example was done using $f_r = 7.5\sqrt{f_c'}$ as in ACI 318 (1995). Walker (1957a,b) and Hover (1984) studied the influence of surface drying on flexural tension strength. The essential results are shown in Figure 8.35, from Walker and Bloem (1957a), in which the modulus of rupture is shown as a function of the time during which the concrete test specimen is exposed to air drying. The modulus of rupture (MOR) of wet concrete is seen to decrease

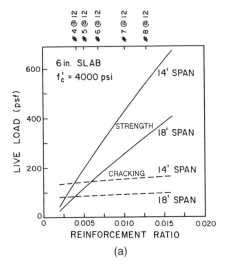

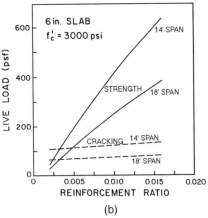

FIGURE 8.34 Allowable live load calculated on the basis of strength versus cracking live load (a) $f_c' = 4000$ psi (28 MPa); (b) $f_c' = 3000$ psi (21 MPa), *Source:* Courtesy of American Concrete Institute; Hover, K.C. 1988. The effects of drying and form and shore removal on flexural cracking in beams and slabs. In *Forming Economical Concrete Buildings: Proceedings of the Third International Conference*, Publication SP107, pp. 169–184. American Concrete Institute, Farmington Hill, MI.

immediately upon exposure to air. This decrease continues for several days of air drying until a minimum is reached after 3 to 6 d, followed by a slow restoration of the original wet MOR. Continued slow drying eventually leads to a modulus of rupture that is greater than the original "wet-tested" value. It was found that the MOR could vary from a high of $16\sqrt{f_c'}$ to an apparent lower limit of approximately $4\sqrt{f_c'}$, depending entirely on moisture conditions at the time of testing. Critically, during the first few days and hours of surface drying, the actual modulus of rupture (and therefore the actual cracking resistance) could be as little as 55% of the commonly expected value ($4\sqrt{f_c'}$ rather than $7.5\sqrt{f_c'}$).

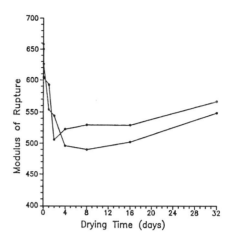

FIGURE 8.35 Data presented by Walker and Bloem (1957b) concerning the influence of drying on the flexural strength of concrete. *Source:* Courtesy of American Concrete Institute; Hover, K.C. 1988. The effects of drying and form and shore removal on flexural cracking in beams and slabs. In *Forming Economical Concrete Buildings: Proceedings of the Third International Conference*, Publication SP107, pp. 169–184. American Concrete Institute, Farmington Hill, MI.

The surface of concrete beams and slabs begins to dry immediately upon the removal of the formwork and the subsequent exposure to dry, moving air. While this surface drying is taking place, the removal of shoring and further construction operations load the members. The removal of forms and shores from flexural members is therefore a delicate issue, not only because of the strength development of early-age concrete but also because the surface drying, which is an almost unavoidable consequence of form removal, can decrease the cracking resistance by almost 50%.

In order to further investigate the phenomenon, beam specimens were cast in forms with steel sides and oiled plywood bottoms. Four days after casting, the forms were removed from three beams, which were then permitted to dry for 4 h in laboratory air. Three other beams were removed from their forms one at a time and immediately tested so as to avoid appreciable air drying. The average MOR of the beams that were tested immediately upon form removal was 573 psi (4.0 MPa), while the average MOR of the beams that had dried for 4 h was 450 psi (3.1 MPa), representing a 25% decrease in cracking resistance due to short-term drying.

It is not unreasonable to envision a situation in which the conditions of this experiment are realized on the job site, where the removal of beam sides may precede the removal of the beam bottom and supporting shoring by several hours. In that case, the self-weight of the slab or beam would be applied during this critical drying period. Similarly, in the case of rapid construction where the weight of falsework, materials, flying forms, etc., is applied to the top of a recently stripped beam or slab, it is likely that superimposed load would be applied during the critical drying period during which the cracking resistance is substantially reduced.

As part of a related experiment, flexural specimens were cast against steel forms with oiled plywood bottoms, with companion specimens cast in similar forms that had been lined with polyethylene. From tests conducted at 4 h, 2 h, and immediately after removal of forms, the modulus of rupture of the beams with the plastic-lined forms was on an average 25% greater than for the beams cast in steel and wood forms. When companion specimens were tested after 28 d of drying in laboratory air, the results for the two types of forms were statistically indisguishable. It was concluded that the influence of the form surface on the cracking resistance was primarily owing to the effects of moisture conditions at early ages.

It was pointed out that timing the removal of forms and shores so as not to load a member while the concrete surface is drying can reduce the incidence of flexural cracking or can increase the applied load necessary to initiate such cracking. The use of smooth form surfaces or nonabsorptive form liners may be beneficial in reducing flexural cracking.

8.4 Strength Consequences of Construction Loads

It would be instructive to begin this section with the following four paragraphs reproduced, with only minor paraphrasing, from Kaminetzky and Stivaros (1994).

> Throughout the history of concrete construction, numerous construction failures have occurred. Many statistical surveys point out that failures and total collapses occur more frequently during construction than during the service life of a structure. Also, it has been well documented that failure of concrete structures during construction most often occur as a result of formwork failure, concrete member failure due to overloading, or lack of concrete strength.
>
> The most common and often devastating failures are punching-shear failures. These are triggered by a localized failure around a single column at an upper-level floor, resulting in progressive collapse going all the way down to the lowest level. Such a domino-effect failure can be stopped by localizing it to a single level. Because punching shear is the weakest link in the chain, the use of large-size columns coupled with continuous reinforcing bars running through the column periphery at both the top and the bottom of the slab will minimize the possibility of progressive collapse.
>
> The most devastating and well-known collapses, during construction, resulting from punching shear, are those that occurred at Bailey's Crossroads, Virginia (Skyline Plaza) [Leyendecker and Fattal, 1977], Commonwealth Avenue, Boston, Massachusetts (1971); and Cocoa Beach, Florida (Harbor Cay Condominium) [Lew et al., 1982a,b]. All of these multistory flat-plate buildings in late stages of construction collapsed vertically in a progressive manner to the ground. The triggering cause was typical: localized punching-shear failure at the slab-column connection. Premature removal of formwork coupled with insufficient concrete strength at the time of the collapse were partially blamed for these disasters. The maturity of concrete was affected by the low temperatures that prevailed during the construction of these projects.
>
> Although the Cocoa Beach collapse was mainly attributed to design and construction errors and not directly to concrete strength, procedures for concrete strength determinations were followed prior to the removal of forms. Construction records showed that laboratory-cured cylinders were used in order to determine the actual field strength of the slabs prior to stripping, when proper procedures and common sense dictate testing of field-cured cylinders.

A reexamination of the Harbour Cay Condominium failure later attributed it more directly to deficiency of punching-shear strength [Stivaros and Halvorsen, 1992]. The Harbour Cay was a five-story flat-plate concrete building. A typical floor framing plan is shown in Figure 8.36. At the time of collapse, flying forms on the fifth floor supported the fresh concrete roof slab. Three levels of wood reshores rested below. Figure 8.36 shows the flying form and reshore layout and indicates the portion of the roof believed to have been in place at the time of collapse.

The National Institute of Standards and Technology (NIST) investigators [Lew et al., 1982a] used the ICES-STRUDL II finite-element program to analyze the Harbor Cay structure. The finite-element model consisted of rectangular plate bending elements for the columns and three-dimensional truss elements for the shoring system. The distribution of loads from the shoring and reshoring procedure was examined by means of the Grundy-Kabaila simplified method. The analysis indicated that when all the floors were reshored to the ground, the removal of the flying forms at the topmost level completely relieved the forces in the reshores. The existing slabs always supported their own dead load and did not share in supporting a portion of the load of the newly cast floor. Only after the first-story reshores were removed, thus removing a direct load path to the ground, did the existing floor slabs share the load of a newly cast slab. These observations stem from Grundy-Kabaila's basic assumption of an infinitely stiff shoring system.

As a result of the above observations, to determine the state of load distribution at the time of collapse, NIST investigators studied two separate load cases (Figure 8.37). In the first, the structure without reshores was analyzed under the dead load of the slabs. In the second, the reshored structure was analyzed under the load of the fresh concrete and the formwork. The final result was

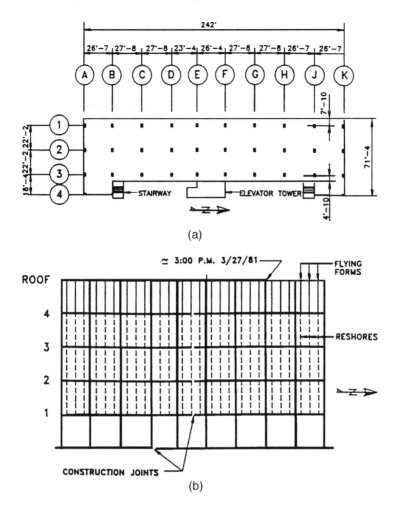

FIGURE 8.36   (a) Typical floor plan of the Harbour Cay Condominium; (b) assumed state of construction at the time of collapse [Stivaros and Halvorsen, 1991].

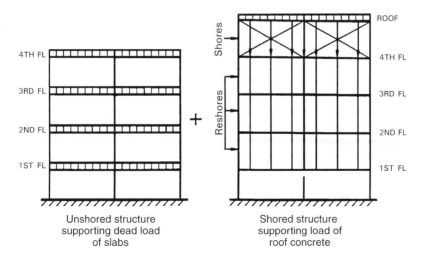

FIGURE 8.37   Two load cases considered in NIST analysis of Harbour Cay Condominium [Stivaros and Halvorsen, 1991].

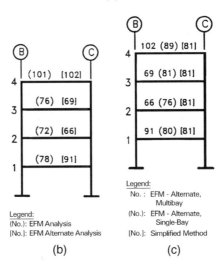

FIGURE 8.38   (a) Punching-shear forces along column line 2 (kips); (b) slab loads on bay B-C from multibay EFM analysis (kips); (c) slab loads on bay B-C from EFM and simplified method, (kips), Harbour Cay Condominium analysis. [Stivaros and Halvorsen, 1991].

obtained by superimposing the results of the two analyses. Given in Figure 8.38a are the combined punching-shear forces, in kips, for the slab-column connections along column line 2 of the building.

Stivaros and Halvorsen (1992) carried out a preliminary equivalent frame analysis of the Harbour Cay failure, using the same structural data and loading conditions as those used by NIST investigators. The preliminary analysis used the plane frame capabilities of ICES-STRUDL II. The results of this analysis are shown in Figure 8.38a in square brackets. After the preliminary analysis indicated reasonable results, Stivaros and Halvorsen developed their special-purpose computer program [Stivaros and Halvorsen, 1990] that recognized time-dependent material properties and time in the construction process. This program was applied to the construction-load distribution of the Harbour Cay Condominium failure in two different ways. First, the structure was analyzed according to the two loading cases and structural details identified by Lew et al. (1982a). The results of this first analysis are shown in parentheses in Figure 8.38a.

An alternative approach assumed more realistic construction-loading conditions. Grundy-Kabaila's simplified method, as mentioned, assumes an infinitely stiff shoring system. However, the Harbour Cay Condominium used compressible wooden reshores. In the course of the alternative analysis, the program "built" the structure from the first floor to the roof and considered the effects of all the cycles of loading and unloading as the construction progressed upward. Thus the effect of compressible reshores could be assessed. Variations in concrete maturity, strength, and stiffness are also incorporated. The punching-shear forces at slab-column joints along the height of column stack C-2 are shown in Figure 8.38b. The results of the alternative analysis, which considered the two separate loading cases, are shown in parentheses.

The accuracy of the Grundy-Kabaila simplified method was tested against observations of the Harbour Cay Condominium collapse. Figure 8.38c shows the punching-shear forces at every slab-column joint along the height of column stack C-2. The first number indicates the force obtained by the alternative equivalent multibay frame, the second number (in parentheses) represents the force obtained by the alternative method using a single-bay model, and the third number (in square brackets) represents the force obtained from the simplified method that invariably uses a single-bay model. A comparison shows that the alternative equivalent single-bay model and the simplified method underestimate the punching-shear forces along column stack C-2 on the top floor by 13 and 21%, respectively, as compared to the results of the equivalent multibay frame model. On the floors below, both the simplified method and the equivalent single-bay frame model differ by smaller percentages from the equivalent multibay frame model.

Based on the simplified method and the equivalent single-bay frame analysis, punching-shear force at the top slab-column joint along column stack C-2 was within the punching shear strength of that joint. However, the analysis based on the equivalent multibay frame model indicated that the available punching shear strength was exceeded at the same joint. According to Stivaros and Halvorsen (1992), "Here is exactly where the significance of the EFM multibay model lies ..."

8.4.1 Safety Analysis

Gardner (1985) was the first to attempt a systematic safety analysis of construction (shoring/reshoring) schemes for multistory buildings, as described in this section.

The strength of a reinforced concrete slab is a function of the design (factored) load, age of the slab, and the critical mode of behavior of the slab: flexure, bond, or shear. The required strength for various live-load/dead-load ratios, in terms of dead load, can be calculated from ACI 318-95 (1995) as follows:

$$U = 1.4D + 1.7L = D\left[1.4 + 1.7\frac{L}{D}\right] \qquad (8.14)$$

Design strength (nominal strength reduced by the appropriate strength-reduction factor) must equal or exceed the required strength and was assumed by Gardner to equal the required strength. It should be noted that in U.S. design practice, the design strength is assumed to be available at 28 d of age. The strength available at the early age at which a slab experiences construction loads is, of course, lower than the 28-d design strength.

Gardner and Poon (1976) had concluded earlier that the tensile and bond strengths of concrete were proportional to the compressive cylinder strength raised to the power 0.8. As shear is governed by tensile strength, it was assumed that shear strength was also related to compressive strength the same way. Data in Table 8.7 were prepared by Gardner (1985) from Klieger's (1958) data, using the 0.8 power criterion to estimate early-age strength as a fraction of the 28-d, 73°F cylinder strength for both Type I normal Portland cement and Type III high-early-strength cement concretes.

Gardner (1985) noted that no existing design code specifies a construction load factor. ANSI A10.9-1983 (1983) recommends a load factor of 1.3 for both dead and live construction loads.

TABLE 8.7   Development of Tensile/Bond/Shear Strength, Percent of 28-d, 73°F (22.8°C) Value

	Type I Cement Temperature			Type III Cement Temperature		
Age, d	73°F (22.8°C)	55°F (12.8°C)	40°F (4.4°C)	73°F (22.8°C)	55°F (12.8°C)	40°F (4.4°C)
1	0.31*	0.15*	0.03*	0.54*	0.33*	0.11*
2	0.47	0.28	0.11	0.65	0.50	0.30
3	0.59*	0.40*	0.18*	0.74*	0.62*	0.43*
4	0.66	0.49	0.24	0.78	0.66	0.54
5	0.72	0.57	0.32	0.81	0.70	0.63
6	0.76	0.63	0.39	0.83	0.73	0.70
7	0.79*	0.68*	0.44*	0.85*	0.75*	0.77*
8	0.81	0.72	0.48	0.86	0.77	0.80
9	0.83	0.75	0.52	0.88	0.79	0.82
10	0.85	0.77	0.56	0.89	0.81	0.84
11	0.86	0.80	0.59	0.90	0.82	0.86
12	0.88	0.82	0.62	0.91	0.84	0.88
13	0.89	0.84	0.64	0.92	0.85	0.89
14	0.90	0.86	0.67	0.92	0.86	0.90
21	0.96	0.94	0.80	0.97	0.93	0.99
28	1.00*	1.02*	0.88*	1.00*	0.96*	1.07*

Source: Gardner, N.J. 1985. Shoring, reshoring, and safety. *Concrete Int.* 7(4):28–34.
*Klieger's (1958) original data.

TABLE 8.8   Factored Construction Loads on Slabs with Two Shores and Two Reshores.

	Age of Slab (cycles after casting)							
	1		2		3		4	
Slab	a	b	a	b	a	b	a	b
1	0*	0*	1.54*	1.54*	1.54*	1.54*	1.94	2.37
2	1.54*	1.69*	2.31*	2.31*	1.94	2.37	1.94	2.37
3	0.77*	0.77*	2.37	2.79	1.94	2.37	1.94	2.37
4	1.52	1.94	2.79	3.21	1.94	2.37	1.94	2.37
5	1.10	1.52	2.60	3.01	1.94	2.37	1.94	2.37

Source: Gardner, N.J. 1985. Shoring, reshoring, and safety. *Concrete Int.* 7(4):28–34.
Note: Operation "a" is removal of lowest level of shores and rshores; operation "b" is casting of next slab.
*Slab supported on grade so construction live load transmission is directly to the ground.

Kaminetzky and Stivaros (1994) indicated preference for the more conservative load factors required by ACI 318 for service conditions, that is, 1.4 and 1.7 for construction dead and live loads, respectively. Gardner used a construction load factor of 1.4 and assumed that the forms have self-weights equal to 10% of the slab self-weight and that the simplified Grundy-Kabaila method is approximately 10% in error. This yielded a factored construction load combination (required strength) of

$$U = \left[1.1 \times 1.1 \times 1.4 \times \text{ load ratio } + \frac{1}{N} \right] \text{ dead load} \qquad (8.15)$$

$$\quad\;\; \text{(a)} \quad\;\; \text{(b)} \quad\;\; \text{(c)} \qquad\quad \text{(d)} \qquad\quad \text{(e)}$$

where (a) is error in simplified method, (b) is weight of formwork, (c) is load factor, (d) is calculated load ratio (Table 8.2), and (e) is an allowance for the construction live load where N is the total number of shore and reshore levels.

Using Eq. (8.15), the factored construction loads at various stages of construction are shown in Table 8.8 for two levels of shoring and two levels of reshoring. A complete version of Table 8.8 for other combinations of shore and reshore levels is shown in Table 8.9.

LE 8.9 Factored Construction Loads with Cycles of Construction

	Slab	Construction Cycle 1 a	1 b	2 a	2 b	3 a	3 b	4 a	4 b	5 a	5 b	6 a	6 b	7 a	7 b
m hores	1	1.54	3.94												
	2	1.54	3.94												
	3	1.54	3.94												
m hore	1	1.54	1.54	2.19	3.04										
	2	2.19	3.04	2.19	3.04										
	3	2.19	3.04	2.19	3.04										
m hores	1	1.54	1.54	1.54	1.54	2.02	2.60								
	2	1.54	1.54	2.02	2.60	2.02	2.60								
	3	2.02	2.60	2.02	2.60	2.02	2.60								
m hores	1	1.54	1.54	1.54	1.54	1.54	1.54	1.94	2.37						
	2	1.54	1.54	1.54	1.54	1.94	2.37	1.94	2.37						
	3	1.54	1.54	1.94	2.37	1.94	2.37	1.94	2.37						
	4	1.94	2.37	1.94	2.37	1.94	2.37	1.94	2.37						
m hores	1	1.54	1.54	1.54	1.54	1.54	1.54	1.54	1.54	1.89	2.23				
	2	1.54	1.54	1.54	1.54	1.54	1.54	1.89	2.23	1.89	2.23				
	3	1.54	1.54	1.54	1.54	1.89	2.23	1.89	2.23	1.89	2.23				
	4	1.54	1.54	1.89	2.23	1.89	2.23	1.89	2.23	1.89	2.23				
	5	1.89	2.23	1.89	2.23	1.89	2.23	1.89	2.23	1.89	2.23				
m hores	1	1.54	1.54	1.54	1.54	1.54	1.54	1.54	1.54	1.54	1.54	1.86	2.15		
	2	1.54	1.54	1.54	1.54	1.54	1.54	1.54	1.54	1.86	2.15	1.86	2.15		
	3	1.54	1.54	1.54	1.54	1.54	1.54	1.86	2.15	1.86	2.15	1.86	2.15		
	4	1.54	1.54	1.54	1.54	1.86	2.15	1.86	2.15	1.86	2.15	1.86	2.15		
	5	1.54	1.54	1.86	2.15	1.86	2.15	1.86	2.15	1.86	2.15	1.86	2.15		
	6	1.86	2.15	1.86	2.15	1.86	2.15	1.86	2.15	1.86	2.15	1.86	2.15		
ms hores	1	0	0	2.19	3.04										
	2	2.19	3.04	3.46	4.31										
	3	0.92	1.77	2.84	3.68										
	4	1.55	2.40	3.14	3.99										
	5	1.25	2.09												
	*	1.35	2.19	3.04	3.89										
ms hore	1	0	0	1.54	1.54	2.03	2.60								
	2	1.54	1.54	2.87	3.43	2.03	2.60								
	3	1.18	1.74	2.74	3.30	2.03	2.60								
	4	1.32	1.87	2.81	3.37	2.03	2.60								
	5	1.25	1.81	2.77	3.33	2.03	2.60								
	6	1.28	1.84	2.79	3.35	2.02	2.60								
	*	1.27	1.82	2.81	3.37	2.03	2.60								
ms hores	1	0	0	1.54	1.54	1.54	1.54	1.94	2.37						
	2	1.54	1.69	2.31	2.31	1.94	2.37	1.94	2.37						
	3	0.77	0.77	2.37	2.79	1.94	2.37	1.94	2.37						
	4	1.52	1.94	2.79	3.21	1.94	2.37	1.94	2.37						
	5	1.10	1.52	2.59	3.01	1.94	2.37	1.94	2.37						
	6	1.30	1.72	2.69	3.11	1.94	2.37	1.94	2.37						
	7	1.20	1.62	2.64	3.06	1.94	2.37	1.94	2.37						
	*	1.23	1.66	2.66	3.08	1.94	2.37	1.94	2.37						
ms hores	1	0	0	1.54	1.54	1.54	1.54	1.54	1.54	1.89	2.23				
	2	1.54	1.54	2.31	2.31	1.54	1.54	1.89	2.23	1.89	2.23				
	3	0.77	0.77	1.93	1.93	1.89	2.23	1.89	2.23	1.89	2.23				
	4	1.16	1.16	2.54	2.88	1.89	2.23	1.89	2.23	1.89	2.23				
	5	1.25	1.59	2.59	2.93	1.89	2.23	1.89	2.23	1.89	2.23				
	6	1.20	1.54	2.57	2.91	1.89	2.23	1.89	2.23	1.89	2.23				
	7	1.22	1.55	2.57	2.91	1.89	2.23	1.89	2.23	1.89	2.23				
	*	1.22	1.55	2.57	2.90	1.89	2.23	1.89	2.23	1.89	2.23				

(Continues)

TABLE 8.9 (*continued*)

	Slab	Construction Cycle												
		1		2		3		4		5		6		7	
		a	b	a	b	a	b	a	b	a	b	a	b	a	b
2 forms	1	0	0	1.54	1.54	1.54	1.54	1.54	1.54	1.54	1.54	1.86	2.14		
4 reshores	2	1.54	1.54	1.54	1.54	1.54	1.54	1.54	1.54	1.86	2.14	1.86	2.14		
	3	0.77	0.77	1.93	1.93	1.54	1.54	1.86	2.14	1.86	2.14	1.86	2.14		
	4	1.16	1.16	2.13	2.13	1.86	2.14	1.86	2.14	1.86	2.14	1.86	2.14		
	5	0.95	0.95	2.39	2.66	1.86	2.14	1.86	2.14	1.86	2.14	1.86	2.14		
	6	1.34	1.61	2.59	2.86	1.86	2.14	1.86	2.14	1.86	2.14	1.86	2.14		
	7	1.13	1.40	2.49	2.76	1.86	2.14	1.86	2.14	1.86	2.14	1.86	2.14		
	*	1.20	1.47	2.52	2.79	1.86	2.14	1.86	2.14	1.86	2.14	1.86	2.14		
2 forms	1	0	0	1.54	1.54	1.54	1.54	1.54	1.54	1.54	1.54	1.54	1.54	1.84	2.07
5 reshores	2	1.54	1.54	1.54	1.54	1.54	1.54	1.54	1.54	1.54	1.54	1.84	2.07	1.84	2.07
	3	0.77	0.77	1.93	1.93	1.54	1.54	1.54	1.54	1.84	2.07	1.84	2.07	1.84	2.07
	4	1.16	1.16	2.13	2.13	1.54	1.54	1.84	2.07	1.84	2.07	1.84	2.07	1.84	2.07
	5	0.95	0.95	2.02	2.02	1.84	2.07	1.84	2.07	1.84	2.07	1.84	2.07	1.84	2.07
	6	1.06	1.06	2.43	2.67	1.84	2.07	1.84	2.07	1.84	2.07	1.84	2.07	1.84	2.07
	7	1.24	1.48	2.51	2.75	1.84	2.07	1.84	2.07	1.84	2.07	1.84	2.07	1.84	2.07
	*	1.19	1.43	2.48	2.72	1.84	2.07	1.84	2.07	1.84	2.07	1.84	2.07	1.84	2.07

Source: Gardner, N.J. 1985. Shoring, reshoring, and safety. *Concrete Int.* 7(4):28–34.
*Converged.

Gardner (1985) provided the following three examples of safety analysis.

8.4.1.1 Example 8.1

Determine if a flat-slab structure designed for a live-load/dead-load ratio of 1.00 by ACI 318 can be constructed using a two-shore two-reshore system with a seven-day casting cycle and form stripping one day before casting. Assume an ambient temperature of 73°F and Type I Portland cement concrete.

$$U = D\left[1.4 + 1.7\frac{L}{D}\right] = 3.1\,D$$

Assuming that the 28-d design strength is equal to the required strength U, the strengths available at various ages of interest are as follows:

Age	Strength Available
6 days	$0.76 \times 3.1\,D = 2.36\,D$
7 days	$0.79 \times 3.1\,D = 2.45\,D$
13 days	$0.89 \times 3.1\,D = 2.76\,D$
14 days	$0.90 \times 3.1\,D = 2.79\,D$
20 days	$0.96 \times 3.1\,D = 2.98\,D$
27 days	$1.00 \times 3.1\,D = 3.1\,D$
28 days	$1.00 \times 3.1\,D = 3.1\,D$

Using the load data from Table 8.8, the adequacy can be checked for every slab at every stage of construction. The results of the check are shown in Table 8.10. It can be seen that the proposed construction scheme overloads slab 4 at ages 13 and 14 d and slab 5 at 14 d. Thus it cannot be used without modification.

8.4.1.2 Example 8.2

Determine the construction cycle for a slab designed for a live-load/dead-load ratio of 0.5, using one shore plus five reshore levels at a temperature of 55°F. Assume Type I Portland cement concrete and a specifed concrete strength of 4000 psi.

TABLE 8.10 Comparison of Strength and Factored Load for Worked Example

Age After Casting	Strength Available	Load Applied				
		Slab 1	Slab 2	Slab 3	Slab 4	Slab 5
6	2.36	0	1.54	0.77	1.52	1.10
7	2.45	0	1.69	0.77	1.94	1.52
13	2.76	1.54	2.31	2.37	2.79	2.60
14	2.79	1.54	2.31	2.79	3.21	3.01
20	2.98	1.54	1.94	1.94	1.94	1.94
21	2.98	1.54	2.37	2.37	2.37	2.37
27	3.10	1.94	1.94	1.94	1.94	1.94
28	3.10	2.37	2.37	2.37	2.37	2.37

Source: Gardner, N.J. 1985. Shoring, reshoring, and safety. *Concrete Int.* 7(4):28-34.

From Table 8.8, for one level of shores and five levels of reshores, the factored construction loads are $1.86D$ at stripping and $2.15D$ at casting of the next slab. The required strength is

$$U = D\left[1.4 + 1.7\frac{L}{D}\right] = 2.25D.$$

Assuming that the 28-d, 73°F design strength is equal to the required strength U, the concrete strength at casting at 55°F needs to be $2.15/2.25 = 96\%$ of the 28-d strength, which occurs at approximately 23 d. The concrete strength at stripping needs to be $1.86D/2.25D = 83\%$, which occurs at 13 d.

In most cases a 23-d cycle would not be economically acceptable.

8.4.1.3 Example 8.3

Redo the above example to achieve a 7-d placing cycle.

The strength of slab needed at casting (7 d) is 0.96×4000 psi $= 3840$ psi; the strength of slab needed at stripping (6 d) is 0.83×4000 psi $= 3320$ psi. To obtain a concrete strength of 3840 psi at 7 d, either the concrete mix must be redesigned and/or heat must be supplied to accelerate the hydration process.

It should be noted that the above safety analysis is probably flawed if punching-shear governs safety, as it usually does for slab systems without column-line beams. This is because of two important reasons. First, as mentioned earlier, simplified Grundy-Kabaila type analysis cannot lead to accurate estimates of punching-shear forces at slab-column joints. Second, the assumption that punching-shear strength varies with the 28-d compressive cylinder strength of concrete raised to the power 0.8 was not verified in later tests by Gardner (1960), who found punching-shear strength to vary with the cube root of the 28-d compressive cylinder strength.

8.4.2 Refined Safety Analysis

A more refined safety analysis was carried out by Kaminetzky and Stivaros (1994) on a test multistory flat-plate office building (Figure 8.39), assumed to have been designed according to ACI 318 with a 50 psf (2.4 kPa) live load and an additional superimposed partition load of 30 psf (1.5 kPa). This example assumed one level of shores and three levels of reshores with a construction rate of two floors per week. It was also assumed that the shores and reshores were removed and relocated a day before casting the top slab. A 50 psf (2.4 kPa) construction live load was assumed present only during the placement of concrete in the top slabs.

The *in situ* concrete strength can be estimated by testing field-cured cylinders in combination with other available nondestructive methods. Figure 8.40 shows the assumed compressive strength-time relationship for two different construction temperatures: 55 and 73°F.

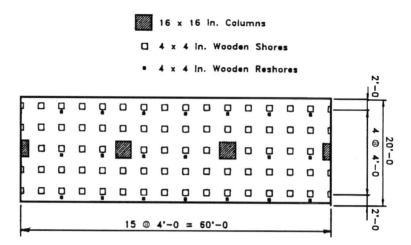

FIGURE 8.39 Typical floor plan of construction example studied by Kaminetzky (1994).

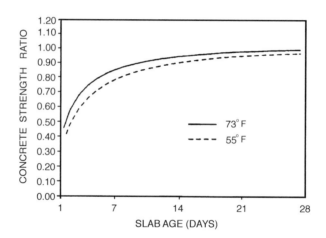

FIGURE 8.40 Compressive strength development versus time assumed by Kaminetzky (1994).

Stivaros' computer program [Stivaros and Halvorsen, 1990], based on the equivalent frame method of analysis, was employed to evaluate the construction load distribution between the shoring system and the concrete frame. The program, as mentioned earlier, "builds" the structure from the first floor to the roof, and evaluates the loads on each floor slab and the support system for every construction stage. The punching-shear stresses due to applied loads are also computed at each column location for every construction step. The program provides the available load-carrying capacities of the slabs and the available punching-shear strengths of slab-column connections at the time of application of the construction loads; thus a safety comparison can be made. The load-carrying capacities are based on flexural considerations, accounting for partially developed concrete compressive strength. A key assumption is that the load factors and strength-reduction factors used for the design of the slabs for service conditions are the same for the construction stage as well. The available punching-shear strength is evaluated according to ACI 318 (1995) and is equal to $4\phi\sqrt{f_{ct}}$, where $\phi = 0.85$ and f_{ct} is the available concrete strength at the time the load is applied.

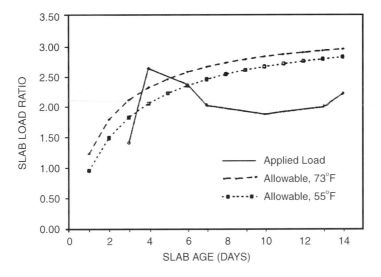

FIGURE 8.41 Allowable and applied slab loads from Kaminetzky's (1994) analysis, live load = 50 psf.

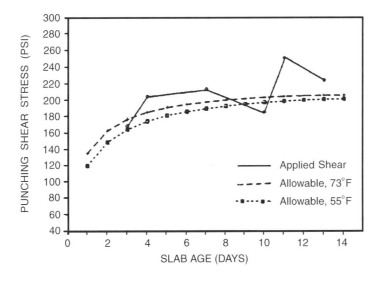

FIGURE 8.42 Allowable and applied punching-shear stress from Kaminetzky's (1994) analysis.

The results of the construction load analysis are shown in Figures 8.41 and 8.42. The figures show the maximum factored slab loads (normalized to the self-weight of the slab) and punching-shear stresses that occurred on the floor slabs during construction. Also shown in the figures are the load-carrying capacities (allowable loads) and allowable punching-shear stresses of the concrete slabs at two temperature conditions: 55 and 73°F.

Figure 8.41 shows that the maximum applied factored load, which occurred on 4-d-old slabs, exceeded the allowable loads for both temperature conditions. This means that the proposed construction scheme cannot be safely executed. Similar unsafe construction conditions are shown in Figure 8.42 where the punching-shear stresses due to the applied loads exceeded the allowable

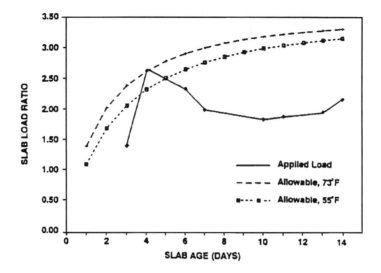

FIGURE 8.43   Allowable and applied slab loads from Kaminetzky's (1994) analysis, live load = 70 psf.

stresses. For this particular example, the available strengths in both flexure and shear have been exceeded. However, because of the possibility of inelastic redistribution of bending moments, and because of the improbabilities of any redistribution of punching-shear stresses, a sudden shear failure is likely to occur prior to flexural failure.

Kaminetzky and Stivaros (1994) pointed out that had some floors of the above building been designed for an additional live load of 20 psf (1.0 kPa), the slabs would have been flexurally adequate to carry the construction loads for the 73°F temperature condition at a rate of two floors per week (Figure 8.43). Also, a construction rate of one floor per week would have been safe even for the 55°F temperature condition. However, the construction operation still could not proceed safely because of the punching-shear deficiency of the slabs.

8.5 Serviceability Consequences of Construction Loads

Floors in residential as well as office buildings are often made of thin, solid concrete slabs with two-way reinforcement. The trend in recent years has been toward a progressive decrease in the ratio of the thickness of slab to length of span. This obviously causes a corresponding reduction in the flexural rigidity of slabs. The consequent deformations have not always received as much attention as preventing collapse. Large permanent deformations of floor slabs have sometimes resulted. While such deformations are often aesthetically unacceptable, their effects may also be seriously disturbing from a practical point of view. Cracking of brick partitions, jamming of doors, and cracking of windows have not been uncommon. At least in one reported case, long-term deflection of an interior span of a floor caused an adjacent balcony to slope inward. Rain falling on the balcony ran inward under the doorway and formed a pool in the middle of the floor, damaging carpets. Excessive long-term deflections of slabs are clearly to be avoided.

8.5.1 Causes of Excessive Deflections

Excessive deflections can sometimes be attributed to a single cause. More commonly, however, the problem may be traced to a combination of several contributing factors. Potential problem areas include design, construction, materials, environmental conditions, and change in occupancy [Scanlon, 1987].

8.5.1.1 Design

The single most common cause of large deflections associated with design is the selection of a slab thickness that is too small for the spans used. Another common problem is inadequate flexural reinforcement; as a result premature yielding may occur, causing a substantial loss of flexural stiffness and leading to excessive deflection.

8.5.1.2 Materials

Higher than normal creep and shrinkage characteristics have been identified as contributing factors in cases of large deflections reported in Australia.

While achieving high concrete strengths, high-shrinkage characteristics have sometimes been produced; the increase in slab warping has more than offset the benefits obtained from the increase in the modulus of elasticity of the concrete. In post-tensioned slabs, higher than normal prestress losses may lead to unanticipated deflections. Alkali-aggregate reactions produced by certain types of aggregate and cement may cause cracking that can adversely affect flexural stiffness.

8.5.1.3 Environmental Conditions

For slab surfaces exposed to daily or seasonal temperature fluctuations, temperature gradients set up through the member thickness may lead to unanticipated deflections.

8.5.1.4 Change in Occupancy

Slabs adequately designed and constructed for the original design loading may suffer deflection problems if the occupancy changes and if the change involves an increase in the sustained load applied to the slab.

8.5.1.5 Construction

The following is a list of factors, as compiled by Taylor and Heiman (1977), contributing to excessive long-term slab deflections:

1. Formwork has not been cambered in cantilevers and large interior panels, so that even early deflections are obvious.
2. Slabs have been supported by props bearing on sole plates that were of inadequate size to prevent appreciable settlement into the ground surface and hence induced slab deformation before stripping.
3. Construction loading from propping or the storage of materials during the early life of the slabs has been severe enough to cause extensive slab cracking and hence loss of stiffness.
4. Top reinforcement at supporting elements has been pushed down during slab construction, substantially reducing its effective depth and hence reducing the contribution made to slab stiffness by continuity at supports.
5. Curing has in many instances been inadequate, and has sometimes led to insufficient strength development and excessive shrinkage cracking.

Item 4 above is a surprisingly common occurrence. Quite often, the top steel over supports is not securely held in place and is displaced toward the neutral axis, greatly reducing the stiffness over the support. If the effective depth is 7 in instead of 8 in, as has been observed, the stiffness is reduced by 23%, and the member tends to behave as if it were simply reinforced.

Discussion here obviously needs to focus on the very important item 3 above. As noted earlier, when the shore/reshore method of multistory building construction is used, high early-age short-duration loads are imposed on the floor slabs. These loads can be comparable in magnitude to, or even exceed, the design service loads and are applied to concrete slabs that have not achieved their specified concrete strength. Owing to the slab concrete being immature with a reduced modulus of elasticity (see preceding section), the immediate deflections due to the construction loads are relatively large. Creep effects may be looked upon as being dependent upon the magnitude of the applied stress relative to the developed concrete strength. Hence creep deflections due to construction loads are also large. Deflection due to concrete shrinkage must also be considered.

8.5.2 Components of Long-Term Deflection

Shrinkage and creep are the two primary contributors to long-term deflections of reinforced concrete members.

Shrinkage produces a shortening of the concrete in a member that is resisted by the reinforcing steel. The magnitude of shrinkage curvature (Figure 8.44b) depends on the amount of non-

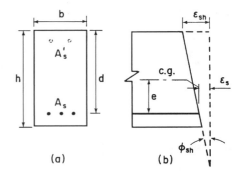

FIGURE 8.44 Shrinkage curvature of a section of a reinforced concrete flexural member.

symmetry of reinforcement and on the relative areas of concrete and steel in a reinforced concrete member. These shrinkage curvatures, which are the greatest for beams with tension reinforcement only, generally have the same sign as those due to moments produced by transverse loads, and thus they magnify the deflection due to load.

Considering creep alone, the increase in strain with time will be as shown by line B in Figure 8.45b, while line A indicates the strain distribution immediately on application of the load. Strain at the extreme fiber in compression increases considerably, while the strain at the level of the steel changes very little. As this occurs, the compressive stresses in the concrete are reduced (Figure 8.45c), since the neutral axis moves toward the reinforcement; the steel stresses increase, since the internal lever arm shortens. It can be seen from Figure 8.45b that the relative increase in curvature caused by creep is less than the relative increase in strain due to creep. Thus the relative increase in deflection also is less than the increase in strain due to creep.

In addition to the difficulty of determining the creep-time history of a particular concrete under constant, uniformly distributed sustained stress, a reinforced concrete flexural member is subjected to a nonuniform stress distribution and often a variable load history. A detailed analysis of the effects of a variable stress history, even for uniformly loaded members, is usually not feasible, and shorter, more appropriate methods are used. One such method uses a reduced or effective modulus called the sustained modulus of elasticity, which refers to concrete under a constant sustained stress:

$$E_{ct} = \frac{\sigma_{\text{constant}}}{\epsilon_{\text{initial}} + \epsilon_{\text{creep},t}} = \frac{\sigma_{\text{constant}}}{(1 + C_t)\epsilon_{\text{initial}}} = \frac{E_c}{1 + C_t} \tag{8.16}$$

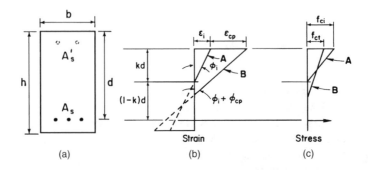

FIGURE 8.45 Strain and stress distribution and curvature due to initial and creep deformation in a section of a reinforced concrete flexural member.

where $C_t = \epsilon_{\text{creep},t}/\epsilon_{\text{initial}}$ is the creep coefficient at time t, the ultimate value of which may be denoted by C_u. It may be parenthetically noted that in considering creep deformations, the use of a specific creep strain δ_t (creep per unit stress), or creep coefficient, C_t (ratio of creep strain to initial strain), yields the same results [ACI Committee 209, 1971], since

$$C_t = \delta_t E_{ci} \tag{8.17}$$

This can be seen from the following relations: creep strain at time $t = (\sigma_{\text{constant}})\delta_t = (\epsilon_{\text{initial}})C_t$, and $E_{ci} = \sigma_{\text{constant}}/\epsilon_{\text{initial}}$. The choice of δ_t or C_t is a matter of convenience, depending on whether applying the creep factor to stress or strain is desired.

With reference to Figure 8.45 and the earlier discussion on that figure

$$\frac{\Delta_{\text{creep},t}}{\Delta_{\text{initial}}} = \frac{\phi_{\text{creep},t}}{\phi_{\text{initial}}} = k_{rc}\frac{\epsilon_{\text{creep},t}}{\epsilon_{\text{initial}}} = k_{rc}C_t \tag{8.18}$$

provided that the same moment of inertia is used in the computation of short- and long-term deflections. The factor k_{rc} accounts for the effect of compression steel and the offsetting effects of a downward movement of the neutral axis due to creep strain distribution and upward movement of the neutral axis due to progressive cracking under creep loading (plus the small effect, if any, of repeated live-load cycles). These offsetting effects normally appear to result in a downward movement of the neutral axis, such that k_{rc} is less than unity.

Since shrinkage and creep deflections are additive, their combined value is often estimated in approximate calculations with a single time-dependent factor applied to the initial deflection:

$$\Delta_t = k_r T_t \Delta_{\text{initial}} \tag{8.19}$$

where T_t is a multiplier for additional long-term deflections due to creep and shrinkage, the ultimate value of which may be denoted by T_u. The factor k_r in Eq. (8.19) and the factor k_{rc} in Eq. (8.18) serve largely the same purpose. The following values have been suggested for the two factors [ACI Committee 209, 1971]:

$$k_{rc} = 0.85 - 0.45(A_s'/A_s) \geq 0.40 \tag{8.20}$$

$$k_r = 1.00 - 0.60(A_s'/A_s) \geq 0.40 \tag{8.21}$$

ACI 318 (1995) basically takes the Eq. (8.19) approach to estimating of long-term deflections.

The opportunity usually does not arise for temperature to influence deflections of floor systems, since these are usually protected from large temperature changes soon after casting. In slabs that are not so protected, temperature changes can be important. Since the thermal coefficients of steel and concrete are roughly the same, little if any curvature results from temperature changes that are uniform across a section of a member. With regard to differential temperatures (such as inside versus outside temperatures of heated or cooled buildings), the deformation of reinforced concrete members is similar to or opposite that due to shrinkage. For example, the deflection would be downward in the case of roofs of heated buildings and upward in the case of roofs of air-conditioned buildings [Branson, 1977].

Since shrinkage is independent of load, it appears illogical for deflection directly caused by shrinkage to be related to load by the initial deflection. There is, however, an indirect way in which shrinkage can influence deflection. The stiffness of a cracked reinforced concrete section is usually less—sometimes much less—than the stiffness of the surrounding gross concrete section. It would appear that this change from an uncracked to a cracked section, which may be brought about and extended over a considerable period by shrinkage (plus creep and thermal) effects, offers the potential for a corresponding continuing increase in deflection. Branson (1987) cited laboratory tests by others to indicate that the formation of new cracks during sustained loading depends on the development of earlier cracks during the initial loading stage. For example, in beams with a low 0.60% steel and a steel stress $f_s = 20$ ksi at initial loading, about half the cracks occurred at initial loading and the remainder during the sustained loading. In beams with the same low steel percentage but $f_s = 30$ ksi, almost all cracks occurred at initial loading. In beams with a higher 2.0% steel, cracks appeared rather early, with little additional cracking occurring during the sustained loading. According to Branson, this effect, which would seem to require the use of a $k_r > 1$ in Eq. (8.19), is for most practical purposes taken into account through the use of a suitable effective moment of inertia in initial deflection calculations and through the use of the k_r factor as given by Eq. (8.21).

8.5.3 Experimental Investigation

Fu and Gardner (1986) fabricated five nominally identical single-span one-way slabs. All the slabs, shown in Figure 8.46, were 8 in (203 mm) wide, 25 in (64 mm) deep, spanned 77 in (1960 mm), were reinforced with 5-mm diameter wires in one direction, and were cast from the same mix. Ready-mixed concrete with a target 28-d strength of 4000 psi (28 MPa) was used.

Only the midspan deflections of the five one-way slabs were measured under load. After the formwork was stripped, measurements were made every day for the first month and then weekly afterward. In addition, shrinkage of four concrete prisms, compressive strength, and modulus of elasticity of the concrete were also measured.

The five slabs were subjected to different load histories during the first 28 d after casting. Slabs 1 and 2 were subjected to simplified load histories, while slabs 3, 4, and 5 were subjected to load histories typical of 1 shore plus 2 reshores, 1 shore plus 2 levels of preshores (this term is discussed later), and 3 levels of shores. The load histories are given in Figure 8.47. The measured deflections are given in Figure 8.48a for the first 28 d and Figure 8.48b for the first 489 d.

An examination of the deflection curves in Figure 8.48a reveals that the slopes of all these curves were virtually the same after day 28; subsequent deflections after removal of the construction loads seemed to be independent of prior loading. In other words, the additional deflections due to creep and shrinkage of concrete during the first 28 d of loading could not be recovered despite removal of the construction loads and even though identical loads were sustained on the slabs afterward.

The net deflections after 28 d, taken as the sum of deflections due to loading (plus) and unloading (minus) alone and excluding deflections due to creep and shrinkage, were measured to be 0.60, 0.63, 0.62, 0.60, and 0.58 mm for slabs 1 to 5, respectively.

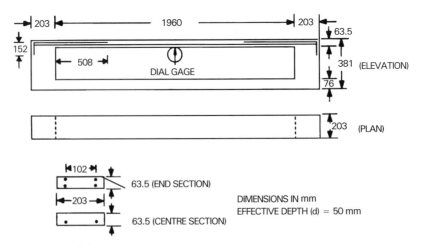

FIGURE 8.46 Details of one-way slab models tested by Fu and Gardner. *Source:* Fu, H.C. and Gardner, N.J. 1986. Effect of high early-age construction loads on the long term behavior of slab structures. In *Properties of Concrete at Early Ages*, Publication SP-95, pp. 173–200. American Concrete Institute, Farmington Hills, MI.

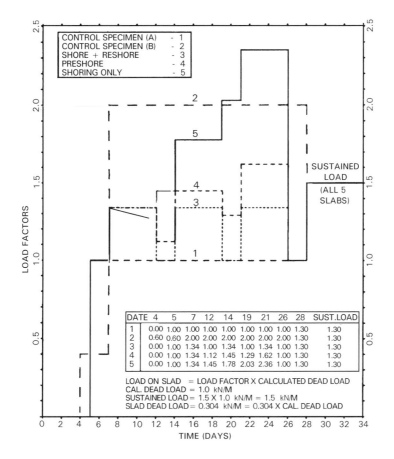

FIGURE 8.47 Loading histories on slab models tested by Fu and Gardner. *Source:* Fu, H.C. and Gardner, N.J. 1986. Effect of high early-age construction loads on the long term behavior of slab structures. In *Properties of Concrete at Early Ages*, Publication SP-95, pp. 173–200. American Concrete Institute, Farmington Hills, MI.

It is equally interesting to note that the increases in deflections from day 28 to the end of the observation period (say 489 d) for all five slabs were very close indeed: 1.60, 1.41, 1.49, 1.50, and 1.57 mm for slabs 1 to 5, respectively. However, owing to the distinct loading histories in the initial stages, the corresponding final deflections after 489 d were 3.26, 4.13, 3.32, 3.51, and 4.31 mm, giving ratios of final deflection to net deflection at 28 d of 5.43, 6.56, 5.35, 5.85, and 7.43.

8.5.4 Control of Slab Deflections

The state-of-the-art for control of two-way slab deflections was reviewed in ACI Committee 435 (1991). The methods of calculating slab deflections were presented, and the effects of two-way action, cracking, creep, shrinkage, and construction loads were considered.

Probably the surest way to control deflections is to carefully sequence the operations of shore removal and reshoring. With a column layout like the one shown in Figure 8.49, stringers would normally be run in the short direction at about 4 ft o/c, supported on shores. Ribs or purlins would then be run in the orthogonal direction, also at about 4 ft o/c, supported on the stringers. Finally, the 4 ft × 8 ft plywood sheets would be supported on the ribs.

It has been suggested [Cantor and Rizzi, 1982], on the basis of successful experience with flat-plate buildings in the New York area, that alternate plywood sheets in both the long and the short direction (shown shaded in Figure 8.49) be supported directly by extra shores (indicated by x in

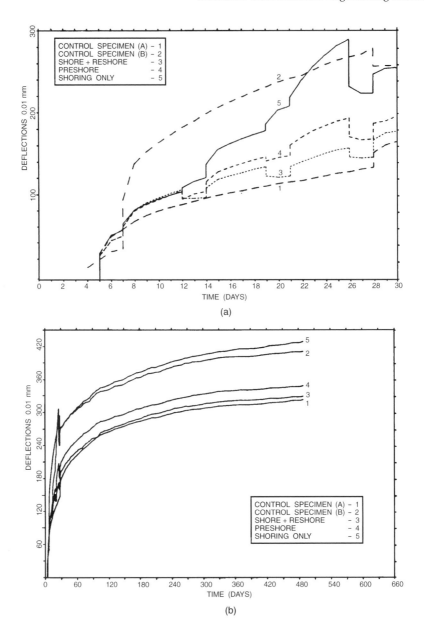

FIGURE 8.48 Deflections due to construction loads measured in slab models tested by Fu and Gardner. *Source:* Fu, H.C. and Gardner, N.J. 1986. Effect of high early-age construction loads on the long term behavior of slab structures. In *Properties of Concrete at Early Ages*, Publication SP-95, pp. 173–200. American Concrete Institute, Farmington Hills, MI.

Figure 8.49). These latter shores (attached to the plywood, rather than to the stringers as the other shores are), also called preshores or permanent shores, may be installed at the time the other shores are installed or just before shore removal.

When the time comes for removal of formwork, a day or two after casting or when concrete strength reaches a certain minimum value, the regular shores, the stringers, the purlins, and the plywood sheets that are not directly held by the extra shores, can all be removed. However, before the extra shores and the plywood sheets held by them are removed, reshores should be installed at

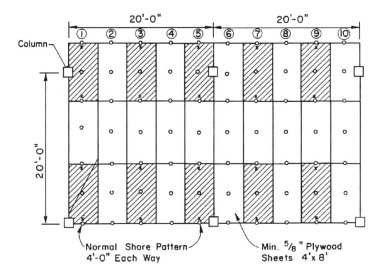

FIGURE 8.49 Sequence of shore removal and reshoring that restricts the slab span left unsupported at an early age [Cantor and Rizzi, 1982].

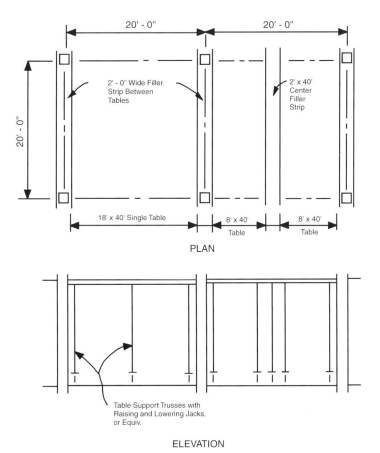

FIGURE 8.50 Restriction on slab span left unsupported at an early age with the use of flying forms [Cantor and Rizzi, 1982].

about 8 ft o/c directly to the concrete slab. The extra shores or preshores should be removed only after the reshores have been installed.

The above scheme of removing the shores and installing the reshores does not permit more than 8 ft of slab span to be left unsupported at any time until the slab is sufficiently mature. With such short unsupported slab spans, slab deflections under usual circumstances cannot assume disturbing proportions, however high the loading may be during construction and however immature the slab may be when it is called upon to support those loads.

The above consideration applies for the use of flying forms also. Figure 8.50 shows a reinforced concrete slab supported on reinforced concrete columns spaced at 20 ft o/c in both directions. With such a column layout, an 18 ft wide form table would normally be used. However, in that case, as soon as the flying form is removed, 18 ft of slab span would be left unsupported. If deflections are of concern, two 8 ft wide form tables, with a filler strip of formwork in between, may be used instead. With the narrower form tables, even when they are removed, no more than 8 ft of slab span would be left unsupported. Admittedly, two 8 ft wide form tables with a filler strip of formwork are significantly more expensive than a single 18 ft wide form table. The added expense has to be weighed carefully against any advantage that is to be gained in terms of reduced deflections.

8.6 Codes and Standards

The following documents have been mentioned in this chapter:

1. *Guide to Formwork for Concrete* (ACI 347R-88), American Concrete Institute, Farmington Hills, MI.
2. *American National Standard for Construction and Demolition Operation–Concrete and Masonry Work–Safety Requirements.* (ANSI A10.9-1983), American National Standards Institute, New York.
3. *Safety and Health Regulations for Construction,* (29 CFR Part 1926), U.S. Department of Labor, Bureau of Labor Standards, Washington, DC.

The first of the items is an ACI committee report, the second is an ANSI standard, and the third is a piece of federal regulation. None of the three is adopted into the three model building codes, one of which typically forms the basis of the legal code of virtually every jurisdiction within the United States. These are the Uniform Building Code (UBC) published by the International Conference of Building Officials, Whittier, California; the BOCA/National Building Code (BOCA/NBC) published by the Building Officials and Code Administrators International, Country Club Hills, Illinois; and the Standard Building Code (SBC), published by the Southern Building Code Congress International (SBCCI), Birmingham, Alabama. All three codes adopt the ACI 318 *Building Code Requirements for Structural Concrete* (1995) for regulations concerning concrete design and construction. The latest (1995) edition of this standard contains interesting and important provisions related to the loading imposed on a concrete structure during construction. These provisions are discussed below. It should also be mentioned that the ASCE Standards Committee on Design Loads During Construction is currently engaged in developing a standard on the subject. As of this writing, the first committee letter ballot on the standard has been concluded.

For the first time, the 1995 edition of the ACI 318 standard contains the following requirements, which are directly quoted from the standard:

ACI 318—6.2—Removal of forms, shores, and reshoring

ACI 6.2.1—Removal of forms
Forms shall be removed in such a manner as not to impair safety and serviceability of the structure. Concrete to be exposed by form removal shall have sufficient strength not to be damaged by removal operation.

ACI 6.2.2—Removal of shores and reshoring

The provisions of ACI 6.2.2.1 through ACI 6.2.2.3 shall apply to slabs and beams except where cast on the ground.

ACI 6.2.2.1—Before starting construction, the contractor shall develop a procedure and schedule for removal of shores and installation of reshores and for calculating the loads transferred to the structure during the process.

(a) The structural analysis and concrete strength data used in planning and implementing form removal and shoring shall be furnished by the contractor to the building official when so requested.

(b) No construction loads shall be supported on, nor any shoring removed from, any part of the structure under construction except when that portion of the structure in combination with remaining forming and shoring system has sufficient strength to support safely its weight and loads placed thereupon.

(c) Sufficient strength shall be demonstrated by structural analysis considering proposed loads, strength of forming and shoring system, and concrete strength data. Concrete strength data shall be based on tests of field-cured cylinders or, when approved by the building official, on other procedures to evaluate concrete strengths.

ACI 6.2.2.2—No construction loads exceeding the combination of superimposed dead load plus specified live load shall be supported on any unshored portion of the structure under construction, unless analysis indicates adequate strength to support such additional loads.

ACI 6.2.2.3—Form supports for prestressed concrete members shall not be removed until sufficient prestressing has been applied to enable prestressed members to carry their dead load and anticipated construction loads.

The ACI 318 Commentary explains:

In determining the time for removal of forms, consideration should be given to the construction loads and to the possibility of deflections. The construction loads are frequently at least as great as the specified live loads. At early ages, a structure may be adequate to support the applied loads but may deflect sufficiently to cause permanent damage.

It is further stated in the commentary that:

The removal of formwork for multistory construction should be a part of a planned procedure considering the temporary support of the whole structure as well as that of each individual member. Such a procedure should be worked out prior to construction and should be based on a structural analysis taking into account the following items, as a minimum:

(a) the structural system that exists at the various stages of construction and the construction loads corrresponding to those stages;

(b) the strength of the concrete at the various ages during construction;

(c) the influence of deformations of the structure and shoring system on the distribution of dead loads and construction loads during the various stages of construction;

(d) the strength and spacing of shores or shoring systems used, as well as the method of shoring, bracing, shore removal, and reshoring including the minimum time intervals between the various operations; and

(e) Any other loading or condition that affects the safety or serviceability of the structure during construction.

For multistory construction, the strength of the concrete during the various stages of construction should be substantiated by field-cured test specimens or other approved methods.

It should be noted that the above requirements of ACI 318 and the accompanying suggestions in the commentary, while fairly comprehensive and meaningful, do not reflect current practice. How much influence these requirements are going to have on practice remains to be seen.

It is also of interest to note here that prescriptive detailing requirements for structural integrity were added for the first time to the 1989 edition of the ACI 318 standard for cast-in-place concrete structures. Similar requirements have now been added to ACI 318-95 for precast concrete structures. The requirements were prompted by experience showing that the overall integrity of a structure can be substantially enhanced by minor changes in the detailing of reinforcement. It is the intent of Section 7.13 of ACI 318-95 (requirements for structural integrity) to improve the redundancy and ductility in structures so that in the event of damage to a major supporting element or an abnormal loading event, the resulting damage may be confined to a relatively small area and the structure may have a better chance to maintain overall stability.

Specifically for flat-plate and flat-slab construction, ACI 318-89 required that at least two of the column strip bottom bars or wires in each direction be made continuous, pass within the column core, and be anchored at exterior supports. ACI 318-95 contains the following expanded requirements for flat plates or flat slabs:

> **ACI 318—13.3.8.5**—All bottom bars or wires within the column strip, in each direction, shall be continuous or spliced with Class A splices. . .At least two of the column strip bottom bars or wires in each direction shall pass within the column core and shall be anchored at exterior supports.

The continuous column strip bottom reinforcement provides the slab with some residual ability to span to the adjacent supports, should a single support be damaged. A structural integrity provision has now also been added for other two-way slabs without beams:

> **ACI 13.3.8.6**—In slabs with shearheads and in lift-slab construction, at least two bonded bottom bars or wires in each direction shall pass through the shearhead or lifting collar as close to the column as practicable and be continuous or spliced with a Class A splice. At exterior columns, the reinforcement shall be anchored at the shearhead or lifting collar.

References

ACI Committee 209. 1971. Prediction of creep, shrinkage and temperature effects in concrete structures. In *Designing for Creep, Shrinkage, and Temperature in Concrete Structures*, Publication SP-27, pp. 51–93. American Concrete Institute, Farmington Hills, MI.

ACI Committee 347. 1988. *Guide to Formwork for Concrete*, ACI 347R-88, 33 pp. American Concrete Institute, Farmington Hills, MI.

ACI Committee 435, 1991. *State-of-the-Art Report on Control of Two-Way Slab Deflections*, ACI 435-9R-91, 14 pp. American Concrete Institute, Farmington Hills, MI.

ACI Committee 318. 1995. *Building Code Requirements for Structural Concrete and Commentary*. ACI 318-95 and ACI 318R-95, 369 pp. American Concrete Institute, Farmington Hills, MI.

Agarwal, R.K. and Gardner, N.J. 1974. Form and shore requirements for multistory flat slab type buildings. *ACI J. Proc.* 71(11):559–569.

American National Standards Institute. 1983. *American National Standard for Construction and Demolition Operation—Concrete and Masonry Work—Safety Requirements*, ANSI A10.9-1983, 22 pp. Washington, DC.

Beresford, F.D. 1964. An analytical examination of propped floors in multistory flat plate construction. *Construction Rev.* 37(11):16–20.

Beresford, F.D. 1971. Shoring and reshoring of floors in multistory buildings. *Symp. on Formwork*, pp. 559–569. Concrete Institute of Australia, North Sydney.

Blakey, F.A. and Beresford, F.D. 1965. Stripping of formwork for concrete in buildings in relation to structural design. *Civil Eng. Trans. Inst. Eng. Australia* 7(2):92–96.

Branson, D.E. 1977. *Deformation of Concrete Structures.* McGraw–Hill, New York.

Byfors, J. 1980. *Plain Concrete at Early Ages*, Fo. 3, No. 80, 566 pp. Swedish Cement and Concrete Research Institute, Stockholm, Sweden.

Cantor, I.G. and Rizzi, A.V. 1982. Shore and reshore procedures for flat slabs. *Proc. Int. Conf. Forming Economical Buildings*, Chicago, pp. 18.1–18.12. Portland Cement Association, Skokie, IL.

Carino, N.J. and Lew, H.S. 1982. Re-examination of the relation between splitting tensile and compressive strength of normal weight concrete. *ACI J. Proc.* 79(3):214–219.

CEB. 1972. *Manual: Structural Effects of Time-Dependent Behavior of Concrete.* Bulletin D'Information, No. 80, Comite Europeen du Beton (CEB), Paris.

City of Boston. 1971. *The Building Collapse of 2000 Commonwealth Avenue, Boston, Massachusetts*, Report, Mayor's Investigating Commission, 159 pp. Boston, MA.

Fattal, S.G. 1983. *Evaluation of Construction Loads in Multistory Concrete Buildings.* NBS Building Science Series No. 146, 130 pp. National Bureau of Standards, Washington, DC.

Fintel, M. Iyengar, S.H., and Ghosh, S.K. 1987. Column shortening in tall structures—prediction and compensation. Publication EB108-01D, Portland Cement Association, Skokie, IL.

Fu, H.C. and Gardner, N.J. 1986. Effect of high early-age construction loads on the long term behavior of slab structures. *Properties of Concrete at Early Ages*, Publication SP-95, pp. 173–200. American Concrete Institute, Farmington Hills, MI.

Gardner, N.J. 1960. Relationship of the punching shear capacity of reinforced concrete slabs with concrete strength. *ACI J. Proc.* 87(1):66–71.

Gardner, N.J. 1985. Shoring, reshoring, and safety. *Concrete Int.* 7(4):28–34.

Gardner, N.J. and Poon, S.M. 1976. Time and temperature effects on tensile, bond and compressive strengths. *ACI J. Proc.* 73(7):405–409.

Grundy, P. and Kabaila, A. 1963. Construction loads on slabs with shored formwork in multistory buildings. *ACI J. Proc.* 60(12):1729–1738.

Hansen, T.C. and Mattock, A.H. 1966. Influence of size and shape of member on the shrinkage and creep of concrete. *ACI J. Proc.* 63(2):267-289.

Hickey, K.B. 1968. *Creep of Concrete Predicted from Elastic Modulus Tests*, Report No. C-1242. United States Department of the Interior, Bureau of Reclamation, Denver, CO.

Hover, K.C. 1984. The Effects of Moisture on the Physical Properties of Hardened Concrete and Mortar. Ph.D. Thesis. Cornell University, Ithaca, NY.

Hover, K.C. 1988. The effects of drying and form and shore removal on flexural cracking in beams and slabs. In *Forming Economical Concrete Buildings: Proceedings of the Third International Conference*, Publication SP107, pp. 169–184. American Concrete Institute, Farmington Hills, MI.

Hurd, M.K. 1989. *Formwork for Concrete*, Publication SP-4, 5th ed., 475 pp. American Concrete Institute, Farmington Hills, MI.

Kaminetzky, D. and Stivaros, P.C. 1994. Early-age concrete: Construction loads, behavior and failures. *Concrete Int.* 16(1):58–63.

Klieger, P. 1958. Effect of mixing and curing temperature on concrete strength. *ACI J. Proc.* 29(12):1063–1081.

Lasisi, M.Y. and Ng, S.F. 1979. Construction loads imposed on high-rise floor slabs. *Concrete Int.* 1(2):24–29.

Lew, H.S. 1985. Construction loads and load effects in concrete building construction. *Concrete Int.* 7(4):20–23.

Lew, H.S. and Reichard, T.W. 1978. Mechanical properties of concrete at early ages. *ACI J. Proc.* 75(10):533–542.

Lew, H.S., Carino, N.J., Fattal, S.G., and Batts, M.E. 1982a. Investigation of the construction failure of Harbour Cay Condominium in Cocoa Beach, FL. NBS Building Science Series No. 145, National Bureau of Standrads, Washington, DC.

Lew, H.S., Carino, N.J., and Fattal, S.G. 1982b. Cause of the condominium collapse in Cocoa Beach, FL. *Concrete Int.* 4(8):64–73.

Leyendecker, E.V. and Fattal, S.G. 1977. *Investigation of the Skyline Plaza Collapse in Fairfax County, Virginia*, Building Science Series No. 94, 88 pp. National Bureau of Standards, Washington, DC.

Liu, X.-L., Chen, W.-F., and Bowman, M.D. 1985a. Construction loads on supporting floors. *Concrete Int.* 7(12):21–26.

Liu, X., Chen, W.-F., and Bowman, M.D. 1985b. Construction load analysis for concrete structures. *J. Struct. Eng. ASCE.* 111(5):1019–1036.

Marosszeky, M. 1972. Construction loads in multistory structures. *Civil Eng. Trans. Inst. of Eng. Australia* 14(1):91–93.

Nawy, E.G. 1996. *Reinforced Concrete—A Fundamental Approach*, 3rd ed., 850 pp. Prentice Hall, Upper Saddle River, NJ.

Neville, A.M. 1981. *Properties of Concrete*, 3rd ed. Pitman, London.

Nielsen, K.E.C. 1952. Loads on reinforced concrete floor slabs and their deformations during construction. *Proc.* No. 15, 113 pp. Swedish Cement and Concrete Research Institute, Stockholm.

Pauw, A. 1960. Static modulus of elasticity of concrete as affected by density. *ACI J. Proc.* 57(6):679–687.

Pfeifer, D.W., Magura, D.D., Russell, H.G., and Corley, W.G. 1971. Time-dependent deformations in a 70-story structure. In *Designing for Creep, Shrinkage, and Temperature in Concrete Structures*, Publication SP-27, pp. 159–185. American Concrete Institute, Farmington Hills, MI.

Price, W.H. 1951. Factors influencing concrete strength. *ACI J. Proc.* 22(6):417–432.

Recommendations for an International Code of Practice for Reinforced Concrete. American Concrete Institute, Farmington Hills, MI, and Concrete Association, London.

RILEM Commission 42-CEA. 1981. Properties of set concrete at early ages: State-of-the-art report. *Mater. Struct.* 14(84):399–450.

Russell, H.G. and Corley, W.G. 1977. *Time-Dependent Behavior of Columns in Water Tower Place*, Publication RD025B, 10 pp. Portland Cement Association, Skokie, IL.

Sbarounis, J.A. 1984. Multistory flat plate buildings—Construction loads and immediate deflections. *Concrete Int.* 6(2):70–77.

Scaffolding and Shoring Institute. 1977. *Recommended Safety Requirements for Shoring Concrete Formwork*, 13 pp. Cleveland, OH.

Scanlon, A. 1987. Excessive slab deflection—A serviceability failure. *J. Forensic Eng.* 1(2):21–29.

Stivaros, P.C. and Halvorsen, G.T. 1990. Shoring/reshoring operations for multistory buildings. *ACI Struct. J.* 87(5):589–596.

Stivaros, P.C. and Halvorsen, G.T. 1991. Equivalent frame analysis of concrete buildings during construction. *Concrete Int.* 13(8):57–62.

Stivaros, P.C. and Halvorsen, G.T. 1992. Construction load analysis of slabs and shores using microcomputers. *Concrete Int.* 14(8):27–32.

Taylor, P.J. 1967. Effects of formwork stripping time on deflections of flat slabs and plates. *Australian Civil Eng. Const.* 8(2):31–35.

Taylor, P.J. and Heiman, J.L. 1977. Long-term deflections of reinforced concrete flat slabs and plates. *ACI J. Proc.* 74(11):556–561.

Troxell, G. E. Davis, H.E., and Kelly, J.W. 1968. *Composition and Properties of Concrete*, 2nd ed. McGraw–Hill, New York.

U.S. Department of Labor. 1972. *Safety and Health Regulations for Construction*, 29 CFR Part 1926, 97 pp. Bureau of Labor Standards, Washington, DC.

Walker, S. and Bloem, D.L. 1957a. Effects of curing and moisture distribution on measured strength of concrete. *Highway Res. Rec.* 36:334–346.

Walker, S. and Bloem, D.L. 1957b. Studies of flexural strength of concrete, Part 3: Effects of variations in testing procedures. *Proc. Am. Soc. Test. Mater.* 57:1122–1142.

Wheen, R.J. 1982. An invention to control construction floor loads in tall concrete buildings. *Concrete Int.* 4(5):56–62.

Winter, G. and Nilson, A.H. 1991. *Design of Concrete Structures*, 11th ed. McGraw–Hill, New York.

(a)

(b)

(a) Deflection test of a prestressed beam (Courtesy Portland Cement Association). (b) Deflection of a prestressed concrete beam prior to failure (Courtesy Dr. Edward G. Nawy, Rutgers University).

9

Deflection of Concrete Members

by
Russell S. Fling, P.E.
A practicing, consulting structural engineer since graduating from Ohio State University in 1949. He has served on many technical committees of the American Concrete Institute and as President of the Institute in 1976.

9.1 Introduction

When applying strength design procedures, engineers can obtain building structures that have adequate strength but unsatisfactory serviceability, that is, they exhibit excessive deflection. Thus the size of many flexural members is determined by deflection response rather than by strength. The purpose of this chapter is to outline efficient procedures for estimating deflection, discuss factors affecting the variability of deflections, suggest procedures for use in the design process to reduce the expected deflection, and enable design engineers to proportion building structures closer to both strength and serviceability requirements. The result could be more economical structures compared to those designed too conservatively because of concerns about deflection. Throughout this chapter, the discussion assumes that a competent design is prepared according to ACI 318 Building Code [ACI 318, 1995] and that construction follows good practices.

0-8493-2666-4/97/$0.00+$.50

9.2 Elastic Calculation Methods

9.2.1 Selection of Methods

Perhaps the most important step in computing deflection is to sketch the deflected shape of the structure, especially if its geometry or loading is somewhat complicated. Computations of deflection magnitude will be meaningless if the engineer has the wrong concept of deflection response. Horizontal members can deflect upward as well as downward, and vertical members can deflect in either direction. Sometimes a member is in double curvature and deflects in both directions. One load may cause a member to deflect in one direction, and another load may cause the same member to deflect in the opposite direction. If an engineer has difficulty visualizing the direction of deflection, experimentation with a simple model of heavy paper, balsa wood, plastic, or other flexible material should clarify the deflection response.

The labor in preparing deflection calculations can be considerably reduced by the judicious selection of a few critical members in a structure for which deflection calculations will be made and disregarding all other members. The success of this approach depends on the skill of the engineer in selecting critical members. Labor in preparing deflection calculations can also be reduced by first deciding the reason for limiting the deflection (see Section 9.6) and directing the calculations to that end.

Finally, deflection calculations can be minimized by determining the deflection limit and span and then selecting an appropriate calculation method by referring to Figure 9.1. There are no precise lines of demarcation between methods. Experienced engineers will consider computation time available, the importance of the member and its deflection response, and the importance that owners and users of the structure will assign to proper deflection behavior before selecting a calculation method. Some details of the normal or extended calculation methods can be used in a simpler method as the situation warrants. For these reasons, an engineer may want to start with the simplest calculation method and extend it if results of the first calculation indicate a potential deflection problem.

9.2.2 Indirect Method (Minimum-Thickness Tables)

Deflection can be limited indirectly by limiting the span-to-depth ratio of a member or by limiting the stress level. For beams and one-way slabs, deflection need not be further calculated if the minimum thickness given in the ACI Code is met and if members do not support and are not attached to partitions or other construction likely to be damaged by large deflections. This is an important qualification as many members do support or are attached to fragile building elements. Likewise, deflection of flat slabs and flat plates need not be calculated if the thickness is limited to values given in the ACI Code. A satisfactory deflection response can normally be expected if the

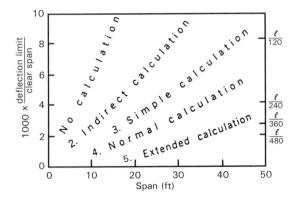

FIGURE 9.1 Recommended calculation procedures.

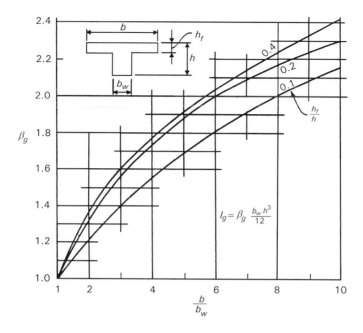

FIGURE 9.2 Moment of inertia I_g of uncracked T-beams, in^4.

superimposed load is small in relation to the self-weight of the concrete, as is usually the case with buildings intended for human residence.

Alternatively, experience indicates that flexural members remaining essentially uncracked at service loads will generally not have excessive deflection. This condition can be easily checked using structural mechanics. That is, $f_r \leq M_{cr}/S$. The section modulus S of T-beams can be approximated by increasing the section modulus by half as much as the moment of inertia is increased by the flanges for tensile stress at the bottom of the stem $[1 + (\beta_g - 1)/2]$ and increasing the section modulus twice as much as the moment of inertia for tensile stress at the top of the flange $[1 + 2(\beta_g - 1)]$. The factor β_g can be taken from Figure 9.2. The error introduced by this approximation is normally less than 10%.

Use of indirect methods should be limited to members with a span no more than about 25 or 30 ft and normal allowable deflections. Reasons for these limitations are further discussed in Sections 9.4 and 9.6.

9.2.3 Simplified Method (Use of Graphs to Estimate Stiffnesses)

For a quick estimate of deflection, use the midspan moment at service loads as calculated for strength design (or maximum moment in cantilevers). If only factored moment is available, divide it by the estimated average load factor to obtain service moment. Alternatively, service moments may be computed directly. If concentrated loads or variable uniform loads are present, use an average uniform load, taking care to make due allowance for loads concentrated near the center of the span. If end moments are not equal to zero or to fixed end moments, use an appropriate moment coefficient from standard references.

Estimate whether or not the beam is cracked at midspan by using Eq. (9.1) and calculate either $E_c I_g$ or $E_c I_{cr}$. Use Figure 9.2 to assist in calculating I_g and Figure 9.3 to assist in calculating $E_c I_{cr}$. Do not include the effects of compression reinforcement. Use the appropriate flexural stiffness in the usual equations for deflection of indeterminate structures to determine the immediate deflection.

$$M_{cr} = f_r I_g / y_t \tag{9.1}$$

where M_{cr}, f_r, I_g, E_c, I_{cr}, and y_t are defined in the ACI Code.

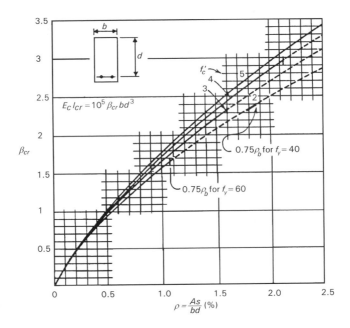

FIGURE 9.3 Flexural stiffness $E_c I_{cr}$ of cracked beams, pounds per square inch.

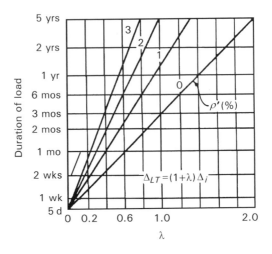

FIGURE 9.4 Values of multiplier λ for sustained load.

Estimate the additional long-term deflection due to creep and shrinkage by multiplying the immediate deflection by a factor λ, taken from Figure 9.4. Do not calculate incremental deflection. Incremental deflection is that portion of the total deflection that occurs after the installation of deflection-sensitive elements of the construction and that stops when they are removed.

9.2.4 Normal Method

For a more careful estimate of deflection, use actual service loads and moments and calculate deflection using normal methods of structural mechanics. Calculate I_g using the normal procedures of structural mechanics, then calculate the cracking moment for the midspan section. If service moment M_a is less than the cracking moment, use I_g in the deflection calculations. If service moment is more than the cracking moment, calculate I_{cr} and I_e and use the latter in the calculations. Computation of I_g, I_{cr}, and I_e is developed in Chapters 4 and 27 of this Handbook, andefer the resulting equations are given below. Neglecting compression reinforcement (and tension reinforcement in the gross section) will meet ACI Code requirements.

For the cracked moment of inertia I_{cr} use Eq. (9.2)

$$I_{cr} = nA_s(1 - k)\,jd^2 \tag{9.2}$$

Alternatively, for the cracked flexural stiffness $E_c I_{cr}$ use Eq. (9.2a).

$$E_c I_{cr} = E_s A_s(1 - k)\,jd^2 \tag{9.2a}$$

where the depth of the compression block kd by elastic theory is computed using

$$k = [(\rho n)^2 + 2\rho n]^{0.5} - \rho n \tag{9.3}$$

and the internal lever arm by elastic theory jd is computed using

$$j = (1 - k/3) \tag{9.4}$$

For the effective moment of inertia I_e, use Eq. (9.5)

$$I_e = I_{cr} + (M_{cr}/M_a)^3(I_g - I_{cr}) \tag{9.5}$$

The labor in making these calculations can be considerably eased, and the chance for error reduced, by using a spreadsheet program prepared by the user.

Deflection of concrete increases with time, see Section 9.3.1, below. Also see Chapter 8 of Fling (1987) and Fling (1996).

Incremental deflection, which is less than the total long-term deflection, may need to be compared to the allowable deflection. See Section 9.6 for a discussion of incremental deflection.

9.2.5 Extended Method

Extended methods involve consideration of all relevant factors affecting deflection as listed in Section 9.4. By carefully allowing for each relevant factor, an engineer can significantly improve the accuracy of deflection calculations. Extended calculations may result in better estimates of deflection, but they do not necessarily meet ACI Code requirements.

9.3 Other Calculation Considerations

9.3.1 Long-Term Deflection

ACI Code procedures using a single multiplier for additional long-term deflection may be used for simplified method and normal method calculation procedures. The initial deflection is multiplied by a single factor that depends only on the duration of the load and the quantity of compressive reinforcement. No other factor, such as ambient conditions, materials, or construction conditions, need be considered, yet these and many other factors do affect long-term deflection. This observation alone should raise a warning flag about the accuracy of the procedure under all possible conditions.

When more accurate deflection estimates are needed, using extended method procedures, consider the six parameters that affect the magnitude of the creep coefficient and shrinkage strain, listed by Branson and Christiason (1971), ACI Committee 209 (1971), and ACI Committee 435 (1995).

TABLE 9.1   Variations in Creep Coefficient from Standard Conditions

Parameter	Standard Conditions	Variations,* %
Age at loading	7 d	−23
Relative humidity	40%	−40
Minimum thickness	6 in	−38
Slump	3 in	+36
Cement content	7 bags/cy	+5
Fines content	50% passing No. 4 sieve	+56

Source: ACI 435 (1995).

*Variations in the ultimate creep coefficient. A positive variation indicates deflection would be increased, and a negative variation indicates deflection would be decreased.

The numerical values they assigned to each factor for potential variations from standard conditions for the creep coefficient have been summarized in Table 9.1. In addition, it is known that temperature, coarse aggregate, and the stress/strength ratio affect creep [Smadi et al., 1987; ACI 209, 1971].

The volume/surface ratio has an important effect on modulus of rupture f_r. In thin members (small V/S ratio), excess mixing water has a shorter average distance to the surface where it can evaporate, leaving the concrete more prone to shrinkage and a lower flexural strength. Conversely, thicker members (large V/S ratio) take longer to dry out, maintaining the flexural strength as the concrete gains strength with time. Experience indicates that thin members with a V/S ratio of 2 may have a flexural strength of only $5(f_c')^{0.5}$, whereas thick members with a V/S ratio of 6 or more may have a flexural strength of $10(f_c')^{0.5}$. While this is only a one-third variation from the ACI Code value for flexural strength, the effect on deflection can be more dramatic because it may determine whether a member is cracked in flexure or not and to what extent it is cracked. When flanges are a significant part of a member, they should be included in the V/S calculation.

Current computation procedures do not consider the amount and location of tensile reinforcement or the location of compressive reinforcement with respect to the centroid of the gross concrete section, even though it is likely that location of reinforcement has some affect on shrinkage warping and long-term creep. Indeed, experience indicates that ribbed slabs do deflect more than thicker beams that are loaded to the same stress level and have the same span as the slabs. This can be attributed to thin flanges with potentially higher creep and shrinkage due to their lower volume/surface ratio and to flange location as far as possible from tensile reinforcement. The distance from tension steel to centroid of the compression zone in ribbed slabs is often 0.8 to 0.9 times the total depth, whereas in rectangular beams it is usually about 0.6 to 0.7 times the depth.

The factors for additional long-term deflection contained in the ACI Code make no distinction between members that are fully cracked and those that are uncracked. If a beam is **uncracked**, the creep deflection should be proportional to the creep coefficient. If, however, a beam is **cracked**, it is more likely that creep deflection is proportional to some fraction of the creep coefficient.

The long-term creep factors given in the ACI Code and illustrated graphically in Figure 9.4 normally give satisfactory results. This may be because most concrete flexural members are partially cracked and the ultimate creep coefficient is normally about 2.0, rather than because the procedure is valid over a wide range of conditions. For conditions other than "normal," the procedure may be flawed and contribute to erratic fluctuation in the ratio of calculated to measured deflection.

9.3.2 Continuous Members

For continuous members, it is essential that an accurate assessment of the distribution of moments be used in calculating deflection. Use of inaccurate moment values is the largest single source of error [Fling, 1996].

Deflection calculations may proceed in the normal manner for structural mechanics, however, most continuous members are not prismatic if there is flexural cracking somewhere in the span. Thus the most important decision is what flexural stiffness value *EI* to use. For simplified or normal calculation methods (see Section 9.2), it is satisfactory to use the midspan flexural stiffness because it has the greatest effect on deflection. For greater accuracy, an average of midspan and end-region flexural stiffness can be used, taking care to give greater weight to the midspan stiffness than to the end-region stiffnesses. Guidance on averaging the stiffnesses is given ACI 435 (1995). For cantilevers, flexural stiffness at the end support should be used for calculations.

9.3.3 Two-Way Construction

When two-way construction consists of beams and slabs of normal proportions, deflection is considered for each direction (beams or slabs) independently of the other. When two-way slab systems, with or without beams in one or both directions, fall into the limitations of Chapter 13 of the ACI Code [ACI 318, 1995], deflection can still be considered for each direction independently of the other, but deflection of that portion of the slab on the column line (column strip) will be greater than that portion spanning in midbay (middle strip). Appropriate allowance for this phenomenon can be made by adjusting the moments used in the deflection calculations in proportion to the positive moments used for design of the slab system. The procedure is known as the Crossing-Beam Method and is discussed in ACI 435 (1995). For all types of two-way construction, the maximum deflection at midbay will be the sum of the deflections in the two directions.

9.3.4 Prestressed Members

Normally, prestressing tends to balance the dead load. With perfect balance, there would be no immediate deflection under dead loads and no long-term deflection. Only live-load deflections would occur. Such balance can rarely be achieved because of construction tolerances and the difficulty of placing tendons in the theoretically correct location. Also, prestressing cannot simultaneously balance concrete self-weight, superimposed dead loads, and semipermanent live loads. The engineer must make a choice of which loads to balance. Nevertheless, prestressed members can generally be made thinner with lower deflection than nonprestressed members under similar conditions. Chapter 3 of the latest ACI Committee 435 report [ACI 435, 1995] gives extensive guidance on calculation of deflection of prestressed members.

9.3.5 Torsional Deflection

Sometimes it is necessary to calculate the rotation of a member with torsional moment or the deflection of an element supported by the member but eccentric to is center of gravity. In such cases, the twist *q* over a length l_t of a member can be calculated as

$$\theta = \sum T / GK \tag{9.6}$$

where

$\sum T$ = summation of the service torsional moment along length l_t of the member (see Figure 9.5)

GK = torsional stiffness similar to the flexural stiffness *EI*

For an uncracked member it is sufficiently accurate to take

$$G = 0.3 E_c$$

$$K_g = \sum x^3 y / 3 \tag{9.7}$$

$$GK_g = 0.1 E_c \sum x^3 y$$

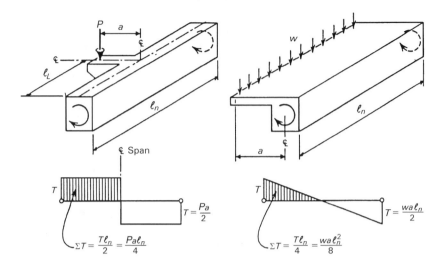

FIGURE 9.5   Twist on an isolated beam due to eccentric loads.

where

  x = shorter overall dimension of the rectangular part of the cross section
  y = longer overall dimension of the rectangular part of the cross section

Component rectangles of flanged sections should be selected to yield the largest value of $\sum x^3 y$ as in strength design for torsion.

For a torsionally cracked member, it is sufficiently accurate to use an equation developed by Lampert (1973).

$$GK_{cr} = \frac{E_s (x_0 y_0)^2 A_n (1 + m)}{2(x_0 + y_0)s} \tag{9.8}$$

where

$$
\begin{aligned}
E_s &= \text{modulus of elasticity of steel} \\
x_0 \text{ and } y_0 &= \text{shorter and longer dimension, respectively, between longitudinal corner bars} \\
s &= \text{stirrup spacing} \\
A_n &= \text{Area of one stirrup leg} \\
m &= \frac{(A_s + A'_s)s}{2 A_n (x_0 + y_0)}
\end{aligned}
$$

where

  A_s = area of longitudinal tension steel
  A'_s = area of longitudinal compression steel

Deflection of an element supported by a twisting member can be calculated by multiplying the angle of twist in the member at the element by the distance from the center of gravity of the member to the point in question. See Figure 9.6.

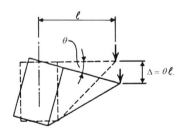

For most cases in which twist or deflection due to twisting members is significant, the member will be at least partially torsionally cracked, and torsional stiffness given by Eq. (9.8) should be used, as it will give an upper bound to the deflection.

FIGURE 9.6   Deflection due to beam rotation.

Torsional stiffness of an uncracked member is usually much larger than that of the same member torsionally cracked, but there is little research available to indicate the proper transition from uncracked to cracked stiffnesses. Procedures for calculating torsional rotation are not intended for application to moment-distribution procedures.

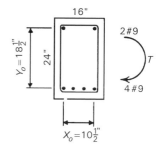

9.3.5.1 Example 9.1

What is the twist in an 11-ft length of a rectangular beam 16 in wide by 24 in deep with a uniform service torsional moment of 50 ft-kips? The beam is reinforced with four #9 bottom and two #9 top bars, as shown in Figure 9.7. Shear reinforcement is #4 bars at 5 in center to center. Assume $f'_c = 3$ ksi.

FIGURE 9.7 Beam subject to rotation.

Solution. Using Eq. (9.8), $2(x_0 + y_0) = 2(10.5 + 18.5) = 58$, and cracked torsional stiffness is

$$GK_{cr} = \frac{29 \times 10^6 (10.5 \times 18.5)^2 \times 0.20}{58 \times 5} \times \left(1 + \frac{6.00 \times 5}{0.2 \times 58}\right) = 2.7 \times 10^9 \text{ psi}$$

Using Eq. (9.6), torsional rotation or twist is

$$\theta = \frac{50 \times 10^3 \times 12 \times (11 \times 12)}{2.7 \times 10^9} = 2.9 \times 10^{-2} \text{ rad} \approx 1.7°$$

Actual twist may be less than that calculated if the beam remains partially uncracked.

9.3.6 Temperature Deflection

When there is a temperature gradient across a flexural member from one side to the other, temperature warping will result. In most structures, the temperature gradient is small and therefore the movement is small. In some structures, such as the top level of a parking garage, the gradient may be large but the resulting movement is inconsequential because it does not affect either the serviceability of the structure or nonstructural elements. Occasionally, a structure will have large thermal gradients and serviceability or nonstructural elements will be affected. For example, tall buildings with exposed exterior columns may experience thermal lengthening and shortening relative to internal columns that are at a nearly constant temperature. In other cases, an exposed roof may deflect up and down with seasonal or daily changes in temperature and affect partitions and ceilings below. In such cases, ACI Committee 435 (1995) offers guidance on temperature deflection.

9.4 Factors Affecting Deflection

In order to properly assess the variability of deflections, it is important to understand and consider all parameters that have an effect on the variability. Table 9.2 lists most of the parameters that are known or that might be determined before construction starts. Table 9.3 lists parameters that, by their very nature, cannot be known before construction starts. Parameters are separated between Tables 9.2 and 9.3 in order to illustrate the difficulties facing a design engineer. The value of the parameters that have the largest effect on the actual deflection of a structure are unknown at the time of design (those in Table 9.3). Indeed, they cannot be known. While it is theoretically possible to determine the value of parameters in Table 9.2 before construction starts, it is frequently impractical to obtain all necessary information. Thus an even higher proportion of variables

TABLE 9.2 Parameters Affecting Deflection: Known Before Construction

1. Loading
   1.1 Specified maximum live loads
   1.2 Dead load based on assumed unit weights and specified sizes
   1.3 Location and arrangement of loads

2. Flexural Stiffness
   2.1 Member sizes (including flange width) and bar placement
   2.2 Presence of compression and tension bars in uncracked concrete
   2.3 Tolerances on concrete outlines and bar locations
   2.4 Relation of steel reinforcement to centroid of concrete section
   2.5 Location of compression reinforcement at end sections, midspan, and in columns
   2.6 Specified concrete compressive strength f_c', flexural strength f_r, and modulus of elasticity E_c
   2.7 Modulus of rupture f_r based on specified minimum compressive strength f_c', versus the actual average f_c'.

3. Moment Distribution
   3.1 Frame completion before loads are applied.
   3.2 Homogeneous, isotropic, monolithic material
   3.3 Rigid joints
   3.4 Pattern loading

4. Time-Dependent Factors
   4.1 Multiplier based on Yu and Winter's (1960) work that ignores relevant parameters such as cracked versus uncracked sections, volume/surface ratio (V/S), absolute size of member, and shape of member (e.g., T-beam versus rectangular beam)
   4.2 Creep coefficient and shrinkage coefficient
   4.3 Properties of concrete proportional to its compressive strength (or its square root); relationships assumed constant for all ages and not interrelated (e.g., shrinkage is assumed not to affect effective modulus of rupture f_r)
   4.4 Constant normal ambient conditions of humidity and temperature
   4.5 Calculations not separated for creep deflection using reduced E_c and shrinkage warping

5. Criteria for Deflection Limits (considering only span/deflection ratios and ignoring other relevant parameters such as the following)
   5.1 Spans over 30 ft or under 10 ft
   5.2 Absolute values of deflection limits
   5.3 Dynamic (vibration) characteristics

cannot be determined by the design engineer before construction. When the parameters that can be anticipated before construction are separated from the parameters that cannot be anticipated, the effect of unanticipated variations can be more than four times the effect of anticipated variations.

Design engineers attempting to accurately assess probable deflection should consider all parameters affecting deflection and assign realistic values to each parameter, not "conservative" values as is done for strength design. That is, in computing deflections, loads should not be higher and concrete strength lower than expected. The potential variability of deflection can be estimated by computing deflections using realistic maximum and minimum values for parameters. While the length of the lists in Tables 9.2 and 9.3 may seem daunting, design engineers can increase the accuracy of their computations by carefully selecting the values of parameters used in the calculations.

It is unlikely that the actual deflection would reach either the extreme minimum or extreme maximum computed deflection, assuming the maximum variation is in the same direction for all parameters. For example, if the probability of reaching one of the extreme values in each of 9 steps is 20%, the probability of the actual deflection reaching either extreme of the computed deflections is less than one chance in a million. Note, however, that the probability of variation in some parameters could be higher than 20% in actual practice. Furthermore, the probability that actual deflection will be less than computed is about the same as the probability it will be more than computed, even though the latter event gets more attention in actual construction. Factors considered in more precise calculations can be categorized under computational errors, loading, flexural stiffness, fixity, creep and shrinkage, and construction variations.

TABLE 9.3   Parameters Affecting Deflection: Not Known Before Construction

1. Loading
   1.1 Actual live load
   1.2 Actual superimposed dead loads
   1.3 Concrete dead load due to actual dimensions and weight of concrete
   1.4 Construction loads due to shoring and reshoring procedures
   1.5 Loading history: magnitude, duration, and frequency of application
   1.6 Arching of masonry and fabricated walls, thus reducing load on supporting beams

2. Flexural Stiffness
   2.1 Bar sizes and quantities actually installed
   2.2 In-tolerance and out-of-tolerance construction of member sizes and bar placement
   2.3 Actual concrete properties f_c', f_r, E_c

3. Moment Distribution
   3.1 Moment redistribution due to incomplete frame (e.g., upper columns or adjacent beams not yet constructed)
   3.2 Moment redistribution due to creep and shrinkage
   3.3 Moment distribution due to nonprismatic members (prismatic outlines may become nonprismatic members because of cracking and location of reinforcement)
   3.4 Moment redistribution due to construction overloads
   3.5 Cracking of supporting columns, especially those with light reinforcement and low axial load levels
   3.6 Transverse (deflection) and torsional (rotation) movements of supporting girders
   3.7 Joint stiffness less than infinite

4. Time-Dependent Factors
   4.1 Actual shrinkage coefficient and its time variation
   4.2 Actual creep coefficient and its time variation
   4.3 Stress/strength ratio f_c/f_c' ratio at time of loading
   4.4 History of actual ambient conditions of humidity and temperature
   4.5 Location of compression reinforcement with respect to centroid of section

5. Criteria for Deflection Limits
   5.1 Changed or refined criteria

9.4.1 Computational Errors

Because computations are usually made by humans, computational errors become a significant source of discrepancy between the calculated and actual values of deflection. Some of these computational errors are outlined below.

- There are many steps in the calculation of deflection. If the probability of error in each step is 1%, then the probability of error in the final calculation could be almost 10%. If the probability of error in each step increases to 5%, then the probability of error in the final calculation could be as much as 40%. The magnitude and type of individual errors is such that even an experienced engineer might not notice errors of 1% to 5% on casual inspection. Calculation errors of 25 to 50% are not uncommon.

  The length and intricacy of deflection computations can be alleviated and the chances of error minimized if a computer program is used. The program should be capable of taking into account most of the known parameters affecting deflection. The program, no matter how sophisticated, should credibly predict the deflection of actual structures (not just laboratory specimens) with reasonable accuracy over a wide range of conditions. These are high standards, and whether such a program exists is open to question. Practicing design engineers should make a habit of comparing deflection performance with calculated deflection and thereby sharpen their judgment skills.

- Factored moments or loads determined in the design for strength might be inadvertently used in deflection computations instead of the actual service moments or loads.

- Maximum moments from pattern loading or from moment coefficients might be used rather than actual moments for the loading condition under consideration. Unrealistically large moments result in unrealistically large computed deflections.

9.4.2 Loading

- Consider the loading history, including the age at loading and unloading for each increment, as it will affect the immediate deflection because concrete strength and modulus of elasticity may be different at different ages. The degree of cracking could affect the flexural stiffness under first, partial loading but this is usually not a factor in multistoried buildings because construction loading frequently equals or even exceeds design loading.

- Use actual loads rather that those assumed in strength design. It is not unusual to find that live loads specified by the building code having jurisdiction are never reached in the actual structure. On the other hand, numerous studies indicate that reshoring loads on lower floors in multistoried buildings can be as much as twice the self-weight of the floor construction. Concrete dead load is frequently more than the specified live load plus superimposed dead load.

- Consider the proportion of long-term versus transient loading. Permanent live load will cause more creep deflection than will transient live load because creep occurs only under loads lasting for some time. Some live loads may remain in place for significant periods of time and contribute to creep deflection, while others are in place such a short time that they have no measurable effect on long-term deflection.

- Correctly assess the live load. If actual service live load is smaller than the design live load, the ratio of applied service moment M_a to cracking moment M_{cr} is lower, probably resulting in higher effective moment of inertia I_e. Thus both dead-load as well as live-load deflection would be lower.

- Consider redundancy. For example, walls may arch from end to end of beams and carry their own weight. Also, concrete members transverse to the main span might carry some load, thus reducing the moment and hence the deflection of the member under consideration.

- Calculate the service load moments separately for live and dead loads rather than proportion the moments at critical sections for every loading stage to the moments at the final loading stage. For many beams, dead load moments will be nearly equal to moments in a fixed-end beam. But with live load on just one of the spans (pattern loading) live-load moments compared to fixed-end moments are proportionally larger at midspan and smaller at the ends.

    If deflections are calculated using separate moments for dead and live loads, the computed total long-term deflection and the computed incremental deflection may be 5% to 10% different from those calculated using the same moment pattern. A separate distribution of moments for dead and live loads may not be justified for simplified and normal calculation methods but should be considered for extended-method calculations.

- Consider the support that members transverse to the span may provide. Such support has the effect of reducing the load. For example, slabs nominally designed as spanning one way may actually receive some support from parallel side members.

9.4.3 Flexural Stiffness

- Use the actual, measured modulus of elasticity of concrete E_c, if available.

- Even if preconstruction tests are required, use the measured modulus of rupture of concrete f_r because it is one of the more important parameters affecting deflection. The ACI Code specifies a ratio of $f_r/(f_c')^{0.5} = 7.5$, whereas research indicates the ratio usually varies from a low value of 7.5 to a high of about 10. A one-third increase in f_r, hence M_{cr}, will increase the effective moment of inertia I_e by as much as 75%.[1] The ACI Code value is "conservative," which means the actual deflection will frequently be less than the computed deflection. On the other hand, if there is significant restraint to shortening of flexural members by shrinkage,

---

[1] For example, assume $I_{cr}/I_g = 0.25$. When $M_{cr}/M_a = 7.5/10 = 0.75$, then $I_e = (0.75)^3 I_g + [1 - (0.75)^3] 0.25 I_g = 0.57 I_g$ and $1/0.57 = 1.76$.

the resulting tensile stresses may reduce the effective or apparent modulus of rupture f_r. For this to occur, there must be massive walls parallel to the anticipated shrinkage movement or other similarly stiff members resisting the shrinkage shortening of the flexural members. When interpreting the results of tests for modulus of rupture, make allowance for the relative humidity of ambient conditions around the test specimen and for its size. See the discussion in Section 9.3.1.

- If construction loads are not allowed to crack the concrete prematurely, consider the effective moment of inertia at every loading stage based on the amount of cracking at that stage. If the maximum load on a structure occurs during construction, only one cracking condition need be considered and one flexural stiffness calculation need be made, that is, for the amount of cracking that occurs at full load. This assumption results in simpler calculations and is usually justified by the field observation that the most severe loading condition frequently occurs during construction when material stockpiles, shoring loads from the floor above, and other construction loads are on the structure. In recent years, there have been many papers published on this subject. [Grundy and Kabaila, 1963; Sbarounis, 1984]

- For investigation of completed structures, use the actual location of reinforcement as built, especially if there is significant deviation from the specified location.

- Use the amount and location of compression reinforcement in computing I_g and I_{cr}. The transformed area of the steel can contribute significantly to the flexural stiffness of the cross section, even though it is not part of the ACI Code requirements.

- Use the amount and location of tension reinforcement in computing I_g. As with compression reinforcement, it can contribute significantly to the gross flexural stiffness. For cracked-section flexural stiffness, tension reinforcement is specifically considered in the calculation.

- For deflection calculations based on curvature along the span, use variation in flexural stiffness EI along the span due to change in cross section, change in amount of reinforcement, and degree of cracking.

- Include the effect of flanges, even though they may be small. If a rectangular beam instead of a T-beam (flanges are usually present) is used in computations, both the cracked and un-cracked (or gross) moment of inertia will be unrealistically low and the computed deflection unrealistically high.

  For almost 50 years (1947–1996) the ACI Code (currently Section 8.10) has specified essentially the same criteria for determining the width of the flange. While the rationale for these criteria may be lost, the criteria have served well for strength design because few, if any, failures due to inadequate flange width have been reported. For deflection computations, the situation is not so good. The design engineer is concerned with an accurate estimate of deflection, neither high nor low, as contrasted with an effort to estimate minimum strength, not necessarily the actual strength. ACI Code T-beam procedures are inconsistent and probably conservative.

- Make a realistic assessment of the contribution of end-region stiffness to the overall stiffness rather than a simple averaging of end and midspan region stiffnesses. A simple average gives far more weight to the end-region stiffnesses than is justified. Using midspan stiffness is satisfactory for simple- and normal-calculation procedures, but accuracy can be increased somewhat for extended procedures by including the effect of stiffness in the end regions.

  For beams continuous on two ends, ACI 435 (1995) recommends the following:

$$I_e = 0.70 I_{e(m)} + 0.15(I_{e(1)} + I_{e(2)}) \qquad (9.9a)$$

For beams continuous on one end only:

$$I_e = 0.85 I_{e(m)} + 0.15(I_{e(1)}) \qquad (9.9b)$$

where subscript (*m*) indicates a midspan location and subscripts (1) and (2) indicate locations at ends one and two.

9.4.4 Factors Affecting Fixity

- Consider the support rotation for cantilevers because movement due to support rotation at the end of the cantilever could be greater than the flexural deflection of the member itself. Furthermore, rotation could result in the end rising or lowering, depending on the back-span dimensions and loading.
- Consider the restraint that nearby, unloaded, parallel members provide through the torsional stiffness of supporting beams. In a recent load test of a large beam and slab roof, it was found that parallel members as far as 200 ft away helped restrain end rotation of a loaded member [Fling, 1996]. Of course, this is a consideration primarily when individual members are loaded. If the entire structure is loaded simultaneously, restraint from parallel members will be minimized.
- Base the moment distribution on actual conditions of loading and stiffness of members rather than on assumed prismatic members. Until the advent of computers, such a consideration was beyond the capacity of manual computations. Depending on the physical conditions, a moment analysis based on nonprismatic members may give small to significant results for deflection considerations.
- Make allowance for stiffness of joints if they are weak or have inadequately anchored reinforcement. The effect will be similar to rotation of the support for cantilevers. At present, there are no satisfactory analytical tools for dealing with this consideration.
- Analyze end spans carefully because they are especially sensitive to assumptions of moment at critical sections. The less stiffness the end support is assumed to have the higher the positive moment in the end span and the higher the computed deflection, even though more positive-moment steel is provided to resist the higher moment. For this reason, an engineer might use wider beams and more reinforcement in end spans simply to control deflection rather than as a necessary requirement for strength. Edge columns in the upper stories of multistoried buildings are frequently cracked from the high moment induced by the beams framing into them. Thus their stiffnesses will be less than the gross uncracked section assumed in the moment analysis.
- Carefully control shoring and reshoring procedures because moment distribution has the largest effect on deflection variability. Unsuitable procedures, especially near construction joints, may impose moments during construction that are more adverse than those assumed in design.

9.4.5 Construction Variations

In design, there is little an engineer can do about construction variations beyond taking care to specify reasonable tolerances and procedures. One might also make a study of the effect certain tolerable variations have on the deflection of a critical member. Of course, in an investigation of an existing member, some construction variations can be measured and allowance made in the deflection calculations.

ACI Committee 117 has recommended tolerances on concrete outlines, steel placement, and material properties that are considered reasonable and achievable [ACI 117, 1990]. If deflection is computed using the maximum variation in one direction, the effect on deflection can be dramatic [Fling, 1992]. Fortunately, variations tend to cancel each other, but if a number of variables each affected deflection in the same direction, the effect could be significant.

Following is a list of the most serious variations.

- Tolerances on concrete outlines may result in a larger or smaller member than specified.
- Concrete cover on bottom bars may be smaller than specified owing to gravity. The increase in effective depth will increase the cracked moment of inertia. Likewise, the cover on top bars may be greater than specified with a reduction in effective depth and moment of inertia, but end-region stiffness has a much smaller effect on deflection than does the midspan stiffness.
- Concrete modulus of rupture is more variable than compressive strength. However, variability could be along the length of a member and thus tend to reach an average in its effect on deflection. There are few studies on this variation.
- Concrete compressive strength may average 15% higher than the specified strength, with a consequent increase in modulus of elasticity E_c of 7%. Depending on job conditions, the average could diverge from the specified strength even more than 15%.

  When a structure is loaded before it has reached its design strength, the adverse effects on deflection may be even more severe than indicated by a lower compressive strength because the creep coefficient can be up to 50% higher for stress/strength ratios f_c/f_c' greater than 0.50 than for f_c/f_c' less than 0.5. For this reason, engineers should specify procedures that avoid loading deflection-sensitive structures before concrete reaches its design strength.
- Concrete modulus of elasticity E_c depends on many parameters other than compressive strength, such as, for example, the aggregates. Lightweight aggregates in the concrete will generally result a lower modulus. The ACI Code gives an average value, but specific aggregates could vary significantly. In addition, certain very hard aggregates could result in a much higher modulus than assumed for the strength of the concrete.
- If the number of bottom tension bars is more than or less than specified, the effect on deflection will be proportional, if the section is cracked.

9.4.6 Creep and Shrinkage

Factors affecting creep and shrinkage have been discussed by ACI Committee 209 (1971) and some are noted in Section 9.3.1. These factors include age at loading, relative humidity, minimum thickness, volume/surface ratio, slump, cement content, air content, aggregates, admixtures, and ambient temperature.

9.5 Reducing Deflection of Concrete Members

Building structures designed by the limit-states approach alone may have adequate strength but unsatisfactory serviceability response. That is, they may exhibit excessive deflection. Thus the size of flexural members is in many cases determined by deflection response rather than by strength. This section gives design procedures for reducing the expected deflection that will enable design engineers to proportion building structures to meet both strength and serviceability requirements. The result could be more economical structures compared to those designed with unnecessarily conservative deflection response. The discussion assumes that a competent design is prepared in accordance with the ACI Code and construction follows good practices.

To properly evaluate options for reducing deflection, a design engineer must know the level of stress in the member under consideration. That is, is the member uncracked, partially cracked, or fully cracked, and does light or heavy reinforcement indicate low or high stresses? Heavily reinforced members tend to be fully cracked because they are highly stressed. This chapter considers only two limiting conditions, uncracked members and fully cracked members. For the purposes of this chapter, if the applied moment in the positive region is more than twice the cracking moment, considering the effect of flanges, the member may be considered to be fully cracked. Frequently, a member is only partially cracked ($M_{cr} < M_a < 2M_{cr}$) and the statements about both limiting conditions are not strictly applicable. Instead, an engineer must use his judgment or make appropriate calculations to interpolate between the limiting conditions. See Section 9.2.4, for methods for computing the degree of cracking in a member.

In addition to the stress conditions, there may be physical or nonstructural constraints on the use of some options, such as limits on increasing concrete outlines. For all options, there are financial implications that must be evaluated for each situation. Some options may increase the cost and some may have offsetting considerations that reduce the cost, while still others may have little effect on cost.

Except as noted, only non-prestressed, reinforced concrete members are discussed.

For each option presented, there is a discussion on the effect of implementation on deflection, the approximate range of potential reduction of deflection, and appropriate situations in which the option should be considered. Options are arranged in three groups: design techniques, construction techniques, and materials selection.

9.5.1 Design Techniques

Option 1: Make the Member Deeper

While increasing the depth may not be possible after schematic design of the building has been established because it affects the architectural and mechanical work, there are many instances where beam depth can be increased. The reduction in deflection is approximately equal to the square of the ratio of effective depths d for cracked sections[2] and approximately equal to the cube of the ratio of total concrete depths for uncracked sections.[3] For example, if an 18 in deep, rectangular beam with an effective depth of 15.5 in is increased to 20 in deep, and all other parameters kept the same, the cracked stiffness will increase by $(17.5/15.5)^2 = 1.27$ and the uncracked stiffness will increase by $(20/18)^3 = 1.37$. For heavily reinforced members, if the amount of steel is reduced when the depth is increased, the cracked stiffness is increased only in proportion to the increase in depth, or by 13% for this example.[4] The increase in stiffness of an uncracked T-beam when it is made deeper will be less than that for a rectangular beam because the flanges do not change. Flanges tend to have a fixed influence rather than a proportional influence on uncracked stiffness.

If, by increasing the depth, the concrete tensile stress in a member is reduced sufficiently so that it changes from a cracked, or partially cracked, to an uncracked member, the stiffness could increase dramatically. The uncracked stiffness can be as much as three times the partially cracked stiffness [Grossman, 1981].

Option 2: Make the Member Wider

This option is not applicable to slabs or other members with physical constraints on their width. However, where beams cannot be made deeper because of floor to floor height limitations, but can be made wider, the increase in stiffness is proportional to the increase in width if the member is uncracked. If the member is cracked and remains cracked after increasing the width, the increase in stiffness is very small. However, if a cracked member will be uncracked after the width is increased, the stiffness will increase dramatically, possibly by as much as a factor of three [Grossman, 1981].

Option 3: Add Compression Reinforcement

Using ACI Code procedures, compression reinforcement has no effect on immediate deflection but can reduce additional long-term or incremental deflection by a factor of up to about 50% [Branson, 1971]. The effect on total deflection is somewhat less. For example, if immediate deflection is

---

[2] The moment of inertia of cracked section $I = nA_s(1 - k)jd^2$ and $I \approx d^2$.
[3] The moment of inertia of uncracked sections $I = bh^3/12$ and $I \approx h^3$.
[4] In cracked sections, $A_s = M/f_s jd$. Substituting in the equation in footnote 2, $I \sim d$.

0.50 in and additional long-term deflection is 1.00 in, the total deflection is 1.50 in. The addition of 2% compression reinforcement reduces the additional long-term deflection to 0.50 in or by 50% and the total deflection to 1.00 in or by 33%.

Long-term deflection has two components, creep deflection and shrinkage warping. Compression reinforcement reduces deflection because concrete creep tends to transfer the compression force to the compression steel, which does not itself creep. The closer the steel is to the compression face of the member, the more effective the steel is in reducing long-term creep deflection. Thus compression steel will be more effective in deeper beams than in shallower beams or slabs if concrete cover to the compression face is a constant value. Indeed, in some very shallow members, owing to the requirements of minimum bar cover, compression steel could be at or near the neutral axis and almost totally ineffective in reducing long-term creep deflection.

Shrinkage warping occurs where the centroids of steel and concrete do not coincide and the shrinkage of concrete, combined with the dimensional stability of steel, warps the member, similar to a piece of bimetal subject to temperature variations. Compression steel reduces shrinkage warping because it brings the centroid of all reinforcement (tension plus compression) closer to the concrete neutral axis. While this is true of all flexural members, compression steel is especially effective for T-beams where the neutral axis is close to the compression face and far from the tension steel. If the T-beam has a thin slab subject to higher than normal shrinkage because of its high surface/volume ratio, then compression reinforcement will be more effective than for a rectangular beam. This will be true for ribbed slabs or joist systems as well.

Option 4: Add Tension Reinforcement

For uncracked members, addition of tension steel has hardly any effect on deflection. For fully cracked members, addition of tension steel will reduce both immediate and long-term deflection almost in proportion to the increase in steel.[5] For example, if the total deflection of a cracked member is 1.50 in, as in the example above, increasing the tension steel by 50% will reduce the deflection to about 1.10 in. Of course, the increased steel area should still be less than the maximum permitted by the ACI Code or $0.75\rho_b$ (rho balanced). This option is most useful for lightly reinforced solid and ribbed slabs. The option of adding more tension reinforcement is not available or is limited for heavily reinforced beams unless compression reinforcement is also added to balance the increase in tension-bar area in excess of $0.75\rho_b$. If this option is selected, excessive bar congestion may result.

Option 5: Add, or Increase Prestressing

Many prestressed members are designed for load balancing, that is, the upward reaction of the prestressing tendons approximately equals the downward force of dead and other sustained loads. Deflection due to live-load and other transient loads will be the same as for an ordinary reinforced concrete member, subject to two qualifications. If prestressing keeps the member in an uncracked state, whereas otherwise it would be cracked, the live-load deflection will be smaller. On the other hand, if the member size is reduced to take advantage of the prestressing, the live load deflection could be larger. For this reason, the span/depth ratio is usually limited to 48 for floor slabs with light live loads and 52 for roof slabs. See the ACI Code Commentary, Section R18.12.3 [ACI 318, 1995]. If the member has a high ratio of live-load to dead load, the span/depth ratio must be proportionally smaller to give satisfactory deflection performance.

If the sole purpose of prestressing a member is to achieve satisfactory deflection response, then prestressing to achieve full dead-load balancing is not necessary. Only a prestressing force sufficient to produce satisfactory deflection response is required. In this situation, the member may be partially cracked.

---

[5]The moment of inertia of cracked sections $I = nA_s(1 - k)jd^2$ and $I \approx 0.9A_s$ because $(1 - k)j$ does not vary much with changes in A_s.

Option 6: Revise Structure Geometry

Common solutions include increasing the number of columns to reduce the length of the spans, adding cross members to create two-way systems, and increasing the size of columns to provide more end restraint to flexural members (this is especially effective for end spans).

Option 7: Revise Deflection Limit Criteria

If deflection of a member is "excessive," the deflection limits may be reexamined to determine if they are unnecessarily tight. If experience or analysis indicates the limits (see Section 9.6) can be loosened, then other action might not be required. Many building codes do not set absolute restrictions on deflection. An engineer might decide that the building occupancy or construction conditions such as a sloping roof do not require the normal deflection limits.

9.5.2 Construction Techniques

Option 8: Cure the Concrete to Allow It to Gain Strength

Deflection response is determined by concrete strength at first loading, not by its final strength. If the construction schedule makes early loading of the concrete likely or desirable, then measures to ensure high strength at first loading can be effective. For example, if design compressive strength f_c' is 4 ksi and the member would be uncracked as designed, and if it is loaded when concrete strength is 2.5 ksi, it could be highly cracked owing to a lower modulus of rupture even though its load-carrying ability is satisfactory. A cracked member will deflect several times as much as the same member in an uncracked state. Furthermore, the modulus of elasticity of 4 ksi concrete is higher than that of 2.5 ksi concrete. See the discussion in options 19 and 20 for the effect of these parameters on deflection.

Option 9: Cure the Concrete to Reduce Shrinkage and Creep

Good curing will not affect the immediate deflection, but additional long-term deflection will be reduced. Assuming that the long-term component of deflection is evenly split between shrinkage and creep, if shrinkage is reduced 20% by good curing, the additional long-term deflection due to shrinkage will be reduced by 10%. The effect will be most pronounced on members subject to high shrinkage such as those with a low volume/surface ratio (small members), those with thin flanges, structures in arid atmospheres, or members that are restrained. The effect of good curing on creep is similar to the effect on shrinkage.

Option 10: Control Shoring and Reshoring Procedures

Many studies indicate that the shoring load on floors of multistoried buildings can be up to two times the dead weight of the concrete slab itself [Sbarounis, 1984]. Because the design superimposed load is frequently less than the concrete dead load, the slab may be seriously overstressed and cracked owing to shoring loads instead of remaining uncracked as assumed by calculations based on design loads. Thus the flexural stiffness could be reduced to as little as one third of the value calculated by assuming design loads only. Furthermore, the shoring loads may be imposed on the floor slabs before the concrete has reached its design strength. See the discussion in option 8.

Construction of formwork and shoring should ensure that a sag or negative camber is not built into the slab. Experience indicates that frequently the apparent deflection varies widely between slabs of identical design and construction. Some reasons for this may be that such slabs were not all built level or at specified grade or that the method and timing of form stripping was not uniformly applied. Also, construction loads may not have been applied uniformly.

Option 11: Delay the First Loading

This allows the concrete to gain more strength than specified or helps insure that it reaches its design strength. Both the modulus of elasticity E_c and the modulus of rupture f_r will be increased. An increase in E_c will increase the flexural stiffness as noted in option 19. An increase in f_r will reduce the amount of cracking or even allow the member to remain uncracked with an increase in flexural stiffness EI as noted in option 20.

Option 12: Delay Installation of Deflection-Sensitive Elements or Equipment

Such delay will have no effect on immediate or total deflection, except as noted in option 11, but incremental deflection (that occurring from the time a deflection-sensitive component is installed until it is removed or deflection reaches its final value) will be reduced. For example, if the additional long-term deflection is 1.00 in, and installation of partitions is delayed for 3 months, the incremental deflection will be approximately 0.50 in or one-half as much as the total (see Section 9.5.2.5 of the ACI Code).

Option 13: Locate Deflection-Sensitive Equipment to Avoid Deflection Problems

For example, printing presses, scientific equipment, and the like that must remain level should be located at midspan where the change in slope is very small with an increase in deflection. On the other hand, because the amplitude of vibration is highest at midspan, vibration-sensitive equipment may be best located near the supports.

Option 14: Provide Architectural Details to Accommodate the Expected Deflection

For example, partitions that abut columns may show the effect of deflection by separating horizontally from the column near the top, even though the partition is not cracked or otherwise damaged. Architectural details should accommodate such movements. Likewise, windows, walls, partitions, and other nonstructural elements supported by or located under deflecting concrete members can be provided with slip joints to accommodate the expected deflections or differential deflections between concrete members above and below the nonstructural elements.

Option 15: Build Camber Into the Floor Slab

Camber will have no effect on computed deflection of the slab; however, cambering is effective if the objective is to have a level floor slab, after deflection takes place, for installation of partitions and equipment. For best results, deflection must be carefully calculated (not overestimated), the pattern of cambering specified (not just a single value), results monitored during construction, and procedures revised as necessary for slabs constructed later.

Option 16: Ensure Top Bars Are Not Displaced Downward

Downward displacement will always reduce strength. The effect on deflection in uncracked members will be minimal. The effect on cracked members (those that are heavily loaded) is in proportion to the square of the ratio of change in effective depths for cantilevers but much less for continuous spans because the flexural stiffness (and resulting deflection) of the member is determined primarily by member stiffness at the midspan section. Thus the deflection of cantilevers is especially sensitive to misplacement of top bars. In continuous members, if the reduction in strength at negative moment regions results in redistribution of moments, deflection could be increased.

9.5.3 Materials Selection

Option 17: Select Materials That Reduce Shrinkage and Creep or Increase the Modulii of Elasticity and Rupture

Materials having an effect on these properties include aggregates, cement, silica fume, and admixtures. See options 19 and 20.

Option 18: Use a Mix Design That Will Reduce Shrinkage and Creep or Increase the Modulii of Elasticity and Rupture

For example, use a lower water/cement ratio and/or a lower slump, and/or change the proportions of materials. See options 9, 19, and 20 for a discussion of the effect of these properties on deflection.

Option 19: Use a Concrete With a Higher Modulus of Elasticity

Using ACI Code procedures, the stiffness of an uncracked member will increase in proportion to the elastic modulus or in proportion to the square root of the cylinder strength. (ACI Code Sections 9.5.2.2 and 8.5) The stiffness of a cracked section is affected very little by a change in modulus of elasticity.

Option 20: Use Concrete With a Higher Modulus of Rupture

Stiffness of uncracked members and highly cracked members will not be affected. Stiffness of partially cracked members will increase because the degree of cracking will be reduced. The increase in stiffness (decrease in deflection) will depend on steel percentage, the increase in modulus of rupture, and magnitude of applied moment.

TABLE 9.4   Deflection-Reducing Options

	Effect on Stiffness of Section	
Description	Uncracked	Cracked
Design Techniques		
1. Deeper members	$(h^*/h)^3$ for rectangular beams less for T-beams	$(d^*/d)^2$ or (d^*/d) if change to uncracked section, up to 300%
2. Wider members	(b^*/b)	Small, or change to uncracked section
3. Add A'_s	Up to 50% for D_{LT} No effect for D_i	Up to 50% for D_{LT} No effect for D_i
4. Add A_s	No effect	A_s^*/A_s
5. Add prestress	Reduces dead load deflection to nearly zero	Reduces dead load deflection to nearly zero and changes member to uncracked
6. Structure geometry	Large effect	Large effect
7. Revise criteria	See text	See text
Construction Techniques		
8. Cure: f'_c	Same as higher E_c and f_r	Same as higher E_c and f_r and could change to uncracked section
9. Cure: ϵ_{sh} and ϵ_{cr}	For long-term deflection $(\epsilon_{sh}^*/\epsilon_{sh})$ and $(\epsilon_{cr}^*/\epsilon_{cr})$	For long-term deflection $(\epsilon_{sh}^*/\epsilon_{sh})$ and $(\epsilon_{cr}^*/\epsilon_{cr})$
10. Shoring	Large effect, see text	Large effect, see text
11. Delay 1st loading	Similar to options 19 and 20	Similar to options 19 and 20
12. Delay installation	Up to 50% + depending on time delay	Up to 50% + depending on time delay
13. Locate equipment	See text	See text
14. Architectural details	See text	See text
15. Camber	See text	See text
16. Top bars	No effect	Up to $(d^*/d)^2$ for cantilevers
Materials		
17. Materials	See options 19 and 20	See options 19 and 20
18. Mix design	See options 9, 19, and 20	See options 9, 19, and 20
19. Higher E_c	(E_c^*/E_c) or $(f_c'^*/f_c')^{0.5}$	Small
20. Higher f_r	None	Significant
21. Use fiber reinforcement	See options 9 and 20	See options 9 and 20

Note: Here * = a superscript denoting parameters that have been changed to reduce deflection. D = (delta) = deflection. ϵ = (epsilon) = strains; ϵ_{sh} = shrinkage strain, ϵ_{cr} = creep strain. Other symbols are the same as in ACI 318 (1995) Building Code. Options are given in text, Section 9.4, Design Techniques.

Option 21: Add Short Discrete Fibers to the Concrete Mix

The effect is to reduce shrinkage (see option 9) and increase the cracking strength (see option 20), both of which will reduce deflection, but the increase in cost may be substantial.

Table 9.4 summarizes some of the preventive measures needed to reduce or control deflection. This table can serve as a general guide for the design engineer but is not all inclusive, and engineering judgment has to be exercised in the choice of the most effective parameters to control deflection behavior.

9.6 Allowable Deflections

Allowable deflection is the other side of the equation to which computed deflection should be compared. If there are no deflection limits, then the computed deflection is irrelevant. When the computed deflection is more than the presumed allowable deflection, it is appropriate to carefully consider the reason for limiting deflection. In some cases, the deflection limit might be increased while in other cases, it might be tightened.

Incremental deflection is that portion of the total deflection that occurs after the construction or installation of a part of the facility, until it is removed. In many cases, incremental deflection controls the design, not the live-load deflection or the total deflection.

As spans exceed 25 to 30 ft, the absolute deflection becomes more critical than span/deflection ratio. For example, a span/deflection ratio of 360 for a 30 ft span is 1 in. Deflections of this amount and greater can be seen with the naked eye and many owners and users will object. Also, standard tolerances for many construction components have stated or assumed values that do not exceed about 1 in.

Deflection limitations can be categorized into four groups, Sensory acceptability, serviceability of the structure, effect on nonstructural elements, and effect on structural elements. Each group is discussed below. For a more complete discussion see ACI 435 (1968).

9.6.1 Sensory Acceptability

- Droopy cantilevers and excessive sag in long-span beams may be unacceptable to the public. The total deflection is relevant. Cambering can alleviate this concern almost regardless of the amount of deflection.

- Floors that vibrate can distress occupants. The rate of change in the acceleration is the relevant parameter, but live-load deflection is commonly used to judge the acceptability of floors. More sophisticated analyses consider amplitude, frequency, and damping of vibration. While vibration is not a problem for most structures, it should be considered if the structure supports, for example, vibration-sensitive laboratory equipment or a long-span open floor in an important public building.

9.6.2 Serviceability of the Structure

- Roofs and outdoor decks that should drain water must do so. The total deflection is relevant. Cambering and other construction such as topping fills or tapering insulation must also be considered. Tolerances on levelness of floor-finishing generally require a slope of 1% or greater to assure complete water drainage.

- Floors for gymnasia and bowling alleys must remain plane and level. Both total deflection and incremental deflection are relevant. Cambering may compensate for some of the total deflection, but it is difficult to build in most structures. Incremental deflection occurring after the floor has been installed should be limited to an acceptable amount for the anticipated use.

- Members supporting sensitive equipment such as printing presses and certain building mechanical equipment must not deflect more than the amount permitted for the equipment.

The incremental deflection occurring after the equipment has been installed is relevant. In some cases, deflection between two points on the span, not the ends, is relevant and limits can be measured in thousandths of an inch.

9.6.3 Effect on Nonstructural Elements

- Partition walls built of brittle materials such as masonry and plaster cannot tolerate large deflections. Even an incremental deflection of span/360 after installation of the partition may not be tolerable. More forgiving materials such a gypsum wallboard partitions (drywall) can tolerate larger incremental deflections. Prefabricated metal partitions may not be damaged by incremental deflections, but the total deflection may cause problems if the partitions are set to a laser-leveled ceiling and must meet a floor that is not level.

- Most ceilings installed today are not damaged by deflections that are acceptable by other criteria.

- Curtain walls, windows, and certain other nonstructural building elements can be damaged by incremental deflection that exceeds the tolerance built into the element in question. In some cases, such as windows or cabinetwork that must fit between concrete floors above and below, the total deflection must be limited. In such cases, tolerances and perhaps other construction details must also be considered.

9.6.4 Effect on Structural Elements

- Occasionally, deflections can cause instability of the primary structure, such as arches, shells, and long columns, for example.

- Deflection can cause higher stresses in another element as, for example, in masonry supporting a beam that rotates at the support. Mutually perpendicular cantilevers meeting at their tips must have the same deflection but may have different deflection properties. Thus deflection becomes a problem in moment analysis. In other cases, a slab supported on a deflecting beam may rest at the other end on an unyielding support, thus causing a moment pattern in the slab that is different from that computed using two unyielding supports. Engineers should be alert to special situations of this type that may distress the structure as well as cause damage to nonstructural elements.

References

ACI Committee 117. 1990. Standard tolerances for concrete construction and materials (ACI 117-90). In *Concrete International: Design & Construction*, Vol. 2, No. 8, pp. 38–46. American Concrete Institute, Farmington Hills, MI.

ACI Committee 209. 1971. Effects of concrete constituents, environment, and stress on creep and shrinkage of concrete. In *Designing for Effects of Creep, Shrinkage, Temperature in Concrete Structures*, Publication SP-27, pp. 1–42. American Concrete Institute, Farmington Hills, MI.

ACI Committee 318. 1995. Building code requirements for structural concrete (ACI 318-95) and Commentary (ACI 318R-95). American Concrete Institute, Farmington Hills, MI.

ACI Committee 435. 1984. Allowable deflections, ACI 435.3R-68. *ACI J. Proc.* 65(6):433–444.

ACI Committee 435. 1984. Deflections of continuous concrete beams. ACI 435.5R-73. *ACI J. Proc.* 70(12):784–789.

ACI Committee 435. 1995. Control of deflection in concrete structures. ACI 435-95, pp. 1–77. American Concrete Institute, Farmington Hills, MI.

Branson, D.E. and Christiason, M.L. 1971. Time dependent concrete properties related to design-strength and elastic properties, creep, and shrinkage. In *Designing for Effects of Creep, Shrinkage, Temperature in Concrete Structures*, Publication SP-27, pp. 257–278. American Concrete Institute, Farmington Hills, MI.

Branson, D.E. 1971. Compression steel effect on long-term deflections. *ACI J. Proc.* 68(8):555–559.

Fling, R.S. 1987. *Practical Design of Reinforced Concrete.* John Wiley & Sons, New York.

Fling, R.S. 1992. Practical consideration in computing deflection of reinforced concrete. In *Designing Concrete Structures for Serviceability and Safety*, SP-133. American Concrete Institute, Farmington Hills, MI.

Fling, R.S. 1996. Deflection of a concrete beam and slab roof. In *Recent Developments in Deflection Evaluation of Concrete*, SP-161, E.G. Nawy, ed., pp. 1–24. American Concrete Institute, Farmington Hills, MI.

Grossman, J.S. 1981. Simplified computations for effective moment of inertia, I_e and minimum thickness to avoid deflection computations. *ACI J. Proc.* 78(6):423–439.

Grundy, P. and Kabaila, A. 1963. Construction loads on slabs with shored formwork in multi-story buildings. *J. Am. Concrete Inst. Proc.* 60(12):1729–1738.

Lampert, P. 1973. Postcracking stiffness of reinforced concrete beams in torsion and bending. In *Analysis of Structural Systems for Torsion*, SP-35, pp. 385–433. American Concrete Institute, Farmington Hills, MI.

Sbarounis, J.A. 1984. Multistory flat plate buildings: Effects of construction loads on long-term deflections. In *Concrete International: Design & Construction*, Vol. 6, No. 4, pp. 62–70. American Concrete Institute, Farmington Hills, MI.

Smadi, M.M., Slate, F.O., and Nilson, A.H. 1987. Shrinkage and creep of high-, medium-, and low-strength concretes, including overloads. *ACI Mater. J.* 84(3):224–234.

Yu, W.W. and Winter, G. 1960. Instantaneous and long-term deflections of reinforced concrete beams under working loads. *ACI J. Proc.* 57(1):29–50.

(a)

(b)

(a) Preload prestressing system for concrete liquid containers (Courtesy Preload Corporation, New York). (b) Structural concrete high rise system (Courtesy Portland Cement Association).

10

Structural Concrete Systems

by
Terry O. Blackburn, P.E., Ph.D.
Principal and Director, Structural Department of Schoor DePalma-Engineers and Design Professionals, Manalapan, New Jersey. Registered engineer in several states and practicing engineering for more than three decades. Served as Visiting Professor at Rutgers University.

10.1 Overview

10.1.1 Introduction

Concrete structural systems must be durable, constructable, economical, and functional. The system selected must be strong and in many cases aesthetically pleasing. The system must have deflections that are within acceptable limits and, in seismic areas, must have the ability to absorb the large amounts of energy generated by seismic events.

Selection of a structural system can sometimes be a difficult process. In many cases, structural steel is the more economical system to use. In general, when the system is to be hidden by architectural finishes, concrete systems are not the systems of choice. When the structure itself becomes an architectural expression, concrete is often the material of choice.

In most cases, the formwork for the concrete system selected represents almost half of the total expense of the structure. Obviously, repetitive systems that reduce the cost of formwork relative to the cost of the concrete and reinforcing are candidates for concrete systems. In fact, under certain conditions concrete systems may be the structural systems of choice, even when the concrete is not exposed or architectural. The plastic nature of the material provides an effective structural solution to any unusual requirements. In such cases, structural steel systems cannot provide the freedom of design that concrete provides.

10.1.2 Durability

Properly proportioned concrete, when placed, finished, and cured in accordance with established standards, provides a durability that is virtually maintenance free and seldom matched by structural steel systems. The weathering steels, often used in bridge design, are the closest rivals to a good dense concrete with proper air entrainment. The problems associated with rust staining have largely limited the use of steel structural systems to bridge structures.

Durability of concrete is directly related to the quality of the concrete. Dense, well-consolidated concrete with a low water/cement ratio and proper amounts of entrained air will be durable in all but the most hostile environments. Varying cement types, additives, and surface treatments are effective in extending the durability of concrete subjected to less than desirable conditions.

10.1.3 Constructability

The choice of the structural system must be based upon the availability of skilled labor to accomplish the design requirements. Concrete construction, unlike steel, is somewhat regional with respect to accepted and common practices. For example, the use of high-strength concrete (f_c' in excess of 5000 psi) is more prevalent in metropolitan areas of the United States. Experience with the construction of ductile moment frames is more widespread in geographic areas traditionally considered to be seismically active. Bridges, highways, water treatment facilities, and buildings each have their own special sets of design requirements, specialized techniques, and selection of materials developed to address these needs.

Localized labor practices and costs have a major influence on the degree to which concrete structural systems are utilized. Cast-in-place concrete is considerably more labor intensive than the fabrication of precast concrete. Conversely, transportation and erection of precast systems are obviously more expensive than those of cast-in-place systems.

In theory, the optimum reinforcement for reinforced concrete would be an extremely large number of very small reinforcing bars to provide a uniform distribution of tensile reinforcing. The reality is that one is forced to use the least number of reinforcing bars and to use large bars to keep the labor costs within reason. Local union agreements, in many cases, place restrictions on which bars may be cut and fabricated in the field versus which may be shop cut and fabricated.

Obviously, shop fabrication is more desirable because of the equipment available and the controlled work environment that generally results in a better product.

The cost of formwork constitutes a major portion of the cost of concrete structures. It is not uncommon for the cost of the formwork to represent 50% of the total cost of the in-place concrete. Systems that require simple, straightforward, reusable forms have a significant cost advantage over systems requiring more complex forms. Many times the effort to reduce the complexity of formwork leads to simpler configurations that unfortunately, require additional concrete along with additional loads, which must be carried by the foundation system. Whenever formwork can be reused, more complex configurations can be utilized. The ability to quickly strip, and reerect forms is a major cost savings in that overall construction time and "general condition" related costs are reduced.

For a structure under construction, the use of concrete and or masonry usually results in the greatest time exposure to the elements. Given this significant length of exposure, the geographic location and the time of year that construction takes place can have a major impact on the costs of concrete construction. Obviously, some steps can be taken to allow concreting operations to continue during weather extremes, but this invariably results in additional job costs. An added concern during periods of weather extremes is quality control.

10.1.4 Appearance

In many cases, concrete is called upon to perform several functions. In addition to providing strength for a structure, it also acts as cladding and must be durable, weather resistant, and have a pleasing appearance. The design of the exposed finish may require a simple rubbed surface to remove minor surface blemishes and form marks or a considerably more expensive architectural surface finish. Great care must be taken to successfully achieve consistent architectural finishes. Elements that must be considered include:

- a workable concrete mix to facilitate placing and finishing activities,
- consistent, controlled, water content in the mix,
- use of cement from a single batch, since ASTM C 150 allows for a wide variation in the coarseness and thus color of the cement powder,
- accurate batching of the components of the cement,
- leak-free formwork to eliminate bleed marks due to loss of paste,
- proper placing techniques to avoid segregation of the mix and cold joints,
- consideration of varying weather conditions,
- proper curing,
- careful formwork removal, and
- protection of the concrete from damage and staining during subsequent construction operations.

Failure to successfully accomplish all of the above will almost certainly lead to less than satisfactory finishes, placing the designer and owner in the unenviable position of deciding to accept blemished concrete or delaying the project while the defective concrete is removed and replaced. Repaired honeycombing of exposed surfaces invariably deteriorates with time (at least with respect to appearance) and is a major contributor to unsightly concrete. In some instances, surface coatings are the only way to obtain an acceptable surface finish.

Structural concrete not required to meet the visual requirements of architectural concrete must nevertheless be durable and weather resistant. The placing and curing requirements above should still be required. The color variations resulting from use of cement from varying batches and manufacturers and some relaxation of moisture control may be acceptable for the application.

10.2 Building Loads

10.2.1 Introduction

Design loads dictated by model building codes reflect the statistically probable maximum loads that can be expected to act upon a structural system. These loads are typically quoted as service loads, meaning that they have not been factored upward. Load factors are employed to account for inconsistencies in material properties, construction and fabrication practices, and the predictability of the load itself. They are, in essence, safety factors. In addition, load factors are employed to conduct analysis of the structural system in the ultimate stress range, as opposed to the service stress range.

Typically, the yield point of commonly used construction materials, such as steel and concrete, is well established. Therefore it is possible to design to the ultimate performance range of these materials. However, it is not desirable to design structures so that they perform in the ultimate range under typical loading conditions. This would not leave a sufficient factor of safety in the event of anomalies in the design assumptions. Therefore the service loads quoted by the model building codes are intended to be applied to materials within their elastic stress ranges. If the designer wishes to analyze a structure by considering the ultimate performance of the materials, then load factors must be applied to the service loads.

As noted above, load factors are the product of both materials analysis and probabilistic theory concerning the construction and fabrication processes. Concrete design employs load factors of 1.4 for dead loads and 1.7 for live loads; other combinations of load factors are employed depending on the code being used. The distinction between these values is primarily based upon the lack of predictability of how live loads will be applied to a structure. How to classify a dead load and a live load is a question to be addressed by the designer. However, certain truisms should be considered.

Material weights are generally predictable and can be specifically calculated as they apply to a given structural member. The weights of large, immoveable objects, such as mechanical equipment, permanent shelving or storage racks, or planters are also fairly predictable. These elements can be classified as dead loads with little risk of inaccuracy.

Human occupancy, furniture, and transient storage that fluctuates in volume and intensity are among the items that are less predictable and thus subject to a higher load factor. In addition, external loads that act on structures are equally unpredictable. Wind, seismic, hydrostatic, and earth loads must be factored as live loads to consider their inherent unpredictable nature.

Maximum load combinations, such as live plus dead plus wind, have the statistical tendency to occur so infrequently that the various model building codes and material-design manuals permit reductions in the overall load applied to the structural system. The reduction factors differ according to the codes and manuals, so the designer should check the applicable references for the particular project.

Model building codes determine the load factor combinations that must be used. Individual jurisdictions must adopt the provisions of a model building code in order for it to be the governing code. Jurisdictions might consist of municipalities, counties, or whole states. Jurisdictions may adopt only certain provisions of a model code and not others. They might supplement the model code with additional design information that is specifically applicable to a given region.

The designer is cautioned to fully understand the applicable building code where his project is to be built. It is the location of the structure that dictates the governing code, not the location of the designer, client, or reviewing agency. The designer must also be aware of local provisions that supersede the model code. For instance, municipalities located close to hurricane zones may adopt more stringent wind loads than are dictated by the adopted model code.

Lastly, there are third-party sources that assemble construction data pertaining to loads and material performance. These resources, such as Factory Mutual and Underwriter's Laboratories, often develop their data to assist insurance companies in establishing rate structures for coverage. As such, they tend to be slightly more conservative in their statement of loads than the model building codes. The designer is obligated to follow these design guidelines only if the jurisdiction has adopted them into the local building code or if the building owner, for whom the designer is producing the design, dictates that the more stringent design standard be used. If these resources are used, specific attention should be paid to the treatment of snow and wind loading and to the fire rating of structural assemblies.

The most frequently referenced model building codes are the Uniform Building Code (UBC) and the Building Officials and Code Administrators National Building Code (BOCA). These model codes have regional appeal and tend to be adopted in specific areas of the United States. The BOCA code is primarily adopted in the northeast and along the Atlantic seaboard. The UBC has been adopted in a wide region covering much of the Midwest, Southwest, and Pacific Coast. It is incumbent on the designer to make certain which model code, and what provisions, have been adopted for the location of the project.

For the purposes of the following discussion, the UBC will be referenced. There are many similarities between the UBC and BOCA codes; however, certain distinctions make interchanging the codes inadvisable. In addition, the requirements of the local building jurisdiction may not permit such an interchange.

10.2.2 Gravity Live Loads

Analytically, design loads can be divided into two primary groups: gravity loads, which predominantly act vertically on the structure, and lateral loads, which predominantly act horizontally on the structure. Gravity loads account for all dead loads and those live loads associated with occupancy of the structure. The designer can refer to several sources for information to provide weight data regarding various building materials. Manufacturers provide tabulated load data for proprietary building materials. Nonproprietary material weights, such as concrete, asphalt, and roofing and flooring materials, can be found in American Society of Civil Engineers publication ASCE-7, Tables C1 and C2.

Live-load data is developed with respect to the use and occupancy of a given structure. UBC Chapter 16 is devoted to establishing the intensity of various live loads as they relate to various uses and types of structures. UBC Tables 16-A and 16-B provide a breakdown of the maximum anticipated design live loads for a wide variety of use conditions. The designer should refer to these tables to establish the design live load for the use that most closely matches the anticipated use of the given structure.

Local building codes do not generally permit interpolation between specified live loads; therefore, larger, multiuse facilities may require subdivision into several analytical pieces for design purposes. It is both permissible and recommended that different live loads be applied to the structural model, as required to satisfy the variety of intended uses. An example of this is the application of a 100 psf load near public means of exit, though the remainder of the building requires only a 50 psf live load to satisfy an office-use criteria.

The designer is reminded that the UBC live-load data for use and occupancy are quoted as services loads. Appropriate load factors must be applied to permit analysis of the ultimate material stress range.

10.2.3 Lateral Live Loads

The second group of live loads that must be considered by the designer are lateral loads. Lateral loads are frequently the least predictable live loads. Such loads include wind, seismic, hydrostatic, and earth loads. It is in the determination of the intensity of these loads that the UBC differs most clearly from other model building codes, including the BOCA code. The designer must first establish the applicable basic wind speed for the locale of the given structure. The basic wind speed is defined as the fastest wind speed in miles per hour for a given locale, associated with an annual probability of 0.02, and measured at a point 33 ft above the ground.

The next design task is to determine the applicable exposure for the given structure. The exposure classifications differentiate between sites that are subjected to high, direct wind forces, owing to terrain characteristics; and those that do not experience such wind forces, owing to shielding effects, building or forest density, or other terrain irregularities. Because subsequent adjustment factors are determined based on the exposure classification, it is crucial that the designer make a thoughtful selection.

The designer is cautioned that wind forces are notoriously unpredictable. Large, concentrated forces can accumulate owing to irregular building geometry or aerodynamic effects around canopies, roofs, balconies, or multiple-story structures. Consequently, it is not recommended that liberties be taken with the exposure classification in an attempt to refine the load analysis. The most prominent physical features of the terrain, surrounding buildings, the given structure, and potential changes over the life span of the structure should be considered when selecting the exposure group.

The design wind pressure is determined by applying several modification factors to the basic wind pressure. These modification factors consider gust effects, the windward or leeward face of the building, the type of structure involved, and the slope of the roof. Special factors are provided for chimney, tower, sign, and flagpole structures. When determining the wind force on an entire structure, the designer is reminded that the combination of both windward and leeward forces must be considered.

The UBC distinguishes between "primary frames and systems" and "elements and components of structures." This distinction is based on the acknowledgment that, while the combination of windward and leeward forces acts only upon a primary system, components often experience intense concentrated forces owing to surface irregularities of the building. Consequently, the determination of forces and gusts acting on components requires that the designer consider both the size and the position of the component in question relative to the entire building.

In practical application, buildings with multiple roof heights and articulated facades often require the development of a wind-pressure chart superimposed over each elevation of the building. This exercise permits the designer to account for major component features on the building exterior. In addition, because the design industry has moved progressively farther away from requiring individual structural engineering consultants to shoulder the entire design responsibility for all building components, the development of a wind-force chart by the primary structural engineer provides appropriate information for subconsultants to design cladding systems and canopies.

Application of wind loads to a structural model is generally considered to occur at the slab-column joints for each floor level. This mode of application addresses the usual facade configuration in which the cladding is connected to the structure at each floor diaphragm. Where the facade treatment attaches to the structure in a different fashion, the designer must follow the load path from the cladding to the framing to determine the most accurate mode of application of wind loads. The designer is reminded that the wind-load data provided in the UBC are service loads and must be factored upward by the appropriate load factor in order to work in the ultimate material stress range.

The second form of lateral load to be considered is seismic load. Seismic load is generated from the movement of the ground and thus acts on the structure in a very different manner than wind loads. Seismic forces are transferred to the structure through the foundation elements. The influence on the supported floors occurs as a function of the weight of each floor. The acceleration of the ground thus causes the mass of each floor to accelerate, resulting in a seismic force in each diaphragm.

There is a delay between the initial occurrence of the seismic force at the foundation level and the influence on the supported floors. Depending upon the configuration of the structural components and the distribution of stiffness, the frequency and period of motion of the structure will vary. These variables, frequency and period, play an important role in establishing the design loads to be applied to the structural system. The stiffer the lateral load resisting system, the shorter the period of motion, and the greater the frequency. Ductile systems tend to absorb load rather than transfer it through to the other structural members. Thus ductile systems result in longer periods of motion and lower frequencies of vibration.

The detailing of reinforcement and the degree of confinement to increase ductility is of major importance. Sufficient reinforcement at the locations of large stress concentrations have to be provided. Beam, or slab, to column connections are particularly susceptible to stress concentrations owing to the large differences in stiffness between the members. In addition, reentrant corners, edges near shearwalls, and openings for stairs and elevators must be carefully detailed to avoid cracking problems. Chapter 21 of the ACI 318 Building Code covers the provisions for seismic proportioning of members and their detailing. The reader is referred to Chapter 26 of this handbook for details of design and proportioning of seismic-resistant concrete structures, their shear-wall components, and the latest provisions on this subject.

It is important to highlight some of the major factors in the context of this general discussion. Typically, shear-wall systems provide satisfactory resistance to seismic loads, owing to their unique performance characteristics. Shear walls tend to be very stiff at the base but gradually become more flexible as they increase in height. At extensions above approximately 120 ft, shear walls tend to deflect more than $h/300$, where h represents the height of the structure. It is advisable that the deflection level *not* exceed $h/500$.

Consequently, for high-rise construction, a dual system is recommended, which employs a combination of both shear-wall elements and moment-resisting frame elements. Frames are flexible throughout but tend to deflect in a regular and predictable manner throughout their height. Thus at lower elevations on the structure, the shear walls act to restrain the frames, while at higher elevations in the structure, the frames act to restrain the deflection of the shear walls. The designer should note that a dual system employing ordinary moment-resisting frames and concrete shear walls provides better resistance than either an independent shear-wall system or an independent ordinary moment-resisting frame system.

Lastly, when considering deflections of structures subjected to seismic loads, the designer must acknowledge that the building codes are developed to prevent catastrophic collapses. This should be understood to mean that conformance with the provisions of the model codes will not assure that a structure will survive seismic activity unharmed. It can be expected that cracking and possibly spalling of concrete near high stress concentration zones may occur. In addition, peak deflections due to maximum anticipated seismic ground accelerations can be more than 10 times greater than those that will be generated using the design seismic base shear. However, adherence to code provisions will dramatically improve the potential for a structure to survive seismic activity. Localized repairs are preferable to the complete demolition of a structure.

10.3 Composite Steel-Concrete Construction

10.3.1 Introduction

Composite construction is the use of two or more building materials or systems that are bonded or interlocked to act as a single unit. In most common building systems, composite construction involves the use of concrete and steel. These materials work extremely well together for beam construction owing to the high compression resistance and stability provided by concrete combined with the good tension-resisting attributes of steel.

In the early part of the twentieth century, composite beam construction existed primarily as a result of fire protection needs, the end result of complete concrete encasement of steel I-shaped beams. In most cases, the composite action of the two building materials was not realized and was completely neglected in design. The composite action existed owing to chemical bonding and friction at the material interface and by the shear strength of the concrete along the shortest failure plane. Although this type of construction and design is still permitted, it is very rarely used.

With the development of welding techniques, a more solid mechanical interlock could be developed to resist the horizontal shear that develops during bending. The use of a mechanical interlock to develop this composite action was first utilized in bridge construction beginning in the 1930s.

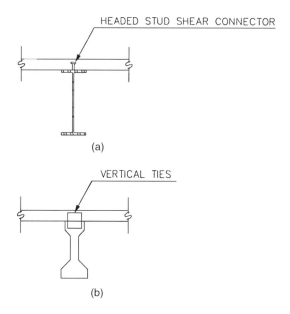

FIGURE 10.1   Composite beam systems.   (a) Composite
steel beam; (b) Composite precast prestressed concrete beam.

Economics prevented the widespread use of composite construction of this nature in buildings until
the 1960s. Today, composite construction has been used worldwide in both bridge and building
construction. Figure 10.1 illustrates two composite beam-slab systems.

10.3.2 Advantages and Disadvantages

Several advantages can be realized with the use of composite construction. As a rule of thumb,
composite action becomes most efficient when loads are heavy, the beams are spaced as far apart
as practical, and spans are relatively long. Typically, a 20 to 30% reduction in steel weight can
be gained, providing better economy and, in many instances, shallower beam depths. Shallower
beam depths may result in substantial savings in high rises, where floor to floor depths can be
reduced. Lower floor to floor depths result in smaller overall building heights, generating savings
from reduced wall materials and reduced lengths of mechanical, electrical, and plumbing risers.

Compared with a noncomposite floor system, the stiffness and overload strength of a composite
floor system is considerably greater. This is because the concrete slab acts as a large cover plate,
shifting the neutral axis upward and allowing more efficient use of the two materials. This often
allows a section that works as a noncomposite member to be used like a composite member for
longer spans without creating deflection concerns.

Although there are some disadvantages to composite construction, the advantages and overall economy favor its use. The cost of providing and placing mechanical shear connectors has
significantly decreased over the years but should be considered, particularly on small jobs where
mobilization of additional trades may completely offset the savings provided by reduced steel weight.
In some cases, although smaller beam sizes may theoretically work as composite members, their
use is not practical. Beams with flange thicknesses less than one quarter inch require careful shear-
stud placement to guard against burn-through failure during welding. The use of projecting shear
connectors also increases safety hazards by keeping workmen from walking on the beams, and for
that reason, the installation of the shear connectors is often delayed until the deck forms or metal
decking is installed.

10.3.3 Other Considerations

The advantages of composite action are largely unrealized in continuous-beam or frame construction owing to the limited continuity in areas of negative bending. Since concrete is unable to effectively resist tension, continuity in these areas is limited to that which is provided by bar reinforcement and is usually neglected. Since the negative moment at a support often exceeds the positive midspan moment, composite action will have little effect other than to reduce midspan deflections.

End reactions for composite beams are almost always larger than those for noncomposite beams of the same size. Engineers and fabricators often design connections based on uniformly loaded, noncomposite beam capacity. Such a practice potentially understates the shear loads and should not be used for composite construction.

Long-term creep deflections are usually small enough to be neglected in composite beams but may warrant special consideration where heavy, sustained loading situations are encountered and when deflection criteria are more stringent.

10.3.4 Composite Action

Composite steel beam and concrete-slab systems behave in a similar manner to reinforced concrete T-beams. The analysis procedure is based on a standard transformed-section methodology. For buildings, AISC limits the portion of the slab which can be considered to participate as a flange for beam action. The AISC limits are very similar in nature to the T-beam construction requirements of the ACI 318 Code. The effective flange width on each side of the beam centerline for an interior beam is taken as the smaller value of one-eighth the beam span, one-half the distance to the nearest adjacent beam, and the distance to the edge of the slab. The slab thickness criteria of ACI 8.10 is not considered in the AISC code.

Although composite construction using steel and concrete has only been discussed thus far, it should be noted that conventional slabs are sometimes designed to act compositely with precast, prestressed concrete beams. In such cases, the T-beam requirements of the ACI 318 Building Code should be followed.

The computation of section properties is based on the principle of transforming all components into a single, homogenous member. This is done on the basis of the ratio of the moduli of elasticity of the two materials. If a steel beam is used, the concrete slab is converted into an equivalent width of steel by dividing the effective flange width by the modular ratio, $n = E_s/E_c$. The calculations necessary for composite design are similar to those for a built-up beam. The composite section must be proportioned to resist the loads, and the shear connectors must be adequate to ensure that the section acts as a solid, single member.

The design of the shear connectors between the slab and the beam is based on the total horizontal shear that exists at the interface. The shear flow is the force per unit length that must be resisted at the interface to achieve composite action. Shear flow is given as

$$f = \frac{VQ}{I}$$

where V is the shear force, Q is the first moment of the effective area above the interface with respect to the neutral axis, and I is the moment of inertia of the composite section. If a given connector has a shear capacity of q, the maximum spacing can then be determined as

$$s \leq \frac{q}{f}$$

In Figure 10.2, the shear flow at the steel-concrete interface is $v_i b_{eff}$. Numerous types of shear connectors have been utilized over the years. They include headed studs, spiral rods, channels, angles, and L-shaped connectors.

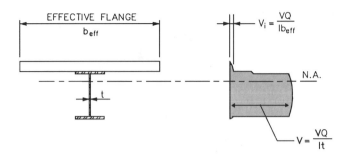

FIGURE 10.2   Shear-sress distribution across a composite section.

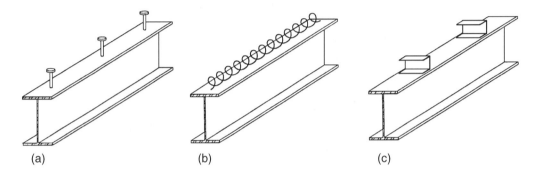

FIGURE 10.3   Types of shear connectors. (a) Headed stud; (b) spirals; and (c) channels.

Figure 10.3 illustrates some of these connectors. In building construction, composite sections usually involve using steel deck as a slab form. When composite metal decking is used, AISC only allows the use of welded studs. The minimum stud length is equal to the rib height of the deck plus 1 1/2 in. The shear studs are commonly fastened in the field with special stud-welding guns. This is done to prevent damage to the connectors during transportation and to allow easier steel erection and deck placement. It should be noted that shear connectors must be capable of resisting both horizontal and vertical forces. Because of the tendency of the slab to separate vertically from the beam, it is good practice to limit the connector spacing to around 2 ft. AISC limits the maximum stud connector spacing to eight times the total slab thickness. The reader is referred to the AISC Code for specifics regarding the use of form deck for composite construction.

Chapter 17 of the ACI 318 Building Code covers the design requirements for composite concrete flexural members. For strength computations, ACI indicates that no distinction shall be made between shored and unshored members. In general, if the composite member is assumed to resist all of the loads, the design shall be based on the standard requirements for a monolithically cast member by using transformed-member properties. To rely on composite action, the factored shear force must be less than the nominal horizontal shear strength multiplied by a strength-reduction factor ϕ such that the factored shear V_u has to satisfy the following expression:

$$V_u \leq \phi V_{nh}$$

where the nominal horizontal shear strength, V_{nh}, is computed as

$$V_{nh} = v_{nh} b_v d$$

and where

  b_v = the width of the contact surface
  d = distance from the extreme compression fiber to the centroid of the tension steel for the
      composite section
  V_{nh} = maximum allowable horizontal shearing stress, in pounds per square inch.

The value of V_{nh} is determined by the following conditions:

1. When the contact surface is clean and intentionally roughened, $V_{nh} \leq 80$ psi.
2. When minimum ties are provided and the surface is clean but not intentionally roughened, $V_{nh} \leq 80$ psi. Minimum ties shall meet the requirements of ACI 11.5 and shall not be spaced farther apart than four times the slab thickness or 24 in.
3. When minimum ties are provided and the surface is clean, free of laitance, and intentionally roughened to an amplitude of about 1/4 inch, $V_{nh} \leq (260 + 0.6\rho_v\, f_y)\lambda$ but not more than 500 psi. In the equation, ρ_v is the ratio of the tie reinforcement area to the contact surface area and λ is the correction factor related to unit weight of concrete as indicated in ACI 11.7.4.3.

10.3.5 Unshored versus Shored Construction

For unshored construction, the beam member must be capable of supporting its own self-weight plus the weight of the wet slab concrete. Once the slab concrete attains about 75% of its 28-day compressive strength, the beam and slab are considered to act as composites. The composite section must then be capable of resisting the live loads and any additional superimposed dead loads. For unshored construction, the compressive stresses in the concrete slab will seldom be critical. For unshored construction, the service-load stresses can be computed as

$$f = \frac{M_c}{S_b} + \frac{M_s}{S_c}$$

where M_c is the moment due to construction loads, M_s is the moment due to superimposed loads after slab curing, S_b is the section modulus of the beam, and S_c is the section modulus of the *composite member*. When shored construction is used, the working stresses are computed as follows:

$$f = \frac{M_c + M_s}{S_c}$$

Although shored construction is usually more costly, dead-load deflections can be significantly reduced. Shored construction also results in higher concrete compressive stresses.

10.3.6 Composite Columns

Composite-column construction usually combines the use of concrete and steel. There are two primary forms of composite column construction: 1) concrete-encased steel sections and 2) steel-encased concrete members. It is important to realize that composite construction implies that there is a shear transfer between the concrete and steel. Simply filling a steel pipe or tube column with concrete will not create a composite member.

Section 10.16.3 of the ACI 318-95 Code requires that any axial load assumed to be resisted by the concrete be transferred to the concrete via lugs or brackets in direct bearing. Similarly, all axial load strength not taken by the concrete must be developed by a direct bearing or shear connection to the structural steel member. The capacity of a composite column is computed on the basis of the same requirements as a conventionally reinforced concrete column.

For structural steel encased concrete sections, the thickness of the wall jacket must be sufficient to reach longitudinal yield stress before buckling outward. To ensure this, ACI 10.16.6.1 stipulates minimum wall thicknesses for both rectangular and circular sections.

For concrete-encased structural steel members, ACI 10.16.7 sets limits on the compressive strength of the concrete and the design yield strength of the structural steel for members with a spirally reinforced steel core. The radial confining pressure provided by the spiral results in sufficient composite action between the concrete, reinforcing bars, and steel core such that the reinforcing bars assist in both stiffening and strengthening the member. ACI therefore permits the inclusion of the longitudinal bars when computing the area and moment of inertia of the steel core for evaluation of slenderness effects.

For structural steel cores confined by tie reinforcement, it is likely that there will not be complete interaction between the concrete, steel, and reinforcing bars. Therefore use of the longitudinal reinforcing bars is permitted for computing the moment of inertia of the steel core since they assist in strengthening the section but do not effectively stiffen it. The tie-spacing requirements are similar to the spacing requirements of ACI 7.10.5, with the exception that tie spacing for composite columns may not exceed one-half the smallest side dimension of the member.

This increased spacing requirement is to help maintain the concrete core, which may separate from the smooth surfaces of the structural steel. ACI 10.16.8 defines the requirements for concrete-encased steel sections with transverse tie reinforcement.

10.4 Foundations

10.4.1 Shallow Foundations

Foundation systems are commonly referred to as shallow or deep, depending upon the depths to which they extend to achieve adequate bearing capacity. Shallow foundations are those foundations that transfer column loads either directly or through relatively short piers, pilasters, or walls to the supporting soil below. The most common types of shallow foundations are strip or wall footings, spread footings, and combined footings. Strip footings are commonly used beneath walls and rely on one-way action as they cantilever a short distance on either side of the wall.

Spread footings are usually square or rectangular pads that act to distribute individual column loads over a soil area large enough to support imposed loads. Combined footings act to distribute loads from two or more columns to the soil. Spread and combined footings rely on two-way distribution of the loads to the soil. In general, these footings are subjected to axial loads, shears, and moments from above that mobilize resisting soil pressures that can be determined by one of the following formulas.

For those footings in which the resultant vertical reaction occurs in the middle third of the footing (Figure 10.4)

$$f_{p\,max} = \frac{P}{A}\left(1 + \frac{6e}{L}\right)$$

For those footings in which the resultant vertical reaction occurs outside of the middle third of the footing (Figure 10.5), equilibrium requires that the total resisting force equal the imposed force. Assuming a triangular pressure distribution,

$$P = f_{p\,max}\frac{Bx}{2}$$

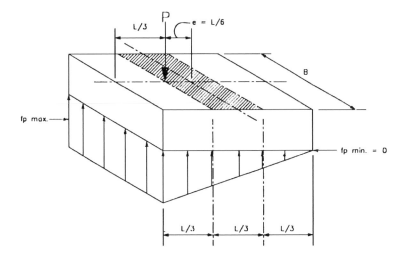

FIGURE 10.4   Footings with resultant vertical reaction of the middle third of the footing.

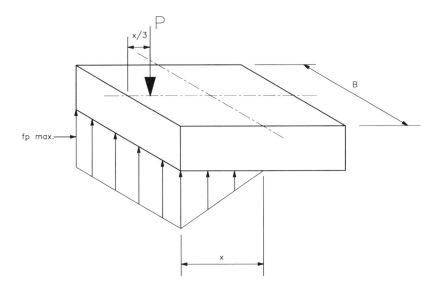

FIGURE 10.5   Footings with resultant vertical reaction outside the middle third of the footing.

where

$$x = 3\left(\frac{L}{2} - e\right)$$

$$f_{p\,\text{max}} = \frac{4P}{3A(1 - 2e)}$$

$$f_{p\,\text{min}} = 0$$

where

  f_p = soil pressure in kips per square foot
  P = vertical load in kips
  e = eccentricity of the vertical load in feet
  A = contact area of the footing in square feet
  L = length of footing, perpendicular to axis of the moment (eP) in feet
  B = width of footing, parallel to axis of the moment (eP) in feet

The design of a shallow footing must consider resistance to flexural and shear forces. ACI 318, Chapter 15 outlines the code requirements for the design of isolated column footings. Section 10.5.4 states that the minimum reinforcement for structural footings must meet the shrinkage and temperature requirements for steel given in Section 7.12. For grade-60 reinforcement, the minimum area of the temperature flexural reinforcement becomes

$$A_{s(\text{min})} = 0.0018(bh)$$

where b is the width and h is the thickness of the footing. The maximum spacing of footing reinforcement must not exceed the smaller of $3h$ or 18 in.

The thickness of a footing is usually governed by shear. This is because it is generally cheaper and easier to increase the depth of a footing than to try and provide web reinforcement in the form of stirrups. The standard requirements of ACI 318, Chapter 11 are used for shear design. For spread footings, both one-way shear based on beam action and two-way, punching shear must be checked. The critical section for one-way shear is located at a distance d from the face of the concrete column or half-way between the face of the column and edge of the base plate for steel columns.

Another form of shallow foundation is a mat foundation, which becomes viable in locations with relatively low bearing conditions and potential water conditions. If the sum of the footing areas exceeds one-half of the total building areas, it is usually preferable to combine the footings into a mat (also referred to as a raft) foundation. Since mat foundations may be used at locations where the bearing capacity can be marginal, it is important to recognize that the excess loads (above those acting on the natural deposit prior to construction) imposed upon the soil are significant. Excess loads can be reduced by increasing the basement depth. This increases the factor of safety with respect to bearing and also reduces settlement.

It should be recognized that the seat of settlement of a mat foundation extends deeper than conventional isolated footings; thus consideration should be given to compressible layers within the depths of concern. Because of the random distribution of compressible zones in subsoil, combined with the stiffening effect of the mat and the superstructure frame, it can safely be assumed that the differential settlement of a mat foundation (per total inches of maximum settlement) will not be more than 0.5 of the corresponding value for buildings supported on isolated footings.

Prior to the advent of sophisticated computer-aided analyses, the analysis of mat foundations involved several simplifying assumptions. As a result, it has been common practice to use twice as much reinforcing as the analysis indicated. If different portions of the mat carry significantly different loadings, it is advisable to use control joints. Irregular shapes such as narrow appendages cause problems and must be carefully designed if they cannot be avoided. Otherwise, cracking and rotation will occur in the vicinity of the junction of the appendage and the main segment of the mat.

10.4.2 Deep Foundations

As the depth to reach suitable support conditions increases, alternate systems must be considered such as drilled piers, caissons, and piles. These systems may receive support from end bearing on high-capacity geological strata such as bedrock or other dense strata, or they may develop their capacity through skin friction with the surrounding soil. Since these systems have significant surface areas exposed to the surrounding soil, their capacity may be reduced as the surrounding soil consolidates. Clusters of friction piles tend to act as a unit rather than as isolated individual members.

Deep foundations are somewhat specialized and require considerable design input from the geotechnical engineer. The selection of a system is usually dictated by geotechnical considerations. Timber and steel piles that support pile caps are the most commonly used deep foundations. The capacities of the piles in these foundations range from relatively low 10-ton values for timber piles to values in excess of 200 tons for steel piles. Piles with 40-ton capacities are the most commonly used steel piles because most building codes require load tests for capacities in excess of 40 tons. The load tests are costly and time consuming. The capacity of piles not subjected to load tests are determined by any of a number of empirical formulas.

Steel wide flange members and open-ended pipe piles, which, when used, have relatively small steel cross sections and displace very little soil as they are driven, are thus suitable for driving through soils with dense strata. Closed-end pipe piles displace soil and densify the soil in the immediate vicinity of the member. As clusters of these pipe piles are driven, the soil contained within the pile array becomes relatively dense and causes the entire cluster to act as a monolithic unit. The structural capacity of closed-end piles may be increased by filling the pile with concrete, however, in many cases the capacity of the pile will be dictated by geotechnical considerations rather than its internal structural capacity. Filling piles with concrete adds considerable stiffness to the member. In cases where concrete fill is used to stiffen the member, proper reinforcement must be provided and the interior of the pile must be cleaned prior to concreting.

Precast prestressed concrete piles are somewhat less commonly used than steel and timber piles in the United States but are popular abroad. When precast piles are used in the United States, they are usually high-capacity, hollow cylindrical piles with large diameters. The diameters are usually considerably larger than steel-pipe piles. Bridges and piers are candidates for this type of pile. The large diameter helps develop a stiff member, which makes placement easier, especially when underwater placement through considerable distances and soft material is involved. As mentioned, while not especially popular in the United States, solid precast prestressed piles are used. These piles usually have cross-sectional dimensions that more closely approximate steel piles.

The stresses resulting from driving must be carefully controlled with precast piles. The hammer introduces a compressive wave that travels down the pile and reflects back as a tension wave. The precompression supplied by the prestressing strands must be adequate to prevent damage to the pile owing to tensile driving stresses. The prestressing tends close any cracks that develop during the life of the pile. If the pile can be maintained in a crack-free condition, its stiffness is greatly increased. Prestressed piles are sometimes coated for protection. Obviously, any coating is susceptible to damage during driving and should not be depended upon solely for the longevity of the pile. Coatings in general extend the useful life of both steel and concrete used in marine environments, since the most hostile environment is in the splash zone, an area in which the coating is less likely to be damaged by the driving operations.

Lateral load resistance of pile foundations requires careful consideration. In a limited number of conditions, the lateral resistance of the pile foundation is developed by soil pressures reacting against the vertical face of the pile cap. A standard technique is battering the piles to utilize the vertical loads available in order to provide a horizontal component that resists the applied lateral loads. In lightly loaded structures with batter piles, uplift forces may develop as a result of the lateral loads. In many cases, the geotechnical engineer can determine a "point of fixity" at which the pile can be considered to have a rigid support. This enables the structural engineer to investigate the feasibility of developing lateral resistance using the flexural capacity of the piles. Flexural stresses introduced into the pilecaps must be considered when the cantilever approach is used.

Special consideration must be given to the proper anchorage of an uplift pile into the pile cap. The problem arises because of the pile's relatively limited depth of penetration into the bottom of the pile cap. The problem is not as pronounced with respect to steel piles as it is with timber piles. Load-resisting lugs can easily be welded onto steel piles. Timber piles have potential problems owing to the parallel orientation of the grain with limited edge distance. In addition to the traditional reinforcing rod inserted through holes drilled through the pile, there are a number of commercially available anchors.

Drilled piers and caissons are concrete foundation elements that may or may not be permanently cased or reinforced. Usually, the decision to case or not is driven by the surrounding soil conditions. Unless the soil has the ability to maintain a vertical cut, casing is usually called for. If any clean out, bottom preparation, or inspection is required, casing is almost certainly required, even if it is removed as the concrete is placed. Since most drilled piers and caissons are high-capacity members, verification of the bearing capacity is a must. If verification requires that personnel be lowered into the hole, a minimum diameter of 30 in should be considered. Verification may include a relatively simple visual confirmation that conditions are as expected or it may entail drilling into the rock to confirm the anticipated properties.

As seismic requirements continue to be refined for almost all geographic regions, there is generally a recognition that lateral loads may be greater than previously anticipated and the ability to absorb large amounts of seismically generated energy is of paramount structural importance. Batter piles, the most common method for resisting lateral loads, depend upon axial transfer of the loads from the structure to the support strata and may experience excessively large loads in a seismic event. It is probable that the design community will move to flexural-resisting elements for substructures as well as superstructures. This will result in large flexural and shear stresses, requiring considerable reinforcing to provide required strength and ductility. Traditionally, piers and caissons have been minimally reinforced, with the reinforcing frequently located in the upper portion of the member to assure adequate connection to the supported structure. The designer should not be surprised to encounter some reluctance from contractors at the amount and detail of the reinforcing required in areas previously considered safe from earthquakes.

Differential settlements must be considered with respect to the type of superstructures supported by the foundations. Simple post and beam construction has more tolerance to movement than structures with shear walls or moment frames. Shear-wall structures usually have several isolated very stiff elements. Excessive differential settlement may cause damage immediately adjacent to the shear walls, as rotations and displacements tend to be concentrated at the perimeter of the shear elements. While moment frames have the inherent ability to accommodate considerable movement, the permanent stresses introduced into the structure must also be considered in the design of overstress during extreme loading events such as seismic activity, which may cause overload and failure.

In most cases, foundations can tolerate minor cracking, and thus standard ultimate-load analyses are commonly used. Thickness of the members is frequently governed by shear considerations. This is especially true of pile-cap design, which requires that large concentrated loads be safely transferred to the support piles. For those structures that cannot tolerate cracking, thickness of the members must also be checked to assure that flexural stresses will be below the modulus of rupture.

For these cases, stresses related to volume changes must also be considered. Since foundations should be placed at elevations where the support soil is not subjected to seasonal (or annual) volume changes, there is the potential that the foundations may be placed at or below water level. The construction documents must clearly define requirements to assure that the concrete is properly placed in these conditions. This is especially true of concrete-placing operations for deep-foundation systems. Consideration must be given to material selection and foundation protection for construction in aggressive soils (such as acid and acid-producing soils) that attack and deteriorate concrete.

Chapter 14, Foundations of Concrete Structures by M. Gunaratne, in this handbook details the geotechnical engineering of foundations in all their categories.

10.5 Structural Frames

10.5.1 Rigid Frames

It is a common in the design of concrete structures to design members as an isolated entity. Once the overall structural system and layout have been determined, a structural analysis is performed to determine the moments, shears, and axial forces in each of the structural members. The individual members are then proportioned to resist these forces.

A structural building frame system relies upon continuity between beam and column members to distribute and resist shears and moments induced by various loadings. As a building material, concrete naturally lends itself to frame-type construction since it can easily be shaped, via formwork, to resist the applied loads in an optimal manner. Continuity is achieved, in part, by providing longitudinal reinforcement through the joint.

For a concrete frame system to perform as intended, particular attention must be given to the design and detailing of the beam-column joints and to proper construction procedures and placement sequences. Frame connections and construction issues are discussed later in this section.

10.5.2 Braced and Unbraced Frames

Building-frame systems can be subdivided into two categories: 1) nonsway or braced frames, and 2) sway or unbraced frames. The majority of concrete building structures fall into the braced-frame category. In most cases, the bracing for frames is accomplished with structural walls placed at stairwells, or elevator shafts, where they serve the secondary purpose of providing a certain level of fire protection. In a general sense, a braced frame is defined as a frame in which the majority of side-sway buckling is prevented by diagonals, shear walls, or other bracing members relative to the restraint provided by the frame itself.

To better develop an understanding of frame behavior, one must first consider the unbraced frame. Stability in an unbraced frame is dependent on the internal stiffness of the beam and column members that comprise the frame system. Lateral deflection in an unbraced frame consists of a displacement component resulting directly from horizontal loads as well as a component caused by unsymmetrical gravity loads, member properties, or frame geometry. When a building deflects laterally, the weight of the structure acts at an eccentricity to the support locations, introducing secondary bending moments in the beam and column members. This phenomena is known as the P-delta effect. Figure 10.6 demonstrates the P-delta effects in a typical unbraced frame. In braced frames, P-delta effects are generally small enough to be neglected.

10.5.3 Column Proportioning

A major factor affecting the design of unbraced frames is the reduction in axial capacity as a result of "slenderness effects." For concrete frames, ACI considers a column to be slender if the column experiences more than a 5% reduction in its axial load capacity due to moments resulting from P-delta effects. Elastic stability of a column exists until a critical load, corresponding to the Euler buckling load, is reached. This critical load is greatly affected by rotational end restraints and lateral bracing, which alter the length and number of half-sine waves in the deflected shape of the column. To account for various end and bracing restraints, most codes have adopted

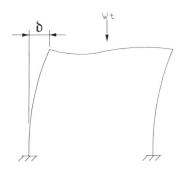

FIGURE 10.6 Sway of portal frame.

the concept of "effective length." The effective length is the actual length of the column multiplied by a modification factor, k, necessary to produce a column with pinned-end restraints having the same buckling-load capacity.

Lateral drift of columns results in an increase in column moments, which reduces the axial capacity of the column. Provided that the axial load is below the critical value, the structure will stabilize with increasing lateral deflection as the load becomes greater. The resulting load versus moment curve is nonlinear, with a stability convergence process that can be described with a second-order differential equation. As a result, the additional forces and moments that result from material nonlinearity, cracking, and P-delta effects are generally considered in what is known as a second-order analysis. The 1995 ACI Code allows the use of such nonlinear, second-order analysis and provides a simplified design method for approximating these slenderness effects. The simplified design combines the forces based on a first-order, elastic analysis with a moment-magnifier approach.

To utilize the moment-magnifier design method, one must first establish whether a column is designated as a sway or nonsway column. Usually, this is readily evident by inspection, provided the column is located within a building level where lateral deflection is limited by stiff bracing members such as shear walls. If there is any doubt, ACI Section 10.11.4.1 permits a column to be considered nonsway if there is less than a 5% end-moment increase in the elastic moments due to second-order effects. Having established the type of frame that the column is a part of, the engineer can then design for the magnified moments given by the approximate equations and methods found in ACI Section 10.12 for nonsway columns or in Section 10.13 for sway columns.

10.5.4 Beam Proportioning

Frames and continuous beams are statically indeterminate members. With the faster and more powerful microcomputers available today, moments and shears in such members can be determined using any one of several frame analysis programs. For smaller, less complex structures, other procedures such as traditional elastic analyses (moment-area, slope-deflection, moment-distribution, etc.), plastic analyses, or approximate methods (such as the portal or cantilever methods) can be employed.

The greatest moments in a frame often result because of pattern loadings, also referred to as skip live-loading or checkerboard loading. Influence lines based on the Mueller-Breslau principle are often used to determine which spans should and should not be loaded to produce the worst-case design moments or shears. An influence line is a graphical representation of a design parameter at a particular point owing to a unit load that moves across the structure. To account for the effects of pattern loadings, ACI Section 8.9.2 requires that continuous-beam members be proportioned to resist loads produced by two cases: (1) the dead load placed on all spans with the live load placed on two adjacent spans—this condition produces the largest negative moment at the support as well as the worst-case shear force; (2) the dead load placed on all spans with the live load positioned on alternate spans—this loading results in the maximum and minimum positive moments at midspan and the maximum negative moment at the exterior support.

For smaller size structures where computer modeling is not warranted, the hand calculations required to produce the moment envelopes for the various loading patterns become quite tedious. To simplify design, ACI has developed the use of moment and shear coefficients that can be used to approximate actual member forces. The approximate analyses permitted by ACI 318 Code Section 8.3 apply only to braced frames where significant moments due to lateral loads do not exist. The following criteria must be met for the simplified moment and shear coefficients to be valid: (1) There must be two or more spans of approximately equal lengths (the larger of two adjacent spans must not exceed the shorter span by more that 20%). (2) Loads must be uniformly distributed. (3) Unfactored live load must not exceed three times the unfactored dead load. (4) The members must be prismatic. Provided that the above criteria are met, the approximate equations give slightly conservative design moments and shears.

Chapter 27 Proportioning Concrete Structural Elements by the ACI 318-95 Building Code by E.G. Nawy, in this handbook outlines the procedures and presents the equations governing the analysis and design of concrete structural members.

10.5.5 Beam-Column Joints

Considering that joints are often the weakest link in a structural system, considerable research on beam-column and slab-column connections has been conducted in recent years and has led to the development of the current ACI-ASCE Committee 352 recommendations for design of monolithic connections. There are several parameters that interact to influence the mechanics of a joint. These parameters include joint shear-stress level, joint confinement, and the bond between the reinforcement and the concrete. ACI Committee 352 has provided design recommendations for ensuring adequate development length and horizontal joint reinforcement and also has set limits on the horizontal shear capacity of the joint, depending on the type and classification. The importance of adequate joint detailing has become increasingly evident in recent years owing to research and better understanding of seismic failure modes. Concrete confinement in the form of transverse closed-tie reinforcement can greatly improve the ductility of concrete, which is a highly brittle material.

10.5.6 Construction Considerations

An increased behavioral understanding of reinforced concrete has resulted in more stringent reinforcement-detailing requirements. These requirements often make construction more difficult, particularly at beam-column joints where significant rebar congestion occurs. For exterior and knee joints, where the primary longitudinal reinforcement cannot be run continuously through the joint, hooked-bar anchorages must be used. This further increases joint congestion and may prevent adequate concrete placement. Many times, geometric limitations prevent the use of larger diameter reinforcing bars owing to lengthy hook extensions and large bend diameters. In such cases, the designer must be cognizant of the construction implications that his/her design may have.

Even with proper design and detailing attention, improper construction techniques can significantly affect the performance of individual members or affect the continuity between them. The concrete placement sequence has significant importance on the behavior of frames. ACI Section 6.4.5 dictates that the column concrete be placed and allowed to set prior to placing any concrete in the floor supported by those columns. This is to ensure that any settlement or bleeding of the column concrete while in the plastic state occurs beforehand, thus preventing any gaps or cracking at the beam-slab and column interface.

10.6 Concrete Slab and Plate Systems

10.6.1 One-Way Beam-Slab Systems

The selection of a beam-slab structural system is most frequently driven by the geometry of a given column bay. Rectangular bays, with an aspect ratio exceeding 2 : 1, will function to distribute nearly 100% of the shear and moments in the short direction. Configuring beams in the long-span direction only, with slabs spanning one-way, perpendicular to the beams, creates a structural system that maximizes the benefits of each element. In addition, the continuity created by casting the slab system integrally across each beam support allows the framing system to redistribute load between the positive and negative moment zones so as to provide redundancy at the ultimate load state. Lastly, deflections are minimized owing to the continuity of the system.

Under standard loading conditions (see Section 10.2), slabs can be kept thin (see Table 10.1), with reinforcing steel provided primarily in one direction only. Nominal transverse temperature and shrinkage reinforcement must always be provided to prevent cracking. It is important to stress that it is more effective to use smaller diameter bars at closer spacing than larger diameter bars at larger spacing. The former is essential to control cracking development in the slabs.

The distribution pattern of the primary reinforcing steel closely follows the pattern of the bending-moment diagram (see Figure 10.7). Where negative moments are greatest, over the beam supports, top-reinforcing steel is provided. The cutoff point for the top steel occurs where the concrete no longer requires steel to resist tension stresses. The ACI Code requires that reinforcement extend beyond this point a distance equal to the greater of the effective depth of the slab or $12d_b$, the diameter of the bar. Also, at least one-third of the total tension reinforcing provided for negative moment must be extended beyond the point of inflection not less than the effective depth of the slab, $12d_b$, or $1/16$ the clear span, whichever is greater. Chapter 27 of the handbook gives the expressions for determining the development length required for the various conditions and categories. In practice, the ACI criteria result in extensions of the top reinforcing steel for distances of $1/3$ (span) beyond each side of the support (see Figure 10.8).

Using the guidelines established in Table 10.1, the resulting slab thicknesses will be sufficiently proportioned to resist shear stresses from typical loadings. Where extremely heavy loads (exceeding 250 psf) are experienced, the slab shear capacity should be checked. In accordance with Chapter 11

TABLE 10.1   Minimum Thickness h of Non-prestressed One-Way Slabs

Loading Condition	h
Simply supported	$L_n/20$
One end continuous	$L_n/24$
Both ends continuous	$L_n/28$
Cantilever	$L_n/10$

Note: L_n = effective span or cantilever arm.

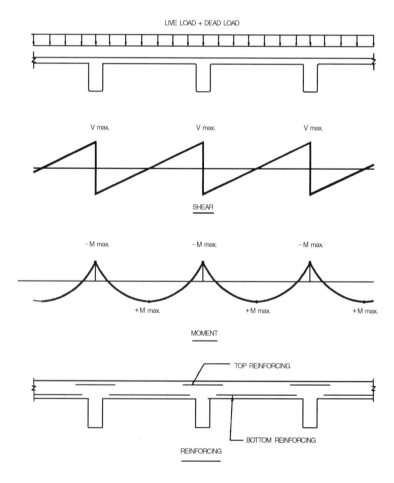

FIGURE 10.7   Moments and reinforcement locations in continuous beams.

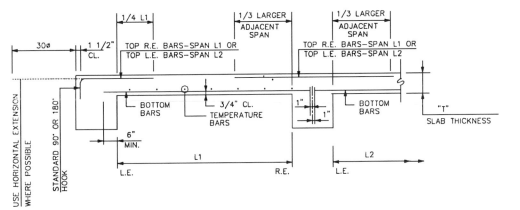

FIGURE 10.8 One-way slab bar bending and placing detail.

of the ACI 318 Building Code, the critical shear plane is located at a dimension d away from the face of the support for one-way slabs.

Special detailing requirements are limited in scope for one-way slab design. As long as the load path to the supporting beam element is maintained, the structure will perform as intended. Thus openings parallel to the slab span are easily accommodated by providing internally reinforced headers along the short edges of the openings. The designer should strive to orient all major openings in a one-way system such that the long dimension is parallel to the span of the slab. No special detailing at columns is required, as the beam element is intended to carry 100% of the load from the slab to the column. One-way slab systems have the following advantages:

1. Long-span capability of the beam elements permits wide column spacings and frame elements for lateral resistance.
2. Predictable slab thicknesses, reinforcing requirements, and deflection performance allow the designer to concentrate design efforts elsewhere.
3. Reinforcing detailing and placement is prioritized in one direction only, reducing complication at the construction site.

One-way concrete-slab systems are frequently used in parking structures, where the predictable traffic patterns require long open-column bays in one direction but permit shorter bays in the other direction.

10.6.2 Flat Plates and Flat Slabs

Where the designer is presented with a fairly regular, essentially square column bay, the most economical concrete structural system available is the two-way column-supported slab. As a monolithic material, the placed concrete naturally spans in two directions. By taking advantage of this natural tendency, the designer can achieve significant economy while maintaining desired deflection control and lateral resistance.

Flat plates are distinguished from flat slabs by the treatment at the columns. Both systems are entirely beam-free. However, flat slabs employ drop caps or column capitals to assist with shear transfer at columns. Flat plates are flat on their underside, and shear transfer is accomplished by proper proportioning of the concrete plate with the appropriate design and use of shear-moment transfer reinforcement.

Two-way column-supported slabs derive redundancy from the fact that 100% of the applied loads must be carried by the structure in each of the two orthogonal directions. Thus for a column bay of widths I_a and I_b, the structure is proportioned and designed such that 100% of the load can at any time be carried in either the I_a or the I_b direction. Load transfer in a two-way system is accomplished by effective widths of slab that serve as shallow beams. For analytical purposes, these

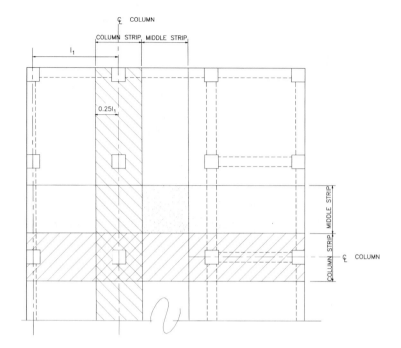

FIGURE 10.9   Column and middle strips in two-way slabs and plates.

effective widths are divided into column strips and middle strips. By definition, column strips are widths of slab centered on the column grid line, in both grid directions. The ACI Code specifies that column strips shall extend 0.25 L_1 or 0.25 L_2 on each side of the column, whichever is less, where L_1 and L_2 are the adjacent slab spans. Middle strips occupy the region between the column strips, at the centers of the column bays, in both directions, as shown in Figure 10.9.

Chapter 13 of the ACI 318 Code addresses the design of two-way slab systems. The code allows that design may be made by any procedure satisfying conditions of equilibrium and geometric compatibility. The code then describes two alternative design approaches: the direct-design method and the equivalent-frame method. The direct-design method is an empirical method developed essentially from the elastic theory of the distribution of moments in continuous slabs. Strict conformance to the limitations must be maintained or the reliability of the method will be compromised. The slab system must operate within the following constraints if the direct-design method is used:

1. There shall be a minimum of three continuous spans in each direction.
2. Panels shall be rectangular, with a ratio of longer to shorter span, center-to-center of supports, not greater than 2.
3. Successive span lengths center-to-center of supports in each direction shall not differ by more than one-third the longer span.
4. Offset of columns by a maximum of 10% of the span (in the direction of the offset) from either axis between center lines of successive columns shall be permitted.
5. All loads shall be due to gravity only and uniformly distributed over the entire panel. Live load shall not exceed two times the dead load.
6. For a panel with beams between supports on all sides, the relative stiffness of beams in two perpendicular directions shall not be less than 0.2 or greater than 5.0.

Using factors provided in the ACI 318 Code, Chapter 13, the total gravity load acting on the slab system is distributed into the column strip and middle strip zones. This method acknowledges that the summation of the negative and positive moments for any given span must equal the simple

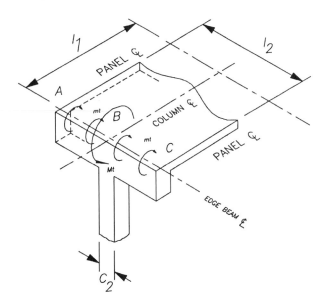

FIGURE 10.10 Torsional beam element in two-way action.

beam moment for a corresponding span. Consequently, the total static moment acting on any panel is denoted M_o, and the positive and negative moment distribution factors sum to 1.0 M_o. For interior spans, the total static moment M_o is distributed 65% to the negative moment and 35% to the positive moment. End-span conditions require reference to the tabular information provided in the ACI Code. After the moment-distribution factor is derived, additional factors are employed to proportion the moments between the column strips and the middle strips.

The equivalent-frame method is also essentially based on elastic theory. The slab is divided into frames comprised of the column grid and the slab on each side, extending to the middle of the panel. Each slab-column frame can be designed separately at each floor level, assuming the columns to be fixed at the floors above and below. Using stiffness coefficients for the various intersecting components at the column junction, inclusive of the stiffness of the torsional beam, a moment distribution is performed. Figure 10.10 illustrates the torsional moment transferred to the slab from the torsional beam. The moment of inertia may be calculated using the gross area of concrete. Determining the distribution of moment through the slabs and columns resolves the flexural phase of the design.

The second and perhaps more critical aspect in the design of flat plates or slabs by either method is the shear transfer at the column junction. Because neither the flat-plate nor the flat-slab systems have beams that can transfer load into the columns, all of the load carried by the column strips must be transferred through the slab thickness. In a flat-slab system, drop caps and column capitals can be used to supplement the shear capacity of the slab concrete alone. The designer must consider two forms of shear transfer.

As seen in Figure 10.11, a portion of the unbalanced moment is transferred to the column by a vertical balancing couple in the form of vertical shear acting at the face of the column, while another portion is transferred by a horizontal force couple (flexure) occurring within the slab depth. ACI 318, Chapter 11 indicates that approximately 60% of the unbalanced moment is transferred by flexure, with the balance transferred by shear. These percentages have to be calculated for each case. The designer must remember that the vertical reaction due to gravity loads in the slab must be added to the unbalanced moment forces to gain a complete picture of the slab-column interface.

For columns that are essentially square in cross section, such shear failure typically governs the proportioning of the plate or column cap. The shear-failure plane occurs as a 45° crack that forms around the full perimeter of the column. The failure plane is angled upward from the bottom of

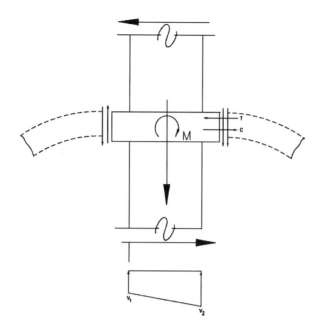

FIGURE 10.11   Shear-moment transfer mechanism.

the column. The critical shear zone for analysis is considered to be a distance $d/2$ from the face of the column, where d is the effective depth of concrete above the reinforcing.

Where the applied shear forces greatly exceed the concrete-slab shear strength, the designer has two primary options. He may provide additional concrete in a region slightly wider than the critical shear zone. This is accomplished by adding a flat drop cap consisting of several additional inches of concrete. Alternatively, a column capital, shaped like an inverted cone, can be provided to widen the critical shear zone and to thicken the concrete in the zone, thereby increasing the concrete shear capacity. Chapter 27 of this handbook lists the maximum allowable shear capacity V_c that can be used in the design for two-way action in order to determine the thickness of the plate or slab as governed by shear.

The other alternative is the use of special steel reinforcing assemblies. Commonly referred to as shear heads, these assemblies can consist of steel beams or channels welded in a cross shape, placed on top of the column, and poured monolithically with the slab. Other shear-head assemblies consist of flat steel bars with welded-head shear studs. These assemblies are placed so that they overlap the column and extend outward, thus increasing the critical shear zone.

Where significant shear-head reinforcing is required, the slab is proportioned too thin for the applied loads or the tributary areas. Shear failure is a sudden failure mode. The designer must exhibit extreme caution in analyzing shear transfer. Although not in favor for many architectural designs owing to the infringement of the ceiling plenum space, drop caps provide the designer with a simple but effective means of increasing slab capacities without resorting to complicated reinforcing schemes that are ultimately only as reliable as the installation method. The designer is encouraged to analyze the perimetric punching shear transfer as early in the design process as possible.

A final means of increasing the slab shear capacity, if the other two alternatives are not economically justifiable, is increasing the size of the columns. In practice, this method is only effective when the differential between the applied shear force and the slab capacity is within 20%. In this case, a 4 inch increase in the column dimensions will increase the slab shear capacity sufficiently to support the shear transfer.

As noted earlier, perimetric shear is typically the governing shear condition for design purposes. However, where rectangular columns with an aspect ratio in excess of 2:1 are provided, one-way beam shear can govern the slab thickness. Where applied shear forces exceed the concrete capacity, drop caps can be employed to increase the critical shear zone and the thickness of the concrete. Column capitals are not used for rectangular columns. The designer may find more success in increasing the long dimension of the column, if this option is available.

10.6.2.1 Post-tensioned Two-Way Plates

Prestressed post-tensioned two-way floors are frequently used today, particularly in apartment buildings. The primary advantages of post-tensioning are the ability to use thinner slabs, crack control derived from the presence of a significant precompression force in the concrete, greater shear capacity as a result of the precompression force, and greater deflection control as a function of reverse cambering.

The designer is cautioned that the advantages realized from post-tensioning in the slab elements can be quickly lost in the column elements. The extremely high compression forces generated by the tensioning strands cause large unbalanced moments in exterior columns. In order to adequately reinforce the columns to resist these forces, much of the available room for shear reinforcing and negative flexural steel is occupied. The designer must carefully detail these conditions to assure that no conflicts exist among the different types of required reinforcement that could jeopardize the design once it is under construction.

The final aspect of two-way column-supported slab design considered here is the transfer of lateral moments between the slab-diaphragm system and the columns. Under uniform gravity loading, flexural stresses are essentially balanced at interior columns, and shear transfer is accomplished via independent analysis. Where unbalanced moments are to be transferred into the columns, the analysis of shear and flexure becomes combined.

Analysis of this moment transfer requires calculation of the polar moment of inertia for the column element. Then, by applying the standard stress analysis theory of shear stress and bending stress, the proportion of the slab or drop cap at the column head can be checked. The designer is cautioned that the analysis of punching shear as a function of unbalanced moment transfer is of utmost importance and should be addressed as early in the design process as possible to allow for changes in the structural approach, if necessary. Because slabs tend to be thinner in post-tensioned systems, the analysis of punching shear and unbalanced moment transfer at the columns takes on even greater significance.

10.7 Liquid Containing Structures

10.7.1 Introduction

Concrete structures for liquid containment consist mainly of water-treatment plants, wastewater-treatment facilities, tanks, and reservoirs. The concrete used for these facilities should be watertight and should be protected against contamination from groundwater. In addition, exposed concrete structures require protection against freeze-thaw damage and environmental effects.

In containment concrete, low water permeability and low shrinkage are essential to prevent damage and cracking of concrete. Although proper design, such as specifying proper control joints and reinforcing steel is important, the most effective way of controlling shrinkage and minimizing water permeability is by designing good quality concrete mixes. In addition, proper quality concrete, concrete batching, delivery to the site, and placing are essential factors for good quality containment concrete.

The compressive strength of concrete used for containment structures should be a minimum of 4000 psi at 28 days of age. The air content should be $6 \pm 1\%$. The amount of mixing water should be enough to produce a slump of 2 to 3 in. The minimum cement content for a 4000 psi concrete should be 6.0 bags. Containment concrete should have a maximum water/cement ratio of 0.45 lb/lb.

The use of a minimum-cement concrete is important to control shrinkage of the concrete. The use of a water-reducing admixture generally helps reduce the amount of mixing water used. The use of high-range water reducers may also be used to reduce the amount of mixing water and increase the slump, easing the placement. Slump after the addition of superplasticizers should not exceed 7 in. In warm weather, a high-range water reducer and retarder may be used to prevent quick setting of the cement. For more information on admixture types and their functions, ASTM-C494 as well as reputable admixture manufacturers may be consulted.

In order to further reduce permeability and increase the density and durability of concrete, the addition of microsilica to the concrete mix may be considered. In general, the addition of 5% microsilica by weight of Portland cement reduces the permeability of concrete considerably. Microsilica-containing concrete is widely used for garage floors to reduce chloride infiltration into the slabs.

The size and amount of coarse aggregate used in concrete is also a governing factor in controlling the shrinkage and cracking of concrete. The size of aggregate should not exceed 1 1/2 in (ASTM-C33, size #57). The amount of coarse aggregate should be between 1750 lb and 1850 lb. The higher the amount of coarse aggregate, the lower the amount of sand, which is a factor in controlling shrinkage. All other concreting practices such as placement techniques, finishing procedures, and curing should be done as per the ACI specifications. The joint type and spacing as well as the reinforcing percentage and the other design details of the project should be designed by the structural engineer.

The most important aspect of concrete tanks is that they be leak-free. This requires that careful consideration be given to the analysis, detailing, and construction of tanks. The requirement that the tanks be free of leaks leads to the rather obvious conclusion that the stresses in the tank must be must be below the modulus of rupture. Since reinforcing becomes effective at strains near those that cause cracking in concrete, reinforcing is present in the tanks to resist the tensile stresses if the concrete ruptures, thus averting a catastrophic failure. The thickness of the walls is therefore selected to keep shear and flexural stresses below the cracking point.

The ideal configuration for tanks is the circular shape. Resisting loads with tensile members is more efficient than doing so with flexural members. Most of the loads generated by the contents of a circular tank are very efficiently resisted by circumferential tensile forces. When space or process constraints must be considered, rectangular tanks may be the proper choice.

10.7.2 Circular Tanks

The analysis and design procedures for circular liquid-containing tanks are not complicated. They follow the elastic theory of shells see [Billington, 1982 and Nawy, 1996]. The method basically consists of a membrane analysis of the tank with the introduction of shears and moments to account for boundary conditions that are noncompliant with the membrane theory. The Portland Cement Association (PCA) publication *Circular Concrete Tanks Without Prestressing* [PCA, 1957] provides procedures and examples for the design of circular tanks. Much of the information and formulas contained herein are extracted from the PCA pamphlet. There are also computer programs that can analyze these tanks. From a practical standpoint, it should be noted that the base slabs for most tanks are relatively thick for several reasons and the nominal amounts of steel to control volumetric changes of the concrete result in capacities greater than those required by the analysis.

The basic approach to design of a circular tank is to select a wall thickness that is strong enough that the stresses in the wall will always be less than the modulus of rupture of the concrete. Circumferential reinforcing is provided to carry all of the ring tension should the concrete crack. Vertical reinforcing is selected to resist the flexural stresses caused by the support conditions. It is interesting to note that using reduced allowable stresses in the reinforcing steel causes higher stresses in the concrete.

The area of reinforcing steel required to resist the ring tension, T, is

$$A_s = \frac{T}{f_s}$$

where A_s is the area of steel required and f_s is the design stress for the reinforcing steel. The stress in the concrete can be expressed as

$$f_c = \frac{(CE_s A_s + T)}{(A_c + nA_s)}$$

where f_c is the stress in the concrete, C is the thermal coefficient of expansion, A_s is the cross-sectional area of the steel, A_c is the cross-sectional area of the concrete, $n = E_s/E_c$, and E_c and E_s are the moduli of elasticity of concrete and steel, respectively. Inserting varying design stresses for the reinforcing confirms that as the steel design stresses decrease the stresses in the concrete increase. Of interest is the revelation that as the design stress of reinforcing approaches infinity, A_s approaches zero and the stress in the concrete approaches T/A_c. In other words, the effects of thermal stresses disappear.

Since volumetric changes cause tensile stresses in the concrete, efforts should made to control shrinkage. Construction joints should be specified on the construction documents. Joint spacings in excess of 30 ft will almost invariably lead to shrinkage cracking. A closer spacing should be considered for relatively low walls because of the circumferential fixity at the base and the unrestrained top surface of the wall. Expansive concrete is an option used by several designers. An alternative to expansive concrete is to attempt to limit those components contributing to shrinkage in the concrete matrix. By limiting the amount of cement, sand, and water in the concrete, major contributors to shrinkage are reduced. The reduction in paste results in reduced workability, which can, at least in part, be addressed by the use of superplasticizing admixtures in the mix. Prestressing is an option for reducing or eliminating tensile stresses in the concrete.

Attention to details is an important part of producing leak-free tanks. In general, it is better to use smaller reinforcing rods with their greater specific area for to developing bond, shorter lap splices, and smaller bend radii. Conversely, it is important that the bar spacing allow for effective concrete placement to avoid honeycombing and inadequately consolidated concrete. Most leaks in concrete tanks occur at joints. Acknowledgment of difficulties in placing concrete should be reflected in the design details. For example, in order to ensure proper water stop installation at the slab-wall intersection, consider lowering the top layer of reinforcing as shown in Figure 10.12.

Also, consideration must be given to joint details that do not utilize "starter walls" as a part of the slab. This is because effective consolidation of concrete, necessary for a leak-free joint, is difficult for these cases. Reinforcing congestion at joints makes proper placement of the concrete difficult and may contribute to honeycombing and leaks.

Prestressed concrete tanks, both pretensioned and post-tensioned, are probably the most efficient leak-proof circular containment vessels. The principles of the shell theory for cylindrical shells applies equally. The design factors used are similar to those for the reinforced concrete already discussed. Two types of prestressing systems are used: (1) circular strand wire-

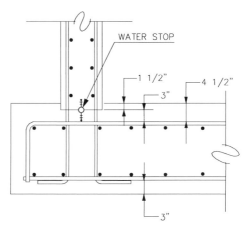

FIGURE 10.12   Detailing of wall-base intersection.

wrapped, such as the preload system used extensively worldwide, and (2) post-tensioned, both horizontally and vertically. ACI Committee 344 Reports [ACI 344, 1970, 1989a,b] give detailed requirements for maximum allowable stresses and other pertinent provisions for design. Nawy (1996) presents a detailed chapter on the design of prestressed circular tanks of both types, with examples, as well as the design of the cylindrical shell roof of a typical circular tank.

10.7.3 Rectangular Tanks

Performance criteria for rectangular reinforced concrete tanks are usually similar to those of circular tanks, namely, no leakage. The wall thickness of relatively shallow tanks is dictated by the need to keep flexural stresses below the modulus of rupture. The wall thicknesses of deep tanks, such as those associated with pump stations, are often dictated by shear stresses; flexural stresses are usually less than the modulus of rupture cracking stresses.

The Portland Cement Association's *Rectangular Concrete Tanks* [PCA, 1994] provides guidelines and design procedures. Extensive additional information is available in the literature [Timoshenko and Winowsky-Krieger, 1959; Young, 1989].

The corners of rectangular tanks tend to be subjected to large stresses. Hence considerable attention should be devoted to detailing and joint placement to limit cracking to an acceptable minimum. Construction joint locations must be incorporated into the design documents to limit the effects of volume changes.

As the horizontal dimensions of the tanks increase, the effects of the corners diminish, and the wall behaves more like a cantilevered retaining wall. Introduction of counterforts reintroduces two-dimensional behavior and should be considered in long walls. This technique can also be utilized for large-diameter circular reinforced concrete tanks in which the circumferential stresses become exceedingly large.

Buried rectangular tanks present a special problem at wall corners. Slabs at the corners of rectangular tanks tend to curl upward; thus the corners must be properly attached to the walls. Since diagonal cracks occur at the corners, additional diagonal reinforcing steel is usually provided.

10.8 Mass Concrete

10.8.1 Introduction

Large structures such as dams, bridge foundations, mat foundations, nuclear plants, and large-span deep beams require the placement of large quantities of concrete that are required to act in a monolithic manner. The successful placement of mass concrete requires careful planning, mixture design, placement, and consolidation. Quality mass concrete, like other concrete, begins with the proper mixture for the job. Volume changes within the concrete can cause internal cracking, with its attendant reduction in strength and degree of watertightness. To keep shrinkage to a minimum, factors that are major contributors to shrinkage must be controlled.

Successful placement of mass concrete requires the prevention of early setting of the cement and shrinkage of the concrete. During the placement of concrete, temperature increases due to the hydration of the cement. Therefore minimizing the increase of heat of hydration is essential to prevent shrinkage and cracking of the concrete. The ACI Committee 224 report [ACI 224, 1989], *Control of Cracking in Concrete Structures*, deals with the preventive measures necessary to control internal temperature in mass concrete.

Concrete used for various large structures has to have excellent control of early setting in order to prevent shrinkage, cracking, and the problems associated with placement and workability of concrete. Among the factors to be considered for mass concrete are the following:

    a. engineering design details of mass concrete; this phase of the work requires considerable knowledge and experience of structural engineering;

b. selection of materials and proportioning of concrete mixes and

c. control of concrete quality during batching and placement.

10.8.2 Materials for Mass Concrete

The following materials are appropriate for mass concrete:

- *Portland cement with low heat of hydration.* It is common to use Type II or Type IV Portland cement (ASTM-C150). The minimum amount of Portland cement should be used to achieve the design strength.

- *Pozzolans (ASTM-C311 and C618).* Addition of a pozzolan such as fly-ash in the range of 20 to 25% by weight of cement will assist in reducing the heat of hydration, workability, and long-term strength gain.

- *Aggregates (ASTM-C33).* Both fine and coarse aggregates should meet ASTM-C33 requirements. For dams, the coarse aggregate should preferably be size No. 1. In case of structural mass concrete for beams, coarse aggregate should have a nominal size of 1 in to 1 1/2 in. The amount of coarse aggregate should be high, approximately 1900 lb/yd^3 of concrete. Especially in concrete dams, for the ratio of fine aggregate to total aggregate by absolute volume should be low; approximately 25%.

- *Mixing water.* The amount of mixing water to be used should be low to provide a slump of 2 to 3 in.

- *Admixtures.* Admixtures should be used to entrain air and reduce the water/cement ratio. The water reducer should preferably be a Type D water reducing and retarding admixture, as specified in ASTM-C494.

- *HRWR.* Mass concrete to be pumped should be treated with a high-range water-reducing admixture (Type F), or a high-range, water-reducing and retarding admixture (Type G) as specified in ASTM-C494 may be used. The slump before the addition of the superplasticizer should be 2 to 3 in; after the addition of superplasticizer the slump should be 5 to 7 in.

- *Coolants.* The temperature of mass concrete for dams should be controlled to be between 40° to 50°F, especially in hot weather conditions. The use of finely chipped ice may be added to concrete to control the temperature. Other methods of cooling the aggregates, such as keeping them damp and under shaded conditions, will reduce the temperature. When steel forms are used, they should be sprayed with cold water if necessary. Curing the concrete with cold water also will help in controlling the temperature of cast concrete. The maximum temperature for mass structural concrete, such as beams, should be 70°F. In this case also, the use of chipped ice may be considered to lower the concrete temperature.

10.8.3 Concrete Mixture Design

Concrete mixtures used for mass concrete should be proportioned in a qualified laboratory in accordance with ACI-211 (1992). The materials used in concrete mix designs should be the same materials intended for use in the project. The compressive strength of concrete used for dams generally is low, 3000 psi to 3500 psi. The full design compressive strength is achieved in 6 to 12 mo after placement. In the case of mass structural concrete, the design compressive strengths are generally between 4000 psi and 6000 psi. The design strength is based on 28 days age.

10.8.4 Quality Control

Thorough quality control and quality assurance is essential for proper execution of a mass concrete project. Selection and quality of ingredients to be used, batching, and the placing of the concrete should be observed, and the necessary tests be performed as per the applicable ACI and ASTM specifications.

Equipment to allow rapid placement of large concrete quantities should be used to reduce the possibility of cold joints. The typical low-water, low-slump concrete used for mass concreting must be properly consolidated to eliminate cold joints and honeycombing. Segregation of aggregates must be avoided. Large-volume delivery systems are less susceptible to segregation than smaller-volume systems. With minimum handling, there is less potential for segregation. Large-volume buckets with their wide discharge openings are effective means of delivering large quantities with minimum segregation. Once placed, the concrete must be properly vibrated to knit successive layers together in a joint-free matrix. Transportation of concrete using vibrators should not be done because of the potential for segregation.

Preplacement planning must include the involvement of the concrete supplier to resolve potential problems such truck as breakdowns or the opposite, backup of trucks leading to excessive mix time. Effective communication is an absolute must. Proper equipment maintenance greatly reduces the potential for breakdown. If a pump-delivery system is utilized, consider of having a spare unit on-site to ensure an uninterrupted delivery to the point of placement.

10.9 On-Site Precasting—Tilt-Up Construction

On-site precasting is used in lift-slab, tilt-up, and miscellaneous general construction. Lift-slab on-site precasting consists of casting the floor and roof slabs on top of one another so that they can be lifted into place once the columns and jacking equipment are in place. Section 10.10 discusses the lift-slab system in more detail, thus the specifics of the system will not be discussed here. Site-casting of the slabs provides economies resulting from

- reduced formwork costs since only edge forms are required and these are reused several times,
- reduced placing costs since all of the concrete placement is at ground level,
- reduced curing and protection costs owing to the limited surface area exposed to the elements,
- reduced reinforcing costs since there is considerable repetition from floor to floor and the materials do not need to be hoisted to elevated floors,
- the soffits of the slabs being uniformly flat thus simplifying design and installation of mechanical and electrical systems, and
- favorable fire resistance ratings that may be obtained without the application of fireproofing to the underside of the slabs.

Several of these advantages can also be realized by site-casting specific structural members. In the northeastern portion of the United States, it is not unusual to site-cast concrete columns to be used with structural steel roof systems. The concrete columns provide favorable fire protection ratings while also providing high-capacity concrete members at a price that compares very favorably with structural steel columns. The site-cast columns can be made as durable as steel columns with respect to impact damage from material-handling equipment.

The majority of site-cast concrete is used in tilt-up construction. In addition to many of the potential cost savings listed above, tilt-up construction can replace exterior structural steel with concrete bearing walls that can be architecturally treated at a relatively low cost. Figure 10.13 shows the various components and their arrangement along a bearing wall of a tilt-up building. Basically, the system consists of the following:

- preparation of the site and casting the footings;
- casting the slab on grade, omitting a strip several feet wide around the perimeter of the building and erosion-control measures such as draping plastic sheeting over the exposed soil

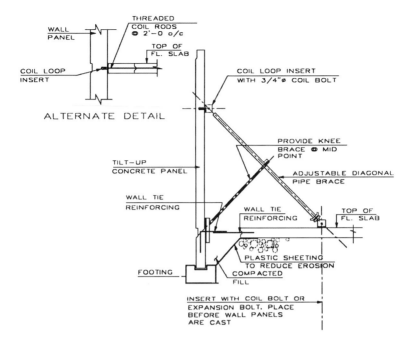

THREADED
COIL RODS
@ 2'—0 o/c

WALL
PANEL

TOP OF
FL. SLAB

COIL LOOP
INSERT

COIL LOOP INSERT
WITH 3/4"⌀ COIL BOLT

ALTERNATE DETAIL

PROVIDE KNEE
BRACE @ MID
POINT

TILT—UP
CONCRETE PANEL

ADJUSTABLE DIAGONAL
PIPE BRACE

WALL TIE
REINFORCING

WALL TIE
REINFORCING

TOP OF
FL. SLAB

PLASTIC SHEETING
TO REDUCE EROSION

FOOTING

COMPACTED
FILL

INSERT WITH COIL BOLT OR
EXPANSION BOLT. PLACE
BEFORE WALL PANELS
ARE CAST

FIGURE 10.13   Typical panel bracing.

at the perimeter of the building between the exterior footings and the edge of the slab on grade;

- placing edge forms for the wall panels directly on top of the slab on grade;
- patch any blemishes on the surface of the slab on grade, and temporarily patch all joints occurring within the area of the wall panels since these will read through onto the wall panels;
- apply a high-quality bond breaker to the slab on grade surface;
- apply any architectural finishes that will be exposed on the surface cast against the slab on grade;
- place reinforcing and tilt-up hardware such as lift and bracing inserts, place concrete, using care to avoid shifting reinforcing, inserts, or architectural finishes;
- apply curing compound or moist cure wall panels;
- install bracing inserts in the slab on grade if they have not been previously placed;
- place leveling pads on footings and mark for transverse and longitudinal alignment;
- tilt up wall panels and lift into position using a suitably sized rubber-tired moving crane;
- brace wall panels to slab on grade using a system of pipe braces and dimensional lumber attached to the wall and slab at the appropriate inserts;
- stabilize the base of the wall using concrete infill at the oversized shear key in the footing;
- erect the structural steel and when the entire structure is plumbed and square weld structure to tilt-up wall panels;
- backfill around the exterior and interior of the perimeter footings, and
- extend the slab connection reinforcing from the wall panels and place the perimeter strip between the wall and edge of slab on grade.

While greater economies can be realized by using the system for bearing-wall construction, the system also deserves consideration when nonbearing walls are contemplated. The system usually becomes economical when the floor area is 40,000 ft² or more. Obviously, this threshold area will vary from region to region. The important requirement is to have adequate room to precast the panels on the floor slab. Local labor rates and work rules will also effect the viability of the system. In northern regions, it would not be prudent to begin a tilt-up project at the onset of winter or other anticipated extended periods of inclement weather. It is especially important not to leave a slab on grade that is exposed to freezing weather unprotected because of the possibility of the slab heaving. Irregularities in the slab such as this are difficult to remove and will "read through" to the wall panels.

The wall panels can be cast so that when lifted, the exterior panel face is either cast against the slab or facing upward. The exterior face-down panel is definitely favored for the majority of applications. The exterior face-up panel will have exposed insert holes on the exterior of the panels that must be patched. The exterior face-up panel also results in a "blind pick" since the crane is invariably operated from the interior of the building. The erection of exterior face-up panels greatly and needlessly complicates the erection process and placement of the various inserts.

The exterior face-down casting position provides an opportunity to economically treat the exterior face of the panel if desired. Since the contact surface of the floor slab will in all probability contain blemishes, treatment of the wall panels may be more effective than the elimination of floor blemishes that might very well be acceptable for the intended floor use. Typical finishes for the panels include graphics formed into the surface, accents, aggregate finishes, or combinations thereof. The author has effectively utilized a combination of accent strips and exposed aggregates on several occasions.

Accent strips are constructed using inexpensive dimensional lumber. The exposed aggregates are placed on a layer of treated sand that is bounded by the accent strips. Prior to placing the surface aggregate the sand receives two applications of bond breaker to prevent the concrete from adhering to the sand or the aggregate. After the sand has been sprayed, the aggregates are carefully spread over the areas to receive the finish. Care must be exercised during the placing operations to avoid displacement of the aggregates. Stainless-steel reinforcing accessories must also be utilized, or rust staining of the surface will occur. During the design, reducing thickness to accommodate the surface treatment of the panels must be considered.

Footings supporting the wall panels may either be continuous or discrete. If continuous footings are not utilized, additional horizontal reinforcement near the bottom of the panels must be provided to ensure adequate flexural capacity to enable the panel to span from footing to footing. In many cases, use of continuous footings is an option that results in ease of construction and reduction of potential for differential settlement.

Panel thicknesses are determined by the economics of reduced concrete and panel lift weight versus additional reinforcing. Panel thickness is also related to the number and arrangement of lift inserts to necessary ensure that the flexural stresses are kept below the cracking level. Allowance for stress increase resulting from partial adhesion of the panel to the slab on grade, if a sand or aggregate layer is not used, should also be incorporated into the design. The panels will almost certainly be considered to be slender compression elements, requiring that the effects of panel deflection (P-delta) under lateral loads be considered in the analysis. If the structural framing for the roof utilizes steel joists, the connection of the joist bottom chord to the panels should be avoided since the deflection of the joists will introduce a potentially large moment into the panels through the fixity of the top and bottom chords. It is more economical to consider the additional unsupported height in the design of the panel. If possible, seat the joists at or near the middepth of the panels to reduce the eccentricity of the loads supported by the panels.

Lateral loads can be effectively resisted through horizontal diaphragm action in the roof structure. The lateral loads can effectively be transferred from the roof deck to the wall panels as shown in Figure 10.14. The angle is also a cost-effective replacement for a joist placed adjacent to the panel and simplifies roofing details through its zero displacement. All welded attachments between adjacent panels should be avoided since the connections invariably occur at the panel edges and will usually crack the panels owing to excessive stresses caused by thermal contraction of the panels. The

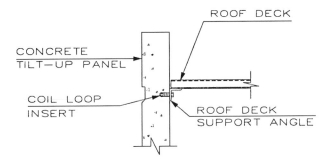

FIGURE 10.14   Alternate detail of transfer of lateral loads.

potential for adjacent panels to transversely displace at joints can be greatly reduced by attaching clip angles that span the joint and attach to the panels with inserts cast into the panels. The bolted connections seem to allow enough movement to keep the edge stress below cracking.

10.10 Lift-Slab Construction

10.10.1 Introduction

The lift-slab method of construction was first introduced in the 1950's by Phillip N. Youtz. He later joined Thomas B. Slick to refine the design concept and construction techniques of his system. From this research, the original concept of lift-slab construction was introduced. The structural components of lift-slab buildings are no different from conventionally constructed buildings. The difference is how the building is erected. The procedure is to cast the concrete floor slabs, one on top of the other, on the slab on grade and lift them to their final elevation with hydraulic jacks.

The first step is to prepare the site for construction and install the foundation system for the building, using normal construction techniques. The first level of columns are then erected on the foundation with the steel lifting collars in place. The columns are usually two stories high for economy. Once all of the columns are in place, the concrete slab on grade is placed, allowed to set and coated with a separating material. The first steel shear collar is then slid down the column and is aligned on the slab on grade. The slab reinforcing is installed, along with embedded electrical conduit and plumbing lines directly on the slab surface. The edge of the slab is formed, and finally the concrete is placed. The above procedure is repeated for each of the supported floors plus the roof.

Once the roof slab is completed, it is loaded with the upper column sections and the slab-lifting equipment. The lifting jacks are placed atop each of the columns and are connected to the roof-lifting collar by threaded rods. The roof slab is then lifted to the top of the column section and temporarily secured. With the roof slab in this position, the tops of the columns are braced and the column capacity is increased for the successive lifts. The upper-level slabs are then lifted to the underside of the roof and are temporarily secured. The lower-level slabs are lifted and are permanently attached to the column sections at their designated elevations. The lifting jacks are removed and the next column sections are erected. The jacks are then placed on top of the next column section and the lifting sequence is repeated until all the floor and roof slabs are in their final position. Once all the slabs are secured to the columns, the exterior facade can be constructed and the floors are ready to be fitted out.

10.10.2 Foundations

The process of lifting the precast concrete slabs does not impose any loads on the foundation system that are different from the design service loads. No special foundation requirements are necessary for this type of construction. The foundation type is selected and proportioned based on the geotechnical characteristics of the project site and the dead loads and service live loads anticipated for the building's occupancy. The concrete slab on grade is to be placed and finished based on the requirements of its end use. Note that the finish of the slab will be reflected on the ceiling above. Any depressions in the slab on grade for floor finishes can be filled with lean concrete to prevent them from being reflected on the ceilings above.

10.10.3 Columns

The columns for lift-slab buildings are usually steel wide-flange shapes but precast and cast-in-place concrete columns have also been used. Our discussion will be limited to steel wide flange shapes, noting that concrete columns are similarly erected. The length of the column sections is usually two stories high and is limited by the design column section and the length of the lifting rods supplied by the contractor. There are no limiting column-spacing requirements imposed by the lift-slab process. It is controlled by architectural layout, span of the slabs, and the loads supported by the slab. The lift-slab contractor may, however, limit the total area to be lifted at a time, thus dividing the slab into sections. This is determined by the number of jacks and support equipment available to the contractor. The boundary between two adjacent slabs is known as a pour strip. After the slabs are lifted into place, these areas are formed and concrete is placed to complete the floor system.

10.10.4 Supported Slabs

The concrete slabs used at the inception were conventionally reinforced two-way flat slabs. These slabs worked well, were easy to construct, and were sized to eliminate drop panels. This limits formwork to block outs for the mechanical chases and perimeter-slab edge. The major drawback with conventional two-way slabs is that they are thick, heavy, and their spans are relatively short. Today, most slabs are two-way flat plates that are post-tensioned. Post-tensioned slabs are thinner for the same loading conditions or have greater capacity for the same span and slab thickness. The slabs are generally the same, whether the building is lift-slab or conventionally constructed. The advantage is obtained by casting all the slabs at grade level, one on top of the other. There is no need for large cranes to hoist material to the elevated floors, a man lift is not required, the time required to move material is reduced, and the fall hazard for the workmen is nearly eliminated.

10.10.5 Lifting Collars

Lifting collars, threaded onto the columns prior to erection , are centered on each column and cast into each of the supported slabs. The original steel castings were found to be expensive and had limited lift capacity. Today, lift collars are exclusively steel sections that are welded with full-penetration welds. The purpose of the lift collar is threefold. It connects the slab to the lifting jack via a threaded rod, attaches the slabs to the columns for load transfer, and acts as shear reinforcement in the slabs (shear heads) to eliminate drop panels in the thinner post-tensioned slabs.

10.10.6 Lifting Jacks

The lifting jacks used to raise the slabs are mounted to the top of each of the columns and are attached to the slab by two threaded rods. The jacks are hydraulic and are driven in such a manner that they lift in unison at a rate of four to ten feet per hour. This is to eliminate damage to the slabs as they are being lifted.

10.10.7 Critical Component Design

The design of the structural components of a lift-slab building is, for the most part, the same as for a conventionally constructed building. The differences occur in the critical components such as the lifting collar, the slab at the column connection, and the columns, which need to be checked for all conditions and end restraints during construction.

10.10.7.1 Columns

The columns of a building utilizing the lift-slab method of construction serve a dual purpose: they support the building in its final position, braced at each floor by the slab, and are used to support the lifting jacks during positioning of the slabs. The columns require analysis for all the load cases proscribed by local building codes as well as for construction loads. During jacking of the roof slab, the columns are free-standing, with base fixity achieved by the base plate detail and encasement of the columns by the slab on grade. The stage during which the roof slab is lifted up along the free-standing columns is the most critical condition. The column sections need to be sized for this condition as well as for conditions during successive lifts. The first load condition is best checked using Euler's column formula for columns fixed at the base and free at the top.

$$P_{(all)} = \frac{\pi^2 E I}{4L^2 (\text{F.S.})}$$

where

$P_{(all)}$ = allowable Load (kips)
E = modules of elasticity (kip per square inch)
I = moment of inertia (in^4)
L = column height (inches)
F.S. = factor of safety

Once the roof slab is at the top of the column section and is secured (temporarily or permanently), the slab connects all of the columns together and fixes the top of the column section. The column condition now becomes fixed at the bottom and fixed against rotation at the top but is allowed to translate laterally. This condition reacts the same way as columns that are pinned at both ends and the following column expression should be used:

$$P_{(all)} = \frac{\pi^2 E I}{L^2 (\text{F.S.})}$$

The capacity of the columns at this stage quadruples. This extra capacity in the columns may allow a number of slabs to be lifted simultaneously.

10.10.7.2 Lift Collar

Lift collars (see Figure 10.15) must be designed to perform adequately for multiple purposes. The collar is sized to transfer the construction loads imposed on the slabs to the lifting jacks and all of the service loads to the column. Finally, the collar must be sized to reinforce the slab to keep the diagonal tension stresses in the slab below the allowable stress dictated by the building code.

The design of the collar also depends on the forces to be transferred from the slab to the column. Depending on the designers assumptions, the forces could be shear only, shear with partial-moment transfer, or shear with full-moment transfer. The design of the lift collar is the responsibility of the lift-slab contractor. The design engineer should note the type of collar desired and anticipated design forces on the design documents.

10.10.8 Applications

The lift-slab method of construction is best suited for buildings whose column layout is ideal for two-way post-tensioned concrete slabs. Although higher structures have been lifted, 3 to 20-story buildings are the best candidates. The exterior facade does not affect the decision

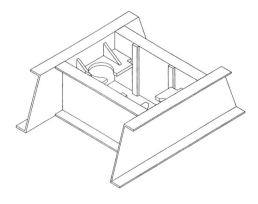

FIGURE 10.15   Lift collar in lift slab construction.

as to whether this system should or should not be used. Constricted construction sites warrant investigation of lift slab because large track-mounted tower cranes are not required. When construction of the superstructure is scheduled to occur during cold seasons, lift slab becomes economical in that once the foundation is completed and the columns erected, a shelter can be constructed and heated to provide a protected work area.

10.11 Slip-Form Construction

10.11.1 Introduction

Vertical slip-form construction is a process of placing concrete continuously with a single form that is constructed on the ground and raised as the concrete is cast. Casting is done at a rate that prevents the formation of a cold joint in previously placed concrete. The result is a continuous placing sequence resulting in a monolithically erected wall with no visible joints. This construction process utilizes lifting jacks located on the ground or on the working platform that elevates the form and the workers scaffolding attached with smooth rods or pipes. These rods or pipes are embedded in the hardened concrete. The construction technique is similar to an extrusion process.

The die (slip form) moves upward as it extrudes the concrete wall. The rate of the extrusion process is controlled by the setting time of the concrete and the crew's ability to prepare the wall for the pour. The average time of lift for any project is 6 to 8 in/h, placing approximately 4 to 10 in layers of concrete per lift. However, the time varies anywhere from 2 to 4 in/h, for the lower levels of a nuclear reactor containment structure, to 20 to 50 in/h, for an underground shaft lining. The project foreman looks for zero slump from the 4 to 10 in layer below the layer being placed.

It is important to note that utilizing this construction procedure, the concrete wall being "slipped through" supports only its own weight. The jacking rods located in the wall support the weight of the slip-form structure. The concrete keeps the jacking rods from buckling. In some cases, the jacking rods become part of the permanent structure. The other alternative is have the jacking rods slip through a thin pipe sleeve as the form slips upward. Engineers designing these walls should not rely on the steel rods for reinforcement owing to lack of bond between the bars and the concrete.

Slip-form construction, even though in a state of continuous motion, can be interrupted by weekends and evenings simply by extruding the form away from the poured wall as it sets. The form should never be allowed to adhere to the concrete. Slip-form construction is used for various applications such as bridge piers, building cores, apartment-house shear walls, chimneys, communication towers, cooling towers, silos, and many other applications. In many cases, the procedure can be used to erect a structure in half the time of conventionally formed work. In addition, the working platforms rise with the form and decrease the labor costs of dismantling and reerecting scaffolds at each floor. Close inspection of the concrete must be performed at all times to ensure elimination of blowout (concrete plastic state) or bonding of concrete to the slip form during the construction process.

10.11.2 Materials and Methods

The major components of a slip-form assembly consist of lifting jacks and rods, yokes, wales, sheathing, a working platform, and suspended scaffolding (see Figure 10.16). All components, with the exception of the lifting jacks and concrete, climb integrally when the form is elevated. The lateral loads of the form are transferred through the sheathing and wales, which are supported by the yokes. The vertical loads of the scaffolding and platforms, in addition to live loads (equipment), are carried by the wales and transmitted to the yokes and into the jacking rods embedded in the concrete. The weight of the concrete acts as a lateral support to the rods.

The **lifting jacks** provide the forces that lift the forms upward. There are three types of jacks, namely, screw, hydraulic, and pneumatic, hydraulic being the most common. Hydraulic jacks range anywhere from 3 to 25 tons/jack in uplift force. Care must be taken to properly position the jacks. The dead and live loads dictate the size and number of lifting jacks required for the application. All jacks must lift equal weight, otherwise distortions in the wall will resulting in excessive stress to the yokes at certain locations. Operations must include quality control to avoid excessive live loads concentrated in an area of the working platform.

Jacking rods require careful design as well. These rods provide the required support of the slip-form structure and are located in the poured concrete wall. If they remain in the wall, they have reduced or minimal bond strength and should not be designed as a typical reinforcing rod. The rods can be salvaged by sliding them through a thin metal sleeve attached to the yoke. The rods must be vertical at all times when placed in the walls. This will prevent buckling of the rods in the concrete as the form rises.

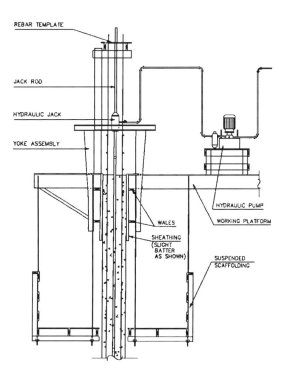

FIGURE 10.16   Lifting jack assembly.

The **yokes** are connected to the lifting jacks and consist of a horizontal member and two yoke legs. Depending on the design of the wall, the yoke should be engineered to compensate for the loads that will be applied to it. The yoke acts as a clamp holding the wales and sheathing in place and transferring the weight of the entire slip-form assembly directly onto the vertical jacking rods. In addition, the yokes support the reinforcing steel of the concrete wall. The yoke has to support the lateral pressure of concrete as per the ACI standard, which is described in the following formula:

$$p = 100 + \frac{6000\,R}{T}$$

where

p = maximum lateral pressure (pounds per square foot)
R = rate of concrete placement (foot per hour)
T = temperature of concrete forms (degrees Farenheit)

Additional frames or false yokes are sometimes located between the yokes. The primary difference between the two is the false yokes do not transfer the vertical loads onto the jacking rods. Instead, the false yokes only support the lateral loads transmitted through the wales.

The **wales** are the longitudinal supports for the sheathing. There is usually an upper and lower wale on each side of the concrete wall. More wales should be provided if the form height warrants the additional support. Normally this is a height exceeding 4 ft. The wales resist the loads and transfer these loads directly to the yokes. The wales must be engineered to withstand the ACI lateral pressure requirement as indicated previously. In addition, ACI Committee 347 also requires that tolerances be maintained for the finished structure. The variation of wall thickness cannot be greater than ±3/8 in for walls up to 8 in thick or ±1/2 in for walls thicker than 8 in. Therefore good engineering practice must be applied in the design of these forms. As per ACI SP-4, timber wales should be a minimum of 2 or 3 ply lumber or at least one ply of 2 in material, and when the span between jacks approaches 10 ft, the wales should be braced to act as a truss in the vertical plane between yokes. For curved walls, the minimum depth of segmental wales should be 4 1/2 in at the center after cutting.

The **sheathing** is the vertical support in direct contact with the concrete. Swelling of sheathing must be controlled because the sheathing is in continuous contact with moist concrete for the duration of the construction. Protecting sheathing prior to construction should be accomplished either by presoaking lumber or with a waterproofing preparation. In many cases, a nonabsorbtive surface that can withstand moisture and temperature conditions is selected. Sheathing should be higher on the exterior face of the form to avoid splashing materials onto workers on the scaffolding below. As per ACI SP-4, forms should be constructed of at least 1-in board, 3/4-in plywood, 10-ga steel sheet, or other approved material. The sheathing, when constructed, should have a slight batter to it. This will enable the forms to self-clear. The batter should be tapered inward at the top of the form and tapered outward at the bottom of the form. The middle area of the sheathing is the finished dimension of the concrete wall. The height of the form is usually 4 ft. The sheathing is kept well oiled as it rises to eliminate excessive drag forces. The construction engineer should compensate for this additional vertical load in the design.

The **working platform** houses all the construction equipment necessary for the form to advance. The platform also keeps the poured walls square. Care should be taken in design of the platform. A minimum of 75 psf is recommended for design. In addition, concentrated buggy loads must be considered.

Suspended scaffolds are located below and on either side of the form. They provide a work area that enables work crews to finish the concrete wall that has been placed. Since the concrete is still wet, it is workable.

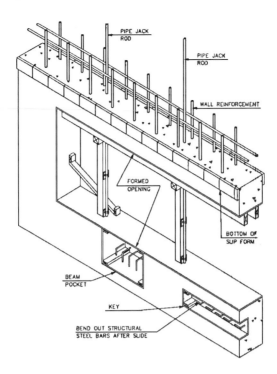

FIGURE 10.17 Blockouts for floors, corbels and openings.

10.11.3 Advantages and Disadvantages

One disadvantage is that the monolithic pour precludes the placement of floors during the extrusion process. Horizontal reinforcing protruding through the walls cannot be placed until the slip form passes through the wall. Provisions can be incorporated to prepare for floors, corbels, or openings (see Figure 10.17). This is accomplished by blockouts. The crews provide openings or pockets by creating forms within the form. Quality control is of considerable importance. Care must be taken to lift the form at the right time. As a result, the construction procedure requires skilled laborers and experienced project management. Placing concrete too quickly can result in a blow out and too slowly can cause the concrete to adhere to the form, which could damage the wall as well as the form. Construction engineering is an important aspect of the overall procedure. Careful preparation and analysis should be undertaken prior to construction. Temperature and climatic conditions are all factors that must be incorporated in the design of slip-form construction. All the above incur additional costs to the project as opposed to using conventional forms. Another disadvantage is the difficulty of inspecting the wall since steel is placed only a few feet above prior to the pour.

The advantage of slip-form construction is the speed at which a wall can be erected. A slip-formed wall can eliminate up to three months construction, compared to conventionally formed work and the wall is free of horizontal joints. The taller the building, the more cost effective the system can be. The level of safety is greatly increased when utilizing the slip form process. Preparation for the process must be carefully carried out by well-trained and skilled individuals. As a result, all staging is eliminated and the working areas only have to be built once.

10.11.4 Economy

Slip-forming tall cores (in excess of 12 stories) can greatly reduce the cost of construction. Only one form has to be built which can often be reused for every level to the top. The cost of forms per contact area is greatly reduced. The overall question of cost, however, is difficult to establish owing to varying costs with regional conditions, available skilled labor, and the complexity of the project. This issue must be evaluated by the conditions that dictate the design and the location of the site.

10.12 Prestressed Concrete

10.12.1 Introduction

When gravity loads are applied to a flatwork element (slab or beam), moments and shears are created within the element. Applied moments are resisted by the development of a force couple within the element section. The couple consists of a tension and a compression force, each occurring on opposite sides of the neutral axis within the section. Shears are resisted by the effective concrete area of the section; at times, supplemental resistance is provided by shear-reinforcing steel.

In conventional concrete design, the moment-resisting couple is essentially the only resistance to bending moments. Thus the proportioning of the section dimensions, in association with the sizing of adequate reinforcing steel, must be sufficient to resist the applied moments, with tension stresses often being the limiting design criterium. By designing so that tension stresses are resisted by mild-steel reinforcement, the failure mode of an element will be ductile and gradual, giving ample warning of precipitating failure. This is ensured by designing all beams, slabs, and plates as underreinforced, namely, the reinforcement percentage does not exceed 75% of the balanced condition, in most cases, not more than 50% in order to prevent congestion of reinforcement.

Pretensioned (precast) and post-tensioned concrete design, generally grouped as prestressed design, involves an additional form of resistance to the applied moments and shears. By externally applying a compressive force to the element, the stress distribution over the section is dramatically altered. The compressive force acting on the section reduces the tensile stresses at the tension zones, thereby preventing cracking and also increasing considerably the diagonal capacity of the section. This permits the section size to be reduced while keeping the total resulting stress within the allowable range. Chapter 27 of the handbook gives the basic expressions for the stresses and the diagrams of stress distribution across the depth of a prestressed beam as affected by the external compressive prestressing force [see also Nawy (1996) for details of design of prestressed concrete structures].

Thus the use of prestressing methods can allow the designer to reduce section proportions, while carrying the same, or larger, load as a conventionally reinforced concrete section. There is a corresponding reduction in cost, in terms of the gross volume of concrete. The designer must acknowledge that the prestressing equipment, materials, and installation can, in some circumstances, cause the total construction cost to exceed that of conventionally reinforced concrete. Preliminary cost/benefit analysis is advisable before deciding to use a prestressing system, provided that other requirements of span and load might necessitate the use of a prestressed-concrete design.

10.12.2 Pretensioned (Precast) Concrete

Prestressed concrete falls generally into two groups: pretensioned or precast concrete and post-tensioned concrete. Precast concrete usually is made in offsite casting plants and may or may not include prestressing. Construction elements ranging from hollow core slabs, to T sections, to architectural wall panels are regularly precast. Precasting offers the controlled environment of a plant, which results in very consistent production in terms of quality and delivery speed. Precast plants work with heated, steel form beds with polished form surfaces. The quality of the concrete is controlled and is usually 5000 psi or higher. The color, consistency, and dimensions of precast elements are usually of better quality than can be produced in the field. Concrete admixtures, to achieve rapid set and also to gain durability advantages, are frequently added to produce higher quality concrete.

As noted earlier, precast concrete can be either prestressed or not. Conventionally reinforced precast concrete is most frequent in wall panels, columns, and miscellaneous elements such as highway dividers. Prestressed elements are most frequently produced for structures carrying high loads, bridging long spans, or subjected to harsh environments. In these cases, the precompression force improves both load-carrying performance, by reducing tension stresses, and durability, by closing surface cracks on the tension face that might allow the intrusion of corrosive salts or chemicals.

Where prestressing is used in precast applications, bonded prestressing is usually used. Bonded prestressing consists of placing high tensile strength (270 ksi) wire strands, or bundles, in the setting beds prior to pouring the concrete. The strands are then tensioned by jacking them tight against a rigid frame at the head and foot of the setting bed. Once the required jacking force has been introduced to the strands, fresh concrete is poured into the bed, in direct contact with the strands. As the concrete sets, the strands become bonded to the concrete.

The initial jacking force introduced to the strand is released into the precast element via the bond between the strand and the concrete. This produces a fairly even distribution of compressive stress along the length of the member. In addition, because the bond is continuous along the member's length, a bonded precast element can be cut into smaller pieces without losing the precompression force.

The designer is cautioned that the latter characteristic of precast concrete is only applicable where the strand drape is flat, thus producing an even precompression force over the entire length of the element. Where the strand drape is not flat (discussed further below), the influence of the jacking process will be to introduce both a precompression force and an internal moment into the element. In such a case, cutting the element could result in the loss of the internal moment, as it would not be resolved by the balancing remainder of the element that has been cut away. In practice, hollow-core plank is frequently cast in long lengths, with an essentially flat strand. The plank is then cut to the appropriate length for project needs. Some forms of wall panel are also produced in this fashion. However, where strict quality control is required to assure proper architectural appearance, each precast member is usually cast alone.

When designing precast concrete sections, the designer must be cognizant of several forces that act on the member during its life. It is not sufficient to simply analyze the service loads to be applied once the member is situated in its final position. Because concrete cures over a period of time, the relative strength of a concrete member changes as the concrete strength increases. Thus loads that are applied to a member shortly after casting could cause stresses that exceed the capacity of the member at that specific moment in its life.

The designer must therefore consider three types of loads that act on every precast member at some point. First are stripping stresses, created when the casting forms are released. This action causes suction forces that can induce large, concentrated tensile stresses in the concrete. Second are transport stresses, created from lifting and stacking the members for transport around the casting yard and to the job site. Lifting lugs are frequently cast into the precast members to provide a location for the attachment of lifting equipment. The use of strategically located lifting lugs, allows the designer to anticipate and design for the stresses that result from moving the members. Last are service stresses, which occur as a result of the design loads that are expected to act on the structure, determined by its use classification. These are the traditional design loads that must always be addressed.

Once precast elements arrive at the job site, they are erected piece by piece and bearing pads are used to separate the elements. Precast concrete shop drawings specifically detail all the connections between the precast pieces and any other structural members. Local stress concentrations must be taken into consideration during the design process to address conditions where ledges or corbels will support adjoining members.

In certain applications of precast concrete, a final cast-in-place concrete topping is applied once the precast superstructure is erected. This is typical in precast plank or double-tee construction, where a consistent, durable wearing surface is required. The topping course, typically 2–3 in thickness, is sometimes reinforced with welded wire fabric, thus providing catenary reinforcement, which aids in the distribution of loads over the entire floor system.

10.12.3 Post-tensioned Concrete

While pretensioning is generally conducted in a plant setting, post-tensioning is mostly conducted in the field. The "post" term is something of a misnomer, insofar as the precompression force introduced into the concrete is added prior to the introduction of any load on the structural system. However, the term has evolved because of the fact that the procedures occur at the construction site, rather than at a precasting plant, and thus are "post" because the members are not brought to the site ready to install.

Post-tensioning design follows the same procedures and expressions used in the design of pretensioned elements except for prestress losses. By introducing a precompression force into the concrete section, the maximum tensile stress is reduced by the compressive stress, thereby permitting the use of smaller sections that support the same loads as conventionally reinforced concrete members. Construction procedures can be considered an extension of those used in traditional cast-in-place (*situ*-cast) concrete. Formwork must be erected to support the fresh, poured concrete; mild reinforcing steel must be placed to provide system ductility; and site curing procedures must be followed to assure that quality, final concrete is produced.

The concrete stresses permissible in flexural prestressed members occur at three load levels: stresses occurring immediately after prestress transfer, stresses at service load, and stresses at the ultimate load level. Section 18.4 of the ACI 318-95 Building Code presents the permissible stresses at the extreme fibers of the members for these loading stages. Also, in this handbook Chapter 11, Construction of Prestressed Concrete Structures by B. Gerwick and Chapter 12, Unbonded Post-Tensioning in Building Construction by F.G. Barth and E.G. Nawy's book, *Prestressed Concrete—A Fundamental Approach* address most of the design and construction aspects pertinent to pretensioning and post-tensioning beams and one-way and two-way slabs and plates.

It is important to recognize that in two-way slab design, attention must be given to reentrant corners, large openings, L-shaped plan configurations, balconies, repetitive perimeter openings, or steps in slab elevations. Each of these conditions poses a difficult design problem if conventionally reinforced concrete is used. The problems are multiplied when the system is precompressed and compelled to shorten. Cracks can rapidly develop at corners, and localized crushing of concrete can occur, particularly due to stress concentration at these locations. To counter this condition, significant mild steel reinforcement has to be provided to prevent cracking propagation at the early stages of post-tensioning or loading.

Acknowledgments

The author wishes to extend deep appreciation to Jeremy Ivers, P.E., Scott McConnell, P.E., Anthony Scenna, P.E., and Anthony Voloninno, his co-workers at Schoor DePalma, and to Dr. Kamil Sor, his friend and colleague for many years, for their significant contributions to this chapter. The generous support of his fellow co-owners at Schoor DePalma and the patience and encouragement offered by his wife Marie are greatly appreciated.

References

ACI Committee 344. 1970. *Design and Construction of Circular Prestressed Concrete Structures*, ACI 344R. American Concrete Institute, Farmington Hills, MI.

ACI Committee 224. 1989. *Control of Cracking in Concrete Structures*. American Concrete Institute, Farmington Hills, MI.

ACI Committee 344. 1989a. *Design and Construction of Circular Wire and Strand Wrapped Prestressed Concrete Structures*, ACI 344 1R. American Concrete Institute, Farmington Hills, MI.

ACI Committee 344. 1989b. *Design and Construction of Circular Prestressed Concrete Structures With Circumferential Tendons,* ACI344 2R. American Concrete Institute, Farmington Hills, MI.

ACI Committee 318. 1995. *Building Code Requirements for Structural Concrete,* ACI 318-95 and ACI 318R-95. American Concrete Institute, Farmington Hills, MI.

ACI Committee 347. 1995. *Formwork for Concrete,* Handbook SP-4, 6th ed. American Concrete Institute, Farmington Hills, MI.

ACI Committee 435. 1995. *Control of Deflection in Concrete Structures.* American Concrete Institute, Farmington Hills, MI.

American Society for Testing and Materials: ASTM Standards C33, C150, C311, C494, C618— referred to in the text. ASTM, Philadelphia, PA.

Billington, D.P. 1982. *Thin Shell Concrete Structures.* McGraw–Hill, New York.

Camellerie, J.F. 1978. Vertical slipforming. *Concrete Const. J.*

Nawy, E.G. 1996. *Prestressed Concrete—A Fundamental Approach,* 2nd ed. Prentice Hall, Upper Saddle River, NJ.

Peurifoy, R.L. *Formwork For Concrete Structures.* McGraw–Hill, New York.

Portland Cement Association. 1957. *Circular Concrete Tanks Without Prestressing.* Skokie, IL.

Portland Cement Association. 1994. *Rectangular Tanks.* Skokie, IL.

Russillo, M.A. 1988. *Lift Slab Construction: Its History, Methodology, Economics and Applications.*

Timoshenko, S. and Woinowsky-Krieger, S. 1959. *Theory of Plates and Shells.* McGraw–Hill, New York.

United States Lift Slab Corp. 1955. *Engineering Manual.*

Young, W.C. 1989. *Roark's Formulas for Stress and Strain.* McGraw–Hill, New York.

(a)

(b)

Hoisting double-T prestressed roof element (Courtesy Professor Ben C. Gerwick, Jr.).

11

Construction of Prestressed Concrete

by
Ben C. Gerwick, Jr.
Chairman, Ben C. Gerwick, Inc., San Francisco, California and Professor Emeritus of Civil Engineering, University of California at Berkeley. An international authority on prestressed concrete and former president of FIP.

0-8493-2666-4/97/$0.00+$.50
© 1997 by CRC Press LLC

11.1 Introduction

The effective construction of prestressed concrete structures requires the practical application and implementation of sophisticated engineering technology in a safe and reliable manner under the constraints of time, budget and the external environment. It incorporates the assembly of materials, equipment, men, and the execution of the work so as to attain the results envisioned in the design.

The key elements in prestressed concrete construction are the following:

1. The production of concrete that has stable, predictable properties, not only of strength but also of creep, shrinkage, elastic modulus, and durability. High strength is essential to attain efficiency of structural performance.
2. The forming (molding) of concrete into the design shape and within the specified tolerances.
3. The incorporation of mild steel reinforcement, accurately placed and held during concreting.
4. The placement of high strength steel wires, strands, or bars to fit the design profile and the stressing and anchoring and corrosion protection of such elements.
5. Installation of the composite structural elements or assemblages described above, in their final positions, whether they are cast-in-place or prefabricated.

During construction, prestressed concrete passes through a number of steps or stages, each of which must be considered. Then, after attaining the designed state, it undergoes a lifetime of relatively subtle but cumulatively important changes. Although conventional reinforced concrete passes through a similar life history, the changes and the importance of stages is generally far less

FIGURE 11.1 Prestressed concrete cantilever segmental bridge. Columbia River Bridge on I-5.

dramatic, and the consequences of oversights are not as severe. Hence in this chapter, the emphasis is on the practical attainment of safety and stability. We are working with an active, rather than a passive structural system.

Environmental conditions will be given considerable emphasis. Important prestressed concrete structures have been completed and now serve in Arctic, temperate, and tropical zones. they are successfully employed in the ocean, in the urban environment of large cities, and in remote deserts. Prestressed concrete is used in buildings, bridges (Figure 11.1), offshore platforms (Figure 11.2), floating structures, towers, poles, tanks, and railroad sleepers. Each of these uses has its own special criteria and practices. Each has its own constraints and limitations. The reliable performance of prestressed concrete, that is, its durability, throughout its design life requires careful consideration during design and construction.

The requirements for such widespread use can be summed up in a few key words:

- reliable, stable material properties;
- full consideration of the stages of construction of a prestressed concrete structure; and
- durability—long-term performance.

FIGURE 11.2 Statfjord B. Offshore Platform with prestressed concrete shafts. *Source: Construction of Prestressed Concrete Structures*, Gerwick, B.C., Jr. © 1996. John Wiley and Sons, Inc. Reprinted by permission of John Wiley and Sons, Inc.

Various authorities have given their precepts or commandments. These generally include the following:

1. When prestress is applied, the concrete element must be allowed to shorten in the direction of the prestress. If the structure is restrained from shortening, it will not be prestressed.
2. When the prestressing tendon is curved, it exerts a radial shear force.
3. Prestressing applied to concrete too soon after casting, before it has gained strength and maturity, will suffer excessive creep. Prestress applied too late after concreting may result in extensive shrinkage cracks.
4. Prestressing eccentric to the centroid of the concrete section produces bending. This bending, for example, camber, tends to grow with age owing to creep.
5. The energy and hence force induced in a prestressing tendon by stressing is very large and results in a temporarily dangerous missile. Safety precautions are essential.

11.2 Concrete and Its Components

The concrete material components must meet the general requirements for construction of reinforced concrete structures and, in addition, must usually conform to more stringent requirements in order to attain the design objectives.

11.2.1 Aggregates

Coarse aggregates are usually siliceous or limestone. Both gravels from riverine deposits and crushed rock are used. Desirable properties are hardness, soundness, nonreactivity, surface roughness, cubic as opposed to plate cleavage, low water absorption, and impermeability. Slightly reactive aggregates may often be utilized in specific applications if proper chemistry of cementitious materials is selected. Some limestones are not only porous, but are also permeable: their use in marine structures requires special actions to preserve durability.

Lightweight (ceramic) aggregates are extensively used. Those of high strength, low creep, and low absorption are most appropriate to prestressed concrete application. Fine aggregates may be from natural deposits or they may be crushed from coarse aggregate. In any event, screening will be needed to ensure a proper proportion of coarse sands and a suitable grading curve, preferably with a low percentage of the fine fractions, since a mix rich in cementitious material will fill the interstices, while a large percentage of very fine particles will have a high water demand.

The aggregates must be clean and free from salt and dirt. They should be properly stored above ground and protected from snow and ice and from excessive heat. Shielding from the sun may be provided by portable sheets of galvanized metal on temporary posts in the aggregate pile. To a large degree, cooling of aggregates may be accomplished by evaporation. A water soaker hose may be effective. Conversely, aggregates should be above freezing when batched. They may be heated by steam. Care must be taken to avoid accumulation of ice in the aggregate, whether by snow or the freezing of condensate, since ice particles may be unintentionally weighed and batched as coarse aggregate.

Aggregates, both fine and coarse, should not contain or be encrusted with salts and silt. Washing is usually required, preferably with fresh water; but if fresh water is not available, even brackish water can be used to remove the condensed salt crystals and silt.

11.2.2 Cementitious Materials

The practical construction of prestressed concrete is based on the use of Portland cement as the principal cementing agent. ASTM cement Types I, II, and III are used. Type I can be used in geographical areas and applications where sulfate attack is not a problem. Type II, with closer control over the constituent compounds, especially tricalcium aluminate (C_3A), is widely employed where durability is a concern.

Type III, high early strength, attains sufficient strength to enable early application of prestress and hence has significant benefits to both manufacturers of precast concrete and constructors of cast-in-place concrete. It does, however, result in greater setting and drying shrinkage and thermal strains. The benefits of Type III cement are high early strength and that it can be largely attained by fine grinding of Type II cement and by the incorporation of microsilica fume.

Fly ash is a pozzolanic material that gains strength over a longer period of time. It can be substituted for up to 20% or more of the cement. The silica of the fly ash binds up the soluble calcium hydroxide, hence substantially increasing impermeability. It also provides resistance to sulfate attack and alkali-aggregate reactivity while lowering the heat of hydration and reducing thermal strains. It, of course requires, a separate silo and should be weigh-batched separately from the cement. In some areas, fly ash is interground with the cement. For proper strength gain, continuous moisture is essential over an extended period of up to 30 d. This is best obtained by wet curing for 3 d followed by sealing with one or two coats of membrane-sealing compound.

Finely ground blast-furnace slag (BFS) can be used with a limited portion of Portland cement to act as an efficient cementitious material with low heat of hydration and low permeability. Mix proportions of BFS to cement range from 50–50% to 70–30% and even 80–20%.

11.2.3 Admixtures

Many revolutionary new admixtures are now commercially available. The most important are water reducing agents, especially high-range water reducers (superplasticizers), air-entraining agents, and microsilica, called silica fume. Corrosion-inhibiting admixtures, retarding admixtures, pumping aids, and thixotropic admixtures are also available. Tests mixes are essential to ensure that the admixtures selected are compatible with the particular cement and with other admixtures.

High-range water reducers (superplasticizers) are the best way to achieve the low water/cementitious material ratios essential for high-performance concrete. The usable "life" of HRWRA is typically 1 1/2 h after mixing, meaning that the concrete must be placed and consolidated within that period to prevent sudden premature stiffening. This life can be extended by adding retarders. Some HRWRA admixtures incorporate a retarder and hence have an extended slump life. HRWRAs are the best way to obtain pumpable concrete.

Air entrainment creates tiny air bubbles of proper distribution throughout the concrete, thus enabling the structure to resist a large number of freeze-thaw cycles. Its use is essential in cold regions. Its efficacy is dependent on in-place characteristics. Unfortunately, these can only be verified by microscopic examination of hardened concrete specimens, although other "real time" tests give valuable indications as to the entrained air and can be used for control once they have been properly calibrated by a petrographic inspection. Day-to-day control can be exercised by use of a meter that measures the total air content (entrapped plus entrained air) in the fresh concrete mix.

Microsilica or silica fume consists of very small particles of silica about 1/100 the size of cement grains, which combine with gypsum to produce high early strength, very low impermeability, improved bond, and enhanced fatigue endurance. Microsilica also increases electrical resistivity. It is used in percentages (by weight of cement) of 3 to 10%. More than 6% makes the concrete quite sticky and difficult to place and consolidate. However, this property of added cohesiveness can be beneficially used for concrete placed underwater. Microsilica can be introduced as a powder or in a slurry, or in some cases, ground with the cement. Although most commercially-available microsilica is a mineral byproduct of silicon production, rice hull ash has recently been used as a source, with similar and possibly enhanced properties. Use of HRWRA is essential to the incorporation of microsilica in the mix.

Proprietary corrosion-preventing admixtures, based on calcium nitrite, are available. Their long term efficacy is clouded by contradictory and inadequate evidence. Similar admixtures that reduce the diffusion of chlorides in concrete are commercially available but are relatively expensive.

Thixotropic agents are available that promote the initial flow of concrete and self-leveling but gel when movement ceases, thus preventing bleed. They are often used for underwater concreting and for grouting post-tensioning ducts.

11.2.4 Water

Water for prestressed concrete is required to have no more than 650 ppm of chlorides and 1000 ppm of sulfate ion; although in many specifications, the latter is increased to 1300 ppm. Water furnishes the essential electrolyte that initiates the chemical reactions and promotes fluidity (workability) by coating the particles. However, excess water is inimical to strength and durability. Therefore the water/cementitious-material ratio (W/CM) should be limited to a maximum of 0.42 and preferably 0.40. In practice, W/CM ratios of 0.32 for precast elements and 0.37 for cast-in-place concrete are attainable. In hot weather, ice may be added to the mix in place of water, to lower the temperature of the fresh concrete.

11.2.5 Batching of Concrete

Concrete is weigh-batched in the same manner as in conventional reinforced concrete practice except that in prestressed concrete construction, there are usually more dry components (cement, coarse aggregate, fine aggregate, fly ash, and microsilica) and more wet components in the form of admixtures, each of which must be accurately dosed. The moisture content of the aggregates must be ascertained from free-moisture meters in the bins and corrections made to the amount of water added.

11.2.6 Mixing of Concrete

It is obvious that mixing must be very thorough in order to blend the many components. For this reason, turbine mixers and shear mixers are favored. Even when the concrete is subsequently transported in ready-mix trucks, premixing at the plant is desirable.

The order in which admixtures are added to the mix is important, in particular that of air entrainment and superplasticizer since the latter may inhibit the proper development of entrained air. Adding half the dosage, then mixing for a brief period before adding the remainder has solved the problem in some cases.

11.2.7 Transporting and Placing

Transporting and placing follows conventional practice; ready mix trucks, buckets, conveyors, or pumps are used. For discharge of stiff mixes from the trucks, screw conveyors may be used. In some high-rise construction using slip forms, where the required rate of concreting at any one location is low, buckets have been used to raise the concrete to elevated hoppers, from which the concrete has been distributed by buggies running on runways. The loss of slump and of entrained air content when pumping concrete to high elevations must be considered.

11.2.8 Consolidation

Internal high-frequency vibration is the typical means of consolidation for cast-in-place concrete. It should also be used to supplement external vibration in all precast concrete elements except very thin (150 mm or 6 in or less) members, since the vibratory energy is not effective at depth.

11.2.9 Curing

Membrane curing is generally considered to be the most reliable for vertical surfaces. For horizontal surfaces, wet curing is best, although sealing with polyethylene sheaths is also widely practiced.

Concrete containing fly ash, and especially that containing microsilica, requires special attention to curing in order to prevent surface crazing.

For precast elements, and occasionally for cast-in-place segments, steam curing is used to accelerate gain in strength. Steam curing at atmospheric pressure provides both moisture and heat and results in generally superior concrete. However, certain practical precautions are required with steam curing.

- Concrete should have its initial set (3–4) h before the temperature is elevated, since otherwise it may be damaged as it expands.
- The maximum curing temperature should be limited to 65°C (150°F).
- Steel forms expand before concrete. Hence changes in section must be detailed to accommodate length change so as not to crack the concrete. Soft rubber gaskets may be used between flanges of forms to absorb their early expansion.
- Concrete, once cured, should not be exposed to cold, drying winds until it has cooled to within about 10° to 15°C (20° to 30°F) above ambient. Not only is it important to prevent sudden cooling of the outer surface while the inner core is still hot but also to prevent rapid evaporation of moisture. The latter may require supplemental water or membrane curing. Membrane-curing compounds evaporate with heat. Hence in hot weather a second coat may be required.

In cold weather, and especially with dry winds, the concrete must be protected by insulated forms or blankets (tarpaulins) until the surface temperature is no more than 15°C (30°F) above that of the core (Figure 11.3). Otherwise, cracking will occur owing to the early contraction of the outer layer.

11.2.10 Testing

Testing of high-strength and high-performance concrete is a demanding technology, requiring trained personnel and sophisticated equipment. Specimens must be taken and prepared carefully if the results are to be meaningful. Cylinders should be vibrated with a pencil vibrator in steel molds

FIGURE 11.3   Spreading rubberized nylon tarpaulins in preparation for steam curing. *Source: Construction of Prestressed Concrete Structures*, Gerwick, B.C., Jr. © 1996. John Wiley and Sons, Inc. Reprinted by permission of John Wiley and Sons, Inc.

since plastic and cardboard molds will give erroneously low results. The cylinders should be subject to the same temperature regime as the structural element, for example, placed under the steam hood near each end and at the center of the bed. When removed, cylinders should be placed in a water bath containing lime and kept at 20°C (70°F) until one day before the test. Then they should be removed to dry out before testing. Ends of cylinders should be rubbed smooth to remove any surface irregularities. Numbers should never be etched in the surface! Cylinders should be carefully capped with new capping compound or ground smooth to test without capping. Similarly, cubes, where used, should be ground and tested without capping. Be sure there is a strong screen around the testing machine as high-strength concrete fails explosively.

Testing of other properties is even more complex. Air entrainment characteristics are determined by microscopic petrographic examination of thin slices cut from a core in the hardened concrete.

Impermeability is very difficult to measure at the upper limits associated with high-performance concrete. The validity of the rapid chloride permeability tests of both ASTM and British Standards have recently been questioned.

On many important projects involving prestressed concrete, various properties are specified. These include permeability and determination of the extent of microcracking. In practice, some of the values attained in the laboratory have proven to be difficult or unattainable in the field. The contractor should therefore start his mix development and testing program at the earliest possible moment, working in cooperation with the engineer and his technicians so as to avoid delays and rejections.

11.3 Reinforcement and Prestressing Systems

11.3.1 Reinforcing Steel

Because in prestressed concrete structures, the role of primary reinforcement has been largely replaced by prestressing steel, there has been an unfortunate tendency to denigrate the importance of supplementary conventional reinforcing. Conventional reinforcement fulfills the essential functions of distribution and resistance in the orthogonal directions to the prestressing, most typically in the transverse direction. It serves to augment prestressing steel in critical areas of concentrated stress. It serves as confinement, resisting the bursting and delamination stresses due to prestressing itself (Figure 11.4). It also resists punching shear under concentrated loads.

FIGURE 11.4 I-headed bars lock against reinforcing steel to provide high shear capacity and confinement.

FIGURE 11.5 Mechanically headed bars used to replace conventional closed stirrups.

Since the role of reinforcing steel is secondary to the structural behavior and hence is concerned with stress distribution and crack prevention, close spacing with small bars is most often employed. Confining reinforcement and shear reinforcement usually consist of stirrups or spirals, although mechanically headed bars (Figure 11.5) have been used on a number of recent major structures to ease placement and reduce congestion.

Subject to the design engineer's approval, bundling of bars may be acceptable. This clustering of three, or four bars together facilitates concrete placement and vibration, although it often complicates steel installation. Splices must be staggered.

Reinforcing bars typically develop anchorage by having an embedment length that extends beyond the location of high stress, preferably terminating in a confined zone or core. Bends, such as those of stirrups, are only about 70% effective in ultimate capacity, since the concrete under the bend crushes at high stress. Mechanical heads typically develop the full strength of the bar and permit the use of larger bars or those with higher yield strength. They are much easier to install and reduce congestion of steel bars.

It is a serious mistake by both designer and constructor to reduce anchorage lengths to the absolute minimum permitted by code or specifications, since practicable placement tolerances frequently result in inadequate length for anchorage.

A typical concern arises with the provision of "cover," that is, the thickness of concrete outside the reinforcing bars. Since cover is a major factor in durability, as well as in mechanical behavior, it is normally specified to a tolerance of −6 mm (−1/4 in). This means that reinforcement has to be installed to the same tolerance.

Main bars should be tied with wire at every intersection. Stirrups need to be tied to the longitudinal bars at at least three intersections so as to prevent displacement along the bars during concreting (a typical problem in concreting that has serious consequences.)

Lap splices in reinforcement are another source of difficulty. These can be more easily achieved if the bar is a little longer than the minimum required by code. The designer and/or the code should specify whether or not the overlapping bars need to be tied together with soft iron wire. In general, this is good practice. Compression splices are rendered most efficient by being well tied together at both ends: otherwise the concrete may locally crush under one or both bar ends. Lap splices, especially those in tension, need to be confined on both orthogonal axes, transverse and through thickness.

Reinforcing bars need support. Both concrete "dobe" blocks and plastic chairs are used and are left in the completed concrete structure.

Reinforcing bars are frequently epoxy-coated in order to protect against corrosion. The electrostatic fusion process is most widely used. Problems sometimes occur with insufficient thickness of epoxy over the lugs of the deformations. Regardless of method of manufacture or whether the bars are prebent before coating or bent afterward, there will be some scratches and some "holidays" due to handling and placement. Fiber slings and pads must be used. Abraded areas must be touched up in the field after placing.

Special care has to be taken to ensure that the reinforcing steel is stored in a salt-free area before concreting, that is, where seawater chlorides cannot be carried by fog or spray to contaminate coated or uncoated bars.

11.3.2 Prestressing Tendons and Assemblies

Prestressing tendons are made of wire, both parallel and stranded, or rods of high yield strength steel. Bars and rods consists of heat-treated alloys, which have been prestretched beyond yield and tempered in the manufacturing process. They are rolled with spiral threads, either cut in upset ends or continuously rolled so that the threads also act as deformations.

Cold-drawn wire is typically low alloy, which has been tempered. Wires may have forged headed ends, as in the BBRV system or may be wound into seven-wire strands and prestretched.

The seven-wire strand, with an outside strand diameter of 12 to 15 mm (0.5 to 0.6 in) is most typical for both the pretensioning and post-tensioning systems discussed in Sections 11.5 and 11.6. For pretensioning, the strands are directly embedded in the concrete, gaining anchorage by bond on the twisted surface. For special applications, such as railroad ties (sleepers), a wire with a deformation is stranded into one outer layer of the strand in order to shorten the stress transfer length. For post-tensioning, bars at the end may be threaded to receive screwed-on end anchorages. Headed wires and rods are directly anchored into anchorage plates, while for strands, the anchorage is developed by a wedged cone, locking the strand between a male cone and female sleeve. Wedges can also be used to anchor bars, while headed rods are anchored by slotted plates and shears. Headed wires and strands may be grouped together in large bundles to concentrate enormous forces, as great as 500 tons or more, in a small cross-sectional area.

Bars, wires, and strands are usually left "black," that is, uncoated. They may be galvanized for corrosion protection: hot-dip process galvanizing appears most reliable. A chromate wash applied at the end of the galvanizing process will passivate the zinc and inhibit the liberation of individual hydrogen atoms that results from reaction with the alkali cement. There is evidence, however, that the protection resulting from galvanizing has a limited life of 10–15 years.

Epoxy coating may be applied to steel strands. Sand is dusted on while the epoxy is still wet to improve bond.

Prestressing steels, which are under high sustained stress, are subject to a long-term plastic rearrangement of molecules known as stress-relaxation. This can produce long-term losses in prestress of the member, up to 13%. By a special tempering process, the stress relaxation may be reduced significantly, to about 6%.

Seven-wire strand has approximately 30% voids as compared to a solid bar of the same external diameter. "Compact strand" is available, in which the wires are shaped so that they fit tightly together. This allows an approximately 15% increase in total force to be obtained within the same external dimensions and is a valuable technique for solving space problems in highly stressed members and in repairs.

Prestressing wire and strand are shipped in coils that are wrapped with heavy, water-resistant paper (Figure 11.6). Bars and rods are similarly shipped in bundles, wrapped for protection from moisture. For protection against corrosion, vapor phase inhibiting crystals may be inserted in the package. More frequently, however, the tendons are coated with water-soluble oil.

FIGURE 11.6 Seven-wire prestressing strand is shipped in coils to the precast concrete plant or jobsite. *Source:* Courtesy of VSL Corporation, Campbell, California.

11.3.3 Anchorages and Splices

The anchorages and splice hardware for prestressing systems consist of cast alloy steels, machined to accurate dimensions with very close tolerances to provide a positive grip to the tendons under high forces.

The anchorage includes, either separately or monolithically, a bearing plate that transfers force to the concrete. The concrete at the anchorage, being under a force that is typically several times its uniaxial compressive strength, must be confined, either by a ring integral with the anchorage or by spiral or both. The anchorage also includes the head and wedges for gripping the tendon. Anchorages must be transported, stored, and handled with extreme care. Some European specifications require storage in a heated, dehumidified, warehouse. Anchorage assemblies must be clean and free from foreign material, except the grease specified by the system manufacturer. Damaged or visually defective anchorages should never be used. A failure may result in the other anchorage and tendon being propelled at high velocity.

Splices of prestressing tendons are similar to anchorages in that they anchor to both ends of the tendons. It is fundamental that the splice be free so as to allow the tendon to elongate upon sressing: thus a recess or opening must be temporarily left in the concrete at the splice location.

Post-tensioned anchorages are either encased (Figure 11.7) or recessed for reasons of durability, fire protection, and appearance. A recessed pocket is formed

FIGURE 11.7 High-strength alloy steel bars are used to prestress this wall to enable it to resist the pressures of Arctic sea ice. *Source: Construction of Prestressed Concrete Structures;* Gerwick, B.C., Jr. © 1996. John Wiley and Sons, Inc. Reprinted by permission of John Wiley and Sons, Inc.

in the end block with protruding steel ties that are temporarily bent aside. The surfaces of the pocket or recess are coated with bonding epoxy or latex. Then a fine (small coarse aggregate) concrete is placed in a manner so as to eliminate a shrinkage and bleed pocket at the top. After the form is stripped, the joint may be sealed with a penetrant epoxy.

11.3.4 Ducts

Ducts are used to preform holes in the concrete structure during casting so that later, after the concrete has hardened, tendons may be inserted and stressed. Several types of ducts have been used, including steel pipe; semirigid steel ducting, made of strips spirally wound to lock to each other; thin flexible metal ducting; and plastic (Figure 11.8). Ducts are usually circular in cross section, although "flat" ducts, with a rectangular cross section are used where the concrete cross section and other steel, etc., constrict the available space.

Flexible metal ducts, made of thin corrugated steel and usually galvanized for temporary corrosion protection, are used for tendons of sharp or reversing curvature. The ducts are readily wound in coils and can be spliced by a short overlapping sleeve of similar material. These ducts are very flexible, so that they develop unwanted local curvature, called "wobble." This can be prevented by insertion of a mandrel of smaller diameter, for example, electrical conduit. After the concrete has been placed, the mandrel is drawn forward and backward to prevent any bond with mortar that may have leaked into the duct, and then removed.

Semirigid metal ducts are bright steel strip, wound spirally with crimped overlaps (Figure 11.9). This ducting has less wobble, but still requires use of a mandrel in those applications where even slight deviations have serious consequences. This material is normally furnished in straight lengths but can still be deviated over large-radius curves, depending on the sheet thickness used.

Semirigid and flexible ducts are neither watertight nor mortar tight. If a mandrel is not used, a wire or single strand should be run back and forth to clean out any in-leaking mortar.

For long straight runs, especially long vertical runs, thin-walled pipe, with screwed splice fittings is employed. Wall thicknesses of 1 to 2 mm are employed.

Splicing of metallic ducts is done by sleeving, with an adequate overlap of at least two diameters and taping with waterproof tape. However, the latter has not proven fully reliable. Heat-shrink tape is now the preferred method.

Ducts may be epoxy coated in zones especially vulnerable to corrosion, such as transverse post-tensioning of bridge decks in geographical areas where salts are applied to prevent icing.

Plastic ducts of polyethylene or polystyrene, with corrugations inside and out, are increasingly employed because they are watertight and grout tight, have controllable flexibility, and give

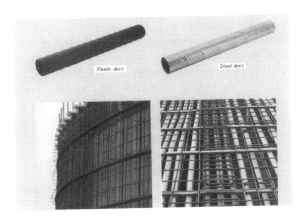

FIGURE 11.8 Ducts for post-tensioning may be of steel or plastic. *Source:* Courtesy of VSL Corporation, Campbell, California.

FIGURE 11.9   Both longitudinal and transverse ducts in place
to receive tendons after concrete has been placed and cured.

long-term corrosion protection to the tendons. They can readily be fabricated to exact length and can be spliced, either by fusion or by use of sleeves and glue or heat-shrink tape. Plastic ducts should have a minimum thickness of 2 mm; plastic ducts are available both as circular tubes and as rectangular "flat" ducts.

When a plastic duct is installed on a profile or plan that is curved, and wires or strands are later pulled in, the duct may be cut by abrasion, resulting in subsequent grout leakage. Ribbed ducts, in which the strand or wire will bear on enlarged ribs, have been successfully used to overcome this problem, as has lubrication with water-soluble oil. The ribs or corrugations enhance the bond between the plastic duct and the surrounding concrete. Use of 3 to 4 mm thick plastic duct will also minimize the problem of abrasive cutting.

The integrity of ducts, after installation in the concrete, is critical to the successful completion of prestressing. The duct location and the tolerances determine the subsequent profile of the tendon. Any blockage, as by in-leakage of mortar during concreting or out-leakage and crossover of grout from one duct to an adjacent one, will prove very costly, time consuming, and in some cases, impracticable to correct.

11.4 Special Provisions for Prestressed Concrete Construction

11.4.1 Concreting in Congested Areas

It is typical of much prestressed concrete construction that certain zones, such as the ends of girders and locations of deviation of prestress, are heavily congested (Figure 11.10). Such areas may include, for example, zones in which post-tensioning anchorages, trumpets, spiral confinement, transverse, and vertical reinforcing steel bars compete for the limited space available. At the same time, such zones may make the highest demand on the concrete for compressive and tensile strength.

Placing concrete of the required quality, without any voids, honeycomb, or segregation requires one or more of the following techniques:

1. bundling of reinforcing steel bars, especially stirrups, or substitution of headed reinforcement to give more clearance for the concrete;
2. changing concrete mix to reduce maximum size of coarse aggregate, for example, to 10 mm, while increasing cement content as necessary to obtain required strength;
3. prior to concreting, insertion of tremie tubes through the reinforcement, gradually withdrawing these as concrete is placed;

FIGURE 11.10 Congestion of conventional reinforcement and multiaxial prestressing ducts creates problems for concrete placement. *Source: Construction of Prestressed Concrete Structures*, Gerwick, B.C., Jr. © 1996. John Wiley and Sons, Inc. Reprinted by permission of John Wiley and Sons, Inc.

FIGURE 11.11 Vibrating concrete in order to ensure thorough consolidation below ducts. *Source: Construction of Prestressed Concrete Structures*, Gerwick, B.C., Jr. © 1996. John Wiley and Sons, Inc. Reprinted by permission of John Wiley and Sons, Inc.

4. prior to concreting, marking the location of reinforcing steel on the forms to enable subsequent insertion of vibrator;
5. prior to concreting, inserting the vibrator within a tube that penetrates through reinforcement (Figure 11.11), gradually withdrawing tube as concrete is placed, allowing vibrator to consolidate concrete below tube; and
6. in extreme cases, precasting of end block in the horizontal position, then setting end block in forms and joining it to the cast-in-place concrete by protruding reinforcing dowels and the axial post-tensioning.

11.4.2 Special Reinforcement at Location of Curvature of Prestressing Tendons

At locations where the centroid of prestressing deviates sharply in relation to the centroid of the concrete section, radial stresses are introduced. These may cause cracking or, in extreme cases, actual pull-out of the concrete in punching shear. It is important to tie this zone to the main concrete section.

Typical zones where this phenomenon occurs and has led to damage in the past are at intermediate anchorages, especially where the anchorages are in bolsters or blisters protruding from the concrete section, and at deviation points for deflected-strand pretensioning. Similar conditions exist where external post-tensioning is constructed alongside an existing concrete section but is eccentric to it.

In all of the above cases, "U" stirrups or equal are required. Their legs must be anchored deeply within the concrete core, preferably in a compressive zone.

11.4.3 Special Reinforcement to Distribute Anchorage Zone Strains

It has long been recognized that the concentrated forces at anchorages of post-tensioning tendons creates radial bursting forces that lead to cracking in the surrounding concrete. All commercial producers of prestressing systems include appropriate spiral reinforcement or the equivalent.

Not so well recognized is that in the zone between two anchorages, or between the ends of two groups of pretensioned strands, splitting tension is developed. This requires distributed reinforcement to prevent detrimental cracking.

When prestressing tendons are anchored at locations other than the ends of the member, for example, the intermediate anchorages of continuity tendons or dead-end anchorages, the concrete behind the anchorage is subject to tension as the concrete in front of the anchorage tries to shorten. Either adjacent prestressing tendons or conventional reinforcing bars are required to distribute these stresses back into the body of the concrete behind.

In vertical walls and webs, looped U tendons are often used to provide prestress against shear. These situations occur in the vertical walls of silos, offshore platforms and bridge piers, and in the web of deep girders. Unless adjacent Us overlap at the bottom, that is, if the adjacent Us are separated, then high-tensile strains exist between them that can lead to cracking. Orthogonal reinforcing steel bars are needed.

Additional bars for the above cases should be well distributed. One or more bars, or loops, as the case may be, should be located as close as practicable to the location where the potential crack will initiate.

The number and size of bars, while calculable by finite-element methods, is often determined on an empirical basis, for example, by provision of enough conventional reinforcement to transfer 50% of the prestress force to the zone behind. Failure to counter the tensile forces and strains can lead not only to cracking but to a serious reduction in shear capacity.

It can be argued that the above problems are the responsibility of the design engineer. However, the counterargument is that these local strains are associated with the anchorages of the prestressing system, which are typically selected by the contractor or his subcontractor and hence, to some degree, are his responsibility. In any event, the occurrence of the cracking, etc., described above may initially be blamed on the constructor and may have serious cost consequences for him.

Conversely, the extra reinforcing required is minimal compared with the overall project and therefore it appears prudent for the constructor to ensure that it is provided. The matter should be resolved before the concrete is cast.

11.4.4 Embedments

Steel embedment plates are frequently installed in the sides or surfaces of prestressed concrete members. Frequently, these are square plates. As a result of at least three phenomena, cracks often originate at the corners.

1. **Steam curing.** The steel plate expands rapidly as the temperature rises, before the concrete expands and before it gains strength.
2. **Concrete shrinkage.** The concrete surrounding the embedment is subject to both setting and drying shrinkage.
3. **Strains at sharp corners.** Such strains arise due to prestress.

Solutions that will prevent or minimize such cracks are as follows:
A. Place soft wood or a neoprene strip around the embedment, which will later be removed and filled with epoxy, latex, cement mortar, or an equal material.
B. Round the corners of embedment plates.
C. Install short diagonal bars across corners.

11.4.5 High-Performance Concrete

The criteria for high-quality concrete has been presented in general terms in Section 11.2. High performance, above the range normally associated with high-quality reinforced concrete, is often required for prestressed concrete members (Figure 11.12). Typical high-performance requirements specify very high strength (above 60 MPa, or 8400 psi), enhanced impermeability (less than 10^{-10} m/s), special requirements limiting microcracks, high abrasion resistance, and high tensile strength (above 4 MPa, or 600 psi). The attainment of these requires advanced concrete technology and strict control of field construction practices.

Typical means of attaining these properties include:

1. rescreening and rewashing of aggregates;
2. selection of aggregates with surface roughness, for example, crushed rock and crushed sand;
3. smaller coarse aggregate sizes;
4. higher sand content;
5. higher cement content;
6. replacement or addition of fly ash;
7. use of blast-furnace-slag cement;
8. addition of silica fume (microsilica);
9. use of a high-range water reducing agent or higher than normal doses of conventional water-reduction agents;
10. special admixtures related to specific properties desired;
11. selection of coarse aggregates known for high abrasion resistance, for example, very dense aggregate;
12. precooling of concrete mix; and
13. curing practices.

FIGURE 11.12 Sylan Viaduct in French Alps combines high performance, high-strength concrete truss members with external tendons inside trussed extension.

As stated in Section 11.2.3, development, testing, and approval of special mixture designs takes time and therefore should be started as early as possible. Extensive treatment of this subject is given in Chapter 2 and in Nawy (1996a).

11.4.6 Lightweight Concrete

Prestressed lightweight concrete has a substantial history of very satisfactory performance in highly demanding structures, including bridge girders (Figure 11.13) and offshore structures. In several important cases, beneficial properties other than light weight have been the rationale for selection. These properties include:

- durability,
- insulation,
- fire resistance, and
- in conjunction with silica fume, enhanced fatigue resistance.

Structural lightweight concrete has often been selected to reduce the inertial mass of structures in seismic regions and to reduce draft of floating structures.

The performance of lightweight concrete is highly dependent on the specific aggregates selected, which in turn depend on the raw materials (principally clay), the manufacturing process, and the temperature and duration at which they are fired in the kiln. Unfortunately, because of the previous widespread use of low-quality, lightweight aggregates for less-demanding applications, such as fire proofing, and because of the wide range of properties covering lightweight aggregates most current codes specify increased allowances for creep and shrinkage. However, through careful selection of the highest quality lightweight aggregates and the use of some natural sand, creep and shrinkage can be kept within the same ranges as with natural stone aggregates.

Some properties of these high quality lightweight concretes are, however, less than those of hard-rock (conventional) concrete of the same strength. These include modulus of elasticity, shear strength, and tensile strength.

High-performance lightweight concrete has a unit weight of only 75–80% of conventional hard-rock concrete; however, compressive strengths of up to 62 MPa (9000 psi) can be obtained by the use of natural sand for the fine aggregate and the addition of microsilica.

Microsilica is especially beneficial to lightweight concrete because it chemically bonds with aggregate particles, which in turn gives better fatigue endurance under cyclic loads.

FIGURE 11.13   Prestressed lightweight concrete girder being prepared for shipment by rail. *Source: Construction of Prestressed Concrete Structures*, Gerwick, B.C., Jr. © 1996. John Wiley and Sons, Inc. Reprinted by permission of John Wiley and Sons, Inc.

11.4.7 Modified Density Concrete

By using natural sand to replace the lightweight fine aggregate in part or whole and using 50–60% of hard-rock coarse aggregate (by volume), a concrete having about 90% of the density of all-hard-rock concrete can be produced that has almost the same properties, including modulus, shear, and tensile strength as the comparable all-hard-rock concrete. This mixture has been used on a number of offshore structures where weight was critical for draft and stability.

11.4.8 Composite Construction

Composite steel-concrete construction is widely used for high-speed-railroad bridges in Germany, as well as for major bridges elsewhere. Composite construction has also been proposed for offshore structures where the membrane characteristic of the steel plate, and its two-dimensional reinforcement properties, can be combined with the shell action of the concrete to resist concentrated forces. Connectors used to ensure shear transfer have been welded studs, shear rings, and special shear ribs.

Prestressing has been suggested as a potentially synergistic way in which to utilize composite properties, since the steel is kept from local buckling if it is properly anchored to the concrete.

Composite construction using precast prestressed slabs and beams with structural concrete topping has been widely utilized for floor slabs of buildings and for decks of short-span bridges. Shear transfer may be accomplished by the roughened concrete surface alone in the case of light live loads or by means of reinforcing ties. Cleanliness of the surfaces and proper preparation, for example, saturated or surface damp, are important to ensure bond. The constructor must ensure that the specified degree of roughness of the lower slab or beam is attained. Steel ties are very effective.

Typically, the precast member is longitudinally pretensioned. Conventional reinforcing in the topping slab is used to distribute forces transversely and to provide live-load continuity, although in certain instances, especially bridge and wharf decks, the composite system has also been post-tensioned transversely. The designer of such a transversely post-tensioned structure will presumably have adequately considered the effects of differences in moduli and the inherent eccentricity of two orthogonal layouts of tendons. In areas where significant eccentricity is unavoidable, use of steel ties will prevent delamination.

11.4.9 Scaffolding and Falsework for Post-tensioned Cast-in-Place Construction

A great many building and bridge structures are constructed on scaffolding and falsework. After the cast-in-place concrete has gained strength, the structure is post-tensioned. This prestressing redistributes the dead loads, typically raising the span off of the central scaffolding and transferring it to the end supports.

In the typical case, this redistribution and induced camber are beneficial to the construction process, making it easier to remove the scaffolding. However, the constructor must ensure that any temporary supports at the reaction points are capable of taking the dead load imposed on them.

Prestressing also shortens the span. If one end is on neoprene bearing pads it will be free to move. The intermediate scaffolding will be distorted by the shortening but is usually able to accommodate this because of its height and flexibility. However, this should be verified in each case.

If the ends are fixed, then the act of prestressing pulls the supports toward each other. The stiffer support will take most of the force, and in turn, will counter the prestressing, reducing its efficacy. Owing to creep, the problem gets more severe as time goes on. This situation has unfortunately arisen on a substantial number of building projects where the effects were not adequately foreseen, resulting in spalling at the edges of bearing seats and, in at least one case, splitting a stiff column.

If a permanent connection must be made at both ends, and the supports do not have adequate flexibility to accommodate the shortening with only small opposing force, then it is best to temporarily place one end on a neoprene pad or sliding bearing, then fix it after the elastic shortening and early creep has taken place. Of course, long-term creep may still cause problems, but its magnitude will be less. Neoprene strips should be used at the edges of the bearing surface to allow minor rotation and minimize the tendency to spall.

11.4.10 Architectural Prestressed Concrete

Many applications of prestressed concrete have a criterion for architectural and a esthetic appearance. Prestressing can be used to help attain the criterion by preventing cracking. One- and two-way prestressing has been utilized for this purpose (Figure 11.14). Care must be taken that neither the ends of pretensioned strands nor the anchorages of post-tensioned tendons can rust and stain the surface. Pretensioned strands can be cut back by a small flame torch or by a welding rod and then plugged with latex cement mortar or a light colored epoxy mortar.

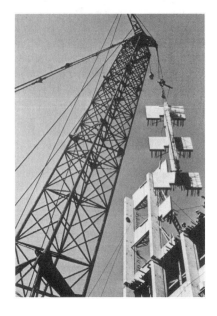

FIGURE 11.14 Two-way pretensioned concrete wall units serve both structural and architectural purposes in the construction of new university building.

Post-tensioned anchors should be recessed and later filled with concrete or mortar. In some cases, epoxy mortar is used. These plugs should be tied to the parent concrete by steel ties, since otherwise cracks may occur around the plugs and allow corrosion products to leach out. Alternatively, in some architectural applications, the architect has elected to emphasize the anchorage, in which case, heavy galvanizing or epoxy coating has been preapplied.

Some architectural panels incorporate ceramic tiles as facing elements with a backing of concrete. Prestressing locks the ceramics in place and prevents cracking, which otherwise might reflect through the face. Great care has to be taken in locating the tendons to offset the difference in modulus of the facing from that of the backing.

Some architectural treatments of concrete involve processes that are corrosive to prestressing steel. Acid washing or etching is especially dangerous as residue may remain in ducts and recesses.

Joints between segments require special attention if they are to be blended into the overall members. Any epoxy residue needs to be removed. Use of white cement as part of the mortar mix will help to prevent discoloration, since patches tend to be darker than the surrounding concrete.

Steam curing may adversely affect colored concrete. Condensation may drip onto colored concrete and leave unsightly marks.

Microsilica or anti-bleed admixtures can be used to prevent unsightly bleed holes. Fly ash or microsilica can be used to prevent efflorescence. These can also be used to reduce permeability and thus minimize the unsightly fungus growth that occurs in semitropical and humid environments.

11.5 Post-tensioning Technology

11.5.1 Principles

With post-tensioning, the concrete sections are cast first with all conventional passive reinforcement. Then after the concrete has gained sufficient strength, tendons are placed, usually through holes formed by ducts. These tendons are stressed so as to react against the concrete and precompress it. The concrete must be free to shorten under the precompression. The tendons are then anchored and corrosion protection, such as grout or grease, is installed. There are many variations on the above procedures, of which the most common and important will be described here.

11.5.2 Storage of Tendons and Anchorages

Tendons and anchorages are high-strength steel, accurately made to very close tolerances. As such they are subject to corrosion in storage, or even more insidious, to contamination that will lead to hidden long-term corrosion.

Tendons and anchorages should be stored in a covered area, fully protected from the weather, and raised off the ground. Some specifications require that the storage facility be dehumidified. Usually, heat is applied to keep the relative humidity well below 50%. In other countries, including the United States and Canada, open storage is acceptable, provided it is up off the ground and dry.

Among the many cases of corrosion of tendons due to inadequate storage are the following:

1. Tendons were stored in mud adjacent to a roadway on which salt was applied to prevent icing.
2. Tendons were stored on a beach and subject to immersion at high tide.

11.5.3 Installation of Ducts and Anchorage Bearing Plates

Ducts, whether of steel strip or plastic, must be carefully placed to true profile and rigidly held so as not to be displaced during concreting. Their location and alignment determines the position of the tendons that are later installed.

Ducts must be kept free from dents and flattening, as otherwise they will not permit the free installation of the tendons. Dents, bulges, and burred ends must not be allowed to occur during the subsequent placing of reinforcement, and care must be taken in the use of the vibrator when consolidating the concrete. Preferably, ducts should be supported on saddles of plastic or steel, rather than bearing directly on the reinforcing steel, which may dent the duct when the load of fresh concrete impinges on it.

It is particularly difficult to align ducts accurately across splices. A temporary internal mandrel may be used to ensure proper alignment across splices. Sleeves, overlapping at least two diameters on each end, should be sealed with heat-shrink tape. Screw fittings can be used on ducts comprised of pipe. While waterproof tape and epoxy have been used to seal overlapping sleeves, tests and experience have shown that these are not as reliable as the heat-shrink tape.

The duct cross section should be about twice the gross cross section of the tendons and should allow 6 mm (1/4 in) clear all around in order to permit proper encasement in grout.

The preferred anchorage for post-tensioning tendons is a recess pocket formed in the end of the concrete member. This is "boxed out" with forms attached to the end form. When the duct is placed, the trumpet and bearing plate are attached by screws and sealed to prevent mortar inflow. This rigid attachment serves to ensure that the bearing plate is truly normal to the design tendon axis.

Vent tubes and grout tubes must be properly affixed to the duct and taped securely to prevent rupture or leakage. Vents are installed at the high points of upward-curving tendons or at about 20 m intervals on near-horizontal ducts. A tube is also installed at each end just behind the bearing plate to act as an inlet when injecting grout and as a vent at the far end.

Drains are seldom used at the present time since water flushing of ducts is not normally employed. However, they should be used at the bottom of U ducts installed vertically, as for example, in the webs of girders (Figures 11.15 and 11.16). The drain tube may also function as a grout tube.

Ducts should preferably be delivered to the job site with plastic covers already fitted. If not, covers should be fitted during installation in order to keep foreign material, including rain, from entering the duct.

For the primary prestressing of major structures, such as bridge girders, provision of at least one extra longitudinal duct on each side will facilitate corrective action in case of excessive creep or blockage of a duct.

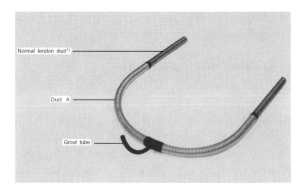

FIGURE 11.15 U-duct is specially fabricated to hold vertical tendons in webs of bridge girders. *Source:* Courtesy of VSL Corporation, Campbell, California.

11.5.4 Installing and Stressing Tendons

Groups of steel strands and bundles of steel wires can be preassembled with a nose piece attached. A pull-wire is fed through the duct; one means to feed it through the duct is to blow it by compressed air acting on a rubber ball. A line is attached to the ball which can then be used to pull in the pull wire. The nose piece is often a Chinese finger grip of spirally wound wire. The bundle is then pulled through the duct.

However, it is usually more efficient to push strands through one at a time, using a pushing feeder of mechanically driven rolls. The end of the strand is fused together by a torch so that the individual wires won't separate. Water-soluble oil is brushed onto the strand as it enters the duct.

Once all strands have been pushed through the duct and through holes in the anchorages, they are pulled to a nominal tension by a single-strand jack and temporarily wedged. This ensures that all strands are of the same length. See Figure 11.17 for a graphic representations of the sequence.

Then a multistrand jack is fitted so as to extend the tendons to their designed stressing level, which is usually about 72% of the ultimate strength. The wedges are hydraulically pushed home (Figure 11.18). When there is considerable curvature in the tendon, the tendon is stressed from both ends. Cycling it by pulling from first one jack and then the other reduces frictional loss.

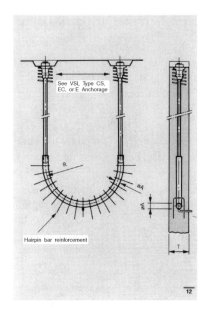

FIGURE 11.16 Duct, tendons, and anchorages for vertical prestress of deep webs of long-span bridges. *Source:* Courtesy of VSL Corporation, Campbell, California.

Final stress in the tendon should match theoretical elongation, using calculations based on the modulus of elasticity as furnished by the manufacturer. The modulus varies with the lay of the strand. Allowable tolerance is usually set at 5%. Where this value is exceeded, the cause should be determined. It may be excessive friction, in which case, adding water-soluble oil and cycling helps, or the cause may be in the assumed modulus. This data is recorded for each tendon. After the jack is removed, the ends should now be protected from rain and spray by a covering of polyethylene or similar material.

11.5.5 Corrosion Protection of Tendons

The standard method of providing corrosion protection of tendons is by injection of cement grout. If properly done, this encapsulates the strands and penetrates between the wires of the strands.

In the case of ducts no larger than two times the gross area of the tendon, cement and water are the principal components of the grout. Sand or other fines are incorporated only in the rare cases of very large ducts.

Water reducing admixtures are necessary for corrosion protection. Fly ash may be used to replace up to 20% of the cement. Thixotropic admixtures are very valuable in reducing bleed which leads to voids at high spots in the profile. Thixotropic grouts require a positive-displacement pump for injection. Special grouting admixtures have proven very successful for promoting full filling of ducts and preventing bleed pockets.

Expansive admixtures, such as aluminum powder have been used, but have sometimes created problems with excessive pressure and with the control of expansion. A thixotropic antibleed admixture is believed to be more effective.

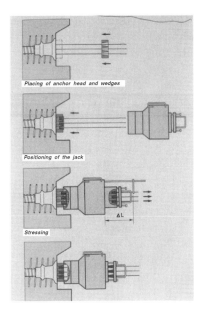

FIGURE 11.17 Sequence for stressing post-tensioning tendon. *Source:* Courtesy of VSL Corporation, Campbell, California.

The grout should be injected from one end. The first grout ejected from a vent is dark owing to the water-soluble oil. Grout should be wasted until the ejected material has the same color and consistency as the grout to be injected.

When the grout ejected from the first vent is deemed satisfactory, that vent is closed. When the end vent is discharging satisfactory grout, that vent is closed. This forces the grout through the tendon anchorages and seals the strands at these critical locations.

Vertical ducts present a special problem in that the strands act as wicks, promoting bleed. Use of a thixotropic admixture helps but is not always 100% effective. One solution is to have an extra hole in the top anchorage plate, through which a tube leads to a small tank above. This tank or receptacle is filled during grouting; the grout then can feed down to fill the duct as the column of grout subsides.

Silica fume, as an admixture, also reduces bleed. However, the combination of silica fume and thixotropic or antiwashout admixture is frequently too viscous for effective injection.

FIGURE 11.18 The stressing operation. *Source:* Courtesy of VSL Corporation, Campbell, California.

The anchorage pockets then need to be completed. Preferably, small sized bars have been placed that can be bent down into the pocket. The anchorage pockets should preferably be filled with a "fine" concrete, that is, one made with small size coarse aggregate, say 8–10 mm (pea gravel), and a mix rich in cement. Silica fume may be added to minimize bleed.

Frequently, the surface of the anchor and the pocket is coated with bonding epoxy. In other cases, latex-modified concrete is used to improve bond.

Shrinkage should be prevented by using the "window box" technique, in which the concrete is filled to an elevation higher than the pocket, so that if bleed occurs, the grout will feed down into the space. Any excess concrete can be easily chipped off after a day or so. The joint around the pocket may be painted with an epoxy that has high capillarity and can be sucked into any shrinkage crack.

In a number of cases, especially nuclear reactor containments, the tendons and the anchorages are encapsulated in grease instead of cement grout. The grease is injected in much the same manner as grout and periodically examined and, where necessary, additional grease is injected.

This technology is now giving way to an advanced system in which each strand of the tendon is sheathed in polyethylene and grease is injected, so that the strand is free to move within the sheath. This protection is very positive. A cluster of such sheathed strands is fed into a duct. They are then stressed and the duct is grouted. Thus a multiple protection system is achieved.

The critical zone in such a system is at the anchorage, where the strands must be bared in order for the wedges to grip them. Caps with grease or grout fittings are provided that screw onto the anchorage so that they too may be injected after stressing.

11.5.6 External Tendons

The term external tendons is used to describe post-tensioning tendons that are not directly incorporated into ducts in the concrete. Usually, these tendons are placed inside the box of a trapezoidal box girder, but they may also be outside. External tendons are unbonded. They are anchored at blisters or bolsters, usually near the ends of the member, and may be deviated at intermediate points along the profile.

The tendons are typically encased in a heavy polyethylene sheath. At deviation points, a steel sleeve may be used to prevent the tendon from cutting into the polyethylene. The polyethylene sheath may be continued through the sted sleeve. The steel sleeve is preformed to the design radius. It should have belled ends to prevent stress concentrations (Figure 11.19).

There are, of course, large radial forces at the deviators that must be resisted by proper reinforcing details. Deviators must be adequately anchored to the webs and/or flanges.

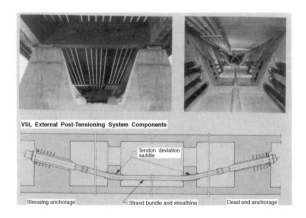

FIGURE 11.19   External tendons. *Source:* Courtesy of VSL Corporation, Campbell, California.

One frequent problem encountered with external tendons is the failure to leave enough space in which to place a large multistrand jack on the prescribed angle. This must be considered in the working drawings.

11.6 Pretensioning Technology

11.6.1 General

Pretensioning denotes the process by which tensioned high-strength steel wires or strands are incorporated in a concrete segment. The process is relatively simple in concept, economical, and technically efficient. However, it requires a major plant facility, able to temporarily restrain the forces in the tensioned tendons until the concrete cast around them has gained sufficient strength to effectively bond the tendons and transfer their force to the concrete.

The process lends itself to prefabrication, that is, precasting. Most precast pretensioned concrete is produced in permanent plants. However, for large projects, especially in remote areas, contractors have found it practicable to set up job-site plants. Although the process is adaptable to a wide range of shapes and configurations, economy favors the use of standardized members fabricated on a repetitive process with a fixed cycle.

The principal standardized members produced are piles, bridge girders, building floor slabs, roof slabs and wall panels, and poles. Nonstandardized but repetitive segments that lend themselves to pretensioning include stadium seats.

11.6.2 Description of Process

In its simplest form, the pretensioning facility consists of a casting slab or bed on which the segments will be fabricated, reaction frames or stands at the end to temporarily resist the tendon forces, hydraulic jacks for tensioning, tarpaulins or hoods to cover the segments during curing, and lifting equipment to remove the completed segments for storage and shipment.

11.6.3 Tendon Installation and Stressing

The tendons most commonly employed are seven-wire strands with a nominal diameter of 15 mm (0.6 in), having an ultimate tensile strength of 1900 MPa (270,000 psi) and a 0.2% offset yield strength, which is 80% of ultimate. These strands are spaced apart to develop bond for stress transfer, and arranged in appropriate patterns to develop the design compressive stress in the concrete cross section.

Multiwedge strand anchors are used to temporarily grip the ends of each strand so that the jacks can impart the desired force and elongation. Both single-strand and multistrand jacks are used. After the strands are properly stretched, they are anchored to the stands at each end of the bed (Figures 11.20 and 11.21).

When strands are stretched with a multiple jack, very large forces are stored. To prevent serious accidents, nuts should be progressively run up on the jack's extension arms, so as to limit travel in the event of loss of hydraulic pressure.

FIGURE 11.20 Stretching strands down pretensioning bed.

FIGURE 11.21   Fully extended jacks impart prestress to pretensioning strands. *Source: Construction of Prestressed Concrete Structures*, Gerwick, B.C., Jr. © 1996. John Wiley and Sons, Inc. Reprinted by permission of John Wiley and Sons, Inc.

The achievement of the proper prestress can be determined both by gauge pressure readings of calibrated jacks and by the elongation. Most specifications require that these deviate not more than 5%. Variations are caused by difference in moduli of the strands or by frictional restraint at the ends and intermediate supports.

It is important to keep foreign materials such as oil and grease off of the strands. Welding should never be permitted on or near the tendons since it will destroy the special properties that have been induced by cold drawing of the wires.

Unstressed reinforcing steel can usually be most easily placed after the strands are tensioned. The entire reinforcing assembly can then be held in proper position by tying it to the strands or by using plastic chairs and dobe block or by hanging it from overhead spreader bars with plastic wires.

11.6.4 Forms

The forms for concrete segments are typically made of steel. Since the forms are standardized members, they minimize the effort and time required to set them in place and subsequently strip them.

For some configurations, the forms may be fixed and tapered so that the segment can be removed without moving the forms. For other shapes, mechanical means may be used to close and open the forms.

The end gates of the forms are especially critical, since they must allow the strands, and perhaps some reinforcing bars, to pass through without leakage. Hence they may have to made up in small pieces. Rubber gaskets can be used to seal around the strands and at the sides.

For some products, such as piles, it is important that the head be normal or "square" to the longitudinal axis. This can best be achieved by making the end gates up in pairs, spread apart, and rigidly held in a frame.

11.6.5 Concreting

The concrete mix is designed to gain strength rapidly so that the tendons may be released, transferring their force into the concrete; the segment is then removed to storage. The strength required for release is controlled by two factors: adequate bond strength to limit the transfer length at the ends of the member, and adequate strength to minimize the creep under sustained stress. Typical release strengths range from 25 MPa (3500 psi) to 35 MPa (5000 psi).

FIGURE 11.22   Concreting and finishing operations.

Concreting follows normal practices of placement (Figure 11.22). External vibrators are often used; they may be moved progressively to brackets on the steel form, or permanently mounted on them. They produce excellent consolidation and drive water and air bubbles from the outer exterior 100 to 150 mm (4–6 in) of concrete thickness. For members thicker than about 200 mm (8 in), internal vibration is also required, even with so-called flowing concrete. Otherwise, in densely reinforced and congested zones, such as the end blocks of girders and the heads and tips of piles, honeycomb and rock pockets may occur.

11.6.6 Curing

To gain strength rapidly, accelerated curing is usually applied, which provides heat and moisture. Most commonly, such curing will consist of low-pressure steam. Ideally, adequate strength will be gained in 8 to 12 h, enabling a daily cycle of production.

Forms must be free when the tendons are released, since the concrete will shorten. If the forms are locked to the concrete, for example, by change in cross section, they also will be forced to shorten.

A variety of means are used to prevent damage to the forms. Safe removal can be effected by lifting, sliding, or hinging. If tapered and smooth, of constant cross-section, and sufficiently rugged, then the concrete member may partially slide in the form. Multiple short forms may be designed to slide on the bed.

11.6.7 Release of Prestress

After curing is complete, the tendons are released from the stands, transferring force into the concrete. The concrete shortens under compressive stress. In order to behave as a prestressed concrete member, the shortening must not be restrained. In practice, when heavy segments are cast, the friction between the segment and the soffit of the forms may restrict the shortening. In many cases, this problem may be overcome by initially lifting one end of the segment before the other, thus allowing the member to shorten and become prestressed before the full load is realized. Any discontinuity in the cross section, whether inherent in the design or accidental, such as a fin due to leakage from the forms, may prevent shortening.

Since at this early stage the concrete segment is subject to drying shrinkage and thermal shrinkage, cracks may occur. Also, steel forms expand under steam or heat curing more rapidly than concrete, forcing a crack. To prevent cracks at changes in cross section, soft neoprene or rubber gaskets may be installed to accommodate the dimensional changes.

Insulation of the surfaces after curing will postpone drying and thermal shrinkage strains until the concrete has more tensile strength.

Release of the pretensioning into the segments is preferably accomplished by use of a hydraulic jack. This requires reinstalling the jack, exerting a slight additional pull to loosen the wedges, then slowly backing down on the force, allowing the strands to slacken. Even then the friction of heavy concrete segments on the bed or in the forms may prevent their movement, and hence prevent release of prestress into the individual segments.

The release then is accomplished by burning the strands. Preferably, this should be done in a balanced pattern, by using low heat (yellow flame), then cutting with the blue flame. The intention is to reduce the stress gradually. A sudden shock release increases the development length over which the prestress force is transferred to the concrete.

Over this transfer length, particularly near the ends of the segment, the prestress compression is not uniform over the concrete cross section. Tension develops between strands and especially between groups of strands. Orthogonal reinforcement, in the form of short bars or mesh, is often needed at the ends in order to prevent splitting cracks.

While this supplemental reinforcement should be part of the design, in practice, it is often left up to the contractor and hence needs to be considered in preparing shop drawings for fabrication.

11.6.8 Cycle

The typical cycle in precast pretensioned fabrication is as follows:

4 am	Terminate steam curing cycle.
5 am	Remove test cylinders from under tarpaulins for testing; replace tarpaulins.
6 am	Test cylinders. Verify that required release strength has been obtained.
6:30 am	Remove tarpaulins.
7 am	Detension tendons. Cut strands between segments.
8 am	Lift products out to storage, store on timber sleepers at correct points to minimize deflection due to creep.
8:30 am	Clean forms.
9 am	Stretch new strand tendons the length of the bed.
10 am	Equalize lengths with single-strand jack.
10:30 am	Stress strands.
11 am	Place reinforcing steel. Place end stops or gates.
1 pm	Place concrete, consolidate, and finish.
2 pm	Cover progressively to prevent premature drying out.
6 pm	Turn on low pressure steam, raising temperature at 1°C every 2 min, to 60°C (140°F). Hold temperature constant until 4 am.

Obviously, times and operations have to be adjusted to fit individual products and reinforcing steel patterns.

11.6.9 Tendon Profile

The efficient design of many segments, such as slabs, beams, and girders, requires that the profile of the tendons follow a path other than a straight line. This means that the strands must be deflected. Deflection has been accomplished in a number of ingenious ways. One such way is described in the following.

Strands are initially stressed to a precalculated reduced stress. Then they are pulled down or up at specific points, forcing the strands into a series of chords approximating the design path (Figure 11.23). This of course raises the stress in the strands to near the desired value. Then the jack brings the stress to the design level. By this method, frictional losses and variations of stresses between segments are minimized.

Another method is to pull the tendons over a series of rollers, set at selected deviation points. Unfortunately, the fractional losses with this method are cumulative and result in unequal tendon stresses in several segments.

FIGURE 11.23   Deflecting pretensioned strands with special hold-down devices.

The hardware at the deviation points can be likened to the arrow in a stretched bow. Thus they are points of danger to personnel during placement of reinforcement and subsequent concreting until the concrete has at least reached final set. These points should be clearly marked with red paint on the tops of the forms and all personnel kept clear. This is especially important for the vibrator operator.

The release of a member made with deflected strands requires special consideration. If the deflected strands are released first, the pull-down devices hold the segment tightly against the casting bed, preventing shortening and hence preventing effective transfer of prestress. Conversely, if the deviation points are released first, the resulting upward force in an unstressed member could break it. In practice, the first method is used, that is, first releasing the longitudinal strands, followed by release of the pull-downs at the deviation points. In some advanced installations, provisions have been incorporated to allow limited longitudinal movement at the deviation points.

11.7 Prestressed Concrete Buildings

11.7.1 General

Buildings represent perhaps the largest overall use of prestressed concrete and certainly the most diversified use of precast pretensioned concrete segments. Conversely, standardized modular configurations have been extensively employed for roof slabs, floor panels, beams, and wall panels. Cast-in-place post-tensioned concrete has been widely used for the floor slabs of lift-slab construction for heavy beams, and girders. Post-tensioning permits the full integration of slabs, beams, and girders. Similar monolithic construction is attained with precast pretensioned construction by jointing and cast-in-place infill and topping.

Since the construction process differs significantly between the two processes, they will be treated separately in the following sections. However, the two systems may be combined in any particular building.

11.7.2 Precast Pretensioned Concrete

The most widely employed precast concrete segment is the double tee (Figure 11.24). It is used for roof slabs and wall panels. In combination with cast-in-place topping, double tees are used for floor slabs. Widths are modular: in the United States, this means either 4 or 8 ft (1.22 m or 2.44 m).

FIGURE 11.24   Double-tee precast, pretensioned concrete slab. *Source: Construction of Prestressed Concrete Structures,* Gerwick, B.C., Jr. © 1996. John Wiley and Sons, Inc. Reprinted by permission of John Wiley and Sons, Inc.

11.7.2.1 Manufacture

The precast segments are cast in steel forms with tapered legs. The taper plus the slight flexibility of the steel forms permit the segments to be lifted out or "stripped" without adjustment to the forms. Inserts can be placed in the bottom of the legs in order to reduce their depth for shorter spans.

Straight strands are employed for short spans, while deflected strands are employed for longer and more heavily loaded spans. Since the typical deflecting forces are low, much less than with bridge girders, the system of pushing up at the ends while holding down at the middle or third points is commonly employed.

Shear reinforcement may be in the form of bent reinforcing bars or welded wire mesh. A few widely spaced strands may be run in the top slab to prevent shrinkage cracking. Transverse bars and/or diagonal bars will provide cross-slab reinforcement and can anchor embedments for connection between flanges of adjacent slabs. Where topping is to be placed for composite behavior, the top surface is roughened with a transverse broom finish.

When double-tee segments are employed for walkways and small bridges, the legs are usually wider and deeper to accommodate the necessary strands and shear reinforcement. For longer spans, single tees with deeper and thicker stems are used (Figure 11.25). Another form of slab, the hollow-

FIGURE 11.25   Single-tee slab for longer spans.

FIGURE 11.26 Hollow-core floor slabs.

core slab (Figure 11.26), is made by several proprietary processes, resulting in a flat top and bottom with multiple longitudinal holes.

Both lightweight and conventional concrete is used in manufacture. For lightweight concrete, the replacement of part or all of the fines with natural sand will reduce problems of creep and shrinkage.

11.7.2.2 Erection

Precast segments are typically transported by truck and erected by a large truck crane with extended outriggers. Proper procedures and safety practices must be followed in order to ensure safe and efficient operations. The Prestressed Concrete Institute (1995) has published a manual, *Erection Safety for Precast and Prestressed Concrete*, that includes sections on the following important aspects:

- preplanning the erection,
- site conditions,
- cranes,
- equipment,
- rigging,
- tools,
- unloading,
- lifting,
- fall protection, and
- setting/connecting/releasing.

Preplanning is proving especially valuable to contractors, who use it to determine access for cranes and trucks, the swing of the crane and segment in three-dimensional space, and the sequence of erection and means for accurate positioning. Temporary bracing and staying may be required. Marking of the exact seats of segments beforehand will save valuable time in final positioning.

Connecting or jointing details are quite critical. Tolerances must be considered, both relative to adjacent segments and cumulative. Details of welding and bolting have proved to be very important, especially when the structures are subjected to dynamic loads such as earthquake, hurricane, and tornado.

Large truck cranes with outriggers are the most commonly used means of erection (Figure 11.27). It is essential that the outriggers be properly supported by timber pads or the equivalent to prevent settlement during the swing of the load. Care must be taken to ensure that neither the boom nor the

FIGURE 11.27   Erecting a tapered double-tee floor slab in circular parking garage.

rigging swing into already erected elements. Adequate room must be available for the counterweight during swing.

Tower cranes are also employed. The weight of units determines their capacity and reach.

Because of the large number of relatively light segments that are typically erected in a building, attention must be given to the detailing and construction of the lifting inserts. These must be compatible with the hooks and slings employed by the rigger. They must be properly anchored into the member to resist inclined as well as vertical forces.

Local pull out can produce a serious accident. Adequate factors of safety must be provided to take care of the dynamic amplification of lifting and minor accidental lateral impact.

When a cast-in-place concrete topping slab is placed on top of precast-pretensioned slabs, the latter often require temporary shoring in order to sustain the dead load of the fresh concrete without excessive deflection (Figure 11.28). This shoring is typically made tight and the grade adjusted either by the use of wedges or screw jacks. These must be set in such a manner that they do not become loose or tip while the top concrete is being placed.

The shores react against a lower slab, which may have been placed only a week before, in which case the topping may not have adequate strength. Serious progressive collapses have occurred when the shoring has not been carried far enough or was not strong enough to carry the accumulated loads.

FIGURE 11.28   All precast, pretensioned building nears completion. Marin County Civic Center, California.

Other serious collapses have occurred when the shoring was removed too early. The in-place strength of concrete should be ascertained by companion cylinders carried alongside or through nondestructive testing (Schmidt hammer). Particular care must be taken during winter when the low temperatures delay the strength gain.

Creep must be considered by both the designer and erector, since lengths and camber will change with time. For that reason, it is desirable to let the precast segments mature a reasonable length of time before erection, for example, one to three months. Rigid welded connections at both ends are to be avoided. Bearing slabs must be of adequate length and be adequately reinforced. Neoprene bearing pads will allow minor rotation and shortening.

Shear at the ends of slabs becomes increasingly critical with time owing to the tensile stresses induced by creep and the increase in the transfer length of pretensioned strands with time. The erector needs to consider this when preparing his shop drawings.

Camber is affected by a number of factors: eccentricity of prestressing, the modulus of elasticity of the concrete, creep, differential temperature from top side to bottom, and differential moisture. The cumulative effect of these factors and of various tolerances may create difficulties, especially in thin, highly stressed slabs such as those most often employed in roofs. These difficulties may in turn cause problems in waterproofing. In extreme cases, loss of camber may lead to ponding of rainwater, which in turn increases downward deflection.

The camber of adjacent segments may not fully match. Adjacent sections can be pulled to match with minimal force, provided the differences in camber and the stiffness of the members are not too great. Where excessive force is required, typically more than 100 kg (220 lb), the design engineer should be consulted.

11.7.3 Cast-in-Place Post-tensioned Buildings

Both bonded and unbonded tendons are employed for cast-in-place post-tensioned buildings. Unbonded tendons are extensively used in flat slabs, both lift-slabs and cast-in-place slabs. The advent of polyethylene-sheathed and greased strands has made this process reliable and durable. The details of this type of construction are presented in Chapters 10 and 12.

Conventional post-tensioning in ducts, with subsequent encasement of the strands in grout, is typically employed for deep beams and girders. In parking garages, this method is often employed for the ramp (Figure 11.29).

For lift-slab construction and wherever precast columns are used, post-tensioning through the columns is desirable to ensure shear friction transfer from the slab, so as to offset subsequent shrinkage.

FIGURE 11.29 Cast-in-place concrete floor slab will be post-tensioned in both directions. *Source:* Courtesy of VSL Corporation, Campbell, California.

A key detail in post-tensioned building construction is the protection of anchorages to prevent moisture in leakage and to safeguard against fire. Both conventional and polyethylene-sheathed tendons have bare strands through anchorage zones. Thus this is the most vulnerable area for potential corrosion. Recessed anchorages, filled with concrete and tied to the structural slab, are the most reliable.

11.8 Prestressed Concrete Bridges

11.8.1 General

Prestressed concrete has been quite successfully used in the field of bridges, ranging from low to medium-span precast pretensioned bridges to post-tensioned girder spans of 250 m length and, beyond that, to cable-stayed concrete bridges of 600 m; the latter are a derivative form of post-tensioning. Prestressed concrete bridges are now used worldwide in remote developing countries as well as in the urban centers of highly developed countries. Prestressing has proven very versatile in its application and can be used to solve many complex problems of curvature and skewness. It has become the standard by which alternative materials and systems are measured.

This success, however, has not come without some problems. Some of these are common to all reinforced concrete, for example, the corrosion of steel in bridge decks on which salts are applied to prevent icing. Many of the problems peculiar to prestressing are due to the widespread adoption of this new technology, sometimes without proper attention to details of design and workmanship. Though in most cases, these problems were resolved on the site, they have proven costly and cause delays to the constructor. Fortunately, most problems are now in the past, in that we understand their causes and have adequate solutions available.

11.8.2 Precast Pretensioned Bridge Girders

The precast pretensioned I-beam girder is widely used for trestle-type bridges, for viaducts, and for overcrossings (Figure 11.30). This element has proven to be adaptable to mass production, with resultant economies and reductions in cost.

The girders are manufactured on long-line pretensioning beds, with deflected strands as described in Section 11.6. They are then transported, principally by truck and dolly, so that the support is within 3 ft (1 m) of each end. Long and deep-webbed girders are hog-rodded to prevent buckling during transport.

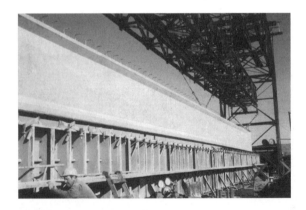

FIGURE 11.30 Precast, pretensioned I-girder for high-level bridge across Napa River, California.

FIGURE 11.31   Erecting precast girder using trussed spreader beam. *Source: Construction of Prestressed Concrete Structures,* Gerwick, B.C., Jr. © 1996. John Wiley and Sons, Inc. Reprinted by permission of John Wiley and Sons, Inc.

Girders must be positively restrained against tipping. At this stage, there is no additional dead load from the slab acting, and hence the girder stresses are usually within safe limits only when they are near vertical. If tipped too far, they may fail explosively owing to overcompression in the top flange. Blocking and chaining are used to prevent tipping. (Note that wire rope stretches under sustained and repeated loads, and hence chains or structural members are the only safe means for providing restraint.)

The girders are usually erected by crane. When one crane is used, slings leading at 45° to the horizontal or more (60° is preferable) are employed. This of course limits the height to which the girders can be lifted. These slings develop a horizontal thrust owing to their angle, which may overload the compression flange of the girder, especially because of the low buckling resistance of the girder in the Y-Y (transverse) direction.

To overcome this, a spreader beam with vertical slings from the girder to the beam is often used (Figure 11.31). Then the bending strength of the beam and its column strength may both be utilized. A pipe strut is usually used for the spreader beam, but trusses are employed for very long girders.

Lifting and setting of precast girders is a demanding erection process in that there are dynamic forces due to inertia and swinging as well as to long and high lifts. Most erection is carried out by crawler and truck cranes; the latter use their outriggers.

The picking inserts in the girders must be properly anchored deep into a compressive zone and must be capable of taking the force at an angle. Slings must be properly designed to meet the amplified load due to angle. Both must be adequate for the dynamics of inertia force and acceleration. The latter is especially critical when floating equipment is used for lifting.

As described in Chapter 6, Section 6.2, the rules of crane and rigging safety must be carefully followed.

When bridge girders are erected, they are initially vulnerable to tipping. Tipping may be caused by girders being set on a superelevation, by wind, or by being contacted by a line from a second girder while it is being lifted. For this reason, long and deep I-girders should be braced as soon as they are set.

Cast-in-place decks are usually designed to act compositely with the precast girders. To ensure full transfer of horizontal and vertical shear across the joint, the top flange of the precast girder is roughened and multiple stirrups are employed to tie the girder and deck together.

Transverse diaphragms are cast between the end blocks of the girders and sometimes at intermediate locations, although the latter has proven to be less prevalent than was necessary in the early years of this new technology. Experience and tests have shown that the deck slab provides adequate transfer in most cases.

Girders are usually set on neoprene bearing pads. To facilitate accurate and rapid setting, it is helpful to mark a line on the pad for the planned ends of the girder, so that no further measurements in either direction need to be made.

The cantilever suspended span concept has been used to extend the span of precast prestressed girders. The two side spans are continued over the main piers and are haunched and prestressed for negative moment. Either pretensioning or post-tensioning may be used to provide the required moment capacity. Then the suspended span, designed for simple span behavior, is installed.

At the interim stage, when the side spans are set but do not have the relieving load of the suspended span, they may be overstressed by the prestressing. This can be countered by addition of mild steel reinforcement at the bottom of the haunch or by the use of temporary counter prestress.

In the typical bridge, the deck acts as the compressive flange, while the web transmits the shear to the bottom flange. The bottom flange has to have enough prestress to resist both superimposed dead load and design live loads. The top flange has to be large enough to sustain the dead-load compression. In longer girders, this frequently results in temporary excessive precompression in the bottom flange. Enlarging both flanges results in a "bulb tee," which has the desired stress characteristics at all stages and provides the necessary room for the prestressing tendons.

Other cross-sectional shapes have been employed for shorter span bridges. These include the double tee and the triple tee or "M" shape. The typical double tee used in buildings has a rather narrow web, which may not provide adequate cover for durability or adequate strength for shear. Hence special forms with a wider web may be used.

Precast, prestressed slabs are also employed for short spans and have the advantage of minimum depth and a flat underside. For slightly longer spans, hollow cores may be formed, so as to reduce weight. These slabs may also have a cast-in-place composite topping.

To temporarily support the load of fresh concrete, the precast member may require shoring at its midspan or third points.

11.8.3 Post-tensioned Girders, Cast-in-Place on Falsework

This is a widespread application of prestressed concrete. Falsework shoring is set up, adequately supported on the ground, or on the new foundations or on piling, and capable of resisting the dead load of the concrete with minimal deflection. Provision should be made to offset the calculated deflection, whether it be elastic, bending of falsework beams, or elasto-plastic deformations of the ground support. It is important that the shores be adequately braced for both transverse and longitudinal movement during casting of the concrete (Figure 11.32).

For continuous structures, the concrete is usually cast in progression from the center of the span to each end, so as to reach its deflected profile prior to casting the concrete over the piers, thus preventing cracking in the negative moment zone.

As the cast-in-place sections are prepared, the ducts are placed to the required profile and alignment. At vertical construction joints, the sections are spliced. Since splices and construction joints are a principal source of the problems that occurred in the early years of this technology, attention must be paid to details, as described in Section 11.3.4. This includes accurate prolongation of the profile across the joint, as by the use of mandrels, the application of heat shrink tape to the ends of splice sleeves, and proper preparation of the construction joint surfaces (Figure 11.33).

In the negative-moment zone, over the piers, the design often calls for several ducts, with their tendons, to be lined up, one above the other in groups of two, three, or more. Space between these ducts may be very limited, so that the potential exists for an upper duct to "pull through" into one below. To prevent this, small reinforcing bar "hair-pins" may be placed transversely, between each pair of ducts. In extreme cases, the lower tendon must be stressed first and grouted, then the second tendon is stressed and grouted, and so on.

FIGURE 11.32 Structural steel falsework supports high-level cast-in-place post-tensioned bridge. Ma Wan Viaduct, Hong Kong.

FIGURE 11.33 Draping ducts for cast-in-place post-tensioned bridge. I-205 north approach, Columbia River Bridge between Washington and Oregon. *Source: Construction of Prestressed Concrete Structures*, Gerwick, B.C., Jr. © 1996. John Wiley and Sons, Inc. Reprinted by permission of John Wiley and Sons, Inc.

It generally proves economical and practicable to space the tendon so that the shear strength (in double shear) of the concrete and the dowel effect of the transverse bar are adequate to resist the radial component of the tendon force.

As cautioned in Section 11.4.9, the design of supporting falsework must consider not only the stage when the concrete is cast, and hence imposes its dead load on the supports, but also the stage when the prestressing force has transferred the dead load, distributing it with concentrated forces under the zones of concave curvature downward, while relieving the load on the zones of convex curvature.

The cumulative effect of shortening between joints of multispan continuous girders must be considered. Intermediate supports must be able to accept the curvature imposed by this shortening, or better, be temporarily freed from it by temporary devices that allow the girders to move the short increment of prestress deformation. Stressing from each end will minimize the actual movements over the intermediate piers.

Some very important bridges of moderate span have been cast-in-place and post-tensioned with temporary support on a steel deck truss. When the span is post-tensioned, it rises up off the truss so that it is now independently supported directly on the piers. The truss can now be slid forward past the pier to the adjacent pier and the process repeated for the adjacent span. This results in a series of simple spans that can subsequently be made continuous for live load by post-tensioning for negative moment over the supports.

To facilitate the truss's sliding from one span to the next, the pier must have an appropriate configuration, such as a division into two shafts, between which the truss slides, or the provision of steps on the outside of the piers on which individual trusses may be slid forward.

Transverse post-tensioning is increasingly employed, especially for box girders with cantilevered deck flanges. Since space is limited, ducts must be kept small. Rectangular ducts encasing four side-by-side strands are frequently employed. Tendons may have a dead-end anchorage in the concrete at one end and a stressing anchorage in a pocket at the other end.

To provide additional corrosion protection, plastic ducts may be used. In other cases, plastic-sheathed and greased strands are employed, with special care taken to protect the anchorages.

The primary longitudinal post-tensioning tendons are located in the upper flange in the negative moment region near the piers. This means that the transverse tendons should go over them, above the webs, and under them, near the transverse centerline. This is a three-dimensional problem of location, and some eccentricity is unavoidable.

11.8.4 Post-tensioned Precast Segmental Bridges

The concept here is that short segments of the full cross section of the bridge are cast in a prefabrication site or casting yard. These segments are then transported to the site and erected on falsework or on a truss, as described in the previous section. The segments are then jointed and post-tensioned.

Because the segments are cast as relatively small, discrete units, it is possible to obtain close tolerances for reinforcing and duct placement, as well as for finished concrete dimensions. The ability to obtain high-strength and high-performance concrete is enhanced. At the same time, segments are kept to a reasonable size for transport and erection.

After erection, segments are carefully aligned to the correct profile. Thus dead-load deflection is taken care of prior to stressing.

Joints may be constructed by a variety of means. Cast-in-place joints, typically 500 mm to 1500 mm in width can be constructed, in which the reinforcing steel is made continuous by lap splices, welded splices, or couplers. Ducts can be readily spliced by sleeves and sealed by heat-shrink tape. Thin concreted joints, typically 75 mm in thickness, were used in the past but have not proven to be successful because of difficulty with duct continuity.

Vertical joints usually require shear keys. While in the past, trapezoidal keys were widely used, experience has shown them to be subject to diagonal cracking at the corners. Installing a short diagonal bar across the potential crack is one solution. More recently, curved shear keys, corrugated on a large pitch, have been used for both cast-in-place and precast concrete segments.

Dry joints have been used for many bridges in which the precast concrete deck girder segments have been match cast. Match casting consists of constructing each successive segment with its trailing edge cast against the leading edge of the preceding segment (Figures 11.34 and 11.35). Thus a perfect fit is ensured. Ducts are extended from the first segment to the second by use of a mandrel.

Some problems with fit up have occurred with match casting when steam curing has been employed. The segments tend to warp in unequal fashion with the heat. To prevent this, it is best to steam cure the first and second segment together, prior to separation. A bond breaker must be used on the common joint surface. After separation, a light sand blasting, wire brushing, or water-jet blasting can be employed to clean the mating surface.

With dry joints, a thin o-ring seal can be placed in a recess around each duct to seal against grout leakage and cross over of grout between ducts.

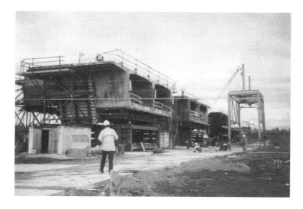

FIGURE 11.34 Precast concrete segments are match-cast in elevated cells to facilitate their alignment and movement.

FIGURE 11.35 Precast match-cast segment for viaduct in Riyadh, Saudi Arabia. Note shear keys and anchor blocks (blisters).

Dry or match cast joints have been greatly improved by the use of epoxy glue on the mating surfaces. Typically, during erection, the second segment of the joint is raised. It is positioned by the use of two mating dowels, either cast in the segment or temporarily affixed to the top. After verifying proper fit, the segments are moved apart and epoxy glue is applied with a gloved hand. Since epoxy is incompatible with free moisture, the surface must be protected from rain, etc., during this process.

The second segment is again pulled into contact with the first and a temporary precompression is applied by stressing bars at the top, bottom, and both sides. The precompression on the concrete surface of the joint should be 0.3 MPa (50 psi). The epoxy sets under the applied pressure. Many such matings have been successfully made using only epoxy glue to seal the ducts at the joint. However, use of a thin o-ring gasket is believed to be a conservative and justified step to prevent inadvertent blockages in later grouting.

Assembly of complete and multiple spans has been successfully carried out both by erection on falsework and by lifting up a completed span with one or two crane barges. In some instances, the full span, less than two pier-head segments, has been assembled, jointed, and prestressed on a barge, then lifted up from the preset pier heads.

Prestressing such an assembly on a barge requires the consideration of the extreme loads imposed on the barge deck at the girder ends owing to the post-tensioning. Additional posting or support of the deck may be required.

11.8.5 Cast-in-Place Cantilever Segmental Bridge Construction

This process has been successfully employed on spans up to 200 m and more (Figure 11.36). With cable-stayed concrete segmental bridges that employ the external tendon principle, much longer spans, 500 m and longer, have been attained.

The concept is to cast two segments, one on each side of a pier. The casting is followed by prestressing the segments over the pier to resist the negative moment in cantilever. Then two more segments can be cast and post-tensioned, and so on. Each stage must be carefully analyzed to ensure adequate prestressing for the dead load of that segment plus the forms and the weight of the next segment during its casting.

Since it is not feasible to cast two segments simultaneously, one on each side of the pier, the pier shaft must be adequately reinforced to withstand the temporary bending that is induced.

These cantilevered decks are typically extended to near midspan. After a waiting period, to allow as much of the creep as practicable to take place, the two extended arms are locked together and a closure pour is made.

At the initial stage of this process, a pier-cap segment is first constructed which is temporarily locked by vertical post-tensioned bars to the pier shaft. Then a small skid derrick is erected on the pier-cap segment. The derrick lifts up a prefabricated form for the first cantilevered segment. Reinforcing steel and ducts are placed, and the segment concreted. The concrete mix is designed to attain adequate strength for prestressing within 1 to 3 d. This segment is then post-tensioned to the pier cap segment, and the skid derrick is moved out onto it and temporarily bolted to the deck. This leaves room on the pier cap to install a second small skid derrick headed in the other direction. It now raises the forms, and constructs and stresses its cantilevered segment. Now the process can proceed in parallel on each arm of the deck. A typical cycle requires 4 to 7 d for each pair of segments.

In many designs for continuous bridges, a hinge is required near the quarter point. In order to permit the cantilever segmental process to continue past the hinge, it is temporarily locked with removable post-tensioning bars.

In other designs, a hinge is placed at midspan. To prevent undesirable movements as heavy live loads cross the hinge, complex devices that transfer shear but no moment are installed.

FIGURE 11.36 Cast-in-place cantilever segmental construction. I-205, Columbia River Bridge.

The cantilever segmental process is very demanding for the contractor; accurate and thorough quality control are essential. Prestressing ducts must be accurately placed to close tolerances and rigidly held in position so as not to be displaced during subsequent concreting and vibrating. At the ends of each segment, the ducts are held in exact location by the end bulkhead of the form. To ensure that the profile of a duct is continuous across the joint, without a small, sharp bend, a pipe mandrel should be inserted, which can be extended into the duct in the next segment when it is formed. Pipe mandrels also serve the secondary purpose of ensuring that the duct remains clear and open when adjacent tendons are grouted.

Each time a form is extended from the preceding segment, it is purposely set sufficiently high at its leading end to counter the deflection under the weight of the fresh concrete. Concrete placement should proceed from the outer end back toward the previously completed segment.

Segments are typically very congested by conventional reinforcement in three directions and longitudinal and transverse post-tensioning tendons. Near the piers, vertical post-tensioning tendons are often installed as well. Thus congestion makes concreting difficult. Fortunately, the use of HRWRA (superplasticizers) makes obtaining a very workable mix feasible while still achieving high early strength and high long-term strength.

Because of the congestion in the deck near the pier, the horizontal section through the middle of the top deck may have a greatly diminished concrete section; the area is largely occupied by closely spaced longitudinal and transverse ducts. Thus horizontal shear transfer across this section is diminished and may result in laminar cracking due to the unavoidable eccentricities in the prestressing centroid in the two directions. This problem has been successfully prevented in a number of cases by provision of small-diameter "hairpin" stirrups in the deck at a nominal spacing of about 0.6 m (24 in) in each direction.

While the above should be considered for design, it should be considered by also the contractor since laminar cracking is not readily apparent at early ages and repairs are very difficult, usually requiring stitch bolting.

Creep and shrinkage, as well as elastic deformations, affect the profile of the completed bridge. Thus it is necessary to take accurate readings at the same time each day, preferably early in the morning before the sun's heat distorts the structure. This enables corrections to be calculated by the engineer and implemented on the next segment.

In lieu of the temporary support of forms by a small skid derrick or overhead frame, a gantry may be used, extending over at least one and one-half spans. This gantry is moved forward after the two cantilever arms are completed on one pier. When its leading edge reaches the next pier, temporary supports are placed, enabling the gantry to move forward one-half span farther. The rear end of the gantry, now over the first pier, is locked to it by post-tensioning bars. The gantry can now support the forms and the fresh concrete of the cantilever segments as they are constructed.

Earlier, the installation of near-vertical tendons in the webs was mentioned. These tendons resist the high shear forces near the piers. Frequently, these are installed in U-shaped ducts, with the two anchors in the deck slab. During construction, these ducts must be covered to prevent rain and curing water from entering as well as to protect them from being clogged by debris. Even soda bottles have been found wedged in the U at duct bottoms! Grouting of these vertical ducts must employ the special procedures discussed in Section 11.5.4, to ensure complete filling of the duct.

Continuity tendons are installed after the closure has been completed. They are designed to provide positive moment capacity over the midportion of the span. Typically, their anchorages are in bolsters (blisters) on the webs or bottom flange and the tendons are installed in the bottom flange.

Anchorage bolsters should be staggered in balanced pairs, with adequate longitudinal spacing and reinforcing to distribute the tensile strains behind them. Proper reinforcing details are also required to resist the pull-out forces of the sharp curvature at these anchorages.

11.8.6 Precast Cantilevered Segmental Construction

In this type of construction, the segments are prefabricated in a casting yard, using the match-cast process. Care is taken to ensure the correct positioning of the longitudinal ducts and their extension into the next segment through the use of pipe mandrels (Figure 11.37).

The erection process in the field may proceed in a manner similar to that described for cast-in-place segments. In this case, the skid derricks or frames must lift a segment that is positioned directly underneath and raise it into position. Then epoxy glue is applied to the faces of the segments. The tendons are then installed and stressed.

The process is inherently very rapid, in that a pair of segments can be completed each day. In some cases, using multiple shifts, two pairs of segments have been completed per day.

The gantry scheme of erection is especially well adapted for this concept. The precast segments can be transported along the completed deck, oriented at right angles to their final position. They can be lifted by the rear end of the gantry and run forward to their final location, where they are turned 90° for erection.

Usually, a longer gantry is used, extending over 2 1/2 or even 3 spans. This enables the work to procede farther ahead, while previous spans are being post-tensioned and grouted and closures are being effected.

Provided the gantry has adequate capacity, several segments may be suspended from it; thus minimizing the amount of post-tensioning required. If the gantry is of adequate length and strength, so that it can support the entire double cantilever midspan to midspan, then the required post-tensioning will be reduced to that required for the permanent structure.

Because the segments are precast and therefore the concrete has attained greater maturity, the creep and shrinkage are reduced. Nevertheless, the profile and alignment have to be checked daily. Minor corrections can be made as needed by the insertion of wire mesh shims. Alternatively, corrections may be accomplished by additional post-tensioning. A few more strands may be inserted in an existing duct, compact strands may be used, or an additional tendon may be placed in a spare duct.

One significant disadvantage of precast segmental construction is the lack of mild steel reinforcement across the joints, which would minimize crack width under ultimate load conditions as well as improve stress distribution. This is especially needed at the ends of transversely cantilevered deck flanges, where shear lag may reduce the effective prestress.

The latter problem can be resolved by including a high-strength bar in the end of each flange that is stressed at each segment. Couplers can be used to extend the bar from one segment to the next.

FIGURE 11.37 Erecting precast match-cast segment in cantilever construction.

Perhaps the best means of providing the desired monolithic behavior is to include additional ducts, spaced out over the contact surfaces where no primary tendons cross, and to install unstressed strands and grout them, so as to provide a nominal steel area across the joint. Another method is to provide mild steel dowels in slots or holes across each joint that are subsequently grouted.

At the present time the majority of bridges have not used such unstressed steel, but instead have taken the alternative route of using extra primary prestress to ensure that no tensile stresses ever arise during the design loading conditions. Thus precast segmental construction often requires some additional prestress as compared with cast-in-place segmental construction.

11.8.7 Incremental Launching

The incremental launching method is suitable for straight bridges (on a tangent) and for bridges of a constant circular alignment and superelevation. Unfortunately, it is unsuitable in its present state of development for constructing spirals. The bridge may be on a straight-line profile or a constant vertical curve.

With this process, all work is carried out from one end, in a job-site "factory." A segment, typically 6 m long, is cast to its full cross section. Then a second segment is cast immediately behind and post-tensioned to the first. Both segments are seated on bearing plates of stainless steel that slide on teflon with a friction coefficient of 0.03 to 0.05. Large hydraulic jacks are installed below the first segments and arranged so that they can pull the two segments forward.

As each succeeding segment is cast, it is post-tensioned to the preceding assemblage and they are pulled forward as a group. Obviously, since the leading segments overhang the abutment, they must be counterbalanced by the segments still at the casting site.

A trussed steel "nose" extension is attached to the leading segment. When the nose reaches the next pier, it is guided onto teflon and stainless-steel bearings. The nose is slightly tapered so that it can engage the sliding bearings. Despite the elastic downward deflection, the "nose" raises the bridge to grade as it progressively slides past the second pier.

Side guides should be provided at each pier bearing to prevent the bridge from creeping laterally. This is especially important when the bridge is superelevated or on a curved alignment.

Because all work is carried out at one location, it is feasible to provide an enclosure so that work can be carried out regardless of weather conditions.

Stage post-tensioning is critical to this method. During launching, since the girders are alternating between positive and negative moment, the post-tensioning must be more or less concentric. For the final service condition, after the bridge is in final position, the centroid of prestress must be raised over the piers and deflected downward over the central portion. To physically accomplish this, several methods have been employed.

1. Using external tendons located inside the box of a box girder the tendons are jacked to their required profile, then secured to the webs with stirrups and concreted in place. This requires an accurate computation of the increase in the stress due to deflection. The process is conceptually similar to the deflection of pretensioned strands.
2. Additional prestressing tendons with exaggerated profile are added to move the centroid of the total prestress to the desired location. These additional tendons can be internal tendons, that is, in preplaced ducts, or external tendons inside the central opening of the box girder.

11.8.8 Lift-In and Float-In Erection

Lift-in and float-in erection refers to processes in which the span is preassembled and post-tensioned, either on a barge or on shore, then transported to the specific site and positioned beneath the piers.

FIGURE 11.38 Full-width, full-length bridge span is prefabricated on land. Great Belt Western Bridge, Denmark.

For the lift-in process, heavy lift jacks can be placed on the pier caps so as to raise the complete span. When the span is to be erected one girder at a time, the lifts may be made by skid derricks on top or by floating cranes below. On overland projects, the bridge span may be constructed at ground level and then raised by lift jacks to the pier tops. Then a cast-in-place closure is constructed. This closure should incorporate a shear key or keys. The lifted-in span is connected to the pier caps is by post-tensioning through the pier cap.

Float-in spans are similar, except that in this case, a falling tide or deballasting may be used to lower the span onto the pier caps. This concept is most suitable to low-level bridges and trestles. Float-in may also be used with a large crane on a catamaran-type barge (Figure 11.38). The crane picks the girder; the barge then moves to straddle the pier. The crane then lowers the girder onto its bearing. This process has been used for very large spans over 100 m in length and up to 8000 t in weight (Figures 11.39, 11.40, and 11.41).

FIGURE 11.39 Erecting one of the 124 spans of the Great Belt Western Bridge. Each is 100 m long and weighs 7000 t.

11.8.9 External Tendons

This relatively new technology was introduced in Section 11.5.6. External tendons are tendons located outside the concrete cross section of the bridge girder, although they may be inside the box of a box girder.

External tendons may constitute the entire primary post-tensioning or a portion of the total, the remainder being internal, grouted tendons. The external tendons are anchored at or near the ends in typical recessed pockets. The tendons themselves are multistrand cables encased in a polyethylene sheath. At several locations along the span, either at third or quarter points, the tendons are deviated to the correct profile by structural concrete deviators, that is, partial diaphragms anchored to the webs. Since the forces on these deviators are concentrated, the deviators must be heavily reinforced themselves as must their ties to the concrete webs.

The bending at the deviator must be over a curve of substantial radius in order to minimize the normal bearing stress on the strands that acts concurrently with the axial elongation changes under varying live load. A steel sleeve with belled ends is used, which is prebent to the required radius. The polyethylene duct is run

FIGURE 11.40   Construction of one of the five bridges of the King Fahd Causeway. Saudi Arabia–Bahrain.

through this sleeve. The strands are grouted in the duct. The duct is usually left free in the steel sleeve.

One of the advantages of using external tendons in bridges is the ability to remove and replace one or two tendons at a time. This is feasible only if the anchorages are accessible for removal and subsequent replacement and restressing. Anchorages may have screwed-on, grease-filled caps. Although such replacement is frequently a design criterion for bridges that will be treated with salts in the winter to prevent icing, and therefore exposed to a very corrosive environment, the protection afforded by polyethylene sleeves plus grout is believed to be essentially permanent.

The use of external tendons has also been applied to concrete truss spans, in which the truss members can be used as the deviators. In this case, since the tendons are exposed to daylight, UV degradation of the polyethylene may occur over time.

FIGURE 11.41   Prestressed concrete girders being erected at Prince Edward Island Bridge, Eastern Canada.

11.9 Prestressed Concrete Piling

11.9.1 General

Prestressed concrete piling are used on a very large scale as bearing piles for building foundations, as bridge supstructures, and as support for wharves and quays and tanks and towers. They have been used for offshore terminals and for fender piles and sheet piles for quays and bridge piers (Figure 11.42).

Piles up to 600 mm (24 in) in cross section and, occasionally, to 800 mm (36 in) are typically pretensioned, using standard pretensioning practices as described in Section 11.6. Further specifics for manufacture of piles will be given later in this section.

Cylinder piles, 800 to 1650 mm (36–66 in) in diameter have been produced both by pretensioning and by post-tensioning, the latter again using standard post-tensioning processes.

Fender piles and sheet piles are usually pretensioned. Sheet piles may incorporate sheathed-and-greased tendons arranged in a vertical profile to suit the final service conditions. They are stressed from the top only after the sheet pile is at final grade.

11.9.2 Durability of Piles

Piling for most land foundations, such as for the support buildings, are generally fully embedded in a benign environment of soil below the water table with little oxygen. Hence corrosion of steel has not been a problem. Exceptions may arise when piles are driven through garbage dumps or highly corrosive soils.

For piling that extends above the permanent water table, durability must be considered. In desert or dry arid conditions near the sea, chlorides pose a threat to the steel: Corrosion has occurred in serious proportions in the Middle East for piles in both water and land foundations.

Preventing corrosion of prestressing steels and conventional reinforcing in marine environments requires a dense, highly impermeable mix, at least 350 kg/m³ (600 lbs/yd³) of cement containing at least 5–10% C3A, and a low water/cementitious material ratio, preferably less than 0.43. Fly ash may be used to replace 15–20% of the cement.

In marine installations subject to freezing, the problems of freeze-thaw attack may be especially severe. High tide and waves saturate the concrete, while low air temperature cause the concrete to freeze during low tide. Some sections of the pile may see two cycles of freeze-thaw a day throughout the depth of winter.

Substantial air entrainment is required, with the proper spacing factor and specific surface, as shown by petrographic analysis of hardened specimens or cores. Coatings, other than those that "breathe" water vapor, can be counterproductive, since the moisture migrates to the cold face and is trapped behind the impermeable coating, where it then freezes, popping the cover off. So high-strength, high-impermeability, minimum water absorption by aggregates, low W/CM ratios, and adequate air entrainment are all required for this environment.

FIGURE 11.42 Prestressed concrete cylinder pile for pier of Napa River Bridge, California. *Source: Construction of Prestressed Concrete Structures*, Gerwick, B.C., Jr. © 1996. John Wiley and Sons, Inc. Reprinted by permission of John Wiley and Sons, Inc.

11.9.3 Manufacture

The manufacture of pretensioned piling is a mass-production process, typically involving a small number of standardized cross sections. Hence economy of production is of great importance. Forms may be double or quadruple so that two to four lines can be manufactured simultaneously.

Preferably, the forms are designed so as to require minimal effort for setting and release. Flexible steel side panels may be held in position during concreting by intermittent straps across the top, these being easily removed for lift out of the hardened pile. Or the side forms may be stiff enough to hold the fresh concrete without significant deflection, yet be tapered so that the member may be lifted out directly. A draft of 1 on 50 is adequate for steel forms of moderate stiffness.

For octagonal and cylinder piles, the two upper sloping surfaces may be formed by portable panels that clamp to the fixed form and use a neoprene gasket between (Figure 11.45). Alternatively, the upper sloping forms may be hinged.

The internal voids of cylinder piles have been formed in many ways. All the methods used face the problem of keeping the void form accurately centered so that it won't float up when the concrete is vibrated. One method that has proven capable of consistently producing high-quality cylinder piles is using an internal steel form that is collapsible. A form segment is typically half the length of a pile, so that after lift out of the completed pile, each half-length segment may be retracted.

In a typical manufacturing sequence, the internal void forms are set up in the lower half of the fixed outer form and are then blocked up. Coiled spreading spirals of appropriate amount are then set at the ends of the piles (Figure 11.43). The strands are pulled down the bed, through the coiled spiral but outside the inner void form. The strands are now properly spaced around the circle and tensioned with a single strand jack. They may be fully tensioned to the desired value (70 to 75% of ultimate) or only partially tensioned, with the remaining tension applied by a large jack. Now the spirals are distributed to their correct spacing and tied to the strand as needed.

At appropriate stages, end forms (end gates) are set in place: these are usually fabricated in several segments. Rubber or plastic stops are used to seal against mortar leakage where the strands penetrate the end gates (Figure 11.46).

The upper side forms are set in place and the pile is ready for concrete. In a properly designed pile, the spirals will be very closely spaced at the head and tip. Additional longitudinal bars or tubes may also be inserted at the pile head. The head and tip are zones where the highest concrete quality is essential, so vibration must be thorough. To permit this, the spirals may be bundled and a pencil vibrator used to penetrate between them.

After the piles in a line have been cast and cured, the upper forms are laid back, the strand tension is released, and the strands are cut between piles. This is an especially critical time for

FIGURE 11.43 Spreading spirals along tensioned strands of prestressed concrete piling.

FIGURE 11.44   Placing inserts for lifting piles in pretensioned concrete piles.

FIGURE 11.45   Closing up forms during manufacture of pretensioned cylinder piles.

FIGURE 11.46   View of end-gates separating piles during manufacture. Note heavy spiral to resist driving stresses as well as plastic hinging during earthquakes.

cylinder piles, as well as for all large piles, in that the upper surface of the pile is now exposed to both rapid drying shrinkage and thermal shrinkage as the top cools off while the warm bottom half is still protected by the forms. Application of membrane curing compounds and blanketing the pile will prevent the longitudinal cracks that often occur, especially in winter and on windy days.

Cylinder piles are also manufactured by sliding a long internal mandrel along the bed, so as to form the void. The mandrel may be coated so as to slide with minimal friction. The rate of progress is matched by the stiffening of the concrete so as to prevent slumping of the top as the mandrel's tail end passes. Internal heat has been used to accelerate the rate of stiffening.

During the main part of the mandrel slide or 'slip', the cement paste from the concrete will lubricate the mandrel, but at the start it is necessary to coat the mandrel with fluid paste or at least to wet it. Particular attention has to be given to the ends of the piles to prevent the drag of the mandrel from causing cracks or spalls.

The bottom strands must be supported in such a way as to maintain their cover. Slumping of the top and displacement of the strands will cause eccentricity in the cross section and even delamination, which may then lead to breakage or spalling during driving, as well as reduce the piles' performance under lateral loads.

Cylinder piles are also manufactured by assembly of short concrete "pipe" segments. Jointing can be by epoxy or by a 200 mm wide cast-in-place joint. The pile is then post-tensioned through preformed holes and the tendons are grouted. This process has been used for cylinder piles from 1 m to 3.5 m in diameter.

Piles are lifted from the forms by one of several means. Except in the splash zone of marine piles, lifting inserts or bundled loops of strand may be used (Figure 11.44). They must be adequately embedded in the pile. Spacing of these inserts may be designed to minimize negative bending moments.

Since these negative moments peak at the lifting points and are augmented by friction in the forms for large and heavy piles, short lengths of mild steel bars may be incorporated in the top of the pile.

Where inserts are not permitted, a thin band may be preplaced in the forms, enabling the pile to be raised a short distance by a jacking frame so that it can be progressively blocked about 50 mm (2 in) above the soffit. Then a lifting wire rope "choker" can be inserted so that slings from an overhead gantry or crane may lift the pile clear. Piles should be blocked at the same picking points while they are in storage, so as not to develop a sweep due to creep.

While in storage, the projecting ends of the strands, especially those at the pile head, should be burnt back into the concrete (Figure 11.47). A daub of epoxy mortar may be applied if desired.

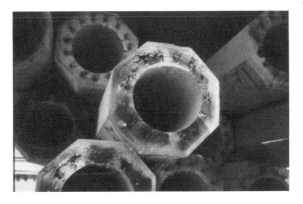

FIGURE 11.47   Pretensioned hollow-core piles in storage.

FIGURE 11.48   Loading heavy prestressed cylinder piles on barge by rolling.

During load out and transport, piles should be supported so that they do not develop a net tension in the concrete (Figure 11.48). For truck transport, the dynamic amplification must be considered, especially for the overhanging ends.

11.9.4 Pile Installation

Although one of the primary benefits of prestressing for piles is that they can be driven in a wide range of soil conditions and achieve the required penetrations, even though this requires prolonged hard driving, there are a number of very specific requirements to prevent pile damage or breakage.

Lofting the pile, that is, lifting it from the horizontal to the vertical, typically requires two or more lifting points (Figure 11.49). The slings or lines to these points will typically be at an angle and this angle will vary during the lift. This means that the vertical components of the force in the lines will be different. Thus the situation differs from that in the manufacturing yard where all lifting forces are normal to the pile.

FIGURE 11.49   Lifting prestressed concrete pile into position for driving.

Fortunately, it turns out that in most cases the maximum moments in the pile occur during initial lofting, so that the calculation of angles and resultant moments and stresses is relatively simple. A dynamic amplification factor should be applied to the pile's dead load. For very long piles, the moments and stresses should be checked through the several stages of lofting from horizontal to vertical.

Most cases of damage to prestressed concrete piles occur during the initial and early phases of driving. This is a consequence of the fact that the tension capacity of the pile is the sum of the precompression due to prestress and the tensile strength of the concrete; the latter is only 10% of the compressive strength.

High tensile stresses during driving are due to a rebound wave of the pile hammer blow from the pile tip in soft driving, when the tip has little or no resistance. Consider the following three typical cases.

The pile is driven by a diesel hammer (Figure 11.50). Initially, the pile tip is in soft clay or mud. To start the diesel hammer, it has to be raised to almost full stroke, typically about 2 1/2 m (8 ft).

Thereafter, in soft driving, the diesel hammer automatically delivers soft blows, but that first blow at full stroke is the one that cracks the pile. The tensile rebound wave is greater than the pile's tension capacity.

This cracking may or may not be noticed, so driving continues. Eventually, the pile encounters hard driving, and the hammer delivers maximum compressive stresses across the cracks. This leads to local crushing at the crack and eventually to fatigue of the strands. Finally, the pile plunges, obviously broken somewhere below ground. Typically, the hard driving is blamed, but the initial damage was done on the first blow.

A second case arises when the pile is driven through compacted soil that overlies softer materials such as mud. The pile is subjected to hard driving when in the compacted soils. Then it suddenly breaks through and plunges 3 m or more. The single hammer blow, with little tip resistance, is what causes the cracking.

A somewhat similar situation arises when the pile is required to penetrate a hard stratum. The

FIGURE 11.50 Driving long prestressed concrete piling for major naval facility in Puget Sound, Washington.

contractor employs jetting to reduce the end bearing. Prolonged jetting erodes a hole. The pile drives through with relative ease owing to the low tip resistance. In this case, the tension is exacerbated by the fact that the pile is gripped by skin friction along its sides while there is essentially no resistance at the tip, so that the compressive wave from the hammer blow is literally driving the end off.

In all these cases, the actual breakage may not occur until there is subsequent sustained hard driving.

Prevention of damage due to tensile rebound stresses starts with a cushion block placed on top of the concrete pile. This block is contained within the driving head or helmet on which the pile hammer ram delivers its blow. The material and thickness of this pile cushion block is selected so as to attenuate the peak compressive force, which lasts only a few milliseconds during each blow. When properly selected, this cushion block can actually aid penetration by lengthening the period of application of compressive force while reducing its the peak.

The best material, after thousands of trials, has proven to be softwood, such as pine or spruce, laminated in multiple layers. To hold this softwood block together, plywood sheets may be nailed top and bottom and inserted in between every third or fourth lamination. The required thickness varies with the pile, the soil, and the hammer. Thicknesses of 150 to 350 mm (6 to 14 in) are typical.

The cushion block compresses during driving. Except when it breaks into pieces and falls out, or catches fire and burns up, it is not necessary to replace the block during the driving of any one pile. Exceptions occur for piles in highly stratified ground, where soft soils lie below hard soils.

A new cushion block should be used for each pile. Even if the cushion block appears undamaged, its attenuating compressibility has been greatly reduced.

For the typical case of a diesel hammer driving a pile whose tip is in soft mud, the pile may be driven down to moderate tip resistance by using the diesel hammer as a drop hammer, that is, raising the ram half distance and dropping it so as to prevent the damage resulting from a high starting blow. For the second case, predrilling through the fill is indicated. In both cases, the pile may be pulled down through the soft soil by a line rigged to the hoist engine.

The third case is more complex, especially where penetration of the intermediate hard stratum proves difficult and requires extensive jetting. Steps that may be used are as follows:

  1. Increase prestress; however, this is costly and piles may have already been cast.

2. Limit the jetting at the tip, but increase the jetting higher up along the sides in order to reduce skin friction.
3. Install a new, thick cushion block when driving through the hard stratum and before redriving the pile.
4. Prebore by jet or augur to break up the hard stratum.

Hundreds of thousands of prestressed concrete piles have been driven successfully without breakage, provided they had the required design and a suitable cushion block was used.

The design prestress should ensure that if a crack does occur for any reason, the steel strand will not be stressed beyond yield. Cracking will occur when the residual prestress in the concrete, after losses that have occurred to that date, plus the effective tensile strength of the concrete exceed the yield strength of the strand. This requires both adequate prestress and an adequate area of steel.

Values of 5 MPa (725 psi) have been successfully used for hundreds of thousands of prestressed concrete piles driven in "normal" soils, where moderate to hard resistances were encountered throughout the driving cycle.

A value of 7 MPa (1000 psi) has been used for piles subject to potentially severe tensile rebound stresses. However, the calculated value is extremely sensitive to the assumed tensile strength of the concrete, which is progressively degraded by repeated hammer blows in soft soil.

Prestress as high as 8 MPa (1200 psi) may be employed for sensitive conditions and piles with high importance, for instance, large-diameter cylinder piles.

To prevent the pile head from spalling, additional spiral confinement should be placed close to the head. This can be a circular hoop placed just beneath the chamfer at the corners of the head, perhaps 50 mm from the head, and of adequate diameter to extend to the sides, since cover is not normally a requirement at this location.

The tip of the pile should be square, not pointed or wedge shaped since these shapes tend to wedge the pile out of vertical and, in the case of batter piles, out of proper inclination. Extra spiral, similar to that for the pile head, will prevent local spalling when riprap or boulders may be encountered.

11.10 Tanks and Other Circular Structures

The first application of prestressing technology was circular tanks, for which circumferential wrapping with high strength wire under tension was used to create precompression in the walls of the tank. Walls were cast in place using jump forms or slip forms.

Today, such walls are often made of precast concrete staves, set vertically and joined by grout infill or, with wider openings, concrete infill. The wires or strands are then wound under tension. These are then encased by mortar, usually applied pneumatically by a wet shotcrete process (Figure 11.51).

The placing and tensioning of the wires is almost always performed by a specialist contractor using a proprietary process. Both black and galvanized wires are used. Therefore in this section, only those elements of work performed by the general contractor will be discussed.

The circular wall is usually set on neoprene pads to permit inward deformation under prestressing and hence the development of a state of prestress in the concrete at the base.

In seismic regions, special shear keys and/or restrainer strands connect the base slab and the walls. The details must be executed with great care, not only to ensure safety under earthquakes but also to ensure proper performance and leak tightness under normal service conditions.

When precast staves are used, they are erected to true position and verticality and temporarily supported by "tilt-up" braces. By marking the location of each stave or panel beforehand, minor errors in position will be minimized and prevented from accumulating.

Tanks may also be prestressed by internal post-tensioning. If the joints between panels are relatively wide, 200 mm or more, they may be treated like similar joints between bridge segments, with overlapping reinforcing and sleeving of ducts so that they may be internally post-tensioned.

FIGURE 11.51   Prestressed concrete digester tanks under construction. Los Angeles, California.

Anchorages are in blisters, spaced at 120°, so that each tendon covers 240° and alternate tendons overlap. Blisters must be adequately tied to the wall by stirrups in order to resist the radial stresses. Tendons are usually stressed from both ends in order to offset the friction due to curvature.

The joints between panels are filled with high-strength grout or concrete. This grouting is very critical since it must fill the complete joint with a high-strength material that will be able to withstand 70% or so of the ultimate strength of the wall panels, applied within a short period after grouting.

The encasement with wet shotcrete is then applied, first by washing the wall with fresh water to remove any contaminants, such as wire-drawing lubricant or water-soluble oil, then by applying a flash coat of neat cement grout, and finally by placing several coats of shotcrete. Where the tanks are to be backfilled, that is buried tanks, a coating of epoxy-asphalt is also applied.

The long-term performance of shotcrete is determined by its impermeability and is dependent to a large degree on the expertise of the nozzle man, who must direct his spray at a slight angle so that rebound is not entrapped.

A critical location in tank construction lies at the top of the wall, where any small crack, even a microcrack, between wall concrete and shotcrete can result in progressive corrosion due to moisture infiltration and freeze-thaw attack. This location should be covered by monolithic concrete or epoxy asphalt, as specified or approved by the design engineer.

The walls of tanks that will contain sewage or organic materials, hot chemicals, etc., must be free from cracks such as those due to drying shrinkage or thermal strains. Coating the interior surface with a suitable polyurethane material, durable to the contained fluids and able to span minor cracks, should be considered.

Where a circular structure such as a penstock is in contact with the soil, or backfilled, epoxy-asphalt coatings may provide protection against contamination and penetration of salts from the soil.

In freeze-thaw or deep-freeze environments, a special problem arises from the fact that the concrete is saturated and the freeze front is in the middle of the wall. This can lead to delamination and the spalling of the exterior concrete. An internal coating or membrane appears to be essential.

Containments for nuclear power reactors and for high-temperature gas reactors are usually thick-walled structures with either internal post-tensioning or exterior wrapping. The construction of these structures is similar to that of tanks, but the walls are much thicker and, of higher strength concrete, the reinforcing is much denser and more complex, and the post-tensioning ducts and anchorages are very thickly congested. Since thermal strains in the thick concrete walls need to be minimized, the fresh concrete mix should be precooled by ice or liquid nitrogen, and walls should be protected or insulated after the forms are stripped.

11.11 Prestressed Concrete Pavements

Prestressed concrete pavements have been used for major airport runways, taxiways, heavy industrial floors, container terminal yards, etc., especially where it is desirable to minimize irregularities and provide a truly level surface during service.

Post-tensioning is usually employed and is centered in the concrete cross section. Since the pavement is subject to water collection, major changes in temperature, etc., provision of a crack-free structure is important. Cracks prior to stressing must be avoided by such means as precooling the mix, placing concrete at night, and covering the concrete for several days with ponding, wet burlap, or polyethylene, etc.

To permit shortening under prestress, the slab is usually placed on a 50 mm thick bed of sand and covered by a polyethylene sheet.

Splices in post-tensioning ducts and couplers in bar tendons are made at the longitudinal joints. These construction joints are an obvious location of leakage and subsequent corrosion and so must be made with great care.

The edges of slabs tend to warp upward owing to thermal strains. Many designs incorporate an edge beam so that the weight of the concrete offsets the thermal strains and any design eccentricities in the centroid of prestressing.

11.12 Maintenance, Repair, and Strengthening of Existing Prestressed Concrete Structures

Although prestressed concrete structures, properly designed and constructed, have proven highly durable in a wide range of environments, there have been numerous cases where corrosion has seriously damaged both the conventional reinforcing steel and the prestressing tendons. The most prevalent cases are the following:

- decks of prestressed concrete bridges, to which deicing salts have been applied during winter;
- decks of parking structures where tires have carried salts from the adjoining streets;
- piling and the underside of wharf decks of coastal marine structures in tidal and splash zones; and
- unbonded tendons, wrapped in paper and bitumastic, in an area near the seacoast where chlorides are present in the fog.

In the above microenvironments, chloride contamination is the principal cause of corrosion. Chloride ions penetrate the concrete to the reinforcing and prestressing steel and depassivate it. With the presence of water and the permeation of oxygen, electrochemical corrosion proceeds.

Carbonation from CO_2 in the atmosphere can cause similar corrosion attacks but penetrates more slowly and is more easily prevented. Prevention of these attacks must be primarily focused on initial design and construction, where steps can be implemented to make the concrete more impermeable, to give adequate cover over the steel, and to limit cracking.

The lives of existing prestressed concrete structures that have not been significantly damaged may be prolonged by a rigorous maintenance program.

11.12.1 Maintenance Program

- Wash down concrete surfaces on which salt has been deposited by spray, accident, or intention.
- Treat the surfaces with silane to render them relatively impermeable to water that contains chlorides.
- Seal static cracks with epoxy injection or coatings.

- Seal active cracks with flexible membranes, such as polyurethane, which have crack-spanning ability.
- Where feasible, reduce the relative humidity in enclosures to below 50%. This is applicable to such structures as seawater pump rooms.
- Apply penetrating corrosion inhibitors. These are a relatively new development that may work in special cases, particularly where the concrete can be dried so that water in the pores does not impede penetration.
- Where applicable, flood the structure with water, even seawater, to keep the concrete fully saturated. The oxygen content of seawater is only a few percent of the oxygen content in the atmosphere and oxygen does not penetrate saturated concrete very rapidly. This applies to prestressed concrete penstocks, conduits, sewers and piping, especially during the period between completion of construction and initiation of service.
- Repair and seal laminar cracks by stitch bolting and epoxy injection.
- Where freeze-thaw attack is eroding the cover of concrete over the reinforcement, provide insulation and coating or covering.
- Install a cathodic protection system, connected to both the reinforcing steel and the prestressing steel. This step is normally instituted after serious corrosion of the reinforcement or prestressing has developed.

11.12.2 Repairs

In many cases, serious damage to a prestressed concrete structure will have only penetrated to the conventional mild steel, causing delamination, spalling, and cracking. Thus conventional repairs can be instituted. These include removal of the damaged concrete, cleaning of existing rebar, replacement of badly corroded rebar, and patching with new concrete or mortar.

The problem that often arises is that the new salt-free concrete becomes cathodic to the adjoining chloride-contaminated concrete, leading to accelerated corrosion at the periphery of the patch. Various techniques have been developed to prevent or minimize this process.

This section primarily addresses the repair of those structures in which the prestressing tendons have become corroded.

Prior to undertaking repairs, the existing structure should be adequately shored. The shores must extend down to firm and adequate support. In cases where the damage is due to corrosion, after the structural situation is rectified, cathodic protection may be applied in order to prevent further corrosion. In the case of exposed beams, new external tendons can be installed on each side. They are tied to the existing beam by drilled and grouted dowels and then encased in new concrete.

Bolting steel plates on the bottom and sides of existing beams or gluing on carbon sheets provides external strengthening sufficient to offset the loss of prestress. The structure then becomes essentially a reinforced concrete structure instead of a prestressed concrete structure. However, if the steel plates can be tensioned, these become in effect, external tendons and impart precompression in the concrete.

For temporary repairs to badly corroded structures, structural steel beams may be placed to give additional support in critical areas. Wedging or shimming may be used to ensure contact and predeflect the steel. Cathodic protection and/or coatings may be applied for corrosion protection of the steel. Overlays of concrete with new tendons, such as polyethylene-encased and-greased strands, may be used in certain cases.

In cases where the damage to the tendons is localized, new short tendons, such as bars, may be placed in adjacent slots cut in the concrete, anchored at the ends, and stressed at the center by special center-pull jacks.

Where tendons in a deck are corroded but the concrete is still sound, slots may be cut and new tendons installed, stressed, and encased in concrete one at a time.

Existing tanks may be repaired by additional circumferential prestressing, followed by shotcrete to give corrosion protection to the new tendons.

11.12.3 Strengthening Existing Structures

This is the arena where prestressing has a major role to play, since it is ideally adapted to providing greater strength and correcting unacceptable deflections.

Post-tensioning by means of external tendons has been used to correct excessive sag in long-span bridges and to provide additional load capacity. It has been used to transversely tie cantilevered additions to existing bridges in order to widen them. When new structural elements are added adjacent to existing structures, post-tensioning can tie them so that both new and old deform together, sharing the load.

In these cases, post-tensioning has been applied either by drilling holes through the existing structure or placing external tendons. The external tendons must be adequately tied to the existing concrete by dowels.

Post-tensioning is extensively employed to strengthen existing buildings and bridges for earthquake resistance. It has been used effectively to transfer the loads from existing columns to new underpinning.

Construction follows the general guidelines for post-tensioning with the following special provisions:

1. Seats for anchorages must be carefully prepared by mortar or grout to ensure the anchor plates can be fully seated on a plane normal to the tendon's trajectory. The concrete beneath must have the ability to accept high compressive stress without bursting or cracking owing to the concentrated load. Thus it may be necessary to remove and replace defective concrete. It may be necessary to preinstall stitch bolts transversely and through thickness to provide confinement.
2. Where external tendons are placed and are to be bonded to the existing concrete, the surface should be roughened and cleaned. A bonding epoxy may be applied just ahead of the new concrete.
3. Adequate corrosion protection and, where applicable, fireproofing, must be provided for the anchorages.
4. In planning for strengthening, consideration must be given to access for the jacks and bearings for them to react against.
5. Ensure that the retrofitted structure can accommodate the elastic and plastic shortening due to prestress.

11.13 Demolition of Prestressed Concrete Structures

Buildings and bridges must eventually be demolished and removed in order to make room for larger structures or other uses. Such demolition can be piecemeal, using the time-honored methods of headache ball, concrete saw, drill and burning torch, or the more modern approach that uses controlled explosives, toppling the entire structure within a matter of seconds and turning it into rubble.

Prestressed concrete structures require special considerations, owing to the fact that tremendous amounts of energy are stored in the stressed tendons. Sudden release can produce a missile that can fly 100 m or more with a velocity sufficient to penetrate a board fence. More serious is the potential for injury to workmen or the public.

When removing a structure in stages, it is usually necessary to shore under the span being demolished in order to prevent its collapse. The shores should be wedged up to fit tightly against the member so as to prevent erratic loading as the tension is released. Heavy timbers or blasting mats should be placed behind the end anchorages to absorb impact energy. Unbonded tendons should be distressed gradually. One method is to heat the tendon with the yellow flame of a burning torch, allowing it to slowly elongate. The other, applicable to large post-tensioning tendons is to cut one wire at a time. Controlled heating is the most reliable method.

When the precompression in the structural element is fully released, the element will transfer its load onto the shores. They must be designed to take the resultant loads.

Bonded tendons are more difficult to release. Since their transfer length is typically short, they have to be cut at frequent intervals. However, the projectile phenomenon is not a factor.

When demolishing by explosives, millisecond delays should be used to ensure that unbonded tendons are cut close behind the anchorages prior to the release of the main spans.

Demolition of a small area in order to repair a damaged zone should follow isolation of the concrete by a saw cut, so that spalling, etc., will not extend farther than intended. Stitch bolts through the thickness may be installed around the periphery. Then the concrete may be removed by a small chipping gun or jet blast of water, taking care not to seriously damage the tendons. The tendons may be clamped at their ends or bonded by epoxy injection if they have not already been bonded by grout. They are then heated to relieve tension and cut for splicing or other remedial repair.

Acknowledgments

Many of the pictures of post-tensioning operations were made available by VSL Corporation. The author's firm, Ben C. Gerwick, Inc. provided many pictures of prestressed concrete manufacture and erection.

References

Gerwick, B.C., Jr. 1996. *Construction of Prestressed Concrete Structures*, 2nd ed. Wiley, New York.

Libby, J.R. *Modern Prestressed Concrete*, 3rd ed. Van Nostrand–Reinhold.

Nawy, E.G. 1996a. *Fundamentals of High Strength High Performance Concrete*. Addison Wesley Longman, London and California.

Nawy, E.G. 1996b. *Prestressed Concrete—A Fundamental Approach*, 2nd ed., Prentice Hall, Upper Saddle River, New Jersey.

Podolny, W., Jr. and Muller, J. 1982. *Construction and Design of Prestressed Concrete Segmental Bridges*. Wiley, New York.

Precast/Prestressed Concrete Institute (PCI). 1995. *Erection Safety for Precast and Prestressed Concrete*. Chicago.

(a)

(b)

(a) Unbonded post-tensioning reinforcement (Courtesy FBA, Inc., Belmont, CA). (b) High rise building with prestressed post-tensioned floors (Courtesy Portland Cement Association).

12

Unbonded Post-tensioning in Building Construction

by
Florian G. Barth, P.E.
Principal, FBA, Inc., Belmont, California. Expert in design of post-tensioned concrete structures and in the design and rehabilitation of buildings in seismic regions.

12.1 Developments in Unbonded Post-tensioning

12.1.1 Introduction

During the last three decades, unbonded post-tensioning has progressively become the predominant construction choice for commercial concrete buildings in the United States. The utilization of unbonded tendons is now standard practice for concrete structures. Owing to excellent performance records combined with economical and versatile application, the total use of unbonded tendons increased from approximately 7,500 tons in 1976 to over 40,000 tons of prestressing steel in 1990, adding up to over one billion square feet of slab constructed with unbonded post-tensioning.

Early applications of unbonded tendons and their limitations in design and construction were soon succeeded by well-established analysis, design, detailing, and field procedures. Driven by durability considerations, post-tensioning manufacturers have implemented product improvements essential for corrosive environments. The most meaningful testimony for unbonded post-tensioned construction is the successful performance of over one billion square feet of concrete construction.

There are four common types of tendon systems, namely monostrand tendons, single-bar tendons, multistrand tendons, and multiwire tendons. During the 1960s and early 1970s, several proprietary tendon systems were used for concrete building structures. Unbonded single strand or monostrand (1/2-in diameter, seven wire) tendons with various types of sheathing applications were used predominantly over construction using unbonded single-bar tendons in plastic (or metal) ducts and unbonded button-head wire tendon. The application of unbonded monostrand tendons successfully offers an economical and versatile application for post-tensioning thin slabs and narrow beams. Positioning monostrand tendons horizontally, side by side (maximum bundle of four), together with the small tendon diameter, provides the maximum possible eccentricity within a member cross section. In addition, the monostrand post-tensioning system utilizes compact anchorage devices and anchor recess pocket formers, together with small, lightweight stressing jacks, thus permitting the stressing operation to be executed with hand-carried equipment.

The use of 5/8 in diameter single-bar tendons (smooth rods) was discontinued owing to the labor and material costs involved in splicing the standard bar length. In contrast, unbonded multiwire tendons (button-head wire tendons) were used until the mid 1970s for beams and transfer girders, which required large localized forces. The economical application of multistrand tendons assumed this application until today. Multistrand tendons, originally intended as bonded tendon, but kept ungrouted for surveillance (nuclear), are specifically excluded and not part of this review.

12.1.2 Unbonded Post-tensioning System Technology

In the United States, the first unbonded monostrand tendons were used in the mid 1950s for building construction using greased and paper-wrapped seven-wire strand. The spirally applied continuous paper strip was intended to be bond breaker between the strand and concrete and the grease coating took the role of corrosion protection. Plastic sheathing introduced during the mid to late 1960s assumed the role of (i) bond breaker, (ii) protection against damage by mechanical handling, and (iii) a barrier against intrusion of moisture and chemicals. The strand coating, commonly referred to as grease, (i) reduces the friction between the strand and the sheathing and (ii) provides added protection against corrosion (Figure 12.1). Three principal polyethylene coating applications were used for several years, namely a plastic tube into which the grease-coated strands were pushed (stuffed or push-through tendons); a continuous polyethylene strip positioned parallel with the strand, wrapped around the coated strand, and sealed at its seam (heat sealed); and the extrusion of polyethylene over the coated strand. Primarily used in Canada, the stuffed or push-through tendons were discontinued during the early 1970s in the United States and remained in use in Canada until approximately 1990. Heat-sealed fabrication of monostrand sheathing was supplied until the early to mid 1990s. In corrosive environments, stuffed and heat-sealed tendons, have the inherent shortcoming of either trapping or the allowing access of corrosive substances in the oversized sheathing.

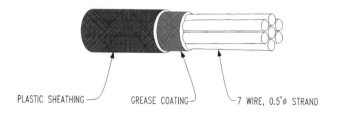

PLASTIC SHEATHING   GREASE COATING   7 WIRE, 0.5"⌀ STRAND

FIGURE 12.1   Unbonded monostrand tendon.

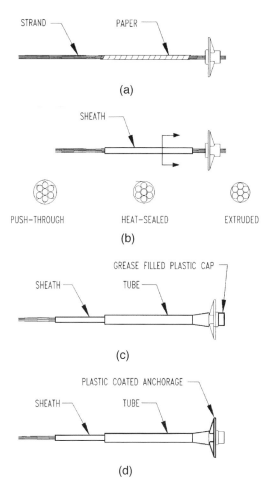

FIGURE 12.2 Unbonded tendon evolution. (a) Paper-wrapped, 1995–1970±; (b) plastic sheath types, 1980–present; (c) encapsulated system, 1985; and (d) electrically isolated tendon.

Extruded polyethylene sheathing application has shown excellent performance since its first introduction in 1969. The extrusion process applies the sheathing, eliminating voids between the sheathing and the grease coating. The extrusion sheathing application, with encapsulation of anchor zones for corrosive environments, is now used almost exclusively, in the United States. Again, the sheathing progression for most parts of the United States was paper wrapped, stuffed or push-through, heat-sealed, extruded, encapsulated, and for special applications, electrically isolated (Figure 12.2). Electrically isolated monostrand tendons have had limited application so far [Schupack, 1980].

12.1.2.1 Material Specifications

Prestressing Steel. Prestressing steel properties have essentially remained unchanged, conforming to ASTM A-416 Grade 250 or 270 k. The seven-wire low-relaxation strand, with a nominal diameter of 0.5 in, and a specified tensile strength of 270 ksi is typically anchored near 70% of its ultimate strength. The material shall be packaged at the source in a manner that prevents physical damage to the strand during transportation and protects the material from deleterious corrosion during transit and storage.

Anchorages. Tendon anchorages and couplings shall be designed to develop the static and dynamic strength requirements of Section 3.1.6(1) and Section 3.1.8(1) and (2) of the Post Tensioning Institute's (PTI) Guide Specifications for Post-Tensioning Materials, *Post Tensioning Manual*, 5th edition, hereafter referred to as the PTI Manual [PTI, 1990]. Castings shall be nonporous and free of sand, blow holes, voids, and other defects. The average compressive concrete bearing stress of anchorages shall not exceed the limits set forth in Section 3.1.7 of the PTI Manual. For wedge-type anchorages, the wedge grippers shall be designed to preclude premature failure of the prestressing steel due to notch or pinching effects under the static and/or dynamic test load conditions stipulated under Section 3.2.3(1) of the PTI 1990 Manual, for both stress-relieved and low-relaxation prestressing steel materials.

Anchors intended for use in corrosive environments shall include design features permitting a watertight connection of the sheathing to the anchorage and a watertight closing of the wedge cavity for stressing and nonstressing (fixed) anchorages. Intermediate stressing anchorages shall be designed to permit complete watertight encapsulation of the unbonded tendon.

Sheathing. Sheathing for unbonded single-strand tendons shall be made of a material with the following properties:

- sufficient strength to withstand unrepairable damage during fabrication, transport, installation, concrete placement, and tensioning;
- watertightness over the entire sheathing length;
- chemical stability, without embrittlement or softening over the anticipated exposure temperature range and the service life of the structure; and
- nonreactivity with concrete, steel, and the tendon corrosion preventive coating.

Minimum thickness of the sheathing used in normal (non-corrosive) environments shall not be less than 0.025 in for medium- or high-density polyethylene or polypropylene. Sheathing thickness for tendons used in corrosive environments shall not be less than 0.040 inches for medium- or high-density polyethylene or polypropylene.

Corrosion Preventive Coating. The corrosion preventive coating material shall have the following functions and properties:

- provide corrosion protection to the prestressing steel;
- provide lubrication between the strand and the sheathing;
- resist flow from the sheathing within the anticipated temperature range of exposure;
- provide a continuous nonbrittle film at the lowest anticipated temperature of exposure; and
- be chemically stable and nonreactive with the prestressing steel, the sheathing material, and the concrete.

The film shall be an organic coating with appropriate polar, moisture displacing, and corrosion-preventive additives. The minimum weight of coating material on the prestressing strand shall be not less than 2.5 lb of coating material per 100 ft of 0.5 in diameter strand. The amount of coating material used shall be sufficient to ensure essentially complete filling of the annular space between the strand and the sheathing. The coating shall extend over the entire tendon length.

12.1.3 Durability of Unbonded Tendons

Post-tensioned concrete members inherently provide enhanced durability owing to the limitation of cracks that provide access to reinforcement for corrosive agents. The use of post-tensioning typically eliminates most slab joints, which if not eliminated, may provide access for corrosive agents to beams and columns. The excellent durability performance of selected concrete structures in corrosive environments reflects the durability potential of unbonded monostrand tendons. Most of these structures were built without specific durability design and construction considerations. The visual inspection of a 15-year-old parking structure in Baltimore following demolition confirmed that no significant corrosion had occurred on the unbonded tendons over the 15-year service life [Suarez and Posten, 1990].

The distinguished characteristic of an unbonded tendon is that, by design, it does not form a bond along its length with the surrounding concrete. The axial force in the stressed tendon is transferred to the concrete primarily by the anchors provided at each end. Since the force of an unbonded tendon is primarily resisted by the anchors at each end, the long-term integrity of tendons and anchors throughout the service life of an unbonded tendon are of concern.

PTI published specifications for unbonded post-tensioning tendons in 1985 that included special durability requirements for tendons used in structures exposed to aggressive environments. These specifications provide for watertight encapsulation of the strand within the tendon sheathing over its entire length, including watertight enclosure of the tendon anchorage device (Figure 12.3). The specifications further require the use of a specially formulated corrosion-inhibiting grease and include many other secondary provisions to insure enhanced durability.

Nevertheless, some buildings constructed with unbonded tendons have suffered durability short-comings. A survey of 215 concrete structures in Toronto, Ottawa, and Montreal concludes the

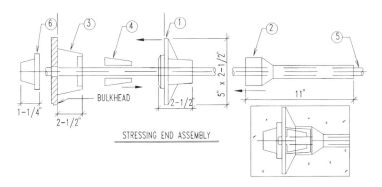

NOTES:
1. LOCATE ANCHOR AT BULKHEAD PER PROJECT PLANS
2. INSTALL GROMMET FLUSH BETWEEN BULKHEAD AND ANCHOR FOR TIGHT SEAL.
3. SLIDE SLEEVE TIGHT AGAINST ANCHOR. BE SURE NO BARE STRAND IS EXPOSED. TAPE IF NECESSARY.
4. AFTER POURING, AT TIME OF STRESSING, REMOVE GROMMET AND INSERT WEDGES.
5. AFTER STRESSING, CUT STRAND TO WITHIN 1/8" OF END OF END CAP AND GREASE END CAP PRIOR TO INSERTING IT TIGHT AGAINST ANCHOR
6. PATCH STRESSING POCKET PER PROJECT PLANS.

ITEM	DESCRIPTION
1	ANCHORAGE
2	PROTECTION SLEEVE
3	GROMMET
4	MONOWEDGES TYPE 1.5
5	STRAND (GREASED AND COATED)
6	END CAP

FIGURE 12.3 Encapsulated system for tendons in corrosive environments (stressing end).

following: "The evidence indicates that durable structures can be built, and that poor performance must be attributed to design and construction practices whose effectiveness falls short of that required by the environment" [Litvan and Bickley, 1987]. To understand and size environmental considerations for the selection of an appropriate unbonded tendon system, a map is available that divides the United States into five principal environmental zones. The selection of zones was in part based on the geographical use of deicing salts and presence of airborne salts from oceans. Walker (1990) offers detailed information on the zoning requirements and related system recommendations.

12.2 General Notes and Standard Details

12.2.1 General Notes

This section is intended to offer a sample layout of general notes for the construction of an unbonded post-tensioned concrete structure, typically outlined on structural drawings. The sample notes are limited to the post-tensioning activity of the concrete frame or member only. The notes are neither considered complete nor applicable for every project that includes the construction of unbonded post-tensioned members. However, with minor revisions and additions/deletions the notes have been used by the author Watry (1996) on over 500 concrete structures.

The post-tensioning notes presented below are grouped into sections titled General, Material, Installation, Concrete Pour, Stressing, Inspection, Other Related Notes, and Allowances.

The sections "Other Related Notes and Allowances" address fundamental postconstruction concrete frame behavior. The design engineer should be aware that the concrete frame may develop cracking as a result of concrete creep, shrinkage, temperature deformation, and elastic shortening on members that are partially or fully restrained from movement. For this reason, notes should be added on the drawings that disclose the likelihood of potential concrete cracking and allow for a funding mechanism for crack repair in the form of a postconstruction material maintenance allowance.

12.2.1.1 Post-tensioning

General.

Material, installation, stressing, and finishing specifications. As a minimum, any part of the tendon assembly and the tendon shall meet requirements set forth in Chapter 3 "Specifications" of the PTI Manual.

Marking of tendon location. If desired by the owner, the tendons may be marked using the dye-transfer method or by paint marking the formwork along the tendon lines just prior to placement of concrete. The paint transfers to the concrete soffit to permanently locate tendons [PTI Manual, 1990].

Power-driven fasteners. No power-driven fasteners or inserts shall be shot or drilled into the post-tensioned slab after concrete is placed without the written authorization of the engineer.

Openings. All openings, penetrations, and inserts shall be preplanned to the fullest extent possible. No changes shall be made in the field without prior approval of the engineer.

Formwork. For multilevel structures, the formwork shall extend beyond the slab edge, or scaffolding shall be provided to allow adequate room for the stressing operation.

Shop drawings. The contractor shall submit shop drawings showing tendon layout, dead-end and stressing-end locations, and tendon support layouts with details necessary for installation to the engineer for approval. The contractor shall supply the engineer with two blueline sets of shop drawings a minimum of one week prior to placement. The review of shop drawings by the engineer is only for general compliance with the structural drawings and specifications. A set of approved shop drawings must be filed with the City Engineer by the contractor.

Post-tensioned slab review.   The tendon and mild reinforcement layout of a post-tensioned slab shall be reviewed by the engineer or his/her designated representative prior to concrete pour. The engineer shall be notified at least 48 hours in advance.

Field foreman.   The field foreman responsible for the placement, stressing, and finishing of all post-tensioning material shall have a minimum of five (5) years specialized experience in this capacity for this type of construction.

Materials.

Strand quality.   One sample of each reel or heat shall be tested by an approved laboratory. Test results or mill certificates shall be submitted to the engineer before stressing of tendons. Post-tensioning tendons shall be stress-relieved or be of low-relaxation quality and shall conform to the following:

seven-wire strand ASTM designation	A-416
1/2 in diameter tendon area	0.153 in^2
ultimate strength	270 ksi

PT hardware quality.   All anchorages, couplers, and miscellaneous hardware shall be standard products and approved by governing agencies and the engineer.

Tendons.   Unbonded strands shall be encased in slippage sheathing that shall consist of a sealed durable waterproof plastic tubing (minimum 0.04-in thickness) capable of preventing the penetration of moisture and cement paste and that will contain a rust-inhibiting grease coating. Tears in the sheathing shall be repaired to restore the watertightness of the sheathing. The sheathing application shall be limited to the extrusion process.

Tendons shall be secured during shipping and supported during handling to avoid damage to the tendon sheathing. During shipping and storage, the tendons shall be covered or protected to avoid moisture access to post-tensioning material.

Installation.

Installation of unbonded tendons.   If the post-tensioning supplier does not install the material supplied, detailed instructions for the installation and stressing of tendons shall be furnished. The contractor responsible for hiring the independent post-tensioning placer shall insure that the installation crew meets the standards set forth above. The supplier shall provide technical assistance necessary to properly install, stress, and finish all post-tensioning material.

Tendons.   Tendons shall be shop fabricated with preassembled fixed-end anchorages. Anchor casting with plastic pocket formers shall be used at all stressing ends to recess the anchor in enough concrete to achieve required cover.

Banded layout.   The "banded" tendon placement procedure shall be used for this project.

Tendon placement.   Care shall be taken that tendons are located and held in their designated positions. Tolerances for the location of the prestressing steel shall not be more than +1/8 in vertically, except as noted or approved by the engineer. Access to stressing ends shall be maintained where shown.

Strand bundles.   The maximum allowable number of strands per bundle is four (4) for slabs and six (6) for beams.

Tendons over columns.   A minimum of two (2) tendons in each orthogonal direction shall be placed directly over the supporting column.

Tendon adjustments.   Small deviations in the horizontal spacing of the slab tendons will be permitted when required to avoid openings, inserts, and dowels with specific locations. Where locations of tendons seem to interfere with each other, one tendon may be moved horizontally in order to avoid the interference.

Twisting.   Twisting or entwining of individual wires or strands within a bundle or a beam shall not be permitted.

Vertical profiles.   Profiles shall conform to controlling points shown on the drawings and should be in approximate parabolic drape between supports, unless noted otherwise. Low points are at midspan unless noted otherwise. Harped tendons shall be straight between high and low point controls.

Horizontal profiles.   Should the tendons be horizontally curved to miss an opening or other obstructions, tendon bundles shall be flared with a minimum distance of two inches between each individual tendon while horizontally curved. In addition, #3 hairpins at 12 in on center for each tendon shall be installed, transferring the horizontal radial force via the hairpin mild reinforcing to the concrete.

Prestress cover.   All dimensions showing the location of prestressing tendons are to the center of gravity of the tendon (CGS) unless noted otherwise.

Minimum chairing.   Tendons shall be secured to a sufficient number of positioning devices to ensure correct location during and after the placing of concrete and shall be supported at a maximum of 3 ft 6 in on center. Chairs greater than 2.5 in shall be stapled to the formwork.

Support bars.   Support bars located at the face of drop panels shall be #6 or greater. Drop panels greater than 4 ft in width shall have additional #6 or greater support bars at the center, with a maximum support bar spacing of 4 ft for larger panels. All other support bars shall be minimum #4 bars. Continuous support bar lap splices shall be 24 in minimum.

Anchors.   Anchorages shall be recessed a minimum of two (2) in. Place two (2) #4 bars continuous behind all anchorages unless otherwise noted. Splices shall be 24 in minimum and staggered. Special anchorage zone reinforcement shall be provided for groups of six or more anchors for 1/2-diameter strand tendons spaced at 12 in or less on center.

Blockouts.   All pockets or blockouts required for anchorage shall be adequately reinforced so as not to decrease the strength of the structure. All pockets should be waterproofed to eliminate water leakage through or into the pocket.

Pipes.   Plastic or metal conduits may be embedded in the slab providing that the following criteria are met:

   A. The diameter does not exceed one-quarter of the slab thickness or 2 in, which ever is less.
   B. Conduits greater than or equal to 1-in diameter shall be located within the middle third of the slab.
   C. Conduits smaller than 1-in diameter may be located anywhere within the slab as long as the minimum cover requirements are observed.
   D. Center-to-center spacing of the conduits is not less than three (3) times the diameter of the largest conduit.
   E. Conduits must not interrupt the post-tensioned cables.
   F. No conduit may be placed within the column shear cone.
   G. It is undesirable to have excess amounts of conduit entering the slab from one location. If this condition exists, the conduits must be fanned out immediately.

Penetrations.   Penetrations shall not be permitted in beams or drop caps unless permitted on post-tensioning drawings or typical details.

Concrete Placement.

Concrete consolidation. The contractor shall take precautions to assure complete consolidation and densification of concrete behind all post-tensioning anchorages.

Concrete placement. When concrete is placed in post-tensioned slabs, special care shall be taken at all column drop caps (panels). Insert the pump hose into the column drop panel below reinforcement and fill until concrete reaches the top reinforcing layer. Monitor concrete elevation to avoid floatation of top reinforcing. After the drop panel is full of concrete, place concrete over the top reinforcing layer to specified slab thickness. Vibrate adequately in and around column drop panels.

Pumped concrete. If concrete is placed by the pump method, horses shall be provided to support the hose. The hose shall not be allowed to ride on the tendons.

Chlorides. Grout or concrete containing chlorides shall not be used.

Stressing of Tendons.

Tendon stresses. Such stresses shall conform to the following:

maximum tendon jacking stress	216 ksi
maximum tendon stress at anchorage immediately after prestress transfer	189 ksi

Effective force. Forces shown on structural drawings are effective forces after all losses. All losses (short- and long-term losses) due to creep, shrinkage, tendon relaxation, and elastic shortening, including friction losses and losses due to wedge seating, may be assumed as 14 ksi. Thus the effective force per tendon may be assumed to be 24.8 kips for stress-relieved tendons, and 26.8 kips for low-relaxation tendons, when tendon length is less than 100 ft. For variance from this value or for tendons over 100 ft, the post-tensioning supplier shall provide friction and long-term loss calculations for the engineer's approval. Friction losses may not be averaged or assumed to redistribute along the tendon length. The available effective force shall be established at a location along the tendon length where the force demand meets the minimum effective tendon force.

Concrete strength at stressing. Prior to transfer of prestress, concrete shall reach a minimum compressive strength of $f'_c = 2,750$ psi. Minimum concrete strength shall be established by breaking concrete test cylinders. The stressing shall not commence until concrete reaches the specified strength. However, tendons shall be stressed within 72 h after concrete reaches the minimum specified strength to mitigate early age concrete cracking. This may not apply to stage stressing of transfer floor or mat foundations.

Calibration. The ram and attendant gauge used shall have been calibrated within sixty (60) days of use.

Tendon stressing. The stressing operation shall be done by jacking under the immediate control of a person experienced in this type of work. Continuous inspection and recording of elongations is required during all stressing operations.

Stressing sequence. In general, uniformly distributed tendons shall be stressed before concentrated beam strip (banded) tendons, and slab tendons shall be stressed before beam tendons. Additional stressing sequence requirements shall be as specified below.

Two-way slab sequence:
1) Stress through distributed tendons
2) Stress banded tendons
3) Stress added distributed tendons

Beam and slab sequence:
1) Stress temperature tendons
2) Stress through slab tendons
3) Stress beam tendons
4) Stress transfer girder tendons
5) Stress added slab tendons

Elongation.   Individual tendon field readings of elongations and/or stressing forces shall not vary by more than ±7% from the calculated required values shown on the shop drawings. If the measured elongations vary from calculated values by more than ±7%, the contractor shall provide friction calculations and/or other justification to the satisfaction of the engineer.

Member forces.   The post-tensioned force provided in the field for each structural member shall not be less than the values noted on the structural drawings. In this context, structural members are beams or slabs, whether with banded or distributed tendons, each serving their respective tributary.

Tendon ends.   Do not burn off tendon ends until the entire floor system has been satisfactorily stressed and the engineer's approval is obtained.

Anchor painting.   The stressing end anchors and wedges shall be spray painted with rust-inhibiting paint or a similar coating for corrosion protection prior to grouting of recess pockets.

Grouting of stressing pockets.   Stressing pockets shall be filled with nonshrink grout as soon as practical after stressing to limit moisture access to the tendon.

Deshoring.   Slabs or beams may be deshored when all tendons have been satisfactorily stressed and the engineer's approval is obtained, unless shoring is required to carry floors on above levels. In areas supporting a partial span, such as near pour strip or construction joint, the shoring in the partial span and immediate back span shall stay in place until the remaining section of the span has been poured and stressed.

Inspection for Prestressing Steel.   Continuous special inspection shall be provided during the placing of reinforcing steel, tendons, and prestressing steel for all structural concrete. Tendon placement and integrity of the protective wrapping for post-tensioned tendons shall be inspected prior to placement of concrete. During all stressing of post-tensioned concrete, the special inspection shall include recording of field-measured elongation and jacking force for each tendon.

Admixtures.   No admixtures shall be added to the concrete mix without the approval of the engineer, unless noted otherwise. Admixtures concrete containing chlorides shall not be used in post-tensioned slabs.

Special Notes to the Owner.   Under normal conditions, and for conventional buildings, reinforced concrete as well as post-tensioned concrete develops cracks. The cracks are due to inherent shrinkage of concrete, creep, and the restraining effects of walls and other structural elements to which the beams/slabs are tied.

The early age concrete cracks that may develop are usually of cosmetic nature. The slab typically retains its serviceability and strength capability. Owing to special features of unbonded post-tensioning, it is possible that a number of hair cracks, which would normally spread over a wide area, will integrate into a single crack with a width exceeding 0.01 in. It is emphasized that although special efforts are made to reduce the potential causes and number of such cracks, it is not practical to provide total articulation between the floor system and its supports and thereby achieve complete inhibition of all cracks.

Most early-age cracks develop during the first two years after construction of the floor system is complete. Cracks that are wider than 0.01 in may need to be pressure epoxied. Refer to the notes under "Allowances."

The objective of providing joints is to allow movement. Movements due to creep and shrinkage may be noticeable at joints for up to two years after construction, beyond which movements due to variations in temperature will persist. In aggressive environments, cracks should be repaired at the time they are first noticed.

Material Allowances.

Reinforcement allowance.   The contractor shall provide "additional supply" reinforcement for the engineer to use at the engineer's discretion during construction.

Post-tensioning allowance.   The contractor shall provide "additional supply" of post-tensioning for the engineer to use at the engineer's discretion during construction.

Pressure epoxy allowance.   The contractor shall include the cost of $0.10 per square foot of elevated concrete slab for pressure epoxy injection of cracks that may develop in the structure during the first two years.

12.2.2 Standard Details

In accordance with standard industry procedures for most construction-related work, details for un-bonded post-tensioned members are first developed by the design engineer or architect. The details typically show the member geometry, reinforcing layout (location of tendons and mild reinforcing), and other embedments. After the construction contract is awarded, the post-tensioning supplier commonly prepares more project-specific drawings, called shop drawings. The shop drawings for post-tensioning materials are normally prepared in much more detail than the design drawings. Typically, they are submitted for review and approval to the design engineer/agency before the fabrication of tendons is initiated. It is essential that details of the post-tensioning tendons, nonpre-stressed reinforcement, ducting for electrical or mechanical service, and other embedment items be reviewed and coordinated during the detailing stage. It is not uncommon that final details for different materials are shown on different shop drawings, indicating incompatible or conflicting layout. In most cases, details can be rather easily adjusted at the shop-drawing stage to accommodate all embedded items. When conflicts do arise during the development of shop drawings, or during construction, the tendon layout should govern over other elements or embodiments conflicts.

Many specialized structural details have been developed for the construction of unbonded post-tensioned members. The following selected details illustrate typical design and detailing practices for most common applications of unbonded post-tensioning in building construction.

12.2.2.1 Tendon Anchorage Zone

The anchorage zone, probably the most critical concrete region, has to retain the tension force of unbonded tendons during the service life of the post-tensioned structure. Figure 12.4 shows typical details of stressing and dead ends. Unless otherwise detailed on the design or post-tensioning installation drawings, banded tendon anchorage zones in normal-weight concrete for groups of six or more 0.5-in diameter single-strand tendons with anchor spacing of 12 in or less should be reinforced in accordance with Figure 12.4b.

For restriction of anchorage zone embedments, see Figure 12.14. The flaring of tendon bundles near anchorage zones, in the horizontal plane of the concrete member, and fixed-end tendon staggering requirements are suggested in Figure 12.5.

12.2.2.2 Two-Way Slab Tendons Over Column Supports

The congested mild-reinforcing arrangement and tendon layout of banded and uniform tendons over column supports must be detailed so the field personnel can understand the various layers of top reinforcement and their support system. Figures 12.6 and 12.7 indicate a typical two-way slab tendon layout over interior and exterior columns. The layout of tendon groups is arranged so that except for the bundle of uniform tendons (minimum two tendons) directly above the supporting column, all tendons in both directions have essentially the same eccentricity within the proposed slab section in the negative moment region. A detailed account of all layers can be found in Figure 12.8.

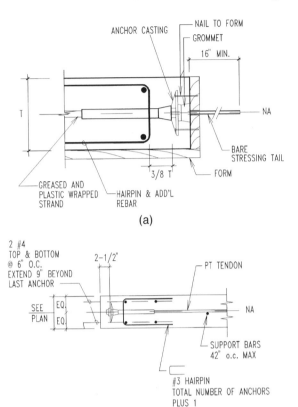

FIGURE 12.4 Stressing and dead ends. (a) Stressing end anchor for banded tendon in a corrosive environment; (b) dead end at banded tendons in a corrosive environment.

12.2.2.3 Beams

A number of field problems can be eliminated if simple installation recommendations are considered during the installation of the beam reinforcement.

Figures 12.9 and 12.10 show typical column-beam joints and beam sections with detailed information on tendon and mild reinforcement layout. Besides showing a detailed bar layout for mild reinforcing, the tendon support system is very critical. First, the tendon support bars must be stable and properly secured to ensure a firm tendon profile during the concrete pour, but second, large tendon bundles, such as multiple bundles of six tendons, that are not properly spaced and layered may result in tendons bunching up at locations of curvature (high and low points) and developing splitting forces in the concrete beam. The use of tendon support bars, which guide the installation crew in the appropriate spacing of tendon bundles, has successfully addressed this concern (Figure 12.11).

12.2.2.4 Joints

Section 12.3 discusses in detail the types and optimum locations for joints to mitigate the restraint effects of post-tensioned members. Besides restraint considerations, joints may have to be located to limit concrete pour size (Figure 12.12a) or allow for intermediate stressing (Figure 12.12b). The selection of pour-strip location and duration of pour-strip opening shall be based on a numerical shortening evaluation of the structure. A sample reinforcing layout for a typical stressing block out and a 3 ft wide pour strip is shown in Figure 12.13.

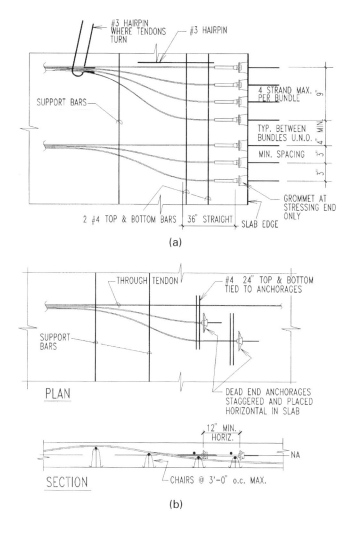

FIGURE 12.5   Flaring of tendon bundles and fixed-end tendon staggering. (a) Flaring of banded tendons at the slab edge in a corrosive environment; (b) placement of added tendons.

The preceding discussion exemplifies that the successful performance of a unbonded post-tensioned structure is directly related to the understanding and extent of detailing of specific performance considerations.

12.2.3 Typical Field Shortcomings—Problems and Solutions

12.2.3.1 Preventing the Most Frequent Problems

Prior to the placing of the concrete, the post-tensioning installation should be checked for the following.

The area behind the anchors (18 in behind the anchor at 45° angles on each side as shown in Figure 12.14) should be free of sleeves, block outs, large conduit, or any other voids or congestion that could allow the concrete to crush or form a void in this high-stress zone. If penetrations need to be positioned within the 45° region, steel pipe inserts must be used as specified by the engineer. Frequently, the electrical, mechanical, and railing contractors place their sleeves just before the pour,

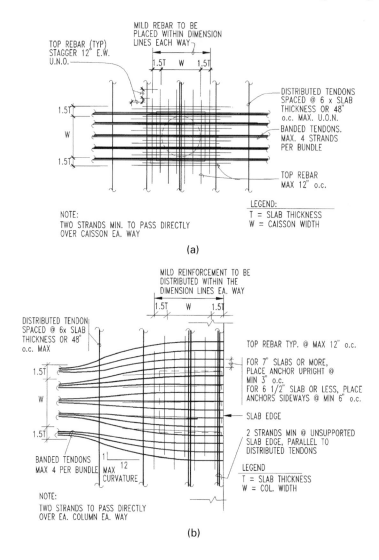

FIGURE 12.6 Typical two-way slab tendon layout. (a) Top reinforcement at interior columns; (b) top reinforcement at exterior columns.

after the tendon-placement inspection. While checking the anchor zones, make sure a sufficient strand tail extension is protruding through the edge form. It is typically much easier to adjust the tendon by a few inches prior to placing the concrete than it is to use splicers and special equipment to stress a short tendon later.

For an encapsulated system, the sheathing should be connected to the anchor according to the manufacturer's recommendation to insure that there will be no exposed strand and to provide continuous protection. However, in normal environments, since there is no connection between the anchor and the sheathing, special care shall be taken to minimize the length of greased strand behind the anchor. The maximum length of greased strand should not exceed 1 in. If concrete is cast against on unsheathed strand, the rifling pattern of the strand will be cast in the concrete, forming spiral grooves that twist the strand when stressed. The spiraling of the strand will cause the stressing jack to spin at the end of the stressing cycle and could injure the person operating the stressing equipment or break the hydraulic supply hoses on the jack. Even if no one is harmed, the twisting motion of the strand through the jack grippers causes premature wear. It is important

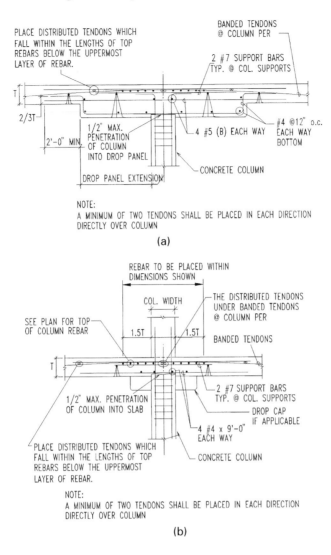

FIGURE 12.7 Typical two-way slab tendon layout. (a) Typical interior column; (b) typical drop-panel section.

to tape wrap exposed strands before the concrete pour, since there is no fix once the concrete has hardened.

If more than 1 in is exposed, repair the sheathing right up to the back of the anchors. If this area is hard to access owing to bursting steel or other obstructions, make a circular cut on the sheathing 18 to 24 in back from the anchor, slide the sheathing forward until it touches the anchor, and then repair the bare spot at a location away from the congestion. The second option has the advantage of leaving the most critical area (that zone 12 in behind the anchor) covered with good sheathing.

The quality of the installation can be jeopardized by people walking on the placed cable before a pour. The pocket former needs to be held tight against the anchors so that concrete slurry does not leak into the anchor cavity. This can happen if the concrete vibrator bounces the edge form and separates the pocket former from the anchors. There is no substitute for having the anchors tightly attached. If a small amount of post-tensioning coating is applied to the tip of the pocket former before inserting it in the anchor cavity, it will make a seal between the two pieces that will keep concrete slurry out even if a small gap develops.

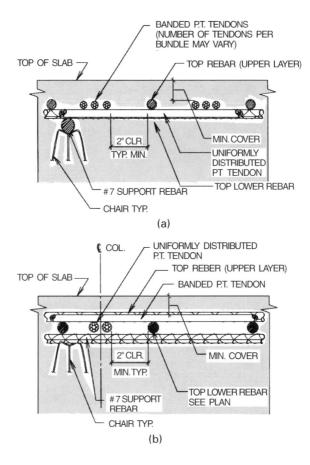

FIGURE 12.8 Layers in typical two-way slab tendon layout. (a) Section through banded tendon at column; (b) section through distributed tendon at column.

12.2.3.2 Slipping Strand and Jack Hang Up

When wedges fail to hold the strand, the most common cause is concrete slurry in the wedge seat of the anchor cavity. If a separation between the pocket former and the anchor occurs, concrete slurry can flow into the anchor cavity and set up in the form of a ring around the strand at the back of the anchor. This will stop the wedges from penetrating into the anchor the proper distance and result in strand slippage.

If the elongation requires more than one cycle of the stressing equipment, it can cause the jack to become locked onto the tendon. On the first cycle, the wedges usually hold because they do not have to be fully seated and the pressure is low. On the second cycle, the wedges bottom out on the concrete slurry and the strand will be free to slide back into the concrete, stripping the teeth on the wedges and hanging up the jack.

Several different methods can be used to detension the tendon, thereby releasing the jack, depending on individual site conditions as follows. A second jack should not be used on the back of the one that is hung up to detension it. Once the jack is hung up, a split troubleshooting anchor (Figure 12.15) is needed to insert behind the nosepiece of the jack bearing on the anchor cast in the concrete but in front of the gripper block of your jack in order to free up the jack.

During this procedure, do not exceed the recommended gauge stressing pressure. Open the jack just enough to insert the troubleshooting split anchor on the strand. Insert the wedges in the

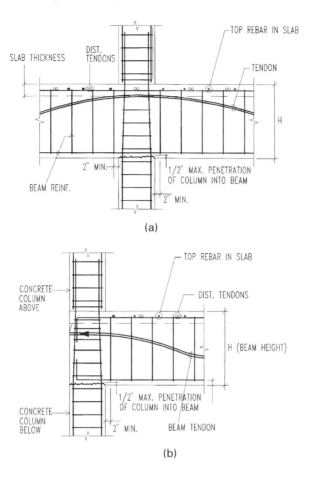

FIGURE 12.9   Typical column-beam joints. (a) Typical column-beam section; (b) exterior beam-column connection.

troubleshooting anchor and slowly release the pressure on the jack until the stress in the strand is taken up by the troubleshooting anchor. Continue closing the jack until the jack grippers in the jack gripper block are released from strand. Extend the jack fully and then retract the jack approximately 2 in. Engage the jack grippers and extend the jack to stress the strand again and release the wedges in the troubleshooting anchor. (Applying a thin coat of Never-Seez or a similar product on the back of the wedges will make them easier to remove during this step.) Slowly release the pressure in the jack and let the strand slide back into the concrete slab until fully relaxed. (If the jack bottoms out before relaxing the strand, simply repeat all steps of the sequence again.) Once the jack is fully released, remove it and the wedges from the anchor in the slab. It is not uncommon to find that the wedge seating was either restricted by a small film of concrete slurry around the tapered sides of the anchor cavity or a ring of concrete formed around the strand at the back of the anchor. If necessary, remove the debris by scraping or chipping the slurry out of the anchor. Since the area is very congested, a small screwdriver may be the proper tool for this procedure. After removal of all debris, clean the wedge seat of loose materials and dust using compressed air. Insert a new pair of wedges and stress the tendon.

12.2.3.3 Honeycomb in Concrete

Rock pockets, sand pockets, or voids should be repaired prior to the stressing operation. Remove all loose material and dust prior to repair. Wet the concrete surface before repair. When patching,

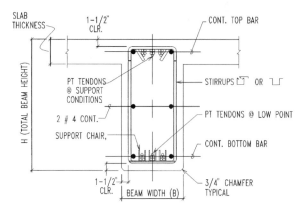

NOTE: KEEP TENDON GROUPS UNIFORMLY SPACED THROUGHOUT
THE LENGTH OF BEAM

(a)

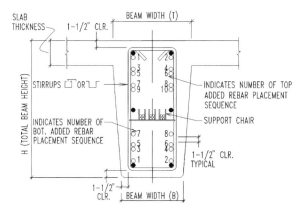

NOTE: AT CONDITIONS WHERE THERE ARE TWO TYPES OF ADDED BARS
(TOP OR BOTTOM), THE LONGER BARS SHALL BE PLACED FIRST

(b)

FIGURE 12.10 Typical beam sections. (a) Placement of tendons in beam; (b) placement of reinforcement in beams.

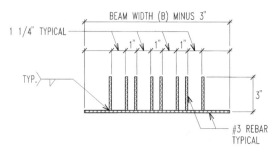

NOTES: 1. SUPPORT BAR @ 42" o.c. MAX.
SPACING IN BEAM WITH MORE THAN
NINE (9) TENDONS

2. FILL SUPPORT CHAIR SLOTS UNIFORMLY
WITH TENDONS

FIGURE 12.11   Tendon support bar.

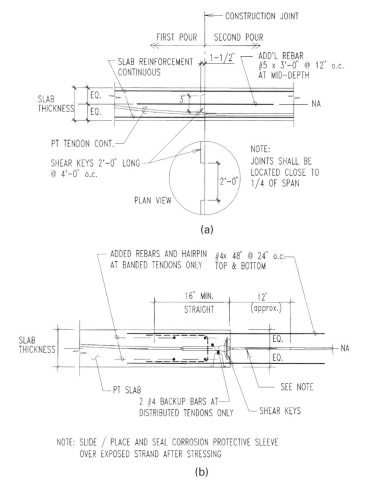

FIGURE 12.12   Joint locations. (a) Construction joint with no intermediate stressing; (b) construction joint with intermediate stressing in a corrosive environment.

use a high-strength non-shrink concrete grout mix with an epoxy binder. Grout strength should equal or exceed specified concrete strength. DO NOT use grout that contains calcium chloride or other materials containing chloride. When the patch has attained proper strength, the stressing operation may proceed.

Honeycomb in anchorage zones should be repaired, as this is essential to avoid blowouts. After detensioning of tendons, all loose material and dust shall be removed until sound concrete surfaces are encountered. Stress tendons after the repaired area reaches the minimum concrete compressive strength specified. Prior to stressing, check the quality of the patch by tapping it with a hammer to sound for voids. A hollow sound indicates a poor patch that is not suitable for stressing.

12.2.3.4 Splicing Tendons

Tendons are sometimes too short to reach an edge form because of misplacement or misfabrication. If the tendon is in one pour only and not continuous, every effort should be made to replace the short tendon with a tendon of proper length instead of using couplers. If tendons are continuous from another pour, thus making tendon couplers necessary, the Engineer of Record and the post-tensioning-material supplier should be notified. The coupler location should be determined by

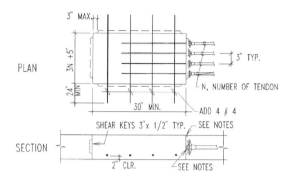

NOTES: – CUT STRESSED STRANDS; GREASED AND PLACE END CAP
        – CONTINUE ALL INTERRUPTED BARS SHOWN IN PLANS THROUGH THE BLOCKOUT
        – FILL BLOCKOUT WITH NON-SHRINK CONCRETE OF MIN. SAME STRENGTH AS SLAB

(a)

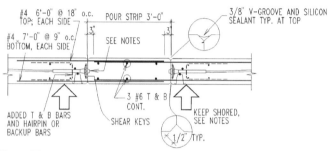

NOTES: 1– RETAIN SHORING UNTIL CLOSURE CONCRETE REACHES SLAB'S DESIGN STRENGTH OR 14 DAYS
        2– CLOSURE STRIP TO BE POURED WITH NON-SHRINK CONCRETE NOT LESS THAN X DAYS
           AFTER STRESSING
        3– ROUGHEN & CLEAN JOINTS & WET PRIOR TO PLACING CONCRETE
        4– ELIMINATE ACCIDENTAL MISALIGNMENT BETWEEN EDGE OF SLABS THAT ARE TO BE JOINED
           WITH A CLOSURE STRIP. USE MECHANICAL METHODS SUCH AS JACKING IF NECESSARY.
        5– RAISE EDGE OF SLAB 1/4" @ CLOSURE TO ALLOW FOR SETTLEMENT.
        6– PROVIDE WATERPROOFING MEMBRANE IF REQUIRED FOR WATER-TIGHTNESS
        7– CUT STRESSED STRAND TAILS; GREASE AND PLACE END CAP

(b)

FIGURE 12.13   Sample reinforcing layouts. (a) Stressing blockout in a corrosive environment; (b) closure strip in a corrosive environment.

the post-tensioning-material supplier such that the coupler is centered in the member and not at a point of tendon curvature. Couplers should not be located side by side. If more than one tendon requires splicing, couplers should be staggered at half bay increments per tendon group.

A PVC pipe of sufficient inside diameter to hold the coupler and of sufficient length to allow for subsequent elongation movement should be used. Also, an additional piece of sheathed strand of sufficient length to reach the edge form is required, along with two pocket formers. P/T coating shall be used to fill the void in the PVC pipe. The tapered tip of the pocket former that normally fits inside the anchor cavity can be cut off when being used for splicing, thereby reducing the length of the PVC pipe needed. The original strand is first cut with a saw or abrasive plate at the coupler location and one pocket former is placed on the strand. The strand should be marked before coupling to make certain that the proper length of strand has been fully inserted into the coupler. The coupler is then coupled to the original strand. The PVC pipe is placed over the coupler. The second pocket former is placed over the new strand (the strand is marked) and inserted into the coupler. A pocket former is taped to one end of the PVC pipe, which is then packed tightly with P/T coating, allowing no air voids. The second pocket former is affixed to the PVC pipe completing a tightly sealed coupler.

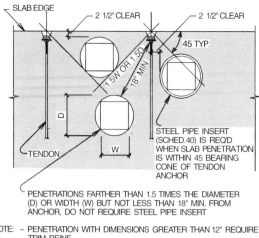

FIGURE 12.14 Openings at post-tensioning anchorage.

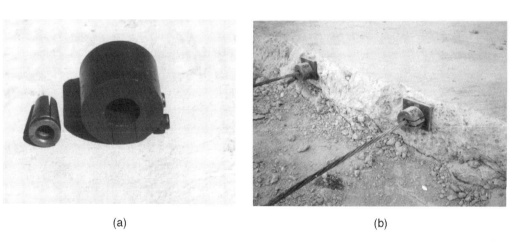

(a)   (b)

FIGURE 12.15 Split troubleshooting anchor.

The tendon coupler's location within the PVC pipe must permit the coupler to move the required elongation amount in the direction of stressing. Allowance for movement in both directions must be provided when the tendon is to be stressed from both ends. Conservatively, a minimum of 1.5 times the total expected elongation at the splice location shall be allowed for. A dark crayon or paint mark on the deck will facilitate locating the coupler after the pour, should that become necessary if the above procedure was not properly followed.

12.2.3.5 Tendons too Short to be Stressed Using Normal Stressing Procedure

Short tendons can result from an incorrect tendon-fabrication cutting list, misfabrication, misplacement, or a job-site mistake such as cutting tendons off prior to stressing.

During stressing, most conditions may be addressed with special equipment that can be obtained from the post-tensioning-material supplier. If a tendon is too short to be stressed using a standard jack, in some cases a short tendon can be stressed simply removing the nose piece and using jack feet. When using jack feet, care shall be taken to center the jack with the tendon before applying pressure. If the tendon is stressed without being centered on the anchor, it will rub on the side of the anchor, and inserting one of the wedges may be impossible. This will cause the other wedge to be drawn all the way to the back of the anchor cavity, breaking or damaging the strand. Without the hydraulic-seating attachment, the wedges will have to be inserted and seated using a hand-seating tool and a hammer. Tendons that are too short for the above procedure will have to be stressed using a coupler with a short piece of strand fixed on one end of the coupler.

Tendons that were cut with a torch prior to stressing have lost some of the temper in the steel owing to the heat. If the jack grippers or the coupler grip near the previously heated area, the tendon may slip at a very low pressure. If this condition exists, make the first pull as short as possible (so the stressing pressure is kept low), install the wedges in the anchor, and regrip the tendon farther away from the end that was heated.

12.2.3.6 Lift-Off Procedures

The purpose of a lift off is to verify the force of a tendon after it has been stressed. A lift off may be required when the recorded elongation is out of code-recommended tolerance. Project specification may call for a selected force verification using the lift-off method. A "lift-off test" may be conducted by use of the standard hydraulic stressing jack on previously stressed and anchored monostrand post-tensioning tendons to determine the residual effective force in the tendon at the anchorage. The lift-off test is preferable and most easily done before the stressing tails of the tendons have been cut off. While it may be possible to conduct a lift-off test after the stressing tails have been cut off, this possibility is determined by the length of tendon protruding beyond the wedges in the stressing pocket as well as the possibility of connecting the hydraulic jack to this length of tendon (this may be dangerous).

When the tendon is initially stressed and anchored, the wedge seating that occurs develops a mechanical-friction force between the strand, wedges, and anchorage casting. During the lift-off test, it is necessary to stress the tendon in excess of the residual effective tendon force at the anchorage by an amount equivalent to this mechanical-friction force in order to "break the wedges loose" and determine the force remaining in the tendon. This process will be reflected during the lift-off test by stressing to a level (reflected on the gauge attached to the ram) sufficient to break the wedges loose and a subsequent reduction in the gage pressure to reflect the residual force in the tendon. It should be understood that the lift-off test determines the residual force in the tendon at the anchorage. Determination of the force level in the tendon at other locations requires detailed consideration of friction and wedge-seating effects.

12.2.3.7 Cracked Wedges

Hairline cracks may appear in the case-hardened surface of wedges owing to deformation of the wedges around the strand at the time of seating. These cracks do not affect the integrity of the post-tensioning system.

12.2.3.8 Shooting Power-Driven Fasteners

The structural designer of an unbonded post-tensioned member should offer detailed information regarding the limited application of power-driven fasteners that may be used on a particular project member. Frequently, developers are concerned about damage to tendons if future plumbing penetrations are added. This problem can be solved for structures in which changes are anticipated. The dye transfer technique relies on the dye color marked on the forms to transfer to the concrete soffit after the member is poured. Alternatively, markers may be installed to visually mark the location of each band and tendon bundle.

12.2.3.9 "Hazardous" Statement

The procedures described in this part of Section 12.2.3 may be hazardous. Only qualified experienced personnel, with a minimum of five years of specialized experience in the installation and repair of unbonded post-tensioned systems should attempt these procedures.

12.3 Crack Mitigation

A comprehensive discussion of restraint cracking and mitigation of restraint cracking is provided in publications available from the American Concrete Institute and the post-tensioning Institute. The discussions and illustrations in this section are extracted with minor modifications from these publications.

12.3.1 Crack Causes and Types

Three factors, when combined, lead to restraint cracks in post-tensioned slabs. First, post-tensioned slabs tend to shorten. Second, walls and columns restrain free movement of a slab. Third, the tension developed in a slab owing to restraint exceeds the slab's tensile capacity.

Factors causing shortening of slab are

- shrinkage of concrete,
- creep in concrete due to precompression,
- elastic shortening due to precompression, and
- fall in temperature.

For a typical parking structure in Southern California with 70% ambient humidity and a moderate temperature variation of 40°F, the contributions of the above factors to slab shortening are as given in Figure 12.16 and Table 12.1. It is noteworthy that two-thirds of slab shortening is typically due to shrinkage of concrete. Axial creep and elastic shortening, which are the only direct consequences of post-tensioning, precipitate about one-sixth of total shortening.

In order to appreciate the magnitude of shortenings that are likely to occur in a post-tensioned slab, consider the example

TABLE 12.1 Contribution of Different Factors to Typical Slab Shortening*

Description	Percentage
Shrinkage	66
Creep	11
Elastic shortening	7
Temperature	16
Total	100

*For a parking structure in southern California.

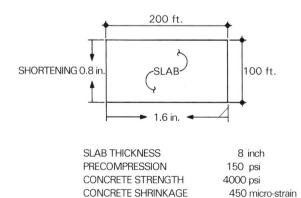

SLAB THICKNESS   8 inch
PRECOMPRESSION   150 psi
CONCRETE STRENGTH   4000 psi
CONCRETE SHRINKAGE   450 micro-strain

FIGURE 12.16 Factors contributing to slab shortening.

shown in Figure 12.17. For the 200 × 100 ft slab shown, the shortenings—if free to take place—are estimated at 0.8 in per 100 ft of slab length. Obviously, this shortening cannot materialize in most cases, since the slabs are commonly tied to supporting structural elements. The interaction of the slab with its restraining structural elements is the crucial factor in the formation of cracks.

Referring to the breakdown of shortenings in Figure 12.17, only 18% of the calculated shortening is due to post-tensioning. The balance is common to nonprestressed as well as post-tensioned slabs. This shows that there is little difference between post-tensioned and nonprestressed slabs as far as crack initiation is concerned. However, crack propagation is fundamentally different between the two types of slabs.

Prominent characteristics of cracks in unbonded post-tensioned slabs as compared to the regular reinforced concrete are the following:

   (i) Cracks are lesser in number. Instead of a multitude of hairline cracks, fewer cracks form.

  (ii) Cracks are generally wider. They are spaced farther apart and generally extend deeper into the slab. In regular reinforced concrete, the spacing between cracks is of the order of slab depth, whereas in post-tensioned slabs it is more related to the span length and the overall dimensions of the slabs. In most cases, crack spacing is more than one quarter of the shorter slab span.

 (iii) Cracks are normally longer and continuous. Continuous cracks may extend over one span and beyond. In nonprestressed concrete, cracks are generally shorter in length.

 (iv) Cracks commonly do not coincide with locations of maximum moments. Restraining cracks do not necessarily develop at the bottom of midspan or the top of supports where the bending moments are maximum.

  (v) Cracks occur at axially weak locations. Axially weak regions are typically found at construction joints, pour strips, cold joints, paths with reduced discontinuities in slab, and finally, where precompression is reduced either owing to termination of tendons or friction losses in tendons. Figure 12.17 compares typical crack patterns on the soffit of an interior panel of a two-way slab construction. For the regular reinforced concrete structure, the shrinkage cracks are shown coinciding with the locations of maximum tension.

Unbonded post-tensioned slabs generally exhibit poorer cracking performance as a result of lesser bonded reinforcement that mobilizes the concrete in the immediate vicinity of a crack. Hence a series of large slab segments separated by wide cracks rather than well-distributed small cracks are produced unless either the unbonded post-tensioning is accompanied by a sufficient nonprestressed reinforcement or in-plane restraining actions are present that result in a similar improvement of the crack distribution.

Examples of common cracks in slabs, columns, and walls due to restrained movement are illustrated below.

12.3.1.1 Slab Cracks

Figure 12.18 shows crack formations in two of many similar slab conditions investigated by the authors. The examples are representative of many slabs having similar crack formation. The slabs are post-tensioned in both directions and designed as a two-way system according to Chapter 18 of ACI 318 [ACI, 1995]. The precompression provided by the tendons in the longitudinal direction is, in both cases, dissipated into the supporting walls, since the primary transverse cracks extend across the entire width of the slab and through its thickness. In

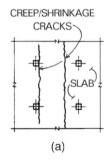
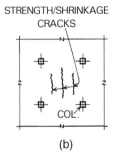

CREEP/SHRINKAGE CRACKS   STRENGTH/SHRINKAGE CRACKS

(a)   (b)

FIGURE 12.17 Reflected ceiling view of slabs. (a) post-tensioned slab; (b) reinforced concrete slab.

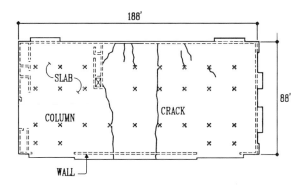

FIGURE 12.18 View of reflected ceiling showing cracks in post-tensioned slab, Village Serramonte, California.

the two cases exemplified in Figure 12.18, the prime cause of cracks is the restraining effects of the perimeter walls.

12.3.1.2 Column Cracks

Short columns at split levels in parking structures, as illustrated in Figure 12.19, can develop severe cracks and spalling of concrete owing to the shortening of the parking decks immediately above and below. The same figure shows a release detail with a central dowel for prevention of such cracks. For simplicity, the stirrups in the short column are not shown.

Columns tied to half-height walls as shown in Figure 12.20a develop cracks similar to those in the short columns described in Figure 12.19. The crack formation is especially severe in beam-slab floor constructions. Provisions of full-height or half-height joints between the walls and the columns, illustrated in Figure 12.20, are effective methods of mitigating such cracks.

End columns of slabs 150 ft or more in length are particularly susceptible to cracks of the type illustrated in Figure 12.21. The moment generated in the column due to this displacement should be accounted for in the design of such columns.

12.3.1.3 Wall Cracks

Figure 12.22 illustrates the most common crack formation due to overall behavior of walls tied to post-tensioned slabs. The diagonal tension cracks shown form at the ends of the walls owing to the movement of the slab and extend over a region having a length of approximately one to two wall heights from the wall end. Such cracks can be reduced or eliminated by design.

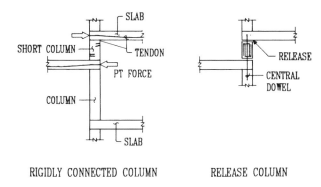

FIGURE 12.19 Cracking in short column at split level of parking structure.

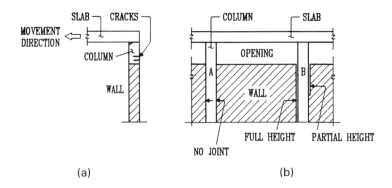

(a)   (b)

FIGURE 12.20   Wall-column release. (a) Side view of column tied to wall; (b) front view.

12.3.2 Crack-Mitigation Measures

The principle techniques of crack mitigation are described in the following sections.

12.3.2.1 Planning the Layout of Restraining Members

The most effective method of restraint-crack prevention is the good selection of wall and column locations during the architectural planning of the building. The equal number and length of walls may be positioned so as to reduce the tendency of crack formation by allowing the slab to move freely

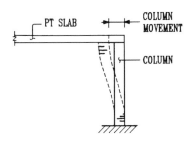

FIGURE 12.21   Cracks in end columns of long buildings.

toward a planned point of zero movement (Figure 12.23a). Figure 12.23b shows examples of unfavorably arranged walls and layouts in which the walls impede the free movement and thus create conditions conducive to crack formation.

12.3.2.2 Structural Separation

Slabs of irregular geometry in plan are particularly susceptible to cracking. Figure 12.24a shows a small slab area appended to a larger rectangular-shaped region. The structural separation shown in the figure between the two post-tensioned slabs consists of a physical gap between the slabs equal to 0.5 to 1 in. For the particular example shown, it is advisable to continue the slab separation through the supporting walls. The major difference between such structural separations and expansion joints is that the structural separation discussed herein loses its significance after a period

of two to three months, during which time the bulk of the slab shortening takes place. The structural separation does not need to be designed to remain serviceable during the lifetime of the structure. An expansion joint that has been designed to accommodate temperature-induced movements must be detailed to remain operational during the in-service life of the structure.

Smaller areas separated by openings or irregular slab geometries, such as the appendix shown in the top right corner of Figure 12.24b cannot generally follow the overall pattern of shortening of the entire slab area. Their connection to the main slab is mostly over short lengths. Stairwells, elevator shafts, and other walls

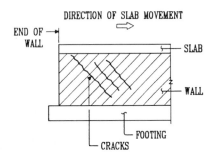

FIGURE 12.22   Cracks in wall due to slab movement.

impart substantial restraint against free movement of small slab areas. Moreover, for most cases, it is neither economical nor practical to effectively posttension small slab areas less than 20 ft in length. The author's practice has been to provide a separation between the two slab areas and construct the detached smaller region as a nonprestressed slab. The structural separation for such conditions need not extend through the supporting walls. Typically, the separation is achieved by placing Styrofoam© sheets 0.5 to 0.75 in thick vertically between the two slabs.

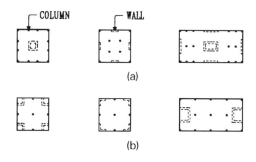

FIGURE 12.23 Planning in layout of shear walls to mitigate slab crack. (a) Favorable arrangement of restraining walls; (b) unfavorable arrangement of restraining walls.

12.3.2.3 Closure Strips, Joints, and Favorable Pour Sequencing

A closure strip, also referred to as a pour strip, is a temporary separation of approximately 30 to 36 in between two regions of slab that will be constructed and post-tensioned separately. Each region is allowed to independently undergo shortening. After a period of typically 30 to 60 d, the gap between the two post-tensioned slab regions—the closure strip—is closed by placing and consolidating nonshrink concrete. The reinforcement that extends from the concrete slab on each side into the closure strip provides the continuity of the slab over the strip.

The width of a closure strip is determined by the net distance required to position a stressing jack between the two sides of the strip and conclude the stressing operation. The reinforcement across the closure strip is designed on the basis of actions (moments and shears) occurring at the location of the strip when the entire slab is combined in a continuum. Between two adjacent supports, the preferred location of a closure strip is, for regular conditions, at a quarter span where the moments are typically small. Other considerations, however, may dictate the location of closure strip. The position of the closure strip in relation to the entire slab is discussed at the end of this section. For corrosion protection, it is emphasized that as a good practice the stressing ends of the tendons terminating in the closure strip should be cut, sealed, and grouted in the same manner as at free edges.

The time necessary to keep a closure strip open is determined by the extent of shortening deemed necessary before the tow-slab regions are tied together. Some engineers specializing in design of post-tensioned slabs use an empirical value of 0.25 in as the hypothetical displacement that can be accommodated in a post-tensioned member without apparent impairment of its serviceability. On

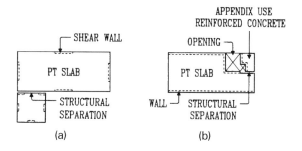

FIGURE 12.24 (a) Separation between large areas forming an irregular shape; (b) separation between a large area and a small appendix.

this premise, a closure concrete should be placed when the calculated balance of shortening on each side of the closure strip is 0.25 in or less. The shortenings are calculated using standard procedures, in which concrete is assumed to be free to move. Obviously, once the closure strip is poured and the two slab regions are tied together, the balance of computed shortening referred to can not take place. This empirical procedure is backed by the satisfactory performance of closure strips in place. It generally leads to closure-strip concreting between 14 to 120 d.

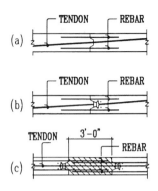

FIGURE 12.25 Details of slab joints. (a) Construction joint with no stressing; (b) construction joint with intermediate stressing; and (c) closure strip.

Construction joints are joints at predetermined locations in the slab between two concrete placements. The joints provide a planned temporary break between two slab regions for the purpose of crack control and construction operations. They are also used to subdivide a larger slab area into constructionally manageable sizes. A construction joint as shown in Figure 12.25 differs from a cold joint in that (i) its location is determined by design as opposed to the location at which a concrete batch is finished, and (ii) there is a time gap of commonly 3 to 7 d between the placement of first pour and the second pour. This time gap is applicable to joints that are designed for crack control.

Construction joints may or may not have intermediate stressing. Intermediate stressing of tendons is carried out for long tendons where friction losses are appreciable.

From the performance experience of post-tensioned slabs, the following guidelines for the provision of closure strips or structural separations may be considered.

(i)   If the slab length is less than 250 ft, no closure strip or structural separations are necessary, unless the supporting walls are unfavorably placed.

(ii) If the slab length is longer than 250 ft, but less than 375 ft, provide one centrally located closure strip.

(iii) If the slab length is longer than 375 ft, provide a structural separation.

12.3.2.4 Released Connections

Released connections are effective means of crack mitigation when favorable layout of supporting structural elements or provision of construction separations and closure strips cannot be fully implemented. Released connections are those in which a joint is detailed and constructed so as to permit a limited movement of the slab relative to its support. Released connections may be used in conjunction with closure strips and structural joints. Released connections with successful results are now common practice for post-tensioned slab construction in California. Released connections are grouped into wall/slab release, slab joints, and wall joints.

Wall/Slab Release. Figure 12.26 shows several types of commonly used wall/slab connections. To facilitate slippage, a slip material is normally provided at the interface of wall and slab. For simplicity in presentation, the connections shown are for the end walls and a terminating roof slab, but they are equally applicable, with appropriate modifications, to interior walls and intermediate slabs.

The connection type, shown in Figure 12.26a, with no ties between the slab and its supporting wall, is the most effective release joint, but its application is restricted by the fact that in many cases, walls in must be designed to transfer shear forces, in addition to gravity loading, at their interface to the slabs. Moreover, the stability of the walls owing to lateral loads may become a governing consideration. Such releases, where possible, are employed at the corners of the slab areas. The maximum length of a "no-tie" release is recommended to be limited to the height of the respective wall.

A permanent release with a dowel encased in a compressible material is shown in Figure 12.26b. The dowel is provided to impede catastrophic movements of the wall, as in the event of an earthquake.

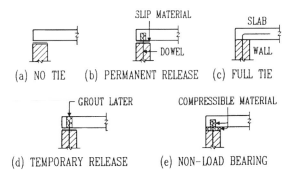

FIGURE 12.26   Typical details of different wall-slab connection types.

This permanent release detail is used more frequently than the no-tie connection. However, it is more costly and requires greater care during construction.

A temporary release as shown in Figure 12.26d is one where the slab is initially constructed released from the wall. After the shortening of the slab has taken place, to the extent that the balance is considered acceptable, the joint is fixed by grouting the pockets.

Slab/Column Release.   Columns may either be designed to withstand the anticipated forces conducive to lateral displacements between their ends without signs of distress or may be released to accommodate relative displacements of slab to column at the joints. The latter option, where applicable, leads to superior slab performance. Several items must be reviewed in arriving at a satisfactory solution.

Maximum displacements are typically at the end columns as shown in Figure 12.27. A detail providing rotational release at the base of the column, as shown in the same figure, may prove adequate. Where columns are excessively bulky, as may be required for architectural reasons, it becomes necessary to provide a detail that accommodates displacements in addition to rotation.

Wall Joints.   Wall joints are vertical separations between adjacent walls that enable the walls to accommodate displacements of slabs/beams supported by walls. Wall joints are very effective in mitigating cracks in slab/beams as well as cracks in the supporting walls themselves. Figure 12.28 shows the plan of a rectangular slab resting on perimeter walls and interior columns. For clarity, the columns are not shown. The wall joints (WJ) provided at the corners of the slab extend through the entire height of the walls. They allow the end wall to move toward the center of the slab without being impeded by the longitudinal walls. Such wall joints perform best when accompanied by a slip joint between the slab and cross walls as shown in Figure 12.29. The detail shows joints with no ties at the corners, which allows the wall shown at left to follow the movement of the slab to the

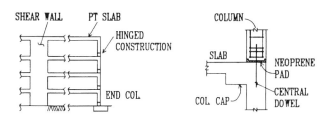

FIGURE 12.27   Hinged construction at base of end columns. (a) Elevation; (b) detail of hinge construction.

right without interference from the cross wall shown in elevation. The size of the gap is estimated to be 0.75 in per 100 ft of slab movement accommodated by the wall. Wall joints need not in all cases extend through the entire height of a wall down to the lower level.

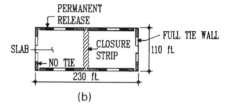

(a)

12.3.2.5 Addition or Improved Layout of Mild Reinforcement

In addition to the well-planned layout of shear walls and supporting structures and provision of releases, it is necessary to place additional mild reinforcement at locations of potential distress to mitigate crack formation. Figures 12.30 through 12.31 illustrate examples of typical cases.

Figure 12.30 shows reinforcement added next to nonreleased exterior walls. Owing to design shear-transfer requirements between a slab and its supp-

(b)

FIGURE 12.28   Wall joints. (a) Plan showing wall joints (WJ) and closure strip above; (b) plan showing arrangement of different wall-slab connections.

orting wall, it might not always be feasible to provide sufficient release details to prevent all cracks. The reinforcement shown in Figure 12.30 is found to be highly effective for such conditions. The steel is placed parallel to the wall over a width equal to approximately 10 ft normal to the wall. The steel area is determined as 0.0015 times the cross-sectional area of the slab over one-third of the transverse span. The bars are spaced alternately at the top and bottom at approximately 1.5 times the slab thickness. Note that this is not a code requirement, but rather a practice found to yield satisfactory results for the elimination of potential restraint cracks.

12.3.2.6 Addition or Improved Layout of Tendons

Figures 12.32 and 12.33 show two conditions where wall restraints can lead to significant losses of precompression in the central region of the slab and consequently lead to formation of cracks. In addition to other measures, such as the releases described in the preceding sections, it is helpful to lay out the tendons so as to deposit additional compression in regions where losses are expected to be highest. Dead ending and overlapping of tendons as illustrated in Figures 12.32 and 12.33 can serve this purpose.

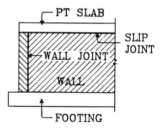

FIGURE 12.29   Elevation of corner wall showing wall joint.

The detailing of strand layout around discontinuities and openings is also of importance. Figure 12.34 illustrates two arrangements for tendon layout at an interior opening. The detail on the right shows a common practice where the sides of the opening are pulled apart. Cracks at the corners of such openings are not uncommon. The detail on the left demonstrates an alternative tendon layout, whereby the opening is provided with an additional precompression ring to counteract crack-precipitating stresses at the corners.

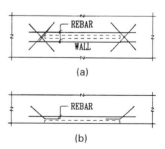

(a)

(b)

FIGURE 12.30   Crack-mitigating rebar next to shear walls. (a) Interior shear wall; (b) exterior shear wall.

12.3.3 Crack Evaluation Summary

From a study of crack formation in post-tensioned structures, a number of general conclusions have been formulated, as follows:

i) Shortening cracks (cracks due to constraint against free movement of the slab) are common in post-tensioned slabs supported on walls and stiff columns.

ii) Shortening cracks can be reduced significantly through crack-mitigation measures. The principle crack reduction procedures are

- planning for layout of constraints,
- structural separations,
- closure strips, joints, and favorable pour sequencing,
- released connections,
- addition or improved layout of mild reinforcement, and
- addition or improved layout of tendons.

  Concerning the implementation of crack mitigation procedures the following guideline is recommended:

- For small and simple slab geometries (10,000 ft² or less) supported on regular-size columns, design the slab to withstand the forces generated by shortening. It is not generally cost effective to implement crack-mitigation measures.

- For slabs with substantial restraint, it is necessary to implement crack-mitigation measures.

iii) Most shortening cracks are not structurally significant.

  The most common cause of shortening cracks is the exposure of reinforcement and post-tensioning to corrosive elements; aesthetics and leakage are the next most common considerations.

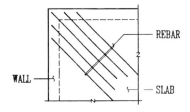

FIGURE 12.31 Reinforcement at slab corners.

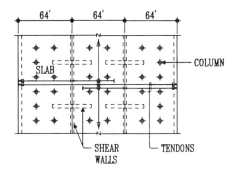

FIGURE 12.32 Tendon arrangement to compensate restraining effects of transverse walls.

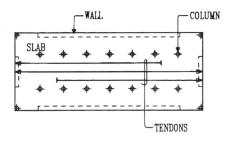

FIGURE 12.33 Tendon arrangement for mitigating cracks in central spans.

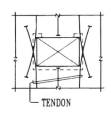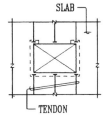

CRACK INHIBITING LAYOUT   CRACK PROMOTING LAYOUT

FIGURE 12.34 Arrangement of tendons at an interior opening.

v) For slabs with significant support restraints, such as perimeter walls, it is often necessary to conduct a one-time maintenance routine to repair shortening cracks. In such cases, notes should be added to the structural drawings indicating the following:

- Shortening cracks are likely to occur.
- Shortening cracks do not normally impair the structural integrity of slabs.
- Slabs should have a one-time crack maintenance operation, which consists of
  - inspecting and evaluating slabs and supporting members two years after construction,
  - determining cracks to be repaired, and
  - repairing cracks.

12.4 Evaluation and Rehabilitation of Building Structures

12.4.1 Evaluation

Even though concrete is considered to be one of the most durable construction materials, building structures need to be evaluated during their useful life for various reasons. Some of the more common reasons may include: deterioration (serviceability shortcomings, loss of strength), change in loading of a structure, building modification, overloading (disaster), or simply as an assurance evaluation.

The evaluation of an existing unbonded post-tensioned elevated floor system may be divided into three principal activities, namely, (i) examination of the existing floor system and structure (conditional survey), (ii) diagnosis, and (iii) prognosis or findings. All three steps are necessary before a repair or retrofit can be outlined. The level of evaluation may vary from simple nondestructive examinations and preliminary calculations for initial reporting to rigorous destructive testing with detailed analytical diagnosis of the structure.

12.4.1.1 Examination of an Existing Unbonded Post-tensioned Floor System

As part of the structural assessment, the engineer should initially complete a detailed survey of the structural floor framing system, geometry, and the "as is" condition of the structure, including environmental impact. This stage includes the collection of existing information about the structure, surveying the floor system for signs of distress, and establishing a listing of proposed nondestructive and destructive testing based on the distress observed and information gathered.

To establish as-built data, the examination should consider (i) material properties, and (ii) the detailed survey of the condition of each material and the material configuration or quantity.

i) The physical testing of concrete may include verification of the concrete compressive strength, material uniformity, mix properties, permeability, and aggregate type. In addition, chemical testing may be performed to establish the concrete constituents, which are used in the evaluation of concrete reactivity or concrete resistivity. The tests should determine the amount of chloride, sulfate, or other materials that may result in a chemical attack on the concrete section of concern. Properties such as the tensile strength of the mild steel and post-tensioning strands may be of interest. Should the tendon anchorage zone be of concern, it may be necessary to test individual elements of the anchorage system.

ii) The material-condition survey of the examination may focus on locations of primary distress but should extend to the comparative performance of the entire structure. Concrete voids (consolidation), delamination, spalling, discolored concrete, chemical attack, excessive air voids, and cracking should be investigated to determine the extent, formation, and amount of concrete damage within the distressed concrete area. In addition to selectively locating the position and amount of reinforcing steel and post-tensioning, the level of deterioration, as a result of corrosion, should be assessed. Specifically, post-tensioned slab edges should be surveyed for exiting tendons or loose grout plugs at stressing ends. The grout plug removal at tendon-stressing ends may allow the visual assessment of the anchorage zone. Along the tendon length, removal of the tendon sheathing may be necessary to view the condition of the high-strength strand wires at high and low points.

Besides the factual material-condition review of structural members, the survey should include a performance survey. Besides the typical serviceability considerations, this would include the performance review of joints, hinges, attachments, locations of movement, and locations of restraint. As part of the material-condition survey, each distress location should be assessed using time as one of the evaluating parameters.

12.4.1.2 Diagnosis

After all project documents have been collected and reviewed and after detailed inspection and testing of the floor system of the structure has been performed, the formal process of analytically assessing the existing unbonded post-tensioned floor member takes place.

First, the floor framing system should be checked for adequate strength. The selected modeling must accurately represent the actual geometry and boundary conditions using the two-dimensional simple beam frame or the equivalent frame slab strip for a two-way slab. The calculated elastic factored-load moments may be increased by the code-permissible plastification, commonly referred to as "redistribution of moments." The elastic support moments may be raised or lowered to the maximum percent of the code-permissible redistribution for each respective support. This band of the factored-load moment demand yields a range of acceptable solutions.

After establishing the band of acceptable moment-demand solutions, determine the capacity of the existing structure at critical points. Select a redistribution based on the capacities of the existing structure and the permissible percentages computed, and, finally, establish whether or not the redistribution made on the basis of the existing capacities falls within the permissible range. The demand-versus-capacity check should include consideration of possible strength loss due to distressed member cross sections. For conditions where the demand design strength exceeds the member's capacity, refer to Section 12.4, Retrofit. For unbonded post-tensioned slabs in particular, extensive destructive testing should be conducted to confirm that calculated member capacity is based on observed tendon's quality and quantity.

Second, serviceability of the unbonded floor system should be reviewed. Sectional stress checks at critical locations and immediate and long-term deflections should be calculated. The observed distress may reveal a direct relationship to the original design, material sections, construction, and/or applied loads and maintenance. In most cases, serviceability limitations, such as durability or deferred maintenance shortcomings, initiate concerns that are typically answered by evaluations. The more common serviceability shortcomings include corrosion of strands and anchorage zones and broken strand wires and tendon concrete cover at high and low points.

12.4.1.3 Findings

Assessing a post-tensioned member, particularly one showing signs of distress, is not simply a matter of conservatively selecting the desired material properties and load flow path, as is the case for a newly designed member. After an analytical model representing the unbonded post-tensioned member has been developed, it should be used to perform a detailed review. The objective is to explore the consequences of variation in tested material properties, assumed loading and load distribution, boundary conditions, extent of distress, and rate of material deterioration. As part of a member evaluation process, it is common to apply different modeling techniques in the process of understanding how the structure behaves and where hidden strength reserves may be available. After the diagnosis of the structural element is completed, the engineer should have developed a clear understanding of how the structure behaves and be clear on the causation of distress, including future behavior, considering time and site-specific adverse environmental conditions as additional dimensions. Based on the preceding description, it is obvious that the prognosis is based on a multitude of variables that may only be understood by a reviewer with extensive experience in the evaluation of unbonded post-tensioned concrete structures. In closing, it should be noted that the review or evaluation of an existing post-tensioned member requires the highest degree of professional expertise, knowledge, and integrity from the assessing engineer.

12.4.2 Repair

Unbonded post-tensioned concrete members are considered one of the more difficult structural members to assess for repair or retrofit. It is not uncommon that engineers who are, not experienced with unbonded post-tensioning, misinterpret the distress observed. The author frequently encounters assessment reports of crack development that conclude that cracks may have developed as a result of member strength deficiency, where, in fact, the cracks may have developed as a result of reverse tendon-profile curvature or restraining effects. Even if the assessing engineer correctly found the cause of concrete cracking of the post-tensioned member, the effects the distress has on the structure and its repair may be grossly misjudged. For this reason, it is not uncommon that proposal requests for evaluation and repair require the contractors to have a minimum of five (5) years of specialized experience in designing, analyzing, and repairing unbonded post-tensioned members. This is one of the fundamental considerations before assessing the repair of a member.

The common repairs outlined below may be used to address distress resulting from poor detailing or construction execution and deterioration of the materials.

12.4.2.1 Blowout

The sudden exiting of a tendon at the slab edge or at tendon profile high or low points, during or after stressing, is typically referred to as "tendon blowout." In contrast, if the concentrated precompression force behind tendon anchors or the reverse tendon-profile curve causes the sudden disintegration of a localized concrete pocket, it is typically referred to as "concrete blowout."

Typical Examples of Tendon Blowout. Tendon blowouts are typically recorded either at the slab edge or on the slab surface or soffit. During the installation of mechanical, electrical, or structural elements, should the contractor partially rupture or cut the unbonded tendon(s), the tendon tail may eject from the slab edge. The amount of existing slab edge is primarily related to the type of sheathing that was supplied. Extruded tendon should typically result in minimal tendon exiting (up to 3 ft) at slab edges (Figure 12.35). Whereas, if unbonded tendons are used, with a stuffed or heat-sealed sheathing application, the tendon exiting during a sudden release of stored energy may be unpredictable. The minimal void between the sheathing and the creased strand (of stuffed or heat-sealed tendons) limits the internal friction, allowing the stored energy to travel past the existing location.

FIGURE 12.35   Tendon exiting at slab edges.

FIGURE 12.36 Vertical loop resulting from exit of seven-wire strand.

During a sudden release of the entire tendon force, the strand may exit vertically (also known as vertical tendon blowout) at locations of minimal concrete cover, typically the high and low points of profiled tendons, owing to the vertical component of the draped profile. The seven-wire strand may exit, resulting in a 1 to 3 ft vertical strand loop (Figure 12.36). Conditions noted above may be a result of the accidental cutting of a strand or due to deterioration of the member.

Horizontal tendon blowouts are typically found where horizontal tendon curves around openings are not detailed and executed correctly. Tendons are typically installed side by side in groups of two to four unbonded tendons. The tension force allows the strands to ride on each other in the horizontal radial plane, creating a splitting force that can result in a tendon blowout (Figure 12.37). Tendon groups should be reduced and limited to groups of two maximum over the length of horizontal curvature. In addition, adequate reinforcing steel (U-pins) should be added to account for and tie back the centrifugal forces.

FIGURE 12.37 Tendon blowout resulting from strands riding on each other.

Typical Examples of Concrete Blowout.   Concrete blowouts are most frequently recognized during the tendon-stressing operation. The anchoring of tendon forces tests the compressive strength of the concrete pocket immediately behind the tendon anchorage. A simple void, rock pocket, low concrete strength, or lack or reinforcing behind the tendon anchor may cause the concrete to pulverize, resulting in a concrete blowout. Figures 12.38 and 12.39 show pulverized concrete behind the tendon anchor due to low concrete strength and voids, respectively. Most concrete blowouts are recognized during the stressing operation or within several months of stressing the tendons.

In addition, tendons that are stretched over a longer distance near the concrete surface (i.e., with minimal concrete cover) and that have a reverse tendon-profile curvature may split the concrete section over the distance of reverse curvature to allow the tendon to straighten. This may take place during the stressing operation or at a later date if the concrete section experiences additional concrete stresses due to loading.

FIGURE 12.38 Pulverized concrete behind tendon anchor due to low concrete strength.

FIGURE 12.39 Pulverized concrete behind tendon anchor due to voids.

How to Repair Concrete Blowout.   Depending on the location and severity of the blowout, adjacent tendons may have to be detensioned before concrete removal can begin. After detensioning (as necessary), the damaged concrete is removed in sufficient amounts to expose any damaged strands and allow the resetting of the anchorages. In some cases, it may be necessary to use couplers in the repair to increase tendon length owing to strand damage.

It is important that the back side of the opening be cut square and perpendicular to the tendon so that slippage of the concrete patch during stressing will be avoided. Remove all loose debris and clean the surface of dust.

Make sure all the anchor zone reinforcements have been replaced and fill the area to be repaired with a high-strength nonshrink concrete grout mix. Do not use grout that contains calcium chloride or other materials containing chlorides.

Stress the tendons only after the grout patch has attained the required design strength as approved by the engineer of record.

12.4.2.2 Tendon Rupture

Tendons may rupture partially if only one wire is damaged or totally when several wires are damaged, causing the remaining wires to fail under sustained tension load. The cause of tendon rupture for tendons should be determined. One of the most critical actions for unbonded post-tensioned structures is to immediately determine the extent of damage aged. If corrosion is the leading cause of the damage, several tendons should be spot checked to establish the extent of damage to each tendon and the slab area or beams in which damage has been recorded. If preliminary calculation indicates that the loss of the ruptured tendon(s) affects the strength of the unbonded post-tensioned member, the member should immediately be shored.

Destructive testing to recover representative strand samples should be considered in order to record the tendon location, wire condition at the location of rupture, concrete cover, type of sheathing, condition of the crease, remaining cross-sectional area of the seven-wire strand, and other project-specific conditions. Once the extent of damage and type of tendon sheathing has been identified, strand replacement or splicing of tendons may be considered.

Strand rupture or breakage can also occur from misalignment of wedges, the anchor to the strand not being perpendicular, overstressing, and/or internal damage to the tendon. Misalignment of wedges occurs when the two or three parts of the wedges are offset prior to stressing. The wedges can pinch one or more wires owing to different circumstances. Internal damage to the tendon could be caused by nicks in the strand or heating of the strand due to torch cutting of adjacent objects prior to concrete placement. Damage can be caused after concrete placement by drilling, saw cutting, or power actuated studs that are shot into the concrete.

If a strand does not hold, remove the wedges, clean the cavity, install new wedges and restress. Overstressing of a tendon can occur by misreading the pressure gauge or using a jack and gauge that are out of calibration or that are not a matched set. The strand may either break or be stressed beyond yield. If the strand breaks, the Engineer of Record will determine how the structure is affected and whether replacement is necessary. If the wedges hold and the strand does not break, it is usually preferable to leave the tendon in the overstressed condition. Attempts to detension the tendon may damage tendon or break it necessitating replacement of the strand.

Prior to the replacement of existing strands, adjacent strands may have to be detensioned. After exposing stressing and dead ends, it may be advantageous to weld a 1/4-in wire rope to the existing damaged strand tail, allowing the repulling of the new strand. If the existing strand is ruptured, not allowing the rethreading of a wire, the new strand may have to be pushed into the sheathing void by hand. Strand replacement can be effectively executed for unbonded monostrand tendons with heat-sealed or stuffed sheathing. The slightly oversized sheathing typically can accommodate new strands. For long tendons or tendons with extruded sheathing, a smaller size strand, such as 7/16 or 3/8 in diameter strands, may be selected. Strand replacement for tendons with paper wrapped sheathing is nearly impossible. For all strand replacement, adequate creasing application should take place as part of the strand installation procedure.

The substitution of the original 1/2 in diameter strand with a smaller strand may be acceptable for the following reasons. Most older structures used stress-relieved type strands that are typically replaced by new low-relaxation strands. The strain tempering is very effective in improving the stress-strain characteristics and has the additional advantage of substantially reducing time-dependent losses due to relaxation of the strand. Stress-relieved type strands may experience over 10% more loss in stresses due to relaxation, than do low-relaxation strands. The stress loss due to concrete shortening may not have to be compensated for.

12.4.2.3 Cracking of Concrete Members

Crack Development. The most frequent crack development is typically due to restraint of adjoining members. This phenomenon is extensively covered in Section 12.3 above.

Crack development due to the incorrect installation of the tendon profile or anchors and over-balancing of the member self-weight is in most cases misinterpreted.

The incorrect placement of an unbonded tendon profile high point near the quarter span of a beam instead of over the column support has resulted in notable beam cracking approximately two beam depths away from the support (Figure 12.40). In this case, even though the inspection reports

FIGURE 12.40 Beam cracking due to incorrect tendon profile high point placement.

FIGURE 12.41 Misaligned tendon revealed by destructive testing.

and survey of the installation crew confirmed correct placement of the tendons, destructive testing revealed the misaligned tendon profile (Figure 12.41). A simple realignment of the tendon profile reconditioned the beam for its intended life cycle.

The layout of unbonded tendons should incorporate appropriate locations for the tendon's dead ends and stressing ends. The localized concrete zone surrounding groups of added tendons may be subjected to high tension stresses, which, if combined with flexural stresses, may result in crack development. Figure 12.42 illustrates a significant crack that developed in a post-tensioned beam at the location of added tendon dead ends. The dead ends were not staggered, allowing a load distribution as called for in the standard details.

Repair of Cracks. The objectives of crack repair on structures with cracks caused by restraint effects tendon profile or with cracks in anchorage zones, which were surveyed by the author, served the following purposes:

- In most cases, repair was conducted as a precautionary measure to end the exposure of reinforcements and post-tensioning to weather and moisture. In some cases, it was performed to stop leakage.
- It was rarely necessary to carry out repairs in order to restore structural strength.
- Occasionally, repairs were conducted for aesthetic reasons.

FIGURE 12.42　Crack in a post-tensioned beam where tendon dead ends were added.

Which cracks should be repaired.

- Cracks that are determined to be of structural significance should be repaired regardless of width and location. Most such cracks are due to poor design, deficient detailing, or bad workmanship.
- Cracks that affect the serviceability of a structure, such as deflection and local distress, may be left unrepaired if the diminished serviceability is acceptable and the repair is not cost effective.
- Under normal conditions of service, shortcomings due to deterioration may be encountered if cracks exceed 0.01 in width. Such cracks should be sealed to prevent intrusion of moisture and possible oxidization, loss of steel area, and possible spalling.
- Cracks in structures exposed to especially adverse conditions should be sealed, even if they are less than 0.01 in width. Also, cracks that show rust stain should be sealed.

When cracks should be repaired.　Restraint cracks are best repaired after the shrinkage and creep shortenings are essentially complete. Generally a lapse of approximately two years after construction is adequate, after which cracks may be repaired. A time delay in sealing of cracks is only justifiable if corrosion considerations permit. Cracks caused by reverse tendon profile should not be repaired until the reverse profile is relieved, neutralizing the cause of cracking.

How to repair cracks.　There are numerous reports on methods for sealing cracks in prestressed concrete structures. The most common and effective procedure is the injection of an epoxy resin compound under pressure into the cracks in order to fill in the crack voids. For details, consult the manufacturers' literature.

- For cracks that are "nonworking," that is, they no longer move, the best sealing method is to inject the cracks with an epoxy resin of low viscosity. This is done in such a manner that the cracks filled with the resin and the concrete on each side of the cracks are reunited by the "gluing" action of the resin. Another method is to rout a groove along the crack throughout its entire length and fill the groove with an epoxy compound. The latter scheme is not recommended in highly corrosive environments.
- Cracks that are "working," that is, they open and close as a result of loads, temperature, etc., cannot normally be successfully sealed with epoxy compounds but must be sealed with flexible sealant that can withstand the movements to which the cracks are subjected.

12.4.3 Structural Retrofitting Using Unbonded Post-tensioning

From the mid-1980s to the mid-1990s, a series of natural disasters have tested the performance of existing structures on the west and east coasts of the United States. As a result, building-code revisions addressed improved strength and serviceability considerations. The perception of building performance during recent disasters, the ever-changing building codes, and the gradual deterioration of existing structures have sparked increased public interest in the retrofitting of structures. The objective of retrofitting a structure is to modify or improve the existing member's strength and/or serviceability. The options of strengthening an existing concrete floor system for gravity loads include the following:

- added drop caps, drop panels, or beams at slab soffit;
- slab overlay that supports the existing slab dead load;
- increase or jacketing of existing beams, girders, and columns;
- add columns or remove and replace existing columns;
- added gridwork of beams at slab soffit;
- external applied metal plates attached to the existing concrete slab; and
- external prostrating.

The options outlined in the following sections are limited to members being retrofitted with unbonded post-tensioning. It is understood that other elements or connections may need to be strengthened (columns, walls, and foundations and their connections) within the structure as a result of the external tendon retrofit application. The application of external prestressing for nonprestressed or prestressed floors is a widely used retrofit option for gravity-load strengthening and serviceability. The following section differentiates between gravity-load strengthening and serviceability considerations.

12.4.3.1 External Prestressing for Gravity-Load Strengthening

A strength requirement insures that all elements of the building provide an adequate factor of safety against injury or material damage in the event of a code-specified overload. Slabs with inadequate strength or slabs that are subject to overload initially exhibit significant crack formation in tandem with noticeable deflections. If a thorough evaluation indicates inadequate member capacity for code-predicted factored load demand, the member should be subject to strength retrofit.

Using an external unbonded post-tensioning retrofit scheme, the author has used two principal approaches to compensate for the strength shortfall of the existing structure, namely, direct-member strengthening (for one-way slabs and beams) and indirect-member strengthening (for two-way slabs).

Direct-Member Strengthening. Direct-member strengthening is typically used for one-way members such as one-way slabs and beams. First, the engineer should establish the existing member's capacity and scale the strength shortfall by comparing the established capacity with the load demand. The strength-shortfall compensation may be readily supplied by attaching externally stressed unbonded tendons on each side of a beam. The tendons should be profiled (typically, harped with one or two deviators) so as to uplift or unload the beam equal to or more than the amount of load that the existing capacity can not safely sustain. The external tendons are only intended to supplement the existing capacity. It may be difficult to establish the capacity of damaged structures or members with highly deteriorated reinforcing. In such cases, the external post-tensioning may be considered to take all loads. Where anchors are attached to columns or beams, the applied load is retained in the form of precompression in the existing member. The installation of deviators (deflector saddles) or anchors should miss all main member reinforcing.

Indirect-Member Strengthening. Indirect-member strengthening, typically used for two-way slabs, takes advantage of the possibility of alternate load passes by using the existing structure's capacity. Initially, a two-way member may be examined, using code-factored loading, to establish the as-is failure mechanism and locations of hinge formation. The objective is to search for and select an alternate failure mechanism that is capable of safely sustaining the factored loading. Through the addition of external applied post-tensioning upward forces, the failure mechanism may be altered to accomplish this objective.

External prestressing, if used to supplement the member's strength, should be encased in fire-retardant material that meets the fire resistivity or rating requirements for the particular application.

12.4.3.2 External Prestressing for Serviceability

Serviceability describes the in-service functionality of a building for its users. Excessive out-of-level floors, inadequate drainage, perceived vibration, perceived sagging of ceiling lines, exposure of reinforcement to corrosive elements owing to excessive crack widths, and unsightly cracks are the primary serviceability considerations. Serviceability may be influenced by original design, material selection, construction, applied loads, and maintenance.

When a structure reflects signs of serviceability shortcomings, such as excessive deflections or cracking, external prestressing has been effective as a corrective retrofit. For example, the installation of tendons at the underside of a slab or at each side of a beam, profiled to result in upward forces where desired, may be an effective and economical retrofit solution. For cracks or deterioration of members, additional work may be required beyond the application of external post-tensioning. The retrofitting of nonprestressed members, using external forces to counteract excessive elastic deflection and plastic deformation, may need to be analyzed using specialized software to model the time-dependent creep deformation.

If the application of external tendons is used solely for the purpose of improving serviceability shortcomings in a structure with adequate strength to sustain the code-predicted factored loading, the tendons may not require corrosion protection if they are aesthetically acceptable.

12.4.3.3 Retrofit Application of External Unbonded Post-tensioning

The principal considerations for the selection of a retrofit scheme are performance, durability, economy, and appearance. The performance records of external prestressing on hundreds of retrofitted structures across the United States and its versatile and economical application has resulted in today's frequent use of this system.

External tendons may be threaded through existing concrete members (such as walls or beams), directly attached to existing elements, or routed over deflector supports. When selecting a particular tendon layout support system, access, available space, fire protection requirements, and aesthetics need to be considered. The following section is intended to offer selected examples of the external prestressing installation application. The information on member selection and connections, shown in the details, offers project-specific design information prepared by the author, which may not be applicable to other retrofit projects.

Beam Retrofit. During the course of evaluating the nonprestressed beams of a private parking structure in Woodland Hills, California, excessive postelastic deformation and concrete cracking were recorded. The use of bundled unbonded tendons on each side of the beam, with one deviator at midspan, was selected to utilize the upward force component to instantly neutralize elastic deflection and to remove the postconstruction plastic deformation (creep) to near zero within a 10-year period after the retrofit was successfully installed. A time-dependent analysis was performed to establish the deformation of the nonprestressed beams during their predicted useful life. A typical elevation and details of tendon attachments are shown in Figures 12.43 and 12.44. An alternative connection of external unbonded tendons is shown in Figure 12.45.

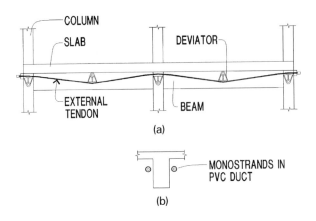

FIGURE 12.43 Schematic of external tendons profiled on each side of the beam. (a) Typical elevation; (b) section.

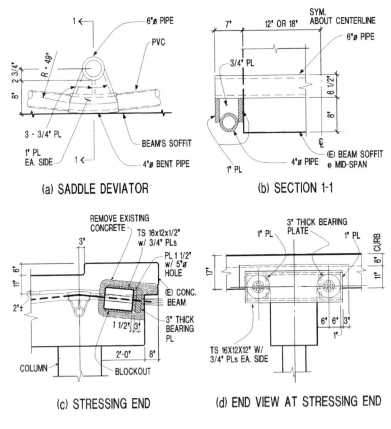

FIGURE 12.44 Retrofit details.

Slab Retrofit. During the final construction stages of a hybrid structure (a three-story residential wood framing over a one-story concrete garage) located in Glendale, California, excessive deflection and cracking of the elevated concrete slab was recorded. An initial document review revealed that inadequate reinforcement was specified in the original design. The nonprestressed cast-in-place concrete slab was supported on an array of orthogonally spaced concrete columns. The first-mode failure mechanism of parallel hinge line formation was altered by the introduction of external upward forces along said hinge line (Figure 12.46). The upward force was calculated, utilizing

FIGURE 12.45 Alternative connection of external unbonded tendons.

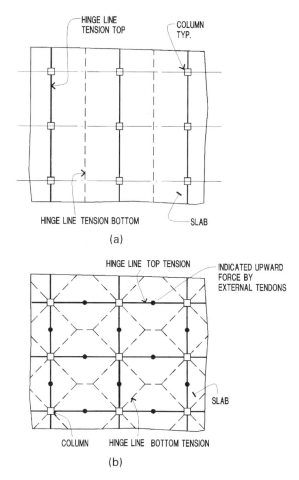

FIGURE 12.46 Simplified failure mechanisms of column-supported slabs. (a) Failure mechanism as constructed; (b) new failure mechanism after retrofitting the slab.

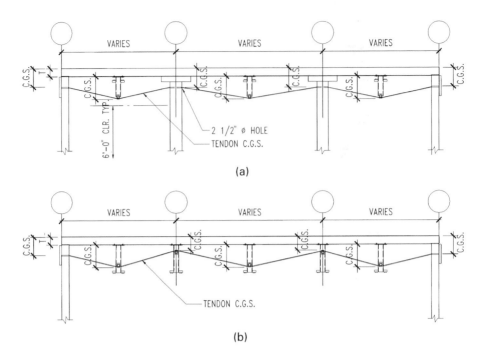

FIGURE 12.47 External tendon layout support systems. (a) Typical harped tendon layout at column lines; (b) typical harped tendon layout at mid-span.

the existing slab's capacity, allowing for an alternate failure mechanism with a significantly higher capacity limit. Three principal external tendon layout support systems were proposed.

The first system consists of simple steel tube deviator or saddle supports anchored at each end of the structure (Figures 12.47 and 12.48). The fabrication and application of this externally applied system can be readily installed around utility pipes and other obstructions. This layout should be considered if adequate headroom is available, which allow for large harped-profile shapes. However, short spans may result in large tendon angular changes over the deviator, which may result in localized damage to the tendons. Aesthetics considerations may be difficult to satisfy, especially if the entire system requires fire-protection coating (Figure 12.49).

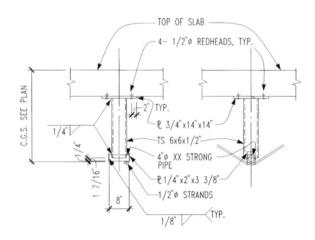

FIGURE 12.48 Typical saddle detail.

FIGURE 12.49 Fire-protection coating.

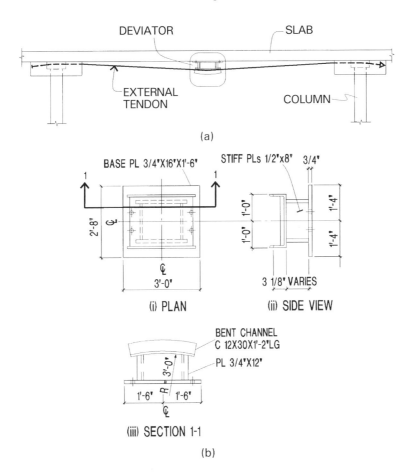

FIGURE 12.50 Schematic of external tendon at column line. (a) Typical elevation; (b) typical tendon deviator detail.

The second option may allow for using more tendons with a less vertically harped profile and tendons anchored in added drop panels at columns (Figure 12.50). The advantage of this option is that tendons may be discontinued at column lines for localized application, and it may be installed in areas with limited headroom. The concrete for the added column capitals may have to be poured

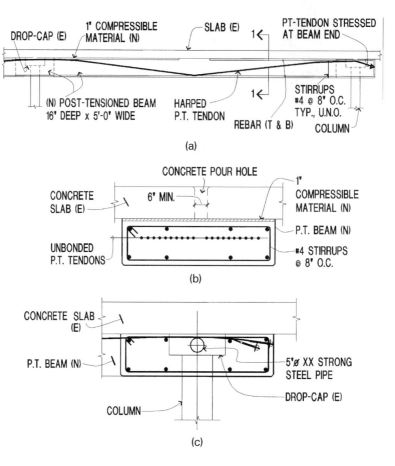

FIGURE 12.51   Retrofit beam and details. (a) Typical beam elevation; (b) section 1-1; and (c) rebar through beam.

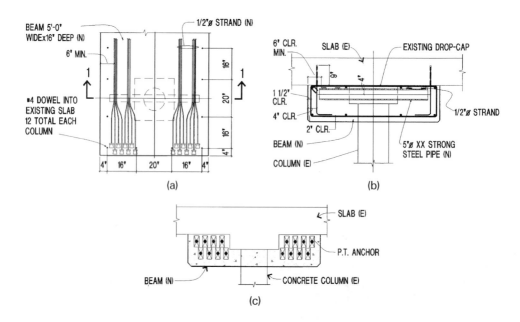

FIGURE 12.52   Details of retrofit using external tendons. (a) Plan; (b) section 1-1; and (c) post-tensioning anchor orientation.

FIGURE 12.53   Tendons installed within a reinforcing cage.

through access holes from above the slab. A fire-protective coating must be applied if external tendons are needed to supplement strength shortcomings.

The third option was added to address fire protection and aesthetic considerations. The objective of this option is to install additional unbonded tendons without the retrofitting of the structure becoming visible. The tendons are installed within a reinforcing cage reflecting the intended vertical profile (Figures 12.51 to 12.53). A concrete beam is poured at each column grid line to encase the tendons, offering fire protection and concealing the retrofit approach in concrete, hiding the afterthought even to a trained eye. Enough tendons must be selected to balance the dead load of the added concrete beam in addition to achieving the required upward force. The beams are isolated from the existing slab with contact at midspan for upward load transfer. The concrete for the added beams may have to be poured through access holes from above the slab. As this option addresses the durability and aesthetic of the retrofit effectively, the economic penalty of adding a grid of concrete beams may not be of great significance.

12.5 Demolition of Post-tensioned Structures

12.5.1 Introduction

The purpose of this section is to address the concerns of the industry, primarily contractors and engineers, regarding the unique properties of prestressed concrete that must be addressed in order to safely demolish structures that contain unbonded prestressing systems and devices.

The scope of this section is limited to the structural engineering considerations that must be made in order to properly evaluate prestressed structures for demolition and not for the purpose of mandating regulatory requirements. The reader is advised to refer to those publications prepared by OSHA (promulgates rules and regulations regarding health and safety in the workplace) and the American National Standards Institute (provides standards for safety during construction and demolition operations.) In addition, the work of the National Association of Demolition Contractors in their publication, *Demolition Safety Manual*, covers the use of demolition equipment and safety procedures to be followed for all forms of structures to be demolished including precast and prestressed concrete structures.

The recommendations are presented for the guidance and information of professional engineers, who must add their engineering judgment to the application of these recommendations.

12.5.2 Structural Evaluation

The demolition of an unbonded post-tensioned structure should be carefully evaluated by an engineer to ensure safety during all phases of the demolition operation. Shoring that is to be used to support the structure during all phases of demolition, including tendon destressing, should be designed to accommodate vertical loads and horizontal loads, including any associated structural deflections.

12.5.2.1 Material Considerations

Wood and steel structures allow the demolition contractor to stage the demolition sequence and procedures based on the member strength and configuration. Concrete structures have inherent uncertainties concerning the location, size, splice, prestressing force, and quantity of tensile reinforcing beyond the information provided by original construction documents. The layout and amount of mild and prestressed reinforcing in the continuous beams of frames must be reviewed, accounting for the altered modeling of a span by a span-demolition sequence.

12.5.2.2 Identification of Structural System Used

It is necessary to review available records relating to the design and construction of the building or structure to be demolished. These include design drawings, shop drawings, project specifications, field reports, repair records, job photographs, and correspondence. These records should be retained by the building owner, and may be available from the Engineer or Architect of Record, or the contractor, and should be on file with the local building department.

12.5.2.3 Condition Survey of Building or Structure

Prior to demolition operations, an engineering condition survey of the premises should be performed to determine the type and condition of the structural elements in order to prevent a collapse of any portion of the structure. This survey should include adjacent structures, overhead and underground utilities, and sidewalk vaults that may be affected by the demolition operations. All necessary permits and notices to adjacent owners and utilities should be obtained. The condition survey should include intrusive probes of the slab, girder, and column connections of critical elements in order to verify both their presence and their condition.

12.5.2.4 Determination of the Condition of Critical Elements

All structural elements reviewed should be to ensure that sufficient load-carrying capacity and stability is maintained for each stage of demolition. Strength capacity may be affected when adjacent supports and lateral load-resisting elements such as shear walls are removed. Strength reduction may already be diminished owing to previous damage from fire, corrosion, age, and improper and undocumented alterations to the structure.

All structural elements reviewed should be for the controlled release of stored energy in cutting tendons, especially the unbonded tendons. Location and direction of anchor wedge seating of intermediate anchors and splices may be of great interest to avoid a progressive collapse in a multi-frame structure.

Stability requirements for each stage of demolition should be evaluated. Locations of stair and elevator towers may be isolated from the primary structure and may require special attention.

A structural analysis may be required to determine unbalanced post-tensioning thrusts on framed elements, shear forces on wall systems, cantilever construction, continuously designed floor elements, and locations of critical tensile stresses in post-tensioned concrete elements owing to temporary loading and unloading conditions. The presence of wind forces and other environmental effects should be considered in the analysis, including temporary crane and equipment loads that may be supported on the structure.

The prime concern to be addressed in all demolition of post-tensioned concrete buildings and/or prestressed elements is the sudden release of stored energy caused by either removing the adjacent concrete cover or cutting or burning the stressed tendon strands, either intentionally or accidentally, which may cause a sudden uncontrolled collapse to occur. In addition, the possibility occurs that anchors and tendons may be released suddenly, resulting in steel ejection at slab edges or blowouts along the tendon length.

12.5.2.5 Preparation of Demolition Plan

A written plan for controlled demolition operations should be prepared that includes the following:

   i) identification of anchorage layout within the structure,
   ii) site protection and barriers,
   iii) sequence of dismantling structural elements and demolition of entire assemblies,
   iv) location of all temporary shores and supports,
   v) location of all equipment loads,
   vi) sequence of cutting strands,
   vii) sequence of cutting partial structural elements,
   viii) control of other site operations,
   ix) preparation of demolition sequence and procedural drawings, and
   x) predemolition survey.

12.5.3 Deconstruction Analysis

Mass concrete demolition prior to the mid-1980s was typically executed by positioning equipment such as the crane and impact ball outside the collapse envelope of a structure. This was considered nonengineered demolition or removal of concrete. The demolition contractor's survey of available documents and field conditions were the primary basis for equipment selection and the removal sequence or procedure.

Engineered demolition involves the analytical evaluation of a concrete structure during all stages of demolition to verify the adequacy of its strength and stability. The Deconstruction analysis is essentially the reverse process of the original construction of a structure, with revised partial framing and alternate load patterns. Engineered demolition is mandatory for post-tensioned structures if the operator is on the structure, or is within the collapse envelope of the structure, to be demolished. The contractor should seek experienced engineering advice. The process of "engineered demolition" should, at a minimum, include a thorough review of the existing condition of a structure, its proximity to other structures, any utilities above and below grade, the preparation of the demolition sequence, the procedures, the assignment of equipment to be used, an analysis of all actual altered structural models, and a stability analysis of the structure considering the member demand (all demolition-load combinations) and the member's capacity (remaining concrete cross section, tendon profile, tendon layout, and location of anchors).

Absent design guidelines for engineered demolition of concrete structures, several state agencies have agreed on minimum considerations. The following guidelines have been used on over 50 major engineered demolition projects since 1990.

The intent of the stability analysis is to confirm that the structure to be demolished has the capacity during all stages of demolition to safely support the weight of equipment or other demolition-related loading such as debris loading. The analysis is limited to a strength review only; no serviceability stress checks are considered. Strength-reduction factors are employed as recommended for new construction. Concrete configuration, reinforcing quantity, size of bars, splice locations, and prestressing forces are obtained from the as-built drawings. To calculate the demand or required strength (U), live loads are magnified by a load factor of 1.3, except as noted below ($U = 1.0D + 1.3L$). The live-load factor is assumed to include adequate increase for impact loading as the equipment live load is a well-defined load. Where supporting members indicate minor visual damage such as cracking at column-beam joints or questionable rebar configuration, the live-load factor is increased to 2.0. For damaged sections of the structure, shoring should be considered and installed to support the dead and live load of its respective tributary area.

Minimum temporary lateral-load capacity of the structures to be demolished is defined as a horizontal force, applied at the center of mass of a section, equal to 2% of its own weight. Pinned joints of supporting elements may require lateral bracing during demolition.

Adequacy of the structure and/or a member of a structure to be demolished is established when the capacity/demand ratio is equal to or exceeds 1.0.

12.5.4 Methods of Demolition

The following demolition methods may be used on prestressed concrete structures where applicable and permitted by local regulations.

12.5.4.1 Ball and Crane

The method requires the controlled swing (choking) or dropping of a steel "headache" ball from a crane. The chief advantage is that the workers are outside of the building collapse envelope during this work. A disadvantage is that it cannot be used on tall structures and on those structures that are close to adjacent ones or in congested urban areas.

12.5.4.2 Explosives

Explosives require experienced personnel to detonate critical building elements in order to effect a proper controlled collapse of the structure. The chief advantage is that this is a fast method of demolition. Disadvantages include ground and air vibration effects that may damage adjacent facilities. Users of this method should consider the stored energy effects of prestressed concrete when determining the charge levels of the explosives being used. Owing to the sudden stored-energy release of unbonded tendons, explosives had limited application for purposes of demolition.

12.5.4.3 Pressure Bursting

This method includes hydraulic bursters, gas expansion bursters, and hydraulic jacks in order to break the concrete into sections.

12.5.4.4 Thermal Lance

This method uses a steel rod in conjunction with oxygen or acetylene to melt concrete aggregates in order to form a series of boreholes that permit further demolition by other methods such as pressure bursting or use of hand tools.

12.5.4.5 Torch

Typically, an acetylene torch is used to burn strands and to cut steel connectors.

12.5.4.6 Diamond Saw and Drills

Regular cuts into a concrete section, using a diamond plate saw, can be made in order to remove sections or to sever concrete to be removed from sections that will remain. Line drilling by using cores can be used in slabs and beams.

12.5.4.7 Hand-Held Percussion Tools

These are typically operated from a compressor and are used locally to remove concrete from tendons and reinforcing steel to permit access for torches and other saws to cut the reinforcement.

12.5.4.8 Pneumatic and Hydraulic Breakers

These are large pieces of equipment such as excavators, special pulverizers, and shear or hammer attachments. These can extract and demolish thick floors with a maximum boom reach of up to 85 ft. New age hydraulic crushers are extensively used of because the pulverizer separates and processes the base materials during the demolition of a floor or wall.

12.5.4.9 Water Jetting

High-pressure water cuts and removes aggregate from concrete. Very high pressures can cut steel. The disadvantage is that large volumes of water are required during the demolition.

12.5.4.10 Other Methods

Other methods include chemical reactions, cutting by lasers, plasma-arc thermal cutting, and variations of methods previously described.

12.5.5 Other Considerations

12.5.5.1 Type of Construction

The type of original framing and erection should be investigated, lift-slab framing or other types of unique erection require special attention.

12.5.5.2 Proximity to Other Buildings

Adjacent buildings and structures must be protected during all phases of the demolition. Adequate dust and noise control should be addressed. If applicable, vibration measurements should be taken.

12.5.5.3 Accessibility of the Exterior Slab Edge and Beam Ends

Exterior wall assemblies should be safely removed in order to provide access to exterior edge strips and beam ends.

12.5.5.4 Interior Closure Strips

These strips should be carefully opened in order to release tendon stresses, if applicable. Controlled cutting of one side of the closure strip at a time is required in order not to adversely affect the adjacent concrete slab area of the other side of the closure strip. Adequate shoring of both sides of the closure strip may be required before any cutting proceeds when structural drawings and shop drawings are not available.

12.5.5.5 Intermediate Stressing Joints or Construction Joints

See the above description for interior closure strips.

12.5.5.6 Height of Structure

Special consideration for high-rise construction containing prestressed elements may be necessary.

12.5.5.7 Condition of Concrete

A strength evaluation of concrete may be required to determine where severe deterioration has occurred in the structure.

12.5.5.8 Condition of Reinforcement

A strength evaluation of floor systems may be required where severe deterioration of reinforcement has occurred.

12.5.5.9 Shoring Requirements

Adequate shoring is required for all phases of demolition.

12.5.5.10 Protection of Personnel and Public

Adequate site supervision, sidewalk bridges, barriers and other protective devices should be employed.

12.5.5.11 Partial Demolition

Where a structure is to undergo partial demolition in order to alter or maintain adjacent parts, special consideration to demolition procedures is required. Lack of consideration for impact loads and sudden stress release may damage existing concrete. Reduction of mechanical energy from pneumatic hammers, in areas to be preserved, may be limited to 1,200 ft-lb or less, in order not to fracture sound concrete unnecessarily.

Defining Terms

Added tendons: Tendons, usually short in length, placed in specific locations, such as end bays, to increase the structural capacity at that location without having to use full-length tendons.

Anchor: For monostrand tendons, normally a ductile iron casting that houses the wedges and is used to transfer the prestressing force to the concrete.

Anchor, trouble-shooting: See troubleshooting anchor.

Anchorage: A mechanical device comprising all components required to anchor the prestressing steel and permanently transmit the prestressing force to the concrete.

Anchorage, dead-end: See dead-end anchorage.

Anchorage, intermediate: See intermediate anchorage.

Anchorage, live end: See live-end anchorage.

Anchorage zone: The region in the concrete adjacent to the anchorage subjected to stresses (forces) resulting from the prestressing force.

Back-up bars: Reinforcing steel used to control the tensile splitting forces in the concrete resulting from the concentrated anchor force developed by the stressed tendons.

Bearing plate: A metal plate that bears directly against concrete and is part of an overall anchorage system.

Blowout: A blowout is a concrete failure that occurs during or after stressing. This may be explosive in character.

Bonded tendon: A tendon in which the annular space(s) around the prestressing steel (strand/s) are grouted after stressing, thereby bonding the tendon to the concrete section.

Bulkhead: See edge form.

Bursting steel: Reinforcing steel used to control the tensile bursting forces developed at the bearing side of the anchor as the concentrated anchor force from the stressed tendon spreads out in all directions.

Cantilever: Any rigid horizontal structural member projecting beyond its vertical support.

Casting: See anchor.

Chair: Hardware used to support or hold prestressing tendons in their proper position to prevent displacement before and during concrete placement.

Chuck, single-use splice: See coupler.

Coating: Material used to protect against corrosion and reduce the friction of the prestressing steel.

Coupler: A device, normally spring loaded, for connecting two strand ends together, thereby transferring the prestressing force from end to end of the tendon.

Creep: The time-dependent deformation (shortening) of prestressing steel or concrete under sustained stress (load).

Dead-end anchorage: The anchorage at the end of the tendon that is usually installed before the tendon arrives at the project site and that is not used for field stressing of the tendon.

Detensioning: A means of releasing the prestressing force from the tendon.

Edge form: Formwork used to limit the horizontal spread of fresh concrete on flat surfaces such as floors.

Effective prestress: The prestressing force at a specific location in a prestressed concrete member after all prestress losses have occurred.

Elastic shortening: The shortening of a member that occurs immediately after the application of the prestressing force.

Elongation: Increase in the length of the prestressing steel (strand) under the applied prestressing force.

Encapsulated System: A system that provides watertight connections at all stressing, intermediate, and dead ends and that has the wedge-cavity side of the anchorage covered by a watertight cap filled with a corrosion-protective coating material.

Fixed-end anchorage: See dead-end anchorage.

Friction loss: The stress (force) loss in a prestressing tendon resulting from friction created between the strand and sheathing due to the wobble and/or profile of the tendon during stressing.

Friction, wobble: See wobble friction.

Jack grippers: Wedges used in the jack to hold the strand during the stressing operation.

Jack-gripper plates: Steel plates designed to hold the jack grippers in place in the jack.

Hand-seating tool: A small, hand-held device used to properly align (seat) the wedges in the anchor prior to attaching the jack to the strand for stressing.

Initial prestress: The stress (force) in the tendon immediately after transferring the prestressing force to the concrete. This occurs after the wedges have been seated in the anchor.

Installation drawings: Drawings furnished by the post-tensioning-material supplier showing information such as the number, size, length, marking, location, elongation, and profile of each tendon to be placed.

Intermediate anchorage: An anchorage, located at any point along the tendon length, that can be used to stress a given length of tendon without the need to cut the tendon. Normally used at concrete pour breaks to facilitate the early stressing and removal of form work.

Jack: A mechanical device (normally hydraulic) used for applying force to the prestressing tendon.

Jacking force: The maximum temporary force exerted by the jack while introducing the prestressing force into the concrete through the tendon.

Kip: One kip = 1000 lb force.

Live end: See stressing end.

Live-end anchorage: The anchorage at the end of a tendon that is used to stress the prestressing steel (strand).

Monostrand: One single strand.

Multi-use splice chuck: A coupler that uses three-piece wedges and is made of heavier material for repeated use.

Nose piece: The front part of the jacking device that fits into the stressing pocket to align the jack with the anchor.

Split donut: See troubleshooting anchor.

Split pocket former: A temporary device used at the intermediate end during casting of concrete to provide an opening in the concrete for access, by the installer and the stressing equipment, to the anchorage area.

Stage stressing: Sequential stressing of tendons in separate steps or stages in lieu of stressing all the tendons during the same stressing operation.

Strand: High-strength steel wires twisted around a center wire. For unbonded tendons, seven-wire strand conforming to ASTM A-416 is almost exclusively used.

Stresses: Internal forces acting on adjacent parts of a body.

Stressing equipment: Consists normally of a jack, pump, hoses, and a pressure gauge.

Stressing end: The end of the tendon at which the prestressing force is applied.

Stressing force: See jacking force.

Tendon: A complete assembly consisting of anchorages, prestressing steel (strand), protective coating, and sheathing. It imparts the prestressing force to the concrete.

Tendon, bonded: See bonded tendon.

Tendon, unbonded: See unbonded tendon.

Tensile stresses: Internal forces directed away from the part of a body on which they act.

Tension: The effect of tensile forces on a body.

Trouble-shooting anchor: A special anchor used for structural modification or repair of existing tendons. The anchor consists of a removable segment that allows it to slide onto an existing strand. The segment is then returned and tightened by a screw or bolt.

Unbonded tendon: A tendon in which the prestressing steel (strand) is prevented from bonding to the concrete and thus is free to move relative to the concrete; therefore the prestressing force is permanently transferred to the concrete by the anchorage only.

Wedges: Pieces of tapered metal with teeth that bite into the prestressing steel (strand) during transfer of the prestressing force. The teeth are beveled at the front end to assure gradual development of the tendon force over the length of the wedge. Two-piece wedges are normally used for monostrand tendons.

Wedge set: The relative movement of the wedges into the anchor cavity during the transfer of the prestressing force to the anchorage, resulting in some loss of prestressing force.

Wobble friction: The friction caused by the unintended horizontal deviation of the tendon.

Note: Local practices, customs and usage may employ terminology, jargon and nicknames different from the terms and definitions set forth in this chapter. Check with your local engineering community or other qualified person to clarify terms and definitions.

References

Aalami, B.O. and Barth, F.B. 1988. *Restraint Cracks and Their Mitigation in Unbonded Post-Tensioned Building Structures.* Post-Tensioning Institute, Phoenix, AZ.

Aalami, B.O. 1990. Developments in Post-Tensioned Floors in Buildings. Paper presented at FIP—XIth International Congress on Prestressed Concrete, June 4–9. 1990, Hamburg, Germany.

Aalami, B.O. 1994. Strength Evaluation of Existing Post-Tensioned Beams and Slabs. *PTI Technical Notes,* Issue 4. Post-Tensioning Institute, Phoenix, AZ.

Aalami, B.O. 1994a. Unbonded and Bonded Post-Tensioning Systems in Building Construction. *PTI Technical Notes,* Issue 5. Post-Tensioning Institute, Phoenix, AZ.

Aalami, B.O. and Chegini, M. 1995. Structural Retrofitting of Cast-in-Place Concrete Parking Structures. Paper presented at Third National Concrete and Masonry Engineering Conference, June 15–17, 1995, San Francisco, CA.

ACI. 1995. *Manual of Concrete Practice.* American Concrete Institute, Farmington Hills, MI.

Barth, F.G. 1993. Engineered demolition of earthquake-damaged bridge structure. *Concrete Constr.* 38(7):480–486.

Barth, F.G. and Aalami B.O. 1992. *Controlled Demolition of an Unbonded Post-Tensioned Concrete Slab.* Post-Tensioning Institute, Phoenix, AZ.

Buchner, S.H. and Lindsell, P. 1987. Testing of Prestressed Concrete Structures During Demolition. In *Proc. 1st Struct. E/BRE Sem. Struct. Assess.*—Based on Full and Large Scale Testing, pp. 46–51.

Buchner, S.H., Lindsell, P., and Robinson, S. 1985. Monitoring of Prestressed Concrete Structures During Demolition. In *Proc. EDA/RILEM Conf. Demo. Techn.,* Rotterdam.

Collins, M.P. and Mitchell, D. 1991. *Prestressed Concrete Structures.* Prentice–Hall, Englewood Cliffs, NJ.

Federation Internationale de la Precontrainte (FIP). 1982. *Guide to Good Practice: Demolition of Reinforced and Prestressed Concrete Structures.* Cement and Concrete Association.

Hom, S. and Kost, G. 1983. Investigation and repair of post-tensioned concrete slabs. A case study. *Concrete Int.* 44–49.

Libby, J.R. 1984. *Modern Prestressed Concrete,* 635 pp. Van Nostrand Reinhold, New York.

Litvan, G. and Bickley, J. 1987. *Durability of Parking Structures, Analysis of Field Survey*, ACI SP 100-76. American Concrete Institute, Farmington Hills, MI.

National Federation of Demolition Contractors. 1975. *The Demolition of Prestressed Concrete Structures*. Report of Joint Liaison Committee, Leicester, UK.

Nawy, E.G. 1996. *Prestressed Concrete—A Fundamental Approach*, 2nd ed., 810 pp. Prentice–Hall, Upper Saddle River, NJ.

Ojha, S. 1986. Rehabilitation of a Parking Structure. *Concrete Int.* 24–28.

PCI. 1985. *PCI Design Handbook for Precast and Prestressed Concrete*. Prestressed Concrete Institute, Chicago, IL.

Podolny, W., Jr. 1986. The cause of cracking in post-tensioned concrete box girder bridges and retrofit procedures. *Portland Concr. J.* 82–139.

Post-Tensioning Institute. 1989a. *Manual for Certification of Plants Producing Unbonded Single Strand Tendons*. Phoenix, AZ.

Post-Tensioning Institute. 1989b. *Field Procedure Manual for Unbonded Single Strand Tendons*. Phoenix, AZ.

Post-Tensioning Institute. 1990. *Post-Tensioning Manual*, 5th ed. Phoenix, AZ.

Post-Tensioning Institute. 1993. *Specification for Unbonded Single Strand Tendons*. Phoenix, AZ.

Price, W.I.J., Lindsell, P., and Buchner, S.H. 1987. Monitoring of post-tensioned bridge during demolition. In *IABSE Colloquium Bergamo—Monitoring of Large Structures and Assessment of Their Safety*, IABSE Report, Vol. 56, pp. 357–365.

Richardson, M.G. 1987. Cracking in reinforced concrete buildings. *Concrete Int.* 21–23.

RILEM/European Demolition Association. 1985. *Demolition Techniques*. RILEM Committee DRC 37.

Schupack, M. et al. 1980. Electrically Isolated Reinforcing Tendon Assembly and Method. United States Patent No. 4,348,844, filed Sept. 25, 1980.

Schupack M. 1989. Unbonded Performance. Civil Engineering, ASCE. 75–77.

Suarez, M.G. and Posten, R.W. 1990. *Evaluation of the Condition of a Post-Tensioned Concrete Parking Structure After 15 Years of Service*. Post-Tensioning Institute, Phoenix, AZ.

Tanaka, Y. et al. 1989. *Ten Years Marine Atmosphere Exposure Test of Unbonded Prestressed Concrete Prisms*. Post-Tensioning Institute, Phoenix, AZ.

Walker, C.H. 1990. *Durability Systems for Concrete Parking Structures*. Carl Walker Engineers, Inc., Kalamazoo, MI.

Hibernia offshore oil platform under construction, Newfoundland (Courtesy Dr. George C. Hoff).

13

Concrete for Offshore Structures

by
George C. Hoff, D.Eng.
Engineering Consultant, Mobil Technology Company, Dallas, Texas. Expert on offshore concrete platforms and marine structures and on concrete behavior under severe conditions. He was president of the American Concrete Institute.

13.1 Introduction

Offshore concrete structures are generally understood to be those structures exposed to an open-sea environment [ACI Committee 357, 1989; Federation Internationale de la Precontrainte, 1985]. They are designed to remain permanently or semipermanently fixed to the sea bed by gravity, piles, or anchors or to remain afloat and moored. They are often associated with the exploration and production of hydrocarbons but may have many other specialized uses.

Like most other types of concretes, those for use in offshore structures are usually made with local materials by local labor in conformance to local guidelines or specifications. Thus they can vary widely in quality. Depending on the particular application, their strengths can vary from 25 to 65 MPa (3600 to 9500 psi). All such concretes must be extremely durable. Once a concrete structure is placed in the sea, maintenance becomes very difficult owing to the hostile environment and is very expensive. Some offshore concrete platforms have design lives of 50 to 70 years.

The use of concrete in marine structures dates back to the ancient Romans and Greeks. The use of concrete as a hull construction material for commercial vessels began at the end of the 19th century

[Harrington, 1987]. Initial applications were generally used worldwide and consisted of concrete barges and pontoons. The first reinforced concrete seagoing ship was the *Namsenfjord*, constructed in Norway in 1917. The first concrete platform for oil and gas production in the Gulf of Mexico was installed in 1950. Since that time, more than 1000 functionally similar concrete structures have been built in that area [Norwegian Contractors, 1991a]. The first concrete gravity-base structure in U.S. waters was installed in 1978 [Hunteman et al., 1979]. The first large offshore concrete platform for the North Sea (Ekofisk Tank) was installed in 1973. Three concrete platforms, functionally similar to those of the North Sea, have been built in Brazil for South American offshore waters [Anon, 1988a]. Concrete has a long history and a significant and successful presence in offshore and marine applications. Throughout the paper, reference will be made to things that are "onshore," "inshore," and "offshore." Onshore is on the land. Inshore means that the location is away from the land but is close enough to the shore to be in protected waters with respect to the open sea. Offshore means that it is located in the open sea. The term "owner" is also frequently used. The owner of an offshore structure can be a single company or a collection of companies who retain varying percentages of the operation but who have designated a single company to operate and maintain the facility.

13.2 Types of Concrete Structures

Offshore structures used in conjunction with hydrocarbon exploration and production can generally be either bottom founded or floating. Many of the bottom-founded structures are also required to float at various stages of their life. The following descriptions of various types of platforms are very brief but are meant to give the reader a feeling for the enormous versatility that can be realized when concrete is used.

13.2.1 Bottom-Founded Structures

Bottom-founded structures can be further identified as

1. gravity-base structures;
2. concrete cylinder pile-supported structures; or
3. floatable/bottom-founded concrete-hull structures.

Examples of each are shown in Figures 13.1, 13.2 and 13.3.

The gravity-base structure (Figure 13.1), commonly called a GBS, maintains its position on the sea bottom because of its very large weight. The sliding force and over-turning moment due to the maximum environmental loads are resisted by the weight of the concrete, the operating weights on the structure, and any additional ballast weight that is contained within the structure. This type of structure is common where produced oil must be temporarily stored before being removed to a tanker or pipeline. The practical range of water depths for these platforms is 40 to 350 m (130 to 1150 ft). These structures are built at onshore or inshore locations and floated out to their final location. They can also be refloated when platform removal is required (Anon, 1990a). More detailed descriptions of these types of platforms can be found in ACI Committee 357 (1989), Federation Internationale de la Precontrainte (1985) and ACI Committee 357 (1990).

FIGURE 13.1 Typical gravity-base structure.

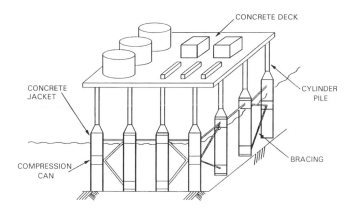

FIGURE 13.2 Typical cylinder pile-supported platform.

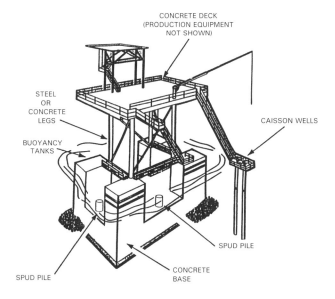

FIGURE 13.3 Floatable/bottom-founded structure. *Source:* Courtesy of Production Management Structural Systems.

Concrete cylinder piled structures (Figure 13.2) were the earliest type of concrete offshore platform used. The first platform of this type in the Gulf of Mexico was installed in 1950. More than 1000 of these platforms have been installed in Lake Maricaibo in Venezuela [Norwegian Contractors, 1991a]. They consist of an array of prestressed concrete piles that are driven into the seabed. The piles are arranged so that a prefabricated template deck can be placed over the array to form the working surface of the platform. The decks can be made of concrete or any other suitable construction material. Concrete jackets are often placed around the piles in the splash zone and boat-impact region of the platform. Steel cross bracing between piles may also be used to stiffen the overall arrangement when the piles become fairly long. The practical range of water depths for these platforms is from 5 to 20 m (16 to 65 ft). The use of concrete cylinder piles is also common for the support of docks, wharves, bridges, and roadways over water.

Floatable/bottom-founded concrete hull platforms generally consist of a bargelike concrete hull that is designed to float. Extending upward from the hull are posts or columns that act as the support frame for the platform (Figures 13.3 and 13.4). These posts or columns can be made of concrete

FIGURE 13.4 Eighteen-year-old floatable/bottom-founded platform after refloating and relocation to wet dock for equipment modifications.

or steel. The hull is floated to its desired location and then water-ballasted down until it sits on the seabed. It is then "pinned" to the seabed by spud piles around its perimeter. These piles maintain the platforms position and help resist sliding and overturning, as the platform does not have sufficient on-bottom weight by itself. Once the hull is piled into position, the topsides deck and equipment are usually added using a crane barge. This type of platform has many variations. It can accommodate some subsea storage of oil in the hull. The practical range of water depths for these platforms is from 4 to 30 m (13 to 98 ft). Platforms of this type that are in use in the Gulf of Mexico have, on numerous occasions, been refloated and reused at different locations. Table 13.1 is a listing of this type of structure constructed by one firm for the Gulf of Mexico and shows typical concrete hull dimensions.

13.2.2 Floating Structures

Floating structures are those structures that will perform their operational function while in a floating mode. These structures require a permanent mooring system. In general, the current family of floating concrete structures includes:

1. concrete tension leg platforms (TLP),
2. deep draft concrete floaters (DDCF), and
3. concrete production/storage barges.

Examples of each are shown in Figures 13.5, 13.6 and 13.7. Large concrete buoy-type floating structures have also been conceptualized.

Concrete tension leg platforms (TLPs) (Figure 13.5) derive their name from the fact that they are fastened to large anchors on the seabed by long tethers that have a predetermined amount of tension in them. These tethers, which originate at the corners of the platform, keep the floating platform in a very precise position. The platform itself can have various configurations but generally resembles the semisubmersible drilling rigs that are common throughout the offshore petroleum industry. It consists of an arrangement of base pontoons, shafts or columns that extend upward from the pontoons and a deck that sits on top of the shafts or columns. The entire hull (pontoons and shafts) and the deck can be made in concrete. The practical water depth for use of this type of platform is from 300 to 1500 m (1000 to 5000 ft). The size of the TLP is generally dictated by the amount of operational weight to be carried. Current designs have ranged as high as 50,000 tonnes (55,000 tons).

TABLE 13.1 Summary of Floatable/Bottom-Founded Concrete Hull Structures for the Gulf of Mexico Constructed by One Contractor Between 1984 and 1989

Year Installed	Location	Water Depth		Hull Dimensions, width × length × height	
		m	ft	m × m × m	ft × ft × ft
1984	Eugene Island Block 45	6.1	(20)	17.7 × 17.7 × 3.7	(58 × 58 × 12)
	Bayou Sorrel	2.4	(8)	9.4 × 11.0 × 3.7	(31 × 36 × 12)
	Cox Bay North	2.1	(7)	19.5 × 46.9 × 3.0	(64 × 154 × 10)
	Cox Bay South	2.1	(7)	19.5 × 33.5 × 3.0	(64 × 110 × 10)
	Vermilion Block 72	7.0	(23)	17.7 × 17.7 × 3.7	(58 × 58 × 12)
	Eugene Island Block 45	7.0	(23)	18.3 × 18.3 × 3.7	(60 × 60 × 12)
1985	Brenton Sound* Block 1	2.7	(9)	18.3 × 26.8 × 3.7	(60 × 88 × 12)
	Lease Platform	NK	NK	8.5 × 30.5 × 3.7	(28 × 100 × 12)
	Eugene Island Block 44	6.4	(21)	17.7 × 17.7 × 3.7	(58 × 58 × 12)
	South Pass	2.4	(8)	21.9 × 49.4 × 3.7	(72 × 162 × 12)
1986	Gordon Island Bay	NA	NA	10.7 × 19.8 × 3.7	(35 × 65 × 12)
	Lease Platform	NK	NK	8.5 × 30.5 × 3.7	(28 × 100 × 12)
	Brenton Sound† Block 2	3.7	(12)	18.3 × 24.4 × 3.7	(60 × 80 × 12)
	Quarantine Bay	NK	NK	21.3 × 38.7 × 3.7	(70 × 127 × 12)
1987	Delta Dock Fld	NA	NA	12.2 × 30.5 × 3.7	(40 × 100 × 12)
	Pt. Ala Hache	NK	NK	21.3 × 40.8 × 3.4	(70 × 134 × 11)
	Pt. Ala Hache	NK	NK	28.0 × 32.9 × 3.4	(59 × 108 × 11)
	W. Lake Verret	NK	NK	15.2 × 29.9 × 3.7	(50 × 98 × 12)
1988	Chandeleur Snd	NA	NA	4.6 × 7.6 × 3.0	(15 × 25 × 10)
	Atchsduya Bay‡	NA	NA	15.5 × 15.5 × 4.3	(51 × 51 × 14)
	S. Marsh Island Block 253	NK	NK	18.3 × 18.3 × 3.4	(60 × 60 × 11)
1989	Main Pass Block 69	NA	NA	12.2 × 18.3 × 3.7	(40 × 60 × 12)
	Eugene Island Block 30	4.3	(14)	21.9 × 25.0 × 4.0	(72 × 82 × 13)
	West Bay	NK	NK	21.9 × 42.7 × 3.7	(72 × 140 × 12)

Source: Norwegian Contractors. 1991a. Durability of concrete in the Gulf of Mexico: Experience from existing marine concrete structures. Prepared by Ben C. Gerwick, Inc., San Francisco for Norwegian Contractors, Joint Industry Project on Concrete for Gulf of Mexico, Floating Production Platforms.

Note: NK, not known; NA, not applicable. Barge structure.

*Designed for 5000 bbl of storage.

†Designed for 4500 bbl of storage.

‡Designed for 2500 bbl of storage.

The deep draft concrete floater (DDCF) (Figure 13.6) is similar in principal to the TLP but uses a conventional mooring system rather than tension tethers. It maintains its position during operations because of its extremely deep draft [greater than 130 m (425 ft)], large weight, low center of gravity, and mooring from the lower portions of the hull. These factors tend to make the structure relatively insensitive to the motions of the sea. Like the TLP, its configuration can have many variations, but in general, it resembles a TLP but with a very deep hull. The pontoons, columns, deck, and any bracing can be made in concrete. The practical water depth for use of a DDCF is from 300 to 900 m (1000 to 3000 ft). Like the TLP, the size of the DDCF is generally dictated by the amount of operational weight to be carried. Current designs have ranged as high as 50,000 tonnes (55,000 tons).

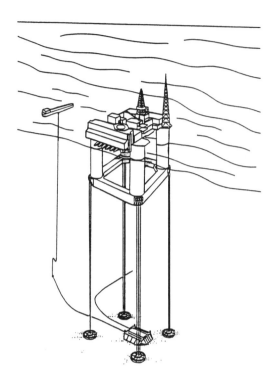

FIGURE 13.5   Concrete tension leg platform (TLP).

FIGURE 13.6   Deep draft concrete floater (DDCF).

FIGURE 13.7   Concrete production barge. *Source:* Courtesy of
Ed. Zublin AG.

A variation of the DDCF is the spar buoy platform (Figure 13.8). Because of the low center of gravity and heavy weight of the concrete, it is relatively insensitive to the motion of the sea. The spar buoy platform can accommodate crude storage, if desired. It also needs a mooring system [Anon, 1991].

Concrete production barges (Figure 13.7) are custombuilt prestressed concrete barges that provide a support surface for the processing equipment, work and storage areas, and living quarters needed for offshore oil and gas production. Drilling is usually not done from these barges but is done from special drilling vessels or jackup rigs. The production wells are usually located on a nearby unmanned fixed platform. The entire barge or selected portions of the barge can be built in concrete. A mooring system must be provided for the barge. Produced oil and other partially processed fluids can be accommodated in the barge. Large floating concrete oil-storage facilities have been built in Japan. The size of the barges is influenced by the sea states in which they must operate and the amount of working area they must provide. The water depth in which a barge can operate is a function of the draft of the barge and the operational sea states. A notable production/storage concrete barge is the Ardjuna Sakti liquified natural gas (LNG) barge currently on station in the Java Sea [Anderson, 1977]. More detailed information on concrete bargelike structures and concrete hulls can be found in ACI Committee 357 (1989), Harrington (1987), and ACI Committee 357 (1988).

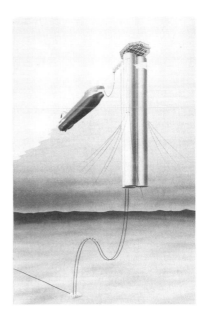

FIGURE 13.8 Spar buoy platform. *Source:* Courtesy of Norwegian Contractors, Stabekk, Norway.

13.2.3 Other Structures

Concrete subsea oil-storage tanks [Anon, 1984, 1986, 1988b] (Figure 13.9) have been proposed for use in water depths ranging from 20 to over 400 m (65 to over 1300 ft). These tanks can be built like the base of a GBS but are fully submerged to the seabed, where they function as a gravity-base structure.

Concrete wall caissons (Figure 13.10) have been used to provide retaining walls for earth-filled islands. These islands provide the working surface for oil and gas exploration and/or production. The caissons are built as floating units, towed to location, joined into a unit, and then ballasted to the seafloor. The framework of the caissons then forms the perimeter of an island. A hydraulic fill is usually used to fill the interior. When the use of the

FIGURE 13.9 Concrete subsea storage tank.

caisson-retained island is complete, the caissons can be refloated and disassembled, and nature is allowed to reclaim the island. A notable application was the Tarsuit Caisson Retained Island [Fitzpatrick and Stenning, 1983] where the caisson was made of lightweight aggregate concrete. Concrete caissons for an artificial island were a strong contender for development of the Wytch

FIGURE 13.10   Concrete   retaining-wall   caissons   [Anon, 1986].

Farm prospect in the offshore southern United Kingdom [Anon, 1990b]. The use of caissons for artificial islands is generally limited to water depths of less than 15 m (50 ft).

Concrete has been used for the base of flare towers and offshore loading buoys. An entire flare boom tower made of concrete was proposed for the Sliepner platform in Norway. Concrete anchors [Anon, 1990c] for the Snorre TLP have been built. The Maureen offshore development uses a concrete offloading buoy. Concrete subsea wellhead protectors for Oseberg II in the North Sea have been built [Norwegian Contractors, 1990]. The potential for concrete use is great and is limited only by the ingenuity of the concrete designer and constructor.

13.3 Concrete Quality

There is a perception that all concrete used in offshore platforms is something unique and special and requires a technology that is beyond "normal practice" for concrete construction. If normal practice means the practice applied to residential construction, this perception is correct. If normal practice means practices applied to any major civil-engineering structure such as a building or bridge, then this perception is wrong. There is nothing unique or special in the application of proper batching, delivery, consolidation, and curing of properly proportioned concrete mixtures. In general, the recommended practices for concrete construction, including materials selection and mixture proportioning, that exist in the various building codes, specifications, and standard practices of most developed countries are entirely sufficient for use in the offshore concrete industry. Somewhat different values for water/cementitious ratio, cementing material content, and concrete cover over reinforcing bars may be required because of marine exposure, but these values are well documented. Examples are given in Tables 13.2 and 13.3.

The concrete provided for offshore North Sea platforms has seen a gradual evolution of cube compressive strength from 50 to 70 MPa (7200 to 10100 psi) [Moksnes, 1989]. Table 13.4 shows the strength development for platforms built by one North Sea contractor. The unique environment in which this concrete is used demands such high quality of concrete. The Ravenspurn North platform [Jackson and Bell, 1990; Roberts, 1990] was built in a more moderate environment in the southern part of the North Sea and required only 50 MPa (7250 psi) concrete, which was delivered from local ready-mix suppliers. The early concrete platforms made in the Gulf of Mexico used concrete with cylinder compressive strengths from 25 to 35 MPa (3600 to 5000 psi). Recent samples from some 33-year-old platforms in the Gulf of Mexico showed an increase in strength from 50 to 69 MPa (7200 to 10,000 psi) over the life of the structure [Hoff, 1991a; Tate and Core, 1989]. The actual strength required for a given structure depends on a large number of factors but is significantly influenced by environmental and operating loads. When these loads are small, the

TABLE 13.2 Summary of Relevant Code Requirements for Water/Cement Ratio, Cement Content, and Compressive Strength

Codes*	Exposure Zones		
	Submerged	Splash	Atmospheric
ACI 357			
Maximum water/cement ratio	0.45	0.40	0.40
Minimum cement content, kg/m^3 (lb/yd^3)	356 (600)	356 (600)	356 (600)
Maximum cement content, kg/m^3 (lb/yd^3)	415 (700)	415 (700)	415 (700)
Minimum 28-d cylinder compressive strength, MPa (psi)	35 (5000)	35 (5000) 42(6100)[†]	35 (5000)
CSA S474			
Maximum water/cement ratio	0.45	0.40	0.40
Minimum quantity of cementing material, kg/m^3 (lb/yd^3)	360 (610)	400 (675)	400 (675)
Minimum quantity of Portland cement, kg/m^3 (lb/yd^3)	300 (510)	300 (510)	300 (510)
Minimum cylinder compressive strength, MPa (psi),			
28 d	30 (4350)	40 (5800)	40 (5800)
91 d	35 (5000)	45 (6500)	45 (6500)
FIP			
Maximum water/cement ratio	0.45 (less than or equal to 0.40 preferred)		
Minimum cement content, kg/m^3 (lb/yd^3)	320 to 360[‡] (540 to 610)	400 (675)	320 to 360[‡] (540 to 610)
Maximum cement content, kg/m^3 (lb/yd^3)	500 (845)	500 (845)	500 (845)
Minimum 28-d cylinder compressive strength, MPa (psi)	32 (4650)	32 (4650) 40(5800)	32 (4650)
DnV			
Maximum water/cement ratio	0.45 (less than or equal to 0.40 preferred)		
Minimum cement content, kg/m^3 (lb/yd^3)	300 (510)	400 (675)	300 (510)
Maximum cement content, kg/m^3 (lb/yd^3)	—	—	—
Minimum 28-d cylinder compressive strength, MPa (psi)	—	—	—
BS 6235			
Maximum water/cement ratio	0.40	0.40	0.40
Minimum cement content, kg/m^3 (lb/yd^3)	320 to 360[‡] (540 to 610)	400 (675)	400 (675)
Maximum cement content, kg/m^3 (lb/yd^3)	—	—	—
Minimum 28-d cylinder compressive strength, MPa (psi)[§]	32 (4650)	32 (4650) 40(5800)[†]	32 (4650)

*See Table 13.3.

[†] if subject to abrasion.

[‡] 320 kg/m^3 (540 lb/yd^3) for a maximum aggregate size of 40 mm (1.57 in); 360 kg/m^3 (610 lb/yd^3) for a maximum aggregate size of 20 mm (0.78 in).

[§] Cylinder strength assumed to be 80% of specified cube strength.

TABLE 13.3 Summary of Relevant Code Requirements for Minimum Concrete Cover

Codes*	Exposure		Other
	Splash or External Atmospheric		
ACI 357			
Untreated reinforcing bars	65 mm (2.6 in)		50 mm (2.0 in)
Prestressing tendons	90 mm (3.5 in)		75 mm (3.0 in)
Cover of stirrups	13 mm (0.5 in)		less than above
CSA S474			
Untreated reinforcing bars	65 mm (2.6 in)		50 mm (2.0 in)
Epoxy coated reinf. bars	50 mm (2.0 in)		35 mm (1.4 in)
Prestressing tendons	90 mm (3.5 in)		75 mm (3.0 in)
	A cover of 75 mm (3.0 in) may be used in atmospheric zone		
Cover of stirrups	15 mm (0.6 in)		less than above
FIP			
Untreated reinforcing bars	65 mm (2.6 in)		50 mm (2.0 in)
Prestressing tendons	90 mm (3.5 in)		75 mm (3.0 in)
DnV			
Untreated reinforcing bars	50 mm (2.0 in)		40 mm (1.6 in)
Prestressing tendons	100 mm (4.0 in)		80 mm (3.2 in)
BS 6235			
Untreated reinforcing bars	75 mm (3.0 in)		60 mm (2.4 in)
Prestressing tendons	100 mm (4.0 in)		75 mm (3.0 in)

*See Table 13.5.

TABLE 13.4 Strength Developments for North Sea Offshore Concrete Structures

Platform (Year)	Concrete in Cell Walls,* m^3 (yd^3)	Specified Concrete Grade	Obtained 28-d Cube Strength, MPa (psi)	Typical Slump, mm (in)
Ekofisk I (1972)	—	C40*	45* (6530)	100 (3.9)
Beryl A (1974)	17100 (22370)	C45	55.0 (7980)	120 (4.7)
Brent B (1974)	40600 (53100)	C45	53.0 (7690)	120 (4.7)
Brent D (1975)	34000 (44470)	C50	54.2 (7860)	120 (4.7)
Statfjord A (1975)	47400 (62000)	C50	54.6 (7920)	120 (4.7)
Statfjord B (1979)	56700 (74160)	C55	62.5 (9070)	160 (6.3)
Statfjord C (1982)	63700 (83320)	C55	67.5 (9790)	210 (8.3)
Gullfaks A (1984)	63400 (82930)	C55	65.2 (9460)	220 (8.7)
Gullfaks B (1985)	45000 (58860)	C55	80.8 (11720)	220 (8.7)
Oseberg A (1986)	43000 (56240)	C60	76.7 (11120)	230 (9.1)
Gullfaks C skirts (1986)	17400 (22760)	C70	83.8 (12150)	240 (9.4)
Gullfaks C (1989)	115000 (150420)	C65	79.0 (11460)	230 (9.1)

Source: Norwegian Contractors. 1990. Concrete subsea structures—well templates. *NC News*, pp. 6–7, 23. Stabekk, Norway.

*Only slipformed concrete in the cell walls except where noted. Does not represent the total concrete in the structure.

strength of the concrete can usually be consistent with that commonly made in the region of the construction.

13.4 Concrete Materials

As noted in Section 13.1, the constituents of the concrete can be local materials. They must be evaluated, however, to ensure that they have the proper concrete-making characteristics and that they will be durable in the environment in which they are used. Most offshore concrete platforms have a service life of 20 years or more. Because of their offshore location, they are not easily accessible for remedial work when problems occur. To eliminate the high cost of future offshore repair work, the materials used and the resulting concrete must be virtually maintenance free for the service life of the structure.

The durability of offshore and marine Portland cement concrete is generally defined as its ability to resist weathering action, chemical attack, abrasion, or any other process of deterioration while retaining its original form, quality, and serviceability when exposed to its environment. This includes resistance to deterioration from freezing and thawing action, chemical attack by the constituents of the seawater, physical abrasion due to wave action, floating or suspended solids and debris, and floating ice, corrosion of steel or other metals imbedded in the concrete, and chemical reactions associated with aggregates in the concrete [Tate and Core, 1989]. When considering these deteriorating actions collectively, it is easily deduced that the most aggressive exposure a concrete can routinely experience is in a tidal zone in freezing weather. For most offshore structures, the most prevalent of the destructive mechanisms is the corrosion of reinforcing bars associated with ingress into the concrete of chlorides from the seawater.

Portland cements should have as low a tricalcium aluminate (C_3A) content as is practical with the local cement production. This helps to reduce the possibility of attacks from sulfates. The total alkali of the cement, calculated as sodium oxide, should not exceed 0.60% to minimize any potential for reactivity with the aggregates. The cement should have some finely divided siliceous material added to it [Hoff, 1991a]. This includes natural pozzolans, fly ash, granulated slag, or condensed silica fume. These products contribute to the formation of a dense binder that inhibits the migration of the seawater into the concrete. They also combine with the alkalis to reduce the amount of available alkalis.

Coarse aggregates can be either normal-density gravel or crushed stone or good-quality lightweight aggregate. The aggregates should be evaluated with respect to their potential for reactivity with the alkalis in the cement. Those aggregates that are potentially reactive should not be used. Aggregates from areas in close proximity to the sea should be checked for concentrations of sea salts. These salts must be washed from the aggregate before it is used. Fine aggregates can be either natural or manufactured sands. They, too, must be nonreactive and free from deleterious materials.

In no instance should seawater or brackish water be used to make the concrete. All mixing water should be potable. Washing of aggregates should also be done with potable water.

Chemical admixtures are essential for the production of durable marine concrete. Air entrainment is needed when cycles of freezing and thawing can occur. High-range water reducing admixtures (HRWRA), commonly called superplasticizers, are required for both consolidation assistance and for improved durability. A HRWRA will allow mixing water reductions of up to 30% without sacrificing workability. This water reduction significantly reduces the permeability of the concrete and contributes to a densification of the binder fraction of the concrete.

13.5 Concrete Properties

Of importance to structural designers are the properties of construction materials at an age when appreciable loads are applied to the structure. For most offshore structures, maximum loadings occur when the structure is put into service. This can vary from 1 to 5 y from the start of construction, depending on the size and complexity of the structure and its ultimate use. The properties of hardened concrete that are used by the designers of offshore concrete platforms are

a) compressive strength,
b) tensile strength,
c) modulus of rupture,
d) modulus of elasticity,
e) Poisson's ratio,
f) stress-strain relationships,
g) fatigue strength,
h) absorption,
i) shear strength,
j) creep and shrinkage,
k) shear-friction capacity,
l) bearing strength, and,
m) thermal properties such as the coefficient of thermal expansion, thermal conductivity, specific heat, and diffusivity.

The numerical value of each of these properties is generally not critical because the design process can usually use whatever values the selected concrete produces. The specific properties may not always be complementary, however. For example, a very high compressive strength concrete [e.g., 65 MPa (9400 psi)] may allow compressive structural members to be reduced in cross section for a given loading. If, however, the corresponding increase in the modulus of elasticity of that concrete allows cracking to occur at lower strain strain levels, then additional reinforcement may be required to reduce the cracking. Because the cross section has now been reduced, the additional reinforcing steel adds to the congestion within the wall and makes concrete placement more difficult. The cost of the in-place reinforcing steel may also be more than the reduction in cost due to using less concrete. Trade-offs between the various properties of the concrete should be attempted, where possible, to achieve the most efficient and cost-effective design.

All the hardened concrete properties should be determined at advanced ages for the specific concrete to be used in an offshore structure. Unfortunately, this is not always possible and early-age properties (e.g., at 28-d) are often used. This gives the design a conservative flavor, but it may add substantial costs to the structure. There is a risk associated with extrapolating early-age data, particularly with high-strength concretes, because the improvement of concrete properties with age may not always follow assumed trends.

Other properties of the concrete are of concern to the constructor rather than the designer. These include

a) workability,
b) pumpability,
c) unit weight,
d) air content,
e) consolidation,
f) thermal gradients, and
g) finishing.

The interrelationship of these properties is a complex problem. Of utmost importance is the unit weight of the concrete. For a structure of given dimensions and configuration, that may also be required to carry a fixed amount of dead load while floating, variations in the concrete unit weight may adversely affect the floating stability of the structure, causing it to sink or overturn. The in-place unit weight, in turn, is affected by the mixture ingredients, their proportions, and the void content, which is both a function of the entrained air content and the entrapped air or voids remaining after consolidation. If the mixture does not have adequate workability to surround the high levels of reinforcing bars that may occur, possibly resulting in additional voids in the concrete. The absorption values determined on the hardened concrete are applied to the hardened density of the concrete to establish what the concrete density is when the structure is in the water. If the actual density varies significantly, so will the actual absorption values, which will be different than those used in the design process.

As described later, the typical structural members in an offshore platform are quite thick. Because most offshore codes require fairly high cement contents (see Table 13.2) for durability purposes, the possibility of significant heat development within the concrete exists. Limiting values for the maximum placing temperature and the maximum heat rise are contained in the codes. Even when meeting these requirements, care must be exercised to minimize thermal gradients so that thermal cracking of the structural members does not occur.

The finish of the concrete surface of an offshore structure may seem like a noncritical item, but a poor finish can have several undesirable effects besides poor appearance. For most offshore structures, the governing design load is caused by the forces from sea waves acting on the surface of the structure. Rough surfaces tend to gather more wave forces and thus reduce the factor of safety planned for a structure. In cold climates, an initially rough surface tends to degrade faster when subjected to cycles of freezing and thawing because there are receptacles in the surface of the concrete where water can collect and freeze. In ice-congested waters, ice moving against and past a structure tends to abrade rough surfaces faster than smooth surfaces [Hoff, 1991a].

Other properties of the concrete that are usually not of concern to either the designer or the constructor are durability properties. These are of concern to the owner, as the offshore structure is usually part of a profit-making venture that has a prescribed lifetime. Some of the durability aspects of the concrete, such as freezing and thawing resistance, are addressed in the code requirements. Matters such as the air-void system in hardened concrete, as defined by the spacing factor, the specific surface, and voids per millimeter (inch), have specific requirements that must be met. Guidance is also provided in the codes to prevent or mitigate such deleterious effects as sulfate attack and alkali-aggregate reactivity through proper materials selection.

Chloride-ion permeability of the concrete should also be evaluated to ensure that a satisfactory concrete is being provided to resist reinforcing-bar corrosion. Although minimum concrete cover over the reinforcing bars is specified for a given exposure zone (see Table 13.3), this may have to be increased if the concrete to be used in the platform does not have adequate resistance to chloride-ion penetration.

The abrasion resistance of the concrete to waterborne sediments, debris, floating objects, and ice is usually not specified as it is a rather site-specific phenomenon. In offshore areas where significant abrasion can occur, such as in ice-congested waters, the resistance of the concrete to the abrading medium must be evaluated and loss rates for the concrete surface determined [Hoff, 1988]. Once these rates are known, measures to accommodate or eliminate the losses, such as additional concrete cover or steel plates in the abrasion zone, respectively, can be implemented.

An evaluation of all the above concrete properties for a specific concrete for a specific structure is the ideal situation, but it has not often been done. When actual numerical values are not available, conservative approximations are chosen, and these result in a satisfactory, but not necessarily cost efficient, design. One study that addressed most of the properties noted above was performed on high-strength lightweight aggregate for use in offshore Arctic structures and is described in ABAM Engineers Inc. (1983), ABAM Engineers Inc. (1984), and ABAM Engineers Inc. (1986).

13.6 Design Considerations

As noted in Section 13.5 concrete offshore structures can be bottom founded or floating. With the exception of a structure that has a base made entirely from prestressed concrete piles, most of the other bottom-founded structures are in a floating mode at some time in their early life. These

TABLE 13.5   Summary of Relevant Codes for Offshore Concrete Structures

1. American Concrete Institute. *Guide for the Design and Construction of Fixed Offshore Concrete Structures.* ACI 357R-84 (revised 1989). Farmington Hills, Michigan. USA.

2. Canadian Standards Association (CSA). *Code for the Design, Construction, and Installation of Fixed Offshore Structures, Part 4; Preliminary Standard S474-M1989: Concrete Structures; and Special Publication S474.1-M1989: Commentary to S474-M1989.* Rexdale, Ontario, Canada.

3. Federation Internationale de la Precontrainte (FIP). *Design and Construction of Concrete Sea Structures,* 4th ed. 1985. The Institution of Structural Engineers, London, England.

4. Det norske Veritas (DnV), *Rules for the Design, Construction, and Inspection of Offshore Structures (with appendices).* 1977. Hovik, Norway.

5. British Standards Institution. *Code of Practice for Fixed Offshore Structures.* BS 6235. 1982. London, England.

6. Det Norske Veritas. *Rules for the Classification of Fixed Offshore Installations Structures.* Classification A/S. 1989 (draft). Hovik, Norway.

7. American Bureau of Shipping (ABS). *Rules for Building and Classing Offshore Installations.* 1983. New York, New York, USA.

8. American Petroleum Institute (API). *Recommended Practice for Planning, Designing and Constructing Fixed Offshore Platforms,* 18th ed. RP 2A. 1989. Washington, DC, USA.

9. American Petroleum Institute (API). *Recommended Practice for Planning, Designing and Constructing Fixed Offshore Structures in Ice Environments.* 2nd ed. RP 2N. 1988. Washington, DC, USA.

10. Federation Internationale de la Precontrainte (FIP). *Design and Construction of Concrete Concrete Ships.* 1986. The Institution of Structural Engineers, London, England.

11. Norwegian Petroleum Directorate (NPD). *Acts, Regulations and Provisions for the Petroleum Activity.* 1985. Oslo, Norway.

12. Norwegian Petroleum Directorate (NPD). *Regulations Concerning Load Bearing Structures in the Petroleum Activity (Draft).* 1990. Oslo, Norway.

structures must then include design provisions for both bottom-founded operational loads and those loads associated with the structure's behavior as a ship.

Design codes and guidelines for offshore concrete structures have been developed by various regulatory agencies and standards groups. A listing of some of the major codes and regulations is given in Table 13.5. These are constantly being upgraded as technology advances. In general, detail design of the individual elements of an offshore concrete structure for such things as shear, tension, flexure, compression, eccentric loads, etc., is not significantly different than for any other type of concrete structure. It is only the types of loads, their frequency and duration, and their magnitude that differs from ordinary civil-engineering structures.

The principal loads the offshore structure encounters are permanent loads, variable functional loads, environmental loads, accidental loads, and deformation loads. These various loads are combined in realistic manners to determine their net effect. Permanent loads include the weight of the structure, any permanent equipment, ballast that will not be removed, and the external hydrostatic sea water up to mean sea level. Variable functional loads are the loads associated with the normal operation of the structure. Loads in this category that are unique to offshore structures include variable ballast, installation and drilling loads, vessel impact, fendering and mooring, weight of petroleum products temporarily stored in the platform, helicopter loads, and crane operations.

Environmental loads include waves, wind, current, ice and snow, and earthquake. Accidental loads include fire, explosion, ship impact, unintentional flooding, unintentional ballast distribution, and changes in presupposed pressure differences. Examples of deformation loads include prestressing, concrete shrinkage, and thermal gradients.

The geotechnical considerations offshore are much more complex than onshore. For bottom-founded structures, this is an extremely critical area of design. The anchors and moorings of floating structures are also significantly influenced by the subsea soil conditions. Seismically active areas warrant special consideration. Specialists in subsea foundation problems, not onshore foundation specialists, should always be used to work on this part of the design problem.

For the first structure in countries or regions that have never used offshore concrete structures before, initially it is best to use the design expertise of companies or firms that have prior experience with these structures. Such firms exist in North America, Europe, Scandinavia, the United Kingdom, and Japan. By involving local design firms in partnerships with these experienced firms, the philosophy and mechanics of the design process can be transferred to the local regions.

13.7 Safety Considerations

Modern offshore concrete platforms are designed with sufficient redundancy to resist major accidental loads. Concrete has exceptionally good impact resistance, and only a few isolated instances of structural damage due to ship impact have been reported. Sufficient ductility can be designed into structural concrete elements to eliminate the problem of progressive collapse. The fire resistance of concrete is well known, with concrete often being used to protect steel from fires in many major structures. A summary of the service record of concrete platforms in the North Sea can be found in Hoff (1986).

Floating concrete structures are designed for one-compartment damage stability which means that any local damage that causes leakage of the sea into the hull will not cause the floating structure to be at risk. Similar criteria are applied to bottom-founded structures so that they are not at risk while in their floating mode. Bottom-founded structures, such as that shown in Figure 13.1, have the unique capability of having each of their shafts or towers operate independently of the others. All living quarters and other major personnel areas can be isolated on top of one shaft and kept removed from the more dangerous areas where drilling and processing of hydrocarbons takes place. All areas can be connected by bridges. This is a significant advantage over a structure where most of the supporting structural members are tied together in some fashion and collectively support all the operations of the platform. In the event of a major fire or explosion in or on shafts where hydrocarbons are present, living quarters would not be affected and successful evacuation of the GBS-type platform could take place.

13.8 Construction Practices

There are no unique construction practices needed to build offshore structures such as those described above. The good practices employed for any major civil-engineering project are sufficient for building an offshore concrete platform. This gives the owner flexibility in selecting construction contractors.

The large offshore concrete platforms of the North Sea have been predominantly constructed using slipforming [Moksnes et al., 1987]. It has been demonstrated that North Sea slipforming techniques can be satisfactorily applied to offshore concrete construction in the hotter climate of the Gulf of Mexico [Norwegian Contractors, 1991b]. The small bargelike platforms in the Gulf of Mexico are cast-in-place [Harrington, 1987]. The Hibernia platform [Michel, 1989] for the east coast of Canada, which was completed in 1996, was originally planned to be jump-formed but was ultimately slipformed. Precast elements have been used in some structures [Yee et al., 1984]. The method of construction can be anything that works and should be left to the discretion of the contractor.

The concrete can be delivered to the form by pump, boom, conveyor, bucket, buggy, barrow, or, again, anything that works. Sophisticated distribution systems normally are not needed. The distribution of the concrete into the slipforms used on North Sea structures has been by wheel-barrow. Proper and sufficient consolidation of the concrete is essential. The equipment and procedures to do this already exist. Adequate curing must be provided. This includes protection from early freezing in regions where this is a possibility.

Two items of offshore concrete construction that differ from most onshore concrete construction are the thickness of the concrete elements and the amount of reinforcing steel that is used. The thinnest concrete walls are usually 350 mm (13 in). The thickest walls can be several meters (more than 6 ft) thick. Typical wall thicknesses are 500–600 mm (20–24 in). Temperature control of the concrete in these thick walls is essential to eliminate thermal cracking problems. Reinforcement densities typically average 400 kg/m^3 (676 lb/yd^3). Extremes in critical areas have been as great as 1100 kg/m^3 (1859 lb/yd^3). The proportioning of the concrete and the consolidation methods must be tailored to ensure the reinforcing steel is completely encapsulated by the concrete. In general, the maximum size aggregates used in these structures has been 19 mm (3/4 in) or less to allow the concrete to move around the large amounts of reinforcing bars. The use of high-range water reducers (superplasticizers) is a necessity.

Prestressing is also used in almost every structure. This is to ensure watertightness of the concrete. The amounts of prestressing are structure and location specific. Standard prestressing materials and practices can be used for these structures, keeping in mind that the work will be done in a marine environment. Special care must be taken to protect the end anchorages of the prestressing from the corrosive environment of the sea.

13.9 Construction Locations

Bottom-founded and floating offshore concrete structures can be constructed, either partially or completely,

1. in dry docks or graving docks,
2. on submersible barges,
3. on skid ways, or
4. in precast facilities onshore.

Precast facilities are used principally for the precast concrete piles and template decks. These facilities can also be used to prefabricate substantial portions of a structure. The prefabricated elements are then transported to and assembled at other locations. The construction of the concrete Arctic drilling structure Glomar Beaufort Sea I, also known as the CIDS (concrete island drilling system), had most of its interior concrete elements made in a prefabrication facility [Yee et al., 1984]. The actual method selected will depend greatly on the existing site facilities where construction is planned and on the economics of the project.

13.9.1 Dry-Dock Construction

The principal method of construction used for large North Sea concrete platforms is to begin the construction in a dry dock or graving dock. Figure 13.11 shows the construction sequence. Once the base of the structure becomes stiff enough and has sufficient buoyancy, the dock is flooded and the base allowed to float. It is then towed from the graving dock to a deeper water location where it is temporarily moored. Construction then continues at that location. Depending on available water depths, this location can be close to the shore or a great distance away. If the structure is located close enough to shore to be reached by a floating or fixed bridge, materials and personnel can reach the structure directly from the shore. Office facilities and concrete production facilities can remain onshore. If the structure is located some distance away from the shore, it will require adjoining barges that support offices, materials laydown areas, concrete batching plants, and other essential equipment. Concrete materials are obtained directly from supporting ships. Work crews are shuttled back and forth from the shore by crew ships.

Once the concrete construction is complete, the structure may be moved to an even deeper water location where it is mated with the topsides equipment and hook up of the equipment is begun.

It generally costs less to construct a GBS entirely in a dry dock rather than partially in the dry dock, at a wet dock or at a jetty mooring. Productivity is generally higher in the dry dock and specialized equipment is minimized. Construction risk is usually lower. For smaller concrete platforms, the

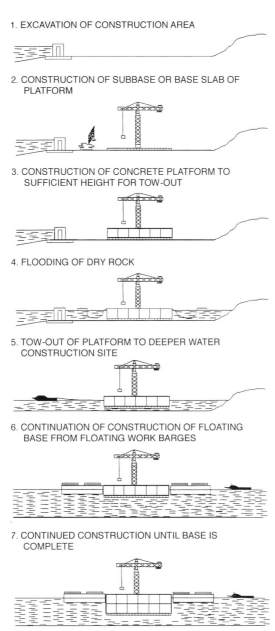

1. EXCAVATION OF CONSTRUCTION AREA

2. CONSTRUCTION OF SUBBASE OR BASE SLAB OF PLATFORM

3. CONSTRUCTION OF CONCRETE PLATFORM TO SUFFICIENT HEIGHT FOR TOW-OUT

4. FLOODING OF DRY ROCK

5. TOW-OUT OF PLATFORM TO DEEPER WATER CONSTRUCTION SITE

6. CONTINUATION OF CONSTRUCTION OF FLOATING BASE FROM FLOATING WORK BARGES

7. CONTINUED CONSTRUCTION UNTIL BASE IS COMPLETE

FIGURE 13.11 Dry dock construction scenario.

platform can be built entirely in the dry dock with sufficient buoyancy to be floated out and towed to its final location. The Ravenspurn North platform [Moksnes, 1989; Roberts, 1990] and all the bargelike platforms (Figures 13.3 and 13.4) built for the Gulf of Mexico are constructed in this manner. Figure 13.12 shows an exploded view of various components of the Ravenspurn North platform [Ove Arup and Partners, 1990]. It is reported [Anon, 1989] that in Europe, concrete GBSs for water depths from 100 to 150 m (330 to 490 ft) can be constructed entirely in a dry dock. The limit is governed by a combination of gate width and sill draft. As noted above, the total support for complete dry dock construction is land based and very similar to constructing

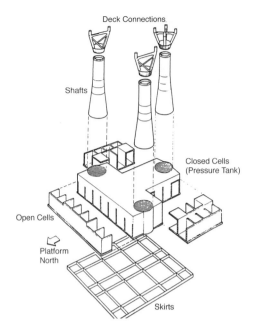

FIGURE 13.12 Principal components of the Ravenspurn North concrete platform. Courtesy of Ove Arup & Partners, London.

a concrete building. Very large platforms can also be built in this manner but require the use of auxiliary buoyancy compartments for floatation. These compartments can be designed to be removed after platform installation or can be left in place. Some small platforms have also used additional buoyancy compartments to satisfy installation requirements.

A major expense in the construction of a concrete platform is the development of a dry dock if one does not exist or the modification of existing dry dock facilities to accommodate the construction. This expense can include land procurement, excavation, cofferdam construction, dewatering systems, dredging of channels for float out, construction of supporting quays, docks, and wharves, and the overall upgrading of the infrastructure to improve project support (roads, bridges, power supply, water supply, sewage treatment, etc.). These costs can easily reach 80% of the project cost in remote areas.

13.9.2 Construction on Barges

To eliminate much of the costs associated with the construction of a dry dock, the construction of some or all of the platform on submersible barges is a practical solution. This eliminates the need for a dry dock and thus greatly expands the potential for construction site locations. Figure 13.13 shows a typical barge construction scenario. Any location with sufficient water depth for barge and supply ship operation is a possible construction site. Preferably, this site should be in sheltered waters and not subject to severe seas. The number of barges required will depend on the size of the structure and the capacity of the barges. Specially built barges are always a solution but it may be more economical to weld together a sufficient number of smaller, standard-size barges to accomplish the same objective. Once the barge or barges have been assembled to provide the working platform, construction then proceeds on the barge(s) as it would on land. Construction support will depend on the water depths available. Initially, construction can probably begin with the barges moored adjacent to the shore so that all construction support can come directly from the land. Figure 13.14 shows the concrete tether anchors for the Snorre tension leg platform (TLP), which were constructed

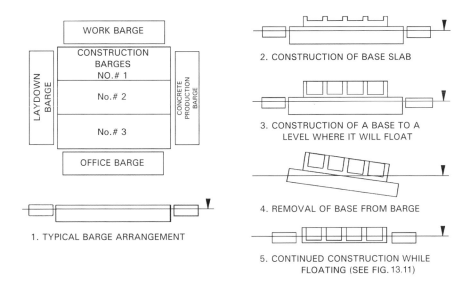

FIGURE 13.13 Barge construction scenario.

(a)   (b)

FIGURE 13.14 Snorre TLP foundation anchors. Courtesy of Norwegian Contractors, Stabekk, Norway.

entirely on barges with direct land support. If the platform is small, the entire platform can be built on the barges at one location. If the platform is large, only a portion would be built on the barges. That portion would be floated off the barges, temporarily moored, and construction completed while the portion is in a floating mode.

13.9.3 Skid Way Construction

Skid ways exist in almost all marine fabrication yards. A skid way is basically a structural slab having a small slope that extends from a construction area down into the sea. Structures that are built on the skid way can be gravity assisted as they are moved down the slope and into the sea. The structures can be self-floating or can be skidded onto barges.

For a concrete GBS, the structure can be built either in part or in its entirety on the skid way [Ben C. Gerwick, Inc., 1989]. Small structures (Figure 13.15) are the most likely candidate for this type of construction. Because the weight of a complete large concrete structure is so great, there is probably

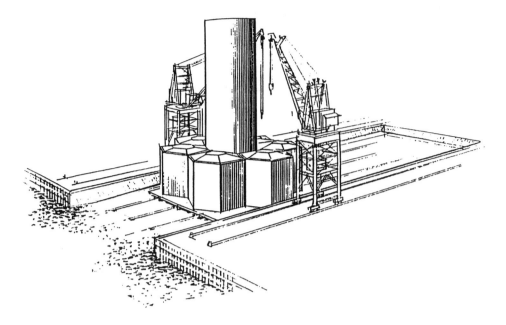

FIGURE 13.15   Skid way construction. Courtesy of Ben C. Gerwick, Inc.

an upper limit where any given skid way could not be used without significant structural upgrading. This upgrading may not be cost effective but does warrant consideration. The base of most concrete structures can probably be constructed on a skid way to a level where it would be self-floating. It can then be skidded into a floating mode and towed to a deeper water site for mooring and completion.

13.9.4 Site Limitations

The principal limitation for any of the construction sites is access to the open sea or to maintained ship channels. As most of the structures will float at some time during their construction and installation phases, water of sufficient depth for them to float and be towed to their final location must exist in close proximity to the construction site. Dredging of new channels is very expensive and may not be possible owing to physical, environmental, or political reasons. Most large structures with deep-drafts have been completed while in a floating mode. When this is done, the deep-water requirement is further constrained because the deep-water location must be in a relatively mild environment with respect to wind, waves, and current. It is also desirable to have the deep-water location close enough to shore that materials supply and construction crew changes can be made without having to stockpile large amounts of material offshore or provide temporary offshore quarters for personnel. Both of these aspects of construction will add greatly to the cost of the project.

If a dry dock is required and must be built, the supporting infrastructure must be carefully studied before a site is selected. If the site is not convenient to major population areas, the site development will have to include such things as accommodations for work crews, dining facilities, water supply, sewage disposal, recreation facilities, communication facilities, transportation, and all the other things required for a small city.

As noted above, the existence of skid ways in operating shipyards does not necessarily mean they can be used for concrete structures because they may be undersized or structurally inadequate. Dredged channels leading away from skid ways most likely will not be of sufficient depth for the draft of larger concrete platforms, and some additional dredging may be required.

13.10 Marine Operations

The subject of marine operations is outside the scope of this paper, but it should be noted that offshore development and marine operations go hand-in-hand. The construction and installation of any type of offshore concrete platform will involve significant marine operations [Federation Internationale de la Preconstrainte, 1991]. The structure itself must float, be towed, and perhaps submerged for final installation. The structure may have its topside equipment installed in a mating operation. In this operation, the structure is ballasted down until it is almost submerged. The assembled topside equipment is then floated over the submerged structure. Deballasting is initiated, and as the structure rises, it picks the topside up off the barges. Deballasting is continued until the structure reaches its final draft for towing operations. Hookup of the topside to the structure is then done. This mating is a critical marine operation.

Marine operations are expensive and may require special vessels. Fortunately, many good specialist firms or consultants with offshore structure installation experience exist around the world, and they should be consulted for this aspect of the use of concrete platforms.

13.11 Cost Considerations

The costs of an offshore structure are usually dictated by the complexity of the structure, the location where it will be built, and the methods of construction used. These structures have the potential for low capital costs because the materials are local and the labor skill level usually does not have to be high. When properly made, concrete can be virtually maintenance free, thus producing an attractive life-cycle cost. Detailed cost discussions are beyond the scope of this paper, but a cost philosophy for these structures is described in des Desert and Gifford (1989) and Hoff (1991b).

13.12 Summary

The information presented in this chapters is intended to provide a general overview of what is required in the use of concrete for offshore structures. A significant amount of information on the subject is distributed throughout the literature but has not yet been assembled in a useful textbook form.

Concrete is the material of choice for permanent constructions in all regions of the world. It can be produced with local labor and local materials. The potential exists for its use in the offshore oil and gas industry in developing regions of the world and in existing offshore regions where steel structures have been predominately used before. The use of concrete for these structures generally must be sold to owners or developers who are used to dealing with steel structures.

Concrete technology is well established for use in offshore structures. No special materials are required, the concrete quality is not unusual, design requirements are well established, competent offshore concrete design firms exist in many regions of the world, no special construction practices are needed, flexibility exists in construction-site selection, capital costs can be low, and life-cycle costs can be very attractive. Because of the severe environment that exists offshore, durability aspects of concrete will require special attention.

The potential for the use of concrete in offshore structures is great and is limited only by the ingenuity of the concrete designer and constructor and their ability to sell their ideas to the owners.

References

ABAM Engineers Inc. 1983. Developmental Design and Testing of High-Strength Lightweight Concretes for Marine Arctic Structures, Program Phase I. Joint Industry Project Report, AOGA Project No. 198. Federal Way, Washington, DC.

ABAM Engineers Inc. 1984. Developmental Design and Testing of High-Strength Lightweight Concretes for Marine Arctic Structures, Program Phase II. Joint Industry Project Report, AOGA Project No. 230. Federal Way, Washington, DC.

ABAM Engineers Inc. 1986. Developmental Design and Testing of High-Strength Lightweight Concretes for Marine Arctic Structures, Program Phase III. Joint Industry Project Report, AOGA Project No. 230. Federal Way, Washington, DC.

ACI Committee 357. 1988. *State-of-the-Art Report on Barge-Like Concrete Structures*, ACI 357.2R-88, 89 pp. American Concrete Institute, Farmington Hills, MI.

ACI Committee 357. 1989. *Guide for the Design and Construction of Fixed Offshore Concrete Structures*, ACI 357R-89, 23 pp. American Concrete Institute, Farmington Hills, MI.

ACI Committee 357. 1990. *State-of-the-Art Report on Offshore Concrete Structures for the Arctic*, ACI 357.1R-90, 117 pp. American Concrete Institute, Farmington Hills, MI.

Anderson, A.R. 1977. World's largest prestressed LPG floating vessel. *J. Prestressed Concrete Inst.* 22(1):21.

Anon. 1984. New idea for tanker loading and crude storage for remote areas. *Ocean Ind.* 19(6):55.

Anon. 1986. Concrete seabed storage. *Ocean Ind.* 21(8):32.

Anon. 1988a. Concrete work for Petrobras provides unique work experience. *Offshore* 48(5):84–87.

Anon. 1988b. Field storage moves to seabed. *Offshore* 48(4):88–89.

Anon. 1989. Concrete advantages in shallow water marginals. *Offshore Eng.*, pp. 143–144.

Anon. 1990a. Anatomy of a concrete platform removal. *Offshore Eng.*, March 1990:28 and 33.

Anon. 1990b. Artificial island finds favour at Wytch Farm. *Offshore Eng.*, July 1990:19.

Anon. 1990c. Concrete gravity foundations. *Offshore Eng.*, August 1990:42.

Anon. 1991. Concrete FPS (floating production system) targets deepwater, marginal fields. *Ocean Ind.* 26(3):78.

Ben C. Gerwick, Inc. 1989. Segmental Construction of Steel-Concrete Composite Base Rafts for Concrete Gravity-Base Structures. San Francisco, CA.

des Desert, L. and Gifford, L.C. 1989. Minimum concrete platform for medium-depth waters. *Ocean Ind.* 24(8):43–48.

Federation Internationale de la Precontrainte. 1985. *FIP Recommendations for the Design and Construction of Concrete Sea Structures*, 4th ed., 29 pp. Thomas Telford, London.

Federation Internationale de la Precontrainte (FIP). 1991. *Marine Practice for Large Offshore Structures*, Cement and Concrete Association, London, England.

Fitzpatrick, J. and Stenning, D.G. 1983. Design and construction of Tarsuit Island in the Canadian Beaufort Sea. *Proc. Offshore Technol. Conf.*, OTC Paper 4517. Houston, TX.

Harrington, K. 1987. *Concrete as a Fabrication Material for Simple Hulls: A Marine Innovation Study*, Ph.D. Dissertation, 454 pp. Sunderland Polytechnic, UK.

Hoff, G.C. 1986. The service record of concrete offshore platforms in the North Sea. *Proc. Int. Conf. Concrete Marine Environ.*, pp. 131–142. Concrete Society, UK.

Hoff, G.C. 1988. Resistance of concrete to ice abrasion—A review. *Proc. 2nd Int. Conf. Concrete Marine Environ.*, ACI SP-109. V. M. Malhotra, ed., pp. 427–455. American Concrete Institute, Farmington Hills, MI.

Hoff, G.C. 1991a. Durability of offshore and marine concrete structures. *Proc. 2nd CANMET/ACI Int. Conf. Durability of Concrete*, Vol. 1. ACI SP-126. V. M. Malhotra, ed., pp. 33–64. American Concrete Institute, Farmington Hills, MI.

Hoff, G.C. 1991b. Considerations in the use of concrete for offshore structures. *Proceedings, Int. Conf. Eval. Rehab. Concrete Structures Innov. Design.* Vol. 2. ACI SP-128. V.M. Malhotra, ed., pp. 749–788. American Concrete Institute, Farmington Hills, MI.

Hunteman, J.E., Anastasio, F.L. Jr., and Deshazer, W.A. 1979. Concrete gravity platform in shallow offshore Louisiana water. *Proc. Offshore Technol. Conf.*, OTC Paper 3473. Houston, TX.

Jackson, G. and Bell, T.A. 1990. The design of the Ravenspurn North concrete gravity substructure: An innovative application of conventional technology. *Proc. Offshore Technol. Conf.*, OTC Paper 6394. Houston, TX.

Michel, D. 1989. Severe environmental loading challenges Hibernia designers. *Offshore* 49(8):73–74.

Moksnes, J. 1989. Oil and gas concrete platforms in the North Sea—Reflections on two decades of experience. *Proc. Int. Expe. Durability Concrete Marine Environ.*, P. Kumar Mehta, ed., pp. 127–146. University of California, Berkeley.

Moksnes, J., Haug, A.K., Modeer, M., and Beravam, T. 1987. Concrete quality in Norwegian offshore structures—15 years of laboratory and in-situ testing of high strength concrete. *Proc. Symp. Utilization High Strength Concrete*, I. Holand et al., eds., pp. 405–416, Stavanger, Norway.

Norwegian Contractors. 1990. Concrete subsea structures—well templates. *NC News*, pp. 6–7, 23. Stabekk, Norway.

Norwegian Contractors. 1991a. Durability of concrete in the Gulf of Mexico: Experience from existing marine concrete structures. Prepared by Ben C. Gerwick, Inc., San Francisco for Norwegian Contractors, Joint Industry Project on Concrete for Gulf of Mexico, Floating Production Platforms.

Norwegian Contractors. 1991b. Concrete for Gulf of Mexico Floating Production Platforms. Joint Industry Study Report No. 948. Vols. 1 and 2. Stabekk, Norway.

Ove Arup & Partners. 1990. The Ravenspurn North Concrete Gravity Substructure (CGS). Information booklet. London, England.

Roberts, J. 1990. Innovation in concrete gravity substructures: The ravenspurn North platform and beyond. *Proc. Offshore Technol. Conf.*, OTC Paper 6347. Houston, TX.

Tate, S.H. and Core, J.A. 1989. Eugene Island 126 concrete refurbishment. *Proc. Offshore Technol. Conf.*, OTC Paper 6146. Houston, TX.

Yee, A.A., Masuda, F.R., Kim, C.N., Doi, D.A., and Daly, L.A. 1984. Concrete module for the global marine concrete island drilling system. *Proc. FIP/CPCI Symp. Vol. 2, Concrete Sea Structures in Arctic Regions*, pp. 23–30.

Foundation preparation for high rise building (Courtesy Portland Cement Association).

14

Foundations for Concrete Structures

by
Manjriker Gunaratne, Ph.D., P.E.
Associate Professor at University of South Florida, Tampa. Specialized in various areas of geotechnical engineering including foundation design, numerical modeling and soil stabilization.

14.1 Foundation Engineering

Geotechnical engineering is a branch of civil engineering in which technology is applied to the design and construction of structures involving earthen materials. Surficial earthen material consists of soil and rock. There are many branches of geotechnical engineering. Soil and rock mechanics are fundamental studies of the properties and mechanics of soil and rock. Foundation engineering is the application of the principles of soil mechanics, rock mechanics, and structural engineering to the design of structures associated with earthen materials. It is generally observed that most common foundation types supported by intact bedrock present no compressibility problems. Hence when designing common foundation types the foundation engineer's primary concerns are the strength and compressibility of the subsurface soil and, whenever applicable, the strength of bedrock.

14.1.1 Soil Classification

14.1.1.1 Mechanical Analysis

According to texture or the "feel," two different soil types can be identified. They are (1) coarse-grained soil (gravel and sand) and (2) fine-grained soil (silt and clay). While the engineering properties (primarily strength and compressibility) of coarse-grained soils depend on the size of individual soil particles, the properties of fine-grained soils are mostly governed by moisture content. Hence it is important to identify the type of soil at a given construction site since effective construction procedures invariably depend on the soil type. Soil engineers use a universal format called the unified soil classification system (USCS) to identify and label soil. The system is based on the results of common laboratory tests of mechanical analysis and Atterberg limits [Bowles, 1986].

In classifying a given soil sample, mechanical analysis is conducted in two stages: (1) sieve analysis for the coarse fraction (gravel and sand) and (2) hydrometer analysis for the fine fraction (silt and clay). Of these, sieve analysis is conducted according to ASTM D421 and D422 procedures, using a set of U.S. standard sieves (Figure 14.1); the most commonly used sieves are numbers 20, 40, 60, 80, 100, 140, and 200, corresponding to sieve openings of 0.85, 0.425, 0.25, 0.18, 0.15, 0.106, and 0.075 mm, respectively. During the test, the percentage (by weight) of the soil sample retained on each sieve is recorded, from which the percentage (R%) passing (or finer than) a given sieve size (D) is determined.

On the other hand, if a substantial portion of the soil sample consists of fine-grained soils ($D < 0.075$ mm), then sieve analysis has to be followed by hydrometer analysis (Figure 14.2). This is performed by first treating the "fine fraction" with a deflocculating agent such as sodium hexametaphosphate (Calgon) or sodium silicate (water glass) for about half a day and then allowing the suspension to settle in a hydrometer jar kept at a constant temperature. As the heavier particles settle, followed by the lighter ones, a calibrated ASTM 152H hydrometer is used to estimate the fraction (percentage, R%) still settling above the hydrometer bottom at any given stage. Further, the particle size (D) that has settled past the hydrometer bottom at that stage in time can be estimated from Stokes' law. It can be seen that R% is the weight percentage of soil finer than D.

Complete details of the above tests are provided in Bowles (1986). For soil samples that have significant coarse and fine fractions, the sieve and hydrometer analysis results (R% and D) can be logically combined to generate grain (particle) size distribution curves such as those indicated in Figure 14.3. From Figure 14.3, it can be seen that 30% of soil type A is finer than 0.075 mm. (U.S. No. 200 sieve), with R% = 30 and $D = 0.075$ mm being the last pair of results obtained from sieve analysis. In combining sieve analysis data with hydrometer analysis data, one has to convert the R% (based on the fine fraction only) and D obtained from hydrometer analysis to R% based on the weight of the entire sample in order to ensure continuity of the curve.

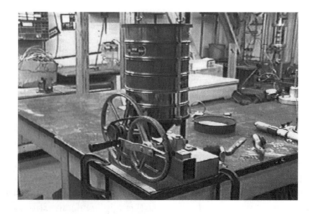

FIGURE 14.1 Equipment for sieve analysis.

FIGURE 14.2 Equipment for hydrometer analysis.

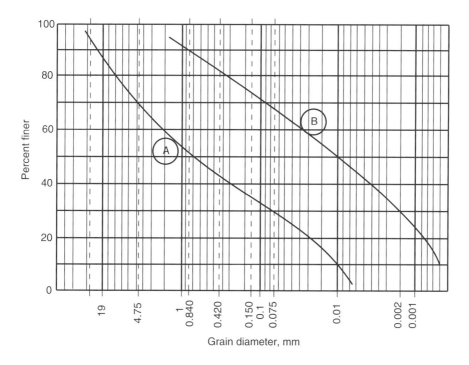

FIGURE 14.3 Grain-size distribution curves.

As an example, let the results from one hydrometer reading of soil sample A be R% = 90 and $D = 0.05$ mm. To plot the curve, one needs the percentage of the entire sample finer than 0.05 mm. Since what is finer that 0.05 mm is 90% of the fine fraction (30%) used for hydrometer analysis, the converted R% for the final plot can be obtained by multiplying 90% by the fine fraction of 30%. Hence the converted data used in Figure 14.3 are R% = 27 and $D = 0.05$ mm.

14.1.1.2 Atterberg Limits

As mentioned earlier, properties of fine-grained soils are governed by water. Hence the effect of water on fine-grained soils has to be considered in soil classification. This is achieved by employing the Atterberg limits or consistency limits. The physical state of a fine-grained soil changes with increasing water content, as shown in Figure 14.4, from a brittle to a liquid state.

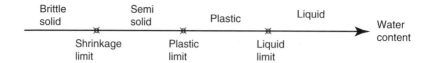

FIGURE 14.4 Variation of the fine-grained soil properties with water content.

Theoretically, the plastic limit (PL) is defined as the water content at which the soil changes from "semisolid" to "plastic" (Figure 14.4). For a given soil sample, this is an inherent property that can be determined by rolling a plastic soil sample into a "worm shape" to gradually reduce its water content by exposing more and more of an area until the soil becomes semisolid. This change can be detected by cracks appearing on the sample. According to ASTM 4318, the plastic limit is the water content at which cracks develop on a rolled soil sample at a diameter of 3 mm. Thus the procedure is one of trial and error. Though the apparatus (ground glass plate and moisture cans) used for the test is shown in Figure 14.5, the reader is referred to Bowles (1986) and Wray (1986) for details.

On the other hand, the liquid limit (LL), which is visualized as the water content at which the state of a soil changes from "plastic" to "liquid" with increasing water content, is determined in the laboratory using the Casagrande liquid limit device (Figure 14.6). This device is specially designed with a standard brass cup where a standard-sized soil paste is laid during testing. In addition, the

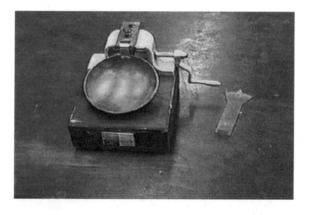

FIGURE 14.5 Equipment for the plastic limit test.

FIGURE 14.6 Equipment for the liquid limit test.

soil paste is grooved in the middle by a standard grooving tool thereby creating a gap with standard dimensions. Then the brass cup is made to drop through a distance of 1 cm on a hard rubber base. The number of drops (blows) required to close the above gap along a distance of 1/2 in is counted. Details of the test procedure can be found in Bowles (1986). ASTM 4318 specifies the liquid limit as the water content at which closing of the standard-sized gap is achieved in 25 drops of the cup. Therefore, one has to repeat the experiment for different trial water contents, each time recording the number of blows required to close the above standard-sized gap. Finally the water content corresponding to 25 blows can be interpolated from the data obtained from all of the trials.

The plasticity index (PI) is defined as follows:

$$PI = LL - PL$$

14.1.1.3 Unified Soil Classification System

In the commonly adopted unified soil classification system (USCS) shown in Table 14.1, the aforementioned soil properties are effectively used to classify soils. Example 14.1 below illustrates the classification of the two soil samples shown in Figure 14.3. Definiing the following two curve parameters is necessary to accomplish the classification:

Coefficient of uniformity $(C_u) = D_{60}/D_{10}$
Coefficient of curvature $(C_c) = D_{30}{}^2/D_{60}.D_{10}$

where D_i is the diameter corresponding to the ith percentage on the grain-size distribution curve.

Example 14.1

Soil A. The percentage of coarse-grained soil is equal to 70%. Hence Soil A is a coarse-grained soil. The percentage of sand in coarse-fraction is equal to $(70 - 30)/70 \times 100 = 57\%$. Thus according to the USCS (Table 14.1) Soil A is a sand. If one assumes clean sand, then

$C_c = (0.075)^2/(2 \times 0.013) = 0.21$ does not meet criterion for SW
$C_u = (2)/(0.013) = 153.85$ meets criterion for SW

Hence Soil A is a poorly graded sand, SP.

Soil B. The percentage of coarse-grained soil is equal to 32%. Hence Soil B is a fine-grained soil. Assuming that LL is equal to 45 and PL is equal to 35 (then PI is equal to 10) and using **Casagrande's plasticity chart** (Table 14.1) it can be concluded that Soil B is a silty sand with clay (ML).

14.1.2 Strength of Soils

The two most important properties of a soil that a foundation engineer must be concerned with are strength and compressibility. Since earthen structures are not designed to sustain tensile loads, the most common mode of soil failure is shear. Hence the shear strength of the foundation medium constitutes a direct input to the design of concrete structures associated with the ground.

14.1.2.1 Determination of Shear Strength

The shear strength of soils is assumed to originate from the strength properties of cohesion (c) and internal friction (ϕ). Using the basic Coulomb's friction principle, the shear strength of a soil can be expressed as

$$\tau_f = c + \sigma \tan \phi \tag{14.1}$$

However, it is also known that the magnitudes of the soil shear strength properties vary with prevailing drainage conditions and to a minor extent with the stress level. Hence it is important to

TABLE 14.1　Unified Soil Classification System

Major Divisions (2)		Group Symbols (τ) (3)	Typical Names (4)	Laboratory Classification Criteria (6)

Coarse-grained Soils — More than half of material is larger than No.200 (75 μm) sieve size. (τ) The No. 200 sieve size is about the smallest parts invisible to the naked eye

- **Gravels** — More than half of coarse fraction is larger than No. 4 sieve size. (4.75 mm)
  - **Clean Gravels** (little or no fines)
    - **GW** — Well-graded gravels, gravel sand mixtures, little or no fines. — $C_u = \dfrac{D_{60}}{D_{10}}$ greater than 4; $C_c = \dfrac{(D_{30})^2}{D_{10} \times D_{60}}$ between 1 and 3
    - **GP** — Poorly graded gravels, gravel-sand mixtures, little or no fines. — Not meeting all gradation requirements for GW
  - **Gravels with Fines** (appreciable amount of fines)
    - **GM** — Silty gravels, gravel-sand-silt mixtures. — Atterberg limits below A-line, or PI less than 4
    - **GC** — Clayey gravels, gravel-sand-clay mixtures. — Atterberg limits above A-line with PI greater than 7
- **Sands** — More than half of coarse fraction is smaller than No. 4 sieve size. (4.75 mm) (for visual classification, 5mm may be used as equivalent to the No. 4 sieve size)
  - **Clean Sands** (little or no fines)
    - **SW** — Well-graded sands, gravelly sands, little or no fines. — $C_u = \dfrac{D_{60}}{D_{10}}$ greater than 6; $C_c = \dfrac{(D_{30})^2}{D_{10} \times D_{60}}$ between 1 and 3
    - **SP** — Poorly graded sands, gravelly sands, little or no fines. — Not meeting all gradation requirements for SW
  - **Sands with Fines** (appreciable amount of fines)
    - **SM** — Silty sands, sand-silt mixtures. — Atterberg limits below A-line, or PI less than 4
    - **SC** — Clayey sands, sand-clay mixtures. — Atterberg limits above A-line with PI greater than 7

(See Sec. 2.5)

Above A-line with PI between 4 and 7 are borderline cases requiring use of dual symbols.

(See Sec. 2.5)

Limits plotting in hatched zone with PI between 4 and 7 are borderline cases requiring use of dual symbols.

Use grain size curve in identifying the fractions as given under field identification.

Determine percentages of gravel and sand from grain size curve. Depending on percentages of fines (fraction smaller than No. 200 sieve size) coarse grained soils are classified as follows:
- Less than 5%: GW, GP, SW, SP
- More than 12%: GM, GC, SM, SC
- 5% to 12%: Borderline cases requiring use of dual symbols.

Fine-grained Soils — More than half of material is smaller than No.200 (75 μm) sieve size.

- **Silts and Clays** — Liquid limit less than 50
  - **ML** — Inorganic silts and very fine sands, rock flour, silty or clayey fine sands or clayey silts with slight plasticity.
  - **CL** — Inorganic clays of low to medium plasticity, gravelly clays, sandy clays, silty clays, lean clays.
  - **OL** — Organic silts and organic silty clays of low plasticity.
- **Silts and Clays** — Liquid limit greater than 50
  - **MH** — Inorganic silts, micaceous or diatomaceous fine sandy or silty soils, elastic silts.
  - **CH** — Inorganic clays of high plasticity, fat clays.
  - **OH** — Organic clays of medium to high plasticity, organic silts.
- **Highly Organic Soils**
  - **Pt** — Peat and other highly organic soils.

Plasticity Chart
For laboratory classification of fine-grained soils

Comparing soils at equal liquid limit: toughness and dry strength increase with increasing plasticity index.

Source: An Introduction to Geotechnical Engineering by Holtz/Kovacs, © 1981. Reprinted by permission of Prentice Hall, Inc., Upper Saddle River, NJ.

characterize the strength properties in terms of the drainage condition (drained or undrained) employed during testing.

A wide variety of laboratory and field methods are used to determine the shear strength parameters c and ϕ of soils. The (1) triaxial test, (2) standard penetration test (SPT), and (3) static cone penetration tests (CPT) are the most common tests used in foundation engineering.

14.1.2.2 Triaxial Tests

In this test, a sample of undisturbed soil retrieved from a site is tested under a range of pressures that encompass the expected field stress conditions due to the building. Figure 14.7 shows the schematic diagram of the important elements of a triaxial setup, and the actual testing apparatus is shown in Figure 14.8.

From the discussion of soil strength, it can be seen that the type of soil and the field-loading rate have a bearing on selection of the laboratory drainage condition and hence the loading rate. Accordingly, three types of triaxial tests are commonly conducted: (1) consolidated drained tests (CD), (2) consolidated undrained tests (CU), and (3) unconsolidated undrained tests (UU). In CU and CD tests, the pressure exerted on the cell fluid is used to consolidate the soil sample back up to the *in situ* stress state before applying the axial compression. On the other hand, in the UU tests the

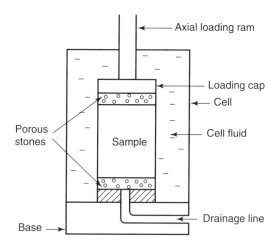

FIGURE 14.7   Schematic diagram of triaxial cell.

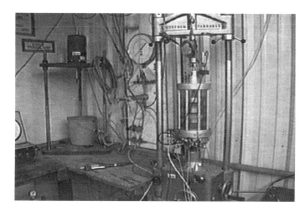

FIGURE 14.8   Triaxial testing apparatus.

cell pressure is applied, with no accompanying consolidation, merely to provide a confining pressure. Computations involving CU and UU tests are given in Examples 14.2 and 14.3, and the reader is referred to Holtz and Kovacs (1981) for more details of the testing procedure.

TABLE 14.2   Measured CU Triaxial Test Data

Test	Cell Pressure, kPa	Deviator Stress at Failure, kPa	Pore Pressure, kPa
1	20	20.2	5.2
2	40	30.4	8.3

Example 14.2

Assume that one conducts two CU triaxial tests on a sandy clay sample from a tentative site in order to determine the strength properties. The applied cell pressures, deviator stresses, and measured pore pressures at failure are given in Table 14.2. The strength parameters can be easily estimated using the Mohr circle method as follows:

Total Strength Parameters.   The total stresses (σ_1 and σ_3) acting on both test samples at failure are indicated in Figure 14.9a. Accordingly, the Mohr circles for the two stress states can be drawn as in Figure 14.10. Then the total strength parameters (sometimes referred to as the undrained strength parameters) can be evaluated from the slope of the direct common tangent, which is the Coulomb envelope (Eq. 14.1) plotted on the Mohr circle diagram, as

$$c_u = 4.0\,\text{kPa} \quad \text{and} \quad \phi_u = 13.2^\circ$$

It is obvious that the generated pore pressure has been ignored in the above solution.

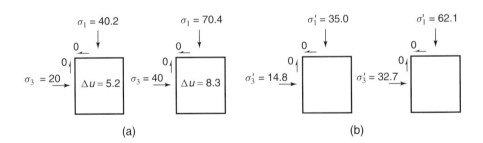

FIGURE 14.9   Stress states at failure. (a) Total stresses (kPa); (b) Effective stresses (kPa).

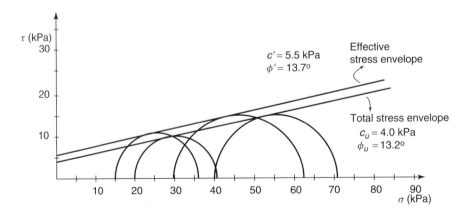

FIGURE 14.10   Mohr circle diagram for a CU test.

Effective Strength Parameters. The effective stresses (σ_1' and σ_3') on both test samples at failure computed by subtracting the pore pressure from the total stress, are indicated in Figure 14.9b. The Mohr circles corresponding to the two stress states are drawn in Figure 14.10. The effective strength parameters (sometimes referred to as the drained strength parameters) can be found from the slope of the Coulomb envelope for effective stresses plotted on the Mohr circle diagram as

TABLE 14.3 Measured UU Triaxial Test Data

Test	Cell Pressure, kPa	Deviator Stress at Failure, kPa	Pore Pressure, kPa
1	40	102.2	N/A
2	60	101.4	N/A

$$c' = 5.5 \, \text{kPa} \quad \text{and} \quad \phi' = 13.7°$$

Example 14.3

Assume that one wishes to determine the strength properties of a medium stiff clayey foundation under short-term (undrained) conditions. An effective method for achieving this is to conduct a UU (Quick) test. The results are presented in Table 14.3; estimate the undrained strength parameters.

Since in these tests, the pore pressure generation is not only measured the total stresses can be plotted, as in Figure 14.11. It can be seen that the deviator stress at failure does not change with the changing cell pressure during this type of test. This is because the soil samples are not consolidated to the corresponding cell pressures during UU (unconsolidated undrained) tests, therefore the soil structure is unaffected by the change in cell pressure.

Hence the following strength parameters can be obtained from Figure 14.11:

$$c_u = 50.6 \, \text{kPa} \quad \text{and} \quad \phi_u = 0°$$

The reader should note that the subscripts u are used to distinguish the UU test parameters.

14.1.2.3 Standard Penetration Test

The standard penetration test (SPT) is the most common field test used to estimate the *in situ* shear strength of foundation soil. In this test, a 140 lb hammer (Figure 14.12) that falls 30 in is used to drive a standard split spoon sampler (Figure 14.13) 18 in into the ground. The number of hammer blows necessary to achieve the last 12 in of penetration is recorded as the blow count (N). Although it is relatively easy to perform, SPT suffers because it is crude and not repeatable. The basic principle underlying the SPT test is the relation between the penetration resistance and shear strength of the soil, which can be visualized as a unique relationship. Since the penetration resistance

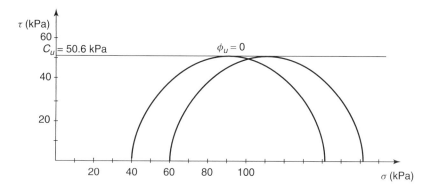

FIGURE 14.11 Mohr circle diagram for a UU test.

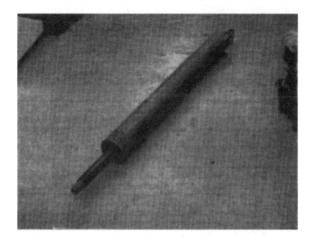

FIGURE 14.12   Standard penetration test hammer.

FIGURE 14.13   Split spoon sampler.

is obviously affected by the overburden, the following correction is applied before determining the soil properties:

$$N' = \sqrt{\frac{1}{\sigma'_v}}\, N \tag{14.2}$$

where σ'_v is the effective overburden stress (in tons per square feet) computed as follows:

$$\sigma'_v = \gamma z - \gamma_w d_w \tag{14.3}$$

where

γ = unit weight of soil
γ_w = unit weight of water

TABLE 14.4 Relation Between SPT Blow Count and Friction Angle of Granular Soils

Description	Very Loose	Loose	Medium	Dense	Very Dense
Relative density D_r	0	0.15	0.35	0.65	0.85
SPT N'_{70}					
Fine	1–2	3–6	7–15	16–30	?
Medium	2–3	4–7	8–20	21–40	>40
Coarse	3–6	5–9	10–25	26–45	>45
ϕ					
Fine	26–28	28–30	30–34	33–38	
Medium	27–28	30–32	32–36	36–42	< 50
Coarse	28–30	30–34	33–40	40–50	
γ_{wet}, kN/m^3	11–16*	14–18	17–20	17–22	20–23

Source: (Bowles, 1995).

Note: Empirical values for ϕ, D_r, and unit weight of granular soils based on a normally consolidated [approximately, $\phi = 28° + 15° D_r (\pm 2°)$] SPT at about 6 m depth.

*Excavated soil or material dumped from a truck has a unit weight of 11 to 14 kN/m^3 and must be quite dense to weigh much over 21 kN/m^3. No existing soil has a $D_r = 0.00$ nor a value of 1.00. Common ranges are from 0.3 to 0.7.

TABLE 14.5 Relation Between SPT Blow Count and Unconfined Compression Strength of Clay

Consistency			N'_{70}	q_u, kPa	Remarks
Very soft			0–2	<25	Squishes between fingers when squeezed
Soft	NC	Young clay	3–5	25–50	Very easily deformed by squeezing
Medium			6–9	50–100	??
Stiff	Increasing OCR	Aged/ cemented	10–16	100–200	Hard to deform by hand squeezing
Very stiff			17–30	200–400	Very hard to deform by hand squeezing
Hard			>30	>400	Nearly impossible to deform by hand

Source: (Bowles, 1995).

*Blow counts and OCR division are for a guide—in clay, "exceptions to the rule" are very common.

z = depth of test location
d_w = depth of test location from the ground water table

Once the corrected blow count (N'_{70}) is determined, one can find the strength parameters based on the empirical correlations shown in Tables 14.4 and 14.5. The subscript 70 indicates 70% efficiency in energy transfer from the hammer to the sampler. This value has been shown to be relevant for the North American practice of SPT. It should be noted that the undrained strength (c_u) of a saturated clay is one-half the unconfined compression strength (q_u).

14.1.2.4 Static Cone Penetration Test

The cone penetration test (CPT) has been gaining popularity as a more reliable and repeatable alternative to SPT. In this test, a standard cone and a sleeve (Figure 14.14) are advanced at a steady rate (1 cm/sec) into the ground while the cone resistance (q_c) and the sleeve friction (f_s) are electronically measured. The entire cone apparatus and the associated computing facilities are usually truck mounted as shown in Figure 14.15. A typical cone profile obtained from a University of South Florida organic soil research site is shown in Figure 14.16. Owing to its ability to measure two parameters (q_c and f_s), CPT is a useful tool for identifying soil type as well as for evaluating soil

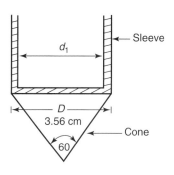

FIGURE 14.14 Cone and sleeve.

FIGURE 14.15 Cone penetration test equipment. From Stinnette, P. 1996. Geotechnical data management and analysis system for organic soils. Ph.D. dissertation. University of South Florida.

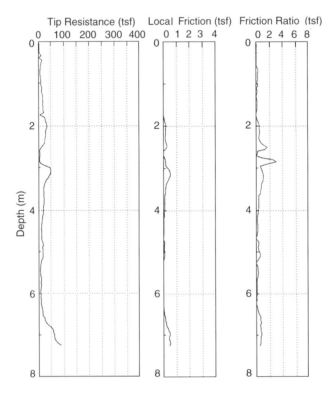

FIGURE 14.16 A typical cone profile. From Mullins, A.G. 1996. Field characterization of dynamic replacement of Florida soils. Ph.D. dissertation. University of South Florida.

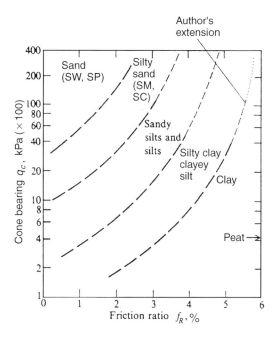

FIGURE 14.17   Soil classification using CPT data. From
J.E. Bowles, 1995.

properties. A convenient parameter termed the friction ratio, F_R, is defined for this purpose as

$$F_R = \frac{f_s}{q_c} \tag{14.4}$$

Figure 14.17 shows a simple chart that can be used for soil classification using CPT data. Currently, it is commonplace to have cone tips fitted with transducers that can produce a continuous record of the ground pore pressures at various depths.

Using CPT data, the undrained strength of a clay can be obtained as

$$s_u = \frac{q_T - p_0}{N_{kT}} \tag{14.5}$$

where

$$q_T = q_c + u_c(1 - a) \tag{14.6}$$

$$N_{kT} = 13 + \frac{5.5}{50} PI \tag{14.7}$$

and p_0 and u_c are the effective overburden pressure and the pore pressure, respectively, measured in the same units as s_u and q_c; a is taken as the approximate diameter ratio $(d_1/D)^2$ (Figure 14.14).

On the other hand, the friction angle of a granular soil can be obtained from q_c (in megapascals) based on the following approximate expression:

$$\phi = 29° + \sqrt{q_c} \tag{14.8}$$

For gravel and silty sand, corrections of $+5°$ and $-5°$, respectively, have to be made.

14.1.3 Compressibility and Settlement

Soils, like any other material, deform under loads. Hence even if the integrity of a structure is satisfied, soil supporting the structure can undergo compression, leading to structural settlement. For most dry soils, this settlement will cease almost immediately after the particles readjust in order to attain an equilibrium with the structural load. This immediate settlement is evaluated using the theory of elasticity. However, if the ground material is wet, fine-grained (low permeability) soil, the settlement will continue for a long period of time with slow drainage of water until the excess pore water pressure completely dissipates. This is usually evaluated by Terzaghi's consolidation theory. In some situations involving very fine clays and organic soils, settlement continues to occur even after the pore water pressure in the foundation vicinity comes to an equilibrium with that of the far field. Secondary compression concepts are needed to estimate this secondary settlement.

14.1.3.1 Estimation of Foundation Settlement in Granular Soils

Very often, settlement of footings founded on granular soils is determined based on the plate load tests discussed in Section 14.2. The most commonly adopted analytical methods for immediate settlement evaluation in granular soils are based on the elastic theory. However, one has to realize that reliable estimates of elastic moduli and Poisson ratio values for soils are not easily obtained. This is mainly because of the sampling difficulty and, particularly, the dependency of the elastic modulus on the stress state. On the other hand, reliable field methods for obtaining elastic moduli are also scarce. The following expressions can be used to find the immediate settlement:

$$s_e = f \frac{Bq_0}{E_s}\left(1 - \mu_s^2\right)\frac{\alpha}{2} \qquad (14.9)$$

where

α = a factor to be determined from Figure 14.18
B = width of foundation
L = length of foundation

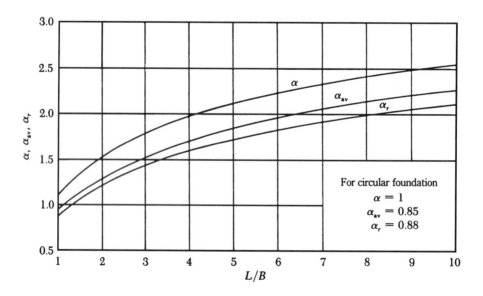

FIGURE 14.18   Chart for obtaining α factor. From Das, B.M. 1995. *Principles of Foundation Engineering.* PWS Publishing.

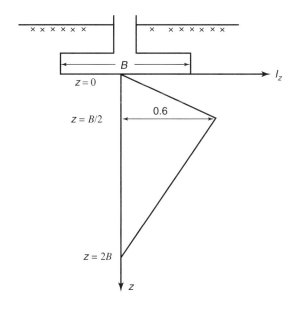

FIGURE 14.19 Strain influence factor. From Schmertmann J.H. and Hartman, J.P. 1978. Improved strain influence factor diagrams. *J. Geotech. Eng. Div., Am. Soc. Civ. Eng.* 104(GT8):1131–1135.

q_o = contact pressure (P/BL)
s_e = immediate (elastic) settlement
E = elastic modulus
f = 0.5 or 1.0 (depending on whether s_e is at the corner or center of the foundation)

Another widely used method for computing granular soil settlements is the Schmertmann and Hartman (1978) method based on the elastic theory.

$$s_e = C_1 C_2 (\overline{q} - q) \sum_0^z \frac{I_z}{E_s} \Delta z \qquad (14.10)$$

where

I_z = strain influence factor in Figure 14.19
C_1 = foundation depth correction factor = $1 - 0.5[q/(\overline{q} - q)]$
C_2 = correction factor for creep of soil = $1 + 0.2 \log$ (time in years/0.1)
$\overline{q}$ = stress at foundation level
q = overburden stress

The elastic properties needed to manipulate the above expressions are provided in Tables 14.6 and 14.7. Furthermore, some useful relationships that can provide the elastic properties from *in situ* test results are given below:

$$E_s \text{ (tsf)} = 8N \qquad (14.11)$$

and

$$E_s = 2q_c \qquad (14.12)$$

A comprehensive example illustrating the use of the above relations is provided in Example 14.4.

14.1.3.2 Estimation of Foundation Settlement in Saturated Clays

The load applied on a saturated fine-grained soil foundation is immediately acquired by the pore water, as illustrated in Figure 14.20a. However, with the dissipation of pore pressure resulting from drainage of water, the applied stress (total stress) is gradually transferred to the soil skeleton as an effective stress (Figure 14.20b). The long-term soil skeleton rearrangement taking place during this process is termed the consolidation settlement.

The soil properties required for estimation of the magnitude and rate of consolidation settlement can be obtained from the laboratory one-dimensional (1-D) consolidation test. Figure 14.21 shows the consolidometer apparatus where a saturated sample (2.5 in diameter and 1.0 in height) is subjected to a constant load while the deformation and sometimes the pore pressure are monitored until consolidation is complete. A detailed description of this is can be found in Bowles (1986). The sample is tested in this manner for a wide range of stresses that encompass the expected foundation pressure.

Using Terzaghi's 1-D consolidation theory, the relationship shown in (Table 14.8 between the degree of consolidation U (settlement at any time t as a percentage of the ultimate settlement) and the time factor, T, can be derived for a clay layer subjected to a constant pressure increment throughout its depth.

Figure 14.22 shows the results of a consolidation test conducted on an organic soil sample. The coefficient of consolidation (C_v) for the soil can be obtained from these results using Casagrande's logarithm-of-time method (Holtz and Kovacs, 1981). Using this method, from Figure 14.22 one can estimate the time for 90% consolidation as 200 sec. Then, by using the following expression for the time factor, one can estimate C_v as 2.5×10^{-4} in²/sec, since $U = 90\%$ when $t = 200$ sec.

$$T = \frac{C_v t}{H_{dr}^2} \tag{14.13}$$

TABLE 14.6 Elastic Properties of Geomaterials

Soil	E_s, MPa
Clay	
Very soft	2–15
Soft	5–25
Medium	15–50
Hard	50–100
Sandy	25–250
Glacial till	
Loose	10–150
Dense	150–720
Very dense	500–1440
Loess	15–60
Sand	
Silty	5–20
Loose	10–25
Dense	50–81
Sand and gravel	
Loose	50–150
Dense	100–200
Shale	150–5000
Silt	2–20

Source: Bowles, 1995.

Note: Value range for the static stress-strain modulus E_s for selected soils (see also Table 5.6). The value range is too large to use an "average" value for design.

Field values depend on stress history, water content, density, and age of deposit.

TABLE 14.7 Poisson Ratios for Geomaterials

Type of Soil	μ
Clay, saturated	0.4–0.5
Clay, unsaturated	0.1–0.3
Sandy clay	0.2–0.3
Silt	0.3–0.35
Sand, gravelly sand	−0.1–1.00
commonly used	0.3–0.4
Rock	0.1–0.4 (depends somewhat on type of rock)
Loess	0.1–0.3
Ice	0.36
Concrete	0.15
Steel	0.33

Source: Bowles, 1995.

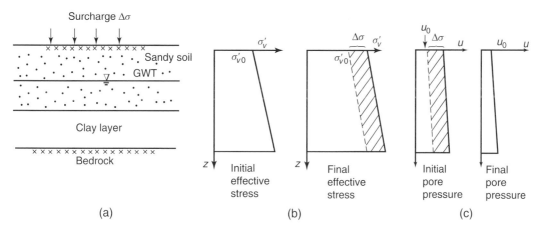

FIGURE 14.20 Illustration of consolidation settlement. (a) Subsurface profile, (b) Effective stress distribution, and (c) Pore pressure distribution.

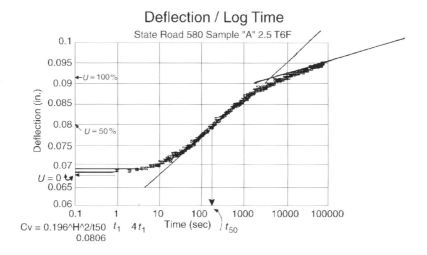

FIGURE 14.21 Laboratory consolidometer apparatus. *Source: Courtesy of the University of South Florida.*

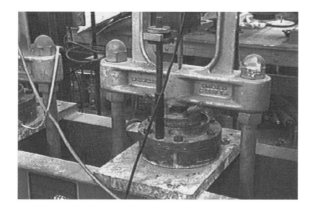

FIGURE 14.22 Settlement versus logarithm-of-time curve. From Stinnette, P. 1992. Engineering properties of Florida organic soils. Masters project. University of South Florida.

where H_{dr} is the longest drainage path in the consolidating soil layer. It should be noted that the water in the laboratory soil sample drains through both sides during consolidation and hence $H_{dr} = 0.5$ in.

When the above consolidation test is repeated for several other pressure increments, doubling the pressure each time, the variation of the post-consolidation (equilibrium) void ratio (e) with pressure (p) can be observed using the following relation between e and the sample strain computed from the monitored sample deformation:

$$\frac{\Delta e}{1 + e_0} = \frac{\Delta H}{H} \tag{14.14}$$

where e_0 and H are the initial void ratio and the sample height, while ΔH and Δe are their respective changes. A typical laboratory consolidation curve (e versus $\log p$) for a clayey soil sample is shown in Figure 14.23. The following important parameters can be obtained from Figure 14.23.

Recompression index, $C_r = (1.095 - 1.045)/(\log 60 - \log 10) = 0.064$

Compression index, $C_c = (1.045 - 0.93)/(\log 120 - \log 60) = 0.382$

Preconsolidation pressure, $p_c = 60$ kPa

All the above information can be used to estimate the ultimate consolidation settlement of a saturated clay layer (of thickness H) due to an average pressure increase of Δp. The ultimate consolidation settlement (s_{con}) can be expressed by the following, depending on the individual case, as illustrated in Figure 14.24.

TABLE 14.8 Degree of consolidation versus Time Factor

U_{avg}	T
0.1	0.008
0.2	0.031
0.3	0.071
0.4	0.126
0.5	0.197
0.6	0.287
0.7	0.403
0.8	0.567
0.9	0.848
0.95	1.163
1.0	∞

Case 1

$$s_{con} = \frac{C_c H}{1 + e_0} \log \frac{p_0 + \Delta p}{p_0} \tag{14.15}$$

Case 2

$$s_{con} = \frac{C_r H}{1 + e_0} \log \frac{p_0 + \Delta p}{p_0} \tag{14.16}$$

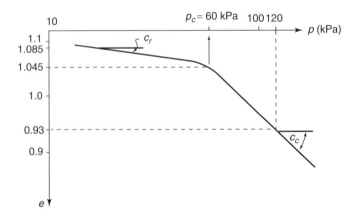

FIGURE 14.23   Laboratory consolidation curve (e versus $\log p$).

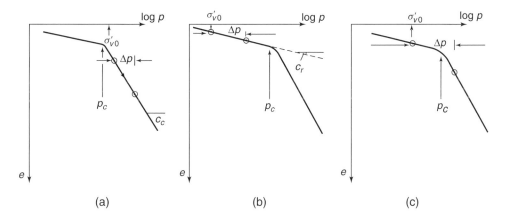

FIGURE 14.24 Illustration of the use of the consolidation equation. (a) Case 1, (b) Case 2, and (c) Case 3.

Case 3

$$s_{con} = \frac{C_r H}{1 + e_0} \log \frac{p_c}{\sigma'_{v_0}} + \frac{C_c H}{1 + e_0} \log \frac{p_0 + \Delta p}{p_c} \qquad (14.17)$$

The average pressure increase in the clay layer can be accurately determined by using Newmark's chart, shown in Figure 14.25. When the footing is drawn on the chart to a scale of $OQ = d_c$ (the depth of the midplane of the clay layer from the footing bottom), Δp can be evaluated by

$$\Delta p = qIM \qquad (14.18)$$

where q, I, and M are the contact pressure, the influence factor (specific to the diagram), and the number of elements of the chart covered by the drawn footing, respectively.

Example 14.4

Assume that it is necessary to compute the maximum differential settlement of the foundation shown in Figure 14.26, which also shows the SPT, elastic moduli (using Eq. (14.11) for sands and using 33% of the estimate for clay), and unit weight profiles as well as the strain influence factor plot. For the above data

> contact pressure ($\overline{q}$) = $200/(1.5)^2$ kPa = 88.89 kPa
> overburden pressure at footing depth (q) = 16.5×1.0 kPa = 16.5 kPa

Immediate Settlement.   Areas of the strain-influence diagram covered by different elastic moduli are

> $A_1 = 0.5(0.75 \times 0.6) + 0.5(0.25)(0.533 + 0.6) = 0.367$ m
> $A_2 = 0.5(1.5)(0.533 + 0.133) = 0.5$ m
> $A_3 = 0.5(0.5)(0.133) = 0.033$ m

Then, by applying Eq. (14.10), one obtains the immediate settlement as

$$s_{center} = [1 - 0.5\{16.5/(88.89 - 16.5)\}][1.0][88.89 - 16.5][0.367(1.0)/(11.5 \times 10^3)$$
$$+ 0.5/(10.7 \times 10^3) + 0.033/(2.57 \times 10^3)] = 5.87 \text{ mm}$$

From Eq. (14.9), s_{corner} can be deduced as $0.5(5.87) = 2.94$ mm.

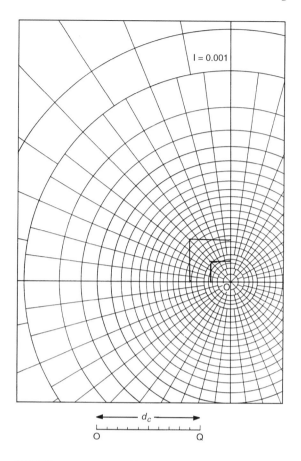

FIGURE 14.25 Newmark's chart. From *An Introduction to Geotechnical Engineering* by Holtz/Kovacs, © 1981. Reprinted by permission of Prentice Hall, Upper Saddle River, NJ.

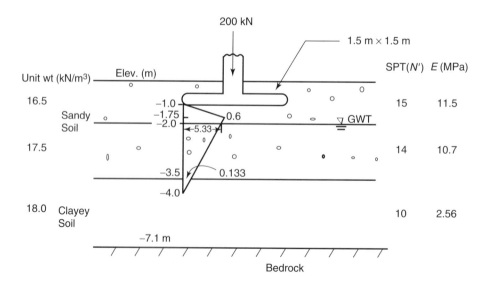

FIGURE 14.26 Settlement computation.

Consolidation Settlement. As for the consolidation settlement, the average stress increase in clay can be obtained as

$$\Delta p_{center} = 4 \times 19 \times 88.89 \times 0.001 = 6.75 \text{ kPa}$$
$$\Delta p_{corner} = 58 \times 88.89 \times 0.001 = 5.2 \text{ kPa}$$

On the other hand, the average overburden pressure at the clay layer is found from Eq. (14.3) as

$$\sigma'_{v0} = 16.5(2) + 17.5(1.5) + 18.0(1.0) - 9.8(2.75) = 54.8 \text{ kPa}$$

From Figure 14.24, one observes that the relevant expression for this situation is Eq. (14.17), and using the above estimates, the consolidation settlement is found as

$$s_{center} = [0.064/(1 + 1.06)](2.5) \log(60/54.8)$$
$$+ [0.382/(1 + 1.06)](2.5) \log[(60 + 6.75)/60] = 0.0245 \text{ m} = 24.52 \text{ mm}$$

As for the corner, the applicable expression from Figure 14.24 is Eq. (14.16). Hence

$$s_{corner} = [0.064/(1 + 1.06)](2.5) \log(54.8 + 5.2/54.8) = 3.06 \text{ mm}$$

Therefore, the total settlement at the center of the footing will be 30.39 mm or 1.12 in, while that at the corner will be 6.0 mm or 0.24 in.

Total Settlement Check. Most building codes stipulate the maximum allowable total settlement to be 1.0 in. Hence, the above value is unacceptable.

Differential Settlement Check. The differential settlement is equal to $(s_{center} - s_{corner})$/distance from center to corner or $(30.39 - 6.00)/(1.06)/1000 = 0.023$. According to most building codes, the maximum allowable differential settlement to prevent structural cracks in concrete is 0.013. Hence the above design fails the differential settlement criterion.

14.1.4 Groundwater and Seepage

Stability analysis of water-retaining concrete structures requires that the uplift forces exerted on them be evaluated. These structures often exist in groundwater flow regimes caused by differential hydraulic heads. Hence an analysis of groundwater seepage invariably has to be performed when estimating the uplift forces.

The most common and the simplest means of seepage analysis is the method of flownets. In this method, two orthogonal families of equipotential and flow lines are sketched in the flow domain (Figure 14.27) using the following basic principles. A flow line is an identified or a visualized flow conduit boundary in the flow domain. On the other hand, an equipotential line is an imaginary line in which the total energy head is the same.

14.1.4.1 Rules Governing the Construction of a Flownet

1. Equipotential lines do not intersect each other.
2. Flow lines do not intersect each other.
3. Equipotential lines and flow lines form two orthogonal families.
4. In order to ensure equal flow in the drawn flow conduits and equal head drop between adjacent equipotential lines, individual flow elements formed by adjacent equipotential lines and flow lines bear the same height/width ratio (typically 1.0).

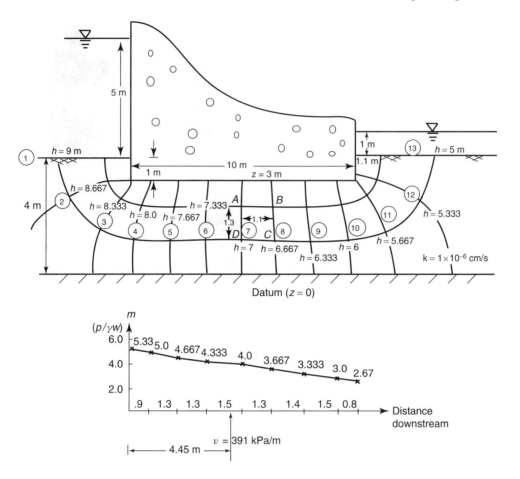

FIGURE 14.27   Seepage under a concrete dam.

With seepage velocities being generally very low, the pressure (p) exerted by seeping water contributes along with the potential energy to the total head (energy/unit weight) of water as

$$h = \frac{p}{\gamma_w} + z \qquad (14.19)$$

The quantity of groundwater flow at any location in a porous medium like soil can be expressed by D'Arcy's law as

$$q = kiA \qquad (14.20)$$

where k is the coefficient of permeability (or hydraulic conductivity) at that location while i, the hydraulic gradient, can be expressed by

$$i = -\frac{dh}{dx} \qquad (14.21)$$

The following example illustrates the flownet method of seepage analysis and evaluation of uplift pressures. For more accurate and rigorous methods, the reader is referred to Harr (1962).

Example 14.5

Assume that it is necessary to establish the pressure distribution on the bottom of the dam shown in Figure 14.27 and the seepage under the dam shown in Figure 14.27.

As the first step in the solution, a flownet has been drawn to scale, following the rules above. Using the bedrock as the datum for the elevation head, total heads have been assigned using Eq. (14.19) for all the equipotential lines as shown. It is noted that the head drop between two adjacent equipotential lines is

$$(9 \text{ m} - 5 \text{ m})/12 = 0.333 \text{ m}$$

Then, by applying Eq. (14.20) to the points where the equipotential lines and the dam bottom (B_i) intersect, the following expression can be obtained for the pressure distribution, which is plotted in Figure 14.27.

$$p = \gamma_w (h - 3.0)$$

Then, the total upthrust can computed from the area of the pressure distribution as 391.34 kPa/m acting at a distance of 4.45 m downstream.

Then, by applying Eq. (14.21) to the element ABCD, one obtains

$$i = (5.333 - 5.0)/1.1 = 0.302$$

FIGURE 14.28 Piezometer probes. From Thilakasiri, H.S. 1996. Numerical simulation of dynamic replacement of Florida organic soils. Ph.D. dissertation. University of South Florida.

Since $k = 1 \times 10^{-6}$ cm/s, one can apply Eq. (14.20) to obtain the quantity of seepage through ABCD as

$$q_1 = 1 \times (10^{-9})(0.302)(1.3)(1) \text{ m}^3/\text{s/m} \quad (\text{since AD} = 1.3 \text{ m})$$

Since all of the conduits must carry equal flow (see rule 4 of the flownet construction)

$$q = 3 \times (10^{-9})(0.302)(1.3)(1) \text{ m}^3/\text{s/m} = 1.18 \times 10^{-9} \text{m}^3/\text{s/m}$$

Note the following important assumptions made in the above analysis:

1. The subgrade soil is homogeneous.
2. The bedrock and concrete dam are intact.
3. There is no free flow under the dam due to **piping** (or erosion).

Hence the design and installation of an adequate pore-pressure monitoring system that can verify the analytical results is an essential part of the design. A piezometer with a geomembrane/sand filter that can be used for monitoring pore pressures is shown in Figure 14.28.

14.1.5 Dewatering of Excavations

Construction in areas of shallow groundwater requires dewatering prior to excavation. Although contractors specialized in such work determine the details of the dewatering program depending on the field performance, a preliminary idea of equipment requirements and feasibility can be obtained by a simplified analysis. Figure 14.29 shows the schematic diagram for such a program

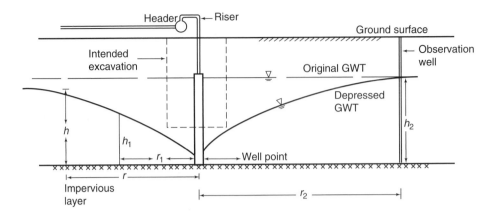

FIGURE 14.29 Dewatering of excavations.

and shows the elevations of the depressed water table at various distances from the center of the well. Observation wells (or bore holes) can be placed at any location, such as those shown at distances of r_1 and r_2, to monitor the water table depression.

In analyzing a seepage situation like this, Dupuit [Harr, 1962] assumed that (1) for a small inclination of the line of seepage, the flow lines are horizontal and (2) the hydraulic gradient is equal to the slope of the free surface and is invariant with depth. Hence, for discharge through any general section such as an observation well, one can write the following expression for the flow by combining Eqs. (14.20) and (14.21):

$$q = k \left(-\frac{dh}{dx} \right) h \qquad (14.22)$$

Noting that q and k are constants throughout the flow regime considered, Eq. (14.22) can be integrated between distances of r_1 and r_2 to obtain

$$q = \frac{\pi k \left(h_1^2 - h_2^2 \right)}{\ln(r_1/r_2)} \qquad (14.23)$$

By defining the extent of dewatering, using parameters r_1, r_2, h_1, and h_2, one can utilize the above expression to determine the capacity requirement of the pump.

14.1.6 Environmental Geotechnology

The amount of solid waste generated in the United States is expected to exceed 510 M tons by the year 2000 [Koerner, 1994]. Thus the immediate need for construction of adequate landfills cannot be overemphasized. Although the construction of landfills involves political and legal issues, properly designed, constructed, and maintained landfills have proven to be secure, especially if they are provided with lined facilities. These are installed on the bottom or sides of a landfill to control groundwater pollution by the liquid mixture (**leachate**) formed by the interaction of rainwater or snowmelt with waste material.

Types of liners for leachate containment are basically (1) clay liners (2) geomembranes, and (3) composite liners consisting of geomembranes and clay liners. Of these, until recently, the most frequently used liners were clay liners, which minimized leachate migration by achieving permeability values as low as 5×10^{-8} to 5×10^{-9} cm/sec. However, owing to the large thickness (0.6–2 m) requirement and chemical activity in the presence of organic-solvent leachates, geomembranes have been increasingly utilized for landfills.

14.1.7 Design of Landfill Liners

As shown in Figures 14.30 and 14.31, the important components of a solid material containment system are (1) a leachate collection/removal system, (2) a primary leachate barrier (3) a leachate detection/removal system, (4) a secondary leachate barrier, and (5) a filter above the collection system to prevent clogging. Some of the design criteria [Koerner, 1994] are as follows:

1. The leachate collection system should be capable of maintaining a leachate head of less than 30 cm.
2. Both collection and detection systems should have 30-cm thick granular drainage layers that are chemically resistant to waste and leachate, and that have permeability coefficient of not less than 1×10^{-2} cm/sec or an equivalent synthetic drainage material.
3. The minimum bottom slope of the facility should be 2%.

Design Considerations for Clay Liners. In the case of clay liners, the U.S. Environmental Protection Agency (EPA) requires that the coefficient of permeability be less than 10^{-7} cm/s. This can be achieved by meeting the following classification criteria:

1. The soil should have at least 20% fines (Section 14.1.1.1., Mechanical Analysis).
2. The plasticity index should be greater than 10 (Section 14.1.1.2, Atterberg Limits).
3. The soil should not have more than 10% gravel-size (>4.75 mm) particles.
4. The soil should not contain any particles or chunks of rock larger than 50 mm.

It is realized that liner criteria can be satisfied by blending available soils with clay minerals like sodium bentonite.

Design Considerations for Geomembrane Liners. Geomembranes are mainly used in geotechnical engineering to perform the functions of (1) separation, (2) filtration, and (3) stabilization. In this

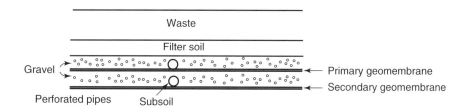

FIGURE 14.30 Typical cross section of a geomembrane-lined landfill. From *Designing with Geosynthetics*, 3rd ed., by Koerner, Robert M., © 1994. Reprinted by permission of Prentice Hall, Upper Saddle River, NJ.

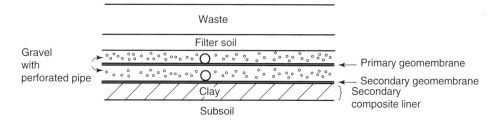

FIGURE 14.31 Typical cross section of a clay/geomembrane-lined landfill. From *Designing with Geosynthetics*, 3rd ed., by Koerner, Robert M., © 1994. Reprinted by permission of Prentice Hall, Upper Saddle River, NJ.

application of geotextiles, the functions of separation and, to a lesser extent, filtration are utilized. Owing to the extreme variation of solid-waste leachate composition from landfill to landfill, the candidate liner should be tested for permeability with the actual or synthesized leachate.

In addition to the permeability criterion, other criteria also play a role in geomembrane material selection. They are as follows:

1. resistance to stress-cracking induced by the soil/waste overburden,
2. different thermal expansion properties in relation to subgrade soil,
3. coefficient of friction developed with the waste material that governs slope stability criteria, and
4. axisymmetry in tensile elongation when the material is installed in a landfill that is founded on compressible subgrade soils.

In selecting a geomembrane material for a liner, serious consideration should also be given to its durability, which is determined by the possibility of leachate reaction with the geomembrane and premature degradation of the geomembrane. For more details on geomembrane durability and relevant testing, the reader is referred to Koerner (1994).

According to U.S. EPA regulations, the required minimum thickness of a geomembrane liner for a hazardous waste pond is 0.75 mm.

14.2 Site Exploration

In addition to screening possible sites, a thorough site study can reveal plenty of vital information regarding the soil and groundwater conditions at a tentative site, leading to more efficient selection of foundation depth and type as well as other construction details. Hence a site investigation that includes a subsurface exploration can certainly aid in economizing the time and cost involved in foundation construction projects.

An exhaustive site study can be separated into two distinct phases: (1) preliminary investigation and (2) detailed investigation. In the preliminary investigation, one would attempt to obtain as much valuable information about the site as possible at the least expense. Useful information regarding the site can often be obtained from the following sources:

1. local department of transportation (DOT) soil manuals,
2. local U.S. Geological Survey (USGS) soil maps,
3. local U.S. Army Corps of Engineers hydrological data,
4. U.S. Department of Agriculture (USDA) agronomy maps, and
5. local university research publications.

A preliminary investigation also involves site visits (or reconnaissance surveys) where one can observe such site details as topography, accessibility, groundwater conditions, and nearby structures (especially in the case of expected pile driving or dynamic ground modification). Firsthand inspection of the performance of existing buildings can also add to this information. A preliminary investigation can be an effective tool for screening all alternative sites for a given installation.

A detailed investigation has to be conducted at a given site only when that site has been chosen for the construction, since the cost of such an investigation is enormous. This stage of the investigation invariably involves heavy equipment for boring. Therefore, at first, it is important to set up a definitive plan for the investigation, especially in terms of the bore hole layout and the depth of boring at each location. Generally, there are rough guidelines for bore hole spacing, as indicated in Table 14.9.

In addition to planning boring locations, it is also prudent on the part of the engineer to search for any subsurface anomalies or possible weak layers that can undermine

TABLE 14.9 Approximate Spacing of Boreholes

Type of Project	Spacing, m
Multistory	10–30
One-story industrial plants	20–60
Highways	250–500
Residential subdivisions	250–500
Dams and dikes	40–80

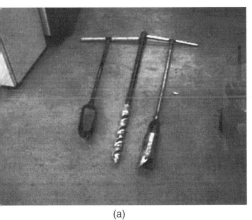

(a)   (b)

GURE 14.32 Drilling equipment. (a) Hand-auger; (b) mechanized auger. Figure courtesy of the University of South ●rida.

construction. As for the depth of boring, one can use the following criteria:

1. If bedrock is in the vicinity, continue boring until sound bedrock is reached, as verified from rock core samples.
2. If bedrock is unreachable, one can seek depth guidelines for specific buildings such as those given by the following expressions [Das, 1995]:
   $D = 3S^{0.7}$ (for light steel and narrow concrete buildings), and
   $D = 6S^{0.7}$ (for heavy steel and wide concrete buildings).
3. If none of the above conditions is applicable, then one can explore up to a depth at which the foundation stress attenuation reduces the applied stress by 90% ($\Delta p/\sigma'_{v_0} = 0.1$ in Example 14.4). This generally occurs around a depth of $2B$, where B is the minimum foundation dimension.

Hand augers and continuous flight augers (Figure 14.32a) can be used for boring up to a depth of about 3 m in loose to moderately dense soil. For extreme depths, a mechanized auger (Figure 14.32b) can be used in loose to medium dense sands or soft clays. When the cut soil is brought to the surface, a technically qualified person should observe the texture, color, and the type of soil found at different depths and prepare a bore-hole log laying out soil types at different depths. This type of boring is called dry sample boring (DSB).

On the other hand, if relatively hard strata are encountered, the investigators have to resort to a technique known as wash boring. **Wash boring** is carried out using a mechanized auger and a water-circulation system that aids in cutting and drawing the cut material to the surface. A schematic diagram of the wash-boring apparatus is shown in Figure 14.33, and the Florida Department of Transportation drill rig, which utilizes the above technique is shown in Figure 14.34.

In addition to visual classification, one has to obtain soil type and strength and deformation properties for a foundation design. Hence the soil at various depths has to be sampled as the bore holes advance.

Easily obtained disturbed samples suffice for classification, index, and compaction properties, while triaxial, and consolidation tests require carefully obtained "undisturbed" samples (samples with the minimum disturbance). Disturbed granular or clayey samples can be obtained by attaching a standard split-spoon sampler (Figure 14.13) to the drill rods. On the other hand, an undisturbed clay sample can be obtained by carefully advancing and retrieving a Shelby tube (Figure 14.35) into a clay layer. However, if one needs to evaluate a granular material for strength, settlement, or permeability, then *in situ* tests have to be performed owing to the difficulty in obtaining undisturbed samples in such soils.

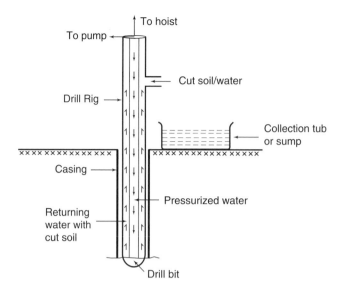

FIGURE 14.33   A schematic diagram of wash boring.

In this regard, the reader is referred to the *in situ* tests shown in Table 14.10.

All of the above tests except the plate load test have been discussed in this chapter. Hence a description of the plate load test is presented in Section 14.2.1.

14.2.1 Plate Load Tests

Plate load apparatus consists of a set of steel plates of standard diameters (12 in, 18 in, etc.), a hydraulic loading and recording mechanism, a reaction frame, and a deflection gage (Figure 14.36). During the test, the plate is laid at the tentative foundation depth and gradually loaded while the magnitude of the load and plate deflection at different stages is recorded. Figure 14.37 shows a typical plot of plate-load results on a sand deposit.

When one scrutinizes Figure 14.37, it can be seen that the ultimate bearing capacity of the plate can be estimated from the change in gradient of the load-deflection curve. Hence the bearing capacity and the settlement of a tentative foundation can be predicted

FIGURE 14.34   Florida department of transportation's CME-75 drill rig.

in the following manner, based on the results of a plate-load test performed on that location. In the following expressions, the subscripts f and p refer to the foundation and the plate, respectively.

Ultimate bearing capacity in clayey soils

$$q_{u(f)} = q_{u(p)} \tag{14.24}$$

Ultimate bearing capacity in sandy soils

$$q_{u(f)} = q_{u(p)} \left(\frac{B_f}{B_p} \right) \tag{14.25}$$

FIGURE 14.35   Shelby tubes.

FIGURE 14.36   Plate load test.

where B_p and B_f refer to the plate diameter and the minimum foundation dimension, respectively.

One can deduce the above expressions based on the basic expression for the bearing capacity of shallow footings (Section 14.3, Eq. (14.28)) when one realizes that predominant contributions for bearing capacity in clay and sand are made by the terms involving N_c and N_γ terms of Eq. (14.28), respectively. On the other hand, the settlement of a footing under a given contact pressure q can be estimated by the corresponding plate settlement s_p (Figure 14.37) using the following expressions:

Immediate settlement in clayey soils

$$s_f = s_p \left(\frac{B_f}{B_p} \right) \tag{14.26}$$

Immediate settlement in sandy soils

$$s_f = s_p \left(\frac{2 B_f}{B_f + B_p} \right)^2 \tag{14.27}$$

TABLE 14.10   Recommended *in situ* Tests

Evaluation Parameter	Test
Permeability	Field pumping test[*]
Settlement	Plate load test[†]
Shear Strength	SPT or CPT[‡]

[*]In Section 14.1, Dewatering of Excavations.
[†]In Section 14.2, Plate Load Tests.
[‡]In Section 14.1, Strength of Soils.

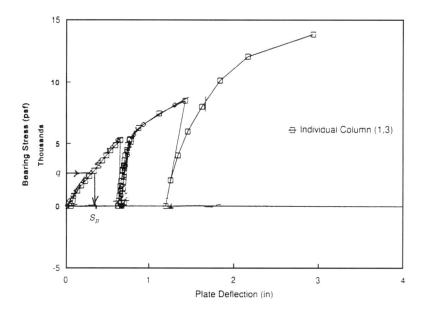

FIGURE 14.37   Typical plate load test results. From A.G. Mullins. 1996. Field characterization of dynamic replacement of Florida organic soils. Ph.D. dissertation. University of South Florida.

14.3 Shallow Footings

A shallow spread footing must be designed for a building column in order to transmit the column load to the ground without exceeding the bearing capacity of the ground and causing an excessive settlement (Figure 14.38). Plate load test results clearly exhibit the existence of a maximum stress (approximately 10 psi in Figure 14.37) that can be imposed on a plate without causing excessive settlement. This maximum stress is termed the bearing capacity of a foundation.

14.3.1 Bearing Capacity of Shallow Footings

In order to avoid catastrophic bearing failures, shallow footings are proportioned based on the bearing-capacity criterion. Two expressions extensively used to evaluate the ground bearing capacity are provided below.

14.3.1.1 Terzaghi's Expression

$$q_{ult} = s_c c N_c + q N_q + s_\gamma (0.5 B \gamma) N_\gamma \qquad (14.28)$$

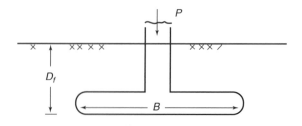

FIGURE 14.38   Schematic diagram of a shallow footing.

14.3.1.2 Hansen's Expression

$$q_{\text{ult}} = s_c d_c i_c c N_c + S_q d_q i_q q N_q + s_\gamma d_\gamma i_\gamma 0.5 B \gamma N_\gamma \qquad (14.29)$$

where

N_c, N_q, and N_γ = bearing capacity factors (Table 14.11)
s coefficients = shape factors based on B/L (Table 14.12)
d coefficients = depth factors based on D_f/B (Table 14.12)
i coefficients = inclination factors based on load inclination θ (Table 14.12)
γ = unit weight of soil in the footing influence zone
c and ϕ = shear strength parameters of the soil

Thus, in order to avoid bearing-capacity failure

$$\frac{q_{n,\text{ult}}}{F} > \frac{P}{A} \qquad (14.30)$$

TABLE 14.11   Bearing Capacity Factors

	Terzaghi's Expression			Hansen's Expression		
ϕ	N_c	N_q	N_γ	N_c	N_q	N_γ
0	5.7	1.0	0.0	5.14	1.0	0.0
5	7.3	1.6	0.5	6.49	1.6	0.1
10	9.6	2.7	1.2	8.34	2.5	0.4
15	12.9	4.4	2.5	11.0	3.9	1.2
20	17.7	7.4	5.0	14.8	6.4	2.9
25	25.1	12.7	9.7	20.7	10.7	6.8
30	37.2	22.5	19.7	30.1	18.4	15.1
35	57.8	41.4	42.4	46.4	33.5	34.4
40	95.7	81.3	100.4	75.25	64.1	79.4
45	172.3	173.3	297.5	133.5	134.7	200.5

TABLE 14.12   Shape, Depth, and Inclination Factors

	Hansen's Expression	Terzaghi's Expression
Shape	$s_c = 1 + (B/L)(N_q/N_c)$	1.0 for strip footings 1.3 for circular footings 1.3 for square footings
	$s_q = 1 + (B/L)\tan\phi$ $s_\gamma = 1 - 0.4(B/L)$	1.0 for strip footings 0.6 for circular footings 0.8 for square footings
Depth	*For $D_f/B < 1$* $d_c = 1 + 0.4(D_f/B)$ $d_q = 1 + 2\tan\phi(1 - \sin\phi)^2(D_f/B)$ $d_\gamma = 1$ *For $D_f/B > 1$* $d_c = 1 + 0.4\tan^{-1}(D_f/B)$ $d_q = 1 + 2\tan\phi(1 - \sin\phi)^2 \tan^{-1}(D_f/B)$ $d_\gamma = 1$	
Inclination	$i_c = i_q = (1 - \beta/90°)^{2*}$ $i_\gamma = (1 - \beta/\phi)^{2*}$	

*Here β is the load inclination to the vertical.

where

$q_{n,\text{ult}}$ = net ultimate bearing capacity based on Eq. (14.31)
     P = structural load
     A = footing area
     F = safety factor

$$q_{n,\text{ult}} = q_{\text{ult}} - q \qquad (14.31)$$

Example 14.6

Proportion a suitable footing for the 1000 kN vertical column load on a sandy ground where the SPT results are as indicated below. Assume that the groundwater table is at a depth of 0.5 m below the ground surface.

Elevation, m	N
1.0	5
2.0	7
3.0	10
4.0	12
5.0	12

An average N value has to be determined from the above data within the influence zone of the footing. For this, one has to assume a footing size, as the influence zone depends on the size of the footing. Hence assume a circular footing of diameter 1.5 m placed at a depth of 1 m from the ground surface. As indicated in Figure 14.39, the influence zone extends from $0.5D_f$ above the footing (i.e., elevation −0.5 m) to 2B below the footing (i.e., −4.0 m). Then, by averaging the corrected N values within this range, one can obtain the average N' as worked out in the table below.

Note that the vertical effective stresses (σ'_v) are obtained using Eq. (14.3) and assuming unit weights of 17.0 kN/m³ and 9.8 kN/m³, respectively, for sand and water, while C_N is obtained using

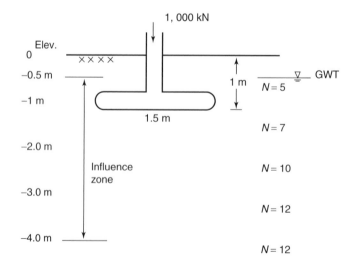

FIGURE 14.39   Foundation influence zone.

Eq. (14.2).

Elevation, m	N	σ'_v (kPa)	C_N	N_r^*
1.0	5	12.1	2.81	14
2.0	7	19.3	2.23	15.6
3.0	10	26.5	1.9	19
4.0	12	33.7	1.69	20.3

*Average $N' = [14(1.0) + 15.0(1.0) + 19(1.0) + 20.3(0.5)]/$
$3.5 = 17$

Then, from Table 14.4, a ϕ of $37°$ can be found. This yields interpolated values $N_c = 60$, $N_q = 49$, and $N_\gamma = 57$ from Table 14.11 (Hansen's factors). The following factors can also be evaluated from Table 14.12.

$s_c = 1.816$, $s_q = 1.657$, $s_\gamma = 0.6$
$d_c = 1.266$, $d_q = 1.176$, $d_\gamma = 1.0$
$i_c = 1.0$, $i_q = 1.0$, $i_\gamma = 1.0$ (since the load is applied vertically)

The following quantities are also needed for Eq. (14.29):

$q = \sigma'_v$ at foundation level $= 12.1$ kPa
$\gamma = \gamma'$ (since the foundation is fully submerged) $= 17.0 - 9.8 = 7.2$ kN/m^3

Finally, by substituting the above values in Eq. (14.29), one obtains the ultimate bearing capacity as

$$q_{n,\text{ult}} = (1.657)(1.176)(12.1)(49 - 1) + (0.5)(0.6)(1.0)(1.5)(7.2)(57) = 1,316.7 \text{ kPa}$$

Note that the cohesion term is dropped owing to negligible cohesion in sandy soils. Then, using Eq. (14.30), one obtains a safety factor of $1316.7/(1000/1.5/1.5) = 2.96$, which provides an adequate design. *Hence a* 1.5 m $\times$ 1.5 m *footing at a depth of* 1.0 m *would suffice.* Note that if the groundwater table was well below the footing (usually greater than $2B$), one would revise the following quantities as

$q = \sigma'_v$ at foundation level $= 17$ kPa
$\gamma = 17.0$ kN/m^3

On the other hand, if the groundwater table was below the footing but still near it (a distance of d below), one can use the following approximation to evaluate the γ term.

$$\gamma = \gamma' + \frac{(\gamma_{\text{dry}} - \gamma')d}{2B} \tag{14.32}$$

As an example, if the groundwater table was 2.0 m below the ground, one can assume a γ_{dry} of 16.5 kN/m^3 to modify the above two quantities as

$q = \sigma'_v$ at foundation level $= 16.5$ kPa
$\gamma = 7.2 + (16.5 - 7.2)1.0/2(1.5) = 10.3$ kN/m^3

14.3.2 Footings with Eccentricity

If a footing has to be designed for a column that carries an axial load (P) as well as a moment (M), or an eccentric axial load, the resulting contact pressure distribution is shown in Figure 14.40a. However, one realizes that this is statically equivalent to the uniform distribution shown in Figure 14.40b. Hence a simple method of computing the bearing capacity is to assume that only the portion of the footing containing the column at its center contributes to bearing capacity. Therefore, in designing

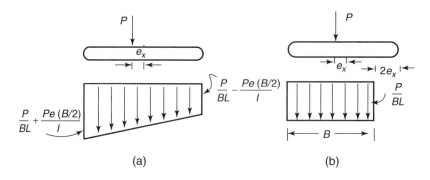

FIGURE 14.40 Simplistic design of an eccentric footing. (a) Pressure distribution due to an eccentric load and (b) Equivalent pressure distribution.

such a footing, modified dimensions (B' and L') have to be used in Eqs. (14.28) or (14.29), where B' and L' are defined as follows:

$$B' = B - 2e_x$$
$$L' = L - 2e_y$$

Example 14.7

Check the adequacy of the footing shown in Figure 14.41 for the soil data obtained from the UU test in Example 14.5 ($c_u = 50.6$ kPa). From Figure 14.41

$$e_x = M_x/P = 50.0\ \text{kNm}/250.0\ \text{kN} = 0.2\ \text{m}$$
$$e_y = M_y/P = 62.5\ \text{kNm}/250.0\ \text{kN} = 0.25\ \text{m}$$

Then

$$B' = 1.1$$
$$L' = 1.1$$

Since $\phi = 0°$, one obtains $N_c = 5.14$, $N_q = 1.0$, and $N_\gamma = 0.0$ from Table 14.11. Hence the only significant term in Eq. (14.29) is the "cohesion" term, and thus only the relevant factors are

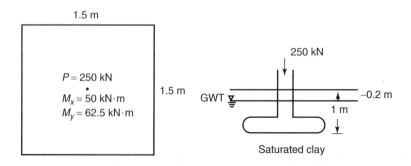

FIGURE 14.41 Design of an eccentric footing. (a) Plan; (b) elevation.

computed as follows:

$$s_c = 1 + 1.0/5.14 = 1.195$$
$$d_c = 1 + 0.4(1.0/1.5) = 1.267$$
$$i_c = 1.0$$
$$q_{ult} = 1.195(1.267)(50.6)(5.14) = 393.78 \text{ kPa}$$

Finally, the safety factor can be computed as

$$F = 393.78(1.1)(1.0)/250 = 1.733$$

Since the safety factor has to be more than 2.5, this is not a safe design. This factor can be improved by increasing the dimensions to about 2.0 m × 2.0 m, depending on the available space.

14.3.3 Presumptive Load-Bearing Capacity

The building codes of some cities suggest bearing capacities for a certain building site based on the classification of the predominant soil type at the site. Table 14.13 presents a comprehensive list of

TABLE 14.13 Presumptive Bearing Capacities

	Presumptive Bearing Capacities from Indicated Building Codes, kPa			
Soil Description	Chicago, 1995	Natl. Board of Fire Underwriters, 1976	BOCA,* 1993	Uniform Bldg. Code, 1991[†]
Clay, very soft	25			
Clay, soft	75	100	100	100
Clay, ordinary	125			
Clay, medium stiff	175	100		100
Clay, stiff	210		140	
Clay, hard	300			
Sand, compact and clean	240	—	140	200
Sand, compact and silty	100	}		
Inorganic silt, compact	125	}		
Sand, loose and fine		}	140	210
Sand, loose and coarse, or sand-gravel mixture, or compact and fine		140 to 400	240	300
Gravel, loose and compact coarse sand	300	} }	240	300
Sand-gravel, compact		—	240	300
Hardpan, cemented sand, cemented gravel	600	950	340	
Soft rock				
Sedimentary layered rock (hard shale, sandstone, siltstone)			6000	1400
Bedrock	9600	9600	6000	9600

Source: Bowles, 1995.

Note: Values converted from pounds per square foot to kilopascals and rounded. Soil descriptions vary widely between codes. The following represents author's interpretations.

*Building Officials and Code Administrators International, Inc.

[†]Bowles (1995) interpretation.

presumptive bearing capacities for various soil types. However, it should be noted that these values do not reflect the foundation shape, depth, load inclination, location of the water table, and the settlements that are associated with the sites. Hence the use of these bearing capacities are primarily advocated in situations where a preliminary idea of the potential foundation size is needed for the subsequent site investigation.

14.4 Mat Footings

Since mat footings are larger in dimension than isolated spread footings, they are commonly used for transferring multiple column loads to the ground in order to prevent bearing-capacity failures.

Thus an ideal application of a mat footing would be on relatively weak ground. However, if the ground has sufficient strength to produce adequate bearing for isolated spread footings, a mat footing will be an economical alternative only if the combined area of the spread footings is less than 50% of the entire building area.

14.4.1 Design of Rigid Mat Footings

14.4.1.1 Bearing Capacity of a Mat Footing

One can use Eq. (14.28) or (14.29) to proportion a mat footing if the strength parameters of the ground are known. However, since the most easily obtained ground strength parameter is the standard penetration blow count, N, an expression is available that uses N to obtain the bearing capacity of a mat footing on a granular subgrade. This is expressed as follows:

$$q_{n,\text{all}} = \frac{N}{0.08}\left(1 + \frac{1}{3.28B}\right)^2\left(1 + \frac{0.33D_f}{B}\right)\left(\frac{s}{25.4}\right) \qquad (14.33)$$

where

$\quad q_{n,\text{all}} =$ net allowable bearing capacity in kilopascals
$\quad\quad B =$ width of footing
$\quad\quad s =$ settlement in millimeters
$\quad\quad D_f =$ depth of footing in meters

Then the following condition has to be satisfied in order to avoid bearing failure:

$$q_{n,\text{all}} > \frac{P}{A} \qquad (14.34)$$

in which the use of a safety factor is precluded by employing an allowable bearing capacity.

Example 14.8

Figure 14.42 shows the plan of a column setup where each column is 0.5 m × 0.5 m in section. Design an adequate footing if the corrected average SPT blow count of the subsurface is 10 and if the allowable settlement is 25.4 mm (1 in). Assume a foundation depth of 0.5 m. Then the bearing capacity can be computed from Eq. (14.33) as

$$q_{n,\text{all}} = 10[1 + 0.33(0.5)/5.0][1 + 0.3/5.0]/(0.08) = 136.87\,\text{kPa}$$

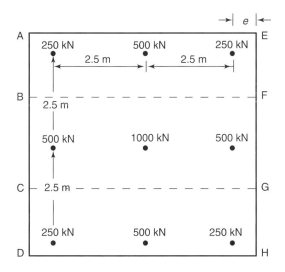

FIGURE 14.42   Illustration of a mat footing.

By applying Eq. (14.34)

$$4000/(5.0 + 2e)^2 < 136.87$$

$$e > 0.2029 \text{ m}$$

Hence the mat can be designed with 0.25 m edge space as shown in Figure 14.42.

For the reinforcement design, one can follow the simple procedure of separating the slab into a number of strips as shown in Figure 14.42. Then each strip (BCGF in Figure 14.42) can be considered as a beam. The uniform soil reaction per unit length (w) can be computed as 4000(2.5)/[(5.5)(5.5) = 330.5 kN/m. Figure 14.43 indicates the free-body diagram of the strip BCGF (Figure 14.42).

It can be seen from the free-body diagram that the vertical equilibrium of each strip is not satisfied because the resultant downward load is 2000 kN, as opposed to the resultant upward load of 1815 kN. This discrepancy results from the arbitrary separation of strips at the midplane between the loads where nonzero shear forces exist. In fact, one realizes that the resultant upward shear at the boundaries BF and CG (Figure 14.42) account for the difference, that is, 185 kN. However, to obtain shear and moment diagrams of the strip BCGF, one can add this modify them as indicated in the figure. This was achieved by reducing the loads by a factor of 0.954 and increasing the reaction by a factor of

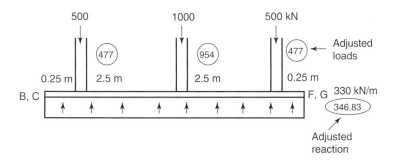

FIGURE 14.43   Free-body diagram for strip BCGF (Figure 14.42).

1.051. The two factors were determined as follows:

> For the loads, $[(2000 + 1815)/2]/2000$
> $= 0.954$
> For the reaction, $1815/[(2000+1815)2]$
> $= 1.051$

The resulting shear and moment diagrams are indicated in Figures 14.44 and 14.45.

Now, using Figures 14.44 and 14.45, one can determine the steel reinforcements as well as the mat thickness. This estimation is not repeated here since it can be found in other chapters of this book.

14.4.1.2 Settlement of Mat Footings

The settlement of mat footings can also be found using the methods that were outlined in Section 14.1, Compressibility and Settlement, assuming that they impart stresses on the ground in a manner similar to that of spread footings.

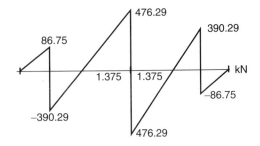

FIGURE 14.44 Distribution of shear on strip BCGF (Figure 14.42).

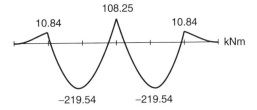

FIGURE 14.45 Distribution of moment on strip BCGF (Figure 14.42).

14.4.2 Design of Flexible Mat Footings

Flexible mat footings are designed based on the principle of slabs on elastic foundations. Owing to their finite size and relatively large thickness, one can expect building foundation mats to generally exhibit rigid footing behavior. Therefore applications of flexible footings are limited to concrete slabs used for highway or runway construction.

The most significant parameter associated with the design of beams on elastic foundations is the radius of relative stiffness $(1/\beta)$ given by the following expression:

$$\frac{1}{\beta} = 4\sqrt{\frac{E h^3}{12(1 - \mu^2)k}} \tag{14.35}$$

where

> E = elastic modulus of concrete
> μ = Poisson ratio of concrete
> k = coefficient of subgrade reaction of the foundation soil usually determined from the plate
> load test (Section 14.2) or Eq. (14.36)
> h = slab thickness

$$k = \frac{E_s}{B\left(1 - \mu_s^2\right)} \tag{14.36}$$

where

> E_s = elastic modulus of subgrade soil
> μ_s = Poisson ratio of subgrade soil

Once β has been evaluated for a particular mat, the shear, moment, and reinforcing requirements can be determined from nondimensional charts that are based on the solution for a concentrated load (P) applied to a slab on an elastic foundation. The following expressions can be used, along

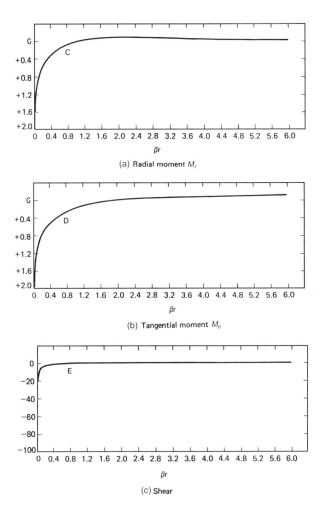

FIGURE 14.46 Radial and tangential moments and shear coefficients in a slab under point load. From *Foundation Analysis*, by Scott, Ronald F., © 1981. Reprinted by permission of Prentice Hall, Upper Saddle River, NJ.

with Figure 14.46, for the evaluation:

$$M_r = -\frac{P}{4}C \tag{14.37a}$$

$$M_\theta = -\frac{P}{4}D \tag{14.37b}$$

$$V = -\frac{P\beta}{4}E \tag{14.38}$$

On the other hand, the moment due to a distributed load can be obtained by drawing the contact area on an influence chart, such as the one shown in Figure 14.47, and then using Eq. (14.39). It should be noted that the scale for the drawing should be selected such that $1/\beta$ is represented by the distance l shown in Figure 14.47.

$$M = \frac{p(1/\beta)^2 N}{10,000} \tag{14.39}$$

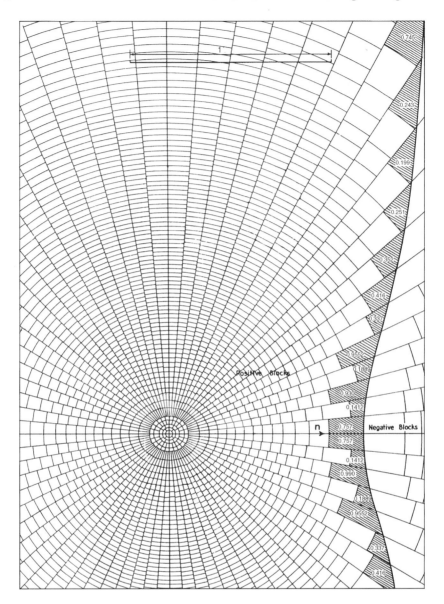

FIGURE 14.47 Influence chart for determining moment at the edge of a slab. *Source:* Pickett, G. and Ray, G.K. 1951. Influence charts for concrete pavements. *Am. Soc. Civ. Eng. Trans.* 116(2425):49–73.

where

    p = distributed load
    N = number of elements covered by the
        loading area drawn

Example 14.9

Plot the shear and moment distribution along the columns A, B, and C of the infinite slab of 8 in thickness shown in Figure 14.48. Consider it to be a flexible footing. Assume a coefficient

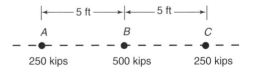

FIGURE 14.48 Illustration of an infinite flexible slab.

TABLE 14.14 Flexible Footing Moments

Distance from A, ft	C Coefficient for Load at A	C Coefficient for Load at B	C Coefficient for Load at C	Moment (kip-ft)
0.0	1.6	0.18	0.0	122.5
1.0	0.8	0.25	0.02	32.25
2.0	0.5	0.4	0.05	84.38
3.0	0.4	0.5	0.08	92.5
4.0	0.25	0.8	0.1	121.88
5.0	0.18	1.6	0.18	222.5
6.0	0.1	0.8	0.25	121.88
7.0	0.08	0.5	0.4	92.5
8.0	0.05	0.4	0.5	84.38
9.0	0.02	0.25	0.8	32.25
10.0	0.0	0.18	1.6	122.5

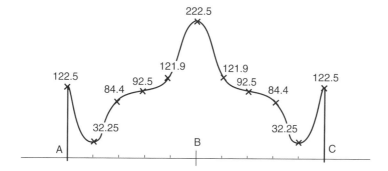

FIGURE 14.49 Moment distribution on a flexible slab (in kip-ft).

of subgrade reaction of 2,600 lb/ft^3. Since $E_c = 5.76 \times 10^8$ psf and $\mu_c = 0.15$, one can apply Eq. (14.35) to obtain:

$$\beta = 0.1156 \text{ ft}^{-1}$$

Using the above results, Figure 14.46a, and Eq. (16.37a), Table 14.14 can be developed for the radial moment (the moment on a cross-section perpendicular to the line ABC in Figure 14.48). These moment values are plotted in Figure 14.49.

14.5 Retaining Walls

When designing a retaining structure, one must ascertain that its structural capacity is adequate to withstand any potential instability that can be caused by the lateral earth pressures of the retained backfill. Hence a major step in the design of a retaining structure is the evaluation of the magnitude, direction, and the line of action of the lateral force. Most of the methods available for analyzing lateral earth pressures assume a yielding soil mass in the vicinity of the retaining structure, and hence the solutions are based on the limit equilibrium.

The magnitude of the lateral force depends on the soil failure mechanism. The mechanism in which the backfill yields with the outward movement of the wall is known as the active failure mechanism (Figure 14.50a), while the yielding of soil due to inward wall movement is termed the passive failure mechanism (Figure 14.50b). Also indicated on Figure 14.50 is the orientation of the failure planes for each condition in the case of a smooth vertical wall supporting a horizontal backfill. The two most widely used analytical methods are illustrated below.

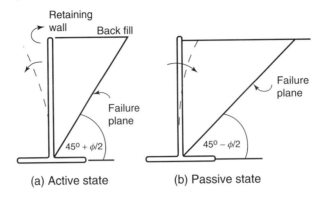

FIGURE 14.50   Illustration of (a) active and (b) passive states.

14.5.1 Determination of Earth Pressures

14.5.1.1 Rankine Method

The Rankine method of analysis can be employed for the relatively simple case of a smooth, vertical wall supporting a homogeneous backfill. A modified form of Rankine's original analysis allows one to obtain the active and passive earth pressure distributions by using the following expressions:

$$\sigma'_a = \sigma'_v K_a - 2c\sqrt{K_a} \tag{14.40}$$

and

$$\sigma'_p = \sigma'_v K_p + 2c\sqrt{K_p} \tag{14.41}$$

where the subscripts a and p stand for active and passive states, respectively, while the coefficients of earth pressure K_a and K_p can be determined from the following expressions:

$$K_a = \cos\beta \frac{\cos\beta - \sqrt{\cos^2\beta - \cos^2\phi}}{\cos\beta + \sqrt{\cos^2\beta - \cos^2\phi}} \tag{14.42}$$

and

$$K_p = \cos\beta \frac{\cos\beta + \sqrt{\cos^2\beta - \cos^2\phi}}{\cos\beta - \sqrt{\cos^2\beta - \cos^2\phi}} \tag{14.43}$$

where

β = inclination of the backfill
c, ϕ = soil strength parameters

The direction of the resultant lateral pressure and the line of action are indicated on Figure 14.51.

Example 14.10

Determine the lateral pressure on the retaining wall shown in Figure 14.51. Assume active conditions. By using Eqs. (14.42), one obtains $K_a = 0.38$. Then, by using Eq. (14.40),

at $z = 0.0$ m, $\sigma'_a = -2(10)(0.616) = -12.32$ kPa;
at $z = 2.0$ m, $\sigma'_a = 0.6$ kPa; and
at $z = 8.0$ m, $\sigma'_a = 19.3$ kPa.

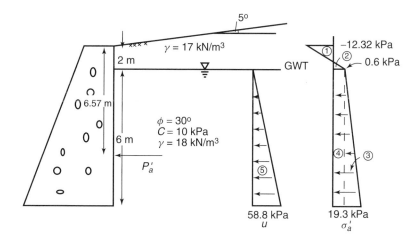

FIGURE 14.51   Illustration of the earth pressure distribution.

The line of action of the effective force can be found by discretizing the pressure diagram into five segments as shown in Figure 14.51.

Segment	Area (A)	Centroidal Distance (z)	Az
1	−11.77	0.64	−7.49
2	0.027	1.97	0.053
3	112.2	6.0	673.2
4	3.6	5.0	18.0

Hence,

$$z = (-7.49 + 0.053 + 673.2 + 18.0)/(-11.77 + 0.027 + 112.2 + 3.6) = 6.57 \text{ m}$$

In obtaining the total force on the wall, one should remember to include the water pressure in segment 5 in addition to the effective force.

14.5.1.2   Coulomb Method

The Coulomb method of analysis is an attractive alternative to the Rankine method owing to its ability to handle more complex cases involving rough, nonvertical retaining walls. The relevant equations for lateral earth pressure coefficients are given below:

$$K_a = \frac{\sin^2(\alpha + \phi)}{\sin^2 \alpha \sin(\alpha - \delta)\left[1 + \sqrt{\frac{\sin(\phi+\delta)\sin(\phi-\beta)}{\sin(\alpha-\delta)\sin(\alpha+\beta)}}\right]^2} \qquad (14.44)$$

$$K_p = \frac{\sin^2(\alpha - \phi)}{\sin^2 \alpha \sin(\alpha + \delta)\left[1 + \sqrt{\frac{\sin(\phi+\delta)\sin(\phi+\beta)}{\sin(\alpha+\delta)\sin(\alpha+\beta)}}\right]^2} \qquad (14.45)$$

where α and δ are the inclination of the wall face to the horizontal and the angle of wall friction, respectively. As illustrated in Figure 14.52, the direction of the resultant lateral force changes when conditions change from active to passive states, with the line of action remaining on the

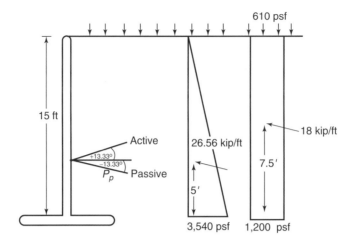

FIGURE 14.52　Illustration of the application of Coulomb's analysis.

generators of the friction cone. The above coefficients can be used in conjuction with Eqs. (14.40) and (14.41).

On the other hand, if $\alpha = 90°$, and $\beta = \delta = 0°$, it can be easily shown that Eqs. (14.44) and (14.42) reduce to

$$K_a = \frac{1 - \sin\phi}{1 + \sin\phi} \tag{14.46}$$

Hence one can see that Rankine's and Coulomb's analytical predictions agree only when earth pressures are predicted on smooth, vertical walls supporting horizontal backfills. In this case, the direction of the lateral force is horizontal.

Example 14.11

Use the Coulomb method to determine the lateral earth pressure on the wall shown; assume passive conditions. Since the angle of wall friction (δ) is not specified in this problem, it would be adequate to assume that $\delta = (2/3)\phi = 13.33°$. Substituting $\alpha = 70°$, $\phi = 20°$, $\delta = 13.33°$, and $\beta = 0°$ in Eq. (14.45), one would obtain $K_p = 1.968$. With $c = 0.0$, and a surcharge of q, Eq. (14.41) reduces to the following.

$$\sigma'_p = \left(\sigma'_v + q\right)K_p \tag{14.47}$$

where $\sigma'_v = 120z$ and $q = 610$ psf.

Then one would obtain the passive pressure distribution as

$$\sigma_p = 236.16z + 1200 \tag{14.48}$$

and the resultant lateral force will be 44.568 kips/ft acting at a distance of 6.0 ft above the base, as shown in Figure 14.52.

14.5.1.3 Trial-Wedge Method

When the backfill is nonhomogeneous or if there is a finite or a concentrated surcharge on the backfill, the above methods cannot be used to determine the earth pressures. In such situations, the lateral force on a retaining wall can be estimated by the graphical construction of the force polygon for a selected potential failure wedge. The following example is provided to illustrate this technique.

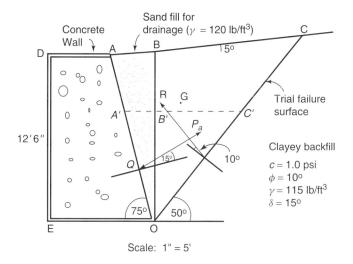

FIGURE 14.53   Trial wedge method.

Example 14.12

Use the trial-wedge solution to estimate the active soil pressure on the retaining wall shown in Figure 14.53.

> **Step 1.**   Select a trial failure surface. An appropriate initial estimate would be the one corresponding to a smooth vertical wall with a horizontal backfill indicated in Figure 14.50. Hence, in this case, select a surface with an inclination of $45° + 10°/2 (= 50°)$ to the horizontal (OC).
>
> **Step 2.**   In the case of active conditions, estimate the depth up to which tension develops $(= 2c/\gamma K_a)$ and demarcate the area of tension cracks. In this case, the vertical depth of tension cracks is 4.4 ft and the tension crack area is indicated by A′ B′ C′. Therefore an effective cohesive force is developed on the OC surface only up to C′. This will have a magnitude of $OC' \times 1 \times 5$ ft $\times 1.5 \times 144$ psf $= 1.526$ kips.
>
> **Step 3.**   Estimate the total weight of the failure wedge. If there is a water table, submerged unit weights must be used under the water table to account for buoyancy. In this case the weight of AOC can be estimated as $(0.5 \times 12.6 \times 3.1 \times 120) + (0.5 \times 12.6 \times 11.5 \times 115) = 10.675$ kips.
>
> **Step 4.**   Draw a force polygon assuming the indicated directions for the two other forces (active force on the wall and the reaction from the intact soil mass). In this case, these are designated as P_a and R on Figure 14.54.
>
> **Step 5.**   Estimate P_a from the force polygon. In this case, $P_a = 5.25$ kips.
>
> **Step 6.**   Repeat the above procedure for a number of different trial failure planes and estimate the minimum lateral force to be the actual P_a and consider the corresponding trial failure plane to be the actual failure plane.
>
> **Step 7.**   If G is the centroid of the actual failure wedge, obtain the line of action of P_a from the point Q on the wall such that QG is parallel to the failure surface OC.

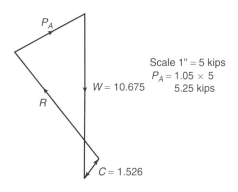

FIGURE 14.54   Force polygon.

14.5.2 Design of Concrete Retaining Walls

There are basically two types of concrete retaining walls: (1) gravity type and (2) cantilever type. The following conditions must be satisfied when designing a concrete retaining wall:

1. safety against overturning due to earth pressure,
2. safety against base sliding due to earth pressure, and
3. prevention of any possible tensile stresses in the base soil.

In addition to these design criteria, the design must also provide for a good drainage system and adequate bearing (Section 14.3.1, Bearing Capacity of Shallow Footings) and no excessive settlements (Section 14.1.3, Compressibility and Settlement).

14.5.2.1 Design of a Cantilever Retaining Wall

The basic design of a cantilever retaining wall is illustrated by the following example.

Example 14.13

Design a suitable cantilever retaining wall to support the backfill shown in Figure 14.55.

Bowles (1995) recommends the tentative dimensions shown in Figure 14.56 for a cantilever retaining wall. Based on Bowles, recommendations, the dimensions shown in Figure 14.55 are assumed for the retaining wall. In computing the lateral pressure on the wall, it is usually assumed that the wall section starts at the cross section CG shown in Figure 14.56. Hence the active wall pressure can be evaluated, using either of the described methods, as 4.496 kips/ft at a height 5 ft from the base. Although there is a passive force owing to the soil berm on the toe, practitioners generally neglect it for a conservative design.

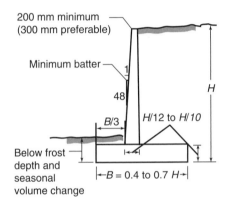

FIGURE 14.55 Tentative design dimensions for a cantilever retaining wall. From J.E. Bowles, 1995.

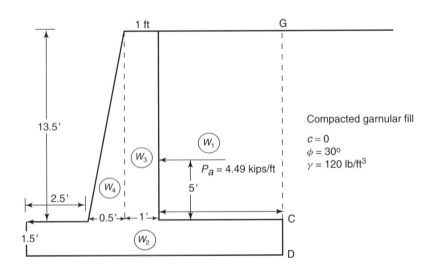

FIGURE 14.56 Cantilever retaining wall design.

The weights of the different retaining wall components are computed as follows:

$$w_1 = (5.0 \times 13.5 \times 120)/1{,}000 = 8.1 \text{ kips/ft}$$
$$w_2 = (9.0 \times 1.5 \times 150)/1{,}000 = 2.025 \text{ kips/ft}$$
$$w_3 = (1.0 \times 13.5 \times 150)/1{,}000 = 2.025 \text{ kips/ft}$$
$$w_4 = (0.5 \times 0.5 \times 13.5 \times 150)/1{,}000 = 0.506 \text{ kips/ft}$$

The locations of the respective centroids from the toe are indicated below:

Element	Weight, w_i	Centroidal distance x_i	$w_i x_i$
1	8.1	6.5	52.65
2	2.025	4.5	9.11
3	2.025	3.5	7.09
4	0.506	2.83	1.43

Stability Against Overturning.   The factor of safety against overturning can be computed as

$$F_1 = \frac{\text{Stabilizing moment}}{\text{Overturning moment}} = \frac{(52.65 + 9.11 + 7.09 + 1.43)}{4.496 \times 5} = 3.13$$

This is adequate since it is greater than 1.5.

Stability Against Sliding.   The normal reaction on the base is equal to $\Sigma w_i = 12.656$. Thus the maximum frictional force on the base is equal to $12.656 \tan \phi = 6.449$ kips/ft. Then the safety factor against sliding can be computed as

$$F_2 = \frac{\text{Stabilizing force}}{\text{Sliding force}} = \frac{6.449}{4.496} = 1.43$$

This is also acceptable since it is greater than 1.25

Check for Tension in the Base Soil.   The eccentricity on the base produced by the above forces is equal to $0.5(8.1 \times 2.0 - 2.025 \times -0.506 \times 1.67 - 4.496 \times 5)/(12.656) = e = -0.723$ m. For tension not to develop in the base soil, $e < B/6$. Since $B/6.0 = 1.5$, the base tension criterion is also satisfied by this design (Figure 14.57).

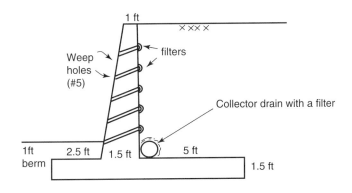

FIGURE 14.57   Design details.

Reinforcement Requirements. In order to determine reinforcement requirements, the shear and bending moment diagrams can be drawn separately for the stem and the base (Figures 14.58 and 14.59).

It should be noted that the maximum shear and moment in the stem occurs at its base and are 3.045 kips/ft. and 16.412 kip-ft/ft, respectively.

The reader should note that in obtaining the shear and moment diagrams for the base, the base soil pressure is represented by a uniform distribution within a distance of B' (equal to $B - 2e$) since this is statically equivalent to an eccentric force of 12.656 kips/ft. It is noted that a maximum shear force of 3.743 kips/ft is produced at the base.

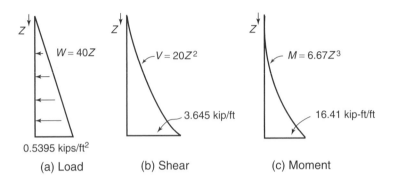

FIGURE 14.58   Load, shear, and moment diagrams for the retaining wall stem.

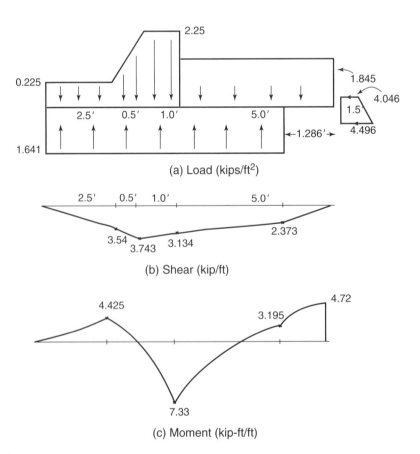

FIGURE 14.59   (a) Load, (b) shear, and (c) moment diagrams for the retaining wall base.

The reader should also note that in the moment diagram, within the stem area of the base, an unbalanced stem moment of 16.512 kips/ft (Figure 14.58) is gradually transfered to the base. Further, the heel of the base will have a moment of 4.2 kip-ft/ft due to the earth pressure distribution on the base thickness as shown in the load distribution diagram (Figure 14.59a). This information can be utilized for reinforcement design.

Drainage Design. In this regard, the interested reader is referred to Bowles (1995) for design details. However, one should remember the following preliminary guidelines:

1. Provide a compacted free-draining soil layer adjacent to the retaining wall if the backfill material is clayey in nature.
2. Provide a perforated collector drain with an appropriate soil or geomembrane filter at the base of the retaining wall.
3. Make provisions for drainage outlets known as **weep holes** in the stem with a soil or geomembrane filter at the wall face to prevent clogging.

14.6 Pile Foundations

A pile foundation can be employed to transfer superstructure loads to stronger soil layers deep underground. Hence it is a viable technique for foundation construction in the presence of undesirable soil conditions near the ground surface. However, owing to the high cost involved in piling, this method is only utilized after other less costly alternatives such as (1) combined footings and (2) ground modification have been considered and ruled out for the particular application. On the other hand, piles may be the only possible foundation construction technique in the case of subgrades that are prone to erosion and in offshore construction.

14.6.1 Advantages of Concrete Piles

Depending on applicability, one of the three different pile types: timber, concrete, or steel, is selected for a given construction situation. Concrete piles are selected for foundation construction under the following circumstances:

1. the need to support heavy loads in maritime areas where steel piles can easily corrode;
2. there are stronger soil types located at a relatively shallower depth that are easily accessible to concrete piles;
3. there are bridge piers and caissons that require large-diameter piles;
4. a large pile group is needed to support a heavy extensive structure so that the total expense can be minimized; and
5. the need for minipiles to support a residential building on a weak and compressible soil.

The disadvantages of concrete piles are that they are damaged by acidic environments (organic soils) and undergo abrasion due to wave action when used to construct offshore structures.

14.6.2 Types of Concrete Piles

The two most common types of concrete piles are (1) precast and (2) cast *in situ*. Of these, precast piles may be constructed to specifications at a casting yard or at the site itself if a large number of piles are needed for the particular construction. In any case, handling and transportation can cause intolerable tensile stresses in precast concrete piles. Hence one should be cautious in handling them so as to minimize bending moments in the pile.

Cast-*in-situ* piles are of two classes:

1. cased type, these are piles that are cast inside a steel casing that is driven into the ground;
2. uncased type, these are piles that are formed by pouring concrete into a drilled hole or into

a driven casing before the casing is gradually withdrawn.

14.6.3 Estimation of Pile Capacity

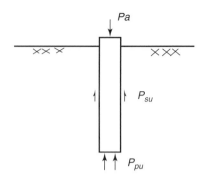

The pile designer should be aware of the capacity of a pile (1) under normal working conditions (**static**) and (2) when it is driven (**dynamic**).

14.6.3.1 Static Pile Capacity

The ultimate working load that can be applied to a given pile depends on the resistance that the pile can produce in terms of side friction and point bearing (Figure 14.60). Hence the expression for the allowable load on a pile will take the following form:

FIGURE 14.60 Load-bearing capacity of a pile.

$$P_a = \frac{P_{pu} + P_{su}}{F} \tag{14.49}$$

where

P_{pu} = ultimate point capacity
P_{su} = ultimate side friction
F = safety factor

The ultimate point capacity component in Eq. (14.49) corresponds to the bearing capacity of a shallow footing expressed by Eq. (14.50), which is a modified form of Eq. (14.28).

$$P_{pu} = A_p\left[cN_c^* + q\left(N_q^* - 1\right)\right] \tag{14.50}$$

where

A_p = area of the pile cross section
q = vertical effective stress at the pile tip
c = cohesion of the bearing layer
N_c^*, N_q^* = bearing capacity factors modified for deep foundations (and a B/L ratio of 1.0)

The bearing capacity factors for deep foundations can be found in Figure 14.61. However, use of the above bearing capacity factors is more involved than in the case of shallow footings as, in the case of deep foundations, the mobilization of shear strength also depends on the extent of the pile's penetration into the bearing layer. In granular soils, the depth ratio at which the maximum strength is mobilized is called the **critical depth ratio** $(L_b/D)_{cr}$. Figure 14.62 shows $(L_b/D)_{cr}$ for the mobilization of N_c^* and N_q^* for different values of ϕ. According to Meyerhoff (1976), the maximum values of N_c^* and N_q^* are usually mobilized at depth ratios of $0.5(L_b/D)_{cr}$.

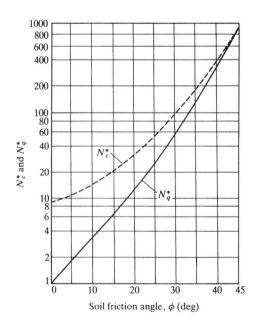

FIGURE 14.61 Bearing capacity factors for deep foundations. From Das, B.M. 1995. *Principles of Foundation Engineering*. PWS Publishing.

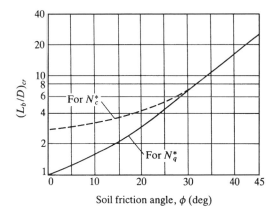

FIGURE 14.62 Variation of critical depth ratio with ϕ. From Das, B.M. 1995. *Principles of Foundation Engineering.* PWS Publishing.

Hence one has to follow an interpolation process to evaluate the bearing-capacity factors if the depth ratio is less than $0.5(L_b/D)_{cr}$. This is illustrated in Example 14.14.

Point Capacity in Sands. In the case of sandy soils where the cohesive resistance is negligible, Eq. (14.50) can be reduced to

$$P_{pu} = A_p\, q\left(N_q^* - 1\right) \tag{14.51}$$

where the limiting point resistance is

$$P_{pu_{max}} = A_p 50\, N_q^* \tan\phi\, kN; \qquad P_{pu_{max}} = A_p N_q^* \tan\phi \text{ kips} \tag{14.52}$$

Point Capacity in Clays. The most critical design condition in clayey soils is the undrained condition where the apparent angle of internal friction is zero. Under these conditions, it can be seen that Eq. (14.50) reduces to

$$P_{pu} = A_p(9.0c_u) \tag{14.53}$$

where c_u is the undrained strength.

Skin-Friction Capacity of Piles. The skin-friction capacity of piles can be evaluated by means of the following expression:

$$P_{sf} = \int_0^L p f dz \tag{14.54}$$

where

p = perimeter of the pile section
z = coordinate axis along the depth direction
f = unit skin-friction at any depth z
L = length of the pile

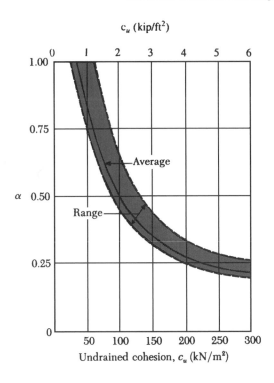

FIGURE 14.63 Variation of α with undrained strength. From Das, B.M. 1995. *Principles of Foundation Engineering.* PWS Publishing.

Unit Skin Friction in Sandy Soils. Since the origin of skin friction in granular soils is due to the frictional interaction between piles and granular material, the unit skin friction can be expressed as

$$f = K\sigma'_v \tan\delta \qquad (14.55)$$

where

K = earth pressure coefficient (K_0 for bored piles and 1.4 K_0 for driven piles)
δ = angle of friction between the soil and pile material (usually assumed to be 2/3 ϕ)
σ'_v = vertical effective stress at the point of interest

It can be seen from the above expression that the unit skin friction can increase linearly with depth. However, practically, a depth of 15B (where B is the cross-sectional dimension) has been found to be the limiting depth for this increase.

Skin Friction in Clayey Soils. In clayey soils, on the other hand, skin friction results from adhesion between soil particles and the pile. Hence the unit skin friction can be simply expressed by

$$f = \alpha c_u \qquad (14.56)$$

where the **adhesion factor** α can be obtained from Figure 14.63.

Example 14.14

Estimate the maximum allowable static load on the 200-mm-square driven pile shown in Figure 14.64.

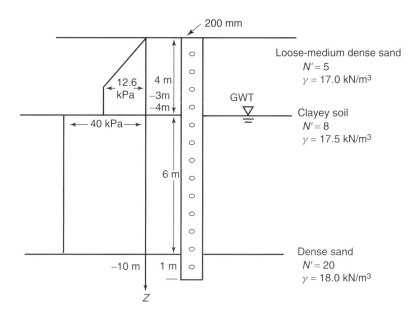

FIGURE 14.64   Illustration of the computation of pile capacity.

From Tables 14.4 and 14.5, the following strength parameters can be obtained, based on the SPT data given.

Soil Layer	Representative, N'	Cohesion, c	Friction ϕ
Loose sand	5	0.0	28°
Clay	8	40 kPa	0°
Medium dense sand	20	0.0	38°

Computation of Skin Friction in Loose Sand.   Applying Eq. (14.55), one would obtain

$$f = 1.4 K_0 (17.0) z \tan \delta$$

up to a depth of -3.0 m (i.e., 15.0×0.2) and constant thereafter.

Assuming $\delta = 2/3(\phi) = 19°$, and $K_0 = (1-\sin\phi)\,\mathrm{OCR} = 0.574$ (since OCR (over consolidation ratio) $= 1.0$ for normally consolidated soils), one obtains

$$f = 4.203z \;\text{kPa} \quad \text{for } z < 3 \text{ m}$$
$$f = 12.6 \;\text{kPa} \qquad \text{for } z > 3 \text{ m}$$

Computation of Skin Friction in Clay.   Applying Eq. (14.56), one obtains

$$f = \alpha(40)$$

where $\alpha = 1.0$ from Figure 14.63;

$$f = 40 \text{ kPa}$$

The dense sand layer can be treated as an end-bearing layer and hence its skin-frictional contribution cannot be included. Since the pile perimeter is constant throughout the depth, the total skin-frictional force (Eq. (14.54)) can be computed by multiplying the area of the skin-friction distribution shown in Figure 14.64 by the pile perimeter of 0.8 m. Hence

$$P_{sf} = (0.8)[0.5(3)(12.6) + 12.6(1) + 40(6)] = 217.2 \text{ kN}$$

Computation of the Point Resistance in Dense Sand. From Figure 14.62, $(L/D)_{cr} = 15$ for $\phi = 30°$. For the current problem, $L/D = 1/0.2 = 5$. Since in this case, $L/D < 0.5(L/D)_{cr}$, N_q^* can be prorated from the maximum N_q^* values given in Figure 14.61. Thus,

$$N_q = [(L/D)/0.5(L/D)_{cr}](N_q^*) = 5/7.5 \times 300 = 208$$

Note that $N_q^* = 300$ was obtained from Figure 14.61. Also

$$A_p = 0.2 \times 0.2 = 0.04 \text{ m}^2$$

$$q = \sigma_v' = 17.0(4) + (17.5 - 9.8)(6) + (18.0 - 9.8)(1) = 122.4 \text{ kPa}$$

Then, by substituting in Eq. (14.51)

$$P_{pu} = 0.04(208 - 1)(122.4) \text{ kN} = 974.3 \text{ kN}$$

But

$$P_{pu\,max} = 0.04(50)(208) \tan 38° = 271.8 \text{ kN}$$

$$\therefore P_{pu} = 271.8 \text{ kN}$$

Finally, by applying Eq. (14.49), one can determine the maximum allowable load as

$$P_{all} = (271.8 + 217.2)/4 = 122.3 \text{ kN}$$

14.6.4 Computation of Pile Settlement

In contrast to shallow footings, a pile foundation settles not only because of the compression the tip load causes on the underlying soil layers, but also because of the compression caused by the skin friction on the surrounding layers. The elastic shortening of the pile itself is another source of settlement. In addition, if an underlying saturated soft clay layer is stressed by the pile, the issue of consolidation settlement will have to be addressed too. In this section, only the immediate settlement will be analytically treated, as a computation of consolidation settlement of a pile group is provided in Example 14.15.

According to Poulos and Davis (1990), the immediate settlement of a single pile can be

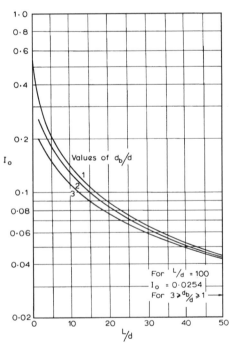

FIGURE 14.65 Influence factor I_0. From Poulos, H.G. and Davis, E.H. 1990. *Pile Foundation Analysis and Design*. Krieger, Melbourne, FL.

estimated from the following expressions:

$$s = \frac{PI}{E_s d} \tag{14.57}$$

and

$$I = I_0 R_k R_h R_v \tag{14.58}$$

where

I_0 = influence factor for an incompressible pile in a semiinfinite medium with $v_s = 0.5$ (Figure 14.65)

R_k = correction factor for pile compressibility $K(= E_p/E_s)$ (Figure 14.66)

R_h = correction factor for a finite medium of thickness h (Figure 14.67)

R_v = correction factor for the Poisson ratio (v_s) of soil (Figure 14.68)

E_s = elastic modulus of soil

E_p = elastic modulus of pile material

d = minimum pile dimension (pile diameter).

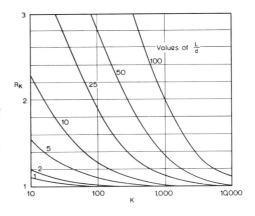

FIGURE 14.66 Correction factor for pile compressibility. From Poulos, H.G. and Davis, E.H. 1990. *Pile Foundation Analysis and Design.* Krieger, Melbourne, FL.

14.6.5 Pile Groups

For purposes of stability, pile foundations are usually constructed of pile groups that transmit the structural load through a pile cap, as shown in Figure 14.69. If the individual piles in a group are not ideally placed, there will be essentially an overlap of the individual influence zones, shown in Figure 14.69. This will be manifested in the following group effects, which have to be considered when designing a pile group:

1. The bearing capacity of the pile group will be different (generally lower) than the sum of the individual capacities, owing to the above interaction.
2. The group settlement will also be different from individual pile settlement owing to additional stresses induced on the piles by the neighboring piles.

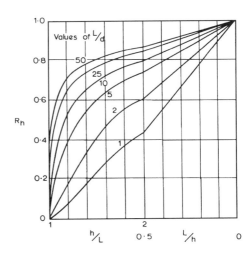

FIGURE 14.67 Correction factor for layer thickness. From Poulos, H.G. and Davis, E.H. 1990. *Pile Foundation Analysis and Design.* Krieger, Melbourne, FL.

Both topics are discussed in the following sections and illustrative examples are provided.

14.6.5.2 Bearing Capacity of Pile Groups

The efficiency of a pile group is defined as

$$\eta = \frac{\text{ultimate capacity of the pile group } \left(P_{g(u)}\right)}{\text{Sum of individual ultimate capacities } \left(\sum P_{(u)i}\right)} \tag{14.59}$$

where

    η = group efficiency

Owing to the complexity of individual pile interaction, the literature does not indicate any definitive methodology for determining the group efficiency in a given situation other than the following common **Converse-Labarre** equation that is applied for clayey soils:

$$\eta = 1 - \frac{\zeta}{90}\left[\frac{(n-1)m + (m-1)n}{mn}\right] \tag{14.60}$$

where

    $\zeta = \tan^{-1}$ (diameter/spacing ratio)
    n = number of rows in the group
    m = number of columns in the group

Although the above expression indicates that the maximum possible efficiency is about 90%, reached at a spacing/diameter ratio of 5, the results from experimental studies (Das, 1995; Bowles, 1995; Poulos and Davis, 1990) have shown group efficiency values of well over 100% under certain conditions, especially in dense sand. This may be explained by possible densification accompanied by pile driving in medium-dense sands. Computation of group capacity will be illustrated in Example 14.15.

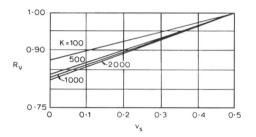

FIGURE 14.68   Correction factor for Poisson ratio. From Poulos, H.G. and Davis, E.H. 1990. *Pile Foundation Analysis and Design*. Krieger, Melbourne, FL.

14.6.5.3 Settlement of Pile Groups

One method for determining the immediate settlement of a pile group is by evaluating the interaction factor (α_F), defined as follows:

$$\alpha_F = \frac{\text{additional settlement caused by adjacent pile}}{\text{settlement of pile under its own load}} \tag{14.61}$$

The settlement of individual piles can be determined on the basis of the method described in Section 14.5.4. Then, once α_F is estimated from Figures 14.70a, 14.70b, or 14.70c one can easily compute the settlement of each pile in a group configuration. At this point, the issue of the flexibility of the pile cap has to be considered. This is because if the pile cap is rigid (thick and relatively small in area), it will ensure equal settlements throughout the group by redistributing the load to accommodate equal settlements. On the other hand, if the cap is flexible (thin and relatively extensive in area), all of the piles will be equally loaded, which results in different settlements.

Under conditions that require the estimation of consolidation settlement under a pile group, one can assume that the pile group acts as a large single footing and use the principles discussed in Section 14.1.3 compressibility and settlement. However, the difference in load attenuation between a shallow footing and a rigid pile group with substantial skin friction

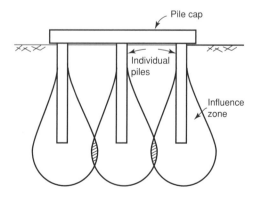

FIGURE 14.69   Illustration of the group effect.

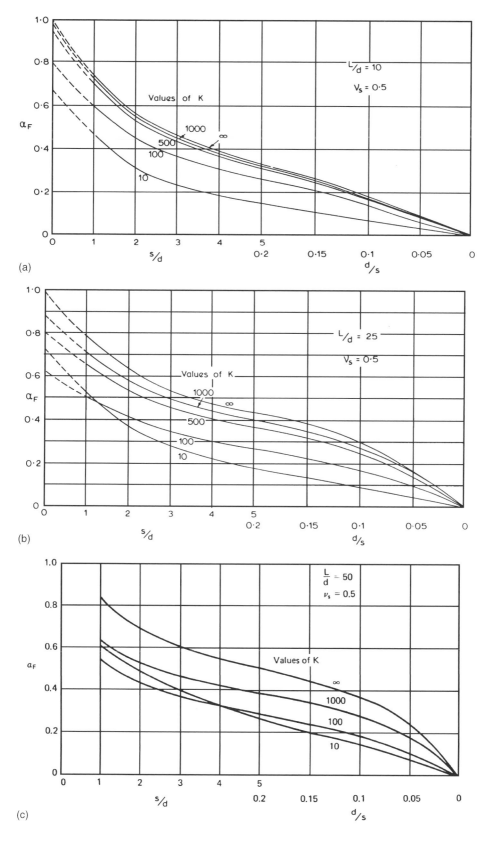

FIGURE 14.70 Determination of α_F factor for $L/D = 10$, (b) for $L/D = 25$ and (c) for $L/D = 50$. L = pile length, D = pile diameter, k and v_s defined in Eq. (14.58). From Poulos, H.G. and Davis, E.H., 1990. *Pile Foundation Analysis and Design.* Krieger, Melbourne, FL.

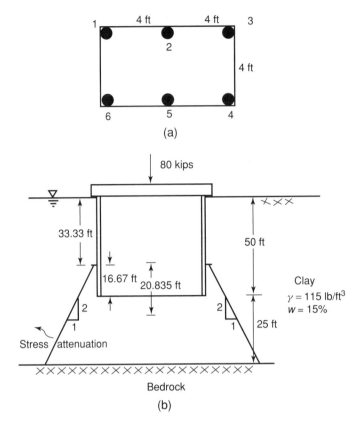

FIGURE 14.71   Illustration of pile group load attenuation.   (a) Plan; (b) elevation.

is accounted for by assuming that the load attenuation originates from the lower middle-third of the pile length, as shown in Figure 14.71.

Example 14.15

The pile group (6, 1 ft × 1 ft piles) shown in Figure 14.71 is subjected to a load of 80 kips. A compression test performed on a representative clay sample at the site yielded an unconfined compression strength of 3 psi and an elastic modulus of 8,000 psi, while a consolidation test indicated no significant over consolidation, with a compression index of 0.3 and a water content of 15%. Estimate the safety of the pile foundation and its settlement. The undrained strength is 0.5 (unconfined compression strength) = 1.5 psi = 216 psf.

Computation of Skin Friction of a Single Pile.   Using Eq. (14.56), $f = 1.0(216) = 216$ psf. From Eq. (14.54), the resultant skin-frictional force = $216(4)(50) = 43.2$ kips.

Computation of End-Bearing of a Single Pile.   Using Eq. (14.53), $P_{pu} = (1)(9)(216) = 1.944$ kips (in fact, one could have easily expected the insignificance of this contribution owing to the "frictional" nature of the pile). Thus, the ultimate capacity of the pile is 45.14 kips.

Estimation of Group Efficiency.   Using Eq. (14.60), $n = 1 - [(3-1)(2) + (2-1)(3)](\zeta)/[90(3)(2)]$ = 0.8. Then, using Eq. (14.59), the group capacity can be obtained as $0.8(45.14)(6) = 216$ kips. Hence the safety factor is 2.7.

Estimation of Single Pile Immediate Settlement. The relative stiffness factor for a rigid pile, K is $E_{concrete}/E_s = 4,000,000/8,000 = 500 L/D = 50/1 = 50$.

From Figure 14.65, $I_o = 0.045$
From Figure 14.66, $R_k = 1.85$
From Figure 14.67, $R_h = 1.00$
From Figure 14.68, $R_v = 1.00$ (undrained $v = 0.5$)

Substituting in Eq. (14.57)

$$s = P(0.045)(1.85)(1.00)(1.00)/(8,000 \times 12) = 0.087 \times 10^{-5} P \text{ in}$$

Analysis of Group Settlement. Assume that the cap is rigid. Therefore the settlement of all six piles must be identical. The total settlement consists of both immediate settlement and consolidation settlement. However, according to the method outlined in Section 14.1.3, Compressibility and Settlement, the consolidation settlement can only be computed for the whole pile group by assuming equal consolidation settlement. Thus one has to assume equal immediate settlements as well.

Owing to their positions with respect to the applied load, it can be seen that piles 1, 3, 4, and 6 can be considered as one type of pile (type 1), while piles 2 and 5 can be categorized as type 2. Thus it will be sufficient to analyze the behavior of pile types 1 and 2 only.

Assume that the load carried by type 1 and type 2 piles are P_1 and P_2, respectively. Then, for equilibrium

$$4P_1 + 2P_2 = 80 \text{ kips} \tag{14.62}$$

Using Figure 14.70, the interaction factors for pile types 1 and 2 due to other piles can be obtained as follows:

Pile Type 1		
Pile i	s/d for pile i	α_F from pile i
1	0	—
2	4	0.4
3	8	0.3
4	4	0.4
5	5.67	0.35
6	8.94	0.25
		$\sum \alpha = 1.7$

Pile Type 2		
Pile i	s/d for pile i	α_F from pile i
1	4	0.4
2	0	—
3	4	0.4
4	5.67	0.35
5	4	0.4
6	5.67	0.35
		$\sum \alpha_F = 1.9$

Using Eq. (14.61), the total settlement of pile type 1 is $2.7(0.087 \times 10^{-5} P_1)$, and the total settlement of pile type 2 is $2.9(0.087 \times 10^{-5} P_2)$.

By equating the settlements of pile types 1 and 2, one obtains

$$2.7\,P_1 = 2.9\,P_2$$
$$P_1 = 1.074\,P_2 \qquad\qquad (14.63)$$

By substituting in Eq. (14.62),

$$2(1.074\,P_2) + P_2 = 40$$
$$P_2 = 12.706 \text{ kips}$$
$$P_1 = 13.646 \text{ kips}$$

Thus the pile group immediate settlement is equal to $2.9(0.087)(10^{-5})(12.706)(10^3) = 0.032$ in.

Computation of Consolidation Settlement.   On the basis of the stress attenuation shown in Figure 14.71, the stress increase on the midplane of the wet clay layer induced by the pile group can be found as

$$\Delta\sigma = 80{,}000/[(8 + 20.835)(4 + 20.835)] = 111.71 \text{ psf}$$

The initial effective stress at the above point is equal to $(115 - 62.4)(54.165) = 2{,}849$ psf. From basic soil mechanics, for a saturated soil sample $e = wG_S = 0.15 \times (2.65) = 0.4$ (assuming the solid specific gravity, G_S, to be 2.65). Then by applying Eq. (14.15), one obtains the consolidation settlement as

$$s_{ult} = (0.3)(41.67)(1/1 + 0.4)\log[1 +$$
$$111.71/2{,}849] = 0.149 \text{ ft} = 1.789 \text{ in}$$

Hence in this case the consolidation settlement is predominant.

14.6.6 Verification of Pile Capacity

There are several methods available to determine the load capacity of piles. The commonly used methods are (1) use of pile-driving equations, (2) use of the wave equation, and (3) full-scale load tests. A brief description of the first two methods will be provided in the next two sections.

14.6.6.1 Use of Pile-Driving Equations

In the case of driven piles, one of the very first methods used to determine the load capacity was using numerous pile-driving equations. Of these equation, one of the more popular is the **Engineering News Record** (ENR) **equation**, which expresses the pile capacity as follows:

$$P_u = \frac{E\,W_h h}{S + C}\,\frac{W_h + n^2\,W_p}{S + C\,W_h + W_p} \qquad\qquad (14.64)$$

where

$n =$ coefficient of restitution between the hammer and the pile (<0.5 and >0.25)
$W_h =$ weight of the hammer
$W_p =$ weight of the pile
$S =$ pile set per blow (in inches)
$C = 0.1$ in

h = hammer fall

E = hammer efficiency (usually estimated by monitoring the free fall)

It is seen how one can easily compute the instant capacity developed at any given stage of driving by knowing the pile set (S), which is usually computed by the reciprocal of the number of blows per inch of driving. To avoid damage to the pile and the equipment, it should be noted that when driving has reached a stage where more than 10 blows are needed for a penetration of 1 in ($S = 0.1$ or **refusal**), further driving is not recommended.

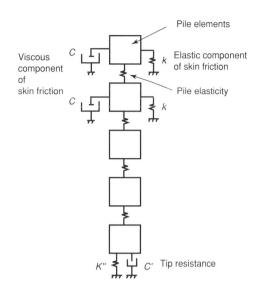

FIGURE 14.72 Idealization of a driven pile.

14.6.6.2 Use of the Wave Equation

With the advent of modern computers, the use of the wave-equation method for pile analysis, introduced by Smith (1960), became popular. Smith's idealization of a driven pile is elaborated in Figure 14.72.

The governing equation for wave propagation can be written as follows:

$$\rho A \frac{\partial^2 u}{\partial t^2} dz = AE \frac{\partial^2 u}{\partial z^2} dz - R(z) \tag{14.65}$$

where

ρ = mass density of pile
A = area of cross section of pile
u = particle displacement
t = time
z = coordinate axis along the pile
$R(z)$ = resistance offered by any pile slice, dz

The above equation can be transformed into the finite-difference form to express the displacement, the force, and the velocity, respectively, of a pile element i at time t as follows:

$$D(i, t) = D(i, t - \Delta t) + V(i, t - \Delta t) \tag{14.66}$$

$$F(i, t) = [D(i, t) - D(i + 1, t)]K \tag{14.67}$$

$$V(i, t) = V(i, t - \Delta t) + \frac{\Delta t\, g}{w(i)}[F(i - 1, t) - F(i, t) - R(i, t)] \tag{14.68}$$

where

$K = EA/\Delta x$
$w = \rho \Delta x A$
Δt = selected time interval at which computations are made

Idealization of Soil Resistance. In Smith's model, the point resistance and the skin friction of the pile are assumed to be viscoelastic and perfectly plastic in nature. Therefore the separate resistances can be expressed by the following equations:

$$P_p = P_{pD}(1 + J V_p) \tag{14.69}$$

and

$$P_s = P_{sD}(1 + J' V_p) \tag{14.70}$$

where

P_{pD} and P_{sD} = resistances at a displacement of D
V_p = velocity of the pile
J and J' = damping factors

The assumed elastic, perfectly plastic behavior for P_{pD} and P_{sD} are illustrated in Figure 14.73.

In implementing this method, the user must assume a value of total resistance (P_u), a suitable distribution of the resistance between the skin friction and point resistance (P_{pD} and P_{sD}), the quake (Q in Figure 14.73), and damping factors J and J'. Then by using Eqs. (14.66)–(14.68), the pile set S can be determined. By repeating this procedure, a useful curve between P_u and S, which can eventually be used to determine the resistance at any given set S, can be obtained.

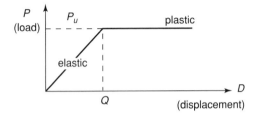

FIGURE 14.73   Assumed viscoelastic perfectly plastic behavior of soil resistance.

Example 14.16

For simplicity, assume that a model pile is driven into the ground using a 1000 lb hammer dropping 1 ft, as shown in Figure 14.74. Assuming the following data, predict the velocity and the displacement of the pile tip after three time steps.

$J = 0.0$ s/ft, $J' = 0.0$ s/ft
$Q = 0.1$ in
$\Delta t = 1/4000$ s
$R_{pu} = R_{su} = 50$ kips
$k = 2 \times 10^6$ lb/in

As shown in Figure 14.74, assume the pile consists of two segments ($i = 2$ and 3) and the time step to be 1/4000 s. Then the following boundary conditions can be written:

$D(1, 0) = D(2, 0) = D(3, 0) = 0$
$F(1, 0) = F(2, 0) = F(3, 0) = 0$
$V(1, 0) = (2gh)_{1/2} = 96.6$ in/s

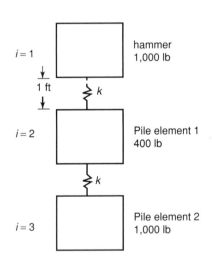

FIGURE 14.74   Application of the wave equation.

After the First Time Step.   From Eq. (14.66),

$$D(1, 1) = D(1, 0) + V(1, 0)\Delta t = 1 + 8.0498(1/4{,}000) = 0.024 \text{ in}$$

$$D(2, 1) = D(3, 1) = 0$$

From Eq. (14.67),

$$F(1, 1) = [D(1, 1) - D(2, 1)]k = (0.024 - 0)(2)(10^6) = 48 \times 10^3 \text{ lb/in}$$
$$F(2, 1) = F(3, 1) = 0$$

From Eq. (14.68),

$$V(1, 1) = V(1, 0) + (1/4,000)(388.8)(0 - 48,000)/1,000 = 91.93 \text{ in/s}$$
$$V(2, 1) = 0 + (1/4,000)(388.8)[48,000 - 0 - R(1, 1)]/400 = 11.664 \text{ in/s}$$
$$V(3, 1) = 0.0$$

After the Second Time Step. By repeating the above procedure, one obtains the following results:

$$D(1, 2) = D(1, 1) + V(1, 1)\Delta t = 0.024 + 91.93(1/4,000) = 0.047 \text{ in}$$
$$D(2, 2) = D(2, 1) + V(2, 1)\Delta t = 0 + 11.664(1/4,000) = 0.0029 \text{ in}$$
$$D(3, 2) = 0$$

$$F(1, 2) = [D(1, 2) - D(2, 2)](2)(10^6) = 88,200 \text{ lb/in}$$
$$F(2, 2) = [D(2, 2) - D(3, 2)](2)(10^6) = 5,900 \text{ lb/in}$$
$$F(3, 2) = [D(3, 2) - D(4, 2)](2)(10^6) = 0$$

$$V(1, 2) = V(1, 1) + 9.72(10^{-5})[F(0, 2) - F(1, 2)]$$
$$= 91.93 + 9.72(10^{-5})(0 - 88,200) = 83.35 \text{ in/s}$$
$$V(2, 2) = V(2, 1) + 24.3(10^{-5})(88,200 - 5,900 - 1,450) = 31.3 \text{ in/s}$$
$$V(3, 2) = 0 + 9.72(10^{-5})(5,900 - 0 - 0) = 0.56 \text{ in/s}$$

After the Third Time Step. By repeating above steps, one obtains the following results:

$$D(1, 3) = D(1, 2) + V(1, 2)\Delta t = 0.047 + 83.35(1/4,000) = 0.0678 \text{ in}$$
$$D(2, 3) = D(2, 2) + V(2, 2)\Delta t = 0.0029 + 31.3(1/4,000) = 0.0078 \text{ in}$$
$$D(3, 3) = 0 + 0.56(1/4, 000) = 0.00014 \text{ in}$$

$$F(1, 3) = [D(1, 3) - D(2, 3)](2)(10^6) = 120,000 \text{ lb/in}$$
$$F(2, 3) = [D(2, 3) - D(3, 3)](2)(10^6) = 15,320 \text{ lb/in}$$
$$F(3, 3) = [D(3, 3) - D(4, 3)](2)(10^6) = 280 \text{ lb/in}$$

$$V(1, 3) = V(1, 2) + 9.72(10^{-5})[F(0, 3) - F(1, 3) - R(1, 3)] = 71.69 \text{ in/s}$$
$$V(2, 3) = V(2, 2) + 24.3(10^{-5})(120,000 - 15,320 - 3,900) = 55.79 \text{ in/s}$$
$$V(3, 3) = 0.56 + 9.72(10^{-5})(15,320 - 70 - 70) = 2.04 \text{ in/s}$$

This computational procedure must be repeated on the computer until the bottom pile segment does not move during a given time step and the velocities of all of the pile segments become zero.

14.6.6.3 Pile-Load Tests

The most reliable method of verifying the capacity of a driven or a bored pile is a full-scale load test. In a load test, a prototype pile is driven at the construction site and gradually loaded while the load and the corresponding deflections up to failure are monitored. Since this is an expensive test, it is usually justified on a few test piles when a pile group is to be installed under similar soil conditions. For further details, the reader is referred to Bowles (1995).

Another frequently used, nondestructive experimental method for estimating the capacity of a pile is the pile-driving analyzer (PDA) method. In this method, an accelerometer and a strain gage are attached to the top of the driven pile in order to monitor the particle velocity and longitudinal stress during shock wave propagation following a hammer blow. By analyzing these records using the wave-equation method (Section 14.6.6.2), one can closely predict the resistance mobilized during a given stage of driving. Details of this method are found in Rausche et al. (1972).

14.7 Caissons and Drilled Piers

Both caissons and drilled piers are very large diameter concrete bored piles that can be used in lieu of pile groups to transmit structural loads to bedrock. The term caisson is reserved for drilled piers involving a waterway. It is known that construction of caissons and drilled piers is more easily accomplished than driving a group of piles in stiff soils like dense sands and stiff clays. An added advantage is that installation of drilled piers and caissons precludes disturbance to nearby structures and ground movement. On the other hand, construction of caissons and drilled piers may involve ground loss and consequent settlements that require careful monitoring and supervision.

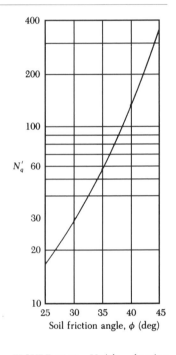

FIGURE 14.75 Vesic's bearing capacity factor N_q' for deep foundations. From Das, B.M., 1995. *Principles of Foundation Engineering.* PWS Publishing.

14.7.1 Estimation of Bearing Capacity

In rare situations when drilled piers are founded on stiff sand or clay, the bearing capacity of the pier can be estimated based on the procedure outlined in Section 14.6.3. However, one realizes that the bearing capacity factors for drilled piers should be somewhat lower than those for driven piles owing to the fact that it is harder for bored piles to mobilize the ultimate shear strength. This is especially the case for drilled piers founded on sand where Eq. (14.51) is applicable. Das (1995) suggests using Vesic's N_q' values, given in Figure 14.75, for computing the bearing capacity using Eq. (14.51).

In most cases however, drilled piers and caissons are embedded in sound bedrock and hence the load bearing capacity criteria have to be adjusted accordingly. The following stepwise procedure can be utilized for drilled pier design in rock.

14.7.1.4 Selection of Diameter

The pier diameter can be estimated using the compressive strength (f_c') of concrete as

$$\beta f_c' = \frac{P_w}{\pi \frac{D^2}{4}} \tag{14.71}$$

where

    β = strength reduction factor (code recommended value is 0.25)
    P_w = allowable (working) load
    D = pier diameter

14.7.1.5 Maximum Embedment Length

The maximum length of embedment (L_{max}) can be estimated by assuming that the overlying soil exerts negligible friction and the pile tip carries zero load. Thus

$$L_{max} = \frac{P_w}{\pi D \tau_{all}} \tag{14.72}$$

where

τ_{all} = allowable shear stress between pile and rock
L_{max} = maximum required length of embedment

The allowable shear stress on the concrete–rock interface can be expressed as (Goodman, 1980)

$$\tau_{all} = \frac{q_u}{20\,F} \tag{14.73}$$

where

q_u = minimum of the unconfined compression strengths of rock and concrete
F = safety factor

Check for Rock Bearing Capacity: The normal stress at a depth of z in a concrete pier embedded in rock (Figure 14.76) can be obtained from the following expression (Das, 1995):

$$\sigma_z = \frac{P_w}{(\pi/4)\,D^2}\,\exp\left[-\frac{2\mu_c\tan\delta}{1-\mu_c+(1+\mu_r)\frac{E_c}{E_r}}\frac{2z}{D}\right] \tag{14.74}$$

where

E_c/E_r = Moduli ratio of concrete rock
δ = Angle of friction between rock and concrete
μ_c and μ_r = Poisson ratios of concrete and rock

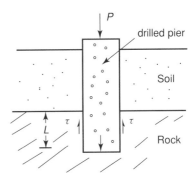

FIGURE 14.76   Drilled pier/rock interface.

For z, one can assume any design embedment depth L ($<L_{max}$) and compare the normal stress at the tip of the pier with the maximum allowable bearing capacity of rock.

Check for Adequate Shear: one can estimate the average shear stress induced in the pier/rock interface by assuming a uniform shear stress distribution τ. Then it is seen that:

$$\tau = (P_w - \sigma_z(\pi/4)\,D^2)\frac{1}{\pi D L} \tag{14.75}$$

Finally, τ can be compared with τ_{all} in Eq. (14.73).

Example 14.17

Design a suitable drilled pier to carry a load of 1000 tons with the following data:

Moduli ratio of concrete–rock = 1.5
friction angle at concrete–rock interface = 35°

allowable compressive strength of rock = 0.65 kips/in^2
allowable compressive strength of concrete = 3 kips/in^2
unconfined compression strength of rock = 4.3 kips/in^2

Determine the Pier Diameter.   From Eq. (14.71)

$$\pi D^2/4 = 1,000(2,000)/(0.25)/(3 \times 1000)$$
$$D = 4.85 \text{ ft}$$

Determine Maximum Pier Length.   From Eq. (14.73)

$$\tau_{\text{all}} = 4.3/20/3 = 0.072 \text{ kips/in}^2 \text{ (Assuming } F = 3.0\text{)}$$

Then, from Eq. (14.72)

$$L_{\text{max}} = 2,000(2,000)/[\pi(4.85 \times 12)(.072 \times 1,000)] = 12.66 \text{ ft}$$

Check for Bearing on Rock.   Assume an embedment depth of 7 ft and the Poisson ratios of rock and concrete to be 0.3 and 0.15, respectively. Then, by applying Eq. (14.74)

$$\sigma_z = 44.78 \text{ tons/ft}^2 = 0.62 \text{ kips/in}^2 < 0.35 \text{ kips/in}^2$$

The bearing is adequate.

Check for Shear.

Shear force on the shaft = $1,000 - 44.78(\pi)(D)^2/4 = 173$ tons
Average shear stress = $173/(\pi D(7)) = 1.622$ tons/ft^2 = 0.0225 kips/in^2 < 0.072 kips/in^2
Thus design a pier of 4.85 ft diameter embedded to 7.0 ft.

References

Bowles, J.E. 1986. *Engineering Properties of Soils and Their Measurements*. McGraw–Hill, New York.

Bowles, J.E. 1995. *Foundation Analysis and Design*. McGraw–Hill, New York.

Das, B.M. 1995. *Principles of Foundation Engineering*. PWS Publishing, Boston, MA.

Goodman, R.E. 1980. *Introduction to Rock Mechanics*. John Wiley, New York.

Harr, M. 1962. *Groundwater and Seepage*. McGraw–Hill, New York.

Holtz R.D. and Kovacs W.D. 1981. *An Introduction to Geotechnical Engineering*. Prentice Hall, Englewood Cliffs, NJ.

Koerner, R. 1994. *Designing with Geosynthetics*. Prentice Hall, Englewood Cliffs, NJ.

Meyerhoff, C.G. 1976. Bearing capacity settlement of pile foundations. *J. Geotech. Eng. Div. ASCE* 102(GT3):197–228.

Mullins, A.G. 1996. *Field Characterization of Dynamic Replacement of Florida Organic Soils*. Ph.D. Dissertation. University of South Florida.

Poulos H.G. and Davis, E.H. 1990. *Pile Foundation Analysis and Design*. Krieger, Melbourne, FL.

Pickett, G. and Ray, G.K. 1951. Influence charts for concrete pavements. *Am. Soc. Civ. Eng. Trans.* 116(2425):49–73.

Rausche, F., Moses, F., and Goble. G.G. 1972. Soil resistance predictions from pile dynamics. *J. Soil Mech. Found. Div., Am. Soc. Civ. Eng.* 98(SM9):917–937.

Schmertmann J.H. and Hartman, J.P. 1978. Improved strain influence factor diagrams. *J. Geotech. Eng. Div., Am. Soc. Civ. Eng.* 104(GT8):1131–1135.

Scott, R.F. 1981. *Foundation Analysis.* Prentice Hall, Englewood Cliffs, NJ.

Smith, E.A.L. 1960. Pile-driving analysis by the wave equation. *J. Soil Mech. Found. Div., Am. Soc. Civ. Eng.* 86(EM4):35–51.

Stinnette, P. 1992. *Engineering Properties of Florida Organic Soils.* Master's Project. University of South Florida.

Stinnette, P. 1996. *Geotechnical Data Management and Analysis System for Organic Soils.* Ph.D. Dissertation. University of South Florida.

Thilakasiri, H.S. 1996. *Numerical Simulation of Dynamic Replacement of Florida Organic Soils.* Ph.D. Dissertation. University of South Florida.

Wray, W. 1986. *Measurement of Engineering Properties of Soil.* Prentice Hall, Englewood Cliffs, NJ.

Sydney opera house, Sydney, Australia (Courtesy Australian Information Service).

15

Specialized Construction Applications

by
John M. Scanlon
Senior Consultant, Wiss, Janney, Elstner Associates, Inc. and formerly, Chief, Concrete Technology Division, US Army Waterways Experiment Station. Vicksburg, Mississippi. Expert in concrete behavior and technology.

15.1 Introduction

Specialized construction applications are considered to be those that contain materials that are not routinely used in conventional structural or mass concrete [ACI 116R], those that are not proportioned using procedures given in the American Concrete Institute Standard Practice 211.1

[ACI 211.1], or those that are placed with equipment or by methods that require additional attention from the contractor to assure the required quality is achieved.

The techniques of mixing, batching, transporting, placing, consolidating, protecting, and placing concrete have drastically changed during the past few decades. The reasons for these drastic changes are as follows:

(1) Owners have demanded that cost escalation of new construction be kept under control.
(2) These demands required that accelerated construction techniques and new materials be developed.
(3) Governmental agencies initiated regulation that required that the environment be better protected.
(4) Construction had to be accomplished with fewer workers; consequently, techniques developed had to use innovative equipment and materials.

Some of the more successful specialized construction techniques, mostly relating to the use of hydraulic-cement concrete, covered in this chapter include:

(1) preplaced-aggregate concrete,
(2) underwater concreting,
(3) conveying concrete by pumping,
(4) vacuum processing, and
(5) cement mortar and plastering.

Research has resulted in many new types of equipment and improved on older types, thereby permitting concrete construction to be accomplished more quickly and, consequently, more economically, with better control and, consequently, superior quality. The uses of admixtures, both mineral and chemical, have greatly expanded concretes use in new construction techniques; extended the life of freshly mixed concrete for as long as the user desires; and allowed concrete to be dropped through water without segregating or separating. They have allowed concrete to be used in corrosive environments without causing the reinforcing steel to corrode and to be used in freezing and thawing environments without the previously experienced rapid deterioration. They have also allowed concrete to attain much greater compressive strength with higher moduli of elasticity for use in high-rise concrete structures.

These vastly improved capabilities have been accomplished in order to keep hydraulic-cement concrete competitive as a primary construction material. This competitiveness has to be maintained for concrete to remain the most cost-effective construction material in the world. Without the progress of concrete technology, the industry will flounder and more exotic construction materials will be developed as alternatives.

15.2 Preplaced-Aggregate Concrete

15.2.1 General

Preplaced-aggregate concrete (PA) [ACI 116R] is concrete produced by placing coarse aggregate in a form and later injecting a hydraulic cement, sand, and fly-ash grout, usually with chemical admixtures, to fill the voids. Smaller-size coarse aggregate (less than 1/2 in) is not used in the mixture in order to facilitate grout injection [ASTM C926]. Much of the information presented on preplaced-aggregate concrete has been taken from the U.S. Army Corps of Engineers' Engineering Manual 1110-2-2000 [U.S. Army Corps of Engineers, 1994].

One of the primary advantages claimed for PA concrete is that it can easily be placed in locations where conventional concrete would be extremely difficult to place. The primary use of PA concrete is the repair of existing concrete structures. PA concrete may be particularly suitable for underwater construction, for placements in areas with closely spaced reinforcing steel and cavities where overhead contact is necessary, and in areas where low-volume change of the hardened concrete is desired. It differs from conventional concrete in that it contains a higher percentage of coarse aggregate since the coarse aggregate is placed directly into the forms with coarse aggregate having point-to-point contact rather than being contained in a flowable plastic mixture.

Therefore hardened PA concrete properties are greatly dependent on the coarse aggregate properties. Drying shrinkage of PA concrete may be less than 50% of that for conventional concrete, which partially accounts for the excellent bond between PA concrete and existing roughened concrete. The compressive strength of PA concrete is dependent on the quality, proportioning, and handling of materials but is generally comparable to that achieved with conventional concrete. The frost resistance of PA concrete is also comparable to conventional air-entrained concrete, assuming that the grout mixture has an air content, as determined by ASTM C231, of approximately 9%.

PA concrete may be particularly applicable to underwater repair of old structures and new underwater construction where dewatering may be difficult, expensive or impractical. Bridge piers and abutments are typical of applications for underwater PA concrete construction or repair. ACI 304.R provides an excellent discussion of PA concrete.

15.2.2 Applications

Preplaced-aggregate concrete has been successfully used on different types of projects including those in the following construction categories:

(1) underwater construction of and/or repair of bridge piers,
(2) resurfacing of lock chambers and guide walls,
(3) massive fills in permanent sheet-piling piers and cofferdams,
(4) construction of atomic-reactor shields, and
(5) resurfacing of dam spillways.

PA concrete is not frequently used. This might be due to the apprehension on the part of the construction industry that the technology exceeds the industry's normal capabilities. Such apprehension is not justified since successful projects can be accomplished by any construction entity that has a knowledgeable concrete technologist on staff.

15.2.3 Materials and Proportioning

Intrusion grout mixtures should be proportioned in accordance with ASTM C938 to obtain the specified consistency, air content, and compressive strength. The grout mixture should also be proportioned such that the maximum water/cement ratio complies with the same ratio that conventional concrete would be required to have for the same environmental exposure and placing requirements. Compressive-strength specimens should be made in accordance with ASTM C943. Compressive-strength testing of the grout alone should not be used to estimate the PA concrete strength because it does not reveal the weakening effect of bleeding. However, such testing may provide useful information on the potential suitability of grout mixtures.

The ratio of cementitious materials to fine aggregate will usually range from about 1 for structural preplaced-aggregate concrete to 0.67 for massive concrete sections. A grout fluidifier meeting the requirements of ASTM C937 is commonly used in the intrusion grout mixtures to offset bleeding, reduce the water/cement ratio and still provide a given consistency, and retard stiffening so that handling times can be extended. Grout fluidifiers typically contain a water-reducing additive or admixture, a suspending agent, aluminum powder, and a chemical buffer to assure timed reaction of the aluminum powder with the alkalies in the hydraulic cement.

Products proposed for use as fluidifiers that have no prior record of successful use in PA concrete can normally be accepted based on successful field use. ASTM C937 requires that intrusion grout, made as prescribed for acceptance testing of fluidifiers, have an expansion within certain specified limits that may be dependent of the alkali content of the cement used in the test. Experience has shown, however, that because of the difference in mixing time and other factors, expansion of the field-mixed grout ordinarily will range from 3 to 5%. If, under field conditions, expansion of less than 2% or more than 6% occurs, adjustments to the fluidifier should be made to bring the expansion to within these limits.

The fluidifier should be tested under field conditions with job materials and equipment as soon as practicable so that sufficient time is available to make adjustments to the fluidifier, if necessary. If the aggregates are potentially reactive, the total alkali content of the hydraulic cement plus fluidifier added to increase expansion should not exceed 0.60%, calculated as equivalent sodium oxide by mass of cement. The grout submitted for use may exhibit excess bleeding if its cementitious material to fine aggregate ratio is different from that of the grout mixture used to evaluate the fluidifier. Expansion of the grout should exceed bleeding, as desired, at the expected in-place temperature. Grouts should be placed in an environment where the temperature will rise above 40°F, since expansion caused by the fluidifier ceases at temperatures below 40°F.

This condition is normally readily obtainable when preplaced-aggregate concrete is placed in massive sections or placements are enclosed by timber forms. If an air-entraining admixture is used in the PA concrete, adjustments in the grout mixture proportions may be necessary to compensate for the significant strength reduction caused by the combined effects of entrained air and the hydrogen generated by the aluminum powder in the fluidifier. However, these adjustments must not reduce the air content of the mixture to a level that compromises its frost resistance. The largest practical nominal maximum size aggregate (NMSA) should be used to increase the economy of the PA concrete. A 1-1/2 in NMSA will typically be used in most PA concrete; however, provisions are made for the use of 3-in NMSA when it is considered appropriate. It is not expected that many situations will arise where the use of aggregate larger than 2 in will be practical and economical. Pozzolan is usually specified to increase flowability of the grout.

15.2.4 Preplacing Aggregate

Since excessive breakage and objectionable segregation are to be avoided, it is necessary to preplace the coarse aggregate in the placement with extreme care. The difficulties are magnified as the nominal maximum size of the aggregate increases, particularly when two or more sizes are blended. Therefore proposed methods of placing aggregate should be carefully established to ensure that satisfactory results will be attained. Coarse aggregate must be prewashed, screened, and saturated immediately prior to placement to remove dust and dirt and to eliminate coatings and undersized particles. Washing aggregate in forms should never be permitted because fines may accumulate at the bottom.

15.2.5 Contaminated Water

Contaminated water is a matter of concern when PA concrete is placed underwater. Contaminants present in the water may coat the aggregate and adversely affect setting of the cement or bonding of the mortar to the coarse aggregate. If contaminants in the water are suspected, the water should be tested before construction begins. If contaminants are present in such quantity or of such character that the harmful effects cannot be eliminated or controlled, or if the construction schedule imposes a long delay between aggregate placement and grout injection, PA concrete should not be used.

15.2.6 Preparation of Underwater Foundation

The cleanup of foundations in underwater construction when the foundation material is glacial till or similar material is always difficult. The difficulty develops when as a result of prior operations, an appreciable quantity of loose, fine material is left on the foundation or in heavy suspension just above the foundation. The fine material is displaced upward into the aggregate as it is being placed.

The dispersed fine material coats the aggregate or settles and becomes concentrated in void spaces in the aggregate just above the foundation, thus precluding proper intrusion and bond. Care must therefore be exercised to ensure that as much loose, fine material is removed as possible before placement of aggregate is allowed to commence.

15.2.7 Pumping

Pumping of grout should be as continuous as practical. However, minor stoppages are permissible and ordinarily will not present any difficulties when proper precautions are taken to avoid plugging of grout lines. The rate of grout rise within the aggregate should be controlled to eliminate cascading of grout and to avoid form pressures greater than those for which the forms were designed. For a particular application, the grout injection rate will depend on form configuration, aggregate grading, and grout fluidity.

15.2.8 Joint Construction

A cold joint is formed in PA concrete when pumping is stopped for longer than the time it takes for the grout to harden. When delays in grouting occur, the insert pipes should be pulled to just above the grout surface before the grout stiffens and then rodded clear. When pumping is ready to resume, the pipes should be worked back to near contact with the hardened grout surface, and then pumping should be resumed slowly for a few minutes. Construction joints are formed similarly by stopping grout rise approximately 12 in below the aggregate surface. Care must be taken to prevent dirt and debris from collecting on the aggregate surface or filtering down to the grout surface.

15.2.9 Grouting Procedure

The two patterns for grout injection are the horizontal layer and the advancing slope. Regardless of the system used, grouting should start from the lowest point in the form.

(1) *Horizontal layer.*   In this method, grout is injected through an insert pipe, that raises the grout until it flows from the next insert hole, 3 to 4 ft above the point of injection. Grout is then injected into the next horizontally adjacent hole, 4 or 5 ft away, and the procedure is repeated sequentially around the member until a layer of coarse aggregate is grouted. Successive layers of aggregate are grouted until all aggregate in the form has been grouted.

(2) *Advancing slope.*   The horizontal-layer method is not practical for construction of slabs having large horizontal dimensions. In such situations it becomes necessary to use an advancing-slope method of injecting grout. In this method, intrusion is started at one end of the form and pumping is continued until the grout emerges on the top of the aggregate for the full width of the form and assumes a slope that is advanced and maintained by pumping grout through successive rows of intrusion pipes until the entire mass is grouted. In advancing the slope, the pumping pattern is started first in the row of holes nearest the toe of the slope and continued row by row up the slope (opposite to the direction of advance of slope) to the final row of pipes. This process is repeated, moving ahead one row of pipes at a time as the intrusion is completed.

(3) *Grout insert pipes and sounding devices.*   The number, location, and arrangement of grout insert pipes will depend on the size and shape of the work being constructed. For most work, grout insert pipes will consist of pipes arranged vertically and at various inclinations to suit the configurations of the work. Grout pipes are generally either 3/4, 1, or 1-1/2 in. Normally, either a 3/4 or 1 in diameter would be needed for structural concrete having a maximum size aggregate of 1-1/2 in or less. If the preplaced aggregate has a maximum size larger than 1-1/2 in, the grout insert pipes should have a diameter of 1-1/2 in.

Intrusion points should be spaced about 6 ft apart; however, spacing wider than 6 ft may be permissible under some circumstances, and spacing closer than 6 ft will be necessary in some situations. Normally, one sounding device should be provided for every four intrusion points; however, fewer sounding devices may be permissible under some circumstances. There should always be enough sounding devices, and these should be so arranged that the level of the grout at all locations can be accurately determined at all times during construction. Accurate knowledge of the grout level is essential in order to accomplish the following tasks:

(a) check the rate of intrusion;
(b) avoid getting the grout too close to the level of the top of the aggregate when placement of the aggregate and intrusion are progressing simultaneously; and
(c) avoid damage to the work that would occur if a plugged intrusion line was washed out while the end of the line was within the grout zone.

Sounding devices usually consist of wells (slotted pipes) through which the level of the grout may be readily and accurately determined. If sounding devices other than wells are considered, conclusive demonstrations should be performed to verify that such devices will readily and accurately indicate the level of the grout at all times.

15.2.10 Finishing Unformed Surfaces

If a screeded or troweled finish is required, the grout should be brought up to flood the aggregate surface, and any diluted grout should be removed. A thin layer of pea gravel or 3/8 to 1/2 in crushed stone should then be worked into the surface by raking and tamping. After the surface has stiffened sufficiently, it may be finished as required. A finished surface may also be obtained on PA concrete by adding a bonded layer of conventional concrete of the prescribed thickness to the surface. The surface should be adequately prepared prior to applying the topping.

15.3 Underwater Concrete

15.3.1 General [ASTM C231, ASTM C943, ASTM C150]

Until recently, underwater concreting was defined as tremie concrete only; but now, owing to research in the use of admixtures, new procedures have been developed so that freshly mixed concrete can now be placed by dropping it through water without the use of a tremie. Although this new system will be discussed here, it is recommended that unless absolutely necessary, the concrete should be deposited through a tremie or possibly, a pumpline.

The placement of concrete underwater requires conveying freshly mixed concrete from the surface of a liquid environment to a location underneath that surface in such a way that the concrete is not damaged by segregation or separation. During the administration of the U.S. Army Corps of Engineers' Repair, Evaluation, Maintenance and Rehabilitation (REMR) Research Program [Scanlon et al., 1983], it was discovered that the cost of dewatering an underwater area to be repaired increased the cost of the repair by 100% when compared to making the same repair in the dry.

Generally many organizations avoid placing concrete underwater; the possibility of major problems occurring is excessive. Also, many owners desire to actually see the concrete that has been placed so that they have a better understanding of the appearance and quality. Underwater concreting normally cannot be observed due to the water. Underwater concreting can be a very successful operation, but it is absolutely necessary for the contractor to pay close attention to details.

At times it is physically or economically impracticable to expose a foundation prior to concrete placement. At such times, suitable underwater placing procedures such as pumping or use of tremies or special concrete buckets should be employed. The recently completed research provides techniques to place freshly mixed concrete through water without the use of a pumpline, tremie, or bucket; new innovative antiwashout chemical admixtures have been developed that permit freshly mixed concrete to be dropped through water without segregation or separation (see Figure 15.1).

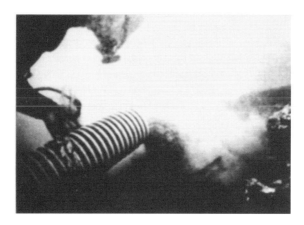

FIGURE 15.1 Clearness of underwater concreting.

Although these are proprietary admixtures, practically all the chemical admixture companies provide versions to the industry.

15.3.2 Structures Conducive to Underwater Placement

Underwater concreting is appropriate for all types of structures: massive sections, walls, slabs, foundations, piers, caissons, stilling basins, cofferdams, conduits, and many others. It is obvious that underwater concreting is most advantageous for relatively large structures such as bridge piers, cofferdams, thick walls, and large foundations. In this chapter, we will discuss the use of tremies, buckets, pumps, aqua valves, and of course, antiwashout admixtures.

15.3.3 Available Methods

The tremie is presently the most often used technique for placing concrete under water. A tremie is defined by the American Concrete Institute in ACI 116R as a pipe or tube through which concrete is deposited under water, having at its upper end a hopper for filling and a bail for moving the assemblage, as shown in Figure 15.2. Underwater-bucket placing consists of lowering a special bucket containing freshly mixed concrete to the bottom of the foundation and opening the bucket slowly to permit the concrete to flow out gently without causing turbulence or mixing with the water.

One of the most troublesome problems when using a tremie is controlling the head pressure on concrete in the tremie pipe so that the pipe does not empty itself too quickly and allow water to enter the pipe from the bottom. Fortunately, a concrete pump can be used to overcome this problem. A concrete pump permits the pumpline to remain full at all times and the end of the pumpline can remain very deep in the fresh concrete, preventing the intrusion of water into the pumpline. We will now discuss underwater-placement techniques in greater detail.

FIGURE 15.2 Tremie concreting operation.

15.3.4 Bucket Placement

The buckets used for underwater placement of freshly mixed concrete should have drop-bottom or roller-gate openings. The gates should be able to be opened from above water. If air is used to open the bucket, the air should discharge through a line to the surface to prevent water disturbance. The top of the bucket must be covered to prevent water from washing the surface of the freshly mixed concrete. One way is to cover the top with canvas or plastic sheets; the covering should be as watertight as possible.

There are special buckets manufactured for underwater placement of concrete; these buckets have sloping tops that minimize the water surge. The first bucket of concrete should be slowly lowered to the bottom of the foundation and allowed to rest on the bottom. The gates should be opened slowly to permit the concrete to flow out gently, without causing turbulence or mixing with the water.

Additional buckets should land on the previously placed concrete and slightly penetrate the surface so that opening the gates and releasing the fresh concrete will result in even less turbulence. The operation is continued until all of the concrete has been placed. An example of concrete placed underwater with buckets is the foundation for the San Francisco-Oakland Bay Bridge with a 240 ft deep foundation.

15.3.5 Tremie

The tremie process is the underwater placing technique used most frequently by contractors. Figure 15.2 depicts a conveyor being used to feed concrete to the tremie hopper for distribution down the tremie pipe in a massive placement. The tremie consists of a vertical pipe through which concrete is placed. The concrete flows from the bottom of the tremie pipe. After the original placement, the end of the tremie pipe is kept submerged in the fresh concrete at all times. Consequently, the first concrete placed is the concrete that will most likely end up on top of the placement in small diameter placements.

It has been observed that in a 36-in caisson the first cubic yard of grout placed ends up on top of the placement and then is allowed to overflow the top of the caisson until the caisson is filled with concrete. Such an occurrence assures the contractor that there will be no voids in the placement and that when the tremie pipe is removed slowly, a good job is almost guaranteed.

The discharge end of the pipe should always remain buried in the freshly placed concrete after the initial placement. A tremie pipe should be made of heavy-gauge steel to withstand all stresses induced by the handling operation. The pipes should have a diameter large enough to assure that aggregate bridging will not occur. Normally, pipes should be between 8 and 12 in in diameter for concrete containing up to 2 in NMSA.

Pipe sections in increments of 10 ft should be used; for deep placements the tremie should be fabricated in sections with joints that permit the upper sections to be removed as the placement progresses, otherwise, the hopper will become too elevated for efficient discharge from a ready-mix truck. The joints between sections of pipe need to be watertight and be capable of being disconnected rapidly so that no major interruptions occur with the placement of concrete.

15.3.6 Basic Tremie Methods

There are two basic methods of initiating concrete placement when using the tremie method un-derwater. These methods are referred to as **wet-pipe** and **dry-pipe** methods.

The wet-pipe method refers to initiating placement with the open-ended tremie pipe on the bottom and with the pipe filled with water. In order to keep the concrete being delivered from being washed out by the water in the pipe, a plug or a go-devil is placed at the upper level of the water in the pipe before the concrete is discharged into the tremie. Therefore, when the concrete is discharged into the tremie, the plug or go-devil is between the water and the fresh concrete, preventing the fresh concrete from being washed out.

The weight of the concrete (approximately 140 lb/ft^3) pushes the water, which weighs approximately 64 lb/ft^3, down the tremie pipe. When the concrete, which has never been in contact with the water, arrives at the bottom, it pushes the plug or go-devil out of the pipe and permits the concrete to be deposited on the bottom of the placement. Many go-devils are designed so that they are lighter than water and float to the top to be used at a later time. Once the fresh concrete begins to flow from the end of the tremie, the end of the pipe should stay submerged in the concrete.

Continuing to deposit fresh concrete into the tremie hopper causes the concrete to continue flowing through the pipe. The end of the tremie pipe should be kept at a depth in the freshly deposited concrete that permits the concrete to flow at a slow speed through the pipe.

If the concrete flows out of the pipe so that the concrete load in the pipe is less than the water load outside the pipe, and if the pipe is not adequately embedded in the freshly placed concrete, the water may refill the pipe. This will not happen if the end of the tremie pipe is adequately embedded into the freshly placed concrete. If the crane operator holding the tremie and hopper feels that the fresh concrete is being completely discharged onto the bottom, he or she should immediately drop the pipe so that it will more deeply embedded. This may prevent the water from getting into the empty pipe.

If water does not get into the empty pipe, concrete placement can continue; but if water does get into the pipe, it would be necessary to restart the placement by reinserting the plug or go-devil as during the placement initiation. An ideal tremie placement is one where the initial concrete placed is the concrete that ends up on top. This will only occur when the placement area is relatively small (say 4 ft^2).

The dry-pipe method requires a tremie pipe that is watertight, including the sectional pipe joints. In this method, a pressure-seal plate is attached to the bottom of the tremie pipe in such a way that the water pressure makes the end completely watertight and the interior of the pipe is completely empty and free of any water. Such a method requires that the walls of the pipe be heavy enough to overcome the buoyancy of the water. Otherwise, it would be impossible to lower the tremie pipe to the bottom of the placement.

Once the end of the tremie pipe with a pressure-seal plate is placed on the bottom, freshly mixed concrete can be introduced into the pipe. After the pipe is sufficiently filled with concrete, the pipe can be slowly raised, which releases the pressure-seal plate owing to the weight of the concrete. After the initial discharge of fresh concrete on the bottom, the placement continues exactly as in the wet-pipe method.

One drawback to the dry-pipe method is that the pressure-seal plate has to be left in the placement. However, it can be retrieved if a rope or cable is attached in such a way that the cable is on the outside of the tremie pipe. An other technical drawback is that should the tremie seal be lost (water gets into the pipe), a plug or go-devil would be needed to reinitiate the start of the placement.

15.3.7 Mixtures for Underwater Placements

The concrete needs to be proportioned for very workable concrete if concrete is to be placed underwater. The slump should be controlled at approximately 7 inches. Normally, the hydraulic cement content should be around seven bags per cubic yard. The maximum size aggregate should be 1 1/2 to 2 in and the fine aggregate (fine) content should be around 45% of the total aggregate content. The concrete should be air entrained at about 6 to 7%.

Any application that improves the workability of concrete should be considered. This includes pozzolans, natural aggregates in lieu of crushed stone, and use of chemical admixtures to extend the setting time and permit additional water reduction.

15.3.8 Use of Antiwashout Admixtures [ACI 304.1R, ASTM C938, ASTM C937, ACI 524R] (Neeley)

Many of the companies dealing with the manufacture of concrete construction materials have developed admixtures for use in concrete that permit the concrete to be placed under water without the use of a tremie. These materials are referred to as antiwashout admixtures. Japan has been a leader in this new concrete technology; although it is believed that Japan obtained its knowledge from a product that was originally used in Germany.

Figure 15.1 depicts the clearness with which concrete containing an antiwashout admixture can be discharged under water. Other terms for this type of concrete are nondispersible concrete and colloidal underwater concrete. The admixture that provides the clearness of this special concrete is known as nondispersible underwater concrete admixture, and in the United States, it is referred to as an antiwashout admixture. This innovative admixture was developed in West Germany around 1981.

The admixture is intended to prevent washout of cementitious material and dispersion of aggregate during underwater placement of concrete. The admixture serves to increase the viscosity and the water retention of the concrete matrix. The antiwashout admixtures presently being marketed in Japan use cellulose or acrylic as the primary ingredient. Admixtures containing acrylic use a polyacrylamide polymer as the primary ingredient.

Admixtures containing cellulose use a nonionic water-soluble cellulose ether, which has an OH (hydroxide ion) base and is almost like water. HEC (hydroxyethylcellulose), HEMC (hydroxyethyl-methylcellulose, and HPMC (hydroxypropylmethylcellulose) are among the various admixtures used. When dissolved, their viscosities differ considerably according to polymerization, molecular weight, and type of substituent. They dissolve in water rapidly when placed in a high-pH environment like concrete. They are also not susceptible to chemical changes within concrete like reaction, gelation, or decomposition.

15.3.9 Characteristics of Antiwashout Underwater Concrete

Antiwashout underwater concretes have slightly different properties than ordinary hydraulic cement concrete because of the effect of the admixture. Fresh antiwashout concrete can be characterized by the following properties.

15.3.9.1 Flowability

Owing to the increased viscosity of antiwashout underwater concrete, the slump transformation takes place over several minutes. The slump is ultimately 8 to 10 in. To have a better understanding of the flowability of this type of concrete, a slump-flow value or a spread value determined by the German Standard DIN 1048 is more suitable than a slump value. The relationship of these values is demonstrated in Figure 15.3. Table 15.1 provides criteria for the relationship between flowability and conditions of execution.

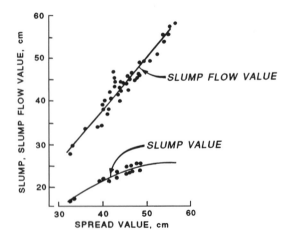

FIGURE 15.3   Relationship of slump, slump-flow value, and spread value.

TABLE 15.1 Criteria of Relationship Between Flowability and Conditions of Execution

Slump Flow Value, cm	Softness	Conditions for Applications	Conditions for Execution
40	Hard consistency	When it is desired to keep the flow small, such as the execution of a slanted path	Concrete pump pressure transmission boundary
45	Medium consistency	General case	Less than 50 m Concrete pump pressure transmission distance
50	Medium soft consistency	When excellent filling capability is needed	Concrete pump pressure transmission distance is 50–200 m
55	Soft consistency (Plastic concrete)	When excellent flowability is especially needed such as in reinforced concrete members of dense fiber and filler for narrow and deep supersoft consistency holes	
	Supersoft consistency		

15.3.9.2 Air Content

Mortar and concrete mixed with cellulose ether have greatly increased air content. Therefore such antiwashout admixtures contain an air-detraining admixture to reduce the air content of the concrete to between 3 and 5%. From a petrographic standpoint, the bubble-spacing factor of concrete containing the antiwashout admixture is about the same as concrete without the admixture, but the freezing and thawing resistance tends to be somewhat low.

15.3.9.3 Bleeding

Concrete containing the antiwashout admixture retains more of the mixing water. Since the normal amount of admixture used is more than double the amount needed to prevent bleeding, very little, if any, bleeding occurs in antiwashout underwater concrete. This lack of bleeding is responsible for the small reduction in quality of the concrete and increases the need for reinforcing steel.

15.3.9.4 Setting Time

The use of antiwashout cellulose admixtures affects the setting time of underwater concrete. When a cellulose antiwashout admixture is used, the setting time [ASTM C 191] is greatly extended. Therefore the antiwashout admixture contains an accelerating admixture. The most common accelerating admixture amounts are adjusted to result in a final setting time from 5 to 12 h. Antiwashout admixtures containing acrylic have no effect on the setting time.

When an air-entraining, water-reducing admixture is added to the antiwashout admixture, the setting time is slightly extended, but the increase in setting time for the normal admixture amounts is less than the 5 h. There are specialty extended setting time admixtures that can extend the setting time for underwater antiwashout concrete by 30 h or more.

15.3.9.5 Underwater Dispersion Resistance

The dispersion resistance of concrete during an underwater placement operation is evaluated by such tests as the cementitious materials outflow rate, the change of water permeation rate, the turbidity of the water, the change of pH value, and the change of composition. The rate of dispersion is decreased as the quantity of antiwashout admixture in the underwater concrete is increased.

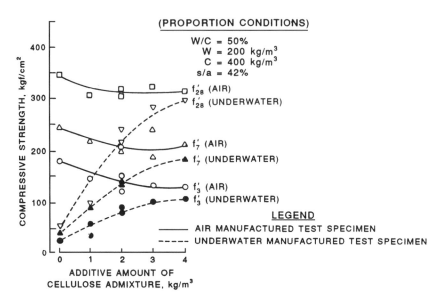

FIGURE 15.4   Relationship of quantity of cellulose admixture and concrete compressive strength.

15.3.10 Characteristics of the Hardened Concrete

15.3.10.1 Compressive Strength

The relationship between the compressive strength of concrete containing no antiwashout admixture and concretes containing various amounts of cellulose antiwashout admixture is shown in Figure 15.4. Generally, the compressive strength of a test specimen fabricated in air is lowered by an increase in the quantity of admixture, but there are instances in which compressive strength has increased slightly. Test specimens fabricated under water are made by placing concrete into water that is 12 to 20 in deep. The compressive strength of such specimens increases with an increase in the quantity of antiwashout admixture used. Consequently, the compressive-strength ratio of test specimens made under water increases compared to those made in air as the quantity of admixture increases.

The amount of admixture to be used is determined by the flowability needed, depth of the underwater placement, horizontal flow distance, desired water cementitious materials ratio, and, of course, the quantity of cementitious materials to be used. In general, the compressive strength ratio referenced above can be expected to range from 0.8 to 0.9.

15.3.10.2 Miscellaneous Strength and Other Characteristics

The ratio of tensile strength and flexural strength to compressive strength of an underwater fabricated test specimen is identical to that of specimens fabricated in ordinary dry concrete. The modulus of elasticity is the same or slightly less than that of ordinary concrete. The unit volume of water in antiwashout underwater concrete is much greater than that of ordinary concrete. Since water retention is so high, drying shrinkage is large at 20 to 35%. Also, creep in air appears to be somewhat greater than in ordinary concrete.

15.3.11 Characteristics of the Horizontal Flow Time of Nondispersible Underwater Concrete

Qualitative changes in antiwashout underwater concrete can be provided by the addition of a water-reducing admixture and cause the concrete to flow longer distances. The underwater concrete with water-reducing admixtures had a slump flow of 50 to 60 cm, a cement content of 364 to 430 kg/m^3, and a water/cement ratio of 0.48 to 0.60. In all of the test results, the final flow gradient was 1/125 to 1/500. Even though the concrete surface was virtually horizontal, qualitative changes were recognizable when the flow distance exceeded 10 m. The area near the edge of the concrete may suffer a drop in unit weight and modulus of elasticity as well as in compressive strength because the quantity of aggregate declines. The greatest flow distance is best determined by fully considering proportions and placement conditions.

TABLE 15.2 Combinations of Antiwashout and Water-Reducing Admixtures

Antiwashout Admixtures	Water-Reducing Admixtures
Cellulose	Melamine sulfonate (triazine)
Acrylic	Naphthalene Sulfonate
	Melamine sulfonate (triazine)
	Acrylic
	Polycarbonic acid

15.3.12 Principal Considerations

(1) Since antiwashout underwater concrete has a high viscosity, the mixer load is increased by 25 to 50%. Therefore the capability of the mixer and the quantity of materials mixed need to be considered.

(2) Use of a water-reducing admixture causes a decline in dispersion resistance and some extension of the time of setting. There are instances wherein a specific flowability cannot be obtained by combining the water-reducing admixture and the antiwashout admixture. Therefore the types of water-reducing admixture and their recommended dosages need to be considered. Table 15.2 illustrates the various combinations of types of antiwashout admixtures and water-reducing admixtures ordinarily used.

(3) Since the dispersion resistance is high, blockage will occur in pump lines only if there is difficulty within the pressure transmission tube during the pumping pressure period. Qualitative changes in the concrete should not occur before or after the pressure is transmitted. However, owing to high viscosity, pressure transmission resistance is 2 to 4 times that of ordinary concrete. It has been reported that the pressure transmission capacity of the squeeze type pumps is inferior to that of the piston type pumps; this therefore has to be considered.

15.3.13 Summary

Antiwashout underwater concrete is being considered for use for many underwater structures and other large-scale projects. Under current conditions, several problems remain. These can be summarized by (1) differences in performance of the more than 10 kinds of antiwashout admixtures presently being marketed, (2) differences in mixing methods and placement methods used by various contractors, and (3) the antiwashout concrete's inappropriateness for use in above-water structures owing to its drying shrinkage and poor resistance to freezing and thawing. Therefore it is recommended that the engineer and contractor fully understand the quality of the antiwashout underwater concrete and the procedures involved in the placement of this relatively new and innovative material.

15.4 Vacuum Processing

15.4.1 General

The ACI report 116R defines vacuum concrete as concrete from which excess water and entrapped air are extracted by a vacuum process before hardening occurs. This process is administered by applying a vacuum to formed or unformed surfaces of ordinary concrete immediately or very soon after the concrete is placed. Additional compaction of the concrete is the primary result of vacuum processing. As the water and entrapped air is removed, the mortar is subjected to consolidation; the concrete becomes denser because approximately 40% of the water close to the surface has been removed.

Entrapped air is also removed, resulting in a concrete surface that is more resistant to freezing and thawing, especially if the concrete originally contained entrained air of at least 9% of the paste fraction. The air, being noncontinuous, is removed from the surface and not from the interior. The depth of water extracted and the amount of water removed depend on the coarseness of the mixture, mixture proportions, and the number of surfaces to which the vacuum is applied.

The depth of water extraction can, under good conditions, extend to 12 in and the amount of water extracted a few inches below the surface can be equal to 1/3 of the mixing water. Removal of an average of 20% of the water down to a depth of 6 in from the surface is common. The best results from vacuum processing occur when (1) the mixture contains a minimum amount of fines, (2) the vacuum can be applied promptly while the concrete is still plastic, and (3) the concrete near the vacuum panel can be vibrated during the first few minutes of the vacuum treatment. Vacuum procedures result in concrete with higher strength and greater durability. At the Bureau of Reclamation, vacuum processing increased the 3-d strength of one concrete from 800 psi to 1,800 psi.

Although vacuum processing improves the surface of hardened concretes, this improved appearance is normally not adequate justification for use of the process. Vacuum processing improves durability, but this, should not be used as justification not to use air entrainment. Vacuum treatment has been used to increase the resistance of concrete surfaces to high velocity liquid flow, but its use should not justify reduced efforts to perfect alinement of flow lines.

15.4.2 Concrete Mixtures

It is not absolutely necessary that special concrete mixtures be used when vacuum processing of the concrete surface is planned. This is not to say that slight changes in the normal mixtures should not be considered. The best results are obtained when the fines are at a minimum; in other words, vacuum processing seems to work best when the mixture is relatively lean and contains a minimum amount of fine aggregate (sand) that is on the coarse side of the grading. Sticky mixtures with an excess of fines do not respond well to vacuum treatment. The treatment is also more effective at low ambient temperatures.

One of the primary goals of concrete construction is to place concrete in a uniform fashion so that the finished results will be uniform. The vacuum should be applied soon after placing the concrete while the concrete is still plastic. The concrete should be highly workable, and during the first few minutes of the vacuuming process the concrete should be slightly vibrated, permitting the small channels left by the water removal to close, thus improving the water tightness.

15.4.3 Early Equipment

During the early use of the vacuum process, the vacuum was applied to the concrete surface by vacuum hoses attached to special vacuum mats or form panels. The mats for unformed surfaces were usually reinforced plywood faced with two layers of screen wire covered by muslin. For unformed curved surfaces, such as the buckets of dams, flexible steel that could adapt to the curved surface was used in lieu of plywood. Sometimes a fiberglass cloth without screen backing was used for the lining of steel forms for concrete pipe. The equipment was very cumbersome and required much preparation time. In addition, it was also expensive and complicated. As a result, the use of the equipment was not very inviting, and very little use was made of the process.

15.4.4 New Equipment

Within the last few years, the vacuum equipment has been simplified and its use, especially in Europe, has greatly increased. The vacuum process is now being used more frequently in the

FIGURE 15.5   Distribution of vacuum pads on newly placed
concrete slab.

United States. The panels have been replaced by vacuum pads that are flexible, light, and easy to
handle (see Figure 15.5). This newer system greatly improves the handling, and the system is much
more efficient and cost effective.

15.4.5 Procedure

After the concrete slab is vibrated, the surface is immediately covered with a base filter pad and a
suction mat connected to a vacuum pump. The pump (see Figure 15.6) creates a vacuum under
the pad. The vacuum causes the atmospheric pressure to compress the concrete and the water is
extracted. The base pad is designed to distribute the vacuum under the entire surface evenly and
permit water to pass through. The vacuum is applied for approximately 4 min for each inch of slab
thickness. After the pads and mats are removed, the surface is normally firm enough to walk on
and finishing is then performed with specially designed low-amplitude, high-frequency disk floats.
Following finishing, the slabs are cured with water, preferably, but practically any method that
prevents the concrete from additional drying can be used. Since internal water has been removed
from the concrete, it is best to assure that adequate water is available for hydration of the cement;
consequently, water curing is best.

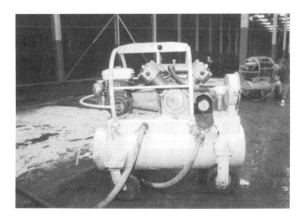

FIGURE 15.6   Vacuum pump used in vacuum processing.

15.4.6 Conclusions

For attaining a highly impervious concrete floor, vacuum processing is a viable method; but like all processing methods, options such as using low water/cement ratios and adequately air-entrained concrete and incorporating proper curing and protection techniques need to be considered during construction. Do not depend 100% on vacuum processing to provide the high-quality, low-permeability, smooth floor that you desire; quality concrete materials must be used, and proven mixing, transporting, placing, finishing, curing, and protection procedures also have to be followed.

15.5 Portland Cement Plaster Construction

15.5.1 General

Portland cement plaster construction, commonly referred to as stucco has been around for many years, but it has normally been considered an art rather than a science. The knowledge necessary to apply Portland cement plaster was previously acquired by learning from others and not from technical literature. The ACI Report 524R, *Guide for Portland Cement Plaster*, was developed in 1992. Previously, little information was available to engineers who wished to include this subject in project specifications. Consequently, the U.S. Army Corps of Engineers initiated worldwide studies to identify the problems associated with defective concretes and Portland cement plaster construction.

15.5.2 Most Frequently Found Problems [ACI 524R]

During visits to the various sites around the world, numerous installations were inspected where the design and the workmanship for stucco were excellent. The recurring problems that were found can be separated into three categories:

(1) questionable design,
(2) incomplete specifications or specifications lacking in detail, and
(3) inadequate inspection and poor workmanship.

Owing to these recurring categories, the following general procedures need to be fully considered when contemplating a Portland cement plaster construction project:

(a) Quality plaster is essential to any successful installation. The plaster must develop adequate tensile strength to resist imposed stress and have sufficient resiliency to accommodate expansion and contraction. Consistency in the batching operation is as important to the development of quality plaster as the ingredients and quantities.
(b) The most important ingredient is the aggregate. Aggregate should conform to specifications. The physical properties of aggregate that have the most pronounced effect on plaster are grading, shape and denseness of the particles, and particle surface characteristics (roughness and porosity).
(c) Curing procedures play a vital role in reducing shrinkage cracking by permitting the plaster to dry slowly and uniformly. Fog curing requires the application of a fine mist at intervals related to job conditions. The purpose of curing is to maintain enough water within the plaster to keep the interior relative humidity above 80% during the specified curing period.
(d) It is acceptable to place a second coat of plaster as soon as the first coat is strong enough to withstand the pressure of the second application. When plaster is applied to a solid backing such as block, concrete, or wire lath backed with rigid sheathing, both base coats can be

applied in one day and the finish coat on the following day. Or successive coats can be applied on consecutive days.

15.5.3 Technical Aspects of Portland Cement Plaster

The desirable properties of fresh Portland cement plaster can be summarized by the following properties:

(a) The ability to stick to the particular substrate. The primary concerns in this area relate to the influence of the aggregate, the water/cement ratio, and the absorptive characteristics of the substrate or base.

(b) The ability of the fresh plaster to stick to itself. To reflect this ability, the plaster should not sag, slough, or separate (delaminate).

(c) The ability to be placed, shaped, floated, and tooled. To reflect this ability, the plaster should already have the first two properties. Plaster without these abilities is generally incorrectly proportioned or possibly, incorrectly mixed.

Hardened Portland cement plaster should have excellent durability against weathering, should be highly impermeable, and should be resistant to temperature changes. Such plaster should also be highly resistant to the action of freezing and thawing. Plaster should be air entrained for better freeze thaw resistance and better impermeability, which provides protection from acid rain and aggressive chemicals. Hardened Portland cement plaster should also be proportioned for high tensile strength. Properly proportioned plasters that have been properly cured should have acceptable tensile strength.

15.5.4 Portland Cement Plaster Materials

The cement used in Portland cement plaster may be practically any type or class of cementitious materials conforming to the various ASTM Standards, such as

(1) Portland cement, ASTM C150,
(2) blended cements, ASTM C595,
(3) masonry cement, ASTM C91, and
(4) plastic cement, ASTM C926.

Should the aggregates be potentially reactive, low-alkali cements should be used, and air-entraining cements should be used, when possible.

Lime conforming to the requirements of Type S, ASTM C206 or C207 should be used along with an air-entraining admixture when possible. Lime is necessary only when regular cement is used.

The aggregates should be either natural of manufactured fine aggregate (sand) complying with the requirements of ASTM C897. Lightweight aggregates such as perlite or vermiculite may be used, but should not be used in base courses when conventional-weight aggregate plaster is to be applied as a finish coat. Perlite or vermiculite aggregates have low resistance to freezing and thawing. Sand should be washed clean, should be free of organic matter, clay, and loam, and should be well graded.

The water used should be as good as water used in hydraulic cement concrete. Drinking water is normally okay. If there is a concern about the water, it should comply with the requirements of ASTM C191 for setting time and ASTM C109 for strength.

Calcium chloride should not be used in Portland cement plaster because of the embedded metals. Should chemical admixtures be considered for use, make sure that they do not contain chlorides and are noncorrosive. Accelerating admixtures that do not contain chlorides or other corrosive materials are available and could be used.

15.5.5 Proportioning

As stated earlier, Portland cement plastering (stucco) is thought of as an art and not a science; therefore, the industry doesn't seem to want to join the twenty-first century; materials are still proportioned by shovel. Measurement of sand is accomplished by counting the number of shovels of sand per bag of cement and seven No. 2 shovels are equated to one cubic foot of sand.

The quantity of water is determined by the appearance of the plaster in the mixer. Some project specifications require the use of a cone, that is, 6 in high by 4 in in diameter at the bottom and 2 in in diameter at the top, for measuring the slump. Many specifications permit a slump of 1 1/2 to 3 in for either hand- or gun-applied plaster. Plasticisers are also normally required, and again the quantity is determined by the appearance. When plastic or masonry cements are used, the addition of plasticisers is not necessary.

When Portland or blended cements are used, it may be necessary to add plasticizer to up to 20% by weight of the cementitious material. Avoid sloppy and overwatered mixtures; they tend to cause segregation and separation of materials. Proportioning Portland cement plaster drastically affects the final quality and serviceability of the hardened plaster. Proportions of the ingredients should be in accordance with project specifications, local building codes, and ASTM C926.

15.5.6 Mixing

Experience dictates that a particular sequence for mixing should be followed: The water should be added first, followed by 50% of the sand, the cement and any admixtures, and finally the remaining 50% of the sand. Normal mixing time is approximately 3 to 10 min. Excessive mixing should be avoided because it could be detrimental to the quality of the plaster.

15.5.7 Bases for Plaster

Metal plaster bases come in three types: There is woven wire plaster base, which is fabricated galvanized steel wire that is reverse twisted into a hexagonal mesh pattern and normally comes in rolls or sheets. Then there is the expanded-metal lath diamond mesh, which is fabricated from coils of steel that are slit and then expanded to form a diamond pattern (chicken wire). The third type is welded-wire lath, which is fabricated from at least 15-gage copper-bearing, cold-drawn galvanized steel wire.

15.5.8 Weather Barrier Backing

There are a number of materials being used as weather bearing backing, including waterproof paper or felt meeting the requirements of Federal Specifications UU-B-790, Type II, Class D. The paper should be free of holes and breaks and should weigh at least 14 lb per 108 ft^2 roll.

A large number of accessories are needed in order to obtain a proper plaster job. Accessories establish plaster grounds and transfer stresses in critical areas of plaster elements. Portland cement plaster should not be considered to be part of the load bearing members; the plastering project should be designed so that the plaster is not placed under stress.

In order to construct a successful plaster project, it is necessary that all locations, that is, corners and joints (expansion and contraction), contain the correct accessories to prevent the plaster from experiencing the normal stress that would occur at these locations. Special accessories are made for corners; they may be expanded flange corner beads, welded or woven wire, vinyl bead, or expanded-metal corner lath. The corner reinforcement must be designed so that plaster can be applied without hollow areas. Inside corner joints must have accessories designed to provide stress relief at internal angles.

Casing beads, often called plaster stops, should be installed wherever plaster terminates or joins a dissimilar material. Plaster screeds are used to establish the thickness of the plaster or to create decorative motifs. Ventilating screeds contain perforated webs that permit air to pass freely from the outside. Additional screeds include drip screeds that are installed on outside plaster ceilings to prevent the water that runs down the face of the structure from penetrating the plaster soffits and the ceiling. Weep screeds are normally installed at the foundation plate line and function as a plaster stop, permitting trapped moisture to escape from the space between the backing paper and the plaster.

15.5.9 Sample Panels

Sample panels or mock-ups should always be constructed, especially on large jobs where all the plastering will not be performed by one crew of plasterers. These panels or mock-ups should include examples of all joints, windows, doors, corners and, in general, all conditions to which the plaster will be exposed. Sample panels should be constructed until the results describe the desired quality and appearance expected of the structure. The panels should be kept close to the project so that workers and supervisors can verify that the previously approved results are being obtained.

15.5.10 Surface Preparation

Concrete surfaces to which plaster will be applied should be straight, true to line, and plane. Concrete surfaces should be cleaned or roughened to increase the likelihood of a good chemical and mechanical bond. The concrete surface should be cleaned with a cleaning agent. Such cleaning should remove most surface contaminates. Other methods of cleaning might include the use of wire brushes, hammers or chisels, water blasting, or light sandblasting.

15.5.11 Application of Plaster

Prior to starting the application of plaster, it is necessary that the substrate be prepared as described above. It is also necessary to verify that the lath and backup paper has been installed along with the necessary accessories. After the required substrate treatment has been verified, proper application procedures need to be followed. The project specifications will normally require or permit the plaster to be applied by hand or by machine. When hand application is permitted, the plasterer applies the plaster to the surface using a trowel.

The plasterer is required to determine the correct amount of water needed for the plaster to be the correct consistency. Plaster pumps are sometimes required to for plaster application. In this case, batches of plaster are prepared in a mixer, and the individual performing the mixing operation is responsible for batching the correct quantities of cement, sand, and water. The quantity of water is determined by the mixer operator. The plaster is placed in a hopper and pumped onto the surface through a hose and nozzle. The nozzle operator controls the spray pattern by adjusting the air jet, air pressure, and nozzle orifice.

15.5.12 Types of Application

Portland cement plaster is normally applied in either two or three coats. The three coats are referred to as the scratch coat, brown coat, and finish coat. The scratch coat should be thick enough to result in a good bond to the substrate, and on substrates containing metal laths, this coat should fully cover the lath. The scratch coat should be scored horizontally so that mechanical bonding is improved. Should the specification require a delay between the application of the scratch coat and the brown coat, it is necessary to cure the scratch coat by moist curing. Should no delay be required, the brown coat should be applied as soon as possible in order to secure good bond. The brown coat should be applied when the scratch coat is rigid enough to receive the brown coat without cracking. The brown coat normally contains more sand then the scratch coat.

The required thicknesses are established either in the local specifications or by construction codes. In some cases, the brown coat is the finish coat. Where a finish coat is specifically required, it is normally proportioned to provide a particular texture, color, or appearance. Prior to applying the finish coat, it is necessary that the brown coat be moist cured for at least 2 d. Finish coats can be applied and finished in numerous aesthetically pleasing patterns. Smooth troweled surfaces are not recommended because of their tendency to crack. Finish coats must not be burnished; burnishing will almost always induce cracking. Moist curing is an absolute necessity.

References

ACI 116R. 1990. *Cement and Concrete Terminology.* American Concrete Institute, Farmington Hills, MI.

ACI 211.1. 1991. *Standard Practice for Selecting Proportions for Normal, Heavyweight, and Mass Concrete.* American Concrete Institute, Farmington Hills, MI.

ACI 304.1R. 1989. *Guide for the Use of Preplaced Aggregate Concrete for Structural and Mass Concrete Applications.* American Concrete Institute, Farmington Hills, MI.

ACI 524R. 1993. *Guide to Portland Cement Plastering.* American Concrete Institute, Farmington Hills, MI.

ASTM C91. 1995c. *Specifications for Masonry Cement.* American Society for Testing and Materials, Philadelphia, PA.

ASTM C109. 1995. *Test Method for Compressive Strength of Hydraulic Cement Mortars Using 2-in Cube Specimens.* American Society for Testing and Materials, Philadelphia, PA.

ASTM C150. 1995a. *Specifications for Portland Cement.* American Society for Testing and Materials, Philadelphia, PA.

ASTM C191. 1992. *Test Method for Time of Setting of Portland Cement by Vicat Needle.* American Society for Testing and Materials, Philadelphia, PA.

ASTM C206. 1984 (1992). *Specification for Finishing Hydraulic Lime.* American Society for Testing and Materials, Philadelphia, PA.

ASTM C207. 1991 (1992). *Specifications for Hydraulic Lime for Masonry Purposes.* American Society for Testing and Materials, Philadelphia, PA.

ASTM C231. 1991b. *Test Method for Air Content of Freshly Mixed Concrete by the Pressure Method.* American Society for Testing and Materials, Philadelphia, PA.

ASTM C595. 1995a. *Specifications for Blended Hydraulic Cement.* American Society for Testing and Materials, Philadelphia, PA.

ASTM C897. 1995a. *Specifications for Aggregate for Job-Mixed Portland Cement-Based Plaster.* American Society for Testing and Materials, Philadelphia, PA.

ASTM C926. 1995a. *Specifications for Application of Portland Cement-Based Plaster.* American Society for Testing and Materials, Philadelphia, PA.

ASTM C937. 1980 (1991). *Specifications for Grout Fluidifiers for Preplaced-Aggregate Concrete.* American Society for Testing and Materials, Philadelphia, PA.

ASTM C938. 1980 (1991). *Practice for Proportioning Grout Mixtures for Preplaced-Aggregate Concrete.* American Society for Testing and Materials, Philadelphia, PA.

ASTM C943. 1980 (1990). *Practice for Making Test Cylinders and Prisms for Determining Strength and Density of Preplaced-Aggregate Concrete in the Laboratory.* American Society for Testing and Materials, Philadelphia, PA.

Bureau of Reclamation. *Concrete Manual,* Revised 8th ed. U.S. Department of the Interior. United States Government Printing Office, Washington, DC.

Gerwick, B.C. 1988. Review of the state of the art for underwater repair using abrasion-resistant concrete. *Technical Report REMR-CS-19.* University of California, Berkeley, CA. Prepared for the U.S. Army Engineer Waterways Experiment Station, Vicksburg, MS.

Kawai, T. 1988. Special underwater concrete admixtures. *Concrete J.* 26(3):45–49.

Khayat, K. and Hester, W. Underwater repair of concrete slab. *Technical Report REMR-CS-13X (Temp. No.)*. University of California, Berkeley, CA.

Neeley, B.D. 1988. Evaluation of concrete mixtures for use in underwater repairs. *Technical Report REMR-CS-18*. U.S. Army Waterways Experiment Station, Vicksburg, MS.

Neeley, B.D. 1989. Antiwashout admixtures for use in underwater concrete placement. *Video Report REMR-CS-3*. U.S. Army Engineer Waterways Experiment Station, Vicksburg, MS.

Neeley, B.D. and Saucier, K.L. 1990. Laboratory evaluation of concrete mixtures and techniques for underwater repairs. *Technical Report REMR-CS-34*. U.S. Army Engineer Waterways Experiment Station, Vicksburg, MS.

Rail, R.D. and Haynes, H.H. Underwater stilling basin repair techniques using precast or prefabricated elements. *Technical Report REMR-CS118 (Temp. No.)*, Naval Civil Engineering Laboratory, Port Hueneme, CA. and Haynes and Associates, Oakland, CA.

Ribar, J.W. and Scanlon, J.M. 1984. How to avoid deficiencies in Portland cement plaster construction. *Technical Report SL-84-10*. Department of the Army, U.S. Army Corps of Engineers. Washington, DC.

Scanlon, J.M., McDonald, J.E., McAnear, C.L., Hart, E.D., Whalin, R.W., Williamson G.R., and Mahloch, J.L. 1983. *REMR Research Program Development Report*. Prepared for Office, Chief of Engineers, U.S. Army, Washington, DC.

U.S. Army Corps of Engineers. 1994. *Standard Practice for Concrete for Civil Works Structures*, EM 1110-2-2000. Washington, DC.

Waddell, J.J. and Dobrowolski, J.A. 1993. Preplaced-aggregate concrete. In *Concrete Construction Handbook*, 3rd ed., pp. 38.1–38.7. McGraw–Hill, New York.

(a)

(b)

(a) Repair of deteriorated beam underside (Courtesy Portland Cement Association). (b) Repair of deteriorated bridge element (Courtesy of Dr. Randall W. Poston).

16

Structural Concrete Repair: General Principles and a Case Study

by
Randall W. Poston, Ph.D., P.E.
Principal of Whitlock Dalrymple Poston & Associates, Inc., Manassas, Virginia. Expert in structural design both in reinforced and prestressed concrete and in retrofit of concrete structures.

16.1 Introduction

Concrete has been used as a construction material in a wide variety of structures ranging from buildings and parking structures to bridges, dams, earth-retaining structures, pressure vessels, tanks, boats, and offshore platforms. Today, it is the most widely used construction material. As with other forms of infrastructure, there is an ever-increasing need to evaluate and repair concrete structures. Repairs may be necessitated for a variety of reasons, ranging from a change in the space requirements of the structure to the deterioration of concrete and corrosion of the steel reinforcement. Although most concrete structures have good long-term performance records, deterioration problems have occurred because of poor construction practices, poor design, lack of quality materials, and aggressive environmental exposure.

Perhaps the most-cited statistic in the technical concrete literature over the past decade concerns the extent of deterioration of this country's roads and bridges, often referred to as the nation's highway infrastructure. The American Association of State Highway and Transportation Officials (AASHTO) rates 40% of U.S. highways as below minimum standards of engineering. According to the U.S. Department of Transportation, 230,000 out of the 575,000 bridges on primary and secondary roads are structurally deficient or functionally obsolete. The value of the concrete-based infrastructure in the United States is estimated to be $6 trillion according to the Civil Engineering Research Foundation (CERF). And as was so simply put in a recent article in Smithsonian [Wolkomir, 1994], "...a lot of that concrete needs fixing."

Repair of structural concrete, that is, repair of structures in which concrete is the principal load-carrying material, can be classified into three major categories: rehabilitation in which repairs are conducted to extend the service life of the structure; restoration in which repairs are effected to upgrade the functionality or possibly change the use of the structure or strengthening to restore the structure's intended load capacity or perhaps, as in the case of restoration, increase its capacity because of an anticipated change in use. As can be surmised, the boundaries between rehabilitation, restoration, and strengthening of concrete structures often overlap. The principal objectives of a repair are to ensure safety and structural integrity, extend useful life, improve aesthetics, change use, and/or to improve serviceability. The design of repairs must therefore address strength, durability, and serviceability analogous to that considered for design of a new structure.

It is not the intent of this chapter to present a comprehensive treatise of all of the aspects of the repair of concrete structures. There is detailed information in the literature about concrete preparation, repair materials, and construction execution. Rather the objective is to present the structural concepts concerning concrete repair that include evaluation procedures and general design concepts. The structural guidelines presented are generally applicable to all types of concrete structures including bridges, buildings, water-containing structures, parking garages, and special structures. A detailed case study of an infrastructure repair project is presented to demonstrate the importance of proper evaluation within the context of a repair project. This comprehensive case study also illustrates the various structural analysis and design techniques that are sometimes necessary to ensure a cost-effective solution in a large-scale infrastructure repair project.

16.2 Limit States Design for Repair

In order to repair a concrete structure, it is important to understand what has "failed." Perhaps one of the simplest definitions of what might constitute "failure" of a structure can be inferred from the definition of limit states given by the structural sage Professor J.G. MacGregor. Professor MacGregor (1976) stated: "When a structure becomes unfit for its intended use, it is said to have reached a limit state." Thus the inference is that when a structure can no longer perform in its intended function, this, in essence, constitutes "failure."

The three generally recognized limit states for concrete structures are

1. serviceability limit state,
2. ultimate limit state, and
3. durability limit state.

Although concrete building codes such as ACI 318-95 [American Concrete Institute, 1995] may not have code provisions and standards that explicitly guard against "failure" for all these limit-state categories, they do so implicitly.

When concrete structures suffer from premature deterioration caused by steel reinforcement, this is generally thought to affect their durability. A concrete structure designed for an intended life of 40 years could be considered to reach the durability limit state if severe corrosion and attendant distress was manifested, for example, in only 15 years. But as conceptually illustrated in Figure 16.1, it is difficult to separate the pernicious effects of corrosion on durability, serviceability, and strength since the deterioration that results from corrosion not only compromises durability, but clearly can also impair the serviceability and strength of the structure. In fact, there is a confluence of limit states with regard to the effects of deterioration in structural concrete. It is, therefore, necessary to design concrete repairs to satisfy all three limit states.

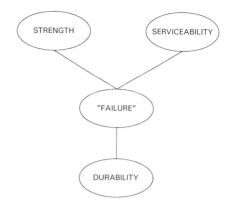

FIGURE 16.1   Limit states in structural concrete repair.

16.3 Evaluation

Prior to undertaking repairs to concrete structures, a condition assessment to determine the existing conditions and nature and causes of observed distress must be conducted. Otherwise, absence of this information could lead to a high risk of failure of the repair. The extent of the condition assessment depends largely on the objectives of the repairs. The factors that affect the repair objectives include safety and structural integrity, the desired service-life extension, change in intended use or loading requirements, serviceability, aesthetics, and cost. Internationally recognized organizations, such as ASCE and ACI, have published guidelines for condition assessment of structures [ASCE, 1990; ACI, 1993].

The Strategic Highway Research Program (SHRP) was established in 1987 by Congress to improve the performance and durability of the nation's highway systems [Transportation Research Board, 1986]. As part of this effort, a number of nondestructive evaluation (NDE) tools were developed that are currently being implemented by states and local government agencies, industry, and private sectors. Some of the methods developed by SHRP include alkali-silica reactivity (ASR) test methods and a rapid method for determining chloride content and rate of corrosion of steel reinforcement in hardened concrete. These along with other existing and emerging NDE techniques aid in diagnosing problems, specifying repairs, and quantifying the extent of adverse conditions and deterioration.

A condition assessment of a concrete structure may be needed even if repairs are not contemplated. In some cases, condition assessment is required for determining the load rating of a structure. Moreover, a condition assessment may be conducted in an on-going inspection program simply to show that a structure is structurally sound. In fact the Federal Highway Administration (FHWA) has mandated biennial inspections for bridges [U.S. Department of Transportation, 1991].

Traditionally, a condition assessment of a concrete structure generally meant a visual examination of the structure along with select coring of various concrete elements for compressive strength testing, perhaps a petrographic analysis, and sounding by chain drag. Although these techniques are still helpful and largely necessary, over the last decade or more other very valuable and practical nondestructive evaluation methods have been developed. These newer-generation nondestructive methods greatly improve and increase the efficiency of the condition assessment process for concrete structures. Some of these recently developed nondestructive evaluation technologies allow for relatively rapid inspection of damage and deterioration of concrete structures.

The objectives of this section are to summarize selected nondestructive evaluation methods that can be used as an aid in conducting condition assessment of concrete structures, highlight some of the more recently developed methods, and discuss the relative merits of various NDE methods.

For our purposes, NDE is defined as any test or method that yields information regarding the quality of a structure or portion thereof that does not impair the serviceability of the element or structure. Therefore destructive testing in the sense of removing samples from a structure or element, such as coring, is considered a nondestructive test for the purposes of this discussion. The techniques and methods discussed are generally applicable to all types of concrete structures.

16.3.1 Information Gathering

Prior to the field-investigation component of the evaluation, it is important to gather as much information as possible related to the original design and construction. This includes gathering structural and shop drawings project specifications, and construction records. Records such as concrete batch tickets, test reports, and weather and daily logs can be particularly useful if there is observed deterioration, cracking, or other types of service-performance problems. These types of original construction documents should be perused to ascertain if existing conditions are related to original construction problems or perhaps to design deficiencies.

16.3.2 Visual Inspection

Despite the advances in nondestructive evaluation methods, visual inspection is still an integral part of a condition assessment and, in fact, is the most widely used form of NDE.

The forms of distress and deterioration that should be recorded in the field evaluation and some of the associated causes are summarized in Table 16.1. In some select cases, the evaluation may be concluded with a simple visual inspection. However, in more cases than not, some form of nondestructive testing (NDT) is required to determine the cause(s) of observed deterioration, assess its implications and the extent to which it affects structural integrity, and develop appropriate repair and maintenance strategies to meet the client's budget.

16.3.3 Testing

A testing program should be developed to determine the types, location, and size of material samples consistent with the size of the structure so that the results are statistically meaningful. For example, determining chloride-ion content at only one or two locations in a bridge deck will not provide sufficient data to draw statistically significant conclusions. Table 16.2 summarizes

TABLE 16.1 Forms of Concrete Distress and Deterioration to be Noted in a Visual Condition Assessment

Description	Typical Causes
Cracking	Plastic shrinkage
	Drying shrinkage
	Restraint
	Subgrade support deficiencies
	Vapor barrier
	Expansion
	Corrosion of reinforcing steel/ prestressing steel or other embedded metal components
	Thermal loading
	Vehicular impact
	Overloading
	Aggregate reaction
Scaling	Inadequate air content
	Finishing problems
	Freeze-thaw cycling
	Chemical deicers
Spalling	Aggregate reaction
	Corrosion
	Freeze-thaw cycling
	Construction
	Poor preparation of construction joints
	Early-age loading
Disintegration	Frozen concrete
	Freeze-thaw cycling
	Low strength
	Chemical attack
	Sulfate attack
Honeycombing and surface voids	Poor placement
	Poor consolidation
	Congested reinforcement
Discoloration and staining	Different cement production
	Different water-cement ratios
	Corrosion
	Aggregates
	Use of calcium chloride
	Curing
	Finishing
	Nonuniform absorption of forms
Efflorescence	Calcium carbonate and other mineral deposits

TABLE 16.2   Selected Tests for Condition Assessment of Concrete Structures

Mechanical/Physical/ Chemical Property	Test Type(s)	Reason for Test
Compressive strength	Swiss hammer (ASTM 805) Windsor probe (ASTM C803) Core for compression testing (ASTM C42) Ultrasonic pulse velocity (UPV) (ASTM 597)	Strength of in-place concrete; comparison of concrete strength in different locations (UPV and Swiss Hammer provide relative differences in strength only)
Reinforcement location	Pachometer X-ray Radar (ASTM D4748)	Steel location and distribution; concrete cover
Corrosion potentials	Half-cell potential (ASTM C876) Linear Polarization (SHRP S-330)	Identification of location of active reinforcement corrosion; corrosion rate
Chloride ion content	Acid-soluble and water-soluble titration (AASHTO T-260) Specific ion probe (SHRP-S/FR-92-108)	Susceptibility of steel reinforcement to chloride-induced corrosion
pH	Phenolphthalein; direct measurement w/pH meter	Assess corrosion protection value of concrete with depth and susceptibility of steel reinforcement to corrosion; depth of carbonation
Air content; Cement and aggregate properties; scaling; Alkali-silica reactivity; and Freeze/thaw susceptibility	Petrographic examination of concrete core removed from structure (ASTM C856) (ASTM C457)	Assist in determination of cause(s) of distress; degree of damage; quality of concrete when originally cast
Permeability	Electrical indication of concretes Ability to resist chloride ion penetration (ASTM C1202) (AASHTO T277) Resistance of concrete to chloride ion penetration (90-day ponding test) (AASHTO T259) Absorption test (ASTM C642)	Establish relative susceptibility of concrete to chloride ion intrusion; assess effectiveness of chemical sealers, membranes, and overlays in repair
Alkali-silica reactivity (ASR)	SHRP rapid test (SHRP-C/FR-91-101)	Establish in field if observed deterioration is due to ASR
Location of delaminations, voids, and other hidden defects	Limited information from sounding (ASTM D4580); impact-echo; Infrared thermography (ASTM D4788); pulse echo radar (ASTM D4748)	Assessment of reduced structural properties; extent and location of unobserved damage and defects
Steel area reduction; defect identification	Invasive probing	Observe and measure rust and area reduction in steel; observe corrosion of embedded post-tensioning components; verify location and extent of deterioration; provide more certainty in capacity calculations.
Concrete component thickness	Impact-echo radar (ASTM D4748); Invasive probing	Verify thickness of concrete; provide more certainty in capacity calculations.
Local or global strength and behavior	Load test Strain measurements Acceleration Deformation Displacement measurements	Uncertainty in integrity and behavior; ascertain acceptability without repair or strengthening; determine accurate load rating

(continues)

TABLE 16.2 *(continued)*

Mechanical/Physical/ Chemical Property	Test Type(s)	Reason for Test
Tensile strength	Pull-off tests (ACI 503R) Splitting tests (ASTM C496) Tension tests Bond test	Assess tensile strength of concrete and steel; relative quality of material
Material property determination	Density (ASTM C642) Moisture content (ASTM C642) Shrinkage (ASTM C341) Dynamic modulus (ASTM C215) Modulus of elasticity (ASTM C469)	Determine mechanical properties of materials and volumetric properties

Note: ASTM refers to American Society for Testing and Materials Testing Standards. AASHTO refers to American Association of State Highway and Transportation Officials Testing Standards. SHRP refers to test methods developed as part of the Strategic Highway Research Program.

some methods of field and laboratory testing that may be conducted to aid in the condition assessment of a concrete structure. Selection of the appropriate test method to determine a particular characteristic depends on several factors, including desired sensitivity and cost. At a minimum, the concrete strength and condition and location of the reinforcement should be assessed in order to generally characterize the structure's strength, safety, and integrity. Other testing is then conducted to determine the extent of deterioration and to establish causes. Establishing the cause of deterioration is an important element in developing appropriate repair strategies within prescribed budget limitations.

Most of the test methods for evaluating concrete structures described in Table 16.2 are well documented in the literature. Because of their relative newness, two of the test methods, impact-echo and the SHRP rapid test for detecting alkali-silica reactivity in concrete, will be briefly described to illustrate the power and versatility of some of the emerging technologies.

16.3.3.1 Impact-Echo

In the impact-echo technique, a transient stress pulse is introduced into a test object by mechanical impact on the surface as illustrated in Figure 16.2. The stress pulse propagates into the object along spherical wave fronts as P- and S-waves and along the surface of the object along a circular wave front as an R-wave. The P- and S-waves are reflected by internal cracks or interfaces and by the external boundaries of the object. The arrival of these reflected waves at the surface where the impact was generated produces displacements that are monitored by a transducer. If the transducer is placed close to the impact point, the waveform is dominated by displacements caused by P-wave arrivals.

Because it is difficult and time consuming to analyze time-domain waveforms in order to determine the arrival times of reflected waves and thereby calculate the depth of reflecting interfaces, waveforms are transformed into the frequency domain using the fast Fourier transform (FFT) technique [Sansalone and Carino, 1986, 1988]. The resulting amplitude spectrum is used to identify the dominant frequencies present in the waveform. Because these frequencies are produced by multiple wave reflections between interfaces, they can be used to determine if a structure is solid or if flaws exist within the structure. Each type of structure—plate (bridge deck, slab), bar (beam, column), hollow cylinder (pipe, shaft line), etc.—exhibits a characteristic frequency response when subjected to impact. The presence of a flaw affects this response.

In a plate, the impact-echo is dominated by reflections between the top and bottom surface of the plate and/or from a defect if one exists. Knowing the P-wave speed (to determine the P-wave speed, a test must be carried out over a portion of the test object of known thickness) in the test object, C_p, the depth T to a reflecting interface can be calculated as

$$T = \frac{C_p}{2f_p} \qquad (16.1)$$

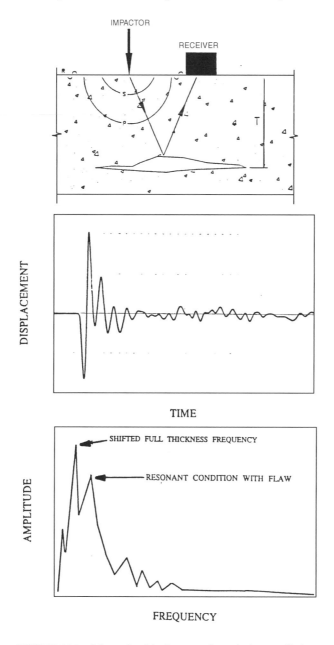

FIGURE 16.2   Schematic of the impact-echo technique applied to
a concrete structure.

where f_p is the frequency of P-wave reflections from the interface. In plates, wave reflections from the side boundaries do not have a significant effect on the response.

An impact-echo test system is composed of three components: an impact source, a receiving transducer, and a waveform analyzer or a portable computer with a data acquisition card that can sample at a frequency of at least 500 kHz. The duration of the impact determines the frequency content of the stress waves generated by the impact. Impacts with shorter durations contain higher frequency (shorter wavelength) components, which are needed for testing thinner structures or detecting smaller or shallow flaws. Hardened steel spheres on spring steel rods are used as impact sources. The impact durations produced by impactors generally range from about 10 to 80 μs.

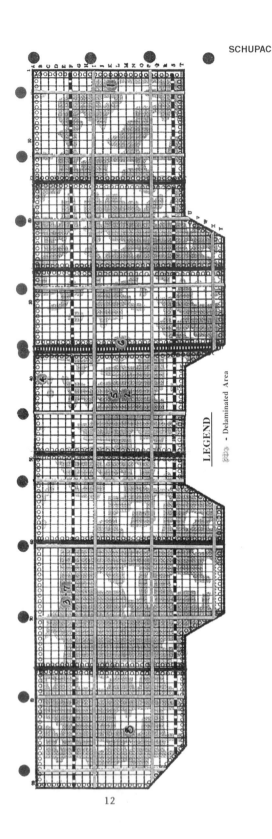

FIGURE 16.3   Results from impact-echo testing on an industrial slab showing areas of delamination.

FIGURE 16.4   Field unit used to detect ASR in concrete.

Figure 16.3 shows the impact-echo test results from a 20,000 ft^2 industrial concrete overlay delaminated from a structural slab at an industrial facility. Impact-echo was used to identify the delaminated repairs, which appear as dark areas in the map. The test results facilitated the selective removal of the delaminated areas, lowering the cost of the repair work. Chapter 19 by Carino details the impact-echo testing system.

16.3.3.2 SHRP ASR Rapid Test

One cause of severe deterioration of concrete exposed to moisture is alkali-silica reactivity (ASR). Alkali, such as the sodium and potassium in cements, reacts with silica or silicates in aggregates, forming a gel-reaction product. The ASR gel-reaction product swells when exposed to moisture. This swelling or expansion in the concrete causes tension stresses that lead to cracking.

The SHRP test [Natesaiyer and Hover, 1988, 1989] detects the existence of gel products. Thus the susceptibility to ASR attack is not determined, only whether the process of ASR has begun. Gel products are detected using uranyl acetate (a uranium salt) sprayed on a concrete surface and viewed under short-wavelength ultraviolet light. Presence of greenish-yellow deposits indicate the presence of ASR gel. Figure 16.4 shows an ASR field unit being used to assess the presence of ASR in bridge-pier concrete.

16.4 Structural Implications

Results from field and laboratory observations and testing must be assessed to determine the cause(s) of observed deterioration. In addition, the implications from the investigation results must be assimilated to determine their effects on strength, serviceability, and durability of the structure. As previously discussed, each of these factors must be considered for a successful concrete repair.

The strength and integrity of concrete structures suffering from deterioration can be assessed using conventional analysis techniques for concrete structures utilizing quantitative assessments of reduced material properties as determined in the investigation. This requires some judgement about the extent of deterioration. It is prudent to reduce the capacity-reduction (phi (ϕ)) factors in the ACI Building Code [ACI, 1995] and AASHTO Specifications [AASHTO, 1992] since they were calibrated for material variations that occur in new concrete construction. Several technical committees within ACI are now examining, through calibration studies, this issue of phi-factor selection for repair and strengthening designs. Depending on the conditions and factors involved, a ϕ factor as low as 0.5 may not be unreasonable.

In extreme cases, where analysis of a structure with reduced properties may be inappropriate or produce questionable results concerning load rating, load testing of the structure is an option. Load testing can be expensive and must be conducted with extreme caution, particularly on deteriorated and suspected grossly deficient structures. However, in some cases, it may be the only viable alternative for demonstrating the integrity of a repaired structure [Poston and Irshad, 1996a,b].

16.5 Repair Principles

Various organizations such as the American Concrete Institute (ACI) and the International Concrete Repair Institute (ICRI) have developed guidelines for the repair of concrete structures. These include technical and construction-related guidelines for, among other things preparing the surfaces of deteriorated concrete and selecting and specifying materials for repair. These are valuable resources that describe in detail the proper construction-preparation techniques for good concrete repair. The reader is referred to the reference list and Emmons (1994) for additional information on these construction-related aspects of concrete repair.

The principles presented in this section are generally applicable for all structural concrete including conventionally reinforced, prestressed, or partially prestressed structures. However, some of the more unique aspects of repairing unbonded post-tensioned concrete structures are presented in Section 16.6.

16.5.1 Material Selection

A general axiom in the repair of concrete structures is to use a material of mechanical characteristics similar to those of the existing parent material. One of the most important mechanical properties to match, but which is often overlooked, is the coefficient of thermal expansion. As an example, epoxy concrete used in a repair exposed to large temperature differentials, such as on a parking garage deck, is not appropriate.

As shown in Figure 16.5, the process of selecting concrete materials for repair must include identification of the causes of deterioration, service conditions, owner requirements, and the application conditions. Thus this somewhat complex selection process must consider constructability and service issues guided by the owner's needs and the engineering requirements of the repair. The final selection of a repair material is based on the relationship between cost and performance. The ICRI Guideline No. 03733 provides in-depth information for the process of selecting repair materials.

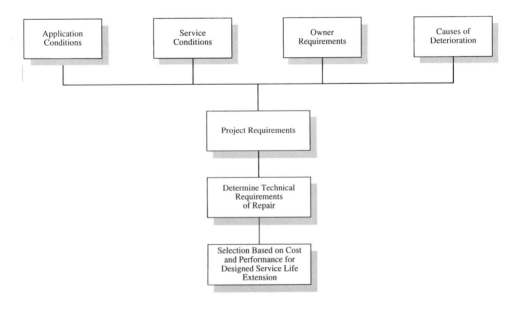

FIGURE 16.5   Representation of material selection process.

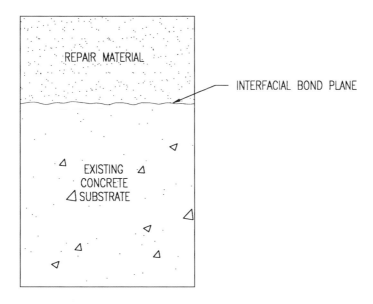

FIGURE 16.6 Bond strength is an important property for the success of repair.

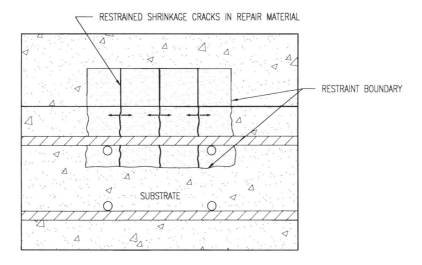

FIGURE 16.7 Dimensional changes in repair material can cause restrained shrinkage cracks.

16.5.1.1 Bond Strength

In almost all cases, one of the most important properties for repair is the bond strength (see Figure 16.6) between the repair material and the existing concrete substrate. Bond failures are generally a result of dimensional incompatibility between the repair material and the existing substrate, such as from drying shrinkage. The failure is generally not a direct result of lack of bond strength.

The relative dimensional changes that can occur, such as those illustrated in Figure 16.7, can affect the repair. Bond can be reduced and shrinkage cracks can appear, and thus the durability of the repair is compromised [Emmons et al., 1994].

16.5.1.2 Dimensional Compatibility

The drying shrinkage of the repair concrete is one of the important factors that influence the dimensional behavior. The existing concrete substrate has already experienced most of its time-dependent effects such as drying shrinkage and creep. The repair concrete that is then added must also undergo its shrinkage. Consequently, it is very important to identify and select a low-shrinkage material.

The time required for a material to achieve dimensional stability is dependent primarily on the ambient temperature and humidity. It is important to adequately cure the repair to mitigate shrinkage cracking. It is recommended that the repair concrete material be limited to less than 0.05% shrinkage at 28 d as measured by ASTM C157.

Other properties of the repair material that are important in achieving dimensional compatibility include the coefficient of thermal expansion, modulus of elasticity, and creep, both compressive and tensile. In general, these properties in the repair material should reasonably match those of the existing concrete in order to achieve dimensional compatibility. By matching these properties, cracking in the repair will be mitigated. This will help ensure long-term durability of the repair.

16.5.1.3 Durability Considerations

The durability of the concrete repair is its ability to resist structural loading and environmental conditions without degradation and deterioration. The environmental conditions that may affect durability include weathering, temperature changes, chemical attack, and abrasion, among others. It is essential that the cause and extent of deterioration of an existing concrete structure be identified. Based on this information, a strategy can be identified to satisfy the durability limit state of the repair.

There are factors and material properties that can affect the durability of a repair. Some of these properties include permeability, water-vapor transmission, freeze-thaw resistance, scaling resistance, sulfate resistance, abrasion resistance, and alkali-silica reaction.

Permeability is the rate of transmission of a liquid through the concrete. In general, the lower the permeability, the better the protection against the ingress of aggressive elements such as chloride ions from chemical deicers and from diffusion of carbon dioxide leading to carbonation.

Water-vapor transmission can lead to failure of a repair that uses an impermeable barrier. The movement of vapor in the concrete can become trapped at the interface of a permeable concrete and an impermeable barrier, such as a membrane, leading to debonding. As shown in Figure 16.8, the repair must make provision for this vapor transmission, such as by adding vents to allow the vapor to escape, or by providing materials that allow vapor transmission such as chemical sealers.

The selected repair material must have adequate freeze-thaw resistance to ensure long-term durability. The cyclic freezing and thawing of concrete that is critically saturated leads to a deterioration of the paste matrix. Repair materials that have entrained air or that proven through testing to provide freeze-thaw resistance should be specified.

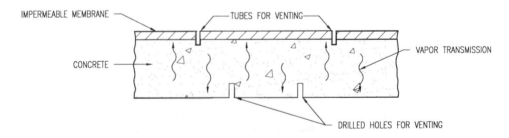

FIGURE 16.8 Provision for vapor transmission in repair.

In industrial applications, abrasion resistance is an important durability consideration. The resistance to abrasion is governed principally by the hardness, quality, and strength of the aggregate.

16.5.1.4 Mechanical Properties

Repair materials must be capable of resisting the internal stresses generated by loading and volume changes. The mechanical properties of the repair material are generally selected to match those of the existing concrete. This ensures compatibility of strains under load.

The principal mechanical properties that should be examined in the repair design include compressive strength, tensile strength, modulus of elasticity, and coefficient of thermal expansion. In some circumstances, Poisson's ratio and flexural strength are also important mechanical properties for repair design.

16.5.1.5 Placement

The placement of the repair material and field conditions can greatly affect the type of repair procedure and material selection. Characteristics such as slump, pumpability, rate of strength gain, heat of hydration, set time, and characteristics in hot or cold weather can affect material selection.

Conventional ready-mixed or machine-mixed concrete is the repair material of choice for existing-concrete deterioration that goes through, or mostly through, the structural element and beyond the reinforcement, or if the area to be repaired is large. Affected areas should be excavated until there is no question as to soundness of concrete. The excavation and repair area should avoid sharp angles that produce stress concentrations. Forms should maintain their shape during casting and maintain existing lines of the structure.

For small repair areas that are relatively deep, rodding a very stiff, low water/cement ratio dry-pack mortar is the method of choice. The rodding should be done in thin lifts, rodding the dry pack between successful lifts.

If large volumes of repair are required in tight places, preplacing aggregate followed by pumping of grout is a good repair method. This usually requires a contractor with some special equipment and experience.

For large-quantity shallow overhead and vertical repairs, shotcrete is an excellent choice. It bonds very well and is capable of supporting itself without a form. The nozzlemen must have good experience to control rebound that can lead to pockets of sandy layers. Mobilization costs require the repair job to be of reasonable size to justify using shotcrete.

Repair of shallow areas of limited size are well suited for proprietary prepackaged repair materials. These prepackaged materials often contain polymers to enhance the bond of the material to the existing concrete substrate. However, it has been observed that some prepackaged repair materials have excessive shrinkage characteristics.

16.5.2 Preparation

In general, deteriorated concrete should be removed of to sound concrete substrate. Partial-depth repairs are desirable because less demolition is required, no formwork is required for horizontal repairs, and they are generally less costly. However, if repair is required on two sides of a slab or wall, then full-depth removal is preferable.

Saw cuts around the perimeter of the repair are advisable, particularly in the case of slabs to eliminate feather edges. After deteriorated concrete is removed, the surfaces to be bonded should be thoroughly cleaned, such as by wet sandblasting. Then all loose particles should be vacuumed or pressure blown from the repair area. The surface should be thoroughly wet. The repair material should be placed when the surface has no standing water. A bonding agent such as a cement slurry coat or epoxy can be advantageous if recommended practices have been followed.

16.5.3 Cleaning and Protecting Reinforcement

Prior to placing the repair material, steel reinforcement and other metal components should be cleaned of corrosion and other scale. Mechanical abrasion can be used, although sandblasting is preferable if environmental conditions permit.

There are various components on the market for coating the reinforcement. These include epoxy, zinc-rich compounds, and cement-based coatings. Epoxy and some of the zinc-rich compounds can act as bond breakers and therefore must be applied carefully. Recent findings regarding epoxy have raised questions about the appropriateness of its use. It has been speculated that the application of epoxy on reinforcing steel in an area under repair results in accelerated corrosion in surrounding areas, sometimes referred to as the "ring effect."

16.5.4 Curing

All concrete repairs must be properly cured to ensure hydration of the cement, strength gain, and to minimize shrinkage and cracking. Curing should follow the same good practices as for new construction.

16.5.5 Crack Repair

Whether a crack in concrete should be repaired to restore structural integrity or merely sealed is dependent on the nature of the structure and the cause of the crack and upon its location and extent. If the stresses that caused the crack have been relieved by its occurrence, the structural integrity can be restored with some expectation of permanency. However, in the case of working cracks, such as cracks that open and close because of temperature changes, the only satisfactory solution is to seal them with a flexible or extensible material.

Cleaning of the crack is essential before any treatment takes place. All loose particles and foreign material must be removed. The method of cleaning is dependent upon the size of the crack and the nature of the contaminants. It may include compressed air, sandblasting, or routing.

Restoration of cracks without restoration of structural integrity requires the use of materials and techniques similar to those used in sealing joints. Typically, the cracks are routed and a urethane sealant is applied. A bond breaker should be used on the bottom of the routed crack so bonding only occurs on two sides.

16.5.6 Corrosion Protection

There are various corrosion-protection strategies that are being used in repair and new construction projects. Their selection depends on the type of structural system, exposure conditions, service life, and, naturally, costs. Selected corrosion protection strategies are summarized in Table 16.3.

16.5.7 Strengthening

A repair that has been conducted for understrength concrete structures, resulting from deterioration or upgrading of load capacity, is the addition of auxiliary bonded reinforcement or addition of members. The strengthening may take one of the following forms: wire mesh or reinforcing bars added to the understrength area and bonded to existing concrete using shotcrete, latex modified concrete or other means; structural steel plates epoxied and bolted through deficient structural concrete to increase capacity, as shown in Figure 16.9; the addition of supplemental steel sub-framing as shown in Figure 16.10; or the use of external post-tensioning as schematically shown in Figure 16.11. Depending on conditions, there may be a need for corrosion protection and fireproofing.

TABLE 16.3　Selected Corrosion Protection Strategies for Concrete Structures

Strategy	Selected Types of Materials	Range of Additional Unit Costs
Reinforcing steel coatings	Epoxy	$0.25–$0.75/lb of steel
Corrosion inhibitors	Admixed in concrete	$5.00–$10.00/yd^3 concrete
Chemical sealers	Methacrylate, silane	$0.50–$1.00/ft^2 of concrete surface
Elastomeric membranes	Urethane, neoprene	$2.50–$3.50/ft^2 of concrete surface
Low-permeability concrete overlays	Fly ash and microsilica	$4.00–$6.00/ft^2
Cathodic protection	Impressed current	$6.00–$15.00/ft^2 of concrete surface
Low-permeability polymer overlays	Epoxy, methacrylate	$1.50–$2.50/ft^2 of concrete surface

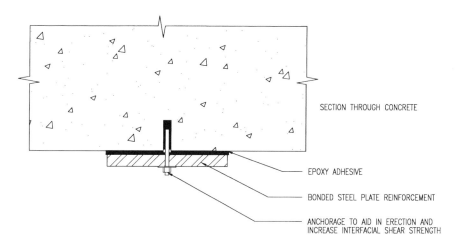

FIGURE 16.9　Strengthening using bolted and epoxied steel plates.

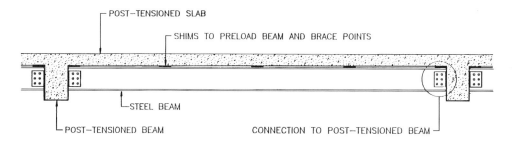

FIGURE 16.10　Strengthening using steel subframing.

The requirements for engaging all or some of the superimposed loads on the structure must be carefully reviewed. Generally, it is preferable to preload the structural member being strengthened to some degree. The amount of preloading depends on whether the deficiency is under service or ultimate loads or both. In some cases, no preloading is required and the strengthening is only effected to ensure the structural element can achieve ultimate strength.

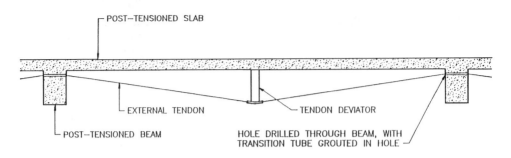

FIGURE 16.11   Strengthening using external post-tensioning.

16.6 Repair of Unbonded Post-Tensioned Concrete Structures

Prestressed concrete is a type of reinforced concrete construction in which steel reinforcement is tensioned so that permanent internal compressive stresses in the concrete improve the response of a member of structure to loading. Compared to conventionally reinforced concrete, prestressed members typically have greater load capacity or deflect less for the same load.

Two different procedures are used for prestressing concrete structures. In pretensioned concrete, the general repair principles presented in Section 16.5 are applicable.

Figure 16.12 illustrates post-tensioning, where the steel is tensioned after the concrete has been cast and gains strength. Metal or plastic ducts are used to prevent bond with multistrand tendons, whereas for single strand tendons, plastic sheathing is used. The prestressing force is transmitted to the concrete through bearing at mechanical anchorages. Tendons are usually stressed from one anchorage, known as the live end, or from both end anchorages if stress losses due to friction and curvature are significant. Pockets for stressing anchors are usually grouted to protect the hardware.

If the duct is filled with a conventional grout, then bonding or partial bonding of the steel to the concrete is achieved. This is referred to as bonded post-tensioned concrete. In unbonded post-tensioned concrete, the duct remains unfilled after anchoring so that the steel is not bonded to the

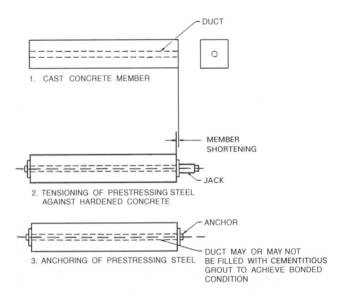

FIGURE 16.12   Post-tensioned concrete.

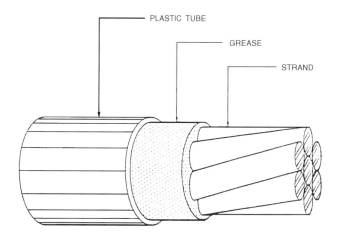

FIGURE 16.13   Plastic sheath filled with grease containing unbonded monostrand.

concrete and is permanently free to move. Unbonded post-tensioning systems are most common in the United States since grouting of tendons can be difficult and costly. Unbonded post-tensioning is used for applications ranging from buildings, bridges, and parking structures to pressure vessels, tanks, and earth-retaining structures.

Steel used for unbonded applications includes high-strength wires, strands, and threaded bars. The most commonly used unbonded tendon is the monostrand system with a 7-wire 0.5 or 0.6 in diameter 270 ksi strand as shown in Figure 16.13. The steel is covered with a corrosion-inhibiting grease or other coating and enclosed in a plastic sheath or a plastic or metal duct.

More than 1.5 billion square feet of unbonded post-tensioned concrete structures have been constructed in the United States since the mid-1960s. There is an ever-increasing need to repair and strengthen these structures to extend their service lives and to restore structural integrity. The need for repair and strengthening is primarily due to one or more of the following reasons: the strand is unintentionally damaged or cut, such as when coring concrete to install a mechanical pipe; there is a change in load or the space requirements of a structure such as adding a stairwell; improper design and/or construction resulting in inadequate structural capacity; and corrosion of the prestressing strand and/or mechanical anchorages.

Prior to repairing an unbonded post-tensioned concrete structure, it is important to conduct a comprehensive engineering evaluation using the techniques previously discussed to determine the causes of observed problems. Because of the potential for tendon release, caution should be exercised during the evaluation and repair phases. The methods for assessing an existing unbonded post-tensioned concrete structure are, for the most part, the same as for a concrete structure constructed with mild steel reinforcement. However, the repair of existing unbonded post-tensioned structures is different in many respects.

The repair of unbonded post-tensioned concrete structures cannot be easily divided into categories where a certain type of repair is best for a certain type of problem. The type of repair depends on the expected cost of the repair, cause of the problem, the need to maintain service of the structure, whether the repair is to a beam, slab, or other structural member, and many other factors. Moreover, once repairs begin, the owner, engineer, and repair contractor must be able to adapt to unobserved latent defects and changed conditions. The most important axiom in repair of unbonded post-tensioned structures is to expect the unexpected.

The general types of repair of unbonded post-tensioning are classified as follows: do nothing; tendon splice or replacement; and strengthening by adding bonded reinforcement, structural members or external post-tensioning to improve or increase structural capacity.

If the repair is due to a corrosion problem, there will likely be other concrete repairs needed because of the deterioration that is generally attendant with corrosion. As with all types of concrete repair, extreme caution should be exercised by all individuals during the investigation, demolition, and repair. The need for shoring and the sequence of repair should be thoroughly thought out prior to starting any work.

16.6.1 Do Nothing

There are some occasions in which nothing needs to be done for a failed post-tensioned strand. If it is known unequivocally that one strand, and possibly two, have been accidentally severed, then these strands can be pulled out and not replaced. Section 18.18.4 of the ACI Building Code [ACI, 1995] allows the total loss of prestress due to unreplaced broken tendons that does not exceed 2% of the total prestress. Thus, for example, in the case of a post-tensioned slab structure that might contain hundreds of tendons in a single floor, the loss of one strand without replacement would be acceptable and would not be considered to have a significant impact on structural integrity.

16.6.2 Tendon Splice or Replacement

Tendon splice or replacement is the most common form of repair to an unbonded concrete structure. This type of repair is generally cost effective when the number of tendons to be spliced or replaced is limited, although there have been several notable examples where the tendons in an entire structure have been replaced [Aalami and Swanson, 1988]. Tendon replacement may include replacement of the strand and/or one or more of the anchorages depending on access for retensioning and anchorage condition. This option is the repair method of choice if the strand has been severed or if corrosion is localized. The cost of tendon splicing and/or replacement can range between $800 to $5000 per strand depending on the difficulty of repair and number of strands needed.

If the strand to be spliced or replaced is already severed, then the repair is somewhat simplified. If the strand is still tensioned, then the following procedure is required:

1. Choose a point in the structure in which the tendon is more accessible. If the location of the problem and thus the splice point is known, then choose that point. Locate the tendon by use of a pachometer. In more complicated repairs, an X ray may be warranted.
2. Make a shallow saw cut in the concrete surrounding the tendon that defines the limits of the concrete to be removed.
3. Remove the concrete around the tendon by light chipping hammer or jackhammer.
4. If the strand is to be respliced, it is preferable to install a temporary anchorage. After placing the temporary anchorage, the strand may then be cut using a grinder. If a temporary anchorage is impractical, then, exercising extreme caution by making sure there is no one in the area of the entire tendon length, the strand may be cut using a grinder. After cutting through just over 2 wires of a 7-wire strand, the strand will yield and then fracture, with an attendant loud noise from the energy release. In some cases, controlled heating of the strand over a length of several feet may be preferred. This is the case when sudden detensioning can push out wedges at inaccessible anchorages, making restressing of the tendons impossible. Heating can result in less shock and more control in the detensioning process.

If the existing strand is to be replaced or partially replaced, the strand piece(s) to be removed can now be pulled from the anchorage. A new strand, coated with a corrosion-inhibiting grease can then be threaded through the existing duct. A special type of splicing hardware allows for retensioning at the point of splice. This is beneficial when access to the end anchorages does not allow the use of a standard hydraulic stressing jack.

With the stressing is completed, the concrete is recast in the excavation. As previously discussed, it is generally preferred to use a concrete of similar strength and modulus as the existing concrete.

FIGURE 16.14   New anchorages and bursting steel.

Exposed reinforcing steel should be cleaned and coated for corrosion protection. Exposed concrete surfaces should be sprayed with water prior to casting concrete. Curing should follow established good practices.

In some cases, the entire post-tensioning strand may need to be replaced because of corrosion. In other cases, such as in buildings where access for splicing may be difficult, anchorages at the edge of a member can become the splice point. If the existing anchorage is not easily accessible for jacking, then a new anchorage can be added for ease of constructability.

It is important to emphasize that if the strands are still tensioned and the anchorages are still engaged, do not try to release the stress by jack-hammering around the anchorage. This is very dangerous. Instead, as previously indicated in the procedure for splicing, cut a slot in the concrete away from the anchorage zone. Making sure no one is in the way of the tendon trajectory at the ends of the structure and above or below, the strands can be detensioned using a grinder or heating at the control point. Again, expect a loud noise once several of the wires of the 7-wire strand have been cut. Now the removal of the anchorages can begin by first providing a shallow saw cut around the boundary of the anchorage zone. The concrete can then be removed by light chipping. The anchorages and strand can then be removed.

Figure 16.14 shows an arrangement of new corrosion-protected anchorages in place. Note that new mild reinforcement for control of anchorage zone bursting stresses has been placed just in front of the anchorages. In this case, the replacement anchorages are epoxy-coated and the plastic transition trumpets are attached integrally to the anchorages and the extruded sheathing to prevent water intrusion. The trumpet is filled with a corrosion-resistant grease.

In order to expedite retensioning of the replaced tendons, a high-early strength concrete may be used. In no circumstances should it contain calcium chloride as an accelerator. Many of these high-early strength concrete materials can gain enough strength (>4000 psi) in 24 h to allow retensioning the next day.

16.7 Construction Issues

The disruption caused by repairs to an existing structure depends on the degree of deterioration, the use type of the structure being repaired, time of day when the construction is being done, expertise of the selected contractor, and other factors. Disruption to the owners and structure users is virtually unavoidable.

Disruption, however, can be minimized with a carefully thought out and detailed schedule that is issued by the contractor to the affected parties. Generally, by breaking the project into workable units that are taken out of service one at a time, completed, then brought back into service is preferable to taking all of the affected areas out of service. Repairs of this nature cost more but allow for continued use of at least a portion of the structure.

Historically, the knowledge level of repair contractors was not what might have been expected. However, in recent years, there has been an increase of competent and knowledgeable contractors for conducting specialized concrete repairs. This has resulted in high-quality, cost-competitive structural concrete repairs.

16.8 Long-Term Repair Performance

The question most often asked by owners relates to the anticipated long-term performance of repair alternatives. If this question can be answered with certainty and repair and maintenance costs are known, then it is a relatively simple matter to conduct a benefit/cost analysis. Of course, how long something will last depends on numerous variables such as quality of repair, exposure conditions, maintenance, and a host of other factors, most of which are difficult to define.

As part of a repair program, costs for a return after the first year to conduct an engineering inspection of the repairs should be included. If the budget permits, some embedded sensors, such as wires to measure half-cell potentials and a corrosion-rate probe should be included to monitor the repairs.

All structures require maintenance. This is even more true with a repaired structure. Depending on the nature of the effected repairs, annual maintenance costs on the order of 2% to 5% of the repair costs may be appropriate for maintaining the repaired structure.

16.9 Case Study

To illustrate some of the general principles involved in structural concrete repair, a detailed case study of a large infrastructure rehabilitation project is presented. This case study of an existing seawall clearly shows the value of conducting an informed evaluation prior to undertaking the repair design. It demonstrates the necessity of determining the cause of deterioration and selecting a multifaceted repair approach to meet the expressed objective of extending the service life for 30 more years. Utilizing a comprehensive approach in this infrastructure repair project ultimately saved the funding public agencies $6 million on a $15 million project.

16.9.1 Background

The 7.5 mile Marina del Rey Seawall was constructed in western Los Angeles, California, during the late 1950s and early 1960s as a means of reclaiming a low-lying swamp area for dry-land uses. Over the last 30 years, some $5 billion of infrastructure has developed in and around this area. The original construction was high quality for the time, as evidenced by its 30-plus years of service life to date. The vertical flexural reinforcement in the cantilevered wall has been subjected to aggressive saltwater exposure, principally through a construction joint located near the base of the wall (see Figure 16.15). The long-term exposure to seawater has led to chloride-induced corrosion of the reinforcing steel crossing the construction joint and attendant concrete deterioration on the hidden (land) side of the wall. Because of the nonredundant nature of the cantilever wall system, loss of primary vertical reinforcement at the wall base by corrosion results in a loss of overall structural integrity.

In February 1986, a failure mechanism (see Figure 16.16) that resulted from severe corrosion of the vertical reinforcement occurred when an isolated 60-ft wall panel collapsed. An engineering investigation into the collapse concluded that the failure was caused by corrosion of the reinforcing steel owing to exposure to natural seawater. No other factors were identified as having contributed to the collapse.

To address concerns about the integrity of the seawall panels, a restoration program with three distinct focuses was developed by various public agencies. The focuses of the restoration program

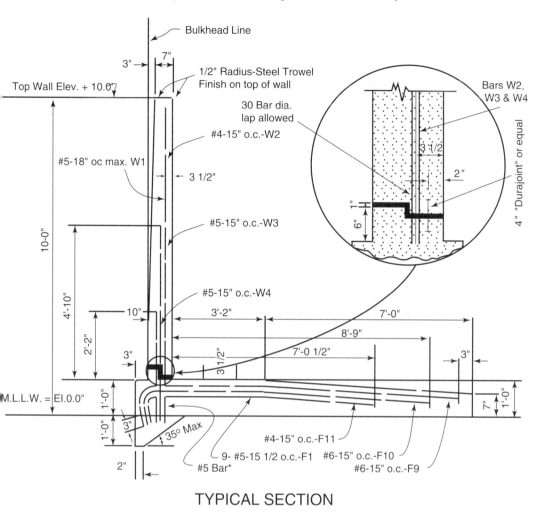

TYPICAL SECTION

FIGURE 16.15   Typical reinforcing details of Marina del Rey Seawall (from original project drawings, October 1959).

FIGURE 16.16   Failure of an isolated 60-ft wall panel due to corrosion.

were as follows:

1. Utilizing an impressed current cathodic protection system to mitigate active corrosion of the seawall reinforcing steel.

2. Utilizing a drilled caisson "strong-back" system to reduce reliance on the vertical reinforcing steel that had been subjected to corrosion and to upgrade the seismic behavior of the system. The strong-back concept was successfully installed during a trial repair in the marina.

3. Conducting nondestructive impact-23 echo testing to locate areas where corrosion damage had occurred. This allowed for categorization of repairs based upon the extent of delamination and concerns for life safety in areas of high public exposure. The impact-echo technique had been proven to be successful in locating corrosion-deteriorated areas in the seawall.

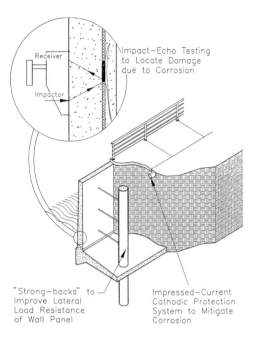

FIGURE 16.17   Schematic representation of proposed rehabilitation strategy for Marina del Rey Seawall.

Figure 16.17 shows a schematic representation of the multifaceted seawall rehabilitation strategy. Each of the three distinct focuses of the restoration program are represented in this figure.

16.9.2 Repair Strategy

Prior to the development of the strong-back concept, several repair strategies were considered. The repair strategies included utilization of rip-rap ballast on the sea side of the wall and installation of precast panels anchored to caissons and use of soil tie backs. These preliminary designs were abandoned because they were cost prohibitive or failed to adequately address the seismic upgrading of the wall.

As part of the strategic planning, it was decided that the rehabilitation must address the need to prevent additional corrosion damage along the base of the seawall, to upgrade the capacity of the seawall due to the loss of strength caused by the corrosion of the primary vertical reinforcing steel, and to provide seismic resistance that was not originally considered in the seawall design.

16.9.2.1 Full Damage Along the Base of the Seawall

On the basis of the results of preliminary impact-echo testing and select concrete coring, significant reinforcing-steel corrosion and attendant concrete deterioration was found to be present along the base of the seawall. Corrosion of reinforcing steel along the base of the seawall was determined to be the cause of the isolated seawall panel collapse in 1986. Because of these factors, all repair strategies had to assume that the vertical reinforcing steel at the vulnerable construction joint had lost capacity to carry tension if corrosion deterioration was found to be present. Therefore, under a fully repaired condition, the connection between the wall and the footing was assumed to be hinged and only capable of transmitting shear across the joint.

16.9.2.2 Service Life of Repairs

The Marina del Rey Seawall had been in service for over 30 years. This was consistent with the expected design life of concrete structures. Much of this nation's concrete infrastructure is now routinely being rehabilitated since replacement is becoming more and more cost prohibitive. At the time of the evaluation, the seawall concrete was not showing any significant signs of visible deterioration due to abrasion, sulfate attack, or other mechanisms. Previous testing had shown levels of chloride-ion penetration sufficient to cause chloride-induced corrosion. However, it was strongly believed that an effective rehabilitation strategy implemented in a timely fashion would extend the life of the seawall by another 30 years or more.

To ensure a minimum 30-year life extension of the seawall, the repair strategy had to mitigate corrosion of the vertical reinforcing steel and provide a load path for forces originally carried by the deteriorated reinforcing steel. Corrosion engineers had indicated that the impressed current cathodic protection system would fully mitigate the corrosion of the vertical reinforcing steel. In essence, it was indicated that the corrosion would virtually stop and the corrosion rate would be essentially zero. The strong-back repair system had been shown to be effective in carrying loads in a panel with existing deterioration along the wall base. With the implementation of these rehabilitation measures the service life the seawall could clearly be extended by 30 years or more.

16.9.2.3 Strategic Ordering of Repairs

Owing to life safety concerns and budget constraints, a strategic ordering of repairs was developed. This logical ordering of repairs was developed to ensure that areas of high public access, such as parks, were given high priority in the ordering of repair to address life safety concerns. Because only a portion of the needed funding was in place at the time, the repairs were categorized to address budget constraints. Impact-echo test results were evaluated to locate wall panels with significant deterioration that would be categorized as the highest priority.

16.9.3 Life-Cycle Cost Analysis

Owing to the complex interaction of concrete corrosion and repair issues, life safety considerations in the marina area, and the budget constraints of the various municipal and state agencies involved, a life-cycle cost analysis of the marina seawall refurbishment project was requested. The principal objective of this analysis was to consider and prioritize various facets of the seawall refurbishment project so that critical decisions could be made with due consideration of the concerns of all agencies and with respect to a logical and strategic ordering of the repairs. Figure 16.18 shows a schematic representation of the interaction of issues that were considered in the Marina del Rey Seawall Rehabilitation Project.

16.9.3.1 Life-Safety Issues

The Marina del Rey area serves as a mixed-use residential and recreational area. Present in the marina are boat yards, restaurants, parks, and residential areas. It was anticipated that the areas of public exposure would be of high priority in the seawall repairs. Use of impact-echo testing in these areas further prioritized the repair sequencing on the basis of the amount of damage detected in panels in public areas. Significant damage detected in areas of high public exposure were naturally categorized as higher priority.

16.9.3.2 Seismic Upgrade

In addition to considering the corrosion-induced deterioration, the engineering design for the Marina del Rey Seawall Refurbishment Project mandated upgrading the seismic capacity of the seawall. The original design did not consider seismic loading. Development of the seismic retrofit included consideration of the degree of corrosion-induced deterioration as determined by impact-echo testing. This deterioration was modeled as a loss of vertical reinforcing-steel capacity at the base of the seawall. As the degree of deterioration increased, additional caissons (strong backs) were required to upgrade the capacity of the seawall. The wall was upgraded to survive the maximum

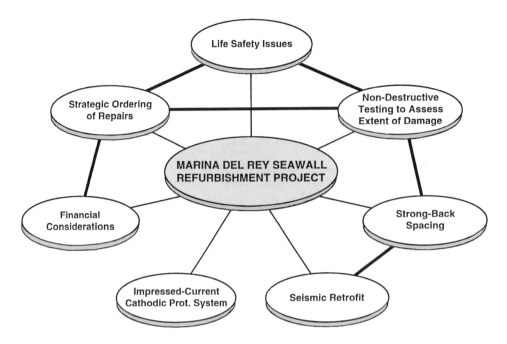

FIGURE 16.18 Interaction of issues being considered in the Marina del Rey Seawall refurbishment project.

credible seismic even though damage would be expected. The proposed strong-back repair strategy minimized the risk of wall collapse during the maximum credible earthquake.

16.9.3.3 Sequence and Ordering of Testing

To provide information on the deterioration level at various locations in the marina, a sequence of impact-echo testing was developed. The sequence and ordering prescribed that a small number of panels initially be tested in all basins. The preliminary testing provided an initial deterioration assessment for the various structural engineering tasks.

16.9.3.4 Deterioration Ratings

Impact-echo testing was used to locate areas where physical manifestations of corrosion damage were present. This information was used to develop deterioration ratings that correlated to the number of caissons required. Both the effects of continuous damage and isolated damage were considered.

Using corrosion damage models, ratings were developed to distinguish the number of caissons required at varying levels of deterioration. The models were based upon estimated lengths of continuous and/or intermittent damage along the construction joint. As impact-echo testing results became available, modifications were made to the preliminary damage models.

Using the corrosion-damage models previously discussed, caisson spacing parameters were developed. The parameters delineated when additional caissons were required owing to the extent of corrosion damage and to minimize the risk of collapse during the maximum credible earthquake.

16.9.3.5 Design Guidelines

The following criteria were agreed upon by the various public agencies involved for developing the life-cycle cost analysis:

- maximum credible seismic event: 0.5 g lateral acceleration,
- surcharge loading: 2 ft of soil, and
- 30 year design life for all rehabilitation measures.

No guidelines were established for a serviceability level seismic event. In essence, the seismic retrofit strategy was to minimize the risk of collapse during the maximum credible seismic event. Damage to the wall during the maximum credible earthquake would be expected and acceptable. After completion of the life-cycle analysis, the parameters were modified to reflect further refinement and development of the analysis models based on the results obtained from the geotechnical investigation and impact-echo testing.

16.9.4 Impact-Echo to Assess Deterioration

Various studies were conducted as part of the strategic planning of the Marina del Rey Seawall Refurbishment Project. Numerous nondestructive test methods were reviewed as candidates for assessing deterioration of the seawall due to corrosion-induced delaminations. The majority of nondestructive testing techniques for assessing reinforcing steel corrosion, such as measurement of half-cell corrosion potentials and chloride-ion content, provide only limited, indirect information about whether or not corrosion is occurring. These testing measures do not provide an assessment of the amount of corrosion damage or the degree of concrete deterioration present.

Stress-wave propagation methods known as pulse-echo and pitch-catch were reviewed as candidates for detecting damage on the soil side of the wall in the vicinity of the construction joint. These methods had limitations that made them impractical for use on the seawall as constructed.

During the project development, the transient stress-wave method known as impact-echo was examined as a possible candidate for nondestructively determining the location of corrosion-induced deterioration. This method was determined to be well suited for finding flaws in concrete. The reader is referred to Section 16.3.3.1 for a description of the impact-echo method.

It is important to note that although impact-echo is effective for finding flaws in concrete structures, such as the corrosion-induced delaminations on the soil side of the wall near the level of the construction joint at the base of the wall, it cannot assess the amount of deterioration in terms of degree of steel-reinforcement corrosion. Impact-echo was, in essence, used to locate the manifestation of corrosion, that is, the attendant delamination of the concrete caused by reinforcing-steel corrosion, but it was not used to assess the degree of corrosion. This implies that when using the impact-echo method, if a delamination or defect is found, it is assumed that the reinforcing steel at this location is in essence completely corroded. There may be some measure of steel area that is still effective, but for condition-assessment purposes, the steel was assumed to be ineffective for developing a tension force and thus incapable of carrying moment at the base of the wall. This was a conservative assumption in the context of the rehabilitation design.

16.9.5 Reliability in Assessing Deterioration

16.9.5.1 Previous Trial Studies on the Seawall

It is generally beneficial in larger projects to conduct trial testing, and in some cases repairs, to prove the veracity of the methods. This provides confidence for all parties concerned prior to spending large sums of money. For the case of the seawall project, previous studies assessing the efficiency of using impact-echo to detect corrosion-induced deterioration on the unobservable soil side of the seawall clearly indicated the veracity of the method. Concrete coring at select locations of delaminations located by the impact-echo method demonstrated the reliability of the procedure for detecting damage.

16.9.5.2 Test Locations

A visual examination in a trial study that sliced a piece of the wall clearly showed that the damage caused by corrosion of the vertical reinforcing was in the range of about 10 to 20 cm (4 to 8 in) above the construction joint. The average of this range is 6 in, which was used to define the height of impact-echo testing in the field-test program.

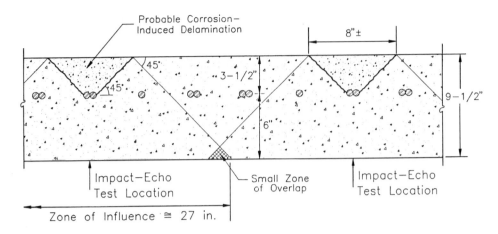

FIGURE 16.19   Influence area of impact-echo test on the seawall for anticipated corrosion-induced delamination.

It is also important to note that a trial study indicated that the corrosion appeared isolated to the local region near the construction joint. All seawall panels were tested about 6 in above the construction joint as part of the field evaluation. However, select panels were tested on a two-way horizontal and vertical grid to confirm that the damage was generally confined to the region near the construction joint.

Figure 16.19 presents a schematic representation of the approximate area of influence during impact-echo testing for the geometry of the wall and assumed type of corrosion-induced delamination. Note that the influence area is on the order of 27 in; thus testing on 25 in centers provided for about a 2 in overlap.

Thus, based on generalized stress-wave propagation principles, there was assurance that most delaminations were located and that finding the delaminations is independent of whether the test is conducted directly over a reinforcing bar. The 25 in spacing used in the evaluation program was determined to be sufficient to detect most of the damage that was present.

16.9.6 Cathodic Protection

The galvanic corrosion of reinforcing steel in concrete is an electrochemical process in which the reinforcing steel reacts with water and oxygen to form iron-oxide products (rust). For galvanic corrosion to occur, four elements must be present. The elements are anodes (locations where corrosion activity occurs), cathodes (areas that are protected by the corrosion activity at anodes), a metallic path between the anodes and cathodes, and an electrolyte. The metallic path is provided by the horizontal and vertical reinforcing steel mat. Moisture in the soil backfill, seawater, and moist concrete all serve as the electrolyte. The corrosion cell is driven by the differences in electrical potential that exist between the anodic and cathodic areas. Figure 16.20 shows a schematic representation of the electrochemical process of corrosion of reinforcing steel.

Cathodic protection (CP) is an electrochemical technique in which external anodes are added to the electrochemical cell present in the seawall. The addition of external anodes to the seawall structure resulted in the shift of electrical potentials at the reinforcing steel from anodic (actively corroding) to cathodic values (noncorroding). Figure 16.21 shows a schematic representation of both sacrificial anode and impressed current cathodic protection systems.

Two types of cathodic protection systems were considered for the Marina del Rey Seawall. A galvanic or sacrificial anode system utilizes metal (typically zinc) anodes attached directly to the reinforcing steel grid. When connected to the reinforcing steel grid, the sacrificial anodes corrode

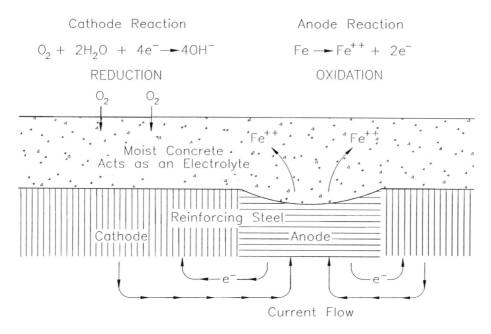

FIGURE 16.20   Schematic of electrochemical corrosion of reinforcing steel in concrete.

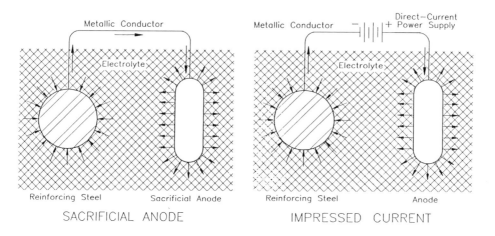

FIGURE 16.21   Schematic representation of cathodic protection systems.

instead of the reinforcing steel. Corrosion of the sacrificial anode results in electrons traveling to the reinforcing steel, which further prevents corrosion of the seawall reinforcing steel. Sacrificial anode systems are best suited to areas of low electrical resistivity and/or to submerged areas. An impressed current cathodic protection system is similar to a sacrificial anode system, except an external power supply is used to provide the electrical potential that drives the corrosion cell.

The seawall rehabilitation concept required that the horizontal (temperature and shrinkage) reinforcing steel carry the redistributed moments that result from caisson installation. This necessitated that both the horizontal and vertical reinforcing steel be protected from long-term corrosion. An impressed current cathodic protection system was determined to be better suited for long-term corrosion control. This is because of the seawall's location above the mean low water level and the measured soil resistivities.

A successful application of an impressed current cathodic protection system must ensure the electrical continuity of all protected areas. Reinforcing steel in both the horizontal and vertical directions were tested to ensure connectivity.

After installation of the impressed current cathodic protection system, the rate of reinforcing steel corrosion is effectively reduced to zero. This has been proven time and again in actual field applications. To mitigate corrosion, cathodic protection must have an electrolyte present. In the seawall, the seawater below the water table and the moist soil acted as the electrolyte. An electrolyte allows the flow of the impressed current from the sacrificial anode to the reinforcing steel in the seawall. In areas where concrete delaminations have occurred due to corrosion, the cathodic protection system will protect the existing reinforcing steel if sufficient electrolyte is present at these locations. The cathodic protection system cannot protect areas where an air gap is present between delaminated concrete and the reinforcing steel. However, this condition was not considered likely to be present to any significant degree.

The Marina del Rey Seawall Refurbishment Project was designed to provide a minimum 30 year life extension to the seawall. Accordingly, all facets of the rehabilitation strategy were designed for the ultimate serviceability and durability limit state to reflect a minimum 30 year life cycle. After installation of the cathodic protection system, no further reinforcing-steel corrosion was expected to occur; thus the 30 year service life was clearly attainable. The steel was, in essence, kept in its existing condition without further corrosion.

16.9.7 Structural Analysis

This section summarizes the results from various structural analyses that were conducted on a typical 60-ft wall panel. Analyses were conducted to assess the integrity of the existing wall without the strong-back system in place for various postulated damage models. Parametric analyses were then conducted for a typical seawall panel that incorporated different combinations of strong backs (caissons) for postulated damage models and included the loads resulting from the assumed maximum credible seismic event.

16.9.7.1 Integrity of the Typical Existing Panel

Analyses were conducted to assess the integrity of a typical 60-ft wall panel in its present condition for various assumed damage models. The threshold damage in terms of the degree of corrosion necessary to cause collapse of a typical wall panel was then established by examining the results of these analyses.

Nonlinear Analysis Model. To assess the integrity of a typical 60 ft seawall panel, as constructed for various assessed-damage models, a nonlinear analysis of the typical wall section was studied. The analytical model used a tangent stiffness formulation. Incremental loads were applied to the structure, then the incremental displacements and forces were calculated and added to the previous displacements and forces to obtain the new displacements and forces. The stiffness method is used to calculate incremental displacements from applied loads, and incremental forces are then computed from the displacements. The model uses a linear formulation that has been modified to include various nonlinear effects. The nonlinearities include second-order deflection effects (P-delta), geometric nonlinearity, and material nonlinearity. Analytically, failure occurs when defined material strength limits are exceeded or by the instability of the structure.

Since the seawall was originally designed principally as a cantilever retaining wall, only a typical 1-ft width of wall was required to represent its behavior under lateral load. Figure 16.22 shows a schematic representation of the nonlinear model used for the wall. The reinforcing steel was terminated (cut off) at the appropriate wall height location as indicated in Figure 16.15. The wall taper was also duly accounted for in this analysis. The base of the wall was assumed to be fixed.

Loading. Figure 16.23 summarizes the load conditions for which a typical seawall panel was analyzed under anticipated actual operating service load conditions. Note that these were not design loads—they represented an assessment of the probable operating loads on the wall as might exist on a typical existing wall panel. For purposes of this analysis, no earthquake loading was considered.

Effect of Reinforcing Steel Loss. Analyses were conducted for the typical seawall panel assuming an average net section loss of reinforcing steel at the construction joint due to corrosion. Incremental lateral loads were applied to the wall for various cases of assumed reinforcing steel loss until a failure occurred. The loads due to the surcharge and water table were applied first; then the lateral earth pressure was incrementally applied until failure occurred. The parameters of interest in this type of analysis are the load necessary to cause failure compared to the probable in-service lateral earth pressure (see Figure 16.23) and the expected horizontal deflection at the tip of wall before collapse occurs (see Figure 16.24).

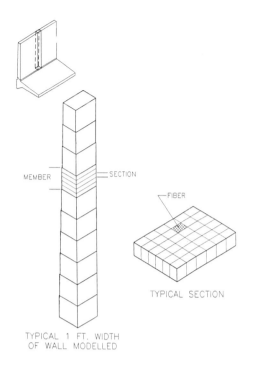

FIGURE 16.22 Representation of the nonlinear model used for analysis of a typical seawall panel.

Table 16.4 summarizes the results of the nonlinear analysis of the wall for various postulated scenarios of the degree of corrosion of the vertical steel at the construction joint. The results of the analysis of the wall as constructed (0% loss of the reinforcing steel) indicated that the wall had adequate integrity for the probable existing operating loads if seismic loads were not considered. This suggested that for wall panels determined to have little or no damage as established by impact-echo testing, some measure of strengthening to upgrade the wall for seismic loading was still needed. The analyses clearly revealed that as the percentage loss of reinforcing steel due to corrosion increases, the load to cause failure and the associated horizontal wall deflection at the top of the wall at failure decreases. Further, the analysis results indicated that an average of about 60% (actual value is 62%) of the reinforcing steel at the construction joint must be lost to corrosion for failure of the wall to occur under current probable operating-load conditions.

This model assumes that the corrosion damage is uniformly distributed along the wall, which may not be the case. Contiguous damage would be even more detrimental than the uniform damage

TABLE 16.4 Summary of Nonlinear Analysis of Typical Wall Panel in Service

Assumed % Loss of Steel Reinforcement Due to Corrosion of Construction Joint	Ratio of Failure Load to Probable In-Service Lateral Earth Pressure	Horizontal Deflection at Top of Wall Before Collapse, in
0%	2.8	$0.74 \approx 3/4$
20%	2.4	$0.74 \approx 3/4$
50%	1.5	$0.45 \approx 1/2$
62%	1.0	$0.41 \approx 3/8$
80%	0.5*	$0.24 \approx 1/4$

*Indicates section fails at about one-half of the present estimated operating service loads.

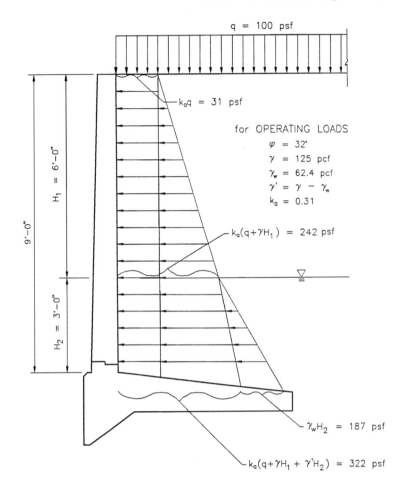

FIGURE 16.23 Estimated operating service wall loads.

implicit in this analysis. This suggests that if impact-echo testing revealed that the wall along the construction joint indicated evidence of randomly distributed corrosion-induced delaminations, say, to about the 40% level, the panel could be approaching collapse conditions and would naturally be of the highest priority in terms of repair sequencing.

The nonlinear analysis results also clearly indicated that there would be very little visual evidence of distress before failure occurred. Table 16.4 indicates that the horizontal deflection at the top of the wall is about 3/4 in just before failure, assuming no corrosion damage. However, for the case of 62% average loss of reinforcement due to corrosion, the calculated horizontal movement at the top of the wall is only about 3/8 in just before collapse occurs. This small amount of relative movement between adjacent wall panels over time would likely not be perceptible owing to the observed construction tolerances and to wall rotations that may have occurred because of differential soil settlement. This analysis clearly indicated that there would be very little warning before failure.

16.9.7.2 Finite-Element Analysis of the Wall With and Without a Strong-Back System

To evaluate the capacity of a typical 60 ft seawall panel for various reinforcing steel corrosion damage models and to assess the effect of strong-back spacing, a parametric study was conducted using linear elastic finite-element analysis techniques.

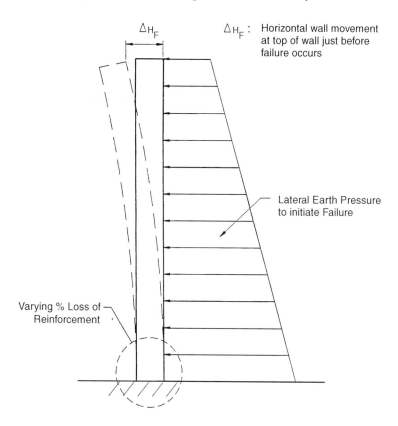

FIGURE 16.24   Horizontal deflection of wall just before failure.

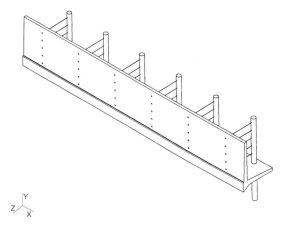

FIGURE 16.25   Seawall with strong-back retrofit.

Model.   A typical 60-ft seawall panel with the proposed strong-back retrofit is shown schematically in Figure 16.25. To evaluate the effects of reinforcing-steel corrosion and strong-back spacing on the seawall stem, a finite-element model was developed for the stem using four node isotropic plate elements. A finite-element model with six caissons installed is shown in Figure 16.26.

It should be noted that the finite-element model took advantage of geometric symmetry about the centerline of the 60-ft seawall panel. For this analysis, it was assumed that loading, strong-back

spacing, and reinforcing-steel corrosion damage is
symmetrical about the centerline of the seawall.
Therefore, for simplicity, only one-half of the seawall was modeled for the parametric study using
appropriate boundary conditions at the plane of
symmetry. For comparison, a typical horizontal
moment dithering for the half-model using symmetry and for a model of the full 60-ft seawall panel
is shown in Figure 16.27. Note that the computed
moments are identical, indicating the validity of
the symmetry model.

A 6 in × 6 in element grid was used for the
model. Element thickness varies to correspond
with the taper in the face of the seawall.

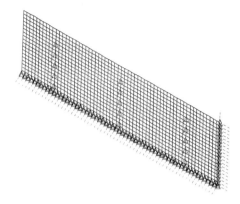

FIGURE 16.26   Finite-element model of a typical wall panel for a case with six caissons.

Since the capacity of a typical seawall panel is
dependent on the strength of the stem, it was not
necessary to model the seawall footing. Instead, boundary elements were used to model the interface between the seawall stem and footing. To model the condition of no loss of reinforcing
steel at a particular node, a boundary condition constraining translation and rotation about the
three axes was assumed. To model the condition of total loss of reinforcing steel at a particular
node, a boundary condition constraining translation, but not rotation, about the three axes was
assumed.

Also, the interface between the seawall and the strong-back system was modeled using translational boundary elements with an assumed spring constant. For this analysis, explicit modeling
of the caisson was not necessary. A tie-back spring constant was assumed to connect the wall to
a rigid element (the caisson). A spring constant of 200 kips/in was assumed. Five tie backs were
assumed for each caisson location, spaced 1 ft 6 in on center starting 2 ft from the top of the
wall.

Loading.   The loads imposed on the stem of the seawall are primarily due to lateral earth pressure.
Using Rankine theory, the active lateral earth pressure diagram shown in Figure 16.28 was developed,
using the following parameters derived from a comprehensive geotechnical investigation:

- soil friction angle, $\phi = 32°$;
- unit weight of soil, $\gamma = 125$ pcf;
- unit weight of water, $\gamma_w = 62.4$ pcf;
- active pressure coefficient, $K_A = 0.31$; and
- 2 ft soil surcharge load.

To determine earthquake loads, a seismic analysis was performed using a procedure by Mononobe
and Okabe, as described in the AASHTO Specifications (1992). This procedure, which uses an
equivalent static loading, replaces the active soil coefficient, K_A, with a seismic active soil coefficient,
K_{AE}. Using a horizontal acceleration of 0.5 g based on a geoseismic study of the area, K_{AE} was
calculated to be 0.62.

The following design load combinations were calculated in accordance with the ACI Building
Code [ACI, 1995]:

$$U = 1.7H \tag{16.2}$$

$$U = 0.75(1.7H + 1.7(1.1E)) \tag{16.3}$$

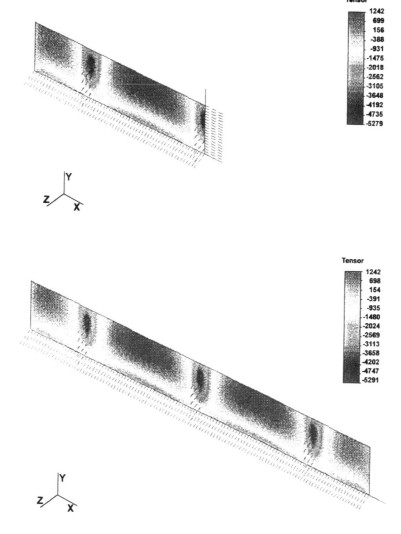

FIGURE 16.27   Horizontal moments from the full and symmetric models (ft-lb/ft).

where

- U = required strength to resist factored loads,
- H = lateral earth pressure, and
- E = lateral earth pressure due to earthquake loading.

The seismic load combination (16.3) resulted in forces approximately 60% larger than load combination (16.2), which considers only the static lateral earth pressure. Thus, load combination (16.3) governed the analysis.

The appropriate load combinations were used in conjunction with the lateral earth pressure diagram shown in Figure 16.28 to determine the pressure loads used in the finite-element analysis.

Wall Capacities.   Moment and shear capacities were calculated for the stem of the seawall in accordance with the ACI Building Code and are summarized in Table 16.5. These capacities were calculated using the reinforcing-steel details shown on the original plans, and assuming

TABLE 16.5 Moment and Shear Capacities of a Typical Undamaged Wall Panel

| Inches | Vertical Moment | | Horizontal Moment | | Shear, lb/ft |
	Negative, ft-lb/ft	Positive, ft-lb/ft	Negative, ft-lb/ft	Positive, ft-lb/ft	
0	−12229	6292	−3636	2534	4190
6	−11899	6292	−3531	2534	3230
12	−11569	6292	−3426	2534	3980
18	−7133	6292	−3321	2534	3870
24	−6928	6292	−3216	2534	3760
30	−6723	4058	−3111	2534	3660
36	−6518	4058	−3006	2534	3550
42	−6313	4058	−2901	2534	3440
48	−2442	4058	−2796	2534	3330
54	−2362	1642	−2691	2534	3230
60	−2282	1642	−2586	2534	3120
66	−2202	1642	−2481	2534	3010
72	−2122	1642	−2376	2534	2900
78	−2042	1642	−2271	2534	2800
84	−1962	1642	−2166	2534	2690
90	−1882	1642	−2061	2534	2580
96	−1802	1642	−1956	2534	2470
102	−1722	1642	−1851	2534	2370
108	−1642	1642	−1746	2534	2260

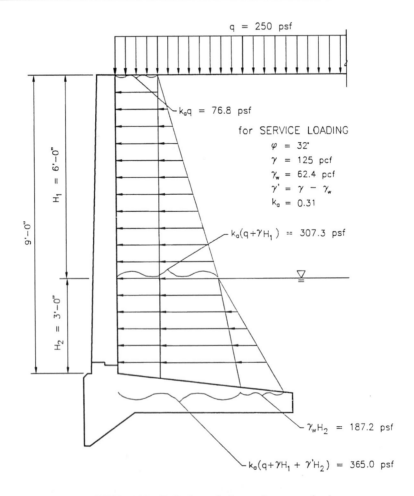

FIGURE 16.28   Preliminary design earth-pressure loads.

TABLE 16.6 Summary of Finite-Element Analysis Cases

Case Number	Corrosion of Vertical Steel at Construction Joint	Load Combination	Number of Strong Backs
1	0%	1	None
2	0%	2	None
3	0%	2	2
4	0%	2	3
5	0%	2	4
6	100%	2	4
7	100%	2	5
8	100%	2	6
9	100%	2	7
10	8% (5 ft at one end)	2	4
11	17% (5 ft between strong backs)	2	4
12	17% (10 ft at one end)	*	None
13	17% (8 ft at one end)	*	None

*Anticipated actual operating service-load conditions.

$f_c' = 4000$ psi and $f_y = 40,000$ psi. Review of the original construction records indicated a concrete strength of 3000 psi was used in the original seawall design. The results of compressive-strength tests conducted during the field investigation indicated an average compressive strength of 4400 psi for the concrete at the time of the evaluation some 35 years later. These test results justified the use of 4000 psi as the *in-situ* concrete compressive strength for the repair design.

Analysis Results. Various combinations of reinforcing-steel corrosion damage scenarios and strong-back spacing were analyzed using the finite-element model and load combinations previously discussed. A summary of the cases analyzed is presented in Table 16.6.

The results of each analysis case were examined and compared with the moment and shear capacities listed in Table 16.5. Vertical and horizontal shears and moments were examined separately, similar to the ACI analysis procedures used for two-way slab design. A sample output of horizontal and vertical moments is shown in Figures 16.29 and 16.30, respectively, for Case 8 (six strong-backs, 100% corrosion of vertical reinforcing steel, and load combination (16.2)).

Negative horizontal moments that occur at strong-back locations were redistributed, as allowed by Section 8.4 of the ACI Building Code. A maximum redistribution limit of 40% was also examined in lieu of the more restrictive 20% limit imposed by ACI. With the lower reinforcement ratio of the horizontal reinforcing steel, it was believed that the wall would exhibit adequate ductility to allow this larger redistribution limit with the caissons in place. For the final design, a detailed moment-curvature analysis of the wall ductility generally confirmed this assumption.

Case 1. The Case 1 analysis examined a typical seawall panel with no corrosion of vertical reinforcing steel, no strong-backs, and no earthquake loading. The average vertical moments were typically slightly below the calculated capacities; however, an 8.5% overstress occurred at the interface with the footing and a 17% overstress occurred 1 ft 6 in above the footing. Considering that a larger value of f_c' was used for the analysis than was originally specified, the overstress indicated that either the load assumptions or design procedures used for the analysis were more conservative than those used for the original design. Hand calculations were also done for this case and were used to verify the results of the finite-element analysis.

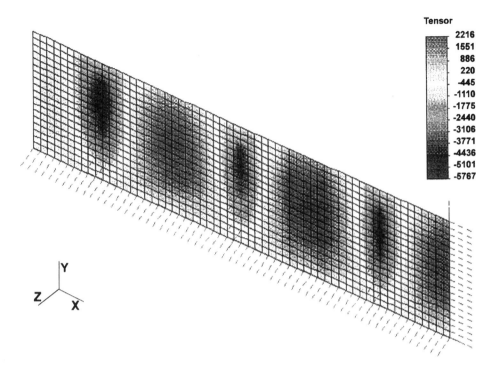

FIGURE 16.29   Horizontal moments for case 8 (ft-lb/ft).

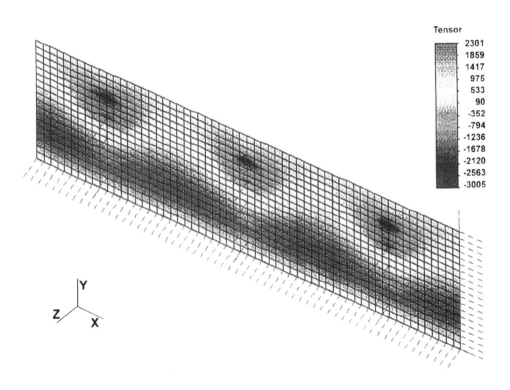

FIGURE 16.30   Vertical moments for case 8 (ft-lb/ft).

Case 2. The Case 2 analysis examined a typical seawall panel with no corrosion of vertical reinforcing steel and no strong backs for earthquake loading. The average vertical moments were larger than the wall capacity from the intersection with the footing to a height of 5 ft 6 in above the footing. The overstress at the interface of the footing was 71%. The results of this analysis indicated the need for a retrofit to address earthquake loading even if there was no corrosion damage.

Using the strong-back retrofit concept, the next three cases were used to determine the minimum number of strong backs required for earthquake loading, assuming no corrosion of vertical reinforcing steel.

Case 3. The Case 3 analysis examined a typical seawall panel for earthquake loading assuming no corrosion of vertical reinforcing steel and strong backs placed 15 ft from each end. By inspection of the horizontal and vertical moments, it was clear that two strong backs would be ineffective for reducing the vertical movement to an acceptable level.

Case 4. The Case 4 analysis examined a typical seawall panel for earthquake loading assuming no corrosion of vertical reinforcing steel and three strong backs placed with a spacing of 20 ft.

The average vertical moment at the base of the seawall for a middle strip equal to half the distance between strong backs is 13,840 ft-lb/ft, which is 13% larger than capacity. In addition, considering a 40% moment redistribution, the average positive and negative moments for the top 4.5 ft of the seawall are 2,710 ft-lb/ft and −2,460 ft-lb/ft, respectively, resulting in an overstress of 7% and 11%.

A more detailed analysis of earthquake loading, soil properties, and surcharge loading provided justification for reducing the load level, and thus it was determined that three strong backs would provide the required lateral-load resistance.

Case 5. The Case 5 analysis examined a typical seawall panel for earthquake loading assuming no corrosion of vertical reinforcing steel and four strong-backs placed with a spacing of 15 ft.

The average vertical moment at the base of the seawall for a middle strip equal to half the distance between strong backs is 13,220 ft-lb/ft, which is 8% larger than capacity. However, considering a 35% moment redistribution, the average positive and negative moments for the top 4 ft 6 in of the seawall are 2430 ft-lb/ft and −2250 ft-lb/ft, respectively, which is less than the average capacities for these regions. Although the vertical moment at the base is slightly overstressed, the horizontal moment capacity was considered adequate when considering a 35% moment redistribution.

Based on analysis of the results for Case 5, the installation of a minimum of four strong backs at each typical panel is required for the assumed earthquake loading.

The next four cases were used to determine the maximum number of strong backs required for earthquake loading assuming 100% corrosion of the vertical reinforcing steel.

Case 6. The Case 6 analysis examined a typical seawall panel for earthquake loading assuming 100% corrosion of the vertical reinforcing steel and four strong backs placed with a spacing of 15 ft. By inspection of the positive and negative horizontal moments, it was clear that both were significantly overstressed.

Case 7. The Case 7 analysis examined a typical seawall panel for earthquake loading assuming 100% corrosion of the vertical reinforcing steel and five strong backs placed with a spacing of 12 ft. As with Case 6, by inspection of the positive and negative horizontal moments, it was clear that both were significantly overstressed.

Case 8. The Case 8 analysis examined a typical seawall panel for earthquake loading assuming 100% corrosion of the vertical reinforcing steel and six strong backs placed with a spacing of 10 ft.

The average positive and negative moments for the top 4 ft 6 in of the seawall assuming a 40% redistribution are 3500 ft-lb/ft and −2340 ft-lb/ft, respectively, resulting in an overstress of 38% and 6%.

Considering the level of overstress of both the horizontal positive and negative moments, the installation of six strong backs would not be sufficient to resist the loadings assumed in the preliminary study. Again, it was later determined that six strong backs was the maximum number required for a given 60 ft long seawall panel when more refined soil parameters became available.

Case 9. The Case 9 analysis examined a typical seawall for earthquake loading assuming 100% corrosion of the vertical reinforcing steel and seven strong backs placed with a spacing of 8 ft 6 in.

The average positive and negative moments for the top 4 ft 6 in of the seawall using a 35% redistribution are 2480 ft-lb/ft and −2000 ft-lb/ft, respectively, which is less than the average capacities for these regions.

Based on analysis of the results for Case 9, a maximum number of seven strong backs is required for the assumed earthquake loading.

Therefore it was determined that for the loading conditions, each seawall panel required the installation of seven strong backs, depending on the amount of vertical reinforcing-steel corrosion. This requirement was relaxed to three to six caissons on the basis of the results from the soil investigation and refinements in the seismic and surcharge loadings.

The next two cases examined the effect of a continuous strip of reinforcing-steel corrosion damage on strong back spacing.

Case 10. The Case 10 analysis examined a typical seawall panel for earthquake loading assuming 5 ft of continuous (8%) corrosion of the vertical reinforcing steel at the end of a panel and four strong backs placed with a spacing of 15 ft.

Based on the analysis of this case, 5 ft of continuous corrosion of the vertical steel at the end of a panel required an additional strong back.

Case 11. The Case 11 analysis examined a typical seawall panel for earthquake loading assuming 5 ft of continuous (8%) corrosion of the vertical reinforcing steel, centered between two strong-backs, assuming four strong-backs placed with a spacing of 15 ft.

The average positive and negative moments for the top 4 ft 6 in of the seawall using a 40% redistribution are 3850 ft-lb/ft and −2380 ft-lb/ft, respectively, resulting in an overstress of 52% and 7%.

Considering this level of overstress of both the horizontal positive and negative moments, it was determined that 5 ft of continuous damage between strong backs required a strong-back spacing of less than 15 ft.

Case 12. The Cases 12 and 13 analyses examined a typical seawall panel with no strong backs using the anticipated actual operating service load conditions to determine the amount of continuous vertical reinforcing steel damage at one end that would lead to a failure ("unzipping") of the panel. In this case, failure would occur by progressive yielding of the uncorroded vertical reinforcing adjacent to the damaged area that continued to the opposite end of the panel. The Case 12 analysis examined 10 ft of continuous damage from one end.

On the basis of the results of this analysis, it was determined that continuous damage of 10 ft or more was sufficient to cause a progressive failure of the seawall panel.

Case 13. The Case 13 analysis used the same assumption as Case 12, except continuous damage of 8 ft from one end was examined.

On the basis of the results of this analysis, it was determined that continuous damage of less than 8 ft would not cause a progressive failure of the seawall panel.

16.9.8 Life-Safety Implications

The Marina del Rey Seawall was at a critical juncture in its 30 year service life. Some $5 billion of infrastructure had developed in and around the marina, making it an integral part of the local economy. Long-term exposure to seawater had resulted in corrosion of the vertical reinforcing steel. Because of the nonredundant nature of a cantilever wall system, continued corrosion of the reinforcing steel could result in the collapse future of seawall panels similar to the incident that occurred in 1986. The probability of failure would increase with time if the corrosion continued unabated. Analysis indicated that a loss of approximately 60% of the vertical reinforcing steel due to corrosion at the construction joint could result in a collapse at the present operating service load levels. The analysis also indicated there would be little visible sign of distress ($\approx$3/8 in horizontal top of wall deflection) prior to collapse.

To address life-safety concerns, a multifaceted restoration program was developed by various local and state agencies. The plan utilized an impressed current cathodic protection system to mitigate active corrosion of the reinforcing steel. A strong-back system was developed to upgrade the seismic behavior of the seawall and the integrity lost by corrosion of the reinforcing steel.

A minimum of three caissons was determined to be necessary to upgrade the seismic capacity of the seawall to satisfy current building-code requirements. The number of additional caissons was directly proportional to the extent of corrosion damage along the base of the seawall. For the case of full damage along the seawall (worst case scenario), analysis indicated that six caissons were required to provide satisfactory capacity.

To determine the extent of the corrosion damage along the base of the seawall, a program of nondestructive impact-echo testing was conducted. Owing to budget constraints of the various municipal and state agencies involved, guidelines for prioritization of the repairs were needed. These guidelines helped to ensure that life-safety considerations were of paramount importance with respect to the prioritization of repairs.

The budget constraints of the municipal and state agencies involved in the Marina del Rey Seawall Restoration Program necessitated a prioritization of repairs based upon two criteria: (1) the extent of damage along the base of the seawall panel and (2) the degree of public exposure of an individual seawall panel. To simplify the categorization of public exposure, use of two categories was proposed. A public exposure categorized as "high" was limited to areas where a seawall panel collapse would result in a high probability of personal injury. All other areas were categorized as "low" exposure.

Analysis indicated that a 60% loss of reinforcing steel might result in the collapse of the seawall panel at its actual operating service load levels. This was based upon the assumption of uniformly distributed corrosion damage across the base of the seawall. Analysis also indicated that contiguous damage from the end of a seawall panel was more severe in nature than uniform damage. For the purposes of categorizing the extent of damage, the adoption of a dual criteria classification was proposed. The classification assigned the panels to one of three categories, as described in Table 16.7.

Wall panels were placed into the appropriate damage category if either criteria in Table 16.7 was satisfied. The use of 40% of total wall damage was recommended as the maximum allowable to provide some additional measure of conservatism. During the impact-echo testing, panels determined to fall in high-damage category were immediately brought to the attention of the public agencies involved.

TABLE 16.7 Prioritization Based on Degree of Wall-Panel Damage

% of Total Wall Damage	Number of Feet of Continuous Damage	Corrosion Damage Category
>40%	>8 ft	High
20–40%	4–8 ft	Intermediate
<20%	<4 ft	Little or no damage (Seismic upgrade required)

TABLE 16.8   Repair Priority

Damage Category	Public Exposure	
	High	Low
High	Highest priority	High priority
Intermediate	Highest priority	High priority
Little or no damage (Seismic upgrade required)	Priority	Priority

Owing to the financial considerations, a phased construction schedule was proposed. To minimize public exposure to the potential collapse of a seawall panel, repairs in areas of high public exposure with high and intermediate damage levels were classified as the highest priority. Other areas were categorized into two additional groups based upon public exposure and degree of deterioration. Table 16.8 specifies the repair prioritization groupings that were made.

Owing to the need to upgrade the seismic capacity of the seawall, no repairs were considered to have a low priority.

16.9.9 Cost Savings

Structural analysis of the Marina del Rey Seawall indicated that the number of caissons required to upgrade the capacity of the seawall for seismic loads was directly related to the amount of corrosion damage present at the construction joint. This indicated that a categorization of seawall panels into logical groupings, based upon the extent of damage detected by impact-echo, could provide a significant cost savings during the construction phase of the project.

Based upon the extent of damage detected during impact-echo testing, the seawall panels were placed into three categories: high damage level, intermediate damage level, and little or no damage present (only seismic upgrade required). The number of caissons required for panels in the high-damage and little-to-no-damage categories was fairly well established. In the intermediate-damage category, a further breakdown of the category was possible. The division of the intermediate-damage category into two groups resulted in four "families" of repairs, categorized in terms of the number of caissons required to provide the upgraded seismic capacity and to address the loss of strength due to corrosion. The categories or "families" of repair are summarized in Table 16.9.

Panels at the intermediate-damage level were placed into separate categories based upon the results of impact-echo testing and further structural analysis that examined the effects of localized corrosion damage and distribution of damage on the repaired system.

After completion of the seawall testing and extensive structural analysis, a construction cost savings estimate was made. This estimate is shown in Table 16.10. The cost savings were based on comparing the number of caissons required to strengthen the walls, assuming 100% corrosion damage, to the number of caissons used in the repair scheme. A savings of $5.8 million was realized by conducting

TABLE 16.9   Caisson Requirements Based on Degree of Damage

Degree of Damage	Number of Caissons Required
High	6
Intermediate–High	5
Intermediate–Low	4
Little or no damage (Seismic upgrade only)	3

TABLE 16.10   Cost Savings Resulting from Impact-Echo Testing

Caissons required assuming 100% damage	3971
Caissons required utilizing impact-echo testing results	2473
Caissons saved as a result of impact-echo testing	1498
Estimated construction cost per caisson	$3,900.00
Construction savings resulting from impact-echo testing	$5,842,000.00

impact-echo testing. On an economic basis, the merits of using nondestructive testing in this infrastructure repair project are clearly seen.

16.9.10 Conclusions

The Marina del Rey Seawall had been in service as a mixed-use residential and recreational area for over 30 years, and some $5 billion of infrastructure had been built around the area. To ensure its continuing service for at least another 30 years or more, an extensive restoration program was developed to mitigate the on-going corrosion of the seawall reinforcing steel, to upgrade the seismic capacity of the seawall, and to restore the integrity of seawall panels that had been lost owing to corrosion.

The following are the more salient conclusions from the presentation of this case study:

1. Analysis of the seawall indicated that with increasing levels of reinforcing steel corrosion, the ductility of the wall decreased. At an assumed 60% loss of reinforcing steel due to corrosion at the seawall base, a failure was possible at the operating service load levels on the seawall. The horizontal deflection of the top of the wall prior to failure was determined to be approximately 3/8 in for the case of 60% reinforcing-steel loss. This would provide little warning of pending failure.

2. The finite-element analysis clearly indicated the need for strong-back installation to seismically upgrade the seawall panels. However, the number of caissons required beyond the number needed for seismic resistance increased as the amount of corrosion damage increased. Between three and six caissons were required to provide adequate lateral load resistance depending on the degree of corrosion damage detected.

3. Threshold criteria were established to prioritize the repairs. The categorization was principally based on life-safety concerns in areas of high public exposure. Seawall panels that were determined to have a high degree of deterioration and that were located in areas of high public exposure were classified as the highest priority.

4. On the basis of the cost estimate for caisson installation and the estimated number of panels that required only seismic-capacity upgrades, a substantial cost savings was realized with the performance of nondestructive impact-echo testing. Some $5.8 million was saved by conducting this testing.

16.9.11 Reports, Standards, and Guidelines

16.9.11.1 ACI Reports

ACI 503R   Standard Specification for Bonding Hardened Concrete, Steel, Wood, Brick, and Other Materials to Hardened Concrete with a Multi-Component Epoxy Adhesive

16.9.11.2 ASTM Standards

ASTM C 39	Test Method for Compressive Strength of Cylindrical Concrete Specimens
ASTM C 42	Test Method for Obtaining and Testing Drilled Cores and Sawed Beams of Concrete
ASTM C 157	Test Method for Length Change of Hardened Hydraulic Cement Mortar and Concrete
ASTM C 805	Test Method for Rebound Number in Concrete
ASTM C 803	Test Method for Penetration Resistance of Hardened Concrete
ASTM C 597	Test Method for Pulse Velocity Through Concrete
ASTM C 876	Test Method for Half-Cell Potentials of Uncoated Reinforcing Steel in Concrete
ASTM C 856	Standard Practice for Petrographic Examination of Hardened Concrete

ASTM C 457	Test Method for Microscopical Determination of the Air-Void System in Hardened Concrete
ASTM C 1084	Test Method for Portland Cement Content of Hardened Hydraulic-Cement Concrete
ASTM C 642	Test Method for Specific Gravity, Absorption, and Voids in Hardened Concrete
ASTM C 496	Test Method for Splitting Tensile Strength of Cylindrical Concrete Specimens
ASTM C 215	Test Method for Fundamental Transverse Longitudinal, and Poisional Frequencies of Concrete Specimens
ASTM C 469	Test Method for Static Modulus of Elasticity and Poisson's Ratio of Concrete in Compression
ASTM C 341	Test Method for Length Change of Drilled or Sawed Specimens of Hydraulic-Cement Mortar and Concrete
ASTM C 1202	Test Method for Electrical Indication of Concrete's Ability to Resist Chloride Ion Penetration
ASTM D 4748	Test Method for Determining the Thickness of Bound Pavement Layers Using Short-Pulse Radar
ASTM D 4580	Practice for Measuring Delaminations in Concrete Bridge Decks by Sounding
ASTM D 4788	Test Method for Deflecting Delaminations in Bridge Decks Using Infrared Thermography

16.9.11.3 ICRI Documents

Guideline No. 03730	Surface Preparation for the Repair of Deteriorated Concrete Resulting from Reinforcing Steel Corrosion
Guideline No. 03732	Guide for Selecting and Specifying Surface Preparation for Sealers, Coatings, and Membranes
Guideline No. 03733	Guide for Selecting and Specifying Materials for Repair of Concrete Surfaces
Guideline No. 03734	Field Guide for Verifying Performance of Epoxy Injection

16.9.11.4 SHRP (Strategic Highway Research Program) Documents

SHRP - C/FR-91-101	Handbook for the Identification of Alkali-Silica Reactivity in Highway Structures
SHRP - S-330	Condition Evaluation of Concrete Bridges Relative to Reinforcement, Corrosion Volume 8: Procedure Manual
SHRP - S-324	Condition Evaluation of Concrete Bridges Relative to Reinforcement, Corrosion Volume 2: Method for Measuring the Corrosion Rate of Reinforcing Steel
SHRP - S/FR-92-108	Condition Evaluation of Concrete Bridges Relative to Reinforcement, Corrosion Volume 6: Method for Field Determination of Total Chloride Content

References

Aalami, B.O. and Swanson, D.T. 1988. Innovative rehabilitation of a parking structure. *Concrete Int.* 10(2):30–35.

ACI Committee 364. 1993. Guide for evaluation of concrete structures prior to rehabilitation. *ACI Mater. J.* 90(5):479–498.

American Association of State Highway and Transportation Officials. 1992. *Standard Specifications for Bridges*, 15th ed.

American Concrete Institute. 1995. *Building Code Requirements for Reinforced Concrete*, ACI 318-95, Commentary, *ACI 318R-95*, pp. 353. Farmington Hills, MI.

American Society of Civil Engineers. 1990. *Guideline for Structural Assessment of Existing Building*, pp. 11–90. New York.

Emmons, P.H. 1994. *Concrete Repair and Maintenance Illustrated*, 295 pp. R.S. Means Company, Inc.

Emmons, P.H., Vaysburd, A.M., and McDonald, J.E. 1994. Concrete repair in the future turn of the century—Any problems? *Concrete Int.* 42–49.

MacGregor, J.G. 1976. Safety and limit state design for reinforced concrete. *Can. J. Civ. Eng.* 3(4).

Natesaiyer, K. and Hover, K.C. 1988. *In situ* identification of ASR products in concrete. *Cement Concrete Res.* 18:455–463.

Natesaiyer, K. and Hover, K.C. 1989. Some field strategies of the new *in situ* method for identification of alkali–silica reaction products in concrete. *Cement Concrete Res.* 19:770–778.

Poston, R.W. and Irshad, M. 1996a. Load tests: Externally post-tensioned bridge girders, Part I 18(8).

Poston, R.W. and Irshad, M. 1996b. Load tests: Externally post-tensioned bridge girders, Part II. 18(9).

Sansalone, M. and Carino, N.J. 1986. Impact-echo: A method for flow detection in concrete using transient stress waves, *NBSIR 86–3452*. 222 pp. National Bureau of Standards, Gaithersburg.

Sansalone, M. and Carino, N.J. 1988. Impact-echo method. *Concrete Int.* 38–46.

Transportation Research Board. 1986. Strategic highway program research plans. Final report, *NCHRP Project 20-20*, TRA 4-1–TRA 4-60. Washington, DC.

U.S. Department of Transportation. 1991. *Bridge Inspectors Training Manual /90*, pp. 1-1–1-3. Federal Highway Administration, Washington, DC.

Wolkomir, R. 1994. Inside the lab and out, concrete is more than it's cracked up to be. *Smithsonian*, 24(10):22–31.

Automation in placement of concrete wall (Courtesy Portland Cement Association).

17

Automation in Concrete Construction

by

Miroslaw J. Skibniewski, Ph.D.
Professor, Civil Engineering, Purdue University, West Lafayette, Indiana. Expert in construction automation and in technology management.

Raghavan Kunigahalli, Ph.D.
R&D Coordinator, Bentley Systems, Inc., Exton, Pennsylvania. Expert in Computer Aided Design and Construction.

17.1 Categories of Construction Automation

Concrete construction automation is a broadly defined planning and technical endeavor that includes two distinct areas:

1) The first is development of programmable, that is, robotic, hardware for the execution of construction work tasks. Significant progress has been achieved in equipment navigation, locomotion systems, and concrete-placement systems.
2) The second is development of computer-based tools for efficient and optimal planning, design, construction, and operation of concrete structures. Of particular importance is the development and practical application of tools for design visualization, quantity takeoff and cost estimation, generation of work schedules and job cost reports, design-construction integration, construction task planning, optimal resource management, and design for constructability and maintainability of concrete structures.

17.2 Automated Construction Equipment and Related Hardware

Construction robotics are now beyond the initial design and academic discourse stages, which was not the case throughout the 1980s and early 1990s [Skibniewski, 1988, 1996]. A number of institutions worldwide developed prototype hardware for various applications that are now in field testing or initial implementation stages.

Currently, the most mature applications of robotics in concrete construction include concrete screeding, surface finishing, scrubbing, and cleaning. A number of Japanese construction firms developed their own prototypes of such machines, including Obayashi, Corp., Taisei Corp., Takenaka Corp., and Shimizu Corp.

An automatic concrete screeder has been developed by Takenaka Corp. (see Figures 17.1a–c) The girder-mounted machine has an automatically controlled screeding tool that operates sequentially along a girder running over a self-shifting rail. When the tool reaches the edge of the rail, the rail is shifted forward automatically after invoking the rail-shifting function on the machine controller. The entire screeder has a modular design and can be manually assembled on the construction site. Assembly and disassembly can be accomplished by two or three workers in about 4 h. The screeding tool includes a screw to perform concrete leveling by transporting surplus concrete to the side. This is done using a vibrating board that ensures an adequately horizontal surface when the concreteis shifted to the side. The maximum weight of the screeding tool is approximately 130 kg. The girder support structure for the trowel is also modular with part ranging from 2 to 3 m in width. The traveling saddle for the trowels is located under the girder. The maximum weight of the girder components is approximately 50 kg. The control system for the screeder includes an inclinometer, a laser leveling device and a rotary encoder [Okuda et al., 1992]. The screeding capacity of the machine is approximately 350 m^2 of a slab area per hour, or 35 m^3 of fresh concrete mix per hour on a slab with a thickness of 18 cm.

An example concrete surface finishing robot, the "Flatkin" by Shimizu Corp., is shown in Figure 17.2. The Flatkin consists of travel rollers, a power trowel, a controller, and a guard frame. A pair of travel rollers are attached at the bottom of the robot's main body. The robot can move back and forth or left and right, using DC motors to drive the rollers. The power-trowel mechanism has three supporting arms. Each arm has a rotating trowel assembly with three trowels each. A gasoline engine is used as a power unit to drive the trowel assemblies, so that the trowel assemblies rotate around their axis and simultaneously rotate around the entire traveling device. The ratio of the rotation speed of the trowels to the revolution speed of the robot around its own axis is approximately 10 and above. The angle between the concrete surface and the trowel can be adjusted by a cam mechanism. This angle is usually changed depending on the hardness of the concrete surface to be finished. The robot is equipped with a guard frame, with a touch sensor mounted as a safety device. In addition to the engine for the trowel, the robot has a small generator that enables the elimination of electric power cables, thereby increasing its mobility. The Flatkin can be operated by radio remote control, which is a useful feature for changing the trowel blade positions depending on the hardness of concrete encountered in a given work area. The robot work output is approximately 600–700 m^2 of concrete surface area per day, utilizing two operators in the process. This makes the robot's productivity four to five times higher than that of human workers utilizing mechanized "walk-behind" trowels and manual tools [Ueno et al., 1988].

A partially automated overhead construction factory system for high-rise reinforced concrete frame buildings had been developed by Obayashi Corp. The "Big Canopy" system integrates a synchronously climbing canopy that houses semiautomated overhead cranes, prefabricated concrete paneling components, computerized management of storage and retrieval of materials on site, and

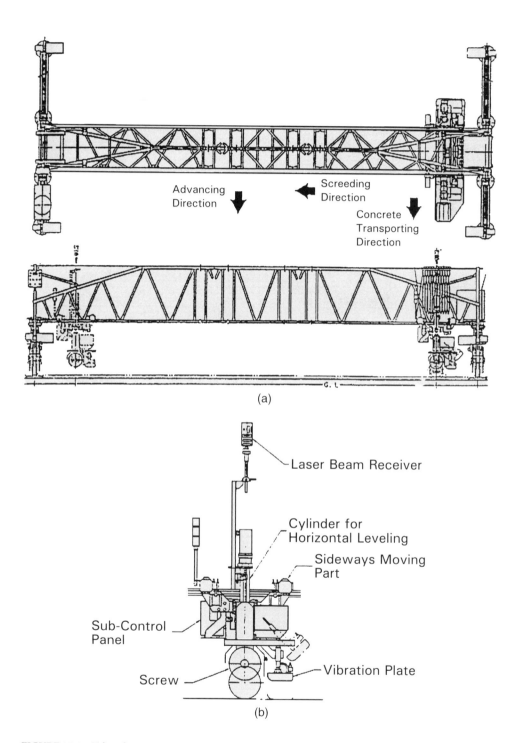

(a)

(b)

FIGURE 17.1   Takenaka Corp. concrete screeder. (a) Top and side views; (b) detail; (c) operational steps.

(*continues*)

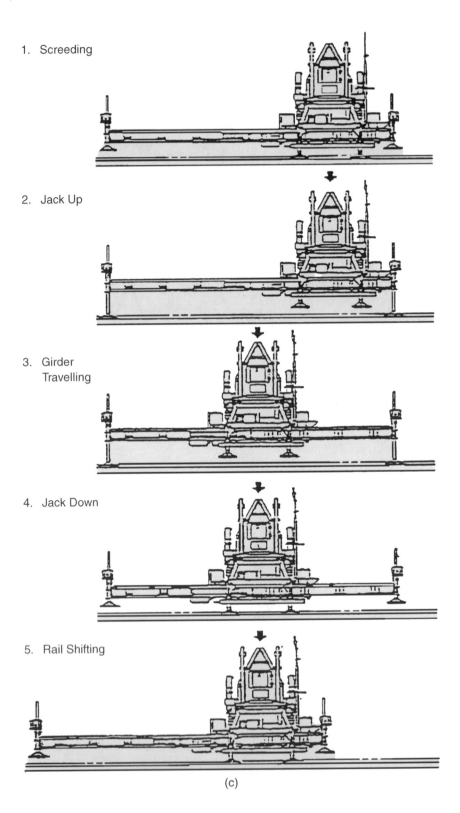

1. Screeding

2. Jack Up

3. Girder Travelling

4. Jack Down

5. Rail Shifting

(c)

FIGURE 17.1 (*continued*)

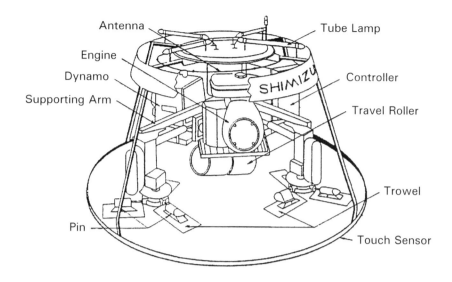

FIGURE 17.2   Concrete surface finishing robot by Shimizu Corp.

a semiautomated structural assembly (see Figure 17.3). The canopy covers the entire story of the building being erected to protect workers from severe weather and to produce a safer work environment. Tower crane posts are used as four columns that support the canopy. Raising of the canopy is performed by the climbing facility of the tower cranes. Safety of motion is maintained by synchronized control. A high-speed construction lift and three hoist cranes are combined to deliver the structural components and materials for assembly. The material is raised to the working floor by the lift and passed to the hoist on the delivery girder. The movement of the hoist is fully automated for maximum work efficiency. Upon the completion of building construction, the canopy is disassembled on the top of the building. The external frame is lowered synchronously and then safely disassembled on the ground.

In automated surveying technologies relevant to concrete construction, the Consortium for Advanced Positioning Systems (CAPS) engineered an application of a laser-based positioning device called Odyssey™, developed by Spatial Positioning Systems, Inc. (SPSi), see Figure 17.4. There are two primary components in the system: transmitters and receivers. At least two transmitters are required to provide positioning signals to a receiver. However, any number of receivers can utilize the positioning signals simultaneously. The transmitters can be set up at convenient locations and generally aimed at the work site. Existing benchmarks are used to calibrate the system using any local coordinate system. Each receiver is composed of two lenses mounted on a pole, a processor, a data entry and retrieval system, and a power supply.

The two lenses form a line, and the position of the lenses and the known geometry of the pole allow the point of position measurement to be projected to the end of the pole. Since the position of the tip of the pole does not change if the pole is slanted, rotated, turned upside down, or sideways, the position of any point that the user touches with the receiver is accurately and "instantly" measured. SPSi system software provides basic functions such as distance between two points, areas, volumes, or angles. However, the integrated site positioning system combines real-time coordinate data from the Odyssey system with CAD design data. The combination of real-time coordinate measurement and CAD representation allows field position and grapical design data to be provided simultaneously to the user. There are a variety of applications for this system in new construction as well as in retrofit projects. For example, in facility characterization, the comparison of 'as designed' with 'as built' physical parameters is a large application area in itself. Other applications inlude industrial plant outage planning and simulation, modular planning, fabrication and construction, consistent site control during construction, providing plant database baselines, and real-time position feedback for automated construction equipment [Beliveau, 1996].

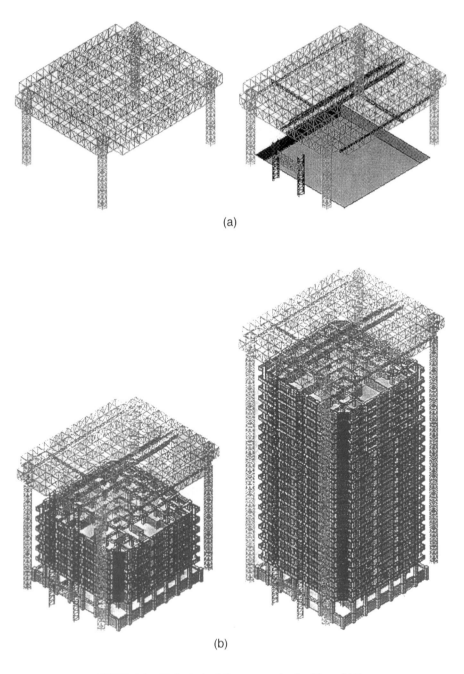

(a)

(b)

FIGURE 17.3 High-rise building automation by Obayashi Corp.

(*continues*)

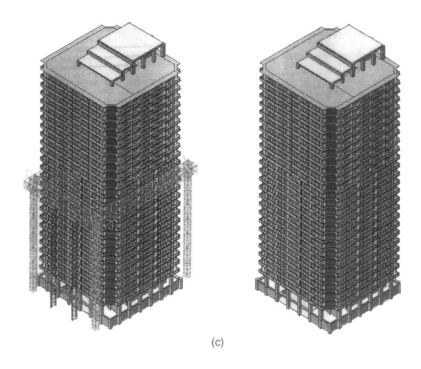

(c)

FIGURE 17.3 (*continued*)

FIGURE 17.4 Laser-based positioning system (Courtesy of SPSi).

17.3 Economics and Management of Robots

The most decisive factor for using a robotic application in construction will be its impact on the overall cost of the concrete construction process [Skibniewski, 1988]. The promising areas of application are in tasks where the work volume, high repetition, and simple control requirements result in a promising robot automation potential. Such tasks include concrete surface treatment (e.g., cleaning, painting, sandblasting, etc.), inspection (e.g., nondestructive testing of concrete and reinforcement steel and assessment of ceramic tile adhesion to concrete surfaces), and concrete placement.

Robot-related costs to be reconciled during the analysis include capital costs and operating costs. The capital costs include research and development expenditures (hardware and software, work-system engineering, calibration, and field hardening). Operating costs include energy, maintenance, downtime, repair, tooling, setup, dismantling, transportation, operator, and other related expenditures.

Robot-related benefits include construction labor and material savings, improved work quality, extension of work activity into additional locations and time periods, and possibly improved productivity. Details of the economic analysis of construction surface finishing tasks can be found in Skibniewski (1988).

As can be expected from the conditions under which the construction industry operates, the most important decision factor for robot implementation will be a short-term profit potential resulting from labor savings through productivity improvement and possibly through increased construction quality. Most conservative construction firms will be unlikely to invest in robot research and development and will rely on robotics technology developed by commercial robotic systems houses. Such robots will then be either sold or leased by commercial vendors operating in the construction equipment market.

A major difficulty that construction firms will initially face is the estimation of robot costs and benefits, as outlined above. This will improve as more experience with particular robot applications is gained. Detailed information on the cost and benefit items for various robot applications in typical job-site settings can accelerate the pace of robotization if it is made available to all interested construction firms. Future construction robot equipment vendors will be well positioned to fulfill this function in cooperation with robot system developers and manufacturers. Once more robotics become available for use on construction sites, significant challenges to both management and technical staff will emerge. For the reasons outlined above (in paragraph 4), robots must be managed wisely and perform at a high quality level in order to ensure maximum economic benefits for the contractor's firm.

Despite a number of advantages over traditional methods of performing construction tasks, robots are currently, and will continue to be in the near future, in short supply in comparison to other construction equipment. Thus robots should be regarded as a scarce resource and their use should be maximized to their full operating potential. Using maximized robot capabilities on as many construction projects as possible, the economic benefits of robot use can be easier to attain. Consequently, robot development costs can be recovered faster, and robot use can spread to other applications and types of construction tasks.

A hypertext-based optimization program called Construction Robotics Management System (CREMS) has been developed for that purpose. As shown in Figure 17.5, it consists of four

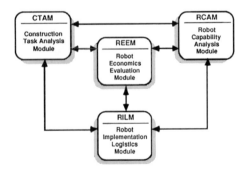

FIGURE 17.5   Construction robotics equipment management system.

modules: Construction Task Analysis, Robot Capability Analysis, Robot Economic Evaluation, and Robot Implementation Logistics [Skibniewski and Russell, 1991].

Automation in construction still constitutes a difficult technical and managerial challenge. However, potential benefits may be significant in this industry. Owing to the lack of investment in research and development by construction firms in most industrialized countries except Japan, in combination with other factors, the industry has been experiencing difficulties in introducing and adopting new technologies at the project site level. To facilitate a more comprehensive impact of automation on the construction industry in the future, further research and development is essential. In particular, more attention should be focused on the redesign of construction sitework environments to enable direct technology transfer from other industries to construction, rather than on development of customized automated construction equipment that would closely resemble the humanlike performance of traditional construction tasks [Skibniewski and Nof, 1989]. Sound methodologies for systematic technology transfer and evaluation are also necessary [Gaultney et al., 1989], particularly those utilizing the recent advances in computing technologies [Hijazi and Skibniewski, 1989]. A better-quality constructed products will increase the competitiveness of concrete construction firms involved and can ultimately lead to greater demand for their services.

17.4 Computer-Aided Design (CAD)

17.4.1 Traditional Architectural CAD Modeling

Reinforced concrete (RC) structural facility delivery processes include the design and drafting of various sectional views, on drawing sheets, as two-dimensional drawings. Even a small design modification necessitates time-consuming redrawing of various sectional views, along with detailed rebar specifications. Emergence of traditional computer-aided design (CAD) systems changed the time-consuming manual design and drafting process into a series of key-strokes and mouse drags-and-clicks on a computer. As basic operations such as drawing and erasing a line, arc, etc., are much faster in a computer, traditional CAD systems reduced the cost of redrawing that resulted from design modifications.

However, traditional CAD systems require manual interpretation of the information pertaining to a given RC structural facility. For instance, a given RC structural component such as beam is represented by several projections along with a set of symbols. This representation process requires manual interpretation not only during the design stage, which provides input of geometric and symbolic information, but also during the construction stage that reproduces the exact shapes and configurations intended during the design stage. Further, the shapes and configurations are manually interpreted during the design and construction stages by different sets of people with widely varying knowledge and skills. In cases of complex reinforcement configurations—resulting from either the shape of the structure or special requirements such as earthquake resistance— total misinterpretation of shapes and configurations is possible during the construction stage, thus necessitating costly redesigns and rework.

More advanced solid-modeling approaches, currently not prevalent in the RC structure domain, employ specialized representation schemes that would facilitate the interpretation process by capturing the shape information pertaining to a given design object. A brief discussion of various solid-modeling approaches is provided below.

17.4.2 Solid Geometric Modeling

A solid geometric model is an unambiguous and informationally complete mathematical representation of the physical shape of an object in a form that a computer can easily process [Mortenson, 1985]. Topology and algebraic geometry provide the mathematical foundation for solid modeling. Computational aspects of solid modeling include data structures and algorithms from computer science and application considerations from the design and construction of engineering projects. The following techniques are available for the solid modeling of civil engineering facilities:

(1) primitive instancing,
(2) cell decompositions,
(3) spatial occupancy enumeration (SOE),
(4) constructive solid geometry (CSG),
(5) sweep representations, and
(6) boundary representation (B-Rep) [Requicha, 1980].

17.4.2.1 Primitive Instancing

The Primitive instancing modeling technique consists of an independent approach to solid-object representation in the context of the group technology (GT) paradigm. The modeling approach is based on the notion of families of objects, each member of the family being distinguishable by a few parameters. Columns, beams, and slabs can be grouped as separate families in the case of general buildings. Each object family is called a **generic primitive** and individual objects within a family are called **primitive instances** [Requicha, 1980].

17.4.2.2 Cell Decompositions

Cell decompositions are generalizations of triangulations. Using the cell decomposition modeling technique, a solid may be represented by decomposing it into cells and representing each cell in the decomposition. This modeling technique can be used for analysis of trusses and frames in industrial and general buildings, bridges, and other civil-engineering structures. In fact, the cell decomposition technique is the basis for finite-element modeling [Mortenson, 1985].

17.4.2.3 Spatial Occupancy Enumeration (SOE)

The spatial occupancy enumeration technique is a special case of the cell decomposition technique. A solid in SOE scheme is represented using a list of spatial cells occupied by the solid. The spatial cells, called **Voxels**, are cubes of a fixed size that lie in a fixed spatial grid. Each cell may be represented by the coordinates of its centroid. Cell size determines the maximum resolution. This modeling technique requires large memory space, thereby leading to inefficient space complexity. However, this technique may be used for motion planning of automated construction equipment under the complete information model [Requicha, 1980].

17.4.2.4 Constructive Solid Geometry (CSG)

Constructive solid geometry (CSG), often referred to as **building-block geometry**, is a modeling technique that defines complex solids as a composition of simpler primitives. Boolean operators are used to execute the composition. CSG concepts include regularized boolean operators, primitives, boundary-evaluation procedures, and point membership classification. CSG representations are ordered binary trees. Operators that may consist of either rigid motion, regularized union, intersection, or difference are represented by nonterminal nodes. Terminal nodes are either primitive leaves that represent subsets of three-dimensional (3D) Euclidean space or transformation leaves that contain the defining arguments of rigid motions. Each subtree that is not a transformation leaf represents a set resulting from applying the motional and combinational operators to the sets represented by the primitive leaves.

The CSG modeling technique can be adopted to develop computer-aided-design and drafting (CADD) systems for civil engineering structures. It can be combined with primitive instancing that incorporates the group technology paradigm to assist the designer. Although the CSG technique is most suitable for design-engineering applications, it is not suitable for construction-engineering applications as it does not store the topological relationships required for construction process planning [Requicha, 1980].

17.4.2.5 Sweep Representation

The sweep representation technique is based on the idea of moving a point, curve, or surface along a given path. The locus of points generated by this process results in one-dimensional (1D), two-dimensional (2D), and 3D objects, respectively. Two basic ingredients are required for sweep representation: an object to be moved and a trajectory to move it along. The object can be a curve, surface, or solid. The trajectory is always an analytically definable path. There are two major types of trajectories: translational and rotational [Mortenson, 1985].

17.4.2.6 Boundary Representation (B-Rep)

The boundary-representation modeling technique involves representing a solid's boundary by decomposing it into a set of faces. Each face is then represented by its bounding edges and the surface on which it lies. Edges are often defined in the 2-D parametric space of the surface as segments of piecewise polynomial curves. A simple enumeration of a solid's faces is sufficient to unambiguously separate the solid from its complement. However, most boundary-representation schemes store additional information to aid feature extraction and determine topological relationships. The additional information enables intelligent evaluation of CAD models for construction process planning and the automated equipment path planning required in Computer-aided design/Computer-aided construction (CAD/CAC) systems [Requicha and Rossignac, 1992; Kunigahalli et al., 1995; Kunigahalli and Russell, 1995a].

A boundary-representation technique that stores topological relationships among geometric entities is most suitable for computer-aided generation of construction process plans. However, primitive instancing, sweep representation, and CSG techniques are useful in developing user-friendly CAD software systems for design of civil-engineering structures. Hence CAD systems that incorporate CSG or the primitive instancing technique during the interactive design process and that employ the boundary-representation technique for internal storage of design information are efficient for use in CAD/CAC systems [Kunigahalli and Russell, 1995b].

17.4.3 Solid Modeling of Reinforcing Elements

17.4.3.1 General

The shape of the boundary of a reinforcing element corresponds to a solid cylindrical primitive. A solid cylindrical primitive consists of three faces, three edges, and two vertices. A boundary-representation scheme must account for storage and manipulation of these topological entities. A boundary-representation scheme, called a rectangle adjacency graph (RAG), supports solid modeling of reinforcing elements for various structural components such as beams, columns, and slabs. A brief description of the RAG modeling approach for reinforcement detail is described next. A more detailed description can be found in Kunigahalli (1997, Vol. 11, No. 2, pp. 92–101, 1997).

17.4.3.2 Description of RAG Scheme for Reinforcement Detail

Beam (or Column) Components. An RC beam or a column component consists of longitudinal bars to resist bending moment and stirrups or ties to resist shear force. The portion of a beam or column having the same configuration of longitudinal reinforcement is called a **region**. The geometric and topological information regarding reinforcement detail in a given region of a beam or column component is stored in a structure called a **Beam_Column_Region**. The information regarding the boundary and reinforcement detail of a beam or column component and the connectivity of a beam or column component to its adjoining structural joints is stored in a structure called **Beam_Column_Component**. Definitions of Beam_Column_Component and Beam_Column_Region structures are provided in the appendix.

The Beam_Column_Component structure contains a beam identification number, pointers to two adjacent structural joints, a pointer to a **Face_Table** structure that stores the boundary representation (B-rep) of a beam or column component itself, and a pointer to a list of Beam_Column_Region structures. A parent pointer included in the Beam_Column_Region structure enables faster identification of the relative location of a region with respect to a complete RC-framed structure. **Long_Circular_List** and **Loop_List** store geometric and topological information pertaining to longitudinal reinforcement and stirrup or tie reinforcement, respectively.

The geometric and topological information regarding an individual reinforcing bar in a given region i is stored using an **Edge-Face** structure that contains a pointer to the parent region and an enumerated type to uniquely identify an individual longitudinal bar in a given region. Geometric and topological information pertaining to lap-spliced bars are stored using a separate structure called **Splicing_Rebar**. The Edge-Face structures of longitudinal bars in a given region i are stored in a circular list. A specification for the ordering of the circular list and labeling of longitudinal bars using the enumerated types has been provided to ensure an unambiguous representation.

There can be more than one loop of stirrup/ties to resist the shear force at a given cross section of a beam or column component. The spacing of loops, and in some cases the configurations of the loops of stirrups/tie bars themselves, may vary along the length of a given region. Geometric information pertaining to the vertices and faces is stored using-floating point numbers. The information regarding the identified edge of a stirrup/tie reinforcing bar is stored using a pointer to an **Edge-Type** structure. The edge-to-edge and edge-to-contact-vertex topological relationships between a stirrup/tie bar and a longitudinal bar at the location of a standard hook can also be stored using a tailored structure designed specifically for storing hook information.

Slab Component. The RAG scheme supports solid modeling of reinforcement detail pertaining to a slab component, designed using one-way and two-way slab theories, that consists of longitudinal reinforcing bars to resist positive bending moment, negative bending moment, and torsion at the four corners. The reinforcement for positive bending moment and torsion are typically provided in two layers of bars, namely, upper and lower, that are placed along the two principal orthogonal directions x and y. There exists a boundary edge-to-edge contact between a given longitudinal bar and every other longitudinal bar placed in the other (orthogonal) direction. Negative bending moment reinforcement for a slab normally results in edge-to-edge contacts with longitudinal bars near the top faces of the beams enclosing the slab.

The geometric and topological information pertaining to a reinforcing bar in a slab component can be stored using an **Edge-Face-Slab** structure of the RAG scheme that stores the bar as an enumerated type. The enumerated-bar type enables identification of appropriate topological relationships that need to be stored. The direction of a given longitudinal bar is also stored using another enumerated type. As only two types of boundary edge-to-edge relationships occur in the case of slab-reinforcing elements, an array of only two elements that contain self-pointers is provided to maintain edge-to-edge topological relationships.

The structures of longitudinal bars in a slab are arranged as lists ordered in the two principal directions x and y. Two such lists in orthogonal directions, which are confined within the boundary of a slab component, give rise to a rectangular grid structure. Thus positive bending moment and torsional reinforcement in a slab component results in a total of five grid structures. Negative bending moment reinforcement for a slab component forms four lists, two in the x-direction and two in the y-direction, respectively.

17.4.4 Computerized Engineering Model (CEM)

Although solid modeling approaches try to capture the shape of a given design object, the focus still remains on the geometric and topological information pertaining to the design object. However, a true engineering model must also account for project-specific information pertinent to the model, the semantic relationships between various components of the model, and context-specific information associated with a given engineering domain. This requirement gave birth to a whole new concept that utilizes the object-oriented paradigm to allow the incorporation of domain-specific expertise into a CAD application [BSI, 1996].

17.4.4.1 Object-Oriented CAD Modeling

An engineering facility-delivery process involves various participants such as owners, architects, structural designers, and construction contractors. These participants often work on different hardware and operating systems. In large engineering organizations, it is not uncommon for different groups, using different hardware and operating systems, to work on a single engineering project. Hence, for a successful implementation of computer-integrated construction (CIC) concepts, it is essential that the CAD data originated in one environment be usable in any other environment without translation. Further, in order to support CIC concepts, a CAD system needs to archive a model in such a way that it can be reactivated after years or decades—for operation and maintenance or renovation purposes—without depending on the hardware or operating systems used during the model-creation process. Such a capability would facilitate smooth progression of users to more cost-effective hardware and operating systems of the future.

An object-oriented CAD system that supports CIC concepts must also be capable of handling large data sets. Further, information pertaining to objects, such as a beam, present in a project model needs to be in consistent state at all times during a CAD session. This is achieved by making sure that the beam information is always accessed and modified by the schema that defined and created the beam object.

The horizontally fragmented nature of the A/E/C industry necessitates simultaneous sharing of a given project model by many users for different purposes. For instance, let us consider a single beam in a project model. A structural designer may need to modify model information such as depth of beam, diameter and spacing of stirrups, and diameter and number of longitudinal bars at the soffit. On the other hand, a rebar subcontractor needs only to query the number and diameter of longitudinal bars, diameter and spacing of stirrups, and grade of concrete and steel. Further, a formwork subcontractor would typically be interested in a query related to the surface area that results from the beam dimension. Apart from allowing tailored intelligent views for different project participants, an object-oriented CEM, which supports CIC concepts, must ensure that modifications to a project model are properly coordinated and the model is kept in a consistent state at all times during the facility-delivery process.

A concrete-engineering project typically involves collaboration between experts in several domains such as design, rebar detailing, formwork installation, rebar placement, and concrete placement. An object-oriented CEM must facilitate integration of the information created by each of the domains and allow easy and consistent access by users in other domains such as HVAC, electrical, and mechanical systems. Hence it should be possible for one schema to reference information defined and maintained by another schema within a given project model that includes schemas that support several disciplines such as architecture, structure, HVAC, construction, and operation and maintenance.

17.4.4.2 Example Object-Oriented CAD System

Objective MicroStation is an example system that addresses the requirements of an object-oriented CAD system and supports the concepts of computer integrated construction (CIC). It includes a schema implementation language called **ProActiveM**. ProActiveM is an object-oriented programming language that allows the engineering application developers to include domain-specific expertise by creating schemas tailored to model domain-specific information. A rebar design and detailing application, for instance, can focus on modeling domain-specific information such as spacing, hook location and type, grade of steel, and lap splicing.

17.4.4.3 Internet CAD

Downloading standard formwork components from a formwork vendor's website and attaching them to an existing CAD model to check if readily available formwork components fit the designed model geometry is not far from reality with the availability of an internet CAD system such as MicroStation Internet. Internet CAD supports an internet programming environment that is simple, is active (i.e., works on all platforms), and possesses web-distribution functionality [Bentley, 1996].

Java is an example Internet programming language that was developed by Sun Microsystems Inc. The Java programming language allowed inline sound and animation in a web page for the first time. Java is not just a Web browser with special features but is a programming language for distributed application. Java allows adding new types of content and the codes necessary to interact with that content. However, Java does not allow two aspects necessary for web-based manipulation of CAD models: maintaining active CAD models and automatic transaction management.

ProActiveM, developed by Bentley Systems, Inc., includes all the functionalities of Java programming language and supports automatic transaction management and maintenance of active CAD models. With CAD systems, such as MicroStation Internet augmented with an Internet programming language such as ProActiveM, civil engineers can create **desktop sites** to work with active CAD models of a given project located at a remote **geographic site** [Bentley, 1996].

17.4.4.4 Example Rebar Modeling Systems

GEOPAK Rebar.   GEOPAK rebar transforms the MicroStation CAD system into a rebar design and detailing system for reinforced concrete structures. Modeling of any regular or irregular reinforcement arrangements that include many configurations of rebar such as straight bars, stirrups, circular and spiral ties, and radial reinforcement is supported. Geopak provides the tight coupling of a given structural component with the rebar model used to reinforce the structural component. For instance, a modification to the model that changes the overall dimension of a structural component will result in an automatic update in the arrangement, spacing, and clearance for the reinforcing steel bars. Automatic scheduling of bar lengths and bar quantities and monitoring of bar marks and shapes is supported for several national reinforced concrete design codes and specifications. Geopak exploits MicroStation's graphical user interface (GUI) capabilities to provide a large range of specialized touch-click-and-drag editing functions to move, copy, or stretch an instance of a reinforcing-bar object. Figures 17.6 and 17.7 show example screens from the Geopak rebar modeling system [Geopak, 1996].

ArtifexPlus.   ArtifexPlus is a rebar and formwork planning software that specializes in complex compound units. Automatic prefabricated generation of completely reinforced structural modules such as staircases and foundations is supported. Reinforcement planning supports the import and export of reinforced areas, thereby enabling complete or partial reuse of reinforcement models for other projects. The mesh-reinforcement program supported by the reinforcement-planning package provides useful routines for laying single meshes or groups of meshes. Detailed representation of the single and double bars of meshes can be created easily, and weight of all meshes is automatically

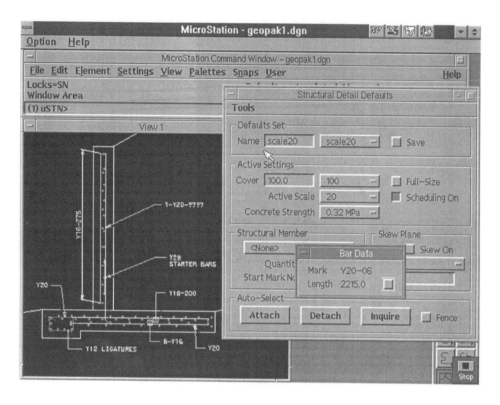

FIGURE 17.6 Geopak rebar modeling.

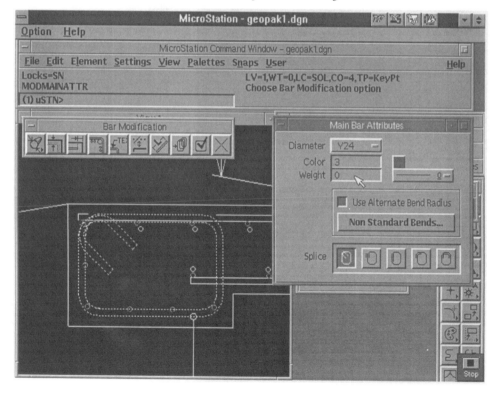

FIGURE 17.7 Geopak rebar modeling detail.

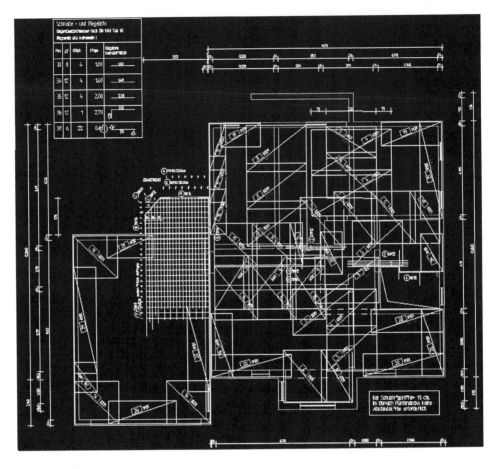

FIGURE 17.8　ArtifexPlus application.

determined and saved. ArtifexPlus supports creation of formwork drawings based on architectural drawings. Figures 17.8 and 17.9 show example screens from an ArtifexPlus application [Computer Unterstutzt, 1996].

17.4.4.5 Automated Manufacturing of Reinforcement Bars

Concrete construction shows a trend toward integrating the achievements of automated design and construction capabilities, at least in off-site production. One example of this trend is the development of an automated rebar manufacturing machine [Navon et al., 1995]. This machine produces bars larger than 16 mm in diameter. The machine receives the raw steel bar in a discrete manner, automatically inputs the bars for processing, cuts and bends the bars, and disposes of the excess material. In addition, the machine deals with the temporary storage of the finished product (see Figure 17.10). The machine was designed and developed with the aid of the ROBCAD graphical simulation system. This made it possible to check, in a 3D environment, the logic of the material flow, the production methodology, relative location of the subsystems, interference between parts of the machine and the product, and as the productivity of the process. The developers of this machine claim a 30% higher productivity compared to traditional rebar-bending machines.

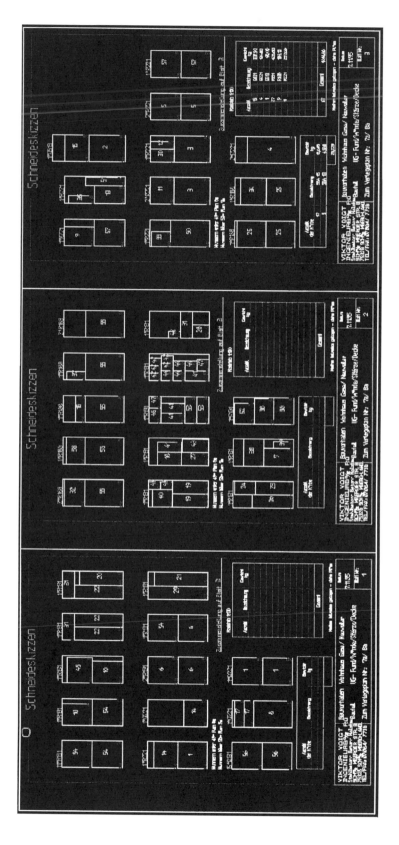

FIGURE 17.9 ArtifexPlus application—cross-sectional sketches.

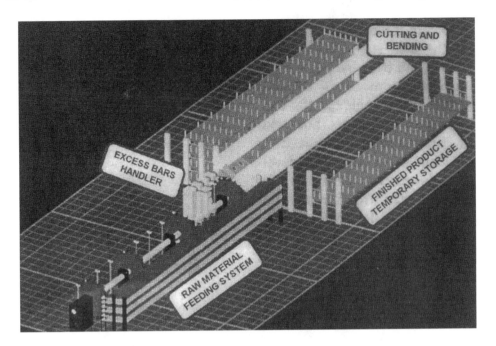

FIGURE 17.10 Rebar cutting and bending machine: general view (Navon et al., 1995).

17.5 Conclusions and Future Activities

Research and development of core technologies for automated construction equipment continues to make remarkable progress. However, much remains to be accomplished in order to bring about the successful automation of today's concrete construction and related work tasks. Concerns for short-term profits, fierce competition among contractors, and lack of top management commitment to technological change are often cited as primary obstacles to the rapid introduction of automation and robotics technology among construction firms.

To date, the primary motivation for the robotization of on-site work was to remove humans from safety and health hazards. Eventually, with shortages of skilled construction labor becoming more acute, the scope of tasks considered for automation will be enlarged to include the simple, high-volume, and repetitive tasks that could be effectively automated. More emphasis on the development of new construction systems will be put on the redesign of traditional work tasks to better match the limited capabilities of automation and robot technologies.

Automation of design and planning of reinforced concrete structures at "A/E/C" firms has become, to a large extent, a reality. The most recent developments in the automation of design-construction integration are taking advantage of the existence of the "information superhighway," allowing company professionals to communicate and transfer data instantly not only within their own organization, but also to clients, subcontractors, and other parties that may be located worldwide. This trend, independent of on-site work automation, is likely to continue.

References

Beliveau, Y. 1996. What can real-time positioning do for construction? *Automation Const.* 5(2).

Bentley, K. 1996. Engineering for the year 2000. Keynote Presentation at the AEC Systems® Conference, June 17–20, Anaheim, CA. *Sponsor*: A/E/C Systems International, Exton, PA.

BSI. 1996. Objective MicroStation, Technical Abstract, Exton, PA.

Computer Unterstutzt. 1996. ArtifexPlus. Technical Description, Berlin, Germany.

Gaultney, L., Skibniewski, M., and Salvendy, G. 1989. A systematic approach to industrial technology transfer: A conceptual framework and a proposed methodology. *J. Inf. Tech.* 4(1):7–16.

Geopak. 1996. GEOPAK Rebar. Technical Description, North Miami Beach, FL.

Hijazi, A. and Skibniewski, M. 1989. Computer simulation of concrete placement operations. *J. Eng. Comp. Appl.* 4(1):19–25.

Kunigahalli, R. 1996. Geometric and topological representation of reinforcement detail for reinforced concrete framed structures. *Proc. 13th Int. Sym. Automat. Robotics Const.*, pp. 521–530. Japan Robot Association, Tokyo.

Kunigahalli, R. 1997. 3-D modeling for computer-integrated construction of R.C. structures. *ASCE J. Comput. Civ. Eng.* 11(2):92–101.

Kunigahalli, R. and Russell, J.S. 1995a. Framework for development of CAD/CAC systems. *Int. J. Autom. Const.* 4(3):327–340.

Kunigahalli, R. and Russell, J.S. 1995b. Sequencing for concrete placement using RAG—A CAD data structure. *ASCE J. Comput. Civ. Eng.* 9(3):216–225.

Kunigahalli, R., Russell, J.S., and Veeramani, D. 1995. Extracting topological relationships from a wire-frame CAD model. *ASCE J. Comput. Civ. Eng.* 9(1):29–42.

Mortenson, M.E. 1985. *Geometric Modeling.* John Wiley & Sons, New York.

Navon, R., Rubinovitz, Y., and Coffler, M. 1995. Development of a fully automated rebar manufacturing machine. *Autom. Const.* 4(3):239–253.

Okuda, K., Yanagawa, Y., Kawamura, T., Ochiai, M., and Aoyagi, H. 1992. Development of the automatic screeding machine mounted on a girder for concrete placing work. *Proc. 9th Int. Symp. Autom. Robotics Const.* Japan Robot Association, Tokyo.

Requicha, A.A.G. 1980. Representation for rigid solids: Theory, methods, and systems. *ACM Comput. Surv.* 12(4):439–465.

Requicha, A.A.G. and Rossignac, J.R. 1992. Solid modeling and beyond. *IEEE Comput. Graphics Appl.* 12:31–44.

Skibniewski, M. 1988. *Robotics in Civil Engineering.* Van Nostrand Reinhold Publishers, New York.

Skibniewski, M. 1996. State of the art in construction automation R&D in the USA. *Proc. 13th Int. Symp. Automa Robotics Const.* Japan Robot Association, Tokyo.

Skibniewski, M. and Nof, S. 1989. A framework for programmable and flexible construction systems. *Robotics Auton. Sys.* 5:135–150.

Skibniewski, M. and Russell, J. 1989. Robotics application in construction. *AACE Cost Eng. J.* 31(6):10–18.

Skibniewski, M. and Russell, J. 1991. Construction robot fleet management system prototype. *ASCE J. Comp. Civ. Eng.* 5(4):444–463.

Skibniewski, M., Arciszewski, T., and Lueprasert, K. 1997. Constructability analysis—A machine learning approach. *ASCE J. Comput. Civ. Eng.* 11(1):8–17.

Ueno, T., Kajioka, Y., Sato, H., Maeda, J., and Okuyama, N. 1988. Research and development of robotic systems for assembly and finishing work. *Proc. 8th Int. Symp. Robotics Const.* Japan Industrial Robot Association, Tokyo.

Yu, W.-D. and Skibniewski, M. 1995. Analyzing construction technology effectiveness with automated constructability review system (ACoRS). *Proc. 1st Canadian Soc. Civ. Eng. Const. Spec. Conf.*, pp. 673–682.

Roller-compacted concrete (RCC) Darling Mills Dam, Australia (Courtesy Ernest K. Schrader).

18

Roller-Compacted Concrete

by
Ernest K. Schrader, P.E.
President, Schrader Consulting Engineers, Walla Walla, Washington. Expert in the theory and application of roller-compacted concrete including planning, design, and construction of more than 65 international projects.

18.1 Introduction

Roller-compacted concrete (RCC) has rapidly become a common material and construction procedure for dams and massive structures. It is also used for overtopping and erosion protection of embankments, and for heavy-duty pavements. This chapter concentrates on the mass applications of RCC, primarily for dams. However, many of the concepts, from testing to material properties and mix designs, apply to all uses of RCC. In a sense, RCC dams can be thought of as a series of bonded pavements or parking lots stacked on top of each other.

This chapter provides an explanation of what RCC is, how it differs from conventional concrete, what its special properties are, and how to use it effectively. The chapter covers specific technical and construction issues, including aggregates and mixture proportioning, laboratory testing, material properties, engineering and design, cost, and construction. Emphasis is placed on areas of controversy and significant interest, such as cost savings, mixture proportions, material properties, watertightness, joint quality, and design options.

18.1.1 What is RCC?

Roller-compacted concrete is concrete, but it is placed by nontraditional methods. It requires a drier or stiffer consistency than conventional concrete. RCC can have a much broader range of material properties than conventionally placed concrete: it can use aggregates not meeting normal requirements; it can be placed at very high production rates; and it can be much less expensive.

By definition [ACI, 1989] RCC is concrete that has a consistency that allows it to be compacted with a vibratory roller. Usually a 10-ton roller intended for asphalt and granular base is used because of its compactive energy with high-frequency and low-amplitude vibration. RCC is usually mixed in a continuous process rather than in batches, delivered with trucks or conveyors, spread in layers using a bulldozer, and given final compaction with a vibratory roller. Figure 18.1 shows a typical application of mass RCC. A condensed summary of the RCC process has been described in earlier literature [Schrader, 1988; Schrader and Namikos, 1988]. More thorough summaries have also been published [ACI, 1989; Jansen, 1989; ICOLD, 1989; Hansen and Reinhardt, 1991; Schrader, 1994, Schrader, 1995].

Freshly mixed uncompacted RCC generally looks like damp gravel that might be used for a road base, although some mixtures that have a wetter consistency look more like a conventional no-slump concrete. Not until the cement has reached a point near final setting or until the hydrated interior is exposed does RCC have the visual appearance of normal concrete.

Portland cement is normally the primary cementing medium, although fly ash or natural **pozzolan** is often used for a major portion of the cementing material. Slag cement has also been used. Low cement content mixtures typically use natural nonplastic fines or rock dust as a filler to compensate for the lack of **paste** that would otherwise exist. At times the fines have cementing abilities.

FIGURE 18.1   Placing RCC with a conveyor, bulldozer, and roller at Concepcion Dam (Courtesy Ernest K. Schrader).

18.1.2 History

The rapid worldwide acceptance of RCC dams is a result of need, success, and economics. Twenty to thirty years ago there were occasional uses of materials that, in hindsight, could be considered to be RCC. These applications were basically stabilized gravel fills and the material was not viewed as an engineered concrete.

In the 1960s, a high-production, no-slump mixture that could be spread with bulldozers was used at Alpe Gere Dam in Italy [Gentile, 1964] and at Manicougan I in Canada [Wallingford, 1970]. A similar process was used as late as the 1980s at Burdekin Falls Dam in Australia. These mixtures were consolidated with groups of large internal vibrators mounted on backhoes or bulldozers.

During the 1970s, a number of organizations were involved with various trials, laboratory evaluations, and the development of philosophies concerning mass RCC. A number of RCC applications for portions of dams and spillways, for temporary structures, and for noncritical uses were completed during this first decade of significant RCC development, including more than a million cubic meters of RCC that were placed at Tarbela dam. In 1974, a preliminary design with extensive laboratory testing was completed by the U.S. Army Corps of Engineers for the Zintel Canyon Dam. This would have been the world's first RCC dam, but owing to funding policy, the dam was not actually constructed until 1992.

The work with RCC in the 1970s formed the basis for RCC dams as they started to appear in the 1980s. Growth and acceptance of this new process was dramatic. In 1983 there was only one major all-RCC dam in the world (Willow Creek, USA) and one completed rolled-concrete dam (RCD) (Shimijagawa, Japan). RCD uses RCC for the interior portion of a concrete dam. In 1996, just 13 years later, we have about 200 large RCC dams worldwide that are either completed or under construction. In the United States alone there now are 115 documented uses of RCC in dams. This includes 26 dams higher than 50 feet, 14 dams lower than 50 feet, 39 uses of RCC to allow overtopping of embankment dams, 9 uses of RCC for added support of existing concrete dams, 3 earth-dam rehabilitation applications, and 24 miscellaneous uses.

Although the United States initially had the greatest number of RCC dams, they are now more prevalent in countries such as China, Spain, and recently, Brazil. However, RCC dams can be found on every continent except Antarctica. RCC dams are either in use, under design, or in planning in countries that have conditions ranging from arctic to tropical and at elevations ranging from sea level to very high mountain regions. Figures 18.2 through 18.5 show examples of completed RCC dams.

FIGURE 18.2 Urugua-I Dam, Argentina, used 105 lb of cement/per cubic yard (no ash) (Courtesy Ernest K. Schrader).

FIGURE 18.3 Monksville Dam, USA, used 100 lb of cement/per cubic yard (no ash) (Courtesy Gannett-Fleming Eng).

FIGURE 18.4 Copperfield Dam, Australia (Courtesy Ernest K. Schrader).

FIGURE 18.5 Echo Lake Dam, USA (Courtesy Ernest K. Schrader).

Rolled-concrete dams (RCD) have been developed and used extensively in Japan. They continue to be very popular there, with over 20 projects completed, under construction, or in various stages of planning and design. However, RCD has not become popular outside of Japan. The process uses a relatively lean RCC for the interior portion of the dam, but encases the entire mass of RCC with about 10 ft of conventional concrete. This includes the upstream and downstream faces, the foundation, and the upper portion of the dam. Traditional **monolith** joint spacing is retained. The result is a very attractive dam that looks and behaves like traditional concrete dams, but the RCD procedure compromises the substantial cost savings and reduction in schedules possible with other RCC dams.

The trend in the United States has gone from using RCC primarily for new dams to using it more for rehabilitation and support of existing dams, and for providing emergency spillway capacity over existing embankment dams. This trend is expected to continue in other countries as they begin to realize the benefits and additional uses of RCC.

As with conventional concrete, there is no known limit to the size or height of dams that can be designed and constructed with RCC. Currently dams on the order of 600 ft high are being planned, designed, or are under construction. One of these very high RCC dams is La Miel-I in Colombia with a height of 620 ft.

18.2 Advantages and Disadvantages

18.2.1 General

The list of RCC dam advantages is extensive, but there also are some disadvantages that should be recognized. Some potential advantages can only be realized with certain types of mixtures, structural designs, production methods, weather, or other conditions. Likewise, some disadvantages only apply to certain conditions and types of mixtures. One condition that stays constant with RCC is that each job must be thoroughly evaluated on its own. What is advantageous for one project with a given set of conditions may not be advantageous for a different project, and what is a problem at one location may actually be a benefit at another location. No single design, mixture, or construction method is ideal for all projects.

Although it is almost routine for efficiently designed RCC dams to be the least costly alternate when compared to other types of dams, there are conditions that may make RCC more costly. A situation in which RCC may not be appropriate is when aggregate material is not reasonably available but there is an abundance of good material for impervious fill.

18.2.2 Cost

The typical reason for using RCC is reduced cost and time. Savings can be dramatic. RCD and more complex RCC designs tend to have less savings, on the order of about 10%, while very simple all-RCC designs can have savings in excess of 50%. Each project must be evaluated on its own. A trade-off in appearance and other characteristics that may be associated with high savings must be acceptable to the owner and compatible with technical requirements.

It is difficult to obtain final actual cost data for RCC dams, although there is an abundance of bid price data for various portions of RCC costs and several reasonable summaries of approximate overall costs can be found in various references [Hansen and Reinhardt, 1991; Schrader, 1988; Forbes, 1988; Schrader, 1995c,d; Schrader and Namikos, 1988]. Apparent discrepancies exist in references and in costs discussed at various meetings for two primary reasons: (1) The work and materials included in the cost may be very inclusive (mobilization, joints, engineering, facings, diversion, spillway, forming, galleries, drains, foundation preparation, etc.) or may only include the very basic costs of RCC production (aggregate, cement, mixing, and placing). (2) Costs are sometimes based on unit bid prices, which can be unbalanced and not the true prices and which do not include subsequent added costs for claims, litigation, time extensions, modifications, and overruns.

Because there are so many interrelated aspects of RCC, a fair way to compare the cost of different designs and projects is to consider the volume and combined cost of applicable mobilization; access and haul roads for RCC and related activities; cement; pozzolan; aggregate; admixtures; mixing; delivery; placing; compaction; grout or bedding mixtures; lift-surface cleaning; curing and protection; upstream and downstream facings; watertightness; spillways; joints; detailed work at the top of the dam; foundation preparation, including leveling concrete if used; galleries; drainage; and cooling or crack mitigation, if required. Consideration should also be given to the required quantity of excavation and the length of diversion and outlet structures.

The reason for this comprehensive comparison is that misunderstandings have occurred without it. An example is when comparing a more massive lower strength design to a less massive higher strength design. The low-strength design may have significantly lower RCC unit costs, but it will probably have more volume, more excavation, and longer conduits. Lower unit costs are usually the result of greater volume, lower strengths, and reduced aggregate, cement, pozzolan, and admixture requirements. High-strength designs have greater unit costs. They may also have significantly more foundation treatment, **gallery** requirements, lift-joint requirements, leveling concrete, monolith joints or crack maintenance, and cooling requirements.

Another example is a design that requires more costly and complex RCC placing with less costly spillway and facing work, whereas a different design may have very simple RCC placing that then requires time-consuming and costly spillway construction.

Even with these problems, general price guidelines can be provided. Typically, the final price of RCC (all ingredients, mixing, delivery, and placing, including joint treatment, drains, cure, facings, and other directly related items) varies from about $22/yd^3$ to $94/yd^3$ (1996 U.S. dollars), depending on the size of the project, strength, and complexity. Table 18.1 provides typical price ranges for different size dams, based on historical records of completed projects and recent estimates for future projects. The table excludes projects in France where RCC tends to be more expensive than elsewhere (but still locally competitive). Another exception is the relatively high final cost of RCC after consideration for all related work, aggregate, crack repair, and settlement of claims and modifications at the Upper Stillwater dam [Parker, 1992]. An exception on the other end of the spectrum is the Zintel Canyon dam, where an uncomplicated project resulted in final costs slightly less than the range indicated in Table 18.1.

Table 18.2 provides a typical price breakdown, showing where the more expensive aspects of the RCC typically lie, namely, in aggregate and **cementitious** materials. The table also demonstrates the increased unit cost of higher cementitious content RCC but also demonstrates the typical lower volume required for a high cementitious content dam compared to a low cementitious content dam.

Historically, and in estimates of future projects, the final cost of an entire dam with all related construction such as outlets, spillway crests and gates, access, foundations, diversion, instrumentation, valves, finishing, and environmental restoration is about 1.5 to 2.5 times the cost of the RCC. The reasons are different from job to job, but the range stays the same. In general, the total project cost for dams is about twice the cost of the RCC. Simple designs with unformed faces and more massive sections tend to be less, while more complex designs and smaller sections tend to be more.

A big part of the savings of RCC dams compared to embankment dams comes from incorporating the spillway into the dam rather than having a separate major excavation and structure. Additional major savings are related to the speed of construction. This often allows the dam to be raised in one dry season, thereby greatly reducing the diversion and cofferdam costs when compared to the cost of providing protection during the wet season for other designs.

TABLE 18.1 Typical RCC Prices in 1996 U.S. Dollars

Volume, yd^3	$/yd^3
1,000–6,000	$42–$94
30,000–100,000	$34–$42
250,000–500,000	$23–$34
1,000,000–7,500,000	$22–$29

Note: Prices include RCC, facings, conventional concrete, and miscellaneous items.

TABLE 18.2   Example RCC Price Breakdown

Item	Lower Strength—More Massive, $/yd^3	Higher Strength—Less Massive, $/yd^3
Aggregate	$11.45	$13.20
Cement + pozzolan	7.25	12.75
Delivery	2.80	3.00
Mixing	1.50	1.60
Cooling	.20	1.40
Place-spread-compact	.55	.60
Cure and protection	.20	.15
Bedding and special mixes	.40	.05
Cleaning and surface prep	.25	.10
Survey-joints-drains-gallery-miscellaneous	.20	.40
Total	$24.80	$33.25
RCC volume, yd^3	1,000,000	850,000
RCC for dam	$24,800,000	$28,262,000
Total dam	$28.60	$38.60
Total volume, yd^3	1,030,000	880,000
Total concrete	$29,461,000	$33,968,000

18.2.3 Schedule

In addition to cofferdam and diversion benefits, being able to schedule RCC placement for dry seasons improves the efficiency of construction. Fast construction reduces interest during construction and provides beneficial use (revenue) at an earlier date. RCC can remove the dam portion of a large project with tunnels and powerhouses from the critical path, and allow a delay in the start of dam construction. Typical placing rates for projects having RCC volumes on the order of 10,000 to 25,000 yd^3 are about 50 to 200 yd^3/h. Medium sized projects of about 50,000 to 200,000 yd^3 generally achieve about 150 to 400 yd^3/h. Large projects of about 400,000 to 2,000,000 can be expected to achieve about 500 to 1,000 yd^3/h. In most cases, especially with medium sized projects, the time required to mobilize, produce aggregates, and prepare foundations exceeds the time required to place the RCC.

Emergency projects have benefited from the speed of RCC construction. Kerville Dam in Texas used RCC to rapidly build a new dam downstream of an embankment dam that was in imminent danger of failure owing to overtopping [ENR, 1986]. Concepcion Dam in Honduras used RCC to rapidly build a water-supply project after declaration of a national emergency in the capitol city of Tegucigalpa [Giovagnoli et al., 1991, 1992].

18.2.4 Equipment, Materials and Manpower

Construction equipment needed for RCC is usually available. Such equipment includes appropriate mixers and conveyors, which are the more difficult pieces of equipment to locate. Depending on the approach taken in design and the location of the project, materials required for RCC can be difficult or very simple to procure. For example, the design of Concepcion Dam took into account the poor-quality of aggregate sources, cement, and pozzolan so that local materials and a low strength mixture could be used in economical construction without delays [Giovagnoli et al., 1992; Gaekel and Schrader, 1992]. These materials could not meet normal requirements for conventional concrete or for RCC requiring medium to high strengths.

RCC dams can be built by labor-intensive methods with reduced equipment requirements, but this should only be considered for smaller projects where economics and political reasons justify it and where relaxed quality can be tolerated. RCC is best suited to high production with large equipment

TABLE 18.3 Shifts Lost Owing to Rain

	Peak Rain Intensity, in/h				
Method	0	0 < 0.04	0.04 < 0.08	0.08 < 0.20	< 0.20
Haul vehicles on dam	0	0*	1*	1*	2
All-conveyor delivery	0	0	0	0*	1*

Note: Assumes a two shift/day work schedule.
*Add one more lost shift if rain duration is greater than 8 h.

and a small labor force. Even where labor is inexpensive and readily available, experience has shown that it is best to use a small work force. It is essential that supervisory staff understand RCC as well as the type of dam construction to be used. Laborers do not require special skills other than those found in the general heavy construction industry, but they should be given an orientation for special concerns when handling RCC. It is common for contractors new to RCC to use the services of a specialist familiar with RCC to assist at the start of placing.

18.2.5 Weather

Rainy and hot weather are arguably the biggest problems in RCC production, but they are not insurmountable. A number of major RCC projects are located in tropical areas. Hot weather causes higher internal peak temperatures, but if the environment remains warm and does not have a severe cold season, thermal cracking can usually be avoided. Hot weather may also reduce the allowable time before the concrete must be delivered, spread, and compacted, and it reduces the time that a lift surface can be exposed before it starts to suffer a significant loss of quality.

Rain can be a significant problem if it is heavy or if equipment drives across the concrete surface when it is wet. Light rain will not significantly damage a compacted surface if conveyor delivery is used so that hauling equipment is not driving over and disturbing the surface. Rain problems are minimized by scheduling RCC placing for the dry season, avoiding placing during hours when rains are common, providing rain protection when appropriate, and using very high production rates with high-speed conveyor delivery. Table 18.3 provides general guidance concerning the probable amount of time that an RCC project will be down as a result of rain. As shown in Table 18.3, down time is a function of rain intensity and duration as well as the method of delivery. If trucks are used for delivery and they must operate on the RCC surface, rain is a much more significant problem compared to when an "all-conveyor" delivery system is used, such as the one shown in Figure 18.1.

18.3 Aggregates and Mixture Proportions

18.3.1 Aggregates

With respect to grading and other properties, aggregates similar to those used in conventional concrete can be used for RCC. However, materials and gradations that would normally be considered totally unacceptable for conventional concrete have been used very successfully in RCC dam construction and can be advantageous.

Although aggregates meeting normal concrete requirements can be used, they are not necessary in RCC dam construction. Monksville and Copperfield Dams used unwashed gravels with minimal processing. The aggregate included friable particles of decomposed granite in the sand sizes. Middle Fork Dam used marlstone oil shale with a specific gravity of about 2.2 and absorption of 12% to 20%. Willow Creek and many other projects have beneficially used dirty overburden or dirty quarry or gravel materials that normally would be wasted or require extensive washing. Because there were no options, Concepcion Dam now has the distinct honor of successfully using probably the worst quality materials in any major concrete dam.

Each project must be evaluated separately to find out how to best use the available aggregate materials. In some cases, such as where a poor foundation controls the design section, a low density may actually be beneficial. In most cases, fines are beneficial, especially if they tend to be pozzolanic, well graded, and nonplastic. Some low-strength aggregates may produce a low but tolerable strength RCC along with a desirable high **creep** and strain capacity. High Los Angeles abrasion losses can usually be accepted because of service conditions, the mixing and compacting equipment, and the typical RCC grading.

RCC mixtures are less affected by particle shape than are conventional concrete mixtures. This is because of the mixing and delivery equipment and the typical grading. The presence of flat and elongated particles is still undesirable, but amounts up to about 40% on any sieve size with an average below about 30% for all sieve sizes have been acceptable. Both crushed and rounded materials work well, but the ideal combination appears to be angular crushed coarse particles with natural rounded sand. When steep unformed slopes are to be constructed, crushed material becomes more important.

Alkali-aggregate reaction in RCC is a subject in itself. A few comments are made here. Potentially reactive aggregates have been used in RCC without problems owing to various reasons, including low cement contents, high pozzolan contents in some mixtures, and the use of low-alkali cements when they were available. A nontraditional concept to consider is that a slight expansion due to alkali-aggregate reaction can actually be beneficial if it offsets thermal contraction.

The key to controlling segregation, minimizing cement contents, and providing a good compactable mixture begins with a grading that is more uniform and contains more material passing the No. 4 sieve than would be common in conventional concrete. Using concepts from conventional mass concrete, earlier projects tended to use 3 in maximum size aggregate (MSA), but the recent trend has been to use smaller MSA, on the order of about 1 1/2 in. This reduces segregation, improves lift-joint quality, and reduces equipment maintenance. The reduced MSA arguably can cause a minor reduction in strength, but such a reduction is minimal and not as significant as would be expected in conventional concrete.

It is inappropriate to provide a single set of upper and lower gradation limits that could be considered correct for all RCC. A wide range of gradations has been used successfully. Table 18.4

TABLE 18.4 Typical Aggregate Gradations for RCC

	Willow Creek	Upper Stillwater	Christian Siegrist	Zintel Canyon	Stagecoach	Concepcion	Elk Creek	Small Project Recommendation
Sieve Size								
4 in						100		
3 in	100					99	100	
2.5 in				100			96	
2.0 in	90	100		98	100	94	86	100
1.5 in	80	95	100	91	95	90	76	97–100
1.0 in	62		99	77	82	80	64	76–90
0.75 in	54	66	91	70	69	72	58	66–80
3/8 in	42	45	60	50	52	56	51	42–56
No. 4	30	35	49	39	40	43	41	34–46
No. 8	23	26	38	25	32	33	34	26–37
No. 16	17	21	23	18	25	25	31	21–31
No. 30	13	17	14	15	15	19	21	15–24
No. 50	9	10	10	12	10	15	15	8–16
No. 100	7	2	6	11	8	9	10	5–12
No. 200	5	0	5	9	5	6	7	3–7
C + P (lb/yd^3)	80 + 32	134 + 291	100 + 70	125 + 0	120 + 130	135	118 + 56	—
Total Paste Volume*	20%	21%	19%	21%	—	21%[†]	21%	—
Workability	Poor	Excellent	Excellent	Excellent	Good	Excellent	Excellent	19–21%

*Total paste is all materials in the full mixture with a particle size smaller than the No. 200 sieve (see text).
[†]Total cementitious materials consisted of pozzolanic cement having approximately 15% natural pozzolan.

shows some example gradations that have been used in RCC. The mixture-proportioning concept that requires fines in the mixture ideally has a grading similar to an impervious gravel with nonplastic fines. Mixes with lower cementitious content generally require more fines. However, the maximum amount of fines that can be added without a reduction in strength, and without causing the RCC to become too sticky to mix, spread, and compact, depends on the plasticity of the fines. Plasticity can be defined in terms of soil mechanics by the liquid limit (LL) and plastic index (PI). Based on the LL and PI, Table 18.5 has been developed and used over the past 15 years as a guide for the maximum amount of fines that can be included for most RCC mixtures.

18.3.2 Mixture Types and Designations

RCC suitable for use in dams can be made with very low cementitious contents, on the order of 60 to 120 lb/yd^3, or it can be made with very high cementitious contents, on the order of 200 to 400 lb/yd^3. Both options, plus intermediate cementitious contents, have been used quite successfully on low and high dams. Both options continue to be popular, and both are expected to be used in the future. The decision-making process for selecting the type of mix is not very dependent on the size or height of the dam. It is often more related to the simple issue of what the designer or owner has used in the past. The decision should be based on factual information related to foundation quality, the degree of reliable inspection expected, facing techniques, climate, cooling and thermal issues, the age at which the reservoir will be filled, and available materials with their associated costs and quality. The best overall option might be a higher strength high cementitious content mix with less mass in one situation but could be a lower strength, low cementitious content, more massive structure in another situation.

Higher cementitious content mixes have less volume but typically have a higher unit cost and more stringent cooling and quality control requirements. Lower cementitious content mixtures typically have lower unit costs and less stringent quality control but more mass. They also may require special attention to achieve watertightness along lift joints. Costs and watertightness are discussed elsewhere in this chapter. Higher cementitious content mixtures tend to result in good cement efficiencies (strength per unit of cementitious material). Lower cementitious content mixtures tend to have even greater efficiencies, on the order of about 10 to 20 psi/lb of cementitious material in a cubic yard of RCC. Efficiencies on the order of 10 psi/lb are typical for excellent quality conventional concrete.

The American Concrete Institute's Subcommittee on RCC has recently completed its proposed third major update of its report on RCC for massive structures. The new report identifies low, medium, and high cementitious content mixes by a combination of cementitious content and strength ranges that are consistent with other publications [Schrader, 1994, 1995a,c,d]. There is a deliberate overlap of the different cementitious content designations, allowing flexibility in judgment in regard to whether the aggregate fines provide some cementitious or pozzolanic value. Nonetheless, the designations, shown in Table 18.6, provide useful guidance.

TABLE 18.5 Maximum Fines Content

Liquid Limit	Plastic Index	Max % Passing No. 200 Sieve
0–25	0–5	10.0
0–25	5–10	8.0
0–25	10–15	6.0
0–25	15–20	3.5
0–25	20–25	2.0
25–35	0–5	8.0
25–35	5–10	6.0
25–35	10–15	4.5
25–35	15–20	3.0
25–35	20–25	1.5
35–45	0–5	7.0
35–45	5–10	5.0
35–45	10–15	4.0
35–45	15–20	2.5
35–45	20–25	1.5
45–55	0–5	5.0
45–55	5–10	4.0
45–55	10–15	3.0
45–55	15–20	2.0
45–55	20–25	1.0

TABLE 18.6 RCC Mixture Designations

Designation	C + P	Strength
Low	85–150 lb/yd^3	700–2100 psi
Medium	140–210 lb/yd^3	1600–3000 psi
High	200–425 lb/yd^3	2500–4500 psi

TABLE 18.7 Types of RCC Mixtures: Typical Extremes

	Dry, Low Cementitious		Wet, High Cementitious	
	lb/yd^3	ft^3	lb/yd^3	ft^3
Cement	100	0.51	150	0.76
Pozzolan (ash)	0	0.00	250	1.60
Water	185	2.96	180	2.88
Air (0.5%)	0	0.14	0	0.14
Aggregate fines (8%)	280	1.80	0	0.00
	——	——	——	——
Total paste	565	5.41	580	5.38
% Paste	14%	20%	14%	20%

Note: Pozzolan and fines may or may not be cementitious in RCC.

At times, past terminology has referred to RCC as being high or low paste, rather than as having high or low cementitious content. This was an inaccurate description of the different types of RCC mixes. Although RCC mixtures have high, medium, and low cementitious contents, they should not have high and low "paste" contents. All RCC mixtures should have about 20% paste by volume (about 14% paste by weight). Paste is considered to be all the materials that constitute the "pasty" portion of the mix. This includes everything smaller than 75 μm—small air bubbles, water, admixtures, pozzolans, cement, and aggregate fines.

If low cementitious content RCC is used, aggregate fines are needed to supplement the otherwise deficient paste content. If a high cementitious content RCC is used, clean aggregates are required in order to keep the paste content from being too high. If the paste content is too low, low strengths, excess water demands, and segregation can be expected. If the paste content is too high, inefficient use of cementitious materials can be expected. Mixes with too much paste can also result in lift surfaces with excess paste and laitance, thereby requiring lift-joint cleaning similar to procedures used for conventional concrete.

Table 18.7 demonstrates the concept of constant paste contents for all types of RCC. It compares two extreme examples of RCC mix types and consistency. These examples are from the very dry, lean Willow Creek Dam project and a mix similar to one of the very high cementitious content, wet consistency mixes at Upper Stillwater Dam.

RCC mixtures can have low or high pozzolan contents. Current use ranges from 0% to about 80% pozzolan. There is no universal optimum. In some cases—such as for Stagecoach dam, using 100% cement resulted in lower strengths than using essentially the same total cementitious content but with 50% of it being pozzolan. In other cases, adding pozzolan while keeping the cement constant actually caused a reduction in strength or no strength change.

The long-term strength gain of RCC is not necessarily determined by the amount of pozzolan that is used, as would normally be expected with traditional concrete. For example, one lean RCC mix at one project using only 92 lb of cement and no pozzolan resulted in one-year strengths that were 380% greater than the 28-d strength, while other mixes with pozzolans at other projects have achieved strength gains of only about 150% over the same time period.

Another observation concerning RCC is that some projects have required a minimum cementitious content in order to get good strength efficiency (strength per pound of cementitious material), while at other projects the efficiency of the cementitious materials decreased with increasing cementitious content.

The important point is that RCC does not follow the same rules and trends as conventional concrete with regard to optimum cementitious and pozzolan content. Each project should be evaluated on its own merits, with its own materials, and with a wide range of options during the initial investigations. Open mindedness on the part of decision makers is also essential, who should not be misled by old traditional concrete experience or guidelines.

18.3.3 Fresh Mixture Consistency

The terms "wet" and "dry" mixture consistency can be a source of confusion. RCC mixtures have no slump. They must be stiff enough to support a large vibratory roller. By traditional concrete standards, all RCC has a dry consistency. However, within the RCC community, mixtures are referred to as being wet or dry on the basis of their appearance and how they behave in the fresh state after compaction.

RCC mixtures that are considered wet generally produce a weaving effect or surface waviness as rollers and trucks drive across the freshly compacted RCC. This is due to an excess of moisture, or more water than is necessary to fill all of the void space. Internal pore pressure develops in the fresh mixture, similar to what occurs in some soils. By contrast, mixtures that are considered to have a dry consistency generally do not weave under traffic after compaction.

The terms wet and dry reasonably describe the behavior of the mixture, but they do not indicate whether the actual water content is high or low. Because of differences in aggregate (texture, shape, and grading), differences in cements and pozzolan, the effects of fines when they are included, and the influences of temperature and humidity, a mixture with a higher water content can actually be drier in appearance than another mixture with a lower water content. This is true when considering water in excess of the water absorbed by the aggregate, and it can be even more exaggerated when total water contents based on oven-dry aggregates are considered. Table 18.7 contains an example.

18.3.4 Water/Cement Ratio and Optimum Water Content

Although it is a somewhat controversial issue, the well-known water/cement ratio rule for traditional concrete does not necessarily apply to RCC. In fact, if what appears to be an excessively high water/cement ratio in a lean mix is decreased too much, it will result in decreased strengths and poor quality material owing to inadequate moisture for compaction. As with embankment materials and soils, there is an optimum moisture for compaction, as shown in Figure 18.6. At this optimum, both strength and density are maximized. The shape of the moisture versus density and moisture versus strength curves help establish the probable type of mix for a project. If the curve is fairly flat, adding more water will have little harmful effect on the strength of the RCC mass while improving the more important lift joint quality. In this case, the wetter mix is desirable. If the curve drops off sharply at moisture contents above optimum, a drier-consistency mix is probably best.

For other reasons, the designer may stipulate in advance that a wet or a dry consistency mix is desired. Typically, dry-consistency mixtures are at or near optimum moisture. They have infinite modified VeBe time (ASTM C 1170) when that test is used for workability. Typically, wet-consistency

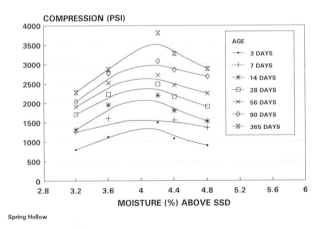

FIGURE 18.6   Effect of moisture content.

TABLE 18.8   Water/Cement Ratio Examples: Dry-Consistency RCC

MSA, in	C + F, lb/yd³	Water, % Above SSD	W/(C + F)	W/C*
3	80 + 32	4.4%	1.63	1.47
3	175 + 00	4.5%	1.06	1.06
3	175 + 80	4.5%	0.73	0.65
1 1/2	315 + 135	4.8%	0.44	0.39

*Water/equivalent cement ratio if the ash volume was cement.

mixtures have modified VeBe times, on the order of about 10 to 20 s, and they are much wetter than the optimum moisture content. Wet-consistency mixtures tend to develop a slight amount of free surface moisture, they leave a pasty lift surface, and they weave under the roller owing to internal pore pressure caused by the extra water.

Because the optimum moisture for RCC is established primarily by the aggregates (there is little or no change in optimum moisture as the cementitious content is adjusted up or down), any major change in water/cement ratio can only be accomplished by increasing or decreasing the cementitious content, as shown in Table 18.8. If, using nonapplicable experience from conventional concrete, a low water/cement ratio requirement is arbitrarily applied to RCC, it will result in higher cementitious contents with associated higher costs, increased heat and thermal stresses, and a more brittle **elastic modulus** with less stress relaxation due to creep. Attempts to change the water/cement ratio by changing the water content have only minor effects on the water/cement ratio, though such changes do alter the mix consistency and cause deviations from the optimum or desired moisture content and compactability.

If the cementitious content is low, the water/cement ratio must be high. Values on the order of 1.0 to 2.0 are common in RCC. This is a major deviation from traditional water/cement ratios in conventional concrete that are more on the order of 0.4 to 0.6. This high water/cement ratio with lean RCC is normal. It does not imply low-quality concrete. It implies a low cementitious content rather than a high water content and wet mix consistency.

Data has been reported implying that a high water/cement ratio automatically causes a major reduction in RCC strength. This erroneous conclusion is the result of incompletely reporting all the data. As the water/cement ratio decreases, strengths were shown to go up dramatically, but this was primarily because the water/cement ratio was decreased by increasing the cementitious content while keeping the water and mix consistency essentially constant. Figure 18.7 shows the complete picture, with the strength shown as both a function of water/cement ratio and as a function of cementitious content.

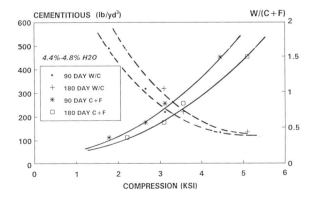

FIGURE 18.7   Strength versus water/cement ratio and cementitious content.

18.3.5 Approaches to Mixture Design and Proportions

RCC mixture proportions should follow the convention used in traditional concrete, that is, identifying the mass of each ingredient contained in a compacted unit volume (cubic yard) of the mixture based on saturated surface dry (SSD) aggregate conditions. Moisture contents are often converted to a percentage of water above SSD aggregate conditions divided by the compacted density of the total RCC mixture. Confusion has occurred when some publications and engineers have based the water content on the oven-dry aggregate mass or some other parameter. Care should be exercised when reviewing the literature.

Mixture proportioning for some projects has relied on making many mortar cubes with various cement, pozzolan, water, and admixture contents in an effort to optimize the mixture without making cylinders of the full mixture. This can be very misleading. Experience at several projects has shown that mortar cubes do not necessarily indicate how the full RCC mixture will perform. In some cases, mortar cubes have actually indicated the opposite of what happened with the full RCC mixture [Gaekel and Schrader, 1992].

There are a number of ways to approach mix designs for RCC [ACI, 1989; Schrader, 1994; Tatro and Hinds, 1992]. One procedure is based on conventional concrete basics and the concept of water/cement ratio. Another procedure uses a series of lab mixtures to pinpoint the optimum cementitious and pozzolan contents for a given set of aggregates. Other procedures are basically variations of these two themes. There is no single approach to mix design that is "best". All approaches to mix designs ultimately require full-scale job trials and tests, with adjustments based on the results of those tests.

The term "soils approach" has been erroneously used to describe a mix-design method developed by the author. This method has nothing to do with soils. The term is misleading. All methods of RCC mix design, including the one mislabeled soils approach, treat the material as a no-slump concrete. They all intend to provide a suitable controlled gradation aggregate. They all intend to optimize the amount of pozzolan (if used). They all produce a consistency suitable for RCC construction. They all determine basic concrete material properties such as density and strength.

18.3.6 Chemical Admixtures

Chemical admixtures have not been very effective in low cementitious content and dry RCC mixes that do not exhibit a fluid paste when subjected to vibration under full consolidation [Schrader, 1984; Gaekel and Schrader, 1992]. If enough water is added to provide a fluid paste, retarders have been useful and water reducers have worked, but this may require very high dosages on the order of five to twelve times the normal rate. The benefit of retarders and water reducers must be balanced against the cost, complication of an additional ingredient, and any strength reduction associated with additional water needed to provide the necessary wetter paste consistency.

Air entraining has been difficult to accomplish with RCC on a routine and consistent basis, but it has recently been done in the United States using special admixtures such as Euco Synthetic Air, and it has been accomplished in China [Chengqian and Chusheng, 1991]. As discussed later under Section 18.4, field experience has shown good resistance to most natural freeze-thaw exposure conditions of mass RCC without air entertainment, but the option of using entrained air has now been demonstrated for some mixtures. Again, the benefits should be balanced against the disadvantages and cost for each application.

18.4 Material Properties

18.4.1 Density and Air Content

The density and air content of RCC depend on the specific gravity of the aggregate, the grading, the moisture content, and the degree of compaction. Fully compacted RCC will typically have less

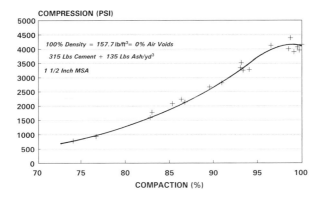

FIGURE 18.8 Twenty-eight-day strength versus compaction.

air content and less water than conventional concrete, so its density will be slightly higher. Because RCC has been made with aggregates having a wide range of specific gravities, the specific densities of compacted RCC at different projects varies greatly from about 2.1 to 2.9.

Fully compacted RCC with reasonable mixture proportions typically has an air content on the order of 0.5% to 1.5%. Specifications commonly require that field densities must obtain an average of 98% of the maximum practical achievable density (MPAD), which is often considered to be 97% of the theoretical air-free density (TAF), providing that no density at any level within a lift of RCC is less than about 93% of the TAF density. The average density is based on measurements taken at the top, middle, and bottom thirds of the lift. Air contents can be obtained in the laboratory using the pressure method for material compacted by tamping or by the vibrating table method, whichever is appropriate for the mixture consistency.

Some of the first RCC projects required greater densities than suggested above, with the thought that densities representing 98% to 100% of the **TAF** would result in significant increases in strength. This is not correct. Unfortunately, some older specifications and concepts continue to be copied. As shown in Figure 18.8, it is important to achieve a reasonably well compacted mix with a density on the order of about 95% of the TAF, but greater densities provide essentially no increase in strength. Excessive compactive effort can actually be harmful. Excessive compaction can cause a reduction in strength if it begins to break the aggregate particles, and, as with most gravel and sands, overcompaction of RCC can also begin to loosen the material and decrease density.

18.4.2 Coefficient of Thermal Expansion

The **coefficient of thermal expansion** for RCC mixes tends to be slightly higher than the thermal expansion coefficient of the aggregate and slightly less than that for a conventional concrete made with the same aggregate but more cement paste. Because of the wide range of aggregate materials that can be used in RCC, the range of thermal expansion coefficients tends to be much wider than would be expected for traditional concrete. Traditional concrete typically has an expansion coefficient somewhere between 4 and 8 millionths per degree Farenheit. Measured expansion coefficients for RCC have varied from about 3.5 to about 18 millionths per degree Farenheit.

18.4.3 Thermal Diffusivity and Conductivity

Thermal diffusivity and conductivity of RCC mixes tend to be similar to the values obtained for the aggregate by itself and similar to conventional mass concrete made with similar aggregate.

18.4.4 Poisson's Ratio

Poisson's ratio for RCC tends be similar to values for conventional mass concrete; typical values are on the order of 0.18 to 0.24, depending on the concrete age, aggregate, and strength. However, at very early ages and very low strengths and for mixtures with soft aggregates and/or very high contents of noncementitious fines, Poisson's ratio can be much higher. Under these conditions, some RCC mixes have had values for Poisson's ratio that are on the order of 0.3 to 0.5.

18.4.5 Autogenous Volume Change

Autogenous volume change (the increase or decrease in size with no applied load or environmental change) cannot be reliably estimated in conventional mass concrete without at least some test data. This is true of RCC also, especially for mixtures that have peculiar cement, pozzolan, and aggregate. It is not unusual to have early expansion followed by a later period of contraction, or vice versa. Typically, the amount of change is minimal and of little consequence, but the potential for expansion or contraction that can be important to large mass structures exists and should be investigated for unusual materials and large projects. Autogenous volume changes should be considered when determining strain and creep properties and when performing a cracking analysis of large RCC structures.

18.4.6 Freeze-Thaw Resistance

Resistance to freezing and thawing of non-air-entrained RCC subjected to natural exposure has been very good, despite the fact that in laboratory tests these materials performed badly. This includes projects such as Monksville, Winchester, and Willow Creek. These projects have unformed and uncompacted downstream faces of exposed RCC with cementitious material contents for the exposed face of 105, 175, and 255 lb/yd^3, respectively. Each project receives almost daily cycles of freezing and thawing during much of the winter. Monksville and Willow Creek have saturation from lift-joint seepage. After more than 10 years, none of these dams has shown any change in the downstream face. Other projects, such as Middle Fork, in severe climates with temperatures that can reach $-25°$F or less and that used about 1 ft of conventional concrete facing have shown no distress.

   Based on test data and observations, the following typical deterioration rates have been developed. These estimates are for low cementitious material contents in the range of 85 to 250 lb/yd^3. The deterioration rate for unformed and uncompacted surfaces subjected to freeze-thaw cycles that penetrate about 1 in is 100 to 250 cycles per inch of erosion. Compacted interior RCC is estimated to deteriorate at about 1000 to 2000 cycles per inch of erosion.

   As discussed in the sections on aggregates and mix proportions, some projects have been able to successfully achieve air-entrainent and good freeze-thaw **durability** even in severe laboratory tests. Early RCC studies indicated that this might be possible only with high cementitious content mixtures with very wet consistencies. However, some lower cementitious content mixtures with a relatively dry consistency have also achieved good air entertainment and durability using synthetic air-entraining admixtures. A recent example is the set of initial mix studies for the Tongue River project that will be exposed to temperatures ranging from about $-40°$F to $+120°$F.

18.4.7 Cavitation and Erosion Resistance

Cavitation and erosion resistance of RCC has been surprisingly good, but experience is still somewhat limited. A historical summary, which includes project plans to use RCC for erosion protection in the future, has been published [Schrader and Stefanakos, 1995]. Very early evaluations of erosion resistance, including full-scale tests at velocities on the order of about 100 ft/s, were documented by the U.S. Army Corps of Engineers (1981). Based on available data, including two high-velocity and high head full-scale trials for a limited duration as well as laboratory and other tests, cavitation and erosion rates have been developed and used with caution to justify exposed RCC spillway surfaces. Assuming good-quality mixtures with cementitious-material contents on the order of 170 to 300 lb/yd^3, an erosion rate of 0.002 psf has been extrapolated for a rolled surface and 0.05 psf for a rough surface. Experience at low-head dams that have been overtopped confirm that these numbers are either reasonable or conservative.

18.4.8 Compressive Strength

The compressive strength of RCC depends on a variety of factors including the aggregate, quantity and quality of fines, quantity and quality of cementitious material, degree of compaction, and moisture content. The efficiency, or strength per pound of cement in a cubic yard of RCC, has been discussed in Section 18.3, Mixture Types and Designations. That section also included general information about the wide disparity of pozzolan or fly-ash performance in RCC and the effect of moisture on compressive strength. The relationship between compressive strength and density or air content is discussed above in Section 18.4.1 Density and Air Content.

Care must be used when reviewing RCC literature that is based only on site-specific aggregates, pozzolans, mixture proportions, moisture contents, and gradations. Publications occasionally offer misleading global statements about RCC strength when, in fact, the experience used as a basis for those statements includes only one project, one set of materials, and one type of RCC mixture. Because of the wide range of materials and mixture proportions that can be used in RCC, it is very difficult to develop global or general statements or rules about RCC strength relationships. Figures 18.9 through 18.14 show examples of the wide range of strengths, different rates of strength gain, and the wide range in performance that can occur with RCC mixes containing fly ash or other pozzolans.

Traditional thinking limits the amount of fines (material passing the No. 200 sieve) in the aggregate to very small amounts, with the common understanding that clean washed aggregates with no fines are best. However, as discussed in Section 18.3, natural or man-made fines are essential in low cementitious RCC mixtures in order to provide adequate paste and to fill void spaces for control of segregation and for compaction. Nonplastic fines typically increase the strength of low cementitious content mixtures. The reasons for this are not entirely clear. In some cases, it is probable that the fines are providing some degree of pozzolanic strength gain, but it also appears that mechanical benefit is also being provided. Because the fines have substantial surface area and they are used primarily in low cementitious content mixes, the traditional thought that sufficient cementitious material is needed to coat all of the aggregate surfaces obviously does not apply to RCC.

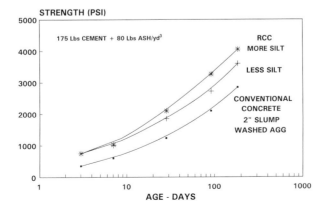

FIGURE 18.9   Effect of fines on strength.

Figure 18.9 shows a case where the addition of fines caused an increase in strength in a medium cementitious content mix. In this situation, the optimum fines content was about 5% to 6%. A slight increase in fines to about 8% resulted in minimal additional strength. At even higher fines contents, the strength could be expected to decrease. The fines were a natural silt with very little plasticity. Figure 18.9 also shows much lower strengths at every age when the aggregate was washed to eliminate all fines, screened to an ideal traditional gradation, and used to make low-slump conventional concrete. Additionally, it shows the strength achieved when the "dirty" RCC aggregate was washed and sorted to an ideal conventional aggregate gradation. The mix made with this gradation was an excellent quality 2-in slump mix with a water-reducing admixture. At all ages, the strength of the conventional concrete was substantially less than the RCC concrete made with the same basic aggregate without the expensive washing.

Figure 18.10 shows another project with a nontypical behavior where an increased fines content of a low cementitious content mix caused a slight reduction in strength. The only way this was detected was by tests. In this case, the fines were siliceous and expected to add substantially to, rather than decrease, the strength.

Figure 18.11 (data courtesy of Gannett-Flemming Engineer, Harrisburg, PA) shows very good strength gain for a mixture that was used in an application where durability and traditional strength levels of about 4000 psi were important. This was achieved with only 270 lb of cement per cubic

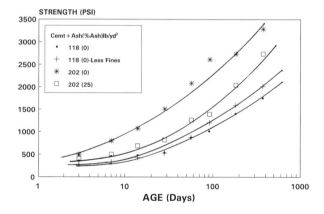

FIGURE 18.10   Agos Dam; compressive strength.

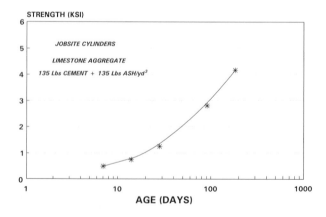

FIGURE 18.11   Compressive strengths, Salt Lick Dam.

yard (135 lb of cement and 135 lb of fly ash). The aggregate was limestone, including the beneficial rock dust which, in the case of limestone, was expected to react somewhat with the siliceous fly ash.

Figure 18.12 shows the benefit of fly ash in a low-strength mass RCC made with lower quality aggregates at Stagecoach Dam. In this case, evaluation of the comparative mixtures at ages up to 28 d indicated that fly ash was not beneficial. However, after 28 d the ash performed extremely well, ultimately achieving a strength greater than the mix made with cementitious material that was 100% Portland cement with no ash. In this case, the aggregates were lower quality and the efficiency of the cementitious material was relatively low by RCC standards.

Although Figures 18.11 and 18.12 indicate that fly ash is very beneficial in RCC and can even be better than adding more Portland cement, this is not always the case. Figures 18.10 and 18.13 show mixtures where adding fly ash increased the cost and complexity of the mix while doing virtually nothing for strength. Figure 18.10 shows that when a mix with 202 lb of cement per cubic yard used fly ash for 25% of the cementitious material, the strength was reduced by about 25% at all ages including the long-term ages up to one year. Figure 18.13 indicates a mix with 175 lb of cement per cubic yard had no significant change in strength when 75 lb of ash per cubic yard was added. In this case, adding silty fines to the mix was found to have the same or better effect as adding fly ash.

Figures 18.11 and 18.13 show RCC mixes that achieved relatively high strengths on the order of 4000 and 6000 psi, respectively. This is comparable to very good quality traditional concrete but with cementitious contents that are about half that used in traditional concrete.

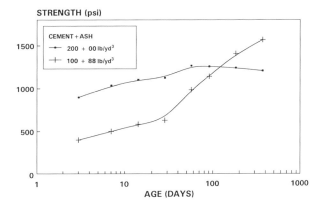

FIGURE 18.12   Strength versus time, Stagecoach.

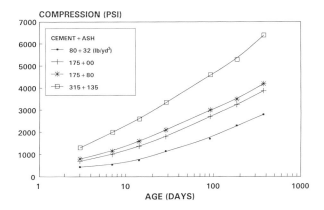

FIGURE 18.13   Strength versus time, Willow.

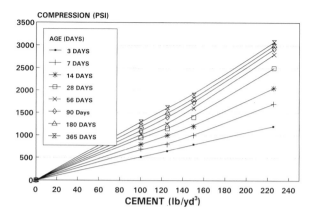

FIGURE 18.14   Compression versus cementitious materials.

Figures 18.9 through 18.13 represent a wide variety of RCC mixtures, with different aggregate types and gradations, different cementitious and fly ash contents, and different maximum aggregate sizes. In all the examples, the individual data points have been plotted to show that although the strength gain trends can be different from mix to mix, the strength gain trend for any given RCC mix plots very predictably with an incredibly small amount of data dispersion. RCC is a reliable engineering material with predictable and dependable properties once a mix design program has been undertaken.

Figures 18.9 through 18.13 plot strength versus time. Another very useful way to present the data is by plotting strength versus cementitious content, with different lines representing different ages as shown in Figure 18.14. From the family of curves shown, the required cementitious content can then be selected to achieve any strength at any age. The shape of these curves is of particular interest. Figure 18.14 shows a generally linear relationship between strength and cementitious content from 0 to 230 lb of cement per cubic yard. However, other RCC mixes have shown no significant strength gain beyond a certain cementitious content. In other cases, no significant strength was achieved until some minimal amount of cementitious material was used.

18.4.9 Tensile Strength

As with compressive strength, the tensile strength of RCC can vary greatly from mixture to mixture. Cementitious-material content is a principal influencing factor, but moisture content and aggregate are also important. Higher-cementitious-material contents, lower moisture, and crushed coarse aggregate tend to increase tensile strength. However, the tensile strain capacity, discussed below, may be more important than the tensile strength. A highly deformable mixture with a low elastic modulus can be more desirable than a stronger mixture that is much more brittle and that cracks earlier under the same amount of deformation.

The Brazilian split-cylinder test (ASTM C496) is a simple and economical method of obtaining an indication of the direct tensile strength of concrete. It is frequently used for RCC, but precautions need to be exercised when deriving the probable direct tensile strength from indirect split-tension results. This applies to traditional concrete as well, but it is more important for RCC where the relationship of direct to indirect strength can vary from project to project as well as within the range of compressive strengths that might be considered for a particular project. Figures 18.15 and 18.16 are examples of the development of split tensile strength with time for a crushed gneiss aggregate with and without fly ash. The mix designations used in Figures 18.15, 18.16, 18.17, and 18.19 indicate the total amount of cement plus fly ash (in pounds per cubic yard), followed by the percentage of cementitious material that is fly ash, followed by the moisture content as a percentage of the full mix.

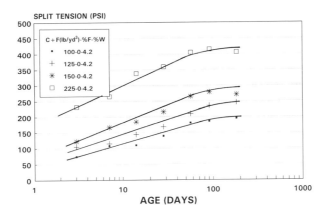

FIGURE 18.15   Tension versus time, Big Haynes.

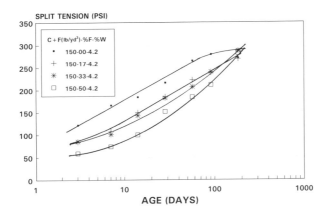

FIGURE 18.16   Tension versus fly ash, Big Haynes.

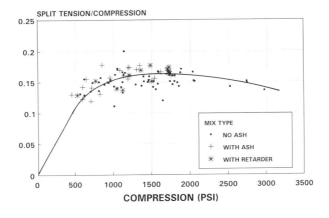

FIGURE 18.17   Ratio (split tension/compression), Big Haynes.

The split tensile strength of RCC mixtures with higher cementitious material contents and higher compressive strengths is typically a lower percentage of the compressive strength than occurs with low-strength mixtures. Some examples of the ratio of split tension to compression strength for various mixes and tests at different projects are 4–7% for Upper Stillwater, 7–12% for Willow Creek, 9–13% for Monksville, 10–18% for Urugua-I, 12–17% for Concepcion, and 13–19% for Middle Fork [Schrader, 1994; 1995]. These have been listed in order from the highest to lowest average strength.

Figure 18.17 indicates the scatter of data and general trend at one project for the ratio of split tensile strength to compressive strength. Although it appears that a straight-line showing one ratio for all strengths could be drawn through the data, it as been demonstrated that the ratio is essentially zero at a strength of essentially zero, and a marked decrease has been demonstrated in the ratio at very low strengths at other projects. Consequently, the trend line shown is considered typical, although the maximum ratio can be much higher or lower for any particular project, as indicated above.

When converting split tensile strengths to direct tensile strengths, a factor should be taken into account based on the compressive strength. Figure 18.18 shows this factor as a function of the logarithm of the compressive strength [Schrader, 1995b]. The factor is almost the same when converting from flexural strengths of beams to direct tensile strength and when converting from split

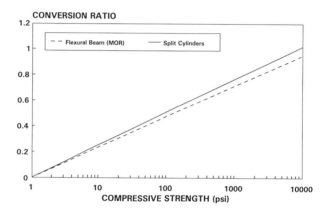

FIGURE 18.18   Tensile strength conversion.

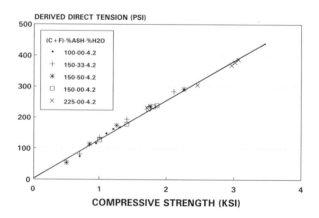

FIGURE 18.19   Direct tension versus compression, Big Haynes.

tensile strengths to direct tensile strengths. The ratio indicated in Figure 18.18 for a particular compressive strength is multiplied by the split-cylinder strength to obtain the derived direct tensile strength.

When all of the adjusting factors are taken into consideration, the final ratio of direct tensile strength to compressive strength typically is a direct linear relationship as indicated in Figure 18.19.

18.4.10 Modulus of Elasticity

The modulus of elasticity and creep for RCC can have an extraordinary range of values. Mixtures with large amounts of cementitious materials and mixtures made by the water/cementitious ratio approach to mixture proportioning usually produce long-term values for the static and sustained elastic modulus and creep similar to conventional mass concrete, namely, modulus values on the order of 3 to 4 million psi, and creep values on the order 0.02 to 0.05 ln (time) for loading ages of about 28 to 90 d. Mixtures made with very high pozzolan contents can be expected to have lower early-age values of static modulus but higher later-age values.

Low cementitious material content mixtures can have very low elastic modulus values and high creep rates. The decrease in modulus and increase in creep tends to be exponentially proportional to the decreases in strength below about 1500 psi. Each incremental decrease in cement content has much more effect than the previous incremental decrease, but each mixture and aggregate can perform differently. Static modulus values on the order of 0.1 to 1.5 million psi are reasonable for cementitious material contents on the order of 100 to 125 lb/yd^3 at ages of 3 to 90 d. Corresponding ultimate values could be on the order of 0.8 to 2.5 million psi. At Burton Gorge Dam, ultimate elastic moduli were on the order of only 0.15 to 0.30 million psi. This was a quality designed into the mixture. It is attributed mostly to the gradation and use of what might normally be considered lower quality aggregates. A "deformable" concrete was desired to avoid thermal stress and because the foundation condition had a low and varying mass modulus that could not be accurately predefined. Figure 18.20 shows the typical range of the modulus of elasticity for traditional concrete and extreme values that occurred with RCC.

Although it is an extreme example, it is important to note the substantial increase of modulus (increased stiffness) at later age for the AGOS mix. This is attributed to pozzolanic activity of the added fly ash, plus some suspected pozzolanic activity due to natural fines. In this case, a similar increase in strength did not develop. The same occurred for some mixes at the La Miel-I project. In both cases, the addition of fly ash caused no noticeable strength gain though it did cause the RCC to become much more brittle. Consequently, the mix in these cases was more susceptible to cracking due to pozzolans.

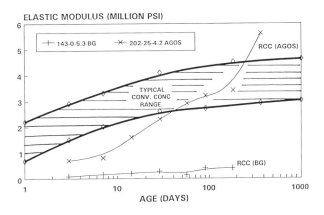

FIGURE 18.20 Modulus of elasticity, typical range.

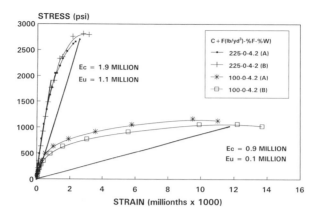

FIGURE 18.21   Modulus of elasticity, elastic and ultimate.

Figure 18.21 shows the derivation of a new and useful value referred to as the "ultimate" modulus [Schrader, 1995b], which is the slope of the secant drawn from the origin to the average peak compressive strength of companion cylinders. It is clear in this typical example that it takes much more deformation and substantial energy to cause the lower strength RCC mixture to fail compared to the higher strength mixture. The mixes were made with identical aggregates and moisture contents and the same source of cement. The only difference was the increased cement content from 100 to 225 lb/yd^3. A lower **ultimate modulus** is associated with more deformation before cracking. This is highly desired in a mass structure but may be a detriment in more traditional structures such as bridges. In a cracking analysis of mass concrete, it is the ultimate condition that is critical and that should be studied, not just the elastic condition. As the mix becomes stronger, the elastic and ultimate modulus come closer to the same value, so it is not as important with higher-strength mixtures. With lower-strength mixtures that are typical of mass placements, using the elastic modulus alone in the cracking analysis can be extraordinarily overconservative. Additional recent work indicates that the ultimate modulus in compression can be a reasonable indicator of the ultimate modulus in tension [Schrader, 1995b,d]. Figure 18.22 shows the relationship between elastic modulus and compressive strength for a typical RCC mixture with reasonable quality aggregates and about 5% fines. Figure 18.23 shows the relationship between ultimate modulus and compression for the same RCC.

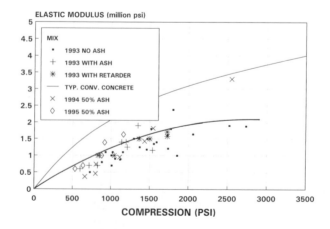

FIGURE 18.22   Elastic modulus versus compression.

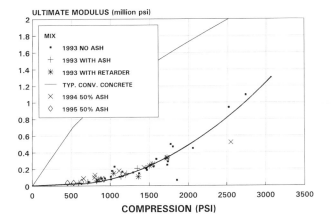

FIGURE 18.23 Ultimate modulus versus compression.

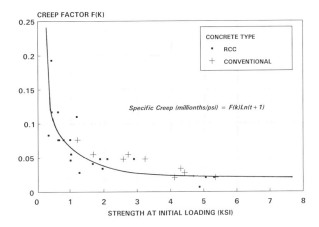

FIGURE 18.24 Specific creep.

18.4.11 Creep

Creep can be thought of in two ways. First, it is the increase in deflection or strain over time due to a sustained load. Second, it is the relaxation of stress over time while maintaining constant deformation or strain. A very high creep rate with associated dramatic reductions in stress over time is possible with low cementitious content RCC mixtures. When combined with the initial elastic modulus, creep reduces the **sustained modulus** [Tatro and Schrader, 1992] that is effective over a long time period, thereby improving the crack resistance of massive placements subjected to thermal stresses. Figure 18.24 shows the relationship between strength at the time of initial loading and creep. The amount of creep over a selected period of time is the creep factor $F(k)$ multiplied by the natural log of the age in days plus 1 d.

18.4.12 Tensile-Strain Capacity and Toughness

Tensile-strain capacity is related to the modulus of elasticity and tensile strength. Fast-load strain capacities (maximum deformability without apparent cracking) are obtained when the load is applied over a time span of seconds or minutes. However, in dams and other massive structures, the primary concern is for slow-load strain capacity, where the strain due to external forces and internal thermal cooling develops over a long period of time. Because of creep, a dam can usually undergo more deformation without cracking if it is slowly stretched versus when it is suddenly deformed. Therefore mixtures, with a low elastic modulus during the period of loading and a high creep rate, such as occurs with low cementitious material content RCC mixtures, can have good slow-load strain properties despite low strengths.

Typical slow-load strain capacities for RCC dam mixtures are on the order of about 90 to 150 microstrain, but values outside of this range are possible. Each mixture should be evaluated, if not by direct slow-load strain capacity tests, then indirectly by extrapolation from static modulus of elasticity and creep studies that can be combined to get a sustained modulus value. The tensile strength divided by the sustained modulus will provide a dependable estimate of the slow load strain capacity over the period of time in question [Ditchy and Schrader, 1988; Tatro and Schrader, 1985; Tatro, 1992]. The problem with extrapolation is that it assumes the creep rate in tension will be similar to compression (as tested). This can be conservative, especially for mixtures that contain a relatively large amount of coarse aggregate. Aggregate-to-aggregate contact can decrease creep in compression.

Because of the low modulus associated with RCC mixtures having lower strength and lower cementitious contents, decreasing, rather than increasing the cementitious content and strength of a mixture can actually increase rather than decrease tensile-strain capacity. This occurs when a decrease in cementitious content results in a greater decrease in modulus than the associated decrease in tensile strength. The ultimate example is the mix that has no cementitious content and insufficient stiffness to define a crack. Obviously, this would also result in an RCC with virtually no strength. Adequate cementitious material is needed to resist the applied loads, but additional strength does not necessarily mean more crack resistance. Additional strength could result in more cracking. Figure 18.25 is an example of tensile-strain capacity and the tensile modulus of elasticity for mixes with different strengths. The mix numbers next to each tensile stress-strain curve represent the mixture that was used; the first number is the cementitious content, the second number is the percentage of fly ash or pozzolan, and the third number is the moisture content of the fresh mix. In this case, the mix with a cementitious content of 405 lb/yd^3 (with 33% of that being fly ash) had a tensile-strain capacity of about 90 millionths of an inch per inch of length. The mix with only 160 lb/yd^3 of cementitious material also used a very poor quality aggregate. Its tensile strength was a low 55 psi, but its tensile-strain capacity was very high at about 340 millionths. The structure built with this mix, Concepcion Dam, has been in service for about eight years with no cracking or distress. Because of the strain capacity of the mix, it was constructed with no cooling to control thermal stresses and contraction.

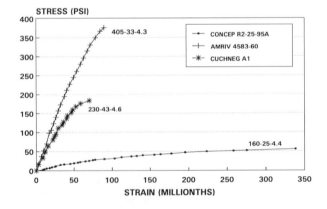

FIGURE 18.25   Tensile-strain capacity; RCC examples.

A very important consideration with lower-strength RCC mixtures is their demonstrated tensile toughness. This is related to ultimate tensile-strain capacity. Substantial toughness is not apparent in higher-strength RCC or most conventional concrete. The indication of substantial toughness in lower-strength mixtures is a relatively new discovery being demonstrated with time, experience, and assemblage of data. Such toughness can contribute to considerable benefits in dam performance and should be considered as an influencing factor in new dam designs and analyses. Eventually, it may lead to more efficient and economical dams that consider fracture toughness in the design. Toughness can be thought of as the ability to absorb energy. Another way to consider it is as a material property that results in the need for substantial energy to be added in order to cause the concrete to fail after it has been stressed beyond its elastic range.

18.4.13 Adiabatic Temperature Rise

Major decisions concerning schedule, cost, construction controls, and cracking potential are based on the results of adiabatic tests. These should not be slighted in any large project. Determinations of adiabatic temperature rise based on calculations from the **heat of hydration** of the cementitious materials and the properties of the aggregates have been reasonably accurate for some RCC mixes and quite inaccurate for others. There does not appear to be any good indicator of when they will be reliable. The only way to assure the proper knowledge of potential peak temperatures as well as the rate of temperature rise for RCC is by proper testing of the full mixture, with large samples that have a volume on the order of about 10 ft^3. Companion tests of the heat of hydration of the cement by itself and the heat of hydration of the cement with pozzolan (if used) are useful for later reference.

A review of adiabatic rise for various RCC projects using Type II moderate-heat cement shows typical ranges (in degrees Farenheit rise per pound of cementitious material in each cubic yard of RCC) of 0.0 to 0.10°F at 1 d, 0.09 to 0.18°F at 7 d, and 0.13 to 0.21°F at 28 d. The adiabatic rise for RCC mixtures with large proportions of fly ash, especially class C ash, should be determined through a later age of 56 to 90 d. Mixtures with lower cement contents tend to produce higher temperature rises per unit of cement. This is even more evident if pozzolan is included.

18.4.14 Thermal Stress Coefficients

Thermal stress coefficients establish how much internal tensile stress will develop in a given mixture for each degree of temperature drop over a specified period of time, assuming the RCC is restrained. RCC can have a broad range of values depending on the mixture proportions and aggregate. It is influenced by creep, coefficient of thermal expansion, and the elastic modulus. Stronger mixtures typically have much higher stress coefficients. For a time period of 28 to 365 d, typical values for low strength mixtures (800 to 1800 psi) with low cementitious material contents are on the order of 4 to 6 psi per degree Farenheit. Higher-strength mixtures (2000 to 4000 psi) have stress coefficients on the order of 8 to 14 psi per degree Farenheit.

18.4.15 Shear Strength and Lift-Joint Quality

The shear strength of an unjointed RCC mass, of RCC containing joints with strengths similar to the mass, and of RCC with the principal load normal to the joints will generally be 10% to 20% of the compressive strength for strengths in excess of 2500 psi, about 15% to 25% for strengths on the order of 2000 psi, and about 25% to 30% for strengths on the order of 1000 psi.

Lift joints, or the layer to layer interfaces, are the weakest part of RCC structures. Special treatments such as a bedding or mortar mix and high cementitious content mixtures can improve this problem considerably, but these techniques are expensive and time consuming. Lift-joint quality as it relates to shear strength is elaborated below. Lift-joint quality in it relates to watertightness is discussed in Section 8.4.16.

The strength of RCC lift joints is dependent on the consistency of the fresh mixture, aggregate characteristics, degree of compaction at the lift surface, cementitious-material content, maturity of the lift surface when it is covered with the next lift, condition of the lift-surface, and lift-surface treatments. A detailed procedure for assessing lift-joint quality in the field, taking into account the many factors that affect the probable lift-joint strength, has been developed [Schrader, 1995c,d]. On the basis of clearly defined criteria, the procedure establishes a numerical value (plus or minus) for each of the following factors that influence the *in situ* shear strength: surface segregation, rain, cure, maturity, surface tightness and condition, surface flatness, the method of RCC delivery, and miscellaneous factors. The sum of the numerical points assigned in each category is referred to as the **lift-joint quality index** (LJQI). The basis of design corresponds to a LJQI of 0.0. Any positive value implies a slight improvement in quality above the basis of design. A negative value indicates a quality less than the basis of design. A graph provided with the procedure indicates the percentage of the design values for **cohesion** and the percentage of the design value for friction that is associated with various values for the LJQI. As a matter of inspection and field control, the contractor may be penalized for slightly negative, but tolerable LJQI work, with a cut-off point at about −4, below which the structure would not perform acceptably and the work is not accepted.

A wide range of possible RCC joint strengths exists, depending on the mix and the above variables, with extremes ranging from about 0 to 300 psi for cohesion and 30° to 60° for friction angles. For example, one summary of strengths for 35 tests including 5 projects, 10 mixtures, 8 joint conditions, and 5 aggregate types resulted in an average cohesion of 155 psi and a friction angle of 50° [Schrader, 1986]. Each project should be carefully examined and preferably tested to determine the shear capacity of its joints. Well-proportioned dry low cementitious material content mixtures with reasonable aggregates and construction controls can be expected to produce friction angles on the order of 45°, with cohesion values on the order of 40 to 110 psi. High cementitious content wetter mixtures typically have similar or slightly better friction angles and cohesion values that may range from about 110 to 300 psi.

Substantial testing of lift joints reported for a variety projects provides useful data for those specific mixtures and conditions [ACI, 1989; Hansen and Reinhardt, 1991; Schrader, 1982a,b, 1984, 1986, 1994, 1995b,c,d; Boggs and Richardson, 1985; Oberholtzer et al., 1988; Gaekel and Schrader, 1992; Dunstan, 1981; Cannon, 1985; Tayabi and Okamoto, 1987; McLean and Pierce, 1988, Dolen and Tayabi, 1988]. Care must be exercised not to use one publication or one set of results that is based on one set of conditions at one project as an absolute basis for what will occur at another project. A general idea can be developed based on a compilation of information from other projects and a knowledge of the mixture and materials proposed for a new project, but absolute values should come from testing the specific mixture and conditions in question.

Every job may have its own peculiarities, but generalized observations typical of most RCC follow: Shear strength increases with age. A wetter consistency at the same cementitious material content can slightly decrease the cohesion value if the moisture is above or below optimum. Higher cementitious content mixtures achieve higher potential cohesion, but the friction is essentially unaffected. Lift joint exposure maturity (exposure time × surface temperature) is a major influencing factor in lift-joint quality or strength. Figure 18.26 shows the general trend or effect of age on cohesion and friction for mixes with both high and low contents of cement plus pozzolan. Figure 18.27 shows the general trend for lift-joint maturity on cohesion and friction for mixes with both high and low cement and pozzolan content. These figures are intended to show typical or general trends; they should not be used as a basis for design without further consideration for the peculiarities of the RCC specific to each project.

Figure 18.28 is an example of site-specific tests for the Miel-I project. The effect of cement content, maturity, admixture, and bedding mix on cohesion is clearly defined for these materials. The mix used only cement, with no fly ash or pozzolan, but it did contain approximately 6% aggregate fines. The bedding mix consisted of a 3/4 in thick layer of high-slump traditional concrete with about 45% sand and a high cement content. It was spread just prior to placing the RCC; the RCC was then

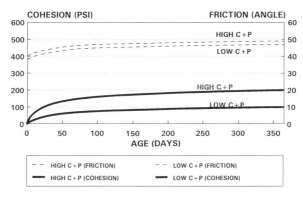

FIGURE 18.26   Friction or cohesion, effect of age.

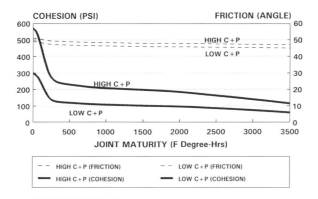

FIGURE 18.27   Friction or cohesion, effect of joint maturity.

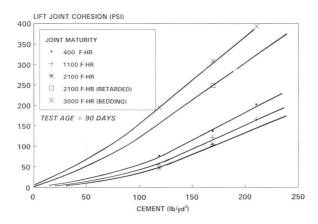

FIGURE 18.28   Miel-I lift-joint cohesion versus cement.

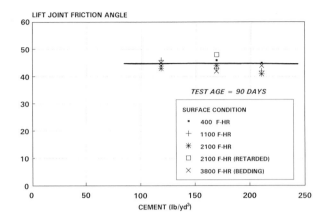

FIGURE 18.29   Miel-I lift-joint friction versus cement.

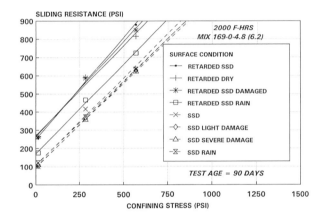

FIGURE 18.30   Miel-I lift-joint shear.

compacted over the bedding. Figure 18.29 shows that the friction angle is essentially independent of maturity, cement content, admixtures, and bedding. It is almost entirely a function of the physical aggregate characteristics.

Figure 18.30 shows a typical plot of total shear stress resistance as a function of the confining stress applied normal to the lift-joint surface. These tests were performed on large blocks that were saw cut from the full RCC mix. The mix contained 169 lb/yd^3 of cement with no fly ash or pozzolan. It had a relatively dry consistency with 4.8% moisture and 6.2% aggregate fines. A surprising result, similar to those of full-scale tests at other projects with other materials, is that allowing the surface to dry just prior to placing the next layer of RCC had virtually no negative effect on the joint strength. This is contrary to traditional practice for conventional concrete. In the field, allowing the surface to dry just prior to placing the next layer allows efficient and very effective cleaning by blowing the surface with compressed air, thereby achieving an even better joint quality.

18.4.16 Seepage, Permeability, and Watertightness

At the time of early RCC projects, confusion developed concerning **permeability** and seepage in RCC dams because they were sometimes discussed or presented and compared using data that had a very different basis. Any one, or any combination, of the following was used: pressure tests and water-loss results from drill holes in dams; recordings of seepage collected from drain holes drilled into dams that may or may not have included holes that extended into the foundation or **abutments**; recordings of discharges from galleries or stilling basins that may have included water from any combination of sources such as foundation drains, drilled drains in the dam, cracks, monolith joints, local runoff, drains in the RCC, and water from construction activities; tests of cores that may or may not have contained joints that may or may not be oriented with or against the flow path; and permeability tests of laboratory prepared cylinders.

Another problem in reporting seepage data is the lack of a full grasp of the details of internal drains, their construction, their purpose, and how they function. For example, Urugua-I Dam originally was supposed to have a grid of face drains behind the upstream membrane so that seepage through the membrane could be detected and isolated if it occurred. Because of the 100% effectiveness of liner systems at other dams and confidence that the liner would have no seepage, this was deleted from the work. Only a single drain line was installed. It was located behind the membrane, just above the foundation. Seepage from this drain has been reported as being due to membrane seepage when, in the opinion of the designers and those who were responsible for construction, the seepage is due to two other causes and not the membrane. One cause is a poor detail where the membrane is connected to the foundation. Water comes under the foundation contact and up behind the membrane to the drain. The other cause is seepage through a portion of the abutment that had questionable grouting. Water probably came into the abutment and then traveled to the dam where some of it migrated upstream to the drain line.

When seepage is present, it typically diminishes naturally with time for both conventional concrete and RCC dams. The reduction is especially dramatic with low cementitious content RCC dams where seepage is primarily along lift joints. Typical reductions are on the order of about 85% to 95% within about 1 to 2 y.

Unlike low cementitious content RCC, which typically experiences initial seepage along lift joints that do not have special treatment, high cementitious content RCC behaves similarly to conventional concrete dams which typically have negligible seepage along the lift joints. As is the case, for example, with Dworshak Dam (conventional) and Upper Stillwater Dam (high cementitious content RCC), watertightness problems with these types of dams are more related to leaking monolith joints and/or leaking cracks. Water loss occurs as high flows or leakage concentrations at fewer isolated locations in high cementitious concrete dams, whereas in the lean RCC dams (without watertight facings or special lift treatment) the unit seepage is smaller but the area of seepage is greater. For example, two years after their respective reservoirs were raised, leakage through one of the cracks at the very high cementitious material content Upper Stillwater Dam was greater than all of the seepage from all sources including foundation drains, local runoff, and lift joints at the very low cementitious material content Willow Creek Dam (it has no monolith joints or through cracks). This is not to say that all high cementitious content RCC dams will have cracks and joint seepage, nor that all low cementitious content RCC dams will have seepage of lift joints. It is meant to point out general differences and where the emphasis should be placed during design for seepage control with different types of mixtures.

Some projects have been designed to allow seepage and let it pass through the structure without being collected or drained away. Seepage in these projects was the most sensible, economical, and appropriate design. Seepage was not a result of a failed design. On the contrary, the design worked. Going back to one of the first RCC dams, the published *Design Memorandum for Willow Creek Dam* [U.S. Army Corps of Engineers, 1981] included a discussion of seepage that could be initially anticipated. Performance of the dam has been almost exactly as predicted. The section or mass of the dam was increased as part of the design in order to offset uplift pressures along lift joints that were allowed to seep without benefit of an internal uplift reduction drain system. The seepage has reduced naturally over time to about 10% of the initial value. Extensive coring soon after construction and again after years of steady seepage show no detrimental effects. This includes tests of joint strength (tension, cohesion, and friction), compression, density, tension, and appearance despite aggressive reservoir water.

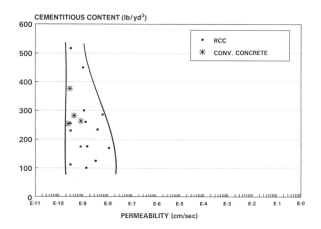

FIGURE 18.31   Sample permeabilities.

Figure 18.31 shows that there is essentially no change in RCC permeability (excluding joints) with changing cementitious contents, including very low cementitious contents of only 100 lb/yd³. The data represent the average permeability values primarily for test samples of RCC actually used on completed dams. For comparison, the figure also shows sample permeability values from tests of conventional mass concrete. The results of the latter are similar to RCC. All of the mixtures have permeability values suitable for dam construction.

Table 18.9 shows specific average test results for different RCC mixtures listed in ascending order of cementitious-material contents. It is clear from comparing the cementitious-material content to the permeability values that there is no defined relationship between cementitious-material content and permeability within the range of 100 to 450 lb/yd³. It is also clear from the table that there is no defined relationship between permeability and the proportion of cementitious material that is pozzolan.

TABLE 18.9   RCC Permeability: Average of Cores

Project	C + P, lb/yd³	Permeability, (cm/s) $\times 10^{-9}$
Urugua-I	101 + 00 = 101	1.4
Willow	80 + 32 = 112	0.3
Zintel	125 + 00 = 125	3.3
Lost Creek	94 + 76 = 170	11.3
Elk Creek	118 + 56 = 174	0.8
Willow	175 + 00 = 175	1.5
Cuchillo Negro	130 + 100 = 230	0.3
Lost Creek	234 + 00 = 234	3.9
Willow	175 + 80 = 255	0.3
Lost Creek	120 + 140 = 260	1.3
Zintel	300 + 00 = 300	1.1
Willow	315 + 135 = 450	1.0
Wes	517 + 00 = 517	0.3

Confusion about seepage can be easily clarified and summarized as follows. Properly designed RCC mixtures all have low permeabilities that are similar to conventional mass concrete, about 1 to 10×10^9 cm/s, regardless of cement and pozzolan content (this statement applies to the compacted mass without joints). A key aspect of properly designed RCC mixtures is that they contain about 20% by volume of material finer than 75 μm. This includes all materials (water, small air bubbles, pozzolan, admixtures, slag, aggregate fines, and cement). If a low cementitious material content RCC is made without adding fines to keep the paste volume at about 20%, the mixture is poorly proportioned and greater permeability can be expected.

Along with these strong statements about permeability of unjointed RCC mass, a companion statement is needed. If special precautions (such as bedding mixture, upstream membrane, or fast lift placement) are not used, low cementitious content RCC dams with a dry consistency will have lift joints that seep. The joints can still achieve good friction and cohesion properties, reasonable tensile strengths, and even look tight, but they will seep. The total quantity of seepage through the joints may be low, but the unit seepage rate along the thin lift joint can be on the order of 1×10^{-3} cm/s.

If this seepage is allowed to penetrate through the dam instead of being intercepted by drains, it may cause uplift and cause the entire mass to look like it is seeping. There are numerous applications where this condition is tolerable and the best overall design, but there are other applications where special precautions to control seepage or a different type of design are appropriate. If a higher cementitious material content mixture with a wetter consistency is used, lift-joint seepage problems will be similar to those encountered with traditional concrete.

A few clarifications are appropriate. Bedding mixture (a retarded sanded grout or high-slump small-aggregate conventional concrete) will provide watertightness if the bedding mixture extends downstream from the upstream face a distance approximately equal to 8% of the hydraulic height above the joint. In the process, the bedding mixture will also provide improved tensile and shear strength. This assumes that the bedding is always placed correctly. In practice, it is reasonable to expect that there will be some less-than-perfect placement of the bedding mixture in any large project. Consequently, it is reasonable to assume that some lift joints may have some seepage, just as occurs with traditionally placed mass concrete. This is usually minimal; it can be picked up by drains; and it will reduce substantially with time. Adding to the width of bedding provides some additional protection, but because of interference with other aspects of the construction, it also makes placement of the bedding and achieving good quality more difficult. If total watertightness is desired, regardless of cracks, joints, failed waterstops, poor-quality lift joints, or low cementitious content mixtures, the impervious upstream membrane system discussed in the next section should be considered.

18.5 Design

18.5.1 General

Efficient RCC designs require balancing considerations including stability of the structure and foundation, stresses from applied loads and dead loads, thermal stresses, construction methods and rates of placement, and mixture proportions including their material properties. Because of the low cost of RCC, the wide range of material properties possible (including properties not possible with conventional concrete), and the ability to use marginal or normally unsuitable materials if necessary there are many more possibilities the designer needs to evaluate than there would be for conventional concrete. Consequently, the designer has a more complex task, but he also has more opportunity to save time and money for the project and maximize the use of available resources and materials.

RCC has been used for a variety of mass applications. Examples include the base on which structural concrete was placed at the Bellefont Nuclear Plant, the base on which hydroelectric power plants were constructed at the Tarbela project, and the base under a number of large spillways. Other examples of mass applications are large buttresses or supporting walls for potential large landslides, as was done along flood channels in El Paso and as was done for the mountainside at Platanovryssi Dam. Still other mass applications involve erosion protection for enormous plunge pools and stilling basins that would otherwise erode from extreme water flows at both low and high velocity [Schrader and Stefanakos, 1995]. The most complex massive RCC structures, and the most common applications, are dams. The remainder of this section on design concentrates on RCC dams, but much of the content applies to other massive applications such as walls and buttresses.

Structural design of RCC dams uses the same basic criteria and procedures used for conventional concrete dams. After the RCC has hardened in place, it is concrete. The concrete is just placed by a more efficient method than has historically been used, and it may have a design section and material properties that are outside the normal range of conventionally placed concrete structures. These attributes, plus the rapid rate of placement and the minimization or elimination of monolith joints, may require a more in-depth study of the structure. General discussions of structural design of RCC dams can be found in ACI (1989), Jansen (1989) and U.S. Army Corps of Engineers (1995). Recent developments have allowed for three-dimensional analyses of RCC structures that simultaneously consider time-dependent material properties, stress-dependent material properties such as the modulus, different mixes used at different locations in the structure, different times of placement for each layer or group of RCC layers, time-dependent development of thermal stresses, dead and applied loads, and seismic loads [Angulo et al., 1995].

RCC dams have been constructed with axes that are straight or curved, with two intersecting straight axes, and with a combination of a curved axis with straight tangent axes. Both vertical and sloped upstream faces have been used. The downstream face can be vertical or at any slope. Within a given section, the downstream slope can be infinitely adjusted, curved, or parabolically shaped. Extensions of the toe, deeper excavations, and keys, have been used to provide stability over poor rock and bad foundation conditions.

RCC dams can be any height ranging, for example, from 6 ft (Ferris dam), to 25 ft (Kerville), 47 ft (Winchester), 125 ft (Copperfield), 160 ft (Monksville), 170 ft (Aoulouz), 210 ft (Concepcion), 270 ft (Urugua-I), 300 ft (Trigomil), 620 ft (La Miel-I), and 660 ft (Longtan–proposed).

The primary areas requiring special attention in RCC dams are overdesign strength requirements for variability, design section options, upstream and downstream facings (facings apply also to rooms in RCC masses, facings of RCC walls, and other nondam applications), and thermal stresses.

18.5.2 Overdesign Strength Requirements

"Overdesign" average strength requirements for variability are required by most codes for general conventional concrete construction. Sometimes the overdesign is achieved through an arbitrary extra factor of safety applied to the required design strength, but more appropriately, specific statistical procedures such as those outlined by ASTM or ACI are used. Until recently, it was not common to apply these specific procedures to dam construction, although various methods have been employed historically to provide a reasonable overdesign in dams. This overdesign is an increase in the average strength requirement to account for the fact that not all cylinders will test at exactly the design strength level. Some will be higher and some will probably be lower. The overdesign factor is a method of limiting the probable number of low strengths to acceptable levels.

Special care is needed with the application of overdesign factors in mass concrete, including RCC, because any extra cement used to increase the average strength may result in an unacceptable increase in thermal stresses. Also, in high-production RCC where the results of meaningful cylinder tests are not available until the structure is essentially complete, testing of variability in the fresh mixture as placed has developed as a practical method of statistical control rather than using cylinder strengths. The overdesign requirement and its application to RCC dams is discussed in detail in the literature [Schrader, 1987].

18.5.3 Design Section Options

Design sections and facing options for large and small RCC dams have been discussed in the literature [Schrader, 1993]. The variety of basic cross-section options that are generally available for RCC dam sections are shown in Figures 18.32 and 18.33. The same options could be applied to conventional concrete, but the cost and construction methods used in conventional construction make most of them impractical.

Figure 18.32 is a typical section for a low dam and/or a dam on a gravel foundation. The extra effort and costs required for a formed vertical upstream face usually are not worth the effort in a low RCC dam. It is easier, cheaper, and faster to simply overbuild the dam at the upstream face without forms. In addition to simpler construction, the extra mass provides additional safety within the RCC and may allow less stringent specifications or inspection. It may also justify using "judgmental" mix designs without prior testing, and local pit-run aggregate.

Compressive and shear stresses in a low RCC dam are so small they are almost meaningless. If the structure is subjected to overtopping, a reasonable level of bond between the top lift joints is necessary. This can be assured by using a bedding mix between the top several RCC layers. Cement

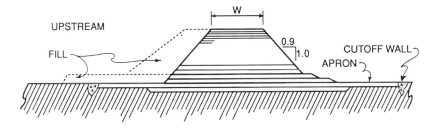

FIGURE 18.32   Small dam section options.

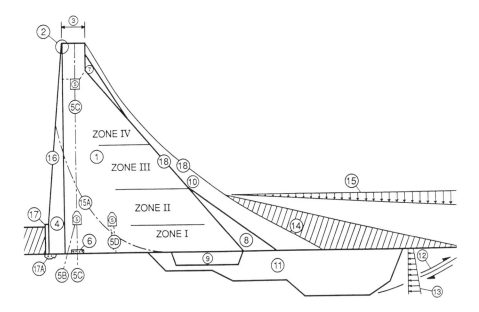

FIGURE 18.33   Large dam section options.

contents are usually dictated by exposure conditions, quality of the mixing equipment, and degree of inspection. Small dams that may need a cement content on the order of only 50 lb/yd^3 for structural loads should have higher cement contents, usually on the order of 150 to 300 lb/yd^3, to account for these factors. For small dams with volumes on the order of several thousand cubic yards, the cost for the extra cement is insignificant.

Extra cement is helpful when placing in cold weather owing to the heat it generates. Thermal cracking in a structure of this size and shape normally is not a structural problem. When needed, watertightness can be provided by using fast construction, a high cementitious content wet mix consistency, an upstream membrane, or a bedding mix between layers at the upstream face. Some structures do not need watertightness.

A downstream slope of 0.9 : 1.0 (0.9 horizontal distance to 1.0 vertical distance) or flatter is suggested for low dams because it is easy to build with any RCC mix. The extra RCC material involved is negligible.

The top width should be selected as the minimum that allows reasonable construction with equipment typical of small projects. The contractor should be allowed to overbuild for his convenience at no additional cost to the owner. A suggested absolute minimum width is 12 ft. Two concerns are required access (if any) after construction and minimum widths for compliance with applicable safety regulations.

Fill material can be used at the upstream (or downstream) face of the dam to steepen the slope, narrow the top width of RCC, and save volume if this becomes economical or if the topography requires it. When placed at the downstream face, the fill hides minor seepage and protects the RCC from exposure. If impervious fill is used at the upstream face, it can provide improved watertightness. When fill is used to steepen the RCC slopes of a dam on a nonrock foundation, consideration should be given to the increase in bearing pressure and sliding that this causes at the base of the dam.

Small RCC dams on pervious foundations may require a cutoff or grouting for foundation stability and/or seepage control, but usually it is faster, cheaper, and easier to spread the foundation of the dam out through wider lower lifts and to use apron slabs of RCC. Settlement studies and a slab/beam analysis of the lower lifts establish how far the extension can go without the slab or apron breaking off. The flow path will also influence the apron width. A simple cutoff wall placed into an excavation without forms is recommended. The downstream cutoff and apron normally should have drain holes.

Figure 18.33 illustrates many of the options for or variations to a typical gravity dam section that are practical with RCC for larger projects. Variations and options are labeled (1)–(17) in the figure. The basic gravity section (1) with a vertical upstream face, constant downstream slope, and a vertical downstream face at the top of the dam has been used for many RCC dams. The low cost of RCC often makes it reasonable to flatten the downstream slope and add more mass than conventional concrete allows. This reduces foundation stress, RCC strength requirements, and lift-joint concerns. Reductions in cement content with related reductions in unit cost and in thermal stresses result. However, the possibility of using higher cementitious contents with higher strengths should also be investigated if the thermal stresses can be tolerated and the volume reduction offsets the increase in cost due to higher unit costs of the RCC. Influencing factors are the length of diversion, the cost and availability of cement and pozzolan, the quality and production costs of the aggregates, and foundation quality.

A parapet wall (2) can reduce costs by reducing the quantity of RCC. The wall can also act as a personnel barrier and curb. Added height or "freeboard" for overtaping waves is not necessary with RCC. Also, by curving the top of the parapet outward, it can return waves by directing them back to the reservoir. The wall can be a continuation of upstream precast panels if that option is used to form the upstream face of the dam. A "breakaway" parapet (fuse plug) designed to fail during overtopping can be designed. One reason to do this is so that water flows over one side of the dam while any downstream powerhouse or access road on the other side remains protected.

The width of the dam (3) should be established after consideration of numerous factors such as the cost of additional RCC and downstream vertical facing; required (not just "nice to have") width for access during operation and construction; inertia (seismic loading) of the laterally unsupported top section of the dam; the effect of the mass on sliding stability due to the added confining load; the effect of the mass on the location of the resultant force for the entire section; the distribution of foundation stresses; and the possibility of causing tensile stress across downstream lift joints for a high dam in the reservoir empty condition.

Adding mass and width to the dam at the base by using a sloped upstream face (4) may efficiently improve stability. An extra benefit is the downward vertical component of the reservoir load on the horizontal projection of the dam face. A condition to check is whether this causes tensile forces to develop or to become unacceptable at the upstream face in both the foundation and lower RCC lifts. Slopes up to about 0.10(H):1.00(V) can be practically built in most RCC dams without noticeable effect on the cost, schedule, or construction practicality.

Tension at the upstream face of both RCC and conventional concrete gravity sections is a controversial issue. Each project should be evaluated for its own set of conditions. What may be acceptable for one location and type of mix may be unacceptable for a different location or mix. The majority of designers and regulating codes in most countries consider that gravity sections should have little or no tension at the upstream face in the normal reservoir or normal operating condition. Minor tension is occasionally allowed for severe flood conditions. However, it is reasonable to provide high cementitious content mixes or bedding mixes across lift joints near the upstream face to accommodate the need for small but sustained tensile stresses on the order of a few percent of the compressive strength for the normal operating condition if this is necessary to achieve an economical design in high dams. Allowing for softening of the foundation with a lower mass modulus at the heel of the dam will also reduce this tensile stress. It is common practice to allow higher tensile stresses on the order of 10% to 20% of the compressive strength for transient load conditions such as maximum flood levels in the reservoir. An additional factor is normally used to allow for about 150% of the static tensile strength of the concrete and lift joints for seismic conditions. With this allowed stress, a factor of safety just greater than 1.0 is typically accepted under earthquake conditions, with higher factors of safety for flood and normal load conditions. Table 18.10 is a simplified summary showing the required safety factors and allowed stresses for different load conditions under different code or agency requirements.

TABLE 18.10 Concrete Dam Factor of Safety Examples*

Agency	Usual	Unusual	Extreme
Sliding within the Dam and at the Foundation Contact			
Corps (1995)	2.0	1.7	1.3
USBR	3.0	2.0	1.0
FERC (General)	3.0	2.0	1.10
FERC (Low hazard)	2.0	1.25	1.0
DIN/DVWK	1.5	1.35	1.2
Sliding within the Foundation			
Corps (Limit equilibrium)	2.0	1.7	1.3
USBR (Shear friction)	4.0	2.7	1.3
USBR (Small dams)	4.0		1.5
USBR (Small-minimal risk)	2.		1.25
FERC (General)	3.0	2.0	1.0
FERC (Low hazard)	2.0	1.25	1.0
DIN/DVWK	2.0	1.5	1.2
Resultant Location at Base			
Corps (1995)	Middle 1/3	Middle 1/2	Within base
USBR	Middle 1/3	Middle 1/3	Within base
FERC (General)	Middle 1/3	Middle 1/3	Middle 1/2
DIN/DVWK	Middle 1/3	Middle 1/2	Within base

Maximum Stress within the Dam

	Comp.	Tension	Comp.	Tension	Comp.	Tension
Corps (1995)	$0.3\,f_c'$	0	$0.5\,f_c'$	$0.6\,f_c'^{2/3}$	$09\,f_c'$	$1.5\,f_c'^{2/3}$
USBR[†]	$0.33\,f_c'$	*	$0.5\,f_c'$	*	$1.0\,f_c'$	*
FERC (General)	$0.33\,f_c'$	$0.10\,f_c'$	$0.5\,f_c'$	$0.10\,f_c'$	$1.0\,f_c'$	$0.10\,f_c'$
FERC (Low hazard)	$0.5\,f_c'$	$0.10\,f_c'$	$0.8\,f_c'$	$0.10\,f_c'$	$1.0\,f_c'$	$0.10\,f_c'$
DIN/DVWK		0		yes		yes

Maximum Stress at the Foundation Contact

Agency	Usual		Unusual		Extreme	
Corps (1995)	Allowable bearing		Allowable bearing		$1.33 \times$ Allowable bearing	
USBR	FS = 4.0		FS = 2.7		FS = 1.3	
FERC (General)	$0.33 \times$ Ultimate bearing	0	$0.5 \times$ Ultimate bearing	0	Ultimate bearing	0
FERC (Low hazard)	$0.5 \times$ Ultimate bearing	0	$0.8 \times$ Ultimate bearing	0	Ultimate bearing	0
DIN/DVWK		0	May open		May open	

*The information shown applies to general conditions. Exceptions may apply or be allowed for special conditions based on the amount of investigation, field conditions, extent of analysis or design, hazard potential or risk, and whether the dam is new or existing.

[†] Max $f_c < 1500$ psi (usual) < 2250 psi (unusual). USBR allows f_t and allows cracked sections for extreme load.

Galleries in RCC dams (5) should be minimized. There has been a tendency to extend galleries beyond where they are actually needed. This causes higher costs, slower production, and lower overall RCC quality. In large open areas such as at the base of a high dam section, galleries typically slow production by about 15% for the uncemented fill method of construction. Conventional forming slows production more. In the upper portions of the dam where there is less room, the decrease in production at the area of a gallery can be 50% or more, and the quality of placement may decline significantly. Where a gallery is actually needed high in a dam for uplift, an open graded rock drain of coarse aggregate should be considered (6). If placed about four lifts high, it is possible to excavate the drain for access if necessary in the future.

In addition to the construction interference that a gallery high up in a dam causes, it also can be a point of weakness in a seismic event (7). Designers of low- and medium-height dams should consider simply overbuilding the dam enough so that galleries and drains are not even necessary. The unit cost of the RCC decreases, while construction, operation, and maintenance are simplified. An example of this is Winchester Dam.

Using a bedding mix between lifts or a high cementitious content RCC is suggested upstream of galleries and between the first three layers in the area above and below the gallery floor and ceiling. This provides watertightness, bond against uplift below the floor, and added sliding resistance against reservoir pressure at the upstream gallery wall.

A **grout curtain** (5B) can be installed prior to the RCC or can be installed afterward from a gallery. The gallery should be large enough to accommodate suitable production equipment, especially at interior corners and intersections.

Internal drains (5C) can be easily drilled with track-mounted rotary percussion equipment. Nominal 3-in holes at spacings of 10 to 15 ft are adequate. These holes can be drilled at high production rates with an accuracy on the order of plus or minus 3 ft in about 120 ft. A very efficient way to drill these holes is immediately after placing the RCC lift that is the gallery floor. When a long gallery with holes starting at the same elevation is called for it is effective to stop RCC for a day while several track drills drive onto the lift and drill the holes. The area is then cleaned, treated as a cold joint, and RCC placing resumes.

High dams, dams with wide bases, dams with high heat (due to high cement contents, high-heat cement, or hot placing conditions), and dams with high elastic modulus values may require longitudinal joints (5D) that can be grouted from a gallery or from outside of the dam. A practical way to make this joint to simply place open graded coarse aggregate at the joint location as each RCC lift is spread. The RCC is then compacted with the aggregate (5D). Grout tubes are installed in the aggregate as it is placed. Before raising the reservoir but after sufficient cooling of the mass, the joint is pressure grouted from the bottom up with expansive grout. A continuous monolithic concrete mass results instead of two masses connected by a thin grout line.

Using a "fillet" (7) in the upper part of the dam at the downstream face provides additional weight that increases sliding stability and offsets some uplift pressures. On a high dam, it moves the resultant force of the entire dam section slightly upstream, whereas on a low dam, it shifts the force downstream. The distribution of stress under the dam, the amount or existence of tensile stress, and the maximum compressive stress are slightly affected. The fillet also reduces the height of the section at the top of the dam that has a vertical downstream face and will eliminate it if extended to the top.

A downstream toe extension (8) can provide additional stability for a high dam where sliding stresses increase significantly with a minor addition in height. It adds both weight and total cohesion, but only in the bottom portion of the dam, which is usually where it is needed. The fillet (7) increases the mass across the full length of the dam, including the upper portions of the foundation where it usually is not needed.

The fillet also adds to foundation bearing and RCC compressive stresses, whereas the toe extension reduces the bearing and maximum RCC stresses. The extended toe requires extra excavation and foundation preparation but only in the deepest section of the dam and not for much of its length.

It is possible that an extended toe in a high dam will result in tension across downstream lift-joint areas at the maximum height for an empty reservoir condition. This can be overcome by an early partial reservoir filling.

A "key" (9) is an effective way of providing additional sliding stability when it is needed in the foundation but not in the RCC. Although adding the key near the upstream face may seem like a good idea because of its potential to act as a cutoff, the downstream location will typically be better. If analyzed for local stresses, an upstream key of a high dam may have tensile forces that could negate sliding friction resistance because there will be little or no vertical stress at the key and a full reservoir. A downstream location has the benefit of maximum vertical confining stresses and the resulting friction. To minimize the width of the key (upstream-downstream) and assure that the required load is transferred to the foundation without slippage across a weak RCC lift surface, bedding mix should be placed between RCC layers in the key or a high paste content mix should be considered. The key provides added foundation stability by extending the foundation failure plane (12) and by the related horizontal component of the downstream foundation-bearing capacity (13). A relatively simple consolidation grouting program in the area downstream of the key may significantly improve stability. A key is usually needed only in the deeper portion of a high dam (if it is on medium to poor-quality rock), at isolated locations where the foundation condition is bad, or for a medium-height dam on an unsuitable foundation.

When the bearing and sliding strengths of a foundation are poor, a conventional concrete dam usually is not economical, but RCC can be a viable option. Using a low-strength and low-cost RCC with a parabolically curved downstream face (10) is one approach. At the PC-1 project, a preliminary design with this concept was prepared for a tailings dam on a clay and weathered-rock foundation. The dam was composed of large monoliths that could undergo significant independent movements caused by time-dependent consolidation of the foundations. Each monolith sat on its own excavated foundation, with steps in the foundation matching the location of monolith joints. The abutments were tied in with embankments that would undergo deformation as required. The foundation for this project was so poor that a massive key was needed to provide sliding resistance and lower the bearing pressure. Because foundation **restraint** is minimal for this type of foundation and cement contents are low owing to the low strength requirements, thermal stresses are minimized. However, the thickness of the key (distance from the downstream surface to the foundation under the key) should be analyzed as a cantilevered beam to assure that it will not break from the rest of the dam. If bearing pressures under the key can accept the added weight (15), a fill (14) can be placed over a portion of the toe to offset some of the cantilever forces. The fill also provides extra sliding stability if it is extended downstream beyond the RCC key.

Regardless of which option is used to widen a dam base, a reduction in bearing pressure and maximum stress occurs in the RCC. Reduced strength requirements allow less cement, less cost, and less thermal stress. Stresses at the lower levels are closer to stresses higher up in the dam, so fewer "zones" requiring different quality RCC at different heights are needed.

Although structural requirements for strength reduce to zero at the top of a dam, some minimum strength is needed for erosion and weathering protection, impermeability, and making the mix cohesive enough to be placed and compacted. The minimum RCC strength should be based on factors such as exposure conditions, function of the dam, risk level, and economics. What is appropriate for one project, owner, and location may not be appropriate for another project, owner, and location. There is some disagreement, but minimum strengths at 1-year values of about 1000 psi have been considered acceptable for the mass.

Early RCC dams used higher-strength mixes for the upstream and downstream regions and lower-strength RCC at the interior. This proved to be a more serious construction and inspection problem than anticipated. The practice is now generally avoided. In addition, other factors have influenced this trend, including the good field performance of low strength RCC exposed at the downstream face under severe weather. If needed, RCC can be protected at the upstream face by constant immersion in the reservoir, by an unbonded impervious membrane (with or without protective precast facing panels), and by conventional concrete placed using one of many possible techniques. The downstream surface can also be protected.

It is usually best to use one mix throughout an entire section for small and medium size dams up to about 120 ft in height. Owing to thermal and economic considerations, higher dams are usually separated into horizontal zones, with higher-strength mixes used in the lower part of the structure. Generally, these zones are on the order of 30 to 60 ft thick, with increases in strength of about 100 to 500 psi per zone. Initial planning for the 340 ft high Binongan Dam used four zones. The current design for the 620 ft high La Miel-I dam uses eight zones.

In addition to the higher compressive strengths needed for higher **principal stresses** in the lower portion of high dams, stronger mixes in the lower zones also provide additional lift-joint tensile strength, added cohesion, and usually a slight increase in friction. When a mix in a lower zone has adequate strength for compression but not for sliding stability, there are several options. Increasing the mass or weight of the dam and widening the base have been discussed. Increasing the paste content of the mixes is another option if it is economical and does not cause thermal cracking due to added heat from hydration and/or a higher elastic modulus. Another option uses bedding between RCC layers. As discussed in Section 18.4, this dramatically increases cohesion and moderately increases friction. This technique is especially useful when "cold joints" occur in low-paste mixes. Contract drawings can simply show where bedding mix is necessary for both cold joint conditions and, in some cases with weak RCC mixes, for "fresh" joints (15A).

18.5.4 Upstream and Vertical Face Options

Design options for the upstream face are detailed in the various sections of Figure 18.34, labeled (16A) through (16M). If the upstream face is sloped (16A), the unformed face may be left exposed when it is aesthetically acceptable and when lift-joint seepage is either tolerable or controlled with bedding or higher cementitious content RCC. If total watertightness is needed and special precautions have not been taken, a flexible geomembrane can be placed over the sloping face (16B). A 0.08 in thick high-density polyethylene (HDPE) membrane was used at Burton Gorge Dam, although more flexible PVC is easier to install and is more common. Where the membrane is exposed, it can be protected with a layer of concrete or shotcrete. Below grade, it is typically protected from damage by a layer of sand.

Reinforced conventional concrete placed after the RCC (16C) uses the same design concept as an upstream face on a rockfill dam. The concrete can provide an attractive and watertight facing when properly designed and constructed. This requires slabs with waterstops in the vertical and horizontal joints. A "plinth" or watertight tie-in to the foundation and abutments is required. Two-way reinforcing distributes shrinkage cracks throughout the slab so that cracks are closely spaced but very small. A drain system is needed between the slabs and the RCC. Anchors are needed to hold the slabs to the dam. These should be designed for the force due to horizontal acceleration in an earthquake. It is possible but difficult to position the anchors in the RCC when it is placed; drilling and grouting them afterward is the alternative. This concrete-facing method is often considered but is seldom designed or used. Though it is expensive, it has merits that can be used. It is a planned option on La Miel-I.

An extension of the concrete-faced option (16D) includes a second facing of porous concrete that acts as a total drain between the RCC and the upstream face. This also isolates the RCC from shrinkage and potential cracking or joint requirements in the facing, and it acts as thermal insulation to reduce gradients near the face. It has been included as an option for the design at Kapachira Dam.

RCC can be placed directly against conventional forms, but without a conventional concrete bedding or facing, the degree of compaction and appearance will be compromised. Threaded anchors to the forms can be compacted into the RCC. After the RCC has been placed high enough that the next form panel can be positioned and anchored, the lower form can be slid out along the anchor and away from the RCC mass (16E). The void between the RCC and form can then be filled with conventional concrete that bonds to the young RCC and is mechanically held by the

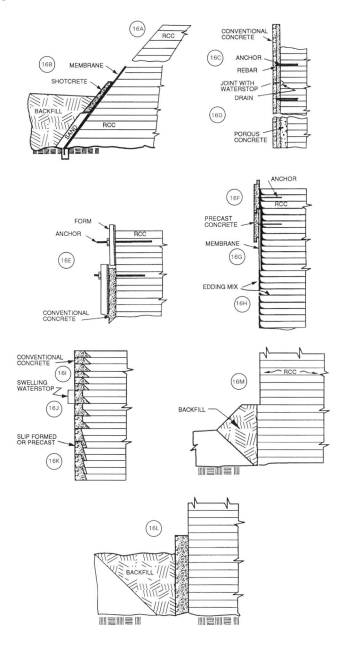

FIGURE 18.34 Upstream face options.

anchor. Instrumentation has shown that by controlling the rate of placement and set time of the concrete with this type of procedure, form pressures can be developed that will stress the anchors and "prestress" the face in place.

Precast panels (16F) make an attractive, economical, and crack-free facing, but the panel joints are not watertight. Anchors are minimal, usually about 75 square feet of panel per square inch of steel anchor area. Watertightness can be provided with a flexible PVC membrane (about 0.08 in thick with welded field seams) attached to the back of the panel. A nut and washer tightened against the membrane with epoxy provides a watertight seal where the anchor penetrates the membrane. This procedure has been very successful in construction, operation, and in tests to a head of 600 ft.

A small amount of bedding is recommended between the membrane and RCC. Drains should be provided behind the membrane to collect seepage if it occurs and to provide additional stability by creating extra uplift reductions.

RCC can also be placed against a conventional form that is later removed. A small amount of bedding against the form has helped to seat the form and provide a better surface. Watertightness can be established with an exposed PVC membrane placed directly against the dam face (16G) or with a high cementitious content mix. The membrane requires an anchored but unbonded procedure specifically developed for concrete dam facings using a special PVC formulation. The CARPI exposed membrane system has been installed with 100% success over more than 10 years on a number of RCC dams and for the rehabilitation of leaking old conventional concrete. Drains between the membrane and RCC improve stability through additional uplift reduction.

Simply extending the bedding mix downstream along the lift joint (16H) for a distance equal to 8% of the hydraulic height will provide watertightness if it is done 100% correctly for 100% of the time. In practice this is not possible. Normal construction with good inspection results in about a 95% reduction in seepage; this may be technically but not aesthetically adequate.

A number of RCC dams, and the facings of rooms and walls on other RCC mass applications, have been built that using the procedure labeled (16I). The procedure results in an attractive conventional concrete face that is completed with the RCC. Usually, the facing has no anchors to the RCC and no reinforcing bars. If a low-water/low-cement/low-shrinkage conventional concrete mix containing a high-range water reducer is carefully used and controlled, a virtually crack-free facing can result— even without vertical joints. The mix should not be thicker (horizontal dimension) than about 1 ft or thermal and shrinkage cracks will probably result. Excellent curing must be provided. Without these precautions, tight cracks at spacings of about 4 to 10 ft can be expected. Normal construction with a reasonable mix will be crack free if joints are provided in the facing about 25 ft apart. The problem with joints in a facing is that it is very difficult to install waterstops. Various projects that have done this have had less than watertight joints. The facing does not make the horizontal lifts watertight. If placing proceeds very quickly (about four to six lifts per day), the fact that the successive layers are placed before the previous layer has fully set will improve watertightness of these joints.

A modified procedure (16J) uses a temporary "blockout" near the upstream face at every other RCC lift. The blockout is removed prior to placing conventional facing and the next RCC lift. Each face placement covers two RCC lifts. Added watertightness can be achieved by using a simple "swelling strip" waterstop that is impregnated with chemical grout and laid along the facing mix lift surface. If seepage penetrates the lift joint, moisture causes the strip to swell and create a watertight pressure seal against the adjacent lift surfaces.

Interlocking upstream-facing elements (16K) have been precast and slipformed. As precast pieces, the upstream area covered by each piece has been only about 10 ft^2 of exposed surface area, so production and placing becomes labor intensive and slow. The small area is a result of the weight of the thick and overlapping shape. The joints are not watertight, and there is concern about stability of the facing if it is not anchored to the RCC. Horizontally slipformed facing can slow production of RCC on dams with a short axis, but the procedure and equipment is better suited to long dams with a large volume of RCC per lift. Careful control of the mix and its delivery are critical, and the facing will develop small shrinkage cracks. RCC can be placed against the facing the same day it is slipped. Consideration should be given to the bond between the unanchored facing and RCC. This may require a high paste RCC mix against the slipformed facing. Sandblasting may be necessary to achieve a bond if the facing is old before the RCC is placed against it. The possibility and consequences of saturation and freezing at the bond line should be evaluated.

Dams in steep canyons, and some large projects, can benefit from an upstream wall (16L) placed across the valley. Such a wall acts as an upstream form for the RCC or as a starting wall for concrete facing and membrane systems; it protects the foundation by containing water and debris; and it allows fill to be placed against the upstream side of the wall, thereby making a practical work area that extends to the face of the dam. Projects such as Copperfield have saved time and money by

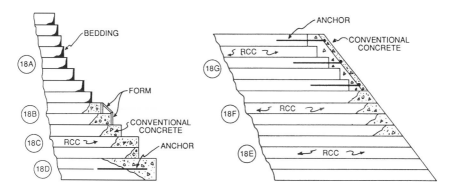

FIGURE 18.35 Downstream face options.

placing backfill lift by lift with the RCC to create a vertical upstream face without forms (16M). This is fast and effective and provides a long level surface upon which subsequent forms can be set when the maximum practical height of backfill is reached.

Regardless of what procedure is used for watertightness, a tight contact is essential at the interface between the upstream face and foundation. Details for this contact and how it varies from one upstream facing and/or water barrier method to another are beyond the scope of this chapter.

18.5.5 Downstream and Sloping Face Options

The downstream face of the dam, or any other sloping face, can be designed using any one of the many options shown in Figure 18.35 and labeled (18A) through (18G). A common, economical, and practical method uses steps with a small amount of bedding placed against reusable-form panels (18A). The panels are one to three lifts high, moveable without equipment, and held by simple methods such as pins hammered into the RCC after compaction. By changing the width of the steps, any average or changing downstream slope can be achieved. If a conventional concrete appearance or protection from weather is desired, conventional concrete can be used for the facing (18B). Larger steps, for example if needed in a spillway, can be built with variations of this method (18C and 18D). Reinforcing steel and anchors are not required with a monolithic construction procedure, but with low-cementitious content mixes, it is essential that the conventional concrete be placed first with a mix that will have lost its slump but not be set by the time the RCC is spread against it and compacted down into it. If the RCC is placed first, the result will look good at the surface, but there will be no reliable contact or interface between the two mixes. Anchorage is then necessary and the RCC should first be compacted at the edge (18D).

Smooth spillways and downstream sloped surfaces have been designed and constructed using no treatment except for hand trimming (18E), unreinforced and unanchored conventional concrete facings with about a 1 ft minimum width (18F), and slip-formed concrete with anchors and two-way reinforcement placed after completion of the RCC and after substantial chipping or preparation of the RCC (18G).

18.5.6 Thermal Considerations

Concrete produces heat when it hydrates and hardens. This is of little consequence to small placements, but it is of major concern to massive placements. If heat from hydration is trapped within the mass, the internal portion of the mass will harden and stiffen at an elevated temperature. Later, as this heat slowly escapes, the mass will try to contract owing to lowering of the temperature. If the mass is restrained, for example, by being bonded to a rock foundation, it is prevented from contracting. The attempted thermal strain is therefore converted into tensile stress. If the thermal stress is greater than the tensile strength, cracking will occur. Because massive structures typically have little, if any, reinforcement, the stability of the structure usually depends on an uncracked section with internal tensile stresses less than the cracking strength of the concrete.

RCC provides several opportunities to reduce internally developed thermal stress. Its lower cement content reduces the total hydration heat in almost a direct proportion to the cement reduction. If the mix has a very low cementitious content, the stiffness or brittleness (modulus) of the concrete can be low, and substantial stress relaxation due to creep can occur.

On the other hand, when conventional forced-cooling methods are needed to control temperatures in RCC, it is not practical or economical to provide forced post cooling; and some of the traditional methods of precooling, as with ice and chilled water, are not very economical or effective. Producing aggregates in cold winter months and storing them in large stockpiles until they are used in warmer months has resulted in reduced placing temperatures. In addition to the controls on aggregate production, the rate of placing, hours of placing (night versus day), and schedule for starting and finishing placing are critical factors for thermal stress analysis. The designer must pay attention to these details and consider them in the thermal evaluation.

Thermal evaluations for RCC mass placements require much more attention to detail than is necessary with traditional mass concrete. There are several reasons for this, but the most important is the large exposed surface area for each thin layer of RCC mass that is placed in each lift. Conventional concrete uses a much smaller exposed surface area with a thick layer of concrete mass for each lift. The heat transfer, either by heat lost to the atmosphere or by heat gained from exposure to the sun, is consequently a much more critical aspect of RCC thermal studies. The time of placement of each layer or lift of RCC can also be a very significant factor, whereas the time of day for placement of conventional concrete is not as important. A low cement content RCC mixture may have an adiabatic temperature rise that is less than the heat absorbed from the sun before the layer is covered with the next layer of RCC. In this case, placing at a faster rate can result in lower temperatures—exactly opposite of what would occur with normal placing of conventional mass concrete.

Details of thermal analyses are beyond the scope of this chapter. Various references contain examples and suggested methods for analysis [Ditchy and Schrader, 1988; Tatro and Schrader, 1985, 1992; Hirose et al., 1988].

When properly accomplished with detailed input of the time-dependent construction schedule, finite element method (FEM) temperature analysis can accurately predict the temperatures of any RCC mix at any location in a structure at any time. This usually requires consideration of the time of day for placement of each layer of RCC, the variable temperature throughout the day, the movement of air across the lift surface, and an accurate determination of the adiabatic temperature rise of each mix. It also requires knowing the physical thermal properties of the materials, considering the type of cure and evaporative cooling, and considering probable interruptions to the placing schedule.

Thermal FEM studies can be simplified by one-dimensional heat-flow studies for the more massive sections of a structure, with two-dimensional analysis being required only in areas of smaller dimensions on the order of about 20 ft wide. The results of the individual studies can then be combined to create a three-dimensional time-dependent result. Using multiple one-dimensional heat flow analysis where possible can actually be better using than two- and three-dimensional studies because it allows better detailing of the FEM mesh, with nodes at the interface of every lift. This typically is every 12 in throughout a structure that extends hundreds of feet in each direction. The exact time of placement of each lift can then be taken into account. Large two- and three-dimensional models that are suited to traditional

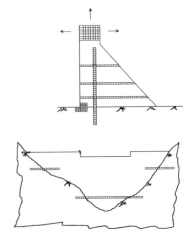

FIGURE 18.36 Sample FEM mesh for thermal studies.

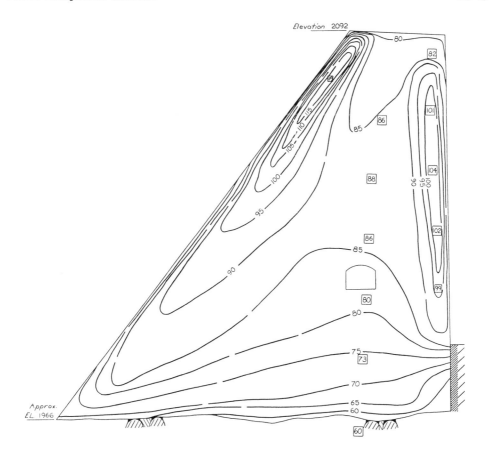

FIGURE 18.37 Predicted versus actual internal temperatures.

concrete placements with a large mass being placed only every few days typically have much larger node spacings and are unable to account for the detailed placing of RCC in 1 ft layers every 3 to 16 h.

Figure 18.36 provides a conceptual approach to using one- and two-dimensional models to properly study a large RCC mass. Figure 18.37 shows the results of a study performed with this methodology, and it gives an idea of the temperature distributions within a structure that used RCC ranging from very lean cement contents in the center of the structure to higher cement contents in the outer 9 ft of the dam. The thermal contours are based on actual field measurements. The temperatures in square boxes indicate the predicted temperature for that mix, location, and time. Predictions were within a degree or two of the actual conditions.

18.6 Construction

18.6.1 General

Because of the interrelationship between mix designs, structural design, material properties, and construction, many aspects of construction have been discussed, at least in part, in the foregoing sections. It is important for the contractor to understand that construction equipment and methods have a direct effect on design and that some procedures, production rates, and equipment may not provide the quality or characteristics needed for a particular design. What is acceptable for one design and set of conditions may not be acceptable for a different design or set of conditions. Construction is discussed in general in ACI (1989), Oury and Schrader (1992), and Schrader (1995a,c,d).

Rapid construction is important to quality, simplicity, and profit in RCC projects. It is more difficult to produce quality construction at low production capacity than at high production rates. In general, RCC mixtures that do not contain a retarder should be delivered within 10 min of when mixing started, spread within 10 min of when they are delivered, and compacted within 10 min of when they are spread. Typical specifications require that the elapsed time from the start of mixing until the end of compaction should not exceed 40 min. The exposed edge of RCC lifts should also be compacted or advanced within about 40 min of when the material was mixed. These time limits are good guides, but consideration should be given to the particular characteristics of each mixture, the temperature, whether the mix is placed in a critical part of the structure, and the practicality of construction.

A common characteristic of profitable and technically successful RCC dams is good communication between the contractor and designer, with decisions being made promptly and with authority in the field. Interruptions and slowdowns result in reduced joint and RCC quality as well as increased costs. Interruptions to a continuous RCC construction operation are bad for both the engineer and the contractor. Direct communications and procedures for on-site problem resolution are essential, without excess personnel, management, or administration. This sounds basic and simple, but it is amazing how many projects have suffered from insufficient authority, responsibility, and responsiveness in the field—from both the contractor and engineers.

When problems develop at the placing area, they must be resolved quickly. In RCC construction, there usually are no alternate monolith blocks where work can progress while the problem is studied. When contingency plans for added monolith joints are approved in advance, it may be possible to raise a portion of a structure or massive placement ahead of the rest of the structure, but this can result in placing difficulties and the design implications must be considered.

Profitable and efficient RCC construction is "machine intensive," with minimal labor per cubic yard of placed RCC. Equipment should do the majority of the work, not labor, even for smaller projects. The rare exceptions are special projects designated to be labor intensive in developing countries with the intent of employing a large unskilled labor force at low wages.

Fueling, formwork, and assembly of embedded items should all be scheduled and planned so that the majority of this work is accomplished off the lifts and during shift changes or scheduled downtime. All unnecessary vehicles and personnel (including visitors and inspector trucks) should be kept out of the placing areas and equipment paths. It is essential that all materials, access, embedded parts, foundation excavation and preparation, and similar work are planned and readied well ahead of time.

18.6.2 Aggregate

The location, size, and withdrawal of aggregate from stockpiles must be coordinated with the RCC plant location and method of feed to minimize segregation and variability. At the very high production rates possible with RCC, several loaders or a conveyor system may be required to keep feed bins full. The length of haul and size of turnarounds need to be considered so that loading equipment can operate rapidly, efficiently, and safely. Aggregate stockpiles and the concrete plant location can be even more important than in the production of conventional concrete. Typically, very large stockpiles that could easily be half the material needed for a season of placement are provided prior to starting RCC placement. Some of the reasons for this are as follows:

- Technical design requirements may require producing aggregate during the winter so that they are stockpiled cold for later use. At Middle Fork and Monksville Dams, winter stockpiling resulted in aggregates that still had occasional frozen areas when the material was withdrawn during the summer though ambient temperatures exceeded 80°F. At Burton Gorge, instrumentation showed that producing RCC aggregate at night resulted in a 10°F lower aggregate stockpile temperature than resulted from producing similar aggregates during the day.

- It may be relatively easy to mobilize aggregate production to full operation while work for the rest of the project is just beginning. It is customary to pay for aggregate in stockpile at the job site if it is tested and found to be acceptable. Payment should be based on an appropriate percentage of the in-place cost of RCC, including some profit.

- The rate of aggregate use during RCC placing may exceed the aggregate production capacity. With large stockpiles, material that occasionally is produced out of specifications may be spread throughout the acceptable material to produce a blend that is within specification. Larger aggregate stockpiles also have the benefit of more-stable moisture contents, which reduces fluctuation in RCC consistency.

Although many RCC dams have been constructed with numerous aggregate size groups and stockpiles, many others have been very successfully constructed with just two size groups. Usually this is plus 3/4 in and minus 3/4 in. Some projects have used a single "all-in" size group. Fewer size groups mean less area for storage and less equipment for loading and transportation. Fewer aggregate bins are required, and less-complicated mix designs are possible. A major benefit that is often overlooked is in the case of a malfunction of an aggregate feed bin. If a plant has four bins and each bin carries a separate material, production stops if one bin malfunctions. If the same-four bin plant is used to carry only two size groups, at least one of the size groups would be fed with more than one bin. When a bin malfunctions, production can proceed, though at a slower pace, with the operating bins while repairs are made to the bin that is down.

When low cementitious content RCC mixes are used, it is necessary to include nonplastic fines (passing 75 μm or No. 200 sieve) in order to compensate for the otherwise low-paste content. When fines are included with a size group of minus 3/4 in, the material is similar to a road base. It has minimal tendency to segregate, especially if it is damp. Also, the moisture content in the pile tends to stay very uniform because the material is not free draining. This is not the case with traditional washed concrete aggregates. Very large and high of minus 3/4 in RCC aggregate containing nonplastic fines stockpiles have been successfully built in layers. The aggregate is later removed without segregation by a front end loader. A potential disadvantage of handling an RCC aggregate with nonplastic fines is its tendency to compact in bins and bridge small gate openings unless special precautions are taken.

18.6.3 Mixing

A comprehensive discussion of RCC mixing and delivery philosophies and specific concerns about handling RCC is available in Oury and Schrader (1992). This reference also includes details and experiences, beyond the scope of this paper, that are pertinent to a wide range of RCC mixing and delivery equipment types.

The concrete plant location should be selected to minimize energy requirements and be appropriate for the terrain, whether the RCC is transported by conveyor or haul vehicle. It should be selected to minimize overall haul distances, vertical lift, and exposure of the fresh mixture to sun and weather. The plant should be located on a raised area so that spillage and wash water drain away without creating a muddy area, especially if vehicular haul is used. The plant location for dams will generally be in the future reservoir area just upstream of the dam and above the cofferdam level or on one of the abutments. The plant should have a bypass or belt discharge that allows wasting rejected RCC without first delivering it onto the dam. This bypass can also be used for sampling, for delivery to trial sections, and for other construction uses.

Both continuous mixers and batch mixers have been used to produce RCC; continuous **pugmill** mixers are the most common. Batch operations with drum mixers tend to cause the most difficulties and concerns. Continuous mixers generally provide greater capacity than batch-type plants. Continuous pugmill mix plants specifically intended for RCC and properly operated and maintained routinely achieve good production rates and uniformity. This applies to plants that operate with volumetric controls as well as to those that operate on weight controls.

Although some RCC has been successfully produced with conventional batch-type plants having drum mixers, problems with low production, bulking, sensitivity to the charging sequence, mixture variability, slow discharge, and buildup in the mixer have been common. This does not mean that acceptable RCC cannot be produced by batch methods and drum mixing, but special attention is needed and low productivity can be expected. Equipment that is well suited to normal "high" production conventional concrete is not necessarily suitable for all RCC mixtures and the typically higher production rates.

RCC mixtures can be very harsh, and some can cause buildup of fines. Drums should be designed or coated to resist buildup that tends to result from the high fines content of some RCC mixtures. Even with these precautions, experience has shown that substantial buildup can develop in drum mixers. If the buildup is not removed, a loss in mixer effectiveness results. Except for special small applications with higher cementitious contents, clean conventional-type aggregates, and aggregate sizes limited to about 3/4 in, transit trucks and mobile batch plants should be avoided. Even with these types of mixtures, slow discharge and high mixture variability should be anticipated.

Pugmill mixers that were originally intended for cement-treated base, asphalt, or moisture conditioning of soils have had difficulties with maintenance and variability when they have been used to construct RCC dams. However, pugmill mixers of both the batch and continuous-mix type have performed well when they are specifically designed and intended for RCC. Typical individual plant capacities range from about 150 to over 400 compact cubic meters per hour. It is generally better to have multiple smaller plants than a single larger plant. If one plant is down, the others can usually continue placing while repairs are made. Also, it is easier to find subsequent uses for smaller plants than for very large specialized plants.

The theoretical or rated peak capacity of the plant should be well above the desired average production. As a general guide, the average sustained placing rate usually does not exceed about 65% of the peak or rated plant capacity when haul vehicles are used for delivery on the dam and 75% when an all-conveyor delivery system is used. These values tend to be lower on smaller projects and higher on uncomplicated larger projects. Table 18.11 shows the average efficiency that can be expected throughout each placement shift for different work schedules and methods of delivery with a properly managed project.

Mixers for RCC need to accomplish two basic functions: They should provide sufficient capacity for the high placing rates typical in RCC, and they should thoroughly blend all ingredients. Typical average placing rates are on the order of about 50 to 150 yd³/h for small projects, 200 to 500 yd³/h for medium projects, and 500 to 1000 yd³/h for large projects. The mixer should operate with little or no downtime. Scheduled maintenance must not be neglected, and repairs should be accomplished rapidly.

TABLE 18.11 Probable Average Sustained Production Through Full Shifts

Shifts Worked	Days Worked	All-Conveyor Delivery	Haul on Dam
2–10 h 2–8 h	6 on—1 off or 12 on—2 off	73%	65%
3–8 h 2–12 h	6 on—1 off or 12 on—2 off	70%	62%
3–8 h 2–10 h 2–12 h	Continuous up to 28 days	68%	60%
3–8 h 2–10 h 2–12 h	Continuous over 28 days	63%	55%

Notes: Assumes Aran continuous pugmill and Rotec or equal proven conveyor. Reduce efficiency by 10% if multiple shifts per day do not overlap. Although Rotec conveyors and Aran mixers (or proven equivalent) may be capable of 80% to 90% efficiency by themselves, the above efficiencies are realistic overall shift values when total maintenance, all types of other equipment breakdowns, aggregate feeding/delivery, raising and moving forms or facings, and slowdowns at abutments are considered.

Variations in free-moisture content of the aggregates can be particularly troublesome when the plant starts up. Some plant operators make the error of overestimating free moisture and provide too little water in the initial mixtures. This is particularly undesirable because most initial mixtures will be used for covering construction joints or foundation areas where the RCC should be slightly on the wet side for improved bond. It is better to start on the high side for moisture and subsequently reduce it to the desired consistency than to start with a mixture that is too dry.

Mixture uniformity must be maintained for all the production rates used. Continuous mixers typically work efficiently above a minimum production rate and up to production levels that are two to three times that of the minimum rate. A consistently uniform mixture must still be provided even when slowing production by say 50% near abutments and when increasing it again after leaving confined areas. Large projects with multiple mixers can simply shut down one or two mixers until the higher production rate is needed again. On smaller projects with one mixer, the mixer itself must be capable of uniform production at varying outputs. Mixture variability is discussed in more detail in Schrader (1987, 1988).

Accurate and consistent control of cement and pozzolan feed is particularly important with continuous-mix plants. This is especially true at lower cement-feed rates. Feeders designed to operate at the high cement-feed rates typical of soil cement often do not operate well at the low cement rates required for some RCC mixes. Maintaining sufficient head in the silos, using air fluffers, and using of vane feeders or positive-displacement cleated belt feeders have been necessary to provide accurate feed.

Proper ribboning or sequencing and feed rates of the aggregates and cementitious material as they are fed into the mixer is critical to minimizing mixing time. The exact timing for adding water to the mixture and the angle of water introduction is critical in drum mixers. Each plant and RCC mixture seems to have its own peculiar requirements that can only be determined by trial and error.

Properly designed pugmills have handled 3 in and larger nominal maximum size aggregate (NMSA), but experience has shown that the amount of material larger than 2 in should never exceed about 8%, and the maximum size should not exceed 4 in. The preferred NMSA for most mass applications of RCC is 2 in.

Accurately introducing the correct quantities of materials into a mixer is only one part of the mixing process. Uniformly distributing and thoroughly blending them throughout the mixture and then discharging them in a continuous and uniform manner is the other part of the process. This can be more troublesome with some RCC mixtures than with conventional concrete mixtures. The accuracy of the concrete plant and methods for control of the mixture during production should be studied for cost effectiveness. If exacting quality control and low variability are necessary, they can be provided in RCC mixtures at increased cost and reduced placing rates. Typical coefficients of variation for RCC compression tests with reasonable weight or volume controls in mass mixtures tend to be about 15% to 20%, with extremes ranging from about 5% to 45%. The recent Platanovryssi project routinely produced excellent variability at about 7%.

18.6.4 Delivery

The volume of material to be placed, access to the placement area, available rental or lease equipment, capital cost for new equipment, and design parameters generally are the controlling factors for selection of equipment and procedures to be used for transporting RCC from the mixing location to the placing area.

Essentially, there are three methods for transporting RCC: (1) by batch, (2) continuously, or (3) by combination of both, typically, using continuous conveyor feed to a hopper on the dam from which vehicles take batches for final delivery to the spreading area. To some extent, transportation may be influenced by the type of mixing equipment used. However, with proper controls and accessories, such as holding hoppers designed to control segregation, continuous mixers can be used with batch transportation and batch mixers can be used with continuous-flow transportation equipment.

The type of transporting equipment used to move RCC from the mixing plant to the placement area will also be influenced by the largest aggregate size in the mixture. Experience indicates that 1 1/2 in NMSA concrete can be transported and placed in nonagitating haul units designed for aggregate hauling and earth moving without substantial uncontrollable segregation. Mixtures with large 3 in NMSA aggregate have more of a tendency to segregate when they are dumped from this type of equipment onto hard surfaces, but with care and proper procedures, these mixtures have also been hauled and dumped successfully. Problems with segregation during the transportation and placing of 6 in NMSA mixtures have been severe.

The entire system of mixing, transporting, spreading, and compacting should be accomplished as rapidly as possible and with as little rehandling as possible. The time lapse between the start of mixing and completion of compaction should be considerably less than the initial set time of the mixture under the conditions in which it is used. A general rule for nonretarded mixtures with little or no pozzolan is that placing (depositing), spreading, and compacting should be accomplished within 40 min of mixing, and preferably, within 30 min of mixing. This time limit is applicable at mix and ambient temperatures of about 70°F. It can be extended for colder weather and should be reduced in warmer weather. It also can be extended for mixtures that are proven to have extended set times because of high pozzolan contents, slags, or effective admixtures with wet RCC consistencies. A simple test to establish the tolerable time for any particular mix and temperature is to compact two cylinders from the same batch every 15 min, starting immediately after mixing. The cylinders are all tested for compressive strength at the same age at about 14 or 28 d. A plot of strength versus time of compaction will quickly indicate the age at which compressive strength becomes seriously affected. Tests have shown that tensile strength and the modulus of elasticity will begin to decrease at about the same time and at the same rate as compressive strength.

The two primary methods of transporting RCC are by conveyor and hauling vehicle. Transport by bucket or dinky has been used, but this slows the rate of production and is more prone to cause segregation. However, if such a system is already available or necessary for large volumes of conventional concrete, it can also be used for the RCC. Pipe delivery has been tried on a few projects with varying degrees of success. If pipe delivery is considered, it should be designed specifically for the mix to be used and potential pressures; it must have a steep slope; the technical issues of moisture loss and temperature gain must be addressed; the length of time in the pipe and its effect on the age of the RCC when compacted must be considered; and sustained satisfactory operation should be proven in full-scale trials.

Transporting RCC by continuous high-speed conveyors directly to the dam is generally preferred. The overall economics, including all direct and indirect costs of alternate delivery systems as well as reliability, the final quality, and schedule should be considered when deciding whether to use or require an all-conveyor delivery system. All aspects of the conveyor system should be specifically designed for RCC of the type used on the project. The advice of personnel experienced with the type of mixture and equipment proposed should be solicited. Conveyor systems that may work well with conventional structural concrete, aggregates, coal, or other materials may not work well with RCC. Conveyor systems that work well with a high-cementitious, wetter, small-aggregate, or no-fines gradation RCC may not work well with a low-cementitious, drier, larger aggregate, or high-fines RCC. Clogged transfers, segregation at the discharge, severe wear at transfers, segregation over rollers, slow belts, not being able to start or stop a loaded belt, drying, loss of paste, and contamination of the RCC lift surface by material dropping off the return side of belts are the most common problems. A detailed discussion of conveyor equipment and methods can be found in the literature [Oury and Schrader, 1992].

Where equipment is readily available for lease or rent, an all-conveyor system can be as economical and practical for small projects as it is for larger projects. The 4000 yd³ Echo Lake Dam in California is an example. The project utilized a highway-mobile all-conveyor delivery system that was driven to the job site and erected in one day, used the next two days for RCC placement where haul units could not possibly work, and then driven off to the next project. It was fed from an equally mobil RCC continuous mix pugmill.

A novel approach has been used recently to assure that, at the time of the design, the engineer can confidently base the design on the better quality, cleaner, and faster lifts placed by specialized RCC conveyors, rather than taking the risk of having the slower production and lower quality lift surfaces that are typical when haul vehicles are used on the surface. A written agreement is made at the time of design, stating that the schedule and design is based on a required predeveloped conveyor delivery system. The supplier guarantees availability of the equipment, the required production capability, and availability of field technicians to assist the contractor. The supplier also guarantees a "not to exceed" price that will be quoted to all qualified bidders. The agreement should address an appropriate allowance for downtime and onsite spare parts.

Allowance is made for contractors to submit a more elaborate conveyor system, whereby, at their option, their costs may be greater but they purchase rather than lease the equipment or they achieve faster production rates. This approach has been used on public projects in at least two countries, with legal documents created that satisfy all public entities as well as the designer, supplier, and contractors. This approach has proven to good for owners, designers, and contractors. It reduces risk, assures a low, preestablished cost, guarantees placement methods that are compatible with design, and helps assure the planned construction schedule.

There are different ways to support and arrange all-conveyor delivery systems. Cuchillo Negro Dam (USA) and Lake Robertson Dam (Canada) are two examples that use vertical posts embedded in and raised with the dam to support and raise the conveyors. Conveyors reach from the posts to essentially anywhere on the dam surface.

Seigrist and Spring Hollow dams (USA) are examples where conveyors were used that were supported from smaller pipe posts anchored to the upstream face of the dam with a long conveyor that extended the full length of the upstream face of the dam and that raised with the dam. The conveyor fed a tripper that ran along the main belt at the upstream face to feed a crawler-mounted mobil conveyor placer on the lift surface. The crawler placer can drive on the surface, extend or retract its conveyor, lift it, or swing it similar to the boom on a hydraulic crane. This allows the operator to deposit the RCC in a layer at its final location. Smoothing the layer with a bulldozer and compaction can follow immediately after the RCC has been deposited from the conveyor. A similar procedure was used for much of San Rafael Dam (Mexico). The crawler placer, with a different setup for the conveyors that fed to it, was used at Concepcion Dam (Honduras) and is being used at Pangue (Chile) and other RCC dams.

At Echo Lake and Big Haynes dams (USA) wheel-mounted conveyors called "creter cranes," and "swingears" reached from outside the dam to the lift the surface and deposit the RCC in layers.

At Burton Gorge Dam (Australia) and others, a partial conveyor system with a belt feeding continuously from a pugmill mixer to a hopper on the dam has been used. Trucks then haul RCC from the hopper to the placing area. Problems with this type of system include continual raising of the hopper, segregation at the edges of each load dumped into and out of the trucks, damage to the surface caused by the truck tires, and insufficient room at the top of the dam for the hopper and trucks.

Exposure time on conveyors should be as short as practical, with 5 min being the desired and 10 min a normal limit. Belt speeds should be on the order of about 10 to 30 ft/s. Covering the conveyor to protect the mixture from drying and from rain should be required for all long sections and preferably for the entire system.

A well-designed conveyor system can also be capable of handling conventional concretes concurrently with RCC. However, this may complicate the placing operation unless separate parallel conveyors for the RCC and conventional concrete are provided. On the Elk Creek project (USA) a larger belt was used for RCC while a parallel smaller one was used for conventional concrete, bedding mixes, facing mixes, and grout.

It is especially important that conveyors do not allow RCC or other material to fall onto the compacted RCC surface along the conveyor path. Because of the rapid rise of RCC dams, conveyor

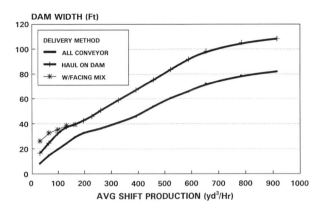

FIGURE 18.38   RCC production in confined work areas.

systems should be designed to be raised quickly. When conveyors are located above the lift surface, provision must be made for clearance of equipment operating under them.

As with conventional mass-concrete conveyor systems, special attention should be given to belt widths, speed, protection, maintenance, incline angles, backup systems, and spare parts. Belt scrapers should be provided to clean the return belt. These typically require frequent attention for adjustment and wear. Properly designed charging and discharge hoppers to prevent segregation at transfer points are essential.

A continuous belt running from the mixer to the final placement area can substantially increase placing rates and significantly reduce other equipment needs and their related labor requirements. Figure 18.38 compares typical average achievable productivities for reduced dam widths when delivery is totally by conveyor and when haul vehicles are used on the dam. Without a conveyor, productivity decreases to very low rates in narrow sections such as at the top of a dam. Figure 18.38 is based on a compilation of actual data at various projects and from computed round-trip delivery times at other projects.

In addition to the benefits of full-conveyor delivery (no hoppers, fast delivery, less congestion, no haul roads, less maintenance, less labor, elimination of lift-surface damage from haul vehicles, and improved productivity), the conveyor can also serve as an access walkway and as a support for lights, water, air, and electric lines to the placement area. A plan has even been developed to use the conveyor as a support for a roof or enclosure to protect the placement area from rain and sun.

When haul units are used to distribute RCC that is conveyed to the lift surface, hopper arrangement that continuously loads them will be necessary at the end of the main conveyor. The objective is to allow the mixers and conveyors to operate and discharge without interruption or waiting for haul vehicles. The recommended minimum size of the hopper is at least twice the size of the haul vehicle. Because of the relatively high unit weight of freshly mixed RCC compared to the loose unit weight of soil, rock, or gravel normally hauled in these vehicles, weight rather than volume normally controls the amount of material hauled per trip.

Bottom-dump trailers and scrapers minimize segregation, spreading requirements, and the distance RCC drops, but they are difficult to use near abutments and obstructions. Scrapers have better mobility than bottom-dump trailers of the highway type. Scrapers and bottom-dumps trailers have the advantage of depositing material in the layer to be spread as they are moving.

Front-end loaders have been used to deliver RCC from a central feed point on the placement to the location where it is spread. This is generally not suitable for large projects. This method has resulted in problems with contamination of joint surfaces, limited productivity, and segregation. However, in special cases where the mixture is not susceptible to segregation, spillage can be avoided, and tire tracking is not a problem, this method may result in the most economical situation that is also technically acceptable. Principle candidates for this approach are smaller dams in tight canyons where the distance for loader travel is minimal. Also, the projects should preferably have a smaller maximum size aggregate, excellent gradation, and a tendency for a higher paste and cementitious content. In addition, each layer should be placed and compacted before the set time of the previous layer. Extra cleaning and/or special grout or bedding mixtures may be appropriate between layers when they are not placed and compacted before the previous layer reaches its final set.

If vehicles are to be used for transporting RCC, a thorough study should be made of the haul route. Problems that may prevent hauling by road include steep and rough terrain, availability of road-building material, plant location, schedule, and environmental considerations. If the concrete plant is located upstream of a dam, the methods used to bring the road through or over the upstream face system may not be practical.

Raising the roads to keep up with the rate of rise of the dam may require so much time that it becomes an inefficient system at higher elevations. To avoid slowing down the mixing and placing operations, raising the roads during a 2 to 4 hr/d shutdown period while maintenance and other work is being done should be considered. Roads must be kept at slopes consistent with the equipment's capabilities and safety requirements.

Haul roads should make the transition onto the lift surface at a shallow angle if possible (plan view) so that turning and damage by tires is minimized. If an immediate right-angle turn is needed (from roads that enter directly onto the dam perpendicular to the face), significant scuffing and lift-surface damage will result. Operators should move slowly while turning and use the largest turning radius possible. The road should be constructed with clean, free-draining rock or gravels if possible.

The last portion of the road prior to entering the lift should be surfaced with large aggregate or clean rock material that minimizes contamination of the RCC surface from truck tires. Simply extending the RCC onto the road will not provide cleaning action. To prevent lift contamination, it may be necessary to use water sprays to wash vehicle wheels before they are allowed on the lift surface, but excess water dripping from the truck and its tires can become a problem. To minimize adverse affects on the surface, hauling equipment should not travel in a concentrated path on the lift. Even with all the above precautions, experience, including observation and cores, has shown that damaged lift surfaces should be expected where haul roads onto the dam are used.

When the RCC is hauled to the placing location and dumped, it should generally be deposited on previously spread but uncompacted material and pushed forward onto the compacted lift surface. This provides remixing action and minimizes clusters of coarse aggregate that otherwise would tend to occur at the lift interface. When RCC is dumped in large piles, larger aggregates tend to roll down the outside of the piles and create clusters. A general rule is to limit the height of a pile to 5 ft. Corecting this kind of segregation is nearly impossible if the rock has already rolled onto a previously compacted lift. Where this condition occurs, the segregated large aggregate must be removed or broadcast onto the RCC layer being spread or wasted.

18.6.5 Placing and Spreading

A preferred technique for placing RCC is to advance each lift from one abutment to the other. An exception is where the distance from abutment to abutment is smaller than the distance from the upstream to the downstream face, such as at the bottom of dams in narrow canyons. In this case, placing can start by working in the upstream-downstream direction. The practice, from embankment construction, of limiting the direction in which rolling equipment can operate does not apply to RCC.

Some early RCC projects considered, recommended, approved, or required placing RCC in paving lanes, typically going from abutment to abutment. This initially seemed like a logical approach, but the practice is now generally discouraged. The problems with it are more serious with lower cementitious content, dryer consistency, and larger aggregate mixtures. Although they may work well with paving mixtures and paving operations, spreader boxes attached to dump trucks, Jersey spreaders attached to dozer equipment, and paving machines lack mobility and occupy important space in narrow areas of the dam. They can be difficult to maneuver at the abutments. Paving lanes tend to leave segregation along the edge of the lanes with dam mixtures. The edges can also become too old to be well mixed and compacted into RCC of the adjacent lane by the time the adjacent lane is placed. The edge also tends to dry out while exposed prior to placing the adjacent lane. This has resulted in concerns over poor quality and weakened or permeable planes through the dam at the interface of paving lanes. If a high cementitious content mix with smaller aggregate, a wetter consistency, and a retarded set time is used, the concerns about placing using pavement lanes are substantially reduced.

Tracked dozer equipment has proven to be best for spreading RCC. It is fast, sufficiently accurate, and contributes to uniformly compacted RCC. By careful spreading, a bulldozer can remix RCC and minimize segregation that results from dumping. Careful attention should be given to assure that remixing is occurring and that the dozer is not simply burying segregated material.

At Elk Creek Dam, retarded RCC mixtures with a Vebe time of 15 to 25 s were end dumped in piles on previously spread but not yet rolled material at least 35 ft from the advancing face. Dozers knocked down the piles and spread the RCC forward into thin unsegregated 6 in thick layers until a full lift thickness of 24 in was reached. The entire surface of each layer was traversed by at least two passes of the dozer tracks. This dozer action produced an average density of 147 pcf. Only the top of the 24 in full thickness of the lift then needed additional compaction by the roller.

Similar results have been achieved with other RCC mixtures having wet mixture consistency. At Nickajack Dam, wet-consistency air-entrained RCC was spread in two 12 in thick layers, with the second layer following behind the first layer. The second layer was compacted before the first reached initial set. The advancing layer was about 100 ft in front of the following layer.

At Burton Gorge Dam in Australia, 100% compaction was achieved using a small dozer in the top portion of the dam by modifying the mixture with retarder, using a wetter consistency, placing rapidly (one lift every 1 to 4 hours), and rigorously tracking the 12 in thick layers as they were spread. This resulted in densities that reached the theoretical air-free density of the mixture. Thorough dozer tracking the same mixture at a dry consistency and without retarder, but while the mixture was less than about 30 min old, achieved densities on the order of about 96% of the theoretical air-free values. This was followed by roller compaction to achieve a higher final density.

Typically, two rollers and one D-6 dozer with a backup smaller dozer can spread and roll nonretarded RCC at a rate of about 300 to 500 yd^3/h in 12 in thick layers. The dozer should operate on fresh RCC that has not been compacted. All turning and crabbing should be done on uncompacted material. Operating the dozer on a compacted surface will damage the surface. When it is necessary for the dozer to drive onto compacted RCC, the operator should limit the movement to straight back and forth travel. Track marks made prior to the mixture reaching initial set can be recompacted by the vibratory roller without significant loss of joint quality. However, damaged surfaces that are recompacted after the mixture has set or if it has dried will have little or no strength, even though they may have acceptable surface appearance. Damaged material can be easily removed by blowing with an air jet, even many hours later. Material that is recompacted early enough will remain cemented in place if blown with an air jet.

Spreading equipment should leave a flat or plane surface of the proper thickness before the roller compacts the lift. Depending on the workability of the mixture, ridges or steps between adjacent passes of the dozer blade can result in uneven compactive efforts and variable quality in the RCC. As a general rule, having a flat surface ready to roll in the least amount of time is more important than having an exact grade with delayed rolling. Typical tolerances for lift thicknesses are on the order of ±2 in. Except for the exposed final lift surface in a structure which typically has a much smaller tolerance.

Where special mixtures are specified, for example at the upstream or downstream face, special procedures are required. If conventional concrete is used against a formed face with a dry consistency RCC mass behind it, the conventional mixture should be placed first and the RCC immediately spread against and on top of the sloping unformed face of the conventional concrete. The conventional mixture should be designed to lose slump rapidly but not set rapidly. This allows the RCC to be compacted into the conventional concrete before either mixture sets. If the conventional concrete does not lose slump soon enough, the roller will sink into it resulting in a variety of construction problems. If rolling is delayed while waiting for the conventional mixture to stiffen, the RCC can become too old for proper compaction. If the roller operator simply stays back from the conventional concrete far enough to avoid sinking into it or shoving it up, the two mixtures will not adequately compact or join together. Using more of the conventional concrete makes the problem worse. Usually, only 12 in of conventional concrete is used, but some projects have been successful with only a few inches of facing. If the facing concrete is wider than about 18 in, the mixture is usually consolidated with immersion-type vibrators while the adjacent RCC is rolled.

If the RCC has a wetter consistency, and especially if it has a delayed set, it is possible to place the conventional concrete mixture after the RCC. The facing concrete still needs to have a relatively low slump when RCC compaction is performed, but it can still be possible to immersion vibrate the interface region of the RCC and conventional concrete. With this procedure, the width of conventional concrete has typically been wider than when conventional facing mixtures are used with drier-consistency RCC. However, experience, coring, and internal destructive investigations have shown that a poor interface of the two mixtures has often resulted with this procedure, even when the two mixtures appear to respond well to immersion vibration and the exposed top surface of the layers looks well consolidated.

The most common compacted lift thickness has been 12 in. Typically, large dual-drum vibratory rollers can develop full compaction of this thickness with only four to five passes. A factor influencing lift thickness is the maximum allowed exposure time before covering one lift with the subsequent lift. Each project should be studied to optimize the benefits of thicker or thinner lifts. Thicker lifts mean longer exposure times but fewer lift joints and fewer potential seepage paths. Thinner lifts result in more potential joints but allow the joints to be covered sooner with better bonds.

At the start of extremely rough foundations and where the foundation has deep holes that have not been filled with dental or leveling concrete, a front-end loader or excavator bucket can sometimes be used to reach the placement site to deposit material. This is a slow operation but may be the only practical solution for some locations. The problem is eliminated with the all-conveyor delivery system or with mobile conveyors that reach across and into areas of rough terrain.

The equivalent of a Cat D-3 or D-4 Case 550 rubber track, or a JD-350 is needed to start the foundation and for tight conditions. With an all-conveyor delivery system, a Cat D-4 is generally capable of spreading RCC at a rate of about 300 yd^3/h. Dozers should have at least hydraulic tilt capability and preferably both tilt and angle hydraulic capability. It is common to underestimate the value of good spreading equipment and operators.

Graders have been used on some RCC projects, but they generally are not necessary and can actually become a problem. They are difficult to maneuver in small areas, and the tires can damage otherwise good compacted surfaces. There also is a tendency to overwork and rework the surface as if it were soil instead of concrete with a limited working time.

18.6.6 Compaction

Maneuverability, compactive force, drum size, frequency, amplitude, operating speed, and required maintenance are all parameters to be considered in selection of a roller. The compactive output in volume of concrete per hour obviously increases with physical size and speed of the roller, but larger size rollers do not necessarily give the same or better density and compactive effort as smaller rollers with a greater dynamic force per unit of drum width. Job size, workability, lift depth, the extent of consolidation due to dozer action, and space limitations will usually dictate roller selection. Large rollers cannot operate closer than about 6 in to vertical formwork or obstacles, so smaller hand-guided compaction equipment is usually needed to compact RCC in these areas. If a slip-formed or precast facing system that has an interior face sloping away from the RCC is used, the large rollers can operate all the way to the facing.

The dynamic force per unit of drum width or per area of impact on tampers is the primary factor that establishes effectiveness of the compaction equipment. Most experience has also shown that rollers with a higher frequency and lower amplitude compact RCC better than rollers with high amplitude and lower frequency, although suitable results have been achieved on some projects using rollers with both high frequency and amplitude. The typical compactor is a 10-ton double or single drum roller with a dynamic force of at least 475 lbs of force per inch of drum width. These rollers are typically used for asphalt and roadway compaction. Larger 15- and 20-ton rollers with more mass and size, typically used with rockfill construction, have been used with RCC, but they usually have larger amplitudes, lower frequency, and are less suited to the aggregate gradings used in RCC. Achieving density and a good lift-joint interface is more difficult with these larger rollers.

In tight areas such as adjacent to forms and next to rock outcrops, large tamping foot-type compactors are most suitable. They are mobile and can provide high-impact energy to produce good density. However, they usually do not leave a smooth surface and they can sink when tamping RCC that has been placed over an excessive thickness of wet bedding mixture, when tamping RCC with excess water, or when tamping next to a conventional concrete mixture that has not lost its slump. One-man vibrating plate compactors intended for sands are not very effective, but the more recent massive plate compactors intended for deep lifts of gravel are effective, though they may require multiple passes. Walk-behind rollers are not very effective in most cases unless they can produce a dynamic force on the order of about 250 lb/in of drum width. Four to six passes of this type roller on fresh RCC not deeper than 12 in usually results in suitable compaction for tight areas, with densities about 98% of that achieved with the large roller. This reduced density is often considered acceptable, except for special areas that are identified as critical and truly in need of greater compaction.

The appearance of fully compacted RCC is dependent on the mixture proportions. Mixtures of wetter consistency usually exhibit a discernible pressure wave in front of the roller. If the paste content is equal to or less than the volume needed to fill all the aggregate voids, rock-to-rock aggregate contact occurs and a pressure wave may not be apparent. This can also occur if the mixture is simply too dry to develop internal pore pressure under the dynamic effect of the roller. Mixtures that have more paste than necessary to fill aggregate voids and a wetter consistency will result in visible paste at the surface that may pick up on the roller drum, depending on the constituents and plasticity of the paste.

The fresh mixture surface should be spread so that the roller drum produces a consistent compactive pressure under the entire width of the drum. If the uncompacted lift surface is not reasonably smooth, the drum may overcompact high spots and undercompact low spots.

Minor damage from scuff marks and unavoidable dozer tears in the surface of a freshly compacted lift can usually be immediately rolled down with the vibratory drum in a static mode or with a rubber tire roller. If the mixture was sufficiently fresh and moist when rerolled, a suitable rehabilitation of the damage will result. If the mixture is too old, severely damaged, etc., the rerolled RCC may look acceptable, but it can and should be easily blown off by an air hose used for general cleanup of loose debris on the lift. For sliding stability, joint tension, and/or watertightness, designs usually require clean and relatively fresh joint surfaces with good bond. This is typically done by suitable large vacuum truck or air blowing. Some tests have shown sandblasting at 24 and 72 h can actually reduce bond.

18.6.7 Joint Treatment and Inspection

Lift joints should be kept from drying or freezing continuously, 24 hours per day, 7 days per week, prior to placing the next lift. The surface should be clean and at or near a saturated surface dry (SSD) condition just prior to placing the next layer of RCC. However, tests and experience have shown that allowing the surface to dry back to just under an SSD condition, as indicated by a change in color from darker to lighter, will greatly facilitate cleaning by air blowing and will not reduce joint quality for most RCC. Some tests have even shown a slight increase in joint strength. However, rewetting the surface after final cleaning and just prior to spreading RCC over it is considered prudent. In addition to the benefit of cure, keeping the surface from drying provides evaporative surface cooling, thereby causing lower internal temperatures and a lower joint maturity value.

If the surface is more than about 2 d old and it has become sufficiently hard, water washing may be necessary if air blowing alone does not adequately clean off any damage, contamination, and general laitance that may be present. Water washing can only be used after the surface has hardened. Sandblasting is generally not advised or necessary.

RCC mixtures generally do not bleed or develop laitance at the surface. An exception is very wet mixtures and some cases of dry mixtures after days of moist cure. If there is no weak laitance or other contamination at the surface, the lift-joint cleaning typically required with conventional concrete is not necessary. Although there is some debate, minor intermittent laitance that may occur in some situations is generally not removed.

If the construction joint is less than about 1500°F degree-hours old and if it has been kept clean and moist throughout its exposure, no joint treatment is required under most conditions. If the surface has been contaminated by dirt, mud, or other foreign elements, the contamination should be removed. The degree-hr maturity is the sum of the temperatures of the lift surface measured every hour during the time that it is exposed. If the surface has been allowed to dry out, exceed about 1500°F degree-hours of maturity, or became damaged, it should be cleaned and may require a full or partial bedding mixture prior to placement of RCC. The 1500°F degree-hours used here is an example. It may appropriately be 900 at one project and 2100 at another. These variables make specifications for RCC lift surfaces difficult to write, and they leave inspection of this most critical item to both the designer and contractor's judgment.

18.6.8 Curing and Weather Protection

After RCC has been placed and compacted, the lift surface must be cured and protected, just as for concrete placed by conventional methods. The surface must be maintained in a moist condition, or at least so that moisture does not escape. It should also be protected from temperature extremes and freezing until it gains sufficient strength.

When haul vehicles are used on the lift surface, RCC placement typically must immediately stop with even the slightest rainfall. Otherwise, the tires will turn the surface into a soft damaged material that then will require waiting for the RCC to harden so that extensive cleanup can be undertaken. When conveyors are used for delivery and little or no vehicular traffic is required on the RCC, construction can continue in slight rainfall. This may require a gradual and very slight decrease in the amount of mixture water used because of the higher humidity and lack of surface drying. Table 18.3, based on data covering various types of mixes and delivery methods throughout the world over the past 15 years, provides guidance with regard to the effect of rain on RCC placing operations.

Immediately after an RCC lift has been compacted, it is essentially impermeable and will not become damaged by light to moderate rain if there is no hauling or traffic on the surface. After a rain, hauling on the lift can resume only after the surface has begun to dry naturally to a SSD condition. A slightly sloped surface will aid in draining free water and speed resumption of placing operations.

Cure during construction has been accomplished with modified water trucks on larger projects and with hand-held hoses for all size projects. Trucks should be equipped with fog nozzles that apply a fine mist that does not wash or erode the surface. Trucks can be augmented with hand-held hoses that reach areas that are inaccessible to the water truck. Provision must be made for maintaining the damp surface while the trucks are fueled, maintained, and refilled with water. Care should be exercised that the trucks do a minimum amount of turning and disruption of the surface. Maintaining access on and off every lift during construction can be a problem that makes trucks impractical. Consequently, the recent trend has been to use hand-held hoses rather then water trucks.

The final lift of RCC should be cured for an appropriate time, generally in excess of 14 d. Curing compound is unsuitable because of the difficulty in achieving 100% coverage on the relatively rough surface, the probable damage to the surface from construction activity, the low initial moisture in the mixture, and the loss of beneficial surface temperature control that is associated with moist curing.

Unformed sloping surfaces such as the downstream face of a dam are very difficult to compact and can be considered sacrificial and unnecessary to cure provided this has been incorporated into the design. Uncompacted exposed RCC will be subject to raveling. While the outside several inches will be incapable of achieving any significant strength or quality, they will serve as protection and a moisture barrier for the curing of the interior RCC.

Defining Terms

Abutment: The foundation along the sides of the valley or gorge against which a dam is constructed.

Autogenous volume change: The change in volume produced by continued hydration of concrete exclusive of effects external forces, water changes, and temperature changes.

Cementitious material: The fine solid-particle material in concrete that reacts in solution with water, and at times other fine solid particles, to create a hardened concrete with strength. Portland cement, sometimes with pozzolan, is usually used.

Coefficient of thermal expansion: The change in linear dimension per unit length divided by the temperature change.

Cohesion: The adhesion of concrete or mortar to other concrete, rock, and other materials.

Creep: Deformation over a long period of time under a continuous sustained load. Also, the relaxation or reduction of stress over a long period of time under a sustained deformation.

Durability: The ability of concrete to resist weathering action, chemical attack, abrasion, and other service conditions.

Gallery: A long, narrow passage in a dam or concrete mass used for access, inspection, grouting, drilling drain holes, and collecting seepage.

Grout curtain: A row of holes filled with grout under pressure near the heel of a dam to control seepage under the dam.

Heat of hydration: Heat generated by chemical reactions of cementitious materials with water, such as the heat evolved during setting and hardening of Portland cement.

Lift joint quality index (LJQI): A numerical designation indicating the quality of a concrete lift joint, derived from evaluation of the factors affecting the quality of the joint.

Modulus of elasticity (elastic modulus): The ratio of stress to strain in the elastic region of behavior prior to development of nonrecoverable microfracturing.

Monolith: A section or block of a large structure such as a dam that is bounded by free faces or contraction joints.

Paste: That portion of a fresh concrete mixture that is composed of a mixture of all particles that will pass through a No. 200 (75 μm) sieve, namely, cement, pozzolan or fly ash, water, small air bubbles, admixtures, and aggregate fines.

Permeability: The rate of flow of water through a unit cross-sectional area under a unit hydrostatic gradient.

Poisson's ratio: The ratio of transverse strain to axial strain resulting from a uniformly distributed axial stress.

Pozzolan: A finely divided powder that is composed of siliceous and other minerals that react with the byproducts of Portland cement hydration to form additional cementitious materials.

Principal stress: Maximum and minimum stress occurring at right angles to a principal plane of stress.

Pugmill: A mixing chamber usually composed of two horizontal rotating shafts to which mixing paddles are attached.

Restraint: Internal or external restriction of free movement of concrete in one or more directions.

Roller-compacted concrete (RCC): A relatively dry concrete mixture that can be compacted by vibratory rolling, usually by a ten-ton roller.

Sustained modulus of elasticity: The modulus of elasticity of concrete that occurs under constant sustained load, including the effects of creep.

Theoretical air-free density: The density corresponding to a concrete that has been compacted to 100% solids with no air.

Ultimate modulus: The ratio of stress to strain that occurs at the maximum load beyond which concrete loses strength and fails.

References

ACI Committee 207. 1989. *Roller Compacted Mass Concrete.* Report 207.5R-89. American Concrete Institute, Farmington Hills, MI.

Angulo, C., Schrader, E.K., Santana, H., Castro, G., Salazar, H., and Lopez, J. 1995. Miel-I Dam. *Proc. Int. Symp. Roller Compacted Concrete*, pp. 443–456.

Boggs, H.L. and Richardson, A.T. 1985. USBR design considerations for roller-compacted concrete dams. In *Roller-Compacted Concrete.* ASCE, New York.

Cannon, R.W. 1985. Design considerations for roller compacted concrete and rollcrete dams. In *Concrete International.* American Concrete Institute, Farmington Hills, MI.

Chengqian, L. and Chusheng, C. 1991. A small naked freeze-resistance RCC dam. In *Roller Compacted Concrete Dams.* Chinese Society of Hydroelectric Engineering, Beijing, China.

Ditchy, E. and Schrader, E.K. 1988. Monksville Dam temperature studies. *ICOLD 16th Cong.*, Vol. III, Q62, pp. 379–396. ICOLD, Paris, France.

Dolen, T.P. and Tayabi, S.D. 1988. Bond strength of roller compacted concrete. Paper presented at ASCE Specialty Conference on Roller Compacted Concrete.

Dunstan, M.R.H. 1981. Rolled concrete for dams—A laboratory study of the properties of high fly ash content concrete. *CIRIA Technical Note 105.* Construction Industry Research and Information Association, London.

Forbes, B.A. 1988. RCC in dams in Australia. In *Roller Compacted Concrete II.* ASCE, New York.

Gaekel, L. and Schrader., E.K. 1992. RCC mixes and properties using poor quality materials—Concepcion Dam. In *Roller-Compacted Concrete III.* ASCE, New York.

Gentile, G. 1964. Study, preparation, and placement of low cement concrete, with special regard to its use in solid gravity dams. *Trans. Int. Cong. Large Dams*, R16 Q 30. ICOLD, Paris, France.

Giovagnoli, M., Schrader, E.K., and Ercoli., F. 1991. Design and construction of Concepcion Dam. *Proc. Int. Symp. RCC Dams.* Beijing.

Giovagnoli M., Ercoli, F., and Schrader., E.K. 1992. Concepcion Dam design and construction. In *Roller-Compacted Concrete III.* ASCE, New York.

Hansen, K. and Reinhardt., W. 1991. *Roller-Compacted Concrete Dams.* McGraw–Hill, New York.

Hirose, T., Nagayama, I., Takemura, K., and Sato, H. 1988. A study of control of temperature cracks in large roller compacted concrete dams. *ICOLD 16 Cong.*, Vol. III, Q 62, pp. 119–135. Paris, France.

International Commission on Large Dams (ICOLD). 1989. Roller-compacted concrete for gravity dams—state-of-the-art. *Bulletin 75.*

Jansen, R.B. 1989. *Advanced Dam Engineering for Design, Construction, and Rehabilitation.* Van Norstrand Reinhold, New York.

McLean, F.G. and Pierce., J.S. 1988. Comparison of joint strengths for conventional and roller-compacted concrete. In *Roller-Compacted Concrete II.* ASCE, New York.

Oberholtzer, G. L., Lorenzo, A., and Schrader, E.K. 1988. Roller-compacted concrete design for Urugua-I Dam. In *Roller-Compacted Concrete II.* ASCE, New York.

Oury R. and Schrader, E.K. 1992. Mixing and delivery of roller compacted concrete. *Roller-Compacted Concrete III.* ASCE, New York.

Parker, J.W. 1992. Economic factors in roller compacted concrete dam construction. In *Roller-Compacted Concrete III*. ASCE, New York.

Schrader, E.K. 1982a. The first concrete gravity dam designed and built for roller-compacted construction methods. In *Concrete International*, pp. 16–24. American Concrete Institute, Farmington Hills, MI.

Schrader, E.K. 1982b. World's first all-rollcrete dam. *Civil Eng. ASCE*, 45–48.

Schrader, E.K. 1984. *Willow Creek Dam Concrete Report*. Final update. Vols. 1 and 2. U.S. Army Corps of Engineers, Walla Walla, WA.

Schrader, E.K. 1986. Discussion of the article Design Considerations for Roller Compacted Concrete & Rollcrete Dams, by R. Cannon. In *Concrete International*, pp. 63–64. American Concrete Institute, Farmington Hills, MI.

Schrader, E.K. 1987. Design for strength variability: Testing and effects on cracking in RCC and conventional concretes. In *Tuthill Symposium*, ACI SP-104, pp. 1–25. American Concrete Institute, Farmington Hills, MI.

Schrader, E.K. 1988. Behavior of completed RCC dams. In *Roller Compacted Concrete II*, pp. 76–91. ASCE, New York.

Schrader, E.K. 1993. Design and facing options for RCC on various foundations. In *Water Power and Dam Construction*, pp. 33–38.

Schrader, E.K. 1994. Roller compacted concrete for dams, The state-of-the-art. *Proc. Int. Conf. Advances Concrete Technol.*, 2nd ed., pp. 371–417. CANMET, Ottawa, Canada.

Schrader, E.K. 1995a. Seepage, permeability, uplift, and upstream facings for RCC dams. *Suppl. Papers, 2nd Int. Symp Advances Concrete Technol.*, pp. 27–43. CANMET, Ottawa, Canada.

Schrader, E.K. 1995b. Strain, cracking, and failure described by an ultimate modulus. *Proc. 2nd Int. Symp. Advances Concrete Technology*, pp. 419–437. Canadian Center for Mineral and Energy Technology, Ottawa, Canada.

Schrader, E.K. 1995c. General report—Construction of roller compacted concrete dams. *Proc. Int. Symp. Roller Compacted Concrete*, pp. 1263–1295.

Schrader, E.K. 1995d. RCC dam overview—Development, current practice, controversies, and options. *Proc. ICOLD Conf.*, pp. 433–452. ICOLD, Paris, France.

Schrader, E.K., and Namikas, D. 1988. Performance of roller compacted concrete dams. *16th ICOLD Cong.*, Vol. III, Q62, pp. 339–364. Paris, France.

Schrader, E.K., and Stefanakos, J. 1995. RCC cavitation and erosion resistance. *Proc. Int. Symp. on Roller Compacted Concrete*, pp. 1175–1188.

Tatro, S. and Schrader, E.K. 1992. Thermal analysis for RCC—A practical approach. *Roller-Compacted Concrete III*. ASCE, New York.

Tatro, S.B. and Hinds, J.K. 1992. Roller compacted concrete mix design. *Roller-Compacted Concrete III*. ASCE, New York.

Tatro, S.B. and Schrader, E.K. 1985. Thermal considerations for roller-compacted concrete. *J. Am. Concrete Inst.*

Tayabi, S.D. and Okamotto, A.S. 1987. *Bonding of Successive Layers of Roller Compacted Concrete*. Construction Technology Laboratories, Skokie, IL.

U.S. Army Corps of Engineers. *Willow Creek Design Memorandum*. Supplement 1 to GDM 2, Phase 2. U.S. Army Corps of Engineers, Walla Walla, WA.

U.S. Army Corps of Engineers. 1995. *Gravity Dam Design*. Engineering Manual 1110-2-2200.

(a)

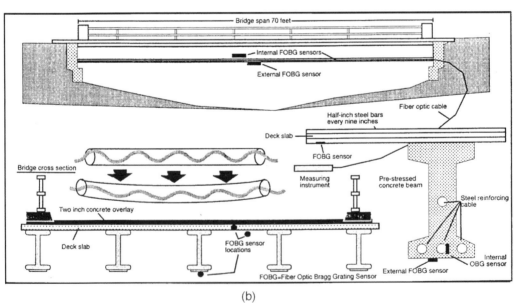

(b)

(a) Nondestructive test of concrete pavement using ground-penetrating radar (Courtesy Portland Cement Association). (b) Schematic of fiber optic instrumentation of bridge deck (Courtesy Dr. Edward G. Nawy).

19

Nondestructive Test Methods*

by
Nicholas J. Carino
Research structural engineer in the Structures Division at the National Institute of Standards and Technology, Gaithersburg MD. Expert in nondestructive test methods for concrete.

19.1 Introduction

Concrete differs from other construction materials in that it can be made from an infinite combination of suitable materials and its final properties are dependent on the treatment it undergoes after it arrives at the job site. The efficiency of the consolidation and the effectiveness of curing procedures are critical for attaining the full potential of a concrete mixture. While concrete is known for its durability, it is susceptible to a range of environmental degradation factors, which can limit its service life. There has always been a need for test methods to measure the in-place properties of concrete for quality assurance and for evaluation of existing conditions. Ideally, these methods should be nondestructive so that they do not impair the function of the structure and permit retesting at the same locations to evaluate changes in properties with time.

Compared with the development of nondestructive test (NDT) methods for steel structures, the development of NDT methods for concrete has progressed at a slower pace, because concrete is inherently more difficult to test than steel. Concrete is highly heterogenous on a macroscopic scale, it is electrically nonconductive but usually contains significant amounts of steel reinforcement, and it is often used in thick members. Thus it has not been easy to transfer the NDT technology developed for steel to the inspection of concrete. In addition, there has been little interest in the

---

*Contribution of National Institute of Standards and Technology and is not subject to copyright in the United States.

traditional NDT community (physicists, electrical engineers, mechanical engineers) to develop test methods for concrete.

There is no standard definition for **nondestructive tests** as applied to concrete. For some people, they are tests that do not alter the concrete. For others, they are simply tests that do not impair the function of a structure, in which case the drilling of cores is considered to be a NDT test. For still others, they are tests that do less damage to the structure than does drilling of cores. This chapter deals with

TABLE 19.1 Nondestructive and In-Place Tests

In-Place Tests to Estimate Strength	Nondestructive Tests for Integrity
Rebound hammer	Visual inspection
Ultrasonic pulse velocity	Stress wave propagation methods
Probe penetration	Ground penetrating radar
Pullout	Electrical/magnetic methods
Break-off	Nuclear methods
Maturity method	Infrared thermography

methods that either do not alter the concrete or that result only in superficial local damage. The author prefers to divide the various methods into two groups: (1) those whose main purpose is to **estimate strength**; and (2) those whose main purpose is evaluate conditions other than strength, that is, to **evaluate integrity**. It will be shown that the most reliable tests for strength are those that result in superficial local damage, and the author prefers the term **in-place tests** for this group. The integrity tests, on the other hand, are nondestructive.

The purpose of this chapter is to provide an introduction to commonly used NDT methods for concrete. Table 19.1 lists the various test methods that will be considered. Emphasis is placed on the principles underlying the various methods so that the reader may understand their advantages and inherent limitations. Additional information on the application of these methods is available in ACI 228.1R (1995), Malhotra and Carino (1991), and Bungey (1989), and the reader is urged to consult these additional references for more in-depth information when necessary. Portions of this chapter are based on previously published works of the author [Carino, 1992a, 1994].

19.2 Methods to Estimate In-Place Strength

19.2.1 Historical Background

Some of the first methods to evaluate the in-place strength of concrete were adaptations of the Brinell hardness[1] test for metals, which involves pushing a high- strength steel ball into the test piece under a given force and measuring the area of the indentation. In the metals test, the load is applied by an hydraulic loading system. Modifications were required to enable this type of test to be made on a concrete structure. In 1934, Professor K. Gaede (Hanover, Germany) reported on the use of a spring-driven impactor to supply the force to drive a steel ball into the concrete [Malhotra, 1976]. A nonlinear, empirical relationship was obtained between cube strength and indentation diameter. In 1936, J.P. Williams (England) reported on a spring-loaded, pistol-shaped device, in which a 4-mm ball was attached to a plunger [Malhotra, 1976]. The spring was compressed by turning a screw, a trigger released the compressed spring, and the plunger was propelled toward the concrete. The diameter of the indentation produced by the ball was measured with a magnifying scale.

In 1938 there appeared a landmark paper by D.G. Skramtajev, of the Central Institute for Industrial Building Research, Moscow [Skramtajev, 1938]. It summarized 14 different techniques, 10 of which were developed in the Soviet Union, for measuring the in-place strength of concrete. This paper should be read by every student of nondestructive testing for its historical content. Skramtajev

---

[1]The term **hardness** is used routinely in the description of a series of tests of metals and concrete, yet this is not a readily quantified mechanical property. If one considers the nature of the hardness test methods that have been developed for metals, it can be concluded that these tests measure the amount of penetration caused by a specific indentor under a specific load. Therefore a more descriptive term for these methods might be **indentation** tests.

divided the test methods into two groups: (1) those that required installation of test hardware prior to placement of concrete, and (2) those that did not require preinstallation of hardware. The methods described by Skramtajev included the following: molds placed in the structure to form in-place test specimens; pullout tests of embedded bars; an in-place punching shear test; an in-place fracture test using a pincer device; penetration with a chisel driven by hammer blows; guns that fired indentors into the concrete; and penetration with a ball powered by a spring-driven apparatus. Readers who are familiar with modern in-place test methods (to be discussed later) will recognize that many of them are variations of methods suggested over one-half century ago.

Skramtajev also commented on the need for in-place testing. For example, he noted that [Skramtajev, 1938]:

- The curing conditions of standard test specimens are not representative of the concrete in the structure.
- The number of standard test specimens is insufficient to assure the adequacy of all members in a structure.
- Standard test specimens that are tested at an age of one month provide no information on the later-age strength of concrete in the structure.
- Surface tests may not provide an indication of the actual concrete strength owing to the effects of carbonation, laitance, and moisture condition.
- Methods requiring preplacement of hardware tend to provide more precise estimates of strength than those that do not require preplacement of hardware, but they lack flexibility for use at any desired location in an existing structure.

It is interesting that 50 years later, the same arguments and limitations are quoted in relation to in-place testing [ACI 228.1R, 1995].

19.2.2 Rebound Hammer

In 1948, Ernst Schmidt, a Swiss engineer, developed a device for testing concrete based upon the rebound principle [Malhotra, 1976, 1991]. As was the case with earlier indentation tests, the motivation for this new device came from tests developed to measure the hardness of metals. In this case, the new device was an outgrowth of the Scleroscope[2] test, which involves measuring the rebound height of a diamond-tipped hammer, or mass, that is dropped from a fixed height above the test surface.

As noted by Kolek (1958), when concrete is struck by a hammer, the degree of rebound is an indicator of the hardness of the concrete. Schmidt standardized the hammer blow by developing a spring-loaded hammer and devised a method to measure the rebound of the hammer. Several different models of the device were built [Greene, 1954], and Figure 19.1 is a schematic of the model that was eventually adopted for field use. The essential parts of the Schmidt rebound hammer are the outer body, the hammer, the plunger, the spring, and the slide indicator. To perform the test, the plunger is extended from the body of the instrument, which causes a latch mechanism to grab hold of the hammer (Figure 19.1a). The body of the instrument is then pushed toward the concrete surface, which stretches the spring attached to the hammer and the body (Figure 19.1b). When the body is pushed to the limit, the latch is released and the hammer is propelled toward the concrete by a combination of gravity and spring forces (Figure 19.1c). The hammer strikes the shoulder of the plunger and it rebounds (Figure 19.1d). The rebound distance is measured on a scale by a slide indicator. The rebound distance is expressed as a rebound number, which is the percentage of the initial extension of the spring [Kolek, 1958]. Currently, different models of the instrument are available, which differ in the mass of the hammer and the stiffness of the spring. Thus different impact energies can be used for different materials.

[2]In Greek, the word "sklero" means "hard."

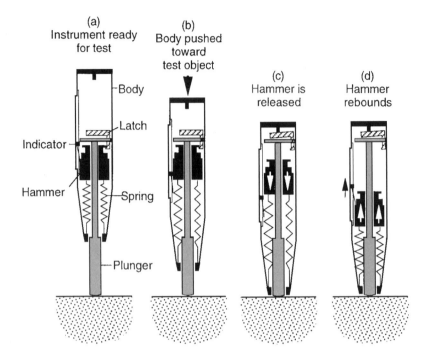

FIGURE 19.1   Schematic cross section of rebound hammer showing principle of operation.

Owing to its simplicity and low cost, the Schmidt rebound hammer is, by far, the most widely used nondestructive test device for concrete. While the test appears simple, there is no simple relationship between the rebound number and the strength of concrete. In principle, the rebound is affected by the movement of the end of the plunger in contact with the concrete. The more the end of the plunger moves, the lower is the rebound. Thus the rebound number is likely to be influenced by the elastic stiffness and the strength of the concrete.

Since the rebound number is indicative of the near-surface properties of the concrete, it may not be indicative of the bulk concrete in a structural member. The report of ACI Committee 228 [ACI 228.1R, 1995] outlines some of the factors that may result in rebound numbers that are not representative of the bulk concrete.

- The moisture condition of the surface concrete affects the rebound number; a dry surface results in a higher rebound number.
- The presence of a surface layer of carbonation increases the rebound number.
- The surface texture affects the rebound number, with smooth hard-troweled surfaces giving higher values than a rough-textured surface.
- The rebound number is affected by the orientation of the instrument in relation to the direction of gravity (approximate correction factors are available).

Because the rebound number is affected by the near-surface conditions, erratic results may occur if the plunger is located directly over a coarse aggregate particle or a subsurface air void. To account for these possibilities, ASTM C 805 requires that 10 rebound numbers be taken for a test. If a reading differs by more than seven units from the average, that reading should be discarded and a new average should be computed based on the remaining readings. If more than two readings differ from the average by seven units, the entire set of readings is discarded.

The rebound hammer was constructed and tested extensively at the Swiss Federal Materials Testing and Experimental Institute in Zurich. A correlation was developed between the compressive strength of standard cubes and the rebound number, and this correlation was provided with the instrument. However, as other investigators began to develop correlations between strength and rebound number, it became evident that there was not a unique relationship between strength and rebound number [Kolek, 1958]. The current recommended practice [ASTM C 805, ACI 228.1R, 1995] is to develop the strength relationship using the same concrete and forming materials as will be used in construction. Without such a correlation, the rebound hammer is useful only for detecting gross changes in concrete quality throughout a structure.

In summary, the rebound number method is recognized as a useful tool for performing quick surveys to assess the uniformity of concrete. However, because of the many factors besides concrete strength than can affect rebound number, it is not generally recommended where accurate strength estimates are needed.

19.2.3 Ultrasonic Pulse Velocity

The ultrasonic pulse velocity method is a stress wave propagation method that involves measuring the travel time, over a known path length, of a pulse of ultrasonic compressional waves (these are waves associated with normal stress). The pulses are introduced into the concrete by a piezoelectric transducer, and a similar transducer acts as receiver to monitor the surface vibration caused by the arrival of the pulse. A timing circuit is used to measure the time it takes for the pulse to travel from the transmitting to the receiving transducers. Figure 19.2 is a schematic of the ultrasonic pulse velocity technique. The speed of compressional waves in a solid is related to the elastic constants (modulus of elasticity and Poisson's ratio) and the density. By conducting tests at various points on a structure, lower quality concrete can be identified by its lower pulse velocity. Naik and Malhotra (1991) provide additional information on the development and application of this method for estimating concrete strength.

The development of a field instrument to measure the pulse velocity occurred nearly simultaneously in Canada and in England [Whitehurst, 1967]. These developments were outgrowths of

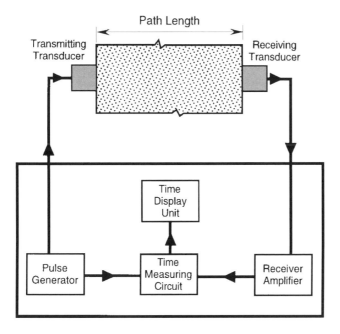

FIGURE 19.2 Schematic of ultrasonic pulse velocity method.

earlier successful work by the U.S. Army Corps of Engineers to measure the speed of a mechanical stress pulse moving through concrete [Long et al., 1945]. The Army Corps of Engineers approach involved two receivers attached to the concrete surface. A horizontal hammer blow was applied in line with the receivers, and a specially designed electronic interval timer was used to measure the time for the pulse to travel from the first to the second receiver. The major purpose of this technique was to calculate the in-place modulus of elasticity.

In 1946 and 1947, engineers at the Hydro-Electric Power Commission of Ontario (Ontario Hydro) worked on the development of a device to investigate the extent of cracking in dams [Leslie and Cheesman, 1949]. The device developed to do so was called the Soniscope. It had a 20-kHz transmitting transducer, was capable of penetrating up to 15 m of concrete, and could measure the travel time with an accuracy of 3%. The stated purposes of the Soniscope were to identify the presence of internal cracking, determine the depth of surface-opening cracks, and determine the dynamic modulus of concrete [Leslie and Cheesman, 1949]. It was further stated that the fundamental measurement was the travel time. The amplitude of the received signal was said to be of secondary importance because the transfer of energy between the transducers and the concrete could not be controlled. It was also emphasized that interpretation of results required knowledge of the history of the structure being investigated.

Early uses of the Soniscope on mass concrete emphasized measuring the pulse velocity rather than estimating strength or calculating the elastic stiffness [Parker, 1953]. Based on velocity readings on a gridwork, the presence of distressed concrete could be easily detected. Parker (1953) reported Ontario Hydro's early attempts to develop relationships between pulse velocity and compressive strength. Forty-six mixtures involving the same aggregate, different cement types, and different admixtures were investigated. The results indicated no significant differences in the velocity-strength relationships for the different mixtures. The results were therefore treated as one group, and the best-fit relationship was determined. Figure 19.3 shows the relationship between estimated strength and pulse velocity and the lower 95% confidence limit for estimated strength. Owing to large scatter, the lower confidence limit was about 45% of the mean strength. Thus the inherent uncertainty in

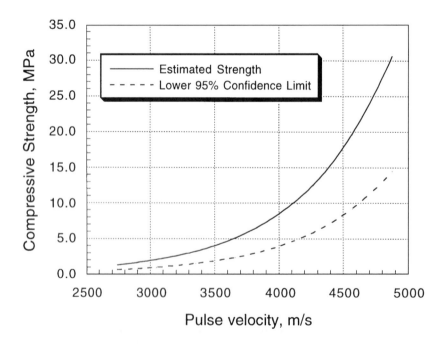

FIGURE 19.3 Compressive strength versus pulse velocity relationship based upon 46 mixtures made with the same aggregate [based on Parker, 1953].

using pulse velocity to estimate strength was established very early. Figure 19.3 also shows that the change in pulse velocity per unit change in strength decreases with increasing strength. This means that pulse velocity is relatively insensitive to strength for mature concrete.

While work on the Soniscope was in progress in Canada, R. Jones and co-workers at the Road Research Laboratory (RRL) in England were involved in independent research to develop an ultrasonic testing apparatus [Jones, 1949a]. The RRL researchers were interested in testing the quality of concrete pavements, which involved shorter path lengths compared with the work at Ontario Hydro. As a result, the apparatus that was developed operated at a higher frequency than the Soniscope, and it was called the ultrasonic concrete tester. Transducers with resonant frequencies from 60 to 200 kHz were used, depending on the desired penetration [Jones, 1953]. Besides using a different operating frequency, the RRL device used a different approach than the Soniscope to measure travel time. This was necessary because of the shorter path lengths in the RRL work. It was reported that the ultrasonic concrete tester could measure travel times to within $\pm 0.2 \, \mu s$.

Jones (1949b) reviewed the test program carried out with the newly developed ultrasonic concrete tester. Among these programs were the following:

- There was an investigation of the variation of pulse velocity with height in standard cube specimens and with depth in slabs. This is one of the first studies to document the top-to-bottom effect that is often mentioned as a problem when planning and interpreting in-place tests [ACI 228.1R, 1995].

- An investigation was performed on the influence of watercement ratio, aggregate type, and aggregate content on pulse velocity. These studies demonstrated the importance of aggregate type and aggregate content on pulse velocity.

- There was an investigation of the relationships between pulse velocity and compressive strength. These studies demonstrated that for a given mixture under uniform conditions there was good correlation between strength and pulse velocity.

Thus Jones established the problems inherent in using the pulse velocity to estimate concrete strength. Despite these early findings, numerous researchers sought to establish correlations between pulse velocity and strength, and many reached the same conclusions as Jones [Sturrup et al., 1984].

In the United States, a Soniscope was developed in 1947 at the Portland Cement Association in cooperation with Ontario Hydro, and field applications were reported by Whitehurst (1951). In his summary of industry's experience in the U.S., Whitehurst published the following tentative classification for using pulse velocity as an indicator of quality:

Pulse Velocity, m/s	Condition
Above 4570	excellent
3660 to 4570	generally good
3050 to 3660	questionable
2130 to 3050	generally poor
Below 2130	very poor

This table was quoted in many subsequent publications. However, Whitehurst warned that these values were established on the basis of tests of normal concrete having a density of about 2400 kg/m^3 and that the boundaries between "conditions" could not be sharply drawn. He mentioned that, rather than using these limits, a better approach would be to compare velocities with the velocity in portions of the structure that are known to be of acceptable quality. Nevertheless, inexperienced investigators often used the above table as the sole basis for interpreting test results.

After the publication of these landmark papers in the late 1940s and early 1950s, a flurry of activity occurred worldwide, and efforts were begun to develop test standards. In the United States, a proposed ASTM test method was published by Leslie (1955), but it was not until 1967 that it finally became a tentative test method [ASTM C 597]. In Europe, the International Union of Testing and Research Laboratories for Materials and Structures (RILEM) organized a working group to study nondestructive testing (R. Jones was appointed chairman), and in 1969, draft recommendations for testing concrete by the ultrasonic pulse method were published [Jones and Făcăoaru, 1969]. In Eastern Europe, the method was used extensively in precast concrete plants.

During the 1960s and 1970s, considerable attention was devoted to gaining more knowledge about the effects of different factors on pulse velocity. Researchers continued to explore the relationship between compressive strength and pulse velocity. However, they appear to have reached a consensus that there is no unique relationship. Numerous studies showed that the type and quantity of aggregate have major effects on pulse velocity but not on strength. Significant effort was also expended to examine whether attenuation measurements could provide additional information about concrete strength. These results were, in general, found to be impractical in field situations because of difficulties in achieving consistent coupling of the transducers, which is critical for measuring attenuation.

Perhaps the most significant advances during this period were in the development of improved field instrumentation. Owing to advances in microelectronic circuitry, the cumbersome instruments developed in the 1940s and 1950s gave way to compact portable devices. In the late 1960s, TNO (Netherlands Organization for Applied Scientific Research) in Delft, Netherlands, developed a portable, battery-operated pulse velocity device that incorporated a digital display of the travel time. In the earlier devices, travel time was measured by examination of oscilloscope displays, which was a time-consuming process. The portable instrument had a resolution of $1 \mu s$, which resulted in low accuracy for short path lengths, and it had limited penetrating ability [Făcăoaru, 1969]. At about the same time, R.H. Elvery of University College, London, developed a similar portable device that was called PUNDIT (Portable Ultrasonic Nondestructive Digital Indicating Tester). It weighed 3.2 kg, had a resolution of $0.5 \mu s$, and could be powered by rechargeable batteries [Malhotra, 1976]. These and other relatively low-cost, portable devices simplified testing and resulted in a worldwide increase in the number of consultants and researchers who could perform this type of testing. Later models of these devices had resolutions of $0.1 \mu s$, and some provided an optional output terminal to allow the received signal to be displayed on an oscilloscope.

In summary, the ultrasonic pulse velocity method is a relatively simple test to perform on site provided it is possible to gain access to both sides of the member. While tests can be performed with the transducers placed on the same surface, the results are not easy to interpret and this method of measurement is not recommended. Care must be exercised to assure that good and consistent coupling with the concrete surfaces is achieved. Other important factors, besides concrete strength, that can affect the measured ultrasonic pulse velocity and that should be considered are discussed in the report of ACI Committee 228 [ACI 228.1R, 1995]. These include:

- moisture content—an increase in moisture content increases the pulse velocity;
- presence of reinforcement oriented parallel to the pulse propagation direction—the pulse may propagate through the bars and result in an apparent pulse velocity that is higher than that propagating through concrete; and
- presence of cracks and voids—these can increase the length of the travel path and result in a longer travel time.

Because of these factors, the ultrasonic pulse velocity should be used for estimating concrete strength only by experienced individuals. Like the rebound number test, the pulse velocity method is very useful for assessing the uniformity of concrete in a structure. It is often used to locate portions of a structure where other tests should be performed or where cores should be drilled.

19.2.4 Probe Penetration

The probe-penetration method involves using a gun to drive a hardened steel rod, or probe, into the concrete and measuring the exposed length of the probe. In principle, as the strength of the concrete increases, the exposed probe length also increases; by means of a suitable correlation, the exposed length can be used to estimate compressive strength. Skramtajev mentioned a similar concept in his 1938 summary paper, and Malhotra (1976) mentions that similar techniques were reported in 1954. Malhotra and Carette (1991) provide an in-depth summary of this technique.

Development of the probe penetration test system began in about 1964 as a joint undertaking by T.R. Cantor of the Port of New York Authority and R. Kopf of the Windsor Machinery Co. [Arni, 1972]. The test system that was eventually commercialized became known as the Windsor probe. The apparatus is supplied with a table that relates exposed probe length to compressive strength for different aggregate hardness as measured by Mohs hardness scale[3] of minerals. The basis for the values in the tables and their uncertainty were not provided [Arni, 1972]. In the late 1960s, independent investigations of the reliability of the Windsor probe system were carried out by the National Ready-Mixed Concrete Association [Gaynor, 1969], the Federal Highway Administration (FHWA) [Arni, 1972], and the Department of Energy, Mines and Resources (Canada) [Malhotra, 1974]. In general, it was found that the probe system had an acceptable within-test variability. However, scatter in the correlation between compressive strength and probe penetration led to rather high uncertainties in the estimated strength. All investigators cautioned against reliance on the manufacturer's correlation tables.

Arni's (1972) study of the uncertainties of the probe penetration and rebound hammer tests is very interesting and worth summarizing. He calculated the number of tests required to detect a strength difference of 1.4 MPa (200 psi) using test cylinders, probe penetration, or rebound number. These estimates were based on the variability of test results and slopes of the correlation equations developed in the FHWA study. For 90% confidence levels, the results were as follows:

Cylinders	8
Rebound	120
Probe	85

Note that these numbers apply for specific data used by Arni. Nevertheless, they point out the inherent inability of in-place tests to detect small differences in concrete strength unless large numbers of tests are performed. This important concept has been largely ignored.

The Windsor probe test method was adopted as a tentative ASTM standard (C 803) in 1975. In 1990, the standard was modified to include the use of a pin penetration device, in which a small pin is forced into the concrete using a spring-loaded driver [ACI 228.1R, 1995; Malhotra and Carette, 1991; Nasser and Al-Manseer, 1987a,b].

The report of ACI Committee 228 [ACI 228.1R, 1995] provides an explanation of the factors affecting probe penetration into concrete. Figure 19.4 is a schematic of the failure zone produced during probe penetration. The probe penetrates until its initial kinetic energy is absorbed by friction and the fracture of the mortar and aggregate. Hence the strength properties of the aggregate affect the penetration depth. As a result, the strength relationship is dependent on the aggregate type. For equal concrete strength, probe penetration would be deeper in a concrete with a soft aggregate than in a concrete with a hard aggregate. See Malhotra (1976), Bungey (1989), and Malhotra and Carette (1991) for additional information of the effects of aggregate type. Probe penetration is not strongly affected by the near-surface conditions and is therefore not as sensitive to surface conditions as the rebound-number method. The direction of penetration is not important provided that the probe is fired perpendicular to the surface. Care must be exercised when testing reinforced concrete to assure that tests are not carried out in the vicinity of the reinforcing steel, especially if the concrete cover is low.

---

[3]A qualitative scale in which the hardness of a mineral is determined by its ability to scratch, or be scratched by, another mineral.

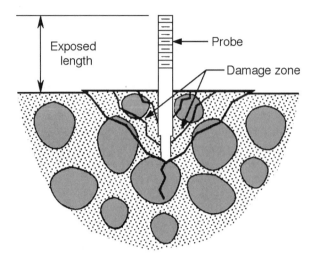

FIGURE 19.4   Schematic of conical failure zone during probe penetration test.

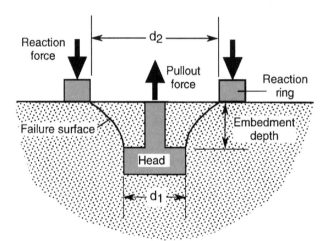

FIGURE 19.5   Schematic of cast-in-place pullout test.

19.2.5 Pullout Test

The cast-in-place pullout test is one of the most reliable techniques for estimating the in-place strength of concrete during construction. In this method, an insert with an enlarged head is cast in the concrete. The insert and the accompanying conical fragment of concrete are extracted by using a tension-loading device reacting against a bearing ring that is concentric with the insert (Figure 19.5). The force required to pull out the insert is an indicator of concrete strength. A comprehensive review of the history and theory of the pullout test is available [Carino, 1991b], and only a brief summary is provided here.

19.2.5.1 History

Ideas for pullout tests originated in the Soviet Union [Skramtajev, 1938]. Tremper (1944) was the first American to report on the correlation between pullout force and companion cylinder strength. An insert developed by Volf (of the Soviet Union) and the one used by Tremper are

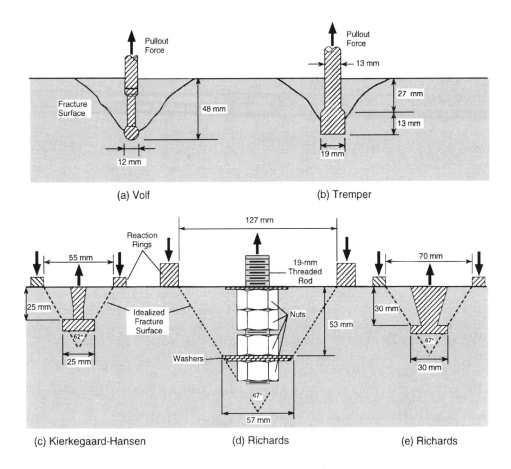

FIGURE 19.6 Various pullout test configurations.

shown in Figures 19.6a and 19.6b, respectively. In both cases, the reaction to the pullout force was applied sufficiently far from the insert that there was negligible interaction between the failure surface and the reaction system. As a result, failure was controlled primarily by the tensile strength of the concrete. This explains why Tremper found that the correlation between pullout force and compressive strength was nonlinear.

Despite Tremper's encouraging results, there was no additional documented work on the pullout test until 1962, when a comprehensive study began in Denmark [Kierkegaard-Hansen, 1975]. The objective was to find the optimum geometry for a field test system that would have a high correlation between pullout load and the compressive strength of concrete. Kierkegaard-Hansen found that the correlation could be improved by constraining the failure surface to follow a predefined path by using a relatively small-diameter reaction ring. The study resulted in the pullout test configuration shown in Figure 19.6c, which was eventually incorporated into the Lok-Test[4] system, the most widely used commercial pullout test system.

Owen Richards, a materials consultant in the United States, carried out independent studies of a pullout test in the late 1960s and early 1970s. The early version of Richards' pullout test configuration was larger than that developed in Denmark. The inserts were manufactured from 19-mm threaded rods, and washers were used to provide the enlarged head. Nuts were used to add

---

[4]As explained by Kierkegaard-Hansen (1975), failure when a small reaction ring is used can be considered a **punching** type of failure. The Danish word for "punching" is **lokning**, so the term **lok-strength** was used to describe the strength measured by the test.

rigidity to the washers and to fix the embedment depth of the washers. The test geometry, which is shown in Figure 19.6d, resulted in an idealized failure surface with an area approximately equal to the area of a standard 152-mm diameter cylinder. Richards preferred to divide the pullout force by the nominal area of the idealized failure to obtain a pullout strength, which was a fictitious quantity because the pullout force was inclined to the surface area.

The first reported pullout tests using Richards' early system were performed at the Bureau of Reclamation [Rutenbeck, 1973]. Strong correlation was obtained between pullout tests performed on slabs and the compressive strength of companion cylinders. Good correlation between pullout strength and compressive strength was also obtained when the inserts were placed in shotcrete panels and compression specimens were cut from the panels. In 1975, Malhotra also reported on the applicability of Richards' pullout test [Malhotra, 1975]. It was found that the coefficient of variation for three replicate tests was less than 5%, which was very encouraging. In a later study [Malhotra and Carette, 1980], it was noted that similar correlations were obtained in different investigations of Richards' system.

Richards' pullout system produced encouraging results, but the large size of the insert required heavy testing equipment and produced significant surface damage. In 1977, a smaller version of the test system was introduced [Richards, 1977], as shown in Figure 19.6e. The apex angle of the conic frustum was maintained at 67°, but the insert was constructed from one piece of steel. The enlarged end of the shank accommodated a pull-rod that passed through a center-hole tension ram.

In the early 1960s, investigations of the pullout test were also conducted in Great Britain [Te'eni, 1970], but the work was apparently never carried to the stage of a practical field test system. A novel feature of the British work was the use of a power function for the correlation equation, rather than a straight line as had been used in Denmark, the United States, and Canada.

The usefulness of the pullout test for evaluating early-age strength was quickly recognized. In 1978, ASTM adopted a tentative test method for the pullout test (C 900-78T). In North America, J. Bickley became an early advocate of the pullout test method for achieving construction safety and economy [Bickley, 1982a].

19.2.5.2 Failure Mechanism

Ever since the test was first described by Skramtajev (1938), there has been an incomplete under-standing of its failure mechanism. Skramtajev correctly noted that the tests subjects the concrete to a combination of tensile and shearing stresses. Kierkegaard-Hansen (1975), the inventor of the widely used Lok-Test system, tried to relate the shape of the extracted conical fragment to the intact cones often observed at the ends of cylinders tested in compression. Jensen and Braestrup (1976) used plasticity theory to relate the ultimate pullout force to the compressive strength of the concrete. Malhotra and Carette (1980) proposed that the pullout strength was related to the direct shear strength of concrete. Recent experimental and analytical studies have tried to gain a better under-standing of the failure process during the pullout test [Ottosen, 1981; Stone and Carino, 1983; Yener, 1994; Ballarini et al., 1986; Krenchel and Shah, 1985; Hellier et al., 1987; Krenchel and Bickley, 1987].

From these independent analytical and experimental studies, it is now understood that the pullout test subjects the concrete to a nonuniform, three-dimensional state of stress. It also has been demonstrated that the failure process involves two circumferential crack systems: a stable system that starts at the insert head at about 1/3 of the ultimate load, propagates into the concrete at a large apex angle, and is arrested as it reaches a tension-free region; and a second system that propagates with increasing load and eventually defines the shape of the extracted cone. Figure 19.7 shows these cracking systems as predicted by Hellier et al. (1987) who used a discrete cracking, finite-element model based on nonlinear fracture mechanics.

Despite general agreement on the cracking process prior to the attainment of ultimate pullout load, there is no consensus on the failure mechanism at the ultimate load. Some believe that ultimate load occurs as a result of compressive failure along a strut running from the bottom of the bearing ring to the insert head [Ottosen, 1981; Krenchel and Shah, 1985; Krenchel and Bickley,

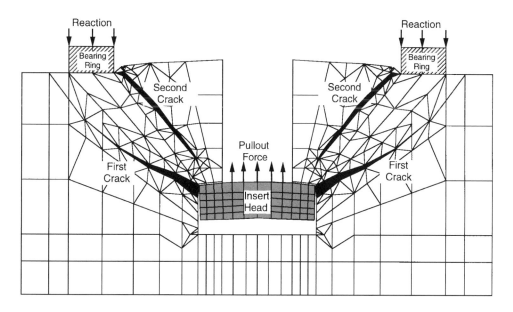

FIGURE 19.7   Crack systems formed during pullout test as predicted from finite-element fracture analysis [Hellier et al., 1987].

1987]. This mechanism has been used to explain the good correlation between pullout strength and compressive strength. Others believe that failure is governed by aggregate interlock across the secondary crack system, and the ultimate load is reached when sufficient aggregate particles have been pulled out of the mortar matrix [Stone and Carino, 1983; Hellier et al., 1987]. In this case, it is argued that there is correlation between pullout strength and compressive strength because both properties are controlled by the tensile strength of the mortar. In the compression test, the ultimate load is associated with the formation and growth of microcracks through the mortar.

While there is no agreement on the exact failure mechanism, it has been shown that the pullout strength has good correlation with the compressive strength of concrete and that the test has good repeatability. In a review of published data, ACI Committee 228 recommends a coefficient of variation (standard deviation divided by the mean) of 8% for the pullout test [ACI 228.1R, 1995]. Recent modifications of ASTM C 900 require a minimum of five individual pullout tests for each 115 m^3 of concrete in a given placement.

19.2.5.3 Post-Installed Tests

A drawback of the standard pullout test is that the locations of the inserts have to be planned in advance of concrete placement and the inserts have to be fastened to the formwork. This limits the applicability of the standard method to new construction. In an effort to extend the application of pullout testing to existing structures, various techniques for performing post-installed pullout tests have been investigated. Some of the more promising approaches are shown in Figure 19.8. However, none of these methods have been standardized by ASTM.

During the 1970s, a need arose In the United Kingdom for in-place tests to evaluate distressed concrete structures built with high alumina cement. Researchers at the Building Research Establishment (BRE) developed a pullout technique using commercial anchor bolts, as shown in Figure 19.8a [Chabowski and Bryden-Smith, 1980]. A 6-mm diameter hole is drilled into the concrete and an anchor bolt is inserted so that the split-sleeve is at a depth of 20 mm. After applying an initial load to expand and engage the sleeve, the bolt is pulled out and the maximum load during the extraction is recorded. Because of the shallow embedment, failure occurs by concrete fracture. Reaction to the pullout load is provided by three feet located along the perimeter of a 80-mm diameter ring. As the bolt is pulled, the sleeve imparts vertical and horizontal forces to the concrete. Hence the fracture surface differs from that in the cast-in-place (CIP) pullout test, and the test has been called an internal fracture test rather than a pullout test. The correlation between ultimate load and compressive strength was found to have a pronounced nonlinearity, indicating that the failure mechanism was probably related to the tensile strength of the concrete. Within-test variability was found to be greater than the CIP pullout test, and the 95% confidence limits of the correlation relationship were found to range between ±30% of the mean curve [Chabowski and Bryden-Smith, 1980]. The relatively low precision of the internal fracture test has been attributed to two principal causes [Bungey, 1981]: (1) the variability in the hole drilling and test preparation; and (2) the influence of aggregate particles on the load-transfer mechanism and the failure-initiation load.

Mailhot et al. (1979) also investigated the feasibility of several post-installed pullout tests. One of these used a split-sleeve and a tapered bolt assembly that was placed in a 19-mm diameter hole drilled into the concrete. As shown in Figure 19.8b, the details differ from the BRE method because the reaction to the pulling force acts through a specially designed high-strength, split-sleeve assembly. Thus the force transmitted to the concrete is predominantly a lateral load because the tapered bolt forces the sleeve to expand laterally. It is likely that failure occurs by indirect tensile splitting, similar to that in a standard splitting-tension test. As with the BRE test, the variability of this test was reported to be rather high. Another successful method involved epoxy-grouting a 16-mm diameter threaded rod into a 19-mm hole to a depth of 38 mm. After the epoxy had cured, the rod was pulled using a tension jack reacting against a bearing ring. This method was also reported to have high variability. The study concluded that these two methods had the potential for assessing the strength in existing construction. However, additional research was recommended to enhance their reliability.

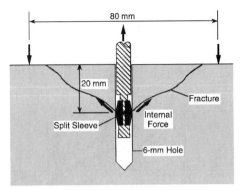

(a) BRE Internal Fracture Test

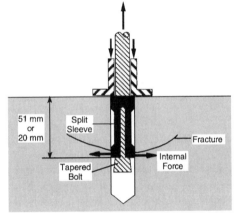

(b) Expanding Sleeve Test

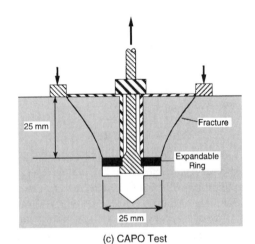

(c) CAPO Test

FIGURE 19.8 Examples of post-installed "pullout tests."

Domone and Castro (1986) also developed a technique similar to the expanding sleeve method shown in Figure 19.8b. However, a torque meter was used to apply the load, and the embedment was 20 mm as in the BRE method. On the basis of a limited number of tests, it was concluded that this method gave better correlations than the BRE method.

Another method was developed by the manufacturer of the Lok-Test[5] system and is referred to as the CAPO test (for cut and pullout) [Petersen, 1984]. The method involves drilling a 18-mm diameter hole into the concrete and using a special milling tool to undercut a 25-mm diameter slot at a depth of 25 mm. An expandable ring is placed in the hole, and the ring is expanded using special hardware. Figure 19.8c shows the ring after expansion. The entire assembly used to expand the ring is pulled out of the concrete using the same loading system as for a CIP pullout test. Unlike the methods discussed above, the CAPO test subjects the concrete to a similar state of stress as the CIP pullout test. The performance of the CAPO test in laboratory evaluations has been reported to be similar to the Lok-Test [Petersen, 1984]. Users of the CAPO test have indicated that the test is cumbersome to perform and care is needed to control the variability.[6] The test surface must be flat and perpendicular to the drilled hole. If these conditions are not achieved, the bearing ring will not seat properly and test results will be erratic.

19.2.6 Break-Off Test

This test measures the force required to break off a cylindrical core from the concrete mass. The method was developed in the early 1970s by R. Johansen at the Cement and Concrete Research Institute in Norway. In cooperation with contractors, Johansen sought a simple, inexpensive, and robust method to measure in-place strength [Johansen, 1977, 1979]. The test method was standardized by ASTM in 1990 (ASTM C 1150). Naik (1991) provides a comprehensive review of research results.

Figure 19.9 is a schematic of the break-off test. For new construction, the core is formed by inserting a plastic sleeve into the fresh concrete. When the in-place strength is to be estimated, the sleeve is removed. Then a special, hand-operated, hydraulic loading jack is placed into the counterbore, and a force is applied to the top of the core until it ruptures from the concrete mass. The hydraulic fluid pressure is monitored with a pressure gage, and the maximum pressure gage reading in units of bars (1 bar $= 0.1$ MPa) is referred to as the break-off number of the concrete.

For new construction, the sleeves are inserted into the top surface of the member after the concrete has been leveled. Alternatively, the sleeves can be attached to the sides of the formwork and filled during concrete placement. For existing construction, a special drill bit can be used to cut the core and the counterbore.

For ease of sleeve insertion into the fresh concrete, the concrete must be workable. In addition, to minimize interference, the maximum aggregate size should be limited to a fraction of the sleeve diameter, which is 55 mm. According to ASTM C 1150, the break-off test is not recommended for concrete having a maximum nominal aggregate size greater than 25 mm. Sleeve insertion must be performed carefully to assure good compaction around the sleeve and a minimum of disturbance at the base of the formed core. Some problems have been reported with keeping the sleeves from floating out of very fluid concrete mixtures [Naik et al., 1987]

The break-off test subjects the concrete to a slowly applied force and measures a static strength property of the concrete. The core is loaded as a cantilever, and the concrete at the base of the core is subjected to a combination of bending and shearing stresses. In early work [Johansen, 1977, 1979], break-off strength was reported as a stress, arrived at by computing the flexural stress at the base of the core corresponding to the rupture force. In later applications (see review by Naik, 1991), the flexural strength was not computed, and the break-off number (pressure gage reading) was related

---

[5]Certain trade names and company products are mentioned to identify specific test equipment. In no case does such identification imply recommendation or endorsement by the National Institute of Standards and Technology, nor does it imply that the products are necessarily the best available for the purpose.

[6]Read, P.H., Bickley, J.A., and Omran, R. Simulated Field Trials. Draft Report for Strategic Highway Research Program (SHRP) Contract 88-C204. January, 1991.

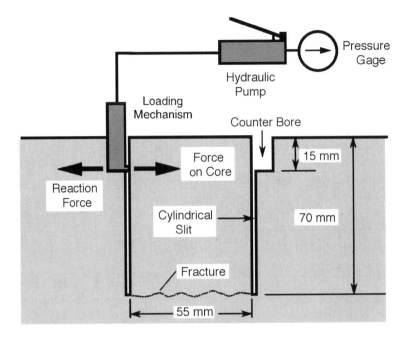

FIGURE 19.9 Schematic of break-off test.

directly to compressive strength. This approach simplifies data analysis, but calibration of the test instrument used in the field is mandatory to assure that the gage readings correspond to the actual forces applied to the test specimen.

The correlations between break-off strength and compressive strength have been found to be nonlinear [Johansen, 1977 and 1979; Barker and Ramirez, 1988], which is in accordance with the usual practice of relating the modulus of rupture of concrete to the square root of compressive strength. It has also been found that the correlation between break-off strength and modulus of rupture may be more uncertain than that between break-off strength and compressive strength [Barker and Ramirez, 1988].

Failure during the break-off test occurs by fracture at the base of the 55-mm diameter core. The crack initiates at the most highly stressed point. It then propagates through the mortar and, in most cases, around coarse aggregate particles located at the base of the core. The particular arrangement of aggregate particles within the failure region would be expected to affect the ultimate load in each test. Because of the relatively small size of the core and the heterogeneous nature of concrete, the distribution of aggregate particles will be different at each test location. Hence one would expect the within-test variability of the break-off test to be higher than that of other standard strength tests that involve larger test specimens. One would also expect that the variability might be affected by maximum aggregate size and aggregate shape. The developer of the break-off test reported a within-test coefficient of variation of about 9% [Johansen, 1979]. This value has, in general, been confirmed by other investigators [Carino, 1992a].

19.2.7 Maturity Method

The **maturity method** is a technique for estimating the strength development of concrete during its curing period by measuring the temperature history of the concrete. Carino (1991a) provides a comprehensive review of the history of the method and some of its applications.

Historically, the maturity method was not classified as a nondestructive test method, but it is now regarded as a useful technique for estimating in-place strength. Its origin can be traced to a series of papers from England dealing with accelerated curing methods [McIntosh, 1949; Nurse, 1949; Saul, 1951]. There was a need for a procedure to account for the combined effects of time and temperature on strength development for different elevated-temperature curing processes. It was proposed that the product of time and temperature could be used for this purpose. These ideas led to the famous **Nurse-Saul maturity function:**

$$M = \sum_{0}^{t} (T - T_0)\Delta t \tag{19.1}$$

where

M = maturity index, degree Celcius-hours (or degree Celcius-days)
T = average concrete temperature, degree Celcius, during the time interval Δt
T_0 = datum temperature (usually taken to be) $10°C$
Δt = time interval

The index computed by Eq. (19.1) was called the **maturity**; however, the current terminology is the **temperature-time factor** [ASTM C 1074]. Saul (1951) presented the following principle, which has become known as the **maturity rule**:

Concrete of the same mix at the same maturity (reckoned in temperature-time) has approximately the same strength whatever combination of temperature and time go to make up that maturity.

Equation (19.1) is based on the assumption that the rate of strength gain is a linear function of temperature; it was soon realized that this approximation may not be valid when curing temperatures vary over a wide range. As a result, a series of alternatives to the Nurse-Saul function were proposed by other researchers [Malhotra, 1971]. However, none of the alternatives received widespread acceptance, and the Nurse-Saul function was used worldwide until an improved function was proposed in the 1970s.

In 1977, a new function was proposed to compute a maturity index from the recorded temperature history of the concrete [Freiesleben Hansen and Pedersen, 1977]. This function was based on the Arrhenius equation [Brown and LeMay, 1988] that is used to describe the effect of temperature on the rate of a chemical reaction. The new function allowed the computation of the equivalent age of concrete as follows:

$$t_e = \sum_{0}^{t} e^{\frac{-E}{R}(\frac{1}{T} - \frac{1}{T_r})}\Delta t \tag{19.2}$$

where

t_e = the equivalent age at the reference temperature
E = apparent activation energy, J/mol
R = universal gas constant, 8.314 J/mol-K
T = average absolute temperature of the concrete during interval Δt, degrees Kelvin
T_r = absolute reference temperature, degrees Kelvin

By using Eq. (19.2), the actual age of the concrete is converted to its equivalent age, in terms of strength gain, at the reference temperature. In European practice, the reference temperature is usually taken to be $20°C$, whereas in North American practice it is usually assumed to be $23°C$. The introduction of this function overcame one of the main limitations of the Nurse-Saul function (Eq. (19.1)) because it allowed for a nonlinear relationship between the rate of strength development and curing temperature. This temperature dependence is described by the value of the apparent activation energy. Comparative studies in the early 1980s showed that this new maturity function was superior to the Nurse-Saul function [Byfors, 1980; Carino, 1982].

19.2.7.1 Effect of Temperature on Strength Gain

The key parameter in Eq. (19.2) is the "**activation energy**" which describes the effect of temperature on the rate of strength development. In the early 1980s, the author began a series of studies to gain a better understanding of the maturity method [Carino, 1984]. From this work, a procedure was developed to obtain the activation energy of a given cementitious mixture. The procedure is based on determining the effect of curing temperature on the rate constant for strength development. The rate constant is related to the curing time needed to reach a certain fraction of the long-term strength, and it is obtained by fitting an appropriate equation to the strength-versus-age data acquired under constant temperature (isothermal) curing. The procedure to determine the activation energy consists of the following steps:

- Cure mortar specimens at different constant temperatures.
- Determine compressive strengths at regularly spaced ages.
- Determine the value of the rate constant at each temperature by fitting a strength-age relationship to each set of strength-age data.
- Determine the best-fit Arrhenius equation (to be explained) to represent the variation of the rate constant with the temperature.

By using the above procedure, the activation energy was determined for concrete and mortar specimens made with different cementitious materials [Tank and Carino, 1991; Carino and Tank, 1992]. It was found that for concrete with a water-cement ratio (W/C) of 0.45, the activation energy ranged from 30 and 64 kJ/mol; while for a W/C of 0.60 it ranged from 31 to 56 kJ/mol, depending on the type of cement and additives.

The significance of the activation energy is explained further here. In Eq. (19.2), the exponential term within the integral converts increments of actual curing time at the concrete temperature to equivalent increments at the reference temperature. Thus the exponential term can be considered as an **age conversion factor**, γ:

$$\gamma = e^{\frac{-E}{R}\left(\frac{1}{T} - \frac{1}{T_r}\right)} \tag{19.3}$$

Figure 19.10 shows how the age conversion factor varies with curing temperature for different values of the activation energy. The reference temperature is taken as 23°C. It is seen that for an activation energy of 30 kJ/mol, the age conversion factor is nearly a linear function of temperature. In this case, the Nurse-Saul equation would be a reasonably accurate maturity function to account for the combined effects of time and temperature, because the Nurse-Saul function assumes that the rate constant varies linearly with temperature [Carino, 1984]. For an activation energy of 60 kJ/mol, the age conversion factor is a highly nonlinear function of the curing temperature. In this instance, the Nurse-Saul function would be an inaccurate maturity function. In summary, Figure 19.10 shows the nature of the error in an age conversion factor if the incorrect value of activation energy were used for a particular concrete mixture. The magnitude of the error would increase with increasing difference of the curing temperature from 23°C.

The reader will have noticed that the term activation energy was introduced within quotation marks. This is because the E-value that is determined when the rate constant is plotted as a function of the curing temperature is not truly an activation energy as implied by the Arrhenius equation. The following discussion is provided for those unfamiliar with the concept of activation energy or the origin of the Arrhenius equation.

The idea of activation energy was proposed by Svante Arrhenius in 1888 to explain why chemical reactions do not occur instantaneously when reactants are brought together, even though the

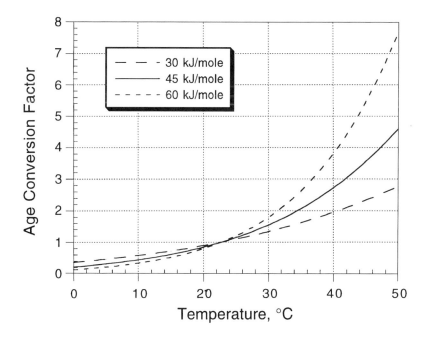

FIGURE 19.10   Effect of activation energy value on the age conversion factor.

reaction products are at a lower energy state [Brown and LeMay, 1988]. Arrhenius proposed that before the lower energy state is achieved, the reactants must have sufficient energy to overcome an energy barrier that separates the unreacted and reacted states. A physical analogy is a brick standing upright. The brick will not instantaneously tip over to its lower, horizontal, energy state. It must be pushed from the higher to the lower energy state. The energy required to push the brick from its upright position to the point of instability, after which the brick falls on its own, is the activation energy for this process.

For molecular systems, the reactant molecules are in constant motion and energy is transferred between them as they collide [Brown and LeMay, 1988]. A certain number of molecules will acquire sufficient energy to surmount the energy barrier and form the lower energy reaction product. As the system is heated, the kinetic energy of the molecules increases and more molecules will surmount the barrier. Thus the rate of reaction increases with increasing temperature. Arrhenius observed that the rate constant, k, of many reactions increased with temperature according to what has since been called the Arrhenius equation, as follows:

$$k = Ae^{-\frac{E}{RT}} \tag{19.4}$$

The term A is called the **frequency factor** and is related to the frequency of collisions and the probability that the molecules will be favorably oriented for reaction [Brown and LeMay, 1988]. It can be seen that the age conversion factor given by Eq. (19.2) is simply the ratio of the rate constants at two different temperatures.

The Arrhenius equation was derived empirically from observations of homogeneous chemical systems undergoing a single reaction. Roy and Idorn (1982) have noted that researchers "... have cautioned that since cement is a multiphase material and also the process of cement hydration is not a simple reaction, homogeneous reaction kinetics cannot be applied." Thus the activation energy obtained from strength-gain data or degree of hydration data is not a true activation energy as originally proposed by Arrhenius.

The author believes that the Arrhenius equation happens to be one of several equations that can be used to describe the variation of the rate constant for strength gain (or degree of hydration) with curing temperature. This has been the motivation for proposing a simpler function than Eq. (19.2) to compute equivalent age [Carino, 1982; Tank and Carino, 1991; Carino and Tank, 1992]. It is suggested that the temperature dependence of the rate constant for strength gain can be represented by the following:

$$k = A_0 e^{BT} \tag{19.5}$$

where

A_0 = the value of the rate constant at $0°C$
B = temperature sensitivity factor, $1/°C$
T = concrete temperature degrees Celcius.

Using Eq. (19.5), the equation for equivalent age at the reference temperature T_r is as follows:

$$t_e = \sum_0^t e^{B(T - T_r)} \Delta t \tag{19.6}$$

where

B = temperature sensitivity factor, $1/°C$
T = average concrete temperature during time interval Δt, degrees Celcius
T_r = reference temperature, degrees Celcius.

It was shown that Eqs. (19.2) and (19.6) would result in similar values of equivalent age [Carino, 1992a]. However, the author believes Eq. (19.6) has the following advantages over Eq. (19.2):

- The temperature sensitivity factor, B, has more physical significance compared with the apparent activation energy: for each temperature increment of $1/B$, the rate constant for strength development increases by a factor of approximately 2.7.
- Temperatures do not have to be converted to the absolute scale.
- Equation (19.6) is a simpler equation.

19.2.7.2 Strength Development Relationships

The key to developing an accurate maturity function for a particular concrete mixture is to determine the variation of the rate constant with curing temperature. Strictly speaking, a rate constant represents the rate at which a chemical reaction occurs at a given temperature. However, in the context of this discussion, the rate constant is related to the rate of strength gain at a constant temperature, and it can be obtained from the equation of strength gain versus age. Thus it is necessary to consider some of the relationships that have been used to represent the strength development of concrete.

The author has successfully used the following hyperbolic equation for strength gain up to equivalent ages at $23°C$ of about 28 d:

$$S = S_u \frac{k(t - t_0)}{1 + k(t - t_0)} \tag{19.7}$$

where

S = strength at age t, d
S_u = "**limiting**" strength
k = rate constant, $1/d$
t_0 = age at start of strength development, d

The basis of this equation has been explained elsewhere [Carino, 1984; Knudsen, 1980]. This model assumes that strength development begins instantaneously at age t_0. Thus the gradual strength development during the setting period is not considered. The parameters S_u, k, and t_0 are obtained by least-squares curve fitting to strength versus age data. The limiting strength, S_u, is the asymptotic value of the strength for the hyperbolic function that fits the data. As will be discussed below, the best-fit value for S_u does not necessarily represent the actual long-term strength of the concrete, and that is why the quotation marks were used in the definition following Eq. (19.7). For the hyperbolic model, the rate constant has the following property: when the age beyond t_0 is equal to $1/k$, the strength equals 50% of the limiting strength, S_u.

An equation similar to Eq. (19.7) was also used by Knudsen (1980) and Geiker (1983) to represent the degree of hydration and development of chemical shrinkage as a function of age. However, Geiker (1983) noted that Eq. (19.7) gave a poor fit for certain cementitious systems. It was found that the following version of the hyperbolic equation gave a better fit to that data than Eq. (19.7) [Knudsen, 1984]:

$$S = S_u \frac{\sqrt{k(t - t_0)}}{1 + \sqrt{k(t - t_0)}} \tag{19.8}$$

Knudsen explained the differences between Eq. (19.7) and (19.8) in terms of the hydration kinetics of individual cement particles. Equation (19.7) is based on **linear kinetics**, which means that the degree of hydration of an individual cement particle is a linear function of the product of time and the rate constant. Equation (19.8) is based on **parabolic kinetics** which means that the degree of hydration is a function of the **square root** of the product of time and the rate constant. Thus Eqs. (19.7) and (19.8) are called the **linear hyperbolic** and **parabolic hyperbolic** models.

Freiesleben Hansen and Pedersen (1985) proposed the following exponential equation to represent the strength development of concrete:

$$S = S_u e^{-(\frac{\tau}{t})^\alpha} \tag{19.9}$$

where

$\tau =$ a time constant
$\alpha =$ a shape parameter

This equation can model the gradual strength development occurring during the setting period, and it is also asymptotic to a limiting strength. The time constant τ represents the age at which the strength has reached $0.37 S_u$. Thus the value of $1/\tau$ is the rate constant for this equation. The shape parameter α affects the slope of the curve during the acceleratory period (following the induction period[7]), and it affects the rate at which the strength approaches the limiting strength.

Figure 19.11 illustrates the performance of these models in representing actual strength development data. Figure 19.11a shows strength data for mortar cubes cured at room temperature and tested at ages from 0.4 to 56 d. Figure 19.11b shows data for standard-cured concrete cylinders tested at ages from 7 d to 3.5 y [Carette and Malhotra, 1991]. The curves are the best-fit curves for Eqs. (19.7), (19.8), and (19.9). For the mortar data, the linear hyperbolic function and the exponential function fit the data well, and these curves in Figure 19.11a are nearly indistinguishable. For the concrete data, the parabolic hyperbolic function and the exponential function fit the data well and these curves cannot be distinguished in Figure 19.11b.

---

[7]After cement and water are mixed together, there is a time delay before strength development begins. This period, is called the induction period. After the induction period, there is rapid strength development, and this is the acceleratory period.

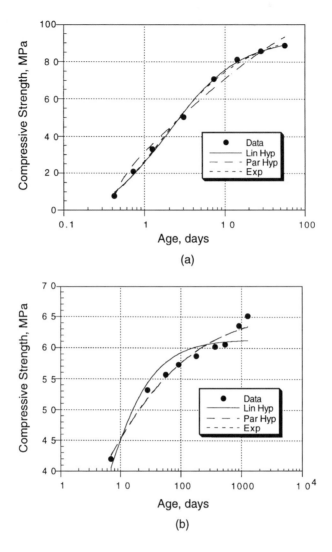

FIGURE 19.11   Fit of strength-age models to data: (a) mortar cubes and (b) concrete cylinders.

The results shown in Figure 19.11 highlight the capabilities of the various strength-age functions. The linear hyperbolic function appears to be a good model for strength development up to about 28 d (equivalent age) but not for later-ages. The parabolic hyperbolic model appears to be better suited for modeling later-age strength gain. The exponential model appears to be capable of modeling strength gain over the full spectrum of ages.

The inherent limitation of the linear hyperbolic function can be understood by considering the ratio of the limiting strength to the 28-day strength. If the t_0 term in Eq. (19.7) is neglected, the following equation is obtained for this ratio:

$$\frac{S_u}{S_{28}} = \beta = 1 + \frac{1}{28k} \qquad (19.10)$$

Thus the value of β is directly related to the rate constant. A higher-value k results in a lower value of β, which means a smaller difference between the limiting strength and the 28-day strength. The rate constant is, in turn, controlled primarily by the initial rate of strength development. The fact is that the ratio of the actual long-term strength of concrete to the 28-day strength does not obey Eq. (19.10). This means that values of S_u obtained by fitting the linear hyperbolic model to strength-age data will be lower than the actual long-term strength of concrete if it is allowed to cure for a long time.

On the basis of the above discussion, one might conclude that the exponential model given by Eq. (19.9) is the best model to use for determining the rate constant at a particular curing tempera-ture. This would be correct if the shape parameter, α, were independent of the curing temperature. Recent test results show that this is not always the case [Carino et al., 1992]. Thus a maturity func-tion based solely upon the variation of the rate constant $(1/\tau)$ with temperature would not be able to account accurately for the combined effects of time and temperature on strength development.

Which strength-gain model should be used? The maturity method is typically used to monitor strength development during construction. Therefore it is not necessary to accurately model the strength gain at later ages. Thus the author believes that the linear hyperbolic model can be used to analyze strength data, up to 28-day equivalent age, to determine the variation of the rate constant with curing temperature. Knudsen[8] suggests that the linear hyperbolic model is suitable up to a degree of hydration of 85%. The suitability of the linear hyperbolic model was also demonstrated in a recent study on the applicability of the maturity method to mortar mixtures with low ratios of water to cementitious materials, typical of those in high-performance concrete [Carino et al., 1992].

19.2.7.3 Estimating Strength

The final discussion deals with estimating strength by the maturity method. The maturity method is generally used to estimate the in-place strength of concrete by using the in-place maturity index and a previously established relationship between maturity index and strength. This assumes that a given concrete possesses a unique relationship between strength and the maturity index. This assumption would be acceptable if the long-term strength of concrete were independent of the curing temperature, but this is not the case. It is known that the long-term strength is affected by the initial temperature of the concrete. Thus if the same concrete mixture were used for a cold-weather placement and a hot-weather placement, the strength would not be the same for a given maturity index. It is proposed that the correct application of the maturity method is to estimate **relative strength**. Tank and Carino (1991) proposed the following **rate constant model** of relative strength development in terms of equivalent age:

$$\frac{S}{S_u} = \frac{k_r(t_e - t_{0r})}{1 + k_r(t_e - t_{0r})} \tag{19.11}$$

where

k_r = value of the rate constant at the reference temperature
t_{0r} = age at start of strength development at the reference temperature

However, the previous discussion has shown that, for the linear hyperbolic model, the ratio S/S_u may not indicate the true fraction of the long-term strength because the calculated value of S_u may not be the actual long-term strength. This deficiency can be overcome by expressing relative strength as a fraction of the strength at an equivalent age of 28 d. By using the definition of β in Eq. (19.10), the relative strength gain equation would be as follows:

$$\frac{S}{S_{28}} = \beta \frac{k_r(t_e - t_{0r})}{1 + k_r(t_e - t_{0r})} \tag{19.12}$$

The value of β would be obtained by fitting Eq. (19.7) to data of strength versus equivalent age.

---

[8]T. Knudsen, the Technical University of Denmark, April 1985, personal communication.

Then the value of S_u is divided by the estimated strength, from the best-fit curve, at an equivalent age of 28 d.

To summarize, the following points represent the author's ideas about the maturity method:

- There is no single maturity function that is applicable to all concrete mixtures. The applicable maturity function for a given concrete can be obtained by determining the variation of the rate constant with the curing temperature.
- The linear hyperbolic function can be used to analyze strength-age data to obtain the rate constants at different curing temperatures.
- Equation (19.5) can be used to represent the variation of the rate constant with curing temperature. The temperature-sensitivity factor governs the rate at which the rate constant increases with temperature and is analogous to the activation energy in the Arrhenius equation.
- The equivalent age can be calculated from the temperature history using Eq. (19.6).
- The maturity method is more reliable for estimating relative strength development than for estimating absolute strength.

19.2.8 Statistical Methods

In 1983, the ACI Code recognized in-place test methods as alternatives to testing field-cured cylinders for assessing concrete strength during construction. The following sentence was added to Section 6.2.1.1, dealing with form removal, of the 1983 Code[9]:

> "Concrete strength data may be based on tests of field-cured cylinders or, when approved by the Building Official, on other procedures to evaluate concrete strength."

The Commentary to the Code listed acceptable alternative procedures and further stated that these alternative methods "require sufficient data using job materials to demonstrate correlation of measurements on the structure with compressive-strength of molded cylinders or drilled cores." Thus, to use the alternative methods, an empirical relationship has to be developed to convert in-place test results to equivalent compressive-strength values. In addition, a procedure is needed to analyze in-place test results so that compressive strength can be estimated with a high degree of confidence. The latter issue is not mentioned in the Code.

The report of ACI Committee 228 [ACI 228.1R, 1995] provides information on developing the **strength relationship**[10] and how it is used to estimate the in-place strength. Basically, the procedure is to perform in-place tests (X values) and standard compressive strength tests (Y values) on companion specimens at different levels of strength and use regression analysis to determine a best-fit curve ($Y = f(X)$) to the data. The relationship and the results of in-place tests on the structure are used to estimate the concrete strength in the structure. In order to obtain a reliable estimate of the in-place strength, statistical methods are needed to account for the various sources of uncertainty. The procedures that can be used for this purpose are reviewed in this section after a brief review of recent developments related to statistical analysis.

19.2.8.1 Background

During the 1960s and 1970s, the traditional method of ordinary least squares (OLS) analysis was used to analyze correlation data and establish the best-fit equation for the strength relationship

---

[9] *Building Code Requirements for Reinforced Concrete*, ACI 318-83. American Concrete Institute, Farmington Hills, MI.

[10] The term "strength relationship" refers to the empirical equation, obtained from a correlation testing program, that relates the compressive strength (or other measure of strength) of concrete to the quantity measured by the in-place test.

and its confidence limits [Natrella, 1963]. In the 1980s, simple procedures were used to estimate lower confidence limits of estimated in-place strength [Bickley, 1982b; Hindo and Bergstrom, 1985]. However, it was recognized that the existing methods were not statistically rigorous and the stated confidence levels were not accurate [Stone et al., 1986]. One of the major deficiencies in using OLS to establish the strength relationship is that OLS assumes the X variable (the in-place test result) has no measurement error. In fact, the within-test coefficient of variation of in-place tests can be two to three times those of compression tests of cores or cylinders. To overcome these deficiencies, a study was undertaken at the National Institute of Standards and Technology (NIST) to develop a rigorous method for obtaining the strength relationship and for estimating the lower confidence limit of the in-place strength [Stone and Reeve, 1986]. The procedure that was developed used a method for least-squares fitting that accounted for error in the X variable [Mandel, 1984]. The rigorous method was discussed in the original report of ACI Committee 228 [ACI 228.1R, 1989], but it has found little use owing to its complexity. Subsequently, a simplification of the rigorous method was proposed, which could be implemented by using a computerized spreadsheet [Carino, 1993]. This simplified method was incorporated into a revision of the ACI 228 report [ACI 228.1R, 1995], and the step-by-step procedure for its implementation is included in the appendices to the report.

During the late 1980s and early 1990s, A. Leshchinsky published a series of papers on statistical methods for in-place tests, which were based largely on work at the Institute for Research of Building Structures in Kiev. These papers compared practices in different countries with those in the former Soviet Union, which had a long history in using in-place methods for quality control in precast plants.

In a review of various national standards, Leshchinsky (1990) concluded that (1) greater numbers of in-place tests are required compared with tests of standard specimens, which makes in-place testing economically unattractive; (2) the number of test specimens necessary to establish the strength relationship and the number of in-place tests on the structure have been established arbitrarily, based on the experiences of those writing the standards; and (3) the number of replicate tests to establish the strength relationship differ from the replications at a test location in a structure. Procedures were presented for selecting in-place tests based on consideration of cost and reliability of the estimated strength, and recommendations were made to reduce the cost of in-place testing.

In another paper, procedures for developing correlations were discussed and criteria were suggested for verifying the correlation at periodic intervals [Leshchinsky and Leshchinsky, 1991]. The notion of a **stable** correlation was introduced. This refers to a strength relationship that is little affected by changes in concrete composition and the construction process. It was noted that methods that have a close connection between concrete strength and the quantity measured by the in-place test tend to have more stable correlations, but they also tend to be more costly than the methods that don't have this close connection. Pullout, break-off, and pull-off tests were identified as possessing stable correlations. It was also shown that the correlations may be affected by the location on the test specimens (top, middle, bottom) where the in-place test is performed [Leshchinsky, 1991a].

Leshchinsky also discussed factors to consider when deciding whether combined methods are justified [Leshchinsky, 1991b]. In addition, the following situations were identified where the combined use of a reliable, expensive method could be combined with a less expensive but less reliable method to achieve an overall cost savings in testing:

- The reliable method is used to **calibrate** the strength correlation of the less reliable method; a correction factor is determined and applied to the strength estimated by the less reliable method.
- The less reliable method is used to identify areas of lower quality concrete where reliable tests should be performed.
- For new construction, a less reliable method is used to determine when tests by the more reliable method should be performed.

There have been few instances in North America where in-place testing has been used for acceptance testing. The major barrier is the lack of a consensus-based, statistical procedure for this application. Leshchinsky (1992) discussed some of the considerations in using in-place testing for acceptance testing. Because the strength of the actual concrete in the structure is measured, the acceptable in-place strength could be less than the design strength. Leshchinsky also reviewed provisions in existing national standards that allow acceptance based on in-place tests. In some of these standards, the required in-place strength depends on the number of tests and the variability of the in-place results. For high variability, the required in-place strength might exceed the design strength. The 1995 revision of the ACI Committee 228 report [ACI 228.1R, 1995] provided the following proposal for acceptance of concrete based on in-place testing:

> The concrete in a structure is acceptable if the estimated average, in-place, compressive strength based on a reliable in-place test procedure equals at least 85 percent of f_c' and no test result estimates the compressive strength to be less than 75 percent of f_c'.

The report states that in order to implement such criteria there needs to be a standard practice for statistical analysis of in-place test results.

19.2.8.2 Correlation Testing

The following is a summary of the guidelines provided in the ACI 228 report for establishing the strength relationship for a specific job [ACI 228.1R, 1995]. The procedure differs depending on whether in-place testing will be used to estimate strength during construction or in an existing structure.

For new construction, the strength relationship is established from a laboratory testing program performed before using the in-place test method in the field. Test specimens are made using the same concreting materials to be used in construction. At regular intervals, measurements are made using the in-place test techniques and the corresponding compressive strengths of standard specimens are also measured. The number of strength levels has a significant effect on the confidence limits of the strength relationship. It is recommended that at least six strength levels be used to establish the strength relationship. More than about nine strength levels may not be economically justified. The range of strengths in the correlation testing must include the range of strengths that are to be estimated in the structure.

For some techniques, such as rebound number and pulse velocity, it is possible to perform the in-place test on standard specimens without damaging them and the specimens can be subsequently tested for compressive strength. For other methods that result in local damage, in-place tests are carried out on separate specimens. It is important that the in-place tests and standard tests are performed on specimens having similar compaction and the same maturity. Curing companion test specimens in the same water bath is a convenient way to assure similar temperature histories. Alternatively, internal temperatures can be recorded and test ages can be adjusted so that the in-place and standard tests are performed at the same maturity index.

For existing construction, the strength relationship is established by performing the in-place tests on the structure and determining the compressive strength from cores taken from adjacent locations. To obtain a wide range of strength, a rebound hammer or pulse velocity survey may be performed first to identify locations with apparently different quality. At each test location, a minimum of two cores should be taken to evaluate the compressive strength. Thus the proper application of in-place testing for existing construction requires taking at least 12 cores to establish the strength relationship. As a result, the procedure may be economical when large volumes of concrete are to be evaluated.

More detailed information on the number of replicate in-place tests and the companion test specimens to use for different test methods may be found in the ACI report [ACI 228.1 R, 1995]. After paired values of in-place test results and concrete compressive strengths are obtained, regression analysis is used to establish the equation of the strength relationship. The ACI 228 report recommends that the natural logarithms of the test results be used so that the following equation is fitted:

$$\ln C = a + b \ln I \tag{19.13}$$

where

$\ln C$ = average of natural logarithms of compressive strengths
$\ln I$ = average of natural logarithms of in-place test results
a = intercept of line
b = slope of line

The report also recommends that the regression analysis to determine the values of a and b be performed using a procedure that accounts for the error on the X variable (in-place test results). This procedure is explained in detail in the appendix of the ACI report. Basically, the ratio of the variances of the in-place test results and compressive strength results are used to define a parameter λ, which is applied to the equations of OLS to obtain the slope and intercept. This approach results in a more reliable estimate of the uncertainty of the strength relationship.

19.2.8.3 In-Place Strength Estimate

In making estimates of the in-place strength, there are several important points to consider:

- Where should the in-place tests be performed?
- How many in-place tests should be performed?
- How should the data be analyzed to obtain a reliable estimate of in-place strength?

These points are also covered in the ACI 228 report. In the case of new construction, a preconstruction meeting should be held to establish where and how many in-place tests should be performed. The ACI report provides guidelines that can be used as a starting point in arriving at these decisions. For existing construction, a pretesting meeting should be held among all parties who share an interest in the test results. Agreement should be reached on the procedures for obtaining, analyzing, and interpreting the test results.

After the in-place test results have been obtained, statistical analysis is used to arrive at a reliable estimate of the in-place concrete strength. The term "reliable estimate" means that there should be a high likelihood that the actual in-place strength exceeds the estimated strength. The statistical procedure that is used should account for the following sources of uncertainty or variability:

- the uncertainty of the strength relationship,
- the variability of the in-place test results, and
- the variability of the in-place concrete strength.

The ACI 228 report includes several approaches that may be used for this purpose. One of these is based on a simplification by Carino (1993) of a rigorous procedure developed earlier by NIST researchers [Stone and Reeve, 1986]. The underlying steps of this procedure are illustrated in Figure 19.12. The average of the in-place test results are used to estimate the average concrete strength using a strength relationship. Next, the lower confidence limit of the average concrete strength is obtained by taking into account the uncertainty of the estimate from the strength relationship. This uncertainty includes a component based on the correlation testing and a component based the variability of the in-place test results. Finally, the variability of the in-place concrete strength is used to obtain the tenth percentile strength, that is, the strength expected to be exceeded by 90% of the concrete in the structure. This variability is estimated using the assumption that the ratio of the standard deviation of compressive strength tests to the standard deviation of in-place test results has the same value in the field as was obtained during the correlation testing [Stone and Reeve, 1986]. The details for applying this procedure are given in the ACI 228 report [ACI 228.1 R, 1995] and additional background information may be found in Carino (1993). This procedure has been implemented in a Windows-based computer program that may be obtained by contacting the author of this chapter.

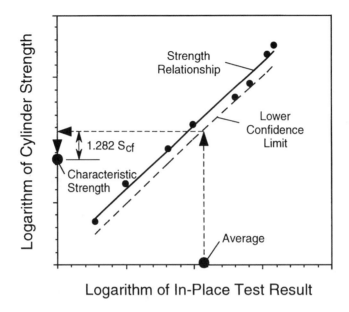

FIGURE 19.12 Statistical procedure to estimate the in-place compressive strength [Carino, 1993].

19.3 Methods for Flaw Detection and Condition Assessment[11]

19.3.1 Introduction

Other types of NDT methods are those used for flaw detection and condition assessment. In this context the term **flaw** can include voids, honeycombing, delaminations, cracks, lack of subbase support, etc. Recent research and development efforts for these methods have far exceeded those for methods to estimate strength. The research impetus has come primarily from the transportation industry, since much of the highway infrastructure is in need of repair as a result of natural aging or the damage resulting from corrosion of reinforcing steel or deterioration of concrete.

The techniques for flaw detection are generally based on the following simple principle: the presence of an internal anomaly interferes with the propagation of certain types of waves. By monitoring the response of the test object when it is subjected to these waves, the presence of the anomaly can be inferred. The interpretation of the results of these types of NDT methods usually requires an individual who is knowledgeable both in concrete technology and in the physics governing the wave propagation.

This section of the chapter reviews the following techniques:

- Visual inspection
- Stress-wave propagation methods
- Infrared thermography
- Ground-penetrating radar
- Electrical/magnetic methods
- Nuclear methods

---

[11]Some of text in this section is based on a draft prepared by the author for a report on NDT methods under preparation by ACI Committee 228.

Emphasis is placed on explaining the underlying principles of the methods, and the reader may find additional information in Malhotra and Carino (1991) and in Bungey (1989).

19.3.2 Visual Inspection

Visual inspection is one of the most versatile and powerful of the NDT methods, and it is typically one of the first steps in the evaluation of a concrete structure [Perenchio, 1989]. Visual inspection can provide a wealth of information that may lead to positive identification of the cause of observed distress. However, its effectiveness depends on the knowledge and experience of the investigator. Broad knowledge in structural engineering, concrete materials, and construction methods is needed to extract the most information from visual inspection. Useful guides are available to assist less experienced individuals in recognizing different types of damage and determining the probable cause of the distress [ACI 201.1R, ACI 207.3R, ACI 224.1R, ACI 362R].

Before performing a detailed visual inspection, the investigator should develop and follow a definite plan to maximize the quality of the recorded data. Various ACI documents should be consulted for additional guidance on planning and carrying out the complete investigation [ACI 207.3R, ACI 224.1R, ACI 362R, ACI 437R]. A typical investigation might involve the following activities:

- perform a walk-through visual inspection to become familiar with the structure;
- gather background documents and information on the design, construction, maintenance, and operation of the structure;
- plan the complete investigation;
- perform a detailed visual inspection; and
- perform any necessary sampling or in-place tests.

19.3.2.1 Supplemental Tools

Visual inspection has the obvious limitation that only visible surfaces can be inspected. Internal defects go unnoticed and no quantitative information is obtained about the properties of the concrete. For these reasons, a visual inspection is usually supplemented by one or more of the other NDT methods discussed in this chapter. The inspector should consider other useful tools that can enhance the power of a visual inspection.

Optical magnification allows a more detailed view of local areas of distress. Available instruments range from simple magnifying glasses to more expensive hand-held microscopes. Some fundamental principles of optical magnification can help in selecting the correct tool. The focal length decreases with increasing magnifying power, which means that the primary lens must be placed closer to the surface being inspected. The field of view also decreases with increasing magnification, making it tedious to inspect a large area at high magnification. The depth of field is the maximum difference in elevation of points on rough textured surface that are simultaneously in focus; this also decreases with increasing magnification of the instrument. To assure that the "hills" and "valleys" are in focus simultaneously, the depth of field has to be greater than the elevation differences in the texture of the surface that is being viewed. Finally, the illumination required to see clearly increases with magnification level, and artificial lighting may be needed at high magnification.

A very useful tool for crack inspection is a small hand-held magnifier with a built-in measuring scale on the lens closest to the surface being viewed [ACI 224.1R]. With such a crack comparator, the width of surface opening cracks can be measured accurately.

A stereo microscope includes two viewing lenses that allow a three-dimensional view of the surface. By calibrating the focus-adjustment screw, the investigator can estimate the elevation differences in surface features.

Fiberscopes and borescopes allow inspection of regions that are otherwise inaccessible to the naked eye. A fiberscope is composed of a flexible bundle of optical fibers and a lens system, and it allows cavities within a structure to be viewed by means of small access holes. The fiberscope is designed so that some fibers transmit light to illuminate the cavity. In some systems, the operator can rotate the viewing head to allow a wide viewing angle from a single access hole. A borescope is composed of a rigid tube with mirrors and lenses and is designed to view straight ahead or at right angles to the tube. The image is clearer using a borescope, while the fiberscope offers more flexibility in the field of view. Use of these scopes requires drilling small holes if other access channels are absent, and the holes must intercept the cavity to be inspected. Some methods discussed in the remainder of the chapter may be used to locate these cavities. Hence the fiberscope or borescope may be used to verify the results of other NDT methods without removing cores.

A recent development that expands the flexibility of visual inspection is the small digital video camera. These are used in a manner similar to borescopes, but they offer the advantage of a video output that can be displayed on a monitor or stored on appropriate recording media. These charge coupled device (CCD) cameras come in a variety of sizes, resolutions, and focal lengths. Miniature versions as small as 12 mm in diameter, with a resolution of 460 scan lines, are available. They can be inserted into holes drilled into the structure for views of internal cavities, or they can be mounted on robotic devices for inspections in pipes or areas with biological hazards.

In summary, visual inspection is a very powerful NDT method. Its efficiency, however, is to a large extent governed by the experience and knowledge of the investigator. A broad knowledge of structural behavior, materials, and construction methods is desirable. Visual inspection is typically one aspect of the total evaluation plan, which will often be supplemented by a series of other NDT methods or invasive procedures.

19.3.3 Stress-Wave Propagation Methods

Tapping an object with a hammer or steel rod (sounding) is one of the oldest forms of nondestructive testing based on stress-wave propagation. Depending on whether the result is a high-pitched **ringing** sound or a low frequency **rattling** sound, the integrity of the member can be assessed. The method is subjective, as it depends on the experience of the operator, and it is limited to detecting near-surface defects. Despite these inherent limitations, **sounding** is a useful method for detecting near-surface delaminations, and it has been standardized by ASTM (D4580).

In NDT of metals, the **ultrasonic pulse-echo** (UP-E) method has proven to be a reliable method for locating small cracks and other defects. The principle of UP-E is similar to sonar. An electromechanical transducer is used to generate a short pulse of ultrasonic stress waves that propagates into the object being inspected. Reflection of the stress pulse occurs at boundaries separating materials with different densities and elastic properties (these determine the acoustic impedance of a material). The reflected pulse travels back to the transducer, which also acts as a receiver. The received signal is displayed on an oscilloscope, and the round-trip travel time of the pulse is measured electronically. By knowing the speed of the stress waves, the distance to the reflecting interface can be determined. If there is no internal defect, the opposite face of the test object is detected.

Attempts to use UP-E equipment designed for metal inspection to test concrete have been unsuccessful because of the heterogeneous nature of concrete. The presence of paste-aggregate interfaces, air voids, and reinforcing steel result in a multitude of echoes that obscure those of real defects. However, in the last 10 to 15 years there has been considerable progress in the development of usable techniques based on the propagation of impact-generated stress waves. This section reviews the basic concepts of stress-wave propagation and reviews the principles of some of the more promising methods. A more comprehensive review is provided by Sansalone and Carino (1991).

19.3.3.1 Basic Relationships

When a disturbance (stress or displacement) is applied suddenly at a point on the surface of a solid, such as by impact, the disturbance propagates through the solid as three different waves: a *P*-wave, an S-wave, and an R-wave. The P-wave and S-wave propagate into the solid along spherical wavefronts. The P-wave is associated with the propagation of normal stress and the S-wave is associated with shear stress. In addition, there is an R-wave that travels away from the disturbance along the surface. In an infinite isotropic, elastic solid, the P-wave speed, C_p, is related to the Young's modulus of elasticity, E, Poisson's ratio, v, and the density, ρ, as follows [Krautkrämer and Krautkrämer, 1990]:

$$C_p = \sqrt{\frac{E(1-v)}{\rho(1+v)(1-2v)}} \tag{19.14}$$

The S-wave propagates at a slower speed, C_s, given by

$$C_s = \sqrt{\frac{G}{\rho}} = \sqrt{\frac{E}{\rho 2(1+v)}} \tag{19.15}$$

where G is the shear modulus of elasticity.

The ratio of S-wave speed to P-wave speed depends on Poisson's ratio, as, follows:

$$\frac{C_s}{C_p} = \sqrt{\frac{1-2v}{2(1-v)}} \tag{19.16}$$

For a Poisson's ratio of 0.2, which is typical of concrete, this ratio equals 0.61. The ratio of the R-wave speed, C_r, to the S-wave speed is given by the following approximate formula:

$$\frac{C_r}{C_s} = \frac{0.87 + 1.12v}{1+v} \tag{19.17}$$

For Poisson's ratio equal to 0.2, the R-wave travels at 92% of the S-wave speed.

In the case of bounded solids, the wave speed is also affected by the geometry of the solid. For rod like solids, the P-wave speed is independent of Poisson's ratio and is given by the following:

$$C_p = \sqrt{\frac{E}{\rho}} \tag{19.18}$$

Thus, for $v = 0.2$, the wave speed in a slender rod is about 5% slower than in a large solid.

When a stress wave traveling through **material 1** is incident on the interface between a dissimilar **material 2**, a portion of the incident wave is reflected. The amplitude of the reflection is a function of the angle of incidence and is a maximum when this angle is 90° (normal incidence). For normal incidence, the reflection coefficient, R, is given by the following:

$$R = \frac{Z_2 - Z_1}{Z_2 + Z_1} \tag{19.19}$$

where

Z_2 = specific acoustic impedance of material 2
Z_1 = specific acoustic impedance of material 1

The specific acoustic impedance is the product of the wave speed and density of the material. The following are approximate Z values for some materials [Sansalone and Carino, 1991]:

Material	Specific Acoustic Impedance, kg/(m²s)
Air	0.4
Water	1.5×10^6
Soil	0.3 to 4×10^6
Concrete	7 to 10×10^6
Steel	47×10^6

Thus it can be shown that when a stress wave traveling through concrete encounters an interface with air, there is almost total reflection at the interface. This is why NDT methods based on stress-wave propagation have proven to be successful for locating defects within concrete.

19.3.3.2 Impact Methods

The greatest success in the practical application of stress-wave methods for testing concrete has been to use mechanical impact to generate the stress pulse. Impact causes a high-energy pulse that results in high penetration of the stress waves. Several techniques have been developed that are similar in principle but that differ in the specific instrumentation and signal-processing methods that are used [Davis and Dunn, 1974; Steinbach and Vey, 1975; Higgs, 1979; Stain, 1982; Sansalone and Carino, 1986; Nazarian and Stokoe, 1986a; Davis and Hertlein, 1991].

Figure 19.13 is a schematic of an impact test. The principle is analogous to other echo methods that have been discussed. Impact on the surface produces a disturbance that travels into the object along spherical wavefronts as P- and S-waves. In addition, a surface wave (R-wave) travels away from the impact point. The P- and S-waves are reflected by internal defects (difference in elastic constants and density) or external boundaries. When the reflected waves, or echoes, return to the surface, they produce displacements that are measured by a receiving transducer. If the transducer is placed close to the impact point, the response is dominated by P-wave echoes [Sansalone and

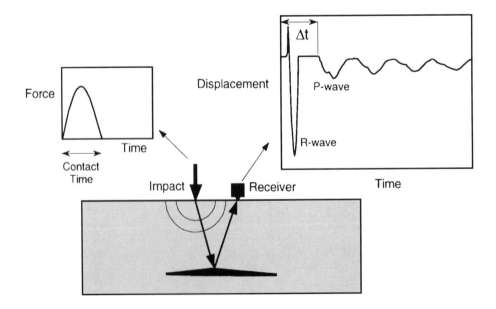

FIGURE 19.13   Schematic of test using impact to generate stress waves.

Carino, 1986]. Using **time-domain** analysis, the time from the start of the impact to the arrival of the P-wave echo is measured, and the depth of the reflecting interface can be determined if the P-wave speed is known.

The first successful applications of impact methods were used in geotechnical engineering to evaluate the integrity of concrete piles and caissons [Davis and Dunn, 1974; Steinbach and Vey, 1975]. The technique became known as the **sonic-echo** or **seismic-echo** method. The long length of the foundation structures allowed sufficient time separation between the generation of the impact and the echo arrival, and determination of round-trip travel times was relatively simple [Lin et al., 1991b; Olson and Church, 1986]. The impact response of thin concrete members, such as slabs and walls, is more complicated than that of long slender members. Work by Sansalone and Carino (1986) has led to the development of a successful technique for flaw detection in relatively thin concrete structures. The technique is known as the **impact-echo method**.

A key development leading to the success of the impact-echo method was the use of **frequency analysis** instead of time-domain analysis of the recorded wave forms [Carino et al., 1986]. The principle of frequency analysis is as follows: The P-wave produced by the impact undergoes multiple reflections between the test surface and the reflecting interface. Each time the P-wave arrives at the test surface, it causes a characteristic displacement. Thus the wave form has a periodic pattern that is dependent on the frequency, f, of the P-wave arrival, which is given by

$$f = \frac{C_p}{2D} \qquad (19.20)$$

where C_p is the P-wave speed[12] and D is the depth of the reflecting interface. This frequency is termed the **thickness frequency**.

To apply frequency analysis, the recorded wave form is transformed into the frequency domain by using the fast Fourier transform technique [Bracewell, 1978]. The computed amplitude spectrum shows the dominant frequencies in the wave form. For slablike structures, the thickness frequency will usually be the dominant peak in the spectrum. The value of the peak frequency in the amplitude spectrum can be used to determine the depth of the reflecting interface by expressing Eq. (19.20) as follows:

$$D = \frac{C_p}{2f} \qquad (19.21)$$

Figure 19.14 illustrates the use of frequency analysis of impact-echo tests. Figure 19.14a shows the amplitude spectrum from a test over a solid portion of a 0.5-m thick concrete slab. There is a frequency peak at 3.42 kHz, which corresponds to multiple P-wave reflections between the bottom and top surfaces of the slab. Using Eq. (19.20) or (19.21) and solving for C_p, the P-wave speed is calculated to be 3420 m/s. Figure 19.14b shows the amplitude spectrum from a test over a portion of the slab containing a disk-shaped void [Carino and Sansalone, 1989b]. The peak at 7.32 kHz results from multiple reflections between the top of the slab and the void. Using Eq. (19.21), the calculated depth of the void is $3420/(2 \times 7320) = 0.23$ m, which compares favorably with the known distance of 0.25 m.

In the initial work leading to the impact-echo method [Sansalone and Carino, 1986], it was noted that the duration of the impact was critical to determining the success of the method. As shown in Figure 19.13, the force-time relationship for the impact may be approximated as a half-cycle sine curve, and the duration of the impact is the **contact time**. The contact time determines the frequency content of the stress pulse generated by the impact [Carino et al., 1986]. As an approximation, the highest frequency component of significant amplitude equals the inverse of the contact time. In

---

[12]Recent studies by Sansalone and co-workers at Cornell University have shown that the P-wave speed in impact-echo testing is approximately 96% of the P-wave speed in an infinite solid.

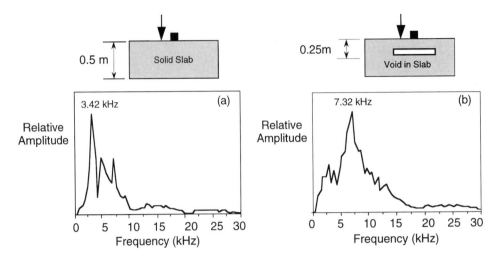

FIGURE 19.14   Examples of amplitude spectra from impact-echo test of concrete slab: (a) solid slab; (b) slab with void.

order to accurately locate shallow defects, the stress pulse must have frequency components greater than the frequency corresponding to the flaw depth (Eq. (19.20)). For example, for a P-wave speed of 4000 m/s, a pulse with a contact time shorter than 100 μs is needed to determine the depth of defects shallower than about 0.2 m. Various impact sources have been used for impact testing. In evaluation of piles, hammers can be used to produce energetic impacts with long contact times (on the order of 1 ms). Such impacts are acceptable for testing long, slender structures but not for slabs or walls. For testing slabs from 0.15 to 0.5 m thick, steel spheres and spring-loaded spherically tipped impactors have been used successfully. Steel spheres are convenient impact sources because the contact time is proportional to the diameter of the sphere [Goldsmith, 1965].

The impact-echo method has been successful in detecting a variety of defects, such as voids and honeycombed concrete in members, delaminations in bare and overlaid slabs, and voids in tendon ducts [Jaeger et al., 1996; Sansalone and Carino, 1986, 1988, 1989a,b; Carino and Sansalone, 1992]. Experimental studies have been supplemented with analytical studies to gain a better understanding of the propagation of transient waves in bounded solids with and without flaws [Cheng and Sansalone 1993a,b, 1995a,b; Sansalone and Carino, 1987; Lin et al., 1991a,b]. Application of the impact-echo method has been extended to prismatic members, such as columns and beams [Lin and Sansalone, 1992a,b,c]. It has been found that reflections from the perimeter of these members cause complex modes of vibration. As a result, the amplitude spectra have many peaks, and the depth of the member is not related to the dominant frequency in the spectrum according to Eq. (19.21). Nevertheless, it has been shown that defects can still be detected within beams and columns, and a successful field application has been reported [Sansalone and Poston, 1992]. The method has also been applied to evaluate the quality of the bond between an overlay and base concrete [Lin and Sansalone 1996a,b].

As with most methods for flaw detection in concrete, experience is required to interpret stress-wave test results. An advance in the interpretation of impact-echo results from tests of slablike structures has been the application of an artificial intelligence technique known as a **neural network** [Sansalone et al., 1991]. In this technique, a computer program is **trained** to recognize amplitude spectra associated with flawed and unflawed structures. After this **training**, the program can be used to classify the results of tests on a structure under investigation. This technique was incorporated into the first commercial impact-echo test system [Pratt and Sansalone, 1992].

Another variant of the impact method is known as **impulse-response, transient dynamic response**, or **impedance testing** [Davis and Dunn, 1974; Higgs, 1979; Stain, 1982; Olson and Wright, 1990; Davis and Hertlein, 1991]. In this approach, the force history of the impact and the response of the structure are measured. Through a signal-processing technique, the measured response and force history are used to compute the characteristic impulse response spectrum of the structure see [Sansalone and Carino, 1991]. The impulse-response spectrum of a structure depends on its geometry, the support conditions, and the existence of flaws or cracks.

Depending on the measured quantity of the structural response (displacement, velocity, or acceleration), the impulse response spectrum has different meanings. Typically, velocity is measured and the resulting impulse-response spectrum has units of velocity/force which are referred to as **mobility**, and the spectrum is called a **mobility plot**. At frequency values corresponding to resonant frequencies of the structure, mobility values are maximum. In the testing of piles, the mobility plot has a series of peaks that correspond to the fundamental and higher longitudinal modes of vibration. The difference between any two adjacent peaks, Δf, is equal to the fundamental longitudinal frequency [Higgs, 1979]. The length of the pile can be calculated by using Δf in place of f in Eq. (19.21).

To illustrate how the method works, impulse-response spectra obtained from two test piles having the same dimensions are shown in Figure 19.15 [Olson and Church, 1986]. Figure 19.15a is the response spectrum obtained from a sound pile. The P-wave speed in this pile was 4140 m/s. The fundamental longitudinal frequency of 138 kHz was calculated by determining the average frequency difference between four successive peaks. The pile length was calculated to be 15.0 m, while the known length was 15.2 m. For comparison, Figure 19.15b shows the response spectrum from a pile that contained a full-width defect at a depth of 9.8 m. The P-wave speed in this pile

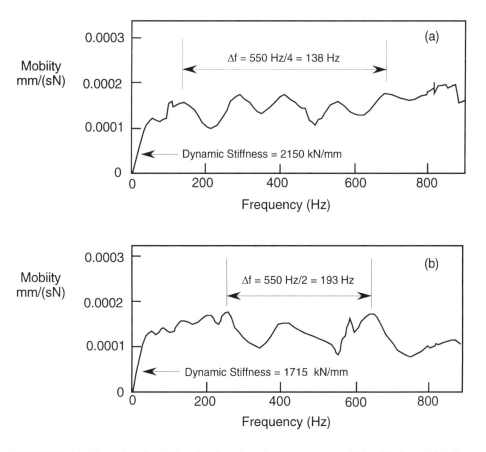

FIGURE 19.15 Examples of mobility plots from impulse-response tests of (a) solid pile and (b) pile with full section defect [adapted from Olson and Church, 1986].

was 4200 m/s. The fundamental frequency of 193 Hz was estimated by averaging the difference between the three successive frequency peaks shown in Figure 19.15b. The depth to the reflecting interface was calculated to be about 11 m. In addition to providing information on the length of a pile, the impulse response function can also indicate the dynamic stiffness of the pile-soil structure. The initial slopes of the spectra in Figure 19.15 are inversely related to the dynamic stiffnesses. Thus the presence of the void is indicated by a reduced pile length and an increased dynamic stiffness.

19.3.3.3 Spectral Analysis of Surface Waves

In the late 1950s and early 1960s, Jones (1955, 1962) reported on the use of **surface waves** to determine the thickness and elastic stiffness of pavement slabs and the underlying layers. The method involved determining the relationship between the wavelength and velocity of surface vibrations as the vibration frequency was varied. Apart from the studies by Jones there seems to have been little use of this technique for testing concrete pavements. In the early 1980s, however, researchers at The University of Texas at Austin began studies of a surface-wave technique that involved an impactor instead of a steady-state vibrator. Digital signal processing was used to develop the relationship between wavelength and velocity. The technique was called **spectral analysis of surface waves** (SASW) [Heisey et al., 1982; Nazarian et al., 1983].

Figure 19.16 shows the configuration for SASW testing [Nazarian and Stokoe, 1986a]. Two receivers are used to monitor the movement of the surface due to the R-wave produced by the impact. The received signals are processed, and a complex calculation scheme is used to infer the stiffnesses of the underlying layers.

Just as the impact is composed of a range of frequency components, the R-wave also contains a range of components of different frequencies or wavelengths. (Note: the product of frequency and wavelength equals wave speed.) This range depends on the contact time of the impact; a shorter contact time results in a broader range. The longer-wavelength (lower-frequency) components penetrate more deeply, and this is the key to using the R-wave to gain information about the properties of the underlying layers [Rix and Stokoe, 1989]. In a layered system, the propagation speed of these different components is affected by the wave speed in the layers through which the components propagate. A layered system is a **dispersive** medium for R-waves, which means that different frequency components in the R-wave propagate with different speeds, which are called **phase velocities** [Krstulovic-Opara et al., 1996].

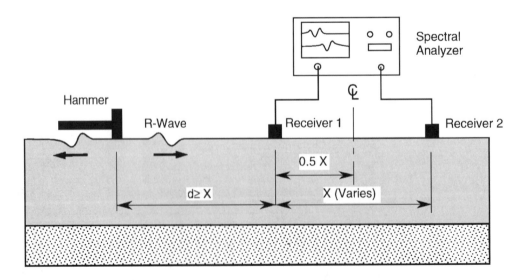

FIGURE 19.16   Schematic of testing configuration for SASW test.

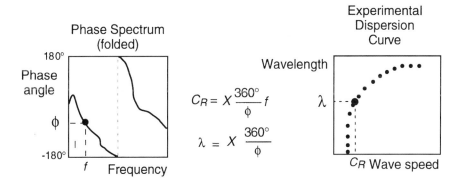

FIGURE 19.17   Schematic of phase spectrum obtained from cross-power spectrum of the receiver waveforms in SASW testing and the dispersion curve relating wavelength and wave speed.

Phase velocities are calculated by determining the time it takes for each frequency (or wavelength) component to travel between the two receivers. These travel times are determined from the phase difference of the frequency components arriving at the receivers [Nazarian and Stokoe, 1986b; Sansalone and Carino, 1991]. The phase differences are obtained by computing the **cross-power spectrum** of the signals recorded by the two receivers. The phase portion of the cross-power spectrum gives phase differences (in degrees) as a function of frequency. A schematic of such a phase spectrum is shown in Figure 19.17.[13] The phase velocities are determined as follows:

$$C_{R(f)} = X \frac{360}{\phi_f} f \tag{19.22}$$

where

$\quad C_{R(f)}$ = surface wave speed of component with frequency f
$\qquad X$ = distance between receivers (see Figure 19.16)
$\qquad \phi_f$ = phase angle of component with frequency f

The wavelength, λ_f, corresponding to a component frequency is calculated using the following equation:

$$\lambda_f = X \frac{360}{\phi_f} \tag{19.23}$$

By repeating the calculations in Eqs. (19.22) and (19.23) for each component frequency, a plot of wavelength versus phase velocity is obtained. Such a plot is called a **dispersion curve**, and is shown on the right side of Figure 19.17.

After the experimental dispersion curve is obtained, a process called **inversion** is used to obtain the stiffness profile at the test site; [Krstulovic-Opara et al., 1996; Nazarian and Desai, 1993; Yuan and Nazarian, 1993; Nazarian and Stokoe, 1986b]. The site is modeled as layers of varying thickness. Each layer is assigned density and elastic constants. Using these assumed properties, the surface motion at the location of the receivers is calculated using surface-wave propagation theory. The calculated responses are subjected to the same signal-processing technique as used for the test data, and a theoretical dispersion curve is obtained. The theoretical and experimental dispersion curves are compared. If the curves match, the problem is solved and the assumed stiffness profile

---

[13]A phase spectrum is usually plotted so that the phase angle axis ranges from $-180°$ to $180°$. Hence the spectrum "folds" over when the phase angle reaches $-180°$, giving the phase spectrum a "sawtooth" pattern.

is correct. If there are significant discrepancies, the properties of the assumed layered system are changed, and a new theoretical dispersion curve is calculated. This process continues until there is agreement between the theoretical and experimental curves. The user should be experienced in selecting plausible starting values of the elastic constants and have the ability to recognize whether the final values are reasonable. Convergence cannot be assumed to indicate that the correct values have been determined, because it is possible for different combinations of layer thicknesses and elastic moduli to result in similar dispersion curves.

The general configuration for SASW testing was shown in Figure 19.16. For reliable results, tests are repeated with different receiver spacings [Nazarian and Stokoe, 1986a]. The receivers are first located close together, and the spacing is increased by a factor of two for subsequent tests. As a check on the measured phase information for each receiver spacing, a second series of tests is carried out by reversing the position of the source. Typically, five receiver spacings are used at each test site. For tests of concrete pavements, the closest spacing is usually about 150 mm.

The SASW method has been used to determine the stiffness profiles of soil sites and of flexible and rigid pavement systems [Nazarian and Stokoe, 1986b; Rix and Stokoe, 1989]. The method has been extended to the measurement of changes in the elastic properties of concrete slabs during curing [Rix et al., 1990].

19.3.3.4 Summary

The impact techniques discussed above offer great potential as reliable methods for flaw detection. While they appear similar in terms of the physical test procedure, different information about the test object can be obtained by using the correct instrumentation and signal-processing methods. Each method is best suited for particular applications. Persons interested in using NDT methods based on stress-wave propagation should develop the ability to use all the methods, so that the most appropriate one can be used for a particular situation.

19.3.4 Infrared Thermography

Infrared thermography is a technique for locating near-surface defects by measuring surface temperature. It is based on two principles. The first principle is that a surface emits electromagnetic radiation with an intensity that depends on its temperature. At about room temperature, the radiation is in the infrared region of the electromagnetic spectrum. The second principle is that the presence of an anomaly having a lower thermal conductivity than the surrounding material will interfere with the flow of heat and alter the surface-temperature distribution. As a result, the surface temperature will not be uniform. Thus, by measuring the surface temperature, the presence of the defect can be inferred. In practice, the surface temperature is measured with an infrared scanner that works in a manner similar to a video camera [Manning and Holt, 1980]. The output of the scanner is a **thermographic image** of temperature differences.

The following are values of thermal-conductivity coefficients for different materials [Halliday and Resnik, 1978]:

Material	Thermal Conductivity, $J/s \cdot m \cdot {}^\circ C$
Steel	46
Ice	1.7
Concrete	0.8
Air	0.024

It can be seen that air has a much lower thermal conductivity than concrete, and this explains why the presence of air voids within concrete can affect the surface temperature distribution when there is heat flow through the concrete.

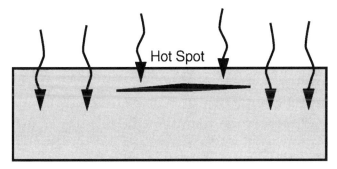

Heat flow inward results in "Hot" spot above flaw

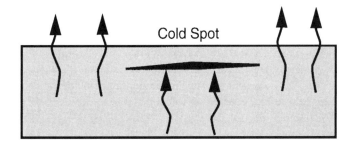

Heat flow outward results in "Cold" spot above flaw

FIGURE 19.18   Effect of a void on the heat flow through a concrete slab.

In civil engineering applications, the method is used primarily to detect corrosion-induced delaminations in reinforced concrete bridge decks. In North America, early research on this application was performed independently in the late 1970s by the Virginia Highway and Transportation Research Council [Clemeña and McKeel, 1978] and by the Ontario Ministry of Transportation and Communication [Manning and Holt, 1983]. Initial studies involved hand-held scanners and photographic cameras to record the thermographic images. This was followed by scanning from a boom attached to a truck and by airborne scanning using a helicopter. Although infrared thermography allowed more rapid surveys than the chain-drag technique [ASTM D 4580], it was not as accurate as chain dragging for determining the extent of the delaminations [Manning and Holt, 1983]. In 1988, ASTM published a standard test method [ASTM D 4788] on the use of the infrared thermography to locate delaminations in exposed and overlaid concrete bridge decks. Additional information on considerations for performing an infrared survey and representative case histories are provided by Weil (1991).

The principle of infrared thermography is illustrated in Figure 19.18. The presence of a void in the concrete slab has a local insulating effect that disrupts the heat flow through the slab and affects the surface temperature. When heat flows into the slab, the area above the void is warmer than the surrounding area; and when heat flows out of the slab, the area above the void is cooler. By measuring the surface temperature distribution, one can infer the presence of the void. Hence, to apply infrared thermography, there must be a heat-flow condition through the test object and a means for measuring small differences in surface temperature.

The required heat-flow condition can be created artificially by using heating lamps, or it can occur naturally through solar heating (heat flow into structure) and nighttime cooling (heat flow out of structure). The latter method is obviously the economical approach. The best time for doing infrared surveys is two to three hours after sunrise or after sunset [Weil, 1991]. Heat flow becomes very low and the surface temperatures become uniform as time elapses following sunrise or sunset.

Even with the proper heat-flow conditions, not all delaminations are detectable. Maser and Roddis (1990) performed analytical studies to gain an understanding of the factors affecting the differences in the surface temperature of a solid concrete slab and a slab with a delamination. It was found that the maximum differential temperature decreased as the depth of the delamination increased and as the width decreased. Also, a water-filled delamination resulted in nearly identical surface temperatures as a solid slab.

In infrared thermography, differences in surface temperature are measured by using an imaging infrared scanner, a device similar to a video camera, which measures the amount of infrared radiation emitted by a surface. As mentioned, the underlying principle of this measurement method is that an object at a temperature above absolute zero emits electromagnetic radiation, whose wavelength depends on the temperature. As the temperature increases, wavelengths become shorter, and, at sufficiently high temperature, the radiation is in the visible spectrum. This is the operating principle of incandescent light bulbs. At room temperature range, the wavelength of the radiation emitted by surfaces is on the order of 10 μm, which is in the infrared region of the spectrum. This radiation cannot be detected by the naked eye. The infrared scanner has a detector that **sees** the infrared radiation. The detector output is related to the amount of incident radiant energy.

The energy emitted by a surface is related to its temperature according to the Stefan-Boltzman law [Halliday and Resnik, 1978]:

$$R = e\sigma\, T^4 \tag{19.24}$$

where

- R = rate of energy radiation per unit area of surface, W/m^2
- e = the emissivity of the surface
- σ = the Stefan-Boltzman constant, 5.67×10^{-8} W/(m$^2 \cdot$ K^4)
- T = absolute temperature of the surface, Kelvin

By proper calibration, the output from the infrared sensor can be converted to a temperature. The emissivity is characteristic of the material and surface texture and has a value ranging from 0 to 1. Since the rate of energy radiation depends on temperature and emissivity, care must be exercised in interpreting thermographic images to assure that apparent temperature differences are not caused by differences in emissivity. The emissivity of bridge deck surfaces can be affected by the type of texturing, oil spots, tire marks, paint, and loose debris.

Equipment for performing a thermographic survey includes an infrared scanner with associated hardware capable of producing a video image of the temperature distribution of the scanned surface; a conventional video camera to provide a visual record for comparison with the infrared record; video recorders to record infrared and conventional video images; analog-to-digital converters to transform the video images into digital data; a computer system and software for data storage and signal enhancement; and a distance-referencing system to correlate the infrared scan with position on the bridge deck [Weil, 1991; Kunz and Eales, 1985]. The equipment is typically contained in a mobile van that travels along the roadway while data are recorded. The resolution of the infrared scanner is improved by lowering its temperature; therefore, a liquid nitrogen cooling system is used to cool the sensor. Available equipment allows resolution of differences in surface temperature as low as 0.1°C.

The following is a summary of the procedure given in ASTM D 4788 for performing an infrared thermographic survey to detect delaminations in bridge decks:

- Remove debris from the surface.
- Allow the surface to dry for at least 24 h before testing.

- There must be at least 0.5°C difference between the surface temperatures in areas above delaminations and in sound concrete. A minimum of 3 h of direct sunlight are generally sufficient to establish this temperature difference. A contact thermometer with a minimum resolution of 0.1°C is used to determine whether the minimum temperature difference has been established.

- Do not test when the wind speed exceeds 50 km/h because the surface temperature will be affected.

- Do not test when the ambient air temperature is less than 0°C because ice in delaminations will give false indications of sound concrete. As a guide, an ambient temperature rise of 10°C, 4 h of sunshine, and a wind speed below 25 km/h should result in accurate data on bare concrete surfaces during winter. For asphalt-covered concrete, at least 6 h of sunshine are necessary during winter.

- Collect data with the van moving at speeds not greater than 15 km/h.

The results of the inspection are usually reported in terms of delaminated area and percentage of delaminated area. After the delaminated areas are identified in the infrared images, the visible video images should be compared to assure that apparent temperature differences were not due to emissivity changes [Kunz and Eales, 1985]. The ASTM standard states that 80 to 90% of the delaminations in a bare bridge deck can be located with this method. It has also been found that the inspection of the same deck by four different operators resulted in a variation of ±5% of the known delaminated area.

In summary, instrumentation and computer software have been developed so that inspection of bridge decks is a fairly routine procedure [Kunz and Eales, 1985]. Trained individuals are required to assure that meaningful data are collected and that the data are correctly interpreted. Infrared thermography is a **global** inspection method. This permits large surface areas to be scanned in a short period of time, which is an advantage over other methods that have been discussed.

19.3.5 Ground-Penetrating Radar (GPR)

Radar is analogous to the ultrasonic pulse-echo technique previously discussed, except that pulses of electromagnetic waves (short radio waves or microwaves) are used instead of stress waves. While the early uses of the technique were for military applications, radar techniques are now used in a variety of fields, such as weather, aerial mapping, and civil-engineering applications. The earliest civil-engineering applications for radar were probing into soil to detect buried pipelines and tanks. This was followed by studies to detect cavities below airfield pavements and more recently for determining concrete thickness, locating voids and reinforcing bars, and identifying deterioration [Bungey and Millard, 1993; Cantor, 1984; Carter et al., 1986; Clemeña, 1983; Kunz and Eales, 1985; Maser, 1986; Maser and Roddis, 1990; Alongi et al., 1982; Cantor and Kneeter, 1982; Steinway et al., 1981; Ulriksen, 1983]. Clemeña (1991) provides a comprehensive review of GPR.

In civil-engineering applications, relatively short distances are involved compared with other uses of radar. As a result, devices for these applications emit very short pulses of electromagnetic waves (microwaves). For this reason, the technique is often called **short-pulse radar** or **impulse radar**. Others call it **ground-penetrating radar** (GPR). In this chapter it will be called GPR. This section discusses the principles of GPR, the instrumentation that is used, and some of the inherent difficulties in using the method.

Propagation of electromagnetic waves is complex. The following presentation is simplified based on assumptions suitable for civil-engineering applications. More detailed treatments are available [Daniels et al., 1988; Halabe et al., 1993, 1995]. The operating principle of GPR is illustrated in Figure 19.19. An antenna above the test object sends out a short-duration pulse (on the order of nanoseconds) of electromagnetic waves. The pulse travels through the test object and when it

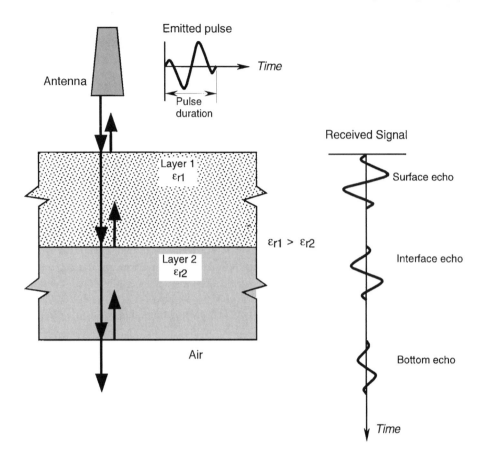

FIGURE 19.19 Reflection of radar pulse at interfaces between materials with different relative dielectric constants and antenna signal caused by arrival of echoes.

encounters an interface between dissimilar materials, some of the energy is reflected back toward the antenna as an **echo**. The antenna receives the echo and generates an output signal, as shown to the right of Figure 19.19. By measuring the time from the start of the pulse until the reception of the echo, one can determine the depth of the interface if the propagation speed through the material is known.

The amplitude of reflection at an interface depends on the difference between the relative dielectric constants[14] of the two materials and is given by the following equation [Clemeña, 1991; Bungey and Millard, 1993]:

$$\rho_{1,2} = \frac{\sqrt{\epsilon_{r1}} - \sqrt{\epsilon_{r2}}}{\sqrt{\epsilon_{r1}} + \sqrt{\epsilon_{r2}}} \qquad (19.25)$$

where

  $\rho_{1,2}$ = reflection coefficient
  ϵ_{r1} = relative dielectric constant of material 1
  ϵ_{r2} = relative dielectric constant of material 2

---

[14]The relative dielectric constant is related to the alignment of charges that occurs in an insulating material when placed in an electric field.

By definition, the relative dielectric constant of air equals 1, and typical values for other materials are as follows [ASTM D 4748]:

Material	Range of Relative Dielectric Constant
Portland cement concrete	6 to 11
Asphalt-cement concrete	3 to 5
Gravel	5 to 9
Sand	2 to 6
Rock	6 to 12
Water	8

The relative dielectric constants for materials such as concrete and soil depend on the moisture content and ionic concentrations [Morey, 1974]. Note that the dielectric constant of water is much higher than the other listed materials. This makes water the most significant dielectric contributor to construction materials and explains why radar is highly sensitive to moisture. As the moisture content increases, the dielectric constant of the material, such as concrete, also increases. Eq. (19.25) shows that when the value of ϵ_r of material 2 is greater than that of material 1, the reflection coefficient is negative. This signifies that there is a phase reversal in the reflected wave, which means that the positive part of the wave is reflected as a negative part. At a metal interface, such as between concrete and steel reinforcement, there is complete reflection, and the reflected wave has reversed polarity. This makes GPR very effective for locating metallic embedments. On the other hand, strong reflections from embedded metals can obscure weaker reflections from other reflecting interfaces that may be present, and reflections from reinforcing bars may mask signals from greater depths.

An important difference between GPR and stress-wave methods, such as the impact-echo method, is the amplitude of the reflections at a concrete-air interface. For stress waves, the reflection is almost 100% because the acoustic impedance of air is negligible compared with concrete. On the other hand, the mismatch in dielectric constants at a concrete-air interface is not as drastic, and only about 50% of the incident energy is reflected at a concrete-air interface. This results in two significant differences between GPR and stress-wave methods. GPR is not as sensitive to the detection of concrete-air interfaces as are stress-wave methods. However, because not all the energy is reflected at a concrete-air interface, GPR is able to penetrate beyond such an interface and "see" features below the interface.

The depth of the reflecting interface is obtained from the measured round trip travel time and the speed of the electromagnetic wave, C, which is dependent upon the relative dielectric constant:

$$C = \frac{C_0}{\sqrt{\epsilon_r}} \tag{19.26}$$

where

C_0 = speed of light in air (3×10^8 m/s)
ϵ_r = relative dielectric constant

If the round trip travel time is Δt, the depth, D, would be

$$D = \frac{C \Delta t}{2} \tag{19.27}$$

Equations (19.25) through (19.27) form the basis for using GPR to inspect concrete structures.

Typical instrumentation for GPR includes the following main components: an antenna unit, a control unit, a display device, and a storage device. The antenna emits the electromagnetic pulse and receives the echoes. The length of the pulse is largely controlled by the antenna design. Longer pulses are associated with longer wavelengths (or lower frequency) and have more penetrating ability, but poorer resolution (poorer ability to detect small objects), than shorter pulses. Typically, an antenna with a predominant frequency of about 1 GHz is used to inspect pavements and bridge decks, and the pulse length is about 1 ns. In air, such a pulse is about 0.3 m long and in concrete the length would depend on the value of ϵ_r. For $\epsilon_r = 6$, the pulse length would be about 120 mm. To be able to measure depths accurately, the echo must arrive after the initial pulse has ceased. Therefore the round trip travel path must exceed the pulse length. For $\epsilon_r = 6$, the minimum depth that can be measured accurately is about 60 mm. As ϵ_r increases, the minimum measurable depth decreases. The pulse is attenuated as it travels through the test object, and there is a limit to the thickness that can be inspected. For concrete, the depth of penetration would depend on the characteristics of the GPR system, the moisture content, and the amount of reinforcement. With increasing moisture content and amount of reinforcement, the penetration decreases. For dry unreinforced concrete, the maximum penetration of the pulse produced by a 1-GHz antenna is about 0.6 m [Clemeña, 1991]. In Figure 19.19, the antenna is in contact with the test objects. It is also possible to use a noncontact **horn** antenna. In this case, the received signal includes an echo from the concrete surface.

The control unit is the heart of a GPR system. It controls the repetition frequency of the pulse, provides the power to emit the pulse, acquires and amplifies the received signal, and provides output to a display device. Data are usually stored in an analog recorder and played back for later analysis and interpretation.

Display devices include oscillographs, which plot the recorded wave forms as a **waterfall plot**, or graphic-facsimile recorders. As an example, Figure 19.20 shows a waterfall plot obtained by plotting the received wave forms next to each other. The plot takes on a topographic appearance, and changes in the pattern of the received signals are relatively easy to identify [Cantor, 1984]. Computer software is also available that permits sophisticated signal processing of the data to aid in interpretation. The operation of the graphic recorder is discussed further here because it is commonly used in the field. Figure 19.21a shows an antenna emitting a radar pulse into a test object

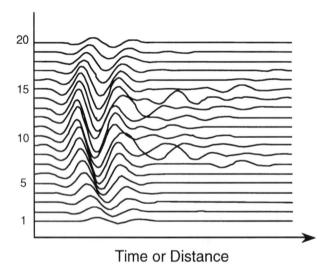

Time or Distance

FIGURE 19.20 Waterfall plot of radar wave forms as on antenna is scanned across the surface of the test object [adapted from Cantor, 1984].

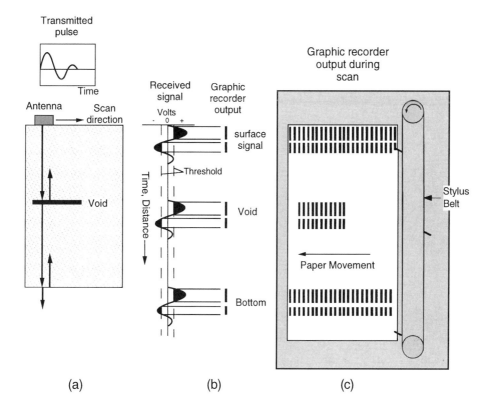

FIGURE 19.21 Schematic to illustrate peak-plotting technique used with graphic recorder: (a) reflections of electromagnetic pulse in test object; (b) received wave form threshold limits; (c) graphic recorder output during surface scan of test object.

containing a void. The shape of the emitted pulse is sketched above the antenna. Figure 19.21b shows the wave form from the receiving side of the antenna. The vertical axis represents time, or it can be transformed to depth by knowing ϵ_r and using Eq. (19.27). The received signal at the start of the waveform represents the transmitted pulse, which is picked up directly by the receiving side of the antenna. The second received signal is the echo from the void, and the third is the echo from the bottom boundary of the test object.

The output of the graphic recorder is obtained by a technique known as **peak plotting**, which is illustrated in Figure 19.21b. First, the operator selects a threshold voltage range. When the amplitude of the received signal goes beyond the threshold range, the pen of the graphic recorder plots a solid line on recording paper. The line is plotted in varying shades of gray, depending on the actual amplitude of the signal. Thus the antenna output is represented on the graphic recorder as a series of dashes as shown in Figure 19.21b. Note that each echo is associated with two dashes. The actual number of dashes depends on the number of cycles in the emitted pulse and the threshold level. This is an important point to understand for proper interpretation of GPR results. As the antenna is moved along the surface, the output is displayed on the graphic recorder. The paper on the recorder moves at a constant speed that is independent of the speed of the antenna motion. The resulting **picture** on the graphic recorder represents a cross-sectional view of the test object, as illustrated in Figure 19.21c. The test equipment provides a means for correlating the position of the antenna during the scan with the location on the paper record. Thus it is possible to determine the depth and approximate size of the reflecting interface.

As was already mentioned, metals are strong reflectors of electromagnetic waves. This makes GPR very effective for locating buried metal objects such as reinforcing bars and conduit. Reinforcing

Scan Direction

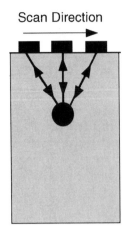

Graphic Recorder Output

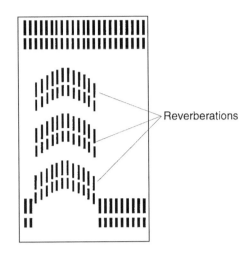

FIGURE 19.22 Schematic to illustrate characteristic graphic recorder output during scan over a reinforcing bar.

bars result in characteristic patterns in the graphic-recorder output, which make them relatively easy to locate. There are two main factors leading to these patterns, which are discussed here with the aid of Figure 19.22. First of all, an antenna behaves like a flashlight in that the beam of radiation has a conical shape. Thus a reinforcing bar produces echoes before the antenna passes directly over the bar; however, the apparent depth of the bar will be larger than the actual cover because of the inclined travel path. Secondly, the pulse undergoes multiple reflections, or reverberations, between the reinforcing bar and the surface, and the output will show multiple echoes from the same bar. The resulting characteristic pattern is shown on the right side of Figure 19.22. The top of the bar is associated with the uppermost part of the arched bands.

When multiple reinforcing bars are present, there will be multiple arch patterns. As the bars become closer together, the arch patterns overlap. Below a certain spacing, the individual bars can no longer be discerned, and the echo pattern is similar to the case of a solid embedded steel plate. The ability to discern individual bars depends on bar size, bar spacing, cover depth, and the configuration of the antenna [Bungey et al., 1994]. Closely spaced bars will also prevent detection of features below the layer of reinforcement. This **masking** effect depends on the wavelength of the electromagnetic waves, the bar size, and cover depth. It has been found that for the 1-GHz, hand-held antenna, 32-mm bars at cover depths of 25 to 50 mm will prevent detection of underlying features when bar spacings are less than 200 mm [Bungey et al., 1994].

Simple methods do not exist to determine bar size from graphic-recorder output. Researchers have attempted to better understand the interactions of GPR with cracks, voids, and reinforcing bars [Mast et al., 1990]. The objective of these studies is to develop procedures to use the recorded data to construct an image of the interior of the concrete. Some advances in this direction have occurred in France, where a prototype system has been developed that uses tomographic techniques to reconstruct the two-dimensional layout of reinforcement in concrete [Pichot and Trouillet, 1990]. In the United States, successful three-dimensional reconstruction of artificial defects and reinforcement embedded in a concrete slab have been demonstrated [Mast and Johansson, 1994]. These imaging methods rely on an antenna array to make multiple measurements and require extensive computational time to reconstruct the internal image.

The detection of delaminations in reinforced bridge decks using GPR is not straightforward. In studies by Maser and Roddis (1990), it was found that a 3-mm air gap in concrete produced little noticeable effect in the received wave form. However, the addition of moisture to the simulated crack resulted in stronger reflections that were noticeable in the wave form. It was also found that the presence of chlorides in moist concrete resulted in high attenuation, because of the increased relative dielectric constant. Thus the ability to detect delaminations will depend on the **in situ** conditions at the time of testing. In addition, the reflections from reinforcing bars are much stronger than from a delamination, and it is difficult to "see" the delamination.

Owing to difficulties in using GPR in reinforced concrete, standardized test procedures for flaw detection do not exist. However, an ASTM standard has been developed (Test Method D 4748) to measure the thickness of the upper layer of a multilayer pavement system. Basically, the technique involves measuring the transit time of the pulse though the pavement layer, and using relationships similar to Eqs. (19.26) and (19.27) to calculate the layer thickness. The procedure is based on using a noncontact horn antenna, and some modifications are required for measurements with a contact antenna. The calculated depth depends on the value of the relative dielectric constant. Errors in the assumed value of the relative dielectric constant can lead to substantial inaccuracies in depth estimations [Bungey et al., 1994]. For data obtained with a horn antenna, the relative dielectric constant of the concrete can be computed directly from the radar signals. For data obtained using a contact antenna, it is necessary to take occasional cores to determine the appropriate value for the pavement materials. The user is cautioned against using the method on saturated concrete because of the high attenuation and limited penetration of the pulse. In ASTM D 4748, it is stated that interoperator testing of the same materials resulted in thickness measurements within ± 5 mm of the actual thickness. Finally, it is noted that reliable interpretation of received signals can only be performed by an experienced data analyst.

While the majority of the applications of GPR have dealt with locating reinforcing bars in structures, locating delaminations in bridge decks, and measuring the thickness of pavement layers, there are other potential uses. Since the dielectric properties of a material like concrete are strongly dependent on the moisture content, microwave measurements can be used to monitor the progress of hydration [Otto et al., 1990; Clemeña, 1991]. This is made possible because the relative dielectric constant of free water is much higher than that of chemically bound water. Clemeña (1991) has also reported on potential applications of microwave measurements to determine water content of fresh concrete.

19.3.6 Electrical and Magnetic Methods for Reinforcement

Information about the quantity, location, and condition of reinforcement is needed to evaluate the strength of reinforced concrete members. This section discusses some of the magnetic and electrical methods that are used to gain information about embedded-steel reinforcement. Additional information may be found in the following references: Malhotra (1976), Bungey (1989), and Lauer (1991).

19.3.6.1 Covermeters

Devices to locate reinforcing bars and estimate the diameter and depth of cover are known as **covermeters**. These devices are based on interactions between the bars and low-frequency electromagnetic fields. The physical principle that is employed is that of **electromagnetic induction**, whereby an alternating magnetic field induces an electrical potential in an electrical circuit intersected by the field. Commercial covermeters can be divided into two classes: those based on the principle of **magnetic reluctance** and those based on **eddy currents**. These differences are summarized below [Carino, 1992b].

19.3.6.2 Magnetic-Reluctance Meters

When current flows through an electrical coil, a magnetic field is created and there is a flow of magnetic flux lines between the magnetic poles. This leads to a magnetic circuit, in which the flow of magnetic flux between poles is analogous to the flow of current in an electrical circuit [Fitzgerald

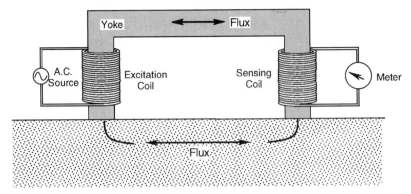

(a) Small current induced in sensing coil when no bar is present

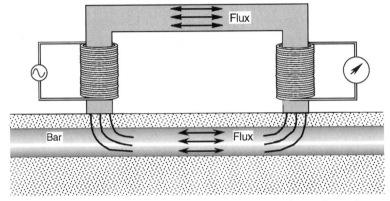

(b) Presence of bar increases flux and increases current in sensing coil

FIGURE 19.23 Covermeter based on the principle of magnetic reluctance [adapted from Carino, 1992b].

et al., 1967]. The resistance to flow of magnetic flux is called **reluctance**, which is analogous to the resistance to flow of current in an electrical circuit. Figure 19.23 is a schematic of a covermeter that is based on changes in the reluctance of a magnetic circuit caused by the presence or absence of a bar within the vicinity of the search head. The search head is composed of a ferromagnetic U-shaped core (yoke), an excitation coil, and a sensing coil. When alternating current (less than 100 Hz) is applied to the excitation coil, an alternating magnetic field is created and magnetic flux flows between the poles of the yoke. In the absence of a bar (Figure 19.23a), the magnetic circuit, composed of the yoke and the concrete between the ends of the yoke, has a high reluctance and the alternating magnetic flux flowing between the poles will be small. The alternating flux induces a small, secondary current in the sensing coil. If a ferromagnetic bar is present (Figure 19.23b), the reluctance decreases, the magnetic flux amplitude increases, and the sensing-coil current increases. Thus the presence of the bar is indicated by a change in the output from the sensing coil. For a given reinforcing bar, the reluctance of the magnetic circuit depends strongly on the distance between the bar and the poles of the yoke. An increase in concrete cover increases the reluctance and reduces the current in the sensing coil. If the meter output were plotted as a function of the cover, a calibration relationship would be established that could be used to measure the cover. Since the size of the bar affects the reluctance of the magnetic circuit, there would be a separate relationship for each bar size.

19.3.6.3 Eddy-Current Meters

If a coil carrying an alternating current is brought near an electrical conductor, the changing magnetic field induces circulating currents in the conductor. These are known as **eddy currents**.

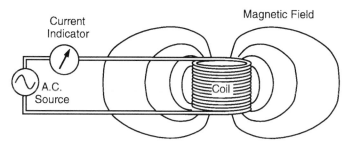

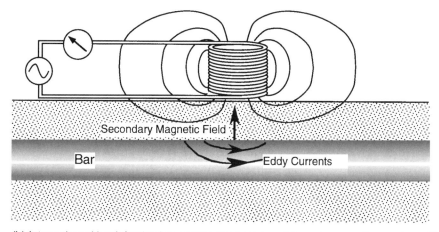

(a) Coil in air results in a characteristic current amplitude

(b) Interaction with reinforcing bar causes changes in coil impedance and current amplitude

FIGURE 19.24   Covermeter based on eddy-current principle [adapted from Carino, 1992b].

Because any current flow gives rise to a magnetic field, eddy currents produce a secondary magnetic field that interacts with the field of the coil. The second class of covermeters is based on monitoring the effects of the eddy currents induced in a reinforcing bar. Figure 19.24 is a schematic of a continuous eddy-current covermeter. In the absence of a reinforcing bar, the magnitude of the alternating current (usually at about 1 kHz) in the coil depends on the coil impedance.[15] If the coil is brought near a reinforcing bar, alternating eddy currents are established within the surface of the bar. The eddy currents give rise to an alternating secondary magnetic field that induces a secondary current in the coil. In accordance with Lenz's law [Serway, 1983], the secondary current opposes the primary current. As a result, the net current flowing through the coil is reduced, and the apparent impedance of the coil increases [Hagemaier, 1990]. Thus the presence of the bar is inferred by monitoring the change in current flowing through the coil.

19.3.6.4 Characteristics

A reinforcing bar is detected by a covermeter when the bar lies within the zone of influence of the search head (yoke or coil). The response is maximum when the search head lies directly above the reinforcing bar. An important characteristic of a covermeter is the relationship between meter amplitude and the horizontal distance from the center of the bar to the center of the search head. The variation has approximately the same shape as the bell-shaped curve of a normal probability

---

[15] When direct current is applied to a circuit, the amount of current equals the voltage divided by the electrical resistance of the circuit. When alternating current is applied to the coil, the amount of current is governed by the value of the applied voltage, the resistance, and another quantity called **inductance**. The vector sum of resistance and inductance defines the **impedance** of the coil.

distribution. The width of the curve defines the **zone of influence** of the search head. A search head with a smaller zone of influence is better able to discern individual bars when they are closely spaced than is a search head with a wider zone of influence. However, focused search heads generally have less penetrating ability. The influence zone of the search head also affects the accuracy when trying to detect the end of a reinforcing bar [Carino, 1992b].

An important distinction between covermeters is the directionality characteristics of the search heads. Owing to the shape of the yoke, a magnetic reluctance meter is directional compared with a continuous eddy-current meter with a symmetrical coil. Maximum response occurs when the yoke is aligned with the axis of the bar. This directionality can be used to advantage when testing a structure with an orthogonal grid of reinforcing bars [Tam et al., 1977].

For a given covermeter, there are unique relationships between meter amplitude and depth of cover. Figures 19.25a and 19.25b show these relationships for a magnetic reluctance and for an eddy-current meter, respectively. These relationships illustrate a basic limitation of covermeters. Since the amplitude is a function of bar diameter and depth of cover, one cannot determine both parameters from a single measurement. As a result, dual measurements are needed to be able to estimate both depth of cover and diameter [BS 1881; Das Gupta and Tam, 1983]. This is done by first

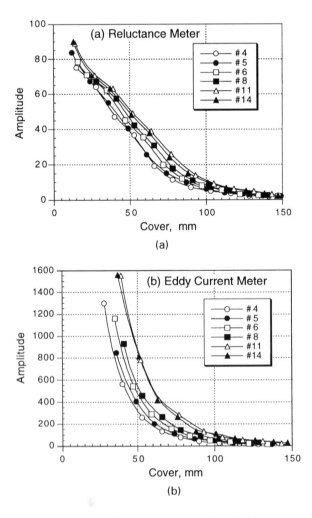

FIGURE 19.25 Amplitude versus cover: (a) results for magnetic reluctance meter, and (b) results for eddy-current meter [adapted from Carino, 1992b].

recording the meter amplitude with the search head in contact with the concrete and then when the search head is located a known distance above the concrete. The difference in amplitudes and the amplitude-cover relationships are used to estimate the cover and bar diameter. The accuracy of this spacer technique depends on how distinct the amplitude-cover relationships are for the different bar sizes. Because these relationships are generally similar for adjacent bar sizes, it is generally only possible to estimate bar diameter within two sizes [Bungey, 1989].

The single-bar, amplitude-cover relationships are only valid when the bars are sufficiently far apart that there is little interference by adjacent bars. For multiple, closely spaced bars, the amplitude may exceed the amplitude for a single bar at the same cover depth. If they are closer than a critical amount, the individual bars cannot be discerned. The critical spacing depends on the type of covermeter and the cover depth. In general, as cover increases, the critical spacing also increases [Carino, 1992b]. Since the response of a covermeter to the presence of multiple, closely spaced bars depends on its design, the user should follow the manufacturer's recommendations regarding minimum bar spacings.

The presence of two layers of reinforcement within the zone of influence cannot generally be identified with ordinary covermeters [Bungey, 1989; Carino, 1992b]. The upper layer produces a much stronger signal than the deeper second layer so that the presence of the second layer cannot be discerned. However, it has been shown that it may be possible to determine lap length when bars are in contact [Carino, 1992b].

In summary, covermeters are effective for locating individual bars, provided that the spacing exceeds a critical value that is dependent on the meter design and the cover depth. By using multiple measurement methods, bar diameter can generally be estimated to within two adjacent bar sizes if the spacing exceeds certain limits that are also dependent on the particular meter. Meters are available that can estimate bar diameter without using spacers to make multiple measurements. Again, the accuracy of these estimates decreases as bar spacing decreases. To obtain reliable measurements, it is advisable to prepare mock-ups of the expected reinforcement configuration to establish whether the desired accuracy is feasible. These mock-ups can be made without using concrete [Carino, 1992b; BS 1881], provided the in-place concrete does not contain significant amounts of iron-bearing aggregates.

19.3.6.5 Corrosion Activity

Electrical methods are used to evaluate corrosion activity of steel reinforcement. As is the case with other NDT methods, an understanding of the underlying principles of these electrical methods is needed to obtain meaningful results. In addition, an understanding of the factors involved in the corrosion mechanism is essential for reliable interpretation of data. The subsequent sections provide basic information about these methods. However, because of the complex interaction of factors, a corrosion specialist should be consulted in planning an investigation.

Corrosion is an electrochemical process involving the flow of charges (electrons and ions). At active sites on the bar, called **anodes**, iron atoms lose electrons and move into the surrounding concrete as ferrous ions. This process is called a **half-cell oxidation reaction**, or the anodic reaction, and is represented as follows:

$$Fe \rightarrow Fe^{2+} + 2e^- \tag{19.28}$$

The electrons remain in the bar and flow to sites, called **cathodes**, where they combine with water and oxygen that are present in the concrete. The reaction at the cathode is called a **half-cell reduction reaction** and is represented as follows:

$$2H_2O + O_2 + 4e^- \rightarrow 4OH^- \tag{19.29}$$

To maintain electrical neutrality, the ferrous ions migrate through the concrete to these cathodic sites where they combine with water and oxygen to form hydrated iron oxide, or rust. Thus, when the bar is corroding, there is a flow of electrons through the bar and a flow of ions through the concrete. When the bar is not corroding, there is no flow of electrons and ions.

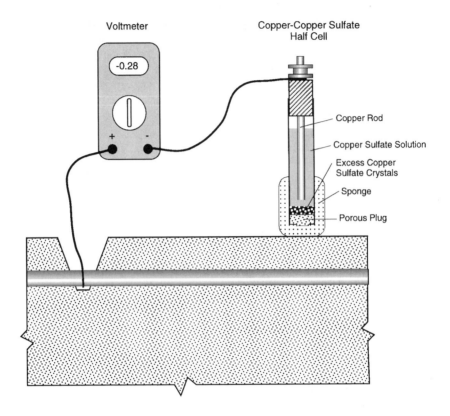

FIGURE 19.26 Apparatus for half-cell potential method described in ASTM C 876.

As the ferrous ions move into the surrounding concrete, the electrons that are left behind in the bar give the bar a negative charge. The **half-cell potential method** is used to detect this negative charge and thereby provide an indication of corrosion activity.

19.3.6.6 Half-Cell Potential Method

Figure 19.26 is a schematic of the apparatus given in ASTM C 876. The apparatus includes a copper-copper sulfate half cell[16], connecting wires, and a high-impedance voltmeter. The positive terminal of the voltmeter is attached to the reinforcement and the negative terminal is attached to the copper-copper sulfate half cell. A high-impedance voltmeter is used so that very little current flows through the circuit. The half-cell makes electrical contact with the concrete by means of a porous plug and a sponge that is moistened with a wetting solution (such as liquid detergent).

If the bar is corroding, electrons would tend to flow from the bar to the half-cell. At the half-cell, the electrons are consumed in a reduction reaction that transforms copper ions in the copper sulfate solution into copper atoms deposited on the rod. Because of the way the terminals of the voltmeter are connected, the voltmeter would indicate a negative value. The more negative the voltage reading, the higher the likelihood that the bar is corroding. The half-cell potential is also called the **corrosion potential** and it is an open-circuit potential because it is measured under the condition of no current in the measuring circuit [ASTM G 15].

---

[16]This half-cell is composed of a copper bar immersed in a saturated copper sulfate solution. It is one of many half-cells that can be used as a reference to measure the electrical potential of embedded bars. The measured voltage depends on the type of half-cell, and conversion factors are available to convert readings obtained with other references cells to the copper-copper sulfate half-cell.

The half-cell potential readings are indicative of the probability of corrosion activity of reinforcement located beneath the reference cell. However, this is true only if all the reinforcement is electrically connected. To assure that this condition exists, electrical resistance measurements between widely separated reinforcing bars should be carried out [ASTM C 876]. This means that access to the reinforcement has to be provided. The method cannot be applied to concrete with epoxy-coated reinforcement.

A key aspect of the test is assuring that the concrete is sufficiently moist. If the measured potential at a test point does not change by more than ± 20 mV within a 5-min period [ASTM C 876], the concrete is sufficiently moist. If this condition is not satisfied, the concrete surface must be wetted; two methods are given in ASTM C 876.

According to ASTM C 876, two techniques can be used to evaluate the results: (1) the **numeric** technique or (2) the **potential-difference** technique. In the numeric technique, the value of the potential is used as an indicator of the likelihood of corrosion activity. If the potential is more positive than -200 mV, there is a high likelihood that no corrosion is occurring at the time of the measurement. If the potential is more negative than -350 mV, there is a high likelihood that there is active corrosion. Corrosion activity is uncertain when the voltage is in the range of -200 to -350 mV. ASTM C 876 also states that, unless there is positive evidence to suggest their applicability, these numeric criteria should **not** be used under the following conditions:

- carbonation extends to the level of the reinforcement,
- to evaluate indoor concrete that has not been subjected to frequent wetting,
- to compare corrosion activity in outdoor concrete with highly variable moisture or oxygen content, and
- to formulate conclusions about changes in corrosion activity due to repairs that changed the moisture or oxygen content at the level of the steel.

In the potential-difference technique, the areas of active corrosion are identified on the basis of half-cell potential gradients. An equipotential contour map is created by locating the test locations on a scaled plan view of the test area. The half-cell voltage readings at each test point are marked on the plan, and contours of equal voltage values are sketched. Regions of corrosion activity are indicated by closely spaced contours.

As has been stated, valid potential readings are possible only if the concrete is sufficiently moist, and the user must understand how to recognize when there is insufficient moisture. Because of the factors involved in corrosion testing, a corrosion specialist is recommended to properly interpret half-cell potential surveys under the following conditions [ASTM C 876]:

- the concrete is saturated with water,
- the concrete is carbonated to the depth of the reinforcement, or
- the steel is coated (galvanized).

In addition, potential surveys should be supplemented with tests for carbonation and water-soluble chloride content. A major limitation of the half-cell potential method is that it does **not** measure the **rate of corrosion**. It only provides an indication of the **likelihood** of corrosion activity at the time the measurement is made. The corrosion rate of the reinforcement depends on the availability of oxygen that is needed for the cathodic reaction. It also depends on the electrical resistance of the concrete that controls the ease with which ions can move through the concrete. The electrical resistance depends on the microstructure of the paste and the moisture content of the concrete.

19.3.6.7 Linear Polarization

The major drawback of the half-cell potential method has lead to the development of several techniques to measure the rate of corrosion [Rodriguez et al., 1994]. The **linear polarization method** is the approach used most frequently in the field [Flis et al., 1992], and efforts were begun for standardization [Cady and Gannon, 1992]. This section provides an overview of the method.

In the field of corrosion science, the term **polarization** refers to the change in the open-circuit potential as a result of the passage of current [ASTM G 15]. In the polarization resistance test, the current necessary to cause a small change in the value of the half-cell potential of the corroding bar is measured. For a small perturbation about the open-circuit potential, there is a linear relationship between the change in voltage, ΔE, and the change in **current per unit area** of bar surface, Δi. This ratio is called the **polarization resistance**, R_p:

$$R_p = \frac{\Delta E}{\Delta i} \tag{19.30}$$

Because the current is expressed per unit surface area of bar that is polarized, the units of R_p are ohms times area. It has been pointed out that R_p is not a resistance in the usual sense of the term [Stern and Roth, 1957], but the term is widely used [ASTM G 15]. The underlying relationships between the corrosion rate of the bar and the polarization resistance were established by Stern and Geary (1957). No attempt is made to explain these relationships, but in simple terms, the corrosion rate is inversely related to the polarization resistance. The corrosion rate is usually expressed as the a corrosion current per unit area of bar, and it is determined as follows:

$$i_{corr} = \frac{B}{R_p} \tag{19.31}$$

where

i_{corr} = corrosion rate in ampere per square centimeter
B = a constant in volts
R_p = polarization resistance in ohms square centimeter

The constant B is a characteristic of the corrosion system and a value of 0.026 V is commonly used for corrosion of steel in concrete [Feliu et al., 1989]. It is possible to convert the corrosion rate into the mass of steel that corrodes per unit of time, and if the bar size is known, it can be converted to a loss in diameter of the bar [Clear, 1989].

Figure 19.27 is a schematic of basic apparatus for measuring the polarization resistance [Escalante, 1989; Clear, 1989]. It is composed of three electrodes. One electrode is composed of a reference half-cell, and the reinforcement is a second electrode called the **working electrode**. The third electrode is called the **counter electrode**, and it supplies the polarization current to the bar. Supplementary instrumentation measures the voltages and currents during different stages of the test. Such a device can be operated in the **potentiostatic** mode, in which the current is varied to maintain constant potential of the working electrode; or it can be operated in the **galvanostatic** mode, in which the potential is varied to maintain constant current from the counter electrode to the working electrode.

The procedure for using such a three-electrode device to obtain the polarization resistance was provided by Cady and Gannon (1992). The basic steps are as follows:

- Locate the reinforcing-steel grid with a covermeter and mark it on the concrete surface.
- Make an electrical connection to the reinforcement (the working electrode).
- Locate the bar whose corrosion rate is to be measured, wet the surface, and locate the device over the center of the bar.
- Measure the half-cell potential of the reinforcement relative to the reference electrode (Figure 19.27b).
- Measure the current from the counter electrode to the working electrode that is necessary to produce a −4 mV change in the potential of the working electrode (Figure 19.27b).
- Repeat the previous step for values of potential of −8 and −12 mV beyond the corrosion potential.

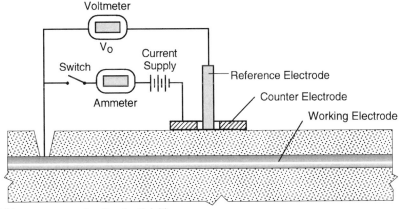

(a) Measure open-circuit potential, V_O

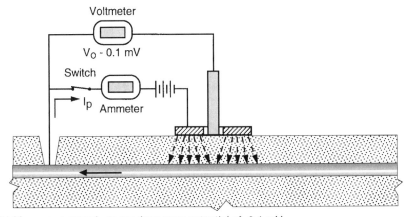

(b) Measure current I_D to produce over potential of -0.1 mV

FIGURE 19.27   Three-electrode, linear polarization method to measure corrosion current.

- Determine the surface area of bar that is affected by the measurement (perimeter of bar multiplied by the length below the counter electrode).
- Plot the potential versus the current per unit surface area of the bar, and determine the slope of the best-fit straight line. This is the polarization resistance.

A major uncertainty in obtaining the polarization resistance is the area of the steel bar that is affected by the current flowing from the counter electrode. In the application of the three-electrode device, it is assumed that current flows in straight lines perpendicular to the bar (working electrode) and the counter electrode. Thus the bar area affected during the tests is the bar circumference multiplied by the length of the bar below the counter electrode. However, numerical simulations show that the above assumption is incorrect and that the current lines are not confined to the region directly below the counter electrode [Feliu et al., 1989; Flis et al., 1992]. In an effort to better control the current path from the counter electrode to the bar, a device has been developed that includes a fourth electrode, called a **guard** or **auxiliary electrode**, that surrounds the counter electrode [Feliu et al., 1990a,b]. The guard electrode is maintained at the same potential as the counter electrode and, as a result the current flowing to the working electrode, is confined to the region below the counter electrode.

The corrosion rate based on measuring the polarization resistance represents the corrosion rate at the time of the test. The corrosion rate at a particular point in a structure is expected to depend on several factors, such as the moisture content of the concrete, the availability of oxygen, and the temperature. Thus the corrosion rate at any point in an exposed structure would be expected to have seasonal variations. Such variations were observed during multiple measurements that extended over a period of more than one year [Clemeña et al., 1992]. To project the amount of corrosion that would occur after an extended period, it is necessary to repeat the corrosion-rate measurements at different times of the year.

At this time, there are no standard procedures for interpreting corrosion-rate measurements obtained with different devices, and a qualified corrosion specialist should be consulted. For example, based on years of experience from laboratory and field testing, Clear (1989) developed guidelines for interpreting measurements obtained with a corrosion-rate device.

There are other limitations that should be considered when planning corrosion rate testing. Some of these have been outlined by Cady and Gannon (1992):

- The concrete surface has to be smooth (not cracked, scarred, or uneven).
- The concrete surface has to be free of water-impermeable coatings or overlays.
- The cover depth has to be less than 100 mm.
- The reinforcing steel can not be epoxy coated or galvanized.
- The steel to be monitored has to be in direct contact with the concrete.
- The reinforcement can not be cathodically protected.
- The reinforced concrete must not be near areas of stray electric currents or strong magnetic fields.
- The ambient temperature must be between 5 and 40°C.
- The concrete surface at the test location must be free of visible moisture.
- Test locations must not be closer than 300 mm to discontinuities, such as edges and joints.

19.3.7 Nuclear (Radioactive) Methods

Nuclear (or radioactive) methods for nondestructive evaluation of concrete involve the use of high-energy electromagnetic radiation to gain information about the internal structure of the test object. These involve a source of penetrating electromagnetic radiation and a sensor to measure the intensity of the radiation after it has traveled through the object. If the sensor is in the form of special photographic film, the technique is called **radiography**. If the sensor is an electronic device that converts the incident radiation into electrical pulses, the technique is called **radiometry**. A review of the early developments in the use of nuclear methods was presented by Malhotra (1976), and more recent developments were reviewed by Mitchell (1991).

Initial work in the late 1940s focused on the use of X rays to produce radiographs that revealed the internal structure of concrete elements, but in the 1950s attention turned to the use of gamma rays. The fundamental differences between these two forms of penetrating radiation are the sources used to generate them and their penetrating ability. X rays are produced by high-voltage electronic devices, and gamma rays are produced by the byproducts of the disintegration of radioactive isotopes. The penetrating ability of gamma rays depends on the radioactive isotope and its age, while the penetration of X rays depends on the voltage of the generating instrument.

Some of the earliest reported work using gamma rays was at Ontario Hydro [de Hass, 1954]. Slabs were constructed with artificial flaws, a pipe containing a radioactive isotope of cobalt was placed beneath the slab, and a Geiger-Müller tube was placed on the top surface of the slab to measure the intensity of radiation. Other early efforts at using gamma ray methods took place in Great Britain during the 1950s, where they were used to locate reinforcing bars, measure density, and locate voids in grouted posttensioning ducts [Forrester, 1970]. Eastern Europe and the Soviet

Union also conducted early studies that eventually led to the development of portable density meters for concrete and soil.

19.3.7.1 Radiometric Methods

There are two basic radiometric methods that use X-rays and gamma rays in nondestructive testing of concrete. In the **transmission** method, the amplitude of the radiation passing through a member is measured. As the radiation passes through a member, the attenuation is dependent on the density of the material and the path length from the source to the sensor. Direct transmission techniques can be used to detect reinforcement. However, the main use of the technique is to measure the in-place density, both in fresh and hardened concrete. Structures of heavyweight and roller-compacted concretes are cases where this method is of particular value. For such applications, the radioactive source is contained in a tube that is pushed into the fresh concrete and the detector is set on the surface of the concrete. The density meter developed at the Technical University of Brno (Czechoslovakia) is an example of such a device. The source can be lowered up to a depth of 200 mm into a hollow steel needle that is pushed into the fresh concrete. A spherical lead shield suppresses the radiation when the source is in its retracted position. Detectors are located beneath the treads used to push the needle into the concrete. It is claimed that the instrument has a resolution of 10 kg/m^3 [Hönig, 1984]. ASTM C 1040 provides procedures for using nuclear methods to measure the in-place density of fresh or hardened concrete. The key element of the procedure is development of the calibration curve for the instrument. This is accomplished by making test specimens of different densities and determining the gauge output for each specimen. The gauge output is plotted as a function of the density, and a best-fit curve is determined.

In the **backscatter method**, a radioactive source is used to supply gamma rays, and a detector close to the source is used to measure the backscattered rays. The scattered rays are lower in energy than the transmitted ones and are produced when a photon collides with an electron in an atom. Part of the photon energy is imparted to the electron, and a new photon emerges, traveling in a new direction with lower energy. This process is known as Compton scattering [Mitchell, 1991]. Backscatter techniques are particularly suitable for applications where a large number of *in situ* measurements are required. Since backscatter measurements are affected by the top 40 to 100 mm, the method is best suited for measurement of the surface zone of a concrete element. A good example of the use of this method is the monitoring of the density of bridge deck overlays. Noncontacting equipment has been developed that is used for continuous monitoring of concrete pavement density during slip-form operations [Mitchell et al., 1979].

Procedures for using backscatter methods to measure concrete density are given in ASTM C 1040. As is the case with direct transmission measurements, it is necessary to establish a calibration curve prior to using a nuclear backscatter gauge to measure in-place density. The inherent precision of backscatter density gauges is less than that of direct transmission devices. ASTM C 1040 requires that a suitable backscatter gauge for density measurement result in a standard deviation of less than 16 kg/m^3, while the standard deviation should be less than 8 kg/m^3 for a direct transmission gauge. According to ASTM C 1040, backscatter gauges are typically influenced by the top 75 to 125 mm of material. The top 25 mm determines 50 to 70% of the count rate, and the top 50 mm determines 80 to 95% of the count rate.

19.3.7.2 Radiographic Methods

Radiography provides a radiation-based photograph of the interior of concrete. From this photograph, the location of reinforcement, voids in concrete, or voids in grouting of posttensioning ducts can be identified. A radiation source is placed on one side of the test object and a beam of radiation is emitted. As the radiation passes through the member, it is attenuated by different amounts, depending on the density and thickness of the material that is traversed. The radiation that emerges from the opposite side of the object strikes a special photographic film that is exposed in proportion to the intensity of the incident radiation. When the film is developed, a two-dimensional visualization (a photograph) of the interior structure of the object is obtained. The presence of high-density material, such as reinforcement, is shown on the developed film as a light area, and a region of low density, such as a void, is shown as a dark area.

The British Standards Institute has adopted a standard for radiographic testing of concrete, BS 1881: Part 205 (Recommendations for radiography of concrete). The standard provides recommendations for investigators considering radiographic examinations of concrete [Mitchell, 1991].

In X-radiography, the radiation is produced by an X-ray tube [Mitchell, 1991]. The penetrating ability of the X rays is dependent on the operating voltage of the X-ray tube. In gamma radiography, a radioactive isotope is used as the radiation source. The selection of a source depends on the density and thickness of the test object and on the exposure time that can be tolerated. The most intense source is cobalt-60, which can be used to penetrate up to 500 mm of concrete. For members with thickness of 150 mm or less, iridium-192 or cesium-137 can be used [Mitchell, 1991]. The film type will depend on the thickness and density of the member being tested.

Most field applications have used radioactive sources because of their greater penetrating ability (higher energy radiation) compared with X rays. However, a system, known as Scorpion II, was developed in France that uses a linear accelerator to produce very high energy X rays than can penetrate up to 1 m of concrete. This system was developed for the inspection of prestressed members to establish the condition and location of prestressing strands and to determine the quality of grouting in tendon ducts [Mitchell, 1991].

19.3.7.3 Summary

While nuclear methods have the ability to "see" into concrete, they are cumbersome and require trained and licensed personnel [Mitchell, 1991]. Testing across the full thickness of a concrete element is particularly hazardous and requires extensive precautions, skilled personnel, and highly specialized equipment. Radiographic procedures are costly and require evacuation of the structure by persons not involved in the actual testing. The use of X-ray equipment poses an additional danger owing to the high voltages that are used. There are limits on the thicknesses of the members that can be tested by radiographic methods. For gamma-ray radiography the maximum thickness is about 500 mm, because thick members require unacceptably long exposure times. Radiography is not very useful for locating crack planes perpendicular to the radiation beam. Because of these major drawbacks, radiographic methods are not used routinely for flaw detection. However, there may be situations where the ability to see the internal structure of the member surpasses these drawbacks.

19.4 Concluding Remarks

This chapter has summarized the available nondestructive techniques for assessing the properties or condition of concrete in structures. The techniques have been divided into two groups:

- those used for estimating the in-place strength, and
- those used for flaw detection and condition assessment.

Emphasis has been placed on describing their underlying principles and highlighting some of their inherent limitations. The user is referred to applicable publications of the American Concrete Institute and relevant ASTM standards for additional information on using these methods.

The key feature of the methods for estimating in-place strength is the strength relationship that correlates the concrete strength to the results of the in-place tests. The strength relationship should be developed experimentally before using the test method to estimate in-place strength. For new construction, test specimens should be made of concrete similar to what will be used in the structure. Care must be exercised to ensure that the companion in-place tests and standard strength tests are carried out on specimens of the same maturity at each strength level. For existing construction, it is necessary to perform in-place tests and obtain cores at different locations so that a wide range of concrete strengths can be used to develop the strength relationship. After the strength relationship has been established, in-place tests are done on the structure, and statistical methods are used to convert the average of in-place test results to a reliable estimate of in-place strength. Generally, in-place test methods that result in local failure of the concrete are more reliable than those that are totally nondestructive.

A variety of methods are available for flaw detection and condition assessment. Most of these methods are based on monitoring the response of the structure when it is subjected to some type of disturbance. Two broad classes of nondestructive methods are those based on stress-wave propagation and those based on electromagnetic-wave propagation. Except for visual inspection, these methods generally require sophisticated instrumentation. All nondestructive test methods have inherent strengths and weaknesses. It is often advantageous to use more than one method to make the assessment. Methods based on stress-wave propagation are suited for identifying the presence of internal concrete-air interfaces, such as those due to cracking or voids. An understanding of the basics of stress-wave propagation is essential for proper interpretation of test results. Electrical methods are well suited for gaining information about embedded reinforcement, such as location, approximate size, and whether active corrosion exists. Radar is appropriate for finding deep metallic embedments and is also sensitive to the presence of moisture. Radar has the added advantage that large portions of a structure can be scanned in a short time.

The importance of having qualified operators cannot be overemphasized. Nondestructive tests are indirect methods by which the property or characteristic of primary interest is inferred by measuring other properties or characteristics. A lack of understanding of the underlying principles and the interferences associated with the method can lead to incorrect assessments of the concrete. When used by properly trained operators, nondestructive test methods offer technical and economic advantages compared with other destructive sampling techniques.

References

ACI Reports, available from American Concrete Institute, Farmington Hills, MI:

    201.1R: Guide for Making a Condition Survey of Concrete in Service

    207.3R: Practices for Evaluation of Concrete in Existing Massive Structures for Service Conditions

    224.1R: Causes, Evaluation and Repair of Cracks in Concrete Structures

    228.1R(1989): In-Place Methods for Determination of Strength of Concrete

    228.1R(1995): In-Place Methods to Estimate Concrete Strength

    362R: State-of-the-Art Report on Parking Structures

    437R: Strength Evaluation of Existing Concrete Buildings

ASTM Standards, available from ASTM, West Conshohocken, PA

    1995 Annual Book of ASTM Standards, Vol. 04.02

    C 597: Standard Test Method for Pulse Velocity Through Concrete

    C 803: Standard Test Method for Penetration Resistance of Hardened Concrete

    C 805: Standard Test Method for Rebound Number of Hardened Concrete

    C 876: Test Method for Half-Cell Potential of Uncoated Reinforcing Steel in Concrete

    C 900: Standard Test Method for Pullout Strength of Hardened Concrete

    C 1040: Test Methods for Density of Unhardened and Hardened Concrete in Place by Nuclear Methods

    C 1074: Standard Practice for Estimating Concrete Strength by the Maturity Method

    C 1150: Standard Test Method for the Break-Off Number of Concrete

    1995 Annual Book of ASTM Standards, Vol. 04.03

    D 4580: Standard Practice for Measuring Delaminations in Concrete Bridge Decks by Sounding

D 4788: Standard Test Method for Detecting Delaminations in Bridge Decks Using Infrared Thermography

D 4748: Standard Test Method for Determining the Thickness of Bound Pavement Layers Using Short-Pulse Radar

1995 Annual Book of ASTM Standards, Vol. 03.02

G 15: Terminology Relating to Corrosion and Corrosion Testing

British Standard: Available from British Standards Institution, London. BS 1881: Part 204. Recommendations on the Use of Electromagnetic Covermeters.

Alongi, A.V., Cantor, T.R., Kneeter, C.P., and Alongi, A., Jr. 1982. Concrete evaluation by radar theoretical analysis. *Transportation Research Record No. 853*, pp. 31–37. Transportation Research Board, Washington, DC.

Arni, H.T. 1972. Impact and Penetration Tests of Portland Cement Concrete. *Highway Research Record 378*, pp. 55–67. Highway Research Board, Washington, DC.

Ballarini, R., Shah, S.P., and Keer, L.M. 1986. Failure characteristics of short anchor bolts embedded in a brittle material. *Proc. Royal Soc. London* A404:35–54.

Barker, M.G. and Ramirez, J.A. 1988. Determination of concrete strengths using the break-off tester. *ACI Mater. J.* 82(6):221–228.

Bickley, J.A. 1982a. Concrete optimization. *Concrete Int.* 4(6):38–41.

Bickley, J.A. 1982b. Variability of pullout tests and in-place concrete strength. *Concrete Int.* 4(4): 44–51.

Bracewell, R. 1978. *The Fourier Transform and its Applications*, 2nd ed., 444 pp. McGraw–Hill Co., New York.

Brown, T.L. and LeMay, H.E. 1988. Chemistry: *The Central Science*, 4th ed., pp. 494–498. Prentice Hall, Upper Saddle River, NJ.

Bungey, J.H. 1981. Concrete strength determination by pull-out tests on wedge anchor bolts. *Proc. Inst. Civ. Eng.* Part 2 71:379–394.

Bungey, J.H. 1989. *Testing of Concrete in Structures*, 2nd ed., 228 pp. Chapman and Hall, NY.

Bungey, J.H. and Millard, S.G. 1993. Radar inspection of structures. *Proc. Inst. Civ. Eng. Struct. Buildings J.* 99:173–186.

Bungey, J.H., Millard, S.G., and Shaw, M.R. 1994. The influence of reinforcing steel on radar surveys of structural concrete. *Constr. Building Mater.* 8(2):119–126.

Byfors, J. 1980. Plain Concrete at Early Ages. Swedish Cement and Concrete Research Institute Report 3:80, 464 pp. Stockholm, Sweden.

Cady, P.D. and Gannon, E.J. 1992. Condition Evaluation of Concrete Bridges Relative to Reinforcement Corrosion, Volume 8: Procedure Manual. *SHRP-S/FR-92-110*, 124 pp. Strategic Highway Research Program, National Research Council, Washington, DC.

Cantor, T.R. 1984. Review of penetrating radar as applied to nondestructive evaluation of concrete. In *In Situ/Nondestructive Testing of Concrete*, V.M. Malhotra, ed. ACI SP-82, pp. 581–601. American Concrete Institute, Farmington Hills, MI.

Cantor, T. and Kneeter, C. 1982. Radar as applied to evaluation of bridge decks. *Transportation Research Record No. 853*, pp. 37–42. Transportation Research Board, Washington, DC.

Carette, G.G. and Malhotra, V.M. 1991. Long-Term Strength Development of Silica Fume Concrete. Paper presented at CANMET/ACI International Workshop on the Use of Silica Fume. April 7–9, 1991. Washington, DC.

Carino, N. J. 1982. Maturity functions for concrete. In *Proc. RILEM Int. Conf. Concr. Early Ages (Paris)*, Vol. I, pp. 123–128. Ecole Nationale des Ponts et Chausses. Paris.

Carino, N.J. 1984. The maturity method: Theory and application. *J. Cement, Concr. Aggregates* 6(2):61–73.

Carino, N.J. 1991a. The maturity method. In *Handbook on Nondestructive Testing of Concrete*, V.M. Malhotra and N.J. Carino, eds., pp. 101–146. CRC Press, Boca Raton, FL.

Carino, N.J. 1991b. Pullout test. In *Handbook on Nondestructive Testing of Concrete*, V.M. Malhotra and N.J. Carino, eds., pp. 39–82. CRC Press, Boca Raton, FL.

Carino, N.J. 1992a. Recent developments in nondestructive of concrete. In *Advances in Concrete Technology*, V.M. Malhotra, ed. MSL 92-6(R), pp. 281–328. Energy, Mines and Resources, Ottawa.

Carino, N.J. 1992b. Performance of electromagnetic covermeters for nondestructive assessment of steel reinforcement, *NISTIR 4988*, 130 pp. National Institute of Standards and Technology. Gaithersburg, MD.

Carino, N.J. 1993. Statistical methods to evaluate in-place test results. In *New Concrete Technology: Robert E. Philleo Symposium*, ACI SP-141, pp. 39–64. American Concrete Institute, Farmington Hills, MI.

Carino, N.J. 1994. Nondestructive testing of concrete: History and challenges. In *Concrete Technology Past, Present, and Future.* Proceedings of V. Mohan Malhotra Symposium, ACI SP-144, pp. 623–678. American Concrete Institute, Farmington Hills, MI.

Carino, N.J. and Sansalone, M. 1992. Detecting voids in metal tendon ducts using the impact-echo method. *ACI Mater. J.* 89(3):296–303.

Carino, N.J. and Tank, R.C. 1992. Maturity functions for concrete made with various cements and admixtures. *ACI Mater. J.* 89(2):188–196.

Carino, N.J., Sansalone, M., and Hsu, N.N. 1986. Flaw detection in concrete by frequency spectrum analysis of impact-echo waveforms. *In International Advances in Nondestructive Testing*, W.J. McGonnagle, ed., pp. 117–146. Gordon & Breach Science Publishers, New York.

Carino, N.J., Knab, L.I., and Clifton, J.R. 1992. Applicability to the maturity method to high-performance concrete, *NISTIR-4819*, 64 pp. National Institute of Standards and Technology. Available from NTIS, Springfield, VA, 22161. PB93-157451/AS.

Carter, C.R., Chung T., Holt, F.B., and Manning D. 1986. An automated signal processing system for the signature analysis of radar waveforms from bridge decks. *Can. Electr. Eng. J.* 11(3):128–137.

Chabowski, A.J. and Bryden-Smith, D.W. 1980. Assessing the strength of concrete of in-situ Portland cement concrete by internal fracture tests. *Mag. Concr. Res.* 32(112):164–172.

Cheng, C. and Sansalone, M. 1993a. The impact-echo response of concrete plates containing delaminations: Numerical, experimental, and field studies. *Mater. Struct.* 26(159):274–285.

Cheng, C. and Sansalone, M. 1993b. Effects on impact-echo signals caused by steel reinforcing bars and voids around bars. *ACI Mater. J.* 90(5):421–434.

Cheng, C. and Sansalone, M. 1995a. Determining the minimum crack width that can be detected using the impact-echo method, Part 1: Experimental study. *Mater. Struct.* 28(176):74–82.

Cheng, C. and Sansalone, M. 1995b. Determining the minimum crack width that can be detected using the impact-echo method, Part 2: Numerical fracture analyses. *Mater. Struct.* 28(177):125–132.

Clear, K.C. 1989. Measuring rate of corrosion of steel in field concrete structures. *Transportation Research Record 1211*, pp. 28–37. Transportation Research Board, Washington, DC.

Clemeña, G.G. 1983. Nondestructive inspection of overlaid bridge decks with ground-penetrating radar, *Transportation Research Record No. 899*, pp. 21–32. Transportation Research Board, Washington. DC.

Clemeña, G.G. 1991. Short-pulse radar methods. In *Handbook on Nondestructive Testing of Concrete.* V.M. Malhotra and N.J. Carino, eds., pp. 253–274. CRC Press, Boca Raton, FL.

Clemeña, G.G. and McKeel, W.T., Jr. 1978. Detection of delamination in bridge decks with infrared thermography, *Transportation Research Record No. 664*, pp. 180–182. Transportation Research Board, Washington, DC.

Clemeña, G.G., Jackson, D.R., and Crawford, G.C. 1992. Inclusion of rebar corrosion rate measurements in condition surveys of concrete bridge decks, *Transportation Research Record 1347*, pp. 37–45. Transportation Research Board, Washington, DC.

Daniels, D.J., Gunton, D.J., and Scott, H.F. 1988. Introduction to subsurface radar. *Proc. Inst. Electr. Eng.* 135, Part F(4):278–320.

Das Gupta, N.C. and Tam, C.T. 1983. Non-destructive technique for simultaneous detection of size and cover of embedded reinforcement. *Br. J. Non-Destructive Test.* 25(6):301–304.

Davis, A. and Dunn, C. 1974. From theory to field experience with the nondestructive vibration testing of piles. *Proc. Inst. Civ. Eng.* 57, Part 2:571–593.

Davis, A.G. and Hertlein, B.H. 1991. Developments of nondestructive small-strain methods for testing deep foundations: A review. *Transportation Research Record No. 1331*, pp. 15–20. Transportation Research Board, Washington, DC.

de Hass, E. 1954. Letter to editor. *J. Am. Concrete Inst.* 25(10):890–891.

Domone, P.L. and Castro, P.F. 1986. An expandable sleeve test for in-situ concrete strength evaluation. *Concrete* 20(3):24–25.

Escalante, E. 1989. Elimination of IR error in measurements of corrosion in concrete. In *The Measurement and Correction of Electrolyte Resistance in Electrochemical Tests.* ASTM STP 1056, L.L. Scribner and S.R. Taylor, eds., pp. 180–190. American Society for Testing and Materials, Philadelphia, PA.

Făcăoaru, I. 1969. Chairman's Report, Meeting of RILEM TC on Non-Destructive Testing of Concrete, Varna, September, 1968. *Mater. Struct.* 2(10):253–267.

Feliu, S., González, J.A., Andrade, C., and Feliu, V. 1989. Polarization resistance measurements in large concrete specimens: Mathematical solution for unidirectional current distribution. *Mater. Struct. Res. Test.* 22(129):199–205.

Feliu, S., González, J.A., Feliu, S., Jr., and Andrade, M.C. 1990a. Confinement of the electrical signal for in-situ measurement of polarization resistance in reinforced concrete. *ACI Mater. J.* 87(5):457–460.

Feliu, S., González, J.A., Escudero, M.L., Feliu, S. Jr., and Andrade, M.C. 1990b. Possibilities of the guard ring for electrical signal confinement in the polarization measurements of reinforcements. *J. Sci. Eng. Corrosion* 46(12):1015–1020.

Fitzgerald, A.E., Higginbotham, D.E., and Grabel, A. 1967. *Basic Electrical Engineering.* McGraw–Hill Book Co., New York.

Flis, J., Sehgal, A., Li, D., Kho, Y.T., Sabol, S., Pickering, H., Osseo-Asare, K., and Cady, P.P. 1992. Condition evaluation of concrete bridges relative to reinforcement corrosion, Volume 2: Method for measuring the corrosion rate of reinforcing steel. *SHRP-S/FR-92-104*, 105 pp. Strategic Highway Research Program, National Research Council, Washington, DC.

Freiesleben Hansen, P. and Pedersen J. 1977. Maturity computer for controlled curing and hardening of concrete. *Nordisk Betong* 1:19–34.

Freiesleben Hansen, P. and Pedersen, J. 1985. Curing of concrete structures. *CEB Inf. Bull. 166*, 42 pp. Comite Euro. International du Beton, Lausanne, Switzerland.

Forrester, J.A. 1970. Gamma radiography of concrete. *Proc. Symp. Non-Destructive Test. Concr. Timber*, pp. 13–17. Institution of Civil Engineers, London.

Gaynor, R. D. 1969. In-place strength of concrete—A comparison of two test systems. *NRMCA Technical Information Letter No. 272.* National Ready Mixed Concrete Assn., Silver Spring, MD.

Geiker, M. 1983. Studies of Portland cement hydration by measurements of chemical shrinkage and systematic evaluation of hydration curves by means of the dispersion model, Ph.D. Dissertation, 259 pp. Technical University of Denmark, Lyngby, Denmark.

Goldsmith, W. 1965. *Impact: The Theory and Physical Behavior of Colliding Solids*, pp. 24–50. Edward Arnold Press, Ltd., London.

Greene, G.W. 1954. Test hammer provides new method of evaluating hardened concrete. *J. Am. Concr. Inst.* 26(3):249–256.

Hagemaier, D.J. 1990. *Fundamentals of Eddy Current Testing.* American Society for Nondestructive Testing, Inc., Columbus, OH.

Halabe, U.B., Sotoodehnia, A., Maser, K.R., and Kausel, E.A. 1993. Modeling the electromagnetic properties of concrete. *ACI Mater. J.* 90(6):552–563.

Halabe, U.B., Maser, K.R., and Kausel, E.A. 1995. Condition assessment of reinforced concrete structures using electromagnetic waves. *ACI Mater. J.* 92(5):511–523.

Halliday, D. and Resnick, R. 1978. *Physics*, 3rd ed., John Wiley & Sons, New York.

Heisey, J.S., Stokoe, K.H., II, and Meyer, A.H. 1982. Moduli of pavement systems from spectral analysis of surface waves. *Transportation Research Record No. 853*, pp. 22–31. Transportation Research Board, Washington, DC.

Hellier, A.K., Sansalone, M., Carino, N.J., Stone, W.C., and Ingraffea, A.R. 1987. Finite-element analysis of the pullout test using a nonlinear discrete cracking approach. *J. Cement, Concr. Aggregates* 9(1):20–29.

Higgs, J. 1979. Integrity testing of piles by the shock method. *Concrete* 13(10):31–33.

Hindo, K.R. and Bergstrom, W.R. 1985. Statistical evaluation of the in-place strength of concrete. *Concr. Int.* 7(2):44–48.

Holt, F.B. and Eales, J.W. 1987. Nondestructive evaluation of pavements. *Concr. Int.* 9(6): 41–45.

Hönig, A. 1984. Radiometric determination of the density of fresh shielding concrete in situ. In *In Situ/Nondestructive Testing of Concrete*, V.M. Malhotra, ed. ACI SP-82, pp. 603–618. American Concrete Institute. Farmington Hills, MI.

Jaeger, B.J., Sansalone, M.J., and Poston, R.W. 1996. Detecting voids in grouted tendon ducts of post-tensioned concrete structures using the impact-echo method. *ACI Struct. J.* 93(4):462–472.

Jensen, B.C. and Braestrup, M.W. 1976. Lok-tests determine the compressive strength of concrete. *Nordisk Betong* 2:9–11.

Johansen, R. 1977. A new method for the determination of in-place concrete strength at form removal. *Proc. RILEM Int. Symp. Testing In Situ Concr. Struct.*, Part II. Budapest, September 12–15, pp. 276–288, Institut des Sciences de la Construction, Budapest.

Johansen, R. 1979. In situ strength evaluation of concrete—The break-off method. *Concr. Int.* 1(9):45–51.

Jones, R. 1949a. Measurement of the thickness of concrete pavements by dynamic methods: A survey of the difficulties. *Mag. Concr. Res.* 1:31–34.

Jones, R. 1949b. The non-destructive testing of concrete. *Mag. Concr. Res.* 2:67–78.

Jones, R. 1953. Testing of concrete by ultrasonic-pulse technique. *Proc. Highway Res. Board, No. 32*, pp. 259–275. National Research Council, Washington, DC.

Jones, R. 1955. A vibration method for measuring the thickness of concrete road slabs in situ. *Mag. Concr. Res.* 7(20):97–102.

Jones, R. 1962. Surface wave technique for measuring the elastic properties and thickness of roads: Theoretical development. *Br. J. Appl. Phys.* 13:21–29.

Jones, R. and Făcăoaru, I. 1969. Recommendations for testing concrete by the ultrasonic pulse method. *Mater. Struct.* 2(10):275–284.

Kierkegaard-Hansen, P. 1975. Lok-strength. *Nordisk Betong.* 3:19–28.

Kolek, J. 1958. An appreciation of the Schmidt rebound hammer. *Mag. Concr. Res.* 10(28):27–36.

Knudsen, T. 1980. On particle size distribution in cement hydration. *Proc. 7th Int. Congr. Chem. Cement (Paris, 1980)*, Vol. II, I-170–175. Editions Septima, Paris.

Knudsen, T. 1984. The dispersion model for hydration of Portland cement: I. General concepts. *Cement Concr. Res.* 14:622–630.

Krautkrämer, J. and Krautkrämer, H. 1990. *Ultrasonic Testing of Materials*, 4th ed., Springer–Verlag, New York.

Krenchel, H. and Bickley, J.A. 1987. Pullout testing of concrete: Historical background and scientific level today. *Nordisk Betong* 6:155–168.

Krenchel, H. and Shah, S.P. 1985. Fracture analysis of the pullout test. *Mater. Struct.* 18(108):439–446.

Krstulovic-Opara, N., Woods, R.D., and Al-Shayea, N. 1996. Nondestructive testing of concrete structures using the Rayleigh wave dispersion method. *ACI Mater. J.* 93(1):75–86.

Kunz, J.T. and Eales, J.W. 1985. Remote sensing techniques applied to bridge deck evaluation. In *Strength Evaluation of Existing Concrete Bridges*, T.C. Liu, ed., ACI SP-88, pp. 237–258. American Concrete Institute, Farmington Hills, MI.

Lauer, K.R. 1991. Magnetic/electrical methods. In *Handbook on Nondestructive Testing of Concrete*, V.M. Malhotra and N.J. Carino, eds., pp. 203–225. CRC Press, Boca Raton, FL.

Leslie, J.R. and Cheesman, W.J. 1949. An ultrasonic method of studying deterioration and cracking in concrete structures. *J. Am. Concr. Inst.* 21(1):17–36.

Leshchinsky, A.M. 1990. Determination of concrete strength by nondestructive methods. *Cement Concr. Aggregates* 12(2):107–113.

Leshchinsky, A.M. 1991a. Correlation of concrete strength and non-destructive tests. *Indian Concr. J.* 65(4):184–190.

Leshchinsky, A.M. 1991b. Combined methods of determining control measures of concrete quality. *Mater. Struct.* 24(141):177–184.

Leshchinsky, A.M. 1992. Non-destructive testing of concrete strength: Statistical control. *Mater. Struct.* 25(146):70–78.

Leshchinsky, A.M. and Leshchinsky, M. Yu. 1991. Correlation formulae in nondestructive testing for concrete strength. *J. Struct. Eng.* 18(3):99–113.

Lin, J.M. and Sansalone, M.J. 1996a. Impact-echo studies of interfacial bond quality in concrete: Part I—Effects of unbonded fraction of area. *ACI Mater. J.* 93(3):223–232.

Lin, J.M. and Sansalone, M.J. 1996b. Impact-echo studies of interfacial bond quality in concrete: Part II—Effects of bond tensile strength. *ACI Mater. J.* 93(4):318–326.

Lin, Y. and Sansalone, M. 1992a. Detecting flaws in concrete beams and columns using the impact-echo method. *ACI Mater. J.* 89(4):394–405.

Lin, Y. and Sansalone, M. 1992b. Transient response of thick circular and square bars subjected to transverse elastic impact. *J. Acoust. Soc. Am.* 91(2):885–893.

Lin, Y. and Sansalone, M., 1992c. Transient response of thick rectangular bars subjected to transverse elastic impact. *J. Acoust. Soc. Am.* 91(5):2674–2685.

Lin, Y., Sansalone, M., and Carino, N.J. 1991a. Finite element studies of the impact-echo response of plates containing thin layers and voids. *J. Nondestructive Eval.* 9(1):27–47.

Lin, Y., Sansalone, M., and Carino, N.J. 1991b. Impact-echo response of concrete shafts. *ASTM Geotechn. J.* 14(2):121–137.

Long, B.G., Kurtz, H.J., and Sandenaw, T.A. 1945. An instrument and a technic for field determination of elasticity, and flexural strength of concrete (pavements). *J. Am. Concr. Inst.* 16(3):217–231.

Mailhot, G., Bisaillon, G., Carette, G.G., and Malhotra, V.M. 1979. In-place concrete strength: New pullout methods. *ACI J.* 76(12):1267–1282.

Malhotra, V.M. 1975. Evaluation of the pull-out tests to determine strength of in-situ concrete, *Mater. Struct.* 8(43):19–31.

Malhotra, V.M. 1976. *Testing Hardened Concrete: Nondestructive Methods*. American Concrete Institute Monograph No. 9, 204 pp. ACI/Iowa State University Press, Ames, IA.

Malhotra, V.M. 1991. Surface hardness methods. In *Handbook on Nondestructive Testing of Concrete*, V.M. Malhotra and N.J. Carino eds., pp. 1–17. CRC Press, Boca Raton, FL.

Malhotra, V.M. and Carette, G.G. 1980. Comparison of pullout strength of concrete with compression strength of cylinders and cores, pulse velocity, and rebound number. *ACI J.* 77(3):17–31.

Malhotra, V.M. and Carette, G.G. 1991. Penetration resistance methods. In *Handbook on Nondestructive Testing of Concrete*, V.M. Malhotra and N.J. Carino, eds., pp. 19–38. CRC Press, Boca Raton, FL.

Malhotra, V.M. and Carino, N.J., eds. 1991. *Handbook on Nondestructive Testing of Concrete*, 343 pp. CRC Press, Boca Raton, FL.

Malhotra, V.M. 1974. Evaluation of the windsor probe test for estimating compressive strength of concrete. *Mater. Struct.* 7(37):3–15.

Malhotra, V.M. 1971. Maturity concept and the estimation of concrete strength. *Information Circular IC 277*, 43 pp. Department of Energy Mines Resources, Canada.

Mandel, J. 1984. Fitting straight lines when both variables are subject to error. *J. Qual. Technol.* 16(1):1–14.

Manning, D.G. and Holt, F.B. 1983. Detecting deterioration in asphalt-covered bridge decks. *Transportation Research Record No. 899*, pp. 10–20. Transportation Research Board, Washington, DC.

Manning, D.G. and Holt, F.B. 1980. Detecting deterioration in concrete bridge decks. *Concr. Inter.* 2(11):34–41.

Mast, J.E. and Johansson, E.M. 1994. Three-dimensional ground penetrating radar imaging using multi-frequency diffraction tomography. *Proc. SPIE Int. Soc. Opt. Eng. Adv. Microwave Millimeter-Wave Detectors* 2275:196–204.

Mast, J.E., Lee, H., Chew, C., and Murtha, J. 1990. Pulse-echo holographic techniques for microwave subsurface NDE. *Proc. Conf. Nondestructive Eval. Civ. Struct. Mater.*, Oct. 15–17, Boulder, Colo., pp. 177–191. National Science Foundation (sponsor).

Maser, K.R. 1986. Detection of progressive deterioration in bridge decks using ground penetrating radar. In *Experimental Assessment of the Performance of Bridges*. Proceedings of ASCE/EM Division Specialty Conference, pp. 42–57. Boston, MA. Oct.

Maser, K.R. and Roddis, W.M.K. 1990. Principles of thermography and radar for bridge deck assessment. *J. Transp. Eng.* 116(5):583–601.

McIntosh, J. D. 1949. Electrical curing of concrete, *Mag. Concr. Res.* 1(1):21–28.

Mitchell, T.W. 1991. Radioactive/nuclear methods. *In Handbook on Nondestructive Testing of Concrete*, V.M. Malhotra and N.J. Carino, eds., pp. 227–252. CRC Press, Boca Raton, FL.

Mitchell, T.M., Lee, P.L., and Eggert, G.J. 1979. The CMD: A device for the continuous monitoring of the consolidation of plastic concrete. *Public Roads* 42(148).

Morey, R.M. 1974. Continuous subsurface profiling by impulse radar. *Proc. Eng. Found. Conf. Subsurface Explor. Underground Excavation Heavy Const.* N.H. Henniker, Aug., pp. 213–232. Engineering Foundation, New York.

Naik, T.R. 1991. The break-off test method. In *Handbook on Nondestructive Testing of Concrete*, V.M. Malhotra and N.J. Carino, eds., pp. 83–100. CRC Press, Boca Raton, FL.

Naik, T.R. and Malhotra, V.M. 1991. The ultrasonic pulse velocity method. In *Handbook on Nondestructive Testing of Concrete*, V.M. Malhotra and N.J. Carino, eds., pp. 169–188. CRC Press, Boca Raton, FL.

Naik, T.R., Salameh, Z., Hassaballah, A. 1987. *Evaluation of In-Place Strength of Concrete by the Break-Off Method*, 101 pp. Dept. of Civil Engineering and Mechanics, University of Wisconsin, Milwaukee.

Nasser, K.W. and Al-Manseer, A.A. 1987a. A new nondestructive test. *Concr. Int. Design Const.* 9(1):41–44.

Nasser, K.W. and Al-Manseer, A.A. 1987b. Comparison of nondestructive testers of hardened concrete. *ACI Mater. J.* 84(5):374–380.

Natrella, M.G. 1963. National Bureau of Standards. *Experimental Statistics*, Chapter 5. Handbook 91. Washington, DC.

Nazarian, S. and Desai, M.R. 1993. Automated surface wave method: Field testing. *J. Geotech. Eng.* 119(7):1094–1111.

Nazarian, S. and Stokoe, K.H., II. 1986a. In-situ determination of elastic moduli of pavement systems by spectral-analysis-of-surface-waves method (practical aspects). *Research Report 368-1F.* Center for Transportation Research, The University of Texas at Austin.

Nazarian, S. and Stokoe, K.H., II. 1986b. In-situ determination of elastic moduli of pavement systems by spectral-analysis-of-surface-waves method (theoretical aspects). *Research Report 437-2.* Center for Transportation Research, The University of Texas at Austin.

Nazarian, S., Stokoe, K.H., II, and Hudson W.R. 1983. Use of spectral analysis of surface waves method for determination of moduli and thickness of pavement systems. *Transportation Research Record. No. 930.*, pp. 38–45. Transportation Research Board, Washington, DC.

Nurse, R. W. 1949. Steam curing of concrete. *Mag. Concr. Res.* l(2):79–88.

Olson, L. and Church, E. 1986. Survey of nondestructive wave propagation testing methods for the construction industry. *Proc. 37th Ann. Highway Geol. Symp.*, Helena, Mont., Aug.

Olson, L.D. and Wright, C.C. 1990. Seismic, sonic, and vibration methods for quality assurance and forensic investigation of geotechnical, pavement, and structural systems. In *Proceedings of Conference on Nondestructive Testing and Evaluation for Manufacturing and Construction*, H.L.M. dos Reis, ed., pp. 263–277. Hemisphere Publishing, New York.

Otto, G., Chew, W.C., and Young, J.F. 1990. A large open-ended coaxial probe for dielectric measurements of cements and concretes. *Proc. Conf. Nondestructive Eval. Civ. Struct. Mater.*, pp. 193–209. National Science Foundation, Washington, DC.

Ottosen, N.S. 1981. Nonlinear finite element analysis of pullout test. *J. Struct. Div. ASCE.* 107(ST4):591–603.

Parker, W.E. 1953. Pulse velocity testing of concrete. *Proc. Am. Soc. Test. Mater.* 53:1033–1042.

Perenchio, W.F. 1989. The condition survey. *Concr. Int.* 11(1):59–62.

Petersen, C.G. 1984. LOK-test and CAPO-test development and their applications. *Proc. Inst. Civ. Eng.* **76** Part I:539–549.

Pichot, C. and Trouillet, P. 1990. Diagnosis of reinforced structures: An active microwave imaging system. In *Proceedings, NATO Advanced Research Workshop on Bridge Evaluation, Repair and Rehabilitation*, A.S. Nowak, ed., pp. 201–215. Kluwer Academic Publishers, New York.

Pratt, D. and Sansalone, M. 1992. Impact-echo signal interpretation using artificial intelligence. *ACI Mater. J.* 89(2):178–187.

Richards, O. 1977. Pullout strength of concrete. *Reproducibility and Accuracy of Mechanical Tests.* ASTM SP 626, pp. 32–40. American Society for Testing and Materials, Philadelphia, PA.

Rix, G.J., Bay, J.A., and Stokoe, K.H. 1990. Assessing in situ stiffness of curing Portland cement concrete with seismic tests. *Transportation Research Record 1284*, pp. 8–15. Transportation Research Board, Washington, DC.

Rix, G.J. and Stokoe, K.H. 1989. Stiffness profiling of pavement subgrades. *Transportation Research Record 1235*, pp. 1–9. Transportation Research Board, Washington, DC.

Rodríguez, P., Ramírez, E., and González, J.A., 1994. Methods for studying corrosion in reinforced concrete. *Maga. Concr. Res.* 46(167):81–90.

Roy, D.M. and Idorn, G. M. 1982. Hydration, structure, and properties of blast furnace slag cements, mortars and concrete. *J. Am. Concr. Inst.* 79(6):444–457.

Rutenbeck, T. 1973. New developments in-place testing of concrete. In *Use of Shotcrete for Underground Structural Support*, ACI SP-45, pp. 246–262. American Concrete Institute, Farmington Hills, MI.

Sansalone, M. and Carino, N.J. 1986. Impact-echo: A method for flaw detection in concrete using transient stress waves. *NBSIR 86-3452*, 222 pp. National Bureau of Standards. Available from NTIS, Springfield, VA, 22161, PB #87-104444/AS.

Sansalone, M. and Carino, N.J. 1987. Transient impact response of plates containing flaws. *J. Res. Nat. Bur. Stand.* 92(6):369–381.

Sansalone, M. and Carino, N.J. 1988. Impact-echo method: Detecting honeycombing, the depth of surface-opening cracks, and ungrouted ducts. *Concr. Int.* 10(4):38–46.

Sansalone, M. and Carino, N.J. 1991. Stress wave propagation methods. In *Handbook on Nondestructive Testing of Concrete*, V.M. Malhotra and N.J. Carino, eds., pp. 275–304. CRC Press, Boca Raton, FL.

Sansalone, M. and Carino, N.J. 1989a. Laboratory and field study of the impact-echo method for flaw detection in concrete. In *Nondestructive Testing of Concrete*, H.S. Lew, ed., ACI SP-112, pp. 1–20. American Concrete Institute, Farmington Hills, MI.

Sansalone, M. and Carino, N.J. 1989b. Detecting delaminations in concrete slabs with and without overlays using the impact-echo method. *J. Am. Concr. Inst.* 86(2):175–184.

Sansalone, M. and Poston, R. 1992. Detecting cracks in the beams and columns of a post-tensioned parking garage structure using the impact-echo method. *Proc. Conf. Nondestructive Eval. Civ. Struct. Mater.*, pp. 129–143. National Science Foundation, Washington, DC.

Sansalone, M., Lin, Y., Pratt, D., and Cheng, C. 1991. Advancements and new applications in impact-echo testing. In *Proceedings, ACI International Conference on Evaluation and Rehabilitation of Concrete Structures and Innovations in Design, Hong Kong*, V.M. Malhotra, ed., ACI SP-128, pp. 135–150. American Concrete Institute, Farmington Hills, MI.

Saul, A.G.A. 1951. Principles underlying the steam curing of concrete at atmospheric pressure. *Mag. Concr. Res.* 2(6):127–140.

Serway, R.A. 1983. *Physics for Scientists and Engineers/with Modern Physics*. Saunders College Publishing Philadelphia.

Skramtajev, B.G. 1938. Determining concrete strength for control of concrete in structures. *J. Am. Concr. Inst.* 34:285–303.

Stain, R.T. 1982. Integrity testing. *Civ. Eng.* April:53–59.

Stehno, G. and Mall, G. 1977. The tear-off method—A new way to determine the quality of concrete in structures on site. *RILEM Int. Symp. Test. In Situ Concr. Struct.* Budapest, Sept. 12–15, pp. 335–347. Institut des Sciences de la Construction, Budapest.

Steinbach, J. and Vey, E. 1975. Caisson evaluation by stress wave propagation method. *J. Geotech. Eng. Div. ASCE* 101(GT4):361–378.

Steinway, W.J., Echard, J.D., and Luke, C.M. 1981. Locating voids beneath pavements using pulsed electromagnetic waves. *NCHRP Report 237*, 40 pp. Transportation Research Board, Washington, DC.

Stern, M. and Geary, A.L. 1957. Electrochemical polarization: I, A theoretical analysis of the shape of polarization curves. *J. Electrochem. Soc.* 104(1):56–63.

Stern, M. and Roth, R.M. 1957. Anodic behavior of iron in acid solutions. *J. Electrochem. Soc.* 104(6):390–392.

Stone, W.C. and Carino, N.J. 1983. Deformation and failure in large-scale pullout tests. *ACI J.* 80(6):501–513.

Stone, W.C. and Reeve, C.P. 1986. A new statistical method for prediction of concrete strength from in-place tests. *ASTM J. Cement, Concr. Aggregates*, 8(1)3–12.

Stone, W.C., Carino, N.J., and Reeve, C.P. 1986. Statistical methods for in-place strength predictions by the pullout test. *ACI J.* 83(5):745–756.

Sturrup, V.R., Vecchio, F.J., and Caratin, H. 1984. Pulse velocity as a measure of concrete compressive strength. In *In Situ/Nondestructive Testing of Concrete*, V.M. Malhotra, ed., ACI SP-82, pp. 201–227. American Concrete Institute, Farmington Hills, MI.

Tam, C.T., Lai, L.H., and Lam, P.W. 1977. Orthogonal detection technique for determination of size and cover of embedded reinforcement. *J. Inst. Eng.* 22:6–16.

Tank, R.C. and Carino, N.J. 1991. Rate constant functions for strength development of concrete. *ACI Mater. J.* 88(1):74–83.

Te'eni, M. 1970. Discussion of Session B. *Proceedings Symp. Non-Destructive Test. Concr. Timber.* June 11–12, 1969, pp. 33–35. Institution of Civil Engineers.

Tremper, B. 1944. The measurement of concrete strength by embedded pull-out bars. *Proceedings Am. Soc. Test. Mater.* 44:880–887.

Ulriksen, C.P.F. 1983. *Application of Impulse Radar to Civil Engineering*. Ph.D. Thesis. Lund University, Dept. of Engineering Geology, Sweden, 1982. Published by Geophysical Survey Systems, Salem, NH.

Weil, G.J. 1991. Infrared thermographic techniques. In *Handbook on Nondestructive Testing of Concrete*, V.M. Malhotra and N.J. Carino, eds., pp. 305–316. CRC Press, Boca Raton, FL.

Whitehurst, E.A. 1951. Soniscope tests concrete structures. *J. Am. Concr. Inst. Proc.* 47(6):433–448.

Whitehurst, E.A. 1967. *Evaluation of Concrete Properties from Sonic Tests.* American Concrete Institute Monograph, No. 2, 94 pp. ACI/Iowa State University Press, Ames, IA.

Yener, M. 1994. Overview, and progressive finite element analysis of pullout tests. *ACI Struct. J.* 91(1):49–58.

Yuan, D. and Nazarian, S. 1993. Automated surface wave method: Inversion technique. *J. Geotech. Eng.* 119(7):1112–1126.

"Reflections"—precast high strength polymer concrete artwork by the Civil Engineering Class '82 with Profs. R.H. Karol and E.G. Nawy, Rutgers University. Each slice cast separately, slices assembled and epoxied together to form the thinking lady statue. (Courtesy Dr. Edward G. Nawy).

20

Architectural Concrete

by
Allan R. Kenney, P.E.
President, Precast Systems Consultants, Inc. Venice, Florida. Expert on Architectural Concrete. Co-editor of the PCI Handbook on this subject.

Sidney Freedman
Director of Architectural Precast Concrete Services at the Precast/Prestressed Concrete Institute, Chicago. Expert in architectural precast concrete since the 1960's and author of the PCI Manual on this subject.

0-8493-2666-4/97/$0.00+$.50
© 1997 by CRC Press LLC

20.1 History of Architectural Cast-in-Place Concrete

20.1.1 The Beginning (Prior to 1965)

Le Corbusier's Carpenter Center at Harvard University is one of the first recognized uses of architectural concrete in the United States. This daring project, along with several others in Europe, triggered a marked interest and excitement among designers in the United States. As using concrete as an architectural medium became an identifiable trend, the architectural concrete industry evolved with the speed of a fad and left an enduring impression on the construction industry.

If there was a specific turning point in the acceptance and use of architectural concrete in the United States, it seems to have come with I.M. Pei's Kips Bay apartment complex in New York city, although there were a number of other projects in the planning and building stages at about the same time. Outstanding achievements can be attributed to I.M. Pei, Eero Saarinen, Minoru Yamasaki, Sert, Jackson, and Gourley, Paul Rudolph, Skidmore, Owings and Merrill (SOM), Pier Luigi Nervi, Phillip Johnson, The Architects Collaborative (TAC), and many others.

Of special importance is the fact that the designs created by the architects listed have no common approach. Each is distinctive. Most of our finest examples of architectural concrete take advantage of the free-form plasticity of the material but show complete recognition of the intense discipline it requires. There have been other projects where this discipline and attention to detail was not given proper consideration and the results were disappointing. Unfortunately, the designers frequently attribute this deficiency to the contractor's lack of knowledge. While this may have been part of the problem, it cannot account for some of the early failures.

20.1.2 Prime Years (1965–1990)

During this 25-year period, the use of architectural concrete bloomed. Sculpturing of the forms, better quality form liners, innovative concrete mix designs, and improved placement and consolidation procedures dramatically improved color uniformity and texture. During this period, architects such as I.M. Pei and Partners; SOM; Hellmuth, Obata and Kassabaum (HOK); Roy P. Harrover; Greiner Engineering; Sikes, Jennings, and Kelly (Figure 20.1); Newhouse and Taylor; Rex Whittiker and Allen; Welton Becket; and others led the way.

FIGURE 20.1 First United Tower, Ft. Worth, Texas. Architect—Sikes, Jennings & Kelly, Houston, Texas. Medium sandblast texture.

20.1.3 Current and Future

In spite of many successes in both commercial and monumental types of construction, architectural concrete is undergoing a period of very flat or declining growth. Architectural cast-in-place concrete will continue to have a limited future unless changes are made in how contractors perceive the process. Right now, high labor costs, difficulties in obtaining a reliable finished product, and possible delays in project completion all work against selection of architectural cast-in-place concrete. Early construction planning can reduce all of these negative factors. Contractors have stated that every dollar spent in planning has saved four to six dollars in construction costs. Polsheck and Partners recently completed the Inventors Hall of Fame in Akron, Ohio, using architectural cast-in-place concrete. I.M. Pei and Partner's Architects recently completed the Rock N'Roll Hall of Fame in Cleveland, Ohio, the National Airlines Terminal at J.F.K. International Airport in New York, and the United States Holocaust Memorial Museum in Washington, DC. All three structures used architectural cast-in-place concrete for the exterior facade. In spite of these four major projects, there currently is a lack of other large projects Architectural design in the past has gone in cycles; the cycle could change again, revitalizing the use of architectural cast-in-place concrete in the future. In the meantime, there is a lack of promotion for this type of construction.

Early planning or a prebid conference including the approved contractors, ready-mix suppliers, architect, and owner is essential to familiarize everyone with the drawing details, specifications, sample mock-ups, and quality of concrete work anticipated for the project. Proper budget estimates need to be prepared. Architects and contractors must be willing to produce the structure with a minimum amount of mistakes and surface blemishes. No matter how good the conceptual design, a building can be ruined by dark lines, pour lines, blotchiness, discoloration, or variations in texture. Architects in recent years have had a cautious approach to the use of architectural cast-in-place concrete owing to potential pitfalls, which can result in owner dissatisfaction.

20.2 History of Architectural Precast Concrete

20.2.1 The Beginning

The first documented modern use of architectural precast concrete was in the Cathedral Notre Dame Du Haut in Raincy, France, by Auguste Perret in 1923—though it was used just as screen walls and infill in an otherwise cast-in-place concrete structure. The depression years followed soon after and then the cataclysm of World War II. Not until World War II ended did the architectural use of precast concrete begin to flourish.

Surface finishing techniques for architectural precast concrete, such as water washing and brushing, bushhammering, sandblasting, and acid etching, were initially developed for cast-in-place concrete, as was obtaining colored surfaces by means of pigments and special colored aggregates. These techniques were also well established within the cast stone industry by the 1930s.

In 1932, John J. Earley and his associates at Earley Studio in Rosslyn, Virginia, began work on producing exposed aggregate ornamental elements for the Baha'i Temple in Wilmette, Illinois (1920–53), one of the most beautiful and delicately detailed architectural precast concrete projects in the United States. The panels are white concrete with exposed quartz aggregate. Lack of funds postponed completion of the exterior precast concrete until 11 years after the building was begun. Another early notable use of large precast panels was for the White Horse Barn (1937), constructed for the Minnesota State Fair. This structure had panels 15 ft long, 7 ft high, and 6 in thick. Cast with a smooth finish and cured for seven days in steam-filled rooms, these panels were attached by the structural concrete around the edges of the panels.

A pivotal development occurred in 1938, when administration buildings at the David W. Taylor Model Testing Basin were built near Washington, DC. Panels 2 1/2 in thick and up to 10 ft by 8 ft were used as permanent forms for cast-in-place walls. The project was significant as the first use of the Mo-Sai manufacturing technique, with the units produced by John Earley in collaboration with the Dextone Company of New Haven, Connecticut. It was also the first project in which large-area exposed aggregate panels were adapted to serve as both the exterior form and preinspected facing for reinforced concrete building construction. Earley had patented the idea of using step (gap)—graded aggregate to achieve uniformity and color control for exposed aggregate work.

Working from this background, the Dextone Company refined and obtained patents and copyrights in 1940 for the methods under which the Mo-Sai Associates (later Mo-Sai Institute, Inc.) operated. The Mo-Sai Institute grew to include a number of licensed manufacturing firms in various parts of the United States. Its public relations and advertising activities, highlighting technical achievements, were a major factor underlying general acceptance of architectural precast concrete.

The surface finishes produced by the Mo-Sai process were dense, closely packed mineral aggregate with a minimum of cement/fines matrices: hence the derivation of the name Mo-Sai from the resemblance to mosaic.

In 1958, a new panel-casting method was introduced in the United States under the name of Schokbeton—shocked concrete, and a number of franchised plants were established. The machinery used in this method was patented in Holland in 1932. The process is mainly a means of consolidating a no-slump concrete mixture by raising and dropping the form about 5/16 in some 250 times per minute. This contrasts with conventional methods of consolidation using of high-frequency and low-amplitude vibration. Although the production of large precast panels by this method was relatively new, small concrete units had been produced in the past on so-called drop tables, which followed the same technique without the refinement of modern machinery. Depending on the cement content, the shocking technique produces compressive strengths of 3,000 to 4,000 psi after 24 h and up to 10,000 psi at 28 d.

Architectural precast concrete usage initially was complicated by the lack of mobile cranes and other efficient material-handling equipment. Because of the lack of this equipment and competition from metal and glass curtain-walling, precast concrete was comparatively slow to develop, and the record of its eventual rise to parity, even its dominance in places, is properly that of the 1960s. Reasons for this expanding usage were improved methods of production, better handling and erecting equipment, and realization that precast panels provided a pleasing variety of surface textures, patterns, and exterior designs that generally could not be accomplished as economically in other materials. There were a number of pioneering early postwar uses of precast concrete in architecture, though they were otherwise of limited significance. These included dormitory units at the University of Connecticut (1948) based on load-bearing wall panels; an eight-story office building in Columbia, South Carolina (1949), with window-wall cladding panels (erected by hand winch); and a six-story office building in Miami (1951) where 4 in thick precast panels were suspended from the soffit of a cantilevered cast-in-place floor slab.

The Hilton Hotel, Denver-completed in 1958-9 was one of the early significant uses of window-wall panels fixed to a structural frame. The Police-Administration Building, Philadelphia, completed in 1962 made history as one of the first major buildings to utilize the inherent structural character-istics of architectural precast concrete. Its 5 ft wide, 35 ft high (three-story) exterior panels carry two upper floors and a roof. This structure was an early model for the blending of multiple systems (precasting and post-tensioning) into one building.

20.2.2 Current Practice

Architectural precast concrete can be provided in almost any color, form, or texture, making it an eminently practical and aesthetically pleasing building material. It is difficult to imagine an architectural style that cannot be expressed with precast concrete. By providing the designer com-plete control of the ultimate form of the facade, the precast concrete industry has experienced steady growth since the explosive 1960s, particularly in the ever-widening range of precast-concrete applica-tions. The widespread availability of architectural precast concrete, the nearly universal geographic distribution of the necessary raw materials, and the high construction efficiency of prefabricated components all add to the appeal of architectural precast concrete construction. Established pre-casters have a high level of craftsmanship and ingenuity along with a thorough knowledge of the material and its potential for converting the designer's vision into a finished structure.

20.3 Applications

20.3.1 Cast-in-Place

Architectural cast-in-place concrete has been used in many colleges, libraries, airport terminals, office buildings, museums, public buildings for city, state, and federal government, hospitals, sound-barrier walls, industrial warehouses, and all types of landscaping products.

20.3.2 Precast

The use of nonloadbearing precast concrete cladding has been the most common application of architectural precast concrete. Cladding panels are those precast elements that resist and transfer negligible load from other elements in the structure. Generally, they are normally used only to enclose space and are designed to resist wind, seismic forces generated from their self-weight, and forces required to transfer the weight of the panel to the support. Cladding units include wall panels, window-wall units, spandrels, mullions, and column covers. Their largest dimension may be vertical or horizontal.

Often the most economical application of precast concrete is as a loadbearing element, which resists and transfers loads applied from other elements. With very few modifications, many cladding panels may function as loadbearing members. The steel reinforcement required to physically handle and erect a unit is often more than that necessary for in-place loads. The slight increase in load-bearing wall panel cost (owing to erection and connection requirements) can be offset by the elimination of separate structural framing (beams and columns) from exterior walls or the reduction of interior shear walls. This savings is most apparent in buildings with a large ratio of wall to floor area.

In many structures, it is economical to take advantage of the inherent strength and rigidity of exterior precast concrete wall panels and design them to serve as the lateral load resisting system (shear walls) when combined with the diaphragm action of the floor construction. There is a trend toward the use of precast concrete units as forms for cast-in-place concrete. This system is especially suitable for combining architectural (surface aesthetics) and structural functions in load-bearing facades or for improving ductility in locations of high seismic risk. Because the cost of formwork is a significant part of the overall concrete cost in a structure, substantial savings can usually be achieved by using precast concrete units as formwork.

In addition to functioning as exterior and interior wall units, precast concrete finds expression in a wide variety of aesthetic and functional uses, including 1) art and sculpture; 2) lighting standards and fountains; 3) planters, curbs, and paving slabs; 4) balconies; 5) screen sound barriers and retaining walls; 6) screens, fences, and handrails; 7) street furniture; and 8) ornamental work.

20.4 Planning

20.4.1 Budget

Regardless of the type of building or its intended use, the architect and owner must begin with a budget estimate. Although forming, placement, concrete, and finishing costs will undoubtedly be more expensive for an architectural concrete job, the end cost of the completed building can well be less than the total cost of a structural concrete job when rental income, maintenance costs, additional finishes, architectural treatments and repair costs are factored into the equation. The life cycle costs are what interests the owner. Architectural concrete projects must have an adequate budget to achieve the desired visual impact while maintaining the integrity of the structure. Once the design architect has determined the image and scope of the project, an initial meeting should be held to include the architect, architectural concrete consultant, structural engineer, and landscape architect, if one is associated with the project.

At this time, shape versus structural requirements and precast versus cast-in-place are evaluated. The capability of local precast facilities and availability of knowledgeable concrete contractors often determines the most economical approach to achieve the finish appearance required. The designer must decide at this time whether to follow the possibility of form and continuity inherent in cast-in-place concrete or the building-block logic, time savings, and precise quality control of precast. Relative costs can vary with time and place. Repetition is essential to maximize economy for precast concrete components. Careful planning is necessary to achieve good repetition without sacrificing design freedom.

The type of finish selected can have a large impact on budget. The architect must provide sufficient funds in the budget to allow the special finish and quality workmanship required for the method selected. He must evaluate whether specialized construction techniques are routinely accepted and executed by local contractors. Casting structural concrete against a form liner or sandblasting the surface does not necessarily produce acceptable architectural concrete. A contractor that assumes this is about to have an expensive, time-consuming lesson and perhaps an unhappy client–not receptive to scheduling future projects with the firm. The secret to successful architectural concrete is early and detailed planning.

Final costs and execution do not end with the building designer, but with the contractor. The general contractor must know how to carry out the designer's intentions without increasing costs to himself or the owner. The contractor must have knowledgeable workmen, especially for the forming and concrete superintendents.

Present bidding practices often result in late award of the precast subcontract. In practice, this means that additional molds must be built in order to meet the project deadline. The necessity for extra molds increases costs and partially offsets the use of high repetition.

The secret to good architectural concrete is in the forming. Forms must be strong enough to accomplish concrete placement within specified form joints and rustications without honeycomb, cold joints, or pour lines. Although the superintendent may have placed many thousands of yards of structural concrete where strength and lack of honeycomb is the criteria, he can be in trouble when visual excellence is required if strict attention is not paid to details.

20.4.2 Drawings

Eventually, the conceptual design must be detailed and working drawings completed. Construction elements such as control joints, construction joints, form-tie locations, concrete shrinkage, reinforcement location, reinforcement coverage, special form requirements, mix design, concrete access points, and type of finish must be considered. The contractor should pay special attention to closure technique or concealed joints in formwork. Articulated lines or rustications should be drawn at the control and the construction joints. The closer the architectural concrete construction follows the natural construction sequence, the more economical the costs. In addition to the items discussed under formwork, Section 20.7.1, the contractor must consider form rigidity. Architectural concrete often includes high columns and walls. Forms must be sufficiently reinforced to eliminate wave action or pulsating action as well as form movement during the use of high-frequency vibrators (Figure 20.2). Form height is a function of drawing details involving rustication details and form-tie locations. Repetitive items in design will increase form usage and allow higher quality forms (such as steel or fiberglass) to be used economically.

FIGURE 20.2   Waviness in the concrete surface due to lack of form rigidity during concrete consolidation.

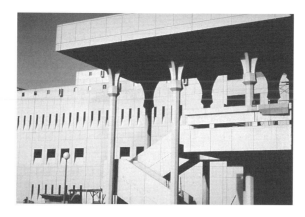

FIGURE 20.3 Matching architectural cast-in-place and pre-cast concrete at King Saud University in Riyadh, Saudi Arabia. Architects—HOK + 4, St. Louis Missouri.

The contractor needs to ensure that the drawing details work. Although the architectural concept of the drawings may be excellent, unless the details work, the building has the potential to be a failure. Good design considers the construction needs as well as the architectural and structural criteria. The drawings set the conditions under which the contractor must work. A wall designed to be 8 in thick, including a double curtain of reinforcing, will not allow proper concrete placement and consolidation because the space is simply too tight. If the wall has rustications and minimum cover requirements on the outside face, space requirements for vibrator placement get even tighter.

Cast-in-place architectural concrete and precast elements may be successfully combined but require detailed effort on the part of the architect and an understanding of the construction process for each system on the part of the contractor. More time will be required for the selection of materials, mix designs, and finishes than if only one system were used. Realistically, large mock-ups and knowledgeable workmen need to be part of the contractor's production process.

The architect has a great deal of freedom of expression in working with architectural concrete. An architect may select thin members or sculptured shapes in precast concrete and combine this with the massiveness and structural strength of cast-in-place concrete to create the curtain wall (Figure 20.3). Today, when costs for different types of facades are compared, precast and/or cast-in-place concrete can be competitive and compare very favorably with glass and metal facades. Design can provide either matching or contrasting colors between architectural precast and cast-in-place concrete.

The precaster's shop drawings (erection and production drawings) should be prepared in general conformance to the *PCI Drafting Handbook,* MNL 119-90. Generally shop (erection) drawings are submitted to the general contractor who, after checking them and making notations, submits them to the architect/engineer for checking and review. Timely review and approval of shop drawings and other pertinent information submitted by the precaster is essential as fabrication should not commence until final approval or an approved-as-noted has been received.

The architect reviews the precaster's erection drawings primarily for conformance to the specifications, then passes them along to the structural engineer so that he can check for conformance to the specified loads and connection locations. This allows the engineer to confirm his own understanding of the forces for the structure at the connection points and the precaster's understanding of the project requirements. Design details, connection locations, and specified loads should not be left to the discretion of the precaster. This is especially important in cladding panels, where the weight of the unit and its torsion and shear must be supported by the frame. Steel frames are more sensitive to the eccentricity of a panel unit causing deformation of the structure and may require bracing.

20.4.3 Specifications

Specifications can be a trap for the unwary contractor. There may be allusions to architectural concrete requirements in the general concrete specifications. When projects, including several monumental buildings, have had high budget estimates, the separate section for architectural cast-in-place concrete has sometimes been removed from the specification. All concrete has been covered under the general concrete section. A short added paragraph may read something like this: "The final finish surface should be a uniform light sandblast and match the approved sample panel in Architect's office." There also may be a short paragraph near the end of the concrete section stating: "No patching or repairs will be permitted without the express permission of the Architect. Only patching of minor blemishes will be permitted."

These short paragraphs added to a general concrete specification can be deadly to the contractor. There is no mention of special mix requirements, mockups, form or form-tie locations, placing consolidation, and finish requirements. Architectural cast-in-place concrete specifications should be used and clearly separated from those for general concrete, and the contractor should always prepare a mock-up for approval whether one is specifically called for in the specification or not. The mock-up can consist of a segment that will be incorporated in the finished structure.

Specifications can be both the friend and ally of the owner, architect, and contractor. Architectural concrete is not structural concrete with only a surface finish applied. It is a material that requires a special process leading to the final finish required. Both architectural cast-in-place and precast concrete must be clearly specified, closely supervised, and produced by knowledgeable workers who care throughout the full construction process. Architectural precast should always have a separate section in the contract specifications, as should architectural cast-in-place concrete.

There are basically two types of specifications: performance and prescriptive. A pure performance specification is one that references a physical sample of concrete. It essentially says: "Produce concrete that looks like the referenced sample and meets all other concrete requirements as contained in the general concrete section." Such a specification assumes that this type of concrete finish is not new and has been satisfactorily produced before by all parties concerned. On the other hand, the prescriptive type specification is necessary when the bidding list must be left open or when the finish is a new type developed by the designer. This type of specification relies on detailed investigation of materials, mix designs, formwork, concrete placement procedures, and methods of finishing. It then gives explicit instructions on what procedures to follow. Even though instructions may be clear, a prescriptive specification requires constant supervision to keep everything rolling smoothly.*

The performance type of specification is the most practical. It can be used when 1) a physical example of the concrete finish is available; 2) the contract is negotiated or only prequalified bidders are selected; or 3) craftsmen are available to produce the required performance.* Too often a contractor is asked to comply not only with standards from ASTM, ACI, PCI and, others, but also with references, guides, or practices published by the above because they are incorporated by reference in other documents mentioned in the contract though no copies or details are furnished. In order to clarify this issue, in 1978 the American Concrete Institute published the following policy that is used today on all appropriate documents: "ACI Committee Reports, Guides, Standard Practices, and Commentaries are intended for designing, planning, executing, or inspecting construction, and in preparing specifications. Reference to these documents shall not be made in the Project Documents. If items found in these documents are desired to be part of the project documents, they should be incorporated directly into the project documents."

20.4.4 Quality Control and Quality Assurance

In theory, for commercial grade concrete construction, the contractor simply hires a testing laboratory approved by the owner or architect and sees that the reports are distributed to the responsible parties. The contractor should review ready-mix concrete facilities and local precast manufacturers for service history, reserve capacity, availability of modern trucks, plant silo, bin, and storage

*Bell, L.W. June 1996. *ACI J.* :63–66.

capacity, and availability of knowledgeable and cooperative personnel. Owners are looking for contractors to spend more time and money in construction reviews and in planning to develop more innovative ways to meet schedules and improve quality. More monumental work is being negotiated or has a very limited bid list. Contractors with prior experience in architectural concrete who can meet budget performance, maintain good safety records, and control the project processes will be in high demand.*

Quality control begins when the architect determines the shape, size, color and texture for the architectural concrete for a specific project. Depending upon project size and scope of work, the contractor's assigned personnel must manage quality and verify that the detailed requirements are carried out. On a small project, this may take only a small part of the project engineer's time. On a large contract, several people under the project engineer may be assigned to the quality-control program. Any personnel involved in quality supervision should not be involved with direct production responsibilities. The project manager should give quality control personnel sufficient time and resources to do an adequate job. Information gained through quality control should be reviewed at weekly project meetings, or sooner, if a problem exists.

The basic purpose of quality-control programs is to ensure that the requirements of the contract documents are met in every step of the process, from selection of materials to curing of the concrete. Written records are needed to substantiate that fact. Top management personnel in construction companies need to educate their talented field people in regard to the benefits of quality control (QC) and quality assurance (QA). The contractor works for the owner and is responsible for constructing the project as designed and specified. He must assure that the quality of construction measures up to or exceeds the standards of material and workmanship required by the contract documents. This is the contractors primary responsibility, but he must also emphasize to his employees, material suppliers, and subcontractors the importance of searching for errors in design or specification and selection of inappropriate materials. Although design review is not part of the contractor's responsibility, it may be good insurance to have a second opinion on soils engineering, structural engineering, exterior-facade details, etc. Replacing or redoing problem construction is time-consuming, frustrating, and always expensive.

The contractor cannot cite lack of skilled workers, irresponsible subcontractors, or rushed or incomplete drawings as an excuse for poor quality. In today's construction, the contractor must live with tight budgets, low construction time, drawings that are completed as construction progresses, and lack of skilled workers. What better reason could there be to cry for improved quality control and quality assurance programs? Quality control is directed toward after-the-fact inspection and testing. Quality assurance is a broader concept that involves planning and a program that incorporates feedback from site problems, inspection, and testing. Quality assurance programs incorporate both "feed-forward" and "feedback" controls for each segment of the construction process. Quality assurance is when the contractor:

1. makes sure the field supervisors and subcontractors recognize the obligations of the contract documents;
2. thinks "quality" in every decision made;
3. identifies project risks and relates them to quality factors as well as time and costs;
4. coordinates and communicates with the Architect/Engineer and owner with overall quality in mind;
5. encourages all of the workers to concern themselves with the quality of construction;
6. uses knowledgeable consultants to assist in planning and on-site review;
7. anticipates problems and tries to alleviate them early;
8. learns to listen; and
9. minimizes reworking or rebuilding.

Most profitable projects are those that prevent defects in the first place through planning, communication, and ongoing QA programs.

*Bainbridge, L.R. and Abberger, W.A. May–June 1994. *PCI J.* :96–97.

An architectural precast concrete manufacturing plant should be certified by the Precast/ Prestressed Concrete Institute (PCI) Plant Certification Program at the time of bidding. Certification should be in product group A1 (architectural concrete). The certification of a producing plant by PCI indicates the plant has in place a system of quality, starting with management. Further, the quality system extends to all areas of plant operations including a thorough and well-documented quality control program. Certification indicates that plant practices are in conformance with time-tested industry standards. It means that the plant regularly demonstrates its ability to produce quality products.

All plants are audited at least twice each year. Audits are unannounced and are usually of two days duration. Audits and grading are based on PCI "Manual for Quality Control for Plants and Production of Architectural Precast Concrete Products," MNL-117-96. The audit covers all aspects of production and quality control as well as engineering and general plant practices. The product evaluations performed by in-house quality control personnel are also reviewed to determine if routine monitoring is correct and accurate.

20.4.5 Prebid Conference

A prebid conference should be held at least three weeks prior to bid. This conference should involve all of the professionals in the design team, preselected general contractors, and anticipated precast manufacturers. Precast manufacturers should have submitted their samples, company brochures, technical literature, and proposed materials prior to the meeting. Plans for how the work is to be accomplished in accordance with the requirements of the contract documents would be available at this meeting. The architect should have the building model and any samples showing anticipated color and texture. On exceptionally large projects, an extra mock-up may be constructed, under a special contract, to be available at the prebid conference. This mock-up would help determine the feasibility of various materials, treatments, and construction procedures prior to writing the particular architectural specifications for bid. Contractors would be told what is expected in the way of workmanship, timing, and results. Samples would be discussed, and material sources and adequacy of reserves reviewed. Precast production schedules, plant facilities, and personnel would also be discussed. At this meeting, potential problems would be discussed and the final architectural finishes clearly spelled out. Generally, one prebid conference is held for all anticipated contractors and precast manufacturers.

20.4.6 Preconstruction Conference

Following award of the contract, the architect should schedule a preconstruction conference with the contractor at a time and place that is mutually agreeable to discuss the architectural concrete system. This allows the design team to meet with the contractor, his key personnel, ready-mix concrete supplier, form supplier, reinforcing and placement subcontractors, testing laboratory, and precast supplier (if applicable). Brochures or literature should accompany any alternate construction proposals. Patented form systems may be suggested at this conference. The contractor should be sure such systems not only meet structural needs but are appropriate for architectural purposes. Butt joints or form-tie locations may be a problem with certain form systems. When precast and cast-in-place construction are to match, the precast manufacturer should match the cast-in-place construction. The precaster has more flexibility in matching color and texture than the cast-in-place contractor, who does not have as many options and specialty techniques available. The preconstruction conference is a good place to set ground rules and begin planning quality-assurance programs. At this meeting, the design team should be responsive to suggestions and requests for approvals and additional information by the contractor. The architect should be responsive to alternate plans submitted by the contractor and make recommendations for the contractor's consideration. The contractor must be ready with his plan of construction, timing, and schedule of items that involve the design team such as mock-up timing, shop-drawing approvals, and any remaining questions as to color or texture.

20.4.7 Mock-up Construction

Mock-up construction is where the contractor demonstrates the results he intends to produce. Materials, equipment, and construction methods used are those intended for the actual work. The mock-up is a proving ground to test the design details, shape, construction-joint locations, gasket procedures, and finishes required under the contract documents. In the case of alternate proposals, the mock-up makes sure that the intent of the designer is understood. For the contractor's own protection, every architectural concrete project should include a mock-up or mock-ups. This is true whether one is specified or not.

A cast-in-place concrete project may have only one large mock-up showing a column, beam, or wall section that is full scale in size and uses the design reinforcing, forming system, concrete mix, placement procedures, and consolidation methods to be used in the actual work. After completion and approval of the construction, the mock-up is used to demonstrate cleaning, repair techniques, sealers, if specified, and protection. An architectural precast project might require three mock-up samples of a size sufficient to demonstrate actual planned production conditions. Any mock-up or combination of mock-ups must be able to establish the range of acceptability with respect to color and texture variations, surface blemishes, uniformity of corners or returns, frequency and size of air-void distribution, and overall appearance. Mock-ups also test the planning and training abilities of the contractor. Mock-up usage should shorten the learning process and set the standards necessary to achieve the designer's objective.

The mock-up should be prepared or erected at the job site and be approved before the majority of the formwork for the architectural concrete construction is built or purchased. When forms are to be site made, the mock-up must be finished at least 45 days before the architectural concrete is scheduled to be cast. On larger cast-in-place projects where custom-built forms are to be used, the mock-up should be finished 90 to 120 days before forms are needed for construction. This will allow time to revise the forms or the construction methods used in the construction of the mock-up. This long lead time can be necessary. Do not let the project be delayed by inadequate planning and lead times. Construct as many full size mock-ups as necessary to meet the approval of the architect. When a mock-up is not called for in the specifications and the concrete called for appears to be structural concrete with only a light texture finish, a contractor would be wise to go through a sample and mock-up procedure at his own expense prior to final decisions on mix design and formwork purchase. The samples and mock-up for cast-in-place concrete may need to be scaled down owing to the cost factor but should not be eliminated from the early phase of the construction process. Sometimes future architectural cast-in-place concrete may use part of the early structural concrete as a testing ground for forming and concrete placement. Areas hidden from view such as basement walls or elevator shafts could be used. Some type of mock-up construction will minimize any confusion on the part of all parties involved in the project design and construction.

Approval of any mock-up should be based on the full mock-up, or in the case of precast, of at least three precast full-scale (not necessarily full size) samples. Realistically, all architectural concrete surfaces will have blemishes, irregular surfaces, and differences in color and texture. Samples and mock-ups should be used to establish the range of acceptability prior to the beginning of actual construction. Mock-ups should remain on the job site or at the plant where they are readily available for inspection and comparisons until completion of the work (Figure 20.4).

FIGURE 20.4 Mockup panels for USAA regional office in Tampa, Florida. Architects—Spillis Candela & Partners Inc., Coral Gables, Florida.

Mock-ups can be most effectively assessed when presented in their final orientation: horizontal, vertical, or sloping. Details should be viewed from a distance typical for viewing the architectural concrete on the building but that is not less than 20 ft.

Reference to another project or an adjacent building should never be the basis of the architectural color and texture demanded for your project. Mock-ups should be required for each project. A different building facade may have been produced under special conditions involving different specifications, use of materials no longer available, or more stringent supervision requirements.

20.5 Materials—Mixture Design

20.5.1 General

Certain variations, mix proportions, and procedures are specific to architectural concrete. A change in aggregate proportions, color, or gradation will affect the uniformity of the finish, particularly where the aggregate is exposed. In smooth concrete, the color of the cement (plus pigment) is dominant. A matte finish will result in a different color than a smooth finish. If the concrete surface is progressively removed by sandblasting, acid, tooling, retarders, or other means, the color becomes increasingly dependent on the fine and coarse aggregate. Darker and more colorful fine aggregates have more influence when used with lighter colored cements and especially with white cement. To produce white architectural concrete of consistent quality, the proper choice of aggregates must accompany the use of a white cement.

20.5.2 Cement

White Portland cement is made of selected raw materials containing negligible amounts of iron and manganese oxides since these materials project a grey color. White cements have inherent color and strength-development characteristics depending on their source: some have a buff or cream undertone, others have a blue or green undertone. Therefore cement of the same type and brand, from the same mill, should be used throughout the entire job to minimize color variations. Color variations in a grey cement matrix are generally greater than those in matrices made with white cement. A grey color may be projected by using white cement with a black pigment or a blend of white and grey cement. Uniformity normally increases with increased percentages of white cement. Grey cement is dominant, and will have much greater color variation than white cement with an added black pigment. Plastic or stainless steel trowels should be used for finishing architectural concrete in order to minimize the danger of staining or darkening the concrete. Trowel marks or surface burns can be minimized by troweling while the concrete surface is moist or by keeping the trowel wet during finish operations.*

20.5.3 Aggregates

The choice of aggregates can have a considerable effect on the finish color of white, buff, or light-toned concrete. The choice of fine and coarse aggregate should be based on a visual inspection of the samples and the mock-up. These samples would have the proper matrix and be finished in the same manner as planned for construction. The method used to expose the aggregate in the final product influences the final appearance. Sandblast textures will generally dull the appearance of the coarse aggregate. All coarse and fine aggregates should be from the same source for the entire project. Normally, aggregates will be selected for exposed architectural purposes and specified in the contract documents as to source, size, and color. Aggregates should have proper durability and be free of staining or deleterious materials. They should be nonreactive with cement, have a proven service record or satisfactory results from laboratory testing, and have the particle shapes (rounded

---

*Freedman, S. December 1968. *Modern Concr.* :30–34; Freedman, S. January 1969. *Modern Concr.* :30–35.

rather than slivers) required for good concrete and appearance. When angular aggregates, such as small size granites are specified, the contractor should inform the architect and owner that they will result in higher color variations, some pour lines, and increased surface blemishes such as bug holes.

Where the color depends primarily on the fine aggregates, gradation control is required. For fine aggregates purchased in bulk, the percentage of fine particles passing the No. 100 sieve should be limited to no more than 5%. This may require double washing and a secondary screening, but the premium for this can be justified by the increased uniformity of color. With a light to medium exposure, a uniform color appearance may be obtained by using crushed sand of the same material as the coarse aggregate. However, for reasons of workability, a percentage of natural sand is always desired in a concrete mix. When maximum whiteness is desired, a natural or manufactured opaque white or light yellow sand should be used. Most naturally occurring sands lack the required whiteness, and the contractor or precaster must look to the various manufactured aggregates for the white base desired. Generally, these consist of crushed limestone (dolomite, calcite) or quartz and quartzite sands.

Coarse aggregates are selected on the basis of color, hardness, size, shape, gradation, method of surface exposure, durability, cost, and availability. Colors of natural coarse aggregate vary considerably according to location and geological classifications. Coarse aggregates should be reasonably uniform in color. Light and dark coarse aggregates require caution in blending for good distribution and color uniformity in adjacent concrete. Extreme color differences between fine and coarse aggregate or between the aggregates and matrix should be avoided. Often a pigment can be added to the matrix to more closely match the color of the aggregates. Extreme color differences between the aggregates and matrix will create uniformity problems and possibly lead to aggregate transparency. The use of closely graded fine and coarse aggregate in a gap-graded combination often results in less segregation and produces a more uniform finish. While gap-grading is an established and well-proven practice, it should not be carried to extremes. Extreme gap-grading may lead to separation of the paste and aggregates and thus create uniformity problems. This can be particularly true when dense hard surface forms are used in combination with high-frequency vibration (Figure 20.5).

The amount of fines, cement, and water should be minimized to ensure that shrinkage remains within acceptable limits and that surface absorption will be low enough to maintain good weathering qualities. The durability of the concrete would normally not be affected by any degree of gap-grading as long as proper concrete cover is maintained over the reinforcement.

20.5.4 Water

Mixing water should be clean and free from oil, acid, iron or rust (which may cause staining), and injurious amounts of vegetable matter, alkalis, or other salts. Almost any natural water that is drinkable and has no pronounced taste or odor is satisfactory. Warm water may tend to cause false set or shorten the working time of the mix. Chilled water or use of ice may retard the set or make mixing more difficult.

20.5.5 Admixtures

Water-reducing and set-controlling admixtures have been found to be effective with respect to water reduction and strength increases when used with Portland cements of low tricalcium aluminate (C_3A) and alkali content. Architectural concrete dosages are in excess of

FIGURE 20.5 Separation of the paste and aggregates when using high density forms and high frequency vibrators. Dark paste lines are evident after sandblasting.

FIGURE 20.6 Pour lines on interior beams where the concrete was placed during hot weather. Higher dosages of admixtures could have been used to minimize the possibility of early set leading to pour lines.

those generally used for structural concrete and may require up to 50% more than the usual recommended minimum. Admixtures are important in architectural concrete for workability, to assist in placement, and to minimize the possibility of pour lines occurring in warm weather due to early set (Figure 20.6). An increase in cement fineness or a decrease in cement alkali content generally increases the amount of air-entraining admixture required for a given air content. High-range water-reducing admixtures (HRWRA) may alter the air void system of air-entrained concrete. Typically, the void-spacing factors are higher than in normal air-entrained concrete. This void spacing is caused by an increase in the average bubble size and a decrease in the specific surface compared to an air-entrained concrete without a HRWRA. In such cases, the dosage of the air-entraining agent would be adjusted to compensate for this effect. In some cases, a complete loss of air may occur when using HRWRA. It may be necessary to change to another HRWRA or use a different air-entraining agent. Certain air-entrained admixtures may be more effective with certain HRWRAs, in the production of adequate air entertainment. All admixtures should be tested to ensure their compatibility.

The use of calcium chloride as an admixture may contribute to the corrosion of metals and darkening, mottling, and discoloration of the concrete surface. Its use is not recommended for architectural concrete. In spite of such terms as "chloride-free," no truly chloride-free admixture exists, since admixtures often are made with water that contains small but measurable amounts of chloride ions. If, in using the available information on the admixture and the proposed dosage rate, it is calculated that the chloride ion limitation will be exceeded, alternate admixtures or procedures should be considered.

When coloring pigments are added to the concrete mixture, the amount should be less than 5% of the weight of the cement. Amounts in excess of 5% seldom produce further color intensity, while amounts greater than 10% may be harmful to the concrete quality. For any coloring agent, it is important to have tests or performance records that indicate color stability in concrete.

20.5.6 Mixture Design

Architectural concrete should provide both proper workability required for consolidation and adequate strength for the given type of concrete. The architect should specify the parameters of concrete performance requirements. A design strength for concrete should be determined by the engineer, based on in-service requirements, while not forgetting construction and, in the case of precast, handling and erection considerations. A concrete mix designed for purely structural reasons or for light acid-etched or very light sandblast texture is normally fully (continuously) graded, which means that it contains all the sizes of aggregates (below a given maximum) in amounts that ensure an optimum density of the mix.

Gradation standards for gap-graded mixes vary widely. The use of a one-sieve size or narrow size range for coarse aggregate, with a small percentage of concrete or masonry sand for workability, results in a more uniform distribution of exposed aggregate when texture is desirable. Gap-graded aggregates in which the coarser particles of sand and finer particles of coarse aggregate are omitted can produce a surface with greatly reduced bug holes provided proper consolidation is used. Lower sand content can result in greater color variations and aggregate transparency when dark colored coarse aggregates are used. Good concrete mix design aims to produce uniformity of color, avoid segregation, and minimize other surface blemishes. This type of mix is not necessarily the same as that used in standard structural concrete. When a contractor accepts the concrete mix design for architectural concrete, whether it be from the architect or a local laboratory, he implies that he is capable of using that mix proportion of specific materials to produce concrete with good workability and of adequate strength that will also produce concrete surfaces with an absolute minimum of surface blemishes. This is a big responsibility and should not be assumed by the contractor until after mock-ups are completed and approved.

The ratio of fine aggregate to coarse aggregate by weight should be $1:2:5$ to $1:3:5$ in gap-graded mixes. The fine aggregate is usually masonry sand. With high cement content concretes, coarsely graded sands may be satisfactory because the cement helps provide the needed fines for workability. With low-cement contents (under 564 lb/yd^3), the fine aggregate particles are necessary for good workable mixtures.

Mixture proportions vary depending on the texture desired. In order to minimize surface blemishes in architectural concrete the following proportions are recommended:

20.5.6.1 Smooth Texture

For adequate consolidation of continuously graded concrete, the desirable amount of air, water, cement, and fine aggregate (i.e., the mortar faction) is about 50% to 65% by absolute volume (45% to 60% by weight). Rounded aggregates, such as gravel, require slightly lower values, while crushed aggregate requires slightly higher values. Fine aggregate content is usually 35% to 45% by weight or volume of the total aggregate content. Aggregate fines below a No. 50 (3 mm) screen should not be in excess of 5%. If suitable form materials, release agents, and placing techniques are used, air voids on a surface can be reduced in size and number to an acceptable level.

20.5.6.2 Very Light to Light Texture

An acid wash or very light sandblast texture should use a maximum coarse aggregate size of 3/8 in (9.5 mm). The very light texture is sometimes called an "Indiana limestone finish." Aggregates may be continuous or gap-graded.

20.5.6.3 Medium Texture

During the process of removing the cement/sand matrix from the exposed surfaces, coarse aggregate in the middle size range may not be able to adhere to the remaining concrete surface, particularly if the particle shape is elongated, sharp, or flat. If these sizes are not eliminated from the mix, the percentage of the surface covered by the matrix (sand and cement) may be too large and aggregate distribution too uneven to provide a good surface appearance. The maximum size aggregate in gap-graded mixes should be 1 in (25 mm) since mixes with larger sizes of aggregate are extremely difficult to place. This is because the larger particles seldom move from their original position even with heavy vibrator effort, and they are more prone to segregation.

20.5.6.4 Heavy Texture

For a heavy exposed aggregate finish using an aggregate of 3/4 in (19 mm) maximum size, the No. 4 to 3/8 in (4.75 to 9.5 mm) particles could be omitted without making the concrete unduly harsh or liable to segregation. In order to achieve a gap between the 1/3 in (8 mm) size and the No. 8

size (2.36 mm), it is advisable to limit the amount of material passing the 3/8 in (9.5 mm) sieve to between zero and 10%. A typical fine aggregate should have 100% passing the No. 8 (2.36 mm) sieve and from zero to 10% passing the No. 100 (150 mm) sieve.

While a sand gradation typical for concrete can usually be used, in heavy exposed aggregate mixes it is advantageous in some cases to use a masonry or industrial sand for the fine aggregate where most of the particles pass a No. 8 (2.36 mm) sieve. The sand should have a fineness modulus of 2.4 or less. A greater explanation of mix requirements for different textures may be found in reference material, such as PCI MNL Manual for Quality Control for Plants and Production of Architectural Precast Concrete Products, 117-96.

20.5.7 Wood Sealers

With wood forms, release agents should generally be used over a sealed form surface. Sealers can minimize nonuniformity in surface finishes and extend the number of uses of the wood forms. Types of wood sealers are lacquer, epoxy, polyester, and polyurethane. Wood sealers help maintain a uniform color in the concrete. The lower the water/cement ratio, the darker the concrete surface, so if the absorption of the form is not uniform, the concrete will vary in color uniformity. Sealed, nonabsorbent forms prevent moisture loss and result in finishes that are uniform in color but lighter than those cast in absorbent forms.

Formwork surfaces must be dry and free from dirt, grease, or other impurities before form sealers are applied. The sealers should be applied to wood when it is new and unoiled. Before it can be sealed, the surface must be sanded to remove all raised grain and rough areas, and all holes or imperfections must be filled with a waterproof filler. For some sealers there are minimum temperatures stated below which they must not be applied. An appropriate drying or curing time should be allowed. The manufacturer's instructions regarding application of the sealer must be followed, and the work should be done by a skilled worker.

20.5.8 Release Agents

Release agent selection should include investigation of the following factors:*

1. compatibility of the agent with the form material, form sealer, and admixtures in the concrete mix—because of the rapid loss of slump, most superplasticized concrete needs a smooth "frictionless" surface along which the concrete can easily move;
2. possible interference with the later application of sealants, sealers, or paints to the form contact face;
3. discoloration and staining of the concrete face;
4. amount of time allowed between application and concrete placement and the minimum and maximum time limit for forms to stay in place before stripping—release agents may require a curing period before being used—if the concrete is too fresh when the agent is applied, some of the release agent will become embedded in the concrete,
5. effect of weather and curing conditions on ease of stripping;
6. uniformity of performance; and
7. conformance to local environmental regulations regarding the use of a volatile organic compound (VOC) compliant form release agent.

Release agents or form oils are still probably the leading cause of discoloration on architectural concrete. Today, many suppliers can offer a 100% nonstaining type if the contractor or precast manufacturer requests it. The safest approach is to evaluate the three or four products being considered by casting small-scale trial batches of concrete and applying the various products. Release agents improperly applied will cause as much color variation as any other factor known. Generally, the thinner the amount of the coating, the better the surface finish. Applying too much of the release agent can

*ACI Committee 303. July 1974. *ACI J.* :317–346.

cause excessive surface dusting on the finished concrete. Discoloration may be caused by excessive use or uneven application of the release agent. Bugholes can be expected where release agents are allowed to puddle on the form or mold or allowed to collect in the many nooks and crannies of a mold. Great care should be taken to see that the equipment used for applying the coating is clean.*

Coating thickness can be controlled by removing excess amounts with a clean cloth. A good rule of thumb is that, if rubbing your finger across the surface results in a deposit of the freshly applied release agent, there is too much present. One brand or batch of form-release agents should be used throughout a project. Information should also be obtained from the release agent manufacturer as to the kind of form surface for which the product is intended as well as the rate of spread and the proper method of application.

Generally, water-based or emulsion type release agents cannot be used in very cold weather because they might freeze. Even at temperatures slightly above freezing, some water-based products thicken enough to produce more bugholes and reduce performance. Also, a form face coated with a water-based agent should be protected from rain. Otherwise, some of the agent can wash off and it may be necessary to apply a second coat to a dry surface before placing concrete.*

For steel forms, release agents should contain a rust inhibitor and be free of water. Rough surfaces on steel forms may be conditioned against sticking by rubbing on a liquid solution of paraffin in kerosene, or the molds may be cleaned and oiled with a nondrying oil, then exposed to sunlight for a day or two.*

In order to reduce color changes, a suitable release agent should be applied for the first and all subsequent uses of glass fiber reinforced plastic molds or plastic form liners. An oil-based emulsion or high-quality household wax containing carnauba wax are preferable. Unsaturated oils, ketones, esters, acids, toluol, toluenes, xylenes, or halogenated solvents should be checked for compatibility with the plastic materials. If curing requires high temperatures, a silicone release agent should be used. Concrete molds will require a release agent consisting of light colored petroleum oils or oil emulsions. The concrete surfaces may be coated with one or two coats of epoxy resin and then waxed. A saponifiable oil should not be used as a release agent.*

Rubber or elastomeric liners may be coated with a thin film of castor oil, vegetable oil, lanolin, or water-emulsion wax. Mineral oil, oil-solvent based release agents, or paraffin wax should not be used on rubber or elastomeric liners as the hydrocarbon solvent will soften the rubber. The rubber or elastomeric supplier's recommendation should be carefully followed. Plastic foam molds are generally lightly sprayed with castor oil, petroleum jelly thinned with kerosene, or paraffin oil.*

As the concrete surface texture increases, the influence of the release agent becomes less important. It is desirable that the brushes or spray equipment be clean and that no laitance or build-up of concrete is on the form surface before the release agent is applied.

20.6 Color and Texture

20.6.1 Color

Architectural concrete can be cast in almost any form or texture to meet the aesthetic and practical requirements of modern architecture. Combining color with texture accentuates the natural beauty of the aggregates. Aggregate colors range from white to pastel to red, black, and green. Natural gravels provide a wide range of rich earth colors, as well as shades of grey. Color selection should be made under lighting conditions similar to those under which the architectural concrete will be viewed. Muted colors look best in subdued northern light. In climates with strong sunlight, much harder and brighter colors can be used with success.

The water/cement (w/c) ratio affects the color; the more water, (higher w/c ratio) the lighter the color even with the same cements. Richer mixes are darker than lean mixes. The common slump test is the usual measure of concrete consistency. Poor concrete mix control and wide variations

---

*PCI MNL-117-96:210 pp.

FIGURE 20.7 Precast addition over 15 years later using the same cement, aggregates and mix design.

in slump will produce different shades of color in any project. Variations in aggregate gradation can cause slump variations even when the w/c ratio remains constant. Color differential will follow variations in the w/c ratio regardless of slump. Good field control involves both aggregate gradation and slump control. Superplastisizers can change workability without adding extra water. If superplastisizers are used on one concrete placement, they should be used on all.

Color and color tone represent relative values. They are not absolute and constant, but are affected by light, shadows, density, time, and other surrounding or nearby colors. For instance, a concrete surface with deeply exposed opaque white quartz appears slightly grey. This is due to the fact that the shadows between the particles "mix" with the actual color of the aggregate and produce the graying effect. These shadows in turn affect the apparent color tone of the matrix. Similarly, a smooth concrete surface will change in tone when striated. A white precast concrete window unit with deep mullions will appear to change tone when bronze-colored glass is installed. A change in color tone is constantly going on as the sun travels through the day. A clear sky or one that is overcast will make a difference, as will landscaping and time. And last, but by no means least, in large city and industrial environments, air pollution can cause the tone to change. This is particularly noticeable when additions are planned to existing buildings built 10 or more years previously (Figure 20.7).

The ease of obtaining uniformity in color is directly related to the ingredients supplying the color. Whenever possible, the basic color should be established using colored fine or coarse aggregates (depending on depth of exposure) and pigments to blend the aggregates and matrix. Extreme color differences between aggregates and matrix should be avoided. In all cases, color should be judged from a full-sized sample that has the proper matrix and has been finished in accordance with planned production techniques.

The sample should be assessed for appearance during both wet and dry weather. The difference in tone between wet and dry concrete is normally less when white cement is used. In climates with intermittent dry and wet conditions, drying out periods often produce blotchy appearances on all-gray surfaces. On the other hand, dirt (weathering) will normally be less objectionable on gray concrete. These comparisons are based on similar water absorption or density of white and gray concrete.

20.6.2 Texture—Cast-in-Place and Precast*

20.6.2.1 General

Textures allow the natural attributes of the concrete ingredients to be expressed, provide some scale to the mass, express the plasticity of the concrete, and normally improve its weathering

---

*PCI Architectural Precast Concrete, 2nd ed. 1989. MNL-122-89 PCI:352 pp.

characteristics. A wide variety of textures is possible, ranging from smooth or polished to a deeply exposed tooled or bushhammered finish. As a general rule, a textured surface is aesthetically more satisfactory than a smooth surface because to a very large extent the texture of the surface camouflages subtle differences in texture and color of the concrete. Exposed aggregate surfaces may be achieved by removing surrounding paste through chemical processes (retarders or acid etching) or mechanically (abrasive blasting, honing and polishing, or tooling). Each method will uniquely influence the appearance of the exposed surface. Different degrees of exposure by any of these methods are defined as follows:

Light Exposure. Here, only the surface skin of cement and some sand is removed to expose the sand and the edges of the closest coarse aggregate. Matrix color will greatly influence the overall panel color.

Medium Exposure. Here, further removal of cement and sand has caused the coarse aggregate to visually appear to be approximately equal in area to the matrix.

Deep Exposure. Here cement and fine aggregate have been removed from the surface so that the coarse aggregate becomes the major surface feature.

A recently popular but very difficult texture is one that is much lighter than the "light exposure." This could be called a brush exposure or a texture similar to Indiana limestone. Indiana limestone results can be best obtained by blending yellow and black sands in the concrete mix design to give a veining effect. This very light texture is difficult to obtain with either acid wash (must use a very diluted solution) or a single-pass sandblast (must use a reduced nozzle pressure of 50 to 60 psi). Brush-blast exposure will expose the surface skin of the cement and part of the sand surface. No coarse aggregate is exposed.

20.6.2.2 Smooth

The smooth or as-cast texture is accepted the way it appears when the forms are removed or when the precast unit is stripped from the mold. Construction of any smooth as-cast surfaces should be approached with caution by the contractor. Even the East Building National Gallery of Art in Washington, DC., designed by I.M. Pei & Partners was not purely a smooth as-cast texture. The contractor, Chas H. Tompkins Co. scrubbed the exposed concrete surface with a weak acid solution to remove construction stains and environmental grime. This gave a slightly weathered look to the basically as-cast concrete surfaces.

A smooth off-the-form finish may often seem to be the most economical, but it is perhaps the most difficult finish to do well, as the color uniformity of gray, buff, or pigmented surfaces is extremely hard to achieve. The cement exerts the primary color influence on a smooth finish because it coats the exposed surface. In some instances, the sand may also have some effect. Initially, this is unlikely to be significant unless the sand contains a high percentage of fines or is itself highly colored. As the surface weathers, the sand will become more exposed, causing its color to become more pronounced. The color of the coarse aggregate should not be significant since it normally is not exposed to view unless the particular unit requires extremely heavy consolidation. Under this circumstance, some aggregate transparency may occur, causing a blotchy, nonuniform appearance.*

While a uniform graded concrete mix can be used with an as-cast finish, its water/cement ratio should be controlled. Concrete slumps should be held in the range of 3 in $\pm$ 1/2 in and never over 4 in. Good consolidation techniques and thorough blending of lifts are extremely important. While the use of concrete pumps is sometimes possible with this type of finish, the mix should not be oversanded or given additional water to facilitate the use of a pump. Oversanding greatly increases the possibility of segregation, resulting in noticeable surface variations. Aggregate transparency or

*PCI MNL-117-96:210 pp.

"shadowing" is a condition in which a light colored, formed concrete surface is marked by dark areas similar in size and shape to particles of dark or deeply colored coarse aggregate in the concrete mix. When encountered, it usually appears on smooth surfaces.[*]

The formwork for smooth-surfaced concrete is perhaps the most critical and the most difficult to control of any type of formwork encountered for cast-in-place or precast concrete, particularly where large single-plane surface areas are involved. Any imperfection in the surface of the form or any misalignment is immediately apparent and becomes the predominant factor in the character of the surface. Impervious surfaces such as plastic liners, steel, overlaid plywood, or fiberglass-surfaced plywood will usually result in a lighter color and more uniform appearance if joints have been properly prepared. In general, the joints of the materials used to construct the casting surfaces are difficult to hide unless rustications are used.

The smooth cement film on the concrete may be susceptible to surface crazing (fine and random hairline cracks) when exposed to wetting and drying cycles. This is, in most cases, a surface phenomenon and will not affect structural properties or durability. In some environments, crazing will be accentuated by dirt collecting in the minute cracks. This will be more apparent in white than gray finishes and in horizontal more than vertical surfaces.

When air voids of a reasonable size, 1/8 to 1/4 in (3 to 6 mm), are encountered on return surfaces, it may be desirable to retain them rather than filling and sack rubbing them. Color variations can occur when sacking is performed. If smooth surfaces are "to be produced without additional surface treatment after stripping, except for possible washing and cleaning," the following precautions should be considered:

1. Pay attention to detailing with provisions for ample draft, proper edges and corners, rustications at form edges, and suitable water drips and other weathering details.
2. Construct forms or molds so that imperfections will not be mirrored in the units. The use of plastic molds or liners with a matte finish or fiberglass overlaid plywood, which is smooth but not glossy, will help reduce crazing tendencies.
3. The mold release agent should be the same throughout construction and be applied under as nearly identical conditions as possible each time. (Some release agents help reduce the crazing tendency by breaking the contact with the glossy surface of the form.)
4. Concrete mix designs should be used that combine a minimum cement content and a constant, low water/cement ratio with high density, in order to minimize crazing, entrapped air voids, and color variations. The mix should be fully graded with no more than 5% aggregate fines passing a No. 50 (300 mm) sieve.
5. Proper consolidation and curing should be used to minimize nonuniformity of color, which shows easily on smooth concrete surfaces, particularly with gray concrete mixes. Uniform curing with minimum loss of moisture from the smooth surface will help minimize crazing tendencies.
6. Minimize chipping or other damage because smooth finish repairs are difficult to perform in terms of texture and color match.

Many of the aesthetic limitations of smooth concrete may be minimized by the shadowing and depth provided by profiled surfaces (fluted, sculptured, board finishes, etc.), which subdividing the panel into smaller surface areas by means of vertical and horizontal rustications, or the use of white cement.

20.6.2.3 Sand or Abrasive Blast[†]

Sand or abrasive blasting of surfaces can provide all three degrees of exposure: light, medium, or deep. Generally, the technique is used when a light exposure is desired, because sandblasting costs

[*]Shilstone, J.M. 1972b. *ACI J. Concr. Const.* :526–530.
[†]PCI. "Manual for Quality Control for Plants and Production of Architectural Precast Concrete Products" 1996, 117-96 PCI:210 pp.

increases, with the depth of exposure. Sand or abrasive blasting of surfaces is suitable for exposure of either large or small aggregates. Uniformity of depth of exposure between panels and within panels is essential in abrasive blasting, as in all other exposed-aggregate processes, and is a function of the skill and experience of the operator. As much as possible, the sandblasting crew and equipment used should remain the same throughout the job. Different shadings and, to some extent, color tone will vary with depth of exposure.

The degree of uniformity obtainable in a sandblasted finish is generally in direct proportion to the depth of sandblasting. A light sandblasting may look acceptable on a small sample, but uniformity is rather difficult to achieve in a full size wall or precast unit. A light sandblast will emphasize visible defects, particularly bug holes, and reveal defects previously hidden by the surface skin of the concrete. The lighter the sandblasting, the more critical the skill of the operator, particularly if the units are sculptured. Small variances in concrete strength at time of blasting may further complicate results.

FIGURE 20.8 Heavy abrasive blasted cast-in-place gap-graded concrete texture. Tennessee State Office Building—Memphis, Tennessee. Architect Roy Harover.

Sculptured units will have air voids on the returns that might show strongly in a light-sandblasted texture. If such air holes are of a reasonable size, 1/8 to 1/4 in (3 to 6 mm), it is strongly recommended that they be accepted as part of the texture because filling and sack rubbing may cause color variations.

Blasting will cause some etching of the face of the aggregate, and the softer aggregates will be etched to a greater extent with heavier exposures. Sandblasted aggregates lose their sharp edges. Blasting of the aggregate surface is more noticeable on dark-colored aggregates that have a glossy surface texture. Sand blasting dark-colored aggregates will produce a muted or frosted effect, which tends to lighten the color and subdue the luster of the aggregate (Figure 20.8). Depth of sandblasting should also be adjusted to suit the aggregate hardness. For example, soft aggregates might be eroded at the same rate as the mortar.

For medium or deep exposure with a sandblasted finish, retarders may be used initially and the matrix removed by sandblasting to obtain a matte finish. The selected retarder strength should only give 50 to 75% of the expected reveal. This approach reduces blasting time and lessens the abrasion of softer aggregates. Using sandblasting to achieve the final texture allows for correction of any variations in exposure, so this method can result in a very uniform surface. Care should be taken to avoid nonuniform exposure that may be caused by the presence of soft and hard spots on the retarded surface. This is especially true and more noticeable on large flat surfaces. Small flat areas or surfaces that are divided by means of rustications will tend to call less attention to these texture variations. Since some aggregates change color after sandblasting, trials on sample panels using different abrasive materials are desirable to check the texture and color tone.

As an additional step toward uniformity, the cement and sand color should be chosen to blend with the slightly "bruised" color of the sandblasted coarse aggregate, as the cement-sand matrix color will predominate when a light sandblast finish is desired. With a light sandblasting, only some of the coarse aggregates near the surface will be exposed, so a reasonable uniform distribution of such aggregates is not controllable.

The concrete mix used, and the matrix strength at time of blasting, will affect the final exposure, as will the gradation and hardness of the abrasive. A good general rule in selecting an abrasive is that the particle size of the abrasive will attack a similar size particle in the concrete surface.

For a more uniform texture, spherical or nearly round abrasives with a close gradation should be used. Materials used in the blasting operation are washed silica sand, certain hard angular sands, aluminum carbide, blasting grit such as power plant boiler slag, carbonized hydrocarbon, crushed chat (a waste material from lead mining), and various organic grits such as ground shells, corn cobs, and rice hulls.

For cleaning or light blasting of a surface, any of the abrasives will be adequate. For deep cutting, an abrasive grit should be used because of its speed of attack and cleaner surface appearance. Some types of colored abrasives impart color to the surface of the concrete. With certain gradation combinations, pressure, and volume, impregnation of the abrasive into the surface can occur. If this happens, an abrasive of similar color to the matrix should be used. Impregnation of abrasive can be minimized by a change in the volume of material, its gradation, and/or the pressure being applied.

Sandblasting may be done with dry abrasive in a stream of compressed air or water rings may be used to introduce water into the compressed air-sand stream at the nozzle. Sand also may be introduced into a high-pressure water washer. When using wet sandblasting, the abraded mortar should be continually washed off already sandblasted areas to prevent staining.

The inside diameter of the hose should be no less than 1 1/4 in (32 mm) or four to six times the diameter of the nozzle orifice in order to keep the sand in continuous suspension while it travels through the hose. Too large a hose would reduce the velocity of media and air and would eventually plug the hose. Smaller diameter whip lines may be used for operator ease of handling. The nozzle at the end of the system is the most important element. The diameter of the nozzle and nozzle pressure should be determined by experimentation. A venturi type nozzle should be used to obtain a uniform blast pattern. Carbide or norbide nozzles should be selected for durability. Nozzle life depends on abrasive hardness and volume as well as the pressure and generally varies from 2 to 4 mo.

The time when sandblasting should take place is determined by scheduling, economics, visual appearance desired, and hardness of the aggregate. The timing of blasting is not as critical as for other finish methods. The concrete matrix will be easier to cut in the first 72 h after casting. As the concrete cures and gains strength, it becomes more difficult to blast to any appreciable depth, thus increasing the cost of the operation. Softer aggregates tend to abrade more when concrete strengths are high and the surface will have a duller appearance. In some cases, the higher costs of deferred blasting may be justified to avoid scheduling problems. However, all surfaces should be blasted at approximately the same age or compressive strength for uniformity of appearance.

When blasting, the operator should hold the nozzle perpendicular to the surface being blasted. Some operators will deviate slightly from this position as it seems to provide a better view of the work. The maximum deviation should be less than 15° as too much deviation from the 90° angle will result in undercutting the coarse aggregate particles. Best results are obtained with the nozzle positioned from about 2 to 6 ft (0.6 to 1.8 m) from the element surface. The exact distance depends on the pressure used, the hardness of the concrete matrix, and the cutting ability of the abrasive. An experienced operator can quickly determine the nozzle position to produce the specified surface finish. Using a circular motion during blasting will minimize pattern marking.

20.6.2.4 Acid Etch

Acid etching of concrete surfaces can result in a pleasing fine sandy texture and retention of detail if the concrete mix and its consolidation has produced a uniform distribution of aggregate particles and cement paste at the exposed surface. The contractor should approach this finish very carefully. Vertical cast-in-place concrete exposed with acid can be an environmental problem. New equipment that uses-high-pressure pumps and a controlled combination of acid and hot water has minimized the application problems. When using this equipment, runoff water is close to neutralized. Some local codes or concerned citizens may require complete containment of the acid residue runoff even though the parts per million of acid are insignificant. This may involve construction of earth dams and a holding pond for the water prior to treatment. It has become increasingly difficult to perform any work requiring acid at job sites owing to potential liabilities and environmental concerns.

All personnel working with or near the acid application area will need protective gear and clothing to prevent injury from the spray. Prior to acid etching, all exposed glass and metal surfaces should be protected with acid-resistant coatings. These include vinyl chlorides, chlorinated rubber, styrene butadiene, rubber (not latex), bituminous paints or enamel, and polyester coatings. Architectural surfaces should be thoroughly flushed with large quantities of clean water immediately prior to application of acid.

Acid etching is most commonly used for light or medium exposure. Acid etching of concrete surfaces will result in a fine, sandy textured distribution of aggregates and cement paste at the exposed surfaces. Concentrations of cement paste and under-and-over etching of different parts of a concrete surface, or variation in sand color or content may cause some uniformity problems, particularly when the acid etching is light or used for large, plain surfaces. Carbonate aggregates (e.g., limestones, dolomites, and marbles) may discolor or dissolve owing to their high calcium content. With lighter textures, color compatibility of the cement and the aggregates becomes more important for avoiding blotchy effects. White or light colors are forgiving to the eye and increase the likelihood of a better color match from unit to unit. Gray is the worst color to select for uniformity. An acid-etch finish is more difficult to patch than many of the deeper texture finishes. However, minor air voids are fairly easy to grout and refinish.

There is a minimum depth of etch that is required to obtain a uniform surface. To attempt to go any lighter than this will result in a blotchy panel finish. A minimum depth of etch will expose the sand and the very tip of the coarse aggregate approximately 1/16 in. (1 mm). It is difficult to achieve a totally uniform very light exposure on a panel that is highly sculptured. This is due to the acid spray being deflected to other areas of the panel, particularly at inside corners. This may be acceptable if the sculpturing creates differential shadowing. Prewetting the concrete with water fills the pores and capillaries and prevents the acid from etching too deeply, and also allows all acid to be flushed after etching. Older dried concretes are likely to be more carbonated. Although the reactions of carbonates with the acid might not be much faster than those with other cement compounds, they cause greater efflorescence so that the reaction is far more obvious and seems to go faster. Acid solutions lose their strength quickly once they are in contact with cement paste or mortar; however, even weak, residual solutions can be harmful to concrete owing to possible penetration of chlorides. Failure to completely rinse the acid solution off the surface may result in efflorescence or corrosion.

20.6.2.5 Retarders

Chemical surface retarders provide a nonabrasive process that is very effective in bringing out the full color, texture, and natural beauty of the coarse aggregate. The aggregate is not damaged or changed by this exposure method. If exposing the bright, natural colors of the aggregate is the prime goal, exposing aggregate from retarded surfaces is one of the best ways to achieve this result. Surface retarders that are to be used to expose the aggregate should be thoroughly evaluated prior to use. A sample panel should be made to determine the effects created by the form or mold and concrete materials. This involves using the particular type of cement, aggregate (proportion determined by specified mix design), and selected release agent. Prolonged exposure of the forms coated with the retarder prior to placing the concrete should be avoided. Water should not contact the retarder on the form surface before the concrete is placed in order to prevent activation of the retarder. When using a retarder, the manufacturer's recommendations should be followed. Surface retarders can be applied by roller, brush, or spray, and care must be taken to ensure uniform application to the form surface.

The performance of the retarder will be influenced by chemistry of the individual cement, mix characteristics, temperature of the concrete, humidity, ambient temperature, characteristics of absorption of the forming material, and total water content of the concrete mix. It is important to consider protecting the treated form surfaces from weathering and ultraviolet rays before casting the concrete. Retarders function by delaying, not preventing, the set of a given amount of cement paste so that the aggregate can be easily exposed. This concept will help in analyzing various mix designs for depth of retardation. If more sand or coarse aggregate is added to a mix with proper consolidation, there will be less cement paste per volume of material at the surface, and thus a deeper exposure. Some retarders are effective for long periods of time while others are active for only a few hours. Water contacting the retarder before the concrete is placed activates the retarder's action prematurely and may result in a nonuniform surface.

The retarded concrete should be removed the same day that the forms are stripped using a high-pressure water blast through a fan nozzle. Any delay in removing the matrix will result in a lighter, less uniform texture. The stripping schedule must be coordinated with that of the work force responsible for form removal so that there will not be long periods between form removal and concrete finishing. For a large project, preliminary tests should be performed before planning the casting to determine the most suitable finishing time. The timing of the surface finishing operation should be consistent each day, as some retarders cease to delay the hardening process as the product cures.

20.6.2.6 Tooled or Bushhammered Finish

This finish is usually achieved by casting concrete against smooth or specially textured or patterned formwork. After removal from the form, the hardened concrete is treated mechanically to create the desired effect. Concrete made with most aggregates can be tool finished, but materials that can be cut or bruised without shattering, such as calcareous limestone and igneous rocks, give results that find the widest acceptance. Aggregate particles of 3/8 in and smaller are more important for scaled and tooled surfaces. The coarser aggregate particles are more important for jackhammered surfaces. Some jackhammered surfaces where, considerable depth of finish is desired, may only require forming practices similar to a good structural concrete. Mechanically fractured surfaces are prepared by one of four methods: scaling, bushhammering, jackhammering, or tooling, which includes reeding. These types of surface textures are described as fractured because the surface is prepared by striking the concrete with a tool, thereby mechanically removing part of the surface.

Concrete may be mechanically spalled or chipped with a variety of hand and power tools to produce an exposed-aggregate texture. Pneumatic or electric tools may be fitted with a comb chisel, crandall, or multiple pointed attachments. The type of tool will be determined by the surface effect desired. Hand tools may be used for small areas, corners, and restricted locations where a power tool cannot reach. Basically, all methods of tooling remove a layer of hardened concrete matrix while fracturing the larger aggregates at the surface. Surfaces attained can vary from a light "scaling" to a bold deep texture achieved by jackhammering with a single pointed chisel.

Orientation of equipment and direction of movement should be kept uniform throughout the tooling process as tooling produces a definite pattern on the surface. Variations due to more than one person working on the surface may occur with this finish. Care should be exercised to avoid exerting excessive pressure on the tool, especially when starting, so as not to remove more material than either necessary or desirable.

Bushhammering at outside corners may cause jagged edges. If sharp corners are desired, bush-hammering should be held back from the corner. It is quite feasible to execute tooling along specific lines. If areas near corners are to be tooled, this must normally be done by hand since power tools will not reach into inside corners. Chamfered corners are preferred with tooled surfaces. With care, a 1 in (25 mm) chamfer may be tooled. Scaling is the lightest texture in bushhammered finishes. It is achieved by passing a triple-pronged scaler (originally developed to remove scale from steel prior to pointing) singly or in gangs over the surface to remove only a thin skin. A single-head scaler is lightweight and can be readily manipulated by one person. No texture as such is brought out by this technique, although some aggregate is exposed and fractured in the process. Under certain conditions, almost the same result can be achieved by light abrasive blasting.

Jackhammering should be done when the matrix has reached a strength approximating that of the coarse aggregate in order to fracture both the mortar and the coarse aggregate. If hammering is started too soon, the tool merely removes the matrix and the coarse aggregate does not fracture. Sometimes the coarse particles are knocked out, leaving blank spaces. If jackhammering is to be performed at a time when the matrix is softer than the coarse aggregate, a chisel-type tool should

FIGURE 20.9 Heavy jackhammer surface. Southern Bell Tower, Atlanta, Georgia. Architect, SOM, Chicago, Illinois.

be used. This tool has a tendency to fracture across the aggregate, while a pointed tool has a tendency to dig into the matrix and not fracture the coarse aggregate. On the other hand, when a concrete becomes very hard the pointed tool does a superior job. Since jackhammering accentuates the presence of coarse aggregate, a higher than normal coarse aggregate content may be desirable (Figure 20.9).*

Although a dense, fully graded concrete mix is desirable, bushhammering may be successfully applied to gap-graded concrete. Natural gravels are inclined to shatter, leading to bond failure and loss of aggregate particles when bushhammered. Aggregates such as granite and quartz are difficult to bushhammer uniformly because of their hardness and may fracture into rather than across the concrete surface. Aggregates such as dolomite, marble, calcite, and limestone are softer and more suitable for bushhammered surfaces.*

In order to prevent loosening of the aggregate, a compressive strength of 4000 psi (28 MPa) is recommended. In many cases, better uniformity may be obtained when the concrete is allowed to age for 14 to 21 d and the surface is dry. Exposing the aggregate by tooling requires trained operators in order to produce a uniformly textured surface, especially when large areas are to be textured. A hammered rib (or fractured fin) finish may be produced by casting ribs on the surface of the unit and then using a hammer or bushhammer tool to break the ribs and expose the aggregate. The ribs may be hammered from alternate sides, in bands, to obtain uniformity of cleavage, or randomly, depending on the effect required. There should be a definite plan, even with a so-called random pattern, because an uneven shading effect may be produced unless care is exercised. Tooling and bushhammering remove a certain thickness of material, 3/16 in (5 mm) on average, from the surface of the concrete and may fracture particles of aggregate, causing moisture to penetrate the depth of the aggregate particle. For this reason, the minimum cover for the reinforcement should be somewhat greater than normally required. It is recommended that 2 in (50 mm) of concrete cover be specified (prior to tooling).*

As a cutting head becomes worn, it should be replaced. Since the texture varies with the condition of the cutting surface, a new head should not be worked next to an area tooled with an old one. Small irregularities in the finished surface can be worked out with further tooling.*

20.6.2.7 High-Pressure Water Jet Blasting

Blasting with high-pressure water instead of sand has received limited use in architectural concrete, particularly with precast concrete. Restrictions on sandblasting in many areas of the country are sure to increase the spread of this method of exposing aggregates at the surface. It uses only water—

*PCI. "Manual for Quality Control for Plants and Production of Architectural Precast Concrete Products" 1996, MNL-117-96, PCI:210 pp.

no harsh chemicals—so there is no pollution and no health risk from burns or fume inhalation. Under special situations or where environmental restrictions exist, the water spray can be mixed in the nozzle with a fine spray of sand. Units are available with more than 10,000 psi of pressure, although usually 5000 psi is sufficient for aggregate exposure. Considerable splatter can develop, which requires that adjacent surfaces be protected.

High-pressure water jets are used in combination with air to expose aggregates. Proper time of application must be determined for each concrete and its curing condition to obtain the desired amount of reveal without loosening the aggregate. Strength of concrete for high-pressure water washing is usually 1500 psi. It is important that the strength of concrete be approximately the same when exposing the aggregate on different concrete placements. This method can be used with or without surface retarders.[*] Regardless of whether retarders are used or not, exposure should be started immediately after forms are stripped. Each equipment operator should be trained on a sample test area. All operators should be protected by rubber wet suits and goggles. They should wear Rubber steel tipped boots with clamp-on instep guards. These are required because the jet has enough force to cut through rubber into the flesh. A dead-man trigger on the gun automatically shuts off the water jet if the gun is accidentally dropped.

20.6.2.8 Form Liners

An almost unlimited variety of attractive patterns, shapes and surface textures can be achieved by casting concrete against wood, steel, plaster, elastomeric, plastic, or foam plastic form liners. These form liners can be attached to or incorporated into the form itself. Most form liners can be used with any forming system. They are particularly useful in large heavy-duty gang form sections because the joints can be permanently aligned and sealed. Architects specify patterned concrete surfaces for the aesthetic interest generated by the play of light and shadow or to economically simulate the traditional patterns of brick, stone, and wood. Adding texture with a patterned form liner helps hide color variations and makes bug holes and other surface blemishes less noticeable. Defects due to leakage at form joints and poorly consolidated concrete also are less noticeable in patterned concrete.[†]

Material selection of the form liner should depend on the amount of usage and whether or not the pattern has undercut (negative drafts). The choice of liner materials may depend on whether the work is cast-in-place or precast. Thin sheets that serve well on horizontal forms may wrinkle and sag in vertical forms unless they are carefully attached. Thicker layers of liner material are more suitable for vertical surfaces. Vertical liners also must withstand the lateral pressure of concrete. This may call for control of concrete placing rates and added support for a contoured liner. The method of attaching form liners should be studied for the resulting visual effect. Form liners should be secured in forms by gluing or stapling, rather than by methods that permit impressions from nail heads, screw heads, rivets, or the like to be imparted to the surface of the concrete, unless desired. Where staples are used, they should be driven by a power stapler to ensure sufficient driving force. Staples should be aligned in a direction consistent with the pattern to be imparted to the concrete. Staples should be spaced close enough to hold the liner securely in place. Where adhesives are used, particular attention should be given to following manufacturer's recommendations regarding adhesive cure period, ambient temperature, and moisture.[0] Protection from the weather is essential. Attempts should be made to camouflage inequalities to within the pattern of the texture. The construction cycle, the probable number of reuses of the lined form, and the cost per use are other concerns in selecting a liner material. Trials should be made to determine the best time for stripping so the surface remains intact and liner reuse can be maximized.

Wood liners, whether used as boards, plywood panels, or nailed-on inserts, work well. Wood liner surfaces should be sealed to minimize discoloration of the concrete caused by differential

---

[*]ACI Committee 303. July 1974. *ACI J.* :317–346.
[†]Hurd, M.K. 1993. *Concr. Const.* :331–336.
[0]Ford, J.H. 1982. *ACI J.* :51–57.

absorption of mix water by the liner. Then the liner should be lightly coated with form oil prior to casting. Sandblasted wood, textured plywood, and rough-sawn lumber are often used for board surface textured finishes where concrete color variations and rough edges are acceptable. To prevent bowing of the boards, all sides should be coated with wood sealer by painting or immersion. Even with a sealer, some types of lumber may absorb moisture from the concrete. In other cases, natural sugars found in lumber such as pine may penetrate the sealer coating when the concrete is cast against the mold, retard the set of the cement at the surface, and cause a dusty, dark, blotchy effect. Fir is a preferred choice for board surface finishes due owing its low sugar content. An effective method of sealing wood to eliminate any moisture transfer is to spray a few light coats of surface resin or urethane onto the wood. Care should be taken not to apply the sealer too thick or the wood grain pattern will be lost. For molds with long casting durations, this process may need to be repeated. The weathering of the lumber can also affect the outcome of the concrete finish. If rough sawn lumber is being used, it is important to produce samples to determine the effect the lumber will produce. The lumber selected should be purchased all at one time from one source to minimize the possibility of variations. Moisture leakage between pieces of lumber should be prevented, or a dark line will result from the change in w/c ratio. The joints may be sealed by using tongue-and-groove lumber. Closed-cell gasket material should be used at edges of the mold to prevent leakage. Wood liners may also be set or embedded in resin, which will eliminate the need for connectors that show as well as seal the edges against leakage.*

Some experts believe porous or absorbent form liners offer a better chance to control bug holes than do impervious plastic liners, but the absorbency that provides this advantage can also pose problems. The more absorptive the form face, the darker the color of the concrete. However, as wood forms are reused and reoiled, their absorbency decreases. If patching or adjusting a much-used panel is required, the location of the new board or plywood will show up as a color difference that takes years to weather away (Figure 20.10). To avoid this problem with wood liners, up one or two extra panels or gang forms should be made and kept in form rotation. Then if one form panel or gang form is damaged, it can be replaced with a form section already seasoned with use.

Elastomerics such as urethane and hot-melt vinyl are used to make liners flexible enough to permit vertical sides or some undercut areas. Elastomeric liners greatly facilitate removal from finished concrete surfaces in cases where other materials would be virtually impossible to strip. The design possibilities are almost limitless. Standard sheets are typically 4 × 8 to 4 × 12 ft, (1.2 × 2.5 to 1.2 × 3.7 m) but sizes up to 12 × 36 ft (3.7 × 11 m) can be special ordered to help minimize horizontal butt joints. Elastomeric materials should have a Shore A-2 hardness of 50 to 60 durometer and a minimum ultimate tensile strength of 600 psi (4.14 MPa). Elastomeric and rubber liners display gasketing characteristics and therefore achieve weep-free seamless joints. They also eliminate the need to cover the small slits cut in the liner for the fasteners. Elastomeric form liners may ripple unless there is a good bond to the base form. Edges of liners should be sealed to each other or to divider strips to prevent bleeding of cement paste. The sealant used should not stain the surface. Liners should not be butt-jointed without a demarcation feature to eliminate nonalignment of the texture. Liner size and module should be coordinated with panel joints, rustication strips, and blockout size.

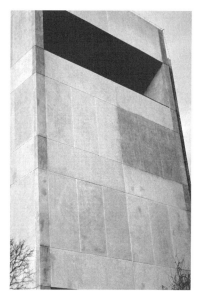

FIGURE 20.10 Wood forms having varying degrees of absorbance as new sections have been added to gang forms.

*PCI MNL-117-96, PCI:210 pp.

Tough and wear-resistant elastomeric liners are relatively heavy and require good vertical support. They are usually attached to the form sheathing with adhesive, but some manufacturers supply these liners prebonded to plywood sheets. Elastomeric liners are sensitive to temperature change and may deform significantly when exposed to surface temperatures above 140°F(60°C). Direct sunlight and the heat of hydration of the concrete can generate this level of temperature.* Shade and temperature control of concrete during curing may be necessary. The liner should be checked for resistance to deterioration caused by oils commonly used as release agents and have rigidity sufficient to resist wrinkling. Solvent-based surface retarders should be checked to be sure they won't degrade the liner and, if necessary, water-based or carried retarders should be used.

Several rigid liner materials such as ABS and polyvinyl chloride (PVC) come in sheets stiff enough to be considered self-supporting. High-impact polystyrene is also used to make rigid form liners. Readily available in 4 × 8 or 4 × 12 ft (1.2 × 2.5 or 1.2 × 3.7 m) sheets, they can also be special ordered in lengths up to 30 ft (9.2 m) or more. Some manufacturers will supply interlocking joints at the edges of the panels that help hide vertical form joints. Rigid liners are well-suited to ribbed patterns, but a detailed closure is needed where the plastic flutes are cut at horizontal joints or openings. Some makers supply prefabricated closure pieces to solve this problem.* This material does not lend itself to intricate patterns, particularly with sharp corners, vertical sides, or undercuts (negative drafts). Basically, the material itself is inflexible. If they are broken, they are difficult to repair at the job site. Sheet plastics may need appropriate backup to resist movement, particularly for wide portions of liners with deep indentations. Movement can occur owing to temperature changes caused by direct sunlight or the mechanical action of concrete placement. It is good policy to maintain a 10° draft on all indentation sides to prevent chipping and spalling during stripping operations. Keep all edges and corners rounded or chamfered. Relief may be more than 1 in (25 mm) deep if the depressed area is sufficiently wide.

Foamed polystyrene or polyurethane liners create deeply revealed designs or blockouts. The preformed foam planks are easily cut to size and readily attached to the form. It is necessary to use a low-solvent contact glue that should dry before contact with the foam plastic is made or the solvent will dissolve the plastic. Liners made of polystyrene foam are used in large sheets like other liners, or in smaller interlocking modules designed to fit together with concealed joints. They typically are single-use, usually destroyed by the stripping process. Because of the one-time use factor, they are not cost efficient for large areas with a repetitive pattern. Repetitive form use requires a different type liner. Rigid plastic forms can be molded to create complex liners with overall patterns or original works of art. Molded patterns have a surface skin that can be coated with a release agent. Strip the foam by hand or with air or water jets. If some of the plastic foam bonds to the concrete, you may need to wirebrush or sandblast at reduced nozzle pressure to remove it.

Metal liners are available in various textures that can be combined with different types of fasteners to achieve an architectural effect. Liner joints should be at rustication strips or mold edges since leakage is difficult to prevent at butt joints. Combination finishes involving the use of more than one finishing method are almost infinite. One common example is the ribbed form liner and sandblasted finish.

20.6.2.9 Applied Coatings

Whenever concrete is to be coated or stained, only form-release agents compatible with the coating should be permitted, unless surface preparation is required to assure good adhesion between the coating and the concrete. Coatings applied to exterior surfaces should be breathable (permeable to liquid water). The coating manufacturer's instructions regarding mixing, thinning, tinting, and application should be strictly followed. Whenever concrete is so smooth that it makes adhesion of the coating difficult to obtain, the surface should be lightly sandblasted, acid etched, or ground with silicon carbon stones to provide a slightly roughened, more bondable surface.

Since there is a vast difference in coating and stain types, brands, prices, and performances, knowledge of coating composition and performance standards is necessary for obtaining a satisfactory

*Hurd, M.K. 1993. *Concr. Const.* :331–336.

concrete coating or stain. In order to select proper coatings, the architect should consult with manufacturers supplying products of known durability and obtain from them, if possible, technical data explaining the chemical composition and types of coatings suitable for the job at hand. For high-performance coatings, proprietary brand-name specifications are recommended. The interior surface of exterior walls should have a vapor barrier (coating or other materials) to prevent water vapor inside the building from entering the wall.

20.6.3 Texture—Precast Only

20.6.3.1 Sand Embedment

Bold, massive, rocklike architectural qualities may be achieved by hand placing 1 to 8 in (25 to 200 mm) diameter stones (cobbles or boulders), fieldstone, or flagstone into a sand bed or other special bedding material. The depth of the bedding material should keep the backup concrete 25 to 35% of the stone's diameter away from the face. Extreme care should be taken to ensure that the aggregate is distributed evenly and densely on all surfaces, particularly around corners, edges, and openings. To achieve uniform distribution and exposure, all aggregate should be of one size gradation. Where facing materials are of mixed colors, their placement in molds should be carefully checked for the formation of unintended patterns or a local high incidence of a particular color. If the intention is to expose a particular facet of the stone, placing should be carefully checked with this in mind before the backup concrete is placed.

The sand-embedment technique reveals the facing material and produces the appearance of a mortar joint on the finished panel. If a white or colored mortar joint is desired, a mortar consisting of one part white cement (plus pigment) to 2-1/2 parts well-graded white or light-colored sand with sufficient water to make a creamy mixture may be placed over the aggregate. If a mortar facing mix is used, the backup mix should be of a low slump with a maximum of 1 in (25 mm) to absorb excess water from the facing mix. Otherwise, the backup mix should be a standard structural concrete with a slump of 2 to 4 in (50 to 100 mm).

The mortar mix or part of the backup concrete should be carefully shoveled onto the stones and further spread and lightly tamped with trowels to ensure that it is worked around all the individual stones. Then it should be screeded to a flat surface before the steel is placed. Care should be taken not to dislodge any of the face stones when placing the first layer of mortar or concrete. Also, care should be exercised during vibration so as not to disturb the sand or large stones, which could cause uneven stone distribution. Upon stripping the precast concrete units, they should be raised, and any clinging sand should be removed by brushing, air blasting, or high-pressure water washing. Some sand bonds to the concrete; therefore the color of the bedding sand should be carefully chosen to harmonize with the exposed stones.

20.6.3.2 Veneer Facing Materials

When natural stone veneer is used, it is recommended that the purchaser of the stone engage someone qualified to be responsible for coordination, which includes delivery, scheduling, and ensuring color uniformity. In evaluating properties of stone, it should be recognized that some natural stones exhibit different properties in different orientations. Also, there may be considerable variation in a given direction of grain for different samples of the same stone. The thin sections of stone are generally more sensitive than thicker sections to strength decreases due to imperfections and inclusions of minerals. Also, a stone that has a crystalline structure with dimensions large enough to approach the thickness of the slab itself will be substantially weakened. In addition, the surface finish, freezing and thawing, and large temperature fluctuations will affect the strength and in turn influence the anchorage system. Nonacid-based masking or plastic tape may be used to keep concrete out of the stone joints so as to avoid limiting stone movement.*

Color control or blending for uniformity should be done in the stone fabricator's plant since ranges of color and shade, finishes, and markings such as veining, seams and intrusions are easily seen

*PCI MNL-117-96, PCI:210 pp.

during the finishing stages. A qualified representative of the owner, who understands the aesthetic appearance requested, or the owner or architect should perform this color control. Acceptable color should be judged for an entire building elevation rather than for individual panels.

All testing to determine the physical properties of the stone veneer should be conducted by the owner prior to the award of the contract. This will reduce the need for potentially costly repairs or replacement should deficiencies in the stone veneer be found after start of fabrication. A contractor, when he buys the stone, should be sure to order an additional 5% to cover breakage and incorrect sizing by the supplier.

A complete bondbreaker between the natural stone veneer and concrete should be used. Connecting the veneer to the concrete should be done with mechanical anchors that can accommodate some relative movement. When using epoxy in anchor holes, 1/2 in long compressible rubber or elastomeric grommets or sleeves should be used on the anchor at the back surface of the stone. Thin brick (1/2 to 1 in thick) rather than whole bricks are generally used in precasting because adequately grouting the thin joints with whole bricks is difficult and results in the use of mechanical anchors. Some bricks are too dimensionally inaccurate for precast concrete applications. They may conform to an ASTM specification for site laid-up application, but they are not manufactured accurately enough to permit their use in the preformed grids that are used to position brick for a precast concrete unit. When both site laid brick and brick precast units are to be used on the same project, the contractor has to be sure the precast tolerance requirements govern. Because variations in brick or tile color will occur, the clay product supplier should preblend any color variations and provide units that fall within the color range selected by the architect. Ceramic glaze units, where required for exterior use, may craze from freeze-thaw cycles or the bond may fail on exposure; therefore, the manufacturer should be consulted for suitable materials and test data.

20.6.3.3 Honed or Polished Surface

Grinding concrete produces smooth, exposed aggregate surfaces. The grinding is called honing or polishing, depending on the degree of smoothness of the finish. Polished exposed aggregate concrete finishes compare favorably with polished natural stone facades, allowing the architect great freedom of design. Honed and polished finishes have gained acceptance because of their appearance and excellent weathering characteristics, which makes them ideal for high traffic areas and polluted environments. In order to produce a good ground or polished finish, it is first necessary to produce a good plain finish. The compressive strength of the concrete should be 5000 psi before the start of any honing or polishing operations. All patches and the fill material on any bug/blow holes or other surface blemishes must also be allowed to reach approximately 5000 psi. It is preferable that the mortar strength of the concrete mix approach the compressive strength of the aggregates, or the surface may not grind evenly or polish smoothly and aggregate particles may be dislodged.

Since aggregates will polish better than the matrix, it is essential to have a minimal matrix area. A continuous graded concrete mix heavy in coarse aggregate is preferred, carefully designed to provide maximum aggregate density on the surface to be polished. In choosing aggregates, special attention should be given to maximum size and hardness. Softer aggregates such as marble or onyx are much easier to grind than either granite or quartz. This will be strongly reflected in the cost of such finishes.

20.6.4 Combination Textures

The combination of a polished or honed surface and acid etching provides a surface on the precast unit which exposes a very high percentage of stone. After the grinding process, the acid removes the cement matrix and fines between the larger aggregate particles. This surface is highly resistant to weathering and is self-cleaning to a high degree. The color of the aggregates predominates in the combined polished acid-etch surface texture. A polished/sandblast finish on a precast unit provides contrast between the smooth polished aggregate and the sandblasted matrix of the concrete. The architect must ensure that the overall design concept includes suitable demarcation between the two textures, as sandblasting overflow will dull the polished surface.

In many cases, stone veneer is used as an accent or feature strip on either precast units or cast-in-place concrete. When stone veneer is done in combination with other finishes, a 1/2 in (12 mm) space is left between the edge of the stone and the adjacent concrete to allow for differential movements of the materials. This space is then caulked as if it were a conventional joint.

Combination finishes involving the use of one or more basic finishes together with form liners are almost infinite. Liners can be used in combination with smooth sandblast, acid, retarded, or tooled textures for either cast-in-place or precast concrete. Care in developing details must be taken to include suitable demarcation between the different textures. Some of the usual combinations include:

1. heavy abrasive blast/light abrasive blast/smooth;
2. striated/abrasive blast/irregular pattern form liner;
3. acid etch/abrasive blast/smooth,
4. ribs or vertical rustication/acid or abrasive blast,
5. tooled/hammered/form liner,
6. tooled/hammered/abrasive blast, and
7. tooled/chiseled/smooth.

20.7 Construction—Cast-in-Place Concrete

20.7.1 Forming

20.7.1.1 General

The selection of forms and forming materials to accomplish a given task is usually limited by the parameters established by the architect in the drawings, specifications, samples, and the contractor's mock-up. Specified forms and surface treatments become vitally important to a successful project when used for architectural purposes. Any material that can contain plastic concrete without deformation is a potential concrete form. Forms must withstand the loads due to both liquid head and compactive effort. This means that the forms must support a full liquid head. With the use of concrete pumps, the concrete can be placed rapidly with a minimum of lift lines. If the forms are not able to fully support the liquid head, bottom movement will occur, resulting in concrete spillage or actual movement of the upper form from the plane of the form below (Figure 20.11). Lightweight form systems are designed for concrete placement at 6 to 8 ft (1.8 to 2.5 m). This

FIGURE 20.11 Movement of upper form from the plane of the form below during placement and consolidation of the concrete.

FIGURE 20.12   Moisture leakage at nail holes and at the edges of duct tape used at form butt joints reduce the w/c ratio at the point of migration resulting in dark areas in the concrete surface.

may be insufficient. The most common forms are made of wood, plywood, concrete, steel, plastic reinforced, and nonreinforced, plaster, or a combination of these materials. For complicated details, forms of plaster, elastomeric rubber, foam plastic, or sculptured sand may be used. These forms are often combined or reinforced with wood or steel depending on the size and complexity of the unit to be produced.

Forms are molds and the cast concrete surfaces will reflect the finest details of the contact surface: wood grain, nail heads, dents, bulges, and the tile smootheners of plastic. All of these can be reproduced with startling—and sometimes distressing—fidelity. This should be remembered when choosing the forming material. Moisture leakage at nail holes and at the edges of the duct tape used at form butt joints reduce the w/c ratio at the point of migration, resulting in dark areas in the concrete surface (Figure 20.12). Overuse of forms can result in wear in areas with variable absorption. Forms that are to be reused should be carefully inspected after each use to ensure that they have not become distorted or damaged and to determine if they have any surface deterioration that may affect their ability to perform. Dark blemishes of this nature may penetrate as much as 1 in (25 mm) into the concrete (Figures 20.13 and 20.14). Once they occur, they cannot be concealed by sandblasting or bushhammering. In fact, due to its low w/c ratio, this dark concrete is actually

FIGURE 20.13   Poor quality form surface where the dense plywood and its surface coating has been worn through repetitive usage.

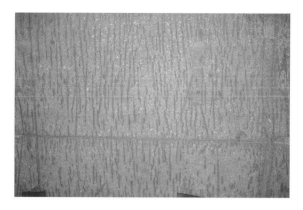

FIGURE 20.14   Result of form with vertical wear marks reflected in the finished concrete surface. Sandblasting made the surface blemishes more apparent.

harder than the adjoining unaffected areas and there will be less erosion from sandblasting—which will tend to accentuate the blemishes. Regardless of the number of reuses, all forms should produce concrete surfaces matching the approved mock-up.

20.7.1.2 Materials

The selection of form materials is the responsibility of the contractor, although the architect may specify certain liners or materials to achieve a particular surface or texture. Selection of materials used for forming must consider the type of surface effect desired, which includes color impact, texture, quality, deflection, and ease of stripping. Economy is of major concern to the contractor and is a function of cost first and then the number of possible reuses while still meeting the criteria of the mock-up.

Deflection, rather than bending or horizontal shear stresses, generally governs the design of the formwork. Forms must be able to support their own weight, the weight of the plastic concrete, and the pressure of the forces from consolidating the concrete. Deflections in the contact surfaces of the formwork reflect directly in the finished concrete surfaces. Forms for architectural concrete must be designed to minimize deflection. The deflection factor as well the as layout of tie-cone holes will affect the number of studs and wales, or the number of studs and size of wales in the case of wooden forms. Limiting deflections may be only 1/16 in (1.5 mm) in 4 ft (1.2 m), which is only half the normally acceptable 1/360 of the span for structural concrete. This limiting deflection is the one that is most restrictive yet still practical. Generally, prefabricated wooden form panels are not suitable for use in architectural concrete owing to the difficulties in making them tight against leakage. Therefore wooden forms for walls are generally built in place. Particular care must be given to alignment, perfection of corners, quality of contact surface, and tightness of joints. Minor defects in any of these items, which may not be objectionable in structural concrete, are not acceptable for architectural concrete.

In any architectural project, the formwork must be superior to that used for structural concrete. Grade B-B plywood is often used for structural concrete construction, but generally is not acceptable for architectural concrete. This plywood absorbs moisture from the concrete and will cause discoloration unless two to three coats of a urethane sealer is applied prior to use. High-density plywood can be used 10 times or more. Today, plywood can be purchased with plant-applied surface treatments that provide a nearly impervious and smooth surface. Birch plywood with a plastic coating is a plant-produced product that is higher in cost, but its high rate of reuse and its availability in sizes larger than 4 × 10 ft (1.2 × 3.1 m) may offset the cost by reducing the number of butt joints. If grain raise is to be transfered to the concrete, impervious coatings should be avoided. Sandblasting the plywood surface will impart a rough grain texture to the concrete.

Steel forms are generally used where high reuse factors or full liquid head construction is required. Steel surfaces are impervious and can provide uniform color to the concrete. The concrete may have some texture or color variation if the steel is rolled in different mills. An epoxy coating on the steel faces can combat the possibility of rust. The steel skin should be thick enough to support the load between its support members and keep deflections within acceptable limits. Quality steel forms should be well braced and manufactured of a heavy enough gauge steel to prevent twisting, buckling, and bending during erection and under the most severe usage. Formwork should incorporate adequate ribbing or channeling to provide rigidity, and when welded to the parent member, the welded areas must exceed 6 in (150 mm) in length to prevent tearing or popping. Skip-welding should be used instead of continuous welding to keep heat out of the plates. If continuous welding must be done on a finish surface, it is recommended that a test section be produced at the joint to determine that the joint area produces an acceptable product without distortion in the concrete. The welds should be ground smooth and coated with epoxy or similar material to hide the joint. Steel plate should come from the same mill as there are different ways of rolling the steel. Variations in the rolling technique can cause differences in the appearance of the concrete cast against metal from two mills. The steel skin should be pickled to remove mill scale. Bluing over welded material has been beneficial for avoiding staining from different surface characteristics. For flat horizontal surfaces, and surfaces to be honed or polished, the form skin should be 3/8 in (10 mm) thick to maintain local flatness after repeated use. Higher plate inertia also imparts a more uniform vibration pattern across the concrete surface during form vibration.

Steel is the preferred forming material for external vibration because it has good structural strength and fatigue properties. Steel forms are well suited for attachment of vibrators and, when properly reinforced, provide good uniform transmission of vibration. Welding at the corners of vibrator-mounted members or brackets should be avoided as this promotes angular crack propagation. Galvanized steel forms may cause concrete to stick and should be avoided. Aluminum and magnesium alloys may be used successfully if they are compatible with concrete. There is no standard test to measure compatibility. Past history of use with the same concrete mixture, forms, and curing conditions is the best known indicator of compatibility.

Fiberglass-reinforced plastic is an excellent solution when it is possible to use one form face piece for two or more adjacent surfaces. Examples are a beam face and soffit or the front and two sides of a column. In this way the support members needed in a fiberglass-reinforced form can be integral with the finish face. Such forms can assure that there will be no leakage at the completely enclosed corners—thus minimizing one of the most objectionable surface blemishes (leakage) in architectural concrete construction. Designers may not like the slightly rounded corners, but no one prefers leakage lines. Fiberglass-reinforced forms must have adequate horizontal and vertical ribs to properly provide rigidity and resistance to deflection.

An appropriate resin must be used on the surface face in order to assure good performance through a reasonable number of reuses, usually over 100. Maintenance of the resin plastic is mandatory for surface uniformity. This can be accomplished by careful cleaning, use of release agents, and occasional touch up of the face surface. Surface conditions, joints, and gel-coat material should be visually inspected prior to each use. Plastic molds should not be used with accelerated curing if concrete temperatures above 140°F (60°C) are anticipated. The susceptibility of the plastic form to attack by the proposed release agent should be determined prior to use. Unreinforced plastics are normally used only as liners for a form system designed to meet all of the structural requirements of concrete containment. Unreinforced plastic is only used to change the characteristics of the surface.

Plaster waste forms are used for custom designs of a sophisticated and detailed nature. The concrete is cast against the mold and the plaster is then broken away from the finished surface. Loose plaster can be removed with high-pressure water washing. Forms are almost always made with ornamental plasters, the methods used being similar to those employed in fibrous plaster work. Plaster forms must be made in sections that can be handled easily; individual pieces should not weigh over 150 lb (68.2 kg). Plaster forms are usually given two coats of white shellac to make them waterproof and nonabsorbent. If this is not done, it is quite certain that there will be a difference in the color

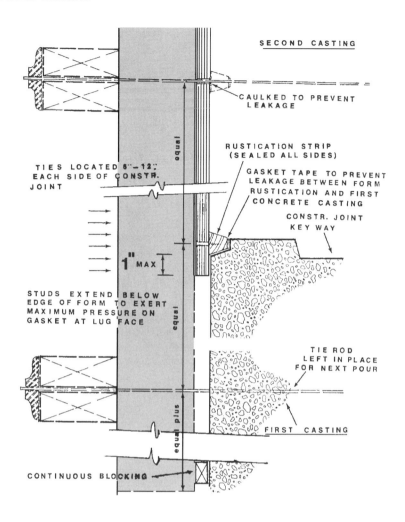

FIGURE 20.15 Typical construction joint; horizontal or vertical—no scale.*

of the concrete as compared with adjacent areas. Obviously one-use forms are relatively expensive and should only be used for nonrepetitive forming or where intricate shapes cannot be formed by more conventional methods. An effective release agent must be used with plaster waste molds.

20.7.1.3 Joints

A surface blemish will result when water is allowed to leak from the form. Leaks of any type in forms will result in either honeycomb (an aggregate-rich surface inconsistent with the normal, dense, adjacent surfaces accompanied by a color change or streaking), mottling, or a dark discoloration surrounding the point of leakage. The darker surface color is a result of less water being available for hydration. Moisture, paste, or grout can never be replaced except by repair of the concrete surface. It is therefore very important that leakage be prevented. Different methods have been advanced to prevent leakage at form joints; none has proved completely successful, especially those attempting to seal the joints flush with one another. Any amount of moisture or grout will create blemishes in exposed surfaces that are difficult and that may even be impossible to remove. For this reason, many designers prefer to acknowledge the joints and mask them with reveals formed by rustication strips. It is good practice to provide rustications at the casting limit atop the preceding cast to define straight level and plumb working lines (Figure 20.15) for construction joints.*

*Shilstone, J.M. 1973. *Architect. Rec.* :161–164.

When concrete is placed for the lift above or for an adjoining wall section, the forms will have a tendency to bulge under the hydrostatic head of the wet concrete and vibratory loads, causing a spill or overpour on the concrete already in place (Figure 20.16). Unless the concrete is to be bushhammered, tooled, or heavily sandblasted, which in the process remove the spill, such an overpour should be prevented. This is best accomplished by attaching a compressible gasket that is set into the reveal at the top of the previous casting and provides a watertight seal beneath the reveal. Tying the form into the previous concrete placement after the gasket is in place helps prevent form deflection. Vertical joints can be handled in a similar manner.*

Control joints may be defined as planes of weakness created by thin sections, encouraging the concrete to crack at predetermined locations. Stresses that cause cracking result from shrinkage, thermal effects, and differential settlement; shrinkage is generally the stress of greatest magnitude. If joints cannot be tolerated, the reinforcement design can attempt to distribute shrinkage and thermal stresses into many small hairline cracks, but it is generally better to weaken the concrete and ensure a controlled crack that can be sealed against water penetration. At times, it is desirable to cut 50% of the temperature steel at the control joint to further weaken the section.

FIGURE 20.16   Spill or overpour at construction joint during placement and consolidation. Tying the form into the previous concrete placement and installing a compressible gasket in the reveal would have prevented this leakage.

The location of control joints should not be determined until a careful study has been made of the structure's geometry, elastic qualities, and steel-reinforcement design. It is an inescapable fact that a large mass of concrete will crack. This must be anticipated in the structural and architectural design. Perhaps the surest way to plan construction joints, and one that reduces their number, is to have them correspond to locations where control joints are required. The relatively weak bond between castings will ensure that cracks develop at those points.

On many occasions chamfers have been used at corners in an attempt to minimize leakage. If the chamfers themselves are not caulked or gasketed, moisture or paste leakage can occur at two points rather than one. It is recommended that all corners, form joints, horizontal and vertical construction joints (sometimes known as cold or pour joints), and control joints be caulked or gasketed to provide positive protection against leakage. Gaskets should be a closed-cell compressible neoprene with adhesive on one side that can be attached to the form. Plastic pressure-sensitive tape (never duct tape) has sometimes been used on the inside of the forms at butt joints when significant texturing of the surface skin is planned. To withstand the stress when the form is filled with concrete, the tape should be flexible and highly adhesive. Tape should be applied prior to the application of a release agent. Care must be taken to prevent displacement of the tape or gaskets during concrete placement. Brush-applied gum adhesive over tape has been used to successfully stabilize the tape or gasket against movement. All joints should be inspected prior to casting to be sure the tape or gaskets have not moved. When tape cannot be used because tape deformation will be visible on the fine texture concrete surface, great care must be given to location of the butt joints. It is recommended that the contractor erect the exposed form face first in order to allow a visual check of all joints and joint treatments to minimize potential leakage.

Other ways to seal plywood butt joints include chamfering form edges and filling them with epoxy, gasketing, and using plywood sheets that have spliced or tongue-and-groove edges that are caulked or glued. Such joints must be backed up and supported with 2 × 4s (50 × 100 mm). The

*Shilstone, J.M. 1973. *Architect. Rec.* :161–164.

FIGURE 20.17 Dark leakage areas around form tie holes where moisture has migrated resulting in localized concrete with a lower w/c ratio.

architect may designate the location of control joints, but the location of construction joints is often left to the contractor. In architectural concrete, it is best if the contractor receives approval from the architect as to where construction joints will be located.

20.7.1.4 Form Ties

Except for small columns and beams, the lateral forces created by the hydraulic pressure of plastic concrete must be resisted by horizontal ties that hold the forms together and maintain their position. Ties may incorporate a spreader device that correctly spaces the distance between forms and acts as a means of removing metal parts to a sufficient depth from the surface, 1-1/2 in (38 mm), to facilitate patching with mortar and prevent corrosion. Form ties significantly influence the visual effect of architectural concrete. They are placed in the forms on a pattern, which should be consistent with the type of form design.* A 4 ft (1.2 m) pattern can be used with a variety of plywood sizes in a gang form. There is a tendency for many contractors to use cone snap ties. The installation of these must be very carefully done. All form-tie systems should incorporate a positive leakage prevention detail. "Bull-eyes" around tie holes are not considered attractive. Leakage can occur around form ties if they are not perfectly seated and sealed (Figure 20.17).

Tie-hole design and spacing is as much an architectural consideration as a structural need for the contractor. Ties for architectural concrete should be planned so they are symmetrical with the member formed and, wherever possible, ties should be located at rustication marks, control joints construction joints, or other points where the visual effect will be minimized. In this way, repair of tie holes will not fall in the flat panel areas. Successfully patching tie holes is not easy because it is hard to match the finished color of the adjacent concrete surfaces. Often, patches tend to accent rather than conceal tie holes. Heavier forms with steel backing and large diameter tapered bolts at a minimum number of locations offer a better approach to architectural concrete than using multiple snap ties. Adjustable, high-strength bolt ties facilitate the installation of subsequent forms by holding them tightly to the concrete of the previous casting, helping prevent leakage.

Ties or bolts that are to be pulled from the form must be coated with nonstaining bond breaker or encased in oiled paper sleeves to facilitate removal. Ties for architectural concrete should have the same safety factor as structural concrete. Form-tie assemblies for architectural concrete should be adjustable so as to permit tightening of the forms. Form tie assemblies fall into one of the following groups:

1. continuous single member ties for specific wall thicknesses and positive breakback characteristics;

*Shilstone, J.M. August 1973b. *Concr. Const.* :363–413.

2. she-bolt ties where an inner male threaded unit is left in the wall concrete and the outer fastening devices are removed and reused;
3. he-bolt ties where the outer fastening devices are reused and an expendable female threaded unit is left in the wall; and
4. tapered high-strength through bolts that can be completely removed from the concrete— these have separate outer units for proper adjustment, when using them, the contractor should place the small-diameter side of the tapered bolt on the exposed face to avoid spalling when the form is stripped; removable ties should be of a design that does not leave holes in the concrete greater than 7/8 in (22 mm), bolts should be tight fitting or holes should be filled with a sealant to prevent leakage at the holes in the form.

All these ties leave round and relatively clean holes, providing proper sealing procedures have been used.* These holes may be left alone, patched flush, or patched with a slight recess for an architectural shadow effect. Snap ties are not suitable for architectural concrete unless a rustic crude look is desired.

20.7.1.5 Form-Tie Removal

Ties should be removed as soon as possible after the formwork has been removed. After forms are removed, uncoated ties or ties that possess staining tendencies should be snapped and the ends should be treated to prevent rust stains. Stainless steel snap ties, when used, present the least trouble with staining, but should still be broken off at least 1 in (25 mm) behind the finished surface. Plastic coated ties should be snapped and the ends treated to prevent rust staining. Holding forms together with twisted wire ties or band iron should not be allowed, as it is nearly impossible to obtain a long-term, stain-free surface without cutting into the concrete surface and cutting back the tie or band iron ends before repairing a rather large nonuniform area.*

Externally braced forms may be used instead of any of the above mentioned form-tie methods in order to avoid objectionable blemishes in the finished surface. This is an expensive alternative unless a high degree of repetitive form use is possible.

20.7.1.6 Form Removal

Assuming that the necessary planning, care, and workmanship have been exercised in producing the quality of work desired, form removal must maintain the results already achieved. Forms should not be removed until the concrete has sufficiently hardened to permit form removal without damaging the concrete surface. Prying against the face of the concrete should not be attempted. When necessary, wooden wedges (not metal), should be used to assist in form removal. Many surfaces are marred by the use of metal wedges to loosen forming. Rough removal of forms may damage them beyond reuse. Elastomeric liners become difficult to remove if the forms are not stripped within 5 to 7 d. Easy removal of forms will reduce labor costs. More importantly, forms that can be stripped easily prevent unnecessary damage to exposed architectural surfaces. Spalled edges require expensive restoration by knowledgeable craftsmen. Sharp corners require a considerable amount of special attention in forming and form removal technique. Sharp edge lines and corners are very vulnerable to chipping or spalls at early ages. Damaged corners require costly repair work.

If forms are removed before the specified curing period is completed, measures must be taken immediately to apply and maintain satisfactory curing. In hot dry climates, wood forms remaining in place should not be considered to be adequate curing. Forms should be removed or loosened so that the concrete surfaces can be kept moist or be coated with a curing agent. In cold weather, architectural concrete should not be allowed to cool more than 40°F (4.4°C) per 24 h following the cessation of heat application. Removal of formwork should be deferred or formwork that is removed should be replaced with insulated blankets to avoid thermal shock and consequent crazing of the concrete surface.

*ACI Committee 303. 1974. *ACI J.* :317–346.

One has to be careful of small projections such as drip lines. Abrupt stripping may crack or break off the projecting tip, making the area vulnerable to moisture and subsequent corrosion of reinforcement. With embedded items such as windows or forms for openings, braces should be stripped first to allow the edge forms to be stripped without spalling the concrete. Intricate details* will require a delay in stripping time to allow the concrete to gain sufficient strength and to shrink from drying. The best procedure calls for stripping to begin away from the intricate area and proceed toward it. If the edges of rustications are to be crisp after sandblasting, they must remain in the face of the concrete. Many reuses call for rustication strips to remain with the forms and new strips to be reinserted prior to sandblasting. Fastening can be by gluing with mucilage or nailing into the form.*

Once formwork is removed, the concrete must be protected to prevent damage from any means including subsequent construction operations and weather. Runoff water from rain or on-site construction should be prevented from running over iron or other materials that can cause staining and then running over finished faces of architectural concrete. Surfaces must be protected from stains, graffiti, and impact damage. Differentiating concrete color hues may be expected between two surfaces where adjacent formwork is stripped at different ages. It is best to strip all adjacent form surfaces at about the same age for greatest uniformity. All forms, regardless of material selected, must be stacked level and horizontal to prevent warping. Even steel forms will warp if stored at an angle against the side of a building.

20.7.2 Reinforcement

20.7.2.1 Reinforcing Steel

Reinforcing steel must be planned in advance so that concrete placement can be carried out without reinforcement reducing the work space needed for consolidation of the concrete. Mechanical, electrical, ductwork, piping, conduit, or other embedments must also be planned as to location (Figure 20.18). If the size of a particular member does not provide adequate work space or otherwise presents difficulties for the contractor, an alternate method of detailing or construction should be considered. Often the suggested solution for a congested member is to use a smaller vibrator. This can be a problem as the smaller vibrator may not be able to do the compactive work necessary to produce the high quality essential for architectural concrete.

Occasionally, special vibrators have been made, such as vibrators with rigid shafts and only a 3 in (76.0 mm) flexible joint. Even special vibrators cannot overcome lack of space. Since the amount of reinforcing required in a member is dependent upon the size of the member, it may be necessary to increase the size of the member. Members cannot always be slim enough to meet the architect's design concept. Concrete members must be of sufficient size to avoid excessive amounts of reinforcing steel. Visualizing the patterns of reinforcing is important at the early stages of design—particularly at member intersections (Figure 20.19). The situation in Figure 20.19 allows little room for placement and consolidation of concrete. Honeycomb and other surface blemishes are a

FIGURE 20.18 Electrical conduit blocking access for placement and consolidation of concrete.

*Dobrowolski, J.A. January 1989. *ACI J.* :35–39.

FIGURE 20.19   Intersection of reinforcing steel allowing little access for placement and consolidation of concrete.

FIGURE 20.20   Bundling of reinforcing bars to allow more working space for vibration.

certainty. For example, beams should be wider than columns that have to go through the floor to avoid the familiar solid maze of vertical and horizontal reinforcing. In other members, bundling bars may provide a solution to afford clear working space for vibration (Figure 20.20).

The assumption that reinforcing steel must be uniformly distributed as detailed on the structural drawings is not always accurate. In many designs, the steel bars can be bundled or otherwise spaced to allow the contractor some space to actually place and consolidate the concrete. Horizontal space available after bundling will allow entry of the vibrating equipment. Full tension welds, cadwelds, or tension splices can be employed where required to eliminate congestion caused by lapping of bars (Figure 20.21). Locating the lapping of bars throughout the beam can reduce the congestion at any one point.

Reinforcing steel for architectural concrete must be accurately positioned and held in place in such a way as to minimize movement during placement and consolidation of the concrete. Wall or beam reinforcement cover can be maintained by use of bolsters or chairs on the inside face of the form to lessen the chance of rust stains or spalling of the concrete.

In reviewing the reinforcing steel design, the contractor should consider the following items in the design:

1. For design purposes, the cover on reinforcing steel for concrete exposed to the weather should be at least 2 in. Since the ACI code allows reinforcing metal to move ±1/2 in from the design location, this sets a minimum of 1 1/2 in of cover over the reinforcement.

FIGURE 20.21 Lapping of bars all occurring at one location causing congestion which can lead to honeycombing.

2. Horizontal bars in walls should be placed toward the outside face of the wall to allow a larger casting space between curtains of reinforcement (Figure 20.22). Single-curtain reinforcement should be positioned at the center of the wall, but again with the horizontal bars outside the vertical.
3. A minimum clear working space of 5 in (design) is necessary to facilitate proper placement and consolidation of low-slump concrete mixes.
4. When computing the space occupied by the reinforcing steel to determine the amount of space available for the worker in the field to do the work, maximum bar diameters should be used instead of the nominal diameters. The effective space occupied by the bars owing to deformations is over 10% larger. Actual space is measured to the outside of the deformations on the rebar surface. These maximum deformations become significant when thin or slender concrete sections are to be cast.
5. Tie wires should be a soft stainless steel specified to be tucked behind the joints and not left dangling adjacent to the exposed face.
6. Chairs should not be used to space reinforcing steel from vertical surfaces. Where it is necessary to support steel in beams, the reinforcing steel should be hung from the forms with stainless steel tie wire, and any chairs should be plastic tipped with sufficient plastic cover to assure that the metal will not rust. If reinforcing steel is hung from the forms, the forms and supporting member must be designed to take the often very heavy weight. This method is generally only practical when steel forms are used. If the surface is to be abrasive blasted, consideration should be given to the material that will resist the abrasive blast. In most cases where there is heavy reinforcement of beams, it may be necessary to use precast concrete blocks of the same concrete used in the construction to support the metal.

Some design codes require use of reinforcement in such quantities that constructibility is markedly reduced. When such conditions are found, the fact that the amount of metal required is within the code is not an acceptable excuse for producing a design that defies proper construction. Double check reinforcing steel details to make sure that casting space is available. For the project manager, this can be the difference between a reasonable project or an impossible task.

20.7.2.2 Placing Accessories

Supporting chairs, spacers, and bolsters should be manufactured of plastic or stainless steel to ensure the absence of surface rust staining, particularly where the concrete member is to be sandblasted. Plastic colors should match the finish surface. Many steel accessory suppliers will custom color their products. Current practices in detailing are considered inadequate to achieve total concealment of accessory feet. The reinforcing steel specification should indicate the need to increase the number

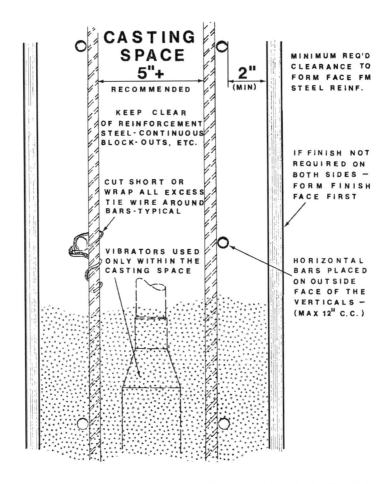

FIGURE 20.22 Double steel curtain wall: may be finished both sides; 12 in usually minimum—not to scale.

of chairs to compensate for steel loads that cannot be tolerated by the plastic-covered chairs, the plastic tips of the chairs, or the form materials used to support the reinforcing cages. Insufficient chairs may cause form indentation. Upon removal of the forms, these plastic or stainless steel tips will project below the ceiling or from the bottom of cast-in-place beams. Any plastic coating tips should be investigated for durability if they will be exposed to weather or sunlight.

Any wire for tying reinforcement should be of soft stainless steel to minimize staining of exposed surfaces. Some stainless steel ties may rust slightly when sandblasted or exposed to severe weather conditions. For this reason, all tie wire should be cut as close as possible to the bars and bent back away from form surfaces. Tie-wire clippings must be removed from any horizontal surfaces (such as beam soffits or ceiling areas) to be exposed to view.

20.7.2.3 Galvanized Reinforcement

Galvanizing reinforcing steel is not recommended for architectural concrete. Adequate minimum cover of well-compacted concrete is the best protection for reinforcing steel. Where galvanizing of reinforcing bars is required, galvanizing is usually performed after fabrication. This may mean several rehandlings for individual bars or partially assembled cages. The ASTM A767/A767/M specification prescribes minimum finished bend diameters for bars that are fabricated before galvanizing. Smaller finished bend diameters are permitted if the bars are stress relieved. Supplementary requirement S1 requires sheared ends of bars to be coated with a zinc-rich paint formulation.

When galvanized reinforcing steel is placed close to nongalvanized metal forms, the concrete may have a tendency to stick to the forms. This may also happen if nongalvanized reinforcement is used close to galvanized forms or form liners. A 2% solution of sodium dichromate or a 5% solution of chromic acid (chromium trioxide) solution applied as a wash to the galvanized surface has satisfactorily reduced the galvanic action of the metal to prevent reaction between the zinc and the alkaline fresh concrete. Addition of chromates to the concrete cannot be recommended, as their effect on concrete performance is not yet fully known. Galvanizing of reinforcement is only recommended when minimum cover requirements cannot be achieved. In these cases, the use of galvanizing should be specifically called for in the contract documents and shown in the shop drawings.

20.7.2.4 Post-Tensioning

Long-span, post-tensioned-beam framing systems have been used in combination with architectural concrete to produce unusual buildings in both the United States and the Middle East. A post-tensioned-beam framing system offers the advantages of both a higher span/depth ratio and a reduction in material quantities. With an architectural cast-in-place facade, considerable planning is required to allow for stressing of the post-tensioned structure without leaving visible signs in the exterior architectural finish. This means using exterior precast concrete panels as covers for exterior stressing pockets, or providing interior stressing pockets, buttresses, and blockouts.

The contractor must be very careful in his placement of anchor pocket formers in the formwork. Improper placement of formed stressing pockets can result in misalignment of jacking equipment and application of an incidental biaxial force to the tendon, which could cause shearing of the tendon at the anchor. Preformed pocket formers can be used and field adjusted to obtain the proper alignment of typical interior shearwall anchorage, as shown in Figure 20.23. An interior girder detail at the core stressing anchorage and at the exterior wall key are shown in Figure 20.24. The 60 ft (18.3 m) single-span beams in the low-rise structure are stressed through to pockets at each end of the beam. A typical top pocket anchor detail is shown in Figure 20.25. The above three details were used at 1515 Poydras in New Orleans. S.O.M. from Chicago provided the architectural and engineering design. The unusual number of unique post-tensioned-beam anchorage configurations required intense review and quality control at the onset of construction. Considerable planning

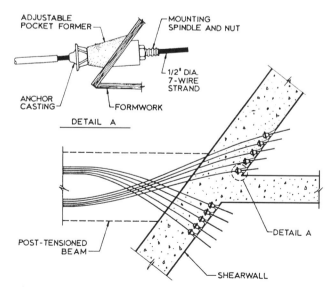

FIGURE 20.23 Plan view of typical interior shearwall anchorage. Pocket formers can be field adjusted for proper tendon alignment.

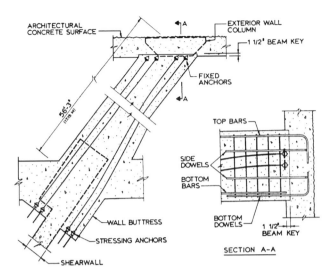

FIGURE 20.24 Interior girder detail at core stressing anchorage and at exterior wall key.

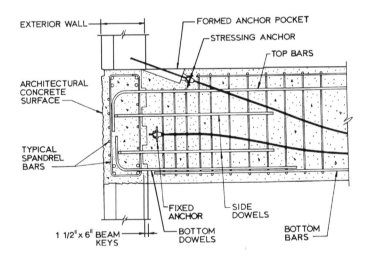

FIGURE 20.25 Typical top pocket anchor detail.

went into materials, formwork sequencing, and special falsework for the doweled reinforcement support. This is a striking example of the integration of structure and form. The exterior concrete finishes range from a heavy hammered texture to a light sandblast. Attention to detailing, finishes, and execution have allowed the architectural design of this building to be accurately expressed through cast-in place and precast structural elements.*

The original design of the post-tensioned channel beams at the roof level of the 800 ft (244 m) long Physical Education building in Saudi Arabia called for massive 4.0 ft (1.2 m) wide by 8.0 ft (2.4 m) deep by 141.6 ft (42.9 m) long cast-in-place concrete beams spanning 120 ft (36.3 m) between cast-in-place lateral supports. Each beam would have a total mass of 170 t (154 Mg) and an architectural surface area of 4670 ft² (432 m²). These channel beams were located 56 ft (17 m) above ground level and spaced at 16.3 ft (4.95 m) on center. The 47 beams were of architectural concrete with a light to medium sandblast finish. Owing to severe schedule restrictions, shoring

*Clark, R.S. and Zils, J. September 1984. *ACI J.* :35–39.

problems, and interference with work at other levels and zones of the building, it was decided to precast and post-tension the channel beams, rather than cast them in place. The precasting would take place at one location at the 56 ft (17 m) level. Caudill Rowlett & Scott (CRS) in Houston, Texas approved the design modification. This change in construction method reduced shoring and forming costs and resulted in a large reduction in labor. The channel beams were cast on a construction platform on a six-day cycle including three days of curing on top of temporary rails set on the cast-in-place lateral support beams. Seven-wire strand was threaded through sheathing to insure minimum friction losses. After six days, the strands were tensioned following a computer program that determined the required gauge pressure and elongation for each strand group. Once all of the strands were properly stressed and anchored, the ends were cut inside the stressing pocket recess (Figure 20.26). Strands were then grouted in the sheathing, and the recess was filled with architectural concrete patch mix, covered with wet burlap, and protected with a wood cover (Figure 20.27). Each channel beam was moved to its final position along the rail using two hydraulic rams. After the beams reached their final position, the hydraulic ram jacking system and rails were removed and the beams were set on thick neoprene bearing pads. This was all done without the use of a crane, only hydraulic jacks were used.*

FIGURE 20.26 After all 7 wire strands were stressed and anchored the ends were cut off inside the stressing pocket recess.

A prestressing strand is required to conform to ASTM A416, Grade 250 (1725) or grade 270 (1860). The most commonly used grade of steel is 270,000 psi (1862 MPa). This strand is 0.5 or 0.6 in (12.7 or 15.2 mm) in diameter and is of low-relaxation type. Unbonded single-strand tendons are composed of steel strand with anchors affixed to each end by toothed wedges coated with corrosion-resistant lubricant and encased in plastic sheathing. For wedge-type anchorages, the wedge grippers should be designed to preclude premature failure of the prestressing steel due to notch or pinching effects under static test load conditions.

FIGURE 20.27 After cutting, the strands were grouted in the sheathing, and the recess filled with architectural concrete patch mix, covered with wet burlap, and protected with a wood cover.

*Kenney, A.R. 1988. *ACI J.* :48–52.

Post-tensioning tendons subject to exposure or condensation and that are not to be grouted should be permanently protected against corrosion by plastic sheathing formed over the coated (greased) strand or by other appropriate means of protection. The sheathing should not be reactive with the concrete, coating, or steel. The material should be watertight and have sufficient strength and durability to resist damage and deterioration during fabrication, transport, storage, installation, concreting, and tensioning. Sheaths should be continuous over the unbonded length of the tendons and should prevent the intrusion of water or cement paste and escape of the coating material.

20.7.3 Concrete Placement

The placement of architectural concrete requires the utmost care on the part of the contractor. Consistent preplanning of each day's concrete placement should be done. The period between the truck and the forms is the most critical time in the life of architectural concrete. The ready-mix truck must transport and mix the concrete to a homogeneous state in spite of low slumps

FIGURE 20.28 Steep-sided bucket with easily movable chute to aid in placement of concrete.

and high coarse-aggregate contents. With some architectural concrete mixes, it may be necessary to load trucks to only 2/3 of their rated mixing volume. The contractor must verify that the ready-mix trucks assigned to deliver architectural concrete are clean, free of hardened buildup, and have less than 20% blade wear. Trucks meeting this criteria should have their numbers recorded or some other means of identification should be used so that only approved trucks will be used for architectural concrete. Trucks should be scheduled so that they arrive at the project site just before the concrete is required, thus avoiding excessive mixing while trucks are waiting or delays in placing successive lifts of concrete. Delays cause nonuniformity of pour line or appearance.

A major decision for the contractor is his method of placing the concrete. Placement can make or break the quality of architectural concrete. The characteristics of certain lay-down buckets or pumps may prevent them from placing the necessary concrete mix. When this happens, other methods or equipment may be required. Steep-sided buckets (Figure 20.28), may be required; conveyors may be used (Figure 20.29); or a different concrete pump may be selected (Figure 20.30). Change the equipment or pump but do not adjust the concrete mix design.

FIGURE 20.29   Conveyors in use at construction site.

FIGURE 20.30 Several concrete pumps in use at construction site.

FIGURE 20.31 Tremie being placed in position for use.

High-quality architectural concrete can be obtained when placing concrete with trucks, tremies (Figure 20.31), buckets, or even wheel barrows, provided proper procedures are followed. When concrete is placed directly from trucks or through tremies, the free fall should be limited to less than 5 ft (1.5 m). Tremies can be made of either sheet metal or reinforced plastic with a rectangular cross section. Minimum inside dimensions would be 4 in (100 mm) × 16 in (400 mm). There must be sufficient space between reinforcing bars if a tremie is to be used. The flow of concrete from a truck or tremie must be directed within the reinforcing cage to ensure minimal bounce or deflection of the concrete off steel, conduit, or other obstructions that cause splatter and/or segregation of the mix. Splatter on the formwork above the level of plastic concrete can harden and cause a surface blemish that is difficult to remove. This is especially serious in finishes having a light or smooth texture.

Conveyors are useful when a low-slump concrete mix with a high coarse-aggregate proportion, 2000 lb (900 kg) of coarse aggregate or more, is required. They are most often used when large floor (ceiling) and beam composite construction is required. Care should be taken with any conveying equipment to prevent contamination by nonarchitectural mixes and to prevent segregation of the mix being transported. If methods of conveyance are varied during the overall construction of architectural concrete, the uniformity of the finish surface color may be affected. When conveyors are used, they should have a high speed and have a discharge end that is easily movable in order to permit proper concrete mix distribution.

Concrete pumps must be rated on their ability to move concrete having a low w/c ratio that meets the project's architectural requirements. There are several pumps today with 5 in (125 mm) or 6 in (150 mm) lines that can handle most architectural concrete mixes. In order to prevent concrete consolidation problems in the walls and minimize bug holes, one contractor elected to pump walls from the bottom up. Pumping ports were installed in the wall forms 12 in (302 mm) from the bottom and on 4 ft (1.2 m) centers along the steel forms. A superplasticizer was used to allow for 8 to 10 in slumps (Figure 20.32). These serpentine architectural cast-in-place walls later had a white coating applied to their surface. In addition to cast-in-place concrete, battery and other vertical forms for precast products have been pumped from the bottom up. Use of a superplasticizer admixture can help the concrete to be pumped while maintaining the required w/c ratio. Variations in w/c ratio will change the color of the exposed architectural surface. When using a concrete pump, the first $1/4$ yd^3 (0.2 m^3) should be wasted or utilized elsewhere, not in the architectural forms. It takes a certain amount of cement paste to coat the inside of the pump and hoses leading to the placement location. An alternative would be to initially charge the pump with a high-cement grout mix prior to placing the regular con-

FIGURE 20.32 Concrete pumped from the bottom up in order to minimize consolidation problems and bug holes. Hospital was designed by Bertram Goldberg of Chicago.

crete. Failure to do one or the other will result in a higher w/c ratio at the point of initial discharge and unsightly discoloration (Figure 20.33). The more uniform the w/c ratio, the greater the uniformity of the color and texture.

The depth of the layers of concrete placed and consolidated depends on the width of the forms, the amount of reinforcing present, the concrete placement method, and the timing required. Each new concrete lift should cover the previously placed concrete within 20 min. Substantial lift or pour lines may result when a 45 min delay in placement of fresh concrete occurs. Fresh concrete is usually placed in lifts of no less than 18 in (450 mm) and no more than 30 in (750 mm). The surface of each

FIGURE 20.33 Adjacent concrete walls from two separate day's placement showing the impact of pumping concrete directly without wasting first 1/4 cubic yard (0.2 cubic meters) to coat the pump and hoses.

lift should be fairly level so that the vibrator does not need to move the concrete laterally. Lateral movement can cause segregation, leading to honeycomb or other surface blemishes. Concrete should be placed continuously at close intervals. A vibrator should be operated at the point of deposit during all concrete placement.

20.7.4 Concrete Consolidation

20.7.4.1 General

The most important single skill in architectural concrete involves the selection and operation of the vibrator. In Europe, vibration operation is recognized as all-important to good concrete work and the skilled vibrator operator is the highest paid man on the concrete crew. All too frequently in the United States, the vibrator operator is one of the most unskilled and untrained men on the concrete crew—often picked for a strong back rather than his knowledge of concrete consolidation. A vibrator not only consolidates the concrete for maximum density but also internally blends the different lifts of concrete together into a single solid mass with few bugholes and no lift lines on the finished exposed surfaces.

20.7.4.2 Equipment*†

All too often vibrators are selected on the basis of their price or maintenance records. Instead, selection should be based on the following vibrator qualities:

1. *Frequency.*   The ability to fluidify the mix through the number of times per minute the head moves from side to side.
2. *Amplitude.*   The ability to kick the mix into place. Amplitude is the measured distance of the head when it goes from side to side as measured from its neutral axis. An increase in amplitude results in an increase in the effective radius of action.
3. *Power.*   The ability to maintain vibration under load.
4. *Size.*   The ability to overcome the resistance of the concrete mix particles to movement. Use the largest diameter head permitted by the form dimensions and the spacing between reinforcement.

High frequency of the vibrator alone will not work. If the power is constant and the frequency is increased, the amplitude decreases because the head does not have the time to travel as far as it did at lower frequency. If the power source is increased, the head has a greater amplitude, provided the frequency is not simultaneously increased. Larger vibrators provide more mass in the head. The larger-mass head is necessary to provide the impulse to drive the concrete effectively into the intricacies of the form. A high-frequency, high-amplitude, high-powered vibrator moves to such a degree that it creates a churning action within the concrete mass. This is the type of vibrator necessary to combine two lifts of concrete into one. Achievement of void-free surfaces favors a higher-frequency vibrator.

Internal vibrators are generally called immersion, poker, or spud vibrators. One of the most commonly used types is the flexible shaft type where the electric motor is outside the vibrator. A flexible shaft leads from the motor into the vibrator head where it turns the eccentric weight. A universal 120 volt, single-phase, 60-(Hz) cycle/s motor is used, and the frequency of this vibrator when operating in air is quite high—in the range of 12,000 to 17,000 vibrations per minute (200 to 283 Hz) (the higher values are for smaller head sizes). Because of this high speed and the heat developed, the vibrator may burn up if operated carelessly outside the concrete. However, when the vibrator is operating in the concrete, the motor is under load and the frequency is generally reduced by about one-fifth.

---

*ACI Committee 309. "Consolidation of Concrete," ACI:40 pp.
†Shilstone, J.M. 1972c. *Concr. Const.* :536–538.

The electric motor-in-head vibrator type has the motor in the vibrator head, as the name implies. Most motor-in-head vibrators are high-cycle units. This means that a 180 high-cycle current is required for their operation instead of the usual 60-cycle commercial power. These vibrators use induction motors and operate at a frequency in concrete of about 10,000 vibrations per minute (vpm); this is only about 5% less than their free speed (speed in air). The high-cycle current is obtained by passing commercial power through a frequency converter or through the use of a special generator. Motor-in-head vibrators operating on a 60-cycle current with a universal motor are also available, but they have less capability than high-cycle units. The electric motor-in-head vibrators are generally at least 2 in (50 mm) in diameter.

When using electric vibrators, all power cables and connectors should be kept in good condition. Vibrators should not be lifted or carried by the electric cables. Sharp bends should be avoided in the shafts of flexible-shaft vibrators. When vibration is finished for the day, vibrators should be cleaned and concrete spills removed. It may be desirable to cover form vibrators to keep concrete from dropping on them, but not in such a way as to interfere with air cooling them by. Whether the con-solidation equipment provides internal or external vibration, or a combination of both, its frequency and amplitude should be designed, tested, and proven for the volume, configuration, and placing technique of the product and the proportioning and consistency of the concrete for that product.

Since internal vibrators are used in wet (conductive) locations, all electric units should be grounded to the power source. Generator sets supplying power should also be grounded in order to maintain continuity of the grounding system. Units operating at less than 50 V, or that are protected by an approved double insulation system, are excepted from the grounding system. In the United States, electric vibrators are subject to the National Electric Code, Article 250-45.

External vibrators may be divided into form vibrators, surface vibrators, and vibrating tables. Form vibrators are external vibrators attached to the outside of the form or mold. They vibrate the form, which in turn transmits the vibration to the concrete. Extremely rugged forms are required where high-amplitude vibration is used. The effective penetration of the vibrations into the concrete ranges from 8 to 24 in (200 to 600 mm). It may be necessary to supplement a form vibrator with internal vibration for sections thicker than 12 in (300 mm) or higher than 4 ft (1.2 m).

Rotary type vibrators are preferred because reciprocating types are very hard on molds. Rotary type vibrators may be operated either pneumatically, electrically, or hydraulically. In the reciprocating type, a pneumatically driven piston is accelerated in one direction, stopped (by stricking against a steel plate), and then accelerated in the opposite direction. These vibrators produce impulses acting perpendicular to the mold with frequencies usually in the range of 1000 to 5000 vpm (20 to 80 Hz).

In the pneumatically and hydraulically driven rotary type models, centrifugal force is developed by a rotating cylinder or revolving eccentric weight. These vibrators generally work at frequencies of 6000 to 12,000 vpm (100 to 200 Hz). The frequency may be varied by changing the air pressure, usually by adjusting the air supply valve or the fluid pressure on the hydraulic models. The amplitude and centrifugal force may be varied by changing the eccentric weight. The hydraulic models are driven by a hydraulic motor that turns an eccentric weight on a shaft. The advantage of this type of vibrator is the combination of an extremely low noise level, similar to an electric motor, with an infinitely variable speed. The unit can be locked into a specific frequency by use of a flow control. The wise contractor also has a spare for each type of vibrator available at the site in case of breakdown. If 180-cycle generators are used to power the vibrators, then two generators should be available at the site.

20.7.4.3 Internal Vibration

Internal vibration is recommended for all standard member cross-sections. The vibrator should be inserted vertically at uniform spacing over the entire area. The vibrator operator should be trained never to force the vibrator into a lift—this produces voids in the earlier lift. Let the vibrator sink rapidly into the concrete to the bottom of the lift, but not less than 6 in (150 mm) into the preceding lift, if there is one. With rapid penetration, the concrete is moved upward and outward, which drives the air and water ahead of the concrete and facilitates its escape to the top of the form surface. Since compaction does not occur below the vibrator's tip, it is imperative for the operator to allow the vibrator to sink rapidly by its own weight into the preceding lift. If the first lift has partially set before the second is placed, it is practically impossible to blend the two layers. Unattractive lift lines and possible honeycomb will result.

The vibrator should be withdrawn slowly upward at the rate of about 3 in/s in a slow pulsating (up and down) motion. Vibrator surge is always upward and generally at a 30° angle from the horizontal. The distance between vibrator insertions should generally be about 18 in (depending on the properties of the mix and vibrator being used); the area visibly affected by the vibrator should overlap the adjacent just-vibrated area by a few inches. Vibrator insertions should be no farther apart than twice the radius of influence. Even the largest head vibrators have a radius of influence of 10 in or less.

For architectural concrete, it is best to have two vibrators and two operators consolidating the concrete. The first operator is stationed at the point of deposit, where that vibrator is operated continuously. The second vibrator is used to slowly blend concrete, eliminate

FIGURE 20.34 First operator is vibrating concrete at point of deposit. Second operator is blending concrete to eliminate lift lines and minimize bugholes.

lift lines, and send entrapped air and water up along the form face and out of the concrete mix (Figure 20.34). When the vibrator head begins to emerge from the concrete, be sure the operator understands that he must remove it from the concrete quickly or else air will be drawn down between the concrete and the form. At no time should the head of the vibrator be half in and half out of the concrete. When that happens, immediately fully extract the vibrator from the concrete. Movement of the vibrator should not cease or pause while it is, in the concrete. Constant manipulation is necessary to prevent the possible formation of harmonic motion, which could distort or deflect the form or cause failure in the form. The very least that can happen under these circumstances is paste leakage, honeycombing, or entrapping air on the form face.

Sometimes it is practical to insert a small vibrator between the reinforcement and the form face. In such cases, the vibrator should be rubber tipped; even so, any contact with the form should be avoided if at all possible because this might mar the form, disfigure the surface, and result in a darker color or leave vibrator paste marks in an exposed aggregate surface (Figure 20.35). This whole procedure is dangerous. Actually, at no time should the head of an internal vibrator come in contact with the forms. The vibrator should be kept just within the steel reinforcing and be inserted no closer than 3 in (75 mm) from the form on the side of the steel away from the form face.

For dry mixes, where the hole around the vibrator does not close during withdrawal, reinserting the vibrator a few inches away may solve the problem. While stiff mixes are to be encouraged, overly dry mixes may result in poor consolidation (honeycomb or excessive entrapped air) and should be avoided. Where air voids in formed surfaces are excessive, the distance between vibrator insertions should be reduced to about 12 to 15 in (300 to 375 mm).

Form liners have more surface area than flat forms, and they require more internal vibration. Architectural concrete will normally require at least twice and maybe three times as much vibration and compaction effect as regular structural concrete. Sometimes very harsh mixes, such as those with gap grading, are used to produce special architectural effects. These also require more powerful vibrators and longer vibration times. The contractor should be sure to chose a sufficiently large vibrator and see that vibration insertions are placed closely enough to consolidate all concrete. The

FIGURE 20.35 Grout lines on exposed face of architectural concrete due to an internal vibrator touching the side of the form during concrete consolidation.

FIGURE 20.36 Concrete wall placed with a superplasticizer in the mixture and no internal or external vibration. Settlement lines occurred at all horizontal reinforcing locations.

vibration should be terminated when the mortar level reaches the top of the aggregate in order to minimize mortar lenses between lifts.

Superplasticized admixtures have changed the vibration techniques used in architectural concrete mixes. Superplasticized mixes require less vibration than standard mixes that do not include a superplasticizer admixture. When superplasticizers were first introduced, the literature stated that no vibration was required. This was a mistake, and several contractors unfortunately believed the technical data. Neither internal nor external vibration was used in placing the concrete. The results were disastrous; settlement cracks occurred at all horizontal reinforcing steel locations (Figure 20.36). It is virtually impossible to overvibrate a well-designed mix. However, it is possible to overvibrate a concrete mix containing a superplasticizer admixture. Overvibration of mixes containing superplasticizers may cause segregation, and the excess mixing action may entrap air or water, which will result in bug holes.

Lightweight concrete behaves differently from hard-rock concrete. Owing to its lack of particle weight, it presents great problems for compaction and has a tendency to float. Considerably more surface blowholes should be expected with lightweight concrete; where these are not acceptable, lightweight concrete should not be used.

20.7.4.4 Form Vibration

Form vibration is recommended in areas inaccessible to internal vibration or where full floor height steel or reinforced plastic forms are used. External vibrators loosen joints in wood forms, causing grout leakage and honeycomb and therefore are not recommended for this use. Forms for external vibration must stand up under the repeated, reversing stresses induced by vibrators attached to the form. External vibration of forms improves surface quality, but the additional stressing of forms and liners can shorten their life. Form vibrators must be capable of transmitting the vibration more or less uniformly over a considerable area.

The form should have adequate skin thickness and suitable stiffeners. The vibrators should be rigidly attached to the form. Special attention should be given to form tightness in order to prevent grout leakage. Trials should be made with form vibrators prior to their large-scale use. These trials should simulate the forming conditions to be encountered on the structure. The size and spacing of form vibrators should be such that the proper intensity of vibration is distributed over the desired area of form. The spacing is a function of the type and shape of the form, depth and thickness of the concrete, force output per vibrator, workability of the mix, and vibrating time. Present knowledge is inadequate to provide an exact solution to this complex problem.*

The recommended approach is to start with a spacing generally in the range of 4 to 8 ft (1.2 to 2.4 m). If this pattern does not produce adequate and uniform vibration, the vibrators should be relocated as necessary until proper results are obtained. Achieving the optimum spacing requires knowledge of the distribution of frequency and amplitude over the form and an understanding of the workability and compactibility of the mix. The frequency can readily be determined by a vibrating reed tachometer. However, the small amplitudes associated with form vibration have been difficult to measure in the past. Inadequate amplitudes mean poor consolidation, while excessive local amplitudes are not only wasteful of vibrator power but can in some cases cause the concrete to roll and tumble so that it does not consolidate properly. Moving one's hand over the form will locate areas of very strong or weak vibration (high or low amplitude) and "dead spots."

Concrete compacted by form vibration should be deposited in layers usually 10 to 15 in (250 to 400 mm) thick. Each layer should be vibrated separately. Vibration times are considerably longer than for internal vibration, frequently as much as 2 min, and possibly as much as 30 min or more in some deep sections. It is desirable to be able to vary the frequency and amplitude of the vibrators. On electrically driven external vibrators, amplitudes can be adjusted to different fixed values quite readily. On air-driven external vibrators, the frequency can be adjusted by varying the air pressure, while the amplitude can be changed by changing the eccentric weight. Since most of the movement imparted by form vibrators is perpendicular to the plane of the form, the form tends to act as a vibrating membrane, with an "oil-can" effect. This is particularly true if the vibration is of the high-amplitude type and the plate is too thin or lacks adequate stiffeners. This in-and-out movement can cause the forms to pump air into the concrete, especially in the top few feet (50 to 100 cm) of a wall or column lift, creating a gap between the concrete and the form. Here there are no subsequent layers of concrete to assist in closing the gap. It is therefore often advisable to use an internal vibrator in this region. Form vibration during stripping is sometimes of benefit. The minute movement of the entire surface of the form helps to loosen it from the concrete and permit easy removal without damage to the concrete surface.

20.7.4.5 Revibration*

The lift should be revibrated after initial consolidation with slow withdrawal of the vibrator head to draw out as many of the air bubbles as possible. The vibrator used to remove air bubbles from the top lift can be of lesser power or a smaller size than that used to initially consolidate the concrete. The usual high-energy vibrators have a tendency to churn air into the top lift of the concrete. This is the area where air and water pockets are most prevalent. If the architectural concrete mix is a

*ACI Committee 309. "Consolidation of Concrete", ACI.

properly designed low-slump mix, do not worry about overvibration. Revibration, after bleeding is substantially complete but before initial set, can be used to further densify the concrete and reduce air and water pockets against the form and exposed concrete face. Revibration can be accomplished any time the running vibrator will sink of its own weight into the concrete and liquify it momentarily. It will accomplish most if it is done as late in the process as possible.

Revibration generally results in improved compressive and bond strength, release of water trapped under horizontal reinforcing bars, minimized leakage under form bolts, and the removal of additional air voids. The greatest benefits are obtained for wetter concrete mixtures. Revibration should not be used where harsh gap-graded mixtures have been used to produce exposed-aggregate surfaces.

20.7.4.6 Spading

Spading may be employed in conjunction with internal or external vibration to improve finish surfaces. A flat, spade-like tool or a large plastic sail batten is repeatedly inserted through the top two lifts of concrete and withdrawn from the concrete adjacent to the form. This forces the coarse aggregate particles away from the form face, and assists the air bubbles in their upward movement toward the top surface.* The spade, regardless of type, must be inserted often enough to cover the entire area of the form surface. Sometimes a wooden 1 × 4 (25 × 100 mm) or 1 in × 6 in (25 × 50 mm) board with a pointed chisel end is used, but better results have been obtained with a plastic sail batten. Spading is usually done in conjunction with revibration by internal or external vibrators. The top lift must be quickly revibrated to close any gaps left by the spading. Although it is a laborious operation, the results can be worthwhile if spading is properly utilized. In addition to spading and revibration, pounding the outside of the forms with wooden mallets or a steel plate attached to an impact hammer, working from the bottom of the form upward, can release air bubbles and improve the concrete surface.

20.7.5 Curing

20.7.5.1 General

The method and period of curing should be consistent to produce a uniform concrete surface without stains, discoloration, drying, or plastic shrinkage cracks.* Curing materials or methods should not allow one section of architectural concrete to cure or dry out faster than other sections. This may produce color variations in the finish surface. Proposed methods should be tried on the site-cast mock-up to determine any adverse effects. Curing involves maintaining a satisfactory moisture content and temperature in the architectural concrete to produce the final qualities desired for both texture and durability of the surface. Unless early stripping is necessary to achieve a specific texture, concrete strength for removal of the forms is usually 2000 psi (13.8 MPa). The stripping strength is generally set by the design structural engineer. Freshly deposited and consolidated concrete should be protected from premature drying and extremes of temperature. The curing period of concrete that is of significant interest is during the early stages of strength development, from initial set until the concrete has reached the design strength appropriate for stripping of the forms. There is no need to wait for the mix water to finish bleeding to the surface before initiating the curing and protection of finish surfaces.

20.7.5.2 Curing in the Forms

Nearly all beams, columns, and undersides of slabs receive their curing by being left in the forms. To prevent staining caused by the type of form material, seal the forms with a liquid sealer prior to use, following the manufacturer's instructions.* For vertical surfaces, the easiest thing is to leave the forms in place and keep them supplied with additional moisture through soaker hoses or other methods. Early removal of forms and application of curing compounds is generally unsuccessful for

---

*ACI Committee 303. July 1974. *ACI J.* :317–346.

architectural concrete, although it is standard procedure for structural concrete. Curing compounds will usually stain the surface, for instance, polyethylene stains the concrete with a whitish color.

Concrete in the form should be maintained at a temperature of not less than 50°F (10°C) during the curing period. In cold climates, the contractor may be required to use insulated forms to use the heat of cement hydration of the in order to maintain the temperature necessary during curing. The contractor may also need to provide heat (if necessary to maintain minimum temperatures and to minimize loss of moisture.)

20.7.5.3 Moist Curing

Flat surfaces can be cured by ponding with the use of perimeter barriers and continuously operated sprinklers. However, the runoff water may create other problems at the project. Curing water should be checked for staining materials such as iron, which may cause rust stains, and the water should be uniformly applied to achieve uniformity of color. Flat surfaces can also be cured with the forms left on and the top surface covered with paper, polyethylene, or clean flannel. Burlap is dirty, can leave residual stains, and is generally not kept continuously wet. It may be necessary to erect wind breaks out of properly supported plastic sheets or temporary plywood walls. For vertical and other formed surfaces, after the concrete has hardened and while the forms are still in place, water should be applied so it runs down the inside of the form if necessary to keep the concrete uniformly wet. Immediately following form removal, the surface should be kept continuously wet with water spray or water-saturated fabric for a period of 5 d. Extending the curing period beyond 7 d does not produce additional beneficial effects, except in areas of very low humidity. Water curing has been found to reduce the possibility of cracks forming and to lighten dark blotchiness and create better uniformity of color. Uniformity of water application is important to obtain uniformity of color. If surface retarding agents are to be used, water curing should be delayed until after the surface treatment has been accomplished. Rapid surface drying can lead to color variations. The contractor must have an adequate crew and all the required equipment prepared so that placement, finishing, and curing can proceed without interruption and as rapidly as possible.

20.7.5.4 Membrane Curing

Liquid curing compounds may cause discoloration or staining and prevent the bond of any repairs or permanent coatings that may be needed. Curing compounds should only be used on the back of walls where the surface is not exposed or where the surface is to be removed by acid etching, sandblasting, or tooling. Manufacturers should be consulted as to the rate of coverage and the effect that their compound has in the above respects. This type of cure should be thoroughly evaluated on the preconstruction mock-up.[*] When used, membrane curing compound should cover the entire surface to be cured with a uniform film that will remain in place without gaps or omissions. Areas with incomplete coating cover should be recoated. If a curing membrane is to be used, it is preferable to use an inorganic material such as sodium silicate. Organic curing compounds discolor after prolonged exposure to sunlight. The briefer the interval between stripping the form and applying the curing compound the better, but the interval should be kept uniform from form to form to minimize color variation in the concrete.

Curing compounds should be applied to the top of a wall as soon as floating is completed. Sodium silicates are especially useful for curing tops of walls because they do not break the bond between concrete and steel if sprayed on reinforcement and they do not interfere with subsequently applied cementitious toppings. Also, the sodium silicates fill the gap at the top of a wall where lower pressure compresses the plastic liner less and results in a slightly thinner wall section. In addition, sodium silicates consume free lime, greatly reducing the potential for efflorescence.

[*]ACI Committee 303. July 1974. *ACI J.* :317–346.

20.7.5.5 Hot Weather Curing

The time between placing of architectural concrete and the start of curing is most critical in hot, dry, and/or windy weather. To minimize variations in color due to nonuniform drying and to prevent plastic shrinkage cracking, curing should commence as soon as practical.* In order to minimize drying shrinkage, it may be necessary to erect wind breaks out of properly supported plastic sheets or temporary plywood walls.

20.7.6 Protection

All freshly placed concrete should be protected from the elements and from any defacements due to subsequent building operations. One of the most common problems is rain or construction runoff water moving across rebar or other stored materials and then flowing down the face of the exposed architectural concrete. Another cause of rust staining is the use of steel scaffolding. On all architectural concrete contracts, nonferrous scaffolding should be used. If steel scaffolding is necessary, it should be maintained with all surfaces painted. Stain removal is discussed in the PCA publication IS214, *Removing Stains and Cleaning Concrete Surfaces.* Corners, edges and other surfaces vulnerable to damage should be protected with suitable guards or barricades until the project is complete.

20.8 Production and Installation of Precast Elements

20.8.1 Production

Production of architectural precast has progressed from reliance on individual craftsmanship to a well controlled and coordinated production-line method with corresponding economic and physical improvements. The contract documents should make reference to the PCI *Manual for Quality Control for Plants and Production of Architectural Precast Concrete Products* (MNL-117-96) as the industry guideline for production. The general objective of this standard is to define the required minimum practices for production and for a program of quality control to monitor the production. Therefore a performance specification based on MNL-117 will ensure a quality product.

20.8.2 Coordination

Even if the quality of the precast concrete components delivered to the job site is good, the success of the entire project is dependent on the quality of workmanship of the erector. A poor erection job can detract not only from the performance of the structure, but also from the appearance of even the best designed precast concrete project. The erector should assist the General Contractor/Construction Manager in taking the lead on a construction site to help solve problems before they occur by coordinating efforts with other participants on the project. The responsibility for the erection of precast concrete may vary as follows:

1. The precast concrete manufacturer supplies the product already erected, either by in-house labor or by an independent erector.
2. The manufacturer is responsible only for supplying the product, Free On Board plant or jobsite. Erection is done either by the general contractor or by an independent erector under a separate agreement with the general contractor and/or owner.
3. The products are purchased by an independent erector, who has a contract with the general contractor and/or owner to furnish the complete precast-concrete package.

On projects where the precast concrete manufacturer does not have the contractual obligation for the erection, it is extremely desirable that the manufacturer assign a representative to observe

---

*ACI Committee 303. July 1974. *ACI J.* :317–346.

and report on planned erection methods. Regardless of contractual obligations, it is recommended that the precast concrete manufacturer maintain adequate contact with the firm(s) responsible for both transportation and erection to ensure that the precast concrete units are properly handled and erected according to the design and project specifications.

The GC/CM is normally responsible for the project schedule and dimensions and coordination with all other construction trades. Relative to erection of precast concrete the GC/CM, in conjunction with the precaster, should carry out the following:

1. The GC/CM should be responsible for coordinating all information necessary to produce the precast concrete erection drawings. A prejob conference should be held as soon as possible after award of the precast concrete contracts. This meeting should consider erection sequencing, weight and size limitations, special rigging and guy and bracing scheme information. This information should be communicated to the precaster's design engineer. Upon receipt of this information, the design engineer should, if necessary, develop a bracing sequence to maintain stability of the structure during erection in conjunction with the erector and engineer of record. Limitations may state, for example, that loading of the structure shall be balanced, requiring that no elevation be erected more than a stated number of floors ahead of the remaining elevations: or limitations may involve the rigidity of the structure, requiring that walls shall not be erected prior to the completion of floors designed to carry the lateral loads.

   In steel frames, it should be determined how far in advance final frame connections must be completed prior to panel erection. Particular consideration should be given to deflection and/or rotation of the supporting structure owing to the precast and other superimposed loads. For concrete frames, what strength of concrete is required should be determined prior to imposing loads of the precast concrete panels. The frame designer should also recognize that connections between panel and frame impose concentrated loads on the frame and that these loads may require supplementary local reinforcing. In the case of multistory concrete frames, consideration should be given to the effects of frame shortening due to shrinkage and creep.

2. The GC/CM should review and approve or obtain approval for all erection drawings and design.

3. The GC/CM should be responsible for the coordination of dimensional interfacing of precast concrete with other trades.

4. The GC/CM should see that proper tolerances are maintained to guarantee accurate fit and overall conformity with precast erection drawings.

5. The GC/CM should be responsible for providing and maintaining clear, level, well drained unloading areas and road access around and into the structure to such a degree that the hauling and erection equipment for the precast concrete units are able to operate under their own power. Erection equipment should be able to handle units directly from the transportation equipment.

Sequencing is of the utmost importance for facilitating erection of the structure. Prior to the manufacture of precast members, the erection and job sequence should be mutually agreed upon by the GC/CM, precaster, and erector. To avoid excessive erection costs, efforts should be made to allow a unit to be handled in one motion from unloading to positioning. Once a sequence is agreed upon, it should be strictly adhered to unless severe unforeseen problems dictate a sequence change.

20.8.3 Field Verification and Layout

Surveys should be required before, during, and after erection:

1. before, so that the starting point is clearly established and any potential difficulties with the support structure are determined early;
2. during, to maintain alignment; and
3. after, to ensure that the products have been erected within tolerances.

The benefits of surveying and laying out the support location prior to erection are as follows: (1) when an error is found it can be corrected prior to starting erection; (2) costly erection delays for crane and crew are avoided if everything fits; and (3) layout crews may find errors on the shop drawings, which, if caught in time, can reduce costly corrections and minimize delays. It is the responsibility of the GC/CM to establish and maintain, at convenient locations, control points, benchmarks, and lines in an undisturbed condition for the erector's use until final completion and acceptance of a project. If the building frame is not precast concrete, then the benchmarks and building lines should be provided by the GC/CM at each floor level. Work points should also be provided for angled and curved building elevations. The erector should be advised by the GC/CM of any known discrepancies in field location, line, or grade and be provided with suitable work areas for layout.

Prior to the start of erection and the scheduling of delivery and handling equipment, a field check of the project should be made by the erector to ensure that foundations, walls, and structural frame are suitably constructed to accept the precast concrete units. This should include a field check of the work affecting the erection contract, including the location, line and grade of bearing surfaces, notches, blockouts, anchor bolts, cast-in-place or contractor's hardware (miscellaneous iron), a check on dimensional tolerances, and a check of all overhead electrical lines. The objective of this survey is to ensure that the areas to receive the precast concrete are ready and accessible to the erector and that the precast units will fit.

Survey notes should be kept of all discrepancies that exceed specified tolerances. Any discrepancies between site conditions and the architectural, structural, and erection drawings, that may cause problems during erection, such as structural steel out of alignment, anchor bolts or dowels improperly installed, errors in bearing elevation or location, and obstructions caused by other trades, should be noted in writing and sent to the precast concrete manufacturer and GC/CM. Erection should not proceed until discrepancies are corrected by the GC/CM, or until erection requirements are modified and reviewed by the design engineer. Verification of remedial work should be the responsibility of the GC/CM. The erector should return to the site for a final survey after all necessary corrections of site conditions have been made by the GC/CM and before any erection is started.

Working from the control points, benchmarks, and lines established by the GC/CM, the erector should establish offset lines and elevation marks as required for use at each floor level. Precast concrete members must be installed to accurate lines and grades to ensure both proper performance and correct relationship to adjoining work. It is the layout man's duty to mark the locations to receive the precast concrete units so that the erector can place them with a minimum of measuring or moving to get them into final position. The erector should establish joint locations prior to actual product installation. This will keep the differential variation in joint width to a minimum, as well as identify problems caused by building frame, columns, or beams being out of dimensional or alignment tolerance. The layout crew should be kept well ahead of erection so if corrections are required, there is adequate time for the design, approval, fabrication, and installation of repairs so as not to hinder the process of erection.

20.8.4 Delivery

Factors to be considered when delivering precast members from the plant to the job site are type, size, shape, and weight of the member; type of finish; weather; road conditions; method of transportation; type of vehicle; routing; distance to the job; and job-site conditions. The shipper not necessarily the precast concrete manufacturer or erector, is responsible for safe delivery of precast members. The precast concrete manufacturer, shipper and erector should develop clearly defined acceptance procedures to avoid later disagreement concerning damage. The precast concrete manufacturer or erector should advise the shipper of any special situations along possible routes and at the job-site for the following reasons:

1. to ensure safe transportation;
2. to meet various governmental transportation regulations;

3. to prevent or minimize in-transit damage with proper supports, frames, blocking, cushioning, and tie-downs. These materials are normally supplied by the precast concrete manufacturer and returned for repeat usage by the shipper. All blocking, packing, and protective materials must be of a type that will not cause damage, staining, or other disfigurement of the units. The blocking points and orientation of the units on the shipping equipment shall be as designated on the shop drawings;

4. to permit their removal for erection from the load in proper sequence and orientation to minimize handling and possible damage to the product;

5. to ensure that the product is oriented and loaded on the trailers so that it can be erected without unnecessary off loading and rehandling—unloading of product on the ground is usually costly, creates site access problems for everybody, and often results in damage to the product; and

6. to ensure that sufficient product is delivered to the site in the prescheduled sequence to allow for orderly, efficient installation in the structure.

20.8.5 Connection Considerations

A certain amount of field adjustment at the connections is normal. Product tolerances make the possibility of a perfect fit in the field impossible. This is true when the precast concrete pieces join to each other, and is even more true, when the precast units must interface with other materials. Each industry has its own recommended erection tolerances that apply when its products are used exclusively. Hardware design for connections should take into account the tolerances for both the precast concrete components and the structure. These considerations may require clip angles and plates with slots or oversize holes to compensate for dimensional variation, field welding, or sufficient shim spaces to allow for elevation variations. Sufficient minimum clearance between precast units and the structure must be provided to allow for product and erection tolerances. Hardware should be designed to compensate for additional stress at the maximum anticipated clearance.

Field adjustments or changes in connections that could create additional stresses in the products or connections should not be permitted without approval by the architect or engineer. Particular care should be taken to prevent damage to the precast concrete unit when adjustments are being made to bring the unit into final position. Whenever possible, the connections should be completed in such a manner to permit operations to take place on the top side of erected members rather than from below where ladders or scaffolds are required, especially for welding. The type of equipment necessary to perform operations such as welding, post-tensioning, or pressure grouting should be considered. Operations that require working under a deck in an overhead position should be avoided, especially for welding. Alternatives to any erection techniques that require temporary scaffolding should be considered. Room to place wrenches on nuts and turn them through a large arc should be provided for bolted connections. Foundation piers should extend above grade so that the anchor bolts can be adjusted during erection and the base plate need not be grouted in a hole, which can fill with water and debris. Properly drypacking column or wall panel bases in a narrow excavation is difficult.

It is desirable to have connections that are designed so that the erector can safely secure the member to the structure in a minimum amount of time without totally completing the connections. If necessary, temporary bracing or connections should be used with final adjustment and alignment in all directions relative to the structure and adjacent components completed independently of crane support. This allows the hoisting device to begin placing the next unit while connections on the first are being completed. The temporary connections should not interfere with or delay the placement of subsequent members. The temporary connections may have to be relieved or cut loose prior to completion of the permanent connections.

Connection details should allow erection to proceed independently of ambient temperatures and without temporary protective measures. Materials such as grout, drypack, cast-in-place concrete, and epoxies need protection or other special provisions when placed in cold weather. Also, welding is slower when the ambient temperature is low. If the connections are designed so that these processes must be completed before erection can continue, the cost of erection is increased and delays may result.

The supply of erection hardware, that is, the loose hardware needed in the field for final connection of precast members, should normally be the responsibility of the precast concrete manufacturer. The responsibility for the supply of contractor's hardware to be placed on or in the structure in order to receive the precast concrete units depends on the type of structure and varies with local practice, but in any case should be clearly defined in the contract documents. Hardware should be incorporated in the structure within the specified tolerances according to a predetermined and agreed upon schedule to avoid delays or interference with the precast erection. Hardware should be maintained and suitably protected (such as capping of inserts and protection of threads) until used by the precast erector. At times, it may be necessary to modify connections in the field to accommodate unforeseen clearance problems or the work of other trades. The GC/CM should clear any modifications through both the precast concrete manufacturer and the designer to ensure proper performance of the connections.

20.8.6 Installation

Hoisting the precast pieces is usually the most expensive and time-critical process of erection. Connections should be designed so that each unit can be lifted, set, connected, and unhooked in the shortest possible time. Before the hoist can be unhooked, the precast piece must be stable and in its final position. Preplanning for the fewest, quickest, and safest possible operations that must be performed before releasing the crane will greatly facilitate erection. Bearing pads, shims, or other devices upon which the piece is to be set should be placed ahead of hoisting. Loose hardware that is required for the connection should be immediately available for quick attachment. In some cases, it may be necessary to provide temporary fasteners or leveling devices, with the permanent connection made after the crane is released. For example, if the permanent connection requires field welding, grouting, drypacking, or cast-in-place concrete, the use of erection bolts, C-clamps, guy lines, pins, or shims should be considered. These temporary devices must be given careful attention to ensure that they will hold the piece in its proper position during the placement of all other pieces erected before the final connection is made.

Shims are often used as spacers or as means of leveling or aligning adjacent components. They serve an interim or temporary load-transferring function. Unless such temporary loading of units has been specifically incorporated or allowed in the contract documents, the erector should be responsible for this temporary loading of the units. High-density plastic and steel shims are commonly used to attain the specified joint dimension. Shims should be placed away from the face of the unit to prevent spalling in case of excessive loading. Also, they should be recessed out of view as they can be unsightly and difficult to remove if exposed to view. Shims should be removed from joints of non-load-bearing units after connections are completed and before applying sealant, unless the shim material itself is readily deformable. If left in place, shims should be noncorrosive or protected so that staining will not occur. For load-bearing units, the use of nonmetallic shims or bearing pads such as neoprene or plastic may be used where bearing pressures do not exceed the allowable concrete stresses and the potential risk of damage to concrete edges is negligible.

The indiscriminate use of shims, particularly steel shims, can sometimes lead to undesirable consequences. For example, steel shims have been used in conjunction with grouted joints in multistory bearing wall construction. Even with well-compacted grout or drypack, the compressive modulus of the steel shim is six times that of the drypack; consequently the grout will deform (compress) more readily than the shim. The principal load-transfer path will remain concentrated through the steel shim rather than distributed along the grout bed. High load concentrations at shims can cause spalling at panel surfaces or crack panels vertically.

Precast concrete units should be erected at locations shown on the erection drawings, within the allowable tolerances. They should be positioned so that cumulative dimensional errors do not exceed allowable tolerances. Horizontal and vertical joints should be correctly aligned and uniform joint width maintained as erection progresses. The installation or prewelding of miscellaneous iron hardware should be performed prior to the start of erection, whenever possible, to minimize the cost of crew and equipment standby. All inserts for lifting and bracing should be filled immediately after units are erected to prevent staining from corrosion-sensitive inserts or water freezing in them. Wall panels should be rigged and hoisted on the structure near the final location and held in place until safely secured. Final alignment and final connections may be made at once or later by follow-up crew, depending on the type of connection.

The connections should allow for easy adjustment in all directions. Panels should be installed on a floor by floor basis, where feasible, to keep loading equal on the structure, or the designer of the structural frame should determine the degree of imbalanced loading permitted (the sequence should be shown on the erection drawings). Limitations may state, for example, that loading of the structure shall be balanced, requiring that no elevation be erected more than a stated number of floors ahead of the remaining elevations; or limitations may involve the rigidity of the structure, requiring that walls not be erected prior to completion of the floors designed to carry lateral loads. Panels should be aligned to predetermined offset lines, established for each floor level. This is important owing to the drift of a high-rise structure caused by sun, wind, and eccentric loads and forces due to construction activity being performed by others. It should be determined how far in advance of precast concrete erection, final connections in steel structures must be completed and what the minimum concrete strength should be for cast-in-place structures prior to loading them with precast concrete units.

Consideration should be given to the number of floors the erector wants to erect in a structure on temporary connections before making final connections. This is especially important for grouted connections and cold-weather conditions. Precast spandrels usually extend from column line to column line at the building perimeter. They connect either at the columns or directly to the perimeter beams. Precautions should be taken against torsional rotation, due to eccentric loadings on perimeter beams, until final connections are made. Shoring may be required to assist in erection until connections are made.

Alignment and all connections should be made, or the spandrels should be safely secured, prior to releasing the load and disconnecting the rigging. On long spandrels, care should be taken to ensure that the spandrel is tied back at the center either through a temporary connection until a slab tie connection is made or with a permanent connection. Spandrels should be installed on a floor by floor basis, where feasible, to keep equal loading on the structure. Spandrels should be aligned to predetermined offset lines established for each floor level. Vertical dimensions between spandrels should be checked to ensure that the opening size is within allowable tolerance. When spandrel panels or column covers and mullions are interspaced with strips of windows to create a layered effect of glazing, precast, and glazing, the general contractor should work closely with the designer to arrive at interface details that can be built within the sequence of construction and that embody all the elements required of the exterior wall design.

20.8.7 Tolerances

Architectural precast concrete product tolerances should comply with the industry tolerances published in the PCI MNL-117. The recommended erection tolerances are shown in Figure 20.37. During wall panel installation, priority is generally given to aligning the exterior face of the units to meet aesthetic requirements. This may result in the interior unit faces not being in a true plane.

A liberal joint width should be allowed if variations in overall building dimensions are to be absorbed in the joints. This may be coupled with a closer tolerance for variations from one joint to the next for appearance purposes. The individual joint-width tolerance should relate to the number of joints over a given building dimension. For example, to accommodate reasonable variations in actual site dimensions, a 3/4 in (19 mm) joint may be specified with tolerances of $\pm 1/4$ in (± 6 mm) but with only a 3/16 in (5 mm) differential variation allowed between joint widths on any one floor

a = Plan location from building grid datum* .... ±1/2 in
a_l = Plan location from centerline of steel** .... ±1/2 in
b = Top elevation from nominal top elevation
    Exposed individual panel .... ±1/4 in
    Nonexposed individual panel .... ±1/2 in
    Exposed relative to adjacent panel .... 1/4 in
    Nonexposed relative to adjacent panel .... 1/2 in
c = Support elevation from nominal elevation
    Maximum low .... 1/2 in
    Maximum high .... 1/4 in
d = Maximum plumb variation over height of structure or 100 ft whichever is less* .... 1 in
e = Plumb in any 10 ft element height .... 1/4 in
f = Maximum jog in alignment of matching edges .... 1/4 in
g = Joint width (governs over joint taper) .... ±1/4 in
h = Joint taper maximum .... 3/8 in
h_{10} = Joint taper over 10 ft length .... 1/4 in
i = Maximum jog in alignment of matching forces .... 1/4 in
j = Differential bowing or camber as erected between adjacent members of the
    same design .... 1/4 in

*For precast buildings in excess of 100 ft tall, tolerances "a" and "d" can increase at the rate of 1/8 in per story to a maximum of 2 in.

**For precast concrete erected on a steel frame building, this tolerance takes precedence over tolerance on dimension "a."

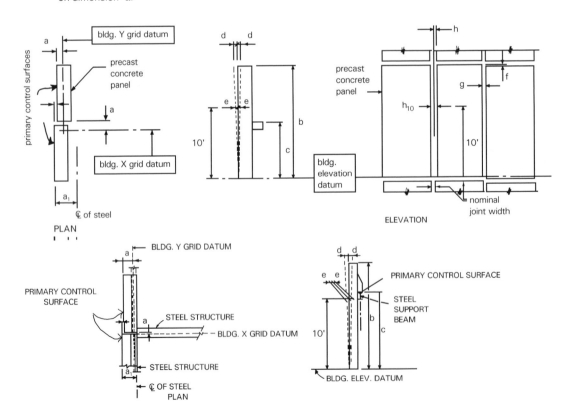

FIGURE 20.37 Erection tolerances for architectural precast concrete wall panels.

or between adjacent floors. Alternatively, a jog in the alignment of an edge may be specified. The performance characteristics of the joint sealant should also be taken into account when selecting a joint size.

Variations from true length or width dimensions of the overall structure are normally accommodated in the joints or, where this is not feasible or desirable, at the corner units, in expansion joints, or in joints adjacent to other wall materials. In a situation where a joint has to match an architectural feature (such as a false joint), a large deviation from the theoretical joint width may not be acceptable and tolerance for building lengths will have to be accommodated at the corner units. Erection tolerances are of necessity largely determined by the actual alignment and dimensional accuracy of the building foundation and frame. The general contractor is responsible for the plumbness, level, and alignment of the cast-in-place concrete foundation and structural frame (except for precast frames) including the location of all bearing surfaces and anchorage points for the precast concrete units. If the precast concrete units are to be installed reasonably "plumb, level, square, and true," the actual location of all surfaces affecting their alignment, including levels of floor slabs and beams, the vertical alignment of floor slab edges, and the plumbness of columns or walls must be known before erection begins.

The architect or engineer should clearly define in the specifications the maximum tolerances permissible in the foundation and building frame alignment and then should see that the general contractor frequently checks to verify that these tolerances are being held. In addition, the architect or engineer should ensure that the details in the contract documents allow for the specified tolerances. Lack of attention to these matters often necessitates changes and adjustments in the field, not only delaying the work but usually resulting in unnecessary extra cost and sometimes impairing the appearance of the units and the completed structure.

With reasonable tolerances for the building frame established, it is equally important that the designer provide adequate clearances (purposely provided space between adjacent members), for example, between the theoretical face of the structure and the back face of a precast concrete panel, in detailing the panel and its relationship to the building structure. If clearances are realistically assessed, they will enable the erector to complete the final assembly without field altering the physical dimensions of the precast units.

A good rule of thumb is that at least 3/4 in (19 mm) clearance should be required between precast members, except for flange-to-flange connections and connection of wall panels to precast members where 1/2 in (13 mm) clearance should be required with a 1 in (25 mm) clearance preferred: 1 in (25 mm) is the minimum clearance between precast members and cast-in-place concrete and 2 in (50 mm) is preferred. For steel structures, 1 in (25 mm) is the minimum clearance between the back of the member and the surface of the fireproofing and 1 1/2 in (38 mm) is preferred. If there is no fireproofing required on the steel, then a 1 in (25 mm) minimum clearance should be maintained. At least a 1 1/2 to 2 in (38 to 50 mm) clearance should be allowed in tall structures regardless of the structural framing materials. The minimum clearance between column covers and columns should be 1 1/2 in (38 mm), with 3 in (75 mm) preferred because of the possibility of columns being out of plumb or a column dimension interfering with completion of the connection. If clearances are realistically assessed, they will solve many tolerance problems. Where large tolerances have been allowed for the supporting structure, or where no tolerances for the structure are given, the clearance must be increased.

All connections should be provided with maximum adjustability in all directions that is structurally or architecturally feasible. Where a 1 in (25 mm) clearance is needed but a 2 in (50 mm) clearance creates no structural or architectural problem, the 2 in (50 mm) clearance should be selected. Closer tolerances are required for bolted connections than for grouted connections. To accommodate any misalignment of the building frame, connections should provide for vertical, horizontal, and lateral adjustments of at least 1 in (25 mm).

Location of hardware items cast into or fastened to the structure by the general contractor, steel fabricator, or other trades should be ±1 in (25 mm) in all directions (vertical and horizontal), plus a slope deviation of no more than ±1/4 in (±6 mm) in 12 in (0.3 m) for the level of critical bearing surfaces. All bearing surfaces are not always level. These tolerances give an acceptable deviation. Connection details should consider the possibility of bearing surfaces being misaligned or warped from the desired plane. Adjustments can be provided with the use of drypack concrete, nonshrink grout, shims, or elastomeric pads if the misalignment from horizontal plane exceeds 1/4 in (6 mm).

20.8.8 Protection of Work

All precast concrete should be furnished to the job site in a clean and acceptable condition and kept in such condition. The erector is normally responsible for any chipping, spalling, cracking, or other damage to the units after they are delivered to the job site and until they are erected and connected. The erector should take necessary precautions to protect the erected precast concrete and the work and materials of other trades from damage during erection.

After the final erection of any portion of precast work to acceptable alignment and appearance, including completion of all connections and joints, the general contractor should assume responsibility for protection of the work. Any cleaning or repair of precast concrete work subsequent to installation and/or acceptance should be done by the erector or precaster, but under the responsibility of the general contractor. Specifications should state this responsibility clearly. It is wholly impractical for the precaster or erector to police the work against damage by others after it is put in place. There should be a carefully established and implemented program of protection and later cleaning for each job that is, under the responsibility of the general contractor, who alone can control all the potential sources of damage.

At the end of each working day, measures may need to be taken to protect the installation from damage. For example, where precast units in partially completed buildings could be damaged by weather, such as water freezing in holes, pipe sleeves, and inserts, adequate temporary protection shall be provided. The GC/CM should provide and maintain temporary protection to prevent damage or staining of exposed precast concrete during construction operations after it has been installed. Rainwater or water from hoses used during the construction of the building can cause discoloration of the precast concrete units by first washing across other building materials (such as steel, concrete, or wood) and then across the precast units. Particular care should be taken to avoid allowing jobsite water to wash the units. Dirt, mortar, and debris from concrete placing should not be allowed to remain on the precast concrete and should be washed off immediately with clean water. The erector should protect adjacent materials, such as glass and aluminum, from damage from field welding or torch cutting operations. Therefore sequencing of work of other trades should be taken into consideration by the GC/CM to prevent such damage.

20.8.9 Sealants

Sealant life and performance are greatly influenced by joint design. For optimum performance and maximum life, the recommendations of the joint-sealant manufacturer should be followed. Joints between precast units must be wide enough to accommodate anticipated wall movements, and particular care must be given to joint tolerance in order for the joint sealant system to perform within its design capabilities. If units cannot be adjusted to allow for proper joint size, saw cutting may be necessary. When joints are too narrow, bond or tensile failure of the joint sealant will occur and/or adjacent units may come in contact and be subjected to unanticipated loading, distortion, cracking and local crushing (spalling). A good general rule is to erect precast units in such a manner as to provide 1/2-in joints between units up to 15 ft long and a minimum of 3/4-in joints for longer units. (Joint width should equal two to four times anticipated movement, depending on the properties of the specified sealant.) Corner joints should be 1 in wide to accommodate the extra movement and bowing often experienced at corners. Sealants should not be installed in joints smaller than 3/8 in wide and 3/8 in deep.

The required sealant depth is dependent on the sealant width at the time of application. The optimum sealant width to depth relationships are best determined by the sealant manufacturer. For a comprehensive discussion of joint sealants used between wall panels, refer to ASTM C962. Sealants used for specific purposes are often installed by different subcontractors. For example, the window subcontractor normally installs sealants around windows, whereas a second subcontractor typically installs sealants around panels. The designer must select and coordinate all of the sealants used on a project for chemical compatibility and adhesion to each other. In general, contact between different sealant types should be minimized. The recommendations of the sealant manufacturer should always be followed regarding mixing, surface preparation, priming, application life, and application procedure. Good workmanship by qualified sealant applicators is the most important factor for satisfactory performance.

20.9 Finish Cleanup

20.9.1 Tie-Hole Repairs*

Tie holes should be plugged to prevent corrosion of the tie and possible staining of the surface, except where stainless steel form ties are used. The holes left in the surface of the concrete as the result of the form tie may be either small or large, depending upon the type of tie used. In a rough-textured surface, small holes can be plugged flush with the surface and concealed. With smooth-surface concrete, the tie holes will be more apparent, and it is better to only partially fill the holes, leaving the holes as a part of the planned appearance. Care must be exercised to avoid smearing the fill material on the surface of the concrete. Materials used for plugging tie holes include Portland cement mortar, epoxy mortar, plastic plugs, precast mortar plugs, and lead plugs. The method should be carefully selected from among those that have shown no staining or discoloration tendencies in actual use. Mortar materials of a dry-tamp consistency and densely tamped into the hole will be less likely to smear on the surface than those of wet consistency.

When Portland cement mortar is used, the tie hole should first be prewet with clean water, and then a neat cement slurry bond coat should be applied to the hole surfaces before they are filled with mortar. If epoxy mortar is used, it should be applied in accordance with the manufacturer's instructions, and a caulking gun should be used to inject it into the tie hole to prevent smearing it on the surface. Cleaning or removal of any fill material is difficult and will usually leave a stain on the surface. Plastic inserts are provided by cone tie manufacturers and can be wedged into the tie hole, leaving a standard predetermined recess. Lead plugs can be wedged into the hole by hammering. Sometimes the removable cone becomes embedded in the concrete owing to form movement or leakage around the cone. It can be removed to produce a neat appearance by drilling out the cone with a diamond bit tool that conforms to the hole size. It may be economical to remove all cones in this manner to ensure neat uniform holes. When tie holes need to be concealed rather than expressed as part of the planned appearance, the repair procedure is similar to that required for blemish repair.

20.9.2 Repairs†

A certain amount of repair of architectural concrete surfaces is to be expected. Blemishes that are beyond the limits established by the quality of the preconstruction mock-up must be repaired. Major repairs of cracks or honeycomb surfaces should not be attempted until an engineering evaluation is made to determine if a sound repair can be made or if concrete removal is required. The repair work should proceed as soon as possible after form removal using the materials and methods already

---

*ACI Committee 303. July 1974. *ACI J.* :317–346
†Kenney, A.R. 1984. *ACI J.* :50–55.

accepted on the approved mock-up. Then the repair and the surrounding concrete will age together, and the chance of color or texture variation will be minimized.

Prior to beginning any repair, the surface blemishes must be evaluated to determine if repairs should be attempted at all. Do not make surface blemishes worse. Too often, indiscriminate repair of architectural concrete surfaces results in accenting blemishes rather than improving them. Much of the skill in repair of surface defects is in knowing what not to attempt. Repair work should be kept to a minimum. Small repairs can be treated by hand tooling or chemical treatment using bleaches, acids, or toners. Hand tooling refers to the use of needle scalers, chisels, bushhammers, or grinders. These tools are useful for eliminating or minimizing hard spots and discolored areas such as the dark lines associated with form-leakage points. Tools can also be effective in blending offsets and cold joints in combination with grout repairs. The importance of establishing a repair method before the need arises cannot be overstressed. Once proven acceptable on the mock-up, immediate repairs can be made without delay and with confidence in the final outcome.

Repair work for architectural concrete requires expert craftsmanship and careful selection and mixing of materials if the end result is to be structurally sound, durable, and aesthetically pleasing. Where adjacent acid etching, sandblasting, tooled, or bushhammering treatments must be matched, experimentation should be performed on unimportant areas. Ingenuity may sometimes be required to establish methods and techniques that are as satisfactory as those in standard use. Excessive deviations in color and texture of repairs from the surrounding surfaces may result in the architectural concrete or precast concrete not being approved until the variation is minimized. Light honeycomb areas, bug holes, pour lines, or cold joints, when sufficient rock is in place, can receive a grout patch mix without a cutting and drypack requirement. All that is necessary is to pre wet the area to be repaired, along with the adjacent concrete, keeping it wet for at least one hour. Then, using a dense sponge float, scrub the predetermined grout mix into the voids between the coarse aggregate. It may be necessary to scrub in a second grout patch mix application to obtain a true flush surface. When the finish has exposed aggregate, just prior to initial set, wash the concrete surface with a soft sponge or bristle brush, exposing the aggregate texture desired. Begin curing procedures immediately to ensure that the repair does not dry out too quickly and develop shrinkage cracks. Cure the repair a minimum of 3 to 7 d before attempting final texturing.

Where the surface blemish cannot be repaired by mechanical, chemical, or grout treatment, the area should be cut out using a pneumatic or hand-held chipping hammer. Edges of the cut area should be square and perpendicular or undercut to the face of the sound adjacent concrete to avoid any feather-edge repairs. In an exposed aggregate area, cut irregularly to conceal the repair. In finishes involving a liner, cut along the texture lines. The procedure for repairs is as follows:

1. After all adjacent finish texture and chipping is complete, remove loose particles and brush or blow all dust from the repair surface.
2. Proportion the patch mix by weight according to the same proportions used in the concrete mix, but substitute 5–50% of the grey or buff cement with white cement. The actual amount is based upon tests to determine what is required to match the finish surface. A pigment or toner is sometimes used for a closer color match or when only white cement was used in the original mix design.
3. Apply a 50% diluted coat of bonding material to the root of the repair area, being careful to avoid brushing or dripping it on any concrete surface to be exposed. This prevents loss of moisture from the patch mix while also improving bond.
4. Fill the repair area with patch mix in layers 1/2 in (12 mm) thick, with a wood or plastic trowel. Vibrate or tamp the patch mix manually to approximately the density of the existing concrete. Strike the area level, and after initial set, if the finish has exposed aggregate, brush the surface to match the surrounding area.
5. Begin curing procedures immediately.
6. Clean the area adjacent to the repair to remove laitance and to restore the original color and texture.
7. After 3 to 7 d, provide the final texture and thoroughly clean the repair area.

8. After 28 d the repair can be evaluated for acceptance when the finished surface is dry. Any repair should match adjacent surfaces in color and texture when viewed at a distance of 20 ft (6 m).

It is preferrable that repairs not be made when direct sunlight is on the repair area. Temporary sunshades should be provided, especially during hot weather. Heat lamps or a small heated enclosure may be necessary to maintain a minimum 50°F (10°C) for curing repairs in cold weather. Repair and patching of precast concrete is an art requiring expert craftsmanship if the end result is to be structurally sound, durable, and pleasing in appearance. Responsibility for repair work is normally resolved between the precaster and the erector. Repairs should be done immediately following occurrence of damage. However, deciding when to perform the repairs should be left to the precaster, who should be responsible for satisfactory final appearance. Since the techniques and materials for repairing precast concrete are affected by a variety of factors including mix ingredients, final finish, size and location of damaged area, temperature conditions, age of member, surface texture, etc., precise methods for repair cannot be detailed here. (See PCI MNL-117-96 for guidance on repair techniques and materials.)

Repairs should be done only when conditions exist to which ensure that the repaired area will conform to the balance of the work with respect to appearance, structural adequacy, and durability. Slight color variations can be expected between the repaired area and the original surface owing to the different age and curing conditions of the repair. Time will tend to blend the repair into the rest of the member so that it will become less noticeable. After all repairs have been completed, the repairs should be coated with a silane or dilute (2 or 3% solids) solution of an acrylic sealer to minimize moisture migration into the repair concrete. The acrylic sealer should not stain the concrete or leave a shiny surface. Even with proper consolidation, the repaired area is more porous than the original hardened concrete and will need this dilute sealer application.

20.9.3 Cracks

Small cracks, under 0.010 in (0.25 mm), may not need repair unless failure to do so will cause corrosion of reinforcement. If repair is required to restore structural integrity for cracks that range in width from 0.003 to 0.015 in, they should be repaired with epoxy injection. Proper preparation of the crack area is very important. Form-release agents, efflorescence, grease, oil, dirt, or fine particles of concrete prevent epoxy penetration and bonding. Preferably, contamination should be removed by vacuuming or clean water flushing or a specially effective solvent, determined by prior experience or testing. The solvent is then blown out using compressed air or adequate time is provided for air drying. Cracks may be pressure injected through entry ports with a low-viscosity, high-modulous, 100% solid, two-component epoxy that will bond to a moist surface. It is usually impossible to completely remove moisture in a crack. Care should be taken to select an epoxy color (amber, white, or grey) that most closely matches the color of the concrete. Toners can be added to the epoxy mixture to more closely match color surfaces for cracks in the exposed face of the architectural concrete. When mixing the low-viscosity epoxy resin, the epoxy hardener, and color toner, do not shake or mix too vigorously, since this can infuse air into the epoxy, which would result in the formation of undesirable voids in the hardened epoxy.

Cracks wider than 0.25 in (6 mm) should be repaired with a gel-type resin system incorporating a mineral filler or a long pot-life material. Some cracks in precast products extending downward from nearly horizontal surfaces may be filled by gravity flow. The minimum width of a crack that can be filled by gravity is a function of the viscosity of the material. If a spalled piece of concrete is available and surfaces still mate, the easiest repair is to simply glue the piece back in place using nonsag epoxy or bonding agents. Broken surfaces on both the original surface and spalled piece should be painted with the epoxy adhesive. Enough epoxy should be applied to the surface that some squeezes out of the joint when mated pieces are bound together. An epoxy with a thick enough consistency should be selected so that it does not sagg or run on a vertical surface. Self-leveling formulations should not be used. In some cases it may be better to secure the repair by inserting pins into epoxy-filled holes drilled in the loose piece and the original surface.

20.9.4 Cleaning

Dirt, mortar, plastic, grout, fireproofing, or debris from concrete placement should not be permitted to remain on the concrete and should be brushed or washed off immediately with clean water. Final cleaning should be performed after caulking is completed and no earlier than 3 d after any repairs have been completed. If at all possible, concrete cleaning should be done when the temperature and humidity allow rapid drying. Slow drying increases the possibility of recurring efflorescence and discoloration. Before cleaning, a small (at least 1 yd^2 (0.8 m^2)) inconspicuous area should be cleaned and checked to be certain that there is no adverse effect on the concrete surface finish or adjacent materials. The effectiveness of the method should not be judged until the surface has dried for at least one week. Materials such as glass, metal, wood, stone, or concrete adjacent to the area to be cleaned should be adequately protected since they can be damaged by contact with some stain removers or by physical cleaning methods. A strip-off plastic that is sprayed on can be used to protect glass and aluminum frames.

The following is a suggested order for testing appropriate procedures for the removal of dirt, stains, and efflorescence (beginning with the least damaging):

1. Dry scrubbing with a stiff fiber (nylon) brush is particularly effective if the surface is brushed shortly after the appearance of efflorescence.
2. Abrasive blasting with industrial baking soda will remove efflorescence without otherwise affecting the concrete surface. Water must not be used to remove any residue on the surface, as salts will be dissolved and carried into the concrete, causing additional efflorescence. Residues should be blown, vacuumed, or brushed from the surface.
3. Wet scrubbing may also be effective in removing efflorescence. This procedure involves, wetting the surface with water, vigorously scrubbing the finish with a stiff fiber brush and thoroughly rinsing the surface with clean water. Low pressure water spraying (water misting), high pressure water jet sprayers, and steam cleaning are alternate methods of wet scrubbing, and
4. Chemical cleaning compounds such as detergents, muriatic or phosphoric acid, or other chemical cleaners should be in accordance with the manufacturer's recommendations. If possible a technical representative of the product manufacturer should be present for the initial test application to ensure it is properly used.

   Areas to be chemically cleaned should be thoroughly saturated with clean water prior to application of the cleaning material to prevent the chemicals from being absorbed deeply into the surface of the concrete. Surfaces should also be thoroughly rinsed with clean water after application so that no traces of acid remain in the surface layers of the concrete. Cleaning solutions should not be allowed to dry on the concrete finish. Residual salts can flake or spall the surface or leave difficult stains. Misapplication of hydrochloric acid can lead to corrosion of embedded metals with shallow cover.

   Care should be taken to use dilute solutions of acid to prevent surface etching that may reveal the aggregate and slightly change the surface color and texture. The entire unit should be treated to avoid a mottled effect. Any of several diluted solutions of acids are effective ways to remove efflorescence:

   a. one part hydrochloric (muriatic) acid in 9 to 19 parts water,
   b. one part phosphoric acid in 9 parts water,
   c. one part phosphoric acid plus one part acetic acid in 19 parts water, or
   d. one part acetic acid (vinegar) in 5 parts water.

   Hydrochloric (muriatic) acid may leave a yellow stain on white concrete. Therefore phosphoric or acetic acid should be used to clean white concrete. Workers using cleaning compounds or acid solutions should be thoroughly trained in their use. The use of proper protective wear should be strictly enforced. Rubber gloves, glasses, and other protective clothing must be worn by workmen using acid solutions or strong detergents. Materials used in chemical cleaning can be highly corrosive and are frequently toxic. All precautions on labels should be observed because these cleaning agents can affect eyes, skin, and breathing. Materials that can produce noxious or flammable fumes should not be used in confined spaces unless adequate ventilation can be provided.

5. Dry or wet abrasive blasting, using sand, ferrous aluminum silicate, or other abrasives, may be considered if this method was originally used in exposing the surface of the unit. An experienced subcontractor should be engaged for sandblasting.

6. Stone veneer-faced precast concrete units should be cleaned with stiff fiber or stainless steel or bronze wire brushes, a mild soap powder or detergent, and clean water using high pressure, if necessary. No acid or other strong chemicals that might damage or stain the veneer should be used. Information should be obtained from stone suppliers on methods for cleaning oil, rust, and dirt stains on the stone.

7. Mortar stains may be removed from brick panels by thoroughly wetting the panel and scrubbing with a stiff fiber brush and a masonry-cleaning solution. A prepared cleaning compound is recommended; however, on red brick, a weak solution of muriatic acid and water (not to exceed a 10% muriatic acid solution) may be used. Acid should be flushed off the panel with large amounts of clean water within 5 to 10 min of application. Buff, gray, or brown brick should be cleaned in accordance with the brick manufacturer's recommendations, using proprietary cleaners rather than acid to prevent green or yellow vanadium stains and brown manganese stains.

    Following the application of the cleaning solution, the panel should be rinsed thoroughly with clean water. High-pressure water cleaning techniques, with a 1000 to 2000 psi (6.9 to 13.8 MPa) washer, may also be used to remove mortar stains.

    Unglazed tile or terra cotta surfaces should be cleaned with a 5% solution of sulfamic acid for gray or white joints, and a more dilute (2%) solution should be used for colored joints. The surface should be thoroughly rinsed with clean water both before and after cleaning. Glazed tile manufacturers generally do not recommend the use of acid for cleaning purposes.

8. For information on removing specific stains from concrete, reference should be made to *Removing Stains and Cleaning Concrete Surfaces*, IS214, published by the Portland Cement Association, Skokie, Illinois.

20.10 Acceptability of Appearance

At the time the visual mock-ups or initial precast production units are approved, the acceptable range in color, texture, and uniformity should be determined. If the procedures determined by the approved mock-up were continued throughout the project, final acceptance should not be a problem. Owing to the inevitable nonuniformity of construction practices, some repairs will normally be required. Their final acceptability will depend on the blending capability and skill of the contractor's or precast manufacturer's restoration personnel. The final product is exposed to view, and faulty work anywhere in the construction process can be easily seen. Acceptance lies with the value judgment of the owner, architect, and/or consultant. For that reason, periodic review during construction and partial acceptance creates good will and confidence for all concerned. Uniformity of texture, intensity of color, and contrast range will determine whether an architectural concrete surface is acceptable.

Acceptability will vary with the characteristics of the architecture and the average viewing distance. Finish texture, color, contrast, aggregate size, and shape all affect appearance. In order to establish a basic criteria for both architectural cast-in-place and precast concrete a definitive rule for acceptability is as follows:

> The finish face surface shall have no obvious imperfections other than minimal color and texture variations from the approved samples or evidence of repairs when viewed in good typical daylight illuminated with the unaided naked eye at a 20 ft (6 m) viewing distance. Appearance of the surface shall not be evaluated when sunlight is illuminating the surface from an extreme angle as this tends to accentuate minor surface irregularities.

After final acceptance, the inspector's records should be completed and filed. If later additions are made or adjoining buildings constructed, these records will be needed for construction.

20.11 Innovations

Today, more and more building contracts are negotiated or have a limited bid list. This will accelerate in the future with significant increases in value engineering. This incentive increases the team member's willingness to offer, approve, and implement ideas quickly. Project control systems will continue to be streamlined and productivity improved as more common systematic problems of information flow are corrected. Involvement of subcontractors and suppliers as equal partners will improve the flow of ideas and total job performance. We forecast use of architectural facades in combination with interior structural concrete will precast concrete to compete with other building systems.

New materials or expanded uses include superplastisizers and 15,000 psi strength concrete. Improved concrete pumps are able to move high aggregate content concrete mixes. Tower cranes will be able to handle heavier loads at a greater reach. Fiberglass and steel forms designed to handle external form vibrators and full liquid head concrete have already begun to be used. Specialty vibrators are currently used to meet specific job performance requirements. Improved sealers, sealants, and specialty coatings are already available. Architectural precast concrete is combining textures, colors, and embedded materials such as clay products or natural stone. The industry is constantly striving to develop improved methods of production, better handling and erection, and new finishing techniques and ways of combining materials. The ability to customize the color, form and texture of concrete allows limitless design possibilities.

Defining Terms

Abrasive blasting: A process in which sand or other materials are used to texture the surface of hardened concrete. The degree of blasting may vary from a light cleaning operation to one that exposes aggregate to a depth of 3/4 in (19 mm) or more.

Admixture: A material other than water, aggregates, or cement used as an ingredient in concrete, mortar, or grout to impart special characteristics.

Aggregate: Granular material, such as sand, gravel, and crushed stone used with a cementing medium to form a hydraulic-cement concrete or mortar.

Architectural concrete: Any concrete with one or more surfaces be permanently exposed to view and where the appearance of these surfaces is important from an architectural standpoint. A precast product with a specified standard of uniform appearance, surface details, color, and texture.

Backup mix: The concrete mix cast into the mold after the face mix has been placed and consolidated.

Bugholes: Small holes on formed concrete surfaces formed by air or water bubbles, sometimes called blowholes.

Bondbreaker: A substance placed on a material to prevent it from bonding to the concrete or between a face material such as natural stone and the concrete backup.

Bonding agent: A substance used to increase the bond between an existing piece of concrete and a subsequent application of concrete such as a patch.

Bushhammering: A process in which pneumatic or hand hammers are used to remove mortar and fracture aggregate at the surface of hardened concrete will produce an attractive varicolored and textured surface.

Coarse aggregate: Aggregate predominately retained on the U.S. Standard No. 4 (4.75 mm) sieve, or that portion of an aggregate retained on the No. 4 (4.75 mm) sieve.

Crazing: A network of visible, fine hairline cracks in random directions breaking the exposed face of a panel into areas from 1/4 to 3 in (6 to 75 mm) across.

Exposed-aggregate concrete: Concrete manufactured so that the aggregate on the face is left protruding.

Face mix: The concrete at the exposed face of a concrete unit used for specific appearance purposes.

Fine aggregate: Aggregate passing the 3/8 in (9.5 mm) sieve and almost entirely passing the No. 4 (4.75 mm) sieve and predominately retained on the No. 200 (75 μm) sieve, or that portion of an aggregate passing the No. 4 (4.75 mm) sieve and predominately retained on the No. 200 (75 μm) sieve.

Form-release agent: A substance applied to the mold for the purpose of preventing bond between the mold and the concrete cast in it.

Gap-graded concrete: A mix with one or a range of normal aggregate sizes eliminated, and/or with a heavier concentration of certain aggregate sizes over and above standard gradation limits. It is used to obtain a specific exposed aggregate finish.

Matrix: The portion of the concrete mix containing only the cement and fine aggregates (sand).

Quality: The appearance, strength and durability that is appropriate for the specific product, its particular application, and its expected performance requirements. The totality of features and characteristics of a product that bear on its ability to satisfy stated or implied needs.

Quality assurance (QA): All those planned or systematic actions necessary to ensure that the final product or service will satisfy given requirements for quality and perform intended function.

Quality control (QC): Those actions related to the physical characteristics of the materials, processes, and services that provide a means to measure and control the characteristics to predetermined quantitative criteria.

References

ACI Committee 303. 1974. *ACI Guide to Cast-in-Place Architectural Concrete Practice.* American Concrete Institute, Farmington Hills, MI. *ACI J.* :317–346

ACI Committee 309. *Consolidation of Concrete.* American Concrete Institute, Farmington Hills, MI.

ACI Committee 533. *Guide for Precast Concrete Wall Panels.* American Concrete Institute, Farmington Hills, MI.

Bainbridge, L.R. and Abberger, W.A. May–June 1994. The trend is towards cost-effective design and construction. *PCI J.* :96–97.

Bell L.W. 1996. Writing specifications for architectural concrete. *ACI J.* :63–66.

Clark, R.S. and Zils J. 1984. Long spans in post-tensioned architectural concrete. *ACI J.* :7–12.

Dobrowolski, J.A. 1989. Formwork removal and architectural concrete. *ACI J.* :35–39.

Ford, J.H. 1982. Reusage of plastic form liners. *ACI J.* :51–57.

Freedman S. 1968. White concrete: Part I. *Modern Concr.* :30–34.

Freedman S. 1969. White concrete: Part II. *Modern Concr.* :30–35.

Friedman, E.L. and Rice, B.P. 1965. Finishes for cast-in-place concrete. *Archit. Eng. News* :26–35.

Heun, R.C. 1985. Imagine the possibilities. *ACI J.* :16–20.

Hurd M.K. 1993. Patterned from liners for architectural concrete. *Concr. Const.* :331–336.

Kenney A.R. 1984. Problems and surface blemishes in architectural cast-in-place concrete. *ACI J.* :50–55.

Kenney, A.R. 1988. Post-tensioning alternate reduces construction time. *ACI J.* :48–52.

PCA. *Removing Stains and Cleaning Concrete Surfaces*, IS214. Portland Cement Association, Skokie, IL.

PCI. *Architectural Precast Concrete*, 2nd ed., MNL-122-89, 352 pp. Precast/Prestressed Concrete Institute, Chicago, IL.

PCI. *Manual for Quality Control for Plants and Production of Architectural Precast Concrete Products*, MNL 117-96, 210 pp. Precast/Prestressed Concrete Institute, Chicago, IL.

PCI. *Recommended Practice for Erection of Precast Concrete*, MNL-127-85, 96 pp. Precast/Prestressed Concrete Institute, Chicago, IL.

PCI. *Erectors Safety Manual for Precast and Prestressed Concrete*, MNL-132-95, 120 pp. Precast/Prestressed Concrete Institute, Chicago, IL.

PCI. *PCI Drafting Handbook*, MNL-119-90, 332 pp. Precast/Prestressed Concrete Institute, Chicago, IL.

Shilstone J.M. 1972a. Ways with architectural concrete. *ACI J.* :17–26.

Shilstone J.M. 1972b. The many faces of architectural concrete: A study of finishes. *Concr. Const.* :526–530

Shilstone, J.M. 1972c. The fine art and hard work of placing and compacting architectural concrete. *Concr. Const.* :536–538.

Shilstone J.M. 1973a. Achieving high quality architectural concrete by understanding details of the construction process. *Architect. Rec.* :161–164.

Shilstone J.M. August 1973b. How to obtain predictable architectural concrete. *Concr. Const.* :363–413.

(a)

(b)

Concrete construction units being placed (Courtesy Ms. Monica Schul-
tese, Precast/Prestressed Concrete Institute (PCI), Mid-Atlantic Region).

21

Masonry Construction

by

Walter L. Dickey, P.E.

Consulting Engineer in Los Angeles, California. Expert in masonry construction and has been consulting since the 1930s. Author of *Reinforced Masonry Design* with R. Schneider.

M.J. Dickey

Consultant, Los Angeles, California. Author and consultant on masonry design and construction.

21.1 Introduction

Masonry is man's oldest permanent construction material, with a long service record—castles, forts, majestic cathedrals reaching to the heavens, and simple home shelter. However, it is a hand-placed material developed by "cut-and-try" or "trial and error." Only recently have we applied true engineering "successive approximation" to its design to provide engineered use rather than simply tradition. Masonry has remained the least understood of our major construction methods, at least as far as its structural behavior is concerned.

One major purpose of this chapter is to describe and clarify the principles of masonry design and construction in a way that is applicable to modern use, taking advantage of the change from old traditional use to modern engineered use. The chapter will emphasize construction as well as design. The many computer programs available will not be discussed because they are rapidly changing and are individual programs suited to different office needs and methodology.

21.1.1 Complexities—Materials

Design of masonry does not warrant great precision. Correctness of principle and schemes or systems is more important. The designer cannot be sure that construction will result in the precise strength specified. There are too many variables that influence the final masonry assembly. The firing of clay units or the manufacture and cure of concrete units, field moisture and temperature conditions, variations in laying by hand, etc., are factors that will influence the end result in sometimes unpredictable and uncontrollable ways. Hence the factors of safety and the allowable design stresses are more conservative than permitted in other concrete methods. An example of such variation is the differing workability and strength of mortar, even though the proportions might remain exactly as specified. Harsh angular sand will give different results than rounded well-graded sand. Some aggregate will be derived from strong stone and some from weak. Workmanship can affect the function of masonry to a considerable degree. Other influencing factors will be mentioned under several subtopics.

21.1.2 Chapter Content

The text will discuss the conventional subjects of stress and strength of the heterogeneous assembly, materials, design, and construction usually covered. In addition, there is discussion of functional factors such as fire rating and the suitability of types of construction for various occupancy requirements. This has become more important as newer shapes and masonry methods have been developed.

The content of this chapter covers average or basic design use of reinforced masonry. It does not cover the complex fringe of the more complicated portions. All these facets of the subject could not be included in one chapter of a concrete manual but are covered in detail in the references on design.

Also, veneer will be discussed. This is commonly designated a "nonstructural" material and hence is generally not part of the engineer's design. However, it is masonry. It also must be designed to support its own weight and to deflect compatibly with the supporting structure. This material will be based on the longstanding veneer chapter of the Uniform Building Code (UBC). It is one of the better code chapters and provides for performance and for prescription. It states that it must be designed and provides the criteria. Then, in lieu of design, it lists several prescriptive methods that may be used as satisfactory alternates.

Garden walls and retaining walls are covered as a separate item because the design policies vary in different locales. One problem is that thorough design is frequently overlooked by professional engineers. The fee for a thorough, precise, accurate design can be a large percentage of the total cost. Paving is also an important facet of masonry, so it is covered, although briefly.

21.2 Historical Perspective

The long history of masonry construction is only briefly mentioned. It leads up to modern masonry, that is, from the 1930s, because the Long Beach earthquake of 1933 represents the birth pains of reinforced masonry as we know it now. Before the '33 earthquake, the masonry used was the traditional unreinforced masonry, largely clay brick.

21.2.1 Ancient Materials

Ancient masonry began with sun-dried mud brick, then fired clay units, which were used to create stone forts for security and stone cathedrals for spiritual inspiration. Many were grand, inspired structures.

21.2.2 Codes

Prior to (and after) 1933 in California and elsewhere, building codes contained prescriptive requirements for unreinforced or so-called plain masonry. This included fired clay and shale, concrete brick, and stone units, with lime mortar or some other filler.

The 1933 Long Beach earthquake occurred on March 10. All the schools and many residential and industrial buildings were destroyed or damaged. Unreinforced masonry had been the major construction method for the institutional buildings. It was imperative that the schools and public buildings be rebuilt immediately. The State of California printed and enacted a code for quake resistance, including masonry, by April 10. The Pacific Coast Building Officials Conference (PCBOC) issued UBC requirements in 1934, and those early provisions spread immediately, first throughout the western building jurisdictions, then to other areas of the world in a continuous expansion of study and application, primarily by the Structural Engineers Association of California. This masonry chapter is derived from the basics of the UBC, that is, basics that have stood the test of time with a good service record since that 1933 start. The principles are sound.

The reinforced masonry of that initial code, Appendix A, was covered, rather thoroughly, by seven 5×7 in pages, which made reference to the lateral force provision. The appendix was developed and expanded to over 60 pages in the 1994 UBC edition. Other codes were developed, with similar provisions, in eastern and southern jurisdictional areas.

As stated elsewhere, earlier traditional masonry had undergone relatively little research and engineering analysis. Hence the field was ripe for study and improvement, so it grew and it grew and it grew. There were many professional groups in the construction industry that devoted time and talent to the improvement of codes for reinforced (modern) masonry. They and some of their contributions are worth mentioning.

The Structural Clay Products Institute (later BIA) and National Concrete Masonry Association (NCMA) were the unit manufacturers' organizations that provided information and that were responsible for code developments for early unreinforced masonry. They did excellent work in the East (BOCA) and the South. However, they contributed little to the use, development, or improvement of the early UBC and its earthquake requirements. They did serve to help distribute the new reinforced and seismic resistance of clay masonry to eastern nonseismic areas. BIA is an excellent source for all brick data, as is NCMA for concrete masonry information.

The Structural Engineers Association of California (SEAOC) was formed in the early 1920s because the prestigious international American Society of Civil Engineers had done nothing to solve California's nagging earth-shaking problems. With that early start, professional training, and constant effort to improve design methods and codes, particularly the UBC, they became leaders in seismic-resistant design. They put the information to good use by applying it and noting the performance of structures after each quake so they could improve the practical use and validity of sound and practical UBC provisions. There have been other codes, such as ACI and ASCE/ACI

530, that have slightly different compositions, but the results are not greatly different from the time-proven UBC.

The Pacific Coast Building Officials Conference (PCBOC) of was formed in the early 1920s to develop the Uniform Building Code so all jurisdiction in the rapidly developing West would have the same provisions. It was also formed to keep the UBC up to date with annual meetings and triannual reprinting. This was to ensure that the UBC would not stagnate or be labeled with the code curse of "out of date." When some Japanese jurisdictions adopted some of the provisions, the organization's name was changed from PCBOC to ICBO—International Conference of Building Officials. The professional caliber of the membership and their activities was excellent and led to the UBC being one of the best design and construction guides for development and distribution of authentic reinforced masonry information. ICBO also provided for the use of improved or new special methods, even those not covered specifically by the code body. They set up a research evaluation procedure whereby new methods or materials could be evaluated by a staff of experienced researchers for detailed recommendations to the ICBO membership. The initial introduction and use of many improvements in masonry design resulted, for example, hollow-brick design, equivalent thickness in fire rating design, fire rating of veneer, and other methods.

Masonry Research was started in 1960 and became the Masonry Institute of America (MIA). It was the first major organization of its type, that is, engineering development and improvement funded by joint contributions from labor and contractors instead of material suppliers. It was so successful that soon there were many chapters in areas throughout the United States. MIA, and others, provide information on masonry and have been instrumental in forming the Western States Clay Products Association (WSCPA) and The Masonry Society (TMS).

The Western States Clay Products Association was formed in the early 1960s to develop codes, materials, and methods for brickwork in the West, where builders were faced with the requirements of seismic activity and the fact that eastern manufacturers, through SCPI and BIA, were not solving the problems. WSCPA initiated test programs and code developments for clay masonry and masonry in general.

The Masonry Society was formed initially by efforts of WSCPA, MIA, and SEAOC, with the purpose of providing a forum for the development and understanding of masonry. The timing was fortuitous. Researchers had discovered that masonry was a complex field ripe for study, a great and interesting source of subjects for master's and doctoral theses, a rich ore to be refined. TMS developed rapidly, attracting great professional talent, and today it provides excellent masonry information.

The American Concrete Institute initiated a code development for concrete masonry, which later became a joint effort with the American Society of Civil Engineers to develop the ASCE/ACI 530 code, which is intended to serve internationally. It is similar to the UBC, with minor variations.

21.3 Definitions

Many terms have developed over the ages of masonry use that have specific meanings and special connotations as applied to masonry construction. For the purpose of this chapter, certain terms are defined as follows.

21.3.1 Areas

Bedded area is the area of the surface of a masonry unit that is in contact with mortar in the plane of the horizontal joint.

Effective area of reinforcement is the cross-sectional area of reinforcement multiplied by the cosine of the angle between the reinforcement and the direction for which the effective area is to be determined.

Gross area is the total cross-sectional area of a specified section.

Net area is the gross cross-sectional area minus the area of ungrouted cores, notches, cells, and unbedded areas. Net area is the actual surface area of a cross section of masonry.

Transformed area is the equivalent area of one material to a second, based on the ratio of moduli of elasticity of the first material to the second.

21.3.2 Bonds

Adhesion bond is the adhesion between masonry units or backing and mortar or grout.

Reinforcing bond is the adhesion between steel reinforcement and mortar or grout.

Bond beam is a horizontal grouted element within masonry in which reinforcement is embedded.

Cell is a void space having a gross cross-sectional area greater than 1 1/2 in^2 (967 mm^2).

Cleanout is an opening to the bottom of a grout space of sufficient size and spacing to allow the removal of debris.

Collar joint is the vertical, longitudinal, mortar or grout joint between wythes.

Column, reinforced is a vertical structural member in which both the steel and masonry resist compression.

Column, unreinforced is a vertical structural member whose horizontal dimension measured at right angles to the thickness does not exceed three times the thickness.

21.3.3 Dimensions

Actual dimensions are the measured dimensions of a designated item. The actual dimension shall not vary from the specified dimension by more than the amount allowed in the appropriate standard of quality.

Nominal dimensions of masonry units are equal to its specified dimensions plus the thickness of the joint with which the unit is laid. For modular, these would be in multiples of 4 in.

Specified dimensions are the dimensions specified for the manufacture or construction of masonry, masonry units, joints, or any other component of a structure.

Grout lift is an increment of grout height poured within the total pour; a pour may consist of one or more grout lifts.

Grout pour is the total height of a masonry wall to be grouted prior to the erection of additional masonry, generally in a continuous sequence. A grout pour will consist of one or more gout lifts.

21.3.4 Grouted Masonry

Grouted hollow-unit masonry is that form of grouted masonry construction in which certain designated cells of hollow units are continuously filled with grout.

Grouted multiwythe masonry is that form of grouted masonry construction in which the space between the wythes is solidly or periodically filled with grout.

21.3.5 Joints

Bed joint is the mortar joint that is nearly horizontal at the time the masonry units are laid on it.

Collar joint is the vertical, longitudinal, mortar, or grouted joint.

Head joint is the mortar joint having a vertical transverse plane.

Masonry unit is clay brick, tile, stone, glass block or concrete block, or brick conforming to the requirements specified in the standards.

Hollow-masonry unit is a masonry unit whose net cross-sectional area in every plane parallel to the bearing surface is less than 75% of the gross cross-sectional area in the same plane.

Solid-masonry unit is a masonry unit whose net cross-sectional area in every plane parallel to the bearing surface is 75% or more of the gross cross-sectional area in the same plane.

Prism is an assemblage of masonry units and mortar, with or without grout, used as a test specimen for determining properties of the masonry that is to be laid.

21.3.6 Reinforced Masonry

Reinforced masonry is that form of masonry construction in which reinforcement acting in conjunction with the masonry is used to resist forces and contains prescriptive amounts of steel.

21.3.7 Shell

A shell is the outer portion of a hollow masonry unit as placed in masonry.

21.3.8 Walls

Bonded wall is a masonry wall in which two or more wythes are bonded to act as a structural unit.

Cavity wall is a wall containing continuous air space with a minimum width of 2 in (51 mm) and a maximum width of 4 1/2 in (114 mm) between wythes that are tied with metal ties. One or more of the wythes may be reinforced.

21.3.9 Veneer

Veneer is a nonstructural facing of brick, concrete, stone, tile, or other approved material attached to a backing for the purpose of ornamentation, protection, or insulation. It does not contribute to the building's support, but only supports itself and local applied wind and seismic loads.

Adhered veneer is a veneer secured and supported through adhesion of an approved bonding material applied over an approved backing.

Anchored veneer is veneer applied to and supported by approved connectors to an approved backing or structural element.

Backing is used in this chapter to denote the surface or assembly to which veneer is attached.

21.3.10 Wall Tie

Wall tie is a mechanical metal fastener that connects wythes of masonry to each other or to other materials.

21.3.11 Web

Web is an interior solid portion of a hollow-masonry unit as placed in masonry.

21.3.12 Wythe

Wythe is the portion of a wall that is one masonry unit in thickness. A collar joint is not considered a wythe.

21.4 Materials

The quality of materials used in masonry is specified in American Society for Testing and Materials (ASTM) standards. The UBC has compiled standards that are in many cases modifications of the ASTM standards. Some UBC standards were developed for additional needs. If no requirements are specified for a material, quality shall be based on generally accepted good practice, subject to the approval of the building official. Reclaimed or previously used units shall meet the same applicable requirements for the intended use as for new units of the same material.

Aggregate

ASTM C 144, Aggregates for Masonry mortar
ASTM C 404, Aggregates for Grout

Cement

UBC Standard 21-11, Masonry Cement (Plastic cement conforming to requirements of UBC Standard 25-1 may be used in lieu of masonry cement when it also conforms to UBC Standard 21-11)
UBC Standard 19-1, Portland Cement and Blended Hydraulic Cements UBC Standard 21-14, Mortar, Cement

Lime

UBC Standard 21-12, Quick Lime for Structural Purposes
UBC Standard 21-13, Hydrated Lime for Masonry Purposes (when Types N and NA hydrated lime are used in masonry mortar, they shall comply with the provisions of UBC Standard 21-15, Section 21.1506.7, excluding the plasticity requirement)

Masonry Units of Clay or Shale

ASTM C34 and C112, Structural Clay Load-Bearing Wall Tile ASTM C56, Structural Clay Nonload-bearing Tile
UBC Standard 21-1, Section 21.101, Building Brick (Solid Units)
ASTM C126, Ceramic Glazed Structural Clay Facing Tile, Facing Brick, and Solid Masonry Units; Load-bearing glazed brick shall conform to the weathering and structural requirements of UBC Standard 21-1, Section 21.106, Facing Brick.
UBC Standard 21-1, Section 21.106, Facing Brick (solid units)
UBC Standard 21-1, Section 21.107, Hollow Brick
ASTM C67, Sampling and Testing Brick
ASTM C212, Structural Clay Facing Tile
ASTM C530, Structural Clay Nonload-bearing Screen Tile

Masonry Units of Concrete

UBC Standard 21-3, Concrete Building Brick
UBC Standard 21-4, Hollow and Solid Load-bearing Concrete Masonry Units
UBC Standard 21.5, Nonload-bearing Concrete Masonry Units
ASTM C140, Sampling and Testing Concrete Masonry Units
ASTM C426, Standard Test Method for Drying Shrinkage of Concrete Block

Masonry Units of Other Materials

UBC Standard 21-2, Calcium Silicate Face Brick (Sand-Lime Brick)
UBC Standard 21-9, Unburned Clay Masonry Units
ACI-704, Cast Stone
UBC Standard 21-17, Test Method for Compressive Strength of Masonry Prisms

Connectors

Wall ties and anchors made from steel wire shall conform to UBC Standard 21-10, Part II, and other steel wall ties and anchors shall conform to A36 in accordance with UBC Standard 22-1. Wall ties and anchors made from copper, brass, or other nonferrous metal shall have a minimum tensile yield strength of 30,000 psi (207 MPa).

All such items not fully embedded in mortar or grout shall either be corrosion resistant or shall be coated after fabrication with copper, zinc, or a metal having at least equivalent corrosion-resistant properties.

Mortar

UBC Standard 21-15, Mortar for Unit Masonry
UBC Standard 21-16, Field Test Specimens for Mortar
UBC Standard 21-20, Standard Test Method for Flexural Bond Strength of Mortar Cement

Grout

UBC Standard 21-14, Grout for Masonry
UBC Standard 21-18, Method of Sampling and Testing Grout

Reinforcement

UBC Standard 21-10, Part I, Joint Reinforcement for Masonry
ASTM A615, A616, A617, A706 and A775, Deformed and Plain Billet-Steel Bars, Rail-Steel Deformed and Plain Bars, Axle-Steel Deformed and Plain Bars, and Deformed Low-Alloy Bars for Concrete Reinforcement
UBC Standard 21-10, Part II, Cold-Drawn Steel Wire for Concrete Reinforcement

21.4.1 Mortar

Mortar has evolved erratically. The original purpose of mortar was to fill in the spaces between irregular rock or cut stones piled on others. Then it became an aid to piling such units, providing a way of laying them more rapidly and accurately with better stress distribution and alignment. Some of the early exotic mixtures included egg whites, clay, urine, or ox blood, but the major development was the addition of lime to sand for workability and for subsequent strength as the lime combined, slowly, with carbon dioxide from the air. Many early structures can be found with such mortar. Later, Portland cement was added for greater plasticity and for early greater strength.

The purpose of mortar is primarily to enhance the strength and the homogeneous character of the masonry. It also provides resistance to moisture penetration and provides good unit alignment through good workability. Recent tests have shown that bond increases the masonry resistance to shear forces. However, reinforced masonry is not designed to utilize tension values. (See later discussion.)

21.4.1.1 Ingredients

Cement. The cements provided for use in mortar and grout in the UBC are Portland cement, masonry cement, and mortar cement. Portland cement is specified as either Type I or II, in accordance with the UBC Standard or ASTM C150. Usually, low alkali is called for to reduce the soluble content of the cement, which is the major contributor to efflorescence, an unsightly deposit of soluble salts left on the surface as the water evaporates. Masonry cements and mortar cements must comply with UBC Standard, and the plastic cements must meet those requirements for Portland cement as set forth in ASTM C-150, except for the limits on soluble residue, air entrainment, and additions subsequent to calcination. There is one approved proprietary material that uses a combination of lime and an air-entraining additive that complies with ASTM C175. Note, then, that there is a "Portland cement" and a "masonry cement" and that they are not interchangeable. Masonry cement is not recommended for mortar in reinforced masonry. The use of certain adhesives or bonding agents or cements of other types may be considered, but they must be approved by such agencies as the UBC through their Evaluation Reports. None of the cementitious materials or additives may contain epoxy resins and derivatives, phenols, asbestos fibers, or fire clays. The simplest method is to use the basic proven Portland cement unless there is good local experience or a service record for the other cements.

Water. The requirements are only that water must come from a domestic supply and that it must be clean and free from injurious amounts of oil, acid, alkali, organic matter, or any other harmful substances. This would apply to both mixing water and to the cleaning or curing water.

Lime. Lime, calcium oxide, (CaO), or hydrated calcium oxide, $Ca(OH)_2$, is probably one of the oldest chemicals made and still in use. It may have been discovered when primitive peoples built fires on limestone rock or in mussel shells and observed what happened. It is one of the oldest mortar ingredients used in modern masonry. In early mortar, it provided needed workability for one of the functions of mortar, that is, to fill up the irregularities between units. It is still used in mortar to provide workability for watertight joints. It actually serves more as a plasticizing agent, while Portland cement provides the basic strength. Lime also increases the water retention or water-holding capacity of the mortar. This action decreases the tendency of the mortar to lose water, a phenomenon known as "bleeding," and reduces separation or segregation of the sand aggregate. While the mineral, lime, is present in many combinations, its conbination with carbon dioxide, which forms limestone (calcium carbonate, $CaCO_3$), is an important source of lime as it is used in mortar. Lime in this form has the marblelike hardness of rock.

Limestone rock, as it comes from the quarry, is changed into a form suitable for use in mortar by crushing, screening, selecting, washing, and grading. The selected stone is then placed in kilns, where it is heated to $2500°F$. This drives off the water of crystallization and also removes certain gases, such as carbon dioxide, from the stone. The result of the calcining and burning is quicklime, which is calcium oxide (CaO). Quicklime is a very caustic material; when it comes in contact with water, a violent reaction occurs, and enough heat is generated to boil the water. It can also cause severe burns, both because of the heat and the chemical action. For this reason, the lime should be carefully added to the water, rather than water being added to the lime. In the past, quicklime was delivered to the job for use in the masonry mortar. The first step in using it was to slake it or hydrate it. This was done by adding enough water so that the oxide became a hydroxide $Ca(OH)_2$. During this slaking process, the caustic quicklime was hot and very hazardous to use. After slaking, the quicklime was screened through a fine mesh and set aside in containers. The result became known as lime putty, and this was what was added to the mortar mix. This quicklime had to conform to UBC Standard.

However, the use of quicklime has given way to a newer substance known as hydrated lime. This product is produced by slaking lime in large tanks at the manufacturing plant, where it is converted to hydrated lime without necessarily saturating it with water. This hydrated lime is actually a dry powder containing just enough water to complete the chemical reaction. After hydration, the lime is pulverized, bagged, and delivered to the job site. This material must conform to UBC Standard.

Sand. The sources of sand can be several. Natural sand is formed by the erosive action of rivers. Such particles are somewhat rounded in shape. They are deposited on flats and on ocean fronts as beach sand. The latter, being subjected to more abrasive wave action, are more rounded. This would make for better workability but probably would have a deleterious effect on strength. Another natural sand source is sand transported by wind, such as blow sand. Sand may also be manufactured. In this case it is a product resulting from crushing stone, gravel, or blast-furnace slag. Manufactured sand is sharper and more angular and thus may require different amounts of fine cementitious particles to provide the lubrication for proper workability. Deleterious substances in sand may be items such as friable particles, lightweight particles, organic impurities, or excess amounts of clay of loam. These materials must be removed at the plant before the sand is sent to the job site.

Unfortunately, too little concern is often shown for the grading of sand. However, the properties of sand have considerable impact on the workability as well as the strength of the mortar. To provide some sort of guide in this respect, grading limits for mortar sand are spelled out in both ASTM C144 and the UBC Standard. The grading limits from the latter are listed in Table 21.1

TABLE 21.1 Sand Grading Requirements (UBC Standard)

% Passing Sieve No.	Limit
4	100
8	95–100
100	25 max.
200	10 max.

TABLE 21.2 Mortar Proportions: Parts by Volume

Mortar Type	Portland Cement	Hydrated Limes or Lime Putty* Min.	Max.	Masonry Cements	Shovel Count† (at 7–8/ft³) 2 1/4 to 3		Parts† ft³ 2 1/4 to 3		Damp Loose Aggregate
M	1	—	1/4	—	21	28	2.81	3.75	Not less than
	1	—	—	1	34	45	4.5	6	2 1/4 and not
S	1	1/4	1/2	—	21–25	28–34	2.8–3.4	3.7–4.5	more than
	1/2	—	—	1	25	34	3.4	4.5	3 times
N	1	1/2	1 1/4	—	25–38	34–51	3.4–5.1	4.5–6.7	the sum
	—	—	—	1	17	22	2.25	3	of the volumes
O	1	1 1/4	2 1/2	—	38–59	51–79	5.1–7.9	6.7–10.5	of the cement and lime used

Source: UBC Table 24, 1988 edition.

*When plastic or waterproof cement is used, hydrated lime or putty may be added, but not in excess of one-tenth the volume of cement.

†Not a part of UBC table.

Actually, it is recommended that the sand for mortar be on the finer side of the ranges permitted in the ASTM and UBC specifications. Those limits are rather broad, and if gradings on the coarse side of the range are used, the mortar will be harsh and not as satisfactory as it could be. (See Table 21.2 on proportions.)

Admixtures. In general, admixtures are neither necessary nor desirable in masonry construction and should not be included unless specifically approved and recommended by the designer or building official. There are many different types of admixtures. They may be used for water reduction, for water retention to prevent bleeding, and for waterproofing. Whatever their purpose, they must be used carefully and in the proper manner and proportions.

There are numerous materials that must not be used in Portland cement mortars under any circumstances, and one of these is gypsum. Gypsum has sometimes been used to accelerate the set, but this introduces the hazard of later deterioration. Calcium chloride has sometimes been used for the same purpose. It is especially undesirable in reinforced masonry because calcium chloride tends to cause corrosion in the reinforcing bars or joint reinforcing.

21.4.1.2 Proportion Limits

Requirements for mortar have developed in the past by trial and error and have resulted in gradual changes in specifications. Now, greatly simplified, the latest edition of the UBC calls for a mortar (1) composed of proportions as listed in Table 21.2 or (2) whose proportions can be demonstrated by laboratory or field experience to produce a mortar achieving the specified design compressive strength for the masonry assemblage, f'_m. In other words, if a mix of any specific proportions of the ingredients has been proven to provide proper strength and workability with the particular masonry units on the job, then it may be used. Under these circumstances, it is not necessary to conform to those proportions precisely spelled out in Table 21.2 for mortar Types M or S.

Table 21.2 evolved from "proportion" limits—not the "property" specification. These proportions provide for a range of aggregate amounts. This is to allow for any field adjustment to proportions depending on mortar workability and water absorption of local aggregates, stemming from particle grading, type, and shape. The indefinable "workability" can only be determined in the field, since it depends primarily on the available sand. Thus the proportions cannot always be precisely stated beforehand. Shovel measurements are adequate and can be easily adjusted to suit job conditions. These shovel counts were added to Table 21.2, since they were not a part of the UBC table.

In contrast, the property specifications, still retained by ASTM, establish mortar acceptability by requiring (1) conformance of the individual ingredients (cement, sand, lime) with appropriate

ASTM laboratory specifications, and (2) that the properties of the proposed mortar mix, as mixed in the laboratory with a specified flow or stiffness, conform to ASTM limits for water retention and cube compression strength. The cube strength is not representative of the field mix in place. Since these acceptance tests are based entirely on laboratory and not job-site conditions, these property specifications are not very reflective of mortar usability in the field and thus are no longer used extensively.

One final note with respect to the origin of the UBC mortar-mix proportions: If sand amounts to approximately three times the combined volumes of cement and lime, then, theoretically at least, the voids in the sand should be completely filled, thereby providing a dense mix. This is a necessity if shrinkage and porosity are to be minimized. So the typical mortar mix specification of 1 part cement, 1/4 to 1/2 part lime, and 4 1/2 parts sand (1 : 1/2 : 4 1/2), based on damp, loose volume, has evolved. Note that the table permits a range between 2 1/4 and 3 for the sand volume to allow for differences in the grading of the sand and the roundness of its particles, which in turn affect workability. And as previously described, these proportions must be adjusted in the field to achieve the desired workability. In the final analysis, the mason must have the last word regarding the mix proportions, not a laboratory technician. After all, the mason is certainly the best judge of how workable the mortar actually is and will literally "sense" whether it is adequate or not.

Lime is a cementing agent whose presence improves the plasticity, workability, and water retention of the mix, all of which are highly important for the attainment of maximum bond strength as well as water tightness in the masonry mortar. Further, it should be noted that the strength of the masonry assemblage in shear and flexure depends on the mortar bond strength plus the contact area between the masonry unit and the mortar, and not directly on the compressive strength of the mortar itself. Thus the deleterious effect of lime in reducing compression strength must not be viewed as a fatal flaw.

21.4.1.3 Mixing

The basic requirement for mixing is simply that the material be thoroughly mixed. Machine mixing is ordinarily called for and is in common use; however, hand mixing is permitted for small batches. The order of placing the ingredients in the mixer is important. Should the cement be added first, it is likely to lump or cling to the damp sides of the paddle blades and therefore not be uniformly dispersed throughout the mix. For this reason, water and sand are put into the mixer first, to scour the blade surfaces and prevent any lumping in the cement. Lime is frequently added last, supposedly to prevent it from coating the sand particles before the cement gel can coat them. No proof seems to exist that this really occurs, so this factor is often neglected.

Mixes are generally measured simply by shovel count in commercial work. This is deemed adequate in view of other existing variables, such as the harshness of field aggregates, which are uncontrollable in the field. The shovel count variation permits the tender to adjust the mix for workability. However, California requires that box measurements be used for accurate proportioning on school projects.

Mixing time, according to UBC, is set at three minutes minimum with whatever amount of water is required to produce the desired workability.

Workability is important for many reasons. For one thing, it results in a good mortar bed into which the units can be placed for proper alignment. Also, it provides for the sound, tight bond so necessary for resistance to water penetration, as well as strength.

Hand mixing can be an effective method for smaller jobs, expeditiously saving equipment, time, etc. The simple method is to spread layers of the dry material in one portion, for example, sand, then cement, then lime, then sand. The layers are cut through with the hoe and the vertical stack is pulled out horizontally along a portion of the mortar box (Figure 21.1).

21.4.1.4 Tempering

Mortar that has stiffened owing to water evaporation may be tempered as frequently as necessary to restore the desired consistency. UBC standards do not place limits on the length of time a mortar

FIGURE 21.1   Hand mixing of mortar. Sequence of adding ingredients in a mortar box for good uniform distribution, that is, horizontal layer of sand, layer of cement, layer of sand, layer of lime with sand over it. The face of the layers is cut with a hoe and pulled into horizontal layers.

may be used after its initial mixing; however, ASTM C270 and many other specifications require that the mortar be in place within 2 1/2 h after initial mixing. This is not a valid limitation, in itself, for the simple reason that it makes no reference to the temperature at the time of placing. If the day is cool and moist and if the mortar is always kept plastic, it may be used satisfactorily for a much longer period. But if the day is hot and dry, 2 1/2 h might be too long. A better limit on retempering could be when the mortar has begun its initial set and hardened enough to become harsh. Further tempering may then result in a weakened mortar that should not be used.

The tempering should not be done by simply splashing water on the surface because it will immediately run off, washing away the fines with it. Rather, the water should be dropped into a basin formed in the mortar and then thoroughly worked into the mix.

21.4.1.5 Testing

Since the individual masonry units are bound together by the mortar, the strength of the resulting assemblage is somewhat dependent on the physical properties of the mortar. As previously noted, the strength of masonry assemblages in shear and flexure depends primarily on the mortar bond strength and not on the compressive strength of the mortar itself. There are no code requirements for bond strength, but the UBC does have a provision for field control of strength adopted from the then California State Division of Architecture procedure. One function of this state agency is the assurance of quality construction in schools and hospitals. So, for this agency, control of field execution is of paramount importance, for example, assuring the actual mortar strength in a wall. It was recognized that (1) different mixes and units of different absorption qualities would result in various strengths of unknown degree, (2) different masonry units absorb varying amounts of water, which changes the water/cement ratio, and (3) the moisture content of mortar as tested in the laboratory for ASTM usually did not result in a fluid consistency suitable for field use.

Therefore, to meet the goal of field control of construction, the agency developed a field sampling technique to measure the condition of the mortar as finally placed in the wall. This consists of (1) mixing the mortar to proper fluid consistency for use, (2) placing a mortar bed for one minute on units intended for the masonry construction, (3) placing that mortar sample into a 2 × 4-in cylinder mold, and (4) obtaining the compressive strengths of the cylinders. The principle presumes that the mortar used, workable and wet, is subjected to the absorption of the masonry units in place and thus resembles, to some extent, the actual condition of the mortar as it exists in the wall. It is noted that the exposure of the test sample to loss of water by absorption is rather short compared to the long-term exposure in the wall. Consequently, the water/cement ratio of the sample is subjected to less water reduction than the actual mortar bed joint placed in the wall. Therefore the mortar in the wall will have a lower water/cement ratio, and correspondingly higher strength than the specimens. For this reason, the specimen was not required to have as high a compressive test strength (1500 psi minimum) as the desired or anticipated wall mortar strength. In essence, this test specimen is influenced in the same manner as the wall mortar, although to a lesser degree, so that the test method should serve as a partial indicator of the condition of the in-the-wall mortar. This method was later adopted into the UBC Standard.

21.4.1.6 Adhesives

Various adhesives have been proposed and used as substitutes for Portland cement mortar. These should not be used, however, unless they have a good test or service record, because many factors affect the quality of the masonry, such as weathering, adhesive strength, durability, heat resistance, and fire resistance. Some of these adhesives have been approved for use in mortars by the ICBO Research Committee; however, their usage is not widespread. Their methods of application and design provisions are specifically outlined in ICBO Evaluation Reports, since their performance and application is different from that of conventional masonry mortar.

21.4.1.7 Fire Clay

Fire clay is a mortar used for fire brick and generally develops its strength during firing in a furnace. It has no place in the mortar used in reinforced masonry. Some have used it as an additive because of the workability it adds to the mortar, but it does not add to the strength. As a matter of fact, it may even weaken the mortar. In time, it may wash out because it will not be subjected to the high refractory temperatures that are needed to make it set and achieve an adequate strength.

21.4.2 Grout

Grout is simply a high-slump concrete made with small-size aggregate, such as sand and pea gravel. It derives its name from the Swedish word "groot" meaning "porridge," which provides some indication of its pouring consistency. It must be fluid enough to fill all voids in the grout space without segregating and to completely encase the reinforcement. It serves to (1) bond the wythes together into a composite element of masonry construction, (2) bond the reinforcement to the masonry so that the two materials will act as a homogeneous material, and (3) increase the masonry volume for bearing and fire resistance. The second bonding function is absolutely essential if the desired flexural resistance, as well as ductility, is to be achieved, since the tensile forces can only be developed by the reinforcement.

21.4.2.1 Ingredients

Cement. Portland cement, Type I or II (UBC Standard 26-1 or ASTM C150), is typically used. Also, Portland blast-furnace slag cement (ASTM C595) may be employed in lieu of a portion of the Portland cement.

Fine Aggregate. The same specifications apply as for masonry mortars.

Coarse Aggregate. The basic ASTM specification for coarse aggregates to be used in masonry is C404, which in turn refers to C33. The UBC Standard covers grout aggregates also. In that specification, a table establishing grading requirements for this material is provided as shown in Table 21.3. The coarse grout aggregate normally used, referred to as pea gravel, has a maximum size of about 3/8-in diameter. It should be noted, however, that in large grout spaces it is permissable to use up to a 3/4-in aggregate. As with sand, it is desirable to keep the grading on the finer side of the range rather than on the coarser side. Also, the grading of the combined sand and gravel mixture should be considered.

There has been some use of lightweight aggregate in grout to reduce the weight of the structure. It has also been used because it improves the insulating properties of the grout. This material is governed by ASTM C330 and the UBC Standard. It should be noted that lightweight aggregates from different sources will perform differently, for some may contain fines obtained by crushing larger particles, producing a harsher aggregate. Lightweight aggregate should be prewet before use to minimize absorption of water from the mix. This is especially important for grout that is to be pumped into the wall. Pumping pressures will force water from the matrix into the pores of the aggregate, tending to stiffen the mix.

TABLE 21.3   Aggregate Grading Requirements

	Amounts Finer than Each Laboratory Sieve, Square Openings, wt%				
	Fine Aggregate			Coarse Aggregate	
	Size 1	Size 2			
Sieve Size		Natural	Manufactured	Size 8	Size 9*
1/2 in	—	—	—	100	100
3/8 in	100	—	—	85–100	90–100
No. 4 (4.76 mm)	95–100	100	100	10–30	20–55
No. 8 (2.38 mm)	80–100	95–100	95–100	0–10	5–30
No. 16 (1.19 mm)	50–85	60–100	60–100	0–5	0–10
No. 30 (595 μm)	25–60	35–70	35–70	—	0–5
No. 50 (297 μm)	10–30	15–35	20–40	—	—
No. 100 (149 μm)	2–10	2–15	10–25	—	—
No. 200 (74 μm)	—	—	0–10	—	—

Source: UBC Standard.

*This is an aggregate group trade name furnished by specific suppliers.

Two basic types of lightweight aggregates are used, natural and manufactured. The natural lightweight aggregates are mined from natural deposits, usually volcanic in nature. Pumice is the most common, and it consists of a volcanic glass full of minute cavities, which cause its light weight. Scoria is a frothy, basaltic magma or lava, or possibly a burned clay or clinker deposit. The aggregate may also be slag refuse produced from the reduction or smelting of ores. Volcanic ash consists of porous particles roughly the size of peas or shot, which look like ashes. Tuff or tufa is a porous, fine-grained volcanic rock of light weight.

Manufactured lightweight aggregates consist of expanded slag, clay, or shale. Expanded slag is a frothy, solidified rock and waste that contains a large number of cavities. Expanded clay or shale has been fired in a kiln. The heat causes the material to bubble and then to become firm by vitrification. Some aggregates are fired in large lumps and then broken and graded to size. Others are formed into pellets of various sizes and fired.

21.4.2.2 Proportions

The mix proportions shall be according to Table 21.4, excerpted from UBC Standards. Water is added to ensure that the grout will flow intimately into place without segregation.

Grout used in high-lift masonry construction, where fluidity is especially important, must be richer and wetter than grout used in low-lift operations. As an alternative, it may have an approved admixture (Grout Aid®) to improve this fluidity and water retention and to minimize volume loss due to absorption.

TABLE 21.4   Grout Proportions by Volume

Type	Parts by Volume of Portland Cement or Blended Cement	Parts by Volume of Hydrated Lime or Lime Putty	Aggregate Measured in a Damp, Loose Condition	
			Fine	Coarse
Fine grout	1	0 to 1/10	2 1/4 to 3 times the sum of the volumes of the cementitious materials	—
Coarse grout	1	0 to 1/10	2 1/4 times the sum of the volumes of the cementitious materials	1 to 2 times the sum of the volumes of the cementitious materials

Source: UBC Standards, Table.

Note: Grout used in masonry construction must attain a minimum compressive strength at 28 days of 2000 psi. The building official may require a compressive field strength test of grout made in accordance with UBC Standard.

21.4.2.3 Mixing

On smaller jobs mixing may be done in a box or wheelbarrow or in a mixer similar to a small concrete mixer. The ingredients must be added to a mixer in a sequence that avoids "balling" of the cement powder, which prevents uniform distribution in the mix. However, on larger jobs and where high-lift grouting operations are employed, the mixing of the grout is done in transit-mix trucks, because better control is provided. In such cases, the grout is pumped directly from the truck into the grout space in the wall. With both methods, care must be taken to ensure that there is no segregation of materials during placing.

One additive used in high-lift grouting is a proprietary material called Grout Aid®. The volume of wet grout used in high-lift operations is composed partially of the excess water needed for fluid placing. When that water is removed by suction or absorption by the adjacent masonry units, the loss of water volume causes a reduction in grout volume. This in turn produces settlement and numerous cracks. Further, there may be arching as the material tends to bridge between the two wythes. Separations may occur as the grout bonds to the two wythes and separates at the center, or the separation may occur at one wythe as the grout in the grout core becomes smaller. Thus a reconsolidation is needed to prevent this volume loss from taking place. Grout Aid® performs this function adequately. It is an admixture that both increases fluidity and water retention and causes a volume expansion within the grout mix, thus offsetting the decrease due to water loss.

21.4.2.4 Testing

Field testing of grout for compressive strength is provided for by a UBC Standard. This standard further establishes a way of obtaining samples that is somewhat representative of the strength of the grout as it exists in the wall, where it has been subjected to absorption by the masonry units, thereby lowering its water/cement ratio. This procedure entails making a mold comprised of the type of units (clay brick or concrete units) used in the wall. The sampling method is shown in Figure 21.2.

21.4.2.5 Pouring Limits

The pouring limits of grout of different mixes are listed in Table 21.5 from the UBC. This may be used as a guide to achieving good grouting. There is a paradox in that field materials and conditions will govern and invalidate the theoretical table. Accordingly, some grout, well mixed with well-

TABLE 21.5 Grouting Limitations

Grout Type	Grout Pour Maximum Height, feet* (×304.8 for mm)	Minimum Dimensions of the Total Clear Areas within Grout Spaces and Cells,*† in (×25.4 for mm)	
		Multiwythe Masonry	Hollow-Unit Masonry
Fine	1	3/4	1 1/2 × 2
Fine	5	1 1/2	1 1/2 × 2
Fine	8	1 1/2	1 1/2 × 3
Fine	12	1 1/2	1 3/4 × 3
Fine	24	2	3 × 3
Coarse	1	1 1/2	1 1/2 × 3
Coarse	5	2	2 1/2 × 3
Coarse	8	2	3 × 3
Coarse	12	2 1/2	3 × 3
Coarse	24	3	3 × 4

Source: UBC Standard.

*The actual grout space or grout cell dimensions must be larger than the sum of the following items: (1) The required minimum dimensions of total clear areas, (2) the width of any mortar projections within the space, and (3) the horizontal projections of the diameters of the horizontal reinforcing bars within a cross section of the grout space or cell.

†The minimum dimensions of the total clear areas shall be made up of one or more open areas, with at least one area being 3/4 in (19 mm) or greater in width.

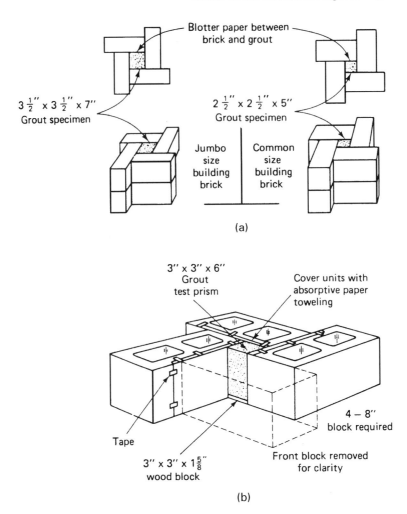

FIGURE 21.2   Grout field-test specimens. (a) The general method using brick, showing two sizes of specimens, both with a height/width ratio of 2:1; (b) use of a hollow block mold to provide a specimen of 2:1 ratio.

graded smooth aggregate, can be poured to heights much higher than the limitation provided by the table. On the other hand, grout with harsh aggregate, not well graded, under dry conditions, might not be pourable to the heights given, that is, the table may falsely indicate that such grouting is feasible.

21.4.3 Concrete Masonry Units

The concrete units most commonly used are hollow load-bearing units and solid bricks. These are manufactured in many sizes, colors, surface textures, and weights, and not all are available economically in all locales. Local availability should be checked before construction begins. Some of the more common shapes and patterns of use are shown in Figures 21.3 through 21.6.

21.4.4 Clay Masonry Units

Earlier masonry units were fired clay or shale, solid "common" brick (ASTM C62) or "face" brick (ASTM C216). As reinforced masonry developed, the use of hollow brick (WSCPA standard or

8" WIDE WALL

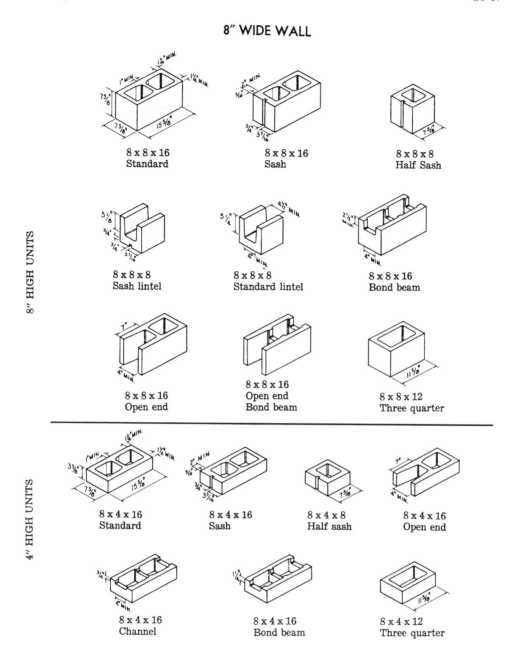

FIGURE 21.3 Typical masonry unit shapes.

ASTM C652) increased because it was easier to reinforce and better engineering of lighter masonry increased. Some typical shapes are shown in Figure 21.7, and bonding details are illustrated in Figure 21.8. Local availability must be checked for various geographic areas. Special shapes can be produced as required by some manufacturers, but the costs are appropriate.

Pavers may be any of the solid concrete or brick shapes as well as the thin or tile shapes listed in various standards. These are not part of the structural function (providing only surface and dead load) but are frequently included in masonry projects. Newer patterns are provided by certain local manufacturers, as discussed later in this chapter.

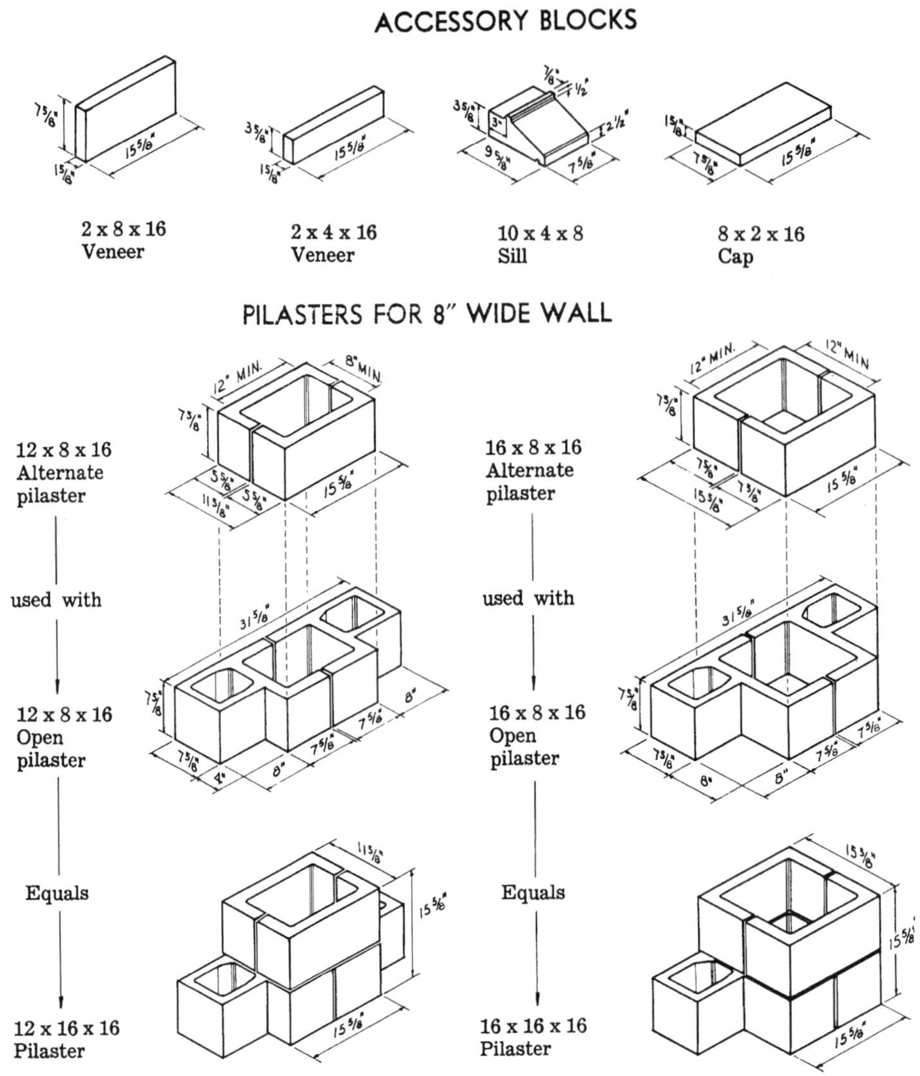
ACCESSORY BLOCKS
2 x 8 x 16
Veneer
2 x 4 x 16
Veneer
10 x 4 x 8
Sill
8 x 2 x 16
Cap
PILASTERS FOR 8″ WIDE WALL
12 x 8 x 16
Alternate
pilaster
used with
12 x 8 x 16
Open
pilaster
Equals
12 x 16 x 16
Pilaster
16 x 8 x 16
Alternate
pilaster
used with
16 x 8 x 16
Open
pilaster
Equals
16 x 16 x 16
Pilaster

FIGURE 21.4 Special shapes.

Units for anchored veneer may be solid concrete or clay, cored units at least 2 in thick, as listed in the standards, or stone units. The limit of 2 in was set simply because it is difficult to lay and anchor thinner units.

Units for adhered veneer may be concrete, clay, or stone not over 15 psf. These may be up to 36 in or 720 psi (unless lighter than 3 psf). They shall have surfaces that will develop at least 50 psi of shear bond.

21.4.5 Reinforcing

Reinforcing may be as permitted for reinforced concrete (i.e., grouted spaces) (see standards). Also, it may be joint reinforcing, that is, specifically fabricated rod or wire. Shapes are shown in Figure 21.9 and in evaluation service reports.

Reinforcing bars shall be deformed bars except that bars 1/4 in or less in diameter may be smooth. Joint reinforcing of 1/4 in or less in diameter may be placed in bed joints as shown in Figure 21.9.

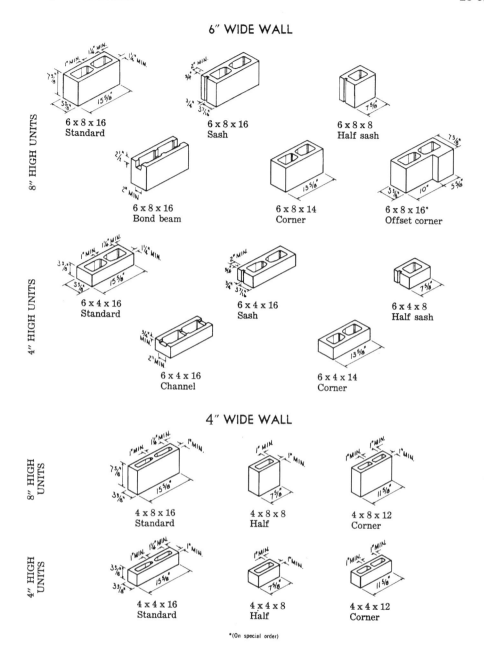

FIGURE 21.5 Hollow-unit shapes.

It is preferably fabricated by welding and either composed of corrosion resistant material or coated for corrosion resistance because it will be placed in mortar near the surface or face of the masonry. Prestressing reinforcing shall be as in the standards for prestressed concrete reinforcing.

Placement of reinforcing is important because it is a necessary ingredient, and it is hand-placed in the masonry assembly. Early placement was frequently in wet concrete or grout and simply aligned by hand, with little assurance of final position. Reinforcing must be securely held in proper position, wired in place, or held by positioners.

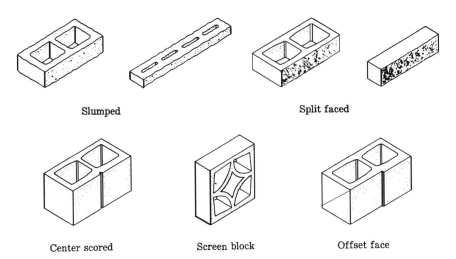

Slumped   Split faced

Center scored   Screen block   Offset face

FIGURE 21.6 Architectural shapes.

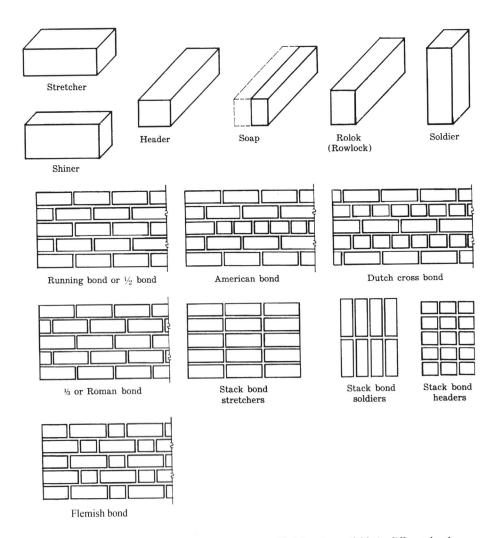

Stretcher

Shiner

Header   Soap   Rolok   Soldier
                          (Rowlock)

Running bond or ½ bond   American bond   Dutch cross bond

⅓ or Roman bond   Stack bond   Stack bond   Stack bond
                    stretchers   soldiers   headers

Flemish bond

FIGURE 21.7 Brick patterns. The actual dimensions of brick units available in different locales may vary slightly so should be checked for size, module, and economic feasibility.

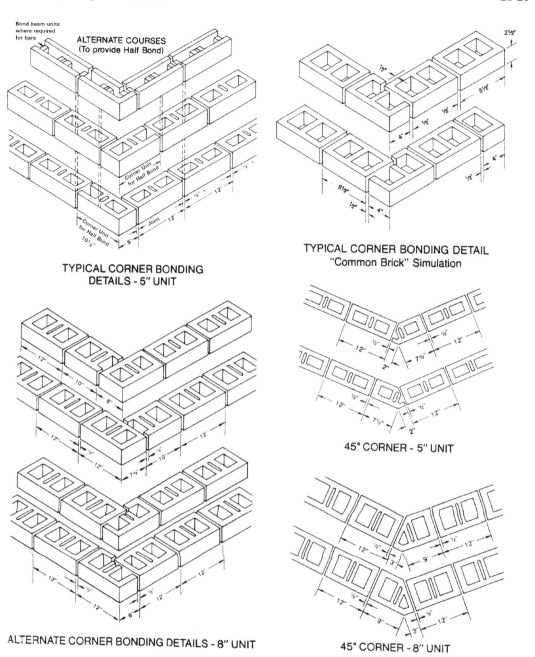

FIGURE 21.8 Typical hollow brick bonding details. The typical bonding details shown for 5″ and 8″ hollow brick may be modified by variations, for example, as indicated in the modified shape to simulate "common brick," but the principles of structural design will remain the same.

21.4.6 Admixtures

Admixtures are not recommended for general use in mortar or grout, except for high-lift grouting in which Grout Aid® must be used. This additive contains a material that increases the flowability with less water and also one that provides volumetric expansion to compensate for the loss of water volume that causes shrinkage cracks and separation of grout from reinforcement and masonry unit faces.

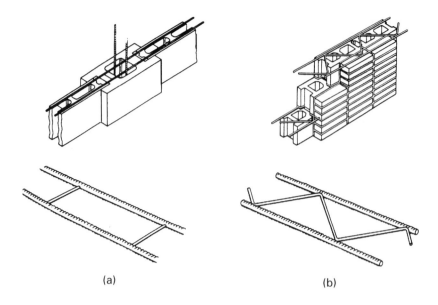

(a)   (b)

FIGURE 21.9 Wall joint reinforcing. (a) Ladder-type joint reinforcing; (b) truss-type joint reinforcing.

21.4.7 Testing

There are Test Standards in UBC and in ASTM that describe the various materials and the tests necessary to assure field compliance. Practice of an application depends on locale, but the designer must be assured of field compliance. That may be by manufacturer or supplier certification or by laboratory testing, inspection and reporting. Payment for the testing must be clearly agreed upon in the bid or contract documents, for example, by the owner or by the contractor.

21.5 Types of Construction

Types of masonry construction commonly used are outlined as follows:

Walls were the initial and most important use of masonry.

Arches make good use of the excellent compressive capacity of masonry. Early on, they provided the possibility of increasing spans beyond the capacity of unreinforced lintels, which were developed by the Egyptions and Greeks to span openings.

The Roman semicircular arch was a logical easily built development. Stresses would flow around the massive adjacent wall material with its great compressive capacity. Modern masonry engineering calculations, were neither required nor appropriate. These arches are also effective for the greater spans and loads of roadways. One interesting development of stone arches was the cutting of the stone so that intricate centering would not be needed. The elements would support themselves, as construction approached the center, known as the key stone (Figure 21.10).

The Gothic arch consisted of two curved soffits coming to a point. This, incidentally, created a member these early engineers thought approached the shape of the resultant forces more closely (i.e., parabola or hyperbola). More important, it allowed the spring lines and crowns or peaks to be located architecturally for different span widths as required. These arches made possible the gloriously high cathedrals, by providing the lateral reactions of flying buttresses. The earlier trial-and-error method had indicated the need for a horizontal reaction at the arch ends, a basic requirement that was sometimes omitted at the end of a row of arches.

The flat arch was an arch with a low rise to span aspect ratio. Some were even flat on the bottom with the arch pattern provided by varying joint or member thickness as shown in Figure 21.10.

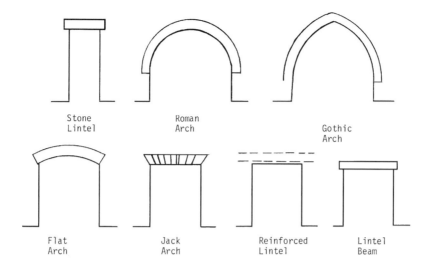

FIGURE 21.10 Masonry bridging openings, from short-span stone lintel to long-span beam.

The **wood lintel** of early masonry construction actually acted as a form until the masonry set and took its load by arching over. Lintels may be provided to make a straight soffit over openings. Lintels may be structural steel, precast concrete, or reinforced masonry. There may be only a triangular loading immediately above the lintel with material above in effect arching over to jambs.

Piers are the masonry sections between the openings. For low load intensity and height they may be considered and designed as short walls. For heavier loads they must be designed as columns.

Columns are vertical structural members that are reinforced for greater capacity, and both the masonry and the larger amount of reinforcing carry the load. The steel increases the masonry capacity. The horizontal dimension at right angles to the thickness is generally limited by codes to not more than three times the thickness. However, that is not a valid structural limitation, and the designer may actually provide whatever shape and size is necessary to do the job.

Pilasters are spaced projections of the wall face. They may be used as elements to carry concentrated loads such as trusses or girders. They may overcome eccentric application of the load and have area for increased wall capacity, or they may have reinforcing for column capacity. They also provide stiffening to the wall and decrease spans so that the vertical and horizontal capacity may be increased for greater spans. They also provide for architectural treatment of walls.

Garden walls are low walls, 6 ft or less in height. Being low, the normal code wind pressures for design are validly assumed to be rather low, say 10 lb or less per square foot. Seismic loads are frequently more critical. Standard details are provided for simple use because through foundation analysis and precise structural design is costly when related to the field construction costs. These walls are valuable landscaping elements, as demonstrated by the single wythe serpentine walls of Thomas Jefferson's estate.

Retaining walls are also treated as separate items. A foundation data definition is adopted from the Los Angeles City Code. It was developed by the Foundation Committee of the SEAOSC and is a conservative method of load determination and resistance capacity, without the expensive foundation investigation, research, and analysis that would be necessary otherwise.

21.5.1 Types of Assemblies

Hollow concrete block walls have been adapted very well as substitutes for the older cut stone and fired-clay masonry. They are manufactured from a relatively lean concrete mix of many shapes, as shown in typical drawings. Surface textures may be smooth, split faced, slumped, scored, glazed, fluted, or exposed aggregate, and they come in several shades, colors, or strengths, as well as load bearing or nonload bearing. Most manufacturers issue catalogs showing the variations and methods of design.

Reinforced block is an economical method of providing increased capacity and wind and seismic resistance. These provide economical walls, such as exist in Southern California, some built in the late '20s and early '30s. The early units were frequently rather crude and made in small plants, but modern units are higher quality and comply with ASTM C-90 for hollow and solid units. These are easily designed and built by traditional methods, with either low-lift or high-lift grouting.

Some special shapes are made in some areas to eliminate the use of mortar joints, for example, special shapes of head joints that interlock, depending on grout for integrity, or in which both head and bed joints interlock, or bed joints are ground for fit or for epoxy.

Traditional solid concrete brick units now comply with UBC Standard 21-3, which is based on ASTM C55. These are similar to concrete brick according to ASTM C90, Typical patterns of coursing in walls are shown in Figure 21.7. Reinforced concrete masonry adds reinforcing to these traditional units. The reinforcing may consist of pencil rod, fabricated wire joint reinforcing complying with UBC Standard 21.10, or concrete reinforcing bars in accord with ASTM Standards.

Traditional clay brick evolved from the sun-dried mud brick and mortar to early fired clay or shale and then to the development of "modern shapes, shown in figures in compliance with ASTM C62 for "common" brick or "building" brick and ASTM C216 for more precise "face" brick. Types of patterns and elements are identified in figures and assemblies in Figure 21.11. These were "solid" if the core holes were not greater than 25% in area. A more recent improvement for more efficient reinforced masonry is the hollow fired-clay brick, ASTM Standard C652, with assembly and shapes as shown in Figure 21.11.

Reinforced Brick Masonry (clay) consists of assemblies of solid brick and grout and hollow brick and grout, reinforced. (See Figure 21.11.)

Prestressing is an effective assemblage method for masonry construction that is strong in compression although weak in tension. Clay masonry has been used effectively, utilizing the beneficial fact that there is little creep to cause loss of prestress. Owing to the high strength of the fired clay; long-span spandrels or bridg-

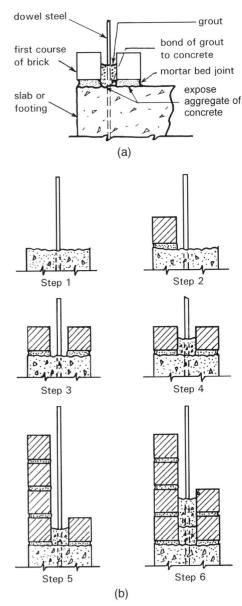

FIGURE 21.11   Steps in reinforced grouted brick masonry.

ing of openings have been done effectively. Also, multistory structures have been built of unreinforced masonry incorporating vertical prestress rods with coupling nuts and final post-tensioning from top to bottom. These must be carefully designed, considering the field procedures, because the method has not had wide usage and every step must be carefully considered.

Veneer has been defined and regarded as a nonstructural facing for ornamentation or insulation, so there are many traditional methods developed by successive approximations or trial and error. However, newer engineering methods and regard for earthquake effects warrants engineering principles.

The UBC Veneer portion (from the 1967 edition) is actually one of the best chapters in the code, but it must be read carefully. In view of the increased recognition of veneer potential, for example in the WSCPA booklets on *Reinforced Veneer* and *Veneer on Steel Studs*, this subject is herein simplified and given greater attention, such as is warranted by other types of masonry. It rationally permits performance or prescription.

The UBC chapter may be outlined simply as follows:

1. Limitations
2. Two types of veneer, both to be designed
   a) Adhered
   b) Anchored
3. Design, to following criteria (performance)
   a) Adhered, develop 50 psi
   b) Anchored, develop for $2\times$ the weight
4. In lieu of design, use proven standards (prescription).

The weakness of the UBC chapter was that it did not explain simple design and should have contained more standard alternates as practiced in different locales.

21.5.2 Limitations of Veneer

Applications not over 4 ft are exempt from UBC standards, as not being hazardous to life. Setting a maximum height of 30 ft above the lower support when veneer is laterally supported by wood prevents excessive differential deflection between the wood construction and the nonshrinking veneer, which might cause failure in the connections, ties, or anchors. This can be waived by specific design solutions. The requirement for support at story heights provides for the differential drift or deflection between stories. However, one can provide design for free movement or control joints so such movement does not cause distress. For example, one five-story, parking deck, with spiral floors, provides lateral ties that permit free in-plane motion but that form lateral support. The vertical support of the five-story deck is provided by the foundations.

Another scheme consists of multi-story masonry resting on the foundation, receiving out-of-plane support by flexible strut-ties at floors, and simply rocking to accommodate inplane building movement.

Independent panels may be supported on high buildings, with differential movement permitted by proper joints and connections.

Incombustible supports are required to minimize the hazard to fire fighters and also to limit the use of wood, a less durable material.

The size of adhered units was limited because of the manual difficulty of placing large pieces of adhered veneer.

The arbitrary weight limit of 15 psi is simply a conservative requirement. Adequate finish surfaces can be provided with such weight. It represents about a 1 1/2 in thick unit. Thicker or heavier units may be laid as brick facing and tied as anchored veneer. Anchored veneer provides a greater assurance of safe field installation.

21.5.2.1 General Design Requirements

Some of the general design issues that must be considered are as follows:

> differential movement between veneer and support;
> shrinkage of concrete areas or concrete supports;
> deflections of supporting structures, both long and short term;
> creep, as it occurs in prestressed or heavily loaded concrete;
> deflection and shrinkage of wood structures;
> deflection of limber steel frame structures and of stud supports and ties; and
> differential temperature-volume change.

Control joints have been effective in permitting the small movements necessary to release stress that such movements or deflections may induce. They must be located in the veneer over any control joints located in the underlying supporting structure.

Proper consideration of differential movements is a subject involving many factors, and its precise solution is impossible because of the paucity of information available. Common sense, experience with material performance, and observations of performance of previous installations provide the best guides. For example, exterior veneers should have expansion joints to relieve stress accumulation. How far apart? Somewhere between 12 and 20 ft seems to be a reasonable maximum.

21.5.2.2 Design Criteria

The UBC requirements for the design of **adhered veneer** include the provision that the bond between veneer and backing develop a shearing strength of 50 psi (or 7220 psf). This ultimate strength is greater than might seem necessary, since the veneer material generally weighs only 10 to 15 psf. But that factor of safety stems from a practical awareness that the workmanship may not always be high quality, that there may be indeterminable internal differential thermal effects, and that absorption and drying volume changes are likely to occur. Also, the possible hazard to public safety must be considered, since veneers are frequently installed over public ways. Fortunately, many adhesive materials and methods are available that assure code compliance.

Design requirements for **anchored veneer** stipulate that both the veneer and its connections resist a horizontal force of not less than twice the weight. The reasons for this very conservative factor of safety are similar to those offered for adhered veneer. This basic design requirement is not difficult to achieve, as shown in simple design examples.

One very important factor to consider when working with veneer is the relative differential movement between building and facing. These distortions are caused by temperature differentials plus the building drift due to wind or seismic forces. An equally important aspect is the control of moisture penetration, which can be minimized or prevented with the use of proper flashing details.

21.5.2.3 Veneer Attachment and Span: Design Example

Assume a 3-in-thick modular brick in running bond veneer attached to a steel stud backing. Check the adequacy of spacing anchors 24 in cc horizontally and 12 in cc vertically.

Since the veneer is much stronger in the horizontal direction, assume that all moment caused by perpendicular out-of-plane loading is carried in this direction. This is somewhat conservative, since there will be some plate action whereby the load could be transmitted in two directions.

$$\text{load required by code} = 2 \times 3 \text{ in} \times 10 \text{ lb} = 60 \text{ psf}$$

$$\text{tie or anchor load} = 2 \text{ ft} \times 60 \text{ lb} = 120 \text{ lb}$$

$$\text{moment horizontally} = \frac{120 \times 24}{12} = 240 \text{ in/lb}$$

$$\text{flexural tension in masonry} = \frac{240 \times 6}{12 \times 3 \times 3} = 13.3 \text{ psi}$$

Therefore, the masonry will function adequately for the code-specified load of two times its own weight. This loading results in low-level stresses, which can be readily carried by the unreinforced portions of the masonry. The 120-lb anchor load is easily carried by No. 9 wire, straps, or manufactured ties and attached with sheet metal screws.

We observe that the design results in construction similar to that specified in UBC Standards. However, note that the design does not require steel-wire reinforcing, which might be costly and difficult to place, thereby causing delays in construction. This wire is totally unnecessary, since that portion of any masonry wall falling between the vertical rebars is unreinforced. However, it is a conservative feature for seismic areas.

The use of stiff brick veneer connected to relatively flexible steel studs is an economically feasible system, but this type of installation requires careful consideration of the deflection of the veneer, steel studs, and ties, and also of the resulting load distribution.

Factors that should receive engineering consideration include the following:

1. When the wind blows or earthquake forces are imposed, the stud framing and the uncracked brick walls deflect almost equally, as forced by the action of the ties.
2. The ties impose the load on the studs, with the top and bottom ties (single or in pairs) providing the span end reactions from the brick span. Since the uncracked brick and the steel studs deflect equally, they share the total resisting moments in proportion to their stiffnesses, until the brick cracks, following which the studs carry the total load, as indicated in the following expressions:

$$\text{stud moment} = \text{total } M \times \frac{E\,I\,(\text{stud})}{E\,I\,(\text{stud}) + E\,I\,(\text{brick})}$$

$$\text{brick moment} = \frac{fr \times bd^2}{6}$$

$$\text{total moment} = \frac{wl^2}{8}$$

Some facts influence the detailing of the wall. The brick wall will crack, and hence there must be a waterproof barrier over the studs. The ties and connections must be corrosion resistant. Also, drainage must be provided for removal of water. It is structurally important that tie strengths be adequate to force the distribution of the loads imposed without failure.

The use of the traditional 2 × veneer weight by the traditional prescriptive design load would be $2 \times 35\# \times 16/12 = 93\#$. This might cause an overstress until the brick cracks. Therefore double ties or heavier ties are recommended for the reactions. Detail drawings are shown for standard installations in Figure 21.12a.

In addition, it is to be recognized that the ties may be spaced by design to provide wider spans that the standard 2 ft^2 per tie. The connections may be spaced to tie to appropriate structural elements with the veneer reinforced by design to span accordingly.

This latter principle can also provide for prefabricated panel installation methods.

Fire rating of veneer has been established for 1- and 2-hour endurance periods. These were established by tests and by an evaluation or research approval of the UBC, ES 5058 (see Figure 21.12b). The essential principle was based on the fact that 5/8 in gypsum wall board or plaster on each side of either wood or steel studs would provide one hour's resistance, and two such wall boards on each side would provide two hours' endurance. Early tests and code acceptance had permitted 1.5 in of masonry to be equivalent to 1 in of gypsum wall board or gypsum plaster. Therefore 1 in of masonry veneer could be substituted for the .625 in of gypsum to give 1-hour endurance. Also 2 in of masonry thickness in lieu of 1.25 in of wall board on either face would provide 2-hour endurance. This increases the potential use of veneer as shown in the required 2-hour rated walls and partitions of Table 6-A of the UBC. (See Table 21.6.) Although 1 in of masonry on a side could provide the resistance required, anchored veneer should be not less than 2 in thick because of the difficulty of laying the masonry.

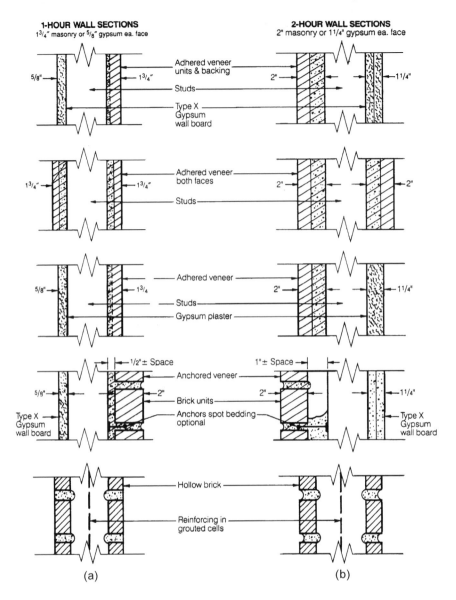

FIGURE 21.12a Fire ratings of brick veneer walls. The 1-hour walls require not less than 5/8″ gypsum board or plaster, or 1 3/4″ of adhered veneer and backing on each face. The 2-hour walls require not less than 1 1/4″ (two 5/8″ boards) wall board or gypsum plaster, or not less than 2″ of brick veneer and backing on each face. The thickness of brick veneer includes the unit, the cement bond coat, and cement backing surface. Walls of hollow brick construction serving as reinforced veneer may be rated for fire endurance.

The UBC Evaluation Service approval showed some sample applications or installations, but many more methods comply with the functional requirements.

21.5.2.4 Paving

Paving is a nonstructural masonry installation. Since the early stone sand-bedded cobblestone paving, there have been many masonry systems used so paving is justifiably a masonry construction method. The development of newer, more expressive patterns is illustrated in Figure 21.13.

ICBO Evaluation Service, Inc.

A *subsidiary corporation of the* International Conference of Building Officials

EVALUATION REPORT
Copyright © 1994 ICBO Evaluation Service, Inc.

Report No. 5058
January, 1994

Filing Category: FIRE-RESISTIVE CONSTRUCTION—Other Fire-resistive Construction   (080)

BRICK VENEER FIRE ENDURANCE RATINGS OF WALL ASSEMBLIES
WESTERN STATES CLAY PRODUCTS ASSOCIATION (WSCPA)
3130 LA SELVA, SUITE 302
SAN MATEO, CALIFORNIA 94403

I. **Subject:** Brick Veneer Fire Endurance Ratings of Wall Assemblies.

II. **Description: A. General:** The brick veneer may be adhered or anchored in compliance with Chapter 30 of the code, and applied to one or two faces of nonbearing wall assemblies for one- or two-hour fire-resistive assembly ratings.

B. **Materials: 1. Thin Veneer Brick Units:** Units are produced from clay or shale in thicknesses from $1/2$ to $1^1/2$ inches and comply with ASTM C 1088, Grade TBS or better.

2. **Brick Units:** Units are made from clay or shale in thicknesses exceeding $1^1/2$ inches and comply with U.B.C. Standard No. 24-1 in the appropriate grade for exposure as described in Section 24.101 (c) or 24.106 (b) of the code.

3. **Mortar:** Type S as set forth Table No. 24-A of the code.

4. **Plaster Backing:** Portland cement plaster complying with Section 4708 of the code.

5. **Lath:** Minimum 3.4 pounds per square yard metal lath complying with Section 4706 of the code.

C. **Installation:** Details of one- and two-hour nonbearing walls are noted in Figures Nos. 1a and 1b. For symmetrical one-hour fire resistance, each face shall have not less than one layer of $5/8$-inch-thick Type X gypsum wallboard, or equivalent gypsum plaster, or $1^3/4$ inch thickness of masonry veneer. For two-hour fire resistance, each face shall have not less than two layers of $5/8$-inch-thick Type X gypsum wallboard, or equivalent gypsum plaster, or 2 inch thickness of masonry veneer.

1. **Steel Framing:** Framing and either gypsum plaster or gypsum wallboard must comply with Items Nos. 13-1.1, 13-1.2, 13-1.3, 13-1.4, 15-1.1, 15-1.2 or 15-1.3, Table No. 43-B of the code.

2. **Wood Framing:** Framing and either gypsum wallboard or gypsum plaster must comply with Items Nos. 14-1.1, 14-1.2, 14-1.3, 14-1.4, 16-1.1, 16-1.2, 16-1.3, 16-1.4, 16-1.5 or 16-1.6, Table No. 43-B of the code.

3. **Adhered Veneer:** Metal lath is installed in compliance with Section 4706 of the code. Where lath is attached to steel framing, minimum 1 inch long, No. 6 drywall screws are used. For exterior walls, a weather-resistive barrier described in Section 4706 (d) of the code is required. As an alternative, paperbacked metal lath recognized in a current evaluation report with paper complying with Section 1708 (a) of the code may be used. The portland cement plaster is applied in compliance with Section 4708 of the code to a minimum $3/4$-inch thickness. The thin veneer units are applied in compliance with Section 3005 (d) 1 of the code in running bond. For one-hour fire resistance, the total thickness of plaster, mortar and brick veneer shall be at least $1^3/4$ inches. For two-hour fire resistance, the total thickness of plaster, mortar and thin brick shall be at least 2 inches.

4. **Anchored Veneer:** Anchored veneer is installed in compliance with Section 3006 (d) 1 of the code for 2-to-5-inch-thick units and Section 3006 (d) 2 of the code for units up to 10 inches thick. Stud spacing is limited to 16 inches on center and a weather-resistive barrier complying with Section 1708 of the code is required on the exterior side of exterior walls. Anchored units may be used for one- or two-hour fire-resistive assemblies.

D. **Identification:** The materials are identified as specified in the code or current evaluation report. Thin veneer units are identified with the manufacturer's name and address, size, type and grade in evidence of compliance with ASTM C 1088.

III. **Evidence Submitted:** Reports of fire-resistive tests conducted in accordance with U.B.C. Standard No. 43-1 and descriptive details.

Findings

IV. **Findings:** That the Brick Veneer Fire Endurance Ratings of Wall Assemblies comply with the 1991 *Uniform Building Code*™, subject to the following conditions:

1. **Materials and installation comply with this report and the code.**

2. **Wall systems are limited to nonbearing assemblies.**

1993 Accumulative Supplement to the U.B.C.: This report is unaffected by the supplement.

This report is subject to re-examination in one year.

FIGURE 21.12b   E.R. 5058.

Concrete masonry paving units, provide wider choices for usage. A so-called "concrete block" pavement consists of a wearing surface of concrete units embedded in a layer of sand over a suitable subbase. The NCMA has developed detailed method and unit specifications to ensure that the wearing surface will be durable. Prior to development by ASTM, NCMA developed a specification for solid concrete block paving units for vehicular traffic, designated NCMA A-10-7 and described in their TEK 87, 115, and others.

Paving brick may be laid in a variety of patterns for different visual effects as shown in Figure 21.14. The effect of unit size will be an important factor in the layout as will the use or nonuse of mortar joints. For example, a basket-weave pattern may be achieved by using modular 3-3/8 in × 11-3/8 in

TABLE 21.6 Types of Construction—Fire-Resistive Requirements in Hours

Building Element	Noncombustible				Combustible				
	Type I	Type II			Type III		Type IV	Type V	
	Fire-resistive	Fire-resistive	1-hr.	N	1-hr.	N	H.T.	1-hr.	N
1. Bearing walls— exterior	4 Sec. 602.3.1	4 Sec. 603.3.1	1*	N	4 Sec. 604.3.1	4 Sec. 604.3.1	4 Sec. 605.3.1	1*	N
2. Bearing walls— interior	3	2*	1*	N	1*	N	1*	1*	N
3. Nonbearing walls— exterior	4 Sec. 602.3.1	4 Sec. 603.3.1	1 Sec. 603.3.1	N	4 Sec. 604.3.1	4 Sec. 604.3.1	4 Sec. 605.3.1	1	N
4. Structural frame[1]	3	2	1	N	1	N	1 or H.T.	1	N
5. Partitions— permanent	1[2]*	1[2]*	1[2]*	N	1*	N	1* or H.T.	1*	N
6. Shaft enclosures[3]	2*	2*	1*	1*	1*	1*	1*	1*	1*
7. Floors and floor-ceilings	2	2	1	N	1	N	H.T.	1	N
8. Roofs and roof-ceilings	2 Sec. 602.5	1 Sec. 603.5	1 Sec. 603.5	N	1	N	H.T.	1	N
9. Exterior doors and windows	Sec. 602.3.2	Sec. 603.3.2	Sec. 603.3.2	Sec. 603.3.2	Sec. 604.3.2	Sec. 604.3.2	Sec. 605.3.2	Sec. 606.3	Sec. 606.3
10. Stairway construction	Sec. 602.4	Sec. 603.4	Sec. 603.4	Sec. 603.2	Sec. 604.4	Sec. 604.4	Sec. 605.4	Sec. 606.4	Sec. 606.4

Source: This table is an excerpt of Table 6-A of the UBC.

Note: For details, see occupancy section in Chapter 3 of the UBC, as well as sections on types of construction and sections referenced in the table.

*Fire-rated brick veneer may be used as 1 and 2 hour walls in the various types of construction by use of the IBCO Evaluation Report No. 5058.

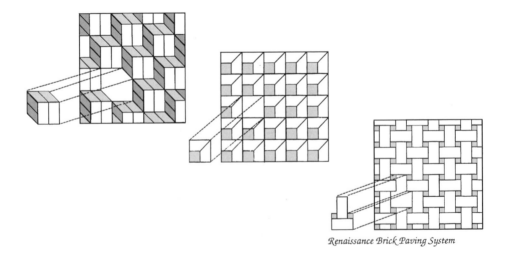

Renaissance Brick Paving System

FIGURE 21.13 Various paving patterns.

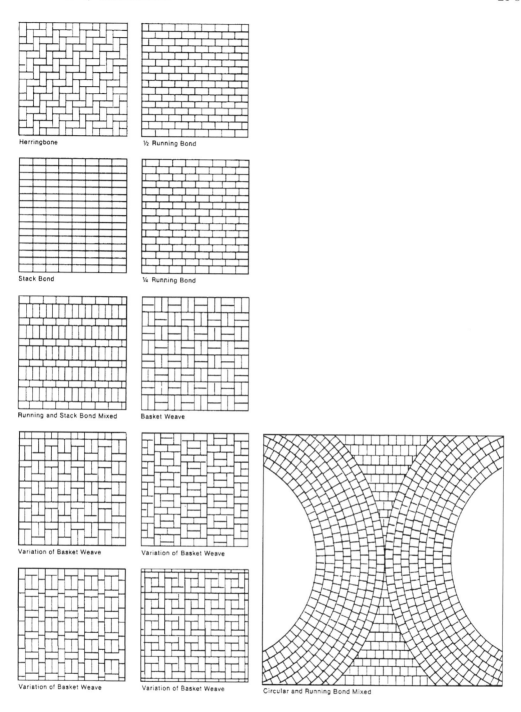

FIGURE 21.14 Brick patterns.

units with normal-size mortar joints resulting in 12 in × 12 in elements. However, laid without mortar joints, the elements would measure approximately 11 1/2 in in one dimension and only 10 1/2 in in the other, resulting in a gap (see Figure 21.15). This gap would not occur in a herringbone or stretcher bond pattern.

The presence of elements such as the hose bibs or connections, planters, curbs, edges, driveways, varying dimensions, and corners will also influence pattern development. These elements may inspire the use of many different patterns within the overall installation. Several examples of how these elements may be handled and how brick can be combined with other surfacing materials are as shown in Figure 21.16. Also note installations of expansion joints as in Figures 21.16–21.18 and 21.21.

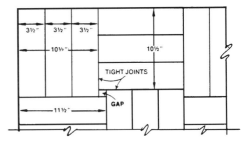

FIGURE 21.15   Paving elements laid without mortar joints.

Site. The site planning must include considerations for traffic, underground utilities, user convenience, and the base support. Successful installation will depend on proper subgrade preparation. Vegetation and organic matter

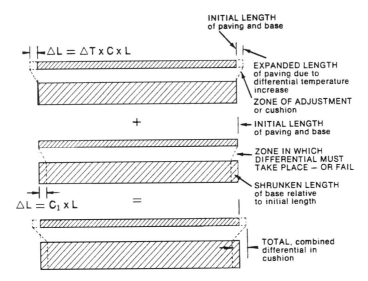

FIGURE 21.16   The cumulative differential effect of paving expansion and base shrinkage. This results in a double shearing stress and tendency to force curvature. Both these factors will have an additional tendency to destroy bond between the surface paving and the base, with consequent possibility of distress. The advantage of close spacing of expansion joint is the reduction of stress accumulation.

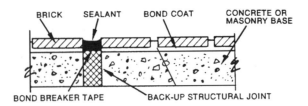

FIGURE 21.17   Expansion joint widths: Exterior—minimum 3/8″ for joints 12′ on center, minimum 1/2″ for joints 16′ on center. Minimum widths should be increased 1/16″ for each 15°F of the actual temperature range greater than 100°F between summer high and winter low. Joints through brick and mortar directly over any structural joints in the backing must never be narrower than the structural joint.

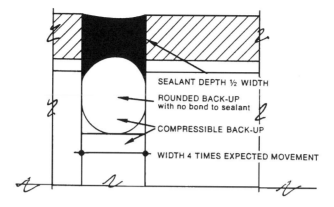

FIGURE 21.18 Expansion joint installation: Set compressible backup strip when mortar is placed or utilize removable wood strip to provide space for backup after mortar has cured. Install sealant after paving and grout are dry. Follow sealant manufacturer's recommendations. Refer to sealant section in ANSI Installation Specifications, A108.

subject to decay must be removed. Soft or poorly compacted sub-base must be removed and replaced with properly compacted material. Expansive clays must be removed, or provision made for the differential expansion that may occur. Plantings must not extend under the sub-base.

Design and Construction Considerations. **Unit paving** installations may consist of several combinations of rigid or flexible bases and pavings.

Rigid paving consists of either thick or thin brick with mortar joints laid in mortar beds. It is stiff and subject to cracking if deformed excessively.

Flexible paving consists of either thick or thin brick without mortars used as either a bed or a joint. It is not subject to cracking due to deformation.

Rigid base is a reinforced concrete slab on grade or a concrete or steel structural slab. It is not subject to large deflection.

Flexible base consists of compacted gravel or a damp, loose, sand-cement mixture tamped into place and grade. It permits considerable deformation. Wood support is also considered to be a flexible base, as is asphalt paving. Asphalt is considered flexible because of the flow characteristics of tar. Only flexible types of paving should be placed over the asphalt paving, unless it is a thick and strong rigid installation that will resist the possible differential deflections.

Traffic loading and exposure to wear or weather determine the kinds of paving and base that should be specified. Heavy vehicular traffic will generally require rigid paving and a rigid base. Light vehicular traffic such as encountered in residential driveways and service areas may be carried by flexible paving supported by a flexible base. Pedestrian traffic is appropriate over any combination of sound paving and base.

Cushion. **Cushion material** may be between the paving and the base and serves as a leveling layer to improve the accuracy of the finished grade and to compensate for irregularities of the base and of the paving units. In some instances it may serve to provide for differential motion between the top surface and the substrate. Cushion material may be sand, pea gravel, or stone screenings, and in wet conditions should not be uniformly graded. "Skip grading" will provide better drainage and inhibit capillary action. Use of one ungraded size will provide the greatest pore drain space, whereas uniform size gradation will provide less pore space, with maximum density. Cushion material may consist of sand and cement mixes of one part Portland cement to 3 to 6 parts damp loose sand.

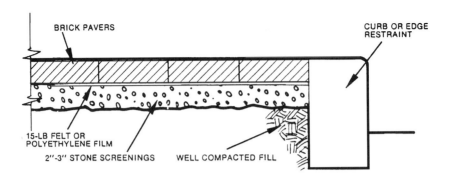

FIGURE 21.19   Gravel base/stone screening bed: "dry" joints. This is effective for patio and pedestrian traffic leading to heavier traffic areas.

Roofing felt may serve as a cushion between the paving and the base (see Figure 21.19). It is installed rather rapidly and will impart some resilience to the support as well as control subsurface drainage. Serving as a membrane, it will stop the flow of ground water into the paving, also of surface water to the substrate and through supported slabs.

Ecology and the flexible brick pavement.   *Flexible type pavement* permits water passage through the paving assembly and the subgrade. This alleviates combined sewer overflow pollution, returns rain water to the ground water, minimizes flash flooding, and reduces rain puddles in parking lots and pedestrian areas. Sand bedding is most effective for the Renaissance pattern shown in Figure 21.13, providing for easy maintenance repair or alteration.

Mortar Joints.   Basic methods for installing mortar joints, with many modifications, are as follows:

One method is the conventional use of mortar placed with a trowel. Brick pavers are buttered with mortar and shoved into a leveling bed of mortar. A second method involves placing each brick unit on a mortar leveling bed with 3/8 to 1/2 in of joint space between the units, followed by placing a grout mixture between the units. Generally, grout proportions of Portland cement and sand are the same as for mortar with the exception that lime may be omitted. When grout is poured into the joints, special care must be taken or the units protected to facilitate cleaning. One clean technique is to pour grout from a coffee pot, or pinched can, in about 3/4 of the joint depth. This settles as water is absorbed. Then a second pour is made, leaving a slight crown that is tooled after stiffening.

A third method involves a dry mixture of Portland cement and sand, using the same proportions as for grout. Brick pavers are installed on a damp cushion comprised of this mixture, followed by the same mixture between the paving units. After cleaning excess material from the paving surface, the paving is sprayed with a fine mist of water until the joints are adequately wet for hydration. The pavement should be maintained in a damp condition for a period of two or three days.

The latter two methods tend to permit the units to become very dirty and stained with the mortar or grout, which is almost impossible to remove. Some improvement may be realized by the use of surface bond breaker applications such as wax coatings, etc. These may be applied with a roller to the brick surface prior to grouting the joints. The use of a bag applicator similar to a cake frosting decorator, and careful tooling, is recommended as a good method to provide tight joints and a clean surface.

Brick paving without mortar joints may be swept with plain dry sand or a mixture of Portland cement and sand to fill the joints. For the proper proportions of Portland cement and sand, refer to the discussion on grout type mortars. Many variations are illustrated in Figure 21.20.

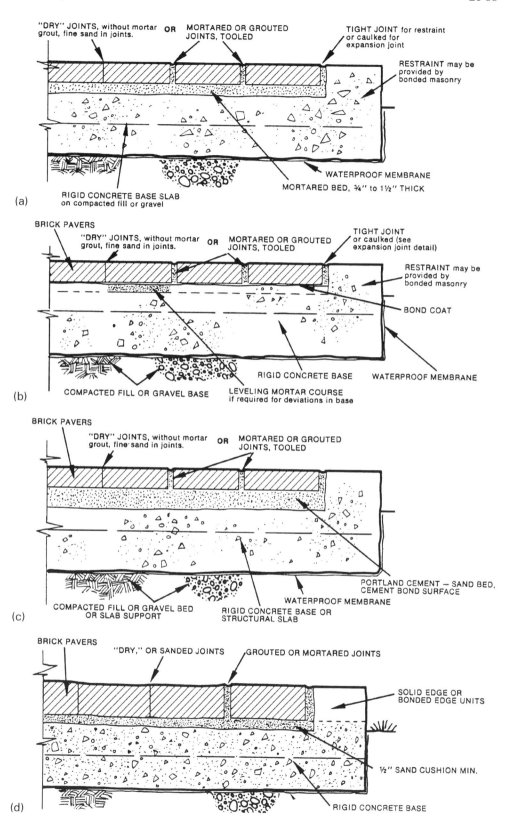

FIGURE 21.20 (*continues*)

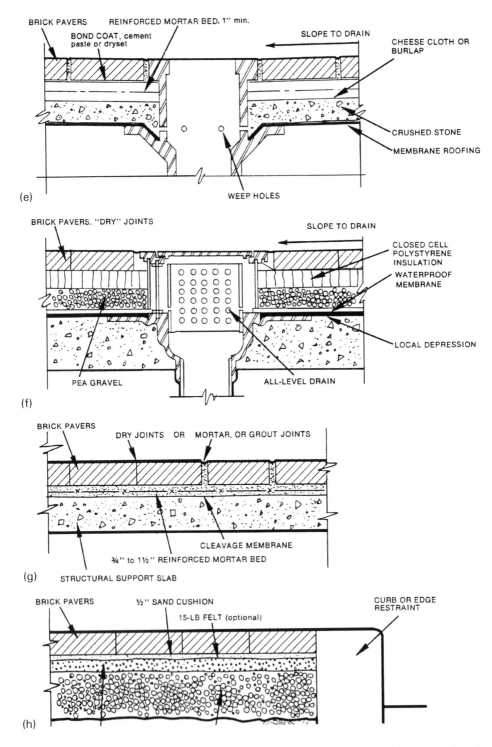

FIGURE 21.20 (*continued*) Examples of paving. (a) Concrete base/mortared bed; (b) concrete base/bond coat bed; (c) concrete base/sand cement bed; (d) concrete base/sand bed; (e) concrete base/roof membrane; (f) concrete base/insulation; (g) concrete slab/mortar bed, reinforced; (h) gravel base/sand bed: "dry" joints.

Thinset Bonding. This is a Portland cement with additives that improve bond and water retention. It may be used where the surfaces are relatively true, but it does not have enough body to provide for large deviations. There is an ANSI A118 specification for the material and an ANSI A108 specification for the application as shown. The manufacturers' recommendations should be followed carefully.

Expansion Joints or Control Joints. Moisture changes and temperature differential allowances must be made for differential movements in material. The paving surface, especially a dark one, may increase in temperature much more than the shielded base. Temperature measurements have shown differentials on the order of 100°F. In addition, a freshly poured concrete base will have shrinkage changes as great as some temperature volume changes, adding to the differential in length. Also, the brick will expand in time. This is shown in Figure 21.16.

Spacing of Joints. Some calculations indicate that control joints in exterior exposed surfaces should be not more than about 12 to 6 ft apart. However, this is dependent upon many factors and local conditions, and some practice has indicated that 16–20 ft is a satisfactory spacing. More important perhaps is regard for joint location relative to shape of area. The joints should be large enough to permit the joint material to accommodate the change in joint dimension.

Expansion Joint Widths. Exterior widths should be a minimum of 3/8 in for joints 12 ft on center and a minimum of 1/2 in for joints 16 ft on center. Minimum widths should be increased 1/16 in for each 15°F of the actual temperature range greater than 100°F between summer high and winter low. Joints through brick and mortar directly over any structural joints in the backing must never be narrower than the structural joint.

Preparation. Brick edges to which the sealant will bond must be clean and dry. Sanding or grinding of these edges is recommended to obtain optimum sealant bond. Primer on brick edges is mandatory when recommended by the sealant manufacturer. Care must be taken to keep primer off brick faces.

Installation. Set a compressible backup strip when mortar is placed, or utilize a removable wood strip to provide space for the backup after mortar has cured. Install sealant after the paving and grout are dry. Follow the sealant manufacturer's recommendations. Refer to the sealant section in ANSI Installation "for further detail" information; Specifications, A108.

Materials. Single-component sealant (not trafficked areas) shall be a nonsag type complying with Federal Specification TT-S-001543 or TT-S-00230c. Two-component sealant shall comply with Federal Specification TT-S00227e; use Type 1 (self-leveling). The backup strip shall be a flexible and compressible type of closed-cell foam, polyethylene, or butyl rubber, rounded at the surface to contact the sealant, as shown in Figure 21.21 and as recommended by sealant manufacturers. It must fit neatly into the joint without compacting and to a height that will allow a sealant depth of 1/2 the width of the joint. Sealant must not bond to the backup material.

21.5.2.5 Efflorescence

Occasionally, efflorescence may occur on paving or masonry exposed to moisture. The phenomena and methods to prevent or minimize it are paradoxically simple yet complicated by many unknown factors.

The basic or simplified mechanism of efflorescence is as follows:

Water enters the masonry through absorption, through cracks, through joints, or through improper details.
Soluble materials are dissolved, either before or after the water enters.

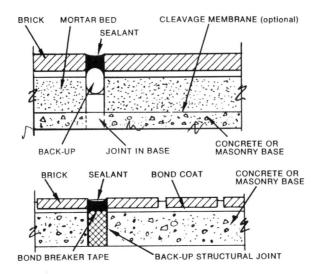

FIGURE 21.21   Control joints alternates.

Water migrates to a surface through the internal porosity and evaporates as vapor, leaving behind whatever it has transported. This may show as a crystalline, powdery, or amorphic deposit at the surface, either visibly on the exterior or in the interior pores of the material. In the one case it is undesirably visible, in the other it may cause interior crystals to grow within the pores and break down the structure of the masonry surface, causing spalling.

There are two basic ways to stop (or minimize) the phenomenon:

1. Eliminate the source of the soluble material.
2. Stop the water migration mechanism.

The sources of efflorescence material are many. One potential source is the ground. Impermeable membranes laid on the ground will prevent penetration. Other sources that may contribute to efflorescence are the mixing water—generally a minute contribution, especially where properly potable; the units—negligible or slight in fired clay or shale and controllable; the lime—very slight; the aggregate—slight since it is generally washed; the cement—which contributes the major portion, as shown by chemical analysis, and observation of the migration from the joints. Type II low-alkali cement apparently causes less efflorescence than others.

The mechanism of efflorescence can be reduced by reduction of porosity, tooling of joints, use of flashing and membranes, and use of coatings. The last is dangerous, especially with "water-retardant" coatings such as repellants. These permit water to enter through cracks or because of pressure, and then prevent the exit of water as liquid through the surface. However, they permit water vapor to exit, which leaves behind deposits of dissolved material. The result is occasionally the growth of deposits in the pores, beneath the repellant, until the pore pressure exceeds the tension value of the masonry material, which may then spall or powder.

Recommendations.

Where suitable, provide a membrane.
Use dense units, and tight tooled joints.
Use low-alkali cement.
Use strong units to resist the pore pressure.
Use a coating that reduces the migration.
Provide flashing and drainage.

Freeze Thaw. This may cause spalling that can be confused with efflorescence spalling. The effect is frequently similar. Standard freeze-thaw tests also are recommended for determination of local suitability.

21.6 Garden Walls and Retaining Walls

As mentioned previously, the loading applied to walls is uncertain, as is the foundation resistance capacity. However, although the loads may be uncertain, they are not great and the assumption of 10 psf on garden walls is adequate and conservative, except for specific areas where high wind is known to be a hazard, as in some of the flat San Bernardino valley (California) stretches. Lower assumed wind loads have negligible reduction of costs. The resistance of the foundation material is also a variable.

The design of typical fences up to 6 ft high is not generally required by most codes, such as UBC, for permit, so arbitrary requirements for steel amounts and spacings may be waived. Some jurisdictions, though, have limits of 4 ft in height.

However, the construction of fences must provide structural validity with due regard for good judgment and good construction. Therefore, a sample of appropriate designs for 5 in hollow unit walls will be considered. See Figures 21.22 through 21.26.

The two basic types of walls considered are shown in Figure 21.22 and are as follows:

1. The simple cantilever above a continuous footing;
2. The simple wall spanning horizontally between pilasters, strengthened portions, or projections of the wall; and
3. Several optional pilasters are shown in Figure 21.25.

The footing requirements are also approximate and conservative for most foundation materials. However, one must be sure that the soil support is sound and firm, with no soft material or potential decay of material.

There are several alternate footing construction types shown that are suitable for any of the several types of wall:

1. continuous concentric concrete strip;
2. continuous eccentric concrete strip, to provide for property line observance;
3. concentric pad footings, as at pilasters;

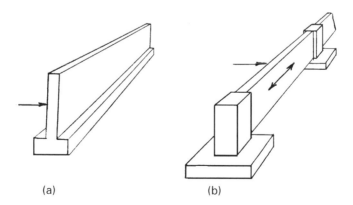

(a) (b)

FIGURE 21.22 (a) A simple cantilever of the fence portion above the footing level as a continuous cantilever. (b) A series of pilasters cantilevering above the footing with the body of the wall spanning horizontally between those pilasters.

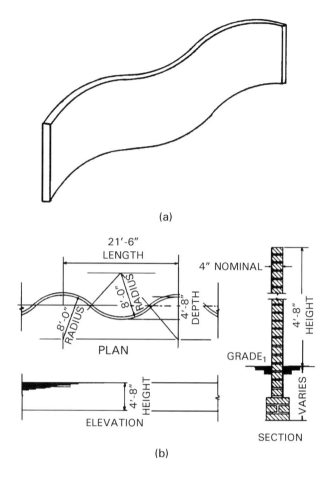

FIGURE 21.23   (a) Serpentine garden wall. (b) As a general rule, the radius of curvature of a 4-in. wall should be no more than twice the height above finished grade, and the depth of curvature should be no less than 1/2 of the height.

4. eccentric pad footings, to provide for construction at property lines;
5. post hole, drilled or pile type of footing; and
6. continuous vertical sheet of filled trench type, as with a narrow trencher.

A typical short form specification is as follows, which is adequate for these small jobs.

Garden Walls

Short Form Specifications

Materials

Units:	Higgins 5 in Hollow Brick, color and texture as approved.
Mortar:	1 part Portland cement, 1/2 part lime, and 3 to 4 parts sand, (i.e., approximately 25 to 30 shovels.)
Grout:	1 part Portland cement to 3 parts sand, mixed to a fluid consistency; Grout Aid® per manufacturer instruction.
Reinforcing:	Wires or deformed bars conforming to ASTM standards.
Concrete:	2000 psi batches, or 1 part Portland cement, not more than 2 1/2 parts sand, and not more than 4 parts gravel.

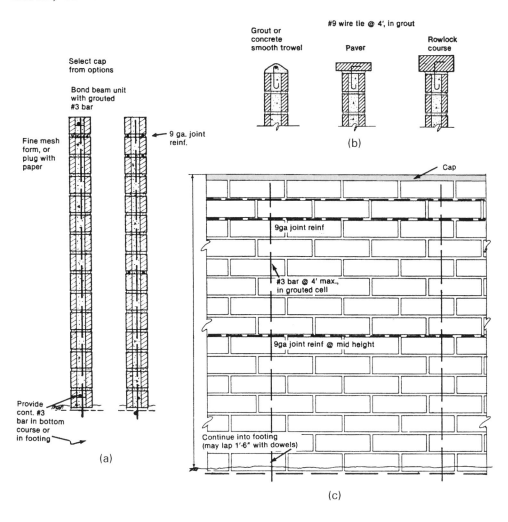

FIGURE 21.24 Continuous cantilever wall on footing.

Laying: Mortar face shells only on bed joint, and in head joints for thickness of the face shells. Units shoved firmly into place, without any tapping or movement after course is laid. Joints to be tooled firmly, round or vee. Grout puddled or vibrated during pouring and then reconsolidated after excess water has been absorbed. Joint reinforcing to be continuous or to be lapped at least 12 in in alternate courses. All vertical and horizontal rebars to be in solid grout.

Retaining walls may be of many types discussed as follows and shown in Figures 21.27 through 21.29.

21.6.1 Gravity

The gravity wall type depends on the massiveness of the masonry itself to achieve lateral stability against earth pressure through dead weight alone. Typically, it is unreinforced, so it must be sized and shaped in such a way that tensile stresses are low or nonexistent. Usually, this calls for a trapezoidal cross-sectional form. Since the advent of modern reinforced masonry, however, this type is not often used.

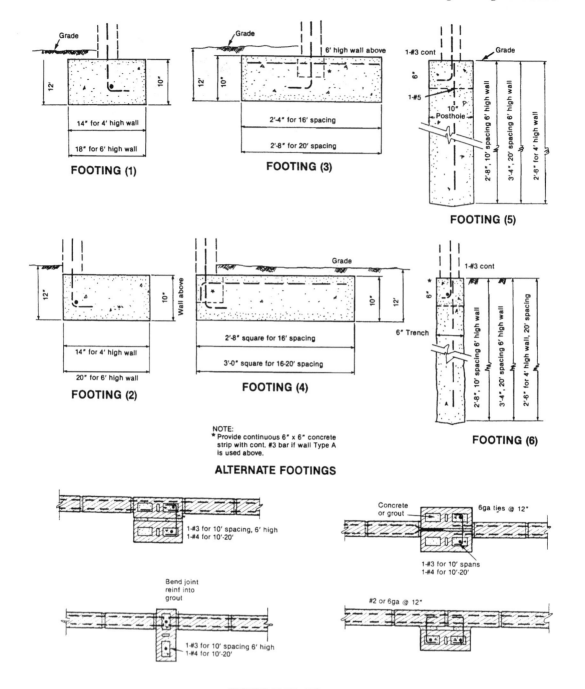

FIGURE 21.25 Pilasters.

21.6.2 Cantilever

The cross section of the cantilever type consists of an L or an inverted T shape. The vertical stem wall acts structurally as a cantilever and, as such, must be reinforced. This stem wall can be readily constructed with hollow masonry units (brick or concrete block), properly reinforced. The base slab (reinforced concrete) essentially maintains lateral stability by restraining the wall against both sliding and overturning. Normally, the L-shaped retaining wall is used when it is located on a property line where the base slab cannot extend beyond the stem face. Otherwise, it is preferable to use the inverted T shape, to minimize flexural stresses within the footing slab.

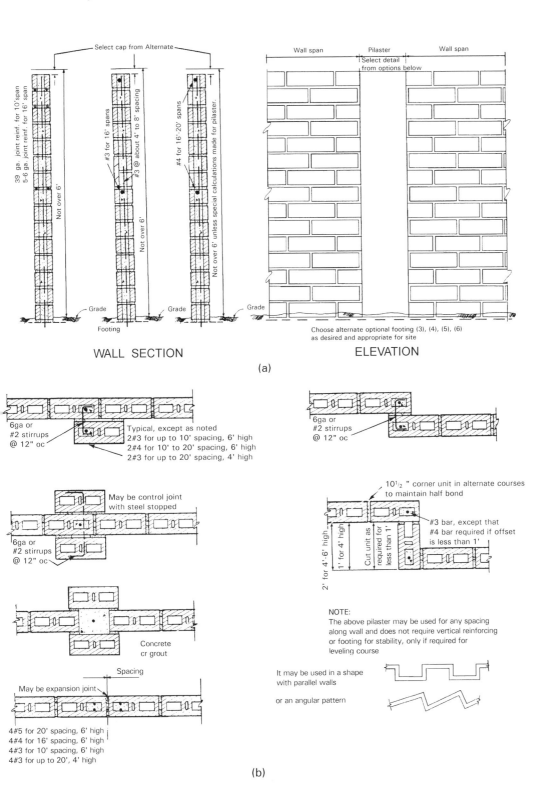

FIGURE 21.26 Wall between pilasters.

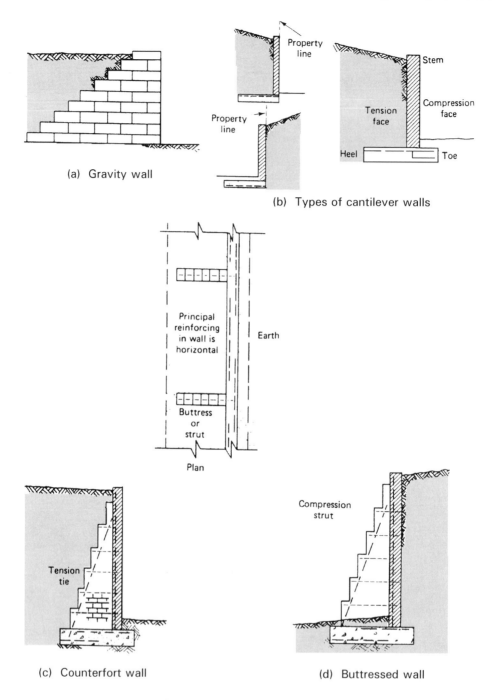

(a) Gravity wall

(b) Types of cantilever walls

(c) Counterfort wall

(d) Buttressed wall

FIGURE 21.27 Basic retaining-wall types.

21.6.3 Counterfort or Buttress

The counterfort or buttress wall types behaves very differently. At intervals along the length of the wall are vertical elements to provide reactions for horizontal spans between these supports. The wall stem reinforcement is horizontal. If the supports are located behind the wall (within the retained earth), they are called **counterforts** and serve as tension ties, whereas if they occur on the exposed face, they are called **buttresses**, in which case they act as compression struts. The wall stem may be assumed to act as a continuous beam or slab. Actually, this type of retaining wall would more than likely be employed for higher walls, where the size or thickness of a cantilever stem would become prohibitive. Refer to Figure 21.27 for sketches of these three types of walls.

21.6.4 Supported Walls

Actually, a basement or subterranean garage wall is a type of retaining wall. In some cases, the wall spans vertically between floors, acting as a continuous beam where several levels exist, or from floor to footing for a single-level basement, as portrayed in Figure 21.28. In others, the wall may be reinforced to span horizontally between intersecting walls, as seen in Figure 21.29. A more refined analysis would recognize that the wall actually acts as a flat plate supported on four sides under various edge-restraint conditions. Note that the subterranean wall generally will carry at least a vertical dead, and very possibly a live, load as well. When the wall spans vertically, the combined effects of vertical axial compressive stress and flexural stress should be analyzed.

21.6.5 Foundation Pressures and Resistance Strengths

Soil pressures and strengths for design purposes depend on the granularity, internal angle of friction, and other physical properties that would be part of a proper soil investigation and consequent calculation based on various theories.

It is recognized that many rather minor retaining walls can be designed for safety using approximate values. One such method was introduced into the Los Angeles City Building Code by the SEAOSC for walls to 15 ft high. (A foundation investigation and report are required for higher or questionable sites.)

The approach in that Arbitrary design method is as follows: Estimation of backfill pressure is based on an empirical equivalent fluid pressure expression, where

$$P = KWH \times H/2$$
P = total lateral earth force
K = weight of backfill, psf
H = height of wall above base in feet

The term KW becomes the equivalent fluid pressure factor as shown in Table 21.7, which lists backfill pressures for various surcharge slopes.

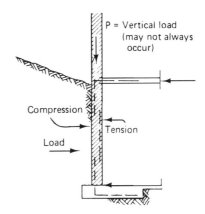

FIGURE 21.28 Supported retaining wall.

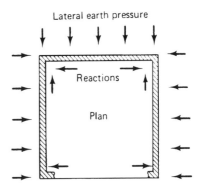

FIGURE 21.29 Intersecting effect of retaining walls.

TABLE 21.7

Surface Slope of Retained Material*, Horizontal to Vertical	Equivalent Fluid Weight, psf
Level	30
5 to 1	32
4 to 1	35
3 to 1	38
2 to 1	43
1 1/2 to 1	55
1 to 1	80

Source: City of Los Angeles Building Code Table 23-E.

*Where the surface slope of the retained earth varies, the design slope shall be obtained by connecting a line from the top of the wall to the highest point on the slope whose limits are within the horizontal distance from the stem equal to the stem height of the wall.

TABLE 21.8 Allowance Foundation Pressure

Class of Material						
Rock: depth of embedment shall be to a fresh unweathered surface except as noted	Value for Increase at Minimum Depth					Maximum Value
Massive crystalline bedrock; basalt, granite and diorite in sound condition*	20					20
Foliated rocks; schist and slate, in sound condition*	8					8
Sedimentary rocks; hard shales, dense siltstones and sandstones, thoroughly cemented conglomerates*	6					6
Soft, or broken bedrocks; soft shales, shattered slates, diatomaceous shales; other badly jointed (fractured) or weathered rock; 12 in minimum embedment	2					2

Soils: minimum depth of embedment shall be one foot below the adjacent undisturbed ground surface*	Kips/ft²				Increase for Depth	Maximum Value
	Lose	Compact	Soft	Stiff		
Gravel, well graded. Well-graded gravels or gravel–sand mixtures, little or no fines	1.33	2.0			20	8
Gravel, poorly graded. Poorly graded gravels or gravel–sand mixtures, little or no fines	1.33	2.0			20	8
Gravel, silty. Silty gravels or poorly graded gravel–sand–silt mixtures	1.0	2.0			20	8
Gravel, clayey. Clayey gravels or gravel–sand–clay mixtures	1.0	2.0			20	8
Sand, well graded. Well-graded sands or gravelly sands, little or no fines	1.0	2.0			20	6
Sand, poorly graded. Poorly graded sand or gravelly sands, little or no fines	1.0	2.0			20	6
Sand, silty. Silty sand, or poorly graded sand–silt mixtures	0.5	1.5			20	4
Sand, clayey. Clayey sands or sand–clay mixtures	1.0	2.0			20	4
Silt. Inorganic silts and very fine sands, rock flour, silty or clayey fine sands with slight plasticity	0.5	1.0			20	3
Silt, organic. Organic silts and organic silt–clays of low plasticity	0.5	1.0	0.5	1.0	10	2
Silt, elastic. Very compressible silts, micaceous or diatomaceous fine sandy or silty soils	0.5	1.0			10	1.5
Clay, lean. Inorganic clays of low to medium plasticity, silty clays, lean clays	1.0	2.0	1.0	2.0	20	3
Clay, fat. Very compressive clays, inorganic clays of high plasticity			0.5	1.0	10	1.5
Clay, organic. Organic clays of medium to high plasticity, very compressible			0.5			0.5
Peat. Peat and other highly organic swamp soils			0			0

Source: The City of Los Angeles Building Code, Table 28-A.

Notes: Kips per square foot; 1 Kip = 1000 lb.

1. Value for gravels and sand given are footings one foot in width and may be increased in direct proportion to footing width to maximum of three times the maximum value, or to the designated maximum value, whichever is the least.
2. Where the bearing values in the above table are used, it should be noted that increased width or unit load will cause increase in settlement.
3. Special attention should be given to the effect of increase in moisture in establishing soil classifications.
4. Minimum depth for highly expansive soils to be one and one half feet.
5. Increases for depth are given in percentage of minimum value for each additional foot below the minimum required depth.

   *The above values apply only where the strata are level or nearly so and/or where the area has ample lateral support. Tilted strata and the relationship to nearby slopes should receive special consideration. These values may be increased one-third to maximum of two times the assigned value for each foot of penetration below fresh, unweathered surface.

TABLE 21.9 Allowable Lateral Bearing per Foot of Depth Below Natural Ground Surface (psf): Natural Soils or Approved Compacted Fill

Soil Type	Loose or Soft	Compact or Stiff	Maximum Values
Gravel, well graded	200	400	8000
Gravel, poorly graded	200	400	8000
Gravel, silty	167	333	8000
Gravel, clayey	167	333	8000
Sand, well graded	183	367	6000
Sand, poorly graded	77	200	6000
Sand, silty	100	233	4000
Sand, clayey	133	300	4000
Silt, inorganic	67	133	3000
Silt, organic	33	67	2000
Silt, elastic	33	67	1500
Clay, lean	267	667	3000
Clay, fat	33	167	1500
Clay, organic	33	—	500
Peat	0	0	0

Source: City of Los Angeles Building Code, Table.

Notes: General Conditions of Use:

1. Frictional and lateral resistance of soils may be combined, provided the lateral bearing resistance does not exceed 2/3 of allowable lateral bearing.
2. A 1/3 increase in frictional and lateral bearing values will be permitted to resist loads caused by wind pressure or earthquake forces.
3. Isolated poles such as flag poles or signs may be designed using lateral bearing values equal to two times the tabulated values.
4. Lateral bearing values are permitted only when concrete is deposited against natural ground or compacted fill, approved by the Superintendent of Building.

TABLE 21.10a Allowable Frictional and Bearing Values for Rock*

Type	Friction Coefficient	Allowable Lateral Bearing, psf	Maximum Values, psf
Massive crystalline bedrock	1.0	4000	20,000
Foliated rocks	0.8	1600	8,000
Sedimentary rocks	0.6	1200	6,000
Soft or broken bedrocks	0.4	400	2,000

Source: Tables 21.10a–21.10c are from the City of Los Angeles Building Code, Table 28-B.

The allowable values for frictional resistant and lateral bearing strengths for fill of various descriptions and depths are given in Tables 21.8 and 21.9, which are taken from the Los Angeles Code (1972 edition).

The allowable assumed frictional and bearing values are shown in the following Tables 21.10a–21.10c.

21.6.6 Design Example

A cantilever retaining wall 8 ft high has a backfill surcharge slope of 2 : 1. Use the Arbitrary code

TABLE 21.10b Allowable Frictional and Lateral Bearing Values for Soils

Frictional Resistance: Gravels and Sands*	
Soil Type	Friction Coefficient
Gravel, well graded	0.6
Gravel, poorly graded	0.6
Gravel, silty	0.5
Gravel, clayey	0.5
Sand, well graded	0.4
Sand, poorly graded	0.4
Sand, silty	0.4
Sand, clayey	0.4

*Coefficient to be multiplied by the dead load.

TABLE 21.10c Allowable Frictional Resistance (psf): Clay and Silt*

Soil Type	Loose or Soft	Compact or Stiff
Silt, inorganic	250	500
Silt, organic	250	500
Silt, elastic	200	400
Clay, lean	500	1000
Clay, fat	200	400
Clay, organic	150	300
Peat	0	0

*Frictional values to be multiplied by the width of footing subjected to positive soil pressure. In no case shall the frictional resistance exceed 1/2 the dead load on the area under consideration.

provisions (since H is less than 15 ft). Assume a brick wall and concrete footing as follows and as illustrated in Figure 21.30:

Design criteria

Solid-grouted brick:

> $f'_m = 1500$ psi. Use 1/2 stresses, since no continuous inspection required; therefore,
> $f'_m = 750$ psi
> $f_s = 20,000$ psi
> Concrete footing $f'_c = 2000$ psi
> Lateral earth pressure $K_h = 43$ psf (Table)
> Vertical earth pressure $K_v = 1/3K_h = 14.3$ psf
> Passive earth pressure $P_h = 300$ psf
> Allowable soil bearing = 2000 psf
> Weight of soil = 100 psf
> Determine lateral loading (1) as $F = wh \times h/2$.
> Determine weight of stem (2) of assumed shape. Calculate moment and reinforcing required on stem.
> Verify stability by summation of rotation of forces that must not exceed allowables.
> Calculate moment on footing necessary to resist overturning and proper reinforcing.
> Verify sliding resistance from allowable resistance.

The result may be seen in Figure 21.31.

21.6.7 Concrete Masonry Retaining Wall

The magnitude of concrete masonry for similar conditions is shown in a typical table and details for various wall heights as presented in Figure 21.32.

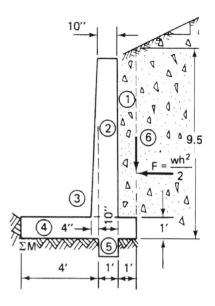

FIGURE 21.30 Loads and critical design steps for design example.

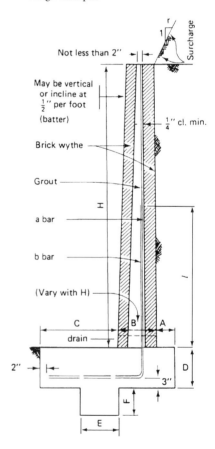

FIGURE 21.31 Typical brick retaining-wall section.

(H)	(t)	(B)	(X) BARS	(Y) BARS
3'	6"	1'-10"	#3 AT 32" O.C.	——
4'	8"	2'-6"	#4 AT 48" O.C.	——
5'	8"	3'-0"	#4 AT 24" O.C.	——
6'	12"	3'-8"	——	#4 AT 24" O.C.
7'	12"	4'-6"	——	#4 AT 16" O.C.
8'	12"	5'-3"	——	#5 AT 16" O.C.

Design for level grade above wall

(H)	(t)	(B)	(X) BARS	(Y) BARS
3'	6"	2'-9"	#3 AT 24" O.C.	——
4'	8"	3'-6"	#4 AT 24" O.C.	——
5'	8"	4'-0"	#5 AT 16 O.C.	——
6'	12"	4'-9"	——	#5 AT 24" O.C.
7'	12"	5'-6"	——	#6 AT 16" O.C.
8'	12"	6'-9"	——	#7 AT 16" O.C.

Design for sloping grade above grade

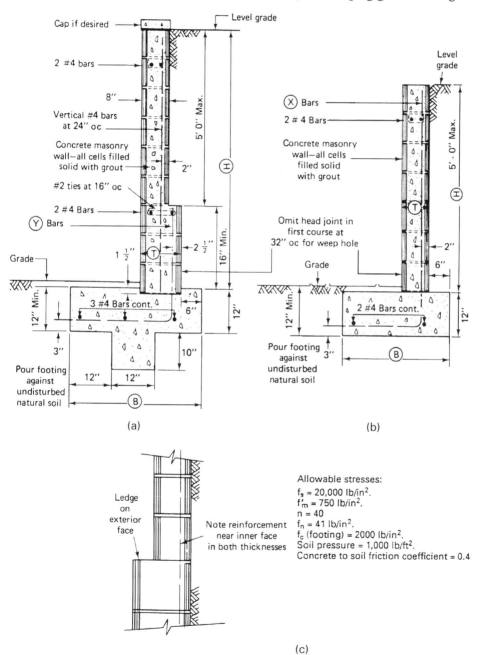

Allowable stresses:
f_s = 20,000 lb/in^2.
f'_m = 750 lb/in^2.
n = 40
f_n = 41 lb/in^2.
f_c (footing) = 2000 lb/in^2.
Soil pressure = 1,000 lb/ft^2.
Concrete to soil friction coefficient = 0.4

FIGURE 21.32 Concrete masonry retaining-wall details. (a) Typical section over 5'–0" 1/2 = 1'–0"; (b) typical section 5'–0" Max. 1/2 = 1'0"; (c) .

21.7 Fire Resistance and Occupancy

Fire resistance, or fire endurance, is a valuable and important property of concrete and masonry and should be used correctly and effectively, such as through the use of newer research and equivalent thickness, with proof. One entirely new provision is the recognition of the fire endurance of masonry veneer.

The International Conference of Building Officials developed a good balance among the various hazards of occupancy as they related to the provisions of fire resistance. It had a good acceptance and service record, but it was rather difficult to apply, even by building officials or plan checkers. The need for clarification brought about the UBC page entitled, Effective Use of the Uniform Building Code (see Figure 21.33). This outlined the steps a plan checker would follow to verify that the occupancy needs were met by the types of construction or fire resistance provided by the project drawings. That requires a slight modification of steps for the designer, summarized as follows:

1. Classify the building.
2. Determine location of the building on the property and the building zone. The distance to property lines determines the minimum rating.
3. Determine (from UBC tables) the occupancy type or use of the facility.
4. Calculate the floor area.
5. Establish the number of stories.
6. Determine the type of construction.
7. Select the appropriate ratings of walls and partitions from the UBC table "Type of Construction."
8. Select the walls to provide the correct rating as given in the table for ratings of walls.

Masonry and concrete have advantages beyond simple fire resistance. They maintain their integrity while providing the required resistance. After exposure, as for 1- and 2-hour endurance, they are generally undamaged. This is opposed to the sacrificial gypsum resistance. A goodly portion of the gypsum resistance is provided by water of crystallization changing to moisture and the heat of vaporization as the moisture changes to steam. The resulting amorphous material has no structural capacity. It may be washed away by a hose stream.

Another advantage to masonry fireproofing is that it does not cause water damage to the building or its contents as do sprinkler systems. Also, the masonry is sure. It is not subject to the hazards of failure from broken pipes, faulty valves, or loss of water supply. Also, it is a more durable material, not subject to puncture or damage from rough usage, and later, repainting costs. Their endurance can be improved easily by coatings such as plaster and wallboard.

Solid brick walls of 8-in, 12-in, 16-in thicknesses, as used in earlier days, had more endurance, intrinsically and by rating, than code requirements, so they were unquestioned. However, thinner walls of 4, 5, and 6-inches were reinforced for more effective use and the evaluation rating became more important.

Concrete block hollow units and hollow clay were excellent for reinforced masonry but did not provide the mass adequate for certain ratings. Other provisions were necessary in some cases.

The *equivalent thickness method* was developed for hollow concrete and hollow clay units and was added to the UBC table. In Addition, a long-time provision became valuable. Applications of cement plaster, gypsum plaster, or wallboard were recognized as important techniques for adjusting the fire endurance of masonry. One and a half times as much masonry was required for equivalence with gypsum. This increased the design flexibility.

Brick veneer 1- and 2-hour ratings were established, verified by test, and approved in ICBO Evaluation Report #5058. Essentially, 2 in of masonry or two gypsum wallboards on each side of

EFFECTIVE USE OF THE UNIFORM BUILDING CODE

The following procedure may be helpful in using the Uniform Building Code:

1. Classify the building:

   A. **OCCUPANCY CLASSIFICATION:** Compute the floor area and occupant load of the building or portion thereof. See Sections 407, 3302 and Table No. 33-A. Determine the occupancy group which the use of the building or portion thereof most nearly resembles. See the '01 sections of Chapters 5 through 12. See Section 503 for buildings with mixed occupancies.

   B. **TYPE OF CONSTRUCTION:** Determine the type of construction of the building by the building materials used and the fire resistance of the parts of the building. See Chapters 17 through 22.

   C. **LOCATION ON PROPERTY:** Determine the location of the building on the site and clearances to property lines and other buildings from the plot plan. See Table No. 5-A and '03 sections of Chapters 18 through 22 for fire resistance of exterior walls and wall opening requirements based on proximity to property lines. See Section 504.

   D. **ALLOWABLE FLOOR AREA:** Determine the allowable floor area of the building. See Table No. 5-C for basic allowable floor area based on occupancy group and type of construction. See Section 506 for allowable increases based on location on property and installation of an approved automatic fire-sprinkler system. See Section 505 (b) for allowable floor area of multistory buildings.

   E. **HEIGHT AND NUMBER OF STORIES:** Compute the height of the building, Section 409 and determine the number of stories, Section 420. See Table No. 5-D for the maximum height and number of stories permitted based on occupancy group and type of construction. See Section 507 for allowable story increase based on the installation of an approved automatic fire-sprinkler system.

2. Review the building for conformity with the occupancy requirements in Chapters 6 through 12.

3. Review the building for conformity with the type of construction requirements in Chapters 17 through 22.

4. Review the building for conformity with the exiting requirements in Chapter 33.

5. Review the building for other detailed code regulations in Chapters 29 through 54, Chapter 56, and the Appendix.

6. Review the building for conformity with engineering regulations and requirements for materials of construction. See Chapters 23 through 28.

FIGURE 21.33 This page was developed to aid building officials and plan checkers verify that occupancy requirements of the UBC were met by plans.

studs, steel, or wood would provide 2-hour endurance. One-hour resistance could be developed by one gypsum board as masonry on each face. The tests proved that 1 in of masonry would provide the resistance required, but there might be difficulty in applying veneer that thin, so E.S. No. 5058 requires thicker veneer.

Some of the possible combinations are shown on the E.S. 5058 (see Figure 21.12), but many others are acceptable. This encourages flexibility and originality in design.

21.8 Field Construction

It is important that the engineer be knowledgable regarding this subject. This is the goal of the concept, the design, the plans, the drawings, and the specifications. The design entity must verify that the intent is accomplished by the end result, the final field construction. Some of the execution of field work items is clarified as follows to help the designer ensure the successful accomplishment of the functional design intent, recognizing that the majority of design personnel may not be highly skilled in hands-on masonry applications.

21.8.1 Mortar

Proportions, mixing, and use are covered under "Materials" (Section 21.4). Proper proportions and good workability will ensure a good tight masonry element. Application specifications have evolved over years of use, but on occasion there may be some deviations. The mortar must be shoved tightly against the masonry unit surfaces, filling the joints of solid unit masonry. It is important that the mortar joints be tooled to make a strong compacted surface and to reduce possibility of leakage.

21.8.2 Grout

Grout must be poured and puddled or vibrated to flow intimately into position without segregation. This requires care in the water content and the handling. Too much water and too much vibration with careless pouring can cause segregation and voids as the water is absorbed by the masonry units.

Grout Aid® must be added for high-lift pour placement just before pouring. If it is added at a mix plant and there is subsequent delay before placing, the expansive property of the admixture may be lost. Excess vibration of the mix may also decrease the effectiveness.

Pouring of grout in high-lift procedures provides good grouting. However, the type of aggregate will influence the workability and fluidity and flow. It is not feasible to control the fine grading and the angularity of the aggregate, the sand, and gravel, so adjustments might be required at the site. The heights limited or permitted for various aggregate sizes by code tables are only guides. (See Table 21.5.)

21.8.3 Hollow Block Units

These are shown under materials. So-called low-lift pouring has been done with 4-ft lifts for many years. This was not determined through engineering study but rather through field usage early on. It is effective for scaffold height, and it may be adjusted with little effect on quality.

Grouting of open-end block must be done carefully, with time allowed for the mortar joints to set fully to prevent a bursting failure due to the hydraulic pressure. One factor often overlooked in hollow concrete block is that the cells are tapered, hence there is a top and a bottom as laid.

21.8.4 Joints

Joints must be tooled. This is especially important in hollow unit construction with raked joints. The tooling compacts the mortar in the joints, making for greater strength and greater water-penetration resistance.

Weather-exposed faces must be waterproofed since the units are generally not adequately resistant and the joints, especially when raked, are not resistant to driven rain.

21.8.5 Solid Concrete Brick

This should be laid with solid "shoved" joints for strength, appearance, tightness, and durability.

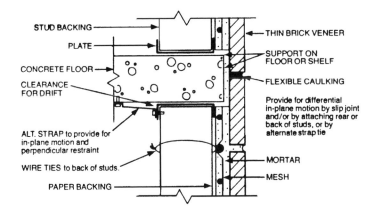

FIGURE 21.34   Adhered veneer.

21.8.6 Solid Clay Brick

This has a long, successful service record. It must be laid with a slight shoving of the units to enhance the mortar-to-unit bond. Highly absorbent brick must be wetted before laying. This not only provides good bond but prevents the dehydration of the mortar before it sets properly.

The units must not be moved after being placed in position because of the hazard of losing bond; the mortar loses plasticity quickly after contact with the unit.

21.8.7 Hollow Clay Brick

This is the newer use of clay brick for reinforced masonry. The laying is similar to hollow concrete block masonry. Since brick is generally several times stronger than concrete block, it is frequently used for structures that require higher strength and more attention to detail. ASTM C-652 arbitrarily considers two types of unit based on amount of void; however, that is meaningless because the design and actual capacity depend on face shell area, unit net strength, reinforcing, and amount of grout added.

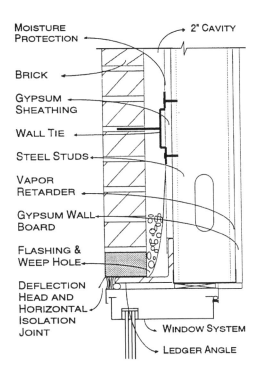

FIGURE 21.35   This drawing shows how this system is explained in the WSCPA design guide on Anchored Brick Veneer Over Steel Studs.

21.8.8 Veneer

Veneer is a method of masonry construction that is frequently overlooked by engineering and control. However, for buildings subject to seismic forces and building deformations it should be observed, for example, to assure that ties are adequate, or even embedded, and that control joints and flashings are installed according to details. (See Figures 21.34 and 21.35.)

21.8.9 Paving

Paving is dependent on the integrity of the supporting submaterial. There must be no material that will decay, disintegrate, or consolidate with time. Slopes for drainage must be at least those shown in the plot, say, one quarter inch per foot.

There must be solid curb-type edge restraints, especially for adjacent planting areas, to prevent root growth under portions of the paving. Field checks should be made of the compaction of the subgrade and foundation material. It should be well compacted at proper moisture content for maximum density. Power tampers are recommended.

21.8.10 Cold Weather

Construction during severe cold is not recommended. However, if schedules require it, proper consideration should be made. Chemical set of cement mortars becomes very slow. They might be over-loaded before setting. This might permit freezing of the contained moisture and a lessening of the final crystalline structure. This is a subject that must be determined for each season in each locale.

21.8.11 Hot Weather

Construction will introduce faster set in hot weather than in normal weather. The initial hazard would be the installation of concrete block that is too hot to handle. This is the only time concrete block should be wet during laying. The wetting is not only to cool the block but to reduce the dehydration of the mortar prior to its initial or chemical set. The masonry might require wetting the first day or two for curing of the mortar or grout before dehydration during set.

21.9 Weather Proofing

Weather proofing is primarily water proofing, that is, water penetration resistance, although insulation for heat flow resistance is a valid weathering condition.

21.9.1 Water Penetration Resistance

This is part of the basic function of exterior walls, and different walls require different detail considerations. Workmanship is an important variable in the effectiveness of treatments.

Solid clay brick walls do not generally fail as barriers and do not generally require coatings or membranes. Water may soak into such a wall and migrate toward the interior but seldom runs through and down the face, especially when joints are full, solid, and well tooled.

Flashing details at openings are important. Parapet walls introduce some hazards. The back faces or surfaces are frequently overlooked, with poor workmanship, multitudinous flashing, and deck connections. The caps must be carefully detailed to prevent entry of driving rain or of water standing on top. Metal caps must be carried far enough down to prevent entry of wind driven rain as it sweeps up and over the top (see Figure 21.36). The provision of a well-troweled mortar or grout type will inhibit such action and may be aesthetically unobtrusive. It must be well coated, however, because it may be subject to cracking due to rich mix and extremes of weather fluctuations.

Flashings and counter flashings must be thoroughly detailed and carefully installed.

For low parapets a simple sound method is to continue the roofing membrane over a cant and up the back and over the top of the low parapet.

Hollow clay brick carries a risk of greater leakage, such as due to, perhaps, flaws in workmanship. A coating will minimize these potential penetrations.

Hollow concrete block should be coated, unless there is a record of good, local performance.

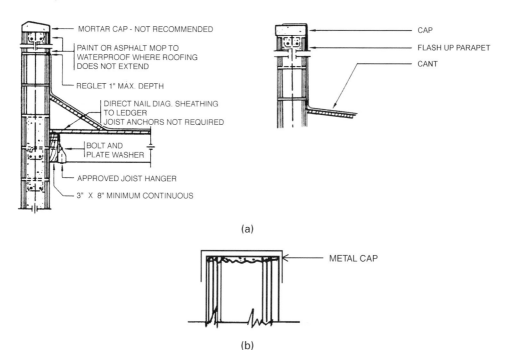

FIGURE 21.36   Parapet cap details.

The increasing use of cost-effective brick veneer on steel studs (BV/SS) utilizes a more sophisticated consideration, as a rainscreen, and space for equalization of pressure and the secondary barrier or drainage wall, the BV/SS system is basically similar to other curtain wall systems.

Leakage occurs because of the presence of rain water on the surface that it flows in sheets in all directions. Lateral movement is greatest near the windward corners and near the top of the building, and water concentrates at surface irregularities.

Some forces that move water through discontinuities are as follows:

a) Gravity can be stopped by proper cap flashing. It is good practice to eliminate horizontal surfaces, such as sills, raked joints, and ledges where water may accumulate and stand or pond.

b) Impact or kinetic energy can be controlled by elimination of any direct path through masonry. When a path exists, such as at a weep hole, care must be taken to provide that the vertical flashing is adequate to dissipate the kinetic energy.

c) Surface tension may be broken by "drips."

d) Capillary action through small passages such as cracks between the unit and mortar can be controlled by introducing a gap or break in the small passage to prevent migration.

e) Differential pressure and air currents are forces that may cause water intrusion. These are minimized by the rainscreen or pressure-equalization concept. The weep holes at the base of the wall and vents at the top allow for the equalization.

f) Air barrier is provided by the interim wall. It need not be perfect but it must resist the flow of air and air pressure.

Anchored veneer details are given in Figure 21.37 and adhered veneers are shown in Figure 21.38 as a good construction practice.

In summary, leakage is due to presence of water, an opening in the wall, and forces that move the water through the wall. The rain screen principle is most important in eliminating the forces causing leakage and draining to the interior.

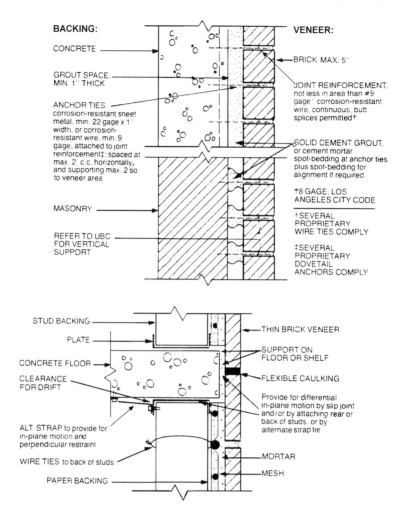

FIGURE 21.37 Anchored veneer details. The Los Angeles City Code permits two veneer features that had not been in the body of the code, but are in special approval. One is that veneer over openings may be supported by bolting to wood lintels, and the other is that the veneer jambs may be considered as structural support for veneer lintels.

Migration or resistance and location of the condensation plane are important in some of the wall material in certain areas. This is generally a detail provision under the responsibility of the architect and mechanical engineer.

21.9.2 Thermal Resistance

This is provided by the calculated amount of insulation and the gypsum wall board. The masonry has a rather low heat-flow resistance but it does have a tendency, through mass, to store or delay heat passage so it is effective in cyclic conditions, for example, hot days and cold nights.

21.9.3 Volume Changes

These are due to temperature change and to moisture expansion and contraction. Joints, called control joints, expansion joints, contraction joints, motion joints, or seismic joints must be provided to relieve the forces that result from restraint. Unless motion is permitted, the forces will be high.

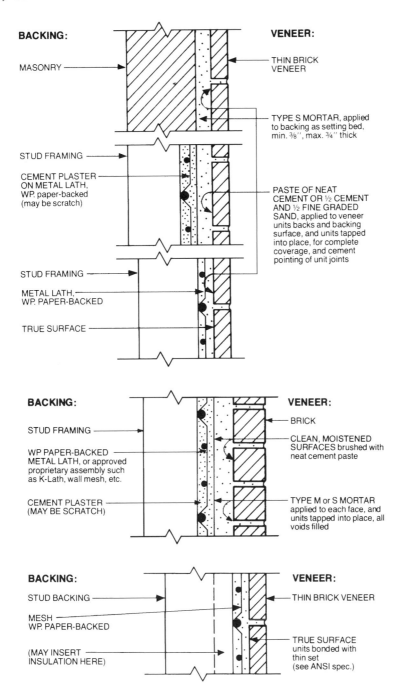

FIGURE 21.38 Adhered veneer.

For example, a bar, free to move, will increase in length with an increase in temperature and decrease in length with a decrease in temperature. The change in length will be $et\ell$, where e is the coefficient of linear expansion, t the change in temperature and ℓ is the length. If the ends of the bar are fixed against motion, a change of temperature, t, will cause a change in the unit stress of Eet and a total of $AEet$, where A is the cross-sectional area of the bar and E is the modulus of elasticity. The examples below indicate the magnitude of the effect.

Example 21.1

What is the change in length of a reinforcing bar 40 ft long due to a 30°F change in temperature, if it is free to move?

$$\text{Change of length} = et\ell = .0000067 \times 30 \times 40$$
$$= .00804 \text{ ft}$$

If the bar is fixed, what is the change in stress?

$$\text{Change in unit stress} = Eet = 30,000,000 \times .0000067 \times 30$$
$$= 6050 \text{ psi}$$

If the masonry were restrained similarly what is the change in stress?

$$\text{Change in stress} = Eet = 1,500,000 \times 0000035 \times 30$$
$$= 1580 \text{ psi}$$

21.9.4 Shoring and Bracing

These factors in masonry are not generally within the scope or responsibility of the design function, that is, the architect or engineer, except for unusual conditions of structure or local wind, etc. This responsibility must be clarified before the start of construction.

21.9.5 Cleaning

Cleaning is an item included in specifications and is important in masonry because of the difficulty. Mortar and grout will bond thoroughly to the units and adjacent construction. The best way to clean is to keep things clean from the outset. Some coatings serve to help keep the desired clean surfaces.

Acid is sometimes used to remove stains. It is effective, but must be used very sparingly, as it will damage tooled mortar joints and is difficult to remove.

21.10 Design

21.10.1 Design Principles

Design principles, or assumptions, can be classed functionally into four types:

Empirical or traditional types are based on ratios of spans for various portions of buildings and some other arbitrary requirements. These are rules that have been practiced for centuries. An example is shown in the city wall of Rothenberg, built in the early 1600s. The thickness, measured through a sally port, compared to the height is the same h/t currently permitted for unreinforced walls.

Incidentally, another modern engineering technique was noted in that old wall. The bottom courses of stone were slightly arched from pilaster pad to pad. As the courses of the upper portion were added, they placed those lower courses in heavy compression, that is, they were prestressed, able to resist the enemy battering ram attack. Prestressing is currently being rediscovered! Empirical design is not currently being used in the western United States.

Arbitrary or prescriptive requirements are types that have been found to be desirable or necessary. One example of this is the arbitrary amount of wall reinforcing required for "reinforced masonry" and "partially reinforced" masonry. The amounts, in both directions, were judicially deemed necessary when reinforced masonry was initiated in the '30s for quake-area construction.

Working stress design is based on the principle of determining the stress in various portions under load, and then limiting the stresses to certain allowable stresses for different material. This is the original method of design used since the development of reinforced masonry in 1933 after the Long Beach earthquake and the widespread UBC adoption and is used in this chapter.

Strength design is a newer method, based on the principle of determining the ultimate strength of a section and then limiting the design capacity to a certain fraction of that by numerical factors. The magnitude of results are similar to those obtained by the working stress design.

21.10.2 Inspection

Inspection is given great emphasis because masonry is a hand-, placed material, and there are so many factors that influence its capacity.

The allowable stress of masonry in use is rather conservative, that is, low fractions of the ultimate strength of the material. If the workmanship is not to be continuously inspected for quality assurance, permitted conservative values are reduced by one-half. This may not be a rational precise theoretical factor, but it is a practical, safe, and perhaps redundant factor.

Tables are provided to aid the selection of working design stresses for various assembly strengths. One of the functions of inspection is the determination of assembly strength. The mortar strength, the grout strength, or the unit strength is not the assembly or masonry strength (see Table 21.11).

The f'_m, the basic design strength, is the strength of the masonry at 28 days. It may be determined by (a) the previous record of strength, (b) assumed from tables giving the probable f'_m given the

TABLE 21.11 Specified Compressive Strength of Masonry, f'_m (psi)* Based on Specifying the Compressive Strength of Masonry Units**

	Specified Compressive Strength of Masonry, f'_m	
Compressive Strength of Clay Masonry Units,*,† psi	Type M or S Mortar,‡ psi (×6.89 for kPa)	Type N Mortar,‡ psi
14,000 or more	5,300	4,400
12,000	4,700	3,800
10,000	4,000	3,300
8,000	3,350	2,700
6,000	2,700	2,200
4,000	2,000	1,600
	Specified Compressive Strength of Masonry, f'_m	
Compressive Strength of Concrete Masonry Units,†,§ psi	Type M or S Mortar,‡ psi (×6.89 for kPa)	Type N Mortar,‡ psi
4,800 or more	3,000	2,800
3,750	2,500	2,350
2,800	2,000	1,850
1,900	1,500	1,350
1,250	1,000	950

*Compressive strength of solid clay masonry units is based on gross area. Compressive strength of hollow clay masonry units is based on minimum net area. Values may be interpolated. When hollow clay masonry units are grouted, the grout shall conform to the proportion in Table 21-B, [in the UBC].

†Assume assemblage. The specified compressive strength of masonry f'_m is based on gross area strength when using solid units or solid grouted masonry and net area strength when using ungrouted hollow units.

‡Mortar for units masonry, proportion specification, as specified in Table 21-B, [in the UBC]. These values apply to portland cement-lime mortars without added air-entraining materials.

§Values may be interpolated. In grouted concrete masonry, the compressive strength of grout shall be equal to or greater than the compressive of the concrete masonry units.

**The above table is the UBC table that relates f'_m to masonry unit strength.

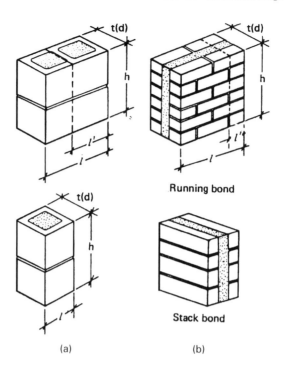

FIGURE 21.39 Masonry assemblage test prisms. These are to be built with the same masonry materials, using similar methods of construction and workmanship as in the masonry project, except that stack bond may be used for hollow unit prisms. The test prisms may be shortened to a length of not less than the wall thickness so as to adjust to a smaller-capacity testing machine. (a) Hollow concrete or clay block; (b) two-wythe clay brick.

combinations of unit strength and type of mortar, or (c) verification by test prisms of the masonry intended for the job. This requires that specimens be made at least 28 days early. Some manufacturers will have reports of previous tests that may be a guide to the f'_m available. Otherwise, the engineer designs to a specific f'_m and requires the prism tests to verify that strength of the assembly being installed is "not less than ..."

The tests of mortar and grout are not necessary if the prism tests are adequate. However, they are desirable for further f'_m verification.

Prism testing is an approximate method of determining f'_m of a specific masonry assemblage. The prism should be built at the site, preferably by the workman in the same manner as he will construct the wall. Its shape will be as shown in Figure 21.39 and Table 21.12.

TABLE 21.12 Compressive Strength of Masonry

Prisms h/t_p*	1.30	1.50	2.00	2.50	3.00	4.00	5.00
Correction factor	0.75	0.86	1.00	1.04	1.07	1.15	1.22

*h/t_p—ratio of prism height to least actual lateral dimension of prism.

Note: The compressive strength of masonry [psi (kPa)] for each set of prisms shall be the lesser of the average strength of the prisms in the set, or 1.25 times the least prism strength multiplied by the prism height-to-thickness correction factor from Table 21-17-A, [in the UBC]. Where a set of grouted and nongrouted prisms are tested, the compressive strength of masonry shall be determined for the grouted set and for the nongrouted set separately. Where a set of prisms is tested for each wythe of a multiwythe wall, the compressive strength of masonry shall be determined for each wythe separately.

The accepted correction owing to height variation is shown, for conversion of test results to consistent design f'_m.

There is a large scatter of results when testing the presumably identical material samples, so groups of five prism specimens are made. Tests should be made for each story and not less than one set for each 5000 square feet laid.

A large amount of masonry is used at low levels of stress so a great deal is installed under design at half stress, that is, without continuous inspection.

Inspection is a very complex and important aspect of masonry, requiring a diverse understanding. The inspector should not only know basic engineering facts but also understand the field operations for a hand-placed material, including basic variable ingredients, mixing, incorporating, combination, weather, curing, etc. Some of field operations have changed with the introduction of seismic engineering into the longtime customary procedures.

21.10.3 Arbitrary, Seismic Provisions

Arbitrary items were introduced in the almost hysterical birth of reinforced masonry, many based on "judgment" from the early studies and observations of quake damage. It was recognized that masonry must be reinforced. Reasonable amounts were determined and "reinforced masonry" was defined as in the 1994 edition of the UBC as part of Special Provisions for Seismic zones 3 and 4.

Reinforcement. The portion of the reinforcement required to resist shear shall be uniformly distributed and shall be joint reinforcement, deformed bars or a combination thereof. The spacing of reinforcement in each direction shall not exceed one half the length of the element, nor one half the height of the element, nor 48 inches (1219 mm).

Joint reinforcement used in exterior walls and considered in the determination of the shear strength of the member shall be hot-dipped galvanized in accordance with U.B.C. Standard 21-10.

Reinforcement required to resist in-plane shear shall be terminated with a standard hook as defined in Section 2107.2.2.5 or with an extension of proper embedment length beyond the reinforcement at the end of the wall section. The hook or extension may be turned up, down or horizontally. Provisions shall be made not to obstruct grout placement. Wall reinforcement terminating in columns or beams shall be fully anchored into these elements.

Bond. Multiwythe grouted masonry shear walls shall be designed with consideration of the adhesion bond strength between the grout and masonry units. When bond strengths are not known from previous tests, the bond strength shall be determined by tests.

Wall reinforcement. All walls shall be reinforced with both vertical and horizontal reinforcement. The sum of the areas of horizontal and vertical reinforcement shall be at least 0.002 times the gross cross-sectional area of the wall, and the minimum area of reinforcement in either direction shall be not less than 0.0007 times the gross cross-sectional area of the wall. The minimum steel requirements for Seismic Zone 2 in Section 2106.1.12.3, Items 2 and 3, may be included in the sum. The spacing of reinforcement shall not exceed 4 feet (1219 mm). The diameter of reinforcement shall not be less than 3/8 inch (9.5 mm) except that joint reinforcement may be considered as a part or all of the requirement for minimum reinforcement. Reinforcement shall be continuous around wall corners and through intersections. Only reinforcement which is continuous in the wall or element shall be considered in computing the minimum area of reinforcement. Reinforcement with splices conforming to Section 2107.2.2.6 shall be considered as continuous reinforcement.

Stack bond. Where stack bond is used, the minimum horizontal reinforcement ratio shall be $0.0015bt$. Where open-end units are used and grouted solid, the minimum horizontal reinforcement ratio shall be $0.0007bt$.

Reinforced hollow-unit stacked bond construction which is part of the seismic-resisting system shall use open-end units so that all head joints are made solid, shall use bond beam units to facilitate the flow of grout and shall be grouted solid.

A new term was introduced, "partially reinforced masonry." The intent was to enable the use of reinforcing in otherwise unreinforced masonry for special loadings without having to use all the requirements of reinforced masonry.

The many special provisions now contained in the pages of the UBC masonry section require continual reference. They contain many detail requirements compared to the initial seven pages of the initial reinforced masonry portions of the 1930s.

21.10.4 Loadings

Loads on a structure are wind, gravity, live and dead, seismic, impacts, volume changes, all listed in the appropriate applicable code sections, Design resistance is outlined herein.

21.10.5 Allowable Stresses

Allowable stresses for elastic or working stress design of the masonry due to the loads are listed as fractions of the masonry strength, or f'_m.

These allowable stresses are permitted a 1/3 increase for resistance to wind or seismic loads. Higher values may be used for the 1994 UBC. (See Table 21.13.) The allowable stresses for unreinforced masonry are as follows:

Axial compression

$$Fa = .25 f'_m \left[1 - \frac{(h')^2}{140r} \right] \quad \text{for } h'/r \leq 99$$

$$Fa = .25 f'_m \left(\frac{70r}{h'} \right)^2 \quad \text{for } h'/r > 99$$

Flexural compression

$$F_b - 0.33 f'_m, 2000 \text{ psi maximum}$$

Combined axial and flexural stress, unity formula

$$\frac{fa}{Fa} + \frac{fb}{Fb} \leq 1 \text{ (see refinement of this basic formula)}$$

21.10.6 Flexural Design

Flexural design is based on the following assumptions:

1. Plane sections before bending remain plane during and after bending.
2. Stress is proportional to strain.
3. Masonry elements combine to form a homogeneous material.
4. No tension is carried by masonry.
5. Composite masonry design is based on elastic transformed section.
6. Summation H, V, and M equal 0.

These concepts are represented in Figure 21.40. Mathematically, this is demonstrated by the drawing of stress–strain force patterns in Figure 21.41.

The assumption that the masonry carries no tensile stresses implies a cracked section. This sometimes leads to erroneous conclusions. For instance, where the steel is located in the center of a wall, the wall would necessarily have to crack through more than one-half the wall thickness in order for the steel to resist the tensile stresses. It is not likely that cracking occurs to this extent. Therefore it would seem somewhat more logical to assume that the masonry does take some tensile stress. However, owing to the uncertainty involved in measuring whatever tensile strength masonry may possess, it is not depended on in the design process. The principle of the transformed section is resorted to in analyzing the internal mechanics at a section. Figure 21.41 shows the straight-line stress and strain distribution patterns assumed to exist in the cracked section. From the internal force system shown in Figure 21.40, assuming that equilibrium obtains throughout, the necessary design and analysis formulas are developed. As previously noted, it is assumed that all the masonry on the tensile side of the neutral axis has cracked and is therefore structurally nonexistent. Thus the transformed section area consists of masonry in compression on one side of the neutral axis and n times the steel area on the other side.

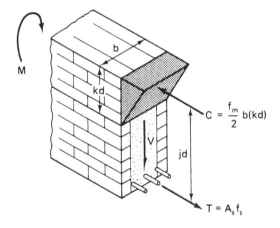

FIGURE 21.40 Internal force system for a cracked system.

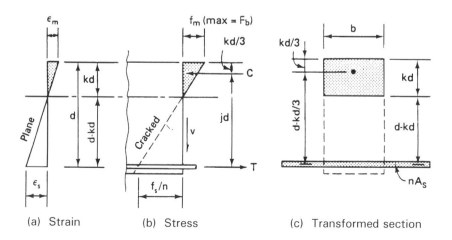

FIGURE 21.41 Stress-strain patterns for cracked section design assumption.

TABLE 21.13 Maximum Allowable Working Stresses for Reinforced Solid and Hollow Unit Masonry (psi) Based on Editions Subsequent to the 1985 Edition UBC*

Type of Stress	Coefficients f'_m — No	Coefficients f'_m — Yes	1350 No	1350 Yes	1500 No	1500 Yes	1800 No	1800 Yes	2000	2500	3000‡	3500	4000	4500	5000	5300§	6000
Special Inspection Required:	No	Yes	No	Yes	No	Yes	No	Yes	Yes	Yes	Yes	Yes	Yes	Yes	Yes	Yes	Yes
Compression: Axial																	
Walls	$0.1 f'_m$	$0.2 f'_m$	135	270	150	300	180	360	400	500	600	700	800	900	1000	1060	1200
Columns	$0.1 f'_m$	$0.2 f'_m$	135	270	150	300	180	360	400	500	600	700	800	900	1000	1060	1200
Compression: Flexural	$0.165 f'_m$	$0.33 f'_m$	225	450	250	500	300	600	667	833	1000	1167	1333	1500	1667	1767	2000
Shear																	
No shear reinforcement																	
Flexural††	$(\times\frac{1}{2})$ 25 max	$1.0\sqrt{f'_m}$ 50 max	18	36	19	39	21	42	45	50**	50	50	50	50	50	50	50
Shear walls‡‡																	
§§ $M/Vd < 1$	$(\times\frac{1}{2})\begin{cases}\frac{1}{3}(4 - M/Vd)\sqrt{f'_m}\\(80 - 45M/Vd)\text{ max}\end{cases}$	$\begin{cases}\frac{1}{3}(4 - M/Vd)\sqrt{f'_m}\\(80 - 45M/Vd)\text{ max}\end{cases}$	See charts for numerical values, or calculate, checking for max														
§§ $M/Vd \geq 1$	$(\times\frac{1}{2})$ 17 max**	$1.0\sqrt{f'_m}$ 35 max	17	35	17	35	17	35	35	35	35	35	35	35	35	35	35
Reinforcing all shear																	
Flexural††	$1.5\sqrt{f'_m}$ 75 max	$3.0\sqrt{f'_m}$ 150 max	55	110	58	116	64	127	134	150**	150	150	150	150	150	150	150
Shear walls††																	
§§ $M/Vd < 1$	$(\times\frac{1}{2})\ \frac{1}{2}(4 - M/Vd)\sqrt{f'_m}\ (120 - 45M/Vd)\text{ max}$		See charts for numerical values, or calculate, checking for max														
§§ $M/Vd \geq 1$	$(\times\frac{1}{2})$ 37 max	$1.5\sqrt{f'_m}$ 75 max	28	55	29	58	32	64	75**	75	75	75	75	75	75	75	75

Modulus of elasticity***	$750 f'_m$	$750 f'_m$	1012	1012	1125	1125	1350	1350	1500	1875	2250	2625	3000	3000	3000	3000	3000
Modular ratio, $n\ E_s/E_m$	$38667 f'_m$	$38667 f'_m$	28.6	28.6	25.8	25.8	21.5	21.5	19.3	15.5	12.9	11.0	9.7	9.7	9.7	9.7	9.7
Modulus of rigidity***	$400 f'_m$	$400 f'_m$	540	540	600	600	720	720	800	1000	1200	1400	1500	1800	2000	2100	2400
Bearing on full area†††	$0.13 f'_m$	$0.26 f'_m$	176	351	195	390	234	468	520	650	780	910	1040	1170	1300	1378	1560
Bearing on $\frac{1}{3}$ or less of area†††	$0.19 f'_m$	$0.38 f'_m$	257	513	285	570	342	684	760	950	1140	1330	1520	1710	1900	2014	2280
Bond Plain bars	30 psi	60 psi**	30	60	30	60	30	60	60	60	60	60	60	60	60	60	60
Deformed	100 psi	200 psi	100	200	100	200	100	200	200	200	200	200	200	200	200	200	200

Note: The allowable reinforcing values are: Deformed bars, $0.5 f_y$, 24,000 psi maxim um; Wire reinforcement $0.5 f_y$, 30,000 psi maximum; Ties, anchors and smooth bars $0.4 f_y$, 20,000 maximum; Compressive stress F_s in columns $.4 f_y$, 24,000 maximum; Compressive stress, deformed bars in flexure, $F_s = 0.5 f_y$ 24,000 psi maximum; The above allowable stresses may be increased by 1/3 for stresses due to wind or seismic.

*Stresses for hollow unit masonry are based on net section.

†Special testing to establish f_m shall be in accordance with either (1) the prism test method, (2) prism test record, or (3) the unit strength method.

‡Ultimate compressive strength may be assumed from the UBC table up to an $f'_m = 3,000$ psi for concrete masonry units. Where half stresses are used, the f'_m is limited to 1500 psi.

§Ultimate compressive strength may be assumed from the UBC table up to an $f'_m = 5,300$ psi for clay masonry units. Where half stresses are used, the f'_m is limited to 2600 psi.

**A heavy vertical line indicates the limit of allowable stresses by the UBC. Values may be exceeded if local jurisdiction permits.

††Web reinforcement shall be provided to carry the entire shear in excess of 20 psi whenever there is required negative reinforcement and for a distance of one-sixteenth the clear span beyond the point of inflection.

‡‡When calculating shear or diagonal tensile stresses in shear walls that resist seismic forces, use 1.5 times the force required by UBC chapter.

§§ M is the maximum bending moment occurring simultaneously with the shear load V at the section under consideration. Interpolate by straight line for M/Vd values between 0 and 1.

***Where determinations involve rigidity considerations in combination with other materials or where deflections are involved, the moduli of elasticity and rigidity noted may be used, but it is preferable that tests be made to determine the correct values. Although the limit of 3000 kips/in² may be exceeded, a higher value is not recommended unless substantiated by tests.

†††This increase shall be permitted only when the least distance between the edges of the loaded and unloaded areas is a minimum of one-fourth of the parallel side dimension of the loaded area. The allowable bearing stress on a reasonable concentric area greater than one-third, but less than the full area, shall be interpolated between the values given.

Design of rectangular sections consists of estimating the section and analyzing it using the demonstrated equations with mechanical aids. These are shown in tables for clarity, although there are many computer programs that permit one to guess and then check very rapidly. Charts, with the precomputed data and curves, actually give the answer as shown in the following outline.

21.10.6.1 Procedure for Design of Reinforced Flexural Stresses

1. Determine the applied load diagrams for P, V, and M.
2. Select the material, its usable strength, f'_m, and consequent allowable stresses.

$$E = 750\, f'_m; \qquad n = 30,000,000/E\,m; \qquad fm = .2\, f'_m.$$

See Table 21.14.

$$fs = \quad ; \qquad v = \quad ; \qquad u = \quad ;$$

$$k = \frac{1}{1 + f_s/nFm}; \qquad j = 1 - k/3; \qquad K = fmkj/2$$

For preliminary design a close estimate is $j = .9$ and $k = .3$.

3. Estimate the size of member

$$M = Kbd^2 \quad \text{or} \quad bd^2 = M/K$$

4. Estimate amount of reinforcing steel

$$As = M/fsjd \quad (j = .9)$$

5. Design member for shear

$$v = V/bjd, \qquad s = f_s A_v/vb \quad \text{or} \quad A_v = vbs/fs$$

For tee sections, replace b with b' (stem).

6. Determine pn

$$p = A/bd; \qquad np = \left(k = \sqrt{2np + (np)^2} - np;\ j = 1 - k/3\right)$$

Select k and j and $2/kj$ from Table 21.14 for flexural coefficients for the np value.

7. Calculate consequent stresses fm in masonry and fs in steel.

$$fm = 2M/bd^2kj \qquad f_s = M/A_s\, jd$$

8. Design tension steel for bond, anchorage, and spacing.

$$u = V/E_0\, jd$$

9. Provide the detail drawings.

TABLE 21.14 Table of Flexure Coefficients

$$p = \frac{A_s}{bd} \qquad n = \frac{E_s}{E_m} \qquad k = \sqrt{2np + (np)^2} - np$$

$$j = 1 - \frac{k}{3} \qquad f_m = \frac{M}{bd^2}\left(\frac{2}{kj}\right) \qquad f_s = \frac{M}{A_s\, jd}$$

np	k	j	2/kj	np	k	j	2/kj	np	k	j	2/kj
0.010	0.132	0.956	15.93	0.115	0.378	0.874	6.04	0.190	0.455	0.848	5.18
0.015	0.159	0.947	13.30	0.120	0.384	0.872	5.97	0.195	0.459	0.847	5.14
0.020	0.181	0.940	11.76	0.125	0.390	0.870	5.89	0.200	0.463	0.846	5.11
0.025	0.200	0.933	10.73	0.130	0.396	0.868	5.82	0.220	0.479	0.840	4.96
0.030	0.217	0.928	9.95	0.135	0.402	0.866	5.75	0.240	0.493	0.835	4.85
0.035	0.232	0.923	9.36	0.140	0.407	0.864	5.68	0.260	0.507	0.831	4.75
0.040	0.246	0.918	8.89	0.145	0.413	0.862	5.62	0.280	0.519	0.827	4.66
0.050	0.270	0.910	8.14	0.150	0.418	0.861	5.56	0.300	0.531	0.823	4.58
0.060	0.291	0.903	7.60	0.155	0.423	0.859	5.51	0.350	0.557	0.814	4.41
0.070	0.311	0.897	7.19	0.160	0.428	0.857	5.46	0.400	0.580	0.807	4.27
0.080	0.328	0.891	6.85	0.165	0.433	0.856	5.41	0.450	0.600	0.800	4.17
0.090	0.344	0.885	6.58	0.170	0.437	0.854	5.36	0.500	0.618	0.794	4.07
0.100	0.358	0.880	6.34	0.175	0.442	0.853	5.31	0.550	0.634	0.788	4.00
0.105	0.365	0.878	6.24	0.180	0.446	0.851	5.26	0.600	0.649	0.784	3.94
0.110	0.372	0.876	6.14	0.185	0.451	0.850	5.22	0.700	0.675	0.775	3.82

21.10.7 Compression

Compression design of masonry is the use of one of the properties of concrete and masonry construction: good values for axial stress resistance. However, it is a property or capacity with poor correlation between theoretical calculations and actual field capacities. Small differences in material and accidental eccentricities have appreciable effects.

Wall bearing capacity is represented by the formula

$$F_a = 0.25\, f'_m \left[1 - \left(\frac{h'}{140r}\right)^2 \right] \quad \text{for } h'/r \le 99$$

$$F_a = 0.25\, f'_m \left(\frac{70r}{h'}\right)^2 \quad \text{for } h'/r > 99$$

This uses a conservative .25 f'_m factor for stress, that is, 1/4 of the crushing strength for the masonry, and omits the rather small contribution of the wall steel. Then the allowable is reduced by a reduction factor, intended to approach the reduction of Euler's equation, that is

$$\frac{P}{A} = \frac{\pi^2 E_m}{(kh/r)^2}$$

There are code changes pending for the inaccurate reduction factors in several codes. Some of the earlier ones were as shown in Figure 21.42.

These compressive reduction factors refer to effective height and provision is made, such as:

However the values in Figure 21.42 depend on a definite location of a point of inflection. These values are difficult to use and to predict for actual structures unless the location of the point of inflection is forced by a construction detail, for example, deeply raked joints or special steel splices (not recommended).

K values

	(a)	(b)	(c)	(d)	(e)	(f)
Buckled shape of column is shown by dashed line	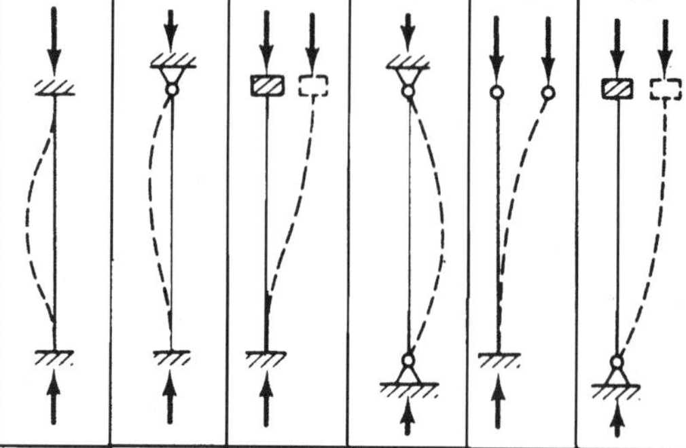					
Theoretical K value	0.5	0.7	1.0	1.0	2.0	2.0
Recommended design value when ideal conditions are approximated	0.65	0.80	1.2	1.0	2.10	2.0

FIGURE 21.42   Effective column length.

Column Loads are limited by the following equation. For reinforced masonry columns, the allowable axial compressive force P_a shall be determined as follows:

$$P_a = \left[0.25\, f'_m A_e + 0.65\, A_s F_{sc}\right]\left[1 - \left(\frac{h'}{140r}\right)^2\right] \quad \text{for } h'/r \leq 99$$

$$P_a = \left[0.25\, f'_m A_e + 0.65\, A_s F_{sc}\right]\left(\frac{70r}{h'}\right)^2 \quad \text{for } h'/r > 99$$

The column formula of the UBC, 1994 edition, combines the allowable stress capacity of the masonry with the allowable steel capacity, all reduced by a hypothetical reduction factor when steel is tied by confining standard ties, both with an arbitrary so-called buckling reduction factor; that is, for reinforced masonry columns, the allowable axial compressive force P_a shall be determined as follows:

Determine the reduction factor R, using tables of radius of gyration. Divide applied load P by R to determine the unreduced capacity required. Select suitable size and reinforcing (between 1% and 4%).

The range of area of column reinforcing is from 1% to 4%. A simple old-fashioned time saver is to use a table such as Table 21.15, which has been used in textbooks, design manuals, and design method publications. The table gives precomputed sizes of columns, the range of steel capacity, and the capacity of masonry for various strengths, f'_m, of some standard column sizes.

Although the masonry column dimensions are stated to be limited by a ratio of 1 to 3, they may be made greater by simply considering several columns side by side and reinforcing accordingly.

21.10.8 Combined Loading

Elements subjected to combined axial and flexural stresses shall be designed in accordance with accepted principles of mechanics or in accordance with the unity formula

$$\frac{fa}{Fa} + \frac{fb}{Fb} \leq 1$$

TABLE 21.15 Masonry Column Capacities and Reinforcement Areas

Load Carried by Reinforcing (at 0.65 × 20,000)				Nominal Sizes (in) Within $\frac{1}{2}$ in of Actual Sizes	A_g Net Area, in	Load Carried by Masonry Area, f'_m Values at .25 f'_m, kips									
Max., at 4%		Min., at $\frac{1}{2}$%				675	750	1350	1500	2000	2500	3000	3500	4000	5000
A_s at 4%	P, kips	A_s at $\frac{1}{2}$%	P, kips			169	188	337	375	500	625	750	875	1000	1250
3.2	42	0.40	5.2	8 × 12	80	13	15	27	30	40	50	60	70	800	100
4.0	52	0.50	6.5		100[a]	16.9	18.8	33.7	37.5	50	62.5	75.0	87.5	100	125
4.6	60	0.57	7.7	8 × 16	115	19	21	39	43	51	72	86	100	115	143
4.8	62	0.60	7.8		120	20	22	40	45	53	75	90	105	120	150
5.2	67	0.65	8.4	12 × 12	130	22	24	44	49	58	82	97	114	130	162
6.2	78	0.75	9.7		150	25	28	50	56	67	94	112	131	150	187
7.2	93	0.90	11.7	12 × 16 or 8 × 24	180	30	34	61	68	80	112	135	157	180	225
8.0	104	1.0	13.0		200	34	37	67	75	89	125	150	175	200	250
9.6	124	1.2	15.6	16 × 16	240	41	45	81	90	107	150	180	210	240	300
10.8	140	1.3	17.5	12 × 24	270	46	51	91	100	120	169	202	237	270	337
12.0	156	1.5	19.5	16 × 20	300	51	56	100	112	134	187	225	263	300	375
14.4	187	1.8	23.4	16 × 24	360	61	67	120	135	160	225	270	315	360	450
15.2	197	1.9	24.7	20 × 20	380	64	71	128	142	169	237	284	333	380	475
16.0	208	2.0	26.0		400	67	75	135	150	178	250	300	350	400	500
18.4	239	2.3	30.0	20 × 24	460										
20.0	260	2.5	32.5		500	84	94	169	188	223	313	374	438	500	625
22.0	286	2.7	35.7	24 × 24	550	93	100	186	206	245	343	411	482	550	688

[a]The values for 100 in^2 are carried past the decimal point, but such precision is not valid when considering the design values compared to true column

in which

> fa = computed axial stress due to design axial load
> Fa = allowable average axial compressive stress for centroidally applied axial load only
> fb = computed bending stress in the extreme fiber due to design bending loads only
> Fb = allowable flexural compressive stress if member were carrying bending only

This is a simple conservative statement, that is, the percent the section is developed in compression plus the percent it is developed in bending shall not be more than 100% developed. However, the code rationally provides that design may be according to "accepted principles of mechanics," a commendably sound engineering provision.

This leaves open the basic design method, that is, guess and then analyze, or "successive approximation" to verify suitability or use precomputed tables or computer methods by estimated calculation or by the example for an 8″ clay brick wall, which follows in Table 21.16.

The unity equation would result in a straight line chart from the value of full compression and zero moment to the point of full moment capacity with zero compression, as in Figure 21.43. However, the assumptions and methods of mechanics can be used to verify greater capacities and obtain a curve such as in Figure 21.43.

The change in UBC 1994 edition values for the buckling factor from the rather simple factors of earlier editions has introduced use of r, the radius of gyration. It may have a theoretical advantage, but it introduces more, rather meaningless, calculation. In order to simplify these, the following tables are included in the 1994 UBC as Table 21-H-1, 21-H-2, and 21-H-3. (See Tables 21.17, 21.18, and 21.19.)

Figure 21.44 illustrates some of the crudities of the attempts to create an axial compression reduction or buckling equations. The 1967 and 1985 UBC editions' equations were far under Euler's

TABLE 21.16   Interaction Curve Calculations for 8-in Clay Block

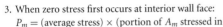

$$f'_m = 2500 \text{ psi}, \ F_b = 8.33 \text{ psi}, \ F_a = 500 \text{ psi}, \ A_m = 90 \text{ in}^2/\text{ft}, \ A_s = 0 \ 20 \text{ in}^2/\text{ft}$$

1. Maximum allowable axial stress—no bending
   (ave. stress = 500 psi):
   $P_m = A_m \times F_a = 90 \times 500$ = 45,000 lb
   M_m = 0

2. Maximum axial stress + bending sufficient to achieve
   maximum allowable flexural compressive stress (ave. stress
   still = 500 psi):
   $P_m =$ = 45,000 lb
   $M_m = (F_b - F_a)\dfrac{bd^2}{6} = 333 \times 12 \times \dfrac{7.5^2}{6}$ = 38,461 in-lb

3. When zero stress first occurs at interior wall face:
   $P_m = (\text{average stress}) \times (\text{portion of } A_m \text{ stressed in}$
   $\text{compression}) = (833/2)(90 \times 1)$ = 37,485 lb
   $M_m = P_m \times (\text{distance from wall } C_L \text{ to}$
   $P_m = 37,485(2/3 \times 7.5 - 7.5/2)$ = 46,855 in-lb

4. When zero stress occurs midway between wall C_L and interior face:
   $P_m = 833/2 \times (90 \times 3/4)$ = 28,113 lb
   $M_m = 28113 \times \left(2/3\,(7.5 - 7.5/4) - \dfrac{7.5}{4}\right)$ = 52,572 in-lb

5. When zero stress occurs at the wall C_L:
   $P_m = 833/2\,(90 \times 1/2)$ = 18,740 lb
   $M_m = 18,740 \times (2/3 \times 7.5/2)$ = 46,855 in-lb

6. When zero stress occurs 1/3 of distance between wall C_L
   and exterior face. From this loading on, steel stressed
   in tension:
   $P_m = 833/2\,(90 \times 1/3)$ = 12,370 lb
   $P_s = (\text{masonry stress }(a\ C_L)\ (n)\ (\text{steel area})$
   $\quad = 833/2 \times 12 \times 0.20$ = −1,000
   $\therefore P =$ $\overline{11,370 \text{ lb}}$
   $M_m = 12,370\left(1/3 \times \dfrac{7.5}{2} + 2/3\,(2/3 \times \dfrac{7.5}{2})\right)$ = 36,120 in-lb

7. When zero stress occurs midway between wall C_L and
   exterior face:
   $P_m = \dfrac{833}{2} \times 90 \times 1/4$ = 9,371
   $P_s = 833 \times 12 \times 0.20$ = −1,999
   $\therefore P =$ $\overline{7,372 \text{ lb}}$
   $M_m = 9371\left(\dfrac{7.5}{4} + 2/3\dfrac{7.5}{4}\right)$ = 29,285 in-lb

8. When zero stress occurs at 2/3 of distance between wall
   C_L and exterior face:
   $P_m = \dfrac{833}{2} \times 90 \times 1/6$ = 6,247 lb
   $P_s = 1667 \times 12 \times 0.20$ = −3,998
   $\therefore P =$ $\overline{2,249 \text{ lb}}$
   $M_m = 6247\left(\dfrac{2}{3} \times \dfrac{7.5}{2}\right) + \dfrac{2}{3}\left(\dfrac{1}{3}\dfrac{7.5}{2}\right)$ = 20,804 in-lb

9. Maximum allowable moment—no axial load:
   $\rho = \dfrac{0.20}{12 \times 3.75} = 0.0044, \ \rho n = 0.0533$
   $\therefore j = 0.91$ (from Table C-1)
   $M = 0.20 \times 20\,(0.91 \times 3.75)$ = 13.65 in kips

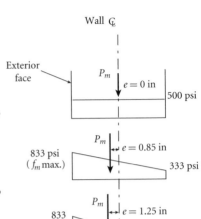

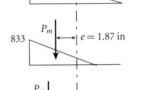

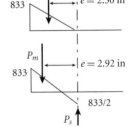

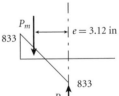

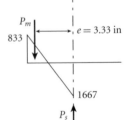

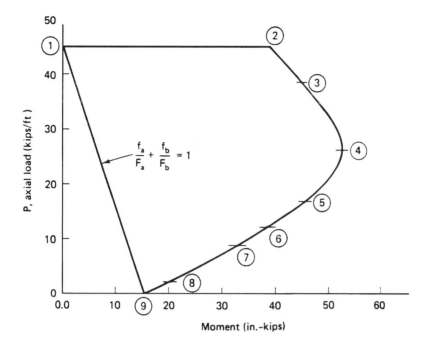

FIGURE 21.43 Interaction curve for 8-in hollow clay block wall. Working stress design.

curve, with no regard for E or exponential variation. The 1994 equation was an attempt, theoretical and mathematical, to approach a consideration of radius of gyration in lieu of merely considering thickness, because the radius of gyration changes in hollow units. On the plot in Figure 21.44 the effect of the differences between hollow and solid units are schematically shown for $f'_m = 2000$ on the h/t coordinate, still without consideration of E, which can vary by code from 1,000,000 to 3,000,000. The curve for Euler's probable buckling is shown for $f'_m = 2000$ and 4000 psi.

TABLE 21.17 Radius of Gyration* for Concrete Masonry Units[†]

Grout Spacing (inches) ×25.4 for mm	Nominal Width of Wall (inches) ×25.4 for mm				
	4	6	8	10	12
Solid grouted	1.04	1.62	2.19	2.77	3.34
16	1.16	1.79	2.43	3.04	3.67
24	1.21	1.87	2.53	3.17	3.82
32	1.24	1.91	2.59	3.25	3.91
40	1.26	1.94	2.63	3.30	3.97
48	1.27	1.96	2.66	3.33	4.02
56	1.28	1.98	2.68	3.36	4.05
64	1.29	1.99	2.70	3.38	4.08
72	1.30	2.00	2.71	3.40	4.10
No grout	1.35	2.08	2.84	3.55	4.29

Note: This table corresponds to Table 21-H-1 of the UBC.

*For single-wythe masonry or for an individual wythe of a cavity wall.

$r = \sqrt{I/A_e}$

[†]The radius of gyration shall be based on the specified dimensions of the masonry units or shall be in accordance with the values shown which are based on the minimum dimensions of hollow concrete masonry unit face shells and webs in accordance with U.B.C. Standard 21-4 for two cell units.

TABLE 21.18 Radius of Gyration* for Clay Masonry Unit Length, 16 Inches[†]

Grout Spacing (inches) ×25.4 for mm	Nominal Width of Wall (inches) ×25.4 for mm				
	4	6	8	10	12
Solid grouted	1.06	1.64	2.23	2.81	3.39
16	1.16	1.78	2.42	3.03	3.65
24	1.20	1.85	2.51	3.13	3.77
32	1.23	1.88	2.56	3.19	3.85
40	1.25	1.91	2.59	3.23	3.90
48	1.26	1.93	2.61	3.26	3.93
56	1.27	1.94	2.63	3.28	3.95
64	1.27	1.95	2.64	3.30	3.97
72	1.28	1.95	2.65	3.31	3.99
No grout	1.32	2.02	2.75	3.42	4.13

Note: This table corresponds to Table 21-H-2 of the UBC.

TABLE 21.19 Radius of Gyration* for Clay Masonry Unit Length, 12 Inches[†]

Grout Spacing (inches) ×25.4 for mm	Nominal Width of Wall (inches) ×25.4 for mm				
	4	6	8	10	12
Solid Grouted	1.06	1.65	2.24	2.82	3.41
12	1.15	1.77	2.40	3.00	3.61
18	1.19	1.82	2.47	3.08	3.71
24	1.21	1.85	2.51	3.12	3.76
30	1.23	1.87	2.53	3.15	3.80
36	1.24	1.88	2.55	3.17	3.82
42	1.24	1.89	2.56	3.19	3.84
48	1.25	1.90	2.57	3.20	3.85
54	1.25	1.90	2.58	3.21	3.86
60	1.26	1.91	2.59	3.21	3.87
66	1.26	1.91	2.59	3.22	3.88
72	1.26	1.91	2.59	3.22	3.88

Note: This table corresponds to Table 21-H-3 of the UBC.

However, the code equations give a method of calculation to serve as a conservative tangible guide; untested, cumbersome, and unproven though they are. It might also be pointed out that the use of the hollow unit cross section may not provide a true consideration because the face shells only are mortared to transmit stress. The cross webs would be non-functionally existent. However, the effect of this would be to show an increased radius of gyration for ungrouted hollow units.

Curves show the effect of h/t on Euler's curve, which suggests the probable buckling stress related to h/t for a compressive element with $f'_m = 2000$ psi. If the f'_m were 4000 psi, the curve would be located as shown by the dotted line. The R, or reduced value, for the element permitted in earlier codes, for example, 1967, do not approach these, even approximately. This emphasizes the lack of information on the compression capacities of masonry walls and columns.

The 1994 UBC approximate curves show an improved theoretical mathematical consideration for R. The capacity of elements beyond the old traditional h/t limit of .25 is extended to show capacities of taller elements. The 1994 curves also increase the allowable usable stress to .25 f/m and indicate the effect of radius of gyration on the reduction factor R as in hollow masonry or in solid masonry, rather than simply h/t (although R is plotted here on basis of h/t as the coordinate.)

In this plot the R factor values are shown by the allowable stress ordinate, showing the relation to h/t (radius of gyration tables converted) for solid and for hollow units. Paradoxically, it shows hollow units masonry to be stronger than solid masonry.

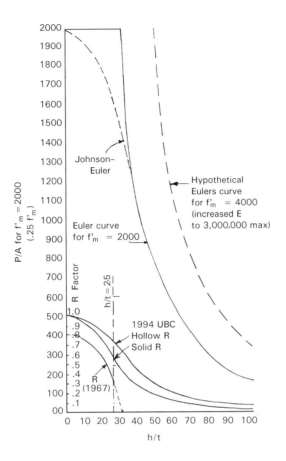

FIGURE 21.44

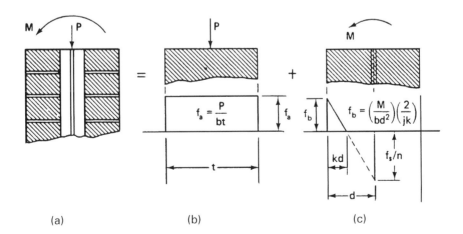

FIGURE 21.45   Combined bending plus direct stress distribution for a wall.

The principle is to assume a plane section rotation that provides stresses limited by the allowable. This principle was used for many chart compilations for reinforced concrete columns subjected to bending and compression. It has been used since the late 1930s for charts for columns developed by the ACI Joint Committee.

A sample calculation of an 8 in wall section is included in Figure 21.45 to illustrate application to a masonry section.

A drawing of a 5 in section is shown in Figure 21.46 to illustrate how greater capacity may be shown for the section.

21.10.9 Allowable Concentric Load on Bearing Walls "P" kips per foot

The curves in Figure 21.47 show maximum P, or axial concentric vertical load, that may be imposed on walls for various stress conditions.

The dashed boundary line shows where the tension stress due to bending force of wind is overcome or exceeded by the compressive axial load stress, so there is no tension on wall.

Combinations of the P and h to the right of the boundary are for compression over the entire section so no tension steel is required for moment, merely minimum arbitrary percentage of steel is required.

In the chart area where tension occurs, normal arbitrary steel is adequate for most conditions and special calculation need be made for only a small area of combinations, i.e., above the capacity provided by arbitrary minimum steel.

The capacity is reduced according to the 1979 UBC reduction factor for height. Due to the fact that a 1/3 increase is applied to the permitted stresses for short time seismic or wind loads when they are combined with vertical load the calculated capacities exceed those indicated for static loads. Therefore the capacity shown is for static or dynamic conditions.

21.10.10 Bolts

Bolt values permitted are rather conservative. Installation is an important facet of bolt capacity and must be carefully detailed with good assurance of proper method. Some of the methods are shown in the following figures; capacities are shown in UBC Tables 21-E-1, 21-E-2, 21-F, which must be followed for projects under UBC. (See Tables 21.20, 21.21, and 21.22.)

21.10.11 Shear

Shear may be a critical capacity for masonry. There are two considerations. One is longitudinal shear such as occurs in beam action, that is

$$v = \frac{VQ}{Ib}$$

The other is in-plane shear or diagonal tension

$$fv = \frac{V}{bjd}$$

Stirrup reinforcing is necessary when computed f_v exceeds the allowable shear stress in masonry, Fv. Web reinforcing shall be provided that is designed to carry the total shear force.

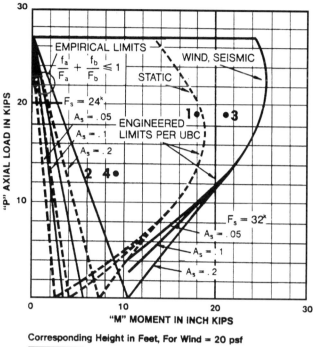

Corresponding Height in Feet, For Wind = 20 psf

8 10 12 14

FIGURE 21.46   The curves are based on allowable stresses for static loads and also for the condition of 1/3 increase for bending allowables for wind or seismic. They also include the effect of reinforcing, for areas of .05, .1, and .2 square inches per foot. These represent percentages of .0009, .0019, and .0037 for comparison with UBC minimum percentages of .0007 and .002. These steel areas might be provided by # 4 @ 48″, #6 @ 48″, and #6 @ 24″ respectively. The straight lines from $M @ P = 0$ and $P @ M = 0$ are the representation of the UBC interaction equation of $\frac{f_a}{F_a} \pm \frac{f_b}{F_b} \leq 1$ while the wider curves indicate the increased capacity based on UBC and strength of material design assumptions.

Examples:

Given:

5″ Hollow Brick wall, grouted, $h = 10'$, wind = 20 psf, $M_W = 3''k$
$f'_m = 2500$ psi, inspected masonry
$F_a = 500$ psi $n = 12$
$F_b = 833 (\times 4/3 = 1111)$
$F_s = 24,000$ $(\times 4/3 = 32,000)$, $A_s = .05^{\Box''}/\text{ft}, .1^{\Box''}, .2^{\Box''}$
Vertical Live Load = 5 kpf @ $e = 2.25''$ $(M = 11.25)$.
Vertical Dead Load = 7 kpf @ $e = 0$ $(M = 0)$
Floor Dead Load = 3 kpf @ $e = 2.25$ $\frac{(M=6.75)}{18.00''}$

Solution:

$h/t = 120''/5 = 24$; $R = .78$

For Max. imposed load, $P = 15$; (equivalent $P = 19.2$) $M = 18'' k$
(See 1)   This is within the capacity for static load.
For Dead Load only $P = 10$ (eq. $P = 12.8$) $M = 6.75''k$
(See 2)   This is within the static load capacity.
For Max. Load plus wind $P = 15$ $(P = 19.2)$ $M = 21''k$
(See 3)   This is beyond the static load capacity but
within the 4/3 increased capacity permitted.

TABLE 21.20   Allowable Tension, B_t, for Embedded Anchor Bolts for Clay and Concrete Masonry, pounds*,†,‡

f'_m psi (×6.89 for kPa)	Embedment Length, l_b, or Edge Distance, l_{be}, inches, (×25.4 for mm; ×4.45 for N)						
	2	3	4	5	6	8	10
1,500	240	550	970	1,520	2,190	3,890	6,080
1,800	270	600	1,070	1,670	2,400	4,260	6,660
2,000	280	630	1,120	1,760	2,520	4,500	7,020
2,500	310	710	1,260	1,960	2,830	5,030	7,850
3,000	340	770	1,380	2,150	3,100	5,510	8,600
4,000	400	890	1,590	2,480	3,580	6,360	9,930
5,000	440	1,000	1,780	2,780	4,000	7,110	11,100
6,000	480	1,090	1,950	3,040	4,380	7,790	12,200

Note: This table corresponds to Table 21-E-1 of the UBC.

*The allowable tension values in Table 21-E-1 are based on compressive strength of masonry assemblages. Where yield strength of anchor bolt steel governs, the allowable tension in pounds is given in Table 21-E-2.

†Values are for bolts of at least A 307 quality. Bolts shall be those specified in Section 2106.2.14.1.

‡Values shown are for work with or without special inspection.

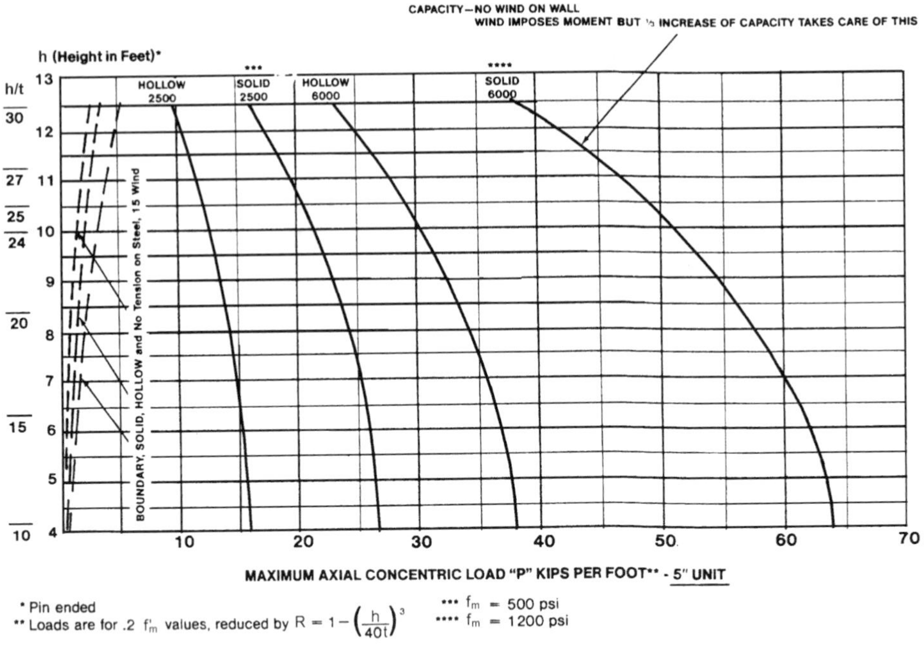

FIGURE 21.47   Where tensions may occur, use the following UBC equation for conservative calculations:

$$\frac{f_a}{F_a} \pm \frac{f_b}{F_b} \leq 1.$$

However this may be simplified and expressed as:

$$\frac{P(\text{imposed})}{P(\text{allowable})} + \frac{M(\text{imposed})}{M(\text{allowable})} \leq 1.$$

The value for P(allowable) may be read from the curves above, which are shown as an example of range of capacity.

TABLE 21.21 Allowable Tension, B_t, for Embedded Anchor Bolts for Clay and Concrete Masonry, pounds[*,†]

Bent bar Anchor Bolt Diameter, inches ($\times 25.4$ for mm; $\times 4.45$ for N)							
1/4	3/8	1/2	5/8	3/4	7/8	1	1 1/8
350	790	1,410	2,210	3,180	4,330	5,650	7,160

Note: This table correspond to Table 21-E-2 of the UBC.

[*]Values are for bolts of atleast A 307 quality. Bolts shall be those specified in Section 2106.2.14.1.

[†]Values shown are for work with or without special inspection.

TABLE 21.22 Allowable Shear, B_v, for Embedded Anchor Bolts for Clay and Concrete Masonry, pounds[*,†]

f'_m, psi	Bent Bar Anchor Bolt Diameter, inches ($\times 25.4$ for mm; $\times 4.45$ for N)						
	3/8	1/2	5/8	3/4	7/8	1	1 1/8
1,500	480	850	1,330	1,780	1,920	2,050	2,170
1,800	480	850	1,330	1,860	2,010	2,150	2,280
2,000	480	850	1,330	1,900	2,060	2,200	2,340
2,500	480	850	1,330	1,900	2,180	2,330	2,470
3,000	480	850	1,330	1,900	2,280	2,440	2,590
4,000	480	850	1,330	1,900	2,450	2,620	2,780
5,000	480	850	1,330	1,900	2,590	2,770	2,940
6,000	480	850	1,330	1,900	2,600	2,900	3,080

Note: This table corresponds to Table 21-F-1 of the UBC.

[*]Values are for bolts of at least A 307 quality. Bolts shall be those specified in Section 2106.2.14.1.

[†]Values shown are for work with or without special inspection.

The area of steel for shear reinforcement placed perpendicular to the longitudinal reinforcement shall not be less than

$$A_v = \frac{s\,V}{F_s\,d}$$

It shall be spaced so that every $45°$ line extending from a point at $d/2$ of the beam to the longitudinal tension bars shall be crossed by at least one line of bars. (See Figure 21.48.)

Bolts and connections between the masonry and other parts of the structure are vital to the proper performance of the system. However, in the past they were not evaluated. Recent research and testing have resulted in better understanding and determination of capacities, with resultant code revisions and potential changes. Field installation methods have a large influence on the actual capacities, as do variations in material and direction of load and edge distance. It is recommended that conservative design be used for bolts, for example, the material cost of a 5/8-in bolt over a 1/2-in bolt is slight, but the capacity may be much greater.

The national concrete masonry industry and the clay brick industry have reworked the long-standing allowable bolt value determination in the UBC. Therefore a conservative approach is to use the UBC tables for capacity.

A more important factor is to provide details and methods of installation that assure proper installation. One great hazard is improper installation in the nonhomogeneous masonry assemblage.

Some of the allowable capacity owing to masonry is based on the resistance of an assumed core of material, as shown in Figures 21.49 through 21.51 in which edge distance, depth, and end distances are important, as well as the tension value and shear value of the bolt. The lateral, or shear, capacity can be improved by adding a nut or plate at the masonry wall face. This will increase the lateral bearing area and impose confining compressive restraint on the masonry outer edge.

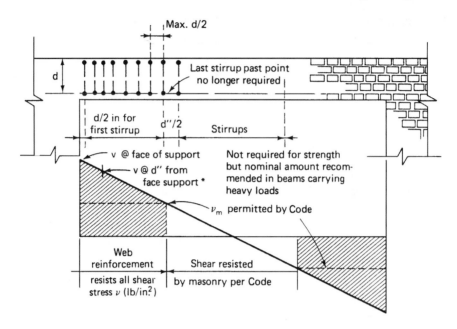

FIGURE 21.48　Location of stirrups.

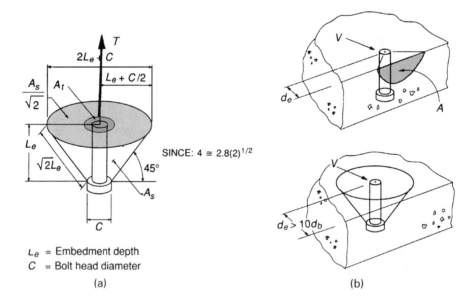

L_e = Embedment depth
C = Bolt head diameter

(a)　　　　　　　　　　　　　　　　(b)

FIGURE 21.49　Bolt stress assumptions. (a) Tensile strength; (b) shear strength. The 1991 U.B.C. required steel reinforcement for bolts located near an edge and located as shown.

The advantage of Figure 21.50 is that the plate or nut not only increases the bolt value to full bolt value but also works well to hold the bolt aligned against a grout form held against the wall face. The bolt shown in Figure 21.51 can be placed after the hollow units are in place and while the grout is being poured or mortar pointed around the bolt.

Shear may be one of the more critical capacities of masonry. If the capacity is exceeded, masonry is assumed to have failed and it does not contribute anything. All the shear capacity is provided by the reinforcement.

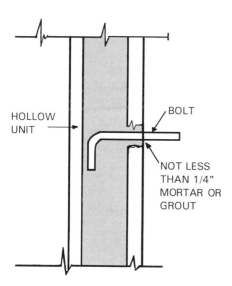

HOLLOW
UNIT

BOLT

NOT LESS
THAN 1/4"
MORTAR OR
GROUT

FIGURE 21.50 Typical bolt installation.

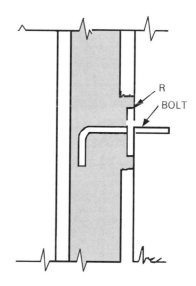

R

BOLT

FIGURE 21.51 Alternate bolt installation.

The flexural member is different from that of shear walls. The shear stress fv due to applied shear V to the section is

$$fv = \frac{V}{bjd}$$

The allowable stress, where there is no shear reinforcement is

$$F_v = 1.0\sqrt{f'_m}, 50 \text{ psi maximum}$$

With shear reinforcement taking all the shear it is

$$F_v = 3.0\sqrt{f'_m}, 150 \text{ psi maximum}$$

Where the required shear reinforcing is

$$Av = \frac{Vs}{Fsd}$$

and the spacing "s" may not exceed $d/2$.

Stirrup bars are more effective if hooked or bent. They may be developed for bond capacity beyond the point of maximum stress, that is 40 bar diameters for full stress.

The shear wall design is based on

$$f_v = \frac{V}{bjd}$$

If V is due to seismic loads, the value of V shall be 1.5 times the values indicated in the seismic force chapter of UBC.

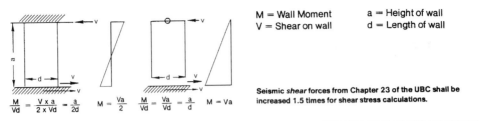

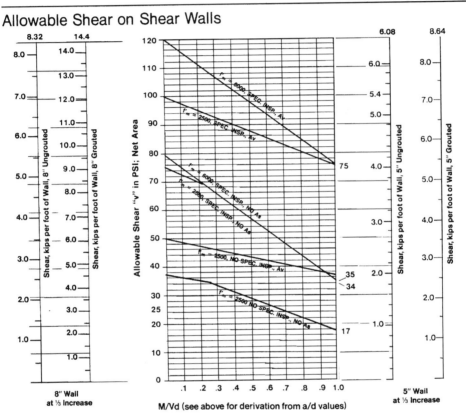

FIGURE 21.52 Example shear curves of 5″ hollow brick. This chart clarifies the variation of allowable unit shear stress for variation of M/Vd from 0 to 1 for f'_m of 2500 and 6000. Also two scales are added to show the shear capacity per ft. of the Hollow unit, ungrouted, and grouted, for those unit shear values. The UBC states the values of M/Vd are related to the a, height, and d, length of walls depending on the fixity at top and bottom of the wall or pier, i.e., for a shear of V on a pier fixed top and bottom, $M/Vd = a/2d$; if fixed at bottom only, $M/Vd = a/d$. This is shown in the figure. Enter the chart, allowable shear, with that M/Vd value, find allowable shear for the appropriate masonry curve, also the capacity per foot of the Hollow wall, grouted or ungrouted, for stress increased by 1/3 for wind or seismic loading. From Higgens Brick Company Redondo Beach, California.

The allowable shear value depends on the fixity as expressed by the following diagrams for various M/Vd, related to a, the height, and d, the length of walls. Figure 21.52 is excerpted from the brochure of the Higgins Brick Company of Redondo Beach, California, one of the earlier users of hollow brick.

Without shear reinforcement, that is, with flexural reinforcement only, the allowable shear stress is:

for $M/Vd < 1$

$$F_v = 1/3\left(4 - \frac{M}{Vd}\right)\sqrt{f'_m}, \left(80 - 45\frac{M}{Vd}\right) \text{ maximum}$$

for $M/Vd \geq 1$,

$$F_v = 1.0\sqrt{f'_m}, 35 \text{ psi (240 kPa) maximum}$$

Where shear reinforcement is designed to take all the shear, the allowable shear stress is:

for $M/Vd < 1$,

$$F_v = 1/2\left(4 - \frac{M}{Vd}\right)\sqrt{f'_m}, \left(120 - 45\frac{M}{Vd}\right) \text{ maximum}$$

for $M/Vd \geq 1$

$$F_v = 1.5\sqrt{f'_m}, 75 \text{ psi maximum}$$

$$A_v = \frac{sV}{F_s d}$$

The maximums for Fv may be increased by one third for wind or seismic resistance. M is the bending moment occurring in the element where V occurs. Fv and the maximum allowable values are one half for the condition of no special inspection.

Shear Friction is not covered specifically but is a recommended concept, such as for the design of connections such as beam seats, where the shear capacity of the masonry perpendicular to the reinforcement is required.

A recommended method of calculation of shear friction reinforcing is

$$Av = \frac{Vu}{\theta Fy}$$

The following values are recommended:

$Vu = 2\times$ applied shear
$\theta = .40$
$Fy =$ steel yield strength

When a member bears on the full area of a masonry element, the allowable bearing stress F_{br} is

$$F_{br} = 0.26 f'_m$$

When a member bears on one third or less of a masonry element, the allowable bearing stress F_{br} is

$$F_{br} = 0.38 f'_m$$

The above formula applies only when the least dimension between the edges of the loaded and unloaded areas is a minimum of one fourth of the parallel side dimension of the loaded area. The allowable bearing stress on a reasonably concentric area greater than one third but less than the full area shall be interpolated.

Note that added jamb bars will function as "hold downs" for wood-sheathed shear walls.

21.10.12 System and Seismic Interaction

System interaction of vertical and lateral forces is of greater importance than determining the precise magnitude of stress and strain in the masonry elements. The actual loads such as wind, seismic, and vertical loads are probabilities, as are the estimated strengths of materials. The basic and most important factor is the correctness of the system or the complete path of the force from point of application to resistance, and compliance with the basic static law, that summation of H, V, $M = 0$.

One example of system function is the consideration of loads on a lintel. A common load and resistance system is the common assumption that the load resisted by a lintel member is a triangular wall area, immediately above the lintel element. The other loads are presumed to "arch" over. However, this arch must have a resisting abutment force that is equal and opposite to the arch thrust, that is, adequate wall plane area. If the lintel were at the end of a wall plane, say, at the end or corner of the building, there would not be additional wall to provide the arch thrust resistance, only a vertical reaction. The moment calculations should then be based on simple $WL/8$ with the total wall load applied, not merely the small triangular area assumed for lintel load.

Figure 21.53 illustrates the common assumption of wall and floor load, but with the wall returned at a corner. The internal load approximate path is shown to illustrate why the assumed end abutment thrust might not exist to provide validity to the arching. The end wall would simply tilt outward, since it could provide vertical support, but not lateral resistance.

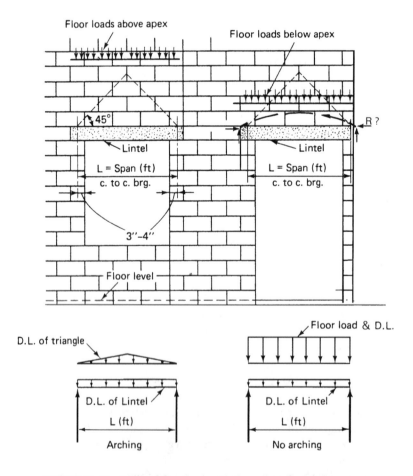

FIGURE 21.53 Wall and floor loads with the wall retained at the corner.

21.10.13 Empirical Design

Empirical Design is still used in some areas, especially for simpler structures. It contains methods that have proven satisfactory for some time. The UBC (1994 edition) lists such requirements in a specification type of code, which is a statement of the law of many jurisdictions, and WSCPA is providing an Evaluation Service Report for residential construction that will serve in more detail for more areas. In this way, some of the simpler structures may be constructed of masonry without the engineering calculations. Modified excerpts from UBC Section 2109 follow in order to provide this well-studied summary for presumed ease and security of masonry design and construction. Other jurisdictions may have similar prescriptive design provisions.

General. The design of masonry structures using empirical design located in those portions of Seismic Zones 0 and 1 as defined in Part III of Chapter 16 where the basic wind speed is less than 80 miles per hour as defined in Part II of Chapter 16 shall comply with the the provisions of this section, subject to approval of the building official.

Height. Buildings relying on masonry walls for lateral load resistance shall not exceed 35 ft (10668 mm) in height.

Lateral Stability. Shear walls shall be provided parallel to the direction of the lateral forces resisted.

Minimum nominal thickness of masonry shear walls shall not be less than 8 in (203 mm).

In each direction in which shear walls are required for lateral stability, the minimum cumulative length of shear walls provided shall be 0.4 times the long dimension of the building. The cumulative length of shear walls shall not include openings, and shall be distributed symmetrically about the center.

The maximum spacing of shear walls shall not exceed the ratio listed in Table 21-L of 1994 Edition of UBC.

Compressive Stresses

General. Compressive stresses in masonry due to vertical loads plus live loads, excluding wind or seismic loads, shall be determined. Dead and live loads shall be in accordance with this code with permitted live load reductions.

Allowable Stresses. The compressive stresses in masonry shall not exceed the values set forth in Table 21-M. The allowable stresses given in Table 21-M for the weakest combination of the units and mortar used in any load wythe shall be used for all loaded wythes of multiwythe walls.

Stress calculations. Stresses shall be calculated based on specified rather than nominal dimensions. Calculated compressive stresses shall be determined by dividing the design load by the gross cross-sectional area of the member. The area of openings, chases or recesses in walls shall not be included in the gross cross-sectional area of the wall.

Anchor bolts. Bolt values shall not exceed those set forth in Table 21-N.

Lateral Support. Masonry walls shall be laterally supported in either the horizontal or vertical direction not exceeding the intervals set forth in Table 21-O.

Lateral support shall be provided by cross walls, pilasters, buttresses or structural framing members horizontally, or by floors, roof or structural framing members vertically.

Except for parapet walls, the ratio of height to nominal thickness for cantilever walls shall not exceed 6 for solid masonry or 4 for hollow masonry.

In computing the ratio for cavity walls, the value of thickness shall be the sums of the nominal thickness of the inner and outer wythes of the masonry. In walls composed of different classes of units and mortars, the ratio of height or length to thickness shall not exceed that allowed for the weakest of the combinations of units and mortar of which the member is composed.

Minimum Thickness

General. The nominal thickness of masonry bearing walls in buildings more than one story in height shall not be less than 8 in (203 mm). Solid masonry walls in one-story buildings may be of 6 in nominal thickness when not over 9 ft (2743 mm) in height, provided that when gable construction is used, an additional 6 ft (1829 mm) is permitted to the peak of the gable.
Exception. The thickness of *unreinforced grouted brick* masonry walls may be 2 in (51 mm) less than required by this section, but in no case less than 6 in (153 mm).
Variations in thickness. Where a change in thickness due to minimum thickness occurs between floor levels, the greater thickness shall be carried up to the higher floor level.
Decrease in thickness. Where walls of masonry of hollow units or masonry-bonded hollow walls are decreased in thickness, a course or courses of solid masonry shall be constructed between the walls below and the thinner wall above, or special units or construction shall be used to transmit the loads from face shells or wythes to the walls below.
Parapets. Parapet walls shall be at least 8 in (203 mm) in thickness and their height shall not exceed three times their thickness. The parapet wall shall not be thinner than the wall below.
Foundation walls. Foundation walls shall be constructed with Type M or S mortar. Where the height of unbalanced fill (height of finished grade above basement floor or inside grade) and the height of the wall between lateral support does not exceed 8 ft (2438 mm), and when the equivalent fluid weight of unbalanced fill does not exceed 30 pounds per cubic foot (480 kg/m^2), the minimum thickness of foundation walls shall be as set forth in Table 21-D. Maximum depths of unbalanced fill permitted in Table 21-P may be varied with the approval of the building official when local soil conditions warrant.

Where the height of unbalanced fill height between lateral supports or equivalent fluid weight of unbalanced fill exceeds that set forth above, foundation walls shall be designed.

Bond

General. The facing and backing of multiwythe masonry walls shall be bonded in accordance with this section.
Masonry headers. Where the facing and backing of solid masonry construction are bonded by masonry headers, not less than 4 percent of the wall surface of each face shall be composed of headers extending not less than 3 in (76 mm) into the backing. The distance between adjacent full-length headers shall not exceed 24 in (610 mm) either vertically or horizontally. In walls in which a single header does not extend through the wall, headers from opposite sides shall overlap at least 3 in (76 mm), or headers from opposite sides shall be covered with another header course overlapping the header below at least 3 in (76 mm).

Where two or more hollow units are used to make up the thickness of the wall, the stretcher courses shall be bonded at vertical intervals not exceeding 34 in (864 mm) by lapping at least 3 in (76 mm) over the unit below, or by lapping at vertical intervals not exceeding 17 in (432 mm) with units which are at least 50 percent greater in thickness than the units below.
Wall ties. Where the facing and backing of masonry walls are bonded with 3/16-in diameter (4.8 mm) wall ties or metal ties of equivalent stiffness embedded in the horizontal mortar joints, there shall be at least one metal tie for each 4 1/2 square feet (0.42 m^2) of wall area. Ties in alternate courses shall be staggered, the maximum vertical distance between ties shall not exceed 24 in (610 mm), and the maximum horizontal distance shall not exceed 36 in (914 mm). Rods bent to rectangular shape shall be used with hollow-masonry units laid with the cells vertical. In other walls, the ends of ties shall be bent to 90-degree angles to provide hooks not less than 2 in (51 mm) long. Additional ties shall be provided at all openings, spaced not more than 3 ft (914 mm) apart around the perimeter and within 12 in (305 mm) of the opening.

The facing and backing of masonry walls may be bonded with prefabricated joint rein-forcement. There shall be at least one cross wire serving as a tie for each 2-2/3 square feet

(0.25 m²) of wall area. The vertical spacing of the joint reinforcement shall not exceed 16 in (406 mm). Cross wires of prefabricated joint reinforcement shall be at least No. 9 gage wire. The longitudinal wire shall be embedded in mortar.

Longitudinal bond. In each wythe of masonry, head joints in successive courses shall be offset at least one fourth of the unit length or the walls shall be reinforced longitudinally as required for stack bond.

Anchorage

Intersecting walls. Masonry walls depending on one another for lateral support shall be anchored or bonded at locations where they meet or intersect by one of the following methods:
1. Fifty percent of the units at the intersection shall be laid in an overlapping pattern, with alternating units having a bearing of not less than 3 in (76 mm) on the unit below.
2. Walls shall be anchored by steel connectors having a minimum section of 1/4 in by 1 1/2 in (6.4 mm by 38 mm) with ends bent up at least 2 in (51 mm), or with cross pins to form anchorage. Such anchors shall be at least 24 in (610 mm) long and the maximum spacing shall be 4 ft (1219 mm) vertically.
3. Walls shall be anchored by joint reinforcement spaced at a maximum distance of 8 in (203 mm) vertically. Longitudinal rods of such reinforcement shall be at least No. 9 gage and shall extend at least 30 in (762 mm) in each direction at the intersection.
4. Interior nonbearing walls may be anchored at their intersection, at vertical spacing of not more than 16 in (406 mm) with joint reinforcement or 1/4 in (6.4 mm) mesh galvanized hardware cloth.
5. Other metal ties, joint reinforcement or anchors may be used, provided they are spaced to provide equivalent area of anchorage to that required by this section.

Floor and roof anchorage. Floor and roof diaphragms providing lateral support to masonry walls shall be connected to the masonry walls by one of the following methods:
1. Wood floor joists bearing on masonry walls shall be anchored to the wall by approved metal strap anchors at intervals not exceeding 6 ft (1829 mm). Joists parallel to the wall shall be anchored with metal straps spaced not more than 6 ft (1829 mm) on center extending over and under and secured to at least three joists. Blocking shall be provided between joists at each strap anchor.
2. Steel floor joists shall be anchored to masonry walls with No. 3 bars, or their equivalent, spaced not more than 6 ft (1829 mm) on center. Where joists are parallel to the wall, anchors shall be located at joist cross bridging.
3. Roof structures shall be anchored to masonry walls with not less than 1/2 in-diameter (13 mm) bolts at 6 ft (1829 mm) on center or their equivalent. Bolts shall extend and be embedded at least 15 in (381 mm) into the masonry, or be hooked or welded to not less than 0.2 square inch (129 mm²) of bond beam reinforcement placed not less than 6 in (153 mm) from the top of the wall.

Walls adjoining structural framing. Where walls are dependent on the structural frame for lateral support, they shall be anchored to the structural members with metal anchors or keyed to the structural members. Metal anchors shall consist of 1/2 in-diameter (13 mm) bolts spaced at a maximum of 4 ft (1219 mm) on center and embedded at least 4 in (102 mm) into the masonry, or their equivalent area.

Unburned Clay Masonry

General. Masonry of stabilized unburned clay units shall not be used in any building more than one story in height. The unsupported height of every wall of unburned clay units shall not be more than 10 times the thickness of such walls. Bearing walls shall in no case be less than 16 in (406 mm) in thickness. All footing walls which support masonry of unburned clay units shall extend to an elevation not less than 6 in (153 mm) above the adjacent ground at all points.

Bolts. Bolt values shall not exceed those set forth in Table 21-O.

Stone Masonry

General. Stone masonry is that form of construction made with natural or cast stone in which the units are laid and set in mortar with all joints filled.

TABLE 21.23 Shear Wall Spacing Requirements for Empirical Design of Masonry

	Maximum Ratio
Floor or Roof Construction	Shear Wall Spacing to Shear Wall Length
Cast-in-place concrete	5 : 1
Precast concrete	4 : 1
Metal deck with concrete fill	3 : 1
Metal deck with no fill	2 : 1
Wood diaphragm	2 : 1

Note: This table corresponds to Table 21-L in the UBC.

TABLE 21.24 Allowable Compressive Stresses for Empirical Design of Masonry

Construction: Compressive Strength of Unit, Gross Area, psi (×6.89 for kPa)	Allowable Compressive Stresses* Gross Cross-Sectional Area, psi (×6.89 for kPa)	
	Type M or S Mortar	Type N Mortar
Solid masonry of brick and other solid units of clay or shale; sand-lime or concrete brick:		
8,000 plus, psi	350	300
4,500 psi	225	200
2,500 psi	160	140
1,500 psi	115	100
Grouted masonry, of clay or shale; sand-lime or concrete:		
4,500 plus, psi	275	200
2,500 psi	215	140
1,500 psi	175	100
Solid masonry of solid concrete masonry units:		
3,000 plus, psi	225	200
2,000 psi	160	140
1,200 psi	115	100
Masonry of hollow load-bearing units:		
2,000 plus, psi	140	120
1,500 psi	115	100
1,000 psi	75	70
700 psi	60	55
Hollow walls (cavity or masonry bonded)† units:		
2,500 plus, psi	160	140
1,500 psi	115	100
Hollow units	75	70
Stone ashlar masonry:		
Granite	720	640
Limestone or marble	450	400
Sandstone or cast stone	360	320
Rubble stone masonry		
Coarse, rough or random	120	100
Unburned clay masonry	30	—

Note: This table corresponds to Table 21-M in the UBC.

*Linear interpolation may be used for determining allowable stresses for masonry units having compressive strengths which are intermediate between those given in the table.

†Where floor and roof loads are carried upon one wythe, the gross cross-sectional area is that of the wythe under load. If both wythes are loaded, the gross cross-sectional area is that of the wall minus the area of the cavity between the wythes.

TABLE 21.25 Allowable Shear on Bolts for Empirically Designed Masonry Except Unburned Clay Units

Diameter Bolt, inches (×25.4 for mm)	Embedment,* inches (×25.4 for mm)	Solid Masonry, Shear in pounds (×4.45 for N)	Grouted Masonry, Shear in pounds (×4.45 for N)
1/2	4	350	550
5/8	4	500	750
3/4	5	750	1,100
7/8	6	1,000	1,500
1	7	1,250	1,850[†]
1 1/8	8	1,500	2,250[†]

Note: This table corresponds to Table 21-N of the UBC.

*An additional 2 inches of embedment shall be provided for anchor bolts located in the top of columns for buildings located in Seismic Zones 2, 3 and 4.

[†] Permitted only with not less than 2,500 pounds per square inch (17.24 MPa) units.

TABLE 21.26 Wall Lateral Support Requirements for Empirical Design of Masonry

Construction	Maximum l/t or h/t
Bearing walls	
Solid or solid grouted	20
All other	18
Nonbearing walls	
Exterior	18
Interior	36

Note: This table corresponds to Table 21-O of the UBC.

TABLE 21.27 Thickness of Foundation Walls for Empirical Design of Masonry

Foundation Wall Construction	Nominal Thickness, inches (×25.4 for mm)	Maximum Depth of Unbalanced Fill, feet (×304.8 for mm)
Masonry of hollow units, ungrouted	8	4
	10	5
	12	6
Masonry of solid units	8	5
	10	6
	12	7
Masonry of hollow or solid units, fully grouted	8	7
	10	8
	12	8
Masonry of hollow units reinforced vertically with No. 4 bars and grout at 24″ o.c. Bars located not less than 4 1/2″ from pressure side of wall.	8	7

Note: This table corresponds to Table 21-P of the UBC.

Construction. In ashlar masonry, bond stones uniformly distributed shall be provided to the extent of not less than 10 percent of the area of exposed faces. Rubble stone masonry 24 in (610 mm) or less in thickness shall have bond stones with a maximum spacing of 3 ft (914 mm) vertically and 3 ft (914 mm) horizontally and, if the masonry is of greater thickness than 24 in (610 mm), shall have one bond stone for each 6 square feet (0.56 m^2) of wall surface on both sides.

Minimum thickness. The thickness of stone masonry bearing walls shall not be less than 16 in (406 mm).

TABLE 21.28   Allowable Shear on Bolts for Masonry of Unburned Clay Units

Diameter of Bolts inches (×25.4 for mm)	Embedments, inches (×25.4 for mm)	Shear, pounds (×4.45 for N)
1/2	—	—
5/8	12	200
3/4	15	300
7/8	18	400
1	21	500
1 1/8	24	600

Note: This table corresponds to Table 21-Q of the UBC.

Tables 21.23 and 21.24 show how the UBC limits empirical design.

Tables 21.25 through 21.29 illustrate how left values are considered in the UBC.

21.11 Specifications

Specifications for masonry have the same need for clarity and specifics as other material methods. Standard specifications, such as ASTM Standards, may be used for materials, but the specification for masonry construction must be a guide specification, one which is studied and adapted to the specific job.

21.11.1 Shop Drawings

Shop drawings are not generally part of masonry construction except when there are complicated reinforcing details. However, the "sample panel" can be considered to serve a similar function. The sample panel should be built before the start of masonry construction. It will provide a basis for approval of the unit color range. It can also incidentally serve for approval of workmanship, method of laying and grouting, joinery or tooling, size variation of units, and wall thickness. It should be built where it can be preserved until after the construction of the masonry is completed. It can frequently be of help if the project results in litigation. Sometimes it may be built in place and left as part of the final construction.

An unusual use was made of this latter method. A hostile designer was shown several in-place samples by a knowledgable contractor, who was anticipating difficulty in obtaining approval. He photographed and plotted the odd-shaped approved sample on drawings. He left the irregular shaped sample in place, and built out away from it with identical construction, exactly like the sample. As expected, the designer disapproved the finished wall, instructing that it be torn out. The contractor agreed to do it but "did not want to remove the portion the designer had officially approved" and asked the designer to show him just what portion should be torn out. It could not be distinguished, so the unfounded disapproval was withdrawn.

21.11.2 Short Form

Short Form specifications may be used for simple projects such as garden walls, retaining walls, etc.

21.11.3 Guide Specifications

Guide Specifications must be thorough to provide for a variety of field conditions. A sample Guide Specification by the Higgins Brick Company for veneer is included here as Figure 21.54 to be a sample because engineers have not generally designed veneer. In view of WSCPA and BIA research and study, engineers will become more frequent designers of BV/SS, reinforced veneer, etc.

Specifications must cover the material and method selected for a specific project, hence should be written for that project. Although frequently ignored in the field, they become very important in litigation. The language need not be "legalese," but it must be complete. The sample concrete

TABLE 21.29   Area of Steel Reinforcement per foot

Size steel	Diameter	Area	Steel Spacing																
			8 in. (0 ft 8 in.)	12 in. (1 ft 0 in.)	16 in. (1 ft 4 in.)	20 in. (1 ft 8 in.)	24 in. (2 ft 0 in.)	28 in. (2 ft 4 in.)	32 in. (2 ft 8 in.)	36 in. (3 ft 0 in.)	40 in. (3 ft 4 in.)	44 in. (3 ft 8 in.)	48 in.* (4 ft 0 in.)	56 in. (4 ft 8 in.)	64 in. (5 ft 4 in.)	72 in. (6 ft 0 in.)	80 in. (6 ft 8 in.)	88 in. (7 ft 4 in.)	96 in. (8 ft 0 in.)
2 No. 9	0.148	0.0345	0.052	0.034	0.026	0.021	0.017	0.015	0.013	0.012	0.010	0.009	0.0086	—	—	—	—	—	—
2 No. 8	0.162	0.0412	0.062	0.041	0.031	0.025	0.021	0.018	0.015	0.014	0.012	0.011	0.010	—	—	—	—	—	—
23/16	0.1875	0.0552	0.083	0.055	0.041	0.033	0.028	0.024	0.021	0.018	0.017	0.015	0.014	—	—	—	—	—	—
21/4	0.250	0.098	0.147	0.098	0.073	0.059	0.049	0.042	0.037	0.033	0.029	0.027	0.024	—	—	—	—	—	—
25/16	0.312	0.0152	0.229	0.15	0.114	0.092	0.076	0.065	0.057	0.051	0.046	0.042	0.038	—	—	—	—	—	—
No. 2	1/4	0.049	0.073	0.05	0.036	0.029	0.024	0.021	0.018	0.016	0.015	0.013	0.012						
No. 3	3/8	0.010	0.165	0.11	0.083	0.066	0.055	0.047	0.041	0.037	0.033	0.030	0.027	0.024	0.021	0.018	0.016	0.015	0.014
No. 4	1/2	0.196	0.293	0.20	0.147	0.118	0.098	0.084	0.073	0.065	0.059	0.054	0.049	0.042	0.037	0.033	0.029	0.027	0.024
No. 5	5/8	0.307	0.460	0.31	0.230	0.184	0.154	0.132	0.115	0.102	0.092	0.084	0.077	0.066	0.057	0.051	0.046	0.042	0.038
No. 6	3/4	0.442	0.663	0.44	0.332	0.265	0.221	0.189	0.166	0.147	0.133	0.120	0.110	0.095	0.083	0.074	0.066	0.060	0.055
No. 7	7/8	0.601	0.900	0.60	0.450	0.361	0.300	0.258	0.226	0.200	0.180	0.164	0.150	0.129	0.112	0.100	0.090	0.082	0.075
No. 8	1.0	0.786	1.180	0.79	0.590	0.471	0.392	0.337	0.295	0.261	0.236	0.214	0.196	0.168	0.147	0.131	0.118	0.107	0.098
No. 9	1.128	1.000	1.50	1.00	0.750	0.600	0.500	0.428	0.375	0.333	0.300	0.273	0.250	0.214	0.187	0.167	0.150	0.136	0.125

*Recommended spacing

Guide Specification	Notes to Specifier

4.01 SCOPE

a) Installation of veneer in the areas indicated by the drawings and specified herein

4.02 WORK INCLUDED

a) Installing, pointing and cleaning of the veneer on surfaces constructed by others

b) Building in of all vents, conduits, inserts, and flashings, as furnished, set and braced by others

c) Removal of surplus veneer material and waste after completion of the veneer work

The detail and listing of "work included" and "work not included" is subject to considerable discussion and is listed here with the intent of clarifying the work included by consideration of the jurisdiction that may be involved.

The specifier should check that such items are included in other appropriate sections, especially item d.

4.03 WORK NOT INCLUDED

a) Embedding or attaching anchoring devices

b) Shoring and bracing

c) Furnishing and fabricating of steel reinforcing

d) Furnishing scratch coat backing or support for veneer

e) Cleaning due to paint, plaster, cement finish and other trades

f) Protection of aluminum frames

g) Preparation of backing surface to receive veneer

4.04 MATERIAL

a) **Water.** Must be clean and potable

b) **Sand** shall be according to ASTM designation C-144, except that not less than 5% shall pass the No. 100 sieve.

a) Domestic water is satisfactory.
b) The sand should be uniformly graded with emphasis on the finer materials, to produce a more dense and workable mortar or grout. Sands on the coarse side of the range do not provide a waterproof or neat-looking job and are difficult to use in the field.

c) **Portland cement** shall conform to ASTM C-150 Type I or Type II low alkali.

c) Specify "low alkali" to reduce efflorescence tendency. Cement is a source of soluble materials that contribute to efflorescence, although there are many other contributors.

d) **Lime** Hydrated lime shall conform to ASTM designation C-207, Type S.

e) **Color.** Sufficient lime-proof colorfast mineral pigment shall be added as approved by the architect.

d) Quicklime may also be used.

e) Proprietary factory mixed-colored pointing mortars are available. Some colors can be achieved only by the use of white cement.

f) **Veneer Units.** Brick shall be as manufactured by Higgins Brick Company, of color and texture as approved by the architect.

g) **Ties and anchors** shall be corrosion resistant. Manufacturers' proven capacity must be submitted.

h) **Storage and Handling.** Material shall be stored and handled in such a manner as to prevent deterioration, chipping, breakage or intrusion of foreign material.

f) The size, color and texture shall be selected and specified by the architect. They may be specified by reference to specific jobs or buildings.

h) The intent here is that the materials be kept in an acceptable condition during and after job delivery.

4.05 MORTAR

a) **Mortar** shall consist of one part portland cement, one-half part hydrated lime, and 4 1/2 parts clean, well graded sand, measured damp loose. Alternate mixes with bond enhancing additives may be used if approved by the architect.

b) **Mortar** shall be mixed long enough for thorough intimate mixing of all ingredients.

c) **Mortar Retempering.** Retempering mortar on boards shall be done by adding water within a basin formed with the mortar and reworked.

d) **Dry Set Mortar** in compliance with ANSI 118.5 may be used for bond coat, as an alternate to the mortar. It shall be installed in compliance with ANSI 108.5.

a) This mortar ratio is in compliance with the uniform code. Up to five or six parts of damp, loose sand might be used to compensate for bulking and to provide for a leaner mix with less shrinking. Latex mortars have proven satisfactory for bond and ease of application.

d) This provides for an economical thin bond coat by a specifically prepared mix and unit application. It may be used if the backing surface is true and accurate and the units are true.

FIGURE 21.54 (*continues*).

masonry specification that follows contains some items that should be included in projects. It will serve as a sample check list for the field accomplishment.

Concrete Masonry Specification

General

The masonry contractor shall examine all drawings and specifications and note all conditions that may affect his work and performance in fulfilling the contract.

Guide Specification

Notes to Specifier

The specifier may ask for a sample area or panel, suggested as follows: " A sample panel not less than 3' x 3' shall be constructed for the architect's approval. It must show the method, jointing, range of completion and approval of the veneer." This will assure a meeting of the minds regarding final appearance and acceptance. The quality and cleanliness of workmanship is more important to the final veneer appearance and acceptance than is the unit. This quality must be mutually agreed upon prior to the start of work.

4-06. SAMPLE AND TESTS
a) **Submit** samples of units, including range of colors, and tolerances for approval.
b) **Provide** sample area for approval.

c) The designer must choose if standard or special tolerance is to be met.

c) The quality shall be as per ASTM C216. Facing Brick Dimensional tolerance deviation shall be not greater than the limits of ASTM C216.

d) This is a minimum statement and should be studied and amplified for specific jobs. It is suggested the specifier list the tests he requires, the number of tests, and who pays for the tests.

d) Tests of material are to be in accord with current applicable ASTM or ANSI.
e) Ties must be substantiated for capacity.

4-07 BACKING
Surface shall be clean and damp, but not wet.

4-08 WORKMANSHIP
a) **Wetting.** The veneer units shall be wetted at least one hour before laying and shall be noticeably damp but free from surface water at the time of laying.
b) **Laying.** Spread a slurry, then mortar approximately three-eights (3/8") inch thick over the backing area, by troweling firmly.Then spread mortar over the adhering face of the unit, sufficient to create a slight excess which will be forced out at the edges of the unit. The units shall be tapped into place so as to eliminate voids in the mortar. As an alternate installation, the "thin-set" or "dry-set" method may be used. The recommendation of the manufacturer on the cement shall be followed, in compliance with ANSI specifications.
c) **Ties.** The ties must be attached and anchored to develop required capacity.

a) Care must be exercised in freezing weather. If the work is subjected to hot dry winds moisture must be added.
b) The code provision for adhered veneer by the use of portland cement has been satisfactory for many years. However, newer methods have been developed through ANSI to improve installation procedures, e.g.:

ANSI A118.1 Dry-set mortar
ANSI A118.2 Conductive mortar
ANSI A118.3 Chemical Resistant
ANSI A118.4 Latex-portland cement
These provide for installation procedures such as:
ANSI A108.1 with portland cement
ANSI A108.2 with portland cement
ANSI A108.3 with portland cement
ANSI A108.5 Dry-set mortar
ANSI A108.7 Conductive dry-set mortar
These also have been used satisfactorily.
However, true surfaces and strict adherence to those specifications must be enforced.

d) **Jointing.** The exposed joints shall be filled as the work progresses. After joints are thumb print firm, they shall be tooled to a smooth concave surface with an approved tool.

d) The specifier shall state the desired effect for tooled joints, or state, "as based on the approved sample," or for example, "as in (designated) site."

e) **Control or expansion** joint shall coincide exactly with any control joints in the backing or support and be continuous through the mortar and veneer and as shown or called for in the details.

e) This subject is a complex one and the specifier must check carefully to see that the intent of the designer is realized by the details and the augmenting specs. Different types of support and veneer may require different joint spacings for crack control by relief of stress accumulation.

4-09. CLEANING

a) **Mortar** stains shall be removed with clear water promptly as work progresses. Keep it clean.

b) Sealing and waterproofing shall be taken into consideration as a deterrent to efflorescence and other problems associated with water penetration.

a) The best way to clean masonry is to KEEP IT CLEAN, avoid mortar contact on the wall surface. Stains must be removed before penetration and before start of set. Acid cleaning or other may be necessary for adequate cleaning if the veneer is dirtied. Specific sealing and cleaning methods should be considered, as recommended by Higgins, Brick &Weatherguard or others.

FIGURE 21.54 (*continued*)   Example guide specification. From Higgens Brick Company, Redondo Beach, California.

Where any deviation is to be made from the plans and specifications, the engineer or architect shall be notified and his written approval obtained before proceeding with the work.

Scope of Work

The work under this section shall include—
The furnishing of all labor, materials, and appliances necessary to complete the masonry construction shown on the drawings and specifications.

All preparations and all masonry work necessary to receive and adjoin other work. Others shall furnish all inserts and attachments as noted in the plans and specifications for installation under this section.

Cooperation with all other trades in laying out his work.

The masonry contractor shall give the work his personal supervision and keep a competent foreman on the job at all times.

Arranging for adequate bracing, forming, and shoring required in conjunction with and in the course of constructing the masonry and not provided for under other sections. Scaffolding for masonry shall be under this section unless arranged for otherwise.

Advising the general contractor as to the position of all dowels to masonry. The general contractor shall be responsible for the placement of dowels in adjoining construction.

Placing of all reinforcing steel and furnishing all reinforcing steel for concrete masonry not provided for under other sections.

Arranging for the necessary storage space and protection for materials at the job site.

Calling for all inspections as required in the course of his work by the engineer, architect and/or building department.

Arranging for and furnishing test specimens, samples of materials, and sample panels as may be required.

Materials

All materials shall conform to the following current standards:

Concrete Masonry Units.   Hollow Load-Bearing Masonry Units shall be units conforming to the ASTM Designation C90.

Masonry units having a dry density of not more than 105 pounds per cubic foot of concrete shall be known a lightweight masonry units.

All masonry units shall have a minimum net tensile strength of not less than 125 psi. (This is desirable although not essential.)

Masonry units shall have cured for not less than 28 days prior to placement in the structure.

All masonry units shall have a maximum linear shrinkage of .06 of 1% from the saturated to the oven dry condition.

Cement.   Cement shall be low-alkali Type I (or Type II, Type III, or Type V) Portland cement conforming to ASTM Designation C150.

Plastic cement shall have less than 12% of the total volume in approved types of plasticizing agents and shall conform to all requirements for Portland cement in ASTM Designation C150, except in respect to the limitations on soluble residue, air-entrainment, and additions subsequent to calcination.

Masonry cement and Mortar cement may also be used if local service record is satisfactory.

Mortar.   Mortar shall be freshly prepared and uniformly mixed in the ratio of 1 part Portland cement, 1/4 part minimum to 1/2 part maximum lime putty or hydrated lime, damp loose sand not less than 2 1/2 and not more than 3 times the sum of the volumes of the cement and lime used, and shall conform to ASTM Designation C270.

Grout.   Grout for pouring shall be of fluid consistency and mixed in the ratio by volume, 1 part Portland cement, 2 1/4 parts minimum to 3 parts maximum damp loose sand where the grout space is less than 3 in in its least dimension.

Grout for pouring shall be of fluid consistency and mixed in the ratio by volume, 1 part Portland cement, 2 parts minimum to 3 parts maximum damp loose sand, 2 parts coarse aggregate where the grout space is 3 in or more in its least dimension.

Grout for pumping shall be of fluid consistency and shall have not less than seven sacks of cement in each cubic yard of grout. The mix design shall be approved by the engineer or architect.

Fluid consistency shall mean that consistency is as fluid as possible for pouring without segregation of the constituent parts.

Lime. Hydrated lime shall conform to ASTM Designation C207. Quicklime shall conform to ASTM Designation C5.

Quicklime shall be slaked and then screened through a 16-mesh sieve. After slaking, screening, and before using, it shall be stored and protected for not less than 10 days.

Aggregate. Aggregate shall be clean, sharp, and well graded, and free from injurious amounts of dust, lumps, shale, alkali, surface coatings, and organic matter.

Aggregate for mortar shall conform to ASTM Designation C144 with preference for maximum amounts of fines.

Aggregate for grout shall conform to ASTM Designation C404.

Admixtures. The use of admixtures shall not be permitted in mortar or grout unless substantiating data are submitted to and approved by the engineer or the architect.

The use of admixtures shall not be permitted in mortar without reducing the lime content. Proportions of admixture shall be as approved by the engineer or architect. Grout Aid® shall be used for high-lift grouting, added at the site immediately before pour.

Inert coloring pigments may be added but are not to exceed 6% by weight of the cement. The formula shall be approved by the engineer or the architect.

The inclusion of fire clay, dirt and other deleterious materials is prohibited.

Water. Water shall be potable, free from deleterious quantities of acids, alkalies, and organic materials.

Reinforcing Steel. Deformed steel bar reinforcement shall conform to ASTM Designation A305.

Steel bar reinforcement shall conform to ASTM Designation A15 or A16. Wire reinforcement shall conform to ASTM Designation A82.

Reinforcement shall be clean and free from loose rust, scale, and any coatings that reduce bond.

Construction

Workmanship. Masonry work shall not be started when the horizontal or vertical alignment of the foundation is a maximum of one inch or more in error.

All masonry shall be laid true, level, plumb, and neatly in accordance with the plans.

Units shall be cut accurately to fit all plumbing ducts, openings, electrical work, etc., and all holes shall be neatly patched.

Extreme care shall be taken to prevent visible grout or mortar stains.

No construction supports shall be attached to the wall except where specifically permitted by the Architect or Engineer.

Reinforcing shall be held firmly in position prior to and during pouring.

Masonry Units. Masonry units shall be sound, dry, clean, and free from cracks when placed in the structure.

All masonry units should be stored so that they are kept off the ground and protected from rain. Wetting the units shall not be permitted except when hot dry weather exists, causing the units to be warm to the touch. The surface only may be wetted with a light fog spray.

Proper masonry units shall be used to provide for all windows, doors, bond beams, lintels, pilasters, etc., with a minimum of unit cutting.

Where masonry unit cutting is necessary, all cuts shall be neat and true.

Mixing of Mortar and Grout. Mortar shall be mixed by placing one-half of the water and sand in the operating mixer to avoid balling of the cement. Then the cement, lime, and the remainder of the sand and water shall be added.

After all ingredients are in the batch mixer, they shall be mechanically mixed for not less than three minutes. Hand mixing shall not be employed unless procedures are specifically approved.

Bonding. For bonding the masonry to the foundation, the top surface of the concrete foundation shall be clean, with laitance removed, and aggregate exposed before starting the masonry construction.

The wall shall be laid up in straight uniform courses with regular running bond where no bond pattern is shown.

Intersecting masonry walls and partitions shall be bonded by the use of steel ties at 24 in on center maximum. Corners shall have a standard masonry bond by overlapping units and shall be solid grouted.

Where stack bond is indicated on the plans, approved horizontal reinforcing shall be provided at 24 in on center maximum.

Veneer shall be bonded to the wall in an approved manner. It may be adhered or anchored as per details.

Joints. The starting joint on foundations shall be laid with full mortar coverage on the bed joint except that the area where grout occurs shall be free from mortar so that the grout will contact the foundation.

Mortar joints shall be straight, clean, and uniform in thickness and shall be tooled as shown on the plans.

All walls shall have joints tooled with a round bar (or v-shaped bar) to produce a dense, slightly concave surface well bonded to the block at the edges, unless specifically detailed.

Tooling shall be done when the mortar is partially set but still sufficiently plastic to bond. All tooling shall be done with a tool that compacts the mortar, pressing the excess mortar out of the joint rather than dragging it out.

Raked joints shall be not more than 1/2 deep and where exposed to the weather shall be tooled.

Where walls are to receive plaster, the joints shall be struck flush.

Where joints are to be concealed under paint, these joints shall be filled and then sacked to produce a dense surface without sheen.

Joints that are not tight at the time of tooling shall be raked out, pointed, and then tooled.

Unless otherwise specified or detailed on the plans, in hollow unit masonry the horizontal and vertical mortar joints shall be 3/8 in thick with full mortar coverage on the face shell and on the webs surrounding the cells to be filled.

Vertical head joints shall be buttered well for a thickness equal to the face shell of the unit, and these joints shall be shoved tightly so that the mortar bonds well to both units. Joints shall be solidly filled from the face of the block to at least the depth of the face shell.

If it is necessary to move a unit after it has been set in place, the unit shall be removed from the wall, cleaned, and set in fresh mortar.

Concrete building bricks shall be laid with full head and bed joints.

Lintels, capping units, and all bearing plates set by the mason shall be set in a full bed of mortar.

When control joints are required, they shall be as detailed on the plans. Verify whether installation of portions are included in this section or under a separate waterproofing section.

Reinforcing

When a foundation dowel does not line up with a vertical core, it shall not be sloped more than one horizontal in six vertical. Dowels shall be grouted into a core in vertical alignment, even though it is in an adjacent cell to the vertical wall reinforcing.

Reinforcing bars shall be straight except for bends around corners and where bends or hooks are detailed on the plans.

Reinforcing steel shall be lapped 30 bar diameters minimum where spliced and shall be separated by one bar diameter or wired together.

Vertical bars shall be held in position at top and bottom and at intervals not exceeding 192 diameters of the reinforcement.

Horizontal reinforcing bars shall be laid on the webs of the units in continuous masonry courses, consisting of bond-beam or channel units, and shall be solidly grouted in place.

Vertical reinforcing steel shall have a minimum clearance of 1/4 in from the masonry, and not less than one bar diameter between bars.

Wire reinforcement shall be completely embedded in mortar or grout. Joints with wire reinforcement shall be at least twice the thickness of the wire.

Wire reinforcement shall be lapped at least 6″ at splices and shall contain at least one cross wire of each piece of reinforcement in the lapped distance.

Grouting: General. Reinforcing steel shall be secured in place and inspected before grouting starts.

Mortar droppings should be kept out of the grout space.

All grout shall be puddled or vibrated in place, with care taken to avoid segregation.

Vertical cells to be filled shall have vertical alignment to maintain a continuous unobstructed cell area not less than 2 in × 3 in.

Cells containing reinforcement shall be solidly filled with grout, and pours shall be stopped 1 1/2 in below the top of a course to form a key at pour joints.

Grouting of beams over openings shall be done in one continuous operation.

The tops of unfilled cell columns under a horizontal masonry beam or bond beam shall be formed with metal lath or special units shall be used to confine the grout fill to the beam section.

All bolts, anchors, etc., inserted in the wall shall be solid grouted or mortared in place.

Spaces around metal door frames and other built-in items shall be filled solidly with grout or mortar.

Low-Lift Grouting. In hollow unit masonry low-lift grouting, the structure shall be grouted in heights of less than four feet. In two-wythe masonry low-lift grouting, the wall shall be grouted in heights of less than 8 or six times the grout joint thickness, whichever is lesser.

In two-wythe masonry walls, one wythe may be carried up 16 in maximum before grouting. The grout joint shall be at least 1 in wide and shall be filled solidly with grout. (See Figure 21.11.)

High-Lift Grouting. Cleanout holes shall be provided at the bottom of all cores containing vertical reinforcement in hollow units masonry and in two-wythe masonry shall be provided by omitting alternate units on the first course of one wythe.

Mortar projections and mortar droppings shall be washed out of the grout space and off the reinforcing steel with a jet stream of water as required to clean the space.

All grout shall be consolidated at time of pouring by puddling or vibrating and then reconsolidated by later puddling before plasticity is lost.

The minimum dimension of the grout space shall be 3 in.

Two-wythe masonry shall cure at least three days and hollow unit masonry shall cure at least 24 hours before grouting.

Grout shall be poured to not more than 4-ft, depths then approximately one hour allowed to elapse before another 4 depth is poured. The full height in each section of the wall shall be poured in one day.

Vertical grout barriers or dams shall be built across the grout space of two-wythe masonry the entire height of the wall to control the flow of the grout horizontally. These barriers shall be less than 25 ft on center.

All reinforcing steel shall be inspected in place before grouting, and there shall be continuous inspection during the grouting operation.

In two-wythe masonry, wire ties, consisting of # 9 wire rectangles, shall connect the wythes and shall be spaced not more than 12 in on center vertically for stacked bond, not more than 24 in on center vertically for running bond, and not more than 32 in on center horizontally.

Wall Cleaning and Protection. Concrete scum and grout stains on the wall shall be removed immediately.

After the wall is constructed it should not be saturated with water for curing or any other purposes.

In desert areas and where the atmosphere is dry, the wall shall have its surface dampened with a very light fog spray during a mortar curing period of three days.

At the conclusion of the masonry work, the masonry contractor shall clean all masonry, remove all scaffolding and equipment used in the work, and remove all debris, refuse, and surplus masonry material, and remove them from the premises.

Recommended Waterproofing

Parapets and masonry projections above the roof provide a metal cap at the top of the wall or a dense capping unit sloped inward.

Where cast-in-place concrete caps are specified, they shall be a minimum of 2 in thick and shall be reinforced.

The inside face of the parapet and all masonry above the roof shall be waterproofed by hot mopping with roofing asphalt, covering with roofing paper, or other approved waterproofing.

No through-wall flashing shall be allowed.

Flashing and counter flashing shall be set 1/2 in into the mortar joint and the portion of wall above the flashing shall be hot mopped with roofing asphalt or other approved waterproofing.

The exterior face of masonry above the roof shall be waterproofed in the same manner as the exterior wall.

Exposed Walls Above Grade. All exterior masonry walls shall be waterproofed.

At the time of waterproofing the wall shall have been completed for at least one month and shall be in a dry condition.

All joints shall be checked for tightness and where cracks are visible, mortar shall be chipped out, tuck pointed, and tooled.

All flashing, heads, jambs, sills, inserts, and similar points shall be thoroughly caulked.

Waterproofing shall be compatible with the masonry units and shall be guaranteed by the applicator and waterproofing manufacturer with full knowledge of the units used and the condition of the wall at the time of the application of the waterproofing.

Waterproofing shall consist of a minimum of two coats with at least 24 hours between applications.

Walls shall be waterproofed as frequently as recommended in the guarantee by the waterproofing manufacturer.

The first coat of waterproofing shall be a filler coat unless otherwise specified.

Walls Below Grade. All walls below grade shall be waterproofed on the exterior surface extending from the foundation pad to above grade.

Waterproofing shall be either asphalt conforming to ASTM D449, Type A, or coal-tar pitch conforming to ASTM D450, Type B, or as otherwise approved.

Surfaces to be waterproofed shall be clean, dry and shall be given either a priming coat of creosote oil conforming to ASTM D43 and two mop coats of hot coal-tar pitch or a priming coat of asphalt primer conforming to ASTM D41 and two mop coats of hot asphalt. Mop coats shall be applied uniformly using not less than 20 lb of tar or asphalt per 100 ft^2 per coat and shall provide a continuous impervious coating, free from pinholes or other voids.

Where known water is present, membrane waterproofing shall be used as specifically noted on the plans.

21.12 Summary

In general, masonry is a concrete, that is, aggregate particles bonded by cement matrix. Its use and design are related to concrete methods, but are also derived from trial and success of methods known from antiquity. As stated before, mortar is a more complex material than concrete as it is known today. It is impossible to cover the design and construction completely in a few handbook pages.

However, the general principles have been presented to help the engineer understand and use mortar.

Emphasis has been given to some of the aspects not usually covered in engineering considerations for masonry but that have the potential to develop from a negligible part of engineering design to an important part.

Veneer, long considered to be nonstructural, is considered in several aspects, for example, the good UBC provision for performance as well as an alternate prescriptive code. Design of performance is clarified for newer uses of masonry veneer such as reinforced veneer, which may be expanded into prefabrication, and improved details of veneer on steel studs are added.

The recognition of the fire endurance value is shown in UBC Evaluation Reports. This provides for potentially greater use in 1- and 2-hour endurance protection as required by code occupancy requirements.

The newer use of hollow clay is also presented with its newer equivalent thickness fire-endurance ratings, as developed by evaluation reports and in code provisions.

Some smaller masonry structures are treated rather thoroughly because the design required is disproportionately large for the cost of the structural element, for example, garden walls, retaining walls, paving, and veneer. Also empirical design has been included because it is acceptable for much of the smaller, less technical masonry uses. Some sample specifications are given in rather complete detail because engineers unfamiliar with masonry frequently are not adept with specification writing.

For further information, pertinent references are included to particularly augment this partial general coverage. Since the subject is voluminous only some of the most pertinent sources are mentioned in references.

References

Dickey, W.L. *Higgins 5″ and 8″ Hollow Brick.* Brochure. Higgins Brick Co., Redondo Beach, CA.

Dickey, W.L. *Masonry Veneer,* Masonry Institute of America.

Dickey, W.L. *Brick Veneer,* Higgins Brick Co., Redondo Beach, CA.

IBCO. 1994. *Uniform Building Code.* International Conference of Building Officials, Whittier, CA.

Masonry Codes and Specifications. Masonry Institute of America. Los Angeles, CA.

Masonry Institute of America. *Masonry Design Manual.* Los Angeles, CA.

Schneider and Dickey. *Reinforced Masonry Design.* Prentice Hall, Englewood Cliffs, NJ.

Western States Clay Products Association. *Design Guide for Anchored Brick Veneer Over Steel Studs,* Orinda, CA.

WSCPA. *Reinforced Veneer.*

WSCPA. *Notes on the Selection, Design and Construction of Reinforced Hollow Clay Masonry.*

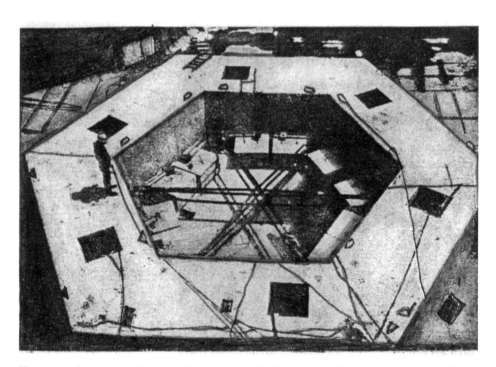

Hexagon marine structure, Kanagawa, Japan: concrete floating block reinforced with glass fiber reinforced plastic tendons. The tendons consist of nine multi-cables, each made of eight 0.3 in (8 mm) diameter Glass Fiber Reinforced Plastic (GFRP) rods. The concrete floating blocks were connected by post-tensioning GFRP tendons and the anchorage was provided by using multi-type anchor heads and wedges. Construction completed in 1993 (Courtesy American Concrete Institute).

22

Part A: Fiber-Reinforced Concrete (FRC)

by
Edward G. Nawy, D.Eng., P.E., C.Eng.
Distinguished Professor, Civil Engineering, Rutgers University—The State University of New Jersey. Expert in concrete structures and materials.

22.1 Historical Development

Fibers have been used to reinforce brittle materials from time immemorial, dating back to the Egyptian and Babylonian eras if not earlier. Straws were used to reinforce sun-baked bricks and mud-hut walls, horse hair was used to reinforce plaster, and asbestos fibers have been used to reinforce Portland cement mortars. Research by Romualdi and Batson (1963) and Romualdi and Mandel (1964) on closely spaced random fibers in the late 1950s and early 1960s, primarily on steel fibers, heralded the era of using the fiber composite concretes we know today. In addition, Shah and Rangan (1971), Swamy (1975), and several other researchers in the United States, United Kingdom, and Russia embarked on extensive investigations in this area, exploring other fibers in addition to steel. By the 1960s, steel-fiber concrete started to be used in pavements in particular. Other developments using bundled fiber glass as the main composite reinforcement in concrete beams and slabs were introduced by Nawy et al. (1971) and Nawy and Neuwerth (1977), as discussed in Section 22.8 of this chapter. From the 1970s to the present, the use of steel fibers has been well established as a complimentary reinforcement to increase cracking resistance, flexural and shear strength, and impact resistance of reinforced concrete elements both *situ* cast and precast.

22.2 General Characteristics

Concrete is weak in tension. Microcracks start to generate in the matrix of a structural element at about 10–15% of the ultimate load, propagating into macrocracks at 25–30% of the ultimate load. Consequently, plain concrete members cannot be expected to sustain large transverse loading without the addition of continuous-bar reinforcing elements in the tensile zone of supported members such as beams or slabs. But the developing microcracking and macrocracking still cannot be arrested or slowed by the sole use of continuous reinforcement. The function of such reinforcement is to replace the function of the tensile zone of a section and assume the tension equilibrium force in the section.

Consequently, the addition of randomly spaced discontinuous fiber elements should aid in arresting the development or propagation of the microcracks that are known to generate at the early stages of loading history. While fibers have been used to reinforce brittle materials such as concrete since time immemorial, newly developed fibers have been extensively used worldwide in the past three decades. Different types are commercially available such as those made from steel, glass, polypropylene, or graphite. They have proven that they can improve the mechanical properties of the concrete, both as a structure and a material, not as a replacement for continuous-bar reinforcement when it is needed, but in addition to it.

Concrete fiber composites are concrete elements made from a mixture comprising hydraulic cements, fine and coarse aggregates, pozzolanic cementitious materials, admixtures commonly used with conventional concrete, and a dispersion of discontinuous, small fibers made from steel, glass, organic polymers, or graphites. The fibers could also be vegetable fibers such as sisal or jute.

Generally, if the fibers are made from steel, the fiber length varies from 0.5 to 2.5 in (12.7–63.5 mm). They are either round, produced by cutting or chopping wire, or flat, having typical cross sections of 0.006–0.016 in (0.15–0.41 mm) in thickness and 0.01–0.035 in (0.25–0.90 mm) in width, and produced by shearing sheets or flattening wire. The most common diameters of the round wires are in the range of 0.017–0.040 in (0.45–1.0 mm) [ACI Committee 544, 1988, 1991, 1993]. The wires are usually crimped or deformed or have small heads on them for better bond within the matrix, and some are crescent shaped in cross section.

The fiber content in a mixture, where steel fibers are used, usually varies from one quarter to two percent by volume, namely, from 33 to 265 lb/yd^3 (20–165 kg/m^3). A fiber content of 50–60 lb/yd^3 is common in lightly loaded slabs on grade, precast elements, and composite steel deck topping. The upper end of the range, more difficult to apply, is used for security applications such as vaults, safes, and impact-resisting structures.

The introduction of fiber additions to concrete as of the early 1900s was aimed principally at enhancing the tensile strength of concrete. As is well known, the tensile strength is 8–14% of the compressive strength of normal concretes with resulting cracking at low stress levels. Such a weakness is partially overcome by the addition of reinforcing bars, which can be either steel or

TABLE 22.1 Typical Properties of Fibers

Type of Fiber (1)	Diameter, in $\times 10^3$ (mm) (2)	Specific Gravity (3)*	Tensile Strength, psi $\times 10^3$ (GPa) (4)	Young's Modulus, psi $\times 10^6$ (GPa) (5)	Ultimate Elongation, % (6)
Acrylic	0.6–0.13 (0.02–0.35)	1.1	30–60 (0.2–0.4)	0.3 (2)	1.1
Asbestos	0.05–0.80 (0.0015–0.02)	3.2	80–140 (0.6–1.0)	12–20 (83–138)	1–2
Cotton	6–24 (0.2–0.6)	1.5	60–100 (0.4–0.7)	0.7 (4.8)	3–10
Glass	0.2–0.6 (0.005–0.15)	2.5	150–380 (1.0–2.6)	10–11.5 (70–80)	1.5–3.5
Graphite	0.3–0.36 (0.008–0.009)	1.9	190–380 (1.0–2.6)	34–60 (230–415)	0.5–1.0
Kevlar	0.4 (0.010)	1.45	505–520 (3.5–3.6)	9.4 (65–133)	2.1–4.0
Nylon (High Tenacity)	0.6–16 (0.02–0.40)	1.1	110–120 (0.76–0.82)	0.6 (4.1)	16–20
Polyester (High Tenacity)	0.6–16 (0.02–0.40)	1.4	105–125 (0.72–0.86)	1.2 (8.3)	11–13
Polypropylene	0.6–16 (0.02–0.40)	0.95	80–110 (0.55–0.76)	0.5 (3.5)	15–25
Rayon (High Tenacity)	0.8–15 (0.02–0.38)	1.5	60–90 (0.4–0.6)	1.0 (6.9)	10–25
Rock Wool (Scandinavian)	0.5–30 (0.01–0.8)	2.7	70–110 (0.5–0.76)	~0.6	0.5–0.7
Sisal	0.4–4 (0.01–0.10)	1.5	115 (0.8)	—	3.0
Steel	4–40 (0.1–1.0)	7.84	50–300 (03–2.0)	29.0 (200)	0.5–3.5
Cement Matrix	—	1.5–2.5	0.4–1.0 (0.003–0.007)	1.5–6.5 (10–45)	0.02

Note: GPa $\times$ 0.145 = 10^6 psi.

*Density = Col. 3 $\times$ 62.4 lb/ft^3 = Col. 3 $\times 10^3$ kg/m^3.

Source: Fundamentals of High Strength High Performance Concrete by Edward G. Nawy, 1996, Addison Wesley Longman, pp. 350.

fiberglass, as **main continuous reinforcement** in beams and one-way and two-way structural slabs or slabs on grade [Nawy and Neuwerth, 1977; Nawy et al., 1971]. As indicated earlier, the continuous reinforcing elements cannot stop the development of microcracks. Fibers, on the other hand, are discontinuous and randomly distributed in the matrix, both in the tensile and compressive zones of a structural element. They are able to add to the stiffness and crack control performance by preventing the microcracks from propagating and widening and also by increasing ductility owing to their energy absorption capacity. Common applications of fiber-reinforced concrete include overlays in bridge decks, industrial floors, shotcrete applications, highway and airport pavements, thin shell structures, seismic and explosion-resisting structures, super flat surface slabs on grade in warehouses and for reduction of expansion joints.

Table 22.1, compiled from several sources including [ACI Committee 544, 1991], describes the geometry and mechanical properties of various types of fibers that can be used as randomly dispersed filaments in a concrete matrix. As there is a wide range of properties in each type of fiber, the designer should be guided by the manufacturer's data on each particular product and the experience with it before a fiber type selected.

22.3 Mixture Proportioning

Mixing the fibers with the other mix constituents can be done by several methods. The method selected depends on the facilities available and the job requirements, namely, plant batching, ready-mixed concrete, or hand mixing in the laboratory. The most important factor is to ensure **uniform dispersion** of the fibers and to prevent segregation or balling of the fibers during mixing. Segregation or balling during mixing is affected by many factors that can be summarized as follows:

1. aspect ratio ℓ/d_f—most important,
2. volume percentage of the fiber,
3. coarse aggregate size, gradation, and quantity, and
4. water/cementitious ratio and method of mixing.

TABLE 22.2   Typical Proportions for Normal Weight Fiber Reinforced Concrete

Material	Range
Cement	550–950 lb/yd^3
W/C ratio	0.4–0.6
Percentage of sand to aggregate	50 to 100%
Maximum aggregate	3/8 in
Air content	6 to 9%
Fiber content	0.5 to 2.5% by volume of mix (steel—1% = 132 lb/yd^3 glass—1% = 42 lb/yd^3 nylon—1% = 19 lb/yd^3)

Source: ACI Committee 544. 1991. State-of-the-art on fiber reinforced concrete. In *Manual of Concrete Practice*, Vol. 5. ACI Committee Report 544.1R-82. American Concrete Institute, Farmington Hills, MI.
Note: 1 lb/yd^3 = 0.5933 kg/m^3; 1 in = 2.54 cm.

A maximum aspect ratio of ℓ/d_f and a steel fiber content in excess of 2% by volume make it difficult to have a uniform mix. While conventional mixing procedures can be used, it is advisable to use a 3/8 in (9.7 mm) maximum aggregate size. The water requirement will vary from that of concrete without fibers depending on the type of cement replacement cementitious pozzolans used and their percent by volume of the matrix. Tables 22.2 and 22.3 give typical mixture proportions for normal-weight fibrous reinforced concrete and fly-ash fibrous concrete mixes, respectively.

A workable method for mixing in a step-by-step chronological procedure can be summarized as follows:

1. Blend part of the fiber and aggregate before charging into the mixer.
2. Blend the fine and coarse aggregate in the mixer, then add more fibers at mixing speed. Lastly, add cement and water simultaneously, or add the cement immediately followed by water and additives.
3. Add the balance of the fiber to the previously charged constituents. Add the remaining cementitious materials and water.
4. Continue mixing as required by normal practice.
5. Place the fibrous concrete in the forms. Use of fibers requires more vibrating than required in nonfibrous concrete. While internal vibration is acceptable if carefully applied, external vibration of the formwork and the surface is preferable in order to prevent segregation of the fibers.

TABLE 22.3   Typical Fly-Ash Fibrous Concrete Mix

Material	Quantity
Cement	490 lb/yd^3
Fly ash	225 lb/yd^3
W/C ratio	0.54
Percentage of sand to aggregate	50%
Maximum size coarse aggregate	3/8 in
Steel fiber content (0.010 × 0.022 × 1.0 in)	1.5% by volume
Air-entraining agent	Manufacture's recommendation
Water-reducing agent	Manufacture's recommendation
Slump	5 to 6 in

Source: ACI Committee 544. 1991. State-of-the-art on fiber reinforced concrete. In *Manual of Concrete Practice*, Vol. 5. ACI Committee Report 544.1R-82. American Concrete Institute, Farmington Hills, MI.
Note: 1 lb/yd^3 = 0.5933 kg/m^3; 1 in = 2.54 cm.

22.4 Mechanics of Fiber Reinforcement

22.4.1 First Cracking Load

Fiber-reinforced concrete in flexure essentially undergoes a trilinear deformation behavior as shown in Figure 22.1. Point A on the load-deflection diagram represents the first cracking load, which can be termed the first-crack strength [Mindess and Young, 1981]. Normally, this is the same load level at which a nonreinforced element cracks. Hence, segment OA in the diagram would be the same and essentially have the same slope for both plain and fiber-reinforced concrete.

Once the matrix is cracked, the applied load is transferred to the fibers that bridge and tie the crack to keep it from opening farther. As the fibers deform, additional narrow cracks develop and continued cracking of the matrix takes place until the maximum load reaches point B of the load-deflection diagram. During this stage, debonding and pullout of some of the fibers occur. But the yield strength in most of the fibers is not reached.

In the falling branch BC of the load-deflection diagram, matrix cracking and fiber pullout continue. If the fibers are long enough to maintain their bond with the surrounding gel, they may fail by yielding or by fracture of the fiber element depending on their size and spacing.

22.4.2 Critical Fiber Length: Length Factor

If ℓ_c is the critical length of a fiber above which the fiber fractures instead of pulling out when the crack intersects the fiber at its midpoint, it can be approximated by Mindess and Young (1981)

$$\ell_c = \frac{d_f}{2v_b}\sigma_f \tag{22.1}$$

where

d_f = fiber diameter
v_b = interfacial bond strength
σ_f = fiber strength

Bentur and Mindess (1990) developed an expression to relate the average pullout work and the fiber matrix interfacial bond strength in terms of the critical fiber length, demonstrating that the strength of a composite increases continuously with the fiber length. This is of significance as it indicates that pullout work may go through a maximum and then decreases as bond strength increases over a critical value. This loss of pull-out work would be reduced to a typical range of $\ell = 10$ mm in the cement-based composites discussed in Section 22.6. If a critical v_b value of 1.0 MPa is chosen and a small diameter fiber $d_f = 20\,\mu$m, namely, a cementitious system with such a small diameter fiber, an increase in bond may result in reduced toughness.

22.4.3 Critical Fiber Spacing: Space Factor

The spacing of the fibers considerably affects cracking development in the matrix. The closer the spacing the higher the first cracking load of the matrix. This is owing to the fact that the fibers reduce the stress intensity factor that

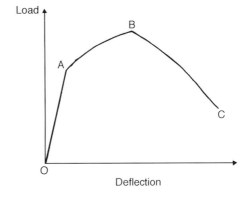

FIGURE 22.1 Schematic load-deflection relationship of fiber-reinforced concrete.

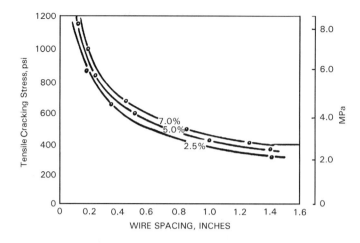

FIGURE 22.2 Effect of steel-fiber spacing on the tensile cracking stress in fibrous concrete for ρ = 2.5, 5.9, and 7.5%. *Source:* Romualdi, J.P. and Batson, G.B. 1963. Mechanics of crack arrest in concrete. *Proc. ASCE Eng. Mech. J.* 89(EM3):147–168.

FIGURE 22.3 Various shapes and sizes of steel fibers.

controls fracture. The approach taken by Romualdi and Batson (1963) to increase the tensile strength of the mortar was to increase the stress-intensity factor through decreasing the spacing of the fibers acting as crack arresters. Figure 22.2, due to Romualdi and Batson, relates the tensile cracking stress to the spacing of the fibers for various volumetric percentages. Figure 22.4 compares the theoretical and experimental values of the ratio of the first cracking load to the cracking strength of plain concrete (strength ratio). Both diagrams demonstrate that the closer the spacing of the fibers the higher the strength ratio, namely, the higher the tensile strength of the concrete up to the practical workability and cost-effectiveness limits. Various shapes and sizes of steel fibers are shown in Figure 22.3.

Several expressions to define the spacing of the fibers have been developed. If s is the spacing of the fibers, one expression from Romualdi and Batson (1963) gives

$$s = 13.8 d_f \sqrt{\frac{1.0}{\rho}} \tag{22.2}$$

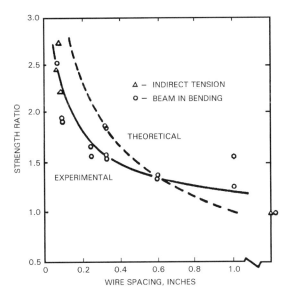

FIGURE 22.4 Effect of fiber spacing on the strength ratio. (Ratio = first cracking load of fibrous concrete divided by strength of plain concrete). *Source:* Courtesy of American Concrete Institute; Romualdi, J.P. and Mandel, J.A. 1964. Tensile strength of concrete affected by uniformly distributed closely spaced short lengths of wire reinforcement. *Proceedings, ACI Journal,* Vol. 61, No. 6.

where

d_f = diameter of the fiber
ρ = fiber percent by volume of the matrix

Another expression due to McKee gives

$$s = 3\sqrt{\frac{V}{\rho}} \tag{22.3}$$

where V is the volume of one fiber element. Another expression, due to Mindess and Young (1981), also taking into account the length of the fiber, gives

$$s = 13.8 d_f \frac{\sqrt{\ell}}{\rho} \tag{22.4}$$

22.4.4 Fiber Orientation: Fiber Efficiency Factor

The orientation of the fibers with respect to load determines the efficiency with which the randomly oriented fibers can resist the tensile forces in their directions. This is synonymous with the contribution of bent bars and vertical shear stirrups provided in beams to resist the inclined diagonal tension stress. If one assumes perfect randomness, the efficiency factor is 0.41ℓ, but it can vary between 0.33ℓ and 0.65ℓ close to the surface of the specimen as trowling or leveling can modify the orientation of the fibers [Mindess and Young, 1981].

22.4.5 Static Flexural Strength Prediction: Beams with Fibers Only

To predict flexural strength, several methods could be applied depending on the type of fiber, the type of matrix, whether empirical data from laboratory experiments is used, or whether the design is based on the bonded area of the fiber or the law of mixtures. An empirical expression for the composite flexural strength based on a composite-material [Bentur and Mindess, 1990] approach can be:

$$\sigma_c = A\sigma_m(1 - V_f) + BV_f\frac{\ell}{d} \tag{22.5}$$

where

σ_c = composite flexural strength
σ_m = ultimate strength of the matrix
V = volume fraction of the fibers adjusted for the effect of randomness
A, B = constants
ℓ/d = aspect ratio of the fiber, where ℓ is the length and d is the diameter of the fiber

The constants A and B obtained from $4 \times 4 \times 12$ in ($100 \times 100 \times 305$ mm) model beam tests by Swamy et al. (1974) and adopted by ACI Committee 544 (1993) produced the following expression:
First crack composite flexural strength, psi

$$\sigma_f = 0.843\, f_r V_m + 425\, V_f\frac{\ell}{d_f} \tag{22.6}$$

where

f_r = stress in the matrix (modulus of rupture of the plane mortar or concrete), pounds per square inch
V_m = volume fraction of the matrix = $1 - V_f$
V_f = volume fraction of the fibers = $1 - V_m$
ℓ/d_f = ratio of length to diameter of the fibers, namely, the aspect ratio

Ultimate composite flexural strength, psi

$$\sigma_{cu} = 0.97\, f_r V_m + 494\, V_f\frac{\ell}{d_f} \tag{22.7}$$

22.5 Mechanical Properties of Fibrous Concrete Structural Elements

22.5.1 Controlling Factors

From Section 22.4, it can be seen that the mechanical properties of fiber-reinforced concretes are influenced by several factors. The major properties are

1. the type of fiber, namely, the fiber material and its shape;
2. the aspect ratio ℓ/d_f, namely, the ratio of the fiber length to its nominal diameter;
3. the amount of fiber in percentage by volume, ρ;
4. the spacing of the fiber, s;

5. the strength of the concrete or mortar matrix; and
6. the size, shape, and preparation of the specimen.

Hence it is important to conduct laboratory tests to failure on the mixtures using specimen models similar in form to the elements being designed. As the fibers affect the performance of the end product in all material-resistance capacities such as in flexure, shear, direct tension, and impact, it is important to evaluate the test specimen performance with regard to those parameters.

The contribution of the fiber to tensile strength, as discussed in Section 22.3, is due to its ability to act as reinforcement and assume the stress from the matrix when it cracks through the interface shear-friction interlock between the fiber and the matrix. This phenomena is analogous to the shear-friction interlock hypothesis presented in Nawy (1996b) on the mechanism of shear friction interlock. Hence deformed or crimped fibers have a greater influence than smooth and straight ones. The pullout resistance in zone AB of Figure 22.1 is proportional to the interfacial surface area [ACI Committee 544, 1993]. The nonround fiber cross sections and the smaller diameter round fibers induce a larger resistance per unit volume than the larger diameter fibers. This is also analogous to the crack-control behavior in traditionally reinforced structural members. There, a larger number of smaller diameter bars, that are more closely spaced are more effective than a smaller number of large diameter bars for the same reinforcement volume percentage [Nawy and Blair, 1971]. One reason is the larger surface interaction area between the fibers and the surrounding matrix resulting in a higher bond and shear-friction resistance.

22.5.2 Strength in Compression

The effect of the contribution of the fibers to the compressive strength of the concrete seems to be minor, as seen in Figure 22.5 [Hsu and Hsu, 1994] for tests using steel fibers. However, the ductility and toughness are considerably enhanced as a function of the increase in the volume fractions and aspect ratios of the fibers used. In Figure 22.5, Hsu and Hsu (1994) show the effect of the increase in volume fraction on the stress-strain relationship of the fibrous concrete through increasing

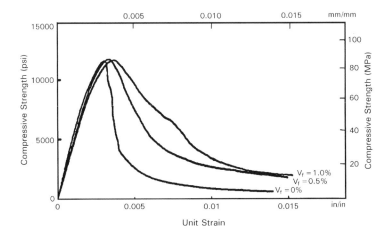

FIGURE 22.5 Influence of volume fraction of steel fibers on stress-strain behavior for 13,000 psi concrete. *Source:* Courtesy of American Concrete Institute; Shah, S.P. and Rangan, B.V. 1971. Fiber reinforced concrete properties. *Proceedings, ACI Journal*, Vol. 68, No. 2.

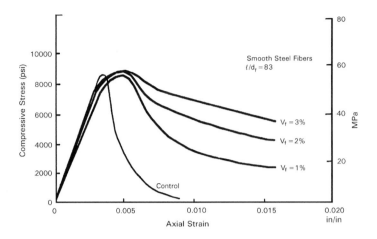

FIGURE 22.6   Influence of volume fraction of steel fibers on stress-strain behavior for 9000 psi concrete. *Source:* Courtesy of American Concrete Institute; Fanella, D.A. and Naaman, A.E. 1994. Stress-strain properties of fiber reinforced concrete in compression. *Proceeding, ACI Journal*, Vol. 91, No. 4.

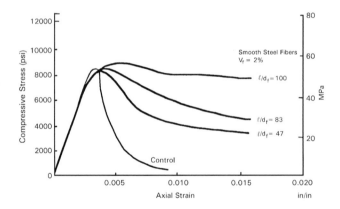

FIGURE 22.7   Influence of aspect ratio of steel fibers on stress-strain behavior. *Source:* Courtesy of American Concrete Institute; Fanella, D.A. and Naaman, A.E. 1994. Stress-strain properties of fiber reinforced concrete in compression. *Proceeding, ACI Journal*, Vol. 91, No. 4.

the fiber volume from zero to 1.5% for concretes having a compressive strength of 13,100 psi (90.3 MPa). Figures 22.6 and 22.7 from Fanella and Naaman (1985) depict a similar trend with respect to both a volume fraction ratio up to 3% and an aspect ratio in the range of 47–100. Figure 22.8, from Shah and Rangan (1971), also demonstrates the influence of the increase in fiber content on the relative toughness of reinforced concrete members.

Toughness is a measure of the ability to absorb energy during deformation. It can be estimated from the area under the stress-strain or load-deformation diagrams. A toughness index (*TI*) expression proposed in Hsu and Hsu (1994) gives

$$TI = 1.421RI + 1.035 \qquad\qquad (22.8)$$

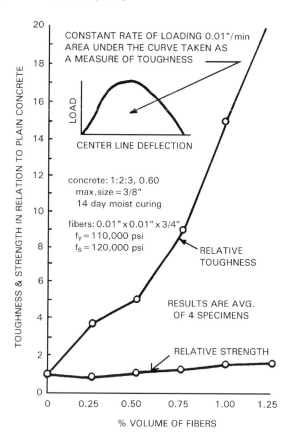

FIGURE 22.8 Relative toughness and strength versus fiber volume ratio. *Source:* Courtesy of American Concrete Institute; Shah, S.P. and Rangan, B.V. 1971. Fiber reinforced concrete properties. *Proceedings, ACI Journal,* Vol. 68, No. 2.

where

RI = reinforcing index = $V_f(\ell/d_f)$
V_f = volume fraction
ℓ/d_f = aspect ratio

Figure 22.9 gives the relationship of the toughness index to the reinforcing index of fibrous high-strength concretes within the limitation of the type, aspect ratio, and volume fractions of the steel fibers used in those tests. In short, by increasing the volume fraction, both ductility and toughness have been shown to increase significantly within the practical limits of workable volume content of fiber in a concrete mix.

22.5.3 Strength in Direct Tension

The effect of different shapes of the fiber filaments on the tensile stress behavior of steel fiber reinforced mortars in direct tension is demonstrated in Figure 22.10. The descending portion of the plots show that the fibers-reinforced with better anchorage quality increase the tensile resistance of the fiber-reinforced concrete beyond the first cracking load.

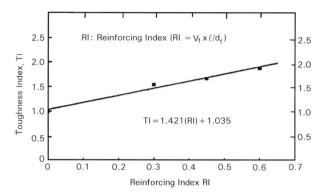

FIGURE 22.9   Toughness index versus reinforcing index of fibrous concrete. *Source:* Courtesy of American Concrete Institute; Shah, S.P. and Rangan, B.V. 1971. Fiber reinforced concrete properties. *Proceedings, ACI Journal,* Vol. 68, No. 2.

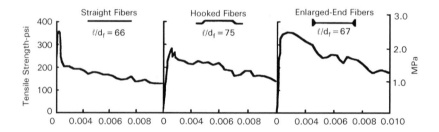

FIGURE 22.10   Effect of the shape of steel fibers on tensile stress in mortar specimens loaded in direct tension. *Source:* Courtesy of American Concrete Institute; Shah, S.P. and Rangan, B.V. 1971. Fiber reinforced concrete properties. *Proceedings, ACI Journal,* Vol. 68, No. 2.

22.5.4 Flexural Strength

Fibers seem to affect the magnitude of flexural strength in concrete and mortar elements to a much greater extent than they affect the strength of comparable elements subjected to direct tension or compression [ACI Committee 544, 1993]. Two stages of loading portray the behavior. The first controlling stage is the **first cracking load stage** in the load-deflection diagram, and the second controlling stage is the ultimate load stage. Both the first cracking load and the ultimate flexural capacity are affected as a function of the product of the fiber volume concentration, ρ, and the aspect ratio, ℓ/d_f. Fiber concentrations less than 1/2% of volume of the matrix and with an aspect ratio less 50 seem to have a small effect on the flexural strength, although they can still have a pronounced effect on the toughness of the concrete element as seen in Figure 22.8.

The flexural strength of plain concrete beams containing steel fibers was defined in Section 22.4, Eqs. (22.6) and (22.7). For structural beams reinforced with both normal reinforcing bars and fibers added to the matrix, a modification of the standard expression for the nominal moment strength, $M_n = A_s f_y (d - a/2)$, has to be made in order to account for the shear-friction interaction of the fibers in preventing the flexural macrocracks from opening and propagating in the tensile zone of the concrete section, as seen in Figure 22.11 [Henager and Doherty, 1976]. In this diagram, the standard hypothesis where the area of concrete in the tensile zone is neglected, therefore modified so that an additional equilibrium tensile force, T_{fc}, is added to the section. This moves the neutral

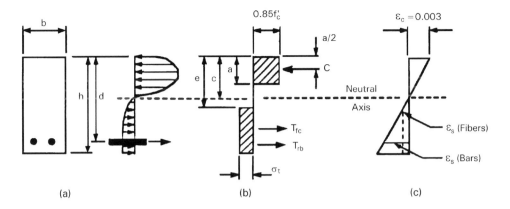

FIGURE 22.11 Stress and strain distribution across depth of singly reinforced fibrous concrete beams. (a) Assumed stress distribution; (b) equivalent stress block distribution; and (c) strain distribution.

axis down, leading to a higher nominal moment strength, M_n. The resulting expression [Henager and Doherty, 1976], for M_n becomes

$$M_n = A_s f_y \left(d - \frac{a}{2} \right) + \sigma_t b(h - e)\left(\frac{h}{2} + \frac{e}{2} - \frac{a}{2} \right) \tag{22.9}$$

$$e = [\epsilon \text{ (fibers)} + 0.003]\frac{c}{0.003} \tag{22.10}$$

$$\sigma_t \text{ (psi)} = \frac{1.12\ell}{d_f \rho_f F_{bc}} \tag{22.11a}$$

$$\sigma_1 \text{ (MPa)} = \frac{0.00772\ell}{d_f \rho_f F_{be}} \tag{22.11b}$$

where

ℓ = fiber length
d_f = fiber diameter
ρ_f = percent by volume of the fibers
F_{be} = bond efficiency of the steel fiber depending on its characteristics; varies from 1.0 to 1.2.
a = depth of the equivalent rectangular block
b = width of beam
c = depth to the neutral axis
d = effective depth of the beam to the center of the main tensile bar reinforcement
e = distance from the extreme compression fibers to the top of the tensile stress block of the fibrous concrete
$\epsilon_s = f_y/E_s$ of the bar reinforcement
$\epsilon_f = \sigma_f/E_c$ of the fibers developed at pullout at a dynamic bond stress of 333 psi
σ_t = tensile yield stress in the fiber
T_{fc} = tensile yield of the fibrous concrete in Figure 22.11
  $= \sigma_t b(h - e)$
T_{rb} = tensile yield force of the bar reinforcement in Figure 22.11
  $= A_s f_y$

22.5.5 Shear Strength

A combination of vertical stirrups and randomly distributed fibers in the matrix enhances the diagonal tension capacity of concrete beams. The degree of increase in the diagonal tension capacity is a function of the shear span/depth ratio of a beam. This ratio determines the mode of failure in normal beams that do not fall in the category of deep beams and brackets as detailed by Nawy (1996a). Williamson (1978) found that when 1.66% by volume of straight steel fibers are used instead of stirrups, the shear capacity increased by 45% over beams without stirrups. When steel fibers with deformed ends were used at a volume ratio of 1.1%, the shear capacity increased by 45–67% and the beams failed by flexure. Using crimped-end fibers increased the shear capacity by almost 100%.

In general, as the shear span/depth ratio, a/d, decreased and the fiber volume increased, the shear strength increased proportionally. Tests by Sharma (1986) resulted in the following expression in the ACI Committee 544 report (1993) for the average shear stress, v_c, for beams in which steel fibers were added:

$$v_{cf} = \frac{2}{3} f_t' \left(\frac{d}{a} \right)^{\frac{1}{4}}$$

(22.12)

where

f_t' = tensile splitting strength
d = effective depth of a beam
a = shear span; equal to distance from the point of application of the load to the face of the support when concentrated loads are acting or equal to the clear beam span when distributed loads are acting.

22.5.6 Environmental Effects

22.5.6.1 Freezing and Thawing

The addition of fibers to a matrix does not seem to result in an appreciable improvement in the freezing and thawing performance of concrete since its resistance to such an environmental effect is controlled by permeability, void ratio, and freeze-thaw cycles. Fibers tend, however, to hold the scaling concrete pieces together, thereby reducing the extent of apparent scaling.

22.5.6.2 Shrinkage and Creep

No appreciable improvement in the shrinkage and creep performance of concrete results from the addition of fibers but perhaps a slight decrease in shrinkage owing to the need for more paste mortar in the mixture when fibers are also used. Cracking due to drying shrinkage in restrained elements can be slightly improved as the cracks are kept from generating because of the bridging effect of the randomly distributed fibers.

22.5.7 Dynamic Loading Performance

The cracking behavior of fibrous concrete elements under dynamic loading seems to be three to ten times better that of plain concrete. Also, the total energy absorbed by the steel fibrous concrete beams can be 40 to 100 times that for plain concrete beams depending on the type, deformed shape, and percent volume of the fibers [ACI Committee 544, 1993].

22.6 Steel Fiber Reinforced Cement Composites

22.6.1 General Characteristics

Fiber-reinforced concretes are designed to contain a maximum 2% by volume of fibers, using the same mixture design procedures and placement as nonfibrous concretes. Fiber-reinforced cement composites, on the other hand, could contain a volume fraction, namely, a fiber content by volume, as high as 8 to 25%. Consequently, neither the design of the mixture nor the constituent materials in the matrix can be similar to those of conventional fibrous or nonfibrous concretes. Either cement only or cement with sand is used in the mixture, with no coarse aggregate, in order to achieve the high strength, ductility and high performance expected from such composites.

In addition, the 1980s saw the development of cement compositions termed **macro defect free cements** (MDF), which had a high Young's modulus and flexural strengths up to almost 30,000 psi (∼200 MPa), and the **densified cements** (DSP) that had particle size less than 1/20 that of Portland cement (0.5 μm). The void content in any matrix can be reduced to a negligible percentage with the addition of pozzolans such as silica fume.

With these developments as a background, the following are the types of cement-based composites that are being studied today:

1. slurry infiltrated fiber concretes, SIFCON and refractory use composite, SIFCA;
2. densified small particle systems, DSP;
3. compact reinforced composites, CRC;
4. carbon fiber cement based composites; and
5. super strength reactive powder concrete, RPC.

These cement-based composites can achieve a compressive strength in excess of 44,000 psi (300 MPa) in compression and an energy absorption capacity, namely, ductility, that can be up to 1000 times that of plain concrete [Reinhardt and Naaman, 1992].

22.6.2 Slurry Infiltrated Fiber Concretes (SIFCON)

Because of the high volume fraction of steel fibers (8–25%), the mix for a structural member is formulated by sprinkling the fiber into the formwork or over a substratum. Either the substratum is stacked with fibers to a prescribed height or the form is completely or partially filled with the fibers depending on the requirement of the design. After the fibers are placed, a low-viscosity cement slurry is poured or pumped into the fiber bed or into the formwork, infiltrating into the spaces between the fibers. Typical cement/fly-ash/sand proportions can vary from 90/10/0 to 30/20/50 by weight [Schneider, 1992]. The water/cementitious ratio (W/C + F) can range between 0.45 and 0.20 by weight, with a plasticizer content of 10–40 oz per 100 lb of the total cementitious weight (C + F). Batch trials of the slurry mix have to be carefully made with regard to the W/(C + F) ratio in order to arrive at a workable slurry mix that can fully penetrate the fibers' depth.

Figure 22.13 gives a stress-strain diagram for a SIFCON mixture [Naaman, 1992] with a compressive strength close to 18,000 psi but with a very large strain capability in the falling branch of the diagram. Figure 22.14 gives the influence of the matrix compressive strength on the stress-strain response of SIFCON in compression [Schneider, 1992]. The fiber content, $V_f = 11\%$, resulted in total uniaxial strain in excess of 10%.

22.6.3 MDF and DSP Cement Composites

The densified small particle systems (DSP) and the compact reinforced composites (CRC) gain super high strength depending largely on the type of compact density cements used for the cement-based

FIGURE 22.12   Fracture surface of steel fiber reinforced concrete (Courtesy American Concrete Institute).

composites and the proper proportioning used to considerably reduce or practically eliminate most of the voids in the paste.

Figure 22.12 illustrates the fracture surface of a steel fiber reinforced concrete specimen.

22.6.4 Carbon-Fiber Reinforced Cement-Based Composites

Petroleum pitch-based carbon fibers have recently been developed for use as reinforcement for cement-based composites. Their diameter varies from 10 to 18 μm (0.0004–0.0007 in) and their lengths vary from 1/8 to 1/2 in (3–12 mm). They have a typical tensile strength in the range of 60–110 ksi (400–750 MPa). They are incorporated in the cement-based composites in essentially the same manner as steel fibers are in concrete and are uniformly distributed and randomly oriented.

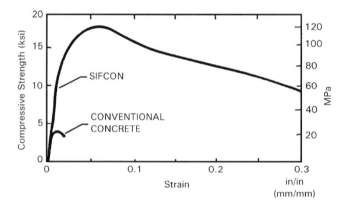

FIGURE 22.13   Stress-strain relationship of SIFCON with rupture strain in the range of 0.45 in/in. *Source:* Naaman, A.E. 1992. SIFCON: Tailored properties for structural performance. In *Proceedings, International RILEM/ACI Workshop*, pp. 18–38. Chapman and Hall, New York.

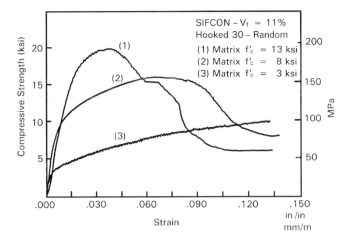

FIGURE 22.14 Influence of matrix compressive strength on the stress-strain response of SIFCON in compression. *Source:* Naaman, A.E. 1992. SIFCON: Tailored properties for structural performance. In *Proceedings, International RILEM/ACI Workshop*, pp. 18–38. Chapman and Hall, New York.

Because of very small size of the carbon fibers and their small diameter, a high fiber count is attained in the cementitious matrix at a typical volume fraction of 0.5 to 3% [Bayasi, 1992]. The spacing between the fibers is approximately 0.004 in (0.1 mm) at a 3% fiber volume fraction. The carbon fibers' function is similar to that of the steel fibers in preventing microcracks from propagating and opening.

22.6.5 Super Strength Reactive Powder Concretes

These concretes have compressive strengths in a range of 30,000–120,000 psi (200–800 MPa). The lower range is used today for construction of structural elements. The higher ranges are used in nonstructural applications such as flooring, safes, and storage compartments for nuclear waste. Concretes in the higher ranges are termed super-high strength concretes and possess the very high ductility necessary for applications in structural systems.

The principal characteristic of such concretes is the use of a powder concrete in which aggregates and traditional sand are replaced by ground quartz less than 300 μm in size [Richard and Cheyrezy, 1994]. In this manner, the homogeneity of the mixture is greatly improved and the distribution in the size of the particles is consequently reduced by almost two orders of magnitude. A major improvement in the properties of the hardened concrete is an increase in the Young's modulus value of the paste by almost a factor of three so that its value can reach to 6×10^6–11×10^6 psi (55–75 GPa), thereby reducing the effects of incompatibility between the moduli of the paste and the quartz powder. Richard and Cheyrezy (1994) have developed the following mechanical characteristics of RPC concrete:

1. improved homogeneity resulting in a Young's modulus up to 11×10^6 psi (75 GPa);
2. increase in dry compact density of the dry solids—while silica fume with its small particle size of 0.1–0.5 μm and an optimum mix content of 25% cement by weight gives excellent dry compact density, additional amounts of precipitated silica improve the dry compact density further;

TABLE 22.4   Mixture Composition and Concrete Mechanical Properties of Super High Strength Reactive Powder Concrete (RPC) Weight–lb/yd³, Stress–psi (MPa)

Mixture Constituents	RPC 200 Concrete	RPC 800 Concrete
Portland cement-type V	1614 (955)	1690 (1000)
Fine sand (150–400 μm)	1775 (1051)	845 (500)
Ground quartz (4 μm)	—	659 (390)
Silica fume (18 m²/g)	387 (229)	389 (230)
Precipitated silica (35 m²/g)	16.9 (10)	—
Superplasticizer (polyacrylate)	22.0 (13)	30.4 (18)
Steel fibers	323 (191)	1065 (630)
Total water	31.2 gal/yd³ (153 ℓ/m³)	36.7 gal/yd³ (180 ℓ/m³)
Cylinder compressive strength	24–33 (170–230)	71–99 (490–680)
Flexural strength	3.6–8.7 (25–60)	6.5–14.8 (45–102)
Fracture energy	15,000–40,000 J/m²	1200–2000 J/m²
Young's modulus	7.8–8.7 ×10³ (54–60 GPa)	(9.8–10.9)10³ (65–75 GPa)

Source: Richard, P. and Cheyrezy, M.H. 1994. Reactive powder concretes with high ductility and 200–800 MPa compressive strength. In *Proceedings, V.M. Malhotra Symposium on Concrete Technology: Past, Present and Future*, P.K. Mehta, ed., ACI SP-144, pp. 507–518. American Concrete Institute, Farmington Hills, MI.
Note: 1 ℓtr/m³ = 0.2 gal/yd³ = 1.69 lb/yd³; 1 ℓtr (water) = 2.204 lb = 0.264 gal; 1000 psi = 6.895 MPa; 1 kg/m³ = 1.69 lb/yd³. Resistant type V cement was used in all the mixtures.

3. increase in the density of concrete by maintaining the fresh concrete under pressure at the placement stage and during setting—this results in the removal of air bubbles, expulsion of excess water, and partial reduction of the plastic shrinkage during final set;

4. improvement in microstructure through hot curing for two days at 194°F (90°C) to speed the activation of the pozzolanic reaction of the silica fume, resulting in a 30% gain in compressive strength; and

5. increase in ductile behavior through the addition of an adequate volume fraction of steel microfibers.

Table 22.4 adapted from Richard and Cheyrezy (1994) fibers gives the mix proportions for type 200 and type 800 RPC concretes. It also lists the major mechanical properties of these concretes.

22.7 Prestressed Concrete Prism Elements as the Main Composite Reinforcement in Concrete Beams

Composite concrete that uses precast pretensioned prisms as its main tension reinforcement has shown its promise as a construction concept for the effective control of cracking and deflection in structural concrete elements, particularly in the negative moment regions of reinforced concrete bridge decks. Most of the earlier studies on this subject were limited to laboratory experimental work (Chen and Nawy, 1994). Introducing highly precompressed prisms as the main reinforcement in the beam tension zone can increase ductility, control and delay the formation and propagation of cracking, reduce deflection, and increase the high-performance characteristics of environmentally exposed structural components, as demonstrated in Chen and Nawy (1994). This work involved several patterns of composite beams, 14,000 psi in concrete strength (97 MPa) as shown in Figure 22.15. All evaluations were based on test-to-failure results. Fiber optic Bragg grating sensors developed for this work were used to monitor strains, deformations, and crack widths in the tested 10 ft (3.3 m) span simple and continuous beams.

The relative mild steel and prestressed prism contents of the beams affect the behavior of these composite members [Chen and Nawy, 1994; Nawy and Chen, 1997]. To account for this influence,

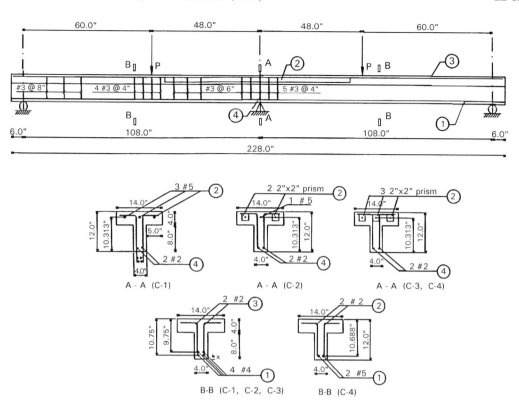

FIGURE 22.15 Prism composite reinforcement geometry. *Source:* Nawy, E.G. and Chen, B. 1998. Fiber optic sensing of the behavior of prestressed prism-reinforced continuous composite concrete beams for bridge deck application. *Proceedings, Transportation Research Board.* National Research Council, Washington, DC.

a combined reinforcement index ω is used in the design of a prism-reinforced structural concrete beam. This index value is obtained from the following expression:

$$\omega = \frac{A_{ps}f_{ps} + A_s m f_y - A'_s f'_y}{bd_p f'_c} \qquad (22.13)$$

where A_{ps}, A_s, and A'_s are the areas of prestressing and nonprestressing reinforcement respectively, f_{ps} is the design stress in the prestressing reinforcement at ultimate load, f_y and f'_y are the yield strengths of deformed bars in tension and in compression, b is the width of the compression face of the member or flange in the case of a T-section, and d_p is the effective depth of the beam cross section.

Other systems of composites can also improve the performance of reinforced concrete beams through the use of two-layer systems, one of which is made out of normal or high-strength concrete and the layer above or below is made of high-strength polymer concrete. In such cases, the beam cross section is built in two-layers, in a similar manner to the SIFCON two-layer system. Several investigations have been conducted that demonstrated that the shear friction interaction of the interface between the two layers of different strength concretes, one of which being polymer concrete, has a resistance to slip superior to the interlock between two layers both made from concrete only [Nawy et al., 1992].

Part B: Fiber-Reinforced Plastic Composites (FRP)

22.8 Historical Development

Use of nonmetallic fibers, particularly fiberglass elements, bundled into continuous reinforcing elements has been considered since the 1950s for prestressing reinforcement [ACI Committee 440, 1996; Nawy, 1996b; Rubinsky and Rubinsky, 1954]. Advances were made in polymer development by using polymer-impregnated bundled fiberglass fibers as rods for anchorages in tunnelling. In the mid-1960s, Nawy and his team [Nawy and Neuwerth, 1977; Nawy et al., 1971; ACI 440R 1996] conducted the original extensive work on the use of bundled and resin-impregnated glass fibers formed into deformed bars as main reinforcement in structural elements. Except for cases where magnetic fields in supporting structures had to be avoided, the commercial applications of the bundled and resin-impregnated type as main reinforcement in structural concrete elements were not recognized until the late 1970s. It is important to state at this juncture that the term "plastic" could be misleading; hence there is general consensus at this time to define FRP as fiber-reinforced polymer composites.

In the 1980s, an increased interest in and use of such glass fiber reinforced plastic (GFRP) reinforcing bars was developing. This was particularly overdue and important for reinforced concrete surrounding or supporting magnetic resonance imaging (MRI) medical equipment. Such equipment includes sensitive magnets and cannot tolerate the presence of any steel reinforcement. Also, where environmental and chemical attacks are present, GFRP reinforcement is more durable and efficient as concrete reinforcement. As stated in ACI Committee 440 (1996), composite rebars have more recently been used in the construction of seawalls, industrial roof decks, base pads for electrical and reactor equipment, and concrete floor slabs in aggressive chemical environments. In 1986, Germany built the world's first highway bridge using composite reinforcement [ACI Committee 440, 1996]. In the United States, significant funds have been expended on product evaluation and further development, and at least nine major companies have been actively marketing this product since the early 1990s.

Additionally, since the 1980s considerable progress has been made in using glass fiber filaments as a supplement to concrete matrices to improve the mechanical properties of concrete, but not as a replacement for the main bar reinforcement in supporting structural components. Glass fibers that are alkali resistant are also gaining wide use. These normally contain zirconium (Z_rO_2) to minimize or eliminate the alkaline corrosive attack on glass present in the cement paste.

Synthetic fibers made from nylon or polypropylene, both loose and woven into geotextile form, have also recently gained some application as more information on their mechanical performance in the matrix is gained and as better understanding of their structural contribution to crack resistance is reached. As other types of nonmetallic fibers were explored during the late 1980s and early 1990s, interest in the use of carbon fibers as possible main reinforcement apart from its use in cement-based composites has increased. It is safe to state now that the science of fibrous concrete and composites has advanced to an extent that justifies its extended use in the years to come.

22.9 Beams and Two-Way Slabs Reinforced with GFRP Bars

In the late 1960s and early 1970s, Nawy and his team at Rutgers University [Nawy and Neuwerth, 1977; Nawy et al., 1971] embarked on researching the use of glass fiber reinforced plastic bars as a substitute for mild-steel reinforcement. Those investigations involved testing to failure a total of 30 beams and 12 two-way slabs. The slabs had an average thickness of 2 1/2 in and overall dimension of 7 ft × 7 ft. The slab panels were 5 ft 6 in × 5 ft 6 in effective spans and were fully restrained along all four boundaries. The GFRP reinforcement was spaced at 3 to 8 in. In the different slabs the reinforcement area varied from 0.196 in²/ft to 0.074 in²/ft in each direction, giving reinforcement percentages of 0.769 and 0.290 percent respectively. The beams were either simply supported or continuous over two spans. The centerline span was 9 ft 11 in. The reinforcement percentages in the beams ranged from 1.045 to 0.696 percent.

These original tests and the analysis results indicated that both the fiberglass-reinforced slabs and the beams behaved similarly in cracking, deflections, and ultimate load to steel reinforced beams. The large number of well-distributed cracks in the GFRP reinforced beams and slabs indicated good mechanical bond developed between the GFRP bar and the surrounding concrete. The research also demonstrated that the equations for flexure accurately predicted the flexural behavior of GFRP reinforced members with the same accuracy as for the mild-steel reinforced beams. A typical stress-strain diagram of the reinforcement is shown in Figure 22.16. Since that time, interest resulted in

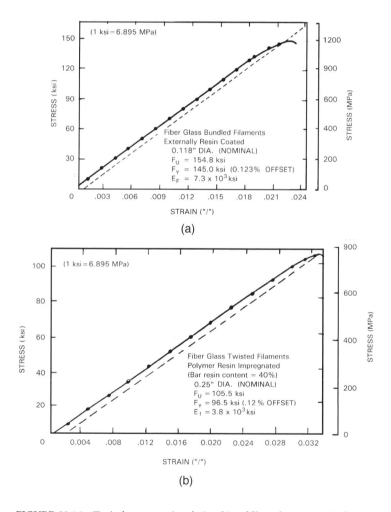

FIGURE 22.16 Typical stress-strain relationship of fiber-glass composite bar reinforcement. (a) Coated filaments. *Source:* Nawy, E.G. and Neuwerth, G.E. 1977. Fiber glass reinforced concrete slabs and beams. *Proc. ASCE J. Struct. Div.* 103(ST2):421–440; (b) impregnated filaments. *Source:* Nawy, E.G., Neuwerth, G.E., and Phillips, C.J. 1971. Behavior of fiber glass reinforced concrete beams. *Proc. ASCE J. Struct. Div.* 97(ST9):2203–2215.

TABLE 22.5 Comparison of Mechanical Properties of GFRP and Steel Reinforcement

	Steel Rebar	Prestressing Steel Tendon	GFRP Bar	GFRP Tendon	CFRP Tendon
Tensile strength					
MPa	483–690	1379–1862	517–1207	1379–1724	1665–2068
ksi	70–100	200–270	75–175	200–250	240–300
Yield strength					
MPa	276–414	—	—	—	—
ksi	40–80	—	—	—	—
Tensile modulus					
GPa	200	200	414–552	48–62	152–165
ksi $\times 10^{-3}$	29	29	6–8	7–9	22–24
Compressive strength					
Mpa	276–414	—	310–482	—	—
ksi	40–80	—	45–70	—	—
Coiff. Therm./°C	11.7	11.7	9.9	9.9	0
Exp. $(\times 10^{-6})$/°F	6.5	6.5	5.5	5.5	0
Specific gravity	7.9	7.9	1.5–2.0	2.4	1.7

Source: ACI Committee 440. 1996. *State-of-the-Art on Fiber Reinforced Plastic (FRP) Reinforcement for Concrete Structures,* ACI Report 440R-96. American Concrete Institute, Farmington Hills, MI.

Note: All strengths are in the longitudinal direction.

investigations by Larralde et al., Satoh et al., Goodspeed et al., Ehsani et al., Zia et al., Bank and Xi, Porter et al., Faza and GangaRao, and Nanni (1993). A summary of their work and publications as well as details of the original Nawy (1971) work are given in the ACI Committee 440 report (1996). Table 22.5 gives a relative comparison of the mechanical properties of GFRP and steel reinforcement.

22.10 Carbon Fibers and Composite Reinforcement

22.10.1 Carbon Fibers

Essentially, there are two types of carbon fibers: high-modulus type I and high-strength type II. The fundamental difference between their properties is the result of the differences in their microstructures, which depend on the arrangement of the hexagonal graphene-layer networks in the graphite [ACI Committee 440, 1996]. The high-modulus type I carbon fibers, in order to attain a modulus of 30×10^6 psi (200 GPa), have their graphine layers aligned approximately parallel to the axis of the fibers. Examples are Kevlar 49 by DuPont and Twaron 1055 by Akzo Nobel. Ultra-high-modulus fiber Kevlar 149 and Twaron 2000 are also available. Table 22.6 lists the minimum average strength values of Kevlar and Twaron reinforcing fibers.

TABLE 22.6 Properties of Kevlar and Twaron Reinforcing Fibers

Property	Kevlar 49	Twaron 1055
Tensile strength, psi (MPa)	525,000 (3,620)	522,000 (3600)
Modulus, psi (MPa)	18×10^6 (124,000)	18.4×10^6 (127,000)
Elongation at break (percent)	2.9	2.5
Density, lb/in^3 (g/cm^3)	0.052 (1.44)	0.52 (1.45)

Source: ACI Committee 440. 1996. *State-of-the-Art on Fiber Reinforced Plastic (FRP) Reinforcement for Concrete Structures,* ACI Report 440R-96. American Concrete Institute, Farmington Hills, MI.

Additionally, hybrid composites made from carbon-glass-polyester are available with a strength up to 115,000 psi (790 MPa), a modulus of 18×10^6 psi (124 MPa), and a density in the range of 0.060 to 0.069 lb/in³. It is important to state that because of the low ductility of carbon fiber in comparison with steel, it is unlikely that it would be used in the form of composite main bar reinforcement. Economically, it would be cost prohibitive. However, it can be used and is being used as prestressing reinforcement and as fabric reinforcement because of its high strength and high modulus as seen from Table 22.6.

22.10.2 Use as Internal Prestressing Reinforcement

FRP carbon tendons made with fiber elements of 0.125 to 0.157 in. (3–4 mm) diameter are used for prestressing. Their ultimate strength is comparable to prestressing strands, ranging between 270,000 and 300,000 psi (1866–2070 MPa).

22.10.3 Use as External Reinforcement

Unidirectional FRP sheets made of carbon (CFRP) or glass fiber (GFRP) bonded with polymer matrix (epoxy, polyster, vinyl, or ester) are being used as protection against corrosion, and they eliminate the need for joints because of the unlimited length of the composite sheets [ACI Committee 440, 1996]. They are useful in increasing the flexural and shear strength of concrete members when these composite plates are epoxy bonded to the exterior facing of the elements. They are also of particular use in the retrofit of deteriorating concrete structures and also retrofitting columns in seismic zones.

Several techniques for wrapping concrete elements with CFRP sheets have been developed by investigators, as detailed in the ACI Committee 440 report (1996) [ACI Committee 440, 1996; Nanni, 1993]. CFRP resin-impregnated strands can be spirally wound onto the surface of an existing concrete element such as a bridge pier or a structural building column. In this manner, owing to the confinement imposed on the member, shear capacity and ductility is improved. Nanni's (1993) work on the effect of wrapping conventional concrete demonstrated that significant enhancement can be achieved in the strength and ductility of the wrapped concrete element.

22.11 Fire Resistance

The resistance of fiber-reinforced polymer plastic composites is relatively lower than that of other systems, with some degradation of the polymer resin content under heat and ultraviolet light that causes some long-term durability problems. The carbon and glass fibers and the fabrics used in the FRP can withstand normal fire exposure and are durable under UV light. But the weak link is the organic polymers used to prepare the fiberglass or carbon used as reinforcing elements through impregnation or wrapping. Recent research [Foden et al., 1996] seems to address this deficiency by substituting an inorganic resin for the organic polymer.

An inorganic resin can be an alkali alumino silicate that can set at moderate temperatures and be able to withstand up to 1000°C. The system is highly impermeable so that it can protect the carbon filaments from oxidation. Tests conducted on carbon, silicon carbide, and glass composites under tension, bending, shear, and fatigue loading [Foden et al., 1996] indicated that the mechanical properties of the nonorganic composite used are comparable to those of organic polymer composites while having the advantage of relatively higher fire resistance.

22.12 Summary

The concretes described in this chapter have demonstrated that the strength, ductility, and performance of concretes and cement-based composites have and will continue to achieve higher plateaus.

A new era in construction materials technology has commenced. It promises a revolutionary impact on the manner in which constructed systems will emerge in the twenty-first century.

Considerable work needs to be done to enhance the practicability of these materials and make them cost effective. It is only with simplicity and practicability in application and the achievement of a cost-effective competitive end product that these developments in the science of materials technology can gain universal acceptance and application.

Acknowledgment

This chapter is based on extensive material taken with permission from *Fundamentals of High Strength High Performance Concrete* by E.G. Nawy, Addison Wesley Longman, London and California, and from *Reinforced Concrete—A Fundamental Approach*, 3rd. ed., and *Prestressed Concrete—A Fundamental Approach*, 2nd. ed., both by E.G. Nawy, Prentice Hall, Upper Saddle River, New Jersey, and from various committee reports and standards of the American Concrete Institute, Farmington Hills, Michigan.

References

ACI Committee 440. 1996. *State-of-the-Art on Fiber Reinforced Plastic (FRP) Reinforcement for Concrete Structures*, ACI Report 440R-96. American Concrete Institute, Farmington Hills, MI.

ACI Committee 544. 1988. *Design Considerations for Steel Fiber Reinforced Concrete*, ACI Committee Report 544.R4-88, pp. 1–16. American Concrete Institute, Farmington Hills, MI.

ACI Committee 544. 1989. *Measurement of Properties of Fiber Reinforced Concrete*, ACI Committee Report 544.2R-89, pp. 1–12. American Concrete Institute, Farmington Hills, MI.

ACI Committee 544. 1991. State-of-the-art on fiber reinforced concrete. In *Manual of Concrete Practice*, Vol. 5. ACI Committee Report 544.1R-82. American Concrete Institute, Farmington Hills, MI.

ACI Committee 544. 1993. *Guide for Specifying, Proportioning, Mixing, Placing and Finishing Steel Fiber Reinforced Concrete*, ACI Committee Report 544.3R-93, pp. 1–10. American Concrete Institute, Farmington Hills, MI.

Bayasi, M.Z. 1992. Application of carbon fiber reinforced mortar in composite slab construction. In *Proceedings, International RILEM/ACI Workshop*, pp. 507–517. Chapman and Hall, New York.

Bentur, A. and Mindess, S. 1990. *Fiber Reinforced Cementitious Deposits*. Elsevier Applied Science, London.

Chen, B. and Nawy, E.G. 1994. Structural behavior evaluation of high strength concrete beams reinforced with prestressed prisms using fiber optic sensors. *Proc. ACI Struct. J.* No. 91-S69, 708–717.

Fanella, D.A. and Naaman, A.E. 1985. Stress-strain properties of fiber reinforced concrete in compression. *Proc. ACI J.* 82(4):475–483.

Foden, A., Lyon, R., and Balaguru, P. 1996. A high temperature inorganic resin for use in fiber reinforced composites. Paper presented at First International Conference on Composites in Infrastructure. Organized by the National Science Foundation. Tucson, AZ, January.

Henager, C.H. and Doherty, T.J. 1976. Analysis of fibrous reinforced concrete beams. *Proc. ASCE Struct. J.* 12(SF1):177–188.

Hsu, L.S. and Hsu, T.C.T. 1994. Stress-strain behavior of steel-fiber high-strength concrete under compression. *Proc. ACI Struct. J.* 91(4):448–457.

Mindess, S. and Young, J.F. 1981. *Concrete*, pp. 670. Prentice Hall, Upper Saddle River, NJ.

Naaman, A.E. 1992. SIFCON: Tailored properties for structural performance. In *Proceedings, International RILEM/ACI Workshop*, pp. 18–38. Chapman and Hall, New York.

Nanni, A. 1993. Flexural behavior and design of R.C. members using FRP reinforcement. *ASCE J. Struct. Eng.* 119(11):3344–3359.

Nawy, E.G. 1996a. *Reinforced Concrete: A Fundamental Approach*, 3rd ed., pp. 848. Prentice Hall, Upper Saddle River, NJ.

Nawy, E.G. 1996b. *Fundamentals of High Strength High Performance Concrete*, 350 pp. Addison Wesley Longman, London and California.

Nawy, E.G. 1996c. *Prestressed Concrete—A Fundamental Approach*, 810 pp. Prentice Hall, Upper Saddle River, NJ.

Nawy, E.G. and Blair, K. 1971. Further studies on flexural crack control in structural slab systems. *Proceedings, International Symposium on Cracking, Deflection and Ultimate Load of Concrete Slab Systems*, ACI SP-30. E.G. Nawy, ed., pp. 1–30 to 1–42. American Concrete Institute, Farmington Hills, MI.

Nawy, E.G. and Chen, B. 1998. Fiber optic sensing of the behavior of prestressed prism-reinforced continuous composite concrete beams for bridge deck application. *Proceedings, Transportation Research Board*. National Research Council, Washington, DC.

Nawy, E.G. and Neuwerth, G.E. 1977. Fiber glass reinforced concrete slabs and beams. *Proc. ASCE J. Struct. Div.* 103(ST2):421–440.

Nawy, E.G., Neuwerth, G.E., and Phillips, C.J. 1971. Behavior of fiber glass reinforced concrete beams. *Proc. ASCE J. Struct. Div.* 97(ST9):2203–2215.

Nawy, E.G., Ukadike, M.M., and Balaguru, P.N. 1992. Investigation of concrete PMC composite. *ASCE J. Struct. Div.* 108(ST5):1049–1063.

Reinhardt, H.W. and Naaman, A.E., Ed. 1992. High performance fiber reinforced cement composite. In *Proceedings, International RILEM/ACI Workshop*, pp. 565. Chapman and Hall, New York.

Richard, P. and Cheyrezy, M.H. 1994. Reactive powder concretes with high ductility and 200–800 MPa compressive strength. In *Proceedings, V.M. Malhotra Symposium on Concrete Technology: Past, Present and Future*, P.K. Mehta, ed., ACI SP-144, pp. 507–518. American Concrete Institute, Farmington Hills, MI.

Romualdi, J.P. and Batson, G.B. 1963. Mechanics of crack arrest in concrete. *Proc. ASCE Eng. Mech. J.* 89(EM3):147–168.

Romualdi, J.P. and Mandel, J.A. 1964. Tensile strength of concrete affected by uniformly distributed closely spaced short lengths of wire reinforcement. *Proc. ACI J.* 61(6):657–671.

Rubinsky, I.A. and Rubinsky A. 1954. An investigation into the use of fiber-glass for prestressed concrete. *Mag. Concr. Res.*

Schneider, B. 1992. Development of SIFCON through application. In *Proceedings, International RILEM/ACI Workshop*, pp. 177–194. Chapman and Hall, NY.

Shah, S.P. 1983. Fiber reinforced concrete. chapt 6. In *Handbook of Structural Concrete*, pp. 6–1 to 6–14. McGraw–Hill, New York.

Shah, S.P. and Rangan B.V. 1971. Fiber reinforced concrete properties. *Proc. ACI J.* 68(2):126–135.

Sharama, A.K. 1986. Shear strength of steel fiber reinforced concrete beam. *Proc. ACI J.* 83(4):624–628.

Swamy, R.N. 1975. Fiber reinforcement of cement and concrete. *J. Mater. Struct.* 8(45):235–254.

Swamy, R.N., Mangat, P.S., and Rao, C.V. 1974. The mechanics of fiber reinforcement of cement matrices. *Fiber Reinforced Concrete*, ACI SP-44, pp. 1–28. American Concrete Institute, Farmington Hills, MI.

Williamson, G.R. 1978. Steel fibers as web reinforcement in reinforced concrete. In *Proc. U.S. Army Service Conference*, Vol. 3, pp. 363–377. West Point.

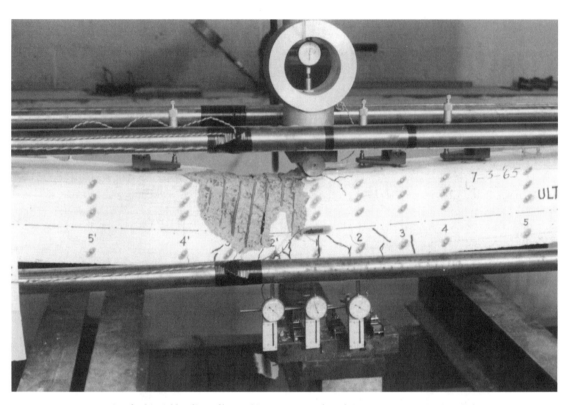

Instrumentation for biaxial loading of beam (Courtesy Dr. Edward G. Nawy, Rutgers University).

23

Performance Evaluation of Structures

by
Richard A. Miller, Ph.D., P.E.
Associate Professor of Civil Engineering, University of Cincinnati, Cincinnati, Ohio. Expert in non-destructive testing and evaluation of existing bridges.

23.1 Introduction

The design process for structures relies heavily on the application of sound structural theory. However, because of the complexities of material behaviors, presence of flaws, possibilities of manufacturing or construction errors, difficulty in accurately estimating environmental effects, and the chance of a catastrophic events, engineering theories are, at best, incomplete simplifications of reality. As a result, an engineer using theory and/or analysis can only assign probabilistic minimum and maximum values to structural responses. Some of these values may be separated from reality by orders of magnitude. There are many cases where much more accurate predictions of structural behaviors are needed. In these cases, performance evaluation is the only reasonable alternative.

0-8493-2666-4/97/$0.00+$.50
© 1997 by CRC Press LLC

Performance evaluation can be defined as the destructive, partially destructive, or nondestructive physical test of a structure, structural system, or structural subsystem to determine critical responses under one or more loading conditions. The purpose of performance testing is the evaluation of service, cracking or ultimate load capacity, degree of safety, reliability, stiffness or flexibility, existence and effect of damage or deterioration, effect of support conditions, deflection under service or ultimate loads, fatigue and cyclic load behavior, and response to impact or vibrational loads. Performance tests can be divided into two groups: tests for quality control (including verification of load capacity) and tests to assess structural performance when existing theories are inadequate or nonexistent.

When most people think of performance evaluation, they usually think of strength evaluation; specifically the method described in the American Concrete Institute's *Building Code Requirements for Structural Concrete* [ACI-318, 1995]. The method, discussed in Chapter 20 of the ACI Code, is intended to evaluate structures that are thought to violate the safety provisions of the code. ACI Committee 437, Strength Evaluation of Structures (1991), presents a more in-depth discussion of structural strength evaluation. However, even the methods presented in that report have limitations. The committee defines several cases where the strength-evaluation techniques discussed in the report may be useful:

1) distressed structures that show signs of damage due to overload, fire, vibration, etc;
2) deteriorated structures that show signs of damage through material degradation such as corrosion, excessive cracking of the concrete, spalling, etc;
3) structures suspected of containing understrength materials or to be substandard in design or construction;
4) structures where the capacity to hold loads applied in the future is in doubt because the original design and/or construction data is not available;
5) change in use of the structure where the new usage would apply loads in excess of the original design loads; and
6) to assess structures undergoing repair, retrofit, or strengthening.

The committee recognizes other areas where performance testing may be valuable, but does not recommend using the strength-evaluation method presented in the report for these areas:

1) structures with unusual design concepts;
2) product development testing used for approval or quality control for mass-produced elements;
3) evaluation of foundations and/or soil conditions; and
4) structural research.

Other possible areas where strength or performance testing may be useful are

1) as an inspection technique to assess the condition of a structure, especially when hidden damage or deterioration is suspected; and
2) to verify the results of analytical modeling.

This chapter attempts to present performance testing in the broadest possible terms.

23.2 ACI 318-95 Provisions on Strength Evaluation of Existing Structures

The method described in Chapter 20 of the ACI Code is the one most commonly used for performance evaluation of concrete structures. It makes sense to begin any discussion of performance evaluation with this method. It is intended to evaluate the safety of a structure whose load carrying capacity is in doubt. If the strength deficiency is well understood and the structural dimensions and material strengths are easily verified, analytical evaluation of the structure is permitted. Otherwise, load testing is required. Load testing is also required to verify the strength of deteriorated structures.

If load testing is to be undertaken, the specification requires the structure to be loaded such that any suspect areas are subjected to maximum stress and deflection. If more than one load placement is needed to achieve this, then multiple load placements must be used. However, only one load intensity is required:

$$U = 0.85(1.4DL + 1.7LL) \tag{23.1}$$

where U is the test load, DL is the unit (unfactored) dead load and LL is the unit live load. The testing sequence is as follows:

1) Obtain an initial measurement of all important structural responses: deflection, strains, crack patterns, crack openings, rotations, etc. Of all these responses, only deflection and cracking are used for acceptance, but the code recognizes that other responses, although not required, may be of interest.
2) The load is applied in four equal increments. Between each increment, structural response is measured. No maximum or minimum time limit between application of the load increments is specified.
3) Once the load is in place, it is left for 24 h. Structural response measurements are taken at the beginning and end of this period.
4) After the 24-h period, all structural responses are measured and the load is immediately removed. Again the structure is left for 24 h and the structural responses are again measured.

There are two acceptance criteria:

1) After removal of the load, the structure must recover at least 75% of the maximum deflection:

$$\Delta_{r\max} = \Delta_{\max}/4 \tag{23.2}$$

where $\Delta_{r\max}$ is the deflection 24 h after removing the load and $\Delta_{\max}$ is the deflection under load after the load is in place 24 h. Because very stiff structures have small deflections, it may be difficult to accurately measure deflection recovery, so the code also accepts a structure if

$$\Delta_{\max} \leq \ell_t^2/20000h \tag{23.3}$$

where ℓ_t is the span of the member (taken as the smaller of center to center of supports or clear span $+h$) and h is the member thickness. For two-way slabs, ℓ_t is the shortest span. Structures that fail the deflection part of the test may be retested after 72 h and accepted if

$$\Delta_{r\max} = \Delta_{f\max}/5 \tag{23.4}$$

where $\Delta_{f\max}$ is the maximum deflection during the second test.
2) The structure cannot exhibit excessive cracking, spalling, or crushing, which may indicate failure. Inclined cracks in members without transverse reinforcement or cracks in anchorage zones must be evaluated.

It is clear that ACI-318 requirements are narrowly tailored to evaluate structural safety against failure. If other structural responses are desired, such as performance and serviceability under working load, changes in structural response due to damage or deterioration, etc., then broader performance-testing techniques are needed.

23.3 Pretest Planning for Reliable Structural Evaluation

The most critical phase of performance evaluation is the pretest planning. In this phase, a detailed plan or critical path should be established not only to assure the tests are performed correctly but to be sure that all critical responses are measured.

Pretest planning can be broken down into several distinct general tasks:

1) Determine the specific purpose of the test and desired final product: This is probably the most overlooked part of pretest planning. Prior to beginning the test, the following items should be evaluated:

   a) Determine the specific desired outcome. Is the engineer looking at service-load behavior? ultimate-load behavior? presence and effect of damage? presence and effect of possible deficiencies in materials or construction? ability to use the structure safely at higher loads than the original design? etc. Before beginning any test, it is important to have a clear picture of the final outcome.

   b) Determine the test method that best meets the goal. When evaluating a structure, engineers often choose load testing as a first, or only, alternative. In fact, there are many ways to evaluate a structure other than just load testing (e.g., modal testing).

   c) If load testing is chosen, be sure the loading method is realistic for the structure. For example, uniform loads are not a good choice for testing bridges as traffic loads tend to be point loads. Also be sure the loading method will produce the desired results, for instance, owing to the problems with creating a moving load, stationary, cyclic loads are often used to simulate traffic loads. However, stationary, cyclic loads may not produce the same response as moving load [Petrou et al., 1994].

   d) Determine the type of data needed to accurately assess the structure. In many performance tests, only deflection is measured. This may not be enough data. Often strains, rotations, differential movements, temperature changes, or accelerations may need to be measured. Instrumentation should be properly chosen to provide the necessary data.

2) Gather all available design and construction information: Once the purpose of the test is established, the next step is to gather all available design and construction information. This provides the basic background of the structure and provides a baseline for comparing actual material test results and actual dimensions. Necessary information on design loads, material specification and strengths, design drawings, construction records, "as-built" drawings, and records of any modifications to the structure must be assembled and carefully examined to form a complete picture of the structure.

   It is important to obtain as much information as possible about the as-built condition of the structure as variations and deviations from the original plans are frequent. Obtaining as-built drawings and/or copies of inspection records is highly recommended.

   Reports on actual material strength are also of great importance, especially in destructive or partially destructive tests. It should be kept in mind that material specifications are usually minimum or maximum properties (as applicable) and the actual materials may have substantially greater or lower properties than required. For example, ASTM A-36 steel has a minimum yield strength of 250 MPa (36 ksi), but in reality the yield strength can be much higher with values of 350 MPa (50 ksi) not being unusual.

3) Pretest inspection of the structure and material sampling: For many structures, the construction records are inadequate. In some cases, the records have been lost or destroyed. In other cases, the urge to save money during construction caused the owner or contractor to use insufficient inspection, so detailed inspection and construction records do not exist. In such cases, an inspection of the structure is necessary.

The structure should first be checked to verify all dimensions. Several measurements of member sizes should be taken at critical sections. Where possible, member support conditions should be checked. At this time, a detailed survey of surface and visible defects should be made. The *Guide for Making a Condition Survey of Concrete in Service* [ACI Committee 201, 1992] provides a good method for making such a survey.

Reinforcing-bar size, spacing, and cover should be documented. Three common nondestructive methods for measuring reinforcing-bar size, cover, and spacing are magnetic, radiographic, and radar [ACI Committee 437, 1991]. Bar size and location can also be found by removing the cover in isolated areas. This last method is often used to calibrate nondestructive testing methods (NDT) methods. Where practical, bar samples should be removed for further evaluation including strength tests and assessment of the degree of corrosion, if present.

Concrete strengths are normally determined by coring (ASTM C42) as this seems to be the only accurate method for determining in-place concrete strength. There are many so-called non-destructive tests, such as the rebound hammer or probe penetration, but these methods do not directly measure strength and show a high degree of variability, especially when different types of concrete are being tested. As a result, the methods require calibration with drilled cores for each type of concrete being tested and, even then, are of questionable accuracy. Tensile strengths can be found by performing the split cylinder test (ASTM C496) or by use of sawed beams (ASTM C42) if practical.

In addition to cores removed for strength testing, cores should also be taken for petrographic and/or chemical testing. Petrographic and chemical testing can reveal potential weakness in the concrete due to alkali-silica reaction (ASR), inhomogeneity, bleed or segregation, poor air-void systems that allow freeze/thaw damage, the potential for reinforcing steel corrosion, abrasion, fire, "d" cracking of aggregates, and weathering.

It is also necessary to sample and test auxiliary materials such as overlays, bearing pads, material in attached structures, etc. This is especially necessary when such materials may affect loading, boundary conditions, and structural or material behaviors.

When sampling the material, the question arises as to how many samples are necessary to yield reliable results. The answer to this question is based on the practicality of sampling the material and the amount of error that can be tolerated. In general, the number of samples is affected by the following:

a) The amount of possible damage that would occur when samples are removed: The attraction of nondestructive testing is that any number of "samples" can be taken without damaging the structure.

b) The cost, in money and time, to perform the test: Often testing must be limited owing to time and budget constraints.

c) The importance of the data: For example, flexural strength of reinforced concrete is only slightly affected by the compressive strength, so less accuracy can be tolerated in determining the compressive strength of a flexural member. On the other hand, the strength of axial members is greatly influenced by compressive strength, so a more accurate determination of compressive strength is needed.

To determine the number of samples needed, any one of a number of statistical techniques can be applied. ASTM C823 provides information on developing a program for material sampling.

1) Perform a pretest analytical investigation: The data collected from records and inspection can be used to create an analytical model of the structure. This model will provide necessary information for designing the test. In the case of load tests, the model will provide information on probable cracking, yield, and ultimate load capacities. The model can also be used to pinpoint critical responses so that instrumentation can be properly placed to record these responses during the actual test. In the case of dynamic tests, frequencies, accelerations, magnification (impact) factors, and vibrational mode shapes can be estimated from the model.

2) Design the testing and instrumentation system: The test can be designed using the results from the model. In the case of load testing, it is necessary to choose type of load, load levels,

load position, and method of loading (i.e., trucks, blocks, hydraulic, etc.). Load-reaction frames for hydraulic-loading cylinders also need to be designed at this point.

Instrumentation grids can be laid out based on the analytical results so that instruments are in place to capture critical responses. The model results are also used to estimate the magnitude of the response so that an instrument with proper range and accuracy is chosen (see Section 23.6 on instrumentation). The data acquisition system(s) should be chosen at this time.

3) Perform the test and evaluate the results: With a properly designed test, performance of the test becomes a relatively simple task. Time spent up front in planning will greatly reduce the chance of problems occurring during the test and lead to a successful outcome.

Unfortunately, evaluation of the results is not an easy task. Even in tests that have excellent pretest planning, there are often ambiguities in the data that must be reconciled. In such cases, it is often useful to create posttest models to assist in data evaluation.

23.4 Nondestructive Testing for Material and Structural Assessment

In pretest evaluation, it is necessary to carefully evaluate the condition of the material and the structure. In the past, this was done by coring. The problem with coring is that it is partially destructive, leaving holes in the structure. Also, coring only permits limited inspection of material; that is, only the material in cored areas is inspected. Since it not practical to take large numbers of cores from the structure, coring leaves large areas uninspected. In spite of these limitations, coring is still very useful for obtaining reasonable estimations of concrete strength and cores can be subjected to petrographic examination, which reveals a wealth of information about microstructure, chemical deterioration, air contents, etc. However, for large-scale detection of flaws, other methods are needed. There are many nondestructive methods for evaluation of concrete structures. Only a few common methods are presented here.

23.4.1 Ultrasonic Testing

Sound waves have been used for testing concrete for many years. Many states still employ the "chain-drag" method for concrete bridge decks. In this method, a large metal chain is dragged across the concrete surface, subjecting the surface to small impacts. Delaminations in the concrete are found when a "hollow" sound is heard during the dragging of the chain. The limitations of this method are obvious: it only detects flaws near the surface, and an experienced person who can differentiate the sounds is needed. More modern methods of testing concrete utilize ultrasonic waves transmitted through the concrete. Equipment to conduct such tests is commercially available.

A simple ultrasonic test, called a pulse-velocity test, consists of creating an ultrasonic pulse, detecting the same pulse at a remote receiver, and measuring the time needed for the pulse to travel the distance. Changes in the pulse transmission time can represent changes in the properties of the concrete. For example, as young concrete gains stiffness, the pulse transmission time decreases. Increases in pulse transmission time often indicate the presence of damage since the pulse cannot transmit through a crack and instead must take a longer path around the damaged area.

The impact-echo technique provides a more quantitative measurement of flaws in the concrete [Carino et al., 1986; Carino and Sansalone, 1992]. In this method, a small impactor is used to create a wave in the concrete. When the wave hits an interface, either a free surface or a crack, it is reflected back to a receiver. By processing the returned signal [Carino and Sansalone, 1992], the depth and size of flaws can be determined.

23.4.2 Infrared Thermographic Testing

The principal of thermographic testing is that concrete containing voids, delamination, or other anomalies will conduct heat at a different rate than sound concrete. When the concrete is heated or cooled, temperature differentials can be measured over large areas using infrared cameras [Weil, 1993; Delahaza, 1996]. Damage is then seen as cooler or hotter areas. Often, heating of the structure by the sun during the day and/or the cooling of structure during night produces sufficient temperature changes to make infrared thermography usable.

23.4.3 Modal Testing

When evaluating a concrete structure, it is beneficial to have information about the actual stiffness or flexibility of the structure. Changes in flexibility over time can be used as a measure of global deterioration, and local changes in flexibility can be used to pinpoint local damage. For modeling purposes, the accuracy of a model can be determined by comparing the actual flexibility of the structure with the flexibility obtained from the model. One method of determining the actual flexibility of a structure is polyreference modal testing.

Modal testing can be used by itself as a nondestructive method of structural evaluation. It can also be used as a pretest evaluation procedure for planning a load test of a structure, as modal testing can identify damaged, deteriorated, or suspect areas of the structure [Aktan et al., 1992]. After the load test, the flexibility matrix obtained from modal testing can be used to verify the results of the load test and to calibrate finite-element models.

In the modal-testing method, structural vibrations are produced by an impact [Aktan et al., 1992; Allbright et al., 1994] (forced vibrations have also been used, but this method is more complex so only the impact method will be discussed here). The structural vibrations resulting from the impact are then measured by seismic accelerometers. By knowing the time histories of the impact (measured with a load cell) and the corresponding accelerations distributed through the structure, the modal frequencies, modal vectors, and damping can be calculated. Modal testing offers the advantage of not having to calculate the mass and stiffness of the member; instead, vibrational characteristics are measured and used directly. From the magnitude and frequency of the vibrations, the natural mode shapes and frequencies can be determined and these can be further used to calculate a flexibility matrix.

Prior to modal testing, an approximate-theoretical or finite-element analysis should be made so that there is a clear understanding of the dynamic structural behavior. The structure can then be discretized by a grid system for the purpose of establishing impact points and accelerometer (reference) points. The results of approximate-theoretical or finite-element analysis are used so that the accelerometers are not placed at grid points that would correspond to the nodes of the first few significant modes. The analysis is also used to establish the bandwidth of expected frequencies, which is necessary so that the dynamic data acquisition system will read only within the band of frequencies of interest.

In theory, the test can be performed with a single accelerometer, but in practice a minimum of three, noncolinear reference points are used, although more reference points will often provide better data [Aktan et al., 1992; Allbright et al., 1994]. Impact to the structure is supplied either by a sledgehammer or a falling weight. In both cases, the impactor must be instrumented with a load cell so that the exact impact load time history is known. Instrumented hammers are commercially available. Each grid point (including the reference points) undergoes impact, and the resulting vibrations are measured by the accelerometers. The impactor load cell and the accelerometers are read using a dynamic analyzer (these are also commercially available). A fast Fourier transform (FFT) is used to transform both the input (impactor) data and the output (accelerometer) data from the time domain into the frequency domain. The frequency response function (FRF), which is basically the ratio of the output FFT to the input FFT, is then calculated. The calculation of the FFT and FRF will normally be performed by the dynamic analyzer, although the operator should have some degree of expertise to assure the quality of the final data.

When impacts are applied to the structure, care must be taken to assure the impactor does not bounce and create multiple impacts while the acceleration data are being taken, as this would corrupt the acceleration data. Also, in some cases the impact is too hard or too soft and the resulting data are outside the range of the data-acquisition system. Such tests should be discarded and redone. This is why it is essential to use a dynamic analyzer capable of processing the results in real time.

Each time a point is struck and acceleration data are taken, a single data set is created for that point. However, one data set per impact point is not enough to yield reliable results, so each point is tested several times to collect multiple data sets, the results of which are then averaged. In general, it is necessary to have at least five good data sets per point.

Once the dynamic data are obtained for all impact points, they can be transferred to a commercial program that will then calculate the modal frequencies, modal vectors (mode shapes), and damping. Anomalies in the frequencies and/or mode shapes can sometime be used to identify damage [Allbright et al., 1994], but the flexibility matrix is usually required for an accurate evaluation. The flexibility matrix can be found from:

$$[F] = [\Psi]\left[\frac{1}{\omega^2}\right][\Psi]^T \tag{23.5}$$

where

$[F] =$ flexibility matrix
$[\Psi] =$ unit-mass scaled modal vectors
$[1/\omega^2] =$ diagonal matrix of ascending natural frequencies

A method for calculating the flexibility matrix from mass-scaled modal vectors can be found in Catbas (1997).

Modal testing does not provide an exact flexibility matrix since an infinite number of modes would be required to calculate the exact flexibility. However, if a sufficient number of modes can be detected, the modal flexibility matrix can be close enough to the actual flexibility to provide highly accurate results.

Once the modal flexibility matrix is obtained, it can be used to evaluate the structure:

$$[P][F] = [\delta] \tag{23.6}$$

where:

$[P] =$ vector of point loads placed at impact points
$[F] =$ modal flexibility matrix
$[\delta] =$ vector of displacements at impact points

Any loading pattern can then be simulated by placing point loads at the corresponding impact points. (A uniform load is simulated by placing point loads at all impact points.) This shows one of the advantages of modal testing. Often, damage to a structure is localized and can only be detected through load testing if the load combination and load placements are ideal to reveal the damage. In a real load-test situation, this would require an unrealistic number of load cases. However, with the modal flexibility matrix a large number of loading patterns can be efficiently checked to see if any damage is detected.

For damage detection, it is best to have a baseline modal-flexibility matrix of the undamaged structure. Damage can then be detected by comparing deflections from subsequent modal-flexibility matrices to deflection found from the baseline matrix. If a baseline matrix is not available, deflections found from the modal-flexibility matrix can be compared to deflection generated from an idealized model of the undamaged structure (e.g., finite-element model), but this comparison is less accurate.

Finally, the modal-flexibility matrix can be used to "tune" the stiffnesses and boundary conditions of a finite-element model. Here, the deflections generated from the model are compared to those generated from the modal-flexibility matrix and the model is adjusted to obtain the best comparison for several load cases. The tuned model can then be used for analysis of the structure.

23.5 Static/Quasistatic Load Testing

Static or quasistatic load testing is the most common form of structural testing. Since many structural, loads are static in nature, a static or monotonic quasistatic load test can provide valuable information on stress, strain, load distribution, and deflection under normal loading conditions. Quasistatic cyclic loading is often used to assess fatigue behavior. For moving or dynamic loads, cyclic quasistatic-load testing is sometimes used as a simpler or less expensive alternative to using actual moving or dynamic loads. However, the validity of using cyclic quasistatic loading to evaluate moving or dynamic loads is questionable [Petrou et al., 1994; Chung and Shah, 1989].

23.5.1 Use of Dead Loads

One method of conducting static-load test consists of placing a dead load on the structure. Typical dead loads are concrete blocks, bricks, sandbags, or containers of gravel, sand, or water. The advantages of using dead loads are as follows:

1) They can easily simulate a uniform load on a structure. A single layer of 200 mm (8 in) hollow core concrete blocks, with 25-mm (1-in) gaps in between to prevent arching, can provide a uniform load of 1.4 kN/m^2 (29 psf).
2) The loading materials are relatively cheap. In many cases, bricks, blocks, crushed stone, or sand can be borrowed from building supply yards or quarries.

The disadvantages of this type of testing are numerous and often relate to safety:

1) Except for exposed structures such as roofs or bridges, the dead load often cannot be placed remotely (as by crane) and must be placed by hand. This presents a danger to workers should the structure fail while load is being applied.
2) Should the structure begin to crack or collapse, the loads cannot be removed quickly or safely enough to prevent additional damage or complete collapse.
3) Should the structure completely or partially collapse, the load may fall freely and perhaps injure workers or unintentionally damage other parts of the structure.
4) The actual load on the structure can only be assessed through the cumbersome process of weighing the individual loading blocks or containers, and there is no easy way to record the load on modern data-acquisition systems.
5) This method cannot be used to simulate moving or fatigue loads.

For safety reasons, dead-load testing should be limited to service-load tests well below damage or collapse limit states. ACI Committee 437 (1991) allows the use of dead-load testing but, because of safety issues, does not recommend it.

When using dead loads, several practices should be observed. All loading blocks or containers should be of the same weight; ACI Committee 437 (1991) allows a ±5% variation between individual units. The loads should be placed with gaps between the individual loading blocks or containers. This is because the structure will deflect under load and if the loading blocks/containers are too close together they may touch and form an arch. The arch formations make the actual loading pattern uncertain.

Loose material (sand, gravel, water) should not be used for load testing as it may gather in low areas of the structure and "pond," again creating an uncertain loading pattern. Any loose material used for loading should be placed in containers and the full containers used to apply the load.

23.5.2 Vehicle Testing

Vehicle testing is preferred for service-load testing of structures such as bridges or parking decks where vehicle loads are the predominant live loads. Vehicles can be empty or loaded with sand or gravel; thus loaded dump trucks are very effective for applying loads [Shahrooz et al., 1994; Aktan et al., 1993]. Vehicle axle or single wheel weights can be obtained using truck scales such as those found at concrete plants, building supply yards, or highway department weigh stations. Many state highway patrols have portable scales that can be used. Vehicular load testing has several advantages:

1) It is the type of load that would exist on bridges, parking decks, garage floors, etc., in service-load conditions.
2) Since trucks can often be supplied by highway departments or local contractors and the sand or gravel to load them can be borrowed from concrete plants or building supply yards, it is relatively inexpensive to use vehicular loading.
3) Vehicular loading can be used to apply static, moving, or impact loads to a structure.

The disadvantages are as follows:

1) As with dead loads, vehicle loading requires workers (drivers in this case) to actually place the load on the structure. Although movable, vehicular loads usually cannot be removed quickly enough to prevent injury to the drivers in the event of sudden cracking or collapse.
2) In the case of sudden collapse, the vehicles would be damaged and the falling vehicles may damage other parts of the structure.
3) It is very hard to apply the load in increments as this would require slowly adding load to the vehicle.
4) Owing to clearance problems, it is difficult to place the loads close together or near barriers.
5) If multiple vehicles are used, it is extremely difficult to get all the trucks to have the same axle loads, even if similar trucks are used. In addition, the loads tend to shift during testing, affecting the axle loads.
6) Unless the vehicles are turned off, engine vibrations will affect the test results.
7) The load cannot be recorded electronically.

23.5.3 Hydraulic Testing

Hydraulic testing consists of applying load through one or more hydraulic cylinders or actuators. This type of loading is preferred [ACI Committee 437, 1991] as it is the safest method. Hydraulic loading can be as simple as using jacks controlled through hand-operated pumps or as complicated as using sophisticated electronic, servohydraulic systems that use electric pumps and computerized controls.

The advantages of hydraulic loading are as follows:

1) Hydraulic tests are safer. Loads can be placed on the structure through a remote-control mechanism, keeping workers out of the way should be structure collapse. Also, electronic hydraulic systems can be equipped with fast unloading circuits and safety shutdowns that will quickly remove all load at the first signs of collapse.
2) In a load-control test, where a structure is subjected to predetermined increasing or decreasing increments of loads, a hydraulic-loading system can precisely control load increment, total load, and rate of load application.

3) Use of servohydraulic, closed-loop control allows the use of displacement or strain-controlled tests, which are more stable than load-controlled tests and less likely to result in sudden collapse. This type of testing will often allow postpeak behaviors to be obtained.
4) Hydraulic testing is the most efficient way to apply cyclic or fatigue loads.

The disadvantages of hydraulic testing are the following:

1) Hydraulic systems apply point loads. Application of a line load or a uniform load over a small area can only be done using loading blocks or a spreader device [Miller et al., 1994; Azizinamini et al., 1992]. Hydraulic loads usually cannot be easily adapted to apply a uniform load over a large area.
2) Computer-controlled servohydraulic systems and the associated control systems are expensive to purchase, maintain, operate, and transport.
3) Skilled technical people are often needed to handle the hydraulics and electrical work to assemble the system.
4) A reaction mechanism is needed for the hydraulic system to push or pull against.

The need for the reaction mechanism is the biggest drawback to hydraulic systems. Sometimes, another part of the structure can be used as a reaction frame, but this presents the possibility of unintentional damage to other structural members. In many cases, a reaction frame tied to a foundation or a self-equilibriating testing frame can be built [Azizinamini et al., 1992]. Another approach, shown in Figure 23.1, is to use post-tensioning tendons grouted into the soil or rock below the structure [Miller et al., 1994].

Hydraulic-loading systems can be precisely controlled through the use of servohydraulic, closed-loop testing. In closed-loop testing, the structure is loaded such that some specific parameter (load, deflection, strain, etc.) increases or decreases at a set rate. In order to understand the usefulness of the closed-loop servohydraulic system and its limitations, it is first necessary to understand how such a system works. A schematic of a typical closed-loop system is shown in Figure 23.2.

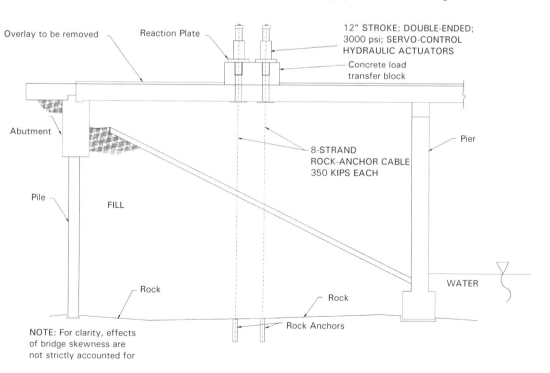

FIGURE 23.1   Use of post-tensioning tendons as a loading system.

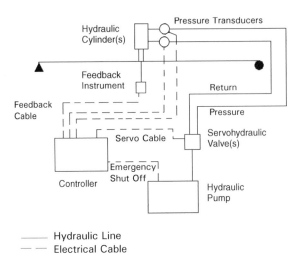

FIGURE 23.2   Closed-loop system.

The system consists of five parts: a pump, a servohydraulic valve, a hydraulic actuator, a feedback instrument, and a controller. The pump supplies hydraulic fluid at set pressure up to a maximum flow rate. The rate of flow of the hydraulic fluid is controlled by a servoelectronic ("servo") valve, which is a high-speed, proportional valve capable of quickly and precisely varying the flow of hydraulic fluid through the system. The flow rate through the valve is directly proportional to the current input to the valve by the controller.

The hydraulic actuator is a hollow cylinder fitted with a piston and rod. To apply load, hydraulic fluid is pumped into a chamber behind the piston. Some actuators are dual action, meaning fluid can be pumped into a chamber on either side of the piston so that the actuator can exert force in either direction. It is important to note that actuator capacities are based on the area of the piston and the maximum pressure supplied by the pump. Quoted actuator capacities are usually based on a 21 MPa (3000 psi) pump pressure. If the pump being used in the test has a maximum pressure different than that given in the actuator manufacturer's specifications, the quoted capacity will be different. Also, since one side of the piston has the rod attached (thus reducing the area), actuator capacities are different in each direction.

During operation, fluid is pumped into the chamber on one side of the piston. If the actuator is dual action, the chamber on the other side of the piston is opened to allow fluid to flow out. If the rod is attached to a flexible structure, pumping fluid into the chamber produces a combination of pressure in the chamber (which translates to load on the structure) and movement of the piston, which translates to deflection of the structure. Unloading is accomplished by releasing fluid and pressure from the active chamber and pumping fluid into the other chamber.

The load applied by the actuator can be accurately measured by measuring the hydraulic fluid pressure and multiplying by the piston area. This method can often be more accurate and less cumbersome than placing a load cell between the actuator and the structure. Pressure can be measured by pressure transducers that can be directly linked into a computer data-acquisition system. In dual-action actuators, there is fluid on both sides of the piston and both chambers have pressure. Therefore, when attempting to calculate load, the differential pressure between the two chambers must be used.

The feedback instrument is any electronic instrument that can read the desired structural response (e.g., deflection) and translate it into a voltage. The controller uses the signal from the feedback instrument to regulate flow through the servovalve. Normally the controller is set to regulate two things, the rate of the structural response and the maximum value of that response.

For the sake of example, assume the system shown in Figure 23.2 is set to be a displacement-controlled test of a beam; that is, the system will control the load such that the beam deflects at a set rate up to some predetermined maximum deflection. Initially, the controller is used to pump some fluid into the actuator and a small load is applied to the beam. An electronic instrument on the beam reads the deflection and converts it to a voltage that is transmitted to the servohydraulic controller. The controller, which is programmed to convert the voltage back to displacement, checks the rate of deflection against the target rate. If the rate of deflection is too slow, the controller will attempt to increase the flow to the hydraulic actuator to increase the loading rate and therefore the deflection rate. If the deflection rate is too fast, the controller will attempt to decrease or, if necessary, reverse the flow to the hydraulic actuator to slow the deflection rate.

When the beam reaches the target maximum deflection, the controller stops the flow to the actuator. However, there is always some electrical and mechanical noise in the system and, as a result, a servohydraulic system will not hold at the specified target but will oscillate about it. This oscillation will introduce noise in the response of the instruments attached to the structure, often giving the responses a "sawtooth" appearance.

Oscillations of the system can also be caused by backlash in the control instrumentation. Backlash is a measure of the change in response needed to elicit a change the instrument reading when the response is reversed (see Section 23.6 on instrumentation). Noise in the system will usually cause the controller to slightly overshoot the intended target response (deflection, strain, etc.), so the controller will unload slightly to correct this condition. However, if the instrument has significant backlash, response under unloading may not be detected immediately, the controller will miss the target and attempt to correct by reloading. Thus instruments with significant backlash will cause large oscillations in the system and should be avoided.

Servohydraulic systems provide an excellent means of conducting a controlled test of structures or elements. However, these systems are extremely complicated and a high level of expertise is required to design, build, and operate such systems. Therefore competent control engineers should be retained to design and operate servohydraulic systems.

23.6 Instrumentation and Data Acquisition

In performance testing, an engineer desires to establish a relationship between load and some structural response. However, the load and structural response must be measured using instruments. Most people tend to assume that the instruments and any associated equipment for data acquisition are transparent; that is, the data that come from the instruments are not affected by the instrumentation itself. This is not true. Each instrument, whether a simple mechanical dial gage or a sophisticated electronic device, has its own particular characteristics that can affect the test results. Unfortunately, instrumentation considerations are frequently overlooked in the pretest planning stage. It is only when the test is complete and the data are found to be nonsensical that any problems with instrumentation are even considered.

23.6.1 Important Issues in Instrumentation Selection

When selecting instrumentation for a performance test, there are several issues that must be addressed.

23.6.1.1 Range

It is useful to speak of the range of the instrument in terms of total range and working range. Total range is the maximum and minimum values of input that the instrument can read. Working range is the anticipated maximum and minimum values of the response that will measured during the test.

It is normally desirable that the working range be about 50% of the total range. Working ranges are usually estimated in the pretest planning stage and will often be inaccurate. If the working range is too close to the total range, the instrument capacity may be exceeded. Also, some instruments exhibit nonlinearities or other errors near the limits of the range. Having a working range that is much smaller than the total range is also undesirable. Since accuracy is normally a function of the total range, trying to make small measurements using an instrument with a large range will usually be inaccurate.

23.6.1.2 Accuracy

Accuracy can be defined as the difference between the instrument reading and the actual quantity being measured. Determination of the accuracy of an instrument is a complex and difficult process that requires careful calibration of the instrument. The instrument accuracy is actually the sum of several instrument effects and errors.

In general, the response of any instrument can be separated into five components (unpublished work by A. Levi, University of Cincinnati), although for a given instrument some of these responses may not be present. The components are given by

$$X_{total} = X_1 + X_2 + X_3 + X_4 + X_5 \qquad (23.7)$$

where

X_{total} = total instrument response
X_1 = response due to the load
X_2 = response due to nonload effects on the structure
X_3 = instrument error
X_4 = error due to external effects on the instrument
X_5 = error due to effects of instrument mounting

In most cases, the total response of the instrument is assumed to be only caused by load and the other responses are ignored. This can be a grievous error. It is necessary to remove, minimize, or at least account for the other responses of the instrument.

The response due to nonload effects, X_2, includes such things as changes in strain and deflection due to temperature. These effects can be significant as changes in strain and deflection of an exposed structure due to temperature can be greater than the response due to load. It is possible to correct for nonload effects either through analytical estimation of the values of these effects or by placing additional instrumentation and/or performing tests under no applied load to specifically measure these effects.

Every instrument has a certain amount of error associated with its measurements, given by the X_3 term in Eq. (23.7). In addition, data-acquisition systems can account for additional sources of error. When considering the sources of error due to the instrumentation and data-acquisition system, it is necessary to consider the following.

1) **Resolution.** Resolution is what most people mean when they speak of accuracy, but is only one component of the total accuracy of the instrument. Resolution is the smallest value of a response that can be registered by the instrument. For mechanical gages, this is the smallest value of a single tick mark. It is possible to interpolate between two marks, but this not objective as it depends on the person reading the gage. For gages with digital displays, the resolution is the smallest value on the display. Resolution of analog electrical instruments is harder to determine. In theory, these instruments have infinite resolution, but in practice the resolution is limited by the resolution of the data-acquisition system, electrical noise, and the behavior of the mechanical components within the gage. Thus for analog electrical instruments the resolution is the smallest response that provides a reliable reading from the gage.

2) **Sensitivity.** Sensitivity is the output of the instrument for a given input and is expressed as the ratio of output to input. Usually, sensitivity is determined by a best-fit straight line. For electrical instrumentation, the output is often influenced by the excitation applied to the

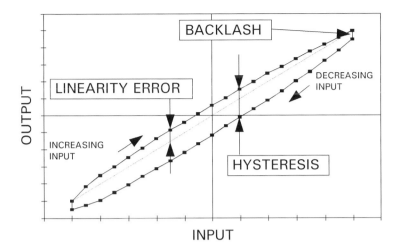

FIGURE 23.3 Definition of sensitivity, linearity, hysteresis, and backlash.

instrument, so this is included in the sensitivity. As an example, consider an instrument that measures displacement, outputs voltage, and is excited by a voltage. The sensitivity may be expressed as X V/V/ mm. This means that a movement of 1 mm will cause the instrument to output X V if a 1-V excitation is used. If a 10-V excitation is used, a 1-mm movement would output $10X$ V.

Sensitivity does not affect accuracy through the instrument itself but rather through the data-acquisition system. If an instrument has a small sensitivity, the output will be small and the data-acquisition system may not be able to read small quantities accurately. Also, small outputs in electrical instruments are often affected by noise.

3) **Linearity.** Ideally, the reading of an instrument has a linear relationship with the response. The linearity error is the measure of how far the data deviates from the best-fit line (Figure 23.3).

4) **Repeatability.** Repeatability is the ability of the device to output the same value for the same response over a number of trials when the response is always increasing or decreasing (e.g., always extending the instrument, always heating the instrument).

5) **Hysteresis and backlash.** Hysteresis is the difference in a reading at a given point depending upon whether the reading was obtained by an increase or decrease in response (e.g., the difference in response when a given point is reached by extending an instrument, as compared to the response at the same point reached by contracting the instrument). Backlash is related to hysteresis and is a measure of the change in response needed to elicit a change in the instrument reading when the response is reversed (e.g., extension to retraction, load to unload, heating to cooling, as seen in Figure 23.3).

Accuracy problems related to resolution and sensitivity can be avoided by proper selection of instruments based on reasonable estimates of the quantities to be measured. Again, this points to the need to perform proper pretest planning. The remaining errors, linearity, hysteresis or backlash, and repeatability, determine the maximum total error of the instrument. Also, it is advisable to calibrate the instrument over the working range ±15%. Smaller ranges tend to have smaller errors, and calibration over the anticipated working range provides a more reasonable estimate of actual error.

Linearity is found by providing a series of inputs over the range and plotting the input against the output. For best results, inputs should be made in both "directions" (e.g., extension and retraction, loading and unloading, etc.). At least three trials in each "direction" should be made and the results plotted on a single graph. The maximum distance from a best-fit straight line to any data point is the maximum total linearity error.

Hysteresis or backlash is found by providing input to the instrument through the range in one direction and then providing input through the same range in the reverse direction. The maximum hysteresis or backlash error is the maximum distance between any two points, one in each direction, corresponding to the same input. Again, at least three trials should be used.

To assess repeatability, input is provided to the instrument through the range but always in the same direction. Maximum repeatability error is the maximum distance between any two points corresponding to the same input but on two different trials.

One common mounting error is the standoff error. Sometimes the mounting bracket for an instrument does not allow the instrument to directly contact the surface. The distance between the instrument and surface is the standoff. If the quantity to be measured has a large gradient (e.g., bending strain at the top in a thin member), the measurement will be in error since the instrument does not measure at the surface but above the surface. This is corrected by minimizing standoff and by correcting for standoff distances in any calculations.

23.6.2 Types of Instrumentation

There are a large number of different instruments that can be used for measuring structural responses. These range from surveying instruments used to measure large deflections to strain gages accurate to 1×10^{-6} strain. Each type of instrument has particular uses, accuracy, and limitations. Clearly, in the limited space available here it is not possible to cover all the possible types of instruments that can be used. A more comprehensive coverage of instruments can be found in Dunnicliff (1993). Although Dunnicliff's work is aimed at geotechnical applications, many of the instruments listed are useable for concrete applications. What follows is a description of some instruments commonly used for concrete.

23.6.2.1 Mechanical Gages

Mechanical gages provide an easy, reliable, and inexpensive method of measuring displacement and strain or distortion. However, the use of these gages is extremely labor intensive and error prone. The gages must be read manually. This exposes the workers to hazard, leads to possible error in recording the data, and reading a large number of gages requires a large work force and/or a large amount of time. Manual gages cannot be directly read by a computer. Thus the data must be either manually input into a computer or manually processed and graphed. Both of these procedures are error prone and time consuming.

There are many types of mechanical gages, but the two most common are the dial gage and the portable strain or distortion gage.

Dial Gages.   Dial gages consist of a rod attached to a dial by a gearing system. As the rod moves, the movement is registered on the dial. These gages have various ranges and accuracies. The accuracy of the dial gage is somewhat difficult to assess. They can be reliably read to the nearest tick mark on the analog dial, but users frequently attempt to interpolate between marks. This adds a degree of subjectivity to the data. Also, analog gages are easy to misread, leading to error. Some more modern "dial gages" replace the mechanical dial with a digital display, thus reducing these errors. Finally, even the best jeweled movement in the gearing system will exhibit some stick (backlash), making small measurements difficult. Although a dial gage may respond to dynamic motions, it is impossible to read accurately under these conditions. It is therefore usable only under static or low-frequency cyclic testing.

The main advantages of these instruments is that they are easy to use, easy to read, and require no auxiliary equipment (data-acquisition systems). They are also cheap, especially if the

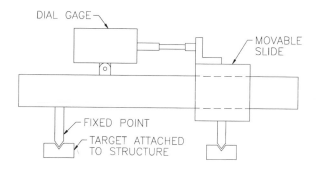

FIGURE 23.4   Portable strain gage.

gages are purchased through a machinery supply house rather then from a scientific supplier.

Portable Strain/Distortion Gage.   The portable distortion or strain gage consists of a metal bar, two points, and dial gage (Figure 23.4). One point is fixed; the other point is attached to the dial gage and is movable. Two targets, pieces of metal with drilled holes, are attached to or embedded in the concrete at the basic length of the gage. A zero reading is established by placing the points of the gage into the holes in the targets and reading the dial gage. Subsequent measurements are taken in the same way and the differential movement is found by subtracting the zero reading. This instrument is particularly useful for measuring over large gage lengths. For smaller gage lengths, vernier calipers or micrometers can be used. The portable distortion or strain gage is also useful in cases where a permanently mounted gage is impractical. Bodocsi et al. (1993) used this type of gage to measure seasonal movements of pavement joints.

   The portable distortion or strain gage has the same problems as the dial gage and has the additional problem that the accuracy is dependant on how well the points sit in the holes in the targets. In general, there is always some play in the interface so the measurements are often not repeatable. For this gage, it is best to take several measurements and average them. Since the gage must be manually placed to take measurements, it is only usable for static or quasistatic tests.

23.6.2.2 Electrical Instrumentation

Electrical instrumentation has greatly improved the ability of the engineer to safely and accurately monitor structural response. These instruments have the advantage that they can be remotely monitored so that the workers never have to enter the structure during the test. This also allows engineers, who cannot be present during the test, to monitor the test results in real time from distant locations.

   Electrical instruments also have the advantage of directly sending all data to computer files, thus reducing data transcription time and errors. Many data-acquisition programs allow the instrument responses or mathematical permutations of these responses to be plotted in real time so that load/deflection, load/strain, or stress/strain graphs can be plotted in during the test.

   One major disadvantage of electrical instrumentation is that these instruments tend to cost much more than similar mechanical instruments. They also require expensive auxiliary equipment such as data-acquisition systems and cables. However, the biggest drawback is the need for electrical power, which is not always available at remote sites. Where power is not available, it is necessary to use a generator. Generators tend to emit a large amount of electrical noise, which affects instrument accuracy; and they are also prone to power surges, which can damage sensitive equipment.

   As with mechanical instruments, in this concise chapter it is not possible to cover all the instruments used, but a few common instruments will be discussed.

Linear Voltage Differential Transformers (LVDT). The LVDT is used to measure displacements (Figure 23.5). Typical LVDTs will have ranges from 2 to 150 mm (0.1 to 6 in). The instrument consists of a primary coil, two secondary coils, and a magnetic core. An AC voltage excites the primary coil, which in turn induces voltage in the secondary coils. The secondary coils are connected series opposing so that only the differential voltage is output. The amount of voltage in each of the secondary coils depends on the position of the magnetic core, and the differential voltage is linear with respect to core position. If the core is attached to the structure (Figure 23.5) so that any deflection of the structure moves the core the same amount, the change in voltage is a direct measure of structural deformation.

FIGURE 23.5 Linear variable differential transformer (DC type).

LVDTs work on alternating current (AC), but when using long lead wires, direct current (DC) is preferred. Many LVDTs contain circuitry so that a DC voltage can be input and a DC voltage is output. This is done by converting the input DC voltage to AC and output AC voltage back to DC. Such LVDTs are frequently called DCLVDTs or DCDTs.

The main error in an LVDT is linearity error, especially near the ends of the range or in the middle when the LVDT passes the null point where the differential voltage changes sign [Hrinko et al., 1994]. Dunnicliff (1993) reports these instruments have no hysteresis, but hysterisis, backlash, and repeatability errors have been found in careful calibration [Hrinko et al., 1994]. The LVDT, theoretically, has infinite resolution, but the actual resolution is limited by cable noise and the resolution of the data-acquisition system.

The instrument responds fast enough to use in any type of testing, including dynamic testing. However, LVDTs have some drift, making them less than ideal for long-term monitoring.

Wire Potentiometers. Wire potentiometers (Figure 23.6) are also used to measure displacement but have much bigger ranges than the LVDT. Typical ranges for a wire potentiometer are 125 to 900 mm (5 in to 30 in). The potentiometer is a contact that moves along a resistor. The resistance in the circuit changes as the contact moves, and this causes a change in voltage in the circuit. The voltage change is directly proportional to the movement of the contact. In the wire potentiometer, a wire is wound around the spindle of a spring-loaded rotational potentiometer. As the wire is pulled, the potentiometer contact moves and registers a voltage change proportional to the wire movement. Spring loading keeps the wire taut at all times. By attaching the wire to the structure, the deformation of the structure can be found from the movement of the wire.

As with the LVDT, the wire potentiometer itself has infinite resolution, and the actual resolution is limited by the data-acquisition system and electrical noise. Wire potentiometers have higher error than an LVDT, but they also have larger ranges so the error, as a percentage of the range, is about the same as for an LVDT [Hrinko et al., 1994].

The main error in a wire potentiometer is backlash and hysteresis [Hrinko et al., 1994]. This will introduce noise in the loading if the wire potentiometer is used the control instrument in a closed-loop test (see the Section 23.5 on closed-loop testing).

Electrical Resistance Strain Gages. An electrical resistance strain gage consists of several loops of wire that are bonded to the structure such that the loops are parallel to the direction of strain.

FIGURE 23.6   Wire potentiometer.

Strain in the structure causes strain in the wires, which is directly proportional to change in the wire resistance. The change in resistance is read by a wheatstone bridge circuit, which is well known and can be found in most standard physics or electric-circuit texts. An excellent discussion of the bridge circuit as it relates to strain gages is found in Dunnicliff (1993). Resistance gages come in several types

1) **Bonded foil gages.**   These gages consist of a grid of thin metal foil bonded to a plastic backing that is in turn bonded to the structure. These gages are not particularly useful for concrete. Surface preparation is critical, and it is often difficult to obtain a smooth, clean surface on concrete, especially in the field. Foil gages are not very robust, requiring that the gages be well protected. Even when well protected, the gages will sometimes fail under extreme field conditions or when embedded.

2) **Weldable gages.**   Weldable gages have the wire loops attached to a metal backing that can then be welded to the structure. In concrete structures, these gages are useful only for attachment to rebar or to metal embedments. As with foil gages, protection of the gage is of primary importance.

3) **Embedment gages.**   These gages are specifically meant to be embedded in concrete. The wire loops are encased in a sheathing that protects the gage during the casting process.

Resistance strain gages are known to be prone to drift, making them unsuitable for long-term monitoring. Also, they are influenced by temperature effects, in spite of the fact that they are made to compensate for temperature effects. Hence they are recommended only for short-term or dynamic monitoring. It is recommended that "zero" readings be taken before the start of each test.

To be read by a data acquisition system, the electrical resistance gage must be wired into a bridge completion circuit. External circuits can be made from unstrained gages or resistors, but such external circuits are cumbersome and difficult to use. Many data-acquisition system manufacturers make special boards with a strain gage completion circuit already wired in. This allows the strain gage to be directly attached to the system. Unfortunately, these boards often have eight or multiples of eight channels, and since the total number of channels in a given system is limited, dedicating all these channels to strain gages may be wasteful if only a few gages are used.

Clip Gage (Omega Strain Gages).   The clip gage (also called an omega strain gage) consists of a metal arch attached to two feet so that the resulting structure resembles the capital Greek letter

FIGURE 23.7 Omega or clip gage.

omega (Figure 23.7). The arch is instrumented with four resistance strain gages wired in a full bridge. The feet of gage are attached to the structure, usually by gluing metal targets to the structure and then attaching the gage to the target with bolts. As the targets move with the structure, strain that is directly proportional to the movement of the targets (and the strain in the structure) is induced in the arch. This strain is read by the strain gages.

Using an arch "desensitizes" the strain gages, and the clip gage can read much larger strains than the strain gages alone. However, the clip gage is less accurate than the strain gages. Clip gages are easy to install and read. They are most useful for measuring fairly large strains (>100 microstrain) over short periods of time under laboratory or otherwise controlled environments. The gages do exhibit considerable drift and noise and are not suitable for long-term field use (unpublished work by A. Levi at the University of Cincinnati).

Vibrating-Wire Strain Gages.   The vibrating-wire gage works on the principle that the frequency of vibration varies with the tension in the wire. A steel wire under an initial tension is placed inside a sealed case. An electromagnetic coil is used to "pluck" the wire. The resulting first mode vibrations induce a voltage in the coil, and the voltage pulses are measured over a set time to determine the frequency of the vibration of the wire. As the gage extends or contracts, the tension in the wire changes and so does the frequency. Thus the change in strain can be determined from the change in the frequency of the wire. Because temperature changes will affect the frequency, the gages have thermistors to measure temperature and a temperature correction is made automatically.

The vibrating-wire gages are quite stable and rugged, making them ideal for long-term field measurements. They also measure temperature as well as strain. However, because the voltage pulses must be measured over time to determine frequency, the gage response is slow, making it suitable only for static tests.

23.6.2.3 Optical Fiber Strain Gages

Fiber-optic sensors represent a different method of measuring strain in concrete structures. Unlike electrical sensors, strain is measured by changes in light waves transmitted through the fibers.

(a) One such sensor is the Fabry-Perot. The Fabry-Perot [Measures, 1995] consists of two mirrors, separated by a distance (gage length) set perpendicular to the axis of the fiber. The mirrors can be internal to the fiber or at the ends of two fibers, separated by a gap and joined in coupler. This device works by sending a coherent (single wavelength) light beam down the fiber. Some of the light is reflected off the first mirror, which is only partially reflective. This forms a reference wave. The remaining light passes through the partially reflective mirror, reflects off the second mirror, and passes back through the partially reflective mirror. This second wave is now out of phase with the reference wave because it has traveled a longer distance. When the two waves are combined, an interference pattern results.

As the fiber is strained, the distance between the two mirrors changes. This changes the path length of the second wave and therefore the interference pattern. Since the interference fringe pattern is dependant on the wavelength of the light (constant) and the path length, strain can be measured by determining the change in the interference pattern and converting this to a change in path length.

(b) A second type of sensor uses a high birefringent optical fiber [Ansari et al., 1996]. Here, a circularly polarized light wave is divided into two modes along the principal axes of the fiber. External deformation causes interference of the two modes, and the output has a sinusoidal variation, the period of which is a fringe. The change in length is related to the number of fringes formed [Ansari et al., 1996].

(c) A third type of sensor uses a Bragg grating [Chen and Nawy, 1994; Nawy and Chen, 1996; Measures, 1995]. The Bragg grating is imprinted on the fiber lead, making a longitudinal, periodic variation in the refractive index of the fiber core. As light passes through the grating, a portion of the light is reflected, but the reflected light has a spectral shift in wavelength. As the grating is strained, the reflected wavelength changes and this change in wavelength can be directly related to strain (Figure 23.8). This sensor has the unique capability of pinpointing the strain at precise locations in the structure rather than detecting average strain in a stressed zone. It was originally developed and used in concrete structural research

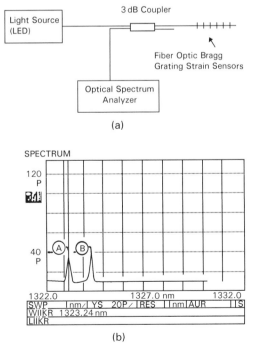

FIGURE 23.8 Fiber-optic Bragg grating sensor (FOBG) system [Nawy and Chen, 1997].

by Nawy et al. (1994) and proved to be effective for on-line remote control sensing of the deformation and cracking of structural elements at their critical locations of high stress, using sensors mounted either internally on the reinforcement or externally on the concrete surface.

23.7 Case Studies in Performance Evaluation of Concrete Structures

In surveying the recent literature on performance evaluation, it was found that the great majority of studies concern bridges. This is not to imply that buildings are not evaluated, only that such evaluations are rarely reported. One reason for this may be that bridge are subjected to more severe loading and environmental conditions and are often required, by law, to have periodic inspections. As a result, performance evaluation of bridge structures may be more frequent. However, the more probable reason for the lack of published material on building structures may be that buildings tend to be privately held structures whose owners may, for legal reasons, be hesitant to publish information on building deficiencies. Although the material presented here is heavily weighted toward bridge structures, the information is applicable to almost any type of concrete structure.

23.7.1 Testing of a Three-Span Slab Bridge

Extensive performance evaluations of a deteriorated, three-span concrete bridge were conducted by Aktan and his team at the University of Cincinnati [Aktan et al., 1992, 1993; Miller et al., 1994]. The bridge, shown in Figure 23.9, had extensive deterioration along the shoulders and sides of the slab. The deterioration was so severe that the top mat rebar on the shoulders was completely exposed.

The testing of this bridge had three main objectives: 1) to determine how severe deterioration affected the elastic response of the bridge; 2) to determine the effect of deterioration on the load capacity and failure mode of the bridge; and 3) to evaluate the effectiveness of nondestructive testing techniques in bridge evaluation.

The first step in the evaluation of this structure was to determine the cause and extent of the damage. Attempts were made to remove core samples from the shoulder area, but the deterioration was so severe that cores could not be removed intact and came out in layers. Some of these samples were used for material evaluation. Petrographic examination showed that the aggregate had extensively "d" cracked. This type of cracking occurs when large pieces of semiporous aggregate

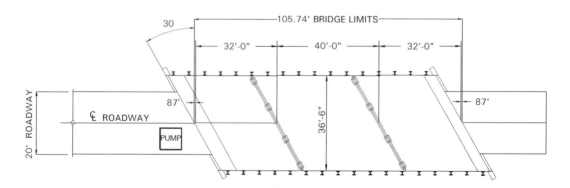

FIGURE 23.9   Three-span concrete slab bridge.

are saturated and then permitted to freeze. Some of the aggregate was also reactive and signs of alkali-silica reaction were found. It is believed that the aggregate cracked and the expansion of the aggregate cracked the surrounding cement paste. This provided a path for water to enter the slab, which then drove the alkali-silica reaction.

There were also signs of carbonation, which lowers the pH of the cement paste and creates an environment where corrosion can occur. Chloride testing revealed that the cracked shoulder areas had high chloride contents from road salt. The chloride, along with the air and water entering the slab through the cracks, caused the steel to corrode, further cracking and damaging the shoulder areas.

It is believed that the reason that the deterioration was limited to the shoulder areas is that chloride-laden ice and snow would have been pushed to the shoulders by traffic and plows. Here the snow melted and either ran over the slab edge (no other drainage was provided) or soaked into the porous asphalt layer and was trapped in between the asphalt and the slab. This would have saturated the shoulder area and provided an environment for the deterioration to occur. It is also of interest to note that chloride intrusion and milder signs of corrosion were found over the piers. It is probable that the chlorides entered through flexural cracking. Owing to the short span of the bridge (the end span was only 30 ft long), it was found that only one axle would actually be at a critical point in the span as a truck passed (Figure 23.10). This rear tandem axle was duplicated by the use of two loading blocks fitted with hydraulic cylinders. The loading blocks were placed in one lane, as this provided a realistic loading pattern and the researchers felt this would also reveal any effect of nonsymmetrical loading on the slab. Estimates of bridge capacity and performance were obtained by running several finite-element analyses [Shahrooz et al., 1994a].

After the finite-element analysis was completed, an extensive instrumentation was instituted (Figure 23.11) for 38 grid points. Two types of instruments were available for measuring deflection, 2-in-range DCDTs and 10-in-range wire potentiometers. The points in the loaded span were instrumented with the 10-in-range wire potentiometers since preliminary finite-element analysis indicated that deflections in excess of 2 in would be expected.

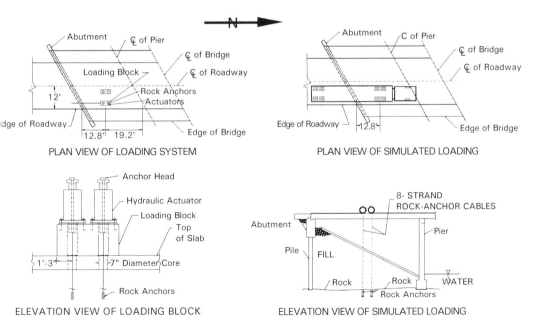

FIGURE 23.10   Load placement to simulate an HS20-44.

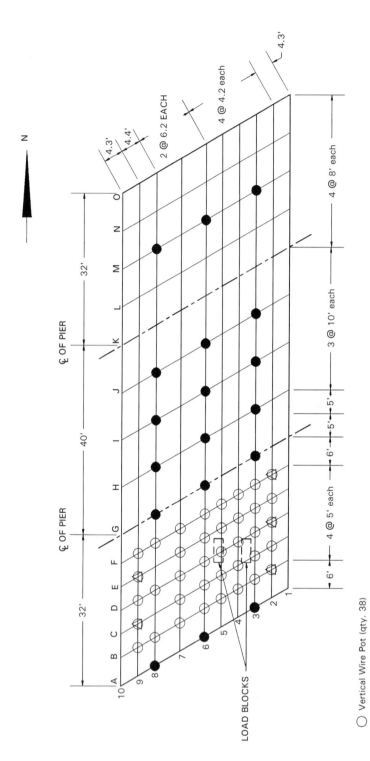

FIGURE 23.11 Instrumentation grid for the slab bridge.

Strains in the reinforcing bars were measured at 11 points on the bottom steel (clustered around the loading blocks) and at 8 points in the top steel (4 over the piers near the loading blocks and 4 at midspan near the loading blocks). Bonded resistance gages were used in these applications. Concrete strains were measured using DCDTs that were attached between two angles glued 6 in apart on the slab. Truck load tests where deflections were measured were conducted. After the truck load tests, the bridge was tested to failure.

Using rock anchors (Figure 23.10), 7 in diameter holes were cored at four points in the deck. A 6 in auger was then passed through the core hole and used to drill a hole 45 ft into a rock layer beneath the bridge. A cable made of several seven-wire prestressing strands was then grouted into the rock. Two concrete loading blocks were then cast on the bridge deck over the core holes. Four 350 kip capacity hydraulic cylinders were then placed into holes cast into the loading block. The shafts of cylinders were hollow and the rock anchor cables were passed through the shafts and tied off into button heads.

The loading cylinders were controlled by an electronic controller. This controller was linked to one of the wire potentiometers monitoring deflection below the loading blocks. The bridge was initially loaded to 64 kips (16 kips/cylinder) and then unloaded. Additional load and unload cycles were made to 112 kips and 124 kips. In each case, the bridge remained linear, although some slight hysteresis was observed on unloading. The bridge was reloaded after several days to 700 kips and then unloaded. It showed some nonlinear behavior and flexural cracking was observed. The bridge failed at a load of 720 kips when it was reloaded the next day. Failure occurred as a circular arc around the loading blocks, resembling a punching shear type of failure (Figure 23.12).

Subsequent analysis of the data revealed several interesting conclusions for the destructive test [Miller et al., 1994]. Initially, the bridge carried the load in a direction parallel to the traffic lanes. When a rotational restraint at the abutment was overcome, the bridge began to carry load perpendicular to the skew. At about 700 kips of total load, the longitudinal reinforcing began to yield and the load path again shifted to parallel to the traffic lanes. This shifted load back to the damaged corner (as evidenced by a sudden increase in top rebar strain at this location). The load was apparently too much for the damaged shoulder and it failed in shear with the failure propagating throughout the entire slab. It is also of interest to note that the top steel in the shoulder areas as completely exposed and corroded, however it was still possible to yield these bars before the failure pulled them out.

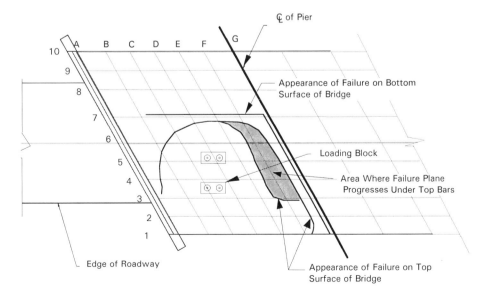

FIGURE 23.12 Final failure of the bridge.

A final objective of this test was to determine the effectiveness of multireference modal testing for real structures. As noted in the Section 23.4.3 on modal testing, the test is performed by using with an instrumented hammer to create impacts at various locations on the structure and reading the subsequent vibrations with accelerometers. The grid for the impacts was exactly the same as that used for instrumentation (see Figure 23.11) except that all points shown on the grid were subject to impact, not just those points that were instrumented for the destructive test. A 12-lb sledgehammer, fitted with a load cell, was used for impact. Accelerometers were placed at points B4, C9, D6, D8, H3, I6, J3, L6, L8, M3, and N7 (Figure 23.11). Modal testing successfully located both visible and hidden damage.

The conclusion that can be drawn from the testing program is that concrete-slab bridges are extremely strong, even if damaged. This particular bridge held the equivalent of 22 HS20-44 trucks before failure. While the damage did not prevent the bridge from safely holding normal traffic loads, it did affect bridge performance since the final failure probably started in the damaged area.

23.7.2 Truck Load Testing of Damaged Concrete Bridges: Testing for Continuity

In the past, continuous concrete slab bridges were a popular choice for highway spans under 100 ft. Although much of this market has now gone to prestressed concrete, many continuous concrete slab bridges still exist and many are still being built, especially in rural areas where labor is less expensive and the cost of shipping precast members may be high.

The top of the slab bridge tends to deteriorate owing to a combination of traffic and road salt. Often the deterioration partially or completely exposes the top reinforcing bars that provide the continuity over the pier supports. When this damage is repaired, deteriorated concrete is removed from the deck and the bars over the piers are further exposed. It is important to determine if the removal of the concrete from around the top bars over the piers could cause a loss of continuity, in effect causing the adjacent spans to behave as simple rather than continuous spans under dead load. In such a case, the dead-load stresses at the midspan of the slabs would increase, decreasing their usefulness.

To test for loss of continuity, three slab bridges were tested before, during, and after repair [Shahrooz et al., 1994b]. The bridges were selected so that one had minor damage, one had moderate damage, and one had severe damage over the piers. All the bridges were three-span, continuous-slab bridges. Spans ranged between 20 and 40 ft. The bridges were repaired in the following manner, were tested prior to repair to establish a base, and were retested after repair with truck loading:

1) The asphalt layer was removed, along with 1/4 in of the top of the concrete deck in two of the bridges.
2) Three feet of the deteriorated shoulder was removed and replaced.
3) All delaminated concrete from the deck was removed, and the deck was patched with microsilica concrete.
4) A 1.25 in thick microsilica concrete wearing surface was placed on the deck.

In order to assess continuity, it was necessary to place the loads on the bridges in three load cases: maximum deflection in the middle span, maximum deflection in the end span, and maximum moment over the piers.

Test results showed the following:

1) For bridge #1, there was little change in deflection or strains during repair, but an increase in stiffness was found after repair. Owing to the low level of damage, this result was expected.
2) For bridge #2, there was little change in stiffness when the damaged shoulder was removed, indicating that the shoulder was so badly damaged that it did not participate in load resistance. After the shoulder was restored, the stiffness of the bridge increased. After the deck was repaired, a further increase in stiffness was noted showing that the repair was very effective.

3) For bridge #3, there was a large loss of stiffness when the shoulder was removed and a large redistribution of moment. The moment transfer was estimated to increase the dead load positive moment by as much as 50%. It was also found that the stiffness after repair was only slightly greater than the before-repair stiffness, indicating that the repair was only marginally effective.

23.7.3 Testing of Shear-Key Cracking in Adjacent Box-Girder Bridges

The adjacent, precast box-girder bridge is used in over 30 states for short-span bridges (under 100 ft). In this type of bridge, precast box girders are placed side by side to form both the superstructure and deck of the bridge (Figure 23.13). Often, the boxes are held together by transverse post-tensioning, but the method of post-tensioning varies widely between states. At one extreme, some states use threaded rods, tightened by hand using a "turn of the nut" method. The rods are often placed only at the quarter points and midspan. At the other extreme, some states use transverse post-tensioning strands that are placed only a few feet apart. Many other intermediate combinations of post-tensioning methods are also used.

Load transfer between adjacent boxes is mostly accomplished through the use of a shear key (Figure 23.14). These keys run the length of the box beam and are usually grouted after the transverse post-tensioning is applied. Many states use a "nonshrink" grout, but this material has often proved unsatisfactory. Some states have switched to magnesium phosphate cements or epoxies.

In this case study, deflectometers developed by El-Esnawi [Hucklebridge et al., 1995] were used to determine the extent of cracking and the load transfer in the shear keys. When one end of the deflectometer is displaced relative to the other, the thin element acts like a fixed beam subject to support displacement. This displacement causes moments at the ends of the "beam" and the resulting strain from these moments is measured by the strain gages.

Tests conducted in this case study showed that the shear keys had cracked extensively allowing large relative displacements of the beams (Figure 23.14). This cracking was found even in relatively

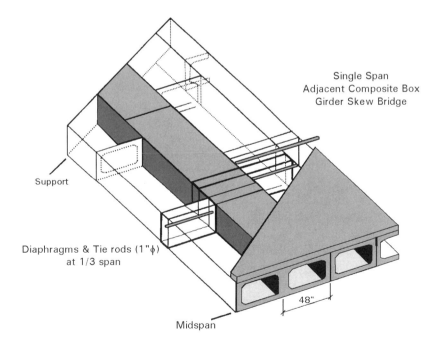

FIGURE 23.13   Adjacent box-girder bridge.

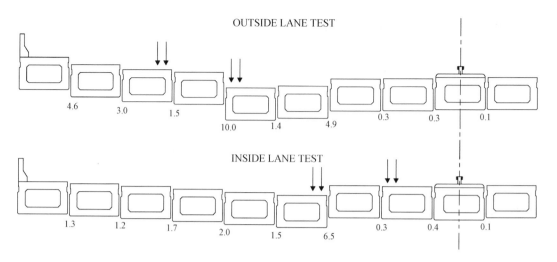

FIGURE 23.14   Relative displacement between adjacent box girders. (Courtesy A. Hucklebridge, Case-Western Reserve University.)

new or recently repaired bridges. However, measured strains in girders showed some load transfer still occured.

23.7.4 Testing of a High-Rise Building After Collapse

One common reason for testing concrete structures is to assure structural safety if the materials of construction or the construction techniques are suspected of being faulty. One such case involved the Skyline Plaza Project at Bailey's Crossroads, Virginia [Dixon and Smith, 1980]. The Skyline Plaza North Project consisted of an office building, apartments, condominiums, and parking structures. In March, 1973, a section of one structure, A-4, collapsed.

This collapse occurred shortly after concrete had been placed on the twenty-fourth floor of what was to be a 26 story structure. The collapse of this floor started a progressive collapse of the lower floors, resulting in the deaths of 14 workers and injury to 30 more. A study was made of the material properties, slab cracking, and the cause of the collapse [Schousboe, 1976]; but of interest here is the load testing that was used to determine the safety and salvageability of the uncollapsed part of the structure.

The A-4 structure was a flat-plate structure 76 ft wide and 389 ft long with irregularly spaced columns. An expansion joint was located 171 ft from the west end of the building. The collapsed area was east of the expansion joint. The investigation of material strengths and slab cracking located 14 areas that were suspect.

ACI 318 Code, Chapter 20, was used to determine the test load. The slab self-weight was 77 psf and an additional 19 psf was calculated as the remaining dead load that would be applied to the structure. The design live load was 40 psf; thus the total load was $TL = 0.85\{1.4(77 + 19) + 1.7(40)\} = 172$ psf.

One interesting aspect of this test was the method of load application. The uniform load was applied by flooding the structure. It should be noted that ACI Committee 437 allows for the use of water as a loading method, but recommends that the water be confined to tanks or containers. The use of flooding is discouraged since the water height may vary owing to variations in the level of the floor and an uncertain load distribution may be created. However, the flooding method was chosen because a large number of areas needed to be tested and this method could be easily and quickly applied.

Before testing began, all mechanical elements (pipes, ducts, etc.) that might interfere with the test were removed and holes in slabs for these elements were sealed. A built-up roofing material was placed over the slab to act as waterproofing. This material was not actually attached to the concrete so that it could be easily removed after the test to allow posttest inspection of the concrete. Bulkheads were built around the area to be tested.

As required by the ACI 318 Code, the acceptance criteria for the slabs was that deflections be under allowable limits. Owing to safety considerations, workers were not allowed in the building during testing so remote monitoring of deflections was needed. This was done using LVDTs mounted on wooden stands and that measured the deflection of the slab from underneath.

The first load applied was 3.5 in of water (approximately 19 psf), which represented the dead load not present on the slab. According to the version of ACI 318 in effect at the time of the test, this load needed to be in place for 48 h before any further testing was done (The current code, ACI 318-95, does not have this requirement). At the end of 48 h, the deflection was measured. The remaining load was then added in three increments (so that the total number of increments was four, as required by the code), with 1 h elapsing between increments. During the hour wait, deflections were monitored. The total load consisted of 18.5 in of water; 3.5 in for the dead load and 15 in for the live load. This load was left for 24 h, after which time deflections were measured. The water was then drained so that only the 3.5 in of water, representing the increment of dead load, remained. This was left for 24 h and deflections were measured again.

Of the original 14 areas tested, 4 did not meet the requirements of ACI 318, Chapter 20. However, the ACI 318 Code allows for retesting parts of a structure that do not pass a load test and this was done. After retesting, two of the areas still failed and one showed an increase in slab cracking. The three areas that did not pass the test were strengthened, but the method of strengthening was not reported.

Another interesting part of this evaluation concerned punching-shear tests around the columns. As previously noted, the Skyline Plaza North building was a flat slab. Flat slabs are always at risk for punching-shear failures. Two columns on the tenth floor in an area of the building scheduled to be demolished were subjected to punching shear tests.

The slabs around the columns were shored from the bottom of the tenth floor to the sixth floor. A portion of the slab around the columns was then cut free (although the reference does not give the size of the cut-out portions, it appears from diagrams and pictures to be about 30 in from each face of the column). Eight holes were cored in the tenth and ninth floor slabs, two on each side of the column. A rectangular testing frame made of back-to-back channels was placed over the holes on the tenth floor. Threaded rods were passed between the channels, through the holes, and attached to post-tensioning jacks anchored to the column two floors below the testing frame.

The loading sequence was as follows:

1) Load all jacks to 3 kips (24 kips total load) and wait 10 min.
2) Add 5 kips per jack (40 kips total) and wait 10 min.
3) Add 5 more kips per jack (40 kips total) and wait 10 min.
4) Add 3 kips per jack (24 kips total). One slab failed in shear before all the load was applied; at a total load of 120 kips.
5) The other slab was loaded at 1.25 kips/jack/min until it failed in shear at 128 kips total load.

Inclined cracks developed from flexural cracks at approximately 60% of the ultimate load. These cracks propagated upward until only a small compression area remained before failure. The shear loads were less than those anticipated, and shearheads were added at many column locations.

As a result of the load-testing program, many parts of the existing structure were strengthened and the building was completed. The authors of the study note that the completed building now performs well and shows no sign of the collapse.

References

ACI Committee 437. 1991. *Strength Evaluation of Concrete Structures*, ACI Report 437R1-91. American Concrete Institute, Farmington Hills, MI.

ACI Committee 201. 1992. *Guide for Making a Condition Survey of Concrete in Service*, ACI Report 201R-92. American Concrete Institute, Farmington Hills, MI.

ACI-318. 1995. *Building Code Requirements for Structural Concrete, ACI 318-95.* American Concrete Institute, Farmington Hills, MI.

Aktan, A.E., Miller, R.A., Shahrooz, B.M., Zwick, M., Heckenmueller, M., Ho, I., Hrinko, W., and Toksoy, T. 1992. *Nondestructive and Destructive Testing of a Reinforced Concrete Slab Bridge and Associated Analytical Studies*, Report #FHWA/OH-93/017. Federal Highway Administration, Washington, DC.

Aktan, A.E., Zwick, M., Miller, R.A., and Shahrooz, B.M. 1993. Nondestructive and destructive testing of a decommissioned RC highway slab bridge and related studies. *Trans. Res. Rec.* 1371:142–153.

Allbright, K., Parekh, K., Miller, R., and Baseheart, T.M. 1994. Modal verification of a destructive test of a damaged prestressed concrete beam. *Exp. Mech.* 34(4):389–396.

Ansari, F., Libo, Y., Lee, I., and Ding, H. 1996. A fiber optic embedded crack opening displacement sensor for cementitious composites. *Proc. 2nd Int'l Conf. NDT Concr. Infrastructure*, pp. 268–277. Society for Experimental Mechanics, Bethel, CT.

Azizinamini, A., Shekar, Y., Boothby, T.E., and Branhill, G. 1992. Load carrying capacity of old concrete slab bridges, *Proc. Third NSF Workshop Bridge Res. Prog.*, pp. 113–116. National Science Foundation, Washington, DC.

Bodocsi, A., Minkarah, I., and Arudi, R. 1993. Analysis of horizontal movements of joints and cracks in Portland cement concrete pavements. *Trans. Res. Rec.* 1392:43–52.

Carino, N., Sansalone, M., and Hsu, N. 1986. Technique for flaw detection in concrete. *J. ACI* 83(3):199–208.

Carino, N. and Sansalone, M. 1992. Detection of voids in grouted ducts using the impact echo method. *ACI Mater. J.* 89(3):296–303.

Catbas, F.N., Lenett, M., Brown, D.L., Doebling, S.W., Farrar, C.R., and Turer, A. 1997. Modal analysis of multi-reference impact test data for steel stringer bridges. *Proc. 15th Int'l Modal Analysis Conf.*, pp. 381–391. Society for Experimental Mechanics, Bethel, CT.

Chung, L. and Shah, S.P. 1989. Effect of loading rate on anchorage bond and beam column joints. *ACI Structural J.* 86(2):132–143.

Delahaza, A. 1996. Nondestructive testing of the concrete roof shell at the Kingdome in Seattle, Washington. *Proc. 2nd Int'l Conf. NDT Concr. Infrastructure*, pp. 256–267. Society for Experimental Mechanics, Bethel, CT.

Dixon, D.E. and Smith, J.R. 1980. Skyline Plaza North (Building A-4)—A case study. In *Full-Scale Load Testing of Structures*, W.R. Schriever, ed., STP 702, pp. 182–199. ASTM, Philadelphia, PA.

Dunnicliff, J. 1993. *Geotechnical Instrumentation for Monitoring Field Performance.* John Wiley and Sons, New York.

Hrinko, W., Miller, R., Young, C., Shahrooz, B., and Aktan, A.E. 1994. Understanding errors and accuracies in DCDTs and wire potentiometers for field testing applications. *Exp. Tech.* 18(2):29–33.

Hucklebridge, A., El-Esnawi, H., and Moses, F. 1995. Shear key performance in multibeam box girder bridges. *J. Performance Const. Fac.* 9(4):271–285.

Measures, R.M. 1995. Fiber optic strain sensing. In *Fiber Optic Smart Structures*, E. Udd, ed., pp. 171–247. Wiley and Sons, New York.

Miller, R.A., Aktan, A.E., and Shahrooz, B.M. 1994. Destructive testing of a decommissioned concrete slab bridge. *ASCE J. of Struct.* 120(7):2176–2198.

Chen, B. and Nawy, E.G. 1994. Structural behavior evaluation of high strength concrete beams reinforced with prestressed prisms using fiber optic sensors. *Proc. ACI Struct. J.* 91(6):708–717.

Nawy, E. and Chen, B. 1996. Bragg grating fiber optic sensing the structural behavior of continuous composite concrete beams reinforced with prestressed prisms. *Proc. 2nd Int'l Conf. NDT Conc. Infrastructure*, pp. 1–10. Society for Experimental Mechanics, Bethel, CT.

Nawy, E.G. and Chen, B. 1997. Fiber optic sensing the behavior of prestressed prism-reinforced continuous composite concrete beams for bridge deck application. *Proceedings, Transportation Research Board*. National Research Council, Washington, DC.

Petrou, M.F., Perdikaris, P.C., and Wang, A. 1994. Fatigue behavior of noncomposite reinforced concrete bridge deck models. *Trans. Res. Rec.* 1460:73–80.

Schousboe, I. 1976. Bailey's Crossroads collapse reviewed. *ASCE J. Const. Div.* 102, Reston, VA.

Shahrooz, B.M., Ho, I.K., Aktan, A.E., deBorst, R., Blaauwendraad, J., Van der Veen, C., Iding, R.H., and Miller, R.A. 1994a. Nonlinear finite element analysis of a deteriorated RC slab bridge. *ASCE J. of Str. Eng.* 120(2):422–440.

Shahrooz, B.M., Miller, R.A., Saraf, V.K., and Godbole, B. 1994b. Behavior of reinforced concrete slab bridges during and after repair. *Trans. Res. Rec.* 1442:128–135.

Weil, G.J. 1993. Nondestructive testing of bridge, highway and airport pavements. *Proc. Int'l Conf. NDT Concr. Infrastructure*, pp. 93–105. Society for Experimental Mechanics, Bethel, CT.

Fire test of concrete panel (Courtesy National Concrete Masonry Association).

24

Fire Resistance and Protection of Structures

by
Mark B. Hogan, P.E.
Vice President of Engineering, National Concrete Masonry Association, Herndon, Virginia. Active member of committees in several professional societies including ACI/TMS Committee 216 on Fire Resistance and Fire Protection of Structures.

24.1 Property Protection and Life Safety

Life safety and property protection for structures under fire conditions are critical functions of all structures. These functions are influenced by the design, construction, and maintenance of structures. Key elements of the design, which have an impact on both the life safety and property protection functions, include the principles of balanced design that incorporate compartmentation to limit the spread of fire, early detection to alert occupants when a fire occurs, and automatic suppression to control a fire until it can be extinguished. Concrete and masonry materials are inherently fire resistant, noncombustible, durable, and maintain structural integrity under fire conditions. These features are used in the design of compartments to contain fire and in the design of structural elements to maintain structural integrity during a fire. Concrete and masonry's durability and permanence can be relied on for life safety and property protection throughout the life of the structure, with minimal investment in maintenance or repair. This chapter presents criteria for the design of concrete and masonry elements to ensure both property protection and life safety functions during fire conditions.

24.1.1 Balanced Design for Fire Safety and Property Protection

Fire safety requires an awareness and understanding of the hazards so that both the potential for fire occurrence and the threat to life and property during a fire are minimized. Death and injury from fire is caused by asphyxiation from toxic smoke and fumes, burns from direct exposure to the fire, heart attacks caused by stress and exertion, and impact due to structural collapse, explosion, and falls. Life safety and property protection are influenced by the design of the building, its fire protection features, and the quality of construction materials, building contents, and maintenance.

Balanced design relies on three complementary systems to reduce the risk of death and the threat to property due to fire:

- a detection system to warn occupants of the fire,
- a containment system to limit the extent of the fire, and
- an automatic suppression system to control the fire until it can be extinguished.

Each of these essential systems contributes to lowering the risk of death and injury from fire as well as to protecting property. The three balanced-design components complement each other by providing fire protection features that are not provided by the other components. Some features of each balanced-design component are intended to be redundant so that in the event that one system is breached or fails to perform the other components continue to provide safety.

Although not a tangible element in fire protection, a strong educational program should be an integral part of any good fire protection plan in addition to the physical components of a balanced-design system.

24.1.1.1 Automatic Detection

Accurate early warning is the first line of defense against slow smoldering fires, typical in dwelling units, with low heat release rates that don't activate sprinkler heads. Detectors that respond to light smoke are important from a life-safety standpoint because they alert occupants in the unit of fire origin to evacuate. Other detection or alarm systems may be used to notify the fire department, thus decreasing response time, expediting rescue operations, and limiting the resulting fire spread and property damage. Detectors wired to a central alarm and installed in corridors and common areas notify all building occupants, permit timely and orderly evacuation, and decrease the potential for injury and death.

The most common detector installed is the smoke-sensing fire detector. Detectors should be wired into a continuous power supply. Their location is determined by judgment and in accordance with the requirements of the general building code. Each dwelling unit in residential construction should be equipped with detectors in all sleeping rooms, in areas adjacent to all sleeping rooms, and on each level of the building, including the basement. The amount of air movement, obstructions within the space, such as partitions or ceiling height, and other factors, will guide the fire-protection engineer in proper selection of detector locations.

The performance of detectors is vulnerable to many unpredictable malfunctions, among which are those owing to acts of sabotage by arsonists, lack of maintenance owing to human error and neglect, and faulty power supply. Young children, the handicapped, elderly, and the deaf may not be able to respond to alarms. Tests on smoke detectors [Menzel, 1934] indicate the need for a regularly scheduled maintenance and testing program as well as periodic replacement of some components.

24.1.1.2 Automatic Suppression

The function of automatic sprinkler systems is to control a fire at the point of origin. While not designed to extinguish a fire, residential sprinklers have been shown to be very reliable and effective in controlling a fire in the room of origin until it can be extinguished. Sprinklers reduce the likelihood of flashover. Flashover, the rapid ignition of volatile gasses, is particularly hazardous in exit corridors. Suppression of a fire allows access to the building to permit rescue and fire suppression efforts to proceed. Sprinklers are credited with preventing multiple deaths in fires.

The National Fire Protection Association (NFPA) maintains minimum standards for the design and installation of sprinkler systems. Sprinkler systems for multifamily housing more than four stories in height are covered by NFPA Standard 13 (1996a). NFPA Standard 13R (1996b) covers the installation of sprinkler systems in multifamily housing up to and including four stories in height. When the interior construction or building contents contain a large amount of combustibles, sprinkler systems should meet the NFPA 13 Standard, regardless of height, to ensure protection in attics, closets, and other concealed spaces built with combustible materials, and to provide additional suppression in all areas owing to the higher fuel loadings.

The NFPA standards cover the design, installation, acceptance testing, and maintenance of sprinkler systems. In addition, there are specific spaces that NFPA 13R does not require to be sprinkled. The standards require an adequate water supply and piping system designed to deliver sufficient water to the sprinkler head. Sprinkler head requirements ensure proper water coverage based on the room dimensions, area to be covered, and fuel loading. The standards also list exceptions for specific spaces that are not required to be sprinklered. Once installation is complete, the standards require inspection and acceptance of the system's piping valves, pumps, and tanks. Testing also includes verification of adequate water flow to the sprinkler heads. After the sprinkler system is in use, it must be maintained; however, specific maintenance requirements and frequency of maintenance are not specified by the standards.

Performance of automatic sprinklers can be vulnerable to system failures owing to inadequate maintenance and inspection or inadequate water supply. Sprinklers are not intended to control electrical and mechanical equipment fires or fires of external origin, such as fires from adjacent buildings, trash fires, and brush fires. Fires in concealed spaces, including some attics, closets, flues, shafts, ducts, and other spaces where sprinkler heads are not required to be installed, can compromise life safety owing to the spread of toxic fumes and smoke. Inadequate water supply can occur owing to low pressure in the municipal water system, broken pipes due to earthquakes or excavation equipment, explosions, freezing temperatures, closed valves due to human error, arson or vandalism, corrosion of valves, pump failure due to electrical outage, and lack of system maintenance.

24.1.1.3 Compartmentation

Compartmentation limits the extent of fire by dividing a building into fire compartments enclosed by fire walls or fire separation wall assemblies and by fire-rated floors and ceilings. Compartments also minimize the spread of toxic fumes and smoke to adjacent areas of a building. Conflagrations beyond the fire compartment are prevented by limiting the total fuel load contributing to the fire. Compartmentation provides safe areas of refuge for handicapped, young, elderly, incapacitated, and other occupants who may not be capable of unassisted evacuation. Compartmentation also provides safe areas of refuge for extended periods when evacuation is precluded due to smoke-filled exit ways or blocked exits. Compartmented construction provides protection for fire and rescue operations. Highly hazardous areas, such as mechanical, electrical, or storage rooms, can be isolated from other occupied areas of a building by fire walls. Fire separation walls and floor and ceiling assemblies between dwelling units in multifamily housing afford protection from fires caused by the carelessness of other occupants. Refuge areas within a building provide protection for occupants by allowing fire fighters to concentrate on extinguishing the fire rather than on rescue efforts.

Compartmentation serves to contain a fire until it can be brought under control by fire fighters. Each compartment is enclosed by fire-resistive components. Floor and wall elements forming the boundaries of each compartment should have a fire resistance rating of at least two hours and should be constructed of noncombustible materials that are capable of preserving the structural integrity of the building throughout the duration of the fire. Openings through compartment boundaries should be protected openings. Doors should be self-closing when fire or smoke is detected. In multifamily housing, each dwelling unit should form a separate compartment. In addition, interior exit ways, as well as storage, electrical, and mechanical rooms, should be separate compartments. Exterior walls should be fire rated to form a barrier to the penetration of exterior fires and to contain interior fires.

The value of compartmentation may be reduced when joints between floors and walls, typically exterior curtain walls, or between walls and ceilings, are not properly fire-stopped. Damage caused by equipment, abuse, or the installation of utilities that are not properly sealed can allow the passage of smoke and gas. Unsealed openings around penetrations can also allow the convective spread of smoke. Self-closing mechanisms on doors in compartment walls may fail if not maintained or if blocked open.

24.1.1.4 Property Protection

The initial cost of providing fire safety can be significant; however, balanced design offers advantages that offset costs. The higher level of protection for both the structure and its contents limits the potential loss due to fire. Immediate and long-term savings will be reflected in lower insurance rates for both the building and its furnishings. Balanced design limits both fire and smoke damage to the building's contents to the compartment of fire origin. Noncombustible compartment boundaries limit damage to the structure itself and reduce repair time following a fire. Repair is generally nonstructural but may include the replacement of doors and windows; electrical outlets, switches, and wiring; heating ducts and registers; and floor, wall, and ceiling coverings.

24.1.1.5 State of the Art in Designing for Fire Safety

Fire-protection engineering is as much an art as it is a science. The number of unknowns and potential fire propagation scenarios are numerous. Fire protection is therefore generally based more on risk assessment than on precise calculation. Currently, building code prescriptive criteria, along with an understanding of the science of fire protection, guides the designer in addressing fire safety [ACI Committee 216, 1995a,b; NFPA, 1994; NIST, 1993].

Some of the more significant fire safety issues requiring consideration are listed in Table 24.1, along with the effectiveness of each component of balanced design. As shown by the table, there may be more than one component that is considered effective in mitigating a particular hazard. Since none of the components are fail-safe, overlapping functions are needed to provide the required level

TABLE 24.1 Fire Safety Functions of Balanced-Design Concept

Function	Automatic Detection	Compartmentation	Automatic Suppression
Controls fire/limits fire growth	□	◩	■
Provides smoke, toxic-fume barrier	□	◩	□
Provides fire barrier	□	◩	□
Limits generation of smoke/toxic fumes	□	◩	■
Allows safe egress	◩	◩	◩
Provides refuge	□	◩	□
Assists fire-fighting efforts	□	◩	◩
Reduces fire and rescue response time	■	□	□
Difficult to vandalize/arson	□	■	□
Performance requires little maintenance	□	■	□
Property Protection Functions and Costs of Balanced Design Component			
Limits the extent of contents damage	◩	◩	◩
Limits the extent of structure damage	◩	■	◩
Low installation costs	■	◩	□
Low maintenance costs	◩	■	□
Limits repair time due to fire damage	◩	■	■

Note: ■ Considered to be effective.
　　　　◩ Considered to be partially effective.
　　　　□ Considered to be ineffective or only slightly effective.

of safety. In addition, there are some functions listed in the table that are addressed by only one component of balanced design. The appendix of NFPA Standard 13R (1996b) also recognizes the need for compartmentation and detection, along with sprinklers, to ensure a reasonable degree of life-safety protection.

There is general agreement among the engineering and scientific communities that the desired approach to improving fire safety in buildings is through computer modeling. To date, limited success has been achieved in developing a system [Allen, 1970] that will assess risk and design protection levels of each of the three components of balanced design. Future efforts will be directed through the government and the private sector in a cooperative effort to develop this important tool.

24.1.2 Design, Construction, and Material Requirements

The fire resistance ratings of concrete and masonry assume the design and construction of these elements comply with the provisions of the *Building Code Requirements for Masonry Structures* [ACI Committee 530, 1995a] and the *Building Code Requirements for Reinforced Concrete* [ACI Committee 318, 1995f] for masonry and concrete elements, respectively. These codes stipulate material requirements by reference to ASTM standards and establish quality assurance provisions for the construction of these elements.

24.2 Fire-Resistance Rating

Two major factors have to be considered in ratings. They are fire endurance and fire resistance. The definition of these two terms, as in the ACI Committee 216 Report (1995a) are as follows:

> **Fire endurance** is a measure of the elapsed time during which a material or assembly continues to exhibit fire resistance under specified conditions of test and performance.
>
> **Fire resistance** is the ability of the material or assembly to withstand fire or to give protection from it. As applied to elements of buildings, it is characterized by the ability to confine a fire or to continue to perform a given structural function, or both.

Building codes establish the minimum level of required fire resistance for specific elements within the structure based on the building's type of occupancy, the function of the element, and other fire-protection considerations. Once the required fire resistance rating of a concrete or masonry element is established, this chapter can assist the designer in meeting that requirement.

The fire-resistance rating criteria presented here is based on the provisions of the *Standard Method of Determining Fire Endurance of Concrete and Masonry Construction Assemblies* [ACI Committee 216, 1995c]. This consensus standard is based on current practice in determining fire ratings of concrete and masonry elements. The standard covers two methods of determining the fire-resistance rating of an element. The most common method for calculating the fire-resistance rating is based on fire test research that has established a correlation between the physical properties of the concrete or masonry member and fire endurance by the *Standard Specification for Fire Tests of Building Construction and Materials*, ASTM E119 [1995]. The second method is based on testing in accordance with the ASTM E119 standard. Fire testing of wall assemblies in accordance with ASTM E119 measures four performance criteria; however, only one of these criteria typically governs the fire-resistance rating of concrete and masonry. Concrete and masonry elements are typically governed by the transmission of heat, which is measured by temperature rise on the non-fire-exposed side of the wall.

A topical summary of the ASTM E119 specification for performance criteria involves the following parameters:

- resistance to the transmission of heat through the wall assembly,
- resistance to the passage of hot gases through the wall sufficient to ignite cotton waste,

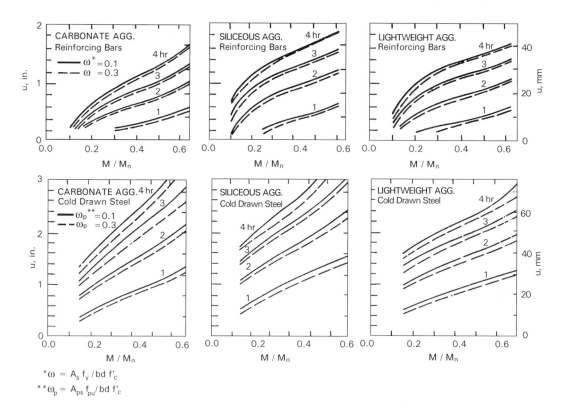

FIGURE 24.1 Fire endurance of concrete slabs as influenced by constituent materials. *Source:* Courtesy of American Concrete Institute; ACI Committee 216, *Guide for Determining the Fire Endurance of Concrete Elements*, ACI 216R-89, 1995.

- load-carrying capacity of load-bearing walls, and
- resistance to the impact, erosion, and cooling effects of a hose stream on the assembly after exposure to the standard fire.

24.2.1 Heat Transmission in Slabs

The structural fire endurance of simply supported concrete slabs as affected by the constituent materials can be interpolated from Figure 24.1 by an effective concrete cover parameter, u, as a function of the moment ratio, M/M_n, where M is the design moment and M_n is the nominal moment strength. In the case of continuous slabs and beams, which is usually the case, a shift in the moment distribution develops, thereby increasing the stresses in the negative reinforcement resulting from the increase in the bending moments at the supports, as seen from Figure 24.2.

However, during a fire, the negative reinforcement remains cooler than the positive reinforcement since it is better protected, being at the top

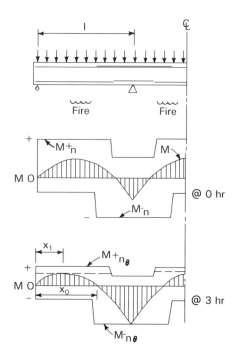

FIGURE 24.2 Moment redistribution of continuous member during fire exposure.

fibers; therefore the increase in the negative moment can be accommodated [ACI Committee 216, 1995a]. Although the moment redistribution that results can be sufficient to result in yielding of the negative reinforcement, the resulting decrease in the positive moment in effect permits the beam or slab span to endure higher temperatures. The negative moment reinforcing bars must be long enough to accommodate the complete redistributed moments and the change of the location of the inflection points. At least 20% of the maximum negative moment reinforcement must be extended throughout the span (CEB/FIP, 1990). One method of increasing the fire endurance of concrete floor slabs is the use of cementitious undercoating as shown in Figure 24.3.

24.2.2 Fire-Resistance Rating of Walls

24.2.2.1 Masonry Walls

The fire-resistance rating of masonry walls, including single-wythe walls, multiwythe walls, and walls with finish treatments, is based on the following criteria, which includes the effect of grouting and the effect of filling the cores of hollow units with specific loose fill materials.

24.2.2.2 Single-Wythe Walls

The calculated fire-resistance ratings of single-wythe masonry walls are determined in accordance with Tables 24.2 and 24.3, for concrete masonry wall assemblies and clay brick wall assemblies, respectively. The equivalent thickness, T_{ea}, of ungrouted and partially grouted walls is based on the equivalent thickness of the masonry unit. The equivalent thickness of the masonry unit is the net volume of the masonry unit divided by the face area of the unit (actual length times actual height). The equivalent thickness, T_e, of solid grouted masonry walls is the actual thickness of the unit. The equivalent thickness, T_e, of hollow masonry unit walls that are completely filled with loose fill is the actual thickness of the unit when loose fill materials are sand, pea gravel, crushed stone, slag,

TABLE 24.2   Fire-Resistance Rating of Concrete Masonry Assemblies

Aggregate Type	Minimum Required Equivalent Thickness, inches, for Fire-Resistance Rating*				
	1 h	1-1/2 h	2 h	3 h	4 h
Calcareous or siliceous gravel (other than limestone)	2.8	3.6	4.2	5.3	6.2
Limestone, cinders, or air-cooled slag	2.7	3.4	4.0	5.0	5.9
Expanded clay, expanded shale, or expanded slate	2.6	3.3	3.6	4.4	5.1
Expanded slag or pumice	2.1	2.7	3.2	4.0	4.7

  *Fire-resistance ratings between the hourly fire-resistance rating periods listed are determined by linear interpolation based on the equivalent thickness value of the concrete masonry wall assembly. Minimum required equivalent thickness corresponding to the fire-resistance rating for units made with a combination of aggregates shall be determined by linear interpolation based on the percent by volume of each aggregate used in the manufacture.

TABLE 24.3   Fire-Resistance Rating of Clay Masonry Walls

Material Type	Minimum Required Equivalent Thickness,* inches, for Fire-Resistance Rating			
	1 h	2 h	3 h	4 h
Solid brick of clay or shale	2.7	3.8	4.9	6.0
Hollow brick or tile of clay or shale, unfilled	2.3	3.4	4.3	5.0
Hollow brick or tile of clay or shale, grouted or filled with perlite, vermiculite or expanded shale aggregate	3.0	4.4	5.5	6.6

  *Fire-resistance ratings between the hourly fire resistance rating periods listed shall be determined by linear interpolation. Where combustible members are framed into the wall, the thickness of solid material between the end of each member and the opposite face of the wall, or between members set in from opposite sides, shall be not less than 93% of the thickness shown.

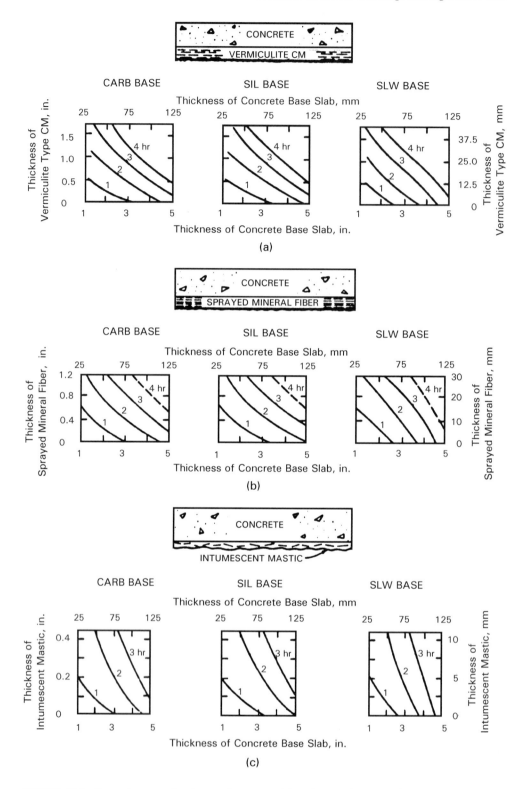

FIGURE 24.3 Fire endurance of undercoated concrete slabs. *Source:* Courtesy of American Concrete Institute; ACI Committee 216, *Guide for Determining the Fire Endurance of Concrete Elements,* ACI 216R-89, 1995.

pumice, scoria, expanded shale, expanded clay, expanded slate, expanded slag, expanded fly ash, cinders, perlite, or vermiculite.

$$T_{ea} = T_e + T_{ef} \tag{24.1}$$

where

T_{ea} = equivalent thickness of masonry wall assembly including finishes, inches
T_e = equivalent thickness of masonry wall assembly excluding finishes, inches
T_{ef} = equivalent thickness of finishes, inches

24.2.2.3 Multiwythe Walls

The fire-resistance rating of multiwythe wall assemblies is determined in accordance with Section 24.2.4, Multiwythe and Multilayer Elements.

24.2.2.4 Expansion or Contraction Joints

The construction of expansion or contraction joints in fire rated masonry wall assemblies are shown in Figure 24.8.

24.2.2.5 Finish Treatments

The effect of finish treatments including gypsum wallboard, plaster or terrazzo on the fire resistance rating are addressed in Section 24.2.6, Finish Treatments.

24.2.3 Concrete Walls, Floors, and Roofs

24.2.3.1 Single-Layer Concrete Elements

The fire-resistance rating of plain and reinforced concrete walls, floors, and roofs that are a single layer in thickness are determined in accordance with Table 24.4 and are based on the equivalent thickness of the element.

24.2.3.2 Equivalent Thickness of Concrete Elements

The equivalent thickness of solid concrete elements with flat surfaces is the actual thickness of the element. The equivalent thickness of hollow core panels with a constant cross section throughout their length is determined by dividing the net cross-sectional area by the panel width. The equivalent thickness of elements in which all of the core spaces are filled with grout or loose fill material, such as perlite, vermiculite, sand or expanded clay, shale, slag or slate, shall be the same as that of a solid wall or slab of the same type of concrete. The equivalent thickness for flanged elements in which the flanges taper is determined at the location of the lesser distance of two times the minimum thickness, or 6 in from the point of the minimum thickness of the flange. The equivalent thickness of elements with ribbed or undulating surfaces is determined as follows:

(a) Where the center-to-center spacing of ribs or undulations is not less than four times the minimum thickness, the equivalent thickness is the minimum thickness of the panel.

TABLE 24.4   Fire-Resistance Rating of Concrete Walls, Floors, and Roofs

Aggregate Type	Fire Resistance Rating for Minimum Equivalent Thickness, inches				
	1 h	1-1/2 h	2 h	3 h	4 h
Siliceous	3.5	4.3	5.0	6.2	7.0
Carbonate	3.2	4.0	4.6	5.7	6.6
Semilightweight	2.7	3.3	3.8	4.6	5.4
Lightweight	2.5	3.1	3.6	4.4	5.1

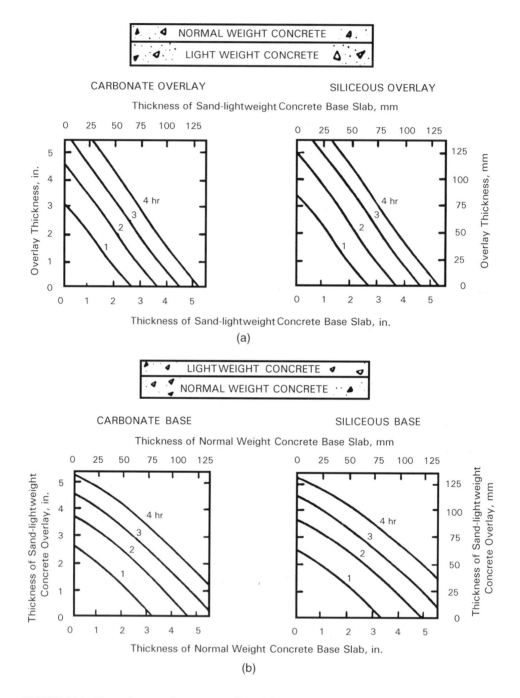

FIGURE 24.4 Fire endurance of concrete two-layer slab system. *Source*: Courtesy of American Concrete Institute; ACI Committee 216, *Guide for Determining the Fire Endurance of Concrete Elements*, ACI 216R-89, 1995. (a) normal weight concrete overlay over lightweight concrete base slab; (b) lightweight concrete overlay over normal concrete base slab.

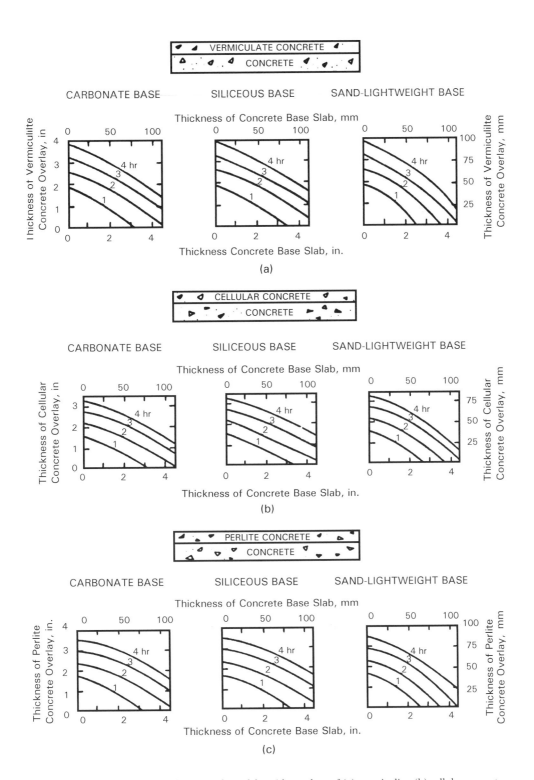

FIGURE 24.5 Fire resistance of concrete base slabs with overlays of (a) vermiculite, (b) cellular concrete, and (c) perlite concrete.

(b) Where the spacing of ribs or undulations is equal to or less than two times the minimum thickness, calculate the equivalent thickness by dividing the net cross-sectional area by the panel width. The maximum thickness used to calculate the net cross-sectional area shall not exceed two times the minimum thickness.

(c) Where the spacing of ribs or undulations exceeds two times the minimum thickness, but is less than four times the minimum thickness, calculate the equivalent thickness from Eq. (24.2).

$$\text{Equivalent thickness} = t + [(4t/s) - 1](t_e - t) \qquad (24.2)$$

where:

s = spacing of ribs or undulations, inches
t = minimum thickness, inches
t_e = equivalent thickness calculated in accordance with (b) above

24.2.3.3 Multiple-Layer Concrete Elements

The fire-resistance rating of walls, floors, and roofs consisting of two or more layers of different types of concrete or masonry are determined in accordance with Section 24.2.4, Multiwythe and Multilayer Elements, or by the following graphical and analytical solution method or numerical solution method. When determining the fire-resistance rating of sandwich panels by Section 24.2.4, Multiwythe and Multilayer Elements, in which the concrete wall panel consists of a layer of foam plastic sandwiched between two layers of concrete that is at least 1 in thick, the following has to be used:

- $R_n^{0.59} = 0.22$ in for foam plastic with a thickness not less than 1 in;
- for foam plastic with a total thickness of less than 1 in, the fire-resistance contribution of the plastic shall be neglected (zero).

24.2.3.4 Graphical and Analytical Solution Method

For solid walls, floors, and roofs consisting of two layers of different types of concrete, use the fire-resistance ratings indicated in Figure 24.4 or as computed from Eq. (24.3) or (24.4). Determine a rating assuming each side of the element is the fire-exposed side. The fire-resistance rating shall be the lower of the two ratings determined, except for floors and roofs, in which the bottom surface shall be assumed to be exposed to fire.

24.2.3.5 Numerical Solution Method

For floor and roof slabs and walls made of one layer of normal-weight concrete and one layer of semilightweight or lightweight concrete where each layer is 1 in or greater in thickness, the combined fire-resistance rating is determined by the following:

- When the fire-exposed layer is of normal-weight concrete

$$R = 0.057(2t^2 - dt + 6/t) \qquad (24.3)$$

- When the fire-exposed layer is of lightweight or semilightweight concrete

$$R = 0.063(t^2 + 2dt - d^2 + 4/t) \qquad (24.4)$$

where

R = fire resistance, hour
t = total thickness of slab, inches
d = thickness of fire-exposed layer, inches

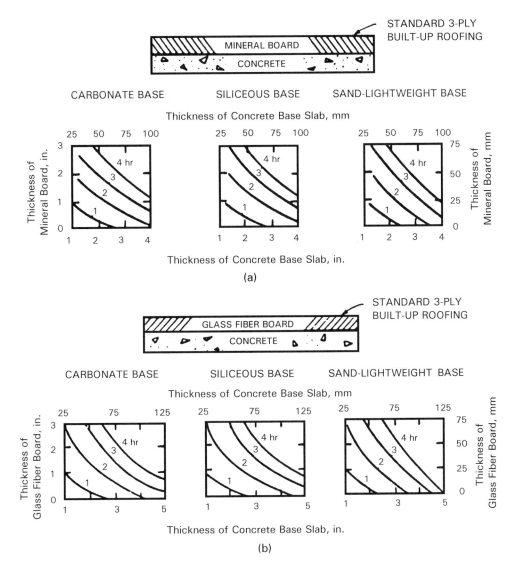

FIGURE 24.6 Fire-retarding insulation on concrete roofs. *Source*: Courtesy of American Concrete Institute; ACI Committee 216, *Guide for Determining the Fire Endurance of Concrete Elements*, ACI 216R-89, 1995. (a) Mineral-board insulation; (b) glass fiber board insulation.

24.2.3.6 Insulated Concrete Roofs

The fire-resistance rating of insulated concrete roofs is based on Figure 24.5 or 24.6 for roofs consisting of a base slab of concrete with a topping (overlay) of cellular, perlite, or vermiculite concrete or of insulation boards and built-up roof. Where a 3-ply built-up roof is installed over a lightweight insulating or sand-lightweight concrete topping, add 10 min to the fire endurance.

24.2.4 Multiwythe and Multilayer Elements

The fire resistance rating, *R*, of multiwythe walls and multilayer elements (Figure 24.7) is determined based on the fire resistance of each wythe or layer and the air space between each wythe or layer in

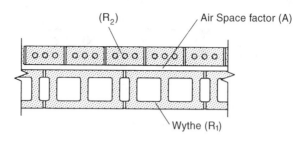

FIGURE 24.7   Multi-wythe walls cross sections.

accordance with Eq. (24.5).

$$R = \left(R_1^{0.59} + R_a^{0.59} + \cdots + R_n^{0.59} + A_1 + A_2 + \cdots + A_n \right)^{1.7} \qquad (24.5)$$

where

$\quad\quad R$ = fire-resistance rating of the multiwythe wall, hours

$R_1, \ldots R_n$ = fire-resistance rating of individual layers or wythes of the element, hours

A_1, A_2, A_n = airspace factor for each continuous airspace within the wall having a width of 1/2 in to 3-1/2 in between layers is taken as 0.03.

24.2.5 Expansion and Contraction Joint Treatment

24.2.5.1 Masonry Elements

Expansion or contraction joints in fire-rated concrete masonry wall assemblies and in clay brick wall assemblies are shown in Figure 24.8.

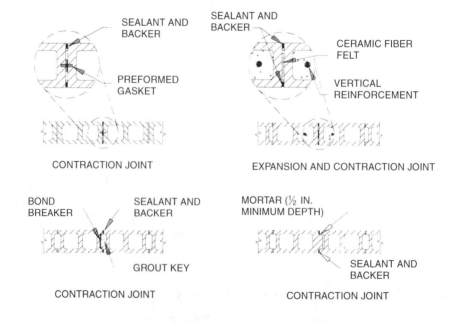

FIGURE 24.8   Expansion and contraction joints for fire-rated masonry walls.

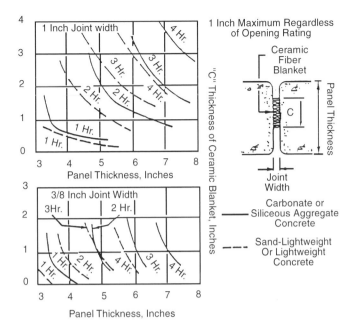

FIGURE 24.9 Ceramic joint fire protection.

24.2.5.2 Precast Concrete Wall Panels and Slabs

In wall panels where openings are not permitted or where it is required that openings be protected, joints must be insulated. In wall panels where permitted or where it is not required that openings be protected, joints do not need be insulated and are considered unprotected openings. Where the percentage of unprotected openings is limited in exterior walls, the area of uninsulated joints is added to the area of other unprotected openings in determining the total area of unprotected openings.

24.2.5.3 Joint Insulation

Fire-rated Concrete Wall Panels. The thickness of ceramic fiber blanket insulation required to insulate joints 3/8 and 1 in wide between fire-rated concrete wall panels are shown in Figure 24.9. For joint widths between 3/8 and 1 in, the thickness of insulation is determined by interpolation. Other approved joint treatment systems that maintain the required fire-resistance rating are also used.

Precast Concrete Slabs. Joints between adjacent precast concrete slabs may be ignored when calculating the equivalent slab thickness, provided that a concrete topping not less than 1 in thick is used. Where a concrete topping is not used, joints should be grouted to a depth of at least one-third the slab thickness at the joint, but not less than 1 in; or the fire-resistance rating of the floor or roof must be maintained by other approved methods.

24.2.6 Finish Treatments

Finish treatments on concrete and masonry elements include gypsum drywall, terrazzo, or plaster. These treatments increase the fire-resistance rating of the element by delaying the temperature rise within the element or through the element that is exposed to fire. The effect of this increase is based on whether the finish is applied to the side of the element being exposed to the fire or to the side that is not exposed to the fire. The fire-resistance rating of elements that may be exposed to fire from either side are determined based on the lower rating determined from assuming the fire exposure is from one side or the other. The fire-resistance rating of the element including the effect of finish treatment(s) is limited to twice the fire rating of the element excluding the effect of finish treatment(s); further, the effect of finish treatments from the non-fire-exposed side of the wall is limited to one-half the fire resistance rating of the element excluding the effect of finish treatment(s).

In order to determine the effect of a specific finish treatment on the non-fire-exposed surface of the element, one has to determine the adjusted thickness of the finish by multiplying the actual

TABLE 24.5 Multiplying Factors for Finishes on the Non-Fire-Exposed Side of Concrete Slabs and Concrete and Masonry Walls

Type of Finish Applied to Slab or Wall	Type of Material Used in Slab or Wall		
	Siliceous- or Carbonate-Aggregate Concrete or Concrete Masonry Unit; Solid Clay Brick Masonry	Semilightweight Concrete; Hollow Clay Brick; Clay Tile	Lightweight Concrete; Concrete Masonry Units of Expanded Shale, Expanded Clay, Expanded Slag, or Pumice Less Than 20% Sand
Portland cement-sand plaster* or terrazzo	1.00	0.75	0.75
Gypsum-sand plaster	1.25	1.00	1.00
Gypsum-vermiculite or perlite plaster	1.75	1.50	1.25
Gypsum wallboard	3.00	2.25	2.25

*For Portland cement-sand plaster 5/8 in, or less, in thickness, and applied directly to concrete or masonry on the non-fire-exposed side of the wall, the multiplying factor shall be 1.0.

TABLE 24.6 Time Assigned to Finish Materials on Fire-Exposed Side of Concrete and Masonry Walls

Finish Description	Time, min
Gypsum wallboard	
3/8 in	10
1/2 in	15
5.8 in	20
2 layers of 3/8 in	25
1 layer 3/8 and 1 layer 1/2 in	35
2 layers 1/2 in	40
Type X gypsum wallboard	
1/2 in	25
5/8 in	40
Direct applied Portland cement-sand plaster	—*
Portland cement-sand plaster on metal lath	
3/4 in	20
7/8 in	25
1 in	30
Gypsum-sand plaster on 3/8 in gypsum lath	
1/2 in	35
5/8 in	40
3/4 in	50
Gypsum-sand plaster on metal lath	
3/4 in	50
7/8 in	60
1 in	80

*The fire-resistance rating of elements with Portland cement-sand plaster finish treatment is determined by adding the actual thickness of the plaster or 5/8 in, whichever is smaller, to the equivalent thickness of the element.

thickness of the finish by the applicable factor from Table 24.5 then adding the adjusted finish thickness to the equivalent thickness of the element, determined from Eq. (24.1). Thereafter the fire-resistance rating of the element can be determined, including the effect of the finish treatment, based on the resulting equivalent thickness, T_a. The effect of a finish treatment on the non-fire-exposed surface of the element is determined by Table 24.5. The fire-resistance rating of the element is the fire-resistance rating of the element excluding the effect of finishes plus the rating obtained from Table 24.6 for the effect of the finish treatment(s).

24.3 Fire Resistance of Columns

24.3.1 Reinforced Masonry Columns

The fire resistance rating of reinforced clay brick or concrete masonry columns are based on the least plan dimension of the column in accordance with the requirements of Table 24.7. The minimum cover for longitudinal reinforcement shall be 2 in.

24.3.2 Reinforced Concrete Columns

Reinforced concrete columns have generally performed well during exposure to fire throughout the history of concrete construction [ACI Committee 216, 1995a]. Circular columns having diameters larger than 12 in (305 mm) and rectangular columns with 12 in minimum dimension are assigned 3 and 4 hours of fire resistance classification, respectively, in most building codes in the United States. The fire-resistance rating of reinforced concrete columns can be determined in accordance with Table 24.8 or 24.9.

TABLE 24.7   Reinforced Masonry Columns

	Minimum Column Dimension, in, for Fire Resistance Rating, h			
Inches	1	2	3	4
Hours	8	10	12	14

TABLE 24.8   Minimum Concrete Column Size

	Fire-Resistance Rating for Minimum Column Dimension, in				
Aggregate Type	1 h	1-1/2 h	2 h	3 h	4 h
Carbonate	8	9	10	11	12
Siliceous	8	9	10	12	14
Semilightweight	8	8-1/2	9	10-1/2	12

TABLE 24.9   Minimum Concrete Column Size when Fire Exposure Conditions are Limited to Two Parallel Sides*

	Fire-Resistance Rating for Minimum Column Dimension, in				
Aggregate Type	1 h	1-1/2 h	2 h	3 h	4 h
Carbonate	8	8	8	8	10
Siliceous	8	8	8	8	10
Semilightweight	8	8	8	8	10

Note: Minimum dimensions are acceptable for rectangular columns with a fire exposure condition on three or four sides, provided that one set of the two parallel sides of the column is at least 36 in long.

24.3.2.1 Minimum Reinforcement Cover for Concrete Columns

The minimum thickness of concrete cover over the main longitudinal reinforcement in columns, regardless of the type of aggregate used in the concrete, must be at least 1 in times the number of hours of required fire resistance but not less than 2 in.

24.3.3 Steel Columns Protected by Concrete and Masonry

The fire-resistance rating of structural steel columns protected by concrete or masonry is determined by Eq. (24.6):

$$R = 0.401(A_s/p_s)^{0.7} + \left[0.285\left(T_{ea}^{1.6}/k^{0.2}\right)\right]$$
$$\times \left[1.0 + 42.7\{(A_s/DT_{ea})/(0.25p + T_{ea})\}^{0.8}\right] \qquad (24.6)$$

where

$\quad R$ = fire-resistance rating of the protected column assembly, hours
$\quad A_s$ = cross-sectional area of the structural steel column, square inches
$\quad D$ = density of the concrete or masonry protection, pounds per cubic foot
$\quad p$ = inner perimeter of the concrete or masonry protection, inches
$\quad p_s$ = heated perimeter of steel column, inches (see Figure 24.10)
$\quad T_{ea}$ = equivalent thickness of the concrete or masonry protection assembly, inches
$\quad k$ = thermal conductivity of the concrete or masonry protection (BTU per hour ft °F)

W section

$$p_s = 2(b_f + d) + 2(b_f - t_w) \qquad (24.7)$$

Pipe section

$$p_s = Xd \qquad (24.8)$$

Square structural tube section

$$p_s = 4d \qquad (24.9)$$

where

$\quad b_f$ = width of flange, inches
$\quad d$ = column dimension, inches (see Figure 24.7)
$\quad t_w$ = thickness of web (in.) (see Figure 24.7, W shape)

The thermal conductivity of concrete masonry is

$$k = 0.0417e^{0.02D}, \text{BTU/hr ft °F} \qquad (24.10)$$

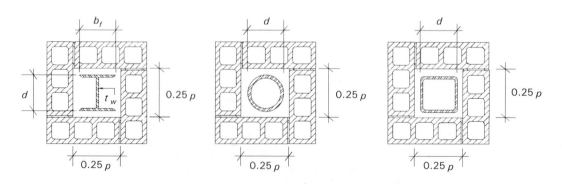

FIGURE 24.10   Masonry column protection for commonly used structural shapes.

TABLE 24.10 Properties of Concrete Masonry Units

Density, d_c		Thermal Conductivity, K	
lb/ft³	kg/m³	Btu/hr ft °F	W/m″K*
80	(1281)	0.207	(0.358)
85	(1362)	0.228	(0.394)
90	(1442)	0.252	(0.436)
95	(1522)	0.278	(0.481)
100	(1602)	0.308	(0.533)
105	(1682)	0.340	(0.588)
110	(1762)	0.376	(0.650)
115	(1842)	0.416	(0.720)
120	(1922)	0.459	(0.749)
125	(2002)	0.508	(0.879)
130	(2082)	0.561	(0.971)
135	(2162)	0.620	(1.073)
140	(2243)	0.685	(1.186)
145	(2323)	0.758	(1.312)
150	(2403)	0.837	(1.449)

*Thermal conductivity at 70°F; °C = (°F-32)(5/9).

TABLE 24.11a Resistance Rating of Steel Columns Protected by Concrete Masonry Units, W-Shape Columns

Column Size	Concrete Masonry Density, lb/ft³	Minimum Required Equivalent Thickness of Concrete Masonry Protection Assembly, T_{ea}, in, for Fire-Resistance Rating				Column Size	Concrete Masonry Density, lb/ft³	Minimum Required Equivalent Thickness of Concrete Masonry Protection Assembly, T_{ea}, in, for Fire-Resistance Rating			
		1 h	2 h	3 h	4 h			1 h	2 h	3 h	4 h
W 14 × 82	80	0.74	1.61	2.36	3.04	W 10 × 68	80	0.72	1.58	2.33	3.01
	100	0.89	1.85	2.67	3.40		100	0.87	1.83	2.65	3.38
	110	0.96	1.97	2.81	3.57		110	0.94	1.95	2.79	3.55
	120	1.03	2.08	2.95	3.73		120	1.01	2.06	2.94	3.72
W 14 × 68	80	0.83	1.70	2.45	3.13	W 10 × 54	80	0.88	1.76	2.53	3.21
	100	0.99	1.95	2.76	3.49		100	1.04	2.01	2.83	3.57
	110	1.06	2.06	2.91	3.66		110	1.11	2.12	2.98	3.73
	120	1.14	2.18	3.05	3.82		120	1.19	2.24	3.12	3.90
W 14 × 53	80	0.91	1.81	2.58	3.27	W 10 × 45	80	0.92	1.83	2.60	3.30
	100	1.07	2.05	2.88	3.62		100	1.08	2.07	2.90	3.64
	110	1.15	2.17	3.02	3.78		110	1.16	2.18	3.04	3.80
	120	1.22	2.28	3.16	3.94		120	1.23	2.29	3.18	3.96
W 14 × 43	80	1.01	1.93	2.71	3.41	W 10 × 33	80	1.06	2.00	2.79	3.49
	100	1.17	2.17	3.00	3.74		100	1.22	2.23	3.07	3.81
	110	1.25	2.28	3.14	3.90		110	1.30	2.34	3.20	3.96
	120	1.32	2.38	3.27	4.05		120	1.37	2.44	3.33	4.12
W 12 × 72	80	0.81	1.66	2.41	3.09	W 8 × 40	80	0.94	1.85	2.63	3.33
	100	1.40	2.42	3.27	4.02		100	1.10	2.10	2.93	3.67
	110	1.46	2.52	3.40	4.16		110	1.18	2.21	3.07	3.83
	120	1.53	2.62	3.52	4.30		120	1.25	2.32	3.20	3.99
W 12 × 58	80	0.88	1.76	2.52	3.21	W 8 × 31	80	1.06	2.00	2.78	3.49
	100	1.04	2.01	2.83	3.56		100	1.29	2.31	3.16	3.90
	110	1.11	2.12	2.97	3.73		110	1.29	2.33	3.20	3.97
	120	1.19	2.23	3.11	3.89		120	1.36	2.44	3.33	4.12
W 12 × 50	80	0.91	1.81	2.58	3.27	W 8 × 24	80	1.14	2.09	2.89	3.59
	100	1.07	2.05	2.88	3.62		100	1.29	2.31	3.16	3.90
	110	1.15	2.17	3.02	3.78		110	1.36	2.42	3.28	4.05
	120	1.22	2.28	3.16	3.94		120	1.43	2.52	3.41	4.20
W 12 × 40	80	1.01	1.94	2.72	3.41	W 8 × 18	80	1.22	2.20	3.01	3.72
	100	1.17	2.17	3.01	3.75		100	1.36	2.40	3.25	4.01
	110	1.25	2.28	3.14	3.90		110	1.42	2.50	3.37	4.14
	120	1.32	2.39	3.27	4.06		120	1.48	2.59	3.49	4.28

where

D = density of concrete masonry, pounds per cubic foot

The thermal conductivity of clay masonry is

$k = 1.25$ BTU/hr ft °F
  for a density of 120 lb/ft³
$k = 2.25$ BTU/hr ft °F
  for a density of 130 lb/ft³

Table 24.10 gives the mechanical properties of concrete masonry units. The minimum required equivalent thickness of masonry units for specific fire-resistance ratings of several commonly used

TABLE 24.11b Resistance Rating of Steel Columns Protected by Concrete Masonry Units, Structural Tube Columns

Tube Nominal Size, in	Concrete Masonry Density, lb/ft³	Minimum Required Equivalent Thickness of Concrete Masonry Protection Assembly, T_{ea}, in, for Fire-Resistance Rating			
		1 h	2 h	3 h	4 h
4 × 4	80	0.93	1.90	2.71	3.43
1/2 in wall	100	1.08	2.13	2.99	3.76
thickness	110	1.16	2.24	3.13	3.91
	120	1.22	2.34	3.26	4.06
4 × 4	80	1.05	2.03	2.84	3.57
3/8 in wall	100	1.20	2.25	3.11	3.88
thickness	110	1.27	2.35	3.24	4.02
	120	1.34	2.45	3.37	4.17
4 × 4	80	1.21	2.20	3.01	3.73
1/4 in wall	100	1.35	2.40	3.26	4.02
thickness	110	1.41	2.50	3.38	4.16
	120	1.48	2.59	3.50	4.30
6 × 6	80	0.82	1.75	2.54	3.25
1/2 in wall	100	0.98	1.99	2.84	3.59
thickness	110	1.05	2.10	2.98	3.75
	120	1.12	2.21	3.11	3.91
6 × 6	80	0.96	1.91	2.71	3.42
3/8 in wall	100	1.12	2.14	3.00	3.75
thickness	110	1.19	2.25	3.13	3.90
	120	1.26	2.35	3.26	4.05
6 × 6	80	1.14	2.11	2.92	3.63
1/4 in wall	100	1.29	2.32	3.18	3.93
thickness	110	1.36	2.43	3.30	4.08
	120	1.42	2.52	3.43	4.22
8 × 8	80	0.77	1.66	2.44	3.13
1/2 in wall	100	0.92	1.91	2.75	3.49
thickness	110	1.00	2.02	2.89	3.66
	120	1.07	2.14	3.03	3.82
8 × 8	80	0.91	1.84	2.63	3.33
3/8 in wall	100	1.07	2.08	2.92	3.67
thickness	110	1.14	2.19	3.06	3.83
	120	1.21	2.29	3.19	3.98
8 × 8	80	1.10	2.06	2.86	3.57
1/4 in wall	100	1.25	2.28	3.13	3.87
thickness	110	1.32	2.38	3.25	4.02
	120	1.39	2.48	3.38	4.17

TABLE 24.11c Resistance Rating of Steel Columns Protected by Concrete Masonry Units, Steel Pipe Columns

Tube Nominal Diameter, in	Concrete Masonry Density, lb/ft^3	Minimum Required Equivalent Thickness of Concrete Masonry Protection Assembly, T_{ea}, in, for Fire-Resistance Rating			
		1 h	2 h	3 h	4 h
4	80	1.26	2.25	3.07	3.79
Standard	100	1.40	2.45	3.31	4.07
0.237 in wall	110	1.46	2.55	3.43	4.21
thickness	120	1.53	2.64	3.54	4.34
4	80	1.12	2.11	2.93	3.65
Extra strong	100	1.26	2.32	3.19	3.95
0.337 in wall	110	1.33	2.42	3.31	4.09
thickness	120	1.40	2.52	3.43	4.23
4	80	0.80	1.75	2.56	3.28
Double extra strong	100	0.95	1.99	2.85	3.62
	110	1.02	2.10	2.99	3.78
0.674 in wall thickness	120	1.09	2.20	3.12	3.93
5	80	1.20	2.19	3.00	3.72
Standard	100	1.34	2.39	3.25	4.00
0.258 in wall	110	1.41	2.49	3.37	4.14
thickness	120	1.47	2.58	3.49	4.28
5	80	1.04	2.01	2.83	3.54
Extra strong	100	1.19	2.23	3.09	3.85
0.375 in wall	110	1.26	2.34	3.22	4.00
thickness	120	1.32	2.44	3.34	4.14
5	80	0.70	1.61	2.40	3.12
Double extra strong	100	0.85	1.86	2.71	3.47
	110	0.91	1.97	2.85	3.63
0.750 in wall thickness	120	0.98	2.08	2.99	3.79
6	80	1.14	2.12	2.93	3.64
Standard	100	1.29	2.33	3.19	3.94
0.280 in wall	110	1.36	2.43	3.31	4.08
thickness	120	1.42	2.53	3.43	4.22
6	80	0.94	1.90	2.70	3.42
Extra strong	100	1.10	2.13	2.98	3.74
0.432 in wall	110	1.17	2.23	3.11	3.89
thickness	120	1.24	2.34	3.24	4.04
6	80	0.59	1.46	2.23	2.92
Double extra strong	100	0.73	1.71	2.54	3.29
	110	0.80	1.82	2.69	3.47
0.864 in wall thickness	120	0.86	1.93	2.83	3.63

column shapes and sizes is shown in Tables 24.11a–c for concrete masonry unit protection and Tables 24.12a–c for clay brick protection.

24.4 Fire Resistance of Lintels

The fire-resistance rating of masonry lintels (beams spanning openings) is based upon the nominal thickness of the beam or lintel and the minimum cover of longitudinal reinforcement, as shown

TABLE 24.12a   Fire Resistance of Clay Masonry Protected Steel Columns, W Shapes

Column Size	Clay Masonry Density, lb/ft³	Minimum Required Equivalent Thickness of Clay Masonry Protection, *T*, in, for Fire-Resistance Rating				Column Size	Clay Masonry Density, lb/ft³	Minimum Required Equivalent Thickness of Clay Masonry Protection Assembly, *T*, in, for Fire-Resistance Rating			
		1 h	2 h	3 h	4 h			1 h	2 h	3 h	4 h
W 14 × 82	120	1.23	2.42	3.41	4.29	W 10 × 68	120	1.27	2.46	3.46	4.35
	130	1.40	2.70	3.78	4.74		130	1.44	2.75	3.83	4.80
W 14 × 68	120	1.34	2.54	3.54	4.43	W 10 × 54	120	1.40	2.61	3.62	4.51
	130	1.51	2.82	3.91	4.87		130	1.58	2.89	3.98	4.95
W 14 × 53	120	1.43	2.65	3.65	4.54	W 10 × 45	120	1.44	2.66	3.67	4.57
	130	1.61	2.93	4.02	4.98		130	1.62	2.95	4.04	5.01
W 14 × 43	120	1.54	2.76	3.77	4.66	W 10 × 33	120	1.59	2.66	3.67	4.53
	130	1.72	3.04	4.13	5.09		130	1.77	3.10	4.20	5.13
W 12 × 72	120	1.32	2.52	3.51	4.40	W 8 × 40	120	1.47	2.70	3.71	4.61
	130	1.50	2.80	3.88	4.84		130	1.65	2.98	4.08	5.04
W 12 × 58	120	1.40	2.61	3.61	4.50	W 8 × 31	120	1.59	2.82	3.84	4.73
	130	1.57	2.89	3.98	4.94		130	1.77	3.10	4.20	5.17
W 12 × 50	120	1.43	2.65	3.6	4.55	W 8 × 24	120	1.66	2.90	3.92	4.82
	130	1.61	2.93	4.02	4.99		130	1.84	3.18	4.28	5.25
W 12 × 40	120	1.54	2.77	3.78	4.67	W 8 × 18	120	1.75	3.00	4.01	4.91
	130	1.72	3.05	4.14	5.10		130	1.93	3.27	4.37	5.34

TABLE 24.12b   Fire Resistance of Clay Masonry Protected Steel Columns, Square Structural Tubing

Tube Size Nominal Size, in	Clay Masonry Density, lb/ft³	Minimum Required Equivalent Thickness of Clay Masonry Protection, *T*, in, for Fire Resistance Rating			
		1 h	2 h	3 h	4 h
4 × 4	120	1.445	2.722	3.764	4.677
1/2 Wall	130	1.616	3.000	4.125	5.108
4 × 4	120	1.565	2.839	3.876	4.785
3/8 Wall	130	1.738	3.116	4.234	5.213
4 × 4	120	1.719	2.986	4.017	4.920
1/4 Wall	130	1.893	3.260	4.371	5.343
6 × 6	120	1.331	2.584	3.615	4.523
1/2 Wall	130	1.500	2.864	3.980	4.959
6 × 6	120	1.478	2.735	3.765	4.669
3/8 Wall	130	1.650	3.014	4.126	5.101
6 × 6	120	1.657	2.913	3.939	4.839
1/4 Wall	130	1.831	3.189	4.296	5.266
8 × 8	120	1.269	2.502	3.522	4.423
1/2 Wall	130	1.436	2.782	3.888	4.862

TABLE 24.12c Fire Resistance of Clay Masonry Protected Steel Columns, Steel Pipe

Pipe Size Nominal Size, in	Clay Masonry Density lb/ft³	Minimum Required Equivalent Thickness of Clay Masonry Protection, *T*, in, for Fire Resistance Rating			
		1 h	2 h	3 h	4 h
4	120	1.744	3.016	4.047	4.950
Standard 0.237 Wall thickness	130	1.915	3.286	4.397	5.369
4	120	1.605	2.886	3.924	4.832
Extra Strong 0.337 Wall thickness	130	1.774	3.158	4.276	5.254
4.2	120	1.255	2.548	3.599	4.520
Extra Strong 0.674 Wall thickness	130	1.415	2.821	3.958	4.950
5	120	1.706	2.970	3.998	4.900
Standard 0.258 Wall thickness	130	1.878	3.243	4.351	5.322
5	120	1.547	2.817	3.851	4.258
Extra Strong 0.375 Wall thickness	130	1.717	3.092	4.208	5.084
5.2	120	1.173	2.441	3.483	4.398
Extra Strong 0.750 Wall thickness	130	1.333	2.717	3.845	4.834
6	120	1.653	2.914	3.941	4.842
Standard 0.280 Wall thickness	130	1.825	3.187	4.295	5.266

TABLE 24.13 Reinforced Masonry Lintels

Nominal Lintel Width, in	Minimum Longitudinal Reinforcement Cover, in, for Fire-Resistance Rating			
	1 h	2 h	3 h	4 h
6	1-1/2	2	NP*	NP*
8	1-1/2	1-1/2	1-3/4	3
10 or more	1-1/2	1-1/2	1-1/2	1-3/4

*NP = not permitted.

in Table 24.13. Cover requirements in excess of 1 1/2 in protect the reinforcement from strength degradation due to excessive temperature during the fire exposure period. Cover requirements may be provided by masonry units, grout, or mortar.

Acknowledgment

The author wishes to acknowledge the assistance of Mark A. Nunn, Senior Building Codes Engineer, Brick Institute of America, for his contributions to this chapter and the development of fire resistance criteria for brick masonry.

References

ACI Committee 216. 1995a. *Guide for Determining the Fire Endurance of Concrete Elements.* ACI 216R-89. American Concrete Institute, Farmington Hills, MI.

ACI Committee 216. 1995b. *ACI Manual for Concrete Practice.* American Concrete Institute, Farmington Hills, MI.

ACI Committee 216. 1995c. *Standard Method for Determining Fire Endurance of Concrete and Masonry Construction Assemblies.* ACI/TMS 216. American Concrete Institute, Farmington Hills, MI.

ACI Committee 318. 1995f. *Building Code Requirements for Structural Concrete.* ACI 318-95. American Concrete Institute, Farmington Hills, MI.

ACI Committee 530. 1995a. *Building Code Requirements for Masonry Structures.* 530-95/ASCE 5-95/TMS 402-95. American Concrete Institute, Farmington Hills, MI.

ACI Committee 530. 1995b. *Specifications for Masonry Structures.* ACI 530.1/ASCE 6-95/MS 602–95. The Masonry Standards Joint Committee, American Concrete Institute, Farmington Hills, MI.

AFMIC. 1921. *Fire Tests of Building Columns.* Associated Factory Mutual Insurance Companies.

Allen, L.W. 1970. *Fire Endurance of Selected Non-Load Bearing Concrete Masonry Walls.* Fire Study No. 25. National Research Council of Canada, Division of Building Research.

Allen, L.W. 1971. *Effect of Sand Replacement on the Fire Endurance of Lightweight Aggregate Masonry Unit.* Fire Study No. 26. National Research Council of Canada, Division of Building Research.

Allen, L.W., Stanzak, W.W., and Galbreath, M. 1974. *Fire Endurance Tests on Unit Masonry Walls with Gypsum Wallboard.* Fire Study No. 32, NRCC 13901. Division of Building Research, National Research Council of Canada.

ASTM. 1995. *Standard Specification for Fire Tests of Building Construction and Materials.* ASTM Standard E 119-95. American Society for Testing and Materials, West Conshohocken.

Blanchard, J.A.C. and Harmathy, T.Z. 1964. *Fire Test of a Load-Bearing Wall Built From Masonry Units (89.1 Percent Solid) of Rotary Kiln, Expanded Shale Aggregate.* National Research Council of Canada, Division of Building Research.

Brick Institute of America. 1996. *BIA Technical Notes.* Reston, VA.

CEB-FIP. 1990. *Model Code for Concrete Structures.* Comite' Euro-International du Beton-Federation Internacionale de Precontraite, Paris.

Cooper, L. 1986. *Why We Need to Test Smoke Detectors. Fire J.*

Foster, H.D., Pinkston, E.R., and Ingberg, S.H. 1950. *Fire Resistance of Walls of Lightweight Aggregate Concrete Masonry Units.* Report BMS 117. National Bureau of Standards, Washington, DC.

Gustaferro, A.H. and Abrams, M.S. 1975. Fire tests of joints between precast concrete wall panels. *Proc. PCI J.* 20(5):

Harmathy, T.Z. and Blanchard, J.A.C. 1962. *Fire Test of a Steel Column of 8-in. H Section, Protected with 4 in. solid Haydite Blocks.* National Research Council of Canada, Division of Building Research.

Hull, A. and Ingberg, H.S. 1925. *Fire Resistance of Concrete Columns.* Technological Paper No. 271. National Bureau of Standards, Washington, DC.

Lie, T.T. and Harmathy, T.Z. 1972. *A Numeral Procedure to Calculate the Temperature of Protected Steel Columns Exposed to Fire.* National Research Council of Canada, Division of Building Research.

Malhotra, M.L. 1982. *Design of Fire-Resisting Structure.* Chapman and Hall, New York.

Menzel, C.A. 1934. *Tests of Fire Resistance and Strength of Walls of Concrete Masonry Units.* Portland Cement Association, Skokie, IL.

National Building Research Institute. 1953. *Fire Resistance of Reinforced Concrete Columns.* National Building Studies, Research Paper No. 18. Her Majesty's Stationary Office, London, England.

Nawy, E.G. 1996. *Fundamentals of High Strength High Performance Concrete.* Addison Wesley Longman, New York.

NBFU. 1973. *Fire Resistance Ratings of Beam, Girder and Truss Protections and Assemblies, Column Protections and Assemblies, Floor-Ceiling Assemblies, Roof-Ceiling Assemblies, Wall and Partition Assemblies.* Report. American Insurance Association Engineering and Safety Service, The National Board of Fire Underwriters.

NBS. 1942. *Building Materials and Structures.* Report BMS 92, and Appendix B. National Bureau of Standards, Washington, DC.

NBS. 1951. *Fire Resistance of Walls of Gravel-Aggregate Concrete Masonry Units.* Report BMS 120. National Bureau of Standards, Washington, DC.

NCMA. 1964. *Estimating Fire Resistance of Concrete Masonry Walls.* National Concrete Masonry Association.

NFPA. 1985. *Standard on Types of Building Construction.* NFPA Report 220. National Fire Protection Association.

NCMA. 1996. *Manual of Facts.* National Concrete Masonry Association, Herndon, VA.

NFPA. 1990. *Standard on Automatic Fire Detectors.* NFPA Report 72E. National Fire Protection Association.

NFPA. *Code for Safety to Life from Fire in Buildings and Structures.* NFPA Report 101. National Fire Protection Association, 1994.

NFPA. 1996a. *Standard for the Installation of Sprinkler System.* NFPA Report 13. National Fire Protection Association.

NFPA. 1996b. *Standard for the Installation of Sprinkler Systems in Residential Occupancies up to Four Stories in Height.* NFPA Report 13R. National Fire Protection Association.

NIST. 1993. *Fire Hazard Assessment Method, Hazard I.* NIST Handbook No. 146. National Institute of Standards and Technology, Gaithersburg, MD.

NRCC. 1963. *Fire Test of Non-Bearing Wall.* Fire Study No. 10. National Research Council of Canada.

Omega Point Laboratories. 1990. *Fire Tests of Various Formulations of Concrete Masonry Units.* National Concrete Masonry Association. Herndon, VA.

(a)

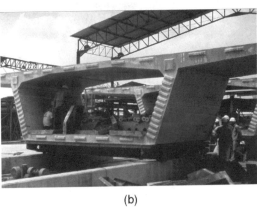

(b)

(c)

(a) Sunshine skyway bridge, Tampa, Florida (Courtesy Portland Cement Association). (b) Prefabricated bridge element (Courtesy Portland Cement Association). (c) Coronado bridge, San Diego, California: 11,720 ft length and 50–200 ft height; girders of precast prestressed lightweight concrete (Steven Simpson, Inc., San Diego, California).

25

Prefabricated Bridge Elements and Systems

by
Michael M. Sprinkel
Research Manager, Virginia Transportation Research Council, Charlottesville, Virginia. Expert since 1970s in materials and construction, particularly public works.

25.1 Practical Applications

A prefabricated concrete bridge element is defined as part of a bridge that is precast away from its final position [Sprinkel, 1985]. A system is a combination of elements. Prefabricated concrete bridge elements and systems are used to construct new bridges and to rehabilitate or replace old ones. Prefabricated elements can reduce design effort, enhance quality, simplify construction, lessen inconvenience to the traveling public, and minimize cost.

Design effort can be reduced when the same design is used on multiple bridge projects. Historically, bridge-design engineers have customized bridge designs for each site, making the prefabrication of elements impractical except for use on major multiple-span bridge projects. Recent efforts have involved making more adjustments to the site to accommodate a standard design and have developed designs that are more versatile.

Fabricating elements in the controlled environment of a prestressed concrete plant enhances quality. Plants are typically certified and well established, although temporary on-site plants are constructed to produce elements for a major bridge project. Plants can use high-quality reusable forms; temperature, relative humidity, and wind can be controlled; the concrete can be batched at the plant; and labor is more effective because tasks are repeated.

Prefabricated elements, set in place at the bridge site, simplify construction by minimizing forming, form removal, and placing and curing concrete in a difficult-to-control environment.

This also minimizes delays and inconvenience to the motorist. Lanes can be open during peak travel periods and closed during off-peak periods for the rapid replacement of roadway sections. Once in place, the fully cured element is ready to receive traffic. Polymer concrete and high-early-strength patching materials now make the rapid connection of prefabricated elements easier.

The most significant reason to use prefabricated elements is the economy realized from the repeated use of forms and the reduction in on-site construction time. Initial construction costs can be lower, depending on the cost for cast-in-place concrete construction. Life-cycle costs will likely be lower because of the higher quality and longer life of the structure. When the cost of delays and inconvenience to the motorist are considered, prefabricated elements that can be assembled and put into use during off-peak traffic periods will almost always be economical.

Prefabricated elements are increasingly popular as highway funds are used to rehabilitate and replace deteriorating bridges. Bridges with high volumes of traffic can usually only be replaced during off-peak traffic periods (at night or on weekends), and prefabricated elements provide an attractive solution. Mass produced, easily assembled elements are just as practical for replacing bridges on low-volume roads.

Prefabricated elements are not the best solution for every bridge construction and replacement project. The demand for a particular shape may be too low to justify an investment in forms. Shipping costs may be too high because the nearest plant is hundreds of miles away. Connection details may cause maintenance problems that result in a higher life-cycle cost. Finally, one of the advantages of concrete is that it can be formed into almost any shape. The architectural and site requirements for a bridge may be so complicated that custom on-site forming is required and the prefabrication of elements is not practical.

A 1984 survey indicated that the use of prefabricated bridge elements was increasing and that the structures were economical in many situations [Hill and Shirole, 1984]. Earlier applications included precast and prestressed slabs and I-beams for simple spans. Later, use expanded to include subdeck panels, deck slabs, parapets, and substructure elements. Currently, all elements in a bridge can be economically constructed or replaced with prefabricated ones.

Recent developments with pozzolans and admixtures have led to the fabrication of elements with concrete compressive strengths in excess of 10,000 psi (68.9 MPa). The higher quality concretes allow smaller cross sections, longer spans, greater girder spacings, and longer service.

25.2 Types of Elements

The most frequently used prefabricated concrete elements and systems are the prestressed I-beam, prestressed box beam, prestressed channel, and slab span [Sprinkel, 1985]. The prestressed subdeck panel was frequently used in the late 70s and 80s, but use has declined in recent years because of reflective cracking in the site-cast overlay concrete. Precast parapets have been used on occasion, but problems with leakage under the parapet have curtailed acceptance. Recent years have seen increased interest in post-tensioned segmental construction for economy in medium and long spans, substructure elements to reduce the environmental impact of construction, and full-depth deck replacement slabs to facilitate the rapid replacement of decks during off-peak traffic periods.

25.2.1 Precast and Prestressed Slab Spans

Slab span elements (Figure 25.1) may be cast in various widths, depths, and lengths to accommodate spans of up to 50 ft (15 m) (Table 25.1). Shorter slabs may be conventionally reinforced and fabricated at simple precast plants. Longer slabs are typically voided and prestressed and/or post-tensioned [PCI, 1975; VTRC, 1980]. Slabs are easy to fabricate, transport, and erect. Department of Transportation (DOT) bridge crews have precast slabs [Sprinkel, 1976].

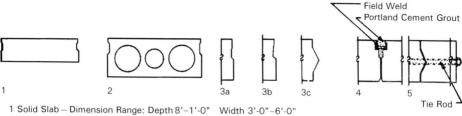

1 Solid Slab – Dimension Range: Depth 8'–1'-0"   Width 3'-0"–6'-0"
2 Voided Slab – Dimension Range: Depth 12"–2'-0"   Width 3'-0"–6'-0"
3a,b,c Typical Keyway Details
4 Typical Connection – Weldplate [Also See Figure 25.2]
5 Typical Connection – Tie Rod

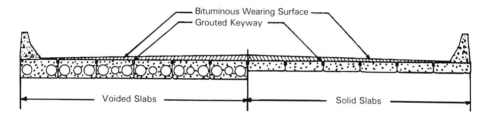

FIGURE 25.1   Slab span. From Sprinkel, M.M. 1985. *Prefabricated Bridge Elements and Systems*, NCHRP Synthesis 119. Transportation Research Board, Washington, DC. With permission.

25.2.2 Multistemmed Beam

Multistemmed beams (Figure 25.2) may be cast in various lengths and increments of width to accommodate short spans. Weld plates and grouted keyways provide shear transfer between beams.

25.2.3 Prestressed Double-Tee and Channel

Most prestressed concrete producers have forms for fabricating double-tees and channels for use in building construction. Additional prestressing, wider webs, and thicker flanges are typically required for bridge loadings (Figure 25.3) [Tokerud, 1975]. Forms have been modified and new forms fabricated to produce members for highway applications when there has been sufficient support provided by a department of transportation (DOT) to justify the investment in forms [Sprinkel and Alcoke, 1977].

Channel beams and double-tees are typically used for medium-length spans, and shear transfer between beams is typically provided by grouted keyways or weld plates. Site-cast concrete is typically placed as an overlay, but channel and double-tee members have been overlaid with asphalt [PCI, 1975].

25.2.4 Prestressed Inverted Channel

The inverted channel (Figure 25.4) may be cast in the inverted position or cast in conventional channel forms and inverted before erection at the bridge site. Longer spans can be achieved

TABLE 25.1   Typical Span Lengths for Elements

Element	Length, ft	Length, m
Precast slab span	10 to 30	3 to 9
Prestressed slab span	20 to 50	6 to 15
Multistemmed beam	25 to 50	8 to 15
Prestressed double-tee and channel	20 to 60	6 to 18
Prestressed inverted channel	30 to 80	9 to 24
Prestressed single-tee	30 to 80	9 to 24
Prestressed I-beam	40 to 100	12 to 30
Prestressed box beam	50 to 100	15 to 30
Prestressed bulb-tee	60 to 80	18 to 24
post-tensioned segmental	50 to 400	15 to 122

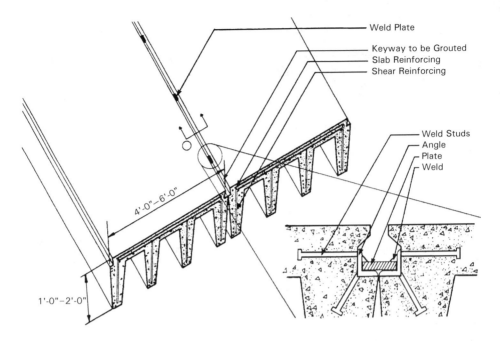

FIGURE 25.2   Multistemmed beam. From Sprinkel, M.M. 1985. *Prefabricated Bridge Elements and Systems*, NCHRP Synthesis 119. Transportation Research Board. Washington, DC. With permission.

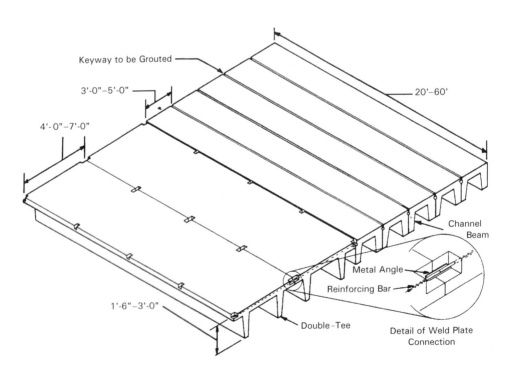

FIGURE 25.3   Double-tee and channel. From Sprinkel, M.M. 1985. *Prefabricated Bridge Elements and Systems*, NCHRP Synthesis 119. Transportation Research Board, Washington, DC. With permission.

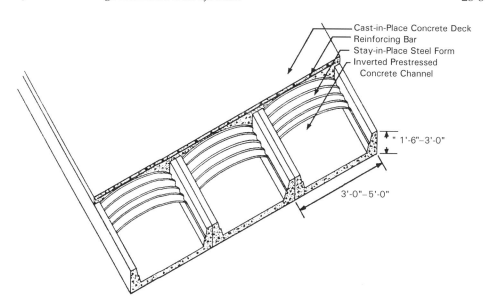

Cast-in-Place Concrete Deck
Reinforcing Bar
Stay-in-Place Steel Form
Inverted Prestressed
Concrete Channel

1'-6"–3'-0"

3'-0"– 5'-0"

FIGURE 25.4   Prestressed inverted channel. From Sprinkel, M.M. 1985. *Prefabricated Bridge Elements and Systems*, NCHRP Synthesis 119. Transportation Research Board, Washington, DC. With permission.

in the inverted position because more prestressing can be placed in the bottom of the beam. The Missouri Department of Transportation used the inverted channel on many bridges [Salmons, 1971]. Site-cast concrete must be placed to connect the channels and provide a deck surface.

25.2.5 Prestressed Single-Tee

Prestressed single-tee beams are used in building construction. Prestressed concrete plants can sometimes fabricate the beams for shorter spans using the same forms as used in building construction with additional prestressing strands to accommodate the heavier loading. With adequate support from a DOT, a precast producer can invest in new forms to produce longer-span beams (Figure 25.5) suitable for highway loadings [Sprinkel and Alcoke, 1977].

The single-tee is unstable by design and must be supported at the bridge site to prevent overturning until the diaphragms can be cast and the keyways grouted. Site-cast concrete is usually placed to connect the tees and to provide a deck [Sprinkel, 1978]. An asphalt wearing surface can be used when the flange of the tee is thick enough to accommodate shear loads.

25.2.6 Prestressed I-Beams

The prestressed I-beam (Figure 25.6) is the prefabricated element most used by DOTs [Sprinkel, 1985]. Many prestressed concrete producers invested in forms during the construction of the interstate system. The standard American Association of State Highway and Transportation Officials (AASHTO) cross sections simplify design and provide for mass production [Panak, 1982]. The beams, cast in a variety of widths and depths, are economical for spans of 40 to 100 ft (12 to 30 m). Spans up to 140 ft (43 m) have been constructed [PCI, 1975; Anderson, 1972]. Longer spans can be achieved by field-connecting the beams end to end and post-tensioning them [Fadl et al., 1977; Oesterle et al., 1989].

The prestressed beams can be positioned more rapidly than a site-cast concrete beam can be constructed. For convenience, other elements are typically constructed with site-cast concrete, limiting the economy of mass production and rapid assembly to the beams. Prestressed concrete subdeck panels have been used, and an ongoing National Cooperative Highway Research Program

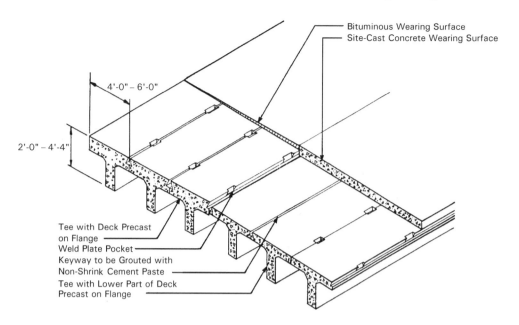

FIGURE 25.5 Prestressed single-tee. From Sprinkel, M.M. 1985. *Prefabricated Bridge Elements and Systems,* NCHRP Synthesis 119, Transportation Research Board. Washington, DC. With permission.

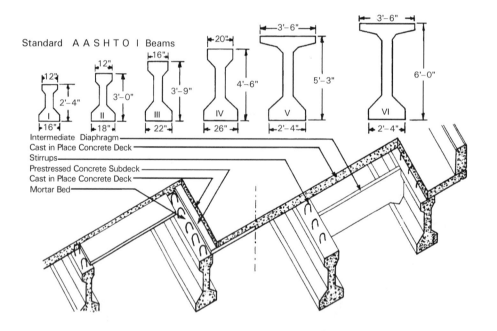

FIGURE 25.6 Prestressed I-beam. From Sprinkel, M.M. 1985. *Prefabricated Bridge Elements and Systems,* NCHRP Synthesis 119, Transportation Research Board. Washington, DC. With permission.

(NCHRP) project, "Rapid Replacement of Bridge Decks," includes the development of designs for prestressed full-depth deck panels to be used with the beams [Yamane et al., in press].

25.2.7 Prestressed Box Beams

The box beam (Figure 25.7) may be precast in a range of widths, depths, and lengths to accommodate spans of approximately 50 to 100 ft (15 to 30 m) [PCI, 1975]. Boxes placed next to each other are

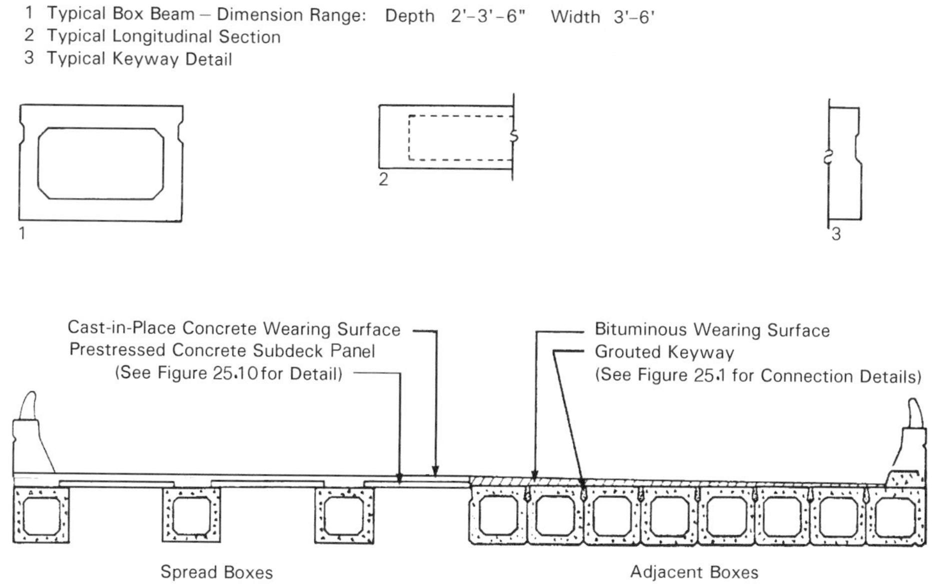

1 Typical Box Beam – Dimension Range: Depth 2'-3'-6" Width 3'-6'
2 Typical Longitudinal Section
3 Typical Keyway Detail

Cast-in-Place Concrete Wearing Surface
Prestressed Concrete Subdeck Panel
(See Figure 25.10 for Detail)

Bituminous Wearing Surface
Grouted Keyway
(See Figure 25.1 for Connection Details)

Spread Boxes   Adjacent Boxes

FIGURE 25.7 Prestressed box beam. From Sprinkel, M.M. 1985. *Prefabricated Bridge Elements and Systems*, NCHRP Synthesis 119, Transportation Research Board, Washington, DC. With permission.

typically tensioned in the transverse direction and covered with a wearing surface of asphalt. Boxes spaced apart are connected with diaphragms. Site-cast concrete is typically used for the diaphragms and deck. Prestressed concrete subdeck panels can also be used with the box beams.

A methodology for the transverse design of concrete box beams without a composite topping has recently been developed [El-Remaily et al., 1996]. The procedure requires that the post-tensioning be applied after the shear keys are grouted, and the design provides a more durable structure.

25.2.8 Prestressed Bulb-Tee

Some DOTs have developed modified versions of the AASHTO girder (Figure 25.8) that are more economical for spans greater than 80 ft [PCI, 1972; Rabbat et al., 1982]. The beams have a high section modulus-to-weight ratio, and spans up to 160 ft have been constructed.

25.2.9 Segmental Construction

Elements (Figure 25.9) are typically full width, match cast, prestressed in the transverse direction, and post-tensioned in the longitudinal direction [VTRC, 1981]. The elements are suitable for use on a wide range of span lengths. For shorter spans, the elements are usually erected on falsework or assembled on a truss supported from pier to pier. For longer spans the elements are erected by balanced cantilever, incremental launching, or progressive placing [Sprinkel, 1985].

A patented segmental concrete overpass system economical for spans of 50 to 115 ft (15 to 35 m) provides at least 2 to 3 ft (0.6 to 0.9 m) of increased underclearance and halves the on-site construction time for a two-span structure [Freyermuth, 1996]. A procedure for the economical replacement of the top slab of a precast post-tensioned segmental bridge has recently been developed, so deck deterioration will not require the replacement of the superstructure [Stelmack and Trapani, 1991].

25.2.10 Prestressed Subdeck Panels

Prestressed subdeck panels are cast in a variety of lengths and widths, typically 4 to 8 ft (1.2 to 2.4 m). The length is a function of the spacing of the supporting beams. The panels are typically

Minimum Width and Depth

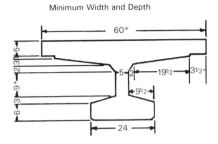

Maximum Width and Depth

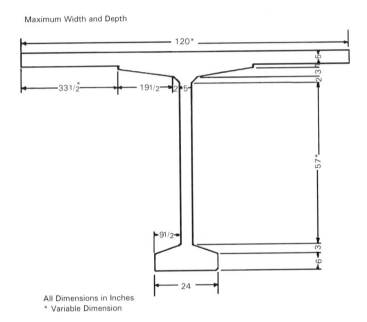

All Dimensions in Inches
* Variable Dimension

FIGURE 25.8 Prestressed bulb-tee. From Sprinkel, M.M. 1985. *Prefabri-cated Bridge Elements and Systems*, NCHRP Synthesis 119. Transportation Research Board, Washington, DC. With permission.

3.5 in (89 mm) thick and are set in a bed of grout about 0.5 in (13 mm) thick. Site-cast concrete is placed over the panels to provide a reinforced deck (Figure 25.10).

The panels are easily installed with a small crane and several laborers and do not need temporary forms or platforms to work from. Cracks usually occur in the site-cast concrete directly above the joints between the panels, and consequently many DOTs have discontinued or restricted the use of the panels. Cracking is less pronounced when the panels are placed on prestressed girders with short spans. Precast concrete subdeck panels can provide an economical and rapidly constructed deck [PCI, 1987].

25.2.11 Prestressed Deck Slabs

The deck is usually the first element in a bridge to deteriorate and to require funds for rehabilitation. In situations where traffic volumes are high, it is often necessary to rehabilitate or replace the deck in sections during off-peak periods. Because of the time required for site-cast concrete to cure, a number of replacement strategies have been developed using prefabricated deck slabs [Issa et al.,

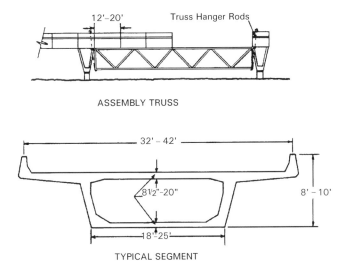

FIGURE 25.9 Post-tensioned segmental construction. From Sprinkel, M.M. 1985. *Prefabricated Bridge Elements and Systems*, NCHRP Synthesis 119. Transportation Research Board, Washington, DC. With permission.

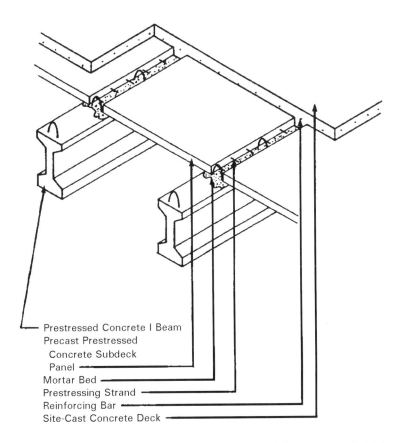

FIGURE 25.10 Prestressed subdeck panels. From Sprinkel, M.M. 1985. *Prefabricated Bridge Elements and Systems*, NCHRP Synthesis 119. Transportation Research Board, Washington, DC. With permission.

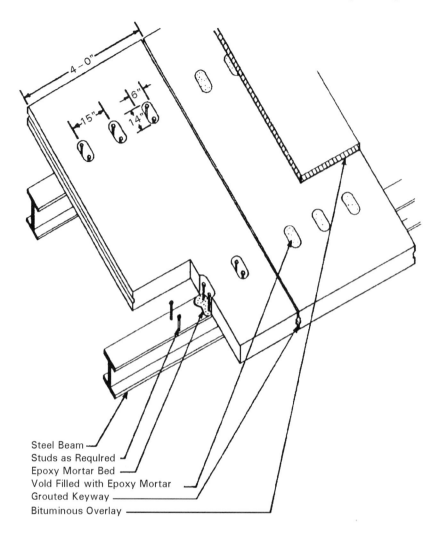

FIGURE 25.11 Prestressed deck slabs. From Sprinkel, M.M. 1985. *Prefabricated Bridge Elements and Systems*, NCHRP Synthesis 119. Transportation Research Board. Washington, DC. With permission.

1995]. Most of the systems involve a transverse segment (Figure 25.11) connected to the supporting beams with a rapid-curing polymer or hydraulic cement concrete. Shear transfer between adjacent slabs is achieved through the use of grouted keyways, site-cast concrete, and post-tensioning. Composite action is achieved through the use of studs on steel beams that extend into voided areas in the slabs that are then filled with polymer or hydraulic cement concrete.

Precast deck slabs can behave in a full-composite manner when connected to steel stringers with studs and epoxy mortar and when keyways are grouted with epoxy mortar [Osegueda et al., 1989]. Improved connection details for use of panels on steel beams and prestressed concrete beams are being developed as part of NCHRP project 12-41 [Yamane et al., in press]. An earlier study identified some suitable connection details and concluded that the deck slabs are more economical than site-cast concrete because of the structural efficiency provided by post-tensioning and prestressing and because of the reduced construction time [Berger, 1983].

A new full-depth precast prestressed concrete bridge deck slab system has been developed that includes stemmed slabs, transverse grouted joints, longitudinal post-tensioning, and welded threaded

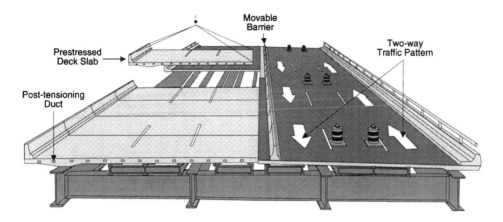

FIGURE 25.12 Prestressed post-tensioned deck slabs were installed at night to replace the deck of the Woodrow Wilson Bridge.

and headless studs [Yamane et al., in press]. The deck slabs are thinner and lighter than a conventional deck and can be constructed faster.

Prestressed deck slabs typically have been used on major bridge deck replacement projects (Figure 25.12) such as the Woodrow Wilson Bridge [Lutz and Scalia, 1984]. Also, most replacements have involved the use of transverse slabs. Longitudinal slabs were successfully used to rehabilitate the Freemont Street Bridge [Smyers, 1984].

25.2.12 Precast Parapet

The precast parapet (Figure 25.13) lends itself to prefabrication because it has a standard shape and can be easily mass produced. Several connection details have been developed to anchor the parapet. The parapet has been used in a number of states but acceptance has been slow because of problems with water and chloride solutions leaking between the base of the parapet and the top of the deck.

25.2.13 Substructure Elements

More time is usually required to construct the substructure than the superstructure, and major reductions in construction time can be achieved by prefabricating the elements of the substructure. Most substructure elements have been prefabricated. Examples include pilings, piers, pier caps, abutments and wing walls. Figure 25.14 shows abutment and wing wall panels placed on temporary pads and anchored with weld plates and a site-cast concrete footing [PCI, 1975]. To simplify erection, abutment and wing wall elements have been precast with the footing and set on a site-cast footing [Sprinkel, 1985]. Prestressed piling has been used for years, but pile caps are usually site cast. Bridges with prefabricated piers,

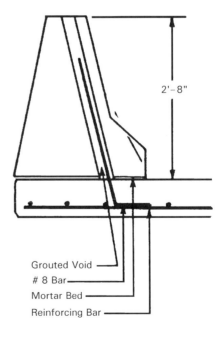

FIGURE 25.13 Precast parapet. From Sprinkel, M.M. 1985. *Prefabricated Bridge Elements and Systems*, NCHRP Synthesis 119, Transportation Research Board, Washington, DC. With permission.

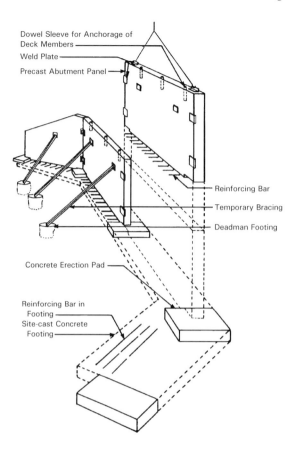

FIGURE 25.14 Precast abutment and wingwall. From Sprinkel, M.M. 1985. *Prefabricated Bridge Elements and Systems*, NCHRP Synthesis 119. Transportation Research Board, Washington, DC. With permission.

pier caps, abutments, and wing walls are limited in number. It is difficult to standardize the elements because of differences between bridge sites and between piers at the same site, which involve soil characteristics, location of bedrock, and the depth at which acceptable bearing can be obtained [Ganga Rao, 1978].

A well-known example of the use of prefabricated piers is the Linn Cove Viaduct [Engineering News Record, 1984]. The entire bridge was prefabricated to minimize environmental impact. Precast segmental superstructure segments were progressively placed and post-tensioned until a pier location was reached. Working from the cantilevered superstructure, holes were drilled into the ground. Prestressed piles were placed in the holes, and precast pier segments were placed and post-tensioned together. Site-cast concrete was placed around the bottom segment (Figure 25.15).

25.2.14 Precast Culverts

Culverts can be used instead of bridges in situations where the cross section will not restrict flow. Culverts are easy to install, do not have a deck to deteriorate, and seldom require extensive plans. Culverts cannot be used on navigable streams.

Precast culvert designs (Figure 25.16) include the pipe, box, inverted U, and arch. Site-cast footings and end walls are typically used with the inverted U and the arch. Concrete pipe is used for spans of 1 to 10 ft (0.3 to 3 m) and concrete boxes are used for spans of 4 to 12 ft (1.2 to 3.7 m) [Concrete

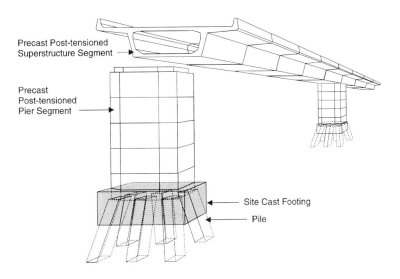

FIGURE 25.15 Prefabricated pier segments.

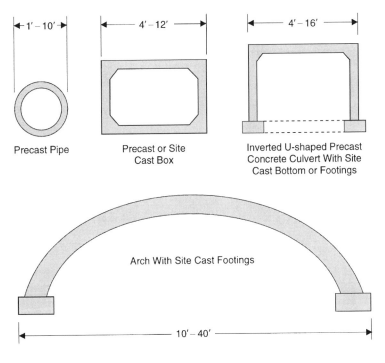

FIGURE 25.16 Precast culverts.

Pipe and Products Company, 1993]. Precast U-shaped culverts have been used for spans up to 16 ft (4.9 m) and the arch shape has been used for 40-ft (12 m) spans [Lambert, 1982; Conspan, 1995].

25.3 Construction Considerations

On-site construction time is typically reduced when prefabricated elements are used because the concrete forming, casting, and curing occurs at a precast plant. Quality elements are typically produced under controlled conditions. Elements are typically inspected at the plant and approved for shipment. Elements should fit together at the site when they are fabricated to the tolerances prescribe by the Prestressed Concrete Institute [PCI, 1977; PCI, 1978]. Precasting operations should be organized to minimize the number of times an element must be moved. Excessive handling is costly and time-consuming and increases the chances for damage [Waddell, 1974].

The contractor should have an approved erection plan. Proper communication between the fabricator and contractor is essential. Elements should be delivered in the order in which they are to be assembled. Each element should be checked for damage that might have occurred during delivery and the plant stamp of approval should be verified.

The hardware, rigging, and equipment required for handling the elements and the lifting locations should be preapproved before lifting an element. Handling and erection stresses can be greater than in-service stresses. Care should be taken to keep the stresses to a minimum. When feasible, elements should be supported during erection as they are during delivery and storage. When lifting equipment is to be placed on the structure, the design should be checked and approved to ensure that the structure is not overstressed. Lifting equipment should be large enough to handle the elements. It is better to have equipment that is too large than too small.

Before placing elements, bearing areas should be properly prepared. Elements that fit properly can be assembled in a few minutes. Additional time is required to make corrections for improperly fitting elements. The advantage of match casting elements is good fit. A variety of mortars and grouts of hydraulic cement and polymer concrete have been developed to facilitate the erection and connection of prefabricated elements [Gulyas et al., 1995]. Temporary shims may be used. High early strength mortars and grouts can anchor the elements in a short time.

Once the elements are assembled, a wearing surface is usually installed. Asphalt is popular because of its low cost, but it should be used in connection with a properly installed membrane to prevent the infiltration of water and chloride ions into the prefabricated elements. Hydraulic cement concrete overlays can be installed to provide the final wearing surface. Overlays are not easy to install and construction should be done according to recommended practice. The recent failure of much of the overlay on the precast, post-tensioned segmental bridge near Honolulu, Hawaii (the two-mile windward viaduct) before opening to traffic illustrates the difficulties associated with constructing a successful overlay. Finally, deck elements can be precast with the final wearing surface, and irregularities can be removed by shotblasting or grinding the surface to provide good ride quality.

25.4 Looking Ahead

The use of prefabricated bridge elements and systems will continue to increase for many reasons. With prefabrication, the work force can be more productive and can produce a better product in the controlled environment of a precast plant, compared to forming and placing reinforcement and concrete outdoors. The enhanced productivity and quality promote economy.

The replacement of bridges and bridge elements is growing as our transportation system ages. The number of structures subjected to high volumes of traffic also continues to increase. Element replacement during off-peak traffic periods is increasingly required, and replacement with prefabricated elements is one of the few feasible options. Reducing delay for the traveling public is an additional economic incentive to use prefabricated elements.

In recent years, the connection details that have caused maintenance problems and reduced the service life of elements have been improved. Better designs, enhanced materials, and more post-tensioning are allowing the construction of bridges with prefabricated elements that are more economical on a life-cycle basis than bridges constructed with site-cast concrete. There will always be a place for site-cast concrete, because concrete can take the shape of any form it is placed in. This flexibility and versatility is needed to satisfy many construction needs. It would be foolish to try to prefabricate concrete for every situation. Even so, the outlook for prefabricated bridge elements and systems has never been better.

References

Anderson, A.R. 1972. Systems Concepts for Precast and Prestressed Concrete Bridge Construction. *HRB Special Report 132: Systems Building for Bridges*, Transportation Research Board, Washington, DC.

Berger, R.H. 1983. Full-Depth Modular Precast, Prestressed Bridge Decks. *Transportation Research Record 903: Bridges and Culverts*, pp. 52–59. Transportation Research Board, Washington, DC.

Concrete Pipe and Products Company, Inc. 1993. *Pipe Design Concepts*. Vienna, Virginia.

Conspan Bridge Systems, Inc. 1995. Dayton, Ohio.

El-Ramaily, A., Tadros, M.K., Yamane, T., and Krause, G. 1996. Transverse design of adjacent precast prestressed concrete box girder bridges. *PCI J.* 41(4):96–113.

Engineering News Record. 1984. *Concrete Today: Markets, Materials, and Methods*, pp. 6, 8, 19, 20.

Fadl, A.I., Gamble, W.L., and Mohraz, B. 1977. Tests of a Precast Post-Tensioned Composite Bridge Girder Having Two Spans of 124 Feet, *Structural Research Series No. 439*. University of Illinois, Chicago, IL.

Freyermuth, C.L. 1996. *Evaluation Findings: The Segmental Concrete Channel Bridge System*, CERF Report: HITEC 96-01. Washington, DC.

Ganga Rao, H.V.S. 1978. Conceptual substructure systems for short-span bridges. *Transp. Eng. J.* 104(1).

Gulyas, R.J., Wirthlin, G.J., and Champa, J.T. 1995. Evaluation of keyway grout test methods for precast concrete bridges. *PCI J.* 40(1):44–57.

Hill, J.J. and Shirole, A.M. 1984. *Economic and Performance Considerations for Short-Span Bridge Replacement Structures*, TRR 950, pp. 33–38. Transportation Research Board, Washington, DC.

Issa, M.A., Idriss, A., Kaspar, I.I., and Khayyat, S.Y. 1995. Full depth precast and precast, prestressed concrete bridge deck panels. *PCI J.* 40(1):59–80.

Lambert, A.V. 1982. Instant arches—European style. *Concr. Int.* 4(1):44–47.

Lutz, J.G. and Scalia, D.J. 1984. Deck widening and replacement of Woodrow Wilson memorial bridge. *PCI J.* 29(3):74–93.

Oesterle, R.G., Glikin, J.D., and Larson, S.C. 1989. *Design of Precast Prestressed Girders Made Continuous*, NCHRP Report 322. Transportation Research Board, Washington, DC.

Osegueda, R.A., Noel, J.S., and Panak, J.J. 1989. *Verification of Composite Behavior of a Precast Decked Simple Span*, TRR 1211, pp. 72–83. Transportation Research Board, Washington, DC.

Panak, J.J. 1982. *Economical Precast Concrete Bridges*, Research Report 226-1F, pp. 1, 12–18, 27–32. Texas State Department of Highways and Public Transportation. Austin, Texas.

Prestressed Concrete Institute (PCI). 1972. Modern concepts in prestressed concrete bridge design. *PCI Bridge Bull.*, Second Quarter.

Prestressed Concrete Institute (PCI). 1975. *Short Span Bridges*. Chicago, IL.

Prestressed Concrete Institute (PCI). 1977. *Manual for Quality Control for Plants and Production of Precast Prestressed Concrete Products*. Chicago, IL.

Prestressed Concrete Institute (PCI). 1978. Precast prestressed concrete industry code of standard practice for precast concrete. *PCI J.* 23(1):14–31.

Prestressed Concrete Institute (PCI). 1987. Precast prestressed concrete bridge deck panels. *PCI J.* 32(2):26–45.

Rabbat, B.G., Takayanagi, T., and Russell, H.G. 1982. *Optimized Sections for Major Prestressed Concrete Bridge Girders*, Report No. FHWA/RD-82/005. Federal Highway Administration, Washington, DC.

Salmons, J.R. 1971. Structural performance of the composite U-beam bridge superstructure. *PCI J.* 16(4):21–23.

Smyers, W.L. 1984. Rehabilitation of the Fremont Street bridge. *PCI J.* 29(5):34–51.

Sprinkel, M.M. 1976. In-House Fabrication of Precast Concrete Bridge Slabs, VHTRC 77-R33. Virginia Highway and Transportation Research Council, Charlottesville, VA.

Sprinkel, M.M. 1978. Systems construction techniques for short span concrete bridges. In *Transportation Research Record 665: Bridge Engineering*, Vol. 2, pp. 222–227. Transportation Research Board, Washington, DC.

Sprinkel, M.M. 1985. Prefabricated Bridge Elements and Systems. *NCHRP Synthesis of Highway Practice 119*. Transportation Research Board, Washington, DC.

Sprinkel, M.M. and Alcoke, W.H. 1977. Systems Bridge Construction in Virginia. *Am. Transp. Builder.* 11–13. Portland Cement Association, Skokie, Ill.

Stelmack, T.W. and Trapani, R.J. 1991. *Design Provisions For a Replaceable Segmental Bridge Deck*, TRR 1290. Vol. 1, pp. 77–92. Transportation Research Board, Washington, DC.

Tokerud, R. 1975. *Economical Structures for Low-Volume Roads*. Special Report 160: Low-Volume Roads, pp. 273–277. Transportation Research Board, Washington, DC.

Virginia Transportation Research Council (VTRC). 1980. *Bridges on Secondary Highways and Local Roads, Rehabilitation and Replacement*, NCHRP Report 222. Transportation Research Board, Washington, DC.

Virginia Transportation Research Council (VTRC). 1981. *Rehabilitation and Replacement of Bridges on Secondary Highways and Local Roads*, NCHRP Report 243. Transportation Research Board, Washington, DC.

Waddell, J.J. 1974. *Precast Concrete: Handling and Erection*. American Concrete Institute, Farmington Hills, MI.

Yamane, T., Tadros, M.K., and Baishya, M. 1997. Full-depth precast prestressed concrete bridge deck system. *PCI J*. In Press.

(a)

(b)

Kaiser Venice Clinic

(c)

Sherman Oaks Fashion Square Parking Structure

(a) CSU Northridge parking structure—suffered no damage in the 1994 Northridge, CA, earthquake while others in area suffered heavy damage (Courtesy, BFL/Owen, engineers of record. (b) Kaiser Venice Clinic, after earthquake (Courtesy FBA, Inc.). (c) Sherman Oaks fashion square parking structure after earthquake (Courtesy FBA, Inc.).

26

Seismic Resisting Construction

by
Walid M. Naja, S.E.
Principal, W.N. Structural Engineers, Belmont, California. Expert in the design and construction of concrete structures in seismically active regions, particularly low to mid-rise buildings.

Florian G. Barth, P.E.
Principal, FBA, Inc., Belmont, California. Expert in the design of post-tensioned concrete structures and in the design and rehabilitation of buildings in seismic regions.

0-8493-2666-4/97/$0.00+$.50

26.1 Fundamentals of Earthquake Ground Motion

26.1.1 Introduction

When transmitted through a structure, ground acceleration, velocity, and displacements (referred to as ground motion), are in most cases amplified. The amplified motion can produce forces and displacements that may exceed those the structure can sustain. Many factors influence ground motion and its amplification. An understanding of how these factors influence the response of structures and equipment is essential for a safe and economical design. Earthquake ground motion is usually measured by a strong-motion accelerograph that records the acceleration of the ground at a particular location. Record accelerograms, after they are corrected for instrument errors and adjusted for baseline, are integrated to obtain velocity and displacement-time histories.

The maximum values of the ground motion (peak ground acceleration, peak ground velocity, and peak ground displacement) are of interest for seismic analysis and design. These parameters, however, do not by themselves describe the intensity of shaking that structures or equipment experience. Other factors, such as the earthquake magnitude, distance from fault or epicenter, duration of strong shaking, soil condition of the site, and frequency content of the motion, also influence the response of a structure. Some of these effects, such as the amplitude of motion, duration of strong shaking, frequency content, and local soil conditions, are best represented through the response spectrum (1.1 to 1.4), which describes the maximum response of a damped single-degree-of-freedom oscillator to various frequencies or periods. The response spectra from a number of records are often averaged and smoothed to obtain design spectra, which also represent the amplification of ground motion at various frequencies or periods of the structure.

This section discusses earthquake ground motion. The influence of earthquake parameters such as earthquake magnitude, duration of strong motion, soil condition, and epicentral distance on ground motion will be presented and discussed.

26.1.2 Recorded Ground Motion

Ground motion during an earthquake is measured by a strong-motion accelerograph that records the acceleration of the ground at a particular location. Three orthogonal components of the motion, two in the horizontal direction and one in the vertical, are recorded by the instrument. The instruments may be located in a free field or mounted in structures. Accelerograms are generally recorded on photographic paper or film. The records are digitized for engineering applications, and during the digitization process, errors associated with instruments and digitization are removed. The digitization, correction, and processing of accelerograms have been carried out by the Earthquake Engineering Research Laboratory of the California Institute of Technology in the past and are now carried out by the United States Geological Survey.

26.1.3 Characteristics of Earthquake Ground Motion

The characteristics of earthquake ground motion that are important in earthquake engineering applications are

1. peak ground motion (peak ground acceleration, peak ground velocity, and peak ground displacement),
2. duration of strong motion, and
3. frequency content.

Each of these parameters influences the response of a structure. Peak ground motion primarily influences the vibration amplitudes. Duration of strong motion has a pronounced effect on the severity of shaking. A ground motion with a moderate peak acceleration and long duration may cause more damage than a ground motion with a larger acceleration and a shorter duration. Frequency content and spectral shapes relate to frequencies or periods of vibration of a structure. In a structure, ground motion is most amplified when the frequency content of the motion and the vibration frequencies of the structure are close to each other. Each of these characteristics is briefly discussed below.

26.1.3.1 Peak Ground Acceleration

Peak ground acceleration has been widely used to scale earthquake design spectra and acceleration-time histories. As will be discussed later, recent studies recommend that in addition to peak ground acceleration, peak ground velocity and displacement should be used for scaling purposes.

26.1.3.2 Duration of Strong Motion

Several investigators have proposed procedures for computing the strong-motion duration of an accelerogram. Page et al. and Bolt proposed the **bracketed duration**, which is the time interval between the first and the last acceleration peaks greater than a specified value (usually 0.05 g). Trifunac and Brady defined the duration of the strong motion as the time interval in which a significant contribution to the integral of the square of the acceleration ($\partial a^2 dt$)—referred to as the **accelerogram intensity**—takes place. They selected the time interval between the 5% and the 95% contributions as the duration of strong motion.

The noted procedures usually result in different values for the duration of strong motion. This is to be expected, since the procedures are based on different criteria. Since there is no standard definition of strong-motion duration, the selection of a procedure for computing it for a study depends on the purpose of the study. The bracketed duration proposed in Page et al. and Bolt may be more appropriate when computing elastic and inelastic response.

26.1.3.3 Frequency Content

The frequency content of ground motion can be examined by transforming the motion from time domain to frequency domain through a Fourier transform. The Fourier amplitude spectrum and power spectral density, which are based on this transformation, may be used to characterize the frequency content. For further discussion regarding frequency content, the reader is encouraged to consult *The Seismic Design Handbook* (a Van Nostrand Reinhold Publication).

Once the power spectral density of ground motion at a site is established, random-vibration methods may be used to formulate probabilistic procedures for computing the response of structures.

26.1.4 Factors Influencing Ground Motion

Earthquake ground motion and its duration at a particular location are influenced by a number of factors, the most important being (i) earthquake magnitude, (ii) distance of the source of energy release (epicentral distance or distance from causative fault), (iii) local soil conditions, (iv) variation in geology and propagation of velocity along the travel path, and (v) earthquake-source conditions and mechanism (fault type, stress conditions, stress drop). Past earthquake records have been used to study some of these influences. While the effect of some of these parameters, such as local soil conditions and distance from source of energy release, are fairly well understood and documented, the influence of the source mechanism and the variation of geology along the travel path are more complex and difficult to quantify. Several of these influences are interrelated, and consequently it is difficult to discuss them individually without incorporating the others. Some of these influences are discussed below.

26.1.4.1 Distance

The variation of ground motion with distance to the source of energy release has been studied by many investigators. In most studies, peak ground motion (usually peak ground acceleration) is plotted as a function of distance. A smooth curve based on a regression analysis is fitted to the data, and the curve or its equation is used to predict the expected ground motion as a function of distance.

These relationships, referred to as motion attenuation, are sometimes plotted independently of the earthquake magnitude. This was the case in earlier studies because of the lack of sufficient numbers of earthquake records. However, with the availability of a large number of records, particularly during the 1971 San Fernando earthquake and subsequent seismic events, the data base for attenuation studies was increased and a number of investigators reexamined their earlier studies, modified their proposed relationships for estimating peak accelerations, and included earthquake magnitude as a parameter. Donovan (1973) compiled a data base of more than 500 recorded accelerations from seismic events in the United States, Japan, and elsewhere, and later increased it to more than 650. A plot of peak ground acceleration versus fault distance for different earthquake magnitudes from his data base shows that, for a good portion of the data, the peak acceleration decreases as the distance from the source of energy release increases.

It has been reported by Housner, Donovan, and Seed and Idriss (1982) that at distances away from the fault or the source of energy release (far field), the earthquake magnitude influences the attenuation, whereas at distances close to the fault (near field), the attenuation is affected by smaller earthquake magnitudes and not by the larger ones. The majority of attenuation studies and the relationships presented in Table 26.1 are mainly from data in the western United States. It is believed by several seismologists that ground acceleration attenuates more slowly in the eastern United States and eastern Canada.

TABLE 26.1 Typical Attenuation Relationships

Data Source	Relationships	Reference
1. San Fernando earthquake, Feb. 9, 1971	$a = 190/R^{1.83}$	Donovan
2. California earthquake	$a = \dfrac{y_0}{1 + (R'/h)^2}$ where $\log y_0 = -(\bar{b} + 3) + 0.81m - 0.027m^2$ $\bar{b}$ is a site factor	Blume
3. California and Japanese earthquakes	$a = \dfrac{0.0051}{\sqrt{T_G}} 10^{0.61m - P\log R + Q}$ where $Q = 0.167 - 1.83/R$ T_G = fundamental period of site	Kanai
4. Cloud (1963)	$a = \dfrac{0.0069e^{1.64m}}{1.1e^{1.1m} + R^2}$	Milne and Davenport
5. Cloud (1963)	$a = 1.24e^{0.67m}/(R + 25)^2$	Esteva
6. U.S.C. & G.S.	$\log a = \dfrac{6.5 - 2\log(R' + 80)}{981}$	Cloud and Perez
7. 303 instrumental values	$a = 1.325e^{0.67m}/(R + 25)^{1.6}$	Donovan
8. Western U.S. records	$a = 0.093e^{0.8m}/(R^2 + 400)$	Donovan
9. U.S., Japan, etc.	$a = 1.35e^{0.58m}/(R + 25)^{1.52}$	Donovan
10. Western U.S. records; USSR; Iran	$\ln a = 3.99 + 1.28m - 1.75\ln[R + 0.147e^{0.732m}]$	Campbell
11. Western U.S. records; worldwide	$\log a = -1.02 + 0.249m - \log(R^2 + 7.3^2)^{1/2}$ $-0.00255(R^2 + 7.3^2)^{1/2}$	Joyner and Boore
12. Western U.S. records	$\ln a = \ln \alpha(m) - \beta(m)\ln(R + 20)$ where m = surface-wave magnitude, $m \geq 6$ local magnitude, $m < 6$ R = smallest distance to source, $m \geq 6$ hypocentral distance, $m < 6$ $\alpha(m), \beta(m)$ are magnitude dependent coefficients	Idriss
13. Italian records	$\log a = -1.562 + 0.306m - \log(R^2 + 5.8^2)^{1/2} + 0.169S$ where $S = 1.0$ for soft sites and 0 for rock	Sabetta and Pugliese

Acceleration a in g; distance to causative fault, R in kilometers; epicentral distance R' in miles; local depth, h, in miles; magnitude m.

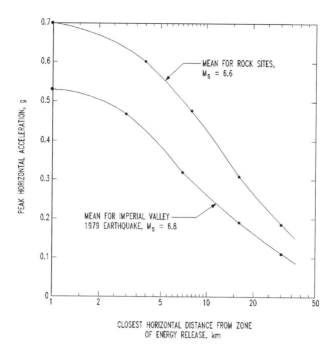

FIGURE 26.1   Comparison of attenuation curves for rock sites and
the Imperial Valley earthquake of 1979. After Seed and Idriss.

The variation of peak ground velocity with distance from the source of energy release (velocity attenuation) has also been studied by several investigators. Velocity attenuation curves have similar shapes and follow similar trends to the acceleration attenuation. However, velocity attenuates somewhat faster than acceleration, and unlike acceleration attenuation, velocity attenuation depends on soil condition.

Distance also influences the duration of strong motion. Correlation of the duration of strong motion with epicentral distance has been studied by Page et al., Trifunac and Brady, Chang and Krinitzsky, and others. Page et al., using the bracketed duration, conclude that for a given magnitude, the duration decreases with an increase in the distance from the source.

26.1.4.2 Site Geology

Soil conditions influence ground motion and its attenuation. Several investigators, such as Boore et al, and Seed and Idriss have presented attenuation curves for soil and rock. According to Boore et al., the peak horizontal acceleration is not appreciably affected by soil condition (it is nearly the same for rock and for soil). Seed and Idriss compare the acceleration attenuation for rock from earthquakes with Richter magnitudes of approximately 6.6 with that for alluvium from the 1979 Imperial Valley earthquake (Richter magnitude 6.8). Their comparison, shown in Figure 26.1 indicates that at a given distance from the source of energy release, peak accelerations on rock are somewhat greater than those on alluvium. The effect of soil condition on peak acceleration is illustrated in Figure 26.2, (after Seed and Idriss, 1982). According to this figure, the difference in acceleration in rock and in stiff soil is practically negligible. There seems to be general agreement among various investigators that soil condition has a pronounced influence on velocities and displacements. Larger peak horizontal velocities are more likely to be expected in soil than in rock.

26.1.4.3 Magnitude

As expected, at a given distance from the source of energy release, larger earthquake magnitudes result in larger peak ground accelerations, velocities, and displacements. Because of the lack of adequate data for earthquake magnitudes greater than approximately 7.5, the effect of magnitude

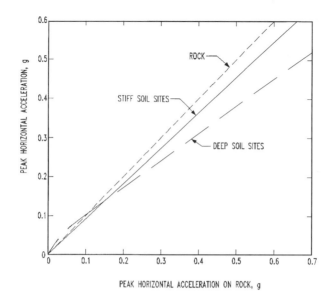

FIGURE 26.2 Relationship between peak accelerations of rock and soil. After Seed and Idriss.

on peak ground motion and duration is generally determined through extrapolation of data from earthquake magnitudes smaller than 7.5.

26.1.5 Evaluating Seismic Risk at a Site

Seismic-risk evaluation is based on information from three sources: (i) the recorded ground motion, (ii) the history of seismic activity in the vicinity of the site, and (iii) the geological data and fault activities of the region. For most regions of the world this information, particularly that from the first source, is very limited and may not be sufficient to predict the size and recurrence intervals of future earthquakes. Nevertheless, earthquake engineers have relied on this limited information to establish some acceptable levels of risk.

Seismic-risk analysis usually begins by developing mathematical models that are used to estimate the recurrence intervals of future earthquakes with certain magnitudes and/or intensities. These models, together with the appropriate attenuation relationships, are commonly utilized to estimate ground-motion parameters such as the peak acceleration and velocity corresponding to a specified probability and return period.

Using the seismic-risk principles of Cornell, Algermissen and Perkins developed isoseismal maps for peak ground accelerations (Figure 26.3) and velocities. The Applied Technology Council has used such maps to develop similar maps for effective peak velocity-related acceleration. The effective peak acceleration and the effective peak velocity-related acceleration are defined by the Applied Technology Council on the basis of a study by McGuire. They are obtained by dividing the spectral accelerations at periods of 0.1–0.5 s and the spectral velocity at a period of approximately 1.0 s by a constant factor.

26.1.6 Estimating Ground Motion

Seismic-risk procedures and attenuation relationships are mostly developed for estimating the expected peak horizontal acceleration at the site.

A statistical summary of the peak ground acceleration ratios for all four soil categories (rock, 30 ft of alluvium underlain by rock, 30–200 ft of alluvium underlain by rock, alluvium) is presented in Table 26.2. The table includes the ratio of the smaller to the larger of the two peak horizontal

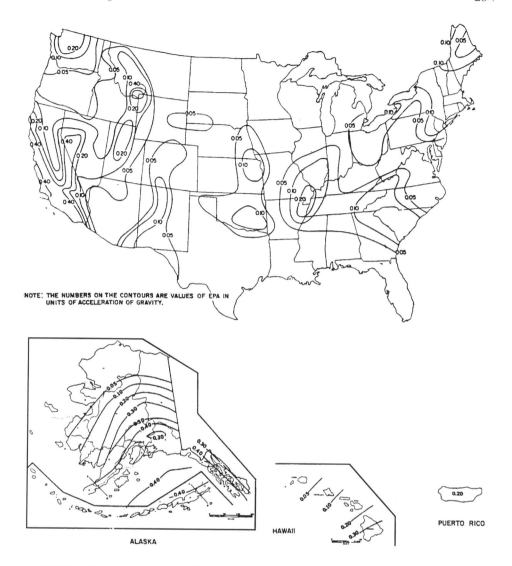

NOTE: THE NUMBERS ON THE CONTOURS ARE VALUES OF EPA IN UNITS OF ACCELERATION OF GRAVITY.

ALASKA   HAWAII   PUERTO RICO

FIGURE 26.3 Contour map for effective peak acceleration. (Applied Technology Council, 2-47).

TABLE 26.2 Summary of Peak Ground Acceleration Ratios*

	a_s/a_l			a_v/a_l		
	Percentile			Percentile		
Soil Category	50	84.1	Mean	50	84.1	Mean
Rock	0.81	0.99	0.82	0.48	0.69	0.52
<30 ft of alluvium underlain by rock	0.89	0.98	0.89	0.47	0.62	0.46
30–200 ft of alluvium underlain by rock	0.82	0.96	0.83	0.40	0.66	0.46
Alluvium	0.75	0.96	0.79	0.42	0.61	0.45

Note: a_l = the larger peak acceleration of the two horizontal components; a_s = the smaller of the two horizontal components; a_v = the peak acceleration of the vertical component.

*Lognormal distribution.

accelerations and the ratio of the vertical to the larger of the two peak horizontal accelerations. In each column, the ratios are generally close to each other, indicating that the soil condition does not influence the acceleration ratios. The 2/3 ratio of the vertical to horizontal acceleration, which has been recommended by Newmark and has been employed in seismic design, is closer to the 84.1-percentile than to the 50-percentile ratio. Although the 2/3 ratio is conservative, its use has been justified as taking into account variations greater than the median and uncertainties in the ground motion in the vertical direction.

26.2 Uniform Building Code

26.2.1 UBC Design Criteria

The Uniform Building Code (UBC) provisions for lateral seismic loads are based on *Recommended Lateral Force Requirements and Commentary* of the Structural Engineers Association of California. The basic guideline of those provisions is that a minor seismic event should cause little or no damage, while a major seismic event should not result in the collapse of the structure. Accordingly, the building is expected to behave elastically when subjected to frequently occurring earthquakes and exhibit inelastic behavior when influenced only by infrequent strong earthquakes. Most low-rise concrete buildings fall into the regular structure type of the Uniform Building Code and accordingly are designed for a loading condition resulting from equivalent static lateral force. This static load depends on the **seismic zone** the structure is located in, the site geology and soil characteristics, the building occupancy, the building configuration and height, and the structural system being used to support the lateral load. It is important to note that code static loads are no more than one quarter to one third of the expected earthquake action. This can be justified in concrete structures where the anticipated cracking results in increased energy absorption only where ductility in each resisting member and between all the resisting members can be maintained.

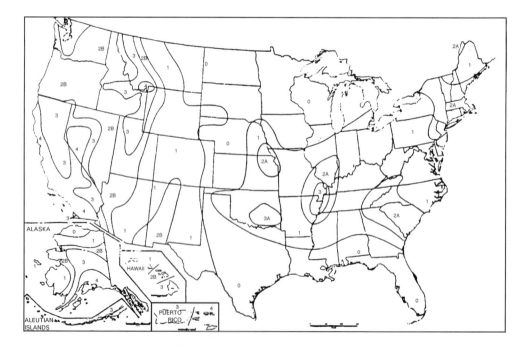

FIGURE 26.4 Seismic zone map of the United States.

26.2.1.1 Seismic Zones

Figure 26.4 shows the current seismic zone map adopted by the Uniform Building Code. Different zones on the map are assigned a different zone coefficient, Z, in accordance with Table 26.3.

TABLE 26.3 Seismic Zone Factor, Z

Zone	1	2A	2B	3	4
Z	0.075	0.15	0.2	0.3	0.4

26.2.1.2 Soil Type, S

Since the soil type a structure is founded on has a great impact on the ground motion at the site, different soil types are assigned values, depending on their stiffness, from 1.0 for rocklike material to 2.0 for a soil profile containing more than 40 ft of soft clay. Table 26.4 lists the site coefficients for different soil types.

26.2.1.3 Importance Factor, *I*

The importance factor, I, depends on the occupancy category of a given structure. The Uniform Building Code distinguishes between five different categories; Essential, Hazardous, Special, Standard, and Miscellaneous as shown in Table 26.4f and condensed Table 26.5. Essential facilities, in addition to having an importance factor of 1.25, require special construction review/inspection/observation procedures. It is argued that the additional measures can improve performance more effectively than solely relying on increased design force levels.

26.2.1.4 Regular and Irregular Structures

The UBC requires that a structure be designated as regular or irregular. Regular structures have no significant physical discontinuities in plan or vertical configurations or in their lateral force-resisting systems. Two assumptions apply for regular structures. First, the equivalent static force distribution provides a conservative envelope of the forces and deformations due to the actual

TABLE 26.4a Seismic Coefficient C_a

Soil Profile Type	Seismic Zone Factor, Z				
	$Z = 0.075$	$Z = 0.15$	$Z = 0.2$	$Z = 0.3$	$Z = 0.4$
S_A	0.06	0.12	0.16	0.24	$0.32 N_a$
S_B	0.08	0.15	0.20	0.30	$0.40 N_a$
S_C	0.09	0.18	0.24	0.33	$0.40 N_a$
S_D	0.12	0.22	0.28	0.36	$0.44 N_a$
S_E	0.19	0.30	0.34	0.36	$0.36 N_a$
S_F			See Footnote 1		

Source: 1997 Uniform Building Code.
Refer to Table 26.4c for soil profile type description.
[1] Site-specific geotechnical investigation and dynamic site response analysis shall be performed to determine seismic coefficients for Soil Profile Type S_F.

TABLE 26.4b Seismic Coefficient C_v

Soil Profile Type	Seismic Zone Factor, Z				
	$Z = 0.075$	$Z = 0.15$	$Z = 0.2$	$Z = 0.3$	$Z = 0.4$
S_A	0.06	0.12	0.16	0.24	$0.32 N_v$
S_B	0.08	0.15	0.20	0.30	$0.40 N_v$
S_C	0.13	0.25	0.32	0.45	$0.56 N_v$
S_D	0.18	0.32	0.40	0.54	$0.64 N_v$
S_E	0.26	0.50	0.64	0.84	$0.96 N_v$
S_F			See Footnote 1		

Source: 1997 Uniform Building Code.
[1] Site-specific geotechnical investigation and dynamic site response analysis shall be performed to determine seismic coefficients for Soil Profile Type S_F.

TABLE 26.4c   Soil Profile Types

Soil Profile Type	Soil Profile Name/Generic Description	Shear Wave Velocity, $\bar{v}_s$ feet/second (m/s)	Standard Penetration Test, $\overline{N}$ [or $\overline{N}_{CH}$ for Cohesionless Soil Layers] (blows/foot)	Undrained Shear Strength, $\overline{S}_u$ psf (kPa)
		Average Soil Properties for Top 100 feet (30480 mm) of Soil Profile		
S_A	Hard rock	>5,000 (1,5000)	—	—
S_B	Rock	2,500 to 5,000 (760 to 1,500)		
S_C	Very dense soil and soft rock	1,200 to 2,500 (360 to 760)	>50	>2,000 (100)
S_D	Stiff soil profile	600 to 1,200 (180 to 360)	15 to 50	1,00 to 2,000 (50 to 100)
$S_E{}^1$	Soft soil profile	<600 (180)	<15	<1,000 (50)
S_F	Soil requiring site-specific evaluation. See Section 1629.3.1.			

Source: 1997 Uniform Building Code.

[1] Soil Profile Type S_E also includes any soil profile with more than 10 feet (3048 mm) of soft clay defined as a soil with a plasticity index, $PI > 20$, $w_{mc} \geq 40$ percent and $S_u < 500$ psf (24 kPa). The Plasticity Index, PI, and the moisture content, w_{mc}, shall be determined in accordance with approved national standards.

TABLE 26.4d   Near-Source Factor $N_v{}^1$

Seismic Source Type	Closest Distance to Known Seismic Source[2,3]			
	≤2 km	5 km	10 km	≥15 km
A	2.0	1.6	1.2	1.0
B	1.6	1.2	1.0	1.0
C	1.0	1.0	1.0	1.0

[1] The Near-Source Factor may be based on the linear interpolation of values for distances other than those shown in the table.

[2] The location and type of seismic sources to be used for design shall be established based on approved geotechnical data (e.g., most recent mapping of active faults by the United States Geological Survey or the California Division of Mines and Geology).

[3] The closest distance to seismic source shall be taken as the minimum distance between the site and the area described by the vertical projection of the source on the surface (i.e., surface projection of fault plane). The surface projection need not include portions of the source at depths of 10 km or greater. The largest value of the Near-Source Factor considering all sources shall be used for design.

TABLE 26.4e   Seismic Source Type[1]

Seismic Source Type	Seismic Source Description	Seismic Source Definition[2]	
		Maximum Moment Magnitude, M	Slip Rate SR (mm/year)
A	Faults that are capable of producing large magnitude events and that have a high rate of seismic activity	$M \geq 7.0$	$SR \geq 5$
B	All faults other than Type A and C	$M \geq 7.0$ $M < 7.0$ $M \geq 6.5$	$SR < 5$ $SR > 2$ $SR < 2$
C	Faults that are not capable of producing large magnitude earthquakes and that have a relatively low rate of seismic activity	$M < 6.5$	$SR \leq 2$

Source: 1997 Uniform Building Code.

[1] Subduction sources shall be evaluated on a site-specific basis.

[2] Both maximum moment magnitude and slip rate conditions must be satisfied concurrently when determining the seismic source type.

TABLE 26.4f Occupancy Category

Occupancy Category	Occupancy or Functions of Structure	Seismic Importance Factor, I	Seismic Importance[1] Factor, I_p	Wind Importance Factor, I_w
1. Essential facilities[2]	Group I, Division 1 Occupancies having surgery and emergency treatment areas Fire and police stations Garages and shelters for emergency vehicles and emergency aircraft Structures and shelters in emergency-preparedness centers Aviation control towers Structures and equipment in government communications centers and other facilities required for emergency response Standby power-generating equipment for Category 1 facilities Tanks or other structures containing housing or supporting water or other fire-suppression material or equipment required for the protection of Category 1, 2 or 3 structures	1.25	1.50	1.15
2. Hazardous facilities	Group H, Divisions 1, 2, 6 and 7 Occupancies and structures therein housing or supporting toxic or explosive chemicals or substances Nonbuilding structures housing, supporting or containing quantities of toxic or explosive substances that, if contained within a building, would cause that building to be classified as a Group H, Divisions 1, 2 or 7 Occupancy	1.25	1.50	1.15
3. Special occupancy structures	Group A, Divisions 1, 2, and 2.1 Occupancies Buildings housing Group E, Divisions 1 and 3 Occupancies with a capacity greater than 300 students Buildings housing Group B Occupancies used for college or adult education with a capacity greater than 500 students Group I, Divisions 1 and 2 Occupancies with 50 or more resident incapacitated patients, but not included in Category 1 Group I, Division 3 Occupancies All structures with an occupancy greater than 5,000 persons Structures and equipment in power-generating stations, and other public utility facilities not included in Category 1 or Category 2 above, and required for continued operation	1.00	1.00	1.00
4. Standard occupancy structures[3]	All structures housing occupancies or having functions not listed in Category 1, 2 or 3 and Group U Occupancy towers	1.00	1.00	1.00
5. Miscellaneous structures	Group U Occupancies except for towers	1.00	1.00	1.00

Source: 1997 Uniform Building Code.

[1] The limitation of I_p for panel connections in Section 1633.2.4 of the UBC Code shall be 1.0 for the entire connector.

[2] Structural observation requirements are given in Section 1702.

[3] For anchorage of machinery and equipment required for life-safety systems, the value of I_p shall be taken as 1.5.

dynamic response. Second, the cyclic inelastic deformation demands during a major level of seismic ground motion will be reasonably uniform in all elements. Irregular structures are those with irregular features as described in Tables 26.6 and 26.7. Vertical irregularities are defined as a distribution of mass, stiffness, or strength that results in lateral forces and/or deformations, over the height of the structure, that are significantly different from the linearly varying distribution obtained from Eqs. (26.4), (26.5), and (26.6). Plan irregularities are encountered where diaphragm characteristics create significant diaphragm deformations and/or stress concentrations. The UBC recognizes two different analysis procedures for the determination of seismic effects on structures. The static lateral procedure is allowed when certain criteria of building height, occupancy, and regularity are met while the dynamic lateral force procedure is always permissible for design. The scope of this chapter is limited to the static lateral force procedure.

26.2.1.5 Structural Systems

The maximum elastic response acceleration of a structure during a severe earthquake can be several times the magnitude of the maximum ground acceleration and depends on the mass and stiffness of the structure and the amplitude of the damping. Since it is unnecessary to design a structure to respond in the elastic range to the maximum seismic inertia forces, it is of utmost importance that a well-designed structure be able to dissipate seismic energy by inelastic deformations in certain localized regions in the lateral resisting system. This translates into accomplishing flexural yielding of the members and avoiding all forms of brittle failure. The code-specified design seismic force recognizes such inelastic behavior and damping and scales down the inertia forces corresponding to a fully elastic response based on the structural system used. This is accounted for in the R factor. The structural systems for buildings and the corresponding R values are listed in Table 26.8 and defined in the following sections.

TABLE 26.5 Occupancy Requirements, I

Occupancy Category	Importance Factor	
	Earthquake	Wind
Essential facilities	1.25	1.15
Hazardous facilities	1.25	1.15
Special occupancy structures	1.00	1.00
Standard occupancy structures	1.00	1.00

TABLE 26.6 Vertical Structural Irregularities

Irregularity Type	Definition
A. Stiffness irregularity—soft story	A soft story is one in which the lateral stiffness is less than 70% of that in the story below or less than 80% of the average stiffness of the three stories above.
B. Weight (mass) irregularity	Mass irregularity shall be considered to exist where the effective mass of any story is more than 150% of the effective mass of an adjacent story. A roof that is lighter than the floor below need not be considered.
C. Vertical geometric irregularity	Vertical geometric irregularity shall be considered to exist where the horizontal dimension of the lateral force-resisting system in any one story is more than 130% of that in an adjacent story. One-story penthouses need not be considered.
D. In-plane Discontinuity in Vertical Lateral Force-Resisting Element	An in-plane offset of the lateral load-resisting elements greater than the length of those elements.
E. Discontinuity in capacity–weak story	A weak story is one in which the story strength is less than 80% of that in the story above. The story strength is the total strength of all seismic-resisting elements sharing the story shear for the direction under consideration.

TABLE 26.7 Plan Structural Irregularities

Irregularity Type	Definition
A. Torsional irregularity—to be considered when diaphragms are not flexible	Torsional irregularity shall be considered to exist when the maximum story drift, computed to include accidental torsion, at one end of the structure transverse to an axis is more than 1.2 times the average of the story drifts of the two ends of the structure.
B. Reentrant corners	Plan configurations of a structure and its lateral force-resisting system contain reentrant corners, where both projections of the structure beyond a reentrant corner are greater than 15% of the plan dimension of the structure in the given dimension.
C. Diaphragm discontinuity	Diaphragms with abrupt discontinuities or variations in stiffness, including those having cutout or open areas greater than 50% of the gross enclosed area of the diaphragm or changes in effective diaphragm stiffness of more than 50% from one story to the next.
D. Out of Plane Offsets	Discontinuities in a lateral force path, such as out-of-plane offsets of the vertical elements.
E. Nonparallel system	The vertical lateral load-resisting elements are not parallel to or symmetric about the major orthogonal axes of the lateral force-resisting system.

TABLE 26.8 Structural Systems, Ω_o, R, and H

Basic Structural System	Lateral Load-Resisting System	Ω_o[a]	R[b]	H[c], ft
A. Bearing wall system	1. Light framed walls with shear panels			
	a. Plywood walls for structures three stories or less.	2.8	5.5	65
	b. All other light-framed walls.	2.8	4.5	65
	2. Shear walls			
	a. Concrete	2.8	4.5	160
	b. Masonry	2.8	4.5	160
	3. Light steel-framed bearing walls with tension only bracing.	2.2	2.8	65
	4. Braced frames where bracing carries gravity loads.			
	a. Steel	2.2	4.4	160
	b. Concrete	2.2	2.8	—
	c. Heavy timber	2.2	2.8	65
B. Building frame system	1. Steel eccentrically braced frame (EBF)	2.8	7	240
	2. Light framed walls with shear panels			
	a. Plywood walls for structures three stories or less.	2.8	6.5	65
	b. All other light framed walls.	2.8	5	65
	3. Shear walls			
	a. Concrete	2.8	5.5	240
	b. Masonry	2.8	5.5	160
	4. Ordinary braced frames			
	a. Steel	2.2	5.6	160
	b. Concrete	2.2	5.6	—
	c. Heavy timber	2.2	5.6	65
C. Moment-resisting frame system	1. Special moment resisting frame (SMRF)			
	a. Steel	2.8	8.5	N.L.
	b. Concrete	2.8	8.5	N.L.
	2. Concrete intermediate moment resisting frames (IMRF)	2.8	5.5	—
	3. Ordinary moment resisting frames (OMRF)			
	a. Steel	2.8	4.5	160
	b. Concrete	2.8	3.5	—
D. Dual System	1. Shear walls			
	a. Concrete with SMRF	2.8	8.5	N.L.
	b. Concrete with steel OMRF	2.8	4.2	160
	c. Concrete with concrete IMRF	2.8	6.5	160
	d. Masonry with SMRF	2.8	5.5	160
	e. Masonry with steel OMRF	2.8	4.2	160
	f. Masonry with concrete IMRF	2.8	4.2	—
	2. Steel EBF			
	a. With steel SMRF	2.8	8.5	N.L.
	b. With steel OMRF	2.8	4.2	160
	3. Ordinary braced frames			
	a. Steel with steel SMRF	2.8	6.5	N.L.
	b. Steel with steel OMRF	2.8	4.2	160
	c. Concrete with concrete SMRF	2.8	6.5	—
	d. Concrete with concrete IMRF	2.8	4.2	—

N.L. No limit

A See Section 1630.4 of UBC Code for combination of structural systems.

B Basic structural systems are defined in Section 1629.6.

C Prohibited in Seismic Zones 3 and 4.

D Includes precast concrete conforming to Section 1921.2.7.

[a] Over-strength factor (seismic force amplification factor).

[b] Coefficient representing overstrength and ductility capacity of lateral-force-resisting system.

[c] Height limit for seismic zones 3 and 4 ($\times$ 304.8 for mm).

Bearing Wall System. A structural system without a complete vertical load-carrying frame. Bearing walls or bracing systems provide support for all or most gravity loads. Resistance to lateral load is provided by shear walls or braced frames. In seismic zones 3 and 4, the shear walls must be specially detailed to satisfy the UBC requirements up to the maximum permissible height, such as 160 ft under certain conditions.

Building Frame System. A structural system with an essentially complete frame providing support for gravity loads. Resistance to lateral load is provided by shear walls or braced frames. Shear walls in seismic zones 3 and 4 are required to be specially designed and detailed to satisfy the UBC and/or the ACI 318 Codes [see Nawy, 1996]. In addition, other structural elements not designated part of the lateral load-resisting system must be able to sustain their gravity load-carrying capacity at a lateral displacement equal to a multiple times the computed elastic displacement of the lateral force-resisting system under code-specified design seismic forces. The UBC restricts the building frame system to a maximum height of 240 ft.

Moment-Resisting Frame System. A structural system with an essentially complete frame providing support for gravity loads. Moment-resisting frames provide resistance to lateral loads primarily by flexural action of members. In seismic zone 1, the moment-resisting frames can be ordinary moment-resisting frames (OMRF) proportioned to satisfy the UBC requirements. In seismic zone 2 (A and B), reinforced-concrete frames resisting forces induced by earthquakes motions must as a minimum be intermediate moment-resisting frames (IMRF). In seismic zones 3 and 4, reinforced-concrete frames resisting forces induced by earthquake motions must be special moment-resisting frames (SMRF).

Dual System. A structural system with the following features:

1. an essentially complete frame that provides support for gravity loads;
2. resistance to lateral load provided by shear walls or braced frames and moment-resisting frames (SMRF, IMRF)—the moment-resisting frames shall be designed to independently resist at least 25% of the design base shear; and
3. designed to resist the total design base shear in proportion to relative rigidities considering the interaction of the dual system at all levels.

26.2.2 Design Base Shear

The Uniform Building Code requires that structures be designed for seismic forces coming from any horizontal direction. Such forces may be assumed to act nonconcurrently in the direction of each principal axis of the structure except for structures with certain plan irregularities.

26.2.2.1 Total Dead Load, W

The dead load of a building, W, is the total dead load and applicable portions of other loads listed below:

1. In occupancies where partition loads are used in the floor design, a load of not less than 10 psf shall be included.
2. In occupancies such as storage and warehouses, a minimum of 25% of the floor live load shall be included.
3. In areas subjected to snow loading greater than 30 psf, the snow load shall be included. Where warranted, such snow load may be reduced up to 75% when approved by the building official.
4. In buildings with permanent equipment, the weight of such equipment shall be included as part of the total dead load, W.

26.2.2.2 Static Force Procedure

The Uniform Building Code equation for calculating the total design lateral force at the base of a building in a given direction is as follows:

$$V = \frac{C_v I}{RT} W \tag{26.1a}$$

The total design base shear need not exceed the following:

$$V = \frac{2.5 C_a I}{R} W \tag{26.1b}$$

The total design base shear shall not be less than the following:

$$V = 0.11 C_a I W \tag{26.1c}$$

In addition, for Seismic Zone 4, the total base shear shall also not be less than the following:

$$V = \frac{0.8 Z N_v I}{R} W \tag{26.2}$$

where

C_v = seismic coefficient in Table 26.4b
C_a = seismic coefficient in Table 26.4a
I = importance factor in Table 26.4f and condensed Table 26.5
R = overstrength and global ductility capacity factor for lateral-force resisting systems in Table 26.8
T = fundamental period of vibration
W = total seismic dead load of the system
V = total design lateral force or shear as the base
Z = seismic zone factor in Table 26.3
N_v = factor used to determine C_v in zone 4 in Table 26.4d

The structure period has to be proportioned using one of the following methods:

Method A. For all buildings, the value T may be approximated from the following formula:

$$T = C_t (h_n)^{3/4} \tag{26.3}$$

where the numerical coefficient C_t has the following values:

$C_t = 0.035$ for steel moment-resisting frames
$C_t = 0.03$ for reinforced concrete moment-resisting frames and eccentrically braced frames
$C_t = 0.02$ for all other buildings

Method B. The fundamental period T may be calculated using the structure properties and deformational characteristics of the resisting elements in a properly substantiated analysis. This requirement may be satisfied by using the following formula:

$$T = 2\pi \sqrt{\left(\sum_{i=1}^{n} w_i \delta_i^2 \right) \div \left(g \sum_{i=1}^{n} f_i \delta_i \right)} \tag{26.4}$$

The values of f_i represent any lateral force distributed approximately using formulas (26.5), (26.6), and (26.7) or any other rational distribution. The elastic deflection δ_i shall be calculated using the applied lateral force f_i. The resulting value from method B shall not be greater 130% of the value of T from mathod A.

26.2.3 Combination of Structural Systems

This section addresses situations where different structural systems are used in different stories, referred to here as "vertical combinations," and situations where different structural systems are used along different axes, referred to here as "combinations along different axes."

26.2.3.1 Vertical Combinations

The Uniform Building Code requires that the value of R used in the design of any story be less than or equal to the value of R used in the given direction for the story above. Where this condition exists, the structure may be designed using either of the following two procedures:

1. The entire structure is designed using the lowest R of the lateral force-resisting systems used.
2. If the structure has a flexible upper portion and a lower rigid portion whose average story stiffness is at least 10 times the average story stiffness of the upper portion and the period of the entire structure is not greater than 1.1 times the period of the upper portion, then it can be designed using the two-stage static analysis procedure as follows:
   a) The flexible upper portion shall be designed as a separate structure using the appropriate value of R and ρ.
   b) The rigid lower portion shall be designed as a separate structure using the appropriate value of R and ρ. The reactions from the upper portion shall be those determined from the analysis of the upper portion amplified by the ratio of the R/ρ of the upper portion over the R/ρ of the lower portion.

26.2.3.2 Combinations Along Different Axes

Any combination of moment-resisting frame systems, dual systems, building frame systems or bearing wall systems may be used to resist seismic forces in structures less than 160 ft in height. However, in seismic zones 3 and 4 where a structure has a bearing wall system in only one direction, the value of R used in the orthogonal direction shall not be greater than that used for the bearing wall system. Also, in seismic zones 3 and 4 and for structures exceeding 160 ft in height, only combinations of dual systems and special moment-resisting frames shall be used to resist seismic forces.

26.2.4 Vertical Distribution of Force

Except where it can be shown using rigorous analysis, the total seismic force shall be distributed over the height of the structure in accordance with Eqs. (26.5), (26.6), and (26.7).

$$V = F_t + \sum_{i=1}^{n} F_i \tag{26.5}$$

where

$$F_t = 0.07\,TV \tag{26.6}$$

The value of the period T used in formula (26.6) may be the period corresponding with the design base shear, V. F_t may be considered to be zero when T is 0.7 seconds or less and need not exceed $0.25\,V$. F_t is provided to account for higher mode response at the top of flexible buildings. The balance of the design base shear shall be distributed linearly over the height of the structure, including the top level, varying from a maximum value at the top to a minimum at the bottom, according to the following formula:

$$F_x = \frac{(V - F_t)w_x h_x}{\sum_{i=1}^{n} w_i h_i} \qquad (26.7)$$

26.2.5 Drift Limitation

The UBC defines story drift as the relative displacement between adjacent stories (above or below) due to the design lateral forces. The calculated drift encompasses both translational and torsional deflections.

The allowable story drift is limited to 0.025 times the story height for buildings having fundamental periods less than 0.7 s. For buildings having fundamental periods equal to or greater than 0.7 s, the allowable drift is limited to 0.020 times the story height. The above drift can be exceeded when it is demonstrated that greater drift can be tolerated by both structural and nonstructural elements that could affect life safety.

26.2.6 Horizontal Distribution of Shear and Horizontal Torsional Moments

The design story shear, V_x (that is, the sum of the forces F_t and F_x above that story) in any story, is distributed to the various elements of the vertical lateral force-resisting system in proportion to their rigidities, considering the rigidity of the diaphragm. Furthermore, the UBC requires that provision be made for the increased shears resulting from horizontal torsion where diaphragms are not flexible. Diaphragms are considered flexible when the maximum lateral deformation of the diaphragm is more than two times the average story drift of the associated story. To account for the uncertainties in load locations, the UBC further requires that the mass at each level be assumed to be displaced from the calculated center of mass in each direction a distance equal to 5% of the building dimension at that level perpendicular to the direction of the force under consideration. This is often referred to as the accidental eccentricity the torsional design moment at a given story is the moment resulting from the combination of this accidental eccentricity and the actual eccentricity between the applied design lateral forces at levels above that story and the center of rigidity of the vertical lateral-resisting elements in that story.

26.2.7 Discontinuous Lateral Force-Resisting Elements

Overturning moments on discontinuous shear-resisting elements are to be carried as loads to the foundation. In seismic zones 3 and 4, columns supporting the overturning forces of discontinuous shear walls are to be designed to resist the axial forces resulting from the dead load, f_1 times the live load and Ω_o (Table 26.8) times the earthquake force, or 0.9 times the dead load $+/- \Omega_o$ times the earthquake force, whichever is more severe. The columns carrying these factored loads may be designed to their axial load strength and must meet special UBC detailing requirements to ensure their ductile behavior under cyclic loading.

$f_1 = 1.0$ for floors in places of public assembly for live load in excess of 100 psf, and for garage live load.
$f_1 = 0.5$ for other live loads.

26.2.8 Story Drift Limitation and P-Delta Effects

The UBC Code specifically requires that member forces and story drifts induced by P-delta effects be considered in the evaluation of overall frame stability. P-delta effects need not be considered when the ratio of secondary moment to primary moment does not exceed 0.10. The ratio may be evaluated for any story as the product of the dead load, floor live load, and snow load (where applicable) above the story times the seismic drift in that story, divided by the product of the seismic shear in that story and the height of that story. It allows, however, that in seismic zones 3 and 4, P-delta need not be considered where the story drift ratio does not exceed $0.02/R$.

26.2.9 Vertical Component of Seismic Forces

In seismic zones 3 and 4, the structural component at horizontal cantilevers has to be designed for a net upward force of $0.7\,C_a\,I\,W_p$. At prestressed members, horizontal components shall be designed using not more than 50% of the dead load for the gravity load, alone or in combination with the lateral force effects in addition to all other applicable load combinations.

26.2.10 Detailed Systems Design Requirements

26.2.10.1 Connections

All connections that resist seismic forces must be designed and detailed on the construction documents.

26.2.10.2 Deformation Compatibility

All framing elements not required by design to be part of the lateral force-resisting system must be investigated and shown to be adequate for vertical-load-carrying capacity when subjected to expected deformation resulting from the required design lateral force and adequate for the P-delta effects. UBC-97 permits occurrences where rigid elements enclosing moment-resistant frames prevent the frame from resisting lateral forces, provided that it can be shown that the action or the failure of the more rigid elements will not impair the vertical and lateral-resisting ability of the frame. Expected deformation shall be determined as the greater of the maximum inelastic response displacement or the displacement induced by a story drift $= 0.0025 \times$ story height.

26.2.10.3 Exterior Elements

Nonbearing, nonshear wall panels or elements that are attached to or enclose the exterior shall be designed to resist the code-prescribed forces and shall accommodate movements of the structure resulting from lateral forces or temperature changes. The subject elements are required to be supported by means of cast-in-place concrete or by mechanical connections in accordance with the following requirements:

1. Connections and panel joints shall allow for a relative movement between stories of not less than two times story drift caused by wind, the elastic story drift caused by design seismic forces, or $1/2$ in, whichever is greater.
2. Connections to permit movement in the plane of the panel for story drift shall be sliding connections using slotted or oversize holes, or other connections providing equivalent sliding and ductility capacity.
3. Bodies of connections should have sufficient ductility and rotation capacity to preclude fracture of the concrete or brittle failures at or near welds and shall be designed for a force F_p, where

$$F_p = C_a I_p \left(1 + \frac{h_x}{h_r} \right) W_p$$

where

$h_x =$ element/component elevation with respect to grade. h_x shall not be taken less than 0.
$h_r =$ structure roof elevation with respect to grade.
F_p shall not be less than $0.7 C_a I_p W_p$ and need not be more than $4 C_a I_p W_p$.

4. All fasteners in the connecting system (inserts, welds, bolts, and dowels) have to be designed for the force F_p. Additionally, fasteners embedded in concrete shall be attached to, or hooked around, reinforcing steel or otherwise terminated so as to effectively transfer forces to the reinforcing steel.

26.2.10.4 Building Separation

All structures have to be separated from adjoining structures a distance equal to Δ_M the displacement due to seismic forces. However, the code allows a smaller separation, when justified by rational analyses, based on maximum expected ground motions. $\Delta_{M_T} = \sqrt{(\Delta_{M_1})^2 + (\Delta_{M_2})^2}$, Δ_{M_1} and Δ_{M_2} are the displacement of adjacent buildings.

26.2.10.5 Ties, Continuity, and Collector Elements

Continuity provisions in the UBC are meant to ensure that all parts of a structure will act together under seismic excitation without localized separation, loss of support, or collapse. As a minimum, any smaller portion of the building shall be tied to the remainder of the building with elements having at least a strength to resist $Z/3$ times the dead load plus live load. Additionally, a positive connection for resisting a horizontal force greater than or equal to $Z/5$ times the dead plus live load, acting parallel to the member, shall be provided for each beam, girder, or truss. Collector elements shall be provided to adequately transfer the seismic forces originating in other portions of the buildings to the members providing the resistance to those forces.

26.2.10.6 Anchorage of Concrete or Masonry Walls

Concrete or masonry walls shall be provided with a positive direct connection to all floors and roofs that provide them lateral support. Such connections shall be capable of resisting the horizontal forces induced by the seismic excitement and as specified in the code. UBC-97 Section 1633 should be consulted for further information.

26.2.10.7 Diaphragms

Floor and roof diaphragms are required to be designed to resist the forces determined in the following formula:

$$F_{px} = \frac{F_t + \sum_{i=x}^{n} F_i}{\sum_{i=x}^{n} w_i} w_{px} \tag{26.8}$$

The force, F_{px}, determined in the above formula need not exceed $1.0 C_a I w_{px}$ but shall not be less than $0.5 I C_a w_{px}$. In structures in seismic zones 3 and 4 containing reentrant corners, where both projections of the structure beyond a reentrant corner are greater than 15% of the plan dimension of the structure in the given direction, diaphragm chords and drag members shall be designed considering independent movement of the projecting wings of the structure. Each of these diaphragm elements shall be designed for the more severe of two assumptions: motion of the projecting wings in the same direction and motion of the projecting wings in opposing directions. This requirement may be deemed satisfied if the dynamic lateral force procedures of the UBC Code in conjunction with a three-dimensional model are used to determine the lateral seismic forces for design.

26.2.11 Proportioning of Members—Load and Strength Reduction Factors

For earthquake loading in seismic zones 3 and 4, the following load factors are to be used:

$$U = 1.4(D + L + E) \tag{26.9}$$

$$U = 0.9D \pm 1.4E \tag{26.10}$$

where

> U = required strength to resist factored loads or related internal moments and forces
> D = dead loads or related internal moments and forces
> L = live loads or related internal moments and forces
> E = load effects of earthquake or related internal moments and forces.

Additionally, in seismic zones 3 and 4, strength-reduction factors shall be modified as follows:

The shear strength reduction factor shall be 0.6 for the design of walls, topping slabs used as diaphragms over precast concrete members, and structural framing members, with the exception of joints, if their nominal strength is less than the shear corresponding to development of their nominal flexural strength. The shear strength-reduction factor for joints is 0.85. Furthermore, the strength-reduction factor for axial compression and flexure shall be 0.5 for all frame members with factored axial compressive force exceeding $A_g f_c'/10$, if the transverse reinforcement does not conform to the UBC provisions.

26.3 Design and Construction of Concrete and Masonry Buildings

26.3.1 Preface

This chapter is intended as guidance to engineers in the design of reinforced concrete and masonry buildings to resist earthquake forces in conformance with the 1997 UBC (refer to Section 26.2). The lateral load-resisting system, including horizontal diaphragms, is addressed. Please note that term definitions, notations, and symbols used in this chapter are defined in the 1997 UBC, Chapters 19 and 21.

26.3.2 Concrete Buildings

26.3.2.1 Introduction

As discussed in Section 26.2, reinforced concrete buildings in general are designed to have adequate strength to resist the equivalent static lateral load and appropriate ductility to allow the structure to displace as a result of dynamic earthquake excitement. The building strength is achieved by selecting the appropriate member sizes, concrete strength, proper placement, and adequate size of reinforcing steel so that the nominal capacity decreased by a reduction factor of each member meets or exceeds the ultimate demands. Ductility is achieved by providing confinement to the concrete section at critical locations by introducing closely spaced and appropriately anchored bars transversely to the longitudinal reinforcement in beams, columns, and boundary members such as at a shearwall's ends (see Figure 26.10).

In designing a building to resist the seismic forces obtained in Section 26.2, the engineer would go through selection of the lateral force-resisting system. This section addresses shear walls and special moment-resisting frames since they are primarily used in seismic zones 3 and 4. The decision of whether to choose shearwalls or moment frames or a combination of the two depends on the architectural layout of the structure. The structural engineer's role in most cases is limited to locating and sizing the lateral resisting elements. The following steps are then followed in the lateral design of the structure:

1. Select the lateral resisting system.
2. Calculate the building weight and base shear (Section 26.2).
3. Distribute lateral load to each floor and calculate the story shear (Section 26.2).
4. Distribute story shear to the various lateral-resisting elements at each floor based on member stiffness (Section 26.3, Horizontal Distribution of Shear and Torsional Moments).

5. Design and detail lateral-resisting components (shearwall or moment frame) for the applied forces (Section 26.3.2.3, design of vertical lateral-resisting elements).
6. Design and detail diaphragm to transfer story shear into shearwall or moment frame.

26.3.2.2 Horizontal Distribution of Shear and Torsional Moments

The following discussion addresses the distribution of shear and torsional moments on the basis that the lateral resisting system consists of shearwalls. The same discussion also applies to buildings with moment frames or combination of shearwalls and moment frames.

The center of mass of the floor is first calculated as follows:

$$X_m = \frac{\sum wx}{\sum w} \tag{26.11}$$

$$Y_m = \frac{\sum wy}{\sum w} \tag{26.12}$$

Figure 26.5 serves as a good illustration for the above formulas. Next, the rigidity of each wall is calculated as follows.

The deflection at the top of the wall is calculated based on the following formula:

$$\Delta_T = \Delta_F + \Delta_S = \frac{Ph^3}{3E_C I} + \frac{1.2Ph}{AE_G} \tag{26.13}$$

where

$r =$ rigidity or stiffness of the wall panel $= 1/$deflection

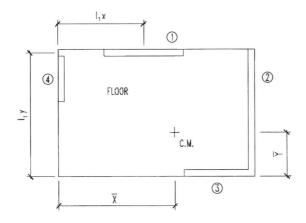

FIGURE 26.5   Schematic center of mass in shear wall.

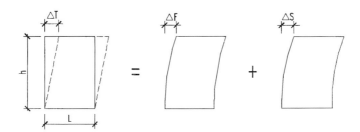

FIGURE 26.6   Cantilever shearwall deflection.

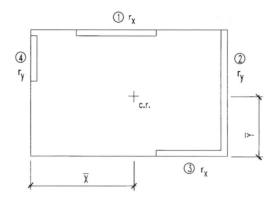

FIGURE 26.7 Schematic center of rigidity in a shear wall.

Then the center of rigidity of the floor is calculated as follows:

$$\overline{x}_r = \frac{r_x x}{\sum r_y} \tag{26.14}$$

$$\overline{y}_r = \frac{r_x y}{\sum r_x} \tag{26.15}$$

Refer to Figure 26.8 for shear distribution formulation.

26.3.2.3 Design of Vertical Lateral-Resisting Elements

This section illustrates the design and reinforcement requirements for shear walls and moment resisting frames through practical examples.

Example 26.1: Shearwall Design and Reinforcement Requirements

(See Figure 26.9): The given values are as follows:

$$f'_c = 4000 \text{ psi}$$
$$\text{wall thickness, } t = 12 \text{ in}$$
$$V = 700 \text{ kips} = \text{shear force}$$
$$M_{\text{base}} = 72,000 \text{ ft-kips} = \text{overturning moment}$$
$$P = 2000 \text{ kips} = \text{total load on wall, includes wall weight}$$

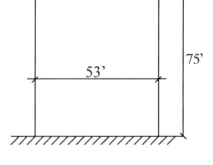

1. a) Note that walls with a height-to-weight (H/W) ratio greater than 2.0 behave primarily as bending members, while walls with H/W below 2.0 resist lateral forces in a predominantly truss-type behavior. The maximum shear stress in walls is limited to

$$2\sqrt{f'_c}$$

   b) The vertical steel in the wall is required to equal the horizontal steel.

   c) At least two curtains of reinforcement shall be used in a wall if the in-plane factored shear force assigned to the wall exceeds $2A_{cv}\sqrt{f'_c}$. A_{cv} is the cross-sectional area bounded by the wall length parallel to the direction of the shear force and the wall thickness.

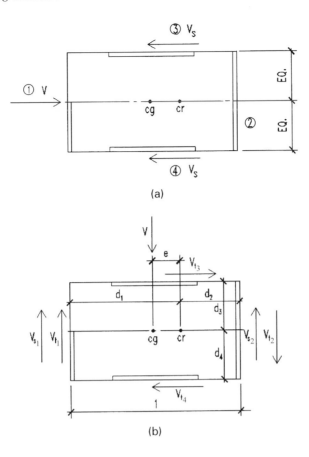

FIGURE 26.8 Shear distribution formulation.

2. a) Check number of curtains required:

$$2\,A_{cv}\sqrt{f_c'} = (2 \times 12 \times (53 \times 12)\sqrt{4000})/1000 = 965.4\,\text{kips}$$

$$V_u = 1.4 \times 700 = 980\,\text{kips}$$

$$V_u > 2\,A_{cv}\sqrt{f_c'} \quad \therefore \text{two curtains of reinforcement are required.}$$

  b) Required vertical and horizontal wall reinforcement:

minimum rebar ratio, $\rho_v = 0.0025$
area of steel in each direction/foot, $\rho_v A_{cv} = 0.0025 \times 12\,\text{in} \times 12\,\text{in} = 0.36\,\text{in}^2/\text{ft}$
try #5 bars in each direction at each face at 18 in on center $\leq S_{max} = 18\,\text{in}$

$$A_s = 2 \times 0.31 \times 12/18 = 0.413\,\text{in}^2$$
$$h_w/l_w = 75\,\text{ft}/53\,\text{ft} = 1.42 < 2$$
$$V_n = A_{cv}(\alpha_c\sqrt{f_c'} + \rho_n f_y)$$
$$A_{cv} = 12 \times (53 \times 12)/18 = 7{,}632\,\text{in}^2$$
$$\alpha_c = 3.0 \ (\alpha_c \text{ varies linearly from 3.0 for } h_w/l_w = 1.5 \text{ to 2.0 for } h_w/l_w = 2.0)$$
$$\rho_n = 2 \times 0.31/(12 \times 18) = 0.0029$$
$$\phi = 0.6$$
$$\phi V_n = 0.6 \times \frac{7632}{1000}(3\sqrt{4000} + 0.0029 \times 60{,}000) = 1657k > 980\,\text{kips} = V_u.\ \text{ok.}$$

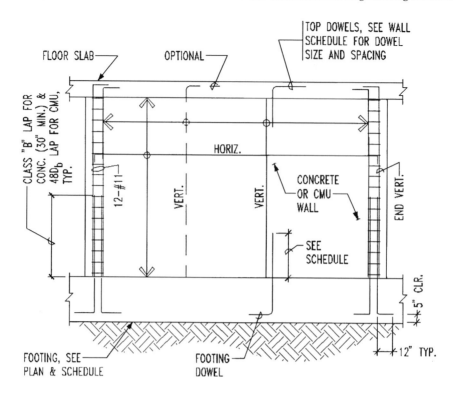

FIGURE 26.9 Typical shearwall elevation.

Check $V_n \le 8A_{cv}\sqrt{f'_c}$

$$V_n = \frac{7632}{1000}(3\sqrt{4000} + 0.0029 \times 60{,}000) = 2{,}776 \text{ kips}$$

$$8A_{cv}\sqrt{f'_c} = 8 \times 7632 \times \sqrt{4000}/1000 = 3861 \text{ kips} \quad \therefore V_n < 8A_{cv}\sqrt{f'_c} \text{ ok}$$

3. Boundary member requirement

$$P_u = 1.4(D + L + E)$$

Boundary members shall be provided where the maximum compressive, extreme-fiber, and stress under factored forces exceeds $0.2 f'_c$.
Gross section properties:

$$l = 53 \text{ ft}, \quad t = 12 \text{ in}$$

$$A_g = 53 \times 12 \times 12 = 7{,}632 \text{ in}^2$$

$$I_y = 12 \times (53 \times 12)^3/12 = 257{,}259{,}456 \text{ in}^4$$

$$f_c = P_U/A_g + M_u(l_w/2)/I_y$$

$$P_U = 1.4 \times 2{,}000 = 2{,}800 \text{ kips}$$

$$M_u = 1.4 \times 42{,}000 = 58{,}800 \text{ kip-ft}$$

$$f_c = \frac{2800}{7632} + \frac{58{,}800 \times 12 \times 53 \times 12/2}{257{,}259{,}456} = 1.24 \text{ ksi}$$

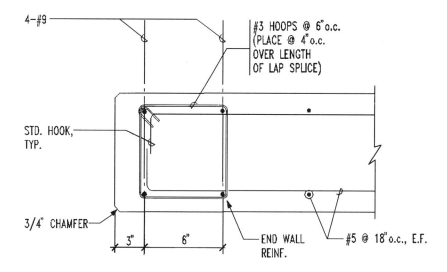

FIGURE 26.10   Typical placement of reinforcement in cast-in-place shear wall.

$$0.2 f'_c = 0.2 \times 4 = 0.8 \text{ ksi} < 1.24 \text{ ksi} \quad \therefore \text{provide boundary member.}$$
$$P_u = 1.4 \times 2000/2 + 1.4 \times 42{,}000/(53 \text{ ft} - 3 \text{ ft}) = 2{,}576 \text{ kips}$$

Check axial capacity of 36 in × 36 in boundary member with 12#11.

$$A_y = 36 \times 36 = 1296 \text{ in}^2$$
$$A_s = 12 \times 1.56 = 18.72 \text{ in}^2$$
$$\rho_g = 18.72/1296 = 0.0144$$
$$\rho_{\min} = 0.01 < \rho_g < \rho_{\max} = 0.06 \text{ ok}$$
$$P_u = \phi P_n = 0.6\big[0.85 f'_c (A_g - A_s) + f_y A_{st}\big]$$
$$= 0.7 P_n = 3{,}061 \text{ kips} > 2576 \text{ kips, O.K.}$$

Use 36 in × 36 in boundary members with 12#11.
Check boundary reinforcement for tension load for load combination:

$$T_u = 1.4 E \pm 0.9 D$$
$$D = 1400 \text{ kips}$$
$$T_u = 1.4 \times 42{,}000/(53 - 3) - 0.9 \times 1400/2 = 546 \text{ kips}$$
$$A_s = T_u/\phi f_y = 546/0.9 \times 60 = 10.11 \text{ in}^2 \text{ O.K.}$$

4. Other requirements:

   a) Required cross-sectional area of confinement reinforcement should be in accordance with
      the UBC and/or ACI 318-95 Codes. Nawy (1996) for a detailed example.
   b) All continuous reinforcement in shearwalls must be anchored or spliced as required by the
      UBC.

26.3.3 Design of Masonry Buildings

26.3.3.1 Introduction

The majority of construction in the masonry industry involves smaller buildings, and these are often designed with the code equivalent static loads. It is the intent of this section to briefly demonstrate how the building must be designed to resist these equivalent static lateral forces. Again, for computational convenience, the effects of ground motion are considered as if the motion acts only in one direction parallel to the perpendicular axes of the building. Several different types of structural systems are used to resist the static lateral forces and carry them into the foundation. Such systems include shearwalls, braced frames, and moment-resisting space frames. Since it has not been possible, so far, to feasibly construct masonry moment-resisting frames, this chapter is limited to the discussion of concrete masonry shearwalls.

26.3.3.2 General Behavior of Box Buildings

A box system does not have an independent vertical load-carrying frame, but rather depends upon the walls to not only carry the vertical loads but also provide the necessary lateral stability. Walls that are perpendicular to the assumed direction of the ground motion must span vertically between the floor diaphragms. The inertial effect of one half the wall height, both above and below the floor level in question, is considered to be transferred to that floor diaphragm. In addition, the inertial effect of the diaphragm dead load itself must be taken by the diaphragm. The diaphragm essentially behaves as a horizontal plate girder, wherein the diaphragm boundary members serve as the girder flanges and the floor functions as the web. The diaphragm therefore spans between the supporting shear walls. Through the diaphragm to wall connection, the total horizontal shear is transferred directly to the shearwall. Depending on the rigidity of the diaphragm, this lateral shear transfer to any shearwall may be based on the adjacent tributary area or the relative rigidities of the various shearwalls. Where rigid diaphragms such as reinforced or post-tensioned concrete floor slabs are used, each wall experiences a torsional shear in addition to the direct shear when the building center of gravity and the center of rigidity of the vertical lateral-resisting elements do not coincide. The total direct lateral shear and the torsional shear are combined so that the sum becomes the total UBC-design force imposed upon the shearwall. Figure 26.8 provides an illustration of the total shear force lateral load combination.

26.3.3.3 Shearwalls

Shearwall Design. Roof and/or floor seismic forces may be carried down to the building foundation by the strength and rigidity of the building walls in a bearing wall system or building frame system. Such walls are called shearwalls. In tall buildings, the design of the shearwalls is generally governed by flexure, whereas in low buildings, shear is often the governing criteria. The shearwall is basically a cantilever member with the shear force resisted by the combined shear strength of the masonry and the horizontal rebar in the wall, while the vertical boundaries (end reinforcement) provide the capacity to resist tension and compression caused by the bending moment and axial forces on the wall. A shearwall by definition is a wall resisting in-plane horizontal shear forces. In addition to horizontal force, the wall is investigated for the effects of the vertical loads it is subjected to. Example 26.2 provides a good illustration for the design of a concrete masonry wall subjected to both gravity and seismic forces.

Shearwall Detailing. In order for a masonry wall to function properly as a shearwall, it must be capable of resisting the diagonal tensile forces imposed by the design shear forces and the vertical compressive or tensile stresses caused by the overturning moment due the same design shear forces. A typical detail for placing reinforcing steel in a masonry wall is shown in Figure 26.11. The horizontal rebar is anchored with a 90° standard hook to ensure that the bar can be fully developed in tension. Also shown in Figure 26.11 is the end reinforcement for a four bar condition to resist tensile forces. Figure 26.9 shows an elevation of the same wall with the wall to floor and wall to footing

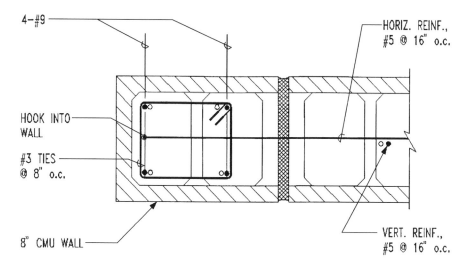

FIGURE 26.11   Typical placement of reinforcement in masonry shearwall.

connections properly shown. It is important to note that the designer must take into account the limited amount of space in the cells when specifying the bar size and spacing so that the placement of reinforcement does not become impossible.

Example 26.2: Masonry Shearwall Design

The given values are as follows:

> Masonry: CMU $f'_m = 1500$ psi
> Grout: type M, 2000 psi
> Steel: Gr 60
>
> Special inspection required
> Wall length: 31 ft
> Wall thickness: 12 in
> Wall height: 12 ft
>
> Shear force: 203.45 kips
> Moment: 3051.7 kip-ft
> Axial loading: 95.00 kips

The following items are required:

1. Stress check
2. Reinforcement requirements

The following calculations are necessary (Calculations are based on design UBC 1991):

> $E_m = 750 \times 1{,}500 = 1{,}125$ ksi
> $E_s = 29{,}000$ ksi
> Ratio of moduli $= 29{,}000/1{,}125 = 25.78$
> Allowable steel stress, $F_s = 24{,}000$ psi
> Increase for seismic design, $1.33 F_s = 1.33 \times 24{,}000 = 31{,}920$ psi
> Allowable bending stress, $F_b = 0.33 f'_m = 0.33 \times 1500 = 495$ psi
> Increase for seismic design, $1.33 F_b = 658.35$ psi
> Allowable axial stress, $F_a / R = 0.2 f'_m = 0.2 \times 1500 = 300$ psi
> Increase for seismic design, $= 399$ psi

Design for bending and axial load is as follows:

$$R = [1 - (h/42t)^3]$$

$$h = 12 \times 12 = 144 \text{ in}$$

$$t = 11.625 \quad \text{for } 12 \text{ in/in} CMU$$

$$R = 1 - \left(\frac{144}{42 \times 11.625}\right)^3 = 0.97$$

$$f_a = \frac{P}{A} = \frac{95 \times 1000}{31 \times 12 \times 11.625} \cong 22 \text{ psi}$$

Allowable axial stress, $F_a = 399 \times 0.97 = 387$ psi
Axial stress ratio, $f_a/F_a = 21.97/387 = 0.057$

$$\frac{f_a}{F_a} + \frac{f_b}{F_b} = 1$$

hence

$$f_b/F_b = 1 - 0.057 = 0.943.$$

Assume masonry stress governs and check for steel stress. If steel stress as calculated exceeds its allowable stress, then discard the assumption and proceed by assuming steel reaches its maximum allowable value first.

(I) Masonry stress governs.

$$f_b = 0.943 \times 658.35 = 621 \text{ psi}$$

$$jk = \frac{2M}{f_b bd^2} = \frac{2 \times 3051.70 \times 12000}{620.98 \times 11.625 \times 354^2} = 0.081$$

where $d = 12 \times 31 \text{ ft} - 18 \text{ in} = 354$ in.

$$k^2 - 3k + 3jk = 0$$

$$k^2 - 3k + 3 \times 0.081 = 0$$

$$k = 0.0833$$

$$f_s = nf_b\left(\frac{1-k}{k}\right) = 25.78 \times 620.98\left(\frac{1 - 0.0833}{0.0833}\right)$$

$$= 176128 \text{ psi} > 31920 \text{ psi} \quad \therefore \text{N.G.}$$

(II) Steel stress governs.

$$A_s = \frac{M}{f_s jd} = \frac{3051 \times 12000}{31920 \times 0.972 \times 354} = 3.33 \text{ in}^2$$

$$= \text{Area of jamb reinforcement}$$

where $j = jk/k = 0.81/0.0833 = 0.972$.

Check compressive stress in masonry.

$$\rho = A_s/bd = 3.33/11.625 \times 354 = 8.09 \times 10^{-4}$$

$$k = \sqrt{2\rho n + (\rho n)^2} - \rho n$$

$$= \sqrt{2 \times 8.09 \times 10^{-4} \times 25.78 + (25.78 \times 8.09 \times 10^{-4})^2}$$

$$- 8.09 \times 10^{-4} \times 25.78 = 0.23$$

$$j = 1 - k/3 = 1 - 0.23/3 = 0.923$$

$$f_b = \frac{2M}{jkbd^2} = \frac{2 \times 3051.70 \times 12000}{0.923 \times 0.162 \times 11.625 \times 354^2}$$

$$= 336 \text{ psi} < 658.35 \text{ psi} \therefore \text{OK}$$

Bending stress ratio:

$$\frac{f_b}{F_b} = \frac{336}{658.35} = 0.51 \therefore \text{OK}$$

Combined bending and axial stress ratio:

$$\frac{f_a}{F_a} + \frac{f_b}{F_b} = 0.06 + 0.51 = 0.57 < 1 \therefore \text{OK}$$

Check shear stresses:

$$\frac{M}{Vd} = \frac{3051.7}{203.45 \times (0.8 \times 31)} = 0.6 < 1 \therefore \text{OK}$$

where $d \cong 0.8 \times 31$.

Allowable range for design with rebar:

for $M/Vd = 1$, $\quad F_v = 1.33 \times 1.5\sqrt{1500} = 77.27 \text{ psi} < 1.33 \times 75 \text{ psi} \therefore \text{OK}$
for $M/Vd = 0$, $\quad F_v = 1.33 \times 2\sqrt{1500} = 103.02 \text{ psi} < 1.33 \times 120 \text{ psi} \therefore \text{OK}$

hence, the design stress is

$$F_v = 77.27 + (1 - 0.6)(103.02 - 77.27) = 87.33 \text{ psi}$$

The governing stress value is $F_v = 87.33$ psi
The actual stress value is

$$F_v = \frac{V_u}{bjd} = \frac{1.5 \times 203.45 \times 1000}{11.625 \times 0.923 \times 354} = 80.34 \text{ psi}$$

The shear stress ratio is

$$\frac{f_v}{F_v} = 80.34/87.33 = 0.92 \therefore \text{OK}$$

Calculate shear reinforcement:

$$\text{Horizontal} \quad A_s = \frac{V_u s}{f_s jd} = \frac{1.5 \times 203.45 \times 1000 \times 12}{31920 \times 0.923 \times 354} = 0.351 \text{ in}^2$$

$$\text{Min} \quad A_s = 0.0007 \times 11.625 \times 12 = 0.1 \text{ in}^2 < 0.351 \text{ in}^2 \therefore \text{OK}$$

$$\rho_n = 0.351/11.625 \times 12 = 0.00252 > 0.002$$

hence min ρ_u governs.

$$A_v = 0.0007 \times 11.625 \times 12 = 0.1 \, \text{in/ft}$$

26.3.4 Shearwall Foundation Analysis and Design

26.3.4.1 Background

The analysis and design presented herein pertains to the sizing of the shearwall foundations and the determination of their structural adequacy. The algorithm adopted is aimed at obtaining economical foundation dimensions with due consideration of all the major parameters such as the actions from the shearwall, the permissible soil pressure, concrete strength, the shearwall foundation, and grade beam dimensions.

This analysis and design is directed toward the seismic-loading case only. The adequacy of the foundation under gravity loading is to be established separately. Under normal conditions, however, the size and reinforcement of a shearwall foundation is governed by the seismic case. The gravity case check will be conducted for unusual cases and dimensions.

The analysis is based on the limit design concept. A probable failure mechanism is assumed (refer to free body diagrams Figure 26.12). By satisfying the requirements of statics, the actions (shears and moments) at critical sections are determined. Based on which the cross-sectional area and the reinforcement at such sections are evaluated. The failure mechanism selected may not be the optimum for all cases, but it is consistently conservative.

The soil pressure is checked by forming the soil stress ratio (SSR). This is the ratio of the calculated unfactored soil pressure to the permissible soil pressure. SSR is to be kept less than 1. The analysis assumes that the foundation is tied to the remainder of the structure through grade beams. If the foundation by itself can resist the applied loading without need for grade beams, the calculated shear transfer (parameter G) between the foundation and the grade beam will become zero or negative. If a shear transfer between the foundation and the assumed grade beams results, this analysis and design must be followed by a grade beam design to ensure that the resulting shear (G) is satisfactorily dissipated into the grade beam and the other structural elements.

The applied vertical loading (P), the seismic moment (M), and the shear (V) from the shearwall are calculated elsewhere and are entered herein (see example 26.3) as input. They act at midlength of the shearwall at the top of the foundation. This design considers the vertical loading and the moment components of the shearwall. The shear component (V) is resisted through friction at the bottom of the foundation, axial loading in the adjoining grade beams, and the connection of the foundation to the slab on grade, as well as by any passive soil pressure that may develop. In common cases, foundation and grade beam sizes, through other considerations, prove to be adequate for V. However, for isolated foundations this may not be the case. If required, this design section is followed by a check for the shear component V.

One type of typical interior foundation is treated in example 26.3. Numerous variations are possible depending on the existence of and location at which a grade beam ties into the foundation. The relationships developed embrace most of the common cases. If a foundation geometry does not fall within the scope of this work and can not be modeled conservatively through the method treated herein, a special analysis and design is required.

In nonsymmetrical cases, different values for the shear transfer force (G) result, depending on the direction of the applied seismic-bending moment. In such conditions, the value yielding the larger shear transfer is calculated and given.

26.3.4.2 Geotechnical Relationships

Soil Pressure. The soil pressure,

$$s = \frac{P_d}{0.25LB} \leq 1.33SBP, \tag{26.16}$$

where P_D is the total ($DL + LL$).

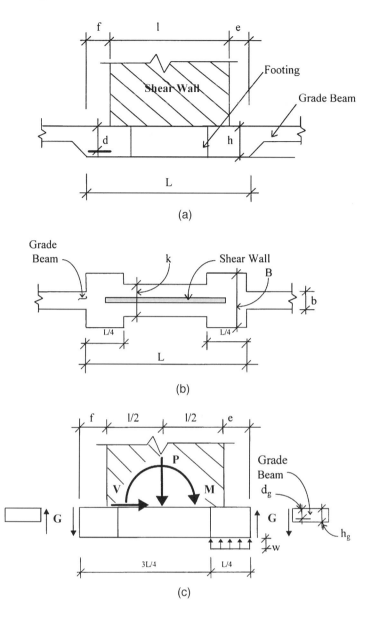

FIGURE 26.12 Shear wall foundation. (a) Elevation—shear wall and footing; (b) plan; and (c) free body diagram of the structural system at limit state.

Soil Stress Ratio. The soil stress ratio,

$$SSR = \frac{s}{1.33SBP} \leq 1. \qquad (26.17)$$

Shear Transfer to Grade Beam, G. (See Figure 26.12). Based on the structural model adopted at the limit state, shear transfer (*G*) to the grade beams occurs if the shearwall footing is not capable, through its geometry and the existing gravity loadings, to resist the overturning moment *M*. The magnitude of the shear *G* depends also on the direction of bending moment *M*. The larger of the two values is considered for design.

$$G = \frac{1}{L}[M - 0.375\,PL + |0.5\,P(e - f)|] \tag{26.18}$$

Factored average soil bearing W

$$W = \frac{4P}{BL} \tag{26.19}$$

26.3.5 Nomenclature

b = width of grade beam;
B = width of foundation pad and its tips;
d = design depth of footing;
d_g = design depth of grade beam;
e = distance from tip of shear wall to tip of footing at bearing end;
f = distance from tip of shear wall to tip of footing at uplift end;
g = distance from tip of shear wall to point of action of shear force G;
G = factored shear force developed between the grade beam and foundation;
h = dimensional depth of footing;
h_g = dimensional depth of grade beam;
j = distance from tip of shear wall to point of action of shear force G;
k = width of footing at midlength of shear wall;
l = length of shear wall;
L = effective length of foundation (equal to $l + e + f$);
M = factored moment at midlength of shear wall (equal to $1.4M_s$);
M_s = unfactored applied seismic moment;
P = factored gravity load (equal to $0.85\,P_d$);
P_a = unfactored gravity loading from shear wall;
P_d = total axial load on footing (equal to $P_a + P_f$);
P_f = total weight of footing;
P_L = unfactored live load from shear wall;
s = unfactored calculated soil pressure;
SBP = allowable soil bearing pressure;
SSR = soil stress ratio (equal to calculated/allowable);
V = factored applied seismic shear;
V_s = unfactored applied seismic shear; and
W = factored calculated soil pressure on soil.

Example 26.3: Shearwall Footing Design Example

(Refer to Figures 26.12 and 26.13).

The following values are given:

from soils report:

$$SBP = 3.0 \text{ kips/ft}^2$$

from lateral analysis:

$$M_s = 10,000 \text{ kips-ft} \qquad P_a = 100 \text{ kips} \qquad P_L = 0 \text{ kips} \qquad V_s = 300 \text{ kips}$$

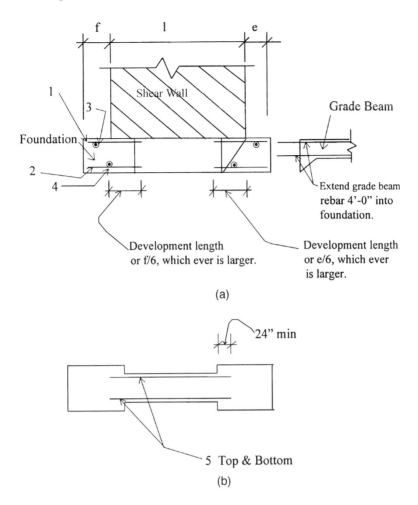

FIGURE 26.13 Reinforcement designation. (a) Elevation and (b) plan. *Note:* The connection between the grade beam and the foundation is modeled as "hinged" in the analysis. It is not necessary to design for a moment across this section.

properties:

$$f'_c = 3000 \text{ psi}$$
$$f_y = 60 \text{ ksi}$$
$$e = 6 \text{ ft}$$
$$f = 6 \text{ ft}$$
$$l = 36 \text{ ft}$$
$$L = l + e + f = 48 \text{ ft}$$

estimated weight of footing:

$$P_f = 60 \text{ kips}$$
$$P_d = P_a + P_f = 100 \text{ kips} + 60 \text{ kips} = 160 \text{ k}$$

$$P = 0.85(P_d - P_L)(160 \text{ kips}) \quad P = 136 \text{ kips}$$

$$M = 1.4M_s = 1.4(10{,}000 \text{ kips-ft}) \quad M = 14{,}000 \text{ kips-ft}$$

from Eq. (26.16)

$$B_{\min} = \frac{P_d}{0.25L(1.33SBP)} = \frac{160 \text{ kips}}{0.25(48 \text{ ft})(1.33 \times 3.0)} = 40.1 \text{ in}$$

assume $B = 48$ in

$$s = \frac{P_d}{0.25LB} = \frac{160 \text{ kips}}{0.25(48 \text{ ft})(4 \text{ ft})} = 3.33 \text{ kips/ft}^2$$

$$SSR = \frac{s}{1.33SBP} = \frac{3.33 \text{ kips/ft}^2}{1.33(3.0 \text{ kips/ft}^2)} = 0.833 < 1.0 \text{ OK}$$

assume $d_g = 26.5$ in; $h_g = d_g + 3.5$ in $= 26.5$ in $+ 3.5$ in; $\therefore h_g = 30$ in.
From Eq. (26.19)

$$W = \frac{4P}{BL} = \frac{4(114 \text{ kips})}{4 \text{ ft}(48 \text{ ft})} = 3.0 \text{ kips/ft}^2 \quad W = 3.0 \text{ kips/ft}^2$$

Reinforcement requirements of foundation.

$$G = \frac{1}{L}[M - 0.375PL + |0.5P(e - f)|]$$

$$= G = \frac{1}{48 \text{ ft}}[14{,}000 \text{ kips-ft} - 0.375(144 \text{ kips})(48 \text{ ft} + |0.5(144 \text{ kips})(6 - 6)|]$$

$$G = 237.67 \text{ kips}$$

see Figure 26.13a, rebar 2D 1

$$M_u = Gf + \frac{Bfh}{12^3}0.15\frac{f}{2} = (237.67 \text{ kips})(6 \text{ ft}) + \frac{(4 \text{ ft})(6 \text{ ft})(3 \text{ ft})}{12^3}0.15\frac{6}{2}$$

$$M_u = 1427.9 \text{ kips-ft}$$

$$A_s = 10.61 \text{ in}^2$$

26.4 Seismic Retrofit of Existing Buildings

26.4.1 Introduction

Historically, the seismic retrofit of existing concrete structures has been executed by introducing concrete shearwalls, new steel or concrete bracing elements, or strengthening existing lateral-resisting members. Recently, new methods such as base isolation and wrapping columns using steel jackets or fiber fabric with epoxy resin have been introduced. The method and the economy of the seismic retrofit depends on the characteristics of the structure. The most critical stage in the retrofit of a concrete structure is the preliminary work for the development of the retrofit scheme. During this phase, the structural components of the structure, the economy, the different retrofit schemes, the timing of the work, and the possibility of loss of occupancy are all investigated so that the appropriate seismic retrofit option may be chosen.

26.4.2 Structure Weakness

Concrete structures observed to suffer the most in the aftermath of major seismic events have exhibited one or more of the following deficiencies:

1. suspect load path or, as is occasionally the case, interrupted load path;
2. lack of compatibility between the vertical elements in the structure; or
3. an inadequate lateral load-resisting system.

For a structure to respond adequately to the applied seismic loads, there must exist a continuous and adequate load path from the point of application of the seismic force to the foundation of the structure. Inadequately designed or detailed connections along the load path can and will lead to local or overall failure, regardless of how well the different components of the lateral-resisting system are designed and detailed.

The issue of deformation compatibility can be seen very clearly in a structure with flexible vertical lateral-resisting elements such as moment-resisting frames. Such frames experience significant horizontal displacements. It is this earthquake imposed deformation that can cause some or all of the lateral and vertical carrying systems to fail should they not have the ability to allow such movement. As building codes evolve on the basis of lessons learned from various seismic events worldwide, higher design loads with a more advanced level of detailing for ductile behavior (not widely practiced before) are demanded. Structures designed according to codes as late as 1967 may be inadequate, based on present standards. Efforts to retrofit such structures to meet present governing codes can be very extensive. Toward achieving our present objectives of ensuring life safety during major seismic events and minimal structural damage during frequently occurring earthquakes, the next two sections address two critical objectives in the seismic retrofit process.

26.4.3 Creation of Adequate Load Path

The structural engineer chooses the load path to support the lateral load from the point of load application to the foundation or soil. The load path or lateral load-resisting system is then analyzed and designed for the corresponding load actions. All components of the system and their connections are verified, and adjustments are made to ensure adequate strength and ductility. In concrete structures, such load paths are achieved by adding new shearwalls or by strengthening existing ones. It is not surprising therefore to observe that the chosen load path is different from the structure's original system. Compatibility of the various vertical structural elements of the structure must be investigated to ensure that the selected load path can be activated.

26.4.4 Establish Deformation Compatibility

As discussed earlier, establishing deformation compatibility is essential to ensuring that all structural elements necessary for stability and load transfer have adequate ductility and the ability to redistribute loading to safely withstand the deformations imposed upon them in response to seismic loads. In parking structures where split level frames are common, postearthquake investigations and studies have shown that most of the distress was due to column-shear failure in short columns. Measures for the mitigation of column-shear failure vary between existing and new structures. The options introduced in the following sections may be considered for each of the construction types.

26.4.4.1 Retrofit Construction

Retrofitting the column implies one of several options:

a) The first is removing the column and replacing it using an adequate section with close tie spacing.
b) The second is jacketing the column so that its shear capacity can be increased. This can be accomplished by placing vertical and hoop reinforcement around the existing column and

by the application of gunite. Another option is to place a two-piece steel jacket and weld it on site. The gap between the existing column and the jacket is then grouted. Epoxy resin and fiber wrapping is another method that is very effective for round or oval columns since the structural contribution of the wrapping is developed through hoop stresses in the wrapping.

26.4.4.2 New Construction

In new construction, short columns can be either designed and detailed as a continuous part of the structure to withstand the higher shear forces or they can be detailed for moment release at their connections to the beam as shown in Figure 26.14.

26.4.5 Establish Adequate Lateral Load Resisting System

It is not possible to achieve our goal of ensuring life safety during major seismic events and minimal structural damage during frequently occurring earthquakes by just ensuring that the structure has an adequate vertical lateral-resisting system. It is imperative that the seismic force be properly collected at the roof and each floor and properly transfered into the vertical lateral-resisting system. Thus the horizontal concrete diaphragms should be analyzed and strengthened as necessary so that this step can be accomplished. The existing vertical resisting system is then strengthened by retrofitting the existing elements or by introducing new elements to supplement the existing capacity. Finally, the existing footing should be investigated and new footings and grade beams should be added where required.

26.4.5.1 Horizontal Diaphragms

When investigating existing buildings for retrofit, especially structures constructed prior to 1970, it is very likely that little attention was given to ensuring that the story force could be adequately transfered into the vertical lateral-resisting system. As part of the seismic retrofit of the building, methods of ensuring that the diaphragm does not fail should be devised. This can be achieved by introducing chord and drag elements at the floor soffit. Attaching thin steel plates onto the concrete soffit using epoxy resin is a method that can cause little disturbance to the function of the structure yet have very positive results. The connection of the diaphragm to the vertical lateral system should be closely examined. The introduction of steel angles bolted to both the floor soffit and the vertical lateral-resisting system is often done with satisfactory results. For diaphragms where a load path into the existing vertical lateral-resisting system cannot be reasonably achieved, the layout of the vertical lateral load-resisting system should be studied. It may be imperative that new vertical members be introduced to allow the establishment of adequate load path.

26.4.5.2 Vertical Load Resisting System

Existing concrete buildings are most likely to have one of three structural lateral systems:

1. space frame system,
2. building frame system, or
3. bearing wall system.

The level of strengthening required to achieve our goal of ensuring public safety and minimal structural damage no doubt differs from one structure to another. However, older buildings with space frames, in general, require more effort for establishing an adequate vertical lateral system. The following options are some of the available methods for strengthening the vertical lateral-resisting systems.

Introduction of New Concrete/Masonry Shearwalls. This is a method frequently used for retrofitting concrete structures. Shearwall placement is simpler, in most cases, than adding moment-

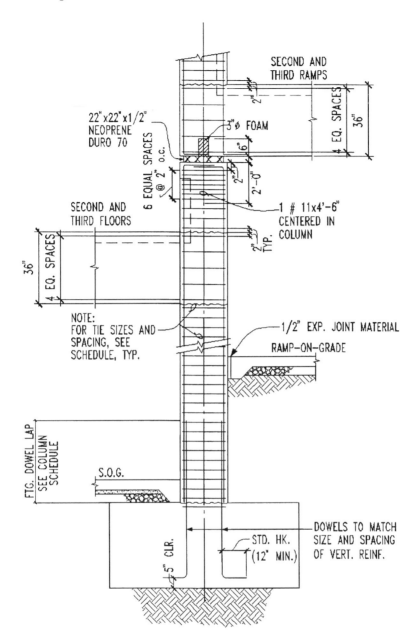

FIGURE 26.14 Slip joint at top of short columns.

resisting frames. Once the size and location of the new shearwalls is established based on the design force, new footings are placed and connected to the various floor systems by way of drilling and bonding reinforcing steel dowels to the existing slab-on-grade. Wall reinforcements are then placed. Walls can then be formed and poured in place or, as is often the case in recent years, a one-sided form can be placed and gunite applied (refer to Figures 26.15 and 26.16).

Retrofitting of Existing Walls. In instances where the building layout or architecture offers no option for new shearwall introduction, existing walls may be thickened to the required adequate dimension (Figure 26.17). Additionally, required wall boundary elements at the end of the wall may

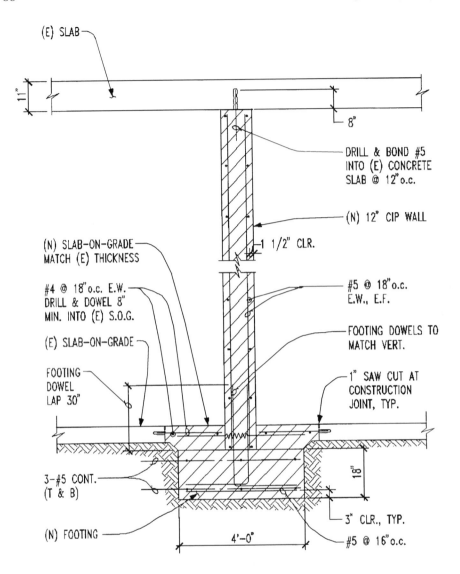

FIGURE 26.15   New concrete wall.

be added if required, as shown in Figure 26.16. Existing shearwalls to be thickened are properly prepared by roughening their surfaces to 0.25 in amplitude. Appropriate dowels are then, drilled and bonded into existing concrete. The required vertical and horizontal reinforcements are tied in place and concrete is placed either conventionally by installing wall forms and pumping the concrete into place or low-slump gunite is shot in place. It is important to note that the existing footings should be verified for their ability to transfer vertical and horizontal reactions to the ground. It is often necessary to enlarge the existing footings at the thickened shearwall. Additionally, grade beams may be introduced to better defuse the lateral force into the ground.

26.4.6 Alternate Methods of Seismic Retrofit

So far, the discussion has been limited to the conventional methods for seismic retrofit of buildings where conventional lateral load-resisting systems are engineered to support the calculated seismic forces that will be developed in a building during severe seismic activity. On the other hand, base

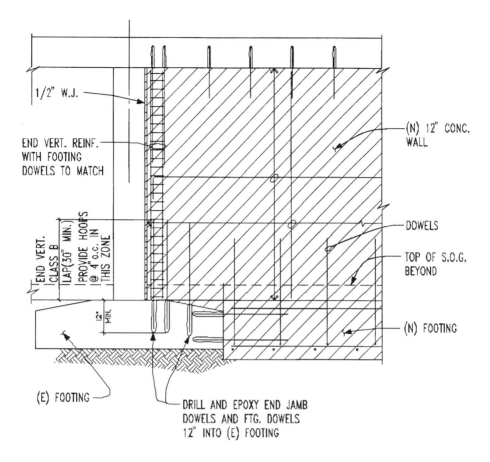

FIGURE 26.16 Wall elevation.

isolation is a means by which the structure's seismic response is significantly reduced by decoupling the building from its seismic forces through

1. increasing the fundamental vibration period of the structure or
2. dissipating the energy delivered to the building.

Base isolation at the present time is far more expensive to implement for retrofitting existing structures than conventional methods. However, structures such as historic landmarks and vital buildings that lack an existing load path for seismic forces and have limited capacity in the existing concrete diaphragm and/or large penetrations in the existing concrete diaphragm that can not be retrofitted conventionally can be successfully strengthened using base isolation. Some of the primary advantages of the base isolation system include:

1. minimum disruption of important historic features,
2. minimum alteration of interior spaces,
3. minimum potential damage to architectural finishes during future earthquakes,
4. greater degree of life safety to building occupants,
5. significantly reduced seismic response, and
6. significantly reduced story drift.

For further information regarding base isolation, the reader is encouraged to consult *Seismic Retrofit of San Francisco City Hall: The Role of Masonry and Concrete*, by Simin Naaseh (1995).

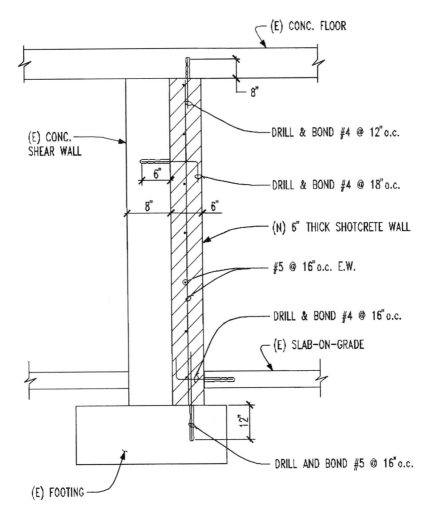

FIGURE 26.17 Retrofit of existing concrete wall.

26.5 Seismic Analysis and Design of Bridge Structures

26.5.1 Seismic Analysis of Bridge Structures

26.5.1.1 Introduction

In this section, a simple method that models a bridge as a single degree of freedom system is presented. As the bridge model becomes more complicated, however, this simple procedure becomes less accurate. In which case, a multimodal dynamic analysis or time history computer analysis is recommended.

There are two basic concepts for bridge seismic analysis. The first is that there is a relationship between a bridge's mass and stiffness and the forces and displacements that affect the structure during an earthquake. Therefore if we can calculate the mass and the stiffness for our structure, we can obtain the earthquake forces acting on it. The second concept is that bridges are designed to behave nonlinearly for large earthquakes. Therefore the engineer is required to make successive estimates of an equivalent linearized stiffness to obtain the seismic forces and displacements of the bridge.

26.5.1.2 Basics

Mass is a measure of a body's resistance to acceleration. It requires a force of one Newton to accelerate one kilogram at a rate of one meter per second squared.

  Stiffness is a measure of a structure's resistance to displacement. It is the force (in Newtons) required to move a structure one meter. The boundary conditions for the bridge need to be carefully studied to determine the stiffness of the structure.

  Period is the time, in seconds, it takes to complete one cycle of movement. A cycle is the trip from the point of zero displacement to the completion of the structure's farthest left and right excursions and back to the point of zero displacement.

  Natural period is the time a single degree of freedom system will vibrate at in the absence of damping or other forces. Natural period (T) has the following relationship to the system's mass (m) and stiffness (k).

$$T = 2\pi \sqrt{\frac{m}{k}} \tag{26.20}$$

  Frequency is the inverse of period and can be measured as the number of cycles per second (f) or the number of radians per second (ω) where one cycle equals 2π radians.

  Damping (viscous damping) is a measure of a structure's resistance to velocity. Bridges are underdamped structures. This means that the displacement of successive cycles becomes smaller. The damping coefficient (c) is the force required to move a structure at a speed of one meter per second. Critical damping (c_c) is the amount of damping that would cause a structure to stop moving after half a cycle. Bridge engineers describe damping using the damping ratio (ξ) where

$$\xi = \frac{c}{c_c} \tag{26.21}$$

A damping ratio of 5% is used for most bridge structures.

26.5.1.3 The Force Equation

The force equation for structural dynamics can be derived from Newton's second law:

$$\sum F = ma \tag{26.22}$$

Thus all the forces acting on a body are equal to its mass times its acceleration.

  When a structure is acted on by a force, Newton's second law becomes:

$$p - f_s - f_D = mu'' \tag{26.23}$$

where

$$f_s = ku \tag{26.24}$$

is the force due to the stiffness of the structure

$$f_D = cu' \tag{26.25}$$

is the force due to damping of the structure and p is the external force acting on the structure (see Figure 26.18). The variables u' and u'' and are the first and second derivatives of the displacement u, k is the force required for a unit displacement of the structure, and c is a measure of the damping in the system.

  Thus Eq. 26.23 can be rearranged as shown in Eq. (26.26).

$$mu'' + cu' + ku = p(t) \tag{26.26}$$

However, for earthquakes, the force is not applied at the mass but at the ground, (Figure 26.19) therefore Eq. (26.26) becomes

$$mu'' + c(u' - z') + k(u - z) = p \tag{26.27}$$

for the relative displacement

$$w = u - z \tag{26.28}$$

and the equation of motion, when there is no external force p being applied, is

$$mw'' + cw' + kw = -mz'' \tag{26.29}$$

In Eq. (26.29), the mass m, the damping factor c, and the stiffness k are all known. The support acceleration z'' can be obtained from accelerogram records of previous earthquakes. Equation (26.29) is a second-order differential equation that can be solved to obtain the relative displacement w, the relative velocity w', and the relative acceleration w'' for a bridge structure due to an earthquake.

26.5.1.4 Caltrans' Response Spectra

Response spectra have been developed so that engineers don't have to solve a differential equation repeatedly to capture the maximum force or displacement of their structure for a given acceleration record. Refer to Figure 26.20 for an example response spectra.

A **response spectra** is a graph of the maximum response (displacement, velocity or acceleration) of different single degree of freedom systems for a given earthquake record. Thus engineers can calculate the structure's period from its mass and stiffness, and use the appropriate 5% damped spectra to obtain the structure's response from the earthquake.

The force equation (26.29) showed three responses that can be obtained from a dynamic analysis; displacement, velocity, and acceleration. We can also obtain these using response spectra. The spectral displacement (Sd) and velocity (Sv) can be obtained from the spectral acceleration using the following relationship:

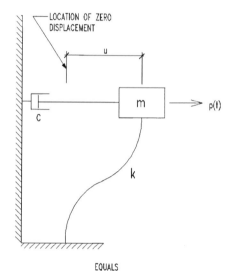

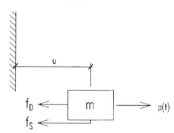

FIGURE 26.18   Forces acting on a structure according to Newton's second law.

$$Sa = \omega Sv = \omega^2 Sd \tag{26.30}$$

Therefore

$$Sd = \frac{Sa}{\omega^2} = \frac{Sa}{(2\pi/T)^2} = \frac{T^2 Sa}{4\pi^2} \tag{26.31}$$

Caltrans developed response spectra using five large California earthquake ground motions on rock. Twenty-eight different spectra were created based on four soil depths and seven peak ground

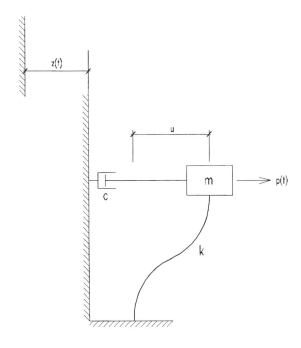

FIGURE 26.19   Earthquake induced displacements.

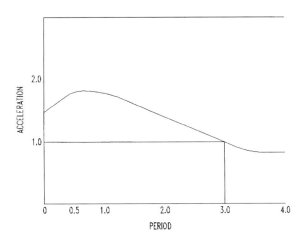

FIGURE 26.20   Example response spectra.

accelerations (PGA). Therefore engineers can obtain the earthquake forces on a bridge by picking the appropriate response spectra based on PGA and soil depth at the bridge site and calculating the natural period of their structure. These response spectra can be found in Caltrans' *Bridge Design Specifications* (1995). However, Caltrans is moving toward using site-specific response spectra for many bridge sites.

26.5.1.5 Nonlinear Behavior

Bridge members change stiffness during earthquakes. A column's stiffness is reduced when the concrete cracks in tension. It is further reduced as the steel begins to yield and plastic hinges form (Figure 26.21). The axial stiffness of a bridge changes in tension and compression as expansion joints open and close. The soil behind the abutment yields for large compressive forces and may

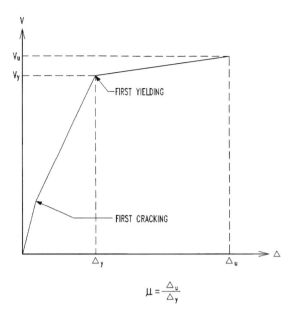

FIGURE 26.21   Nonlinear column stiffness.

not support tension. Engineers must consider all changes of stiffness to accurately obtain force and displacement values for a bridge.

Currently, the standard is to calculate cracked stiffness for bridge columns. A value of $I_{cr} = 0.5(I_{gross})$ can be used unless a moment-curvature analysis is warranted. Also, since bridge columns are designed to yield during large earthquakes, the column force obtained from the analysis is reduced by a ductility factor, and the columns are designed for this smaller force. Caltrans is currently using a ductility factor (μ) of about 5 for designing new columns. However, a moment-curvature analysis of columns should be done when the column's ductility is uncertain.

26.5.1.6 Abutment Stiffness

Longitudinally, the soil behind the backwall is assumed to have a stiffness, which is related to the area of the backwall as shown in Figure 26.22.

$$k_L = (47,000)(W)(h) \text{ (kN/m)} \tag{26.32}$$

Transversely, the stiffness is considered two-thirds effective per length of inside wingwall (assuming the wingwall is designed to take the load) and the outside wingwall is only one-third effective per wingwall length for the resultant stiffness shown in Eq. (26.33).

$$K_T = (102,000)(b) \text{ (kN/m)} \tag{26.33}$$

An additional stiffness of 7,000 kN/m for each pile is added in both directions. Therefore, in the longitudinal direction

$$K_L = (47,000) \, Wh + (7,000)n \text{ (kN/m)} \tag{26.34}$$

In the transverse direction

$$K_T = (102,000)b + (7,000)n \text{ (kN/m)} \tag{26.35}$$

More information on abutment stiffness can be obtained in *Caltrans Bridge Design Aids* (1995).

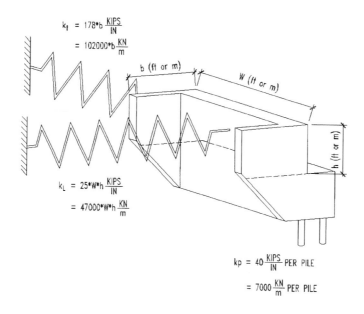

FIGURE 26.22   Abutment stiffness.

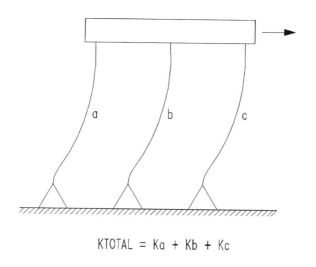

$$KTOTAL = Ka + Kb + Kc$$

FIGURE 26.23   Parallel structural system.

Bridge abutments are only effective in compression. A gap may need to be closed on seat type abutments before the soil stiffness is initiated. The abutment stiffness remains linear until it reaches the ultimate strength of 370 kN/m^2.

26.5.1.7 Parallel and Series Systems

A simplification that allows engineers to analyze by hand many complicated and statically indeterminant structures is the concept of parallel and series structural systems. For a parallel system, all the elements share the same displacement; while for a series system, they all share the same force. Also, their stiffnesses are summed differently. By assuming a rigid superstructure or by making other simplifying assumptions, bridge structures can be analyzed as combinations of parallel and series systems. This concept is particularly useful when evaluating the **longitudinal displacement** of the superstructure. Refer to Figures 26.23 and 26.24.

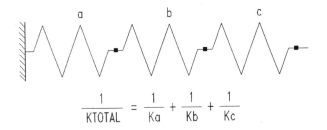

$$\frac{1}{K_{TOTAL}} = \frac{1}{K_a} + \frac{1}{K_b} + \frac{1}{K_c}$$

FIGURE 26.24 Series structural system.

26.5.1.8 Code Requirements

In bridge design, two load cases are considered. Case 1 is for 100% of the transverse force and 30% of the longitudinal force. Case 2 is for 100% of the longitudinal force and 30% of the transverse force. This is to take care of uncertainty as to the earthquake direction and to account for curved and skewed bridges with members that take a vector component of both the longitudinal and transverse force.

26.5.2 Seismic Design of Bridge Structures

26.5.2.1 Introduction

The following collection of guidelines is intended to aid bridge engineers in producing more seismic-resistant bridge designs.

26.5.2.2 Structural System and General Plan

Seismic demands on a structure should be considered conceptually when creating the general plan. First of all, designers can control column demands on superstructures in new bridges by judiciously choosing column sizing, spacing, and flexibility. Secondly, column or frame stiffnesses need to be carefully considered. Column types, shapes, sizes and length are very important in the design process.

Some example concepts that should be considered when establishing the general plan are as follows:

1. Column stiffnesses in a frame should preferably be kept within 20% of each other.
2. Isolators or sliding bearings can be used at the tops of short, stiff columns adjacent to abutments.
3. End diaphragm abutments provide good seismic energy absorbers for short frames (i.e., short length end frame with stiff columns) or short bridges.
4. Frames should not exceed 800 ft as a means of avoiding excessive expansion joint movements.
5. Reduce joint skew angles to control loss of seat support due to frame rotation.
6. Use articulated or single-span bridges in areas of suspected, but poorly defined, faulting.
7. Use double columns at expansion joints.

26.5.2.3 Superstructure Joint Shear

Below is a procedure and example details for determining joint shear reinforcement in box girder bent caps.

Typical flexure and shear reinforcement in bent caps shall be supplemented in the vicinity of columns to resist joint shear. Also, the cap width and/or depth must be sized to limit the joint shear stress to $12 \times (f_c')^{0.5}$. Minimum cap width shall be 2 ft greater than the column dimension in the cap width direction.

The joint shear reinforcement required shall be located within the width of the cap and within a length of the cap equal to (a) 2 times the column width or (b) the column width plus two times the bent cap depth, whichever is less. The effective length of cap shall be reduced appropriately for columns located near the end of the cap and closer than the column width or cap depth.

Following is a procedure for determining the required reinforcement that must be provided within the specified joint shear zone. The reinforcement shall be uniformly distributed.

1. Vertical reinforcement shall be 20% of the column steel. The quantity of vertical steel shall be a combination of cap stirrups and added bars. Added bars shall be hooked around main longitudinal cap bars. All vertical legs shall be well distributed to provide protection for both longitudinal and transverse demand or be supplemented to produce adequate coverage.
2. Horizontal reinforcement shall be stitched across the cap in intermediate layers. The reinforcement shall be hairpin shaped and confine vertical bars in the hairpin bends. The hairpins shall be equivalent in area to 10% of column steel. The hairpins shall be strategically placed in two or more layers, depending on structure depth. Spacing shall be denser outside the column than that used within the column.
3. Side face reinforcement shall be 10% of the main cap reinforcement.
4. Provide #4 J-bars at 6 in on center along longitudinal top deck bars for bent caps skewed greater than 20°. The bars shall be alternately 24 in and 30 in long. Locate the J-bars within a width 'D' either side of the column centerline (D is column dimension).
5. All vertical column bars shall be extended as high as practically possible without interfering with main cap bars.
6. Column spiral extension into the bent cap shall be replaced by equivalent continuous hoops. This substitution facilitates placement of the various horizontal joint shear bars.

    Figure 26.25 shows cap joint shear reinforcement within D distance from the column center line. Figure 26.26 shows cap joint shear reinforcement beyond a distance $D/2$ from column face.

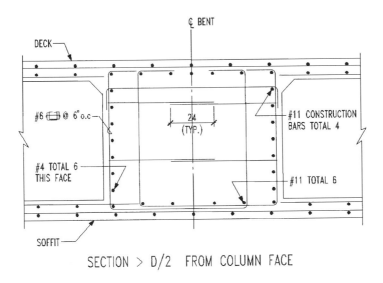

FIGURE 26.25   Example cap joint shear reinforcement skew 0° to 20°.

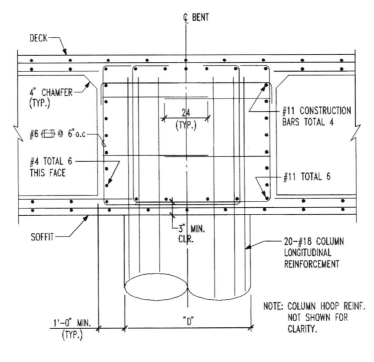

FIGURE 26.26 Example cap joint shear reinforcement skew 0° to 20°.

26.5.2.4 Superstructure Flexural Capacity

Bridge superstructures must be reinforced to resist longitudinal and transverse flexural demands produced by column plastic hinging. Conventional bar reinforcement must be provided in the slabs and bent caps in the designated zone surrounding the column to satisfy such demands. Prestressing bent caps could be an alternative solution for transverse demands, and additionally can supplement joint shear capacity at the column. Pinning column tops in multiple column bents will reduce seismic demand on the superstructure but will require a more expensive fixed foundation.

26.5.2.5 Vertical Accelerations

The 1994 Northridge Earthquake gave evidence of acceleration levels for vertical excitations that were higher than those recorded in the past. The flexural and joint shear reinforcement prescribed so that caps and superstructures can resist column demands should be ample to provide flexural resistance to vertical acceleration demands. However, the vertical capacity of girder stirrups should be checked, and shear-friction reinforcement in the form of girder side-face reinforcement at the cap that is equal to 1.5 gravity load should be provided. Perhaps tiedowns at hinges and abutments will be required.

26.5.2.6 Column Displacement

Displacement demands will generally be determined by STRUDL analysis, although the static method is sufficient for short length bridges (one or two bents). The column displacement capacity can be determined using the "push method" and is dependent on combinations of vertical and horizontal reinforcement. Also, disconnecting random stiff columns from the superstructure by use of sliding bearings or isolation systems is another means of avoiding large load demands on stiff columns.

26.5.2.7 Flared Columns

Flared columns can create some unexpected demands on the superstructure, footings, and columns. Past policy has assumed that lightly reinforced flares will fuse during severe seismic activity. Superstructures, columns, and footings have been designed to resist plastic hinging demands from the column core at top and bottom of column.

However, designers must consider alternative behavior. Developing either the full column section at the top of column or hinging at the base of flares should be investigated. Both conditions will significantly affect shear demands in the column and shear and moment demands in superstructures and footings. Flexural reinforcement for caps and superstructures must be calculated if hinging is assumed at the bases of flares.

26.5.2.8 Bent Cap Designs

The design of bent caps for nonseismic loads must be consistent with assumptions made for supports, especially with respect to column flares. If flares are expected to lose structural integrity during seismic loadings, the cap should be designed to carry all loads without benefit of the flares. However, the bridge must be analyzed for all nonseismic load cases with the stiffness and strength of the in-place flares. As in the case with flared columns, the designer must design the cap for double conditions if any type of structural damage or loss of vertical support is anticipated for other types of supports.

26.5.2.9 Footings

New footing design procedures were tested satisfactorily at University of California, San Diego. Although only one test case was performed, results indicate that serious degradation and bridge collapse will be avoided if standard details are employed. Details included #5 at 12 in o.c. vertical ties and column bar hooks that were turned outward.

Designers should be cautioned that cantilevered footing sections should not exceed $L/d = 2.5$ (L is the cantilevered footing length from the column face; d is the effective depth of footing) in orthogonal directions and $L/d = 3.0$ diagonally to the footing corner with 70-t piles typically spaced at 3 ft with 1.5 ft edge distances.

For higher capacity piles, the designer must assure load paths through connections, vertical ties, development of flexural reinforcement at footing edges (i.e., hooks), etc., for both tensile and compressive demands.

26.5.2.10 Pile-Design Concepts

The designer should strive to prevent damage in foundations. Piles constructed in competent soil, in combination with passive resistance provided by cap and soil interaction and drag forces, will be sufficient to keep piles in the elastic range for flexure or shear under seismic demand.

Problem areas that require guidelines are: (1) piles in soft soils or water, including $P \cdot \Delta$ effects; (2) pile head conditions, pinned (standard) or fixed (fixed places tremendous demands on the cap and requires more than nonstandard cap size and reinforcement); (3) pile demands for pile capacities greater than 70 t (i.e., spacing, edge distance, vertical ties that are more dense or of larger size, flexural steel hooks at footing edges, etc.), and (4) batter piles (use in dense soils, not for use in water, soft or loose soils, capacity in combination with vertical piles, etc.).

These design concepts are not intended to be a complete list of all conditions or problems or a solution to problems. They are intended to alert designers to conditions for which they should ensure that the foundation will have the strength and displacement capacity intended under all conditions of demand.

26.6 Retrofit of Earthquake-Damaged Bridges

26.6.1 Introduction

In California, since the early 1970s, Caltrans has undertaken a major program to ensure the seismic safety of all freeway structures. With a large portion of the work being funded in the aftermath of the Loma Prieta earthquake in 1989, analysis techniques and research to validate such techniques have progressed significantly in the past five years. Early on in the Phase 2 retrofit program, structure over-strength analysis was the primary tool for seismic vulnerability assessment. Hinges and restrainers were always evaluated but none had been examined after a significant earthquake, so knowledge of their performance was based primarily on laboratory tests and hand calculations.

Foundation-strengthening techniques had not been field tested (by real earthquakes) and design methods were based on modified service load tools and simplified assumptions. It was soon clear that these methods, although for the most part very conservative, did not always afford a true picture of a structure's behavior or strength (resistance to collapse) and did not necessarily produce cost-effective designs.

In order to gain a clear picture of the performance of bridges, Caltrans engineers have been focusing on the following:

- load path and limit state analysis,
- input ground motions,
- validity of global models,
- displacement ductility analysis,
- beam and column joint performance,
- hinges and restrainers,
- detailing ductile elements, and
- foundation performance.

The purpose of this chapter is to highlight current thinking on these items and help identify avaliable tools to properly retrofit bridge structures.

26.6.2 Background

The goal of retrofitting is to increase the seismic resistance of a bridge to the point of preventing collapse and thereby avoiding loss of life. It is not practical or economical to retrofit all structures to meet the current seismic design performance criteria for new structures. It is, however, our responsibility to ensure that no structure collapses under the maximum credible event.

Based on the "no collapse" criteria, the primary goal of retrofitting bridges is to ensure that the vertical load-carrying capacity is maintained. Traditionally, this has meant protecting the columns' ability to carry vertical loads while undergoing cyclic lateral loading. Under this kind of loading, some damage is expected to occur in plastic hinge regions of the column. With 25 years of real-life experience, since the San Fernando Earthquake in 1971, we have been provided with a good picture of where damage can be expected to occur on a concrete structure under seismic loads. These areas include

- column plastic hinge regions,
- column shear capacity,
- column to footing and column to superstructure connections,
- expansion joint openings and hinge restrainers,
- abutment backwalls and transverse shear keys,
- pile to footing connections, and
- pile overload.

The following discussions address the seismic analysis and design approaches for retrofitting concrete bridge structures.

26.6.3 Seismic Analysis Approach

26.6.3.1 Load Path and Limit State Analysis

Limit state theory says, roughly, that a structure is only as good as the weakest link (or member) in the structural system. Ensuring that plastic action occurs in a predetermined location is therefore the goal. For concrete bridges, this predetermined location is typically, and preferably, in the column. Controlling the location of the plastic hinge allows the engineer to develop a rational estimate of the damage expected to a structure and therefore its ability to resist collapse. In order for this to happen, all members and components of a structure must be examined to determine whether they can actually carry the forces required to force plastic hinging to occur in the column.

Testing and application of basic engineering principles to the structure's components (column superstructure joints, column footing connections, pile cap and pile connections) have shown that areas of analysis and design once taken for granted as adequate could be vulnerable under extreme seismic loading conditions. Rigorous analysis of the structure may determine that the capacity of a system may likely be controlled by connections to the column and not by the strength of the column itself.

Current analysis techniques commonly used by engineers to estimate reinforced concrete capacity strongly suggest that many superstructures do not have the capacity to force plastic hinging into the column. Since one objective of the retrofit is to ensure that plastic hinging does occur in the column, the superstructure strength and ductility have to be addressed in the retrofit design.

26.6.3.2 Global Behavior

Early retrofit analysis typically relied on global elastic modal models to predict moment and displacement demands. Although this is still often a valid analysis technique, there are situations where global analyses are not the proper tools for seismic-demand prediction. These situations include but are not limited to

- short period ($T < 0.75$ s) structures, and
- very complicated structures (requiring many assumptions).

Other tools that could be used in lieu of or in conjunction with global elastic models are (1) hand calculations with single degree of freedom conditions and (2) pushover analysis to determine the order of plastic hinging in incremental collapse. Hand calculation analysis should be limited to simple structures or structures where the number of assumptions is limited. Pushover analysis can be used for both simple and complex structures.

The value of pushover analyses is that the capacity of the structural system is accounted for while the integrity of each element is determined. For example, if the two end frames of, say, a three-frame structure are designed with substantially more strength or stiffness than the interior frame, a global analysis may well conclude that the interior frame has relatively low seismic demands when in fact it is being supported, or helped, by the stronger adjacent frames. The final result could still be that the system is adequate to avoid collapse; however, it is important that the engineer has a clear idea of where vulnerabilities and strengths exist in the system. Since the goal of retrofit is to prevent the collapse and not just prevent plastic hinging, an efficient retrofit analysis technique would explicitly estimate the system capacity at collapse and not just assume a direct correlation between plastic hinging of a single column and precollapse capacity of the system.

26.6.3.3 Displacement and Curvature Ductility Analysis

An alternative estimate of a members (column) "ductility" can be obtained using ductility analysis techniques proposed by Park et al. The implementation of this approach has been tested extensively and directed explicitly toward bridges by Priestly et al. (1993). The predominant analysis approach until recently has been moment overload — the application of force reduction factors to loads calculated from a dynamic elastic multimodal response spectrum analysis (STRUDL or SAP90 for example) assuming gross moments of inertia I_g. This approach tends to underestimate system capacity.

This displacement ductility approach focuses on the capacity side of the equation and directly addresses the basic premise of the "equal energy theory" — that displacements are equal for a given load whether the structure remains elastic or enters into the plastic state as long as the section has not blown apart (allowable confinement stresses have been exceeded). By directly calculating the plastic moment capacity (instead of factoring up the nominal moment capacity) and the displacement capacity of each member (instead of limiting displacement to a factor of column geometry without regard to the rotational capacity of the section itself), the engineer can make a more accurate estimate of the capacity of the system.

Since the focus in this type of approach is on displacements, it is important to base the analysis on a members effective moments of inertia ($I_{eff} = I_{cr}$) instead of gross moments of inertia. Basing the dynamic analysis on I_{cr} instead of I_g will result in larger estimated displacement demands and smaller estimated moment (force) demands. More refined analysis techniques can be employed that assess a member's performance and estimated damage. This kind of information can be of particular interest when attempting to estimate resistance to collapse.

26.6.3.4 Foundations

Footings typically have required extensive work to increase their capacity to a level consistent with the column plastic moment. Early retrofitting techniques were often conservative (for lateral loads) in ignoring the contribution of existing piles to the capacity of the system. Large-scale testing of existing piles had not yet been done *in situ* and so nominal lateral pile capacities were often assumed when existing pile capacity was analyzed. Construction techniques for increasing the capacity of the footing relied on installing standard piles or prestressed tiedowns (tendons or rods). The technology simply had not been developed to make the leap from standard, new bridge construction to efficient, cost-effective retrofit work.

Since those early days, many new technologies have come into use in the retrofit industry. Field testing of pile foundations at the Cypress structure in Oakland, and the Terminal Separation in San Francisco have shown that the actual lateral capacity of concrete piles is much greater than was originally thought (90 kips/pile versus 40 kips/pile). Note that the assumed tensile capacity of existing piles is most often zero. Pile to footing connections were traditionally designed to develop vertical loads in a downward direction (compression), not upward (tension).

Construction techniques for installing high-capacity piles (up to 1000 kips/pile) in very confined spaces have been field tested with very promising results.

Seismic modeling now takes into account soil springs and group effects instead of simply assuming that all footings are fixed, rigid members. Soil springs in a soft soil may increase the displacement estimate on the column while decreasing the moment demand at the footing. The model (5) may also predict that a footing is moving so much that the piles need to be sized to limit displacement or resist larger loads than would otherwise be expected. Liquefaction and large ground displacements along fault rupture zones continue to be items that elude easy solutions.

26.6.4 Design Approaches

26.6.4.1 Retrofit Approaches

There are three primary concepts used to retrofit structures:

1. reduce loads imposed on the system;
2. increase the capacity of the structural elements; and
3. provide a new or "super" element.

It is not uncommon for a retrofit design to make use of more than one of these concepts. The following section is meant to provide some typical details and ideas that could be used to implement these concepts.

Column Plastic Hinge Regions and Column Shear Capacity.

1. **Steel shells.** Steel shells have been used extensively (and successfully throughout California and in Alaska to add ductility and provide confinement to concrete columns (Figure 26.27).
2. **Composite fiber wrap.** This provides additional confinement and ductility of concrete columns and is an idea that has been under development for several years. Results are starting to yield systems that are economical and structurally equivalent to that of the steel shell (Figure 26.28).
3. **Remove and replace.** Remove and replace portions of the substructure. This option has been used where performance beyond "no collapse" is required.
4. **Superbent.** Providing an additional support system.

Column to Footing Connections and Footings.

1. **Additional top mat on footing.** This is a very common retrofit owing to the fact that top mats of reinforcement were not used in footings until the late 1970s. The top mat adds tensile capacity to the footing (Figure 26.29).
2. **Prestressed footing.** Sometimes used for increased performance.

Column to Superstructure Connections.

1. **Bent cap bolsters.** Used to increase torsional capacity of the bent cap and capacity of the column and cap joint (Figure 26.30).
2. **Prestressed bent cap.** Used to increase torsional capacity of the bent cap and capacity of the column and cap joint.

Expansion Joint Openings and Hinge Restrainers.

1. **Restrainers.** Place additional or replace existing restrainers.

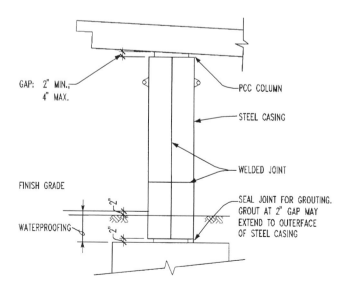

FIGURE 26.27 Steel casing of column.

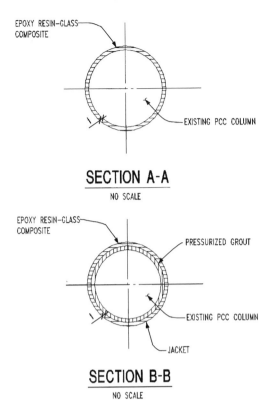

SECTION A-A

NO SCALE

SECTION B-B

NO SCALE

FIGURE 26.28   Composite fiber wrap.

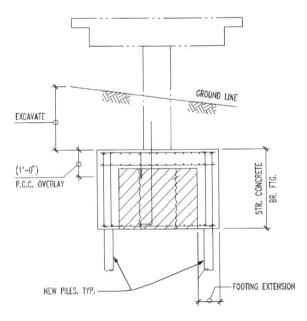

FIGURE 26.29   Footing retrofit with additional top mat.

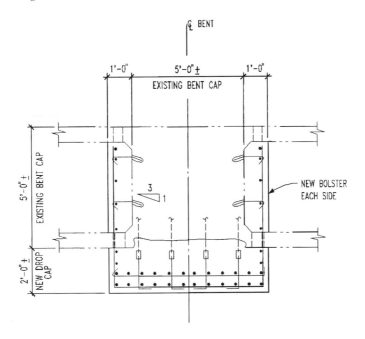

FIGURE 26.30 Bent cap bolsters.

2. **Seat extensions.** (typically pipe) Allows for displacement demands.
3. **Joint lock-up.** Using additional restrainers to reduce joint displacement demands.

Pile to Footing Connections.

1. **Additional piles.** Used for additional tension and compression capacity.
2. **Prestressed tie-downs.** Used for additional footing tensile (only) capacity.

26.6.4.2 Summary

Many more difficulties are encountered in retrofit construction than in standard new bridge construction. However, the field of bridge seismic analysis and retrofit has seen many developments over the past several years. Probably the most significant change in assessing the capacity of a structural system has been the adoption of **Ductility Analysis** and the pseudo nonlinear analysis approach of determining a plastic hinging sequence and ultimately a systematic yield mechanism.

Construction techniques are quickly adapting to the specialized demands of seismic retrofit for bridges. Materials, techniques and equipment have been developed and integrated from other industries, such as installation of high-capacity piles (adopted from the oil industry), and new materials such as carbon fiber are being developed in laboratories.

The field of seismic retrofitting has developed at a tremendous pace over the last several years. Engineers should expect additional changes as research and construction continue. The techniques used on concrete structures in California have proven, under real seismic loads (such as the 1994 Northridge Earthquake), that they are effective.

For additional information on any of the subjects addressed in this chapter, it is strongly suggested that the reader consult the noted references.

Defining Terms

Base shear: The total design lateral force at the base of a structure.
Bearing wall system: The structural system without a complete vertical load-carrying frame.

Building frame system: An essentially complete frame that provides support for dead, live, and snow loads.

Boundary element: Special reinforcement at edges of openings, shearwalls, or diaphragms.

Diaphragm: A horizontal system acting to transmit seismic forces to the vertical lateral-resisting systems.

Dual system: A combination of shearwalls or braced frames and intermediate or special moment-resisting frames proportionally designed to resist seismic forces.

Hoop: A closed tie that may be of several reinforcing elements with 135° hooks having a six-bar diameter (3 in minimum) extension at each end or a continuously wound tie it a 135° hook at each end with a six-bar diameter extension (3 in minimum) that engages the longitudinal reinforcement.

Moment resisting frame: A frame where the frame members and joints are capable of resisting forces primarily by flexure.

Shearwall: A wall designed to resist seismic forces parallel to the plane of the wall.

Story drift: The displacement of one level relative to the level above or below.

Strength: The useable capacity of a structure or its structural members to resist lateral loads within the code allowable deformations.

Structure base: That level in the structure at which the earthquake motions are considered to be imparted to the structure.

References

Algermissen, S.T. and Perkins, D.M. 1972. A Technique for Seismic Risk Zoning, General Considerations and Parameters. *Proc. Microzonation Conf.* 865–877, Seattle, WA.

Algermissen, S.T. and Perkins, D.M. 1976. A Probabilistic Estimate of Maximum Acceleration in Rock in Contiguous United States. USGS Open File Report, 76–416.

Bolt, B.A. 1969. Duration of Strong Motion, *Proc. 4th World Conf. Earthquake Eng.*, 1304–1315, Santiago, Chile.

Boore, D.M., Joyner, W.B., Oliver, A.A., and Page R.A. 1978. Estimation of Ground Motion Parameters. USGS, Circular 795.

Boore, D.M., Joyner, W.B., Oliver, A.A., and Page R.A. 1980. Peak acceleration velocity and displacement from strong motion records. *Bull. Seism. Soc. Am.* 70(1):305–321.

Caltrans Bridge Design Aids Manual. 1995. State of California, Department of Transportation, Sacramento, CA.

Chang and Krinitzsky

Conell, C.A. 1968. Engineering Seismic risk analysis. *Bull. Seism. Soc. Am.* 58(5):1583–1606.

Cooper, T.R. 1995. Seismic Analysis Procedures. *Proc. Third Nat. Concr. Masonry Eng. Conf.*

Donovan, N. C. 1973. Earthquake Hazards for Buildings. Building Practices for Disaster Mitigation, National Bureau of Standards, U.S. Department of Commerce, Building Research Services, 46, 82–111.

Housner, G. W. 1965. Intensity of Earthquake Shaking Near the Causative Fault. *Proc. 3rd. World Conf. Earthquake Eng.*, New Zealand 1(III):94–115.

ICBO. 1994. *Uniform Building Code.* Vol. 2. International Conference of Building Officials, Whittier, CA.

Ingham, J. and Priestly, M.J.N. 1993. Shear Strength of Knee Joints. University of California, San Diego.

McGuire, R.K. 1975. Seismic Structural Response Risk Analysis, Incorporating Peak Response Progressions on Earthquake Magnitude and Distance. Report R74-51, Dept. of Civil Engineering, Mass. Inst. of Technology, Cambridge, MA.

MacRae, Priestly, and Tao. 1993. P-Delta Design in Seismic Regions. SSRP No. 93/05. University of California, San Diego. SC Solutions, Mountain View, CA.

Moehle, J.P. and Soyer, C. 1993. Behavior and Retrofit of Exterior RC Connections. University of California, Berkeley.

Naaseh, S. 1995. Seismic Reftrofit of San Francisco City Hall. *The Third National Concrete and Masonry Enginnering Conferences*, 769–795, San Francisco, California.

Nawy, E.G. 1996. *Reinforced Concrete: A Fundamental Approach*, 3rd ed., 848 pp. Prentice Hall, Upper Saddle River, NJ.

Newark, N.M. and Hall, W.J. 1969. Seismic Design Criteria for Nuclear Reactor Facilities. *Proc. 4th World Conf. Earthquake Eng.* B-4, 37–50, Santiago, Chile.

Newark, N.M. and Hall, W.J. 1973. Procedures and Criteria for Earthquake Resistance Design. Building Practices for Disaster Mitigation, National Bureau of Standards, U.S. Department of Commerce, Building Research Series, 46, 209–236.

Page, R.A., Boore, D.M., Joyner, W.B., and Caulter, H.W. 1972. Ground Motion Values for Use in Seismic Design of the Trans-Alaska Pipeline System, USGS Circular Circular 672.

Priestly, M.J.N. and Seible, F. 1993. Full-Scale Test on the Flexural Integrity of Cap/Column Connections With #18 Column Bars. Report No. TR-93/01. University of California at San Diego.

Priestly, M.J.N. and Seible, F. 1994. Proof Test of Superstructure Capacity in Resisting Longitudinal Seismic Attack for the Terminal Separation Replacement. University of California, San Diego.

Priestly, M.J.N., Seible, F., Xioa, and Hamada. 1993. Proof Test of Circular Column Footing Designed to Current Caltrans Retrofit Standards. *Tests for Rectangular Column Also Performed*. University of California, San Diego.

Seed, H.B., and Idris, I.M. 1982. Ground Motions and Soil Liquefaction During Earthquakes. *Earthquake Engineering Research Inst.*, Berkeley, CA.

Recommended Lateral Force Requirements and Commentary. 1996. Seismology Committee Structural Engineers Association of California.

Seismic Design Book: Edited by Farzad Naeim, published in 1989.

Seyed, M. 1993. Ductility Analysis for Seismic Retrofit of Multi-Column Bridge Structures: Santa Monica Viaduct. Paper presented at ASCE Structures Congress.

Thewalt, C.R. and Stojadinovic, B. 1993. Behavior of Bridge Outriggers: Summary of Test Results. University of California, Berkeley.

Trifunac, M.D. and Brady, A.G. 1975. A study of the duration of strong earthquake ground motion. *Bull. Seism. Soc. Am.* 65, 581–626.

Further Information

For further information, consult CALTRANS "Bridge Design Practice," Section 8, Oct. 1995 and "Seismic Design Memo," Ray Zelinski, CALTRANS Seismic Technical Section, Jan. 1995.

(a)

(b)

(a) Sunshine Skyway Bridge across Tampa Bay, Florida. 4.2 mile long segmental prestressed cable-stayed bridge, one of the longest in the world. (b) Trump Towers, New York City. 12000 psi silica fume concrete (Courtesy Portland Cement Association).

27

Proportioning Concrete Structural Elements by the ACI 318-95 Code

by
Edward G. Nawy, D.Eng., P.E., C.Eng.
Distinguished Professor, Civil Engineering, Rutgers University—The State University of New Jersey.
Expert in concrete structures and materials.

27.1 Structural Concrete

Most structural systems constructed today are made from reinforced, prestressed, or composite concrete having a wide range of characteristics and strengths. Structural concrete, whether normal weight or lightweight, is designed to have a compressive strength in excess of 3000 psi (20 MPa) in concrete structures. When the strength exceeds 6000 psi (42 MPa) such structures are defined today as high-strength concrete structures. Concrete mixtures designed to produce 6000 to 12,000 psi in compressive strength are easily obtainable today when silica fume replaces a portion of the cement content, resulting in lower water/cement (W/C) and water/cementitious materials (W/C + P) ratios. Concretes having cylinder compressive strengths in the range of 20,000 psi (140 MPa) have been used in several buildings in the United States. Such high strength characteristics merit qualifying such concrete as super high strength concrete at this time.

27.1.1 Modulus of Concrete

The ACI 318 Code [ACI Committee 318, 1996; Nawy, 1996a,c] stipulates that the concrete modulus E_c be evaluated from

$$E_c \text{ (psi)} = 33w^{1.5}\sqrt{f_c'} \tag{27.1a}$$

$$E_c \text{ (MPa)} = 0.043w^{1.5}\sqrt{f_c'} \tag{27.1b}$$

The expressions in Eqs. (27.1a) and (27.1b) are applicable to strengths up to 6000 psi (42 MPa). Available research to date for concrete compressive strength up to 12,000 psi (83 MPa) gives the following expressions: [ACI Committee 435, 1995; Nawy, 1996a,c]

$$E_c \text{ (psi)} = \left(40,000\sqrt{f_c'} + 10^6\right)\left(\frac{w_c}{145}\right)^{1.5} \tag{27.2a}$$

$$E_c \text{ (MPa)} = \left(3.32\sqrt{f_c'} + 6895\right)\left(\frac{w_c}{2320}\right)^{1.5} \tag{27.2b}$$

In Eqs. (27.1a) and (27.2a), f_c' is in units of pounds per square inch and w_c ranges between 145 pcf for normal density concrete and 100 pcf for structural lightweight concrete; f_c', in Eqs. (27.1b) and (27.2b) is in units of megapascals and w_c ranges between 2400 kg/m^3 for normal density concrete and 1765 kg/m^3 for lightweight concrete. The modulus of rupture of concrete can be taken as

$$f_r \text{ (psi)} = 7.5\,\lambda\sqrt{f_c'} \tag{27.3a}$$

$$f_r \text{ (MPa)} = 0.632\,\sqrt{f_c'} \tag{27.3b}$$

where

$\lambda = 1.0$ for normal-density stone aggregate concrete
$\lambda = 0.85$ for sand lightweight concrete
$\lambda = 0.75$ for all lightweight concrete.

27.1.2 Creep of Concrete

Concrete creeps under sustained loading owing to transverse flow of the material. The creep coefficient as a function of time can be calculated from the following expression [ACI Committee 435, 1995; ACI Committee 318, 1996]:

$$C_t = \left(\frac{t^{0.6}}{10 + t^{0.6}}\right)C_u \tag{27.4}$$

where the time, t, is in days and C_u, the ultimate creep factor is 2.35. The short-term deflection is multiplied by C_t to get the long-term deflection that is added to the short-term (instantaneous) deflection value in order to get the total deflection.

27.1.3 Shrinkage of Concrete

Concrete shrinks as the absorbed water evaporates and the chemical reaction of cement gel proceeds. For moist-cured concrete, the shrinkage strain that occurs at any time t in days 7 d after placing the concrete can be evaluated from [ACI Committee 435, 1995]

$$(\epsilon_{SH})_t = \left(\frac{t}{35+t}\right)(\epsilon_{SH})_u \tag{27.5a}$$

For steam-cured concrete, the shrinkage strain at any time t in days, 1–3 days after placement of the concrete, is

$$(\epsilon_{SH})_t = \left(\frac{t}{55+t}\right)(\epsilon_{SH})_u \tag{27.5b}$$

where maximum $(\epsilon_{SH})_u$ can be taken as 780×10^{-6} in/in (mm/mm).

Shrinkage and creep due to sustained load can also be evaluated from the ACI expression [ACI Committee 318, 1996]

$$\lambda = \frac{\xi}{1+50\rho'} \tag{27.5c}$$

and Figure 27.1 for the factor ξ that ranges from a value of 2.0 for five years or more to 1.0 for three months of sustained loading.

$\rho' = $ compression steel percentage
$\rho' = A'_s/bd$

27.1.4 Control of Deflection

Serviceability is a major factor in designing structures to sustain acceptable long-term behavior. Serviceability is controlled by limiting deflection and cracking in the members [ACI Committee 435, 1995].

For deflection computation and control, the effective moment of inertia of a cracked section can be evaluated from the Branson equation:

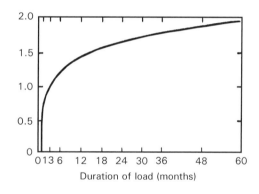

FIGURE 27.1 Long-term deflection multipliers. *Source:* Courtesy of American Concrete Institute; ACI Committee 318. 1996. *Building Code Requirements for Structural Concrete*, ACI 318-95.

$$I_e = \left(\frac{M_{cr}}{M_a}\right)I_g + \left[1 - \left(\frac{M_{cr}}{M_a}\right)^3\right]I_{cr} \leq I_g \tag{27.6}$$

where, for reinforced concrete beams

$M_{cr} = $ cracking moment due to total load $= \dfrac{f_r I_g}{y_t}$

$y_t = $ distance from the neutral axis to the extreme tension fibers
$M_a = $ maximum service load moment at the section under consideration
$I_g = $ gross moment of inertia

In the case of prestressed concrete

$$\left(\frac{M_{cr}}{M_a}\right) = \frac{f_{t\ell} - f_r}{f_L} \tag{27.7}$$

where

M_{cr} = moment due to that portion of the *live* load moment M_a that causes cracking
M_a = maximum service load (unfactored) *live* load moment
$f_{t\ell}$ = total calculated stress in the member
f_L = calculated stress due to live load

For long-term deflection, Figure 27.1 gives the required multipliers as a function of time.

27.1.5 Control of Cracking in Beams

Control of cracking in beams and one-way slabs can be made using the expression [ACI Committee 224, 1989]

$$w_{\max} \text{ (in)} = 0.076\beta f_s \sqrt[3]{d_c A} \times 10^{-3} \tag{27.8}$$

where

$w_{\max}$ = crack width, in (25.4 mm)
$\beta = (h - c)/(d - c)$
d_c = thickness of cover to the first layer of bars, inches
f_s = maximum stress in reinforcement at service load = $0.60\,f_y$, kips per square inch
A = area of concrete in tension divided by number of bars, square inches
  = bt/γ_{bc} where γ_{bc} = number of bars at the tension side

27.2 Structural Design Considerations

27.2.1 General

High-strength concretes have certain characteristics and engineering properties that differ from those of lower strength concretes [ACI Committee 363, 1992]. These differences seem to have larger effects as the strength increases beyond the present 6000 psi (42 MPa) plateau for normal-strength concrete. High-strength concretes are shown to be essentially linearly elastic up to failure, with a steeper declining portion of the stress-strain diagram. In comparison, the stress-strain diagram of lower strength concretes is more parabolic in nature, as seen in Figure 27.2. The stress-strain relationship of the steel reinforcement in this diagram is not to scale in its ordinate value but is intended to show the relative strain following the usual assumption of strain compatibility between the concrete and the steel reinforcement up to yield.

27.2.2 Axially Loaded Columns

The present design practice is to assume adding the contribution of the steel and the concrete in calculating the ultimate state of failure in compression members. For lower strength concretes, when the concrete reaches the nonlinearity load level at a strain of 0.001 in/in, as seen in Figure 27.2, the steel is still in the elastic range, assuming a larger share of the applied load. But, as the strain level approaches 0.002 in/in, the slope of the concrete stress-strain diagram approaches zero while the steel reaches its yield strain that will thereafter be idealized into a constant (horizontal) plateau.

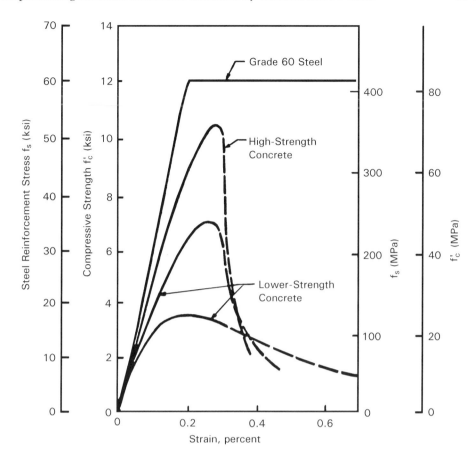

FIGURE 27.2 Concrete and steel stress-strain relationships. *Source:* Courtesy of American Concrete Institute; ACI Committee 363. 1992. *State-of-the-Art on High Strength Concrete,* ACI 363R-92.

The strength of the column using the addition law would then be

$$P_n = 0.85 \, f_c' \, A_c + f_y \, A_s \qquad (27.9)$$

where

f_c' = concrete cylinder compressive strength
f_y = yield strength of the reinforcement
A_c = gross area of the concrete section
A_s = area of the reinforcement

The factor 0.85 representing the adjustment in concrete strength between the cylinder test result and the actual concrete strength in the structural element has been shown by extensive testing to be sufficiently accurate for higher strength concretes [ACI Committee, 363, 1992; Nawy, 1996c].

Confining the concrete in compression members through the use of spirals or closely spaced ties increases its compressive capacity. The increase in concrete strength due to the confining effect of the spirals can be represented by the following expression:

$$f_2' = \frac{1}{4} \big[\overline{f}_c - f_c'' \big] \qquad (27.10a)$$

where

> f'_2 = concrete confining stress due to the spiral
> $\overline{f}_c$ = compressive strength of the confined concrete
> f''_c = compressive strength of the unconfined column concrete

The hoop tension force in the circular spiral is

$$2 A_{sp} f_y = f'_2 D's$$

or

$$f'_2 = \frac{2 A_{sp} f_y}{D's} \tag{27.10b}$$

where

> A_{sp} = cross-sectional area of the spiral
> D' = diameter of concrete core
> s = spiral pitch

Equations (27.10a) and (27.10b) can be improved [ACI Committee, 363, 1992; Nawy, 1996c] leading to the following form for normal-weight concrete

$$\left(\overline{f}_c - f''_c\right) = 4.0\, f'_2 (1 - s/D') \tag{27.11a}$$

For lightweight concrete

$$\left(\overline{f}_c - f''_c\right) = 1.8\, f'_2 (1 - s/D') \tag{27.11b}$$

Figure 27.3 gives the results of peak stress comparisons versus axial strain for spirally reinforced members for low, medium, and high-strength concretes. For higher strength, it shows a lower

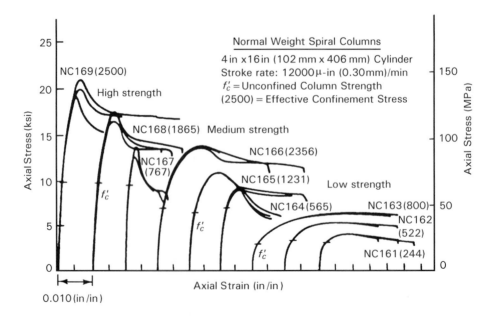

FIGURE 27.3   Stress-strain diagrams of 4 × 6 in normal-weight spirally confined compression prisms. *Source:* Courtesy of American Concrete Institute; Martinez, S. et al. 1984. Spirally reinforced high strength concrete columns. *Proceedings, ACI Journal,* Vol. 81, No. 5.

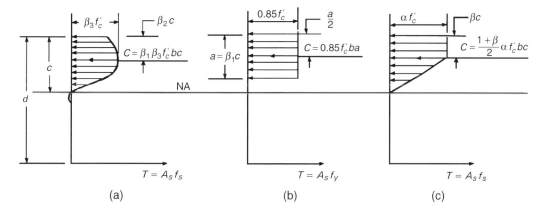

FIGURE 27.4 Concrete compressive stress block. (a) Standard stress block; (b) equivalent rectangular block; and (c) modified trapezoidal block. *Source:* Nawy, E.G. 1996. *Fundamentals of High Strength High Performance Concrete.* Reprinted with permission of Addison Wesley Longman.

strain at peak load and a steeper decline past the peak value. The strength gain in concrete due to confinement seems, however, to be well predicted for high strength concretes in Eq. (27.11).

27.2.3 Beams and Slabs

27.2.3.1 The Compressive Block

Design of concrete structural elements is based on the compressive stress distribution across the depth of the member as determined by the stress-strain diagram of the material. For high-strength concretes, the difference in the shape of the stress-strain relationship discussed in connection with Figure 27.2 results in differences in the shape of the compressive stress block. Figure 27.4 shows possible compressive blocks for use in design. Figure 27.4c could more accurately represent the stress distribution for higher strength concrete.

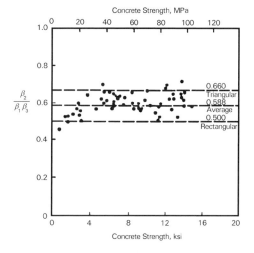

FIGURE 27.5 Stress block parameter $\beta_2/\beta_1\beta_3$ versus concrete compressive strength. *Source:* Courtesy of American Concrete Institute; ACI Committee 363. 1992 *State-of-the-Art on High Strength Concrete*, ACI 363R-92.

However, the computed strength of beams and eccentrically loaded columns depends on the reinforcement ratio. In the ACI 318 Code provisions which use the equivalent rectangular block, the nominal moment strength of a singly reinforced beam is calculated using the following expression:

$$M_n = A_s\,f_y d\left[1 - 0.59\rho\frac{f_y}{f_c'}\right] \tag{27.12}$$

where the coefficient $0.59 = \beta_2/\beta_1\beta_3$. While a detailed evaluation of the factors β_1, β_2, and β_3 indicates a significant difference in their separate values depending on the concrete strength [ACI Committee 363, 1992], Figure 27.5 shows that these differences collectively balance each other and that the combined coefficient $\beta_2/\beta_1\beta_3$ is well represented by the 0.59 value. Consequently, for strengths up to 12,000 psi (42 MPa), the present ACI 318 Code expressions requiring that beams be underreinforced are equally applicable. For considerably higher strengths or for members combining compression and bending or for the over-reinforced members allowed in the codes, some differences in the value of $\beta_2/\beta_1\beta_3$ can be expected.

27.2.3.2 Compressive Limiting Strain

Although high-strength concrete achieves its peak value at a unit strain slightly higher than that of normal-strength concrete (Figure 27.2), the ultimate strain is lower for high-strength concrete unless confinement is provided. A limiting strain value allowed by the ACI 318 Code equals 0.003 in/in (mm/mm). Other codes allow a limiting strain of unconfined concrete of 0.0035 or 0.0038. The conservative ACI value of 0.003 seems to be adequate for high-strength concretes as well, although it is somewhat less conservative than that for lower strength concretes.

27.2.3.3 Confinement and Ductility

As higher strength concrete is more brittle, confinement becomes more important in order to increase its ductility. If μ is the deflection ductility index

$$\mu = \frac{\Delta_u}{\Delta_y} \tag{27.13}$$

where

Δ_u = beam deflection at failure load
Δ_y = beam deflection at the load producing yield of the tensile reinforcement

Table 27.1 shows the ductility index values of concretes in singly reinforced beams ranging in strength from 3700 psi (25 MPa) to 9265 psi (64 MPa). The corresponding reduction in the ductility index ranges from 3.54 to 1.07. Addition of compressive reinforcement and confinement to geometrically similar beams seem to increase the ductility index for $f'_c = 8500$ up to a value of 5.61. Hence the higher the concrete compressive strength the more it becomes necessary to provide for confinement and/or the addition of compression steel, A'_s, while using the same expression for nominal moment strength that is applicable to normal strength concretes. It should be stated that cost would not be affected to any meaningful extent since diagonal tension and torsion stirrups have to be used anyway, and in seismic regions closely spaced confining ties are a requirement. The maximum strain of confined concrete that can be utilized should not exceed 0.01 in/in (mm/mm) in limit design.

27.2.3.4 Shear and Diagonal Tension

Design for shear in accordance with the ACI 318 Code is based on permitting the plain concrete in the web to assume part of the nominal shear V_n. If V_c is the shear strength resistance of the concrete, the web stirrups resists a shear force $V_s = V_n - V_c$. High-strength concrete develops a relatively brittle failure, as previously discussed, with the aggregate interlock decreasing with the increase in the compressive strength. Hence the shear friction and diagonal tension failure capacity in beams might be unconservatively represented by the ACI 318 equations [Ahmed and Lau, 1987]. However, the strength of the diagonal struts in the beam shear truss model is increased through the mobilization of more stirrups and the increased load capacity of the struts themselves. No research data is

TABLE 27.1 Deflection Ductility Index for Singly Reinforced Beams

Beam	f'_c		ρ/ρ_b^*	Ductility Index $\mu = \Delta_u/\Delta_v$
	psi	MPa		
A1	3700	26	0.51	3.54
A2	6500	45	0.52	2.84
A3	8535	59	0.29	2.53
A4	8535	59	0.64	1.75
A5	9264	64	0.87	1.14
A6 (a)	8755	60	1.11	1.07

Source: Pastor, J.A., Nilsen, A.H., and Slate, F.O. 1984. Behavior of high strength concrete beams. *Cornell University Report No. 84–3.* Dept. of Structural Engineering, Cornell University, Ithaca, NY; Nawy, E.G. 1996. *Fundamentals of High Strength High Performance Concrete.* Reprinted with permission of Addison Wesley Longman.

*Ratio of tension reinforcement divided by reinforcement ratio producing balanced strain conditions.

currently available for definitive guidelines on the minimum web steel that can prevent brittle failure. All presently available work shows no unsafe use of the present ACI 318 Code provisions for shear in the design of high-strength concrete members.

27.3 Reinforced Concrete Members

27.3.1 Flexural Strength

27.3.1.1 Singly Reinforced Beams

Flexural strength is determined from the strain and stress distribution across the depth of the concrete section. Figure 27.6 shows the stress and strain distribution and forces. Taking moments of all the forces about tensile steel, A_s, gives, for singly reinforced beams ($A'_s = 0$), a nominal moment strength

$$M_n = A_s f_y \left(d - \frac{a}{2} \right) \tag{27.14a}$$

or

$$M_n = bd^2 \omega (1 - 0.59\omega) \tag{27.14b}$$

where ω = reinforcement index = $A_s/bd \times f_y/f'_c$.
The reinforcement index

$$\rho = \frac{A_s}{bd} \leq 0.75 \overline{\rho}_b \tag{27.15}$$

$$\overline{\rho}_b = \frac{0.85 \beta_1 f'_c}{f_y} \times \frac{0.003 E_s}{0.003 E_s + f_y} \tag{27.16}$$

where $\beta_1 = 0.85$ and reduces at the rate of 0.05 per 1000 psi in excess of 4000 psi, namely,

$$\beta_1 = 0.85 - 0.05 \left(\frac{f'_c - 1000}{1000} \right) \tag{27.17}$$

where β_1 is not to be less than 0.65.

27.3.1.2 Doubly Reinforced Beams

For doubly reinforced sections (which have compression steel)

$$M_n = \left(A_s - A'_s \right) f_y \left(d - \frac{a}{2} \right) + A'_s f_y (d - d') \tag{27.18}$$

if compressive reinforcement is used in a doubly-reinforced section [ACI Committee 318, 1996] (see also Figure 27.7). The reinforcement percentage ρ is

$$\rho \leq 0.75 \overline{\rho}_b + \rho' \frac{f'_s}{f_y} \tag{27.19}$$

where f'_s = stress in the compression.

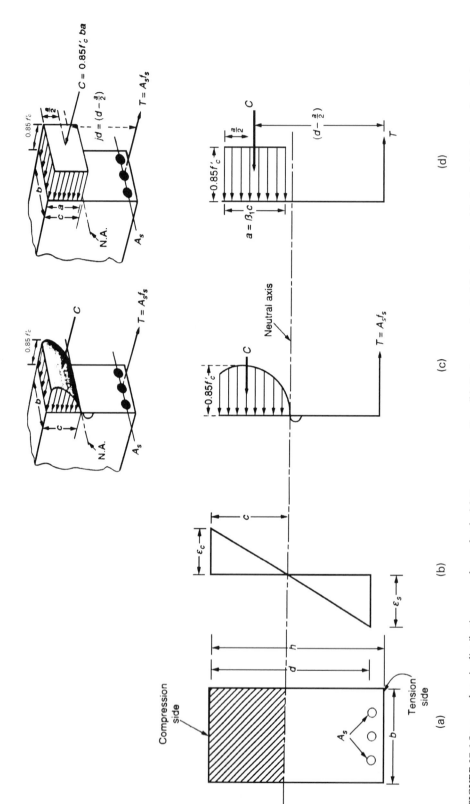

FIGURE 27.6 Stress and strain distribution across beam depth. (a) Beam cross section; (b) Strain across depth; (c) Actual stress block; and (d) Assumed equivalent stress block. *Sources:* Nawy, E.G., *Reinforced Concrete—A Fundamental Approach*, 3rd ed., © 1996. Reprinted by permission of Prentice–Hall, Inc. Upper Saddle River, NJ; Nawy, E.G. 1996. *Fundamentals of High Strength High Performance Concrete.* Reprinted with permission of Addison Wesley Longman.

The depth of the compressive block is

$$a = \frac{A_s f_y - A'_s f'_s}{0.85 f'_c b} \qquad (27.20)$$

where b = width of section at the compression side.

27.3.1.3 Flanged Sections

For flanged sections where the neutral axis falls outside the flange

$$M_n = (A_s - A_{sf}) f_y \left(d - \frac{a}{2} \right) + A_{sf} f_y \left(d - \frac{h_f}{2} \right) \qquad (27.21)$$

where

$$A_{sf} = \frac{0.85 f'_c (b - b_w) h_f}{f_y}$$

b_w = web width
h_f = flange thickness

The depth

$$a = \frac{A_s f_y}{0.85 f'_c b} > h_f$$

$$\rho < \frac{b_w}{b} (\overline{\rho}_b + \rho_f) \qquad (27.22a)$$

where $\overline{\rho}_b$ = reinforcement percentage for a singly reinforced beam

$$\rho_f = 0.85 f'_c (b - b_w) \frac{h_f}{f_y b_w d} \qquad (27.22b)$$

27.3.1.4 Minimum Reinforcement

In Eqs. (27.14c) and (27.19a), the flexural reinforcement percentage ρ has to have a minimum value of $\rho_{min} = 3\sqrt{f'_c}/f_y$ for positive moment reinforcement, and $\rho_{min} = 6\sqrt{f'_c}/f_y$ for negative moment reinforcement but always not less than $200/f_y$ where f_y is in units of pounds per square inch. The factored moment

$$M_u = \phi M_n \qquad (27.23)$$

where $\phi = 0.90$ for flexure.

27.3.2 Shear Strength

External transverse load is resisted by internal shear in order to maintain section equilibrium. As concrete is weak in tension, the principal tensile stress in a beam cannot exceed the tensile strength of the concrete. The principal stress is composed of two components: shear stress and flexural stress. It is important that the beam web be reinforced in order to prevent diagonal shear cracks from opening. The resistance of the plain concrete in the web sustains part of the shear stress and the balance has to be borne by the diagonal tension reinforcement. The shear resistance of the plain concrete is termed as the nominal shear strength, V_c. The nominal V_c is

$$V_c = 2.0\lambda\sqrt{f'_c} b_w d, \text{ lb} \leq 3.5\lambda\sqrt{f'_c} \qquad (27.24a)$$

$$V_c = \left(\frac{\sqrt{f'_c}}{6} \right) b_w d, \text{ N} \qquad (27.24b)$$

or

$$V_c = \left[1.9\lambda\sqrt{f_c'} + 2500\rho_w \frac{V_u d}{M_u} \right] b_w d, \text{ lb} \leq 3.5\lambda\sqrt{f_c'} \tag{27.25a}$$

$$V_c = \left[\left(\sqrt{f_c'} + 120\rho_w \frac{V_u d}{M_u} \right) \div 7 \right] b_w d, \text{ N} \tag{27.25b}$$

Values for λ are given in Section 27.1.

$$\rho_w = \frac{A_s}{b_w d} \quad \text{and} \quad \frac{V_u d}{M_u} \leq 1.0$$

No web steel is needed if $V_u < 1/2 V_c$. For calculating V_n the critical section is at a distance d from the support face. Spacing of the web stirrups is as follows:

$$s = \frac{A_v f_y d}{(V_u/\phi - V_c)} \tag{27.26}$$

where A_v is the cross-sectional area of web steel and ϕ is 0.85 for shear and torsion. The transverse web steel is designed to carry the shear load $V_s = V_n - V_c$. The spacing of the stirrups is governed by the following:

$$V_s \geq 4\sqrt{f_c'} : s = \frac{d}{4}$$

$$V_s \leq 2\sqrt{f_c'} : s = \frac{d}{2}$$

$$V_s \geq 8\sqrt{f_c'} : \text{enlarge section}$$

The minimum sectional area of the stirrups is $A_{v,\min} = 50 b_w s / f_y$ and

$s_{\max} = d/2$ where shear is to be considered
$d =$ effective depth to the center of the tensile reinforcement
$f_y =$ yield strength of the steel in pounds per square inch

27.3.3 Torsional Strength

The space truss analogy theory is used for the analysis and design of concrete members subjected to torsion. It is based on the shear flow in a hollow tube concept and the summation of the forces in the space truss elements [ACI Committee 318, 1996; Hsu, 1993; Nawy, 1996a,b,c]. The ACI 318-95 Code stipulates disregarding the concrete nominal strength T_c in torsion and assigning all the torque to the longitudinal reinforcement, A_ℓ, and the transverse reinforcement, A_t, essentially assuming that the volume of the longitudinal bars is equivalent to the volume of the closed transverse hoops or stirrups. The critical section is taken at a distance d from the face of the support for the purpose of calculating the torque T_u. Sections that are subjected to combined torsion and shear should be designed for torsion if the factored torsional moment T_u exceeds the following value for nonprestressed members:

$$T_u > \phi\sqrt{f_c'}\left(\frac{A_{cp}^2}{P_{cp}} \right) \tag{27.27}$$

where

A_{cp} = area enclosed by the outside perimeter of the concrete cross section
P_{cp} = outside perimeter of the cross-section A_{cp}, inches

Two types of torsion are considered :

(a) equilibrium torsion, where no redistribution of torsional moment is possible—in this case, all the factored torsional moment is designed for;
(b) compatibility torsion, where redistribution of the torsional moment occurs in a continuous floor system—in this case, the maximum torsional moment to be provided for is

$$T_u = \phi 4\sqrt{f_c'}\left(\frac{A_{cp}^2}{P_{cp}}\right) \tag{27.28}$$

The concrete section has to be enlarged if

$$\sqrt{\left(\frac{V_u}{b_w d}\right)^2 + \left(\frac{T_u p_h}{1.7 A_{oh}^2}\right)^2} > \phi\left(\frac{V_c}{b_w d} + 8\sqrt{f_c'}\right) \tag{27.29}$$

where

p_h = perimeter of centerline of outermost closed transverse torsional reinforcement, inches
A_{oh} = area enclosed by centerline of the outermost closed transverse torsional reinforcement, square inches

The transverse torsional reinforcement should be chosen with such size and spacing s that

$$\frac{A_t}{2} = \frac{T_n}{2 A_o f_y \cot \theta} \tag{27.30}$$

where

A_o = gross area enclosed by the shear path = $0.85 A_{oh}$
θ = angle of compression diagonals
  = 45° in reinforced concrete
  = 37 1/2° in prestressed concrete

The longitudinal torsional reinforcement A_ℓ, divided equally along the four faces of the beam, is

$$A_\ell = \frac{A_t}{s} p_h \left(\frac{f_{yv}}{f_{y\ell}}\right) \cot^2 \theta \tag{27.31}$$

$$A_{\ell,\min} = \frac{5\sqrt{f_c'} A_{cp}}{f_{y\ell}} - \left(\frac{A_t}{s}\right) p_h \frac{f_{yv}}{f_{y\ell}} \tag{27.32}$$

where

f_{yv} = yield strength of the transverse reinforcement
$f_{y\ell}$ = yield strength of the longitudinal reinforcement.

The minimum area of transverse reinforcement is

$$A_v + 2 A_t \geq \frac{50 b_w s}{f_y} \tag{27.33}$$

Maximum s is 12 in.

In SI units, the following are equivalent expressions:

Eq. (27.27)

$$\frac{\phi\sqrt{f_c'}}{12}\left(\frac{A_{cp}^2}{p_{cp}}\right)$$

Eq. (27.28)

$$\frac{\phi\sqrt{f_c'}}{3}\left(\frac{A_{cp}^2}{p_{cp}}\right)$$

For Eq. (27.29), the right-hand expression is

$$\phi\left(\frac{V_c}{b_w d}+\frac{8\sqrt{f_c'}}{12}\right)\text{MPa}$$

Eq. (27.32)

$$A_{\ell,\min}=\frac{5\sqrt{f_c'}A_{cp}}{12 f_{y\ell}}-\left(\frac{A_t}{s}\right)p_h\left(\frac{f_{yv}}{f_{y\ell}}\right)$$

Eq. (27.33)

$$A_v+2A_t\geq\frac{0.35 b_w s}{f_y}$$

where f_y is in megapascals. Maximum s is 300 mm.

27.3.4 Compression Members: Columns

27.3.4.1 Nonslender Columns

Flexural members such as beams are designed in such a manner that the reinforcement ratio ρ cannot exceed 75% of the balanced ratio ρ_b and for practical reasons should not exceed 50%. Compression members, on the other hand, are proportioned on the basis of the magnitude of eccentricity. If P_{nb} is the balanced axial load where failure occurs simultaneously by concrete crushing on the compression side and reinforcement yielding on the tension side, then

$P_n < P_{nb}$ is tension failure ($e > e_b$)
$P_n = P_{nb}$ is balanced failure ($e = e_b$)
$P_n > P_{nb}$ is compression failure ($e < e_b$)

$$P_{nb}=0.85 f_c' b a_b + A_s' f_s' - A_s f_y' \tag{27.34}$$

$$M_{nb}=P_{nb}e_b=0.85 f_c b a_b\left(\overline{y}-\frac{a_b}{2}\right)+A_s' f_s'(\overline{y}-d')+A_s f_y(d-\overline{y}) \tag{27.35}$$

where

$$f_s'=E_s\left[\frac{0.003(c-d')}{c}\right]\leq f_y \tag{27.36}$$

The force P_n and the moment M_n at ultimate for any other eccentricity level are

$$P_n = 0.85 f_c' ba + A_s' f_s' - A_s f_y \tag{27.37}$$

$$M_n = P_n e = 0.85 f_c' ba \left(\bar{y} - \frac{a}{2} \right) + A_s' f_s'(\bar{y} - d') + A_s f_s(d - \bar{y}) \tag{27.38}$$

where

$$f_s = E_s \left[\frac{0.003(d - c)}{c} \right] \leq f_y \tag{27.39}$$

where

c = depth to the neutral axis
$\bar{y}$ = distance from the compression extreme fibers to the center of gravity of the section
a = depth of the equivalent rectangular block = $\beta_1 c$, where β_1 is defined in Eq. (27.20b).

The geometry of the compression member section and the forces acting on the section are shown in Figure 27.7. Equations (27.37) and (27.38) are obtained from the equilibrium of forces and moments.

27.3.4.2 Slender Columns

If the compression member is slender, namely, the slenderness ratio $k\ell_u/r$ exceeds 22 for unbraced members and $(34 - 12 \cdot M_1/M_2)$ for braced members, failure will occur by buckling and not by material failure. In such a case, if $k\ell_u/r$ is less than 100, a first-order analysis such as the moment magnification method can be performed. If $k\ell_u/r > 100$, the P-Δ effects have to be considered and

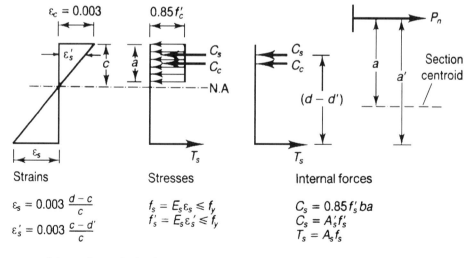

$\varepsilon_s = 0.003 \dfrac{d - c}{c}$

$\varepsilon_s' = 0.003 \dfrac{c - d'}{c}$

$f_s = E_s \varepsilon_s \leq f_y$

$f_s' = E_s \varepsilon_s' \leq f_y$

$C_s = 0.85 f_c' ba$

$C_s = A_s' f_s'$

$T_s = A_s f_s$

c = distance to neutral axis
$\bar{y}$ = distance of section centroid
e = eccentricity of load to plastic centroid
e' = eccentricity of load to tension steel
d' = effective cover of compression steel

FIGURE 27.7 Stresses and forces in columns. *Source:* Nawy, E.G., *Prestressed Concrete—A Fundamental Approach*, 2nd ed., © 1996. Reprinted by permission of Prentice–Hall, Inc. Upper Saddle River, NJ.

a second-order analysis has to be performed. The latter is a lengthy process and is more reasonably executed using readily available computer programs.

Moment Magnification Solution ($k\ell_u/r < 100$). The larger moment M_2 is magnified such that

$$M_c = \delta_{ns} M_2 \qquad (27.40)$$

where δ_{ns} is the magnification factor. The column is then designed for a moment M_c as a nonslender column. The subscript *ns* is non-side-sway; *s* is side sway.

$$\delta_{ns} = \frac{C_m}{1 - (P_u/0.75 P_c)} \geq 1.0 \qquad (27.41)$$

$$P_c = \frac{\pi^2 EI}{(k\ell_u)^2} \qquad (27.42)$$

EI should be taken as

$$EI = \frac{0.2 E_c I_g + E_s I_{se}}{1 + \beta_d} \qquad (27.43a)$$

or

$$EI = \frac{0.4 E_c I_g}{1 + \beta_d} \qquad (27.43b)$$

$$C_m = 0.6 + 0.4\frac{M_1}{M_2} \geq 0.4 \qquad (27.44)$$

If there is side sway

$$M_1 = M_{1ns} + \delta_s M_{1s} \qquad (27.45a)$$

$$M_2 = M_{2ns} + \delta_s M_{2s} \qquad (27.45b)$$

where

$$\delta_s M_s = \frac{M_s}{1 - \dfrac{\sum P_u}{0.75 P_c}} \geq M_s \qquad (27.46)$$

The nonsway moment M_{2ns} is unmagnified provided that the maximum moment is along the column height and not at its ends. Otherwise, its value has to be multiplied by the nonsway magnifier δ_{ns}. The effective length factor *k* when there is single curvature can be obtained from Jackson and Morland's chart, Figure 27.8a. For double curvature, the length factor *k* can be obtained from Figure 27.8b. Discussion of the P-Δ effect and the second-order analysis is given in [Nawy, E.G., 1996a].

27.3.5 Two-Way Slabs and Plates

There are several methods of designing two-way concrete slabs and plates:

(a) the ACI direct design method;
(b) the ACI equivalent frame method where effects of lateral loads can be considered;

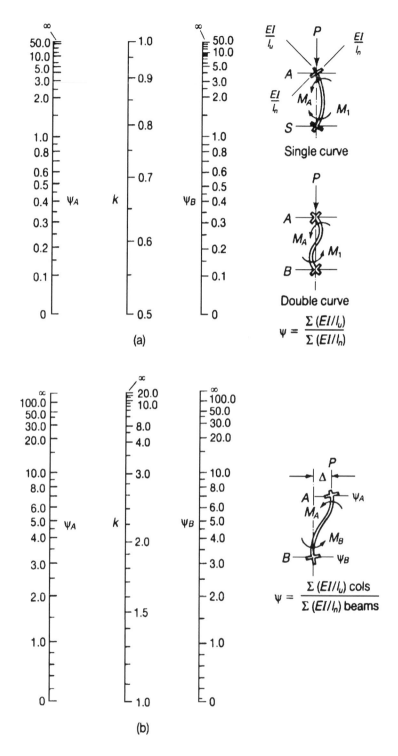

FIGURE 27.8 Slender columns end effect factor *k*. (a) Single curvature; and (b) double curvature.

TABLE 27.2 Minimum Thickness of Slabs Without Interior Beams, $\alpha_m = 0$

Yield Strength, f_y, psi	Without Drop Panels			With Drop Panels		
	Exterior Panels			Exterior Panels		
	Without Edge Beams	With Edge Beams	Interior Panels	Without Edge Beams	With Edge Beams	Interior Panels
40,000	$\ell_n/33$	$\ell_n/36$	$\ell_n/36$	$\ell_n/36$	$\ell_n/40$	$\ell_n/40$
60,000	$\ell_n/30$	$\ell_n/33$	$\ell_n/33$	$\ell_n/33$	$\ell_n/36$	$\ell_n/36$
75,000	$\ell_n/28$	$\ell_n/31$	$\ell_n/31$	$\ell_n/31$	$\ell_n/34$	$\ell_n/34$

Note: ℓ_n = effective span.

(c) the yield line theory;
(d) the strip method; and
(e) elastic solutions.

The subject is too extensive to cover in this overview. However, serviceability as controlled by deflection and cracking limitation, which is of major importance, will be briefly presented.

27.3.5.1 Deflection Control

The thickness of two-way slabs for deflection control should be determined as follows:

Flat Plate.   Use Table 27.2.

Slab on Beams.   If $\alpha_m \leq 0.2$, use

$$\alpha_m > 0.2 < 2.0, \qquad h \geq \frac{\ell_n(0.8 + f_y/200000)}{36 + 5\beta(\alpha_m - 0.2)} \tag{27.47}$$

but slab or plate thickness cannot be less than 5.0 in.

$$\alpha_m > 2.0, \qquad h \geq \frac{(\ell_n0.8 + f_y/200000)}{36 + 9\beta} \tag{27.48}$$

where

α_m = average value of α for all beams on edges of a panel

$\alpha = \dfrac{\text{flexural stiffness of beam section}}{\text{flexural thickness of slab width bounded laterally by centerline of the adjacent panels on each side of beam}}$

β = aspect ratio, long span/short span

27.3.5.2 Crack Control

For crack control in two-way slabs and plates, the maximum computed weighted crack width due to flexural load [ACI Committee 224, 1997; Nawy, 1996a] is as follows where the parameter under the radical is termed as the grid index, G:

$$w_{\max} \text{ (in)} = k\beta f_s \sqrt{\frac{s_1 s_2 d_c}{d_b} \cdot \frac{8}{\pi}} \tag{27.49}$$

For $w_{\max}$ (mm), multiply Eq. (27.49) by 0.145 and use megapascals for f_s

k = fracture coefficient
$k = 2.8 \times 10^{-5}$ for a square uniformly loaded slab

TABLE 27.3 Tolerable Crack Widths

Exposure Condition	Tolerable Crack Width	
	in	mm
Dry air or protective membrane	0.016	0.41
Humidity, moist air, soil	0.012	0.30
Deicing chemicals	0.007	0.18
Seawater and seawater spray; wetting and drying	0.006	0.15
Water-retaining structures (excluding nonpressure pipes)	0.004	0.10

$= 2.1 \times 10^{-5}$ when aspect ratio short span/long span <0.75 but >0.5, or for a concentrated load

$= 1.6 \times 10^{-5}$ for aspect ratio less than 0.5.

$\beta = 1.25 = (h - c)/(d - c)$, where $c =$ depth to neutral axis.

$f_s = 0.40 f_y$ (kip/in^2).

$h =$ total slab or plate thickness.

$s_1 =$ spacing in direction "1" closest to the tensile extreme fibers, inches

$s_2 =$ spacing in the perpendicular direction, inches

$d_c =$ concrete cover to centroid of reinforcement, inches

$d_{b1} =$ diameter of the reinforcement in direction "1" closest to the concrete outer fibers, inches

The tolerable crack widths in concrete elements are given in Table 27.3.
In SI units, Eq. (27.49) therefore becomes

$$w_{\max} \text{ (mm)} = 0.145 k \beta f_s \sqrt{G_I}$$

where f_s is in megapascals, and s_1, s_2, d_c, and d_{b1} are in millimeters.

27.3.6 Development of Reinforcement

27.3.6.1 Development of Deformed Bars in Tension

The full development length ℓ_d for deformed bars or wires is obtained by applying multipliers to a basic theoretical development length ℓ_{db} in terms of the bar diameter d_b and other multipliers as follows. It is obtained from the expression

$$\frac{\ell_d}{d_b} = \frac{3}{40} \frac{f_y}{\sqrt{f_c'}} \frac{\alpha \beta \gamma \lambda}{[(c/d_b) + K_{tr}]} \qquad (27.50)$$

where the term $[(c/d_b) + K_{tr}]$ should not exceed a value of 2.5 but should not be less than 1.5 for usual structures and $\sqrt{f_c'}$ should not exceed 100 psi (≤ 6.9 MPa).

27.3.6.2 Modifying Multipliers of Development Length for Bars in Tension

$\alpha = $ *Bar Location Factor.*

- For horizontal reinforcement α is so placed that more than 12 in of fresh concrete below the development length or splice (top reinforcement) is 1.3;
- For other reinforcement α is 1.0.

$\beta = $ *Coating Factor.*

- For epoxy-coated bars or wires with cover less than $3d_b$ or clear spacing less than $6d_d$ β is 1.5;

- For all other epoxy-coated bars or wires β is 1.2;
- For uncoated reinforcement β is 1.0.

However, the product $\alpha\beta$ should not exceed 1.7.

$\gamma = $ *Bar Size Factor.*

- For No. 6 and smaller bars and deformed wires (No. 20 and smaller, SI) γ is 0.8;
- For No. 7 and larger bars (No. 25 and larger, SI) γ is 1.0.

$c = $ **Spacing or Cover Dimension, Inches.**   Use the smaller of either the distance from the center of the bar to the nearest concrete surface or one-half the center to center spacing of the bars being developed.

$K_{tr} = $ *Transverse Reinforcement Index* $= A_{tr}f_{yt}/1500sn.$

$A_{tr} = $ total cross-sectional area of all transverse reinforcement within ℓ_d that crosses the potential plane of splitting adjacent to the reinforcement being developed, square inches
$f_{yt} = $ specified yield strength of transverse reinforcement, psi (megapascals)
$s = $ maximum spacing of transverse reinforcement within ℓ_d, center to center, inches (millimeters)
$n = $ number of bars or wires being developed along the plane of splitting

The ACI Code permits using $K_{tr} = 0$ as a conservative design simplification even if transverse reinforcement is present.

$\lambda = $ *Lightweight-Aggregate Concrete Factor.*

- When lightweight aggregate concrete is used λ is 1.3.
  However, when f_{ct} is specified, use $\lambda = 6.7\sqrt{f_c'}/f_{ct}$ but not less than 1.0.
- When normal-weight concrete is used λ is 1.0.
  The minimum development length in all cases is 12 in.

$\lambda_s = $ **Excess Reinforcement Factor.**   The ACI Code permits the reduction of ℓ_d if the longitudinal flexural reinforcement is in excess of that required by analysis except where anchorage or development for f_y is specifically required or the reinforcement is designed for seismic effects.

The reduction multiplier $\lambda_s = A_s$ required/ A_s provided and $\lambda_{s2} = f_y/60{,}000$ for cases where $f_y > 60{,}000$ psi. In lieu of using a refined computation for the development length of Eq. (27.50), Tables 27.4a and 27.4b can be utilized for typical construction practices by using a value of

$$\left(\frac{c}{d_b} + K_{tr} \right) = 1.5$$

and $f_c' = 4000$ psi.

Table 27.5 gives minimum development length l_d (inches) in lieu of calculations using Table 27.4. In these two tables the following assumptions are made:

The side cover is 1.5 in on each side;
No. 3 stirrups are used for bars No. 11 or smaller;
No. 4 stirrups are used for bars No. 14 or No. 18; and

TABLE 27.4a   Simplified Development Length ℓ_d Equations

	#6 and Smaller Bars and Deformed Wires	#7 and Larger Bars
Clear spacing of bars being developed or spliced not less than d_b, clear cover not less than d_b, and stirrups or ties throughout ℓ_d not less than the Code minimum	$\dfrac{\ell_d}{d_b} = \dfrac{f_y \alpha \beta \lambda}{25\sqrt{f_c'}}$	$\dfrac{\ell_d}{d_b} = \dfrac{f_y \alpha \beta \lambda}{20\sqrt{f_c'}}$
or	* when $f_c' = 4000$ psi and $\alpha, \beta, \lambda, \lambda_s = 1.0$ $\gamma = 0.8$	** when $f_c' = 4000$ psi and $\alpha, \beta, \lambda, \lambda_s, \gamma = 1.0$
Clear spacing of bars being developed or spliced not less than $2d_b$ and clear cover not less than d_b.	$\ell_d = 38 d_b$	$\ell_d = (38/0.8) d_b$ $= 48 d_b$
Other cases (1.5 times	$\dfrac{\ell_d}{d_b} = \dfrac{3\,f_y \alpha \beta \lambda}{50\sqrt{f_c'}}$	$\dfrac{\ell_d}{d_b} = \dfrac{3\,f_y \alpha \beta \lambda}{40\sqrt{f_c'}}$
the above values)	*$\ell_d = 57 d_b$	**$\ell_d = 72 d_b$

Source: Nawy, E.G., *Reinforced Concrete—A Fundamental Approach*, 3rd ed., © 1996. Reprinted by permission of Prentice–Hall, Inc. Upper Saddle River, NJ.

Note: Table 27.4a is a general table for usual construction conditions giving the required development length for deformed bars of sizes No. 3 to No. 18.

Stirrups are bent around four bar diameters; hence the distance from the centroid of the bar nearest the side face of the beam to the inside face of the No. 3 stirrup is taken as 0.75 in for bars No. 11 or smaller and is equal to the longitudinal bar radius for No. 14 and No. 18 bars.

TABLE 27.4b   SI   Development Length Simplified Expressions

$\leq$ #20	$\geq$ #25
$\dfrac{\ell_d}{d_b} = \dfrac{f_y \alpha \beta \lambda}{2\sqrt{f_c'}}$	$\dfrac{\ell_d}{d_b} = \dfrac{5\,f_y \alpha \beta \lambda}{8\sqrt{f_c'}}$
$\dfrac{\ell_d}{d_b} = \dfrac{3\,f_y \alpha \beta \lambda}{4\sqrt{f_c'}}$	$\dfrac{\ell_d}{d_b} = \dfrac{15\,f_y \alpha \beta \lambda}{16\sqrt{f_c'}}$

27.3.6.3 Development of Deformed Bars in Compression and the Modifying Multipliers

Bars in compression require shorter development length than bars in tension. This is due to the absence of the weakening effect of the tensile cracks. Hence the expression for the basic development length is

$$\ell_{db} = 0.02 \frac{d_b\, f_y}{\sqrt{f_c'}} \tag{27.51a}$$

$$\ell_{db} \geq 0.0003 d_b\, f_y \tag{27.51b}$$

with the modifying multiplier for

(1) excess reinforcement: $\lambda_s = $ required A_s/provided A_s; and
(2) spirally enclosed reinforcement $\lambda_{s1} = 0.75$

27.3.6.4 Development of Bundled Bars in Tension and Compression

If bundled bars are used in tension or compression, ℓ_d has to be increased by 20% for three-bar bundles and 33% for four-bar bundles and f_c' should not be taken greater than 100 psi. A unit of bundled bars is treated as a single bar of a diameter derived from the equivalent total area for the purpose of determining the modifying factors. However, although splice and development lengths of bundled bars are based on the diameter of individual bars increased by 20 or 33% as applicable, it is necessary to use an equivalent diameter of the entire bundle derived from the equivalent total

TABLE 27.5   Tension Reinforcement and Development Length (inches)

Bar Size	Cross Sectional Area, in^2	Bar Diameter, in	Development Length, $\ell_d^{*\dagger}$, in	
			$s \geq 2d_b$ or $d_b^\ddagger$ and Clear Cover $\geq d_b$ $\leq$ #6 : $\ell_d = 38d_b$ $\geq$ #7 : $\ell_d = 48d_b$	Other $\leq$ #6 : $\ell_d = 57d_b$ $\geq$ #7 : $\ell_d = 72d_b$
3	0.11	0.375	15	21
4	0.20	0.500	19	29
5	0.31	0.625	24	36
6	0.44	0.750	29	43
7	0.60	0.875	42	63
8	0.79	1.000	48	72
9	1.00	1.128	54	81
10	1.27	1.270	61	92
11	1.56	1.410	68	102
14	2.25	1.693	82	122
18	4.00	2.257	108	163

Source: Nawy, E.G., *Reinforced Concrete—A Fundamental Approach*, 3rd ed., © 1996. Reprinted by permission of Prentice Hall, Upper Saddle River, NJ. (27, 28)

Note: For $f_c' = 4.000$ psi Normal-Weight Concrete $f_y = 60,000$ psi Steel (α, β, $\lambda = 1.0$, $\gamma = 0.8$ for #6 bars or smaller and $= 1.0$ for #7 bars and larger) For f_c' values different from 4000 psi, multiply table values by ($\sqrt{4000/f_c'}$). For $f_y = 40,000$ psi, multiply by 2/3. $\sqrt{f_c'}$ should not exceed 100.

[*]For Compression development length, ℓ_d = multiplier $\times$ ℓ_{db} where $\ell_{db} = 0.02d_b f_y/\sqrt{f_c'} \geq 0.0003d_b f_y$.

[†]Multiply table values by $\alpha = 1.3$ for top reinforcement; $\lambda = 1.3$ for lightweight aggregate; $\beta = 1.5$ for epoxy-coated bars with cover less than $3d_b$ or clear spacing less than $6 d_b$ and $\beta = 1.2$ for other epoxy-coated bars. Minimum ℓ_d for all cases = 12 in.

area of bars when determining the factors that consider cover and clear spacing and represent the tendency of concrete to split.

27.3.6.5 SI Metric Conversion

Where f_{yt}, is in megapascals, Eq. (27.50) becomes

$$\frac{\ell_d}{d_b} = \frac{15 f_y \alpha \beta \gamma \lambda}{16 \sqrt{f_c'}\left(\dfrac{c}{d_b} + K_{tr}\right)} \quad \text{and} \quad K_{tr} = \frac{A_{tr} f_{yt}}{260 s\, n}$$

27.3.6.6 Development of Welded Deformed Wire Fabric in Tension

The development length, ℓ_d, for deformed welded wire fabric should be taken as the ℓ_d value obtained from Eq. (27.50) or Table 27.4 multiplied by a fabric factor. The fabric factor, with at least one cross wire within the development length and not less than 2 in from the point of the critical section, should be taken as the greater of the following two expressions:

$$(f_y - 35,000)/f_y \tag{27.52a}$$

or

$$(5d_b/s_w) \tag{27.52b}$$

but should not be taken greater than 1.0. Here s_w is the spacing of wire to be developed or spliced, inches.

27.4 Prestressed Concrete

27.4.1 General Principles

Reinforced concrete is weak in tension but strong in compression. To maximize utilization of its material properties, an internal compressive force is induced on the structural element through the use of highly stressed prestressing tendons to precompress the member prior to application of the external gravity live load and superimposed dead load. A typical effect of the prestressing action is shown in Figure 27.9 using a straight tendon, as is usually the case in precast elements [Nawy, 1996b]. In *situ*-cast elements, the tendon can be either harped or, as is usually the case, draped in a parabolic form. Figure 27.10 gives the stress and strain distributions across the beam depth and the forces acting on the section in a prestressed concrete beam.

Stresses due to initial prestressing plus self-weight:

$$f^t = -\frac{P_i}{A_c}\left(1 - \frac{ec_t}{r^2}\right) - \frac{M_D}{S^t} \qquad (27.53a)$$

$$f_b = -\frac{P_i}{A_c}\left(1 + \frac{ec_b}{r^2}\right) + \frac{M_D}{S_b} \qquad (27.53b)$$

Stresses at service load:

$$f^t = -\frac{P_e}{A_c}\left(1 - \frac{ec_t}{r^2}\right) - \frac{M_T}{S^t} \qquad (27.54a)$$

$$f_b = -\frac{P_e}{A_c}\left(1 + \frac{ec_b}{r^2}\right) + \frac{M_T}{S_b} \qquad (27.54b)$$

27.4.2 Minimum Section Modulus for Variable Eccentricity Tendon

$$S^t \geq \frac{(1-\gamma)M_D + M_{SD} + M_L}{\gamma f_{ti} - f_c} \qquad (27.55a)$$

$$S_b \geq \frac{(1-\gamma)M_D + M_{SD} + M_L}{f_t - \gamma f_{ci}} \qquad (27.55b)$$

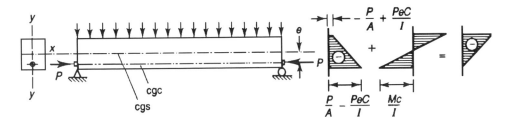

FIGURE 27.9 Stress distribution at service load in prestressed beam with constant tendon eccentricity. *Source:* Nawy, E.G. *Prestressed Concrete—A Fundamental Approach,* 2nd ed., © 1996. Reprinted by permission of Prentice Hall, Upper Saddle River, NJ.

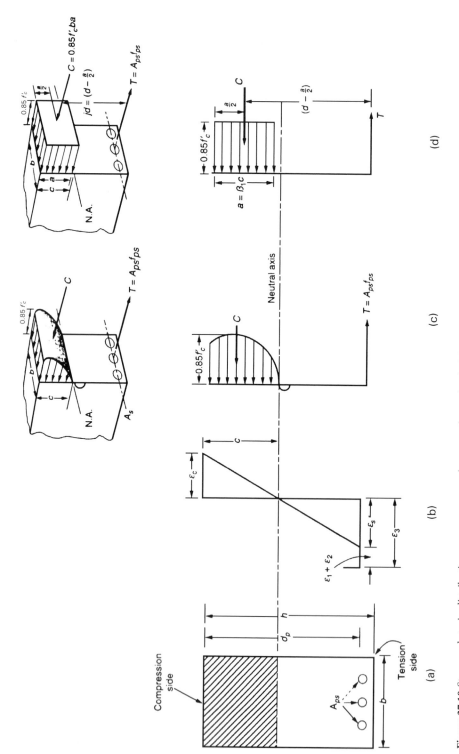

Figure 27.10 Stress and strain distribution across prestressed concrete beam depth. (a) Beam cross section; (b) strain across depth; (c) actual stress block; and (d) assumed equivalent block. *Source:* Nawy, E.G., *Prestressed Concrete—A Fundamental Approach*, 2nd ed., © 1996. Reprinted by permission of Prentice Hall, Upper Saddle River, NJ.

where

γ = percentage loss in prestress
M_D = self-weight moment
M_{SD} = superimposed dead load moment
M_L = live load moment
f_{ti} = initial tensile stress in concrete
f_c = service load concrete compressive strength
f_t = service load concrete tensile strength
f_{ci} = initial compressive stress in concrete
S^t = section modulus at top fibers (simple span)
S_b = section modulus at bottom fibers (simple span)

27.4.3 Minimum Section Modulus for Constant Tendon Eccentricity

$$S^t \geq \frac{M_D + M_{SD} + M_L}{\gamma f_{ti} - f_c} \qquad (27.56a)$$

$$S^b \geq \frac{M_D + M_{SD} + M_L}{f_t - \gamma f_{ci}} \qquad (27.56b)$$

27.4.4 Maximum Allowable Stresses

27.4.4.1 Concrete Stresses (ACI 318)

$f_{ci}' \cong 0.75 f_c'$ psi
$f_{ci} \cong 0.60 f_{ci}'$ psi
$f_{ti} = \sqrt{f_{ci}'}$ psi on span ($\sqrt{f_{ci}'}/4$ MPa)
$\phantom{f_{ti}} = \sqrt{f_{ci}'}$ psi on support ($\sqrt{f_{ci}'}/2$ MPa)
$f_c = 0.45 f_c'$ or $0.60 f_c'$ where permitted by ACI 318.
$f_t = \sqrt{f_c'}$ psi($\sqrt{f_c'}/2$ MPa)
$ = 12\sqrt{f_c'}$ psi if deflection verified ($\sqrt{f_c'}/2$ MPa)

27.4.4.2 Reinforcing Tendon Stresses

Tendon jacking: $f_{ps} = 0.94 f_{py} \leq 0.80 f_{pu}$
Immediately after prestress transfer: $f_{ps} = 0.82 f_{pJ} \leq 0.74 f_{pu}$
Post-tensioned members at anchorage immediately after tendon anchorage: $f_{ps} = 0.70 f_{pu}$
where:
f_{pu} = ultimate strength of tendon
f_{py} = yield strength of tendon
f_{ps} = ultimate design stress allowed in tendon

A prestressed concrete section is designed for both the service load and the ultimate load. A typical distribution of stress at service load at midspan is shown in Figure 27.9. Expressions for the ultimate load evaluation are essentially similar to those of reinforced concrete elements, taking into consideration that both prestressing tendons and mild steel bars are used. Figure 27.10 gives the compressive stress block and the equilibrium forces. Note its similarity to Figure 27.6. For details, refer to a prestressed concrete textbook such as Nawy, 1996b.

27.5 Shear and Torsion in Prestressed Elements

27.5.1 Shear Strength: ACI Short Method if $f_{pe} > 0.40 f_{pu}$

The nominal shear stress of the concrete in the web is

$$V_c \text{ (lb)} = \left(0.60\lambda\sqrt{f_c'} + \frac{700\,V_u d}{M_u}\right) b_w d \tag{27.57a}$$

$$V_c \geq 2\lambda\sqrt{f_c'}\,b_w d \leq 5\lambda\sqrt{f_c'}\,b_w d$$

$$\frac{V_u d}{M_u} \leq 1.0$$

$$V_c \text{ (Newton)} = \left(\frac{\sqrt{f_c'}}{20} + 5\frac{V_u d}{M_u}\right) b_w d \tag{27.57b}$$

$$V_c \geq \frac{\sqrt{f_c'}}{6} b_w d \leq 0.40\sqrt{f_c'}\,b_w d$$

$$\frac{V_u d}{M_u} \leq 1.0$$

27.5.2 Detailed Method

The smaller of the values obtained from flexural shear V_{ce} or web shear V_{cw} has to be used in the design.

27.5.2.1 Flexural Shear

$$V_{ci} \text{ (lb)} = 0.6\sqrt{f_c'}\,b_w d + V_d + \frac{V_i M_{cr}}{M_{\max}} \geq 1.7\sqrt{f_c'}\,b_w d \tag{27.58a}$$

where

$V_{ci} =$ flexural shear force
$M_{cr} = S_b(6\sqrt{f_c'} + f_{ce} - f_d)$
$S_b =$ section modulus at the extreme tensile fibers
$V_d =$ shear force at section due to unfactored dead load
$V_i =$ factored shear force due to externally applied load
$f_{ce} =$ compressive stress in concrete due to effective prestress only at the tension face of the section
$f_d =$ stress due to unfactored dead load at extreme fibers in tension

$$V_{ci} \text{ (Newton)} = \left(\frac{\sqrt{f_c'}}{20}\right) b_w d + V_d + \frac{V_i M_{cr}}{M_{\max}} \tag{27.58b}$$

27.5.2.2 Web Shear

$$V_{CW} = \text{web shear force}$$

$$V_{CW} \text{ (lb)} = \left(3.5\sqrt{f_c'} + 0.3\overline{f}_{ce}\right) b_w d + V_p \tag{27.59a}$$

$$V_{CW} \text{ (Newton)} = \left(0.3\sqrt{f_c'} + \overline{f}_c\right) b_w d + V_p \tag{27.59b}$$

where

$\overline{f}_c$ = compressive stress at center of gravity of section owing to externally applied load
V_p = vertical component of prestressing force

The critical section for calculating V_u and T_u is taken at distance $(h/2)$ from the face of the support.

27.5.3 Minimum Shear Reinforcement

For prestressed members subjected to shear, the minimum transverse web stirrups are the smaller of

$$A_v \text{ (in}^2) = \frac{50 b_w s}{f_y} \tag{27.60a}$$

or

$$A_v = \frac{A_{ps} f_{pu} S}{80 f_y d} \sqrt{\frac{d_p}{b_w}} \tag{27.60b}$$

where f_y is in psi.

27.5.4 Torsional Strength

As discussed in Section 27.2, Beams and Slabs, the nominal torsional strength T_c is disregarded and all the torque is assumed by longitudinal bars and the transverse closed hoops. The expressions used in the case of prestressed concrete elements are essentially the same as those for reinforced concrete elements with the following adjustments for Eqs. (27.27) and (27.27): Multiply the right side by

$$\sqrt{1 + \frac{3 f_{ce}}{\sqrt{f'_c}}}$$

For hollow sections, the left side of Eq. (27.29) becomes

$$\left(\frac{V_u}{b_w d}\right) + \left(\frac{T_u p_h}{1.7 A_{oh}^2}\right)$$

The maximum spacing of the closed hoops is $1/8 \, p_h \leq 12$ in, and the longitudinal bar diameter is not less than $1/16 \, s$, where s is spacing of the hoop steel.

27.6 Walls and Footings

The design of walls and footings should be viewed in the context of designing a one-way or two-way cantilever slab in the case of footings and one-way vertical cantilevers in the case of reinforced concrete walls. The criteria and expressions for proportioning their geometry are the same as those presented in earlier sections of this chapter. Shear V_u in one-way footings is taken at a distance d from the face of the vertical concrete wall or columns and at $d/2$ in the case of two-way footings. The nominal shear strength (capacity) V_c of the one-way slab footing is

$$V_c = 2\sqrt{f'_c} b_w d \tag{27.61}$$

For two-way slab footings, the nominal shear strength V_c should be the smaller of

$$V_c = 4\sqrt{f_c'}\,b_o d \qquad (27.62a)$$

or

$$V_c = \left(2 + \frac{4}{\beta_c}\right)\sqrt{f_c'}\,b_o d \qquad (27.62b)$$

or

$$V_c = \left(\frac{\alpha_s d}{b_0} + 2\right)\sqrt{f_c'}\,b_o d \qquad (27.62c)$$

where b_0 is the perimeter shear failure length at distance $d/2$ from all faces of columns. If the column size is $c_1 \times c_2$, then

$b_0 = 2(c_1 + d/2) + 2(c_2 + d/2)$ for an interior column
β_c = ratio of long side/short side of reaction area
α_s = 40 for interior columns, 30 for end columns, and 20 for corner columns

The same requirement for shear in Eq. (27.62) applies to the shear design of two-way action structural slabs and plates.

Acknowledgment

This chapter is based on extensive material taken with permission from *Fundamentals of High Strength High Performance Concrete* by E.G. Nawy, Addison Wesley Longman, London and California and from *Reinforced Concrete—A Fundamental Approach*, 3rd. ed. and *Prestressed Concrete—A Fundamental Approach*, 2nd ed., both by E.G. Nawy, Prentice Hall, Upper Saddle River, New Jersey, and from various committee reports and standards of the American Concrete Institute, Farmington Hills, Michigan.

References

ACI Committee 363. 1992. *State-of-the-Art on High Strength Concrete*, ACI 363R-92, pp. 1–55. American Concrete Institute, Farmington Hills, MI.
ACI Committee 435. 1995. *Control of Deflection in Concrete Structures.* Committee 435 Report. E.G. Nawy, chairman, pp. 77. American Concrete Institute, Farmington Hills, MI.
ACI Committee 318. 1996. *Building Code Requirements for Structural Concrete*, ACI 318-95. Commentary. ACI 318R-95. American Concrete Institute, Farmington Hills, MI.
ACI Committee 224. 1989. *Control of Cracking in Concrete Structures*, ACI Committee 244 Report. American Concrete Institute, Farmington Hills, MI.
Ahmed, S.H. and Lue, D.M. 1987. Flexure-shear interaction of reinforced high strength concrete beams. *Proc. ACI Struct. J.* 84(4):330–341.
Hsu, T.T.C. 1993. *Unified Theory of Reinforced Concrete*, pp. 312. CRC Press, Boca Raton, FL.
Pastor, J.A., Nilsen, A.H., and Slate, F.O. 1984. Behavior of high strength concrete beams. *Cornell University Report No. 84-3.* Dept. of Structural Engineering, Cornell University, Ithaca, New York.
Martinez, S., Nilsen, A.H., and Slate, F.O. 1984. Spirally-reinforced high strength concrete columns. *Proc., ACI J.* 81(5):431–442.
Nawy, E.G. 1996a. *Reinforced Concrete—A Fundamental Approach*, 3rd ed., pp. 838. Prentice Hall, Upper Saddle River, NJ.

Nawy, E.G. 1994b. *Prestressed Concrete—A Fundamental Approach*, 2nd ed., pp. 810. Prentice Hall, Upper Saddle River, NJ.

Nawy, E.G. 1996c. *Fundamentals of High Strength High Performance Concrete*, pp. 350. Addison Wesley Longman, California and London.

Yong, Y.K., Nour, M.G., and Nawy, E.G. 1988. Behavior of laterally confined high strength concrete under axial loads. *Proc. ASCE J. of Struct. Eng.* 114(2):332–351.

Index

N

X

Y

Z